HANDBOOK FOR DIGITAL SIGNAL PROCESSING

HANDBOOK FOR DIGITAL SIGNAL PROCESSING

Edited by

Sanjit K. Mitra
Department of Electrical and Computer Engineering
University of California, Santa Barbara

James F. Kaiser
Department of Electrical and Computer Engineering
Rutgers University
Piscataway, New Jersey

A WILEY-INTERSCIENCE PUBLICATION

JOHN WILEY & SONS

New York   •   Chichester   •   Brisbane   •   Toronto   •   Singapore

Library of Congress Cataloging in Publication Data:

Handbook for digital signal processing / edited by Sanjit K. Mitra,
  James F. Kaiser.
       p. cm.
    "A Wiley-Interscience publication."
    Includes bibliographical references and index.
    ISBN 0-471-61995-7 (alk. paper)
    1. Signal processing—Digital techniques. I. Mitra, Sanjit
  Kumar.   II. Kaiser, James F.
  TK5102.5.H318   1993
  621.382'2—dc20   92-35700
                                             CIP

Printed in the United States of America

10 9 8 7 6 5 4 3 2

To all the researchers
who contributed to
the growth of digital signal processing

CONTRIBUTORS

RASHID ANSARI, Bell Communications Research, 445 South Street, Morristown, NJ 07960

NIRMAL K. BOSE, Department of Electrical Engineering, The Pennsylvania State University, University Park, PA 16802

C. SIDNEY BURRUS, Department of Electrical and Computer Engineering, Rice University, Houston, TX 77001

YOUN-SHIK BYUN, Department of Electronic Engineering, Incheon University, Incheon 402-749, Korea

JOHN M. CIOFFI, Department of Electrical Engineering, Stanford University, Stanford, CA 94305-4055

JALIL FADAVI-ARDEKANI, AT&T Bell Laboratories, 555 Union Blvd., Allentown, PA 18103

PETRI HAAVISTO, Department of Electrical Engineering, Tampere University of Technology, SF 33101 Tampere, Finland

WILLIAM E. HIGGINS, Department of Electrical Engineering, The Pennsylvania State University, University Park, PA 16802

W. KENNETH JENKINS, Department of Electrical and Computer Engineering, University of Illinois, Urbana, IL 61801

YIH C. JENQ, Department of Electrical Engineering, Portland State University, Portland, OR 97207-0751

JAMES F. KAISER, Department of Electrical and Computer Engineering, Rutgers University, Piscataway, NJ 08855

RAMDAS KUMARESAN, Department of Electrical Engineering, The University of Rhode Island, Kingston, RI 02881

LAWRENCE E. LARSON, Hughes Research Laboratories, 3011 Malibu Canyon Road, Malibu, CA 90265

BEDE LIU, Department of Electrical Engineering, Princeton University, Princeton, NJ 08544

SANJIT K. MITRA, Department of Electrical and Computer Engineering, University of California, Santa Barbara, CA 93106-9560

KALYAN MONDAL, AT&T Bell Laboratories, 555 Union Blvd., Allentown, PA 18103

DAVID C. MUNSON, JR., Department of Electrical and Computer Engineering, University of Illinois, Urbana, IL 61801

YRJÖ NEUVO, Nokia Corporation, Eteläesplanadi 12A, SF-00130 Helsinki, Finland

PHILIP A. REGALIA, Département Electronique et Communications Institut National des Telécommunications, 9 Rue Charles-Fourier, 91011 Evry Cedex, France

TAPIO SARAMÄKI, Department of Electrical Engineering, Tampere University of Technology, SF 33101 Tampere, Finland

OGNJAN V. SHENTOV, Pennie & Edmonds, 1115 Avenue of the Americas, New York, NY 10036

HENRIK V. SORENSEN, Department of Electrical Engineering, University of Pennsylvania, Philadelphia, PA 19104-6390

KENNETH STEIGLITZ, Department of Computer Science, Princeton University, Princeton, NJ 08544

WONGYONG SUNG, Inter-University Semiconductor Research Center, Seoul National University, Shinlim-dong, Gwanak-gu, Seoul 151-742, Korea

GABOR C. TEMES, Department of Electrical and Computer Engineering, Oregon State University, Corvallis, OR 97331

TRAN THONG, Tektronix Federal Systems, Inc., 18700 N.W. Walker Road, Beaverton, OR 97076-4490

P. P. VAIDYANATHAN, Department of Electrical Engineering, California Institute of Technology, Pasadena, CA 91125

PREFACE

The field of digital signal processing (DSP) has been a very active area of research and application for more than 30 years. This broad development has paralleled in time the rapid development of high-speed electronic digital computers, microelectronics, and integrated circuit fabrication technologies. During this period more than 50,000 journal and conference papers and more than 100 textbooks have been published in the general areas of digital signal processing. Almost every electrical engineering department in the country and abroad now offers upper division courses in digital signal processing. Those schools with a graduate electrical engineering program, virtually without exception, offer advanced level courses in digital signal processing. The major electronic instrument manufacturers now routinely incorporate digital signal processing methods in their modern instruments. An ever-increasing assortment of integrated circuit parts specifically tailored to perform common digital signal processing functions is available to the design engineer as system building blocks or parts-in-trade. DSP methodologies have been applied to consumer electronics, communications, automotive electronics, instrumentation, medical electronics, tomography and acoustic imaging, cartography, seismology, speech recognition, and robotics; the list goes on and on. Thus, the practicing engineer is rapidly becoming intimately involved with the many aspects of digital signal processing.

It has now become a virtually hopeless task to be familiar with, or even to become aware of, the majority of the algorithms, implementation strategies, and hardware design techniques that embody the area of digital signal processing. Yet researchers, as well as the active user of DSP methods, must be keenly aware of the most usable and robust methods.

Available textbooks, for the most part are designed to be used as senior or first-year graduate level texts in the engineering schools. Several DSP texts are becoming available that are suitable for advanced graduate level instruction in specialized areas. The practicing engineer or scientist working in the area of signal processing is forced, however, to own a number of different texts and journals to insure a satisfactory coverage of the essential ideas and techniques of the field. The pressing need for a comprehensive handbook on digital signal processing is apparent.

One of this handbook's objectives is a distillation from the extensive literature of the central ideas and primary methods of analysis, design, and implementation of DSP systems. The handbook also points the reader to the primary reference sources that give the finer details of the design and analysis methods. The reference sets are selective rather than exhaustive. The reader is made aware of the significant problems that may be encountered in applying the ideas so that potential pitfalls can be identified and taken into proper consideration.

Software packages for digital signal algorithm design, analysis, and processing are increasingly available as commercial products to design and applications engineers. Some very good DSP software is also available in journal articles and textbooks. This handbook includes a number of selected computer programs and keys, wherever possible, the primary design methods to relevant published software.

This book is comprised of 16 chapters covering both fundamental and advanced topics. Chapter 1 provides an introduction to the basic concepts of signal processing and includes several typical examples of applications of signal processing. The mathematical foundations of signal processing are provided in Chapter 2. Much of this handbook is directed towards the theory and design of linear time-invariant discrete-time systems for the processing of discrete-time signals. Chapter 3 describes several mathematical representations of linear time-invariant discrete-time systems, more commonly called digital filters. Chapters 4, 5, and 13 outline various methods for the design of such filters. Chapter 6 is concerned with the issues that are common to most types of practical implementations of digital filters, such as finite-word length effects. Chapter 7 reviews digital filter structures that minimize such detrimental effects. Chapter 8 delineates some commonly used algorithms for fast computations of discrete Fourier transforms, important DSP tools. It also describes a number of algorithms for fast convolution implementations. Chapter 9 reviews the basic principles of residue number system arithmetic and its DSP applications. The interface structures needed for the digital processing of continuous-time signals are treated in Chapter 10. Chapters 11 and 12 describe various methods for the hardware and software implementations of DSP algorithms. Chapter 14 is devoted to a review of multirate DSP techniques. The subject of adaptive digital filter theory and design is considered in Chapter 15. Finally, Chapter 16 reviews a number of algorithms for spectral analysis.

The authors of this handbook are recognized experts in their respective DSP technologies. Together they have more than 100 years of industrial experience and more than 300 years of teaching and research experience.

The editors thank each of the authors for their timely contributions of high quality texts. The editors also thank Professor Leland Jackson of the University of Rhode Island for carefully reading the complete manuscript and making numerous comments that have improved the style and content of the materials presented. We also acknowledge the comments of Dr. Stanley A. White of Rockwell International who reviewed selected portions of the manuscript. In a book of this size it is almost impossible to ensure 100 percent accuracy in the materials presented. We would appreciate very much if the reader brings to our attention any errors that may have appeared in the final version due to reasons beyond the control of the editors and authors.

SANJIT K. MITRA
JAMES F. KAISER

Santa Barbara, California
Summit, New Jersey

CONTENTS

5   Infinite Impulse Response Digital Filter Design   279

William E. Higgins and David C. Munson, Jr.

6   Digital Filter Implementation Considerations   337

Yrjö Neuvo

7 Robust Digital Filter Structures

P. P. Vaidyanathan

9 Finite Arithmetic Concepts   **611**

W. Kenneth Jenkins

15   Adaptive Filtering   1085

John M. Cioffi and Youn-Shik Byun

16   Spectral Analysis

Ramdas Kumaresan

GLOSSARY OF NOTATIONS AND ABBREVIATIONS

NOTATIONS

Mathematical Notations

$*$	complex conjugation
$\triangleq$	by definition
$\mathbf{k}$	bold lower case letter represents a column or a row vector
$\mathbf{K}$	bold upper case letter represents a matrix
$\mathbf{K}'$	the transpose of the matrix $\mathbf{K}$
$\mathbf{K}^{-1}$	the inverse of the matrix $\mathbf{K}$
$\mathbf{K}^\dagger$	the complex conjugate transpose of $\mathbf{K}$ [i.e., $(\mathbf{K}*)'$]
$\mathbf{A}^{1/2}$	real square root of a positive definite-matrix $\mathbf{A}$
$\mathbf{A}^{-1/2}$	inverse of $\mathbf{A}^{1/2}$
$\mathbf{A}^{t/2}$	transpose of $\mathbf{A}^{1/2}$
$\mathbf{A}^{-t/2}$	inverse transpose $\mathbf{A}^{1/2}$
$\in$	member of a set
$\cap$	intersection of two sets
$\cup$	union of two sets
$\varnothing$	null set
$T_N(x)$	N-th order Chebyshev polynomial in x
$|x|$	magnitude of a function or a number x
$\|x\|$	Euclidean norm or length of x
$\lfloor x \rfloor$	integer part of x
$\lceil x \rceil$	smallest integer larger than or equal to x
$\mathrm{Psgn}(x)$	Periodic signum function of x
$\mathrm{sgn}(x)$	signum function of x
$i|N$	$i = 0, 1, \ldots, N$
$i, j|N$	$i, j = 0, 1, \ldots, N$
$\forall$	for all values of
$-\!\!\bigcirc\!\!-$	element-by-element multiplication of two vectors
$\otimes$	Kronecker product
$\diamond$	replication operation
med	median operation
$O(N)$	on the order of N
$\ni$	such that
$\boxplus$	carry-complement addition

Continuous-time Functions, Laplace Transform and Fourier Transform

t	time variable in seconds
f	frequency variable in hertz (Hz)
s	complex frequency variable: $s = \sigma + j\omega$
$g(t)$	a function of time
$G(s)$	a Laplace transform, a function of s
$\delta(t)$	the impulse or delta function
$\mu(t)$	the unit step function
$g^{\#}(t)$	ideal sampled version of $g(t)$
$G^{\#}(s)$	Laplace transform of $g^{\#}(t)$
$\circledast$	the linear convolution operator
$H(j\Omega)$	frequency response
$\alpha(\Omega)$	$= 20 \log_{10}|H(j\Omega)|$, gain function in dB
$\phi(\Omega)$	$= \arg H(j\Omega)$, phase function
$\tau_g(\Omega)$	$= -(d[\arg H(\Omega)]/d\Omega)$, group delay
$\tau_p(\Omega)$	$= -(\arg H(\Omega)/\Omega)$, phase delay
Ω_p	passband cutoff analog frequency
Ω_s	stopband cutoff analog frequency
Ω_c	3-dB cutoff analog frequency
Ω_0	fundamental angular frequency
δ_p	passband ripple
δ_s	stopband ripple
$\mathbf{K}_*(z)$	matrix obtained from $\mathbf{K}(s)$ by conjugating only its coefficients
$\tilde{\mathbf{K}}(s)$	$= \mathbf{K}_*^t(-s)$

Discrete-time Functions, Sequences, Discrete-time Fourier Transforms, Discrete Fourier Transforms and z-transforms

$g[n]$	a discrete-time sequence
z	z-transform variable or the unit advance operator, $z = e^{sT} = e^{\sigma T + j\omega T}$
T	sampling period in seconds
$G(z)$	z-transform of the sequence $g[n]$
$G(e^{j\omega})$	discrete-time Fourier transform of the sequence $g[n]$
$X[k]$	discrete-Fourier transform coefficients of a sequence $x[n]$
$\tilde{g}[n]$	a periodic sequence
$\langle k \rangle_N$	$= k$ modulo N
W_N	$= e^{-j2\pi/N}$
$\delta[n]$	unit-sample sequence, unit-impulse sequence
$\mu[n]$	unit-step sequence, Heaviside step sequence
ω	$= \Omega T = \Omega/\Omega_s$, normalized digital angular frequency
$\circledast$	the convolution sum, also called aperiodic convolution
$\circledN$	the N-point circular convolution
$H(e^{j\omega})$	frequency response
$\alpha(\omega)$	$= 20 \log_{10}|H(e^{j\omega})|$, gain function in dB

$\phi(\omega)$	$= \arg H(e^{j\omega})$, phase function in radians
$\tau_g(\omega)$	$= -d[\arg H(e^{j\omega})]/d\omega$, group delay
$\tau_p(\omega)$	$= -\arg H(e^{j\omega})/\omega$, phase delay
ω_p	passband cutoff digital frequency
ω_s	stopband cutoff digital frequency
$\|S\|_p$	L_p-norm of $S(e^{j\omega})$
Δ	binary point
$\mathbf{K}_*(z)$	matrix obtained from $\mathbf{K}(z)$ by conjugating only its coefficients
$\tilde{\mathbf{K}}(z)$	$= \mathbf{K}_*^t(z^{-1})$
$\tilde{\mathbf{K}}(e^{j\omega})$	$= \mathbf{K}^\dagger(e^{j\omega})$

ABBREVIATIONS

1-D	one-dimensional (signal)
2-D	two-dimensional (signal)
AAF	anti-aliasing filter
ACU	address computation unit
ADC	analog-to-digital converter
A/D	analog-to-digital
AGU	address generation unit
AIB	analog interface board
AIC	Akaike information-theoretic criterion
ALC	adaptive linear combiner
ALU	arithmetic-logic unit
AM	amplitude modulation
ANSI	American National Standards Institute
AR	autoregressive
ARMA	autoregressive moving average
BFTF	block fast transversal filtering
BIBO	bounded-input–bounded-output
BIFORE	binary Fourier representation
BLMS	block least-mean square
BR	bounded real
CAT	criterion autoregressive transfer function
CAU	control arithmetic unit
CCITT	Comité Consultatif International Télégraphique et Téléphonique (The International Telegraph and Telephone Consultative Committee)
CD	compact disc
CFFT	complex-valued fast Fourier transform
CFM	common factor map
CIRC	cross-interleaved Reed–Solomon coding
cm	$= 10^{-2}$ m, centimeter

CM	connection matrix or coefficient memory
CMOS	complementary metal oxide semiconductor
CPU	central processing unit
CSD	canonical sign-digit
CRT	Chinese remainder theorem
CU	control unit
dB	decibel
DAC	digital-to-analog converter
D/A	digital-to-analog
DAU	data arithmetic unit
DCT	discrete cosine transform
DFE	decision feedback equalizer
DFT	discrete Fourier transform
DHT	discrete Hartley transform
DIF	decimation in frequency
DIT	decimation in time
DM	data memory
DMA	direct memory access
DNL	differential nonlinearity
DOA	direction of arrival
DSA	digital subtraction angiography
DSB–SC	double sideband–suppressed carrier
DSP	digital signal processor(ing)
DST	discrete sine transform
DTFT	discrete-time Fourier transform
ECG	electrocardiography
ECL	emitter-coupled-logic
EEG	electroencephalogram
EFB	error feedback
EFM	eight-to-fourteen modulation
EMSE	excess mean-square error
ESD	energy spectral density
ESS	error spectrum shaping
FDM	frequency division multiplexing
FFT	fast Fourier transform
FHT	fast Hartley transform
FIR	finite impulse response
FLOPS	floating-point operations per second
FM	frequency modulation
FMH	FIR-median hybrid
FNT	Fermat number transform
FPE	final prediction error

GaAs	Gallium-Arsenide
GDFT	generalized discrete Fourier transform
GF	Galois field
GHz	$= 10^9$ Hz, gigahertz
GPR	general purpose register
IC	integrated circuit
ID	instruction decoder
IDFT	inverse discrete Fourier transform
IF	intermediate frequency
i.i.d	independent identically disturbed
IIR	infinite impulse response
INL	integral nonlinearity
IM	instruction memory
I/O	input/output
kHz	$= 10^3$ Hz, kilohertz
KL	Karhunen–Loeve
km	$= 10^3$ m, kilometer
LBR	lossless bounded real
LMS	least-mean square
LPC	linear predictive coding
LPF	lowpass filter
LSB	least significant bit
LSI	large scale integrated
LTI	linear time-invariant
MA	moving average
MAC	multiply-accumulator
Mb/s	megabits per second
MBMSE	minimum block mean-square error
MDL	minimum description length
M-D	multidimensional (signal)
ME	maximum entropy
MESFET	metal semiconductor field effect transistor
MHz	$= 10^6$ Hz, megahertz
min	minute
MIMD	multiple instruction multiple data
MIPS	million instructions per second
MISD	multiple instruction single data
mm	$= 10^{-3}$ m, millimeter
MMSE	minimum mean-square error
MNT	Mersenne number transform
MOS	metal-oxide semiconductor

MOSFET	metal-oxide semiconductor field-effect transistor
MPP	massively parallel processor
MPOS	multiplies per output sample
MPR	McClellan–Parks–Rabiner (algorithm)
MPY	multiplier unit
ms	$= 10^{-3}$ sec, millisecond
MSB	most significant bit
MSI	medium-scale integrated (circuit)
Ms/s	megasamples per second
MU	memory unit
MUSIC	multiple signal classification
MUX	multiplexer
mW	$= 10^{-3}$ w, milliwatt
μs	$= 10^{-6}$ sec, microsecond
μV	$= 10^{-6}$ v, microvolt
NTT	number theoretic transform
NaN	Not a Number
NMR	nuclear magnetic resonance
ns	$= 10^{-9}$ sec, nanosecond
NTE	norm tap error
OSR	over-sampling ratio
PBF	positive Boolean function
PC	program counter
PCM	pulse-code modulation
PE	processing element
PFA	prime factor algorithm
PFM	prime factor map
P-I/O	parallel input–output
PISO	parallel-in serial-out
PM	program memory
PRML	partial response maximum likelihood
PS	program sequencer
ps	$= 10^{-12}$ sec, picosecond
PSD	power spectral density
QMF	quadrature mirror filter
QRNS	quadratic residue number system
RAM	random-access memory
RAR	return address register
RC	resistor-capacitor
REG	register
REM	rapid-eye-movement

RFFT	real-valued fast Fourier transform
RHS	right-hand side
RLC	resistor-inductor-capacitor
RLS	recursive least-squares
RNS	residue number system
ROC	region of convergence
ROM	read-only memory
rpm	revolutions per minute
rms	root-mean square
SAR	successive-approximation register
SC	switched-capacitor
S-M	sign-magnitude
SBM	serial-by-modulus
sec	second
SD	signed digit
SFG	signal flow graph
S/H	sample-and-hold
SIMD	single instruction multiple data
S-I/O	serial input–output
SIPO	serial-in parallel-out
SISD	single instruction single data
SR	shift register
SNR	signal-to-noise ratio
SVD	singular value decomposition
TDM	time-division multiplex
TTL	transistor-transistor logic
ULSI	ultra large scale integration
u.t.	unit of time
VLSI	very large scale integrated (circuit)
WFTA	Winograd–Fourier transform algorithm
WHT	Walsh–Hadamard transform
WSS	wide sense stationary

HANDBOOK FOR
DIGITAL SIGNAL
PROCESSING

1 INTRODUCTION

SANJIT K. MITRA

Department of Electrical and Computer Engineering
University of California, Santa Barbara

A signal is a physical quantity that is a function of one or more independent variables such as time, distance, temperature, or pressure. The variation of the signal amplitude as a function of the independent variable(s) is called its *waveform*. A signal can be naturally occurring or it can be synthetically generated or simulated. The signal can be a continuous or a discrete function of the independent variables. If the signal is a function of a single independent variable, it is called a one-dimensional (1-D) signal. If it is a function of two independent variables, then it is known as a two-dimensional (2-D) signal. A multidimensional (M-D) signal is a function of more than one variable. Examples of some typical signals are described in the following section.

Signals carry information and they need to be processed to extract completely or partially the information contained in them depending on the application of interest. In most practical cases, the signal of interest may have been modified or transformed intentionally or unintentionally. For example, for efficient utilization of a transmission channel, a number of signals are intentionally combined or multiplexed at the transmitting end and transmitted as a single signal over the channel. The multiplexing operation, an example of signal processing, is implemented in such a way that the individual signals can be removed or demultiplexed at the receiving end. The demultiplexing operation is another example of signal processing. A signal may be unintentionally corrupted by another undesirable or interfering signal called *noise* or *interference*. In many applications, the noise component of a corrupted signal needs to be removed or filtered before any other signal processing operation is carried out. The removal of the interfering signal from the signal of interest is a third example of signal processing.

Signal processing is concerned with the mathematical representation of signal in the domain of the original dependent variable(s), or in a transformed domain, and with the algorithmic manipulation of the signal to extract the information being carried.

Handbook for Digital Signal Processing, Edited by Sanjit K. Mitra and James F. Kaiser.
ISBN-0-471-61995-7 © 1993 John Wiley & Sons, Inc.

1-1   THE NEED FOR SIGNAL PROCESSING

To better understand the breadth of the signal processing task, we examine now a number of examples of some typical signals and their subsequent processing in typical applications.[1]

Electrocardiography (ECG) Signal. The electrical activity of the heart is represented by the ECG signal [SH81]. A typical ECG signal trace is shown in Figure 1-1(a). The ECG trace is essentially a periodic waveform. One such period of the ECG waveform as depicted in Figure 1-1(b) represents one cycle of the blood transfer process from the heart to the arteries. This part of the waveform is generated by an electrical impulse originating at the sinoatrial node in the atrium of the heart. The impulse causes the contraction of the atria, forcing the blood in each atrium to squeeze into its corresponding ventricle. The resulting signal is called the P-wave. The atrioventricular node delays the excitation impulse until the blood transfer from the atria to the ventricle is completed, resulting in the P-R interval of the ECG waveform. The excitation impulse then causes the contraction of the ventricle, squeezing the blood into the arteries, and generating the QRS part of the ECG waveform. During this phase the atria are relaxed and filled with blood. The T-wave of the waveform represents the relaxation of the ventricles. The complete process is repeated periodically, generating the ECG trace.

Each portion of the ECG waveform carries various types of information for the physician analyzing a patient's heart condition [SH81]. For example, the amplitude and timing of the P and QRS portions indicate the condition of the cardiac muscle mass. Loss of amplitude indicates muscle damage, whereas increased amplitude indicates abnormal heart rates. Too long a delay in the atrioventricular node is indicated by a very long P-R interval. Likewise, blockage of some or all of the contraction impulses is reflected by intermittent synchronization between P- and QRS-waves. Most of these abnormalities can be treated with various drugs and the effectiveness of the drugs can again be monitored by observing the new ECG waveforms taken after the drug treatment.

In practice, there are various types of externally produced artifacts that appear in the ECG signal [TO81]. Unless these interferences are removed, it is difficult for a physician to make a correct diagnosis. A common source of noise is the 60-Hz power lines, which radiate electric and magnetic fields. These fields are coupled to the ECG instrument through capacitive coupling and/or magnetic induction. Other sources of interference are the electromyographic signals, which are the potentials developed by contracting muscles. These and other interferences can be removed with careful shielding and signal processing techniques.

Electroencephalogram (EEG) Signal. The summation of the electrical activity caused by the random firing of billions of individual neurons in the brain is represented by the EEG signal [CO86; TO81]. In multiple EEG recordings, electrodes

---

[1]A more formal treatment of signals is given in Chapter 2.

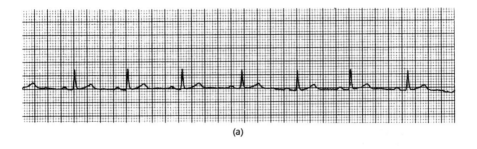

(a)

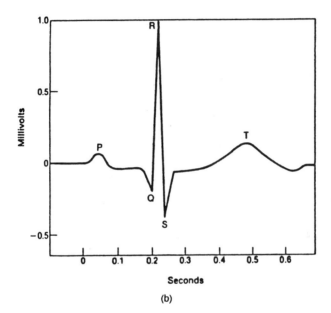

(b)

FIGURE 1-1   (a) A typical ECG trace and (b) one cycle of a ECG waveform.

are placed at various positions on the scalp with two common electrodes placed on the earlobes, and potential differences between the various electrodes are recorded. A typical bandwidth of this type of EEG ranges from 0.5 to about 100 Hz with the amplitudes ranging from 2 to 100 μV. An example of a multiple EEG trace is shown in Figure 1-2.

Both frequency-domain and time-domain analyses of the EEG signal have been used for the diagnosis of epilepsy, sleep disorders, psychiatric malfunctions, and so on. To this end, the EEG spectrum is subdivided into the following five bands: (1) the *delta* range occupying the band from 0.5 to 4 Hz, (2) the *theta* range occupying the band from 4 to 8 Hz, (3) the *alpha* range occupying the band from 8 to 13 Hz, (4) the *beta* range occupying the band from 13 to 22 Hz, and (5) the *gamma* range occupying the band from 22 to 30 Hz.

The delta wave is normal in the EEG signals of children and sleeping adults. As it is not common in alert adults, its presence indicates certain brain diseases.

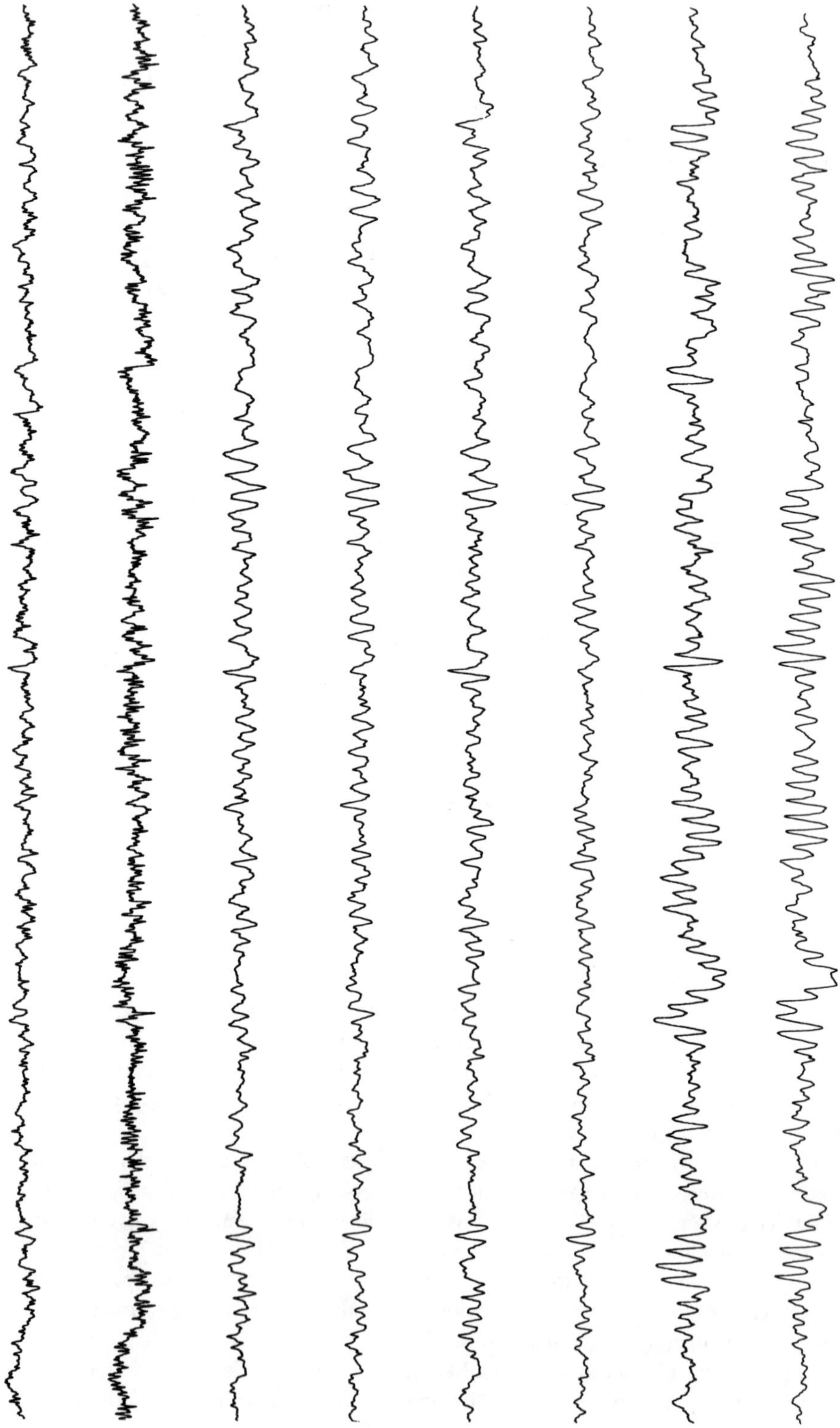

FIGURE 1-2 Multiple EEG signal traces.

The theta wave is usually found in children even though it has been observed in alert adults. The alpha wave is common in all normal humans and is more pronounced in a relaxed and awake subject with closed eyes. Likewise the beta activity is common in normal adults. The EEG exhibits rapid, low-voltage waves, called *rapid-eye-movement* (REM) waves, in a subject dreaming during sleep. Otherwise, in a sleeping subject, the EEG contains bursts of alpha-like waves, called *sleep spindles*. The EEG of an epileptic patient exhibits various types of abnormalities depending on the type of epilepsy, which is caused by uncontrolled neural discharges.

Diesel Engine Signal. Signal processing is playing an important role in the precision adjustment of diesel engines during production [JU81]. Efficient operation of the engine requires the accurate determination of the topmost point of piston travel (called the *top dead center*) inside the cylinder of the engine. Figure 1-3 shows the signals generated by a dual probe inserted into the combustion chamber

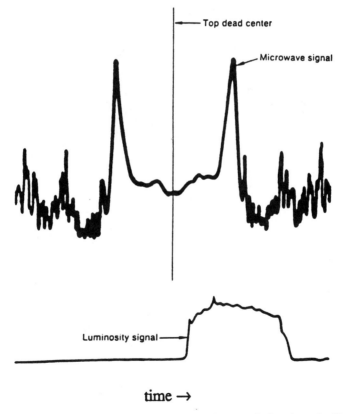

time →

FIGURE 1-3 Diesel engine signals. [Reproduced with permission from R. K. Jurgen, Detroit bets on electronics to stymie Japan. *IEEE Spectrum* **18,** 29–32, (1981) © 1981. IEEE.]

of a diesel engine in place of the glow plug. The probe consists of a microwave antenna and a photodiode detector. The microwave probe captures signals reflected from the cylinder cavity caused by the up and down motion of the piston while the engine is running. Interestingly, the waveforms of these signals exhibit a symmetry around the top dead center independent of the engine speed, temperature, cylinder pressure, or air/fuel ratio. The point of symmetry is determined automatically by a microcomputer and the fuel-injection pump position then adjusted by the computer accurately to within ±0.5° using the luminosity signal sensed by the photodiode detector.

Speech Signal. The linear acoustic theory of speech production has led to mathematical models for the representation of speech signals. A speech signal is formed by exciting the vocal tract and is composed of two types of sounds: *voiced* and *unvoiced* [RA78; SL83]. The voiced sound, which includes, the vowels, and a number of consonants such as B, D, L, M, N, and R, is excited by the pulsatile airflow resulting from the vibration of the vocal folds. On the other hand, the unvoiced sound is produced downstream in the forward part of the oral cavity (mouth) with the vocal cords at rest and includes sounds like F, S, and SH.

Figure 1-4(a) depicts the speech waveform of a male utterance "The boy was mute about his task" [SL83]. The total duration of the speech waveform is about 2.5 s. Expanded versions of the "A" and "S" segments in the word "task" are sketched in Figure 1-4(b) and 1-4(c), respectively. The slowly varying low-frequency voiced waveform of "A" and the high-frequency unvoiced fricative waveform of "S" are evident from the magnified waveforms. The voiced waveform in Figure 1-4(b) is seen to be quasiperiodic and can be modeled by a sum of a finite number of sinusoids. The lowest frequency of oscillation in this representation is called the *pitch frequency*. The unvoiced waveform in Figure 1-4(c) has no regular fine structure and is more noise-like.

One of the major applications of digital signal processing techniques is in the general area of speech processing. Problems in this area are usually divided into three groups: (1) *speech analysis*, (2) *speech synthesis*, and (3) *speech analysis and synthesis* [OP78]. Digital speech analysis methods are used in automatic speech recognition, speaker verification, and speaker identification. Applications of digital speech synthesis techniques include reading machines for the automatic conversion of written text into speech, and retrieval of data from computers in speech form by remote access through terminals or telephones. One example belonging to the third group is scrambling for secure voice transmission. Speech data compression for an efficient use of the transmission medium is another example of the use of speech analysis followed by synthesis. A typical speech signal after conversion into a digital form contains about 64,000 bits per second (bps). Depending on the desired quality of the synthesized speech, the original data can be compressed considerably, for example, down to about 1000 bps.

Musical Sounds. The electronic synthesizer is an example of the use of modern signal processing techniques [LE83; MO77]. The natural sound generated by most musical instruments is generally produced by mechanical vibrations caused by ac-

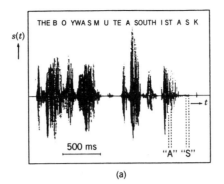

(a)

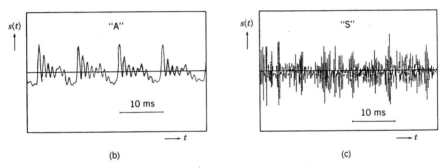

(b)   (c)

FIGURE 1-4 Speech waveform example: (a) a sentence-length segment, (b) expanded version of the voiced segment (the letter A), and (c) expanded version of the unvoiced segment (the letter S). [Reproduced with permission from R. J. Sluyter, Digitization of speech. *Philips Tech. Rev.*, **42**, (7/), 201–223, (1983–1984). © Philips Research Laboratories.]

tivating some form of a mechanical oscillator, which then causes other parts of the instrument to vibrate. All these vibrations in a single instrument together generate the musical sound. For example, in the case of a violin, the primary oscillator is a stretched piece of string (cat gut). Its movement is caused by drawing a bow across; this sets the wooden body of the violin vibrating, which in turn sets up vibrations of the air inside as well as outside the instrument. Likewise, in the case of a piano, the primary oscillator is a stretched steel wire that is set into vibratory motion by the hitting of a hammer, which in turn causes vibrations in the wooden body (sounding board) of the piano. In wind or brass instruments, the vibration occurs in a column of air and a mechanical change in the length of the air column by means of valves or keys regulates the rate of vibration.

The sound of orchestral instruments can be classified into two groups: *quasiperiodic* and *aperiodic*. The quasiperiodic sounds can be described by a sum of a finite number of sinusoids with independently varying amplitudes and frequencies. The sound waveforms of two different instruments, the cello and the bass drum, are indicated in Figure 1-5(a) and 1-5(b), respectively. In each figure, the top waveform is the plot of an entire isolated note, whereas the bottom plot shows an

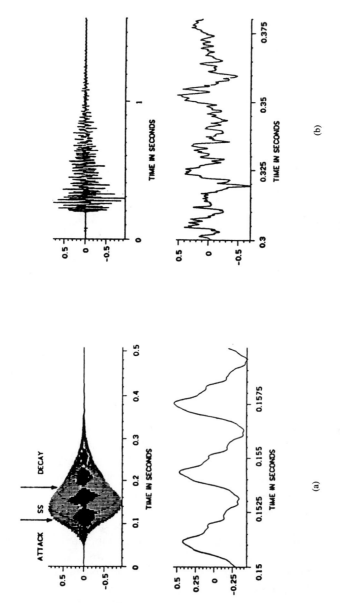

FIGURE 1-5 Waveforms of (a) the cello and (b) the bass drum. [Reproduced with permission from J. A. Moorer, Signal processing aspects of computer music: A survey. *Proc. IEEE* **65**, 1108–1137 (1977). © 1977 IEEE.]

expanded version of a portion of the note, 10 ms of the cello and 80 ms for the bass drum. The waveform of the note from a cello is seen to be quasiperiodic. On the other hand, the bass drum waveform is clearly aperiodic. The tone of an orchestral instrument is commonly divided into three segments called the *attack* part, the *steady-state* part, and the *decay* part. Figure 1-5 illustrates this division for the two tones. Note that the bass drum tone of Figure 1-5(b) shows no steady-state part. A reasonable approximation of many tones is obtained by splicing together these parts. However, high-fidelity reproduction requires a more complex model.

Time Series. The signals described thus far are continuous functions of time as the independent variable. In many cases the signals of interest are naturally discrete functions of the independent variables. Often such signals are of finite duration. Examples of such signals are the yearly average of numbers of sunspots, daily stock prices, the value of total monthly exports of a country, the yearly population of animal species in a certain geographic area, the annual yields per acre of crops in a country, and the monthly totals of international airline passenger, over certain periods. This type of finite extent signals, usually called *time series*, occurs in business, economics, physical sciences, social sciences, engineering, medicine, and many other fields. Plots of some typical time series are shown in Figure 1-6 to 1-8.

There are many reasons for analyzing a particular time series [BO70]. In some applications, there may be a need to develop a model to determine the nature of the dependence of the data on the independent variable and use it to forecast the future behavior of the series. As an example, in business planning reasonably accurate sales forecasts are necessary. Some types of series possess seasonal or periodic components, and it is important to extract these components. The study of sunspot numbers is important for predicting climate variations, for example. Invariably, the time series data are noisy, and their representations require nonstationary models.

Seismic Signals. Reflection seismology methods are usually employed in the geophysical exploration for oil and gas [RO80]. Here regularly placed linear arrays

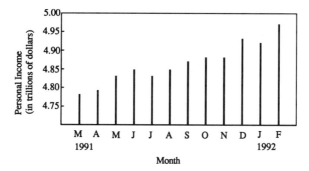

FIGURE 1-6 Seasonally adjusted annual rate of personal income in the United States from March 1991 to February 1992 (From Commerce Department).

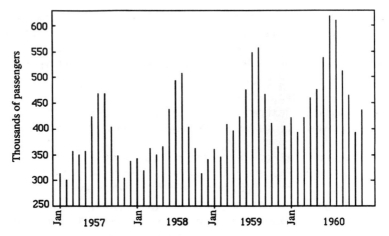

FIGURE 1-7 Totals of international passengers in thousands per month from January 1957 to December 1960. [Adapted from G.E.P. Box and G. M. Jenkins, *Time Series Analysis: Forecasting and Control*. Holden-Day, San Francisco, CA, 1970.]

of seismic sources, such as high-energy explosives, on the ground surface cause seismic waves to propagate through subsurface geological structures. These waves reflect back to the surface from interfaces between geological strata and are picked up by a composite array of geophones laid out in certain patterns. The geophone signals are displayed as a two-dimensional signal that is a function of time and space, called a *trace gather*, as indicated in Figure 1-9. Before these signals are

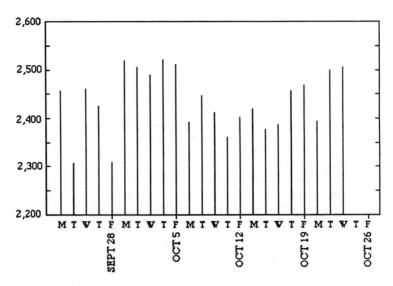

FIGURE 1-8 Daily Dow Jones industrial index from September 24, 1990 to October 24, 1990.

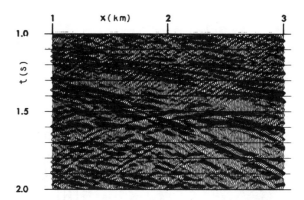

FIGURE 1-9 A typical seismic signal trace gather. [Reprint with permission from V. Barden, Sampling 3-D seismic data. *Int. J. Imag. Syst. Technol.* **1**, 73–77 (1989). © 1989 John Wiley & Sons, Inc.]

analyzed, some preliminary time and amplitude corrections are made on the data to compensate for different physical phenomena. From the corrected data, the time differences between reflected seismic signals are used to map structural deformations, whereas the amplitude changes usually indicate the presence of hydrocarbons. Reflection seismology methods are also used in other applications, such as detection of underground nuclear explosions and earthquake monitoring.

Images. Images in general are two- or three-dimensional signals. Examples of the former are photographs, still video images, radar and sonar images, and chest and dental x-rays. In each case, the image intensity at any point is a function of two spatial variables. For image sequences such as that seen in a television, the image seen at a certain instant, called a *frame*, is essentially a three-dimensional signal for which the image intensity at any point is a function of three variables, the two spatial variables and the time. Figure 1-10(a) shows a photograph of a digital image.

The basic problems in image processing are image signal representation and modeling, enhancement, restoration, reconstruction from projections, analysis, and coding [JA89].

Each picture element in a specific image represents a certain physical quantity; a characterization of the element is called the *image representation*. For example, a photograph represents the luminances of various objects as seen by the camera. An infrared image taken by a satellite or airplane represents the temperature profile of a geographical area. Depending on the type of image and its applications, various types of image models are usually defined. Such models are also based on perception, and local or global characteristics. The nature and the performance of the image processing algorithms depend on the image model being used.

Image enhancement algorithms are used to emphasize specific image features to improve the quality of the image for visual perception or to aid in the analysis of the image for feature extraction. These include methods for contrast enhance-

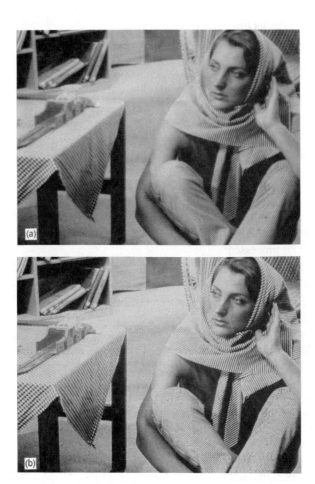

FIGURE 1-10 (a) A digital image and (b) its contrast-enhanced version.

ment, edge detection, sharpening, linear and nonlinear filtering, zooming, and noise removal. Figure 1-10(b) shows the contrast-enhanced version of the image of Figure 1-10(a) developed using a nonlinear filter [MI91].

The algorithms used for the elimination or reduction of degradations in an image, such as blurring and geometric distortion, caused by the imaging system and/or its surroundings are known as *image restoration*. *Image reconstruction* from projections involves the development of a two-dimensional image slice of a three-dimensional object from a number of planar projections obtained from various angles. By creating a number of contiguous slices, a three-dimensional image giving an inside view of the object is developed.

Image analysis methods are employed to develop a quantitative description and classification of one or more desired objects in an image. For digital processing, an image needs to be sampled and quantized using an A/D converter. A reasonable size digital image in its original form takes considerable amount of memory space for storage. For example, an image of size 512×512 samples with 8-bit resolution per sample contains over 2 million bits. *Image coding* methods are used to reduce the total number of bits in an image without any degradation is visual perception quality as in speech coding, for example, down to about 1 bit per sample on the average.

1-2 WHY DIGITAL SIGNAL PROCESSING?

Most naturally occurring signals are continuous functions of the independent variable, which is usually the time, and are commonly called *analog signals*. Digital processing of an analog signal consists basically of three steps: conversion of the analog signal into a digital form, processing of the digital version, and finally conversion of the processed digital signal back into an analog form. Figure 1-11 shows the overall scheme in a block diagram form.

As the amplitude of the analog input signal varies with time, a *sample-and-hold* (S/H) circuit is used first to sample the analog input at periodic intervals and hold the sampled value constant at the input of the *analog-to-digital* (A/D) *converter* to permit accurate digital conversion. The input to the A/D converter is a staircase type analog signal if the S/H circuit holds the sampled value till the next sampling instant. The output of the A/D converter is a binary data stream, which is next processed by the digital processor implementing the desired signal processing algorithm. The output of the digital processor, another binary data stream, is then converted into a staircase type analog signal by the *digital-to-analog* (D/A) *converter*. The lowpass filter at the output of the D/A converter then removes all undesired high-frequency components and delivers at its output the desired processed analog signal. Figure 1-12 illustrates the waveforms of the pertinent signals at various stages in the above process where for clarity for the binary signals the two levels are shown as a positive and a negative pulse, respectively.

In contrast to the above, conceptually, a direct analog processing of an analog signal is much simpler because it involves only a single processor as illustrated in Figure 1-13. It is therefore natural to ask what the advantages are of digital processing of an analog signal.

There are of course many advantages for choosing digital signal processing. The most important ones are as follows [BE84; PR88]:

Unlike analog circuits, the operation of digital circuits does not depend on precise values of the digital signals. As a result, a digital circuit is less sensitive to tolerances of component values and is fairly independent of temperature, aging,

FIGURE 1-11 Scheme for the digital processing of an analog signal.

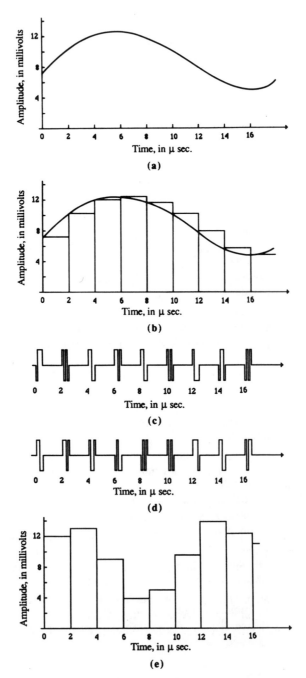

FIGURE 1-12 Typical waveforms of signals appearing at various stages in Figure 1-3: (a) analog input signal, (b) output of the S/H circuit, (c) A/D converter output, (d) output of the digital processor, (e) D/A converter output, and (f) analog output signal. In (c) and (d), the digital HIGH and LOW levels are shown as positive and negative pulses for clarity.

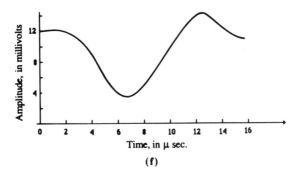

(f)

FIGURE 1-12 (*Continued*)

and most other external parameters. A digital circuit can be reproduced easily in volume quantities and does not require any adjustments either during construction or later while in use. Moreover, it is amenable to full integration and with the recent advances in *very large scale integrated* (VLSI) circuits, it has been possible to integrate highly sophisticated and complex digital signal processing systems on a single chip.

In a digital processor, the signals and the coefficients describing the processing operation are represented as binary words. Thus any desirable accuracy can be achieved by simply increasing the wordlength subject to cost limitation. Moreover, the dynamic ranges for signals and coefficients can be increased still further by using floating-point arithmetic if necessary.

Digital processing allows the sharing of a given processor among a number of signals by time-sharing, thus reducing the cost of processing per signal. Figure 1-14 illustrates the concept of time-sharing, where two digital signals are combined into one by time-division multiplexing. The multiplexed signal can then be fed into a single processor. By switching the processor coefficients prior to the arrival of each signal at the input of the processor, the processor can be made to look like two different systems. Finally, by demultiplexing the output of the processor, the processed signals can be separated.

Digital implementation permits easy adjustment of processor characteristics during processing such as that needed in implementing adaptive filter. Such adjustments can simply be carried out by changing periodically the coefficients of the algorithm representing the processor characteristics. Another application of changing of the coefficients is in the realization of systems with programmable characteristics such as frequency selective filters with adjustable cutoff frequencies. Filter banks with guaranteed complementary frequency response characteristics are easily implemented in digital form.

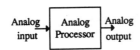

FIGURE 1-13 Analog processing of analog signals.

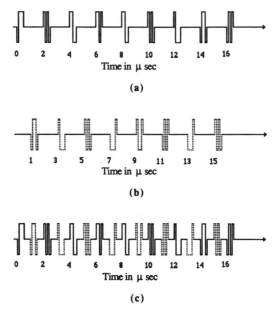

FIGURE 1-14 Illustration of the time-sharing concept. The signal shown in (c) has been obtained by time-multiplexing the signals shown in (a) and (b).

Digital implementation allows the realization of certain characteristics not possible with analog implementation, such as exact linear phase and multirate processing. Digital circuits can be cascaded without any loading problems, unlike analog circuits. Digital signals can be stored almost indefinitely without any loss of information on various storage mediums such as magnetic tapes and disks or optical disks. Such stored signals can later be processed off-line such as in the compact disc player, the digital video disc player, the digital audio tape player, or simply by using a general purpose computer as in seismic data processing.

Another advantage is the applicability of digital processing to very low frequency signals, such as those occurring in seismic applications, where inductors and capacitors needed for analog processing would be physically very large in size.

Digital signal processing is also associated with some disadvantages. One obvious disadvantage is the increased system complexity in the digital processing of analog signals because of the need for additional pre- and post-processing devices such as the A/D and D/A converters, and their associated filters.

The second, but most significant, disadvantage is the available frequency range of operation of a digital signal processor, which is primarily determined by the sample-and-hold (S/H) circuit and the A/D converter, and as result is limited by the state-of-the-art of the technology. As shown later, in general, an analog continuous-time signal must be sampled at a frequency that is at least twice the highest frequency component present in the signal. If this condition is not satisfied then signal components with frequencies above half the sampling frequency appear as signal components below this particular frequency, distorting totally the input analog signal waveform. To avoid this problem, often in many practical applications, the analog signal is passed through an analog lowpass filter with a cutoff frequency less than half of the sampling frequency. The highest sampling frequency reported in the literature presently is around 1 GHz [PO87]. Such high sampling frequencies are not usually employed in practice because the achievable resolution of the A/D converter, given by the wordlength of the digital equivalent of the analog sample, decreases with an increase in the speed of the converter. For example, the reported resolution of an A/D converter operating at 1 GHz is 6 bits [PO87]. On the other hand, in most applications, the required resolution of an A/D converter is from 12 bits to around 16 bits. Consequently, a sampling frequency of at most 10 MHz is presently a practical upper limit. This upper limit, however, is getting larger and larger with advances in the technology.

The third disadvantage stems from the fact that digital systems are constructed using active devices that consume electrical power. For example, the AT&T DSP32C Digital Signal Processor chip contains over 405,000 transistors and dissipates around 1 watt. On the other hand, a variety of analog processing algorithms can be implemented using passive circuits employing inductors, capacitors, and resistors that do not need any power. Moreover, active devices are less reliable than passive components.

However, advantages far outweigh the disadvantages in various applications, and with the continuing decrease in the cost of digital processor hardware, applications of digital signal processing are increasing rapidly.

1.3 APPLICATIONS OF SIGNAL PROCESSING

There are numerous applications of signal processing that we encounter often in our daily life without being aware of them. Due to space limitations, it is not possible to discuss each of these applications. We provide below an overview of a selected few of such applications.

1-3-1 Sound Recording Applications

Today the recording of most musical programs is usually made in an acoustically inert studio. The sound from each instrument is picked up by its own microphone closely placed to the instrument and recorded on a single track in a multitrack tape recorder with as many as 48 tracks. The signals from individual tracks in the master recording are then edited and combined by the sound engineer in a *mix-down* system to develop a two-track stereo recording. There are various reasons for following this approach. First, the closeness of the individual microphones to its assigned instruments provides a high degree of separation between the instruments and minimizes the background noise in the recording. Second, the sound part of one instrument can be rerecorded later if necessary. Third, during the mix-down process the sound engineer can manipulate individual signals using various signal processing devices to alter the musical balances between the sounds generated by the instruments, change the timbre, and add natural room acoustics effects and other special effects [BL78; EA76].

Various types of signal processing techniques are utilized in the mix-down phase. Some of these techniques are used to modify the spectral characteristics of the sound signal and add special effects, whereas others are used to improve the quality of the transmission medium. The signal processing circuits most commonly used are as follows: (1) compressors and limiters, (2) expanders and noise gates, (3) equalizers and filters, (4) noise reduction systems, (5) delay and reverberation systems, and (6) circuits for special effects [BL78; EA76; HU89; WO89]. These operations are usually performed on the original analog audio signals and are implemented using analog circuit components. However, there is a growing trend toward all digital implementation and their use in the processing of the digitized versions of the analog audio signals [BL78; PE86].

Compressors and Limiters. These devices are used for the compression of the dynamic range of an audio signal. The compressor can be considered as an amplifier with two gain levels: the gain is unity for input signal levels below a certain threshold, and less then unity for signals with levels above the threshold. The threshold level is adjustable over a wide range of the input signal. Figure 1-15 shows the transfer characteristic of a typical compressor.

The parameters characterizing a compressor are its compression ratio, threshold level, attack time, and release time, which are illustrated in Figure 1-16.

When the input signal level suddenly rises above a prescribed threshold, the time taken by the compressor to adjust its normal unity gain to the lower value is called the *attack time*. Because of this effect, the output signal exhibits a slight degree of overshoot before the desired output level is reached. A zero attack time is desirable to protect the system from sudden high-level transients. However, in this case, the impact of sharp musical attacks is eliminated, resulting in a dull "lifeless" sound [WO89]. A longer attack time causes the output to sound more percussive than normal.

Similarly, the time taken by the compressor to reach its normal unity gain value when the input level suddenly drops below the threshold is called the *release time* or *recovery time*. If the input signal fluctuates rapidly around the threshold in a small region, the compressor gain also fluctuates up and down. In such a situation,

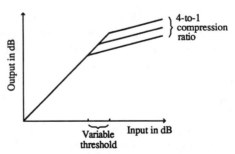

FIGURE 1-15   Transfer characteristic of a typical compressor.

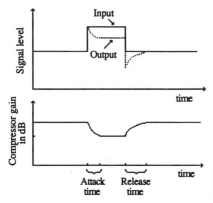

FIGURE 1-16 Parameters characterizing a typical compressor.

the rise and fall of background noise result in an audible effect called *breathing* or *pumping*, which can be minimized with a longer release time for the compressor gain.

There are various applications of the compressor unit in musical recording [EA76]. For example, it can be used to eliminate the variations in the peaks of an electric bass output signal by clamping them to a constant level, thus providing an even and solid bass line. To maintain the original character of the instrument it is necessary to use a compressor with a long recovery time compared to the natural decay rate of the electric bass. The device is also useful to compensate for the wide variations in the signal level produced by a singer who moves frequently, changing the distance from the microphone.

A compressor with a compression ratio of 10-to-1 or greater is called a *limiter* because its output levels are essentially clamped to the threshold level. The limiter is used to prevent overloading of amplifiers and other devices caused by signal peaks exceeding certain levels.

Expanders and Noise Gates. The expander's function is opposite to that of the compressor. It is also an amplifier with two gain levels: the gain is unity for input signal levels above a certain threshold, and less than unity for signals with levels below the threshold. The threshold level is again adjustable over a wide range of the input signal. Figure 1-17 shows the transfer characteristic of a typical expander. The expander is used to expand the dynamic range of an audio signal by

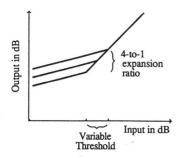

FIGURE 1-17 Transfer characteristic of a typical expander.

boosting the high-level signals and attenuating the low-level signals. The device can also be used to reduce noise below a threshold level.

The expander is characterized by its expansion ratio, threshold level, attack time, and release time. Here, the time taken by the device to reach the normal unity gain for a sudden change in the input signal to a level below the threshold is defined as the *attack time*. Likewise, the time required by the device to lower the gain from its normal value of one for a sudden increase in the input signal level is called the *release time*.

The *noise gate* is a special type of expander that attenuates heavily signals with levels below the threshold. It is used, for example, to totally cut off a microphone during a musical pause so as not to pass the noise being picked up by the microphone.

Equalizers and Filters. Various types of filters are used to modify the frequency response of a recording or the monitoring channel. One such filter, called the *shelving filter*, provides a boost (rise) or cut (drop) in the frequency response either at the low or at the high end of the audio frequency range while not affecting the frequency response in the remaining range of the audio spectrum as shown in Figure 1-18. *Peaking filters* are used for midband equalization and are designed to have either a bandpass response to provide a boost or a bandstop response to provide a cut as indicated in Figure 1-19.

The parameters characterizing a low-frequency shelving filter are the two frequencies f_{1L} and f_{2L}, where the magnitude response begins tapering up or down from a constant level, and the low-frequency gain levels in dB. Likewise, the parameters characterizing a high-frequency shelving filter are the two frequencies f_{1H} and f_{2H}, where the magnitude response begins tapering up or down from a constant level, and the high-frequency gain levels in dB. In the case of a peaking filter, the parameters of interest are the center frequency f_0, the 3-dB bandwidth Δf of the bell-shaped curve, and the gain level at the center frequency. Most often, the quality factor $Q = f_0/\Delta f$ is used to characterize the shape of the frequency response instead of the bandwidth Δf.

A typical equalizer consists of cascade of a low-frequency shelving filter, a high-frequency shelving filter, and three or more peaking filters with adjustable parameters to provide adjustment of the overall equalizer frequency response over a board range of frequencies in the audio spectrum. In a *parametric equalizer*, each individual parameter of its constituent filter blocks can be varied independently without affecting the parameters of the other filters in the equalizer.

The *graphic equalizer* consists of a cascade of peaking filters with fixed center frequencies but adjustable gain levels that are controlled by vertical slides in the front panel. The physical position of the slides reasonably approximates the overall equalizer magnitude response as shown schematically in Figure 1-20.

Other types of filters that also find applications in the musical recording and transfer processes are the *lowpass, highpass*, and *notch filters*. Their corresponding frequency responses are indicated in Figure 1-21. The notch filter is designed to attenuate a particular frequency component and has a narrow width so as not to affect the rest of the musical program.

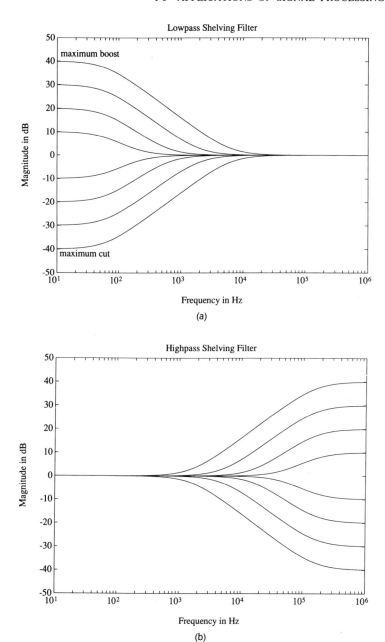

FIGURE 1-18 Frequency responses of (a) low-frequency shelving filter and (b) high-frequency shelving filter.

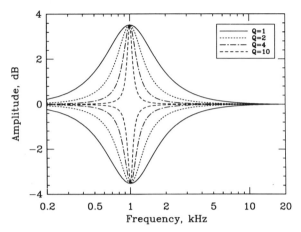

FIGURE 1-19 Peaking filter frequency response.

Two major applications of equalizers and filters in recording are to correct certain types of problems that may have occurred during the recording or the transfer process, and to alter the harmonic or timbral contents of a recorded sound purely for musical or creative purposes [EA76]. For example, a direct transfer of a musical recording from old 78-RPM disks to a wideband playback system will be highly noisy due to the limited bandwidth of the old disks. To reduce this noise, a bandpass filter with a passband matching the bandwidth of the old records is utilized. Often, older recordings are made more pleasing by adding a broad high-frequency peak in the 5–10-kHz range and by shelving out some of the lower

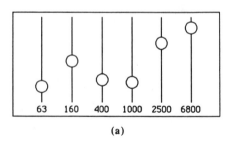

(a)

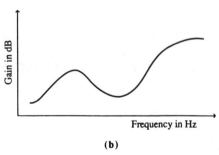

(b)

FIGURE 1-20 Graphic equalizer: (a) control panel settings and (b) corresponding frequency response. [Adapted from J. Eargle, *Sound Recording*. Van Nostrand-Reinhold, New York, 1976.]

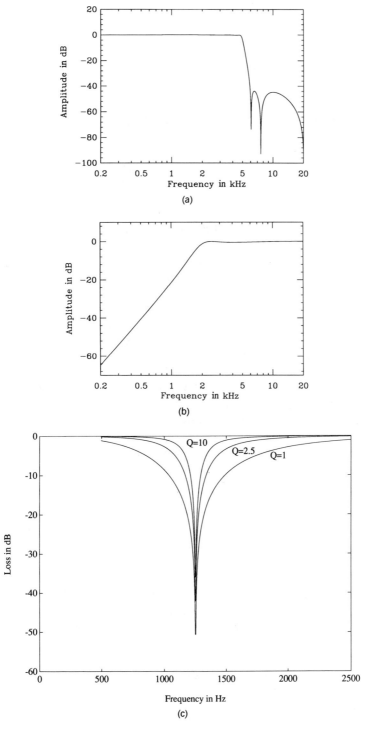

FIGURE 1-21 Frequency responses of other types of filters: (a) lowpass filter, and (b) highpass filter and (c) notch filter.

frequencies. The notch filter is particularly useful in removing 60-Hz power supply hum.

In creating a program by mixing down a multichannel recording, equalization of individual tracks is usually employed by the recording engineer for creative reasons [EA76]. For example, a "fullness" effect can be added to weak instruments such as the acoustical guitar by boosting frequency components in the 100–300-Hz range. Similarly, by boosting the 2–4-kHz range, the transients caused by the fingers against the string of an acoustic guitar can be made more pronounced. A high-frequency shelving boost above the 1–2-kHz range increases the "crispness" in percussion instruments such as the bongo or snare drums.

Noise Reduction System.   The overall dynamic range of human hearing is over 120 dB. However, most recording and transmission media have a much smaller dynamic range. The music to be recorded must be above the sound background or noise. If the background noise is around 30 dB, the dynamic range available for the music is only 90 dB, requiring a dynamic range compression for noise reduction.

A noise reduction system consists of two parts. The first part provides the compression during the recording mode while the second part provides the complementary expansion during the playback mode. To this end, the most popular methods in musical recording are the Dolby noise reduction schemes, of which there are several types [EA76; HU89; WO89].

In the Dolby A-type method, used widely in professional recording, for the recording mode, the audio signal is split into four frequency bands by a bank of four filters; separate compression is provided in each band and the outputs of the compressor are combined, as indicated in Figure 1-22. Moreover, the compression in each band is restricted to a 20-dB input range from -40 to -20 dB. Below the lower threshold (-40 dB), very low-level signals are boosted by 10 dB, and above the upper threshold (-20 dB), the system has unity gain, passing the high-level signals unaffected. The transfer characteristic for the record mode is thus as shown in Figure 1-23.

In the playback mode, the scheme is essentially the same as that is the recording mode, except here the compressors are replaced by expanders with complementary transfer characteristics as indicated in Figure 1-23. Here, the expansion is limited to a 10-dB input range from -30 to -20 dB. Above the upper threshold (-20 dB), very high-level signals are cut by 10 dB, while below the lower threshold (-30 dB), the system has unity gain, passing the low-level signals unaffected.

Note that for each band, a 2-to-1 compression is followed by a 1-to-2 complementary expansion such that the dynamic range of the signal at the input of the compressor is exactly equal to that at the expander output. This type of overall signal processing operation is often called *companding*. Moreover, the companding operation in one band has no effect on the signal in another band and may often be masked by other bands with no companding.

Delay and Reverberation Systems.   Music generated in an inert studio does not sound natural compared to the music performed inside a room, such as a concert

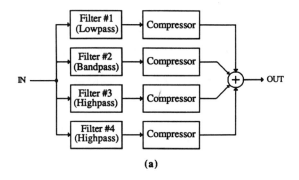

(a)

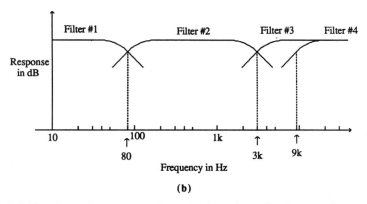

(b)

FIGURE 1-22 The Dolby A-Type noise reduction scheme for the recording mode: (a) block diagram, and (b) frequency responses of the four filters with cutoff frequencies as shown.

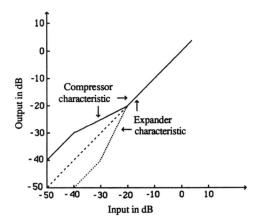

FIGURE 1-23 Compressor and expander transfer characteristic for the Dolby A-type noise reduction scheme.

hall. In the latter case, the sound waves propagate in all directions, and reach the listener from various directions and at various times depending on the distance traveled by the sound from the source to the listener. The sound wave coming directly to the listener, called the *direct sound*, reaches first and determines the listener's perception of the location, size, and nature of the sound source. This is followed by a few closely spaced echoes, called *early reflections*, generated by reflections of sound waves from all sides of the room and reaching the listener at irregular times. These echoes provide the listener subconscious cues as to the size of the room. After these early reflections, more and more densely packed echoes reach the listener due to multiple reflections. The latter group of echoes is referred to as the *reverberation*. The amplitude of the echoes decays exponentially with time as a result of attention at each reflection. Figure 1-24 illustrates this concept. The period of time in which the reverberation falls by 60 dB is called the *reverberation time*. As the absorption characteristics of different materials are not the same at different frequencies, the reverberation time varies from frequency to frequency.

Delay systems with adjustable delay factors are employed to create artificially the early reflections. Electronically generated reverberation combined with the artificial echo reflections is usually added to the recordings made in a studio. The block diagram representation of a typical delay–reverberation system in a monophonic system is depicted in Figure 1-25.

There are various other applications of the electronic delay systems, some of which are described next.

Special Effects. By feeding in the same sound signal through adjustable delay and gain control as indicated in Figure 1-26, it is possible to vary the location of the sound source from the left speaker to the right for a listener located on the plane of symmetry [EA76]. For example, in Figure 1-26, a 0-dB loss in the left channel and a few milliseconds delay in the right channel give the impression of a localization of the sound source at the left. However, lowering of the left-channel signal

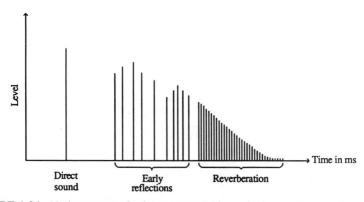

FIGURE 1-24 Various types of echoes generated by a single sound source in a room.

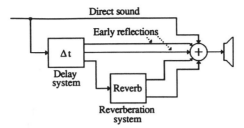

FIGURE 1-25 Block diagram of a complete delay–reverberation system in a monophonic system.

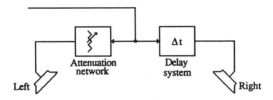

FIGURE 1-26 Localization of sound source using delay systems and attenuation network.

level by a few dB loss results in a phantom image of the sound source moving toward the center. This scheme can be further extended to provide a degree of *image broadening* by phase shifting one channel with respect to the other through allpass networks as shown in Figure 1-27 [EA76].

Another application of the delay–reverberation system is in the processing of a single track into a pseudostereo format while simulating a natural acoustical environment as illustrated in Figure 1-28 [EA76].

The delay system can also be used to generate a chorus effect from the sound of a soloist. The basic scheme used is illustrated in Figure 1-29. Each of the delay units has a variable delay controlled by a low-frequency pseudorandom noise source to provide a random pitch variation [BL78].

It should be pointed out here that additional signal processing is employed to make the stereo submaster developed by the sound engineer more suitable for the record-cutting lathe or the cassette tape duplicator.

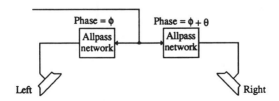

FIGURE 1-27 Image broadening using allpass networks.

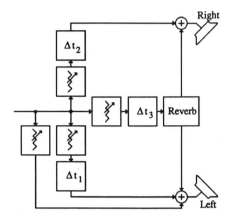

FIGURE 1-28 A possible application of delay systems and reverberation in a stereophonic system.

1-3-2 Telephone Dialing Applications

Signal processing plays a key role in the detection and generation of signaling tones for the push-button telephone dialing [DA76]. In telephones equipped with TOUCH-TONE® dialing, the pressing of each button generates a unique set of two-tone signals, which are processed at the telephone central office to identify the number pressed by determining the two associated tone frequencies. Seven frequencies are used to code the 10 decimal digits and the two special buttons marked with * and #. The lowband frequencies are 697, 770, 852, and 941 Hz. The remaining three frequencies belonging to the highband are 1209, 1336, and 1477 Hz. The fourth highband frequency of 1633 Hz is not presently in use and has been assigned for future applications to permit the use of additional four push-buttons for special services. The frequency assignment used in the TOUCH-TONE® dialing are as shown in Figure 1-30.

The scheme used to identify the two frequencies associated with the button that has been pressed is shown in Figure 1-31. Here, the two tones are first separated

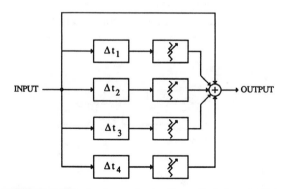

FIGURE 1-29 A scheme for implementing chorus effect.

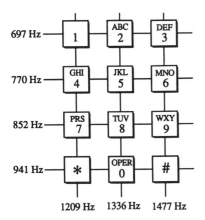

697 Hz
770 Hz
852 Hz
941 Hz

1209 Hz   1336 Hz   1477 Hz

FIGURE 1-30   The tone frequency assignments for TOUCH-TONE® dialing.

by a lowpass and a highpass filter. The passband cutoff frequency of the lowpass filter is slightly above 1000 Hz, whereas that of the highpass filter is slightly below 1200 Hz. The output of each filter is next converted into a square wave by a limiter and then processed by a bank of bandpass filters with a narrow passband. The four bandpass filters in the low-frequency channel have center frequencies at 697, 770, 852, and 941 Hz, respectively. The three bandpass filters in the high-frequency channel have center frequencies at 1209, 1336, and 1477 Hz, respectively. The detector following the bandpass filter develops the necessary dc switching signal if its input voltage is above a certain threshold.

All the signal processing functions described above are invariably implemented in practice in the analog domain. However, the possibility of implementing these functions using digital techniques and digital signal processing has been demonstrated [JA68].

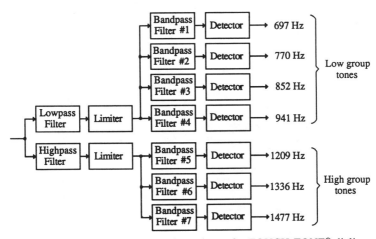

FIGURE 1-31   The tone detection scheme for TOUCH-TONE® dialing.

1-3-3 FM Stereo Applications

For wireless transmission of a signal occupying a low-frequency range, such as an audio signal, it is necessary to transform the signal to a high-frequency range by modulating it on to a high-frequency carrier. At the receiver, the modulated signal is demodulated to recover the low-frequency signal. The signal processing operations used for wireless transmission are modulation, demodulation, and filtering. Two commonly used modulation schemes for radio are *amplitude modulation* (AM) and *frequency modulation* (FM).

We review next the basic idea behind the FM stereo broadcasting and reception scheme as used in the United States [CO83]. An important feature of this scheme is that at the receiving end, the signal can be heard over a standard monoaural FM radio with a single speaker or over a stereo FM radio with two speakers. The system is based on the *frequency-division multiplexing* (FDM) method in which several bandlimited low-frequency signals are first modulated onto several high-frequency subcarriers and combined into a baseband composite signal. The latter is then modulated onto the main carrier developing the FDM signal and transmitted. At the receiving end, the composite baseband signal is first derived from the FDM signal by demodulation. Next, the generated baseband signal is passed through a set of bandpass filters to separate out the individual modulated subcarriers. Finally, the subcarriers are demodulated to recover the original low-frequency signals.

The block diagram representation of the FM stereo transmitter and the receiver are shown in Figure 1-32. At the transmitting end, the sum and the difference of the left and right-channel audio signals, $s_L(t)$ and $s_R(t)$, are first formed. Note that the summed signal $s_L(t) + s_R(t)$ is used in the monoaural FM radio. The difference signal $s_L(t) - s_R(t)$ is modulated using the double sideband suppressed carrier (DSB–SC) scheme using a subcarrier frequency f_{sc} of 38 kHz. The summed signal, the modulated difference signal, and 19-kHz pilot tone signal are then added, developing the composite baseband signal $s_B(t)$. The baseband signal is next modulated onto the main carrier frequency f_c using the frequency modulation (FM) method. At the receiving end, the FM signal is demodulated to derive the baseband signal $s_B(t)$, which is then separated into the low-frequency summed signal and the modulated difference signal using a lowpass filter and a bandpass filter. The cutoff frequency of the lowpass filter is around 15 kHz, whereas the center frequency of the bandpass filter is at 38 kHz. The 19-kHz pilot tone is used in the receiver to develop the 38-kHz reference signal for a coherent subcarrier demodulation for recovering the audio difference signal. The sum and difference of the two audio signals create the desired left audio and the right audio signals.

1-3-4 Electronic Music Synthesis

The generation of the sound of a musical instrument using electronic circuits is another example of the application of signal processing methods [AL80; MO77] The basis of such music synthesis is the following representation of the sound

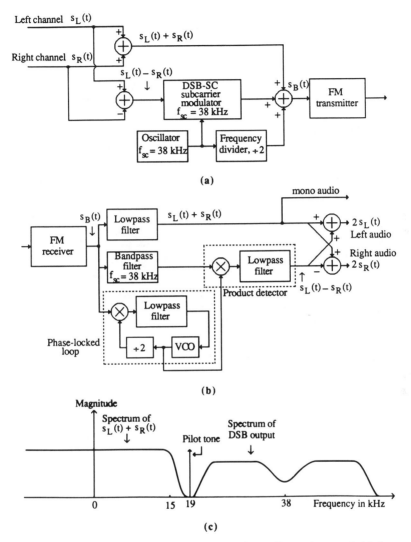

FIGURE 1-32 The FM stereo system: (a) transmitter, (b) receiver, and (c) frequency response of the composite baseband signal $s_B(t)$.

signal $s(t)$:

$$s(t) = \sum_{k=1}^{N} A_k(t) \sin(2\pi f_k(t)t), \qquad (1.1)$$

where $A_k(t)$ and $f_k(t)$ are the time-varying amplitude and the frequency of the kth component of the signal. The frequency function $f_k(t)$ varies slowly with time. For an instrument playing an isolated tone, $f_k(t) = kf_0$, that is,

$$s(t) = A_k(t)\sin(2\pi kf_0). \tag{1.2}$$

f_0 is called the *fundamental frequency*. In a musical sound with many tones, all other frequencies are usually integer multiples of the fundamental and are called *partial frequencies*, also called *harmonics*. Figure 1-33 shows, for example, the perspective plot of the amplitude functions as a function of time of 17 partial frequency components of an actual note from a clarinet. The aim of the synthesis is to produce electronically the $A_k(t)$ and $f_k(t)$ functions. To this end, the two most popular approaches followed are described next.

Subtractive Synthesis. This approach, which nearly duplicates the sound generation mechanism of a musical instrument, is based on the generation of a periodic signal containing all required harmonics and the use of filters to selectively attenuate (i.e., *subtract*) unwanted partial frequency components. The frequency-dependent gain of the filters can be used also to boost certain frequencies. The desired variations in the amplitude functions are generated by an analog multiplier or a voltage-controlled amplifier. Additional variations in the amplitude functions can be provided by adjusting dynamically the frequency response characteristics of the filters.

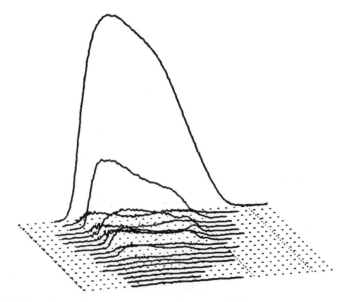

FIGURE 1-33 Perspective plot of the amplitude functions $A_K(t)$ for an actual note from a clarinet. [Reproduced with permission from J. A. Moorer, Signal processing aspects of computer music: A survey. *Proc. IEEE* **65**, 1108–1137 (1977) © 1977 IEEE.]

Additive Synthesis. Here, partial frequency components are generated independently by oscillators with time-varying oscillation frequencies. The amplitudes of the required signals are then individually modified, approximating the actual variations obtained by analysis, and combined (i.e., *added*) to produce the desired sound signal. For example, a piecewise linear approximation of the clarinet note of Figure 1-33 is sketched in Figure 1-34 and can be used to generate a reasonable replica of the note. Usually, some alterations to the amplitude and frequency functions may be needed before the music generated sounds as close as possible to that of the original instrument.

1-3-5 Echo Cancellation in Telephone Networks

In a telephone network the central offices perform the necessary switching to connect two subscribers. For economic reasons, a "two-wire" circuit is used to connect a subscriber to his/her central office, where as the central offices are connected using "four-wire" circuits. A two-wire circuit is bidirectional and carries signals in both directions. The four-wire circuit uses two separate unidirectional paths for signal transmission in both directions. The latter is preferred for long-distance trunk connections because signals at intermediate points in the trunk can be equalized and amplified using repeaters and, if necessary, multiplexed easily. The hybrid coil in the central office provides the interface between a two-wire circuit and

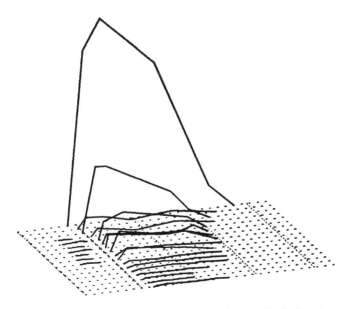

FIGURE 1-34 A piecewise-linear approximation to the amplitude functions of Figure 1-33. [Reproduced with permission from J. A. Moorer, Signal processing aspects of computer music: A survey. *Proc. IEEE* **65,** 1108–1137 (1977). © 1977 IEEE.]

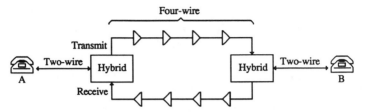

FIGURE 1-35   Basic two-four-wire interconnection scheme.

a four-wire circuit as shown in Figure 1-35. The hybrid circuit ideally should provide a perfect impedance match to the two-wire circuit by a balancing imped-ance so that the incoming four-wire receive signal is passed directly to the two-wire circuit connected to the hybrid with no portion appearing in the four-wire transmit path. However, to save cost, a hybrid coil is shared among several sub-scribers. Thus it is not possible to provide a perfect impedance match in every case because the lengths of the subscriber lines vary. The resulting imbalance causes a large portion of the incoming receive signal from the distance talker to appear in the transmit path and is returned to the talker as an echo. Figure 1-36 illustrates the normal transmission between a talker and a listener, and two possible major echo paths.

The effect of the echo can be annoying to the talker depending on the amplitude and delay of the echo, that is, on the length of the trunk circuit. The effect of the echo is worse for telephone networks involving geostationary satellite circuits, where the echo delay is about 540 ms.

Several methods are followed to reduce the effect of the echo. In trunk circuits of length up to 3000 km, adequate reduction of echo is achieved by introducing additional signal loss in both directions of the four-wire circuit. In this scheme, an improvement in the signal-to-echo ratio is realized as the echo undergoes loss in both directions while the signals are attenuated only once.

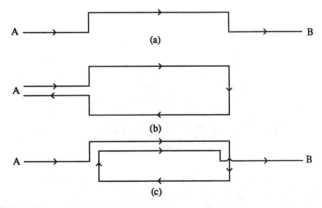

FIGURE 1-36   Various signal paths in a telephone network: (a) transmission path from talker A to listener B, (b) echo path for talker A, and (c) echo path for listener B.

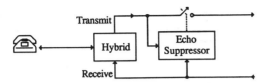

FIGURE 1-37   Echo suppression scheme.

On distances greater than 3000 km, echoes are controlled by means of an echo suppressor inserted in the trunk circuit as indicated in Figure 1-37. The device is essentially a voice-activated switch implementing two functions. It first detects the direction of the conversation and then blocks the opposite path in the four-wire circuit. Even though it introduces distortion when both subscribers are talking by clipping parts of the speech signal, the echo suppressor has provided a reasonably acceptable solution for terrestrial transmission.

For telephone conversation involving satellite circuits, an elegant solution is based on the use of an echo canceler [DU80; FR78; ME82]. The circuit generates a replica of the echo using the signal in the receive path and subtracts it from the signal in the transmit path as indicated in Figure 1-38. Basically, it is an adaptive filter structure who parameters are adjusted using certain adaption algorithms[2] until the residual signal is satisfactorily minimized. Typically, an echo reduction of about 40 dB is considered satisfactory in practice. To eliminate the problem generated when both subscribers are talking, the adaptation algorithm is disabled when the signal in the transmit path contains both the echo and the signal generated by the speaker closest to the hybrid coil.

1-4   TYPICAL APPLICATIONS OF DIGITAL SIGNAL PROCESSING

We now describe several novel applications of digital signal processing techniques.

1-4-1   Compact Disc Digital Audio

One of the best examples for an application of digital signal processing techniques is the compact disc digital audio system [CA82]. On the disc, the signal is recorded

---

[2]See Chapter 15 for a review of adaptive filters.

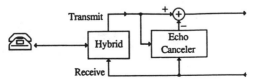

FIGURE 1-38   Echo cancellation scheme.

in a digital form on a spiral track as a succession of *pits* and *lands* as shown in Figure 1-39. These pits and lands represent a series of 1's and O's known as *channel bits*. In fact, each transition from land to pit or vice versa is treated as a 1 with all channel bits in between as O's. The information recorded on the disc consists of the digitized version of the audio signal, along with additional bits added for error correction, and control and display information for the listener.

The left and right channels of an analog stereo audio signal are each sampled at a rate of 44.1 kHz to ensure reproduction of the signal component with frequencies up to 20 kHz. Each channel is coded in standard *pulse-code modulation* (PCM) format using a 16-bit A/D converter. The combined bit rate of the two channels after sampling and digitization is therefore $44.1 \times 10^3 \times 32 = 1.41$ Mb/s. A block of 6 consecutive bits in the audio bit stream is grouped into a *frame*. Blocks of parity bits are then added to successive frames using the *cross-interleaved Reed–Solomon coding* (CIRC) scheme to permit correction of errors during the reproduction of the signal by the compact disc player. In addition, control and display bits are added to each frame to provide information to the listener. The composite bit digital signal, called the *data bit stream*, has a bit rate of 1.94 Mb/s. Next, blocks of eight bits in the data bit stream are translated into blocks of fourteen bits, using the *eight-to-fourteen modulation* (EFM) coding method, which are then linked using three merging bits. Finally, an identical pattern of 27 channel bits is added to each frame resulting in the *channel bit stream* with a bit rate of 4.32 Mb/s. A block diagram of the encoding system in a some-what simplified form is sketched in Figure 1-40 [HE82].

The channel bit stream is then used to turn on and off the light beam of a write laser illuminating the light-sensitive layer on a rotating glass master disc to produce a pattern of pits and lands by means of a photographic developing process. The master disc is used next to produce a nickel impression from which the disc for the player is reproduced.

FIGURE 1-39 Photograph of a part of the compact disc. [Courtesy of Philips Research Laboratories.]

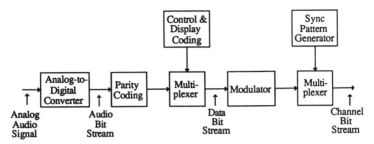

FIGURE 1-40 The compact disc encoding system. Note that in practice there are two audio channels for stereo recording, which together produce the audio bit stream after digitization.

In the reproduction process, the disc is optically scanned by a laser beam, creating a binary data stream that is decoded and converted back into analog signals. The block diagram of the basic components of the player is indicated in Figure 1-41. The inverse of the EFM modulation process is carried out in the demodulator whose output enters the buffer register for temporary storage at a rate depending on the revolution speed of the disc in the player. However, the bit stream leaving the buffer is at a rate synchronized by the clock generator and is therefore independent of the revolution speed of the disc. The output analog signal is thus totally free from wow and flutter commonly encountered in conventional analog record players. The output of the buffer is processed by the error-detection and correction circuit in which the parity bits are used to correct errors or, instead, detect errors if correction is not possible. The errors in the playing disc may have been caused either in the disc manufacturing process and/or through usage. In the latter case, the errors could result from damage from scratches, finger-marks, or dust on the disc. The errors that cannot be corrected are then modified by an error concealment circuit, which masks the errors to make them virtually inaudible either by interpolation or by muting the signal. The output of the error concealment circuit is passed through interpolation filters and digital-to-analog converters generating the two analog stereo outputs.

The digital recording and reproduction of the audio signal in a compact disc have resulted in a signal-to-noise ratio (SNR) and stereo channel separation of

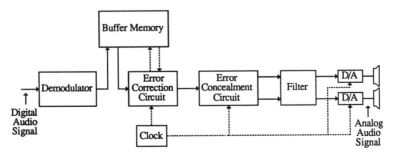

FIGURE 1-41 Block diagram of the signal processing part of the compact disc player.

more than 90 dB each. On the other hand, conventional analog recording and reproduction can offer at most a 60-dB SNR and less than 30-dB channel separation. Moreover, the pits and lands of a compact disc are protected by a transparent layer. Any surface damage and dust are not on the focal plane of the read laser beam scanning the disc and thus have reduced effect. In addition, optical scanning does not cause any physical damage and wear to the disc. Most other errors occurring on the disc can be either corrected or masked by means of error correction and concealment circuits in the player.

In the compact disc audio system and in many other applications involving the digital processing of analog continuous-time signals, the analog-to-digital (A/D) converter and the digital-to-analog (D/A) converter play critical roles.[3] For low-frequency applications such as in digital telephony, the sampling rate used is 8 kHz because the effective bandwidth of the audio signal for conversation purposes is 4 kHz. In such an application, the requirements of both anti-aliasing and reconstruction lowpass filters are not that stringent and relatively simple low-order analog lowpass filters are adequate. However, for high-frequency applications such as in compact disc digital audio, the specifications for the two lowpass filters are quite stringent, making their design and implementation expensive. In such applications, the analog filter design problem has been simplified by an elegant application of multirate digital signal processing techniques.[4] We outline below the basic principles used in both situations.

A simplified block diagram representation of an A/D conversion system for the digital recording of a high-fidelity audio signal is shown in Figure 1-42. The audio band of interest is up to 20 kHz and the basic sampling rate employed is 44.1 kHz. To prevent aliasing distortion, the original analog audio signal should be passed through an analog lowpass anti-aliasing filter to limit the bandwidth to about half the sampling rate before initiating the sampling operation. The necessary specifications for the anti-aliasing lowpass filter are that the magnitude characteristic remain "flat" up to 20 kHz with a "strong" attenuation above 24.1 kHz to prevent aliasing [KA86]. These requirements can be met by a very high-order analog filter with accurate trimming of component values resulting in a very expensive filter. Moreover, such a filter will cause waveform distortion due to excessive phase shifts at some frequencies in the audio signal. There is another difficulty associated with such a system. The A/D converter is designed to generate a 16-bit digital

---

[3]See Chapter 10 for a detailed discussion of these devices.
[4]See Chapter 14 for a discussion of multirate digital signal processing.

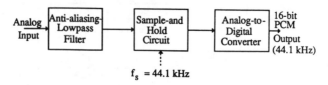

f_s = 44.1 kHz

FIGURE 1-42   Conventional analog-to-digital conversion system.

word in a PCM format for each analog sample generated by the sample-and-hold (S/H) circuit operating at the 44.1-kHz sampling rate. For an A/D converter output in the range -5 volts to $+5$ volts, for example, a 16-bit word implies $2^{16} = 65,536$ quantization levels with a quantization step of 153 μV. The requires an exceptionally stable A/D converter with very high-precision voltage dividers generating the 2^{16} voltage levels.

In practice, the stringent specifications for anti-aliasing filter are avoided by making use of multirate digital signal processing techniques. The modified analog-to-digital conversion scheme as used in compact disc digital audio is sketched in Figure 1-43 [KA86]. Here, the input analog signal is first sampled at a very high rate (3175.2 kHz) and then down-sampled digitally by a factor of 72 to 44.1 kHz. Because of the very high sampling rate, a simple low-order, anti-aliasing filter with a passband flat up to 20 kHz and a very wide transition band can be used, which can be implemented without any accurate trimming. Moreover, the higher sampling rate permits the use of an A/D converter with a very short wordlength, as fewer bits are needed to represent adequately each analog sample. In fact, as indicated in Figure 1-43, a 1-bit coder with a sampling rate of 3175.2 kHz is used. The 1-bit high rate signal is then down-sampled by a factor of 72 and converted into a 16-bit PCM signal with an effective sampling rate of 44.1 kHz. To eliminate the aliasing caused by the down-sampling, a sharp cutoff digital lowpass decimating filter with the same specifications as the analog anti-aliasing filter of Figure 1-42 is inserted before the down-sampler. It is possible to achieve high precision without accurate trimming in the integrated circuit implementation of a digital filter. Moreover, the digital filter can also be designed with exact linear phase, thus eliminating any phase distortion that is generally unavoidable in an analog filter.[5]

In the compact disc player, the digital signal, after decoding and error correction, is a sequence of 16-bit words with a 44.1-kHz sampling rate. It is then converted into a discrete-time sequence of analog samples by means of a D/A converter. The output of the converter is passed through a hold circuit generating an analog staircase signal and then filtered by an analog lowpass filter with a passband of 20 kHz and a transition bandwidth of approximately 2 kHz. The basic block diagram representation of the D/A conversion system is sketched in Figure 1-44. The hold circuit following the D/A converter generates a staircase signal approx-

---

[5]Chapter 3 outlines methods for the design of linear-phase finite impulse response (FIR) digital filters.

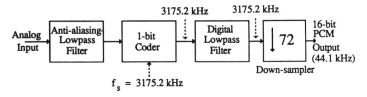

FIGURE 1-43 Oversampling analog-to-digital conversion system.

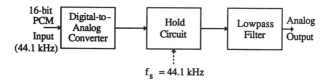

f_s = 44.1 kHz

FIGURE 1-44 Conventional digital-to-analog conversion scheme.

imating the original analog waveform as indicated in Figure 1-45(a). The steps in the staircase signal result in high-frequency components beyond the desired 20-kHz audio bandwidth, which should be suppressed by the analog lowpass filter. The system design requirements dictate that the high-frequency component level be reduced to at least 50 dB below that of the maximum audio signal that can be met by a very high-order expensive analog lowpass filter. Moreover, since it is not possible to design an analog filter with exactly linear phase, use of an analog filter with nonlinear phase results in signal distortion in the desired audio band.

The above problems have again been eliminated by using multirate techniques. The modified D/A conversion system as used in most compact disc players is as shown in Figure 1-46 [GO82]. As indicated in the figure, the 16-bit digital signal is first up-sampled (oversampled) by a factor of four to arrive at a signal with a sampling rate of 176.4 kHz. The up-sampling process is implemented by inserting three zero-valued words in between two consecutive words of the original low-rate digital signal. The lowpass digital filter following the up-sampler "fills in" the missing digital words by interpolation. The specifications of this digital filter are essentially the same as that required for the analog lowpass filter in the direct 16-bit D/A conversion of Figure 1-44. Unlike the analog filter, the digital filter can be designed with exactly linear phase. Moreover, the filter characteristic is insensitive to variations in the rotation speed of the disc as it varies with the 176.4-kHz clock rate.

The multiplication operations carried out by the digital filter increase the word-length to 28 bits. The noise shaper circuit rounds off the oversampled digital signal wordlength to 14 bits and feeds back the rounding-off errors to its input in opposite phase. This latter operation ensures that the SNR in the audio band remains around 96 dB, corresponding to 16-bit quantization. The output of the hold circuit follow-

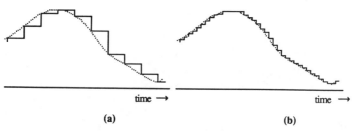

(a)   **(b)**

FIGURE 1-45 Output staircase waveform of the hold circuit: (a) lower sampling rate and (b) higher sampling rate.

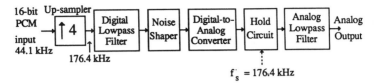

FIGURE 1-46 Factor-of-four oversampling digital-to-analog conversion scheme.

ing the D/A conversion is again a staircase signal but with smaller steps in the waveform (see Figure 1-45(b)). It is now easier to suppress the high-frequency components generated by these steps using a very simple, low complexity analog lowpass filter with less stringent specifications. Moreover, the smaller quantization steps lower the maximum "slew rate" that these circuits need to process, reducing considerably the chances of intermodulation distortion caused when slew rates are exceeded.[6]

1-4-2   Digital Television

Television receivers in use today employ exclusively analog signal processing techniques. A simplified block diagram of a present day TV receiver is as shown in Figure 1-47. However, there is an increasing trend toward the use of digital signal processing techniques in many parts of a TV receiver, and very soon such digital TV receivers are expected to be on the market [AN86]. The parts of a television receiver that are likely to be replaced with digital techniques are encompassed by the dotted lines in Figure 1-47. To make use of digital signal processing techniques, digital TV receivers will include digital field and/or frame memories. There are three main reasons for considering the use of digital signal processing techniques: (1) improved quality, (2) addition of new features, and (3) lower cost.

---

[6]The rate of variation of the output voltage is defined as the slew rate.

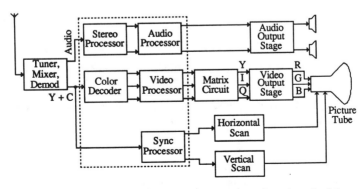

FIGURE 1-47 A simplified block diagram representation of a color television receiver.

To understand the rationale behind improved quality in a digital TV, let us briefly review the operation of a color TV system. There are three different world standards for color television. The standard adopted in the United States and several other countries is called the NTSC system, whereas the other two standards are the PAL and SECAM systems. We restrict our discussion here to the NTSC system [JA89]. In this system, the television camera scans the scene to form an image, called a *frame*, containing 525 scan lines. The scanning rate is 30 frames per second. Each frame is formed by combining two *fields* of $262\frac{1}{2}$ lines each. The first field is composed of even scan lines, and the second contains the odd scan lines. Even and odd fields are transmitted alternately at 60 fields per second and combined in the TV receiver in an interlaced fashion to reduce the flicker.

A color image is composed of three primary color images: red (R), green (G), and blue (B), which are independently generated by the TV camera. These three color signals are first converted into three related signals: a *luminance* or brightness component Y, and two *chrominance* components, I and Q, using the following matrix relation [JA89]:

$$
\begin{bmatrix} Y \\ I \\ Q \end{bmatrix} = \begin{bmatrix} 0.299 & 0.587 & 0.114 \\ 0.596 & -0.274 & -0.322 \\ 0.211 & -0.523 & 0.312 \end{bmatrix} \begin{bmatrix} R \\ G \\ B \end{bmatrix}. \tag{1.3}
$$

The Y component signal is used to produce a black-and-white image in a monochrome TV receiver. The I and Q components are modulated and combined to generate a single chroma signal C, which is then added to the Y component producing the Y + C *composite signal:*

$$
u(t) = Y(t) + I(t) \cos (2\pi f_{sc}t + \phi) + Q(t) \sin (2\pi f_{sc}t + \phi), \tag{1.4}
$$

where $\phi = 33°$, and $f_{sc} \cong 3.58$ MHz, the subcarrier frequency.[7] The relations between the subcarrier frequency f_{sc}, the frame frequency f_P, the field frequency f_F, and the line frequency f_L are as follows:

$$
f_P = 29.97 (\cong 30) \text{ Hz}, \quad f_F = 59.94 (\cong 60) \text{ Hz} = 2f_P,
$$
$$
f_L = 15734 \text{ Hz} = 525f_P, \quad f_{sc} = 3.579545 (\cong 3.58) \text{ MHz} = 227.5f_L. \tag{1.5}
$$

The composite Y + C signal occupies a bandwidth of approximately 4 MHz, which is the bandwidth of a single TV channel. The frequency spectrum of the composite

---

[7]The main difference among the three color television standards, NTSC, PAL, and SECAM, are in the modulation scheme used in generating the chroma component C.

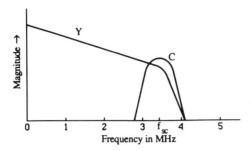

FIGURE 1-48 Frequency spectra of the luminance (Y) and the chrominance (C) signals.

signal is shown in Figure 1-48, from which it can be seen that the spectra of the luminance component Y and that of the chroma component C overlap in the upper part of the available band centered around the subcarrier frequency f_{sc}.

In the TV receiver, the desired composite signal $Y + C$ is obtained after various operations, such as tuning, mixing, and demodulation, are performed on the signal received by the antenna (Figure 1-47). The color decoder then separates the Y and C signals by linear filtering, and recovers the two chroma signals, I and Q, by a second demodulation process. In a conventional TV receiver, an analog one-dimensional lowpass filter with a notch at f_{sc} recovers Y, and an analog one-dimensional bandpass filter centered at f_{sc} recovers C from the composite signal. The analog filtering employed is called *horizontal filtering* because it is based on a combination of signal samples corresponding to picture elements (pixels) on the same line only. Such filters are unable to provide very good separation of the Y and C signals. As a result, some frequency components of the C signal appear at the output of the lowpass filter as a part of the Y signal and show up as a variation of the luminance in the form of dots on the TV screen, which are known as *cross-luminance* effects. Likewise, high-frequency components of Y appear at the output of the bandpass filter as a part of the C signal and result in colored streaks on the screen, called *cross-color* effects.

In a digital TV receiver, improved Y and C signal separation can be achieved by providing *vertical filtering* with the aid of line memories, and *temporal filtering* with the aid of field memories. In the former case, filtering is achieved by combining pixels above or below one another, whereas in the latter case, a combination of pixels from successive fields is used in filtering. A possible scheme for generating the R, G, and B components by digital processing of the $Y + C$ signal is as shown in Figure 1-49. Here, the analog composite signal is first converted into a digital composite sequence $y[n] + c[n]$ by means of an A/D converter operating at a sampling rate f_s, which is typically anywhere from 10 to 20 MHz. The color decoder generates "crude" component signals, $\hat{y}[n]$, $\hat{i}[n]$, and $\hat{q}[n]$, which after additional processing by the video processor block are converted into $y[n]$, $i[n]$, and $q[n]$. These digital signals are converted into their analog equivalents, Y, I, and Q, by means of three D/A converters. The matrix circuit shown converts them

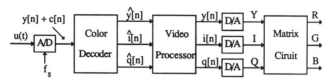

FIGURE 1-49 A possible scheme for digital processing of composite signal.

finally to the R, G, and B signals, using the relation

$$\begin{bmatrix} R \\ G \\ B \end{bmatrix} = \begin{bmatrix} 1.0 & 0.956 & 0.621 \\ 1.0 & -0.273 & -0.647 \\ 1.0 & -1.104 & 1.701 \end{bmatrix} \begin{bmatrix} Y \\ I \\ Q \end{bmatrix}, \tag{1.6}$$

which are used to drive the respective guns in the picture tube.

One possible digital realization of the color decoder is shown in Figure 1-50. The lowpass filter H_1 has a passband width of about 4 MHz with a sharp notch at the subcarrier frequency f_{sc}. The bandpass filter H_2 has a narrow passband centered at f_{sc}. The output $\hat{c}[n]$ of the bandpass filter is demodulated using two quadrature carrier signals of frequency f_{sc}, which are then filtered by two lowpass filters H_3. All the filters shown here perform only horizontal filtering and can be implemented with exact linear phase, an advantage of using digital filters. The video processor block performs the necessary vertical and temporal filtering. The vertical filtering is used to reduce the cross-color and cross-luminance effects and can be employed either before or after the color decoder. On the other hand, temporal filtering is preferable after the color decoding operation to reduce further the cross-effects. Reduction of cross-effects by filtering out interfering frequency components also improves the SNR of the processed video image.

The incorporation of field memories for digital processing also permits the elimination of two annoying imperfections that are unavoidable in a conventional color TV receiver based on analog techniques. One such imperfection, called the *large-*

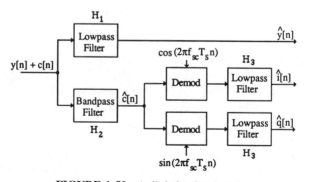

FIGURE 1-50 A digital color decoder.

area flicker, is caused by the periodic variation of the luminance at 60 Hz due to the interlacing of alternate fields and is highly visible in present day receivers with brighter and larger screens. This problem can be remedied completely by storing each field in a field memory and then repeating its display twice, yielding an effective field frequency of 120 Hz. If the image has sharp transitions that are parallel to the scanning direction, these transitions appear at different heights in consecutive fields, causing the second type of imperfection known as the *line flicker*. By employing an additional field memory and a motion detector circuit, both flickers can be reduced by a technique called *frame-doubling*.

The signal received by the TV antenna is the original signal along with its multiple reflections from high structures and/or mountains with the latter appearing after some delay. These delayed signals result in another type of imperfection called "ghost images," which can be eliminated using digital adaptive echo cancellation techniques.[8]

The new features that can be offered in a digital TV receiver are made possible by the availability of field and frame memories. For example, any frame can be frozen to display a still picture. Such still images can be stored for later retrieval for display and/or printing. Images from different TV stations can be combined, offering "picture-in-picture" capability and permitting the viewer to observe transmissions from various TV stations simultaneously. Also, any part of a still image or a video sequence can easily be zoomed to a larger size.

Finally, a digital TV set is expected to have a lower cost for a variety of reasons such as the replacement of expensive analog components by inexpensive digital equivalents, reduced parts count due to an increased use of integrated circuits, increased reliability, and a decrease in the production cost due to improved trimming procedures.

1-4-3 Real-Time Digital Spectrum Analysis

As indicated earlier, in most cases, a signal is a function of time. Such a signal can also be expressed in terms of a superposition of complex sinusoids. The form of such description depends on the type of signal.[9] Often it is useful to study the sinusoids forming the time-domain signal by determining their frequencies and complex amplitudes. A plot of the amplitudes as a function of frequency is called the *spectrum*. This type of signal analysis is called *spectral analysis* and is important in analyzing communication channel characteristics, acoustical phenomena, mechanical vibration, radar and laser pulse characteristics, information encoding schemes, and waveforms from a variety of other diverse applications.[10]

Often, conventional spectral analysis can provide the necessary frequency-domain information. However, in the past few years, there has been an increasing number of applications that demand more complete spectral coverage. Moreover,

[8]See Chapter 15 for a review of adaptive signal processing methods.
[9]Chapter 2 reviews the various forms of frequency-domain description of signals.
[10]See Chapter 16 for a review of various spectral analysis methods.

in many instances, the coverage has to be done in real time. For example, testing or monitoring data encryption schemes and other frequency agile applications almost always require real-time frequency analysis. This is because the frequencies are mobile; they hop around the spectrum. In many other cases, real-time analysis is necessary simply because the event of interest is a short-lived transient—a data channel error, a spurious radar return, or an impact-related acoustic emission, to name a few examples.

One of the first applications of digital signal processing in instrumentation has been in real-time digital spectrum analysis. The main advantage of digital spectrum analysis over conventional spectrum analysis is that a complete view of the whole frequency spectrum (subject to anti-aliasing considerations) is obtained, thus the attribute "real-time," which is often associated with digital spectrum analysis.

The conventional analog spectrum analyzer is based on sweeping a local oscillator over the frequency range of interest, the frequency span, with a certain frequency resolution. A block diagram of an analog swept spectrum analyzer is shown in Figure 1-51. The input signal is first bandpass filtered by the *span filter*. A local oscillator provides a heterodyning signal to translate the frequency span of interest down to an intermediate frequency (IF). The reason for working at an IF as opposed to working at baseband around "dc" is because signal handling (amplification, isolation) is easier and one can look at signals that are not symmetric about the center frequency of interest. The IF signal is then passed through a narrowband bandpass filter, the *resolution filter*. As the local oscillator sweeps through its frequency range, the output of the resolution filter reflects the frequency contents of the spectrum of interest. In a typical spectrum analyzer, a number of resolution filters with different bandwidths are offered. The output of the resolution filter is then passed through a "video" filter that provides time (neighboring frequency) averaging. This is followed by a detector (peak hold) to improve measurements.

With the advent of economical digital technology and computing power, an alternative to analog filter methods became available. This alternative is the *fast Fourier transform* (FFT) approach shown in Figure 1-52. The FFT is a signal processing algorithm that allows complete spectral analysis of digitized waveforms.[11] Its advantage is that any signal that can be digitized can be transformed

---

[11]See Chapter 8 for a review of fast Fourier transform algorithms.

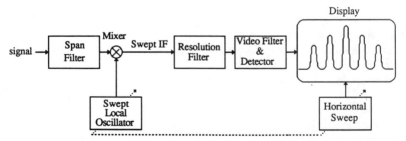

FIGURE 1-51 The block diagram of a conventional analog spectrum analyzer.

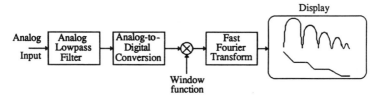

FIGURE 1-52 Spectrum analysis via waveform digitizing and the FFT method.

to a complete description in the complex frequency domain. Even short-lived transients can be transformed if they are captured with sufficient digital points to define the transient.

The signal is first anti-aliased filtered. This analog filter is the equivalent of the analog span filter in Figure 1-51. Following analog-to-digital conversion, the signal is then multiplied by the window function. The window plays the role of the analog resolution filter in Figure 1-51. The fast Fourier transform can be viewed as a bank of parallel heterodyners as shown in Figure 1-53. The output is resampled at the FFT rate since each of the output is a heterodyned signal that is summed over N samples.

From the above model of the FFT, we can easily explain the adjective "real-time" usually associated with digital spectrum analysis. Unlike the analog spectrum analyzer, which can only look at one portion of the spectrum at any instant in time, the digital spectrum analyzer can monitor the complete spectrum of interest at the same time.

The combination of the window and the FFT can also be modeled as a bank of digital identical filters shifted in frequency [TH89]. With the concept of bank of filters [TE89], it becomes possible to relate FFT update rate with resampling, and thus with aliasing control. For example, when abutting FFTs are taken, the resampling rate is 1 to N. With 50% overlap, the resampling rate is 1 to $N/2$. As the overlap is increased, the resampling rate is also increased and each FFT bin output becomes a better anti-aliased signal.

Algorithmic FFT computation of frequency-domain components is complete in that it provides the complex frequency domain (real and imaginary or magnitude

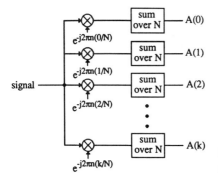

FIGURE 1-53 FFT as a bank of heterodyners.

and phase for each frequency). By contrast, conventional or analog spectrum analyzers only provide frequency magnitude and neglect phase.

In the above we have only discussed digital spectrum analyses that extend from dc to half the sampling rate. Very often one is interested in a high-resolution spectrum for a narrow range of frequency. In this case the basic digital spectrum analyzer is preceded by a heterodyner and a lowpass filter. The heterodyning frequency is set to be the center of the desired passband. An in-phase and quadrature heterodyning is performed, yielding a complex signal that is then filtered to the required bandwidth. The resulting signal is fed into the basic digital spectrum analyzer.

1-5 NOTATIONS

We now summarize the major mathematical definitions and notations that are being followed in this text.

1-5-1 Polynomials

An Nth degree polynomial $P(z) = \sum_{i=0}^{N} p_i z^{-i}$ is said to be *pseudo-Hermitian* if $p_i = c p_{N-i}^*$ for some (possibly complex) c with $|c| = 1$. If $c = 1$, $P(z)$ is then *Hermitian*; and if $c = -1$, $P(z)$ is *skew-Hermitian*. A Hermitian polynomial is simply a *symmetric* polynomial when the coefficients p_i are real. Likewise, a skew-Hermitian polynomial is an *antisymmetric* polynomial when the coefficients p_i are real.

The polynomial obtained by conjugating the coefficients of a polynomial $P(z)$ is denoted by $P_*(z)$. The polynomial obtained by replacing z with z^{-1} and by conjugating the coefficients of a polynomial $P(z)$ is denoted by $\tilde{P}(z)$; that is $\tilde{P}(z) = P_*(z^{-1})$. For an Nth degree polynomial $P(z)$, the polynomial obtained by reversing the coefficients and conjugating them is denoted by $\hat{P}(z)$; that is,

$$\hat{P}(z) = \sum_{i=0}^{N} p_{N-i}^* z^{-i}.$$

Note that for a Hermitian polynomial $P(z) = \hat{P}(z)$. According to these definitions

$$\hat{P}(z) = z^{-N} \tilde{P}(z) = z^{-N} P_*(z^{-1}).$$

1-5-2 Matrices and Vectors

Matrices are being represented by boldfaced uppercase letters, such as $\mathbf{A}$ and $\mathbf{T}(z)$. Column and row vectors are being represented by boldfaced lowercase letters, such as $\mathbf{a}$ and $\mathbf{b}$. The superscripts -1, t, *, and † denote, respectively, the inverse, transpose, conjugate, and transpose conjugate operations. For example, $\mathbf{A}^{-1}$, $\mathbf{A}^t$,

$\mathbf{A}^*$, and $\mathbf{A}^\dagger$ denote, respectively, the inverse, transpose, conjugate, and transpose conjugates of $\mathbf{A}$. Similar notations apply for matrices and vectors that have elements that are functions of z.

$\mathbf{T}_*(z)$ denotes the matrix obtained from $\mathbf{T}(z)$ by conjugating only its coefficients while the variable z remains unchanged. The notation $\tilde{\mathbf{T}}(z)$ denotes $\mathbf{T}'_*(z^{-1})$, which is equivalent to transpose conjugation on the unit circle; that is, $\tilde{\mathbf{T}}(e^{j\omega}) = \mathbf{T}^\dagger(e^{j\omega})$. For a scalar transfer function $H(z)$ with real coefficients, $\tilde{H}(z) = H(z^{-1})$.

Given two positive Hermitian matrices $\mathbf{P}$ and $\mathbf{Q}$ of the same dimensions, the notation $\mathbf{P} < \mathbf{Q}$ means that $\mathbf{Q} - \mathbf{P}$ is positive definite, and $\mathbf{P} \le \mathbf{Q}$ means that $\mathbf{Q} - \mathbf{P}$ is positive semidefinite. For positive definite matrix $\mathbf{P}$ the notation $\mathbf{P}^{1/2}$ stands for a real square root, $\mathbf{P}^{t/2}$ for its transpose, $\mathbf{P}^{-1/2}$ for its inverse, and $\mathbf{P}^{-t/2}$ for its inverse transpose.

1-6 ORGANIZATION AND SCOPE OF THE HANDBOOK

In some sense, the origin of digital signal processing techniques can be traced back to the 17th century when finite difference methods, numerical integration methods, and numerical interpolation methods were developed to solve physical problems involving continuous variables and functions.[12] The more recent interest in digital signal processing arose in the 1950s with the availability of large digital computers. Initial applications were primarily concerned with the simulation of analog signal processing methods. Around the beginning of the 1960s, researchers began to consider digital signal processing as a separate field by itself. Since then, there have been significant and prolific developments and breakthroughs in both the theory and applications of digital signal processing.

It is difficult to include all important topics in any field in one volume, and digital signal processing is no exception. This handbook has therefore been restricted primarily to one-dimensional digital signal processing. Moreover, the emphasis has been placed on the theory and design of signal processing algorithms and their implementation. Very little attention has been given to their applications. To keep the size of the book reasonable, only the topics that are more useful in practice and widely used have been included.

The handbook consists of 16 chapters with Chapter 1 providing an introduction to the field. Chapter 2 is concerned with discrete-time and continuous-time signals, both deterministic and random. In particular, it reviews their time-domain and frequency-domain characterization, and their transformation via linear systems. This chapter also reviews the Laplace transform, Fourier transform, and the z-transform, important tools used for the representation, analysis, and design of linear continuous-time and discrete-time systems, respectively. The effect of sampling a continuous-time signal to generate a discrete-time signal, and the condition for perfect reconstruction of the continuous-time signal from its sampled discrete-time version are treated here.

---

[12]For a brief history of digital signal processing see Section 2-2.

Chapter 3 deals with the time-domain, frequency-domain, and structural representations of linear discrete-time systems. For mathematical convenience, an ideal digital filter can be considered as a linear discrete-time system. Some of the fundamental concepts of such systems including properties and realizability issues are reviewed here. This chapter also discusses block implementation of digital filters in which the input signal is processed in the form of contiguous blocks of samples of equal length, generating an output also in the form of contiguous blocks of samples of equal length.

The following two chapters treat the problem of digital filter design, that is, the development of an appropriate realizable transfer function of the digital filter meeting the prescribed frequency-domain or the time-domain response with some acceptable error. Methods for the design of finite impulse response (FIR) digital filter and the infinite impulse response (IIR) digital filters with piecewise constant magnitude responses in the passbands and stopbands are outlined in Chapters 4 and 5, respectively.[13]

There are various types of implementations of a digital filter structure depending on the application. For example, it may either be a special purpose digital circuit or involve the programming of a general purpose digital signal processor chip. In some applications it may be implemented as a software on a general purpose computer. Chapter 6 is concerned with the issues that are common to most types of implementations. Specifically, it reviews the different types of number representations and arithmetic schemes, certain basic characteristics of filter structures, and transformations for generating new digital filter structures. Other important issues considered here are the analysis of the effects of finite wordlengths on the performance of a practical digital filter.

Chapter 7 outlines the development of digital filter structures that minimizes one or more of the detrimental effects of finite wordlengths. One such effect is the quantization of the filter coefficients causing the filter's actual frequency response to differ from the desired response. This effect is characterized by the sensitivity of the digital filter structure to small changes in the filter coefficients. Structures that are less sensitive to such changes are described. Another effect arises from the quantization of the results of arithmetic operations, in particular, the product. Methods developed for the minimization of these effects are included. This chapter also describes a method developed for increasing the signal-to-noise ratio in the cascade type structures.

The discrete Fourier transform (DFT) is an important tool in the analysis of discrete-time signals and systems and in the design and implementation of digital filters. Chapter 8 is devoted to a review of some commonly used methods for the fast DFT computation and fast finite convolution.

One of the novel concepts considered in recent years for the design of digital signal processors is the residue number system (RNS) arithmetic. Chapter 9 reviews the basic principles of RNS arithmetic and discusses the capabilities and limitations of RNS-based architectures in real-time digital signal processing applications.

---

[13]See Section 3-3 for a definition of FIR and IIR digital filters.

In most applications, signals of interest are continuous-time functions. These need to be converted into digital form for digital filtering and processing, and the processed signal then converted back into continuous-time form. The interface structures between the continuous-time input and output data and the discrete-time digital hardware are the subject of Chapter 10. The interface structures reviewed here are the anti-aliasing and smoothing filters, and the analog-to-digital and digital-to-analog converters.

The next two chapters describe various methods for the implementation of digital filters. Chapter 11 concentrates on the hardware and architecture issues with an emphasis on the increasingly popular digital signal processor (DSP) chips, whereas Chapter 12 is concerned with the software implementation issues, with an emphasis on the programming of some commonly used DSP chips.

Chapter 13 is also on digital filter design. But unlike Chapters 3 and 4, it is concerned with special types of filters, such as Hilbert transformers, integrators and differentiators, smoothing filters, nonintegral delay networks, and median filters. Each of these filters have various applications and are often used in practice.

The topic of multirate digital signal processing is treated in Chapter 14. It includes a review of the basic concepts of sampling rate conversion, design of decimation and interpolation digital filters, multirate filter bank design, and three typical applications.

Chapter 15 is concerned with adaptive filter theory and design. In particular, the most commonly used adaptive filtering methods—LMS, RLS, and frequency-domain or block algorithms—are described here. It also outlines the basic structural characteristics common to adaptive systems and discusses three practical applications.

The final chapter, Chapter 16, deals with spectral analysis, which is used to characterize the frequency contents of a finite length sequence or time-series. It reviews both the classical methods of spectral analysis based on the Fourier transform and the more modern approaches based on parametric models. It also considers techniques developed for processing sinusoidal signals in noise and analysis of spatial time-series data.

ACKNOWLEDGMENTS

I thank my colleagues, Professors Allen Gersho, Hua Lee, and John J. Shynk, who reviewed this chapter and made a number of suggestions for improvement. I also thank Professors Hrvoje Babic and Ulrich Heute for their critical reviews. Thanks are also due to Dr. Tran Thong of Tektronix Federal Systems Inc. for his assistance in the preparation of Section 1-4-3.

REFERENCES

[AL80] H. G. Alles, Music synthesis using real time digital techniques. *Proc. IEEE* **68,** 436–449 (1980).

[AN86] M. J. J. C. Annegarn, A. H. H. J. Nillisen, and J. G. Raven, Digital signal processing in television receivers. *Philips Tech. Rev.* **42**, 183–200 (1986).

[BE84] M. Bellanger, *Digital Processing of Signals*. Wiley, New York, 1984.

[BL78] B. Blesser and J. M. Kates, Digital processing in audio signals. In *Applications of Digital Signal Processing* (A.V. Oppenheim, ed.), Chapter 2, Prentice-Hall, Englewood Cliffs, NJ, 1978.

[BO70] G. E. P. Box and G. M. Jenkins, *Time Series Analysis: Forecasting and Control*. Holden-Day, San Francisco, CA, 1970.

[CA82] M. G. Carasso, J. B. H. Peek, and J. P. Sinjou, The compact disc digital audio system. *Philips Tech. Rev.* **40**(6), 151–156 (1982).

[CO83] L. W. Couch, II, *Digital and Analog Communication Systems*. Macmillan, New York, 1983.

[CO86] A. Cohen, *Biomedical Signal Processing*, Vol. 2. CRC Press, Boca Raton, FL, 1986.

[DA76] G. Daryanani, *Principles of Active Network Synthesis and Design*. Wiley, New York, 1976.

[DU80] D. L. Duttweiler, Bell's echo-killer chip. *IEEE Spectrum* **17**, 34–37 (1980).

[EA76] J. Eargle, *Sound Recording*. Van Nostrand-Reinhold, New York, 1976.

[EN74] M. Engelson and F. Telewski, *Spectrum Analyzer: Theory and Applications*. Artech House, Dedham, MA, 1974.

[FR78] S. L. Freeny, J. F. Kaiser, and H. S. McDonald, Some applications of digital signal processing in telecommunications. In *Applications of Digital Signal Processing* (A.V. Oppenheim, ed.), Chapter 1, pp. 1–28. Prentice-Hall, Englewood Cliffs, NJ, 1978.

[GO82] D. Goedhart, R. J. van de Plassche, and E. F. Stikvoort, Digital-to-analog conversion in playing a compact disc. *Philips Tech. Rev.* **40**(6), 174–179 (1982).

[HE82] J. P. J. Heemskerk and K. A. Schouhamer Immink, Compact disc: System aspects and modulation. *Philips Tech. Rev.* **40**(6), 157–165 (1982).

[HU89] D. M. Huber and R. A. Runstein, *Modern Recording Techniques*, 3rd. ed. Howard W. Sams & Co., Indianapolis, IN, 1989.

[JA68] L. B. Jackson, J. F. Kaiser, and H. S. McDonald, An approach to the implementation of digital filters. *IEEE Trans. Audio Electroacoust.* **AU-16**, 413–421 (1968).

[JA89] A. K. Jain, *Fundamentals of Digital Image Processing*. Prentice-Hall, Englewood Cliffs, NJ, 1989.

[JU81] R. K. Jurgen, Detroit bets on electronics to stymie Japan. *IEEE Spectrum* **18**, 29–32 (1981).

[KA86] J. J. van der Kam, A digital "decimating" filter for analog-to-digital conversion of hi-fi audio signals. *Philips Tech. Rev.* **42**, 230–238 (1986).

[LE83] E. L. Lerner, Electronically synthesized music. *IEEE Spectrum* **17**, 46–51 (1983).

[ME82] D. G. Messerschmitt, Echo cancellation in speech and data transmission. *IEEE J. Select. Areas Commun.* **SAC-2**, 283–297 (1982).

[MI91] S. K. Mitra, H. Li, I.-S. Lin, and T.-H. Yu, A new class of nonlinear filters for image enhancement. *Proc. IEEE Int. Conf. Acoust., Speech, Signal Process.*, Toronto, Canada, pp. 2525–2528 (1991).

[MO77] J. A. Moorer, Signal processing aspects of computer music: A survey. *Proc. IEEE* **65**, 1108–1137 (1977).

[OP78] A. V. Oppenheim, Digital processing of speech. In *Applications of Digital Signal Processing* (A. V. Oppenheim, ed.), Chapter 3. Prentice-Hall, Englewood Cliffs, NJ, 1978.

[PE86] E. H. J. Persoon and C. J. B. Vandenbilcke, Digital audio: Examples of the application of the ASP integrated signal processor. *Phillips Tech. Rev.* **42**, 201–216 (1986).

[PO87] K. Poulton, J. J. Corcoran, and T. Hornak, A 1-GHz 6-bit ADC system. *IEEE J. Solid-State Circuits* **SC-22**, 962–970 (1987).

[PR88] J. G. Proakis and D. G. Manolaklis, *Introduction to Digital Signal Processing.* Macmillan, New York, 1988.

[RA78] L. R. Rabiner and R. W. Schafer, *Digital Processing of Speech Signals.* Prentice-Hall, Englewood Cliffs, NJ, 1978.

[RO80] E. A. Robinson and S. Treitel, *Geophysical Signal Analysis.* Prentice-Hall, Englewood Cliffs, NJ, 1980.

[SH81] A. F. Shackil, Microprocessors and the MD. *IEEE Spectrum* **18**, 45–49 (1981).

[SL83] R. J. Sluyter, Digitization of speech. *Philips Tech. Rev.* **41**(7/8), 201–223 (1983–1984).

[TE89] Tektronix Inc., TEK 3052 Digital Spectrum Analyzer, rev. data sheet. Tektronix Inc., Beverton, OR, 1989.

[TH89] T. Thong, Practical considerations for a continuous time digital spectrum analyzer. *Proc. IEEE Int. Symp. Circuits Syst.*, Portland, Oregon, pp. 1047–1050 (1989).

[TO81] W. J. Tompkins and J. G. Webster, eds., *Design of Microcomputer-Based Medical Instrumentation.* Prentice-Hall, Englewood Cliffs, NJ, 1981.

[WO89] J. M. Worham, *Sound Recording Handbook.* Howard W. Sams & Co., Indianapolis, IN, 1989.

GENERAL REFERENCES

M. H. Ackroyd, *Digital Filters.* Butterworth, London, 1973.

J. K. Aggarwal, ed., *Digital Signal Processing.* Western Periodicals Co., North Hollywood, CA, 1979.

A. Antoniou, *Digital Filters: Analysis and Design.* McGraw-Hill, New York, 1979.

K. G. Beauchamp, *Signal Processing.* Allen & Unwin, London, 1973.

K. G. Beauchamp and C. K. Yuen, *Digital Methods for Signal Analysis.* Allen & Unwin, London, 1979.

R. E. Bogner and A. G. Constantinidas, eds., *Introduction to Digital Filtering.* Wiley (Interscience), New York, 1975.

N. K. Bose, *Digital Filters: Theory and Applications.* North-Holland, New York, 1985.

J. A. Cadzow, *Foundations of Digital Signal Processing and Data Analysis.* Macmillan, New York, 1987.

C.-T. Chen, *One-dimensional Digital Signal Processing.* Dekker, New York, 1979.

D. Childers and A. Durling, *Digital Filtering and Signal Processing.* West Publishing Co., St. Paul, NY, 1975.

D. J. DeFatta, J. G. Lucas, and W. S. Hodgkiss, *Digital Signal Processing: A Systems Design Approach.* Wiley, New York, 1988.

D. E. Dudgeon and R. M. Mersereau, *Two-Dimensional Digital Signal Processing*. Prentice-Hall, Englewood Cliffs, NJ, 1983.

J. L. Flanagan, *Speech Analysis, Synthesis and Perception*, 2nd ed. Springer-Verlag, New York, 1972.

B. Gold and C. M. Rader, *Digital Processing of Signals*. McGraw-Hill, New York, 1969.

U. Grenander and M. Rosenblatt, *Statistical Analysis of Stationary Time Series*. Wiley, New York, 1957.

R. A. Haddad and T. W. Parsons, *Digital Signal Processing: Theory, Applications and Hardware*. Computer Science Press, New York, 1991.

R. W. Hamming, *Digital Filters*, 2nd ed. Prentice-Hall, Englewood Cliffs, NJ, 1983.

L. B. Jackson, *Digital Filters and Signal Processing*. Kluwer Academic Publishers, Higham, MA, 1986.

J. R. Johnson, *Introduction to Digital Signal Processing*. Prentice-Hall, Englewood Cliffs, NJ, 1989.

M. Kunt, *Digital Signal Processing*. Artech House, Norwood, MA, 1986.

H. Y.-F. Lam, *Analog and Digital Filters*. Prentice-Hall, Englewood Cliffs, NJ, 1979.

J. S. Lim and A. V. Oppenheim, *Advanced Topics in Signal Processing*. Prentice-Hall, Englewood Cliffs, NJ, 1988.

A. V. Oppenheim, *Applications of Digital Signal Processing*. Prentice-Hall, Englewood Cliffs, NJ, 1978.

A. V. Oppenheim and R. W. Schafer, *Discrete-Time Signal Processing*. Prentice-Hall, Englewood Cliffs, NJ, 1989.

R. K. Otnes and L. Enochsen, *Digital Time Series Analysis*. Wiley (Interscience), New York, 1972.

A. Peled and B. Liu, *Digital Signal Processing: Theory, Design and Implementation*. Wiley, New York, 1976.

L. R. Rabiner and B. Gold, *Theory and Applications of Digital Signal Processing*. Prentice-Hall, Englewood Cliffs, NJ, 1975.

L. R. Rabiner and R. W. Schafer, *Digital Processing of Speech Signals*. Prentice-Hall, Englewood Cliffs, NJ, 1978.

R. A. Roberts and C. T. Mullis, *Digital Signal Processing*. Addison-Wesley, Reading, MA, 1987.

E. A. Robinson and M. T. Silvia, *Digital Signal Processing and Time Series Analysis*. Holden-Day, San Francisco, CA, 1978.

E. A. Robinson and T. S. Durrani, *Geophysical Signal Processing*. Prentice-Hall, Englewood Cliffs, NJ, 1985.

W. D. Stanley, G. R. Dougherty, and R. A. Dougherty, *Digital Signal Processing*, 2nd ed. Reston Publishing Co., Reston, VA, 1984.

S. D. Stearns and R. A. David, *Signal Processing Algorithms*. Prentice-Hall, Englewood Cliffs, NJ, 1988.

S. D. Stearns and D. R. Hush, *Digital Signal Analysis*. Prentice-Hall, Englewood Cliffs, NJ, 1990.

R. D. Sturm and D. E. Kirk, *First Principles of Discrete Systems and Digital Signal Processing*. Addison-Wesley, Reading, MA, 1988.

F. J. Taylor, *Digital Filter Design Handbook*. Dekker, New York, 1983.

T. J. Terrell, *Introduction to Digital Filters*, 2nd ed. Macmillan Education, London, 1988.

S. A. Tretter, *Introduction to Discrete-Time Signal Processing*. Wiley, New York, 1976.

C. S. Williams, *Designing Digital Filters*. Prentice-Hall, Englewood Cliffs, NJ, 1986.

T. Young, *Linear Systems and Digital Signal Processing*. Prentice-Hall, Englewood Cliffs, NJ, 1985.

2 Mathematical Foundations of Signal Processing

KENNETH STEIGLITZ

Department of Computer Science
Princeton University
Princeton, New Jersey

2-1 SIGNALS

A *signal* is a function, usually of time or space, that conveys information. For example, a speech signal is a representation of the air pressure at a point in space as a function of time. Another familiar example is a picture: here the signal is a representation of the intensity and color as functions of two spatial coordinates x and y. A motion picture film can be thought of as a function of three variables, the x-y coordinates and time.

The independent variable of a signal can be either continuous or discrete. In the first case we say the signal is *continuous-time;* in the second, *discrete-time*. A continuous-time signal is defined at every instant of time, whereas a discrete-time signal is defined only at a set of discrete instants. (For the purposes of discussion, we assume that a signal is a function of the single independent variable *time*, and we make it clear when that is not the case.) In the same way, the dependent variable of a signal, the *values* of a signal, can be either continuous or discrete. Thus there is a natural classification of signals into four categories, as shown in Table 2-1.

When both amplitude and time are continuous we call the signal *analog* (see Figure 2-1). This case corresponds to the kind of signals that are thought of as occurring "naturally" in the physical world; that is, the kinds of signals discussed in introductory physics courses. Such signals cannot be accessed directly by a digital computer but must be converted to digital form for processing by computer.

When the values of a signal are defined only at discrete points in time, and the values themselves are coded with a finite number of bits, so that they can take on only a discrete set of values, the signal is called a *digital* signal (see Figure 2-2).

Handbook for Digital Signal Processing, Edited by Sanjit K. Mitra and James F. Kaiser.
ISBN 0-471-61995-7 © 1993 John Wiley & Sons, Inc.

Table 2-1   Classification of Signals

Time	Amplitude	
	Continuous	Discrete
Continuous	Analog	Quantized boxcar
Discrete	Sampled-data	Digital

These are the kinds of signals that can be stored in the digital storage media used by digital computers: RAMs, ROMs, digital disks, digital tapes, and so on.

The case of discrete-time continuous-amplitude signals, which we call *sampled-data* signals (see Figure 2-3), is especially important from a conceptual point of view, because it is often convenient to think of such signals as the limiting case of digital signals, when we wish to ignore the effect of coding amplitude values with only a finite number of bits. The most successful mathematical techniques for the study of digital signal processing, such as the z-transform and the discrete Fourier transform (DFT), deal with sampled-data signals.

The final case, when time is continuous but the values are discrete, can occur when digital signals are maintained at fixed values between clocking times, as can happen in an electrical circuit (see Figure 2-4). For example, the signal in a cable that carries a signal may (ideally) stay at one of two given voltage levels and change only at discrete times. We can call such signals *quantized boxcar* signals, although there does not seem to be a standard term for this case.

When time is discrete we usually choose the particular values to be equally spaced, and the time between samples is called the *sampling interval* (see Figure 2-2). This not only makes it easier to analyze processing mathematically but is also most convenient in practical implementations. Usually, a digital signal is obtained from an analog signal by sampling at uniform time intervals and then quantizing the amplitude. But even when a signal is generated abstractly, we still refer to its particular values as "samples."

Sometimes signals are classified by the region of support of the time variable. In what follows we use the discrete time variable k to represent time, but the same definitions apply in the continuous-time case as well. If a signal is nonzero only

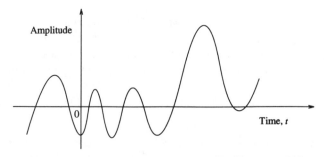

FIGURE 2-1   A continuous-time, continuous-amplitude signal.

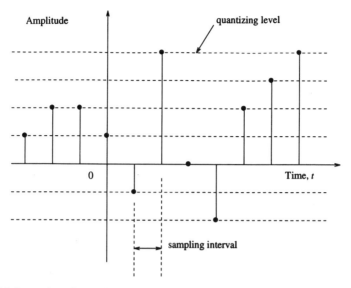

FIGURE 2-2 A *digital signal;* both the amplitude and time are quantized. The sampling interval and quantizing levels are indicated.

over an interval of finite extent, we say the signal is *finite-length*, or *finite-extent*. If it is nonzero only for $k > 0$, we say it is *positive-time*, or *causal;* when nonzero only for $k < 0$ *negative-time*, or *anti-causal*. The term "causal" comes from the fact that a physical system cannot anticipate its input and so responds only after an input is applied.

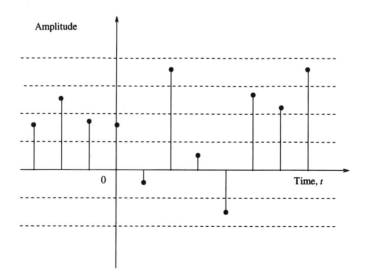

FIGURE 2-3 A *sampled-data signal;* amplitude is continuous, but time is discrete.

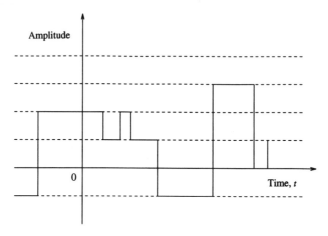

FIGURE 2-4 A *quantized boxcar* signal; the amplitude values are discrete, but time is continuous.

We make a distinction between real- and complex-valued signals. Complex-valued signals arise quite frequently in signal processing—more frequently than might at first be supposed. For example, the real-valued, discrete-time sinusoidal signal

$$x[k] = A \cos (\omega k + \phi)$$

can be thought of as the real part of the complex-valued signal

$$z[k] = Ce^{j\omega k},$$

where $j = \sqrt{-1}$, and $C = Ae^{j\phi}$. In this case $x[k]$ is said to be the *in-phase* component of $z[k]$, and the corresponding imaginary part, the *quadrature* component. We will make good use of complex-valued signals when we come to discuss the frequency domain.

We conclude this section with some definitions describing symmetry relations. If a real-valued signal is symmetric about the time origin, meaning that $x[k] = x[-k]$, we say the signal is *even*. If it is antisymmetric, meaning $x[k] = -x[-k]$, we say it is *odd*. The *even part* of a signal $x[k]$ is defined to be $x_e[k] = (x[k] + x[-k])/2$, and the *odd part* to be $x_o[k] = (x[k] - x[-k])/2$. It is not hard to see that the even and odd parts are in fact even and odd, respectively. Furthermore, the signal is the sum of its even and odd parts, $x[k] = x_e[k] + x_o[k]$.

When the signal is complex, the notions corresponding to even and odd are *Hermitian* and *anti-Hermitian*, respectively. More precisely, the complex signal $z[k]$ is called Hermitian if it obeys the relation $z[k] = z^*[-k]$, where * denotes complex conjugation; and anti-Hermitian if $z[k] = -z^*[-k]$. In a way analogous to the real case, every signal can be written as the sum of its Hermitian and anti-Hermitian parts.

2-2 DIGITAL SIGNAL PROCESSING: DEFINITION AND BRIEF HISTORY

Digital signal processing encompasses:

- The representation of analog signals by digital signals
- The operation on digital signals to transform them to "more desirable" forms and to estimate parameters or make decisions
- The production of analog signals from digital signals

The category of transforming signals to "more desirable" forms includes an enormous variety of tasks: linear and nonlinear filtering, transformation to and from the frequency domain, and changing of effective sampling rate by interpolation or extrapolation, to mention the most common.

The roots of digital signal processing lie in the 17th and 18th century, when astronomers needed to extract the maximum amount of information from their observations of the positions of the stars and planets. The 19th century saw the development of the systematic body of techniques we know as the calculus of finite differences, which laid the foundations for modern mathematical techniques. The work of Boole [BO72], published in 1872, treats difference equations and makes heavy use of operator notation. The next major impetus to the subject came with the increased concentration on signal processing brought about by the technologically sophisticated warfare of World War II. The reader is referred to the book by Wiener [WI48], where much of the analysis is already in terms of sampled-data signals (time series). The problems in the use of digital controllers for feedback systems led to a blossoming of the field in the 1950s. A good view of the state of the field in the late 1950s is provided by the important book by Ragazzini and Franklin [RA58]. By that time the uniform sampling process and the z-transform were well understood and widely used. From that base, and the advances in digital computer technology, the development of digital filter design and implementation techniques was quite natural.

The next important event was the exposition of a fast Fourier transform (FFT) algorithm by Cooley and Tukey [CO65] in 1965. (See also Cooley et al. [CO67] and Heideman et al. [HE84] for an account of the history of such algorithms, which were available but not exploited for many years.) FFT algorithms calculate the Fourier transform of N-point digital signals in time proportional to $N \log N$ instead of N^2, and this makes it possible to use the frequency domain *numerically* as well as theoretically, for signals with thousands of points. The late 1960s and early 1970s saw a rapid expansion and maturing of the field of digital signal processing, both from a practical and a theoretical point of view.

A fundamental change in point of view occurred during the 1960s, a change that affected the way people view the field of digital signal processing as a whole. Through the postwar work on sampled-data control systems, and the development of digital filtering ideas for numerical analysis applications, it was largely assumed that discrete-time signals were samples of analog signals, and the processing that was done digitally was merely an approximation to processing that could, in theory, be done in the analog domain. Thus digital signal processing was viewed as a form of *simulation*. It gradually became clear, however, that discrete-time signals could be viewed as signals in their own right, and their processing could be viewed as more than just a simulation of analog processing. In fact, some operations are quite practical to do digitally and are extremely difficult, if not impossible, to do in the analog domain, such as narrowband filtering with no phase distortion.

We are seeing today another major change in the way people think of digital signal processing: computer technology has reached the point where it is economical to build special purpose processors that can be dedicated to signal processing tasks, and we are seeing the development of custom VLSI chips for many tasks of a more or less specialized nature. Thus digital signal processing implementation has moved from the realm of software for large, general purpose computers, to software for microcomputers, and to the design of the digital hardware itself.

2-3   TIME-DOMAIN REPRESENTATION OF SIGNALS AND FILTERS

This chapter deals with the mathematical techniques that are used in digital signal processing. In many cases the theory in the discrete-time and continuous-time cases are precisely parallel; hence we describe the continuous-time version of a relationship along with the discrete. After that we discuss the meeting ground between discrete- and continuous-time signals: the processes of sampling and reconstruction.

We use $x[n]$ to represent the sample values of a discrete-time signal, where n is an integer variable in the range $-\infty < n < +\infty$, and $x(t)$ the values of a continuous-time signal, where t is a real-valued variable in the range $-\infty < t < +\infty$. When we want to distinguish between a discrete-time signal itself (a sequence) and its sample values, we use $\{x[n]\}$ for the signal. Note that we allow the time variables n and t to vary from $-\infty$ to $+\infty$, so that we deal with "two-sided" signals. One-sided signals are then treated just as special cases of two-sided signals.

Throughout this book we write discrete-time signals, or sequences, with square brackets, as in $x[n]$, and continuous-time signals, or functions of continuous variables, with ordinary parentheses, as in $x(t)$.

Next we discuss some important particular signals. The *unit-sample*, or *unit-impulse* signal (see Figure 2-5), is defined as

$$\delta[n] = \begin{cases} 1 & \text{if } n = 0 \\ 0 & \text{if } n \neq 0. \end{cases} \tag{2.1}$$

The continuous-time counterpart of $\delta[n]$ is the Dirac delta-function $\delta(t)$, which can be defined rigorously using distribution theory. The reader is referred to Zemanian [ZE65] for a rigorous development of generalized functions; for our pur-

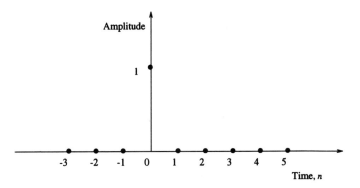

FIGURE 2-5 The special digital signal $\delta[n]$ called the *unit-sample*, or *unit-impulse* signal, the discrete-time counterpart of the Dirac delta-function.

poses it is enough to visualize a function that is zero everywhere except at the origin, where it has a spike of unit area. The cardinal property of the delta-function is that for any function $f(t)$ continuous at the origin,

$$f(0) = \int_{-\infty}^{+\infty} f(t)\delta(t)\, dt. \tag{2.2}$$

That is, the delta-function "punches" out the value of the integrand at the point of its spike. More generally, if $t_1 < t_0 < t_2$,

$$f(t_0) = \int_{t_1}^{t_2} f(t)\delta(t - t_0)\, dt$$

Another fundamental signal is the *unit-step sequence* (see Figure 2-6), sometimes called the *Heaviside step sequence:*

$$\mu[n] = \begin{cases} 1 & \text{if } n \geq 0 \\ 0 & \text{if } n < 0 \end{cases} \tag{2.3}$$

and

$$\mu(t) = \begin{cases} 1 & \text{if } t \geq 0 \\ 0 & \text{if } t < 0. \end{cases} \tag{2.4}$$

The unit-sample and unit-step sequences are related to each other by

$$\mu[n] = \sum_{k=-\infty}^{n} \delta[k], \tag{2.5}$$

$$\delta[n] = \mu[n] - \mu[n - 1], \tag{2.6}$$

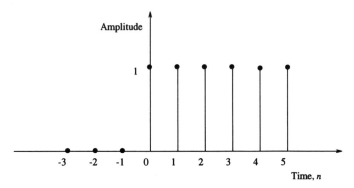

FIGURE 2-6   The special digital signal $\mu[n]$ called the *unit-step sequence*.

corresponding to the following relationships in the continuous-time case:

$$\mu(t) = \int_{-\infty}^{t} \delta(\tau)\, d\tau, \tag{2.7}$$

$$\delta(t) = \frac{d}{dt}\,\mu(t). \tag{2.8}$$

A *filter* is a transformation that maps one signal to another signal: a *sampled-data filter* maps sampled-data signals to sampled-data signals; a *digital filter* maps digital signals to digital, and so on. Most of the time we will actually be dealing mathematically with sampled-data signals and filters, because the signals are real- (or complex-) valued. Finite-wordlength effects are treated as deviations from ideal linear behavior.

More specifically, if x and y are two signals, we express the fact that filter H produces y as output when x is the input by (see Figure 2-7)

$$y = H\{x\}. \tag{2.9}$$

A filter is *linear* if, for all constants a and b and all signals x_1 and x_2, we have

$$H\{ax_1 + bx_2\} = aH\{x_1\} + bH\{x_2\}. \tag{2.10}$$

We next derive a way of characterizing linear filters by their response to the unit impulse. Any discrete-time signal $x[n]$ can be written as a sum of unit impulses as follows:

$$x[n] = \sum_{k=-\infty}^{+\infty} x[k]\delta[n - k]. \tag{2.11}$$

FIGURE 2-7   The digital filter H with input signal $x[n]$ and output $y[n]$.

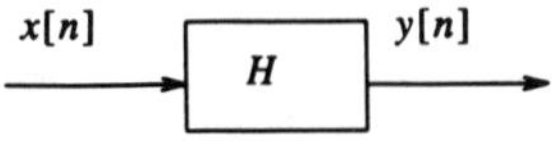

$x[n] \longrightarrow \boxed{H} \longrightarrow y[n]$

If we then apply x to the linear filter H, the output is, by linearity,

$$y[n] = \sum_{k=-\infty}^{+\infty} x[k]H\{\delta[n-k]\}. \tag{2.12}$$

Therefore the effect of the linear filter H is completely determined by the discrete-time signals

$$h_k[n] = H\{\delta[n-k]\}. \tag{2.13}$$

That is, given $h_k[n]$, the output $y[n]$ can be computed explicitly using Eq. (2.12) for any given input $x[n]$. The function $h_k[n]$ is called the *impulse response* of filter H (to an impulse at time k).

The same argument shows that the effect of a continuous-time filter on signal x can be expressed by

$$y(t) = \int_{-\infty}^{+\infty} x(\tau)H\{\delta(t-\tau)\}\, d\tau, \tag{2.14}$$

where

$$h_\tau(t) = H\{\delta(t-\tau)\} \tag{2.15}$$

is the impulse response of linear analog filter H (to an impulse at time τ).

The alert reader will have noticed some sleight of hand here: We defined linearity by the additivity of two items. This can be extended by induction to any finite number of summands, but how do we justify the use of linearity in the case of a countably infinite sum, Eq. (2.12), or an integral, Eq. (2.14)? The answer is that we need to make further assumptions about the behavior of the system, or strengthen our definitions. (See Kailath [KA80] for an interesting discussion of the subtleties in defining linearity.) We will not concern ourselves with such questions of mathematical rigor here. A mathematically rigorous treatment of linear system theory can be carried out using the apparatus of Hilbert space [ST65].

When the impulse response of a linear filter is a function only of the difference between the time of the impulse and the response, we say it is a *linear time-invariant* filter. In this case the filter is oblivious to the time origin; the response to a shifted input is simply a shifted output. The output of a discrete-time linear time-invariant filter can be written in terms of the single-argument impulse response function $h[n]$ as

$$y[n] = \sum_{k=-\infty}^{+\infty} x[k]h[n-k]$$

$$= \sum_{k=-\infty}^{+\infty} h[k]x[n-k]$$

$$= h[n] \circledast x[n] = x[n] \circledast h[n], \tag{2.16}$$

where the second, symmetric form follows from a simple change of variable. This binary operation is called *convolution sum*, or simply *convolution*, and is denoted by the symbol ⊛. Equation (2.16) completely describes a linear time-invariant filter. Thus, given the impulse response sequence $h[n]$ and the input signal $x[n]$, the output signal $y[n]$ can be computed, at least in principle, using this equation. The analogous expressions in the continuous-time case are

$$y(t) = \int_{-\infty}^{+\infty} x(\tau)h(t - \tau)\, d\tau$$

$$= \int_{-\infty}^{+\infty} h(\tau)x(t - \tau)\, d\tau$$

$$= h(t) \circledast x(t) = x(t) \circledast h(t). \tag{2.17}$$

When the impulse response $h[n]$ of a linear time-invariant discrete-time filter is zero for $n < 0$, we say the filter is *causal;* that is, the system does not respond before an input is applied. In the causal case the convolution formulas of Eq. (2.16) reduce to

$$y[n] = \sum_{k=-\infty}^{n} x[k]h[n - k]$$

$$= \sum_{k=0}^{\infty} h[k]x[n - k]. \tag{2.18}$$

Analogously, in the continuous-time case the impulse response $h(t)$ is zero for $t < 0$ and the convolution integrals in Eq. (2.17) become

$$y(t) = \int_{-\infty}^{t} x(\tau)h(t - \tau)\, d\tau$$

$$= \int_{0}^{\infty} h(\tau)x(t - \tau)\, d\tau. \tag{2.19}$$

Figure 2-8 illustrates these formulas, interpreted as a weighting of past values of the input signal.

A further special case arises when the impulse response is of finite duration, in which case all the limits in Eqs. (2.18) and (2.19) are finite, and the filter is called *finite impulse response* (FIR) instead of *infinite impulse response* (IIR) in the general case.

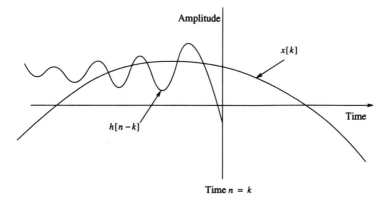

Time $n = k$

FIGURE 2-8 The convolution of a signal $x[n]$ with the causal impulse response $h[n]$, as in Eqs. (2.18) and (2.19). The result at time n is a weighted sum of the signal $x[k]$ times the value of the impulse response k samples in the past, $h[n - k]$.

2-4 FREQUENCY-DOMAIN REPRESENTATION OF SIGNALS AND FILTERS

Besides the unit impulse and step, there is another very important kind of signal, the *complex phasor*, defined by the complex-valued function

$$x[n] = e^{j\omega n} \tag{2.20}$$

in the discrete-time case, or, in the continuous-time case, by

$$x(t) = e^{j\Omega t}. \tag{2.21}$$

The importance of these signals is due to the fact that they are *eigenfunctions* of the shift operator. That is, a shift in the time variable results in the multiplication of a phasor by a complex constant, as can be seen easily from the following (using the discrete-time case to be concrete):

$$x[n - k] = e^{j\omega(n - k)} \tag{2.22}$$
$$= x[n]e^{-j\omega k}.$$

A linear time-invariant filter can be regarded as a linear combination of shift operators, so the effect of such a filter on a phasor is to multiply it by a complex constant that is independent of the time variable (but which in general depends on the frequency variable). This is the basis of the program that we will carry out in this section.

We first mention the issue of the units of the frequency variable ω, which sometimes causes confusion in the discrete-time case. From Eq. (2.20) it is clear that

ω is measured in units of *radians/sampling interval*, and that it can always be taken modulo 2π. Thus it is usual to assume that $-\pi \leq \omega \leq \pi$, and the "highest digital frequency" is π *radians/sampling interval*, called the *Nyquist frequency*. The term *sampling interval* refers here to the increment of the sequence index n and, when the discrete-time signal is an abstract sequence, is, strictly speaking, dimensionless. When the discrete-time signal is thought of as equally spaced samples of a continuous-time signal, the sampling interval is T seconds, and

$$\text{Nyquist frequency} = \pi \text{ radians/sampling interval}$$

$$= \tfrac{1}{2} \text{ revolutions/sampling interval}$$

$$= \pi/T \text{ radians/second}$$

$$= 1/(2T) \text{ Hz.} \tag{2.23}$$

In plots of frequency response one sometimes sees the abscissa labeled "*normalized frequency*" with the Nyquist frequency marked π, in which case the frequency variable is ω in *radians/sampling interval*. Alternatively, one sees the abscissa labeled "*normalized frequency*" with the Nyquist frequency marked 0.5, in which case the frequency variable is $\omega/2\pi$ in *revolutions/sampling interval*. (See Figures 2-16 and 2-17 for examples of the latter convention.) In the continuous-time case there is no inherent limit to the size of the frequency variable Ω, which is measured simply in the units *radians/second*.

The normalized frequency ω can be written as ΩT, where we can now interpret Ω as the unnormalized frequency, having the units of *radians/second*. Thus the Nyquist frequency corresponds to $\omega = \pi$ and $\Omega = \pi/T$. The *sampling frequency* then corresponds to twice the Nyquist frequency and therefore to $\omega = 2\pi$ and $\Omega_s = 2\pi/T$. We reserve the symbol Ω_s for the sampling frequency in *radians per second*.

We return now to the response of a linear time-invariant discrete-time filter to a phasor. If the impulse response is $h[n]$, then, using Eq. (2.22),

$$y[n] = \sum_{k=-\infty}^{+\infty} h[k]x[n-k]$$

$$= \left(\sum_{k=-\infty}^{+\infty} h[k]e^{-jk\omega} \right) \cdot x[n]$$

$$= H(e^{j\omega})x[n], \tag{2.24}$$

where $H(e^{j\omega})$ is called the *frequency response* of the filter. Thus, as we described above informally, the result of filtering a phasor with a linear time-invariant filter is to multiply it by the complex-valued function of frequency, $H(e^{j\omega})$, called the *transfer function* of the filter. The magnitude of this function, $|H(e^{j\omega})|$, is called the *magnitude response*, and the phase angle, $\arg H(e^{j\omega})$, the *phase response* of the filter. Analogously, we have in the continuous-time case

$$y(t) = \int_{-\infty}^{+\infty} h(\tau)x(t - \tau)\, d\tau$$

$$= \left(\int_{-\infty}^{+\infty} h(\tau)e^{-j\tau\Omega}\, d\tau \right) \cdot x(t)$$

$$= H(j\Omega)x(t). \tag{2.25}$$

Note that the signal as a function of time and the transfer function as a function of frequency appear together as a product in these equations. This does not ordinarily happen but is a consequence of the fact that we are dealing here with the very special case when the signal is a phasor.

The expression for the transfer function in terms of the impulse response,

$$H(e^{j\omega}) = \sum_{k=-\infty}^{+\infty} h[k]e^{-jk\omega} \tag{2.26}$$

or

$$H(j\Omega) = \int_{-\infty}^{+\infty} h(\tau)e^{-j\Omega\tau}\, d\tau \tag{2.27}$$

is a Fourier series or Fourier transform, respectively, and can be regarded as a transformation that can be applied to any signal, not necessarily an impulse response. There is a large mathematical literature on the properties of these and related transformations. A classical mathematical treatment can be found in Carslaw [CA30], while a good discussion of the applications of the Fourier integral to engineering problems can be found in Papoulis [PA62]. A very important result in the theory is that the inverse transformation, that from transfer function to impulse response, can be written explicitly as

$$h[n] = \frac{1}{2\pi} \int_{-\pi}^{+\pi} H(e^{j\omega})e^{+j\omega n}\, d\omega \tag{2.28}$$

or

$$h(t) = \frac{1}{2\pi} \int_{-\infty}^{+\infty} H(j\Omega)e^{+j\Omega t}\, d\Omega \tag{2.29}$$

in the discrete- and continuous-time cases, respectively.

What we have done so far is in terms of a phasor of a single frequency, but general signals can be thought of as comprised of a sum of different frequencies. This can be seen clearly from Eq. (2.28), where the impulse response (or any signal) $h[n]$ is written as the linear combination of the phasors $e^{+j\omega n}$, each with weight $H(e^{j\omega})$, and analogously in Eq. (2.29).

So far, frequency is considered to be a real variable: around the circle from $-\pi$ to $+\pi$ in the discrete-time case, and from $-\infty$ to $+\infty$ in the continuous-time case. The next important step is to generalize the mathematics by moving away from the single frequency line into the complex frequency plane. This leads to the z-transform in the discrete case and the Laplace transform in the continuous case, the subjects of the next section.

2-5 THE z- AND LAPLACE TRANSFORMS

Equation (2.26), the expression for the transfer function in terms of the impulse response of a linear time-invariant filter, or the transform of a signal in terms of its sample values,

$$X(e^{j\omega}) = \sum_{k=-\infty}^{+\infty} x[k]e^{-jk\omega} \tag{2.30}$$

can be viewed as a (two-sided) power series in the complex variable $z = e^{j\omega}$:

$$X(z) = \sum_{k=-\infty}^{+\infty} x[k]z^{-k}. \tag{2.31}$$

Such two-sided power series are called *Laurent series* in the mathematical literature, and there is an elegant and extensive theory of such series [CH60; EV66; KN45] that tells us much about existence and the properties of the resulting functions of z.

To be more specific, the existence of the z-transform of a signal can be ensured if we know something about the rate at which the signal decays to zero as n approaches $-\infty$ and $+\infty$. Assume that the absolute value of the signal $x[n]$ at infinity is in fact bounded by exponential functions as follows:

$$|x[n]| \le M_1 K_1^n \quad \text{for } n \ge 0 \tag{2.32}$$

and

$$|x[n]| \le M_2 K_2^n \quad \text{for } n \le 0. \tag{2.33}$$

Then the series Eq. (2.31) converges absolutely in the annular region in the z-plane defined by

$$K_1 < |z| < K_2. \tag{2.34}$$

It is also uniformly convergent on every circle interior to and concentric with this annulus.

As an important example, consider the discrete-time signal

$$x[n] = \begin{cases} a^n & \text{if } n \geq 0 \\ 0 & \text{if } n < 0 \end{cases}$$

$$= a^n \mu[n], \tag{2.35}$$

where a is in general a complex constant. Then the z-transform of $x[n]$, which we conventionally denote by the capital letter, is easily written in closed form because it is a geometric series:

$$X(z) = \sum_{n=0}^{\infty} a^n z^{-n} = \frac{1}{1 - az^{-1}} \quad \text{for } |z| > |a|. \tag{2.36}$$

Note that the z-transform converges for arbitrarily large values of z because $x[n]$ vanishes for negative n, and we can take $K_1 = |a|$ in Eq. (2.32). Figure 2-9 shows the pole and zero of this transform in the complex z-plane.

If we choose for the complex constant a the value $be^{j\theta}$, where b and θ are real, and we then take real and imaginary parts of the transform treating z as a formal variable, we get the useful transforms for damped trigonometric signals:

$$b^n \cos n\theta \, \mu[n] \rightarrow \frac{1 - b \cos \theta \, z^{-1}}{1 - 2 b \cos \theta \, z^{-1} + b^2 z^{-2}} \quad \text{for } |z| > |b| \tag{2.37}$$

and

$$b^n \sin n\theta \, \mu[n] \rightarrow \frac{b \sin \theta \, z^{-1}}{1 - 2b \cos \theta \, z^{-1} + b^2 z^{-2}} \quad \text{for } |z| > |b|, \tag{2.38}$$

where we use the notation $x[n] \rightarrow X(z)$ to mean that the z-transform of $x[n]$ is $X(z)$.

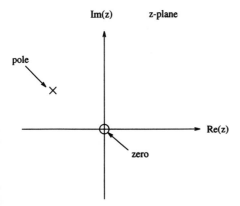

FIGURE 2-9 The complex z-plane, showing the pole at $z = a$ and the zero at the origin of the transform $X(z)$ in Eq. (2.36).

As another example, consider the two-sided signal

$$x[n] = \begin{cases} (\tfrac{1}{2})^n & \text{if } n \geq 0 \\ 3^n & \text{if } n < 0. \end{cases} \tag{2.39}$$

The transform is now the sum of two geometric series, and we can take $K_1 = \tfrac{1}{2}$ and $K_2 = 3$ in Eq. (2.32) and Eq. (2.33) to yield (see Figure 2-10)

$$X(z) = \frac{1}{1 - \tfrac{1}{2}z^{-1}} + \frac{\tfrac{1}{3}z}{1 - \tfrac{1}{3}z} \qquad \text{for } \tfrac{1}{2} < |z| < 3. \tag{2.40}$$

The region of convergence tells us that the first term in the z-transform represents a power series in negative powers of z, while the second term represents a series in positive powers of z. In general, we need to know the region of convergence to perform the inverse z-transform.

The theory of Laurent series tells us that a signal can be recovered from its z-transform by the contour integral

$$x[n] = \frac{1}{2\pi j} \oint_C X(z) z^n \frac{dz}{z}, \tag{2.41}$$

where the integral is over any closed counterclockwise contour C within the region of convergence. If the region of convergence includes the unit circle in the z-plane, which it does in the common situation when the signal decays to zero in both

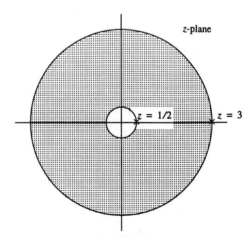

FIGURE 2-10   The complex z-plane, showing the poles of the z-transform of the signal in Eq. (2.39). The region of convergence lies in the shaded annular region between the two circles of convergence.

directions, then we can take the contour to be the unit circle, and Eq. (2.41) reduces to Eq. (2.28). This then gives precise conditions under which the Fourier series and its inverse, Eqs. (2.26) and (2.28), are valid.

The continuous-time theory is exactly analogous, the Fourier transform integral Eq. (2.27) being generalized to the Laplace transform by replacing the variable $j\Omega$ by the complex variables s:

$$X(s) = \int_{-\infty}^{+\infty} x(t)e^{-st}\, dt. \tag{2.42}$$

If we assume that the continuous-time signal $x(t)$ is bounded as

$$|x(t)| \le M_1 e^{\sigma_1 t} \quad \text{for } t \ge 0 \tag{2.43}$$

and

$$|x(t)| \le M_2 e^{\sigma_2 t} \quad \text{for } t \le 0, \tag{2.44}$$

then the Laplace transform converges in the region

$$\sigma_1 < \text{Re}\{s\} < \sigma_2 \tag{2.45}$$

and the inverse Laplace transform is given by

$$x(t) = \frac{1}{2\pi j} \int_{\Gamma} X(s)e^{st}\, ds, \tag{2.46}$$

where the contour Γ goes from $-\infty$ to $+\infty$ within the region of convergence.

2-6 PROPERTIES OF THE z- AND LAPLACE TRANSFORMS

We now summarize some important properties of the z-transform. First, it follows immediately from the form of the transform as a sum that it is linear; that is, if $x[n]$ and $y[n]$ are two discrete-time signals, and a and b are two complex constants, then

$$ax[n] + by[n] \rightarrow aX(z) + bY(z) \quad \text{for } z \in R_x \cap R_y, \tag{2.47}$$

where R_x and R_y are the regions of convergence of the transforms $X(z)$ and $Y(z)$, respectively. (From now it will be convenient to use this notation: R_w will always denote the region of convergence of the transform of the signal w. The notation $z \in R_x \cap R_y$ means simply that z lies in the intersection of the two regions of convergence R_x and R_y.)

The next property relates the transform of a shifted version of a signal to the transform of the signal. The z-transform of $x[n + n_0]$, where n_0 is an integer, is from the definition

$$\sum_{k=-\infty}^{+\infty} x[k + n_0]z^{-k} = \sum_{m=-\infty}^{+\infty} x[m]z^{-(m - n_0)} = z^{n_0}X(z), \qquad (2.48)$$

which gives us the following transform pair:

$$x[n + n_0] \rightarrow z^{n_0}X(z) \quad \text{for } z \in R_x. \qquad (2.49)$$

Clearly, this manipulation of the power series does not change the region of convergence. We can interpret this result in more familiar terms by considering what happens in the case when $x[n]$ is a steady-state phasor of frequency ω. That means that the variable z is on the unit circle at angle ω, and the transform is multiplied by a complex number of unit magnitude and phase angle $n_0\omega$.

There is a strong symmetry between the time and frequency domains, and properties usually come in symmetrical pairs. Equation (2.49) gives us the result of multiplying the transform by an exponential weight; multiplying the time function by an exponential weight a^n yields

$$\sum_{n=-\infty}^{+\infty} a^n x[n]z^{-n} = \sum_{n=-\infty}^{+\infty} x[n](a^{-1}z)^{-n}, \qquad (2.50)$$

so we get the transform pair

$$a^n x[n] \rightarrow X(a^{-1}z) \quad \text{for } z/|a| \in R_x. \qquad (2.51)$$

Note that now the transform is a power series in $a^{-1}z$ instead of z, so the region of convergence is changed accordingly. An especially interesting interpretation of this property is obtained when the constant a is of unit magnitude (say $a = e^{j\theta}$) and $x[n]$ is a phasor. Then Eq. (2.51) becomes

$$e^{j\theta n}x[n] \rightarrow X(e^{j(\omega - \theta)}) \quad \text{for } z \in R_x, \qquad (2.52)$$

so the result of modulating the signal by a complex phasor is to shift the frequency variable in the transform domain by an amount equal to the modulating frequency. This can be called the "heterodyne" principle.

Another property can be derived from the fact that the z-transform converges uniformly within its region of convergence and therefore can be differentiated term by term. Differentiating the definition Eq. (2.31) with respect to z gives

$$\frac{dX(z)}{dz} = \sum_{k=-\infty}^{+\infty} (-k)x[k]z^{-k-1}. \qquad (2.53)$$

Rearranging this gives the transform pair

$$nx[n] \rightarrow -z\frac{dX(z)}{dz} \quad \text{for } z \in R_x. \tag{2.54}$$

The region of convergence is unaffected by the differentiation. This property is useful for deriving z-transforms. For example, differentiating the transform of the unit-step sequence

$$\mu[n] \rightarrow \frac{1}{1 - z^{-1}} \quad \text{for } |z| > 1 \tag{2.55}$$

gives the transform of a linearly increasing signal, the unit ramp signal

$$n\mu[n] \rightarrow \frac{z^{-1}}{(1 - z^{-1})^2} \quad \text{for } |z| > 1. \tag{2.56}$$

The continuous-time case is again perfectly analogous, and the corresponding properties are derived in the same way. The Laplace transform is linear:

$$ax(t) + by(t) \rightarrow aX(s) + bY(s) \quad \text{for } s \in R_x \cap R_y. \tag{2.57}$$

A shift in the time domain results in an exponential weighting in the frequency domain:

$$x(t + t_0) \rightarrow e^{st_0}X(s) \quad \text{for } s \in R_x \tag{2.58}$$

and vice-versa:

$$e^{at}x(t) \rightarrow X(s - a) \quad \text{for } s - \text{Re}\{a\} \in R_x. \tag{2.59}$$

Differentiating the transform with respect to s yields

$$tx(t) \rightarrow -\frac{dX(s)}{ds} \quad \text{for } s \in R_x. \tag{2.60}$$

Finally, there are a number of symmetry properties of the z- and Laplace transforms that are often useful. For example, consider the z-transform on the unit circle when the signal $x[n]$ is an even function of time (i.e., $x[n] = x[-n]$):

$$X(e^{j\omega}) = \sum_{k=-\infty}^{+\infty} x[k]e^{-jk\omega}$$

$$= x(0) + 2\sum_{k=1}^{+\infty} x[k] \cos k\omega. \tag{2.61}$$

Thus the transform on the unit circle of an even signal is a real function of frequency. A signal that is an odd function of time (i.e., $x[n] = -x[-n]$) has a purely imaginary transform on the unit circle. Similarly, suppose that $x[n]$ is a real signal, and take the complex conjugate of its transform:

$$X^*(e^{j\omega}) = \sum_{k=-\infty}^{+\infty} x[k]e^{+jk\omega}$$

$$= X(e^{-j\omega}) \tag{2.62}$$

so that the magnitude of the transform is an even function of frequency, and the phase is odd. As usual, the same results hold in the continuous-time case. More detailed development of these ideas can be found in the texts [JU73; OP75; RA58; TR76], and extensive tables of transforms in Healy [HE67] and Roberts and Kaufman [RO66].

2-7 REAL AND COMPLEX CONVOLUTION

We saw in Eq. (2.16) that the output of a discrete-time filter is the convolution of the input signal and the impulse response of the filter. We also saw from Eq. (2.24) that when the input signal is a pure phasor of frequency ω, the effect of filtering is to multiply it by the transfer function of the filter evaluated at the frequency ω. We now generalize this to arbitrary inputs by using the z-transform. Write the output signal $y[n]$ as a convolution of input signal and impulse response:

$$y[n] = \sum_{k=-\infty}^{+\infty} x[n-k]h[k]. \tag{2.63}$$

Multiplying by z^{-n} and summing yield, after some formal manipulation,

$$Y(z) = \sum_{n=-\infty}^{+\infty} y[n]z^{-n}$$

$$= \sum_{n=-\infty}^{+\infty} \sum_{k=-\infty}^{+\infty} x[n-k]z^{-(n-k)}z^{-k}h[k]$$

$$= \sum_{k=-\infty}^{+\infty} \left[\sum_{n=-\infty}^{+\infty} x[n-k]z^{-(n-k)} \right] z^{-k}h[k]$$

$$= \sum_{k=-\infty}^{+\infty} X(z)z^{-k}h[k]$$

$$= H(z)X(z). \tag{2.64}$$

That is, the z-transform of the output of a linear time-invariant discrete-time filter is the product of the z-transform of the input signal and the z-transform of the impulse response of the filter (its transfer function). While the derivation above was formal, the following result can be established rigorously [KN45, *Problem Book*, Vol. II, Section 6, Problem 5]. If

$$x[n] \rightarrow X(z) \quad \text{for } z \in R_x \tag{2.65}$$

and

$$h[n] \rightarrow H(z) \quad \text{for } z \in R_h \tag{2.66}$$

and $R_x \cap R_h \neq \emptyset$ (the intersection of the regions of convergence is not empty), then $x[n] \circledast h[n]$ converges, and

$$x[n] \circledast h[n] \rightarrow X(z)H(z) \quad \text{for } z \in R_x \cap R_h. \tag{2.67}$$

An analogous result holds for continuous-time systems; the Laplace transform of $x(t) \circledast h(t)$ is $X(s)H(s)$.

It is usually the case that for each result in the time domain there is a corresponding result in the frequency domain. For example, Eq. (2.49) shows that a shift in the time domain corresponds to multiplication by an exponential in the frequency domain, while Eq. (2.52) is the counterpart; it shows that multiplication by an exponential function in the time domain corresponds to a shift in the frequency domain. We thus can expect a frequency domain counterpart to convolution. To find it, consider the product of two signals:

$$w[n] = x[n]y[n]. \tag{2.68}$$

Multiplying by z^{-n} and summing,

$$W(z) = \sum_{n=-\infty}^{+\infty} x[n]y[n]z^{-n}. \tag{2.69}$$

Replacing $x[n]$ by the inverse transform of $X(z)$, we get

$$W(z) = \sum_{n=-\infty}^{+\infty} y[n] \left[\frac{1}{2\pi j} \oint_C X(v)v^n \frac{dv}{v} \right] z^{-n}. \tag{2.70}$$

Finally, interchanging the order of summation and integration,

$$W(z) = \frac{1}{2\pi j} \oint_C X(v) \left[\sum_{n=-\infty}^{+\infty} y[n] \left(\frac{z}{v} \right)^{-n} \right] \frac{dv}{v} \tag{2.71}$$

or

$$W(z) = \frac{1}{2\pi j} \oint_C X(v)Y\left(\frac{z}{v} \right) \frac{dv}{v}, \tag{2.72}$$

which is called the *complex convolution* formula. The region of validity of this formula is determined by the regions of convergence of the transforms of x and y. If $X(z)$ converges in $K_{x_1} < |z| < K_{x_2}$, and $Y(v)$ in $K_{y_1} < |v| < K_{y_2}$, then Eq. (2.72) holds in $K_{x_1}K_{y_1} < |z| < K_{x_2}K_{y_2}$ (see Tretter [TR76], for details). To see that Eq. (2.72) does in fact correspond to convolution in the frequency domain, let $z = e^{j\omega}$, $v = e^{j\theta}$, and take the contour of integration to be the unit circle; then Eq. (2.72) becomes

$$W(e^{j\omega}) = \frac{1}{2\pi} \int_{-\pi}^{+\pi} X(e^{j\theta})Y(e^{j(\omega-\theta)})\, d\theta. \tag{2.73}$$

Note that this convolution is *circular*, in the sense that the functions Y and X are defined on the unit circle, and θ and ω are taken mod 2π.

Setting $\omega = 0$ in Eq. (2.73) yields the useful formula

$$\sum_{n=-\infty}^{+\infty} x(n)y(n) = \frac{1}{2\pi} \int_{-\pi}^{+\pi} X(e^{j\theta})Y(e^{-j\theta})\, d\theta, \tag{2.74}$$

which is a form of *Parseval's relation*. Setting $y(n) = x^*(n)$ (the complex conjugate), we have, from the definition of the z-transform,

$$Y(e^{-j\theta}) = X^*(e^{j\theta}), \tag{2.75}$$

and from Eq. (2.74),

$$\sum_{n=-\infty}^{+\infty} |x(n)|^2 = \frac{1}{2\pi} \int_{-\pi}^{+\pi} |X(e^{j\theta})|^2\, d\theta, \tag{2.76}$$

which can be interpreted as showing that the total energy of the signal in the time domain is equal to the total energy of its frequency components. As usual, completely analogous results hold for continuous-time signals.

Table 2-2 summarizes for reference a few commonly used z-transforms, and the important properties we have derived.

2-8 FINITE-DIMENSIONAL FILTERS

A wide class of linear time-invariant discrete-time filters can be specified by writing a recurrence relation in the time domain that defines the output signal at time n to be a finite linear combination of input and past output values. Specifically, we write the output $y[n]$ as

$$y[n] = \sum_{i=0}^{M} a_i x[n-i] - \sum_{j=1}^{L} b_j y[n-j]. \tag{2.77}$$

Using linearity and the shift property of Eq. (2.49), the z-transform of this equation yields

$$Y(z) = \left(\sum_{i=0}^{M} a_i z^{-i} \right) \cdot X(z) - \left(\sum_{i=1}^{L} b_j z^{-j} \right) \cdot Y(z), \qquad (2.78)$$

which means that the transfer function of this filter, the ratio of output to input z-transforms, is

$$H(z) = \frac{Y(z)}{X(z)} = \frac{\displaystyle\sum_{i=0}^{M} a_i z^{-i}}{1 + \displaystyle\sum_{j=1}^{L} b_j z^{-j}}. \qquad (2.79)$$

(Sometimes a plus sign is used in the definition of Eq. (2.77), which of course results in a minus sign in the denominator of Eq. (2.79).)

The impulse response of the filter, $h[n]$, is then the inverse transform of $H(z)$, by Eq. (2.64). In general, when the denominator in Eq. (2.79) is not a factor of the numerator, the impulse response is infinite in extent, and the filter is an infinite impulse response (IIR) filter. When, on the other hand, the transfer function Eq. (2.79) is a finite-degree polynomial in z^{-1}, the filter output is determined by a

TABLE 2-2 Summary of Important z-Transform Pairs and Properties

Discrete-Time Function	z-Transform
Unit-sample $\delta[n]$	1
Unit-step $\mu[n]$	$1/(1 - z^{-1})$
Exponential $a^n \mu[n]$	$1/(1 - az^{-1})$
Damped cosine wave $b^n \cos(\theta n)$	$\dfrac{1 - b\cos\theta\, z^{-1}}{1 - 2b\cos\theta\, z^{-1} + b^2 z^{-2}}$
Damped sine wave $b^n \sin(\theta n)$	$\dfrac{b\sin\theta\, z^{-1}}{1 - 2b\cos\theta\, z^{-1} + b^2 z^{-2}}$
Linear combination $ax[n] + by[n]$	$aX(z) + bY(z)$
Time shift $x[n + n_0]$	$z^{n_0} X(z)$
Exponential weighting $a^n x[n]$	$X(a^{-1}z)$
Linear weighting $nx[n]$	$-z\,\dfrac{dX(z)}{dz}$
Convolution $x[n] \circledast h[n]$	$X(z)H(z)$
Product $w[n] = x[n]y[n]$	$W(z) = \dfrac{1}{2\pi j} \oint_C X(v) Y\left(\dfrac{z}{v}\right) \dfrac{dv}{v}$
Even $x[n] = x[-n]$	Real $X(e^{j\omega})$
Odd $x[n] = -x[n]$	Imaginary $X(e^{j\omega})$
Real $x[n]$	Even $|X(e^{j\omega})|$ and odd arg $X(e^{j\omega})$

finite linear combination of inputs, and the filter is a finite impulse response (FIR) filter. The interesting and important problems of designing and implementing both IIR and FIR discrete-time filters are discussed in succeeding chapters.

The continuous-time counterpart of Eq. (2.77) is the differential equation

$$y(t) = \sum_{i=0}^{M} a_i \frac{d^i x}{dt^i} - \sum_{j=1}^{L} b_j \frac{d^j y}{dt^j}, \tag{2.80}$$

and the corresponding transfer function is a ratio of polynomials in the Laplace transform variable s. The counterpart of the FIR discrete-time filter is therefore the differential operator

$$y(t) = \sum_{i=0}^{M} a_i \frac{d^i x}{dt^i}, \tag{2.81}$$

which is not practical to implement because differentiation in the continuous-time domain is a very noisy operation. In strong contrast to this, FIR discrete-time filters are very practical and widely used.

2-9 IDEAL SAMPLING

Up to this point we have dealt with discrete-time and continuous-time signals separately. It is now time to consider the transformations from one to the other. We first take up the case of equally spaced sampling, to convert a continuous-time to a discrete-time signal. This can be modeled mathematically by multiplying the continuous-time signal $f(t)$ by a periodic train of ideal impulses with period T (see Figure 2-11):

$$p(t) = \sum_{k=-\infty}^{+\infty} \delta(t - kT). \tag{2.82}$$

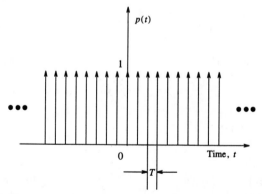

FIGURE 2-11   The periodic train of ideal impulses, $p(t)$.

We represent the resulting impulse train by $f^\#(t)$ (pronounced "f-sharp"), using the felicitous notation of Tretter [TR76].

(Often $f*(t)$ is used to represent the ideally sampled version of $f(t)$, but we reserve ( )* to mean complex conjugate.) Thus

$$f^\#(t) = f(t)p(t) = \sum_{k=-\infty}^{+\infty} f(kT)\delta(t - kT). \tag{2.83}$$

This can be thought of as a continuous-time signal, a sequence of evenly spaced impulses, each weighted by the value of $f(t)$ at the time of the impulse (see Figure 2-12). The Laplace transform of $f^\#(t)$ is, from Eq. (2.83),

$$F^\#(s) = \sum_{k=-\infty}^{+\infty} f(kT)e^{-ksT}, \tag{2.84}$$

which shows that the Laplace transform of $f^\#(t)$ is simply the z-transform of the discrete-time signal $\{f(kT)\}$ evaluated at $z = e^{sT}$:

$$F^\#(s) = [F(z)]_{z=e^{sT}} \tag{2.85}$$

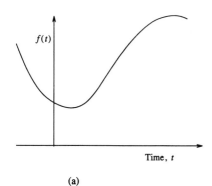

(a)

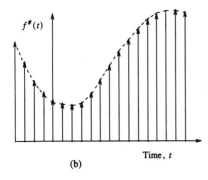

Time, t

(b)

FIGURE 2-12   (a) A continuous-time signal $f(t)$ and (b) the impulse train $f^\#(t)$, representing the ideally sampled version of $f(t)$.

Another form for $F^\#(s)$ can be derived by expanding the ideal impulse train $p(t)$ in a Fourier series. Thus

$$p(t) = \sum_{n=-\infty}^{+\infty} c_n e^{jn(2\pi/T)t}, \tag{2.86}$$

where

$$c_n = \frac{1}{T} \int_{-T/2}^{+T/2} p(t) e^{-jn(2\pi/T)t} \, dt = \frac{1}{T}. \tag{2.87}$$

Therefore

$$f^\#(t) = \left(\frac{1}{T} \sum_{n=-\infty}^{+\infty} e^{jn(2\pi/T)t} \right) \cdot f(t). \tag{2.88}$$

Using the multiplication-shift property of the Laplace transform (Eq. (2.59)), this yields

$$F^\#(s) = \frac{1}{T} \sum_{n=-\infty}^{+\infty} F\left(s + jn\frac{2\pi}{T} \right) \tag{2.89}$$

From this we see that the effect of sampling on the transform is to add versions of the spectrum, shifted by multiples of the sampling frequency $2\pi/T$ (see Figure 2-13). Since components at frequencies $\omega + n2\pi/T$ masquerade as components at ω, these frequencies are called *aliases* of ω, and Eq. (2.89) is called the *aliasing formula*.

2-10 RECONSTRUCTION

From the aliasing formula Eq. (2.89) we can see that if $F(j\Omega)$ is *bandlimited* in the sense that

$$F(j\Omega) = 0 \quad \text{for } |\Omega| \geq \pi/T, \tag{2.90}$$

then there is no overlap between the various shifted versions of the spectrum that are added. Thus in the range $|\Omega| \leq \pi/T$, we can write the true transform of $f(t)$ in terms of the transform of the pulse train $f^\#(t)$:

$$F(j\Omega) = T\{F^\#(j\Omega)\} \quad \text{for } |\Omega| \leq \pi/T. \tag{2.91}$$

This operation can be interpreted as an ideal lowpass filter that produces $f(t)$ from $f^\#(t)$, as illustrated in Figure 2-14. From this it follows that we can recover $f(t)$

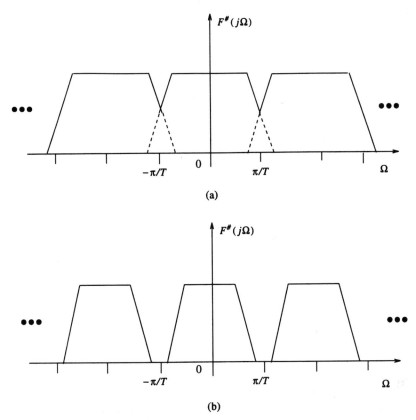

FIGURE 2-13 The transform of a sampled signal, showing that it is the sum of shifted versions of the original continuous-time signal. Original frequency components above the Nyquist frequency π/T radians/second are aliased to frequencies below the Nyquist frequency. Part (a) shows the case when aliasing takes place; part (b) the case without aliasing.

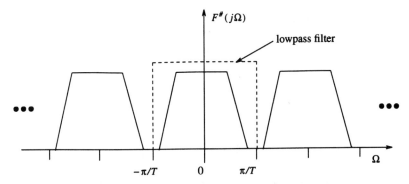

FIGURE 2-14 Transform of a sampled bandlimited signal. The transform of the original continuous-time signal can be recovered exactly with an ideal lowpass filter, shown by dashed lines.

as the inverse Fourier transform of $F(j\Omega)$:

$$
f(t) = \frac{1}{2\pi} \int_{-\pi/T}^{+\pi/T} F(j\Omega) e^{j\Omega t} \, d\Omega
$$

$$
= \frac{1}{2\pi} \int_{-\pi/T}^{+\pi/T} T\{F^{\#}(j\Omega)\} e^{j\Omega t} \, d\Omega. \tag{2.92}
$$

We can now write $F^{\#}(j\Omega)$ in terms of the sample values $f(nT)$; using Eq. (2.84),

$$
f(t) = \frac{T}{2\pi} \int_{-\pi/T}^{+\pi/T} \left(\sum_{n=-\infty}^{+\infty} f(nT) e^{-jn\Omega T} \right) e^{j\Omega t} \, d\Omega
$$

$$
= \sum_{n=-\infty}^{+\infty} f(nT) \left(\frac{T}{2\pi} \int_{-\pi/T}^{+\pi/T} e^{j\Omega(t-nT)} \, d\Omega \right) \tag{2.93}
$$

and, carrying out the integration,

$$
f(t) = \sum_{n=-\infty}^{+\infty} f(nT) \, \frac{\sin\left(\dfrac{\pi}{T}(t - nT)\right)}{\dfrac{\pi}{T}(t - nT)}. \tag{2.94}
$$

This is known as *Shannon's sampling theorem*, or the *cardinal reconstruction formula;* it shows that an ideal bandlimited signal with maximum frequency component $1/T$ Hz is determined completely (ideally) by samples T seconds apart. Equation (2.94) can be written

$$
f(t) = \sum_{n=-\infty}^{+\infty} f(nT) h(t - nT), \tag{2.95}
$$

where $h(t)$ is the impulse response of an ideal reconstruction filter,

$$
h(t) = \frac{\sin(\pi t/T)}{\pi t/T}, \tag{2.96}
$$

which has the property that

$$
h(nT) = \begin{cases} 1 & \text{if } n = 0 \\ 0 & \text{if } n \neq 0. \end{cases} \tag{2.97}
$$

Equation (2.95) is a convolution formula of a special kind; it shows the continuous-time response of a filter to a pulse-train input. The frequency response of this filter, the Fourier transform of $h(t)$, is simply the ideal lowpass response shown in Figure 2-14, passing only frequencies up to π/T radians/second, the Nyquist frequency.

The property expressed by Eq. (2.97) is illustrated in Figure 2-15 and ensures that the reconstructed signal passes precisely through the sample values at times kT, for all integers k; but we know also that when $f(t)$ is bandlimited, the reconstruction is perfect at all values of t.

While of theoretical interest, the cardinal reconstruction formula does not provide a practical way to reconstruct a continuous-time signal from its samples, because it involves an infinite sum and corresponds to an ideally sharp cutoff at the frequency π/T radians/second. A more realistic way to obtain a continuous-time signal from samples is to hold the sample values of the signal constant between sampling instants. This is often what happens by design at the output stage of commercial digital-to-analog converters. This corresponds to the impulse response

$$h(t) = \begin{cases} 1/T & \text{if } 0 \leq t < T \\ 0 & \text{otherwise} \end{cases} \tag{2.98}$$

(normalizing the area to 1). Such a reconstruction scheme is called a *zero-order* or *boxcar* hold and corresponds to the transfer function

$$H(s) = \frac{1 - e^{-sT}}{sT}, \tag{2.99}$$

which can be thought of as an approximation to an ideal lowpass filter. Its magnitude response as a function of frequency Ω is easily computed to be

$$|H(j\Omega)| = \left| \frac{\sin(\Omega T/2)}{\Omega T/2} \right|. \tag{2.100}$$

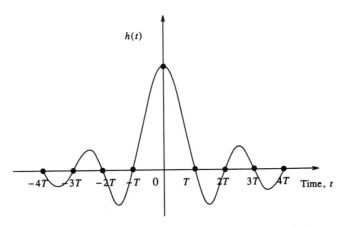

FIGURE 2-15 The impulse response of the ideal lowpass filter.

This magnitude response decreases gradually from 1 at $\Omega = 0$, and at $\Omega = \pi/T$, $|H(j\Omega)| = 2/\pi = 0.637$, so this filter is very far from ideal. A linear, analog postfilter can be used after the zero-order hold to provide a much sharper cutoff, as well as to compensate for the "droop" in the passband, thus approaching ideal behavior more closely. Figure 2-16 shows the magnitude response of the zero-order hold.

A more sophisticated reconstruction filter is the *linear point connector*, which connects sequential sample values with straight-line segments. The impulse response of this filter, normalized to unit area, is

$$h(t) = \begin{cases} (T + t)/T^2 & \text{if } -T \le t \le 0 \\ (T - t)/T^2 & \text{if } 0 \le t \le T \end{cases} \tag{2.101}$$

with the corresponding transfer function

$$H(s) = \frac{e^{sT} - 2 + e^{-sT}}{s^2 T^2} \tag{2.102}$$

and magnitude response

$$|H(j\Omega)| = \left(\frac{\sin (\Omega T/2)}{\Omega T/2}\right)^2, \tag{2.103}$$

which is shown in Figure 2-17. This is just the square of the magnitude response of the zero-order hold, and the linear point connector accordingly provides a closer approximation to an ideal lowpass filter.

In any event, it is usually necessary in practice to follow digital-to-analog conversion with an analog postfilter to remove frequency components above the Nyquist frequency.

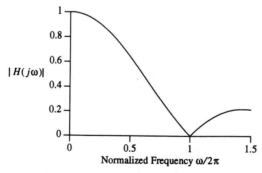

FIGURE 2-16 The magnitude of the frequency response of the zero-order hold. The Nyquist frequency corresponds to $\omega = \pi$.

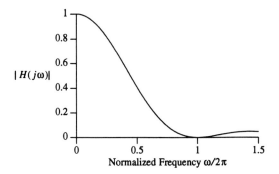

FIGURE 2-17 The magnitude of the frequency response of the linear point connector. The Nyquist frequency corresponds to $\omega = \pi$.

2-11 THE PULSE TRANSFER FUNCTION

The previous section provided an example of a filter with an impulse-train input and a continuous-time output. Any analog filter can be used in this way, and we next summarize the mathematics of this situation.

We have denoted the operation of sampling the continuous-time signal $f(t)$ by writing $f^{\#}(t)$. The corresponding operation on the Laplace transform is denoted in the same way, and we have seen in Eq. (2.89) that

$$F^{\#}(s) = \frac{1}{T} \sum_{n=-\infty}^{+\infty} F(s + jn2\pi/T). \tag{2.104}$$

In general, if $F(s)$ and $G(s)$ are two transforms,

$$[F(s)G(s)]^{\#} \neq F^{\#}(s)G^{\#}(s). \tag{2.105}$$

For example, let $F(s)$ and $G(s) = 1/s$. Then

$$F^{\#}(s) = G^{\#}(s) = \frac{1}{1 - e^{-sT}} \tag{2.106}$$

but (e.g., see the table in Ragazzini and Franklin [RA58])

$$[F(s)G(s)]^{\#} = \left[\frac{1}{s^2} \right]^{\#} = \frac{Te^{-sT}}{(1 - e^{-sT})^2}. \tag{2.107}$$

However, the following result holds, which we call the *pulse transfer function theorem:*

$$[F^{\#}(s)G(s)]^{\#} = F^{\#}(s)G^{\#}(s). \tag{2.108}$$

This has the following interpretation: if a pulse train $f^{\#}(t)$ is applied to an analog filter with transfer function $G(s)$, and the resultant output sampled, the net effect is the same as the digital filter with transfer function $G^{\#}(s)$. $G^{\#}(s)$ is a periodic function of s and can be written as a function of $z = e^{sT}$; it is called the *pulse transfer function*.

To prove Eq. (2.108), first write the continuous-time output $q(t)$ of the analog filter with transfer function $G(s)$ as

$$q(t) = \sum_{k=-\infty}^{+\infty} f(kT)g(t - kT). \tag{2.109}$$

Sampling this at $t = nT$ yields

$$q(nT) = \sum_{k=-\infty}^{+\infty} f(kT)g(nT - kT), \tag{2.110}$$

which is precisely the time-domain version of Eq. (2.108).

The pulse transfer function theorem can be applied to the problem of simulating the behavior of a continuous-time filter with transfer function $G(s)$, say, with a discrete-time filter [RA58]. Imagine that we are going to apply the pulse train $f^{\#}(t)$ in the discrete-time simulation. To approximate what happens when we apply the continuous-time signal $f(t)$, insert an imaginary reconstruction filter with transfer function $R(s)$ and then sample the output of the transfer function $G(s)$. The better the reconstruction filter is at reconstructing $f(t)$, the closer will the output approximate the output of the continuous-time system. The resulting transfer function is

$$D^{\#}(s) = [R(s)G(s)]^{\#} \tag{2.111}$$

and it can be interpreted and implemented as a digital filter.

As a simple example, suppose we want to simulate the integration operator $G(s) = 1/s$. Using the linear point connector as the conceptual input reconstruction device, we get

$$D^{\#}(s) = \left[\frac{e^{sT} - 2 + e^{-sT}}{s^3 T^2}\right]^{\#}$$

$$= \frac{e^{sT} - 2 + e^{-sT}}{T^2}\left[\frac{1}{s^3}\right]^{\#} \tag{2.112}$$

by Eq. (2.108). (That is, any function of e^{sT} can be moved outside the $[\ ]^{\#}$ operator.) Using the transform (e.g., see the table in Ragazzini and Franklin [RA58])

$$\left[\frac{1}{s^3}\right]^{\#} = \frac{T^2}{2}\frac{e^{-sT} + e^{-2sT}}{(1 - e^{-sT})^3}, \tag{2.113}$$

we get

$$D^{\#}(s) = \frac{1}{2} \frac{1 + e^{-sT}}{1 - e^{-sT}}. \tag{2.114}$$

Writing this as a digital filter transfer function, a function of z,

$$D(z) = \frac{1}{2} \frac{1 + z^{-1}}{1 - z^{-1}}. \tag{2.115}$$

We see that this is equivalent to the recursive formula

$$q[n] = q[n - 1] + \tfrac{1}{2} [f[n] + f[n - 1]], \tag{2.116}$$

which is simply the trapezoid rule for numerical integration. Thus the linear point connector connects successive input samples with straight-line segments to form a polygon, the filter G integrates that, and the sampled output corresponds precisely to the cumulative area of the resulting trapezoids. This technique can be used to derive a variety of classical numerical integration and differentiation formulas [HO68] and illustrates the close connections between discrete-time system theory and classical numerical analysis.

2-12 THE DISCRETE FOURIER TRANSFORM

We next consider the case where we have a finite number of samples of a discrete-time signal: in one sense this is the only completely realistic situation. By convention, we start with sample number 0, and such a signal with N points is denoted by $f[0], f[1] \cdots, f[N - 1]$. Sometimes it is convenient to consider f to be periodically extended to all index values, so that $\cdots f[-N] = f[0] = f[N] = f[2N] = \cdots$. Sometimes, however, we consider f to be defined only at the points 0 through $N - 1$. This is a matter of mathematical convenience and makes no difference in practice.

The z-transform of a finite-extent discrete-time signal is perfectly well defined as a finite polynomial in z^{-1}:

$$F(z) = \sum_{n=0}^{N-1} f[n]z^{-n}. \tag{2.117}$$

However, we usually compute it only at the following N equally spaced points on the unit circle:

$$z_k = e^{jk2\pi/N} \quad \text{for } k = 0, 1, \cdots, N - 1. \tag{2.118}$$

The reason for this is that, as we shall see, the values of $F(z)$ at these points are sufficient to determine the original N signal values uniquely. We denote $F(z_k)$ by $F(k)$, $k = 0, 1, \cdots, N - 1$, by an abuse of notation. We therefore have the transformation

$$F(k) = \sum_{n=0}^{N-1} f[n]e^{-jkn2\pi/N} \quad \text{for } k = 0, 1, \cdots, N - 1, \qquad (2.119)$$

which is called the *discrete Fourier transform* (DFT). The DFT is a linear transformation from N-vectors to N-vectors, and Eq. (2.119) can be considered simply multiplication by the matrix with klth element $e^{-jkl2\pi/N}$, $k, l = 0, 1, \cdots, N - 1$. That is, if we consider the signal to be an N-vector f, and the DFT another N-vector F, then $F = Wf$, where W is the $N \times N$ matrix just mentioned. Often, the klth element of W is written W_N^{kl}, where $W_N = e^{-j2\pi/N}$.

To derive the inverse of this transformation, and thereby the inverse of this matrix, multiply Eq. (2.119) by $e^{+jkm2\pi/N}$ and sum on k:

$$\sum_{k=0}^{N-1} F(k)e^{jkm2\pi/N} = \sum_{n=0}^{N-1} f[n] \left(\sum_{k=0}^{N-1} e^{jk(m-n)2\pi/N} \right). \qquad (2.120)$$

The sum in parentheses on the right-hand side is 0 if $m \neq n$, and N if $m = n$, so

$$f[m] = \frac{1}{N} \sum_{k=0}^{N-1} F(k)e^{jkm2\pi/N}. \qquad (2.121)$$

This is the *inverse discrete Fourier transform* (IDFT) and differs from the forward transform only in the normalization factor $1/N$ and the sign in the exponent of the kernel.

The forward and inverse DFT are so similar that programs for one can be used to compute the other, provided that we do a little extra processing. One way to arrange this is to take the complex conjugate of the defining equation for the DFT of $F^*(n)/N$:

$$[DFT[F^*/N]]^* = \frac{1}{N} \sum_{n=0}^{N-1} f[n]e^{jkn2\pi/N} \quad \text{for } k = 0, 1, \cdots, N - 1, \qquad (2.122)$$

from which we see that

$$IDFT[F] = [DFT[F^*/N]]^*. \qquad (2.123)$$

Another such simple relation can be derived by noting that

$$[DFT[F^*(k)]]^* = DFT[F(\langle -k \rangle_N)], \qquad (2.124)$$

where the notation $\langle -k \rangle_N$ means $-k$ mod N; that is, an integer multiple of N is added or subtracted so that the result falls in the range 0 to $N - 1$, inclusive. Thus

Eq. (2.123) can be written

$$\text{IDFT}[F(k)] = \text{DFT}[F(\langle -k \rangle_N)/N].$$ (2.125)

The properties of the z- and Laplace transforms also have their counterparts for the DFT. For example, a shift in the time domain corresponds to a complex-exponential factor in the frequency domain. That is, if we have the DFT transform pair

$$f[n] \rightarrow F(k),$$ (2.126)

then

$$f[n + m] \rightarrow e^{jm2\pi k/N}F(k).$$ (2.127)

Conversely,

$$e^{-jmn2\pi/N}f[n] \rightarrow F(k + m).$$ (2.128)

Note that the shifted functions $f[n + m]$ in Eq. (2.127) and $F(k + m)$ in Eq. (2.128) are defined for their arguments mod N. That is, $f[n + m] = f[\langle n + m \rangle_N]$, and $F[k + m] = F[\langle k + m \rangle_N]$.

The convolution of two signals of finite duration is likewise defined with their arguments mod N; or, what is the same thing, the functions are defined to be periodically extended with period N. It is then a theorem that if we have the DFT pairs

$$f[n] \rightarrow F(k)$$ (2.129)

and

$$g[n] \rightarrow G(k),$$ (2.130)

then

$$\sum_{n=0}^{N-1} f[n]g[m - n] \rightarrow F(k)G(k).$$ (2.131)

This operation is called the *N-point circular convolution* of $f[n]$ and $g[n]$ and is denoted by $f[n] \, \textcircled{N} \, g[n]$. To prove Eq. (2.131), compute the DFT of this convolution, transforming the time variable m to the frequency variable k:

$$\text{DFT}\left[\sum_{n=0}^{N-1} f[n]g[m - n]\right] = \sum_{n=0}^{N-1} f[n] \, \text{DFT}[g[m - n]]$$ (2.132)

$$= \sum_{n=0}^{N-1} f[n]e^{-jnk2\pi/N}G(k)$$

(using the shift property Eq. (2.127))

$$= F(k)G(k).$$

An important fact about the DFT is that it is possible to compute the transform of an N-point signal in time proportional to $N \log N$, instead of N^2, as might be thought necessary from the defining Eq. (2.119). Techniques for doing this, as well as applications, are covered in succeeding chapters.

2-13 TIME-LIMITED SIGNALS

We have seen that sampling a continuous-time signal "aliases" the transform, in the sense described precisely by the aliasing formula Eq. (2.89). This section is devoted to the symmetrical phenomenon: we shall see that sampling on the unit circle in the z-transform domain corresponds to aliasing in the time domain.

Given a discrete-time signal $x[n]$, in general infinite in duration, with z-transform

$$X(z) = \sum_{n=-\infty}^{+\infty} x[n]z^{-n}, \tag{2.133}$$

we ask what finite-duration signal $\tilde{x}[n]$ corresponds to the sampled transform

$$X(e^{jk2\pi/N}) \quad \text{for } k = 0, 1, \cdots, N-1. \tag{2.134}$$

In other words, we need to compute, for $m = 0, 1, \cdots, N-1$, the IDFT

$$\begin{aligned}
\tilde{x}[m] &= \frac{1}{N} \sum_{k=0}^{N-1} X(e^{jk2\pi/N})e^{jkm2\pi/N} \\
&= \frac{1}{N} \sum_{k=0}^{N-1} \sum_{n=-\infty}^{+\infty} x[n]e^{-jkn2\pi/N}e^{jkm2\pi/N} \\
&= \sum_{n=-\infty}^{+\infty} x[n]\left(\frac{1}{N} \sum_{k=0}^{N-1} e^{j(m-n)k2\pi/N}\right).
\end{aligned} \tag{2.135}$$

Again, the sum in parentheses is 0 when $m \neq n \bmod N$, and 1 when $m = n \bmod N$, so

$$\tilde{x}[m] = \sum_{n=-\infty}^{+\infty} x[m+nN]. \tag{2.136}$$

This is as promised; the effect of sampling the transform as in Eq. (2.134) is to add shifted versions of the time function.

From this result we can see that the z-transform of a signal of finite duration is determined uniquely by equally spaced samples of the transform on the unit circle (just as a bandlimited signal is determined uniquely by equally spaced samples in the time domain). More precisely, consider a finite-duration signal $x[n]$, $n = 0$, $1, \cdots, N - 1$. Then the aliased signal $\tilde{x}[n]$ is identical to $x(n)$ in that range, and we can write

$$X(z) = \sum_{n=0}^{N-1} x[n]z^{-n} = \sum_{n=0}^{N-1} \tilde{x}[n]z^{-n}$$

$$= \sum_{n=0}^{N-1} \left(\frac{1}{N} \sum_{k=0}^{N-1} X(e^{jk2\pi/N})e^{jkn2\pi/N} \right) z^{-n}, \qquad (2.137)$$

replacing $\tilde{x}[n]$ by its IDFT. Regrouping, we get

$$X(z) = \sum_{k=0}^{N-1} X(e^{jk2\pi/N}) \left(\frac{1}{N} \sum_{n=0}^{N-1} e^{jkn2\pi/N} z^{-n} \right). \qquad (2.138)$$

Using the closed form for the finite geometric series, we get finally

$$X(z) = \frac{1 - z^{-N}}{N} \sum_{k=0}^{N-1} \frac{X(e^{jk2\pi/N})}{1 - e^{jk2\pi/N} z^{-1}}. \qquad (2.139)$$

This is an explicit interpolation formula that gives the z-transform of a finite-duration signal from samples of its transform; it is analogous to the Shannon sampling formula Eq. (2.94).

Note that the right-hand side of Eq. (2.139) appears to have poles at the points $z = e^{jk2\pi/N}$, but in fact any poles at such points are canceled by the zeros in the factor $(1 - z^{-N})$. The expression is the transfer function of a FIR filter and leads directly to a method of implementing FIR filters, the so-called *frequency-sampling* method. Details can be found in Oppenheim and Schafer [OP75], Rabiner and Gold [RA75], or Tretter [TR76].

2-14 RANDOM SIGNALS

Up to now we have assumed that the signals we have dealt with are completely known. In many practical situations, however, especially in the presence of noise, we have only statistical information about signals. This section is devoted to a brief description of how the transform techniques discussed so far carry over to the random case. We assume that the reader is familiar with the notion of a *random variable*, x, say, and its *expectation*, denoted by $E[x]$. It is important to realize when dealing with functions of time that this operator is not a time average, but rather an *ensemble* average—an average over different realizations of the random variable. Only in certain situations can it be replaced by a time average. For background reading in applied probability theory, the reader is referred to Davenport and Root [DA58] and Thomas [TH71]; for a mathematically rigorous treatment of the foundations of stochastic process theory, see Doob [DO53].

A *stochastic process*, or *random signal*, is a family of random variables, $\{x(t)\}$, in general complex-valued, indexed by a time parameter $t \in T$. When T is the finite set $\{0, 1, \cdots, N - 1\}$, we have a discrete-time random signal of finite duration; when T is the set of all integers, we have a discrete-time random signal; and when T is the real line, we have a continuous-time signal. We discuss mostly the discrete-time case here, but most of what we say carries over to the other cases with little change. A stochastic process is called *wide-sense stationary* if the following two conditions hold:

$$E[|x[n]|^2] < \infty \qquad (2.140)$$

and the function

$$\phi_{xx}[n] = E[x[m + n]x^*[m]], \qquad (2.141)$$

called the *autocorrelation function of* x, does not depend on m, but only on the displacement n. We restrict attention in this section to those random signals that are wide-sense stationary.

The autocorrelation function is a *two-sided* function; that is, it cannot extend for positive values of n without also extending for negative values. To see this, note that

$$\phi_{xx}[-n] = E[x[m - n]x^*[m]] = \phi_{xx}^*[n]. \qquad (2.142)$$

In the special case that $x[n]$ is real, $\phi_{xx}[n]$ is an even function of n.

The *power spectral density* can be defined to be the z-transform of the autocorrelation function:

$$\Phi_{xx}(z) = \sum_{n=-\infty}^{+\infty} \phi_{xx}[n]z^{-n}. \qquad (2.143)$$

This power series may not converge for any value of z, however. Consider the case, for example, when $\phi_{xx}[n]$ is a periodic function. Then the region of guaranteed convergence of the transform for $n > 0$ is $|z| > 1$, while the region of convergence for $n < 0$ is $|z| < 1$. What happens, in fact, is that the transform converges to delta-functions on the unit circle in the z-plane (at the frequency of repetition of the periodic signal and its multiples), and more elaborate mathematical tools, such as the Lebesgue–Stieltjes integral, must be used to deal with such cases [DO53]. We assume here that the power spectral density as defined by the z-transform Eq. (2.143) converges in an annulus in the z-plane that includes the unit circle.

The inverse z-transform Eq. (2.41) tells us that

$$\phi_{xx}[n] = \frac{1}{2\pi j} \oint_C \Phi_{xx}(z) z^n \frac{dz}{z}.$$ (2.144)

Choosing the unit circle as the contour of integration, and letting $z = e^{j\omega}$, this becomes

$$\phi_{xx}[n] = \frac{1}{2\pi} \int_{-\pi}^{+\pi} \Phi_{xx}(e^{j\omega}) e^{jn\omega} \, d\omega.$$ (2.145)

From this we see, by letting $n = 0$, that the mean-square value of the signal x is given by the total integral of the spectral density as a function of frequency ω,

$$E[|x[n]|^2] = \phi_{xx}[0] = \frac{1}{2\pi} \int_{-\pi}^{+\pi} \Phi_{xx}(e^{j\omega}) \, d\omega.$$ (2.146)

The spectral density $\Phi_{xx}(e^{j\omega})$ is a real function of ω, because $\phi_{xx}[n]$ is even, and the variance of x is equal to the average area under one period of it.

An important situation arises when a random signal is passed through a filter. We would like to know the spectral density of the output, which is of course also a random signal, as a function of the spectral density of the input and the transfer function of the filter. Consider then the signal y obtained by filtering x with the filter with impulse response g. The autocorrelation function of y is

$$\phi_{yy}[n] = E[y[m + n] y^*[m]]$$

$$= E\left[\sum_{q=-\infty}^{+\infty} x[m + n - q] g[q] \sum_{r=-\infty}^{+\infty} x^*[m - r] g^*[r] \right].$$ (2.147)

The expectation operator commutes with the convolution sums, so

$$\phi_{yy}[n] = \sum_{q=-\infty}^{+\infty} g[q] \sum_{r=-\infty}^{+\infty} g^*[r] E[x[m + n - q] x^*[m - r]]$$

$$= \sum_{q=-\infty}^{+\infty} g[q] \sum_{r=-\infty}^{+\infty} g^*[r] \phi_{xx}[n - q + r].$$ (2.148)

Taking the z-transform of this and using the shift property Eq. (2.49), we obtain

$$\Phi_{yy}(z) = \Phi_{xx}[z] \sum_{q=-\infty}^{+\infty} g[q] z^{-q} \sum_{r=-\infty}^{+\infty} g^*[r] z^r$$

$$= G(z) G^*(1/z) \Phi_{xx}(z).$$ (2.149)

On the unit circle, this shows that

$$\Phi_{yy}(e^{j\omega}) = |G(e^{j\omega})|^2 \Phi_{xx}(e^{j\omega}). \tag{2.150}$$

This last relation can be used to justify the interpretation of the power spectral density as the *average power* of the signal as a function of frequency. Let G be a bandpass digital filter with center frequency ω_0 and a very narrow passband. Filter the signal x by G, and measure the mean-square value, the average power, of the output signal y. We can think of y as being composed of these components of x at the frequency ω_0. From Eqs. (2.146) and (2.150) we have

$$E[|y[n]|^2] = \phi_{yy}[0] = \frac{1}{2\pi} \int_{-\pi}^{+\pi} |G(e^{j\omega})|^2 \Phi_{xx}(e^{j\omega})\, d\omega. \tag{2.151}$$

Assuming that $\phi_{xx}(e^{j\omega})$ is smooth in the neighborhood near $\omega = \omega_0$, this integral can be approximated by

$$E[|y[n]|^2] \approx \Phi_{xx}(e^{j\omega_0}) \frac{1}{2\pi} \int_{-\pi}^{+\pi} |G(e^{j\omega})|^2\, d\omega. \tag{2.152}$$

Normalizing the filter transfer function by choosing the integral in this last expression to be unity, we have

$$E[|y[n]|^2] \approx \Phi_{xx}(e^{j\omega_0}), \tag{2.153}$$

thus verifying the desired interpretation of the power spectral density as the average power of the random signal at a particular frequency.

In the important special case when a real wide-sense stationary random signal $x[n]$ is uncorrelated from sample to sample, has zero mean, and variance σ^2, it is called *white noise*. Its autocorrelation function is

$$\phi_{xx}[n] = \sigma^2 \delta[n], \tag{2.154}$$

and therefore its power spectral density is

$$\Phi_{xx}(z) = \sigma^2, \tag{2.155}$$

which converges for all z. We can interpret this as meaning that all frequencies are present in equal amounts, which accounts for the term *white*.

A common application of this theory will be seen when we come to consider the effect of arithmetic roundoff noise in a digital filter. If the roundoff noise is considered to be a random signal with a known power spectral density (often white), and if it occurs in the filter at such a position that the transfer function from the place of origin to the output point is known, then the relation Eq. (2.151) can be used to calculate the mean-square value of the output signal caused by the roundoff noise.

The autocorrelation function of a continuous-time random signal $x(t)$ is defined, in analogy with the discrete-time case, as

$$\phi_{xx}(\tau) = E[x(t + \tau)x^*(t)], \qquad (2.156)$$

and its power spectral density as its Laplace transform

$$\Phi_{xx}(s) = \int_{-\infty}^{+\infty} \phi_{xx}(t)e^{-st}\, dt. \qquad (2.157)$$

Because the autocorrelation function of the sampled random signal is the sampled version of the continuous-time autocorrelation function, the power spectral density in the discrete-time case is the aliased version of the continuous-time power spectral density, as in Eq. (2.89),

$$\Phi_{xx}^{\#}(z) = \frac{1}{T} \sum_{n=-\infty}^{+\infty} \Phi_{xx}\left(s + jn\frac{2\pi}{T}\right). \qquad (2.158)$$

Thus components in the power spectral density above the Nyquist frequency are aliased by sampling into components below, just as in the deterministic case, and methods for estimating spectral densities of continuous-time random signals from samples must take this into account.

2-15 SUMMARY: THE SIX DOMAINS OF SIGNAL PROCESSING

The signals and transforms that we have seen in this chapter can be characterized by their domains of definition. By this criterion we have seen three kinds of signals: continuous-time signals of infinite extent; discrete-time signals of infinite extent; and discrete-time signals of finite extent. The Fourier transforms of these signals are also defined on three domains: the Fourier transforms of the continuous-time functions on the continuous frequency axis of infinite extent; the z-transforms of the discrete-time, infinite-extent signals on the unit circle in the z-plane (continuous but finite in extent); and the discrete Fourier transform on the discrete, finite set of points $z_k = e^{jk2\pi/N}$ on the unit circle in the z-plane. We can represent these cases by the diagram in Figure 2-18. We have also indicated that sampling in the time domain corresponds to aliasing in the frequency domain, and vice versa.

Most of signal processing is represented in this simple picture, and as we shall see in the rest of this book, the experienced practitioner of signal processing moves deftly from one to the other of these domains.

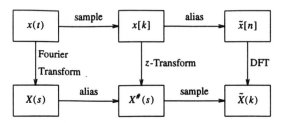

FIGURE 2-18   The six domains of signal processing. The top row is in the time domain; the bottom in the frequency domain. Sampling in one domain corresponds to aliasing in the other.

ACKNOWLEDGMENTS

This work was supported in part by NSF Grant MIP-8912100 and U.S. Army Research Office–Durham Grant DAAL03-89-K-0074.

REFERENCES

[BO72] G. Boole, *A Treatise on the Calculus of Finite Differences*. Dover, New York, 1960; J. F. Moulton, ed., 2nd and last rev. ed., first published in 1872.

[CA30] H. S. Carslaw, *Introduction to the Theory of Fourier's Series and Integrals*, 3rd ed. Dover, New York, 1930.

[CH60] R. V. Churchill, *Complex Variables and Applications*, 2nd ed. McGraw-Hill, New York, 1960.

[CO65] J. W. Cooley and J. W. Tukey, An algorithm for the machine calculation of complex Fourier series. *Math. Comput.* **19**, 297–301 (1965).

[CO67] J. W. Cooley, P. A. W. Lewis, and P. D. Welch, Historical notes on the fast Fourier transform. *Trans. IEEE Audio Electroacoust.* **AU-15**, 76–79 (1967).

[DA58] W. B. Davenport, Jr. and W. L. Root, *An Introduction to the Theory of Random Signals and Noise*. McGraw-Hill, New York, 1958.

[DO53] J. L. Doob, *Stochastic Processes*. Wiley, New York, 1953.

[EV66] H. W. Eves, *Functions of a Complex Variable*, 2 vols. Prindle, Weber and Schmidt, Boston, MA, 1966.

[HE67] M. Healy, *Tables of Laplace, Heaviside, Fourier, and z-Transforms*. W. R. Chambers, Edinburgh, 1967.

[HE84] M. T. Heideman, D. H. Johnson, and C. S. Burrus, Gauss and the history of the fast Fourier transforms. *IEEE ASSP Mag.* October, pp. 14–21, (1984).

[HO68] J. E. Hopcroft and K. Steiglitz, A class of finite memory interpolation filters. *IEEE Trans. Circuit Theory* **CT-15**, 105–111 (1968).

[JU73] E. I. Jury, *Theory and Application of the z-Transform Method*. R. E. Krieger, Publishing Co., Huntington, NY, 1973.

[KA80] T. Kailath, *Linear Systems*. Prentice-Hall, Englewood Cliffs, NJ, 1980.

[KN45] K. Knopp, *Theory of Functions*, Part I. Dover, New York, 1945 (Part II, 1947); *Problem Book*, Vol. I. Dover, New York, 1948 (Vol. II, 1952).

[OP75] A. V. Oppenheim and R. W. Schafer, *Digital Signal Processing*. Prentice-Hall, Englewood Cliffs, NJ, 1975.

[PA62] A. Papoulis, *The Fourier Integral and its Applications*. McGraw-Hill, New York, 1962.

[RA58] J. R. Ragazzini and G. F. Franklin, *Sampled-Data Control Systems*. McGraw-Hill, New York, 1958.

[RA75] L. R. Rabiner and B. Gold, *Theory and Application of Digital Signal Processing*. Prentice-Hall, Englewood Cliffs, NJ, 1975.

[RO66] G. E. Roberts and H. Kaufman, *Table of Laplace Transforms*. Saunders, Philadelphia, PA, 1966.

[ST65] K. Steiglitz, The equivalence of digital and analog signal processing. *Inf. Control* **8,** 455–467 (1965); reprinted in L. R. Rabiner and C. M. Rader, eds., *Digital Signal Processing*. IEEE Press, New York, 1972.

[TH71] J. B. Thomas, *An Introduction to Applied Probability and Random Processes*. Wiley, New York, 1971.

[TR76] S. A. Tretter, *Introduction to Discrete-Time Signal Processing*. Wiley, New York, 1976.

[WI48] N. Wiener, *Extrapolation, Interpolation, and Smoothing of Stationary Time Series*. MIT Press and Wiley, New York, 1948 (2nd ed., 1961).

[ZE65] A. H. Zemanian, *Distribution Theory and Transform Analysis*. McGraw-Hill, New York, 1965.

3 Linear Time-Invariant Discrete-Time Systems

NIRMAL K. BOSE

Department of Electrical and Computer Engineering
The Pennsylvania State University, University Park

3-1 SYSTEM CLASSIFICATION

Systems may be classified in much the same way as signals are classified in the previous chapter. This classification is natural because a system, being a collection of objects united by some form of interaction or interdependence, has the core mission, in signal processing, of transforming a set of signals occurring at its input to another set of signals at its output with more desirable properties. It should be noted, however, that besides processing signals, a system could also serve to generate (e.g., speech and video) as well as to measure, transmit, and receive signals. The range of complexity of systems that are used to process signals is staggering. Such systems are often difficult to model accurately and analyze efficiently. A typical example of an ultracomplex signal processing system is the human visual system. The complexity of created systems ranges from the highly complex supercomputer to ultrasimple ones such as common types of analog or digital low-pass filters. Although systems of prime concern here are those required for processing a signal that is a function of the single independent variable *time*, it should be appreciated that the independent variable in many applications need not be time. In a variety of tasks spanning, typically, medical imaging applications, analysis and visualization of atmospheric turbulence phenomena, and optical and geophysical signal processing, up to three independent spatial variables may be encountered either in conjunction or in the absence of the temporal (or time) variable. Systems required for such tasks are generally called multidimensional systems and, on occasions, have also been referred to as spatiotemporal systems.

3-1-1 Continuous and Discrete Systems

A continuous-time system is one for which the input and output signals, as well as the signals within the system, are continuous-time. An example of a continuous-

Handbook for Digital Signal Processing, Edited by Sanjit K. Mitra and James F. Kaiser.
ISBN 0-471-61995-7 © 1993 John Wiley & Sons, Inc.

time system is an electrical filter built from standard passive circuit elements like resistors, inductors, and capacitors and active components such as operational amplifiers. Similarly, a system is called a discrete-time system when the input and output signals together with the signals within the system are discrete-time. A digital computer is an example of a discrete-time system. For finer classification, it should be noted that a digital computer is a digital system, which is an important subclass of discrete-time systems. A hybrid-time system is one in which both continuous-time and discrete-time signals appear. Examples of such systems are the analog-to-digital and digital-to-analog converters discussed in Chapter 10 of this handbook.

3-1-2   Linear and Nonlinear Systems

Another important system classification is derived from the properties of linearity and its complement—nonlinearity. From the definition advanced in the previous chapter, it is clear that Eq. (2.10), which describes a linear system, translates to the properties of homogeneity and superposition. A system having one input is homogeneous when the multiplication of the input by an arbitrary constant results in the output being multiplied by exactly the same constant. A system satisfies the superposition property when its response (output) to the sum of any two excitations (inputs) equals the sum of its responses to the individual excitations. A system that does not satisfy the properties of homogeneity (or proportionality) and superposition is nonlinear. Linear systems comprise a small but important subclass of systems. Not only are many physical processes modeled with reasonable accuracy by linear systems but, very importantly, analytical tools for analysis of such models are very well developed and widely available. Use of simple models also makes it easier to develop system design methods.

3-1-3   Time-Invariant and Time-Variant Systems

Systems could be classified into categories other than those considered above. An important and analytically tractable class of systems is characterized by the time-invariant property. When the independent variables are either spatial or spatial and temporal, the corresponding property is called shift-invariant. A system is time-invariant when any arbitrary but fixed time shift in its input excitation signal produces exactly the same time shift in its response. A system that is not time-invariant (shift-invariant) is time-variant (shift-variant).

3-1-4   Linear Time-Invariant Systems

The class of systems identified for focus is of the discrete-time *linear time-invariant* (LTI) type. Usually, the signal to be processed begins as a continuous function $u_c(t)$, where t is the time variable. The physical origin of the signal could be diverse and encompasses biological, geophysical, audio, video, speech, sonar, radar, and a variety of environmental phenomena. Usually, the signals are real-valued functions of time or space variables. However, sometimes complex representation of the signal prior to processing is either useful or necessary. For example, when describing narrowband active sonar systems, a complex or quadrature representation of the waveform is convenient for depicting the Doppler phase shifts introduced by the propagation and reflection coefficients. The notion of an analytic signal, which is a complex function of time, has been exploited in communications and signal processing for a long time and is treated in Chapter 13.

Prior to processing a continuous-time signal $u_c(t)$ by a discrete-time LTI system, the continuous-time signal has to be sampled. Assume that the sampling is uniform (equispaced). If $u_c(t)$ is lowpass bandlimited, then, as discussed in Chapter 2, the spectrum of $u_c(t)$ is recoverable from the uniformly sampled sequence $\{u[n]\}$ by lowpass analog filtering, provided the sampling frequency is greater than twice the highest frequency component in $u_c(t)$. In practice, certain lowpass analog filters, called anti-aliasing filters, are also required so that the bandlimitedness constraint imposed on the continuous-time signal might be satisfied.

Since discrete-time signals and systems are of particular interest in this handbook, it will be assumed henceforth that the signal to be operated on by a LTI system is available as a properly sampled sequence. Bear in mind that each element of the sequence has to be quantized (i.e., represented by a specified number of bits) before processing by digital hardware and this restriction of finite machine wordlength is a source of error, which is discussed in Chapter 6. In this chapter, we are concerned only with discrete-time signals and not with digital signals that are obtained after quantization.

Consider the LTI system, denoted by L, in Figure 3-1. The input sequence to L is denoted by $\{u[n]\}$ and the resulting output sequence is $\{y[n]\}$. Since the system L operates on the input $\{u(n)\}$ to produce $\{y(n)\}$, it is convenient to describe the input/output relationship by the equation

$$\{y[n]\} = L\{u[n]\}. \tag{3.1}$$

The system L may be viewed as an operator and a LTI system is characterized by a LTI operator. Since complex sequences occur naturally in sonar and radar applications, as well as in other signal processing tasks [VA87], we allow the sequences to be complex-valued. The LTI system is defined next for the sake of clarity, completeness, and because of its pivotal role in this chapter.

Definition 3.1. Let $\{u_1[n]\}$ and $\{u_2[n]\}$ be two arbitrary input sequences and let c_1 and c_2 be two arbitrary complex constants. A system L is linear provided

$$L[c_1\{u_1[n]\} + c_2\{u_2[n]\}] = c_1L\{u_1[n]\} + c_2L\{u_2[n]\}, \tag{3.2}$$

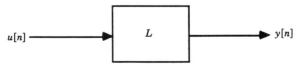

FIGURE 3-1 A linear time-invariant discrete-time system L with a single input $u[n]$ at time index n.

and L is time-invariant provided

$$\{y[n - l]\} = L\{u[n - l]\} \tag{3.3}$$

for any integer l where the excitation/response relationship in Eq. (3.1) is known to hold. The system L is LTI provided Eqs. (3.2) and (3.3) hold.

The examples given next serve to bring out more fully the implications of linearity and time-invariance.

Example 3.1. Assume that the nth output sample $y[n]$ from a discrete-time system is related to the input samples $u[r]$ for values of r shown in the difference equation,

$$y[n] = \sum_{r=n-3}^{n+2} u[r].$$

It is easily established that the system is linear. Furthermore, since for any integer l the system response to the time-shifted input sequence $\{u[n - l]\}$ is

$$\sum_{r=n-3}^{n+2} u[r - l] = \sum_{r=(n-l)-3}^{(n-l)+2} u[r] = y[n - l],$$

the system is also time-invariant.

Example 3.2. The input/output relationship for a discrete-time system, referred to as the accumulator, is

$$y[n] = \sum_{r=0}^{n} u[r].$$

The system is linear. However, it is not time-invariant since a time-shifted input sequence $\{u[n - l]\}$ results in the output

$$\sum_{r=0}^{n} u[r - l] = \sum_{r=-l}^{n-l} u[r] \neq \sum_{r=0}^{n-l} u[r] = y[n - l].$$

Example 3.3. The discrete-time system described by the input/output relationship

$$y[n] = 3u[n] + 5$$

is not linear, even though the points in the plot of $y[n]$ versus $u[n]$, with n viewed as an integer parameter, lie on a straight line. The system is, however, time-invariant.

3-2 TIME-DOMAIN REPRESENTATION OF LINEAR SYSTEMS

3-2-1 Unit-Impulse Response and Convolution Sum

A discrete-time LTI system transforms an input sequence $\{u[n]\}$ to an output sequence $\{y[n]\}$ and it is characterizable completely by a sequence $\{h[n]\}$, which is referred to as the *unit-impulse response*, that is, the response to a unit-impulse input, $u[n] = \delta[n]$. The unit impulse $\delta[n]$ has been defined in Eq. (2.1). An arbitrary input signal $x[n]$ can be expressed as the weighted sum of shifted unit impulses by

$$u[n] = \sum_{l=-\infty}^{\infty} u[l]\delta[n-l]. \tag{3.4}$$

The property of time-invariance implies that if $\{h[n]\}$ is the response to $\{\delta[n]\}$, that is,

$$\{h[n]\} = L\{\delta[n]\}, \tag{3.5}$$

where the operator L characterizes the linear time-invariant system, then $\{h[n-l]\}$ is the response to the shifted input $\{\delta[n-l]\}$. By linearity, the output $\{y[n]\}$ in Eq. (3.1) is obtained from Eq. (3.4), after allowing the operator L to be brought inside the summation sign in Eq. (3.4).[1]

$$y[n] = Lu[n] = L \sum_{l=-\infty}^{\infty} u[l]\delta[n-l]$$

$$= \sum_{l=-\infty}^{\infty} u[l]L\delta[n-l]$$

$$= \sum_{l=-\infty}^{\infty} u[l]h[n-l] \tag{3.6}$$

The preceding equation is called a *convolution sum*; the input sequence $\{u[n]\}$, when convolved with the unit-impulse response sequence, $\{h[n]\}$, of a discrete-time LTI system produces the output sequence $\{y[n]\}$. This fact is expressed notationally as

$$\{y[n]\} = \{u[n]\} \circledast \{h[n]\} = \{h[n]\} \circledast \{u[n]\}. \tag{3.7}$$

---

[1]It is well known that this is not possible to do in the case of some linear operators when an infinite summation is involved. An analysis of cases where the operations described below are permissible has been provided in Borodziewicz et al. [BO83]. The need for caution when faced with the type of situation under discussion has also been pointed out in Zadeh and Desoer [ZA63] and Kailath [KA80].

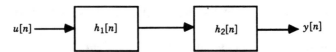

FIGURE 3-2   A cascade connection of two LTI discrete-time systems characterized by unit-impulse responses $h_1[n]$ and $h_2[n]$.

One way of forming a composite LTI system is by cascading two LTI subsystems. In this cascade arrangement, the output from the first LTI subsystem becomes the input to the second LTI subsystem. Consequently, if $\{h_1[n]\}$ and $\{h_2[n]\}$ are the unit-impulse response sequences of the first and second subsystems, then the unit-impulse response sequence $\{h[n]\}$ of the cascaded system shown in Figure 3-2 is

$$\{h[n]\} = \{h_1[n]\} \circledast \{h_2[n]\}. \tag{3.8}$$

Of course, the subsystems in Figure 3-2 could be interchanged without altering the overall unit-impulse response $\{h[n]\}$, by virtue of Eq. (3.7). In Figure 3-3, the two LTI subsystems, whose unit-impulse response sequences are $\{h_1[n]\}$ and $\{h_2[n]\}$, are connected in parallel. The unit-impulse response of this overall system is the sum of the unit-impulse responses of the subsystems.

Example 3.4. Two discrete-time LTI systems whose unit-impulse response sequences are described by

$$h_1[n] = (\tfrac{1}{2})^n, \quad n \geq 0$$

and

$$h_2[n] = \delta[n] - \delta[n-3],$$

as shown in Figures 3-4 and 3-5, are cascaded. The overall unit-impulse response $h[n]$ of the cascaded system is obtained after convolving the two sequences $\{h_1[n]\}$

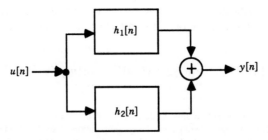

FIGURE 3-3   A parallel connection of two LTI discrete-time systems characterized by unit-impulse responses $h_1[n]$ and $h_2[n]$.

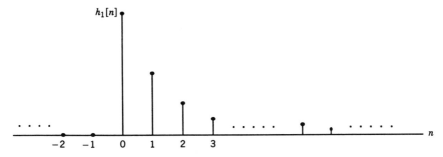

FIGURE 3-4   Plot of $h_1[n]$ versus n in Example 3.4.

and $\{h_2[n]\}$. After noting that $h[n] = h_1[n] - h_1[n - 3]$ for $n \geq 3$, it can be verified that

$$h[0] = 1, \qquad h[1] = \tfrac{1}{2}, \qquad h[2] = \tfrac{1}{4},$$

$$h[3] = -\tfrac{7}{8}, \qquad h[4] = -\tfrac{7}{16}, \qquad h[5] = -\tfrac{7}{32},$$

$$h[6] = -\tfrac{7}{64}, \qquad h[7] = -\tfrac{7}{128} \qquad \text{and so on.}$$

In general, $h[n]$ may be expressed in the form

$$h[n] = \delta[n] + \tfrac{1}{2}\delta[n - 1] + \tfrac{1}{4}\delta[n - 2] - 7 \sum_{l=3}^{\infty} (\tfrac{1}{2})^l \delta[n - l].$$

The sequence $\{h[n]\}$ is sketched in Figure 3-6.

The input/output description of a class of LTI discrete-time systems is often provided by a difference equation,

$$\sum_{k=0}^{N} d_k y[n - k] = \sum_{i=0}^{M} q_i u[n - i], \qquad d_0 \neq 0, \quad n \geq 0, \qquad (3.9)$$

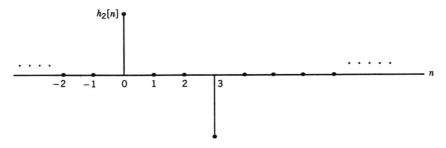

FIGURE 3-5   Plot of $h_2[n]$ versus n in Example 3.4.

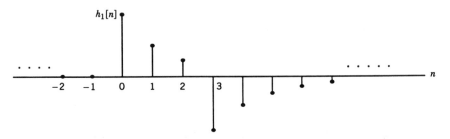

FIGURE 3-6   Plot of $h[n]$ versus n in Example 3.4.

where the coefficients d_k and q_i are constants, and $\{y[n]\}$ and $\{u[n]\}$ are, as before, the output and input sequences. In order to solve for $y[n]$ (without any loss of generality, take $N \geq M$), it is necessary that, in addition to the input, the boundary (initial) conditions,

$$y[-1], y[-2], \cdots, y[-N], \tag{3.10}$$

be specified. An important result for the subclass of LTI systems, which are describable by Eq. (3.9), is summarized next. The justification is provided in Pondelicek [PO81] and Zadeh and Desoer [ZA63].

Let the initial conditions in Eq. (3.10) be all zero. Then there exists exactly one sequence $\{h[n]\}$ that relates the nth sample $y[n]$ of the output to the input sequence $\{u[n]\}$ in the input/output description provided through Eq. (3.9) by the equation

$$y[n] = \sum_{l=-\infty}^{\infty} u[l]h[n-l], \quad n \geq 0. \tag{3.11}$$

The basic content of the result just stated may be succinctly summarized: any difference equation of the form shown in Eq. (3.9) with initial conditions in Eq. (3.10) set to zero has a convolutional solution. Its importance stems from the fact that the concepts of unit-impulse response and transfer function (defined in Section 3-4) are justified, in general, for convolutional systems only. This class of systems can also be analyzed for properties like stability through transform techniques.

The unique sequence $\{h[n]\}$ in Eq. (3.11), which completely characterizes the LTI system described by Eq. (3.9) with the initial conditions in Eq. (3.10) set to zero and with the unit-impulse sequence $\{\delta[n]\}$ as the input, is the unit-impulse response. Therefore $\{h[n]\}$ satisfies the following equations:

$$\sum_{k=0}^{N} d_k h[n-k] = q_n, \quad n \geq 0, \tag{3.12}$$

$$h[-1] = h[-2] = \cdots = h[-N] = 0. \tag{3.13}$$

It is understood that on the right side of Eq. (3.12), $q_n = 0, n > M$.

Knowledge of the unit-impulse response for a LTI system makes possible the calculation of the response of that system to any other input. This is done by convolving the unit-impulse response sequence with the specified input sequence.

Example 3.5. It is required to calculate the response of the cascaded system, whose unit-impulse response $h[n]$ was calculated in Example 3.4, to the input sequence described by

$$u[n] = 4 \cos \left(\tfrac{2}{3}\pi n - \tfrac{1}{8}\pi\right).$$

For the sake of brevity, define $\theta \triangleq \left(\tfrac{2}{3}\pi n - \tfrac{1}{8}\pi\right)$. Then

$$u[n-1] = 4 \cos \left(\theta - \tfrac{2}{3}\pi\right), \quad u[n-2] = 4 \cos \left(\theta - \tfrac{4}{3}\pi\right),$$

$$u[n-3] = u[n] = 4 \cos\theta, \cdots .$$

Therefore, from Eq. (3.11), the response $y[n]$ to the specified input may be grouped as $y[n] = y_A[n] + y_B[n] + y_C[n]$, where

$$y_A[n] = 4 \cos\theta[1 - \tfrac{7}{8} - \tfrac{7}{64} - \tfrac{7}{512} - \cdots] = 0,$$

$$y_B[n] = 4 \cos\left(\theta - \tfrac{2}{3}\pi\right)\left[\tfrac{1}{2} - \tfrac{7}{16} - \tfrac{7}{128} - \tfrac{7}{1024} - \cdots\right] = 0,$$

$$y_C[n] = 4 \cos\left(\theta - \tfrac{4}{3}\pi\right)\left[\tfrac{1}{4} - \tfrac{7}{32} - \tfrac{7}{256} - \tfrac{7}{2048} - \cdots\right] = 0.$$

So $y[n] = 0$. Note that the output sequence need not be zero if the input sequence is changed.

The unit-step sequence $\{\mu[n]\}$ was related to the unit-impulse sequence $\{\delta[n]\}$ through the equation

$$\mu[n] = \mu[n-1] + \delta[n] \tag{3.14}$$

in Chapter 2. The nth sample $s[n]$ of the *unit-step response* sequence for a LTI system L is obtained from use of Eq. (3.14) as

$$s[n] \triangleq L\mu[n] = s[n-1] + h[n]. \tag{3.15}$$

Therefore knowledge of the unit-impulse response allows the unit-step response to be recursively computed from Eq. (3.15) for, say, $n \geq 0$ given also, in that case, the initial value $s[-1]$.

Example 3.6. A discrete-time LTI system is described by the difference equation

$$y[n] = 0.5y[n-1] + 2u[n] - u[n-1], \quad n \geq 0.$$

The system is unexcited before $n = 0$, so that the output sample $y[-1] = 0$. The unit-impulse response $h[n]$ is obtained by solving the above equation after replacing $u[n]$ by $\delta[n]$ and, correspondingly, $y[n]$ by $h[n]$. Setting $h[-1] = 0$ and solving

$$h[n] = 0.5h[n-1] + 2\delta[n] - \delta[n-1], \qquad n \geq 0,$$

one obtains

$$h[0] = 2 \quad \text{and} \quad h[n] = 0 \qquad \text{for } n > 0.$$

Therefore the unit-impulse response of the system is described by

$$h[n] = 2\delta[n], \qquad n \geq 0. \tag{3.16}$$

Setting $s[-1] = 0$, the unit-step response $s[n]$ is obtained from Eqs. (3.15) and (3.16):

$$s[n] = 2\mu[n] \quad \text{or} \quad s[n] = 2 \quad \text{for } n \geq 0.$$

3-2-2 Total Response of a LTI System

The total response of a LTI system is the sum of the responses due to nonzero initial conditions [the LTI system characterized by Eq. (3.9) is said to be at zero-state if the initial conditions in Eq. (3.10) are zero] and a specified input. The two responses are referred to as the zero-input and zero-state responses and their superposition gives the total response by the LTI property. The *zero-input response* is obtained by setting the input sequence samples to zero and then calculating the response of the LTI system to the nonzero initial conditions. The *zero-state response* is obtained by convolving the unit-impulse response sequence with the specified input sequence after setting the initial conditions to zero as explained earlier in this section.

Example 3.7. Let us determine the total response of the system in Example 3.6 with the initial condition $y[-1]$ being nonzero.

The zero-input response $y_2[n]$ is obtained from $y[-1]$ and the difference equation after the input is set to zero.

$$y_2[n] = 0.5y_2[n-1], \qquad y_2[-1] = y[-1].$$

Clearly, for $n \geq 0$,

$$y_2[n] = y[-1](0.5)^{n+1}.$$

Denote the zero-state response by $y_1[n]$. For a step input, $y_1[n] = s[n]$ was computed in Example 3.6 as

$$y_1[n] = 2\mu[n].$$

The total response of the LTI system with nonzero initial condition $y[-1]$ to a step input is then

$$y_1[n] + y_2[n] = 2 + y[-1](0.5)^{n+1}, \quad n \geq 0.$$

3-3 CLASSIFICATION AND PROPERTIES OF LTI DISCRETE-TIME SYSTEMS

A discrete-time LTI system may be classified either as a *finite impulse response* (FIR) or as an *infinite impulse response* (IIR) system. The properties that are important to note in such a classification are causality, recursibility, and stability. As described in Chapter 2, a system for which the output at a certain instant of time does not depend on future inputs or on future outputs is said to be causal. Since $\delta[n] = 0$ for $n < 0$, the unit-impulse response $h[n]$ of a causal discrete-time LTI system must be zero for $n < 0$. It is easy to verify that the LTI system in Example 3.1 is not causal, while the systems in Examples 3.2 and 3.3 are both causal. Obviously, the unit-impulse response sequence for a casual LTI system could be of either finite or infinite extent and this determines the classification into either FIR or IIR type of system.

3-3-1 Classification (IIR and FIR)

For motivation, we initiate the discussion on classification by treating a special case. With reference to the difference equation representation of the subclass of LTI systems in Eq. (3.9), consider the case when

$$d_k = 0 \quad \text{for } k = 1, 2, \cdots, N. \tag{3.17}$$

In that situation, Eq. (3.9) may be rewritten in the form

$$y[n] = \frac{1}{d_0} \sum_{i=0}^{M} q_i u[n - i]. \tag{3.18}$$

On applying the commutative property in Eq. (3.7) to Eq. (3.11), we also have

$$y[n] = \sum_{l=-\infty}^{\infty} h(l)u[n - l]. \tag{3.19}$$

By comparing Eqs. (3.18) and (3.19), we infer that the unit-impulse response of the LTI system described by Eq. (3.18) is given by

$$h[n] = \begin{cases} \dfrac{q_n}{d_0} & \text{for } n = 0, 1, \cdots, M \\ 0 & \text{otherwise.} \end{cases} \tag{3.20}$$

Since the sample $h[n]$ in Eq. (3.20) is nonzero only for finitely many n, the sequence $\{h[n]\}$ is said to have a *finite support* or is of finite length. It is appropriate then to call the LTI system, which is characterized by the unit-impulse response sequence of Eq. (3.20), a finite impulse response (FIR) system. This notion may be generalized to provide the classification alluded to at the beginning of this section. If, on the other hand, the unit-impulse response sequence is of infinite length (nonfinite support), the system is called an infinite impulse response (IIR) system.

3-3-2 Recursive Implementation

Rewrite Eq. (3.9) in the form

$$y[n] = \frac{1}{d_0} \left(\sum_{i=0}^{M} q_i u[n-i] - \sum_{l=1}^{N} d_l y[n-l] \right). \tag{3.21}$$

It is clear from Eq. (3.21) that $y[n]$ is computable from present as well as M past values of the input and N past values of the output. The computation of $y[n]$ based on Eq. (3.21) along with the specified initial conditions in Eq. (3.10) is via the *recursive implementation* of the difference equation in Eq. (3.9) and the filter whose output is being computed is causal. The unit-impulse response sequence $\{h[n]\}$ of this filter is also recursively computable from the equation

$$h[n] = \frac{1}{d_0} \left(q_n - \sum_{l=1}^{N} d_l h[n-l] \right), \tag{3.22}$$

subject to the constraint that $h[n] = 0$ for $n < 0$. Some of the initial terms of the sequence $\{h[n]\}$ computed from Eq. (3.22) are

$$h[0] = \frac{q_0}{d_0},$$

$$h[1] = \frac{(q_1 d_0 - q_0 d_1)}{d_0^2},$$

$$h[2] = \frac{q_2}{d_0} - \left(\frac{1}{d_0} \right) \left(d_1 h[1] + d_2 h[0] \right),$$

$$\vdots$$

$$h[M] = \frac{q_M}{d_0} - \left(\frac{1}{d_0} \right) \left(\sum_{l=1}^{N} d_l h[M-l] \right);$$

and for $n > M$,

$$h[n] = -\left(\frac{1}{d_0} \right) \sum_{l=1}^{N} d_l h[n-l]. \tag{3.23}$$

From Eq. (3.23) it is clear that except in the case when $d_l = 0$ for $l = 1, 2,$ $\cdot \cdot \cdot$, N as in Eq. (3.17), the unit-impulse response sequence $\{h[n]\}$ associated with the LTI system in Eq. (3.9) could, potentially, be an infinite sequence. The recursive mode of computation illustrated in Eq. (3.23) is possible, in general, when the unit-impulse response sequence of a LTI system is one-sided as defined below.

Definition 3.2. A sequence $\{h[n]\}$ is one-sided provided $h[n] = 0$ either for $n < K_1$ or $n > K_2$, where K_1 and K_2 are finite integers. In the first case, $\{h[n]\}$ is a right-sided sequence and in the other case, it is a left-sided sequence.

Obviously, a causal sequence is right-sided. Instead of writing Eq. (3.9) in the form shown in Eq. (3.21), in the more general case we solve for a particular term, say, $y[n - r]$, by pulling it out of the sum:

$$y[n - r] = \frac{1}{d_r} \left(\sum_{i=0}^{M} q_i u[n - i] - \sum_{\substack{l=0 \\ l \neq r}}^{N} d_l y[n - l] \right). \tag{3.24}$$

If the input and values of $y[n - l]$, $l \neq r$, are known either from initial conditions or previous calculations so that $y[n - r]$ is obtainable for any n, then Eq. (3.24) can be implemented recursively. This recursibility feature will depend on which $y[n - r]$ term is pulled out of the sum or, in other words, on certain specific values of r. The case when $r = 0$ was considered in Eq. (3.21). The case when $r = N$ in Eq. (3.24) also provides a recursive implementation of the difference equation in Eq. (3.9). In the latter case, it is possible to compute $y[n]$ sequentially, for $n = 0, -1, -2, \cdot \cdot \cdot$, given $y[N], y[N - 1], \cdot \cdot \cdot, y[2], y[1]$ and the input. The unit-impulse response sequence of the filter implemented by Eq. (3.24) with $r = N$ is clearly nonzero for $n < 0$ and is therefore noncausal. However, it is a one-sided sequence.

In implementing the recursive mode of computation through the use of Eq. (3.24) both when $r = 0$ and $r = N$, attention has to be directed to ensuring conditions that will guarantee stability. The possible occurrence of instability is due to feedback provided through the use of previously computed outputs appropriately weighted by the d_l coefficients. It is to be expected therefore that these d_l coefficients would have a role in the maintenance of any type of stability.

3-3-3 Causality and Stability Conditions

There are various definitions of stability. Probably the most commonly used one is the *bounded-input–bounded-output* (BIBO) type of stability. A discrete-time LTI system is said to be BIBO stable if and only if for any bounded-input sequence the output sequence is also bounded. This implies that if $|u[n]| \leq M_1$, then $|y[n]| \leq M_2$, where M_1 and M_2 are finite numbers, and this condition holds for any such bounded input.

Since a LTI system is characterized by its unit-impulse response sequence, it would be reasonable to expect that BIBO stability information should be obtainable from the unit-impulse response. It can be shown that a discrete-time LTI system is BIBO stable if and only if its unit-impulse response sequence $\{h[n]\}$ is absolutely summable; that is,

$$\sum_{n=-\infty}^{\infty} |h[n]| < \infty. \tag{3.25}$$

For a proof of this result, see Borodziewicz et al. [BO83].

Example 3.8. Consider the cascaded system whose unit-impulse response $h[n]$ was calculated in Example 3.4. Clearly, in the case,

$$\sum_{n=-\infty}^{\infty} |h[n]| = 1 + \tfrac{1}{2} + \tfrac{1}{4} + 7 \sum_{l=3}^{\infty} (\tfrac{1}{2})^l$$

$$= 1.75 - 12.25 + 7 \sum_{l=0}^{\infty} (\tfrac{1}{2})^l$$

$$= 3.5.$$

Since the unit-impulse response sequence is absolutely summable, the cascaded system in Example 3.4 is BIBO stable.

Example 3.9. Consider a discrete-time LTI system having the following input/output description:

$$y[n] = 2y[n-1] + 3u[n], \qquad n \geq 0,$$

and $y[-1] = 0$. The unit-impulse response $h[n]$, for $n \geq 0$, is obtained from

$$h[n] = 2h[n-1] + 3\delta[n], \qquad h[-1] = 0.$$

Clearly, the solution

$$h[n] = 3(2)^n, \qquad n \geq 0,$$

produces an unbounded unit-impulse response sequence. An unbounded sequence cannot be absolutely summable and therefore the LTI system is not BIBO stable.

3-4 TRANSFORM-DOMAIN REPRESENTATION OF LINEAR SYSTEMS

A very convenient tool for analyzing discrete-time LTI systems is the z-transform. This is particularly so when it is possible to provide an input/output description of the system by means of a constant coefficient linear difference equation as in Eqs.

(3.9) and (3.10). The literature on the calculus of z-transforms in extensive. Two books [JU64]; [VI87] have been exclusively devoted to this subject. The definition of the z-transform and its properties are provided in Chapter 2.

3-4-1 Transfer Function, Poles and Zeros

Application of the z-transform to the difference equation in Eq. (3.9), under zero initial conditions, leads to the notion of a transfer function. If $Y(z)$ and $U(z)$ denote, respectively, the z-transforms of the output and input sequences $\{y[n]\}$ and $\{u[n]\}$, then *under zero initial conditions* Eq. (3.9) is transformed to the following equation:

$$Y(z) \sum_{l=0}^{N} d_l z^{-l} = U(z) \sum_{i=0}^{M} q_i z^{-i}. \tag{3.26}$$

The ratio of the output transform and the input transform in Eq. (3.27) defines the *transfer function* $H(z)$:

$$H(z) = \frac{Y(z)}{U(z)} = \frac{\displaystyle\sum_{i=0}^{M} q_i z^{-i}}{\displaystyle\sum_{l=0}^{N} d_l z^{-l}}. \tag{3.27}$$

In Eq. (3.27), $H(z)$ may be viewed as a rational function in the complex variable z^{-1}. Alternatively, the right-hand side of Eq. (3.27) may be recast in the form shown in Eq. (3.28), where $H(z)$ is expressed as a ratio of two polynomials in z:

$$H(z) = z^{N-M} \frac{\displaystyle\sum_{i=0}^{M} q_i z^{M-i}}{\displaystyle\sum_{l=0}^{N} d_l z^{N-l}}. \tag{3.28}$$

Without any loss of generality, it may be assumed that $H(z)$ has been expressed in its *irreducible form*; that is, the numerator and denominator polynomials of $H(z)$ do not have any nonconstant common factor. Equation (3.28) can alternately be expressed in the form

$$H(z) = \frac{q_0}{d_0} z^{N-M} \frac{\displaystyle\prod_{k=1}^{M} (z - \xi_k)}{\displaystyle\prod_{p=1}^{N} (z - \lambda_p)}. \tag{3.29}$$

The set $\{\xi_k\}$ of values of z for which $H(\xi_k) = 0$ are called the zeros of $H(z)$; the set $\{\lambda_p\}$ of values of z for which $H(\lambda_p) = \infty$ are called the poles of $H(z)$.

Thus the roots of the denominator polynomial,

$$D(z) = \sum_{l=0}^{N} d_l z^{N-l}, \tag{3.30}$$

in Eq. (3.28) are the poles of $H(z)$ in the finite part of the z-plane; the roots of the numerator polynomial,

$$Q(z) = \sum_{i=0}^{M} q_i z^{M-i}, \tag{3.31}$$

in Eq. (3.28) are the zeros of $H(z)$ in the finite part of the z-plane. Furthermore, there is a pole (or zero) of multiplicity $(M - N)$ [or $(N - M)$] at $z = 0$ depending on whether $M > N$ (or $M < N$). Consequently, in the extended z-plane, which includes the point at infinity, the number of poles of $H(z)$ equals the number of zeros of $H(z)$.

Example 3.10. The transfer function of a sixth-order digital filter was calculated to be

$$H(z) = \frac{(z + 1)^6}{z^6 + 0.77769595z^4 + 0.11419942z^2 + 0.00175093}. \tag{3.32}$$

The filter has a zero of multiplicity 6 at $z = -1$. The poles $\{\lambda_k\}$ of the filter are given by

$$\lambda_1 = j(0.7673270), \qquad \lambda_2 = j(0.41421356),$$

$$\lambda_3 = j(0.1316525), \qquad \lambda_4 = -j(0.1316525),$$

$$\lambda_5 = -j(0.41421356), \qquad \lambda_6 = -j(0.7673270).$$

3-4-2 Frequency Response

It is often important to calculate the steady-state response of the system characterized either by the difference equation in Eq. (3.9) or by the transfer function $H(z)$ in Eq. (3.27) to a complex sinusoidal input sequence $\{u[n]\}$ defined by

$$u[n] = e^{j\omega nT}, \qquad -\infty < n < \infty. \tag{3.33}$$

In Eq. (3.33) ω is the digital frequency in radians per second and T is the sampling period in seconds. As noted in Chapter 2, the frequency response of the discrete-time LTI system is obtained by evaluating its transfer function $H(z)$ on the unit circle, $|z| = 1$, where $z = e^{j\omega T}$. The magnitude function $G(\omega) = |H(e^{j\omega T})|$ of $H(e^{j\omega T})$ when plotted versus ω gives the *magnitude response*, while a similar plot for the angle $\phi(\omega)$ of $H(e^{j\omega T}) = G(\omega)e^{j\phi(\omega)}$ provides the *phase response*. The magnitude response in decibels (dB) is the plot of $20 \log|H(e^{j\omega T})| = 20 \log G(\omega)$

versus ω. The *group delay function* $\tau(\omega)$ is given by

$$\tau(\omega) = - \frac{d\phi(\omega)}{d\omega}.$$

Although the phase function $\phi(\omega)$ need not be a rational function in ω, the group delay function is. A linear phase response is associated with a constant group delay.

The computations of the magnitude and phase responses are easier using Eq. (3.29). The complex quantity $(z - \xi_i)$ in Eq. (3.29) may be viewed as a vector from the point ξ_i in the z-plane to an arbitrary point z. Likewise, the complex quantity $(z - \lambda_i)$ may be viewed as a vector from the point λ_i in the z-plane to the point z. By introducing the notation

$$(e^{j\omega T} - \xi_i) = Q_i e^{j\theta_i} \tag{3.34}$$

and

$$(e^{j\omega T} - \lambda_l) = D_l e^{j\phi_l}, \tag{3.35}$$

it follows from Eq. (3.29) that

$$G(\omega) = |H(e^{j\omega T})| = \frac{|q_0| \displaystyle\prod_{i=1}^{M} Q_i}{|d_0| \displaystyle\prod_{l=1}^{N} D_l} \tag{3.36}$$

and

$$\arg H(e^{j\omega T}) = \sum_{i=1}^{M} \theta_i - \sum_{l=1}^{N} \phi_l + (N - M)\omega T. \tag{3.37}$$

A typical pole–zero plot and the manner in which the magnitude and phase of the transfer function may be computed at an arbitrary but fixed frequency are indicated in Figure 3-7. The magnitude and phase response of the discrete-time LTI system may then be plotted (Example 3.11 serves to illustrate typical plots) by repeating the computations at different values of frequency corresponding to different points on the unit circle, $|z| = 1$.

Example 3.11. A discrete-time filter transfer function $H_M(z)$ has been designed to satisfy the following specifications:

The order, M, of the filter is 4.
The cutoff frequency is 500 Hz.
The sampling frequency is 10,000 Hz.
The zero frequency gain $H_M(1) = 10$.

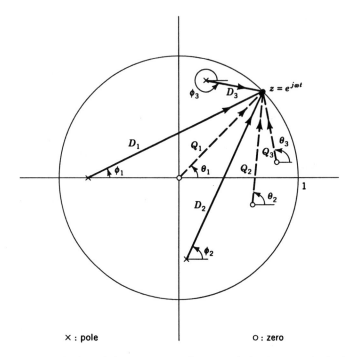

FIGURE 3-7   Magnitude and phase responses from a typical pole–zero plot in the z-plane.

The coefficients of the transfer function may be calculated from the above specifications by using any computer-aided design package available for filter design (see Chapter 5). Use of one such package led to the following set of coefficients: the numerator coefficients in ascending powers of z^{-1} are

$$q_0 = 0.00416580587626, \quad q_1 = 0.016663223505,$$

$$q_2 = 0.0249948352575, \quad q_3 = 0.016663223505, \quad q_4 = 0.00416580587626;$$

the denominator coefficients in ascending powers of z^{-1} are

$$d_0 = 1.0, \quad d_1 = -3.18063855171, \quad d_2 = 3.86119413376,$$

$$d_3 = -2.112155437447, \quad d_4 = 0.438265144825.$$

The magnitude and phase responses are shown in Figures 3-8 and 3-9, respectively.

3-4-3   Stability Test

In the case of discrete-time LTI systems that may be represented, under zero initial conditions, by Eq. (3.9) or, equivalently, by the transfer function in Eq. (3.28), a

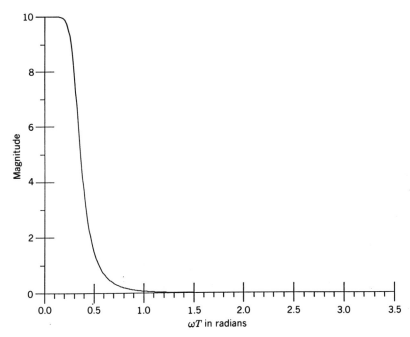

FIGURE 3-8 Plot of magnitude response of filter in Example 3.11. Abscissa is labeled for ωT, where ω is angular frequency in radians per second and T is the sampling period in seconds.

criterion for stability is described next. This criterion can be applied more easily than the one given in Eq. (3.25) to determine stability.

A discrete-time LTI system that is characterizable by a transfer function with relatively prime numerator and denominator polynomials is BIBO stable if and only if the unit circle, $|z| = 1$, falls within the region of convergence of the power series expansion for its transfer function. In the case when the transfer function characterizes a causal LTI system, the stability condition is equivalent to the requirement that the denominator polynomial of the transfer function has all its roots inside the unit circle. A polynomial, which has all its roots inside the unit circle, is called a *Schur polynomial*. When the numerator and denominator polynomials are both Schur polynomials, the transfer function is of the minimum-phase type.

Consider an arbitrary complex coefficient polynomial,

$$Q(z) = \sum_{k=0}^{N} q_k z^k, \qquad q_0 \neq 0, \quad q_N \neq 0, \quad N \neq 0, \tag{3.38}$$

which is devoid of zeros at $z = 0$ (since $q_0 \neq 0$). To determine whether or not $Q(z)$ in Eq. (3.38) is a Schur polynomial, first write the following two rows of coefficients:

$$\begin{matrix} q_N & q_{N-1} & \cdots & q_2 & q_1 & q_0 \\ q_0^* & q_1^* & \cdots & q_{N-2}^* & q_{N-1}^* & q_N^*, \end{matrix} \tag{3.39}$$

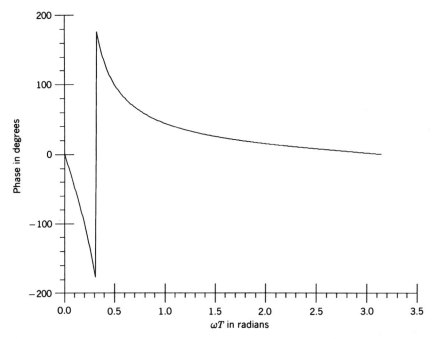

FIGURE 3-9 Plot of phase response for filter in Example 3.11. The abscissa is labeled for ωT, where ω is angular frequency in radians per second and T is the sampling period in seconds.

where the asterisk superscript denotes complex conjugate, and, clearly, there are $N + 1$ columns. The first row is formed from the coefficients of $Q(z)$ ordered (according to descending powers of z) from left to right starting with the coefficient of z^N, the highest power monomial of z; the second row is similarly formed from the coefficients of the *reciprocal polynomial*,

$$Q_r(z) = z^N Q^*((z^*)^{-1}).\qquad(3.40)$$

$Q_r(z)$ is called a reciprocal polynomial because for a zero of $Q(z)$ at $z = \xi_0$, there is a zero of $Q_r(z)$ at $z = (\xi_0^*)^{-1}$. Considering the first column of Eq. (3.39) to be a pivoting column, calculate the determinants of the N (2×2) matrices formed by the pivoting column and each of the remaining columns starting with the last column and ending with the second column after a sequential scan of the columns from right to left. The values of these N complex scalars are written

$$(q_N q_N^* - q_0 q_0^*) \quad (q_N q_{N-1}^* - q_1 q_0^*) \quad \cdots \quad (q_N q_1^* - q_{N-1} q_0^*).$$

From the above coefficients, form a polynomial, $Q_1(z)$, of degree $N - 1$ in the following manner:

$$Q_1(z) = (q_N q_N^* - q_0 q_0^*)z^{N-1} + (q_N q_{N-1}^* - q_1 q_0^*)z^{N-2}$$

$$+ \cdots + (q_N q_1^* - q_{N-1} q_0^*).\qquad(3.41)$$

We denote by $Q_c(z)$ the polynomial obtained by complex conjugating the coefficients of the polynomial $Q(z)$:

$$Q_c(z) = q_N^* z^N + q_{N-1}^* z^{N-1} + \cdots + q_1^* z + q_0^*.$$

The zeros of $Q_c(z)$ are the complex conjugates of the zeros of $Q(z)$. Therefore the zeros of $Q_c(z)$ are in $|z| < 1$ provided the zeros of $Q(z)$ are in $|z| < 1$. The tabular test for determining whether or not $Q(z)$ is a Schur polynomial is based on a result of Schur, which states that $Q(z)$ in Eq. (3.38) is a Schur polynomial if and only if

(a) $|q_0| < |q_N|$ and
(b) $Q_1(z)$ in Eq. (3.41) is a Schur polynomial.

The preceding result, which is proved in Duffin [DU69], when applied successively, generates Table 3-1. After the computations in Table 3-1 are implemented [note that it is unnecessary to compute $Q_N(z)$], the test on $Q(z)$ required for determining whether or not it is a Schur polynomial is summarized. The polynomial

$$Q(z) = \sum_{k=0}^{N} q_k z^k, \qquad q_0 \neq 0, \quad q_N \neq 0, \quad N \neq 0,$$

is a Schur polynomial if and only if in Table 3-1

(a) $\left| \dfrac{q_0}{q_N} \right| < 1$ and

(b) $\left| \dfrac{q_0^{(k)}}{q_{N-k}^{(k)}} \right| < 1, \qquad k = 1, 2, \cdots, N - 1.$

Example 3.12. It is required to determine if the polynomial

$$Q(z) = 20z^4 + 3z^2 + 8z + 10$$

has all its zeros inside the unit circle in the z-plane. The two generating rows (corresponding to rows 1 and 2 in Table 3-1) of the table are

20	0	3	8	10
10	8	3	0	20

Note that here $N = 4$, $q_4 = 20$, and $q_0 = 10$ and hence $|q_0/q_N| < 1$, indicating that the first condition for the stability test is satisfied. The succeeding rows, referred to as the computed rows, of Table 3-1 are computed next. The first computed row is obtained after calculating the determinants of the (2×2) submatrices formed by the first column and each succeeding column in the matrix of the two generating rows above. The determinants are written from left to right in reversed

TABLE 3-1 A Tabular Test for a Schur Polynomial

Polynomial	Row	Coefficients					
$Q(z)$	1	q_N	q_{N-1}	q_2	$\cdots$	q_1	q_0
$Q_r(z)$	2	q_0^*	q_1^*	q_{N-2}^*	$\cdots$	q_{N-1}^*	q_N^*
$Q_1(z)$	3	$q_{N-1}^{(1)}$	$q_{N-2}^{(1)}$	$q_1^{(1)}$	$\cdots$	$q_0^{(1)}$	
$Q_{1r}(z)$	4	$q_0^{*(1)}$	$q_1^{*(1)}$	$q_1^{*(1)}$	$\cdots$	$q_{N-1}^{*(1)}$	
$\cdots$				$\vdots$			
$Q_{(N-1)}(z)$	$2N-1$	$q_1^{(N-1)}$	$q_0^{(N-1)}$				
$Q_{(N-1)r}(z)$	$2N$	$q_0^{*(N-1)}$	$q_1^{*(N-1)}$				
$Q_N(z)$	$2N+1$	$q_0^{(N)}$					
$Q_{Nr}(z)$	$2N+2$	$q_0^{*(N)}$					

It is not necessary to compute $Q_N(z)$

$$Q_j(z) = \sum_{k=0}^{N-j} q_k^{(j)} z^k, \quad j = 0, 1, \ldots, N$$

$$Q_0(z) \triangleq Q(z), \quad q_k^{(0)} \triangleq q_k, \quad k = 0, 1, \ldots, N$$

$$Q_{jr}(z) = z^{N-j} Q_j^*((z^*)^{-1}), \quad j = 0, 1, \ldots, N$$

$$Q_{0r}(z) \triangleq Q_r(z)$$

The coefficients of the polynomial $Q_{(j+1)}(z)$ are obtained from the coefficients of $Q_j(z)$, $j = 0, 1, 2, \ldots, N - 1$, as follows for $k = 0, 1, 2, \ldots, (N - j - 1)$:

$$q_k^{(j+1)} = q_{N-j}^{(j)} q_{(k+1)}^{*(j)} - q_0^{*(j)} q_{N-j-k-1}^{(j)}$$

order of their described computation as illustrated below. Note that after a row is computed, it is desirable to divide each element of the row by the greatest common divisor of the elements of that row. This controls coefficient growth to some extent. The coefficient conditions to be checked for stability are written in a column adjacent to each computed row after division of the elements in that row by their greatest common divisor. The procedure is continued after replacing row 1 above by the computed row following the division of each of its elements by the greatest common divisor of those elements. The number of elements in the successively computed rows decreases and the procedure terminates when a computed row has only two elements, as illustrated below.

$(\div 10)$	300	-80	30	160	
	30	-8	3	16	$\left|\frac{16}{30}\right| < 1$
	16	3	-8	30	
$(\div 2)$	644	-288	218		
	322	-144	109		$\left|\frac{109}{322}\right| < 1$
	109	-144	322		
	91803	-30672			$\left|-\frac{30672}{91803}\right| < 1$

It can now be inferred that $Q(z)$ has all its zeros in the region defined by $|z| < 1$.

3-4-4 Special Types of Transfer Functions

Special transfer functions have interesting properties that are exploited in diverse applications. An *allpass* filter transfer function $H_{al}(z)$ has a magnitude response that is unity at all frequencies; that is,

$$|H_{al}(e^{j\omega T})| = 1 \quad \text{for all } \omega.$$

Such filters are of the IIR type. They are useful for stabilizing unstable filters without appreciable change in the magnitude of frequency response and in phase equalization of IIR designs. An allpass filter transfer function with real coefficients and order N is of the following form:

$$H_{al}(z) = \pm \frac{d_N + d_{N-1}z^{-1} + \cdots + d_0 z^{-N}}{d_0 + d_1 z^{-1} + \cdots + d_N z^{-N}}.$$

The numerator and denominator coefficients are reverse-ordered. This forces the zeros to be reciprocals of the poles. A particularly attractive structure is achieved if $H(z)$ can be expressed as the average of the sum of two allpass transfer functions.

$H_{all1}(z)$ and $H_{al2}(z)$:

$$H(z) = \tfrac{1}{2}[H_{all1}(z) + H_{al2}(z)].$$

This is possible for certain types of transfer functions, such as odd-order Butterworth. For lowpass and highpass cases, the orders of the component allpass functions differ by one, whereas for the bandpass and bandstop cases they differ by two. The phase response of an allpass transfer function of order N exhibits a change of $N\pi$ in the frequency interval $[0, \pi]$. In the passband the allpass sections add up practically in phase and in the stopband they are essentially out of phase. By taking the difference of the component allpass filter transfer functions and multiplying by $\tfrac{1}{2}$, a complementary filter transfer function $G(z)$ is obtained:

$$G(z) = \tfrac{1}{2}[H_{all1}(z) - H_{al2}(z)].$$

The filter characterized by $G(z)$ is said to be *allpass complementary* as $H(z) + G(z) = H_{all1}(z)$. Note also that $H(z) - G(z) = H_{al2}(z)$. The filter pair $H(z)$ and $G(z)$ satisfies also the *power complementary* property because

$$|H(e^{j\omega T})|^2 + |G(e^{j\omega T})|^2 = 1.$$

Filters that satisfy the allpass complementary and power complementary properties are often called *doubly complementary* and find useful practical applications in digital audio [GA80] and as building blocks in tree-structured filter bands [RE87].

Turning to FIR filters, their most important property is their linear phase frequency response, which is satisfied when the filter coefficients are constrained. Specifically, as described in Chapter 4, a FIR filter characterized by the transfer function

$$H(z) = \sum_{i=0}^{M} q_i z^{-i}$$

might satisfy either the conjugate-even symmetry,

$$q_i = q^*_{M-i},$$

or the conjugate-odd symmetry,

$$q_i = -q^*_{M-i},$$

to yield a frequency response that is expressible in the form

$$H(e^{j\omega T}) = H_1(e^{j\omega T})e^{j(a-b\omega T)},$$

where the function $H_1(e^{j\omega T})$ is real-valued and a and b are constants. A filter whose frequency response is expressible as above is said to satisfy the linear phase property.

3-5 STRUCTURAL REPRESENTATIONS OF LINEAR SYSTEMS

There are various types of representations of digital filters that exhibit their internal structural interconnections. Such structural representations are necessary to develop the input/output relations of the filter and often lead to alternate equivalent realizations with better performance characteristics. Three such representations are discussed in this section.

3-5-1 Block Diagram Representation

The basic building blocks in the realization of LTI discrete-time input/output descriptions provided through either a difference equation or a transfer function are adders, multipliers, pick-off nodes, and unit delays. These elements are the discrete-time counterparts of components like resistors, inductors, and capacitors in passive circuit synthesis, or resistors, capacitors, and operational amplifiers in active *RC* circuit synthesis. The symbols for the basic discrete-time filter building blocks are shown in Figure 3-10. Also shown in Figure 3-10 are the respective

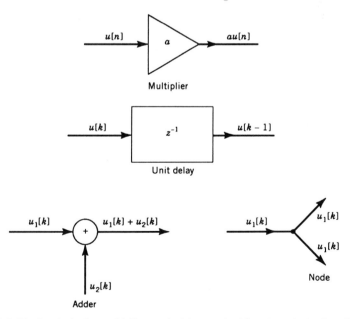

FIGURE 3-10 Symbols for multiplier, unit delay, and adder. A node is also shown for comparison with the adder symbol.

operations performed by each element on inputs to the blocks. In contrast to an adder, which has two or more inputs and a single output, a node has a single signal-carrying input branch incident on it while two or more output branches, each of which carries the same input signal, exit from it. The node in Figure 3-10 shows two such output branches.

One of the most important considerations in the performance evaluation of digital signal (recall that a quantized discrete-time signal is called a digital signal) processing structures is sensitivity under finite arithmetic constraints. The topology of the realization structures significantly influences the sensitivity properties. The various structures that may be used to realize a specified transfer function $H(z)$ shown in Eq. (3.27) are described next. Without any loss of generality, we set

$$d_0 = 1 \qquad\qquad (3.42)$$

in Eq. (3.27).

The direct form 1 realization of the transfer function is shown in Figure 3-11. Note that the element values for the multipliers are obtained directly from the numerator and denominator coefficients of the transfer function.

The number of delays occurring in the direct form realization can be reduced. To do this, we use the simple equivalence [FE83] between realization structures shown in Figure 3-12. This leads to the structure shown in Figure 3-13, from which the canonic structure involving the minimum number of delays is easily obtained in Figure 3-14. Similar to the direct form structure, the multiplier values in Figure 3-14 can be obtained directly from the coefficients of the transfer function. Therefore it is referred to as the canonic direct form 1 structure. In Figure 3-14, the feedforward part containing the multipliers whose values correspond to the numerator coefficients of the transfer function is on the left-hand side of the

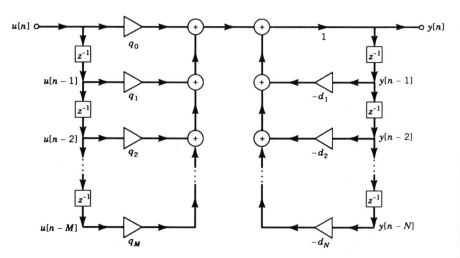

FIGURE 3-11   Direct form 1 realization of Eq. (3.28) with $d_0 = 1$.

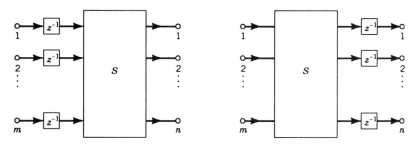

FIGURE 3-12 Equivalence between realization structures.

figure, while the feedback part, which contains the multipliers whose values are the negatives of the denominator coefficients of the transfer function, is on the right.

Figure 3-15 shows the direct form 2 realization obtained from the canonic direct form 1 realization in Figure 3-14 by moving the feedforward part to the right and the feedback part to the left. This structure is referred to, sometimes, as the *transpose* of the structure in Figure 3-14. It is seen that this transposed structure is obtained from the original structure in Figure 3-14 by reversing the direction of signal flows (consequently, the input is made the output and vice versa) and replacing the adders by nodes and the nodes by adders. The transpose operation can be carried out on any structure to yield an equivalent realization with the same transfer function.

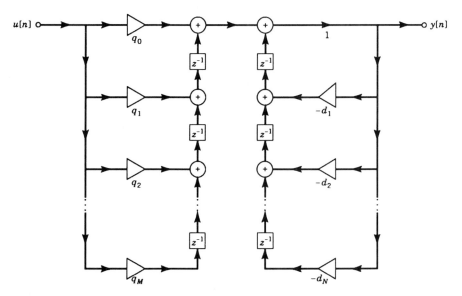

FIGURE 3-13 Structure derived from Figure 3-11 using structure equivalence of Figure 3-12.

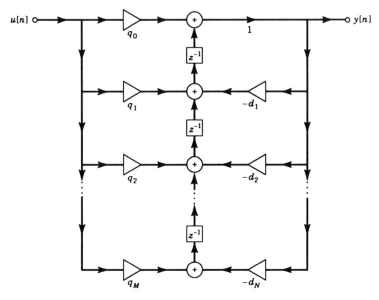

FIGURE 3-14 Canonic direct form 1 structure involving the minimum number of delays obtained from Figure 3-13 with $M = N$.

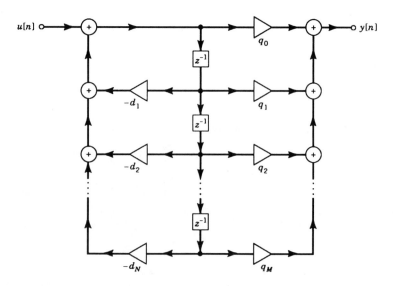

FIGURE 3-15 The direct form 2 structure or the transpose of the canonic direct form 1 structure; $M = N$.

The transfer function $H(z)$ in Eq. (3.27) may be decomposed as a sum of partial fractions. When $M \leq N$, the partial-fraction expansion is of the form

$$H(z) = \beta_{00} + \sum_{i=1}^{K} \frac{\beta_{0i} + \beta_{1i}z^{-1}}{1 + \delta_{1i}z^{-1} + \delta_{2i}z^{-2}}, \qquad (3.43)$$

where all the coefficients are real and K is the least integer that is greater than or equal to $N/2$. Each of the K terms in the summation may be realized in one of the direct forms and connected in parallel along with the multiplier β_{00} as shown in Figure 3-16. It is pointed out that there are two types of parallel forms [SZ77].

The transfer function in Eq. (3.27) may also be factored in the form

$$H(z) = \prod_{i=1}^{K} \frac{\gamma_{0i} + \gamma_{1i}z^{-1} + \gamma_{2i}z^{-2}}{1 + \delta_{1i}z^{-1} + \delta_{2i}z^{-2}}, \qquad (3.44)$$

where each of the coefficients in the decomposition is real. Each of the K biquadratic rational functions under the product sign in Eq. (3.44) may be realized in direct form and then cascaded, as in Figure 3-17.

The realization of lattice structures can easily be programmed as a recursive method for digital filter design. The generic structure is shown in Figure 3-18, where each section, referred to as a *digital two-pair* [BO85; MI73], takes the symmetric lattice form in Figure 3-19.

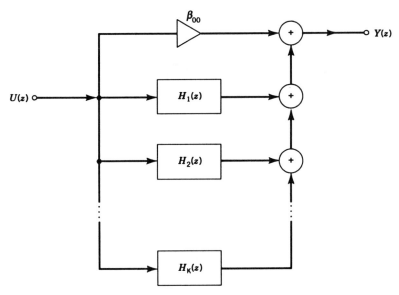

FIGURE 3-16 Parallel realization; $H_i(z) = (\beta_{0i} + \beta_{1i}z^{-1})/(1 + \delta_{1i}z^{-1} + \delta_{2i}z^{-2})$, $i = 1$, $2, \cdots, K$.

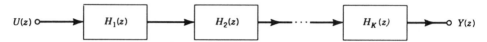

FIGURE 3-17 Cascade realization; $H_i(z) = (\gamma_{0i} + \gamma_{1i}z^{-1} + \gamma_{2i}z^{-2})/(1 + \delta_{1i}z^{-1} + \delta_{2i}z^{-2})$, $i = 1, 2, \cdots, K$.

Essentially, the structure involves a cascade of lattices to which a feed forward part is added by taking as the output a weighted linear combination of signals, tapped as shown in Figure 3-18.

Consider a digital filter transfer function of the form

$$H(z) = Q_N(z)/D_N(z),$$

where $Q_N(z)$ and $D_N(z)$ are Nth order polynomials in z^{-1} of the form

$$Q_N(z) = \sum_{n=0}^{N} q_{N,n} z^{-n} \tag{3.45}$$

and

$$D_N(z) = \sum_{n=0}^{N} d_{N,n} z^{-n}, \qquad d_{N,0} = 1. \tag{3.46}$$

It is assumed, without any loss of generality, that the leading coefficient $d_{N,0}$ of $D_N(z)$ is one. The number of digital two-pairs in Figure 3-18 required to realize the above transfer function is N and the number of taps required is $N + 1$. The ith

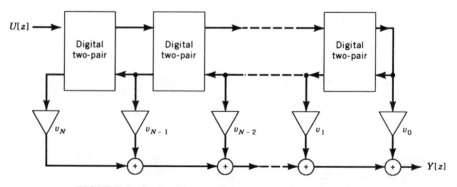

FIGURE 3-18   Structure realizing a rational transfer function.

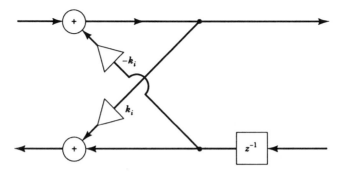

FIGURE 3-19   A typical digital two-pair.

digital two-pair is completely characterized by a parameter k_i, for $i = 1, 2, \cdots, N$. Each of these parameters is referred to as a reflection coefficient and the jth tap is associated with a multiplier value v_j, for $j = 0, 1, \cdots, N$. Consequently, there are $2N + 1$ parameters to be determined. These are the N reflection coefficients $k_1, k_2, \cdots, k_N$ and the $N + 1$ multiplier values $v_0, v_1, \cdots, v_N$. These parameters are recursively obtained by starting from the coefficients of $D_N(z)$ and $Q_N(z)$ as follows. For $n = N, N - 1, \cdots, 1$, the reflection coefficients are first calculated by repeating for each n the loop of computations in the following equations:

$$k_n = d_{n,n}, \tag{3.47}$$

$$d_{n-1,i} = \frac{d_{n,i} - k_n d_{n,n-i}}{1 - k_n^2}, \quad i = 0, 1, 2, \cdots, n - 1. \tag{3.48}$$

The denominator $(1 - k_n^2)$ in Eq. (3.48) has been introduced to preserve monicity (a polynomial in z is monic if the coefficient of its highest power is unity) in the polynomial $D_{n-1}(z)$ akin to polynomial $D_n(z)$. Note that the sequence of monic polynomials, $D_{N-1}(z), D_{N-2}(z), \cdots, D_0(z)$ through the use of Schur's algorithm. It should be noted that the denominator $D_N(z)$ of the transfer function $H(z)$ is completely defined by the reflection coefficients k_i, $i = 1, 2, \cdots, N$. The multiplier values $v_N, v_{N-1}, \cdots, v_0$ in the ladder part are calculated by performing the recursion in the following equations for $n = N, N - 1, \cdots, 0$:

$$c_{N,n} = q_{N,n} - \sum_{l=n+1}^{N} c_{N,l} d_{l,l-n}, \tag{3.49}$$

$$v_n = c_{N,n}. \tag{3.50}$$

Example 3.13. It is required to provide a lattice realization of the filter transfer function of Example 3.11.

Here the filter order is $N = 4$. It may be checked that Eqs. (3.47) and (3.48) generate the following values of the lattice coefficients. The reader is invited to verify each cycle of calculations, which is summarized as

$$k_4 = d_{4,4} = 0.43826514.$$

Furthermore, from the solution to the problem in Example 3.11, and using the notation set in Eq. (3.46),

$$d_{4,3} = -2.11215544, \qquad d_{4,2} = 3.86119413,$$

$$d_{4,1} = -3.18063855, \qquad d_{4,0} = 1.00000000.$$

By making use of Eqs. (3.47) and (3.48), it is possible to calculate the remaining k_i values.

$$d_{3,3} = -0.88893599, \qquad d_{3,2} = 2.68461915,$$

$$d_{3,1} = -2.79104890, \qquad d_{3,0} = 1.00000000,$$

$$k_3 = d_{3,3} = -0.88893599,$$

$$d_{2,2} = 0.97026825, \qquad\qquad d_{2,1} = -1.92853789,$$

$$d_{2,0} = 1.00000000, \qquad\qquad k_2 = d_{2,2} = 0.97026825,$$

$$d_{1,1} = -0.97881989 = k_1.$$

From the solution of the problem in Example 3.11 and using the notation set in Eq. (3.45), it is clear that

$$q_{4,0} = q_{4,4} = 0.00416581,$$

$$q_{4,1} = q_{4,3} = 0.01666322,$$

$$q_{4,2} = 0.02499484.$$

Equation (3.49) and (3.50) can now be used to calculate the multiplier coefficients v_4, v_3, v_2, v_1, and v_0:

$$c_{4,4} = 0.00416581 = v_4,$$

$$c_{4,3} = 0.02991316 = v_3,$$

$$c_{4,2} = 0.09239890 = v_2,$$

$$c_{4,1} = 0.12335185 = v_1,$$

$$c_{4,0} = 0.06001874 = v_0.$$

The lattice realization for this example is shown in Figure 3-20.

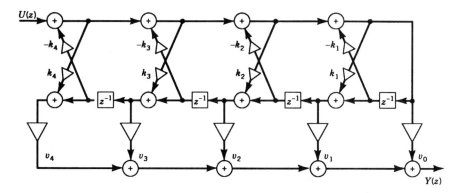

FIGURE 3-20   Fourth-order lattice structure in Example 3.13.

3-5-2   Signal Flow-Graph (SFG) Representation

A signal flow-graph (SFG) consists of junction points called *nodes*, which are connected by directed line segments called *branches*. Signals travel along branches only in the direction described by the arrows of the branches. Each node k is assigned a node value w_k. Branch jk denotes a branch originating at node j and terminating at node k. Each branch has an input signal and an output signal. The input signal from node j to branch jk is the node value w_j and the output signal from branch jk to node w_k is denoted by y_{jk}. The functional dependence of the branch output on the branch input is denoted by the transformation $f_{jk}[.]$. This transformation could be linear or nonlinear and relates y_{jk} to w_j as

$$y_{jk} = f_{jk}[w_j]. \tag{3.51}$$

Two special types of nodes in a SFG are the source nodes and the sink nodes. A source node possesses only outgoing branches; a sink node has only input branches. In the commonly used branch symbol shown in Figure 3-21, the node value w_j at node j is the input to branch jk while the node value w_k at node k had the output y_{jk} from branch jk as one of its components. The other components in w_k are due

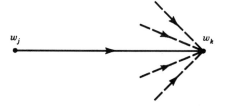

FIGURE 3-21   The nodes j and k, with node values w_j and w_k, are connected by branch jk, whose input w_j and output y_{jk} are related to each other by Eq. (3.51).

to the other branches that may be incident on node k. If the branches incident on node k are numbered $1, 2, \cdots, N$, and if the output of the jk branch is y_{jk} for $j = 1, 2, \cdots, N$, then the constraint equation at node k is

$$w_k = \sum_{j=1}^{N} y_{jk}. \tag{3.52}$$

A SFG is a directed graph because it consists of a set of nodes (including source nodes, sink nodes, and other nodes called internal nodes) and a set of directed branches. As in any directed graph, each branch has two nodes for its end points and the branch exits from one of those nodes and enters into the other. From a discrete-time filter structure containing nodes, adders, multiplier, and unit delay elements, the SFG corresponding to the structure may be derived. Both adders and nodes (also called branch points) in the discrete-time filter structure are represented as nodes in the SFG. A source (sink) node in a SFG has one or more branches exiting (entering) from (into) it and no branch entering (exiting) into (from) it. A node in the SFG corresponding to a branch point (adder) has only one branch entering (exiting) into (from) it and more than one branch exiting (entering) from (into) it. The multiplier and the unit delay elements are both represented in the SFG as oriented branches. When no weight is associated with a branch, then this branch represents a unit multiplier. A nonunity weight associated with a SFG branch indicates the value of the multiplier, while any unit delay element in the structure is explicitly identified as z^{-1} in the corresponding branch of the SFG. The SFG associated with the structure shown in Figure 3-22 is shown in Figure 3-23. The discrete-time structure in Figure 3-22 can also be reconstructed from the SFG in Figure 3-23.

The SFG is useful for computing the transfer function $H(z)$ of a system. To do this, it is convenient to use *Mason's gain formula*. Before introducing this formula, the meanings of feedforward path and feedback loop, path gain, and loop gain should be understood. A feedforward path from input node (or source) to output node (or sink) is a sequence of nodes and branches, starting with the source node and ending with the sink node, in which all nodes (occurring in the sequence) are

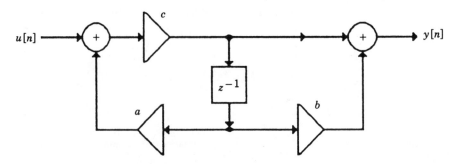

FIGURE 3-22 A first-order digital filter structure.

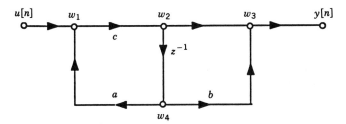

FIGURE 3-23 Signal flow-graph for digital filter structure in Figure 3-22.

traversed only once. The product of all branch values along the path is the path gain. In the feedforward path, the source and sink constitute the terminal nodes. When the terminal nodes of a path coincide, a feedback loop is formed. The product of branch values in a feedback loop gives the loop gain.

According to Mason's gain formula, the transfer function in Eq. (3.27) is given by

$$H(z) = \frac{1}{\Delta} \sum_k H_k \Delta_k, \tag{3.53}$$

where

$$\Delta = 1 - \sum_m P_{m1} + \sum_m P_{m2} - \sum_m P_{m3} + \cdots + (-1)^j \sum P_{mj},$$

with

P_{mk} = product of the loop gains of the mth set of k nontouching loops (loops of order k, where $k = 1, 2, \cdots, j$), where nontouching loops are node-disjoint loops.

Δ_k = the value of Δ for the subgraph of the graph obtained by removing the kth feedforward path along with all those branches touching this path.

H_k = product of the branch gains in the kth feedforward path.

The use of Mason's gain formula is illustrated below. It is evident that P_{m1} is the loop gain or the feedback gain of a first-order loop.

Example 3.14. It is required to calculate the transfer function of the digital filter structure in Figure 3-22.

Consider the SFG representation of the digital filter structure of Figure 3-22 is shown in Figure 3-23. There is only one first-order loop, and two feedforward paths as indicated in Figure 3-24. The loop gain is $P_{11} = cz^{-1}a$. Therefore

$$\Delta = 1 - cz^{-1}a.$$

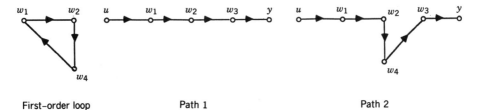

First–order loop   Path 1   Path 2

FIGURE 3-24   The first-order loop and the two paths from input to output in Example 3.14.

The products of the branch gains in the two feedforward paths from the input node to the output node are

$$H_1 = c \quad \text{and} \quad H_2 = cz^{-1}b.$$

It is easily seen that the corresponding values of Δ_k are

$$\Delta_1 = 1 \quad \text{and} \quad \Delta_2 = 1.$$

Applying Eq. (3.53), the transfer function is therefore

$$H(z) = \frac{Y(z)}{U(z)} = \frac{c(1 + z^{-1}b)}{(1 - cz^{-1}a)}.$$

The graphs that were used in the computation of the functions required in the calculation of $H(z)$ are shown in Figure 3-24.

Example 3.15. Consider the SFG in Figure 3-25. This SFG represents a digital filter, which is realized as a cascade of a second-order filter and a first-order filter. Between these two filters is a multiplier of value m, which is used for scaling. In addition to the source node, whose node value at time index n is the input $u[n]$,

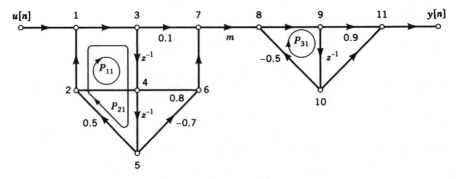

FIGURE 3-25   The signal flow-graph for Example 3.15.

and the sink node, whose node value is $y[n]$, there are 11 other nodes whose node values at time index n are, say, $w_1[n]$, $w_2[n]$, $\cdots$, $w_{10}[n]$, and $w_{11}[n]$. The nodes associated with these node values are labeled sequentially from 1 to 11.

In Figure 3-25, the three first-order loops are identified and their loop gains are calculated to be

$$P_{11} = z^{-1}, \quad P_{21} = 0.5z^{-2}, \quad \text{and} \quad P_{31} = -0.5z^{-1}.$$

Since each of the loops having loop gains P_{11} and P_{21} are not touching the loop with loop gain P_{31}, the second-order loop gains in Figure 3-25 are

$$P_{12} = -0.5z^{-2} \quad \text{and} \quad P_{22} = -0.25z^{-3}.$$

Therefore

$$\Delta = 1 - (z^{-1} + 0.5z^{-2} - 0.5z^{-1}) + (-0.5z^{-2} - 0.25z^{-3})$$
$$= 1 - 0.5z^{-1} - z^{-2} - 0.25z^{-3}.$$

There are six forward paths from the input node to the output node. These paths are shown in Figure 3-26. Each of these paths touches all the loops. Therefore each Δ_k is 1 since the subgraph obtained by removing the kth feedforward path along with all those branches touching this path does not have any loops. It may easily be checked that

$$H_1 = (0.09)m, \quad H_2 = (0.10z^{-1})m, \quad H_3 = (0.72z^{-1})m$$
$$H_4 = (0.80z^{-2})m, \quad H_5 = (-0.63z^{-2})m, \quad \text{and} \quad H_6 = (-0.7z^{-3})m.$$

Therefore the transfer function is

$$H(z) = \frac{Y(z)}{U(z)} = \frac{m(0.09 + 0.82z^{-1} + 0.17z^{-2} - 0.7z^{-3})}{1 - 0.5z^{-1} - z^{-2} - 0.25z^{-3}}.$$

The loop occurring in the SFG of the digital filter structure shown in Figure 3-22 has a z^{-1} branch present in that loop. However, in some structures there may be loops containing no z^{-1} branches. Such loops are called *delay-free loops*. It is shown in the next subsection that the absence of delay-free loops is necessary and sufficient for satisfying an important condition in the realization theory of digital filters. This condition is referred to as the *computability condition*, which guarantees the existence of a scheme for reordering the nodes in a SFG to permit the node variables to be generated in sequence.

3-5-3 Matrix Description

SFGs permit LTI systems to be efficiently modeled. One such model is provided by the matrix description. At the kth node of the SFG, the node value, which is

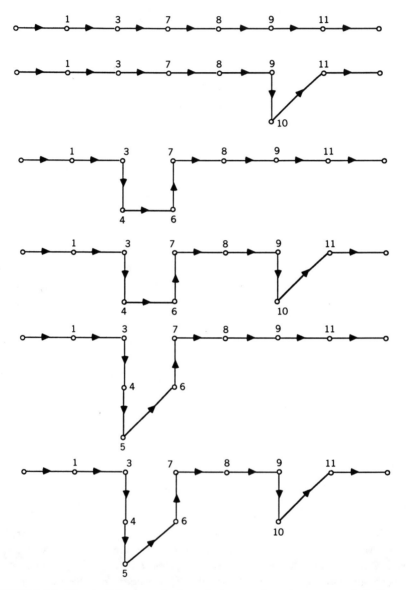

FIGURE 3-26   The six different paths from the input node to the output node in Example 3.15.

given by the sum of the outputs of all the branches entering the node, is denoted by w_k. For the SFG in Figure 3-23, the node values w_1, w_2, w_3, and w_4 associated with each internal node (i.e., nodes other than the source and sink nodes) are indicated. Furthermore, in Figure 3-23, $u[n]$ denotes the input sample while $y[n]$ is the output sample at time index n. The matrix equation linking the node values, input, and output may be written down by writing the constraint equation at each internal node. The resulting four equations are expressed in matrix form:

$$
\begin{bmatrix} w_1[n] \\ w_2[n] \\ w_3[n] \\ w_4[n] \end{bmatrix} = \begin{bmatrix} 0 & 0 & 0 & a \\ c & 0 & 0 & 0 \\ 0 & 1 & 0 & b \\ 0 & 0 & 0 & 0 \end{bmatrix} \begin{bmatrix} w_1[n] \\ w_2[n] \\ w_3[n] \\ w_4[n] \end{bmatrix} + \begin{bmatrix} 0 & 0 & 0 & 0 \\ 0 & 0 & 0 & 0 \\ 0 & 0 & 0 & 0 \\ 0 & 1 & 0 & 0 \end{bmatrix} \begin{bmatrix} w_1[n-1] \\ w_2[n-1] \\ w_3[n-1] \\ w_4[n-1] \end{bmatrix}
$$

$$
+ \begin{bmatrix} 1 \\ 0 \\ 0 \\ 0 \end{bmatrix} u[n], \tag{3.54}
$$

$$
y[n] = [0 \quad 0 \quad 1 \quad 0] \begin{bmatrix} w_1[n] \\ w_2[n] \\ w_3[n] \\ w_4[n] \end{bmatrix}. \tag{3.55}
$$

The SFG in Figure 3-23 may also be constructed from the matrix representations in the two preceding sets of equations. By denoting the column vector of node values by $w[n]$, the preceding two equations may be written in more compact form:

$$
w[n] = F_1 w[n] + F_2 w[n-1] + Bu[n], \tag{3.56}
$$

$$
y[n] = Cw[n]. \tag{3.57}
$$

The matrices F_1, F_2, B, and C in Eqs. (3.56) and (3.57) are easily obtained by comparison with Eqs. (3.54) and (3.55). It is clear that the forms of the matrices F_1 and F_2 depend on the ordering chosen to number the nodes. If this chosen ordering is changed, the matrices would change. From Eq. (3.54), it is clear that the node variables cannot be generated in sequence, since $w_4[n]$ is needed to compute $w_1[n]$. On the other hand, a permutation of the node variables might allow sequential computation of the renumbered node variables. For example, consider the SFG in Figure 3-27. This SFG is identical to the one shown in Figure 3-23

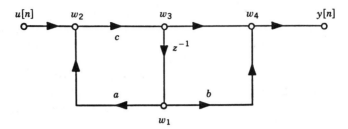

FIGURE 3-27 Signal flow-graph for digital filter structure in Figure 3-23 with nodes renumbered.

except for the fact that the node values have been renumbered. The matrix equations for this renumbered set of nodes in Figure 3-27 are

$$
\begin{bmatrix} w_1[n] \\ w_2[n] \\ w_3[n] \\ w_4[n] \end{bmatrix} = \begin{bmatrix} 0 & 0 & 0 & 0 \\ a & 0 & 0 & 0 \\ 0 & c & 0 & 0 \\ b & 0 & 1 & 0 \end{bmatrix} \begin{bmatrix} w_1[n] \\ w_2[n] \\ w_3[n] \\ w_4[n] \end{bmatrix} + \begin{bmatrix} 0 & 0 & 1 & 0 \\ 0 & 0 & 0 & 0 \\ 0 & 0 & 0 & 0 \\ 0 & 0 & 0 & 0 \end{bmatrix} \begin{bmatrix} w_1[n-1] \\ w_2[n-1] \\ w_3[n-1] \\ w_4[n-1] \end{bmatrix}
$$

$$
+ \begin{bmatrix} 0 \\ 1 \\ 0 \\ 0 \end{bmatrix} u[n], \tag{3.58}
$$

$$
y[n] = \begin{bmatrix} 0 & 0 & 0 & 1 \end{bmatrix} \begin{bmatrix} w_1[n] \\ w_2[n] \\ w_3[n] \\ w_4[n] \end{bmatrix}. \tag{3.59}
$$

Note that unlike Eq. (3.56), the matrix $\mathbf{F}_1$ for Eq. (3.58) has only nonzero elements below its main diagonal. Such a matrix is called lower triangular. The node variables in Figure 3-27 can be computed in sequence. In general, a LTI system may be characterized by equations whose forms are shown in Eqs. (3.56) and (3.57) and the node variables of the SFG for the LTI system can be computed in sequence if and only if the matrix $\mathbf{F}_1$ is lower triangular. In certain situations, no reordering of the nodes in a SFG will permit the node variables to be generated in sequence. Such a SFG is called *noncomputable*. In that case, the SFG will have one or more loops without delay.

In digital signal processing, only computable SFGs are of relevance since it is required to solve Eq. (3.56) by computing successive node values. To fulfill this objective, it is necessary to test for the absence of delay-free loops by relating to the notion of node precedence. An algorithmic procedure for delineating a structure in a form that reveals the precedence relations was advanced by Crochiere ([CR75]). To facilitate exposition, in addition to the constant gain (coefficient) branches, delay branches, and nodes that make up a SFG, the terminology of a coefficient-delay branch is introduced. Such a branch is a cascade of a constant gain (including unity) and a unit delay viewed as one branch. Therefore a unit delay branch belongs to the category of coefficient-delay branch. The procedure begins by searching the original SFG for all nodes that do not have coefficient branches entering them. These nodes are then separated from the network and assigned to node set $\{N_1\}$. At any time index n, the node values in the node set $\{N_1\}$ are clearly computable only from knowledge of past node values for time index $n - 1$ due to coefficient-delay branches, or present source input values at time index n. The subgraph that contains all nodes and branches in the original SFG except those nodes that belong to the node set $\{N_1\}$ and those branches that connect these nodes is then considered. A search is made over this subgraph analogous to the search made over the original SFG with the objective of identifying the set $\{N_2\}$ of all nodes that do not have coefficient branches from other nodes in the subgraph incident on these. Clearly, each node in the node set $\{N_2\}$ must have at least one coefficient branch incident on it, which comes from a node in $\{N_1\}$. The procedure is repeated until at the last stage either none of the remaining nodes in the final subgraph has any coefficient branch incident on it (the original SFG is then computable) or each node in the final subgraph has at least one coefficient branch incident on it (the original SFG is then noncomputable). The next example illustrates how the algorithm outlined above is implemented on a SFG associated with a digital filter structure.

Example 3.16. Consider the SFG in Figure 3-25. On applying the search for all nodes (other than the source and sink nodes) that do not have coefficient branches incident on them, the set $\{N_1\}$ is identified as

$$\{N_1\} = (4, 5, 10).$$

The subgraph with the remaining nodes, which excludes those branches that connect to any of the nodes in set $\{N_1\}$, is shown in Figure 3-28. The second search over this subgraph identifies the set $\{N_2\}$ to be

$$\{N_2\} = (2, 6).$$

The procedure is continued to yield

$$\{N_3\} = (1), \quad \{N_4\} = (3), \quad \{N_5\} = (7),$$
$$\{N_6\} = (8), \quad \{N_7\} = (9), \quad \text{and} \quad \{N_8\} = (11).$$

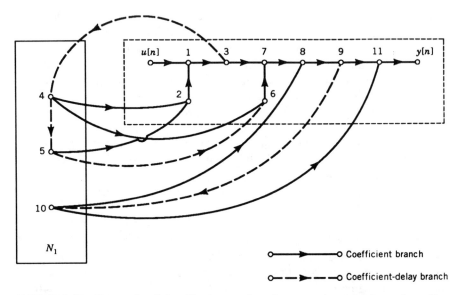

FIGURE 3-28 First cycle of algorithmic procedure for generating precedence form. Remaining subgraph after completion of first cycle is enclosed in dotted box and consists only of the nodes and the directed edges shown by solid lines totally inside this box. Any edge having one terminal node in the set N_1 does not belong to this remaining subgraph.

Since the algorithmic procedure proceeds to completion, the SFG in Figure 3-25 is computable.

The decomposition of the network nodes numbered from 1 to 11 in Example 3.16 into the subsets $\{N_1\}$ to $\{N_8\}$ is called the precedence form because it shows all the precedence requirements of the network in terms of node evaluation. A procedure for numbering the nodes of a computable network so that precedence requirements are satisfied consists of first numbering the nodes in set $\{N_1\}$, then the nodes in set $\{N_2\}$, and so on until all the nodes are numbered.

3-6 STATE-SPACE REPRESENTATION OF LINEAR SYSTEMS

Some important problems involving digital filter structures are convenient to formulate using a state-space description. This description, which has been used extensively in the analysis and design of continuous-time systems and also in estimation and digital control theory, is finding increasing usage in the digital signal processing literature. To facilitate understanding, consider the signal-flow graph in Figure 3-23 for which the representation in Eq. (3.54) was obtained. Rearrange

Eq. (3.54) to obtain

$$
\begin{bmatrix}
1 & 0 & 0 & -a \\
-c & 1 & 0 & 0 \\
0 & -1 & 1 & -b \\
0 & 0 & 0 & 1
\end{bmatrix}
\begin{bmatrix}
w_1[n] \\
w_2[n] \\
w_3[n] \\
w_4[n]
\end{bmatrix}
$$

$$
=
\begin{bmatrix}
0 & 0 & 0 & 0 \\
0 & 0 & 0 & 0 \\
0 & 0 & 0 & 0 \\
0 & 1 & 0 & 0
\end{bmatrix}
\begin{bmatrix}
w_1[n-1] \\
w_2[n-1] \\
w_3[n-1] \\
w_4[n-1]
\end{bmatrix}
+
\begin{bmatrix}
1 \\
0 \\
0 \\
0
\end{bmatrix}
u[n]. \tag{3.60}
$$

Since the matrix on the left-hand side of Eq. (3.60) is nonsingular, it has an inverse. This inverse matrix allows Eq. (3.60) to be recast in the form

$$
\begin{bmatrix}
w_1[n] \\
w_2[n] \\
w_3[n] \\
w_4[n]
\end{bmatrix}
=
\begin{bmatrix}
0 & a & 0 & 0 \\
0 & ac & 0 & 0 \\
0 & ac+b & 0 & 0 \\
0 & 1 & 0 & 0
\end{bmatrix}
\begin{bmatrix}
w_1[n-1] \\
w_2[n-1] \\
w_3[n-1] \\
w_4[n-1]
\end{bmatrix}
+
\begin{bmatrix}
1 \\
c \\
c \\
0
\end{bmatrix}
u[n]. \tag{3.61}
$$

There are a lot of redundancies in Eq. (3.61), because only the node value $w_2[n-1]$ together with the input $u[n]$ is necessary for the calculation of all node values and the output at time index n. Simple manipulations show that the following set of equations carry all necessary information to calculate the output:

$$
w_2[n] = acw_2[n-1] + cu[n], \tag{3.62}
$$

$$
y[n] = (ac + b)w_2[n-1] + cu[n]. \tag{3.63}
$$

Since $w_4[n] = w_2[n-1]$, after defining $x[n] \triangleq w_4[n]$, the above two equations can be written in the form

$$
x[n+1] = acx[n] + cu[n], \tag{3.64}
$$

$$
y[n] = (ac + b)x[n] + cu[n]. \tag{3.65}
$$

Equation (3.64) is the state equation and Eq. (3.65) is the output equation for the filter structure shown in Figure 3-22. Here $x[n]$ is the essential state variable for the first-order structure. The state-space description in the previous two equations involves the minimum number of states and therefore may also be referred to as a minimal state-space realization of the filter in Figure 3-22. For a LTI discrete-time single-input–single-output system, the number of essential state variables is equal to the order of the system, where by order is meant the maximum of the degrees of the numerator and denominator polynomials in the rational transfer function, which characterizes the input/output behavior of the system. Equation (3.61) may also be viewed as a state equation and together with Eq. (3.55) provides a non-minimal realization of the filter in Figure 3-22.

3-6-1 State Equations and Their Solutions

The state update and output equations (for brevity, these two are referred to as state equations) of a discrete-time LTI system, with input $u[n]$ and output $y[n]$, take the form

$$x[n-1] = \mathbf{A}x[n] + \mathbf{B}u[n] \tag{3.66}$$

and

$$y[n] = \mathbf{C}x[n] + \mathbf{D}u[n], \tag{3.67}$$

where $x[n]$ is the state vector at time index n whose elements are the state variables $x_1[n], x_2[n], \cdots, x_N[n]$. In the case of several inputs and outputs, $u[n]$ and $y[n]$ are replaced, respectively, by vectors $\mathbf{u}[n]$ and $\mathbf{y}[n]$. The matrices $\mathbf{A}, \mathbf{B}, \mathbf{C}$, and $\mathbf{D}$ are constant matrices of appropriate dimensions. The vector $x[n]$ in a nonminimal realization could have a dimension equaling the number of internal nodes, and each of its components is a state. The set of essential states has the minimum number of state variables that are necessary and sufficient for the representation.

The state equation (3.66) can be solved for $x[n]$ in terms of a specified initial state $x[0]$ and the input sequence $\{u[n]\}$, and this solution, given below in Eq. (3.68), may be used to compute the state trajectory for $n > 0$.

$$x[n] = \mathbf{A}^n x[0] + \sum_{k=0}^{n-1} \mathbf{A}^{n-k-1}\mathbf{B}u[k]. \tag{3.68}$$

The first term on the right-hand side of Eq. (3.68) is the zero-input response, while the second term involving the summation is the zero-state response.

3-6-2 State Equations from Difference Equations

The state equations [Eqs. (3.66) and (3.67)] may be directly obtained from the difference equation, which provides the input/output description of a discrete-time LTI system. Set $d_0 = 1$ and $N = M$ in Eq. (3.9) and consider the difference equation

$$y[n] = -\sum_{l=1}^{N} d_l y[n-l] + \sum_{i=0}^{N} q_i u[n-i]. \tag{3.69}$$

The matrices in the state equation are identified below, where $\mathbf{A}$, $\mathbf{B}$, $\mathbf{C}$, and $\mathbf{D}$ are of size $(N \times N)$, $(N \times 1)$, $(1 \times N)$ and (1×1), respectively, and we have defined

$$c_i = q_i - d_i q_0$$

for $i = 1, 2, \cdots, N$.

$$\mathbf{A} = \begin{bmatrix} -d_1 & -d_2 & -d_3 & \cdots & -d_{N-1} & -d_N \\ 1 & 0 & 0 & \cdots & 0 & 0 \\ 0 & 1 & 0 & \cdots & 0 & 0 \\ \vdots & & & & & \\ 0 & 0 & 0 & \cdots & 1 & 0 \end{bmatrix}, \quad \mathbf{B} = \begin{bmatrix} 1 \\ 0 \\ 0 \\ \vdots \\ 0 \end{bmatrix} \quad (3.70)$$

$$\mathbf{C} = [c_1 \quad c_2 \quad \cdots \quad c_N], \qquad \mathbf{D} = q_0. \tag{3.71}$$

When $d_0 \neq 1$, the state equations are obtained by replacing the coefficients in $\mathbf{A}$, $\mathbf{B}$, $\mathbf{C}$, and $\mathbf{D}$ matrices according to the following rule:

$$d_i \rightarrow \frac{d_i}{d_0}, \qquad i = 1, 2, \cdots, N; \tag{3.72}$$

$$q_i \rightarrow \frac{q_i}{d_0}, \qquad i = 0, 1, \cdots, N. \tag{3.73}$$

The case when $N \neq M$ may also be handled by setting the appropriate coefficients to zero.

3-6-3 State Equations from Transfer Function

Consider the transfer functions $H(z)$ obtained by applying the z-transform to Eq. (3.69) under zero initial conditions:

$$H(z) = \frac{q_0 + q_1 z^{-1} + \cdots + q_N z^{-N}}{1 + d_1 z^{-1} + \cdots + d_N z^{-N}}.$$

When the transfer function is specified in the form above, the state equation matrices are directly obtained as in Eqs. (3.70) and (3.71).

3-6-4 State Equations from Structures

Consider the direct form 2 realization of a discrete-time LTI filter structure shown in Figure 3-29. Assign state variables $x_1[n]$, $x_2[n]$, $x_3[n]$, and $x_4[n]$ at the outputs

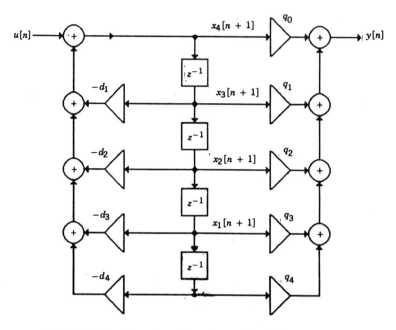

FIGURE 3-29   States for a direct form digital filter structure.

of each unit delay as shown in Figure 3-29. Clearly,

$$x_1[n + 1] = x_2[n],$$

$$x_2[n + 1] = x_3[n],$$

$$x_3[n + 1] = x_4[n],$$

$$x_4[n + 1] = - \sum_{i=1}^{4} d_{5-i} x_i[n] + u[n], \tag{3.74}$$

$$y[n] = \sum_{i=1}^{4} q_{5-i} x_i[n] + q_0 x_4[n + 1]$$

$$= \sum_{i=1}^{4} (q_{5-i} - q_0 d_{5-i}) x_i[n] + q_0 u[n]. \tag{3.75}$$

The preceding set of equations may be recast in the form of state equations as in Eqs. (3.76) and (3.77). The **A**, **B**, **C**, and **D** matrices may then be identified for this problem.

$$\begin{bmatrix} x_1[n+1] \\ x_2[n+1] \\ x_3[n+1] \\ x_4[n+1] \end{bmatrix} = \begin{bmatrix} 0 & 1 & 0 & 0 \\ 0 & 0 & 1 & 0 \\ 0 & 0 & 0 & 1 \\ -d_4 & -d_3 & -d_2 & -d_1 \end{bmatrix} \begin{bmatrix} x_1[n] \\ x_2[n] \\ x_3[n] \\ x_4[n] \end{bmatrix} + \begin{bmatrix} 0 \\ 0 \\ 0 \\ 1 \end{bmatrix} u[n] \tag{3.76}$$

$$y[n] = [q_4 - q_0 d_4 \quad q_3 - q_0 d_3 \quad q_2 - q_0 d_2 \quad q_1 - q_0 d_1]$$

$$\cdot \begin{bmatrix} x_1[n] \\ x_2[n] \\ x_3[n] \\ x_4[n] \end{bmatrix} + q_0 u[n]. \tag{3.77}$$

3-6-5 State Transformation

The state variables of a system are not unique. The state vector $\mathbf{x}[n]$ may be transformed to a new state vector

$$\hat{\mathbf{x}}[n] = \mathbf{T}\mathbf{x}[n],$$

where $\mathbf{T}$ is any nonsingular matrix of compatible order. The state-space descriptions in Eqs. (3.66) and (3.67) then change to the following descriptions:

$$\hat{\mathbf{x}}[n] = \mathbf{TAT}^{-1}\hat{\mathbf{x}}[n] + \mathbf{TB}u[n] \tag{3.78}$$

$$y[n] = \mathbf{CT}^{-1}\hat{\mathbf{x}}[n] + \mathbf{D}u[n]. \tag{3.79}$$

The two different state-space descriptions are associated with two different realizations of the same transfer function. These two realizations will, in general, perform differently under a finite word length constraint. For an extensive discussion on the optimal synthesis of state-space structures for LTI digital filters, see Roberts and Mullis [RO87]. While either the transfer function or the unit-impulse response sequence provides the input/output description of a LTI system, the internal structure of the system may be studied through the state equation representation. The unit-impulse response sequence $\{h[n]\}_{n=0}^{\infty}$ of a causal filter, obtained from the power series expansion about $z^{-1} = 0$ of the transfer function (in the single-input–single-output case)

$$H(z) = \mathbf{D} + \mathbf{C}(z\mathbf{I} - \mathbf{A})^{-1}\mathbf{B} = \sum_{n=0}^{\infty} h[n]z^{-n},$$

is related to the state matrices $\mathbf{A}$, $\mathbf{B}$, $\mathbf{C}$, and $\mathbf{D}$ as follows:

$$h[n] = \begin{cases} h[0], & n = 0 \\ \mathbf{CA}^{n-1}\mathbf{B}, & n \geq 1. \end{cases} \tag{3.80}$$

3-7 BLOCK PROCESSING

Block processing filters are attractive alternatives for linear digital filters because they can be designed for better performance under finite arithmetic constraints and they are responsive to the distributed computational architecture, currently prevalent, for implementation. Basically, block processing filters are multi-input–multi-output filters, which may be recast to be equivalent to another single-input–single-output filter. Block processing is used for high-speed digital filtering and adaptive filtering on super-computers. It also permits the use of FFTs to implement the block equations.

3-7-1 Block Equation from Transfer Function

A matrix representation of convolution was advanced [BU71] in order to obtain an efficient block implementation scheme for Eq. (3.85). To ease notation and explanation, a specific third-order IIR filter is chosen for illustration of the underlying principles in block processing. Let the transfer function of this filter be

$$H(z) = \frac{Y(z)}{U(z)} = \frac{q_0 + q_1 z^{-1} + q_2 z^{-2} + q_3 z^{-3}}{d_0 + d_1 z^{-1} + d_2 z^{-2} + d_3 z^{-3}} = \frac{Q(z)}{D(z)}.$$

The above expression can alternately be written as

$$D(z)Y(z) = Q(z)U(z),$$

or equivalently as

$$d_n \circledast y[n] = q_n \circledast u[n],$$

from which Eq. (3.81) can easily be derived:

$$
\begin{bmatrix}
d_0 & 0 & 0 & 0 & \cdots \\
d_1 & d_0 & 0 & 0 \\
d_2 & d_1 & d_0 & 0 \\
d_3 & d_2 & d_1 & d_0 \\
0 & d_3 & d_2 & d_1 \\
0 & 0 & d_3 & d_2 \\
\vdots
\end{bmatrix}
\begin{bmatrix}
y[0] \\
y[1] \\
y[2] \\
y[3] \\
y[4] \\
\vdots
\end{bmatrix}
$$

$$
=
\begin{bmatrix}
q_0 & 0 & 0 & 0 & \cdots \\
q_1 & q_0 & 0 & 0 \\
q_2 & q_1 & q_0 & 0 \\
q_3 & q_2 & q_1 & q_0 \\
0 & q_3 & q_2 & q_1 \\
0 & 0 & q_3 & q_2 \\
\vdots
\end{bmatrix}
\begin{bmatrix}
u[0] \\
u[1] \\
u[2] \\
u[3] \\
u[4] \\
\vdots
\end{bmatrix}.
\qquad (3.81)
$$

The input, the output, and the coefficient matrices can now be partitioned into blocks or submatrices of dimension $L \geq 3$. For $L = 3$, the submatrices and the input and output blocks are given by

$$\mathbf{U}[0] = \begin{bmatrix} u[0] \\ u[1] \\ u[2] \end{bmatrix}, \quad \mathbf{U}[1] = \begin{bmatrix} u[3] \\ u[4] \\ u[5] \end{bmatrix}, \quad \mathbf{U}[2] = \begin{bmatrix} u[6] \\ u[7] \\ u[8] \end{bmatrix}, \cdots$$

$$\mathbf{Y}[0] = \begin{bmatrix} y[0] \\ y[1] \\ y[2] \end{bmatrix}, \quad \mathbf{Y}[1] = \begin{bmatrix} y[3] \\ y[4] \\ y[5] \end{bmatrix}, \quad \mathbf{Y}[2] = \begin{bmatrix} y[6] \\ y[7] \\ y[8] \end{bmatrix}, \cdots$$

$$\mathbf{Q}_0 = \begin{bmatrix} q_0 & 0 & 0 \\ q_1 & q_0 & 0 \\ q_2 & q_1 & q_0 \end{bmatrix}, \quad \mathbf{Q}_1 = \begin{bmatrix} q_3 & q_2 & q_1 \\ 0 & q_3 & q_2 \\ 0 & 0 & q_3 \end{bmatrix}.$$

$$\mathbf{D}_0 = \begin{bmatrix} d_0 & 0 & 0 \\ d_1 & d_0 & 0 \\ d_2 & d_1 & d_0 \end{bmatrix}, \quad \mathbf{D}_1 = \begin{bmatrix} d_3 & d_2 & d_1 \\ 0 & d_3 & d_2 \\ 0 & 0 & d_3 \end{bmatrix}. \qquad (3.82)$$

$$\begin{bmatrix} \mathbf{D}_0 & 0 & 0 & 0 & \cdots \\ \mathbf{D}_1 & \mathbf{D}_0 & 0 & 0 & \\ 0 & \mathbf{D}_1 & \mathbf{D}_0 & 0 & \\ 0 & 0 & \mathbf{D}_1 & \mathbf{D}_0 & \\ \vdots & & & & \end{bmatrix} \begin{bmatrix} \mathbf{Y}[0] \\ \mathbf{Y}[1] \\ \mathbf{Y}[2] \\ \mathbf{Y}[3] \\ \vdots \end{bmatrix}$$

$$= \begin{bmatrix} \mathbf{Q}_0 & 0 & 0 & 0 & \cdots \\ \mathbf{Q}_1 & \mathbf{Q}_0 & 0 & 0 & \\ 0 & \mathbf{Q}_1 & \mathbf{Q}_0 & 0 & \\ 0 & 0 & \mathbf{Q}_1 & \mathbf{Q}_0 & \\ \vdots & & & & \end{bmatrix} \begin{bmatrix} \mathbf{U}[0] \\ \mathbf{U}[1] \\ \mathbf{U}[2] \\ \mathbf{U}[3] \\ \vdots \end{bmatrix}. \qquad (3.83)$$

From Eq. (3.83) the following recursion equation results:

$$\mathbf{D}_0 \mathbf{Y}[k + 1] + \mathbf{D}_1 \mathbf{Y}[k] = \mathbf{Q}_0 \mathbf{U}[k + 1] + \mathbf{Q}_1 \mathbf{U}[k]$$
$$\mathbf{Y}[k + 1] = -\mathbf{D}_0^{-1} \mathbf{D}_1 \mathbf{Y}[k] + \mathbf{D}_0^{-1} \mathbf{Q}_0 \mathbf{U}[k + 1] + \mathbf{D}_0^{-1} \mathbf{Q}_1 \mathbf{U}[k].$$

$$(3.84)$$

This algorithm states how to calculate a block of output from one prior block of output and present and prior blocks of input. The block implementation is most efficient when the filter order exceeds 25. This implementation is stable, if the original filter is stable. Furthermore, it was shown that the poles of the block filter move toward the origin as the block length is increased. Therefore block implementation reduces error due to finite arithmetic constraints [MI78].

3-7-2   Block-State Implementation

The state-space description in Eqs. (3.66) and (3.67) can be rewritten as follows:

$$\begin{bmatrix} x[n+1] \\ y[n] \end{bmatrix} = \begin{bmatrix} A & B \\ C & D \end{bmatrix} \begin{bmatrix} x[n] \\ u[n] \end{bmatrix}. \tag{3.85}$$

If the state-vector $x[n]$ is of dimension $(N \times 1)$, the updating in Eq. (3.85) appears to require the computation of $(N + 1)$ row matrix–column matrix products (inner products) at each sample interval. However, because the current state and input are sufficient to define the system response for all future time, the system state need only be computed for every Lth sample interval (state decimation) and the input signal may be processed in blocks of L samples. To see how this is achieved, consider this possibility of processing the input vector sequence in blocks of length $L = 3$. To see the development of block-state equations, it is advisable to write the state equations for several values of n:

$$x[n + 1] = Ax[n] + Bu[n],$$
$$x[n + 2] = Ax[n + 1] + Bu[n + 1] = A^2x[n] + ABu[n] + Bu[n + 1],$$
$$x[n + 3] = Ax[n + 2] + Bu[n + 2] = A^3x[n] + A^2Bu[n] + ABu[n + 1]$$
$$+ Bu[n + 2].$$

Likewise, the output equations can be written as

$$y[n] = Cx[n] + Du[n],$$
$$y[n + 1] = Cx[n + 1] + Du[n + 1] = CAx[n] + CBu[n] + Du[n + 1],$$
$$y[n + 2] = Cx[n + 2] + Du[n + 2] = CA^2x[n] + CABu[n]$$
$$+ CBu[n + 1] + Du[n + 2].$$

In matrix form we thus have

$$X[k + 1] = \hat{A}X[k] + \hat{B}U[k],$$
$$Y[k] = \hat{C}X[k] + \hat{D}U[k], \tag{3.86}$$

where

$$\mathbf{X}[k + 1] = \mathbf{x}[n + 3], \quad \mathbf{X}[k] = \mathbf{x}[n];$$

$$\mathbf{U}[k] = \begin{bmatrix} u[n] \\ u[n + 1] \\ u[n + 2] \end{bmatrix}, \quad \mathbf{Y}[k] = \begin{bmatrix} y[n] \\ y[n + 1] \\ y[n + 2] \end{bmatrix};$$

$$\hat{\mathbf{A}} = \mathbf{A}^3, \quad \hat{\mathbf{B}} = [\mathbf{A}^2\mathbf{B} \quad \mathbf{AB} \quad \mathbf{B}];$$

$$\hat{\mathbf{C}} = \begin{bmatrix} \mathbf{C} \\ \mathbf{CA} \\ \mathbf{CA}^2 \end{bmatrix}, \quad \hat{\mathbf{D}} = \begin{bmatrix} \mathbf{D} & 0 & 0 \\ \mathbf{CB} & \mathbf{D} & 0 \\ \mathbf{CAB} & \mathbf{CB} & \mathbf{D} \end{bmatrix}. \tag{3.87}$$

The multi-input–multi-output system described by Eqs. (3.86) and (3.87) is referred to as the block-state realization, $\{\hat{\mathbf{A}}, \hat{\mathbf{B}}, \hat{\mathbf{C}}, \hat{\mathbf{D}}\}$, that is generated by the state-space realization $\{\mathbf{A}, \mathbf{B}, \mathbf{C}, \mathbf{D}\}$ of the single-input–single-output system in Eq. (3.85). Therefore in block form, the transfer function

$$H(z) = \mathbf{C}[z\mathbf{I} - \mathbf{A}]^{-1}\mathbf{B} + \mathbf{D}$$

can be realized by a multi-input–multi-output filter implemented with serial-to-parallel and parallel-to-serial converters as shown in Figure 3-30. The internal structure of the block-state realization is shown in Figure 3-31 for the case where the block length is four.

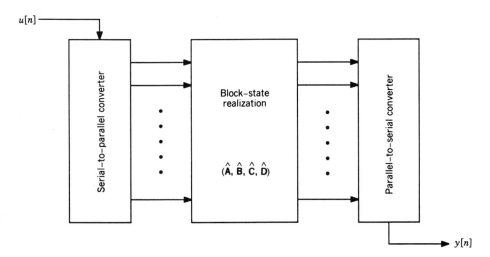

FIGURE 3-30 Block-state realization using serial-in-to-parallel-out and parallel-in-to-serial-out converters.

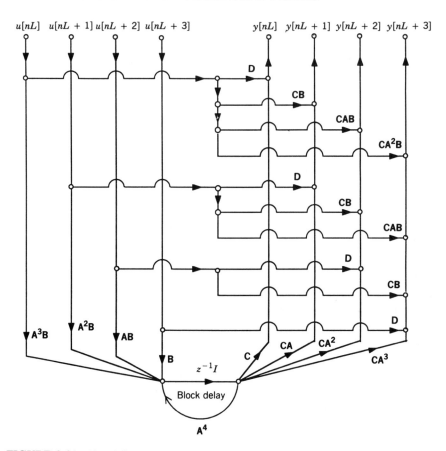

FIGURE 3-31 Signal flow-graph structure of a block-state realization for block length $L = 4$.

The updating of the block-state realization in Eq. (3.86) requires the computation of $(N + L)$ inner products of size $(N + L)$ every L steps or, equivalently, $(1 + N/L)$ inner products at each step. The number of multiplications per output sample is minimized for the block-state realization when the block length is chosen to be $L_{\text{opt}} = \sqrt{2}\, N$ in the case when the state vector $\mathbf{x}[n]$ is of size $(N \times 1)$. Using this optimal block length, the number of multiplications per output sample is $3.41N + 0.5$ and thus increases linearly with the order of the filter. Other realizations, like the cascaded second-order direct forms, might require a lower number of multiplications per output sample; however, the benefits of the block-state implementation arise from low sensitivity properties and high inherent parallelism without significant increase in computational complexity. The high inherent parallelism is ideally suited for use of distributed arithmetic or multimicroprocessor systems [RO87]; [ZE81].

3-8 SUMMARY AND FUTURE TRENDS

The restricted class of discrete-time systems, which are characterized by the properties of linearity and time-invariance, may be characterized in the time domain by its unit-impulse response and in the transform domain by its transfer function. Those LTI systems that can be represented by rational transfer functions are of crucial importance in practice. Such systems may have their input/output relationship characterized in the time domain by constant coefficient linear difference equations with zero initial conditions and in the transform domain by poles and zeros (to within a constant multiplier). Internal properties of the LTI system may be studied through the state-space description.

While considering the problem of synthesis of discrete-time LTI systems having rational transfer functions, attention has to be given to the realization of structures that have low sensitivity to finite arithmetic effects. Other factors that are important include the computational complexity in implementation. With the trend toward distributed computing spurred by the advent of the ultra-large-scale integration (ULSI) era, an effective measure of computational complexity is being redefined in contrast to the traditional one incorporating space/time computational economy applicable to the serial von Neumann type of computational architecture. In order to incorporate massive parallelism in implementation, the current trend is toward developing an information-theoretic computational complexity measure [AB88]. Information-based complexity includes information, partial information, contaminated information, and uncertainty. The intrinsic difficulty as measured by time and space requirements is also considered along with other factors like efficiency of data acquisition and data flow between multiple processors.

ACKNOWLEDGMENTS

This work was supported by HRB-Systems and the Office of Naval Research under the Fundamental Research Initiatives Program. The author thanks Francine Cauffman for patiently typing several versions of the manuscript, KiDoo Kim for his help in drawing the figures, and Jongtae Chun for his very careful reading of the manuscript.

REFERENCES

[AB88] Y. S. Abu-Mostafa, ed., *Complexity in Information Theory*, Springer-Verlag, Berlin and Heidelberg, 1988.

[BO83] W. J. Borodziewicz, K. J. Jaszczak, and M. A. Kowalski, A note on mathematical formulation of discrete-time linear systems. *Signal Process.* **5**(4), 369–375 (1983).

[BO85] N. K. Bose, *Digital Filters: Theory and Applications*. North-Holland/Elsevier, New York, 1985.

[BU71] C. S. Burrus, Block implementation of digital filters. *IEEE Trans. Circuit Theory* **CT-18,** 697–701 (1971).

[CR75] R. E. Crochiere and A. V. Oppenheim, Analysis of linear digital networks. *Proc. IEEE* **65**(4), 581–595 (1975).

[DU69] R. J. Duffin, Algorithms for classical stability problem. *SIAM Rev.* **11**(2), 196–213 (1969).

[FE83] A. Fettweis, K. Meerkötter, and W. Mecklenbräuker, *Digital Signal Processing,* Lect. Notes. VDE-Verlag GmbH, Berlin and Offenbach, 1983.

[GA80] P. Garde, Allpass crossover systems. *J. Audio Eng. Soc.* **28**, 575–584 (1980).

[JU64] E. I. Jury, *Theory and Application of the z-Transform Method.* Wiley, New York, 1964; reprinted by R. E. Krieger Publishing Co., Huntingdon, NY.

[KA80] T. Kailath, *Linear Systems.* Prentice-Hall, Englewood Cliffs, NJ, 1980.

[MI73] S. K. Mitra and R. J. Sherwood, Digital ladder networks. *IEEE Trans. Audio Electroacoust.* **AU-21,** 30–36 (1973).

[MI78] S. K. Mitra and R. Gnanasekaran, Block implementation of recursive digital filters—New structures and properties. *IEEE Trans. Circuits Syst.* **CAS-25,** 200–207 (1978).

[PO81] B. Pondelicek, On compositional and convolutional systems. *Kybernetika* **17**(4), 277–286 (1981).

[RE87] P. A. Regalia, S. K. Mitra, P. P. Vaidyanathan, M. K. Renfors, and Y. Neuvo, Tree-structured complementary filter banks using all pass sections. *IEEE Trans. Circuits Syst.* **CAS-34,** 1470–1484 (1987).

[RO87] R. A. Roberts and C. T. Mullis, *Digital Signal Processing.* Addison-Wesley, Reading, MA, 1987.

[SZ77] J. Szczupak, K. Mondal, and S. K. Mitra, An alternate parallel realization of digital transfer function. *Proc. IEEE (Lett).* **65,** 577–578 (1977).

[VA87] P. P. Vaidyanathan, ed., Special section on complex signal processing. *IEEE Trans. Circuits Syst.* **CAS-34,** 337–399 (1987).

[VI87] R. Vich, *z-Transform Theory and Application.* Reidel, Dordrecht, The Netherlands, 1987.

[ZA63] L. A. Zadeh and C. A. Desoer, *Linear System Theory.* McGraw-Hill, New York, 1963.

[ZE81] J. Zeman and A. G. Lindgren, Fast digital filters with low round-off noise. *IEEE Trans. Circuits Syst.* **CAS-28,** 716–723 (1981).

4 Finite Impulse Response Filter Design

TAPIO SARAMÄKI

Department of Electrical Engineering
Tampere University of Technology
Tampere, Finland

This chapter reviews several design techniques for finite impulse response (FIR) filters along with computationally efficient realization methods. The outline of this chapter is given in Section 4-2 after introducing the filter design problem for both FIR and infinite impulse response (IIR) filters.

4-1 DIGITAL FILTER DESIGN PROBLEM

4-1-1 Digital Filter Design Process

Digital filter design involves usually the following basic steps:

1. Determine a desired response or a set of desired responses (e.g., a desired magnitude response and/or a desired phase response).
2. Select a class of filters for approximating the desired response(s) (e.g., linear-phase FIR filters or IIR filters being implementable as a parallel connection of two allpass filters).
3. Establish a criterion of ''goodness'' for the response(s) of a filter in the selected class compared to the desired response(s).
4. Develop a method for finding the best member in the filter class.
5. Synthesize the best filter using a proper structure and a proper implementation form, for example using a computer program, a signal processor, or a VLSI chip.
6. Analyze the filter performance.

Handbook for Digital Signal Processing, Edited by Sanjit K. Mitra and James F. Kaiser.
ISBN 0-471-61995-7 © 1993 John Wiley & Sons, Inc.

In most cases, the desired response is the given magnitude response or the given phase (delay) response or both. The desired magnitude response is usually specified by determining the frequency region(s) where the input signal components should be preserved and the region(s) where the signal components should be rejected. The phase response, in turn, is often desired to be linear in those frequency intervals where the signal components are preserved. In certain cases, time-domain conditions may be included, for example, in the design of Nyquist filters where some of the impulse-response values are restricted to be zero-valued. There are also applications where constraints on the step response are imposed.

The second step consists of determining a proper class of filters to approximate the given response(s). First, it must be decided whether to use FIR filters or IIR filters. After this, a proper class of FIR or IIR filters is selected. For many computationally efficient or low-sensitivity FIR and IIR filter structures, there are constraints on the transfer function. In these cases, the design of the transfer function and the filter implementation cannot be separated, and the desired filter structure determines the class of filters under consideration.

In order to find the best member in the selected filter class, an error measure is needed by which the nearness of the approximating response(s) to the given response(s) is determined. There are several error measures, as will be seen in Section 4-1-3. In many cases, the maximum allowable value of the error measure, for example, the maximum allowable deviation from the given desired response, is specified. In this case, the problem is to first determine the minimum complexity of a filter (e.g., the minimum filter order) required to meet the criteria. The remaining problem is to find the best member in the class of filters with this complexity to minimize the error measure.

The fourth step is to find or develop a method for finding this best member. This chapter and the following chapter describe several design methods for FIR and IIR digital filters. The fifth step involves synthesizing the filter designed at the previous step. The final step is to test whether the resulting filter meets all the given criteria. Also the performance under finite-precision arithmetic is studied. The last two steps are considered in Chapters 6 and 7.

The above design process is often used iteratively. If the resulting filter does not possess all the desired properties, then the desired response(s), the filter class, or the error measure should be changed and the overall process repeated until a filter is obtained with a satisfactory overall performance.

4-1-2   Filter Specifications

The requirements for a digital filter are normally specified in the frequency domain in terms of the desired magnitude response and/or the desired phase (delay) response. In the lowpass case, the desired magnitude response is usually given by

$$D(\omega) = \begin{cases} 1 & \text{for } \omega \in [0, \omega_p] \\ 0 & \text{for } \omega \in [\omega_s, \pi] \end{cases} \tag{4.1}$$

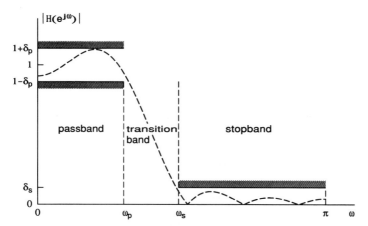

FIGURE 4-1 Tolerance limits for approximation of an ideal lowpass filter magnitude response.

and the specifications are given for the realizable magnitude response $|H(e^{j\omega})|$ as shown in Figure 4-1. It is desired to preserve signal components in the region $[0, \omega_p]$, called the *passband* of the filter, and to reject signal components in the region $[\omega_s, \pi]$, called the *stopband* of the filter. ω_p and ω_s are called, respectively, the passband edge and stopband edge angular frequencies. The permissible errors in the passband and in the stopband are δ_p and δ_s, respectively. The dashed line represents an acceptable magnitude response staying within the limits $1 \pm \delta_p$ in the passband and being less than or equal to δ_s in the stopband. To make it possible to approximate the desired function as close as possible, the specification includes a *transition band* of nonzero width $(\omega_s - \omega_p)$ in which the filter response changes from unity in the passband to zero in the stopband. Note that, because of the symmetry and periodicity of the magnitude response $|H(e^{j\omega})|$, it is sufficient to give the specifications only for $0 \le \omega \le \pi$.

Usually, the amplitudes of the allowable ripples for the magnitude response are given logarithmically (i.e., in decibels) in terms of the maximum passband variation and the minimum stopband attenuation, which are given by

$$A_p = 20 \log_{10} \left(\frac{1 + \delta_p}{1 - \delta_p} \right) \text{ dB} \tag{4.2a}$$

and

$$A_s = -20 \log_{10} (\delta_s) \text{ dB}, \tag{4.2b}$$

respectively. Note that both these quantities are positive. Another commonly encountered passband specification is the peak deviation from unity expressed logarithmically, that is, $A_p = 20 \log_{10} (\delta_p)$; A_s is then specified as $A_s = 20 \log_{10} (\delta_s)$. These quantities are negative.

Above, the passband and stopband edges, ω_p and ω_s, have been given in terms of the angular frequency ω. If the sampling frequency of the filter is f_s, then ω is related to the real frequency f through the equation

$$\omega = 2\pi f/f_s. \tag{4.3}$$

For instance, if the sampling frequency is 20 kHz and the passband and stopband edges of a lowpass filter are 4 and 5 kHz, then the band edges in terms of the angular frequency become $\omega_p = 0.4\pi$ and $\omega_s = 0.5\pi$. The third alternative is to specify the edges in terms of the *normalized frequency* defined by

$$f_{norm} = f/f_s. \tag{4.4}$$

In the above example, the normalized passband and stopband edges are 0.2 and 0.25.

In some applications, it is necessary to preserve the shape of the input signal. This goal is achieved if the phase response of the filter is approximately linear in the passband region $[0, \omega_p]$; that is, arg $H(e^{j\omega})$ approximates on $[0, \omega_p]$ the linear curve

$$\phi(\omega) = -\tau_0\omega + \tau_1, \tag{4.5}$$

where τ_0 and τ_1 can be freely chosen. Instead of the phase response, the criteria for the phase are usually given in terms of the *group delay* response

$$\tau_g(\omega) = -\frac{d \text{ arg } H(e^{j\omega})}{d\omega} \tag{4.6a}$$

or the *phase delay* response

$$\tau_p(\omega) = -\frac{\text{arg } H(e^{j\omega})}{\omega}. \tag{4.6b}$$

These responses have simpler representation forms than the phase response and are often easier to interpret. For instance, the value of the phase delay at a specified frequency point $\omega = \omega_0$ gives directly the delay caused by the filter for a sinusoidal signal of frequency ω_0. If the input signal is periodic or approximately periodic, as an electrocardiogram signal, then the phase delay is required to approximate in the passband a constant τ_0 with the given tolerance δ_d as shown in Figure 4-2. Since the delay of all passband signal components is approximately equal, the signal shape is preserved. If the signal is not periodic, then instead of the phase delay, the group delay may be used. Note that for a constant phase delay, τ_1 is forced to be zero in Eq. (4.5),[1] whereas in the case of a constant group delay, τ_1 can take any value.

---

[1]In the bandpass or highpass case, the shape of a periodic signal is preserved if τ_1 is a multiple of 2π and τ_0 is a constant.

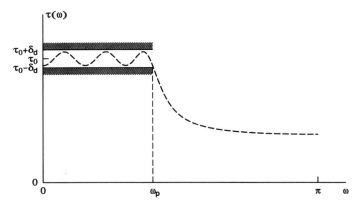

FIGURE 4-2 Tolerance limits for approximation of a constant group or phase delay response.

In the most general case, there are several passbands and stopbands, the desired magnitude response $D(\omega)$ is arbitrary in the passbands, and the allowable approximation error depends on ω in each band. In this case, the specifications can be stated as (see Figure 4-3)

$$D_p(\omega) - e_p(\omega) \le |H(e^{j\omega})| \le D_p(\omega) + e_p(\omega) \quad \text{for } \omega \in X_p, \qquad (4.7a)$$

$$|H(e^{j\omega})| \le e_s(\omega) \quad \text{for } \omega \in X_s, \qquad (4.7b)$$

where X_p and X_s denote the passband and stopband regions of the filter, respectively. $e_p(\omega)$ is the permissible deviation from the desired passband response $D_p(\omega)$, whereas $e_s(\omega)$ is the allowable deviation from zero in the stopband region. These general specifications can be used, for example, in cases where the input signal of the filter is distorted before or after filtering and it is desired to equalize the am-

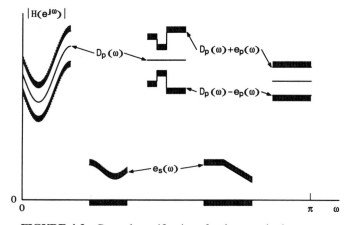

FIGURE 4-3 General specifications for the magnitude response.

plitude distortion. An example of an amplitude distortion occurring after filtering is the one caused by a zero-order hold when reconstructing an analog signal from a discrete-time signal.

The specifications of Eq. (4.7) can be written alternatively in terms of the passband and stopband weighting functions $W_p(\omega)$ and $W_s(\omega)$ as

$$-\delta_p \le W_p(\omega)[|H(e^{j\omega})| - D_p(\omega)] \le \delta_p \quad \text{for } \omega \in X_p, \qquad (4.8a)$$

$$W_s(\omega)\,|H(e^{j\omega})| \le \delta_s \quad \text{for } \omega \in X_s. \qquad (4.8b)$$

$e_p(\omega)$ is related to δ_p and $W_p(\omega)$ through the equation

$$e_p(\omega) = \delta_p/W_p(\omega), \qquad (4.9a)$$

whereas $e_s(\omega)$ is related to δ_s and $W_s(\omega)$ through

$$e_s(\omega) = \delta_s/W_s(\omega). \qquad (4.9b)$$

The specifications of Eq. (4.8) can be combined to give the following form, which is useful in many filter design techniques:

$$|E(\omega)| \le \bar{\epsilon} \quad \text{for } \omega \in X = X_p \cup X_s, \qquad (4.10a)$$

where

$$E(\omega) = W(\omega)\,[|H(e^{j\omega})| - D(\omega)] \qquad (4.10b)$$

with

$$\bar{\epsilon} = \delta_p, \qquad (4.10c)$$

$$D(\omega) = \begin{cases} D_p(\omega) & \text{for } \omega \in X_p \\ 0 & \text{for } \omega \in X_s, \end{cases} \qquad (4.10d)$$

and

$$W(\omega) = \begin{cases} W_p(\omega) & \text{for } \omega \in X_p \\ \dfrac{\delta_p}{\delta_s}\, W_s(\omega) & \text{for } \omega \in X_s. \end{cases} \qquad (4.10e)$$

$D(\omega)$ and $W(\omega)$ are called the *desired function* and the *weighting function*, respectively, and $E(\omega)$ is the *weighted error function*. If the maximum absolute value of this function is less than or equal to $\bar{\epsilon}$ on X, then $|H(e^{j\omega})|$ is guaranteed to meet the given criteria.

For instance, in the bandpass case, the specifications are usually stated as

$$1 - \delta_p \le |H(e^{j\omega})| \le 1 + \delta_p \quad \text{for } \omega \in [\omega_{p1}, \omega_{p2}], \tag{4.11a}$$

$$|H(e^{j\omega})| \le \delta_s \quad \text{for } \omega \in [0, \omega_{s1}] \cup [\omega_{s2}, \pi]. \tag{4.11b}$$

These criteria can be written in the above form using

$$X = [0, \omega_{s1}] \cup [\omega_{p1}, \omega_{p2}] \cup [\omega_{s2}, \pi], \tag{4.12a}$$

$$D(\omega) = \begin{cases} 1 & \text{for } \omega \in [\omega_{p1}, \omega_{p2}] \\ 0 & \text{for } \omega \in [0, \omega_{s1}] \cup [\omega_{s2}, \pi], \end{cases} \tag{4.12b}$$

$$W(\omega) = \begin{cases} 1 & \text{for } \omega \in [\omega_{p1}, \omega_{p2}] \\ \delta_p/\delta_s & \text{for } \omega \in [0, \omega_{s1}] \cup [\omega_{s2}, \pi], \end{cases} \tag{4.12c}$$

and

$$\bar{\epsilon} = \delta_p. \tag{4.12d}$$

In a similar manner, the general specifications for the group or phase delay can be stated in terms of the weighted error function as

$$|E_\tau(\omega)| \le \bar{\epsilon}_\tau \quad \text{for } \omega \in X_p, \tag{4.13a}$$

where

$$E_\tau(\omega) = W_\tau(\omega)[\tau(\omega) - D_\tau(\omega) - \tau_0]. \tag{4.13b}$$

If it is desired to equalize the delay distortion caused by an elliptic filter, then the criteria for an allpass delay equalizer can be written in the above form by selecting $D_\tau(\omega) = -\tau_e(\omega)$, where $\tau_e(\omega)$ is the delay response of the elliptic filter. It should be noted that the actual value of τ_0 in Eq. (4.13b) is not fixed but is an adjustable parameter.

In some cases, it is desired to optimize the frequency-domain behavior of a filter subject to the given time-domain conditions. For instance, in the case of Nyquist or Lth band filters, every Lth impulse-response value is restricted to be zero except for the central value. Furthermore, in some applications, the overshoot of the step response of a digital filter, optimized only in the frequency domain, is too large. In this case, the filter has to be reoptimized with constraints on the ripple of the step response.

4-1-3 Approximation Criteria

Three different error measures are normally used in designing digital filters.

4-1-3-1 Minimax Error Designs. Some applications require that the transfer function coefficients be optimized to minimize the maximum error between the approximating response and the given desired response. The solution minimizing this error function is called a *minimax* or *Chebyshev approximation*. In the case of the weighted error function $E(\omega)$ as given by Eq. (4.10b), the quantity to be minimized is the peak absolute value of $E(\omega)$ on X, that is, the quantity

$$\epsilon = \max_{\omega \in X} |E(\omega)|. \tag{4.14}$$

If the maximum allowable value of ϵ is specified, then the approximation problem is to first find the minimum order of a filter required to meet the given criteria and then optimize the coefficients of a minimum-order transfer function to minimize ϵ. Examples of minimax solutions are elliptic (Cauer) IIR filters and equiripple linear-phase FIR filters.

4-1-3-2 Least-Squared Error Designs. In some cases, instead of the minimax norm, the L_p norm is used. Here, it is desired to minimize the function[2]

$$E_p = \int_X [W(\omega)[|H(e^{j\omega})| - D(\omega)]]^p \, d\omega, \tag{4.15}$$

where p is a positive even integer. It can be shown that as $p \mapsto \infty$, the solution minimizing the above quantity approaches the minimax solution. This fact is exploited in some IIR filter design methods. For FIR filters, L_p error designs are of little practical use since there are efficient algorithms directly available for designing in the minimax sense FIR filters with arbitrary specifications. The exception is the L_2 error or *least-squared error* designs, which can be found very effectively. In this case, the quantity to be minimized is

$$E_2 = \int_X [W(\omega)[|H(e^{j\omega})| - D(\omega)]]^2 \, d\omega. \tag{4.16}$$

4-1-3-3 Maximally Flat Approximations. In the third approach, the approximating response is obtained based on a Taylor series approximation to the given desired response at a certain frequency point and the solution is called a *maximally flat approximation*. In some cases, such as in designing maximally flat (Butterworth) IIR filters, there are two points, one in the passband and one in the stopband, where a Taylor series approximation is applied.

Most of the methods developed for designing digital filters use one of the above approximation criteria. In some synthesis techniques, a combination of these cri-

---

[2]In the literature, the weighting function $W(\omega)$ is not usually raised to the power of p. If it is desired that the solution minimizing E_p approach the minimax solution as $p \mapsto \infty$, then E_p has to be formed as given by Eq. (4.15).

teria is used. For instance, in the case of Chebyshev IIR filters, a Chebyshev approximation is used in the passband and a maximally flat approximation is used in the stopband.

There exist also several simple filter design techniques that do not use directly the above criteria at all. A typical example of such methods is the design of FIR filters using windows, where the Fourier series of an ideal filter is first truncated and then smoothed using a window function.

4-2 WHY FIR FILTERS?

In many digital signal processing applications, FIR filters are preferred over their IIR counterparts. The main advantages of the FIR filter designs over their IIR equivalents (see Chapter 5 for a review of IIR filter design methods) are the following:

1. FIR filters with exactly linear phase can easily be designed.
2. There exist computationally efficient realizations for implementing FIR filters. These include both nonrecursive and recursive realizations.
3. FIR filters realized nonrecursively are inherently stable and free of limit cycle oscillations when implemented on a finite-wordlength digital system.
4. Excellent design methods are available for various kinds of FIR filters with *arbitrary* specifications.[3]
5. The output noise due to multiplication roundoff errors in an FIR filter is usually very low and the sensitivity to variations in the filter coefficients is also low.

The main disadvantage of conventional FIR filter designs is that they require, especially in applications demanding narrow transition bands, considerably more arithmetic operations and hardware components, such as multipliers, adders, and delay elements than do comparable IIR filters. As the transitions bandwidth of an FIR filter is made narrower, the filter order, and correspondingly the arithmetic complexity, increases inversely proportionally to this width. This makes the implementation of narrow transition band FIR filters very costly. The cost of implementation of an FIR filter can, however, be reduced by using multiplier-efficient realizations, fast convolution algorithms (see Chapter 8), and multirate filtering (see Chapter 14).

This chapter reviews some commonly used methods for designing FIR filters along with several computationally efficient realization methods. The outline of the remaining part of this chapter is as follows. Section 4-3 reviews the properties of linear-phase FIR filters. Sections 4-4 through 4-7 are devoted to very fast design

---

[3]The design of arbitrary magnitude IIR filters is usually time-consuming and the convergence to the optimum solution is not always guaranteed.

methods. Section 4-4 considers the design of FIR filters based on windowing, Section 4-5 outlines the design of filters that are optimum in the least-mean-square sense and Section 4-6 treats the design of maximally flat filters. Section 4-7 gives some simple analytic design techniques. Section 4-8 is devoted to the design of FIR linear-phase filters that are optimum in the minimax sense. Section 4-9 shows how the design of conventional minimum-phase filters can be accomplished by using a linear-phase filter as a starting point. In Section 4-10, the design of filters having some additional constraints on the frequency- or time-domain response is considered. Finally, Sections 4-11 and 4-12 are devoted to the design of computationally efficient FIR filters. The filters of these two sections are constructed using subfilters as building blocks. In Section 4-12 these subfilters are identical, whereas in Section 4-11 they are different and are obtained from a conventional transfer function by replacing each unit delay element by multiple delays.

4-3 CHARACTERISTICS OF LINEAR-PHASE FIR FILTERS

Some properties of linear-phase FIR filters are reviewed in this section, such as the conditions for linear phase and the zero locations of these filters as well as different representation forms for the frequency response.

4-3-1 Conditions for Linear Phase

Let $\{h[n]\}$ be the impulse response of a causal FIR filter of length $N + 1$. The transfer function of this filter is

$$H(z) = \sum_{n=0}^{N} h[n]z^{-n}. \tag{4.17}$$

The corresponding frequency response is given by

$$H(e^{j\omega}) = \sum_{n=0}^{N} h[n]e^{-jn\omega}. \tag{4.18}$$

In the above, N is the order of the filter.

For many practical FIR filters, exact linearity of phase is often desired. This goal is achieved if the frequency response of the filter is expressible in the form

$$H(e^{j\omega}) = \overline{H}(\omega)e^{j\phi(\omega)}, \tag{4.19a}$$

where

$$\phi(\omega) = \alpha\omega + \beta \tag{4.19b}$$

and $\overline{H}(\omega)$ is a real and even function of ω. The magnitude and the phase of the above function are, respectively,

$$|H(e^{j\omega})| = |\overline{H}(\omega)| \tag{4.20a}$$

and

$$\arg H(e^{j\omega}) = \begin{cases} \alpha\omega + \beta & \text{for } \overline{H}(\omega) \geq 0 \\ \alpha\omega + \beta - \pi & \text{for } \overline{H}(\omega) < 0. \end{cases} \tag{4.20b}$$

$\overline{H}(\omega)$ is called the *zero-phase frequency response*[4] to distinguish it from the magnitude response $|H(e^{j\omega})|$. To simplify the notation, let $H(\omega)$ represent the zero-phase frequency response. It should always be clear from the context whether H is a function of z, $e^{j\omega}$, or ω, that is, whether the transfer function, the frequency response, or the zero-phase frequency response is considered. The relationships between $H(\omega)$ and $|H(e^{j\omega})|$, and between $\phi(\omega)$ and $\arg H(e^{j\omega})$, are shown in Figure 4-4. Note that the zero-phase frequency response of the filter may take both positive and negative values, whereas the magnitude response is strictly nonnegative.

There are the following four types yielding the phase linearity:

Type I: N is even and $h[N - n] = h[n]$ for all n.
Type II: N is odd and $h[N - n] = h[n]$ for all n.
Type III: N is even and $h[N - n] = -h[n]$ for all n.
Type IV: N is odd and $h[N - n] = -h[n]$ for all n.

[4]Some authors call $\overline{H}(\omega)$ the amplitude response of the filter.

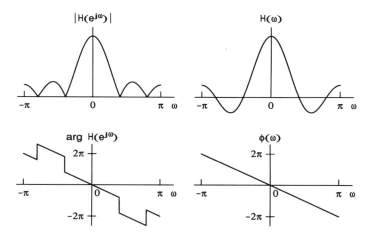

FIGURE 4-4   Relations between the magnitude response $|H(e^{j\omega})|$ and the zero-phase frequency response $H(\omega)$, and between $\arg H(e^{j\omega})$ and $\phi(\omega)$ for an example linear-phase FIR filter.

In each of these four cases, the transfer function is expressible as

$$H(z) = F(z)G(z), \tag{4.21a}$$

where

$$F(z) = \begin{cases} 1 & \text{for Type I} \\ [1 + z^{-1}]/2 & \text{for Type II} \\ [1 - z^{-2}]/2 & \text{for Type III} \\ [1 - z^{-1}]/2 & \text{for Type IV} \end{cases} \tag{4.21b}$$

and

$$G(z) = \sum_{n=0}^{2M} g[n]z^{-n} \tag{4.21c}$$

with

$$g[2M - n] = g[n] \quad \text{for all } n \tag{4.21d}$$

and

$$M = \begin{cases} N/2 & \text{for Type I} \\ (N - 1)/2 & \text{for Type II} \\ (N - 2)/2 & \text{for Type III} \\ (N - 1)/2 & \text{for Type IV.} \end{cases} \tag{4.21e}$$

Hence $H(z)$ can be expressed as a cascade of a fixed term $F(z)$ and a common adjustable term $G(z)$, which itself is a Type I transfer function. The relations between $h[n]$ and $g[n]$ are given in Table 4-1 for the four types. Figure 4-5 shows example impulse responses. In each case, the center of the symmetry is $K = N/2$. For Types I and III, K is an integer and there is an impulse-response sample exactly at this point, whereas for Types II and IV, K is not an integer and it lies between two impulse-response samples. Note that for Type III, the symmetry forces $h[N/2]$ to be equal to zero.

Because of the symmetry property of Eq. (4.21d), $G(z)$ can be expressed as

$$G(z) = z^{-M} [g[M] + g[M - 1](z + z^{-1}) \\ + g[M - 2](z^2 + z^{-2}) + \cdots + g[0](z^M + z^{-M})]. \tag{4.22}$$

TABLE 4-1 Relations Between the Coefficients $h[n]$ and $g[n]$

Coefficient	Type I	Type II	Type III	Type IV
$h[0]$	$g[0]$	$\dfrac{g[0]}{2}$	$\dfrac{g[0]}{2}$	$\dfrac{g[0]}{2}$
$h[1]$	$g[1]$	$\dfrac{g[1] + g[0]}{2}$	$\dfrac{g[1]}{2}$	$\dfrac{g[1] - g[0]}{2}$
$h[n]$	$g[n]$	$\dfrac{g[n] + g[n-1]}{2}$	$\dfrac{g[n] - g[n-2]}{2}$	$\dfrac{g[n] - g[n-1]}{2}$
$h[N-1]$	$g[N-1]$	$\dfrac{g[N-1] + g[N-2]}{2}$	$\dfrac{-g[N-3]}{2}$	$\dfrac{g[N-1] - g[N-2]}{2}$
$h[N]$	$g[N]$	$\dfrac{g[N-1]}{2}$	$\dfrac{g[N-2]}{2}$	$\dfrac{g[N-1]}{2}$

By substituting $z = e^{j\omega}$ in the above equation, the frequency response of $G(z)$ becomes

$$G(e^{j\omega}) = e^{-jM\omega}[g[M] + g[M-1](2 \cos \omega)$$
$$+ g[M-2](2 \cos 2\omega) + \cdots + g[0](2 \cos M\omega)]. \quad (4.23)$$

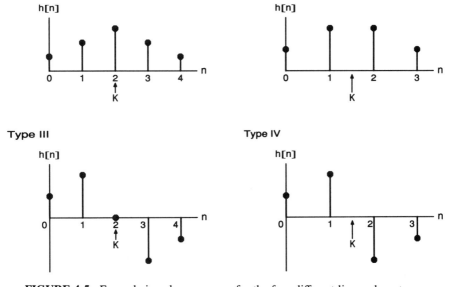

FIGURE 4-5 Example impulse responses for the four different linear-phase types.

Similarly, the frequency response of $F(z)$ can be written, after some manipulations, in the form

$$
F(e^{j\omega}) = \begin{cases} 1 & \text{for Type I} \\ e^{-j\omega/2} \cos(\omega/2) & \text{for Type II} \\ e^{-j(\omega - \pi/2)} \sin \omega & \text{for Type III} \\ e^{-j(\omega/2 - \pi/2)} \sin(\omega/2) & \text{for Type IV.} \end{cases} \tag{4.24}
$$

By combining the above results, the zero-phase frequency response can be expressed as

$$
H(\omega) = F(\omega)G(\omega), \tag{4.25}
$$

where

$$
F(\omega) = \begin{cases} 1 & \text{for Type I} \\ \cos(\omega/2) & \text{for Type II} \\ \sin \omega & \text{for Type III} \\ \sin(\omega/2) & \text{for Type IV,} \end{cases} \tag{4.26}
$$

$$
G(\omega) = \sum_{n=0}^{M} a[n] \cos n\omega, \tag{4.27a}
$$

$$
a[n] = \begin{cases} g[M] & \text{for } n = 0 \\ 2g[M - n] & \text{for } n \neq 0, \end{cases} \tag{4.27b}
$$

and M is given by Eq. (4.21e). The phase term becomes

$$
\phi(\omega) = \begin{cases} -N\omega/2 & \text{for Types I and II} \\ -N\omega/2 + \pi/2 & \text{for Types III and IV.} \end{cases} \tag{4.28}
$$

In Section 4-3-3, some other useful representation forms of $H(\omega)$ are given. Figure 4-6 gives example zero-phase frequency responses for the four types. For Type I, $H(\omega)$ is even about $\omega = 0$ and $\omega = \pi$ and the periodicity is 2π. For Type II, the fixed term $F(\omega) = \cos(\omega/2)$ generates a zero for $H(\omega)$ at $\omega = \pi$, making it odd about this point. The periodicity is 4π. Similarly, for Type IV, the fixed term generates a zero at $\omega = 0$. The resulting $H(\omega)$ is odd about $\omega = 0$ and the periodicity is 4π. For Type III, the fixed term gives a zero at both $\omega = 0$ and $\omega = \pi$, making $H(\omega)$ odd about these points. The periodicity is 2π.

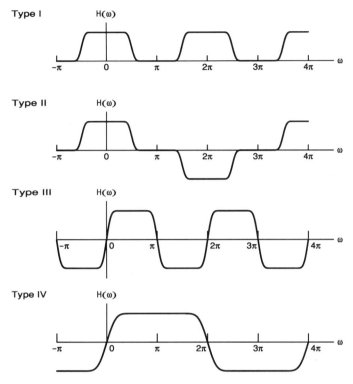

FIGURE 4-6 Example zero-phase frequency responses for the four different linear-phase types.

This chapter concentrates mainly on Types I and II. These filter types are used for conventional filtering applications because, in these cases, the delay caused for sinusoidal signals, $-\phi(\omega)/\omega = N/2$, is independent of the frequency ω. Filters belonging to the remaining two cases have an additional 90° phase shift and they are most suitable for realizing such filters as differentiators and Hilbert transformers (see Chapter 13). In these two cases, the delay caused for sinusoidal signals depends on the frequency. However, the group delay, $-d\phi(\omega)/d\omega$, is a constant (equal to $N/2$ for all linear-phase types).

The above linear-phase filters are also characterized by the property that only $M + 1$ multipliers are needed in the actual implementation because of the symmetry in the filter coefficients. Figure 4-7 gives such an implementation for a Type I filter of order $N = 2M$.

4-3-2 Zero Locations of Linear-Phase FIR Filters

All the poles of an FIR filter lie at the origin. From Eq. (4.21b), it is seen that the fixed term $F(z)$ generates for Type II designs one zero at $z = -1$, for Type III designs one zero at both $z = 1$ and $z = -1$, and for Type IV designs one zero at

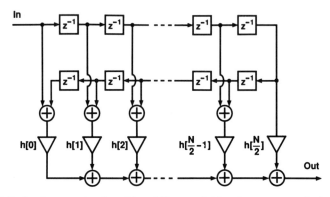

FIGURE 4-7   Implementation for a Type I filter exploiting the symmetry in the filter coefficients.

$z = 1$. What remains is to examine where the zeros of the common adjustable Type I part $G(z)$ are located. From the symmetry condition of Eq. (4.21d), it follows that

$$G(z^{-1}) = z^{2M} G(z).$$ (4.29)

This means that $G(z)$ and $G(z^{-1})$ have identical zeros. The zeros of $G(z)$ thus occur in mirror-image pairs. For $G(z)$ with real coefficients, the zeros are either real or occur in complex conjugate pairs. These conditions imply that $G(z)$ can be factored in the form

$$G(z) = g[0]G_1(z)G_2(z)G_3(z),$$ (4.30a)

where

$$G_1(z) = \prod_{i=1}^{N_1} \left(1 - \left[2\left(r_i + \frac{1}{r_i}\right)\cos\theta_i\right]z^{-1} + \left[r_i^2 + \frac{1}{r_i^2} + 4\cos^2\theta_i\right]z^{-2}\right.$$

$$\left. - \left[2\left(r_i + \frac{1}{r_i}\right)\cos\theta_i\right]z^{-3} + z^{-4}\right),$$ (4.30b)

$$G_2(z) = \prod_{i=1}^{N_2} (1 - [2\cos\bar{\theta}_i]z^{-1} + z^{-2}),$$ (4.30c)

and

$$G_3(z) = \prod_{i=1}^{N_3} \left(1 - \left[\bar{r}_i + \frac{1}{\bar{r}_i}\right]z^{-1} + z^{-2}\right).$$ (4.30d)

Here, $4N_1 + 2(N_2 + N_3) = 2M$ and

1. $G_1(z)$ contains the zeros occurring in quadruplets, that is, in complex conjugate and mirror-image pairs off the unit circle at $z = r_i e^{\pm j\theta_i}$, $(1/r_i)e^{\pm j\theta_i}$ for $i = 1, 2, \cdots, N_1$.

2. $G_2(z)$ contains the zeros occurring in complex conjugate pairs on the unit circle at $z = e^{\pm j\theta_i}$ for $i = 1, 2, \cdots, N_2$.

3. $G_3(z)$ contains the zeros occurring in reciprocal pairs on the real axis at $z = \bar{r}_i, 1/\bar{r}_i$ for $i = 1, 2, \cdots, N_3$.

If $G(z)$ possesses a zero at $z = 1$ or at $z = -1$, then it follows from the symmetry of $G(z)$ and the fact that $G(z)$ is of even order that the number of zeros at this point must be even. If $G(z)$ happens to have k zero-pairs at $z = 1$ (at $z = -1$), then these pairs can be included in $G_3(z)$ by using k terms with $\bar{r}_i = 1$ ($\bar{r}_i = -1$).

For Types II, III, and IV, the locations of the zeros outside the points $z = 1$ and $z = -1$ are similar. The main difference between the four cases is in the number of zeros at $z = 1$ and $z = -1$.

1. Type I designs have either an even number or no zeros at $z = 1$ and at $z = -1$.

2. Type II designs have either an even number or no zeros at $z = 1$, and an odd number of zeros at $z = -1$.

3. Type III designs have an odd number of zeros at $z = 1$ and at $z = -1$.

4. Type IV designs have an odd number of zeros at $z = 1$, and either an even number or no zeros $z = -1$.

Figure 4-8 gives the amplitude response, the impulse response, and the zero locations of a typical Type I filter.

4-3-3 Different Representation Forms for Zero-Phase Frequency Responses

Equations (4.25)–(4.27) give one representation form for the zero-phase frequency response in each of the four different linear-phase cases. This form is used in designing filters in the minimax sense (Section 4-8). Another useful representation form is obtained by expressing $H(\omega)$ directly in terms of the impulse-response coefficients $h[n]$ as follows [OP89; RA75a]:

$$
H(\omega) = \begin{cases}
h[N/2] + \displaystyle\sum_{n=1}^{N/2} h[N/2 - n][2\cos n\omega] & \text{for Type I} \\[2ex]
\displaystyle\sum_{n=0}^{(N-1)/2} h[(N-1)/2 - n][2\cos[(n+1/2)\omega]] & \text{for Type II} \\[2ex]
\displaystyle\sum_{n=0}^{N/2-1} h[N/2 - 1 - n][2\sin[(n+1)\omega]] & \text{for Type III} \\[2ex]
\displaystyle\sum_{n=0}^{(N-1)/2} h[(N-1)/2 - n][2\sin[(n+1/2)\omega]] & \text{for Type IV.}
\end{cases}
\tag{4.31}
$$

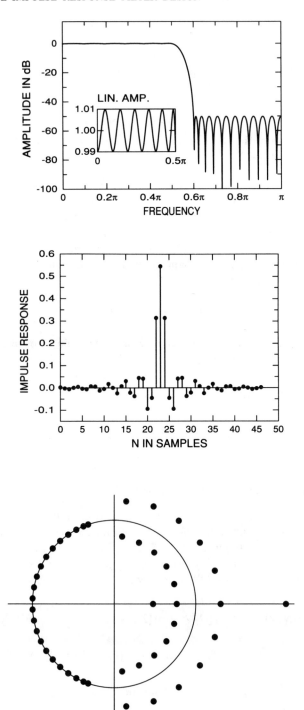

FIGURE 4-8   Amplitude response, impulse response, and zero locations for a typical Type I filter of order $N = 46$.

For later use, we rewrite the above equation in the form

$$H(\omega) = \sum_{n=0}^{M} b[n] \, \text{trig} \, (\omega, n), \tag{4.32a}$$

where

$$\text{trig} \, (\omega, n) = \begin{cases} 1 & \text{for Type I; } n = 0 \\ 2 \cos n\omega & \text{for Type I; } n > 0 \\ 2 \cos [(n + 1/2)\omega] & \text{for Type II} \\ 2 \sin [(n + 1)\omega] & \text{for Type III} \\ 2 \sin [(n + 1/2)\omega] & \text{for Type IV,} \end{cases} \tag{4.32b}$$

$$b[n] = \begin{cases} h[N/2 - n] & \text{for Type I} \\ h[(N - 1)/2 - n] & \text{for Types II and IV} \\ h[N/2 - 1 - n] & \text{for Type III,} \end{cases} \tag{4.32c}$$

and M is related to N through Eq. (4.21e). This representation form is used in Section 4-5 for designing filters in the least-mean-square sense and in Section 4-10 for designing filters based on linear programming.

A very useful representation form for the common adjustable response part $G(\omega)$ as given by Eq. (4.27) or for the overall response $H(\omega)$ for Type I [$H(\omega) = G(\omega)$] can be derived based on the identity

$$\cos n\omega = T_n(\cos \omega), \tag{4.33}$$

where $T_n(x) = \cos (n \cos^{-1} x)$ is the nth degree Chebyshev polynomial. These polynomials can be generated using the following recursion formulas:

$$T_0(x) = 1, \tag{4.34a}$$

$$T_1(x) = x, \tag{4.34b}$$

$$T_n(x) = 2xT_{n-1}(x) - T_{n-2}(x). \tag{4.34c}$$

Using these equivalences, $\cos n\omega$ can be expressed as an nth degree polynomial in $\cos \omega$ and $G(\omega)$ as an Mth degree polynomial in $\cos \omega$:

$$G(\omega) = \sum_{n=0}^{M} \alpha[n] \cos^n \omega. \tag{4.35}$$

This shows that the zero-phase frequency response of a Type I filter of order $2M$ can be determined as an Mth degree polynomial in $\cos \omega$. This fact is exploited in several synthesis techniques.

4-4 FIR FILTER DESIGN BY WINDOWING

The most straightforward approach to designing FIR filters is to determine the infinite-duration impulse response by expanding the frequency response of an ideal filter in a Fourier series and then to truncate and smooth this response using a window function. The main advantage of this design technique is that the impulse-response coefficients can be obtained in closed form and can be determined very fast even using a calculator. The main drawback is that the passband and stopband ripples of the resulting filter are restricted to be approximately equal.

4-4-1 Design Process

FIR filter design based on windowing generally begins by specifying the ideal zero-phase frequency response $H_{id}(\omega)$. Ideal Type I responses in the lowpass, highpass, bandpass, and bandstop cases are shown in Figure 4-9. Since $H_{id}(\omega)$ is even about $\omega = 0$ and periodic in ω with period 2π, it can be expanded in a Fourier series as follows:

$$H_{id}(\omega) = h_{id}^{(0)}[0] + 2 \sum_{n=1}^{\infty} h_{id}^{(0)}[n] \cos n\omega, \qquad (4.36a)$$

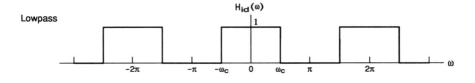

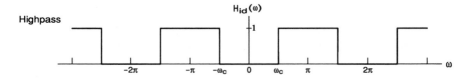

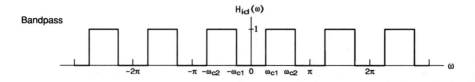

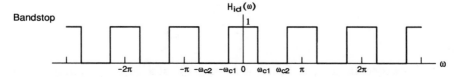

FIGURE 4-9   Zero-phase frequency responses of ideal Type I filters.

where

$$h_{id}^{(0)}[n] = \frac{1}{2\pi} \int_{-\pi}^{\pi} H_{id}(\omega) \cos(n\omega)\, d\omega, \qquad 0 \le n \le \infty. \qquad (4.36b)$$

The corresponding impulse response is of infinite duration and even about $n = 0$; that is, $h_{id}^{(0)}[-n] = h_{id}^{(0)}[n]$ (see Figure 4-10). For instance, in the lowpass case

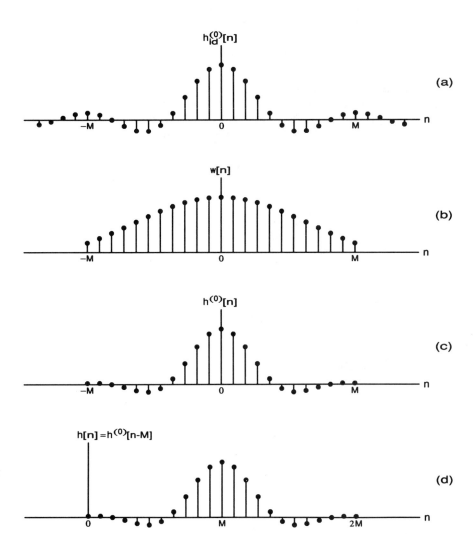

FIGURE 4-10 Impulse responses involved in designing an FIR filter by windowing. (a) Infinite-duration impulse response for an ideal zero-phase filter. (b) Impulse response for a window function. (c) Response obtained by truncating and smoothing the ideal response by the window function. (d) Response for the corresponding causal filter.

with cutoff edge ω_c, the coefficients are

$$h_{\text{id}}^{(0)}[n] = \frac{\omega_c}{\pi}\left(\frac{\sin(\omega_c n)}{\omega_c n}\right) = \begin{cases} \omega_c/\pi, & n = 0 \\ \sin(\omega_c n)/(\pi n), & |n| > 0. \end{cases} \tag{4.37}$$

The coefficients for highpass, bandpass, and bandstop filters are given in Table 4-2.

An approximating finite-duration impulse response is then generated by truncating and smoothing the above response according to

$$h^{(0)}[n] = w[n]h_{\text{id}}^{(0)}[n], \tag{4.38}$$

where $w[n]$ is a *window function* that is nonzero for $-M \le n \le M$ and zero otherwise (see Figure 4-10). Finally, the coefficients of the corresponding causal realizable FIR filter of order $2M$ are obtained by shifting the location of the central impulse-response coefficient from $n = 0$ to $n = M$, giving

$$h[n] = h^{(0)}[n - M], \qquad 0 \le n \le 2M. \tag{4.39}$$

TABLE 4-2   Coefficients of Ideal Zero-Phase Type I Filters

Type	Coefficients
Lowpass filter with edge	$h_{\text{id}}^{(0)}[0] = \dfrac{\omega_c}{\pi}$
angle ω_c	$h_{\text{id}}^{(0)}[n] = \dfrac{\sin(\omega_c n)}{\pi n}, \qquad \|n\| > 0$
Highpass filter with edge	$h_{\text{id}}^{(0)}[0] = 1 - \dfrac{\omega_c}{\pi}$
angle ω_c	$h_{\text{id}}^{(0)}[n] = -\dfrac{\sin(\omega_c n)}{\pi n}, \qquad \|n\| > 0$
Bandpass filter with edge	$h_{\text{id}}^{(0)}[0] = \dfrac{(\omega_{c2} - \omega_{c1})}{\pi}$
angles ω_{c1} and ω_{c2}	$h_{\text{id}}^{(0)}[n] = \dfrac{1}{\pi n}[\sin(\omega_{c2}n) - \sin(\omega_{c1}n)], \qquad \|n\| > 0$
Bandstop filter with edge	$h_{\text{id}}^{(0)}[0] = 1 - \dfrac{(\omega_{c2} - \omega_{c1})}{\pi}$
angles ω_{c1} and ω_{c2}	$h_{\text{id}}^{(0)}[n] = \dfrac{1}{\pi n}[\sin(\omega_{c1}n) - \sin(\omega_{c2}n)], \qquad \|n\| > 0$

For Type III filters, $H_{id}(\omega)$ (e.g., the ideal response of a Hilbert transformer or differentiator) is odd about $\omega = 0$ (cf. Figure 4-6) and the Fourier series contains sine terms, instead of cosine terms, and $h_{id}^{(0)}[0]$ is absent. In this case, $h_{id}^{(0)}[-n]$ $= -h_{id}^{(0)}[n]$ and the terms in the series are $2h_{id}^{(0)}[-n] \sin n\omega$, where $h_{id}^{(0)}[-n]$ can be determined from Eq. (4.36b) by replacing $\cos(n\omega)$ by $\sin(n\omega)$.

For Type II, $H_{id}(\omega)$ is odd about $\omega = \pi$ and the periodicity is 4π (see Figure 4-11(a)). The design of a Type II filter of odd-order N to approximate this response can be performed by first applying the above process with $M = N$ to the response $H_{id}(2\omega)$, which is a Type I response [PA87] (see Figure 4-11(b)). This gives an FIR filter of order $2N$. Since $H_{id}(2\omega)$ is odd about $\omega = \pi/2$, $h_{id}^{(0)}[n] = 0$ for $n = 0, \pm2, \pm4, \cdots$. Correspondingly, $h[N \pm 2r] = 0$ for $r = 0, 1, \cdots, (N - 1)/2$. The Type II filter whose zero-phase frequency response approximates the ideal response of Figure 4-11(a) is then obtained by discarding these zero-valued impulse-response samples, resulting in a filter of order N. The design of Type IV filters can be converted into the design of Type III filters in the same manner [PA87].

4-4-2 Direct Truncation of an Ideal Impulse Response

The simplest window is the *rectangular window* for which

$$w[n] = \begin{cases} 1, & -M \leq n \leq M \\ 0, & \text{otherwise.} \end{cases} \tag{4.40}$$

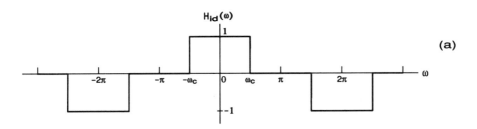

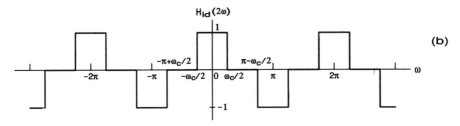

FIGURE 4-11 (a) Ideal Type II lowpass response $H_{id}(\omega)$. (b) Response $H_{id}(2\omega)$, which can be approximated by Type I filters.

The use of this window corresponds to a direct truncation of the infinite-duration impulse response and leads to a solution exhibiting large ripples before and after the discontinuity of the ideal frequency response. This is the well-known *Gibbs phenomenon*. As an example, Figure 4-12 gives the resulting responses $H(\omega)$ for Type I lowpass filters with $\omega_c = 0.4\pi$ for $M = 10$ and $M = 30$. The corresponding filter orders ($N = 2M$) are 20 and 60, respectively. As seen from this figure, the transition bandwidth of $H(\omega)$ becomes narrower when M is increased, but the maximum ripples in the passband and stopband regions remain about the same. In both cases, the first stopband extremum has the value of -0.09 (21-dB attenuation) and the last passband extremum has the value of 1.09.

The Gibbs phenomenon can be explained by the fact that $H(\omega)$ is related to the ideal response $H_{id}(\omega)$ and the frequency response of the window function

$$\Psi(\omega) = \sum_{n=-M}^{M} w[n] e^{-jn\omega} = w[0] + 2 \sum_{n=1}^{M} w[n] \cos n\omega \qquad (4.41)$$

through

$$H(\omega) = \frac{1}{2\pi} \int_{-\pi}^{\pi} H_{id}(\theta) \, \Psi(\omega - \theta) \, d\theta. \qquad (4.42)$$

For the rectangular window,

$$\Psi(\omega) = \sum_{n=-M}^{M} e^{-jn\omega} = \frac{\sin((2M+1)\omega/2)}{\sin(\omega/2)}. \qquad (4.43)$$

This response is depicted in Figure 4-13 for $M = 10$ and $M = 30$. As seen from this figure, $\Psi(\omega)$ appears as a gradually decaying sinusoid starting at a middle

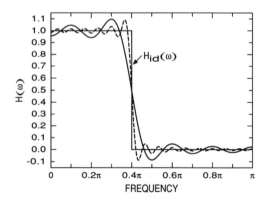

FIGURE 4-12 Responses of Type I lowpass filters designed by truncating the impulse response of an ideal filter with $\omega_c = 0.4\pi$. The solid and dashed lines give the responses for $M = 10$ and $M = 30$, respectively.

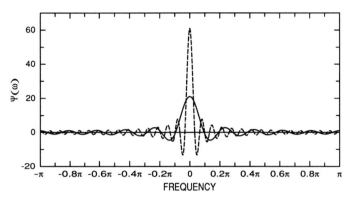

FIGURE 4-13 Frequency responses for the rectangular window for $M = 10$ (solid line) and $M = 30$ (dashed line).

lobe, called the *mainlobe*, whose width is twice that of the *sidelobes* being situated in the intervals between the zeros.

According to Eq. (4.42), the value of $H(\omega)$ at any frequency point ω is obtained in the lowpass case with cutoff edge ω_c by integrating $\Psi(\omega - \theta)$ with respect to θ over the interval $[-\omega_c, \omega_c]$. This is illustrated in Figure 4-14. For $\omega = \pi$, only small ripples of $\Psi(\omega - \theta)$ are inside this interval, resulting in a small value of

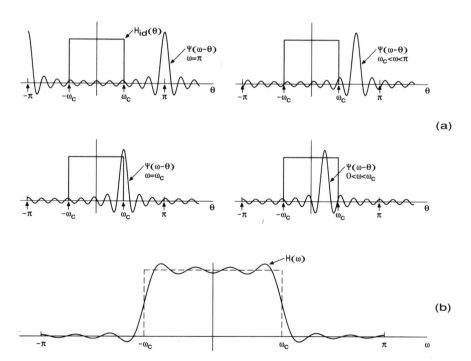

FIGURE 4-14 Explanation of the Gibbs phenomenon. (a) Convolution process. (b) Response for the resulting filter.

$H(\omega)$ at $\omega = \pi$. As ω is made smaller, larger ripples of $\Psi(\omega - \theta)$ are entering into the interval, resulting in larger values in $H(\omega)$ for $\omega < \pi$. The ripples are due to the fact that the area under every second sidelobe of $\Psi(\omega)$ is of opposite sign. For $\omega = \omega_c$, half of the mainlobe is inside the interval $[-\omega_c, \omega_c]$. Since the integral of $\Psi(\omega)$ over the interval $[-\pi, \pi]$ is one and most of the energy is concentrated in the mainlobe, the value of $H(\omega)$ at $\omega = \omega_c$ is approximately $\frac{1}{2}$. When ω is further decreased, the whole mainlobe enters the interval and the area in this interval is approximately one, resulting in the passband response of $H(\omega)$. The ripples around one are due to the fact that the sidelobes of $\Psi(\omega - \theta)$, which are of different heights, go inside the interval $[-\omega_c, \omega_c]$ and leave it as ω varies.

As M is increased, the widths of the mainlobe and the sidelobes decrease. However, the area under each lobe remains the same since at the same time the heights of the lobes increase (see Figure 4-13). This means that as M is increased, the oscillations of the resulting filter response occur more rapidly but do not decrease (see Figure 4-12).

4-4-3 Fixed Window Functions

The Gibbs phenomenon can be reduced by using a less abrupt truncation of the Fourier series. This is achieved by using a window function that tapers smoothly towards zero at both ends. Some of the well-known fixed window functions $w[n]$ [BL58; HA78; HA87; KA63, KA66; RA75a] are summarized in Table 4-3 along with their frequency responses $\Psi(\omega)$.[5] For these fixed window functions, the only adjustable parameter is M, half the filter order. The plots of the frequency responses are given in Figure 4-15 for the last four windows in Table 4-3 for $M = 128$. Also, the responses of the filters resulting when using these windows for $\omega_c = 0.4\pi$ are shown in this figure.

Figure 4-16 depicts, in the lowpass case, a typical relation between $H(\omega)$ and $\Psi(\omega)$, which is given in terms of $\theta - \omega_c$ in order to center the response at the cutoff edge. Note the close similarity to the case where $\Psi(\omega)$, $H_{id}(\omega)$, and $H(\omega)$ correspond to the impulse response, the step excitation, and the response of a continuous-time filter, respectively. As seen from the figure, $H(\omega)$ satisfies approximately $H(\omega_c + \omega) + H(\omega_c - \omega) = 1$ in the vicinity of the cutoff edge ω_c. This means that $H(\omega_c) \approx \frac{1}{2}$. Furthermore, the maximum passband deviation from unity and the maximum stopband deviation from zero are about the same, and the peak passband overshoot $(1 + \delta)$ and the peak negative stopband undershoot $(-\delta)$ occur at the same distance from the discontinuity point ω_c. The distance between these two overshoot points is for most windows approximately equal to the mainlobe width Δ_M. The criteria met by $H(\omega)$ can be given by

$$1 - \delta \leq H(\omega) \leq 1 + \delta \quad \text{for } \omega \in [0, \omega_p], \tag{4.44a}$$

$$-\delta \leq H(\omega) \leq \delta \quad \text{for } \omega \in [\omega_s, \pi], \tag{4.44b}$$

---

[5]The definitions of the window functions of Table 4-3 differ slightly in the literature.

TABLE 4-3   Some Commonly Used Fixed Windows for FIR Filter Design

Window Type	Window Function, $w[n]$, $-M \le n \le M$	Frequency Response, $\Psi(\omega)$
Rectangular	1	$\Psi_R(\omega) \equiv \sin[(2M+1)\omega/2]/\sin(\omega/2)$
Bartlett	$1 - \dfrac{\|n\|}{M+1}$	$\dfrac{1}{M+1}[\sin[(M+1)\omega/2]/\sin(\omega/2)]^2$
Hann	$\dfrac{1}{2}\left[1 + \cos\left[\dfrac{2\pi n}{2M+1}\right]\right]$	$0.5\Psi_R(\omega) + 0.25\Psi_R\left(\omega - \dfrac{2\pi}{2M+1}\right)$ $+ 0.25\Psi_R\left(\omega + \dfrac{2\pi}{2M+1}\right)$
Hamming	$0.54 + 0.46\cos\left[\dfrac{2\pi n}{2M+1}\right]$	$0.54\Psi_R(\omega) + 0.23\Psi_R\left(\omega - \dfrac{2\pi}{2M+1}\right)$ $+ 0.23\Psi_R\left(\omega + \dfrac{2\pi}{2M+1}\right)$
Blackman	$0.42 + 0.5\cos\left[\dfrac{2\pi n}{2M+1}\right]$ $+ 0.08\cos\left[\dfrac{4\pi n}{2M+1}\right]$	$0.42\Psi_R(\omega) + 0.25\Psi_R\left(\omega - \dfrac{2\pi}{2M+1}\right)$ $+ 0.25\Psi_R\left(\omega + \dfrac{2\pi}{2M+1}\right)$ $+ 0.04\Psi_R\left(\omega - \dfrac{4\pi}{2M+1}\right)$ $+ 0.04\Psi_R\left(\omega + \dfrac{4\pi}{2M+1}\right)$

where $\omega_p(\omega_s)$ is defined to be the highest frequency where $H(\omega) \ge 1 - \delta$ (the lowest frequency where $H(\omega) \le \delta$). The width of the transition band, $\Delta\omega = \omega_s - \omega_p$, is thus less than the mainlobe width Δ_M. This means that for a good window function, the mainlobe width has to be as narrow as possible. On the other hand, for a small ripple value δ, it is required that the area under the sidelobes of $\Psi(\omega)$ be as small as possible. These two requirements contradict each other.

For the fixed window functions given in Table 4-3, $H(\omega)$ is characterized by the facts that δ is approximately a constant, regardless of the values of M and ω_c, and the transition bandwidth is inversely proportional to the filter order $N = 2M$; that is,

$$\Delta\omega = \omega_s - \omega_p \approx \gamma/(2M), \qquad (4.45)$$

where γ is also approximately a constant. Some properties of these window functions are summarized in Table 4-4. It gives, for each window, the mainlobe width

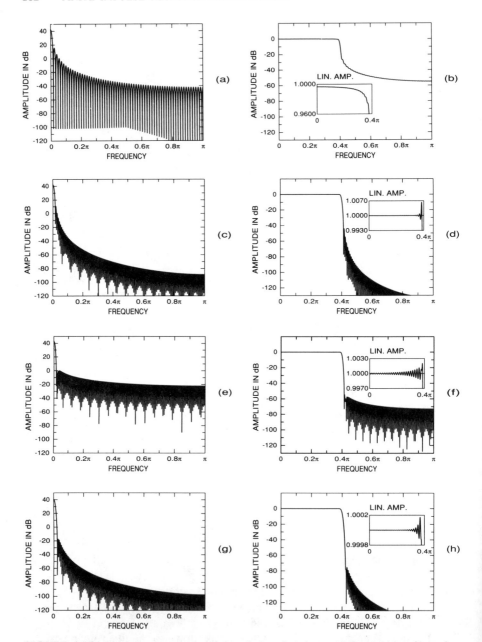

FIGURE 4-15   Frequency responses of the window functions and the resulting filters for the last four windows in Table 4-3 for $M = 128$ and $\omega_c = 0.4\pi$. (a,b) Bartlett window. (c,d) Hann window. (e,f) Hamming window. (g,h) Blackman window.

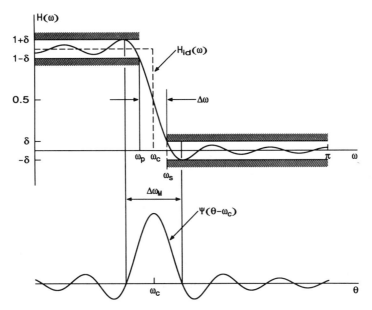

FIGURE 4-16 Typical relations between the frequency response of the window function and the resulting filter response in the lowpass case with cutoff edge ω_c.

Δ_M and the maximum sidelobe ripple in decibels in the case where $\Psi(\omega)$ is scaled to be unity at $\omega = 0$. Furthermore, the minimum stopband attenuation

$$A_s = -20 \log_{10} \delta \qquad (4.46)$$

and the dependence of the transition bandwidth on $2M$ are given.[6] The only exception is the Bartlett window for which the filter response has no zeros on the

---

[6]These values have been determined for the case $\omega_c = 0.4\pi$ and $M = 128$.

TABLE 4-4 Properties of Some Commonly Used Fixed Windows

Window Type	Mainlobe Width Δ_M	Sidelobe Ripple	A_s	$\Delta\omega = \omega_s - \omega_p$
Rectangular	$\dfrac{4\pi}{2M + 1}$	-13.3 dB	20.9 dB	$1.84\pi/(2M)$
Bartlett	$\dfrac{4\pi}{M + 1}$	-26.5 dB	See text	See text
Hann	$\dfrac{8\pi}{2M + 1}$	-31.5 dB	43.9 dB	$6.22\pi/(2M)$
Hamming	$\dfrac{8\pi}{2M + 1}$	-42.7 dB	54.5 dB	$6.64\pi/(2M)$
Blackman	$\dfrac{12\pi}{2M + 1}$	-58.1 dB	75.3 dB	$11.13\pi/(2M)$

unit circle (see Figure 4-15) and it is difficult to locate the stopband edge. The above window functions suffer from the drawback that A_s cannot be varied. Only ω_p and ω_s can be adjusted by properly selecting ω_c and M. Since ω_c is in the center of the transition band, it is selected to be $\omega_c = (\omega_p + \omega_s)/2$. Then M is determined from Eq. (4.45), where the specific value of γ can be found in Table 4-4 for each window type.

4-4-4   Adjustable Window Functions

The above problem can be overcome by using window functions having an additional parameter with which A_s can be varied. There exist three approaches to obtaining good adjustable windows. The first alternative is to minimize the energy in the sidelobes of the frequency response of the window function $w[n]$, whereas the second one is to minimize the peak sidelobe ripple. Both the Kaiser window [KA66, KA74] and the Saramäki window [SA89a, SA91a] provide an approximately optimum solution to the first problem, whereas the Dolph–Chebyshev window [HE68] is the solution to the second problem. The third alternative is to properly combine the first two approaches [SA91a].

The Kaiser window [KA66, KA74] is given by

$$
w[n] = \begin{cases} I_0\left[\alpha\sqrt{1 - \left(\dfrac{n}{M}\right)^2}\right] \Big/ I_0(\alpha), & -M \le n \le M \\ 0, & \text{otherwise,} \end{cases}
\tag{4.47}
$$

where α is the adjustable parameter and $I_0(x)$ is the modified zeroth-order Bessel function of the first kind, which has the simple power series expansion

$$
I_0(x) = 1 + \sum_{r=1}^{\infty} \left[\frac{(x/2)^r}{r!}\right]^2.
\tag{4.48}
$$

For most practical applications, about 20 terms in the above summation are sufficient to arrive at reasonably accurate values of $w[n]$.

For the Saramäki window [SA89a], the frequency response of the unscaled window function ($\overline{w}[0]$ is not equal to unity) can be expressed in the forms

$$
\overline{\Psi}(\omega) = \sum_{n=-M}^{M} \overline{w}[n]e^{-jn\omega} = 1 + \sum_{k=1}^{M} 2T_k[\gamma\cos\omega + (\gamma - 1)]
$$

$$
= \frac{\sin\left[\dfrac{2M+1}{2}\cos^{-1}\{\gamma\cos\omega + (\gamma - 1)\}\right]}{\sin\left[\tfrac{1}{2}\cos^{-1}\{\gamma\cos\omega + (\gamma - 1)\}\right]},
\tag{4.49a}
$$

where $T_k[x]$ is the kth degree Chebyshev polynomial and

$$\gamma = \left(1 + \cos \frac{2\pi}{2M + 1}\right) \Big/ \left(1 + \cos \frac{2\beta\pi}{2M + 1}\right). \tag{4.49b}$$

Here, β is the adjustable parameter, which has been selected such that the mainlobe width is $4\beta\pi/(2M + 1)$. This is β times that of the rectangular window. In the special case $\beta = 1$, $\overline{\Psi}(\omega)$ is the frequency response of the rectangular window.

The desired normalized window function ($w[0] = 1$) is

$$w[n] = \begin{cases} \overline{w}[n]/\overline{w}[0], & -M \le n \le M \\ 0, & \text{otherwise} \end{cases} \tag{4.50}$$

and the corresponding frequency response is $\Psi(\omega) = \overline{\Psi}(\omega)/\overline{w}[0]$.

The unscaled coefficients $\overline{w}[n]$ can be expressed as

$$\overline{w}[n] = v_0[n] + 2 \sum_{k=1}^{M} v_k[n], \tag{4.51}$$

where the $v_k[n]$'s can be calculated using the following recursion formulas:

$$v_0[n] = \begin{cases} 1, & n = 0 \\ 0, & \text{otherwise} \end{cases} \tag{4.52a}$$

$$v_1[n] = \begin{cases} \gamma - 1, & n = 0 \\ \gamma/2, & |n| = 1 \\ 0, & \text{otherwise} \end{cases} \tag{4.52b}$$

$$v_k[n] = \begin{cases} 2(\gamma - 1)v_{k-1}[n] - v_{k-2}[n] \\ \quad +\gamma(v_{k-1}[n - 1] + v_{k-1}[n + 1]), & -k \le n \le k \\ 0, & \text{otherwise.} \end{cases} \tag{4.52c}$$

For the Dolph–Chebyshev window [HE68], the frequency response of the un-scaled window function can be expressed as

$$\overline{\Psi}(\omega) = T_M[\gamma \cos \omega + (\gamma - 1)], \tag{4.53a}$$

where

$$\gamma = \left(1 + \cos \frac{\pi}{2M}\right) \Big/ \left(1 + \cos \frac{2\beta\pi}{2M + 1}\right), \tag{4.53b}$$

and the unscaled coefficients are

$$\overline{w}[n] = v_M[n], \tag{4.54}$$

where $v_M[n]$ can be determined using the recursion relations of Eq. (4.52). Here, β has been selected as for the Saramäki window to make the mainlobe width β times that of the rectangular window.

The transitional windows introduced by Saramäki [SA91a] combine the properties of the Dolph–Chebyshev and Saramäki windows. For the mainlobe width being β times that of the rectangular window, the unscaled frequency response of this window is given by

$$\overline{\Psi}(\omega) = \sum_{n=-M}^{M} \overline{w}(n) e^{-jn\omega} = \prod_{k=1}^{M} (\cos \omega - \cos \omega_k), \tag{4.55a}$$

where

$$\omega_k = \rho \omega_k^{(1)} + (1 - \rho) \omega_k^{(2)} \tag{4.55b}$$

with

$$\omega_k^{(1)} = 2 \cos^{-1} \left[\frac{\cos [\beta \pi / (2M + 1)]}{\cos [\pi / (2M + 1)]} \cos \left(\frac{k \pi}{2M + 1} \right) \right] \tag{4.55c}$$

and

$$\omega_k^{(2)} = 2 \cos^{-1} \left[\frac{\cos [\beta \pi / (2M + 1)]}{\cos [\pi / (4M)]} \cos \left(\frac{(2k - 1) \pi}{4M} \right) \right]. \tag{4.55d}$$

Here, $\omega_k^{(1)}$ and $\omega_k^{(2)}$ for $k = 1, 2, \cdots, M$ are the zero locations of the Saramäki and the Dolph–Chebyshev windows, respectively. For $\rho = 1$ and $\rho = 0$, $\overline{\Psi}(\omega)$ is the unscaled response for the Saramäki and the Dolph–Chebyshev window, respectively. For this transitional window, $0 < \rho < 1$ is an adjustable parameter in addition to β. Accurate values for the unscaled window coefficients $\overline{w}[n]$ are obtained from [PA87]

$$\overline{w}[n] = \frac{1}{2M + 1} \left[\overline{\Psi}(0) + 2 \sum_{k=1}^{M} \overline{\Psi} \left(\frac{2 \pi k}{2M + 1} \right) \cos \left(\frac{2 \pi n k}{2M + 1} \right) \right]. \tag{4.56}$$

Alternatively, the coefficients can be determined by evaluating $\overline{\Psi}(\omega)$ at 2^l ($>2M + 1$) equally spaced frequencies and using the inverse fast Fourier transform (see Chapter 8). With a slightly increased amount of calculation, this window gives a higher attenuation for the resulting filter than the other adjustable windows considered above.

The advantages of the above adjustable windows compared to the fixed windows are their near optimality and flexibility. Given ω_p, ω_s, and the minimum stopband attenuation A_s of the filter, the adjustable parameter (α for the Kaiser window and β for the Saramäki, Dolph–Chebyshev, and transitional windows) can be determined to give the desired value for A_s, whereas M can be determined to give the desired value for the transition bandwidth $\Delta\omega = \omega_s - \omega_p$ of the filter. Experimentally obtained estimation formulas for the adjustable parameter and M are given in Table 4-5 for each window type [KA66, KA74; SA89a, SA91a]. For the transitional window, these equations are for

$$\rho = \begin{cases} 0.4, & A_s \leq 50 \\ 0.5, & 50 < A_s \leq 75 \\ 0.6, & 75 < A_s, \end{cases} \tag{4.57}$$

which has turned out to be a good selection in most cases [SA91a]. Like for fixed window functions, the cutoff edge of the ideal filter is selected to be $\omega_c = (\omega_p + \omega_s)/2$ to center the transition band of the resulting filter at this point.

Because of the characteristics of the Dolph–Chebyshev and transitional windows, the estimation formulas developed for these windows are not as accurate as those for the Kaiser and Saramäki windows.

An informative way to compare the performances of adjustable windows is to design several classes of filters with various values of the adjustable parameter for fixed values of M and ω_c. Based on the resulting filter frequency responses, a plot of the stopband attenuation as a function of the parameter $D = 2M(\omega_s - \omega_p)$ can be generated (D, instead of $\omega_s - \omega_p$, is used to make the plot almost independent of M). Figure 4-17 gives such plots for the above-mentioned adjustable windows for $\omega_c = 0.4\pi$ and $M = 128$. For the Kaiser window and the Saramäki window, the difference in the plots is very small. For comparison purposes, a corresponding plot is also included for filters for which the passband and stopband ripples $\delta_p = \delta_s$ have been minimized in the minimax sense for the given value of D. This plot thus gives an upper limit for the stopband attenuation attainable using window functions. For the Kaiser window and the Saramäki window, the resulting attenuation is 5–7 dB less than this upper limit. The stopband attenuation obtained by the Dolph–Chebyshev window is 1–4 dB worse than that of the Kaiser or Saramäki window. For the transitional window, the improvement is typically 2–4 dB over the Kaiser and Saramäki windows and the resulting attenuation approaches the upper limit.

Example 4.1. It is desired to design with each adjustable window considered above a filter with an 80-dB stopband attenuation for $M = 128$ and $\omega_c = 0.4\pi$. Using the formulas given in Table 4-5, the values for the adjustable parameters for the Kaiser, the Saramäki, the Dolph–Chebyshev, and the transitional windows become $\alpha = 7.857$, $\beta = 2.702$, $\beta = 2.770$, and $\beta = 2.587$, respectively. The resulting attenuations are 79.68, 80.17, 79.29, and 80.75 dB. Figure 4-18 gives the frequency responses of both the window functions and the resulting filters in the case of an exactly 80-dB attenuation.[7] The transition bandwidths for these filters are

---

[7] The value of the adjustable parameter giving exactly the desired attenuation A_s can be obtained in two steps. First, the actual filter attenuation, denoted by A_r, is determined for the estimated value of the adjustable parameter. Then this parameter is reestimated by using, instead of A_s, $A_s - (A_r - A_s)$ in the estimation formula.

TABLE 4-5 Estimation Formulas for the Adjustable Parameter and M for Adjustable Windows to Give the Desired Attenuation A_s and Transition Bandwidth $\omega_s - \omega_p$

Window Type	Adjustable Parameter	M
Kaiser	$\alpha = \begin{cases} 0.1102(A_s - 8.7), & A_s > 50 \\ 0.5842(A_s - 21)^{0.4} \\ \quad + 0.07886(A_s - 21), & 21 < A_s < 50 \\ 0, & A_s < 21 \end{cases}$	$M = \dfrac{A_s - 7.95}{14.36(\omega_s - \omega_p)/\pi}$
Saramäki	$\beta = \begin{cases} 0.000121(A_s - 21)^2 \\ \quad + 0.0224(A_s - 21) + 1, & A_s \leq 65 \\ 0.033A_s + 0.062, & 65 < A_s \leq 110 \\ 0.0345A_s - 0.097, & A_s > 110 \end{cases}$	$M = \dfrac{A_s - 8.15}{14.36(\omega_s - \omega_p)/\pi}$
Dolph–Chebyshev	$\beta = \begin{cases} 0.0000769(A_s)^2 \\ \quad + 0.0248A_s + 0.330, & A_s \leq 60 \\ 0.0000104(A_s)^2 \\ \quad + 0.0328A_s + 0.079, & A_s > 60 \end{cases}$	$M = \dfrac{1.028A_s - 8.4}{14.36(\omega_s - \omega_p)/\pi}$
Transitional	$\beta = \begin{cases} 0.000154(A_s)^2 \\ \quad + 0.0153A_s + 0.465, & A_s \leq 60 \\ 0.0000204(A_s)^2 \\ \quad + 0.0303A_s + 0.032, & A_s > 60 \end{cases}$	$M = \dfrac{0.00036(A_s)^2 + 0.951A_s - 9.4}{14.36(\omega_s - \omega_p)/\pi}$

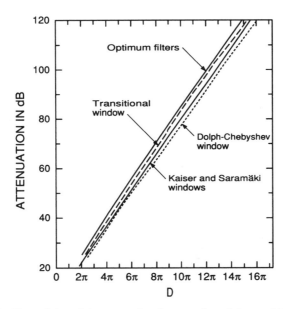

FIGURE 4-17 Plots of the minimum stopband attenuation of the resulting filter versus D = $2M(\omega_s - \omega_p)$ for adjustable windows. $M = 128$ and $\omega_c = 0.4\pi$. For comparison purposes, a corresponding plot for filters for which $\delta_p = \delta_s$ has been minimized in the minimax sense is included.

0.0393π, 0.0390π, 0.0406π, and 0.0373π. From Figure 4-18, it is seen that the responses for the Kaiser window and the Saramäki window as well as the responses of the resulting filters are practically the same. It is interesting to observe from this figure that the filter response for the Dolph–Chebyshev window is flatter than the corresponding responses for the Kaiser and Saramäki windows, whereas the filter response for the transitional window is between those of the Saramäki and Dolph–Chebyshev windows. Also note that the ripples of the sidelobes are of the same height for the Dolph–Chebyshev window. In Section 4-7, this property is utilized in designing filters having an equiripple behavior in the stopband.

4-5 DESIGN OF FIR FILTERS IN THE LEAST-MEAN-SQUARE SENSE

The second straightforward approach for designing FIR filters is based on the use of the least-squared approximation [FA74; KA63, KA66; KE72; LI83c; PA87; TU70; VA87]. In this case, the problem is to find the filter coefficients to minimize

$$E_2 = \int_X [W(\omega)[H(\omega) - D(\omega)]]^2 \, d\omega, \tag{4.58}$$

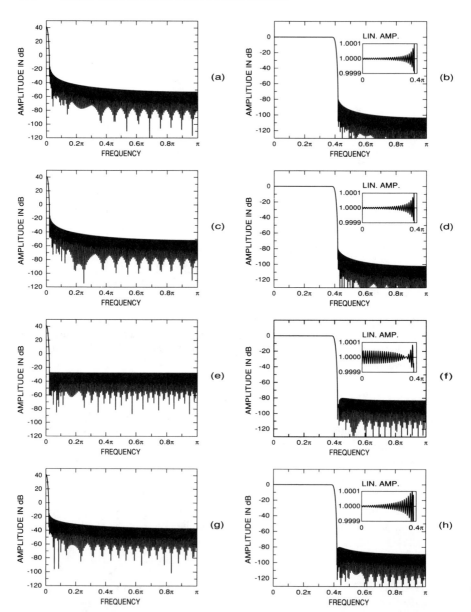

FIGURE 4-18 Frequency responses of the window functions and the resulting filters for adjustable windows giving an 80-dB attenuation for the filter when $M = 128$ and $\omega_c = 0.4\pi$. (a,b) Kaiser window. (c,d) Saramäki window. (e,f) Dolph–Chebyshev window. (g,h) Transitional window.

where X contains the passband and stopband regions, $D(\omega)$ is a desired response, and $W(\omega)$ is a positive weighting function. If $D(\omega)$ and $W(\omega)$ are sampled at a very dense grid of frequencies $\omega_1, \omega_2, \cdots, \omega_K$ on X, then minimization of the above equation may be achieved by minimizing

$$E_2 = \sum_{k=1}^{K} [W(\omega_k)[H(\omega_k) - D(\omega_k)]]^2. \tag{4.59}$$

As shown in Section 4-3-3, $H(\omega)$ can be expressed in the four different linear-phase cases in the form (see Eq. (4.32))

$$H(\omega) = \sum_{n=0}^{M} b[n] \operatorname{trig}(\omega, n). \tag{4.60}$$

By substituting this for $H(\omega_k)$ in Eq. (4.59) and transferring $W(\omega_k)$ inside the parentheses we obtain

$$E_2 = \sum_{k=1}^{K} \left[W(\omega_k) \sum_{n=0}^{M} b[n] \operatorname{trig}(\omega_k, n) - W(\omega_k) D(\omega_k) \right]^2. \tag{4.61}$$

This equation can be written in the following quadratic form

$$E_2 = \mathbf{e}^T \mathbf{e}, \tag{4.62a}$$

where

$$\mathbf{e} = \mathbf{Xb} - \mathbf{d} \tag{4.62b}$$

with

$$\mathbf{X} = \begin{bmatrix} W(\omega_1) \operatorname{trig}(\omega_1, 0) & W(\omega_1) \operatorname{trig}(\omega_1, 1) & \cdots & W(\omega_1) \operatorname{trig}(\omega_1, M) \\ W(\omega_2) \operatorname{trig}(\omega_2, 0) & W(\omega_2) \operatorname{trig}(\omega_2, 1) & \cdots & W(\omega_2) \operatorname{trig}(\omega_2, M) \\ \vdots & \vdots & & \vdots \\ W(\omega_K) \operatorname{trig}(\omega_K, 0) & W(\omega_K) \operatorname{trig}(\omega_K, 1) & \cdots & W(\omega_K) \operatorname{trig}(\omega_K, M) \end{bmatrix},$$

$$\tag{4.62c}$$

$$\mathbf{b} = [b[0], b[1], \ldots, b[M]]^T, \tag{4.62d}$$

and

$$\mathbf{d} = [W(\omega_1) D(\omega_1), W(\omega_2) D(\omega_2), \ldots, W(\omega_K) D(\omega_K)]^T. \tag{4.62e}$$

Here, $\mathbf{e}$ is a K length vector with the kth element being $W(\omega_k)[H(\omega_k) - D(\omega_k)]$. The optimum solution of minimizing E_2 is given by [LI83c; PA87]

$$\mathbf{b} = (\mathbf{X}^T\mathbf{X})^{-1}\mathbf{X}^T\mathbf{d} \qquad (4.63)$$

and it satisfies the "normal equations" [PA87]

$$\mathbf{X}^T\mathbf{X}\mathbf{b} = \mathbf{X}^T\mathbf{d}. \qquad (4.64)$$

It should be noted that if K is much larger than M, then Eq. (4.63) should not be solved directly because it becomes ill conditioned. In this case, the direct solution will probably have large errors. Parks and Burrus [PA87] recommend the use of the software package LINPACK [DO79], which has a special program for solving the above problem.

In the case where both $W(\omega)$ and $D(\omega)$ are piecewise-constant functions, a significantly simpler procedure for finding the optimum solution can be generated [LI83c].

Example 4.2. Consider the design of a Type I lowpass filter of order $N = 46$ ($M = 23$) for $\omega_p = 0.5\pi$ and $\omega_s = 0.6\pi$. $D(\omega) = 1$ on $[0, \omega_p]$ and $D(\omega) = 0$ on $[\omega_s, \pi]$. Figure 4-19 shows the resulting responses in two cases. In both cases, $W(\omega) = 1$ on $[0, \omega_p]$, whereas $W(\omega)$ is unity on $[\omega_s, \pi]$ in the first case and 10 in the second case. The effect of the stopband weighting is clearly seen from the figure. It is also seen that the maximum deviations between the actual and the desired responses are much larger near the passband and stopband edges. This is characteristic of the least-squared-error designs. If the maximum deviations are desired to be minimized, then it is preferred to design the filter in the minimax sense (see Section 4-8). Compare Figure 4-19 to Figure 4-8, which gives a re-

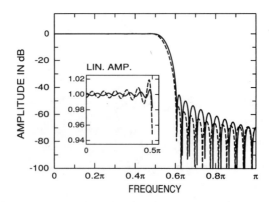

FIGURE 4-19 Amplitude responses for least-squared-error FIR filters of order 46. The solid and dashed lines give the responses for the filters with stopband weighting of 1 and 10, respectively.

sponse for an FIR filter optimized in the minimax sense. The filter orders in these two figures are the same.

4-6 MAXIMALLY FLAT FIR FILTERS

The third straightforward approach for designing FIR filters is to use filters with maximally flat response around $\omega = 0$ and $\omega = \pi$ [HA77; HE71b; KA77b, KA79; VA84]. The advantages of these filters are that the design is extremely simple and they are useful in applications where the signal is desired to be preserved with very small error near the zero frequency. If the maximum deviation from the desired response is required to be minimized, then it is preferred to use filters designed in the minimax sense. These minimax designs meet the given criteria with a significantly reduced filter order.

Consider a Type I filter with transfer function

$$H(z) = \sum_{n=0}^{2M} h[n] z^{-n}, \qquad h[2M - n] = h[n]. \tag{4.65}$$

For maximally flat designs, it is advantageous to express $H(\omega)$ as an Mth degree polynomial in $\cos \omega$ as follows (cf. Section 4-3-3):

$$H(\omega) = \sum_{n=0}^{M} \alpha[n] \cos^n \omega. \tag{4.66}$$

This $H(\omega)$ is determined in such a way that it has $2K$ zeros at $\omega = \pi$ and $H(\omega) - 1$ has $2L = 2(M - K + 1)$ zeros at $\omega = 0$. M is thus related to L and K through $M = K + L - 1$. The above conditions are satisfied if $H(\omega)$ can be written simultaneously in the forms

$$H(\omega) = \left[\frac{1 + \cos \omega}{2}\right]^K \sum_{n=0}^{L-1} d[n] \left[\frac{1 - \cos \omega}{2}\right]^n$$

$$= \cos^{2K}(\omega/2) \sum_{n=0}^{L-1} d[n] \sin^{2n}(\omega/2) \tag{4.67a}$$

and

$$H(\omega) = 1 - \left[\frac{1 - \cos \omega}{2}\right]^L \sum_{n=0}^{K-1} \bar{d}[n] \left[\frac{1 + \cos \omega}{2}\right]^n$$

$$= 1 - \sin^{2L}(\omega/2) \sum_{n=0}^{K-1} \bar{d}[n] \cos^{2n}(\omega/2). \tag{4.67b}$$

The coefficients $d[n]$ and $\bar{d}[n]$ giving the desired solution are

$$d[n] = \frac{(K - 1 + n)!}{(K - 1)!n!}, \qquad \bar{d}[n] = \frac{(L - 1 + n)!}{(L - 1)!n!}. \qquad (4.68)$$

The resulting $H(\omega)$ is characterized by the facts that it achieves the value one at $\omega = 0$ and its first $2L - 1$ derivatives are zero at this point, whereas it achieves the value zero at $\omega = \pi$ with its first $2K - 1$ derivatives being zero at this point. The primary unknowns of the above filters are K and L. Given the filter specifications, the problem is thus to determine these integers such that the criteria are satisfied.

Kaiser [KA79] has stated the specifications for maximally flat filters as shown in Figure 4-20. Here, β is the center of the transition band and γ is the width of the transition band, which is defined as the region where the response varies from 0.95 (passband edge angle) to 0.05 (stopband edge angle). For meaningful specifications, γ has to satisfy $0 \leq \gamma \leq \min(2\beta, 2\pi - 2\beta)$. In the design procedure proposed by Kaiser, the lower estimate for $M = K + L - 1$ (half the filter order) is given by

$$M_{\text{lower}} = (\pi/\gamma)^2. \qquad (4.69)$$

Then ρ is determined by

$$\rho = (1 + \cos \beta)/2. \qquad (4.70)$$

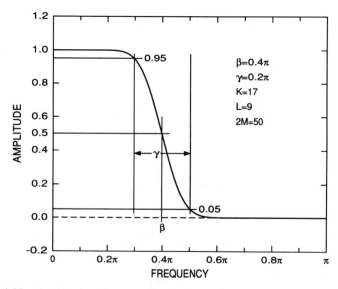

FIGURE 4-20 Specifications for a maximally flat lowpass filter and response for a filter of order 50 meeting the criteria $\beta = 0.4\pi$ and $\gamma = 0.2\pi$.

The next step is to determine

$$K_p = \langle \rho M_p \rangle, \tag{4.71}$$

where $\langle x \rangle$ stands for the nearest integer of x, for the values of M_p in the range $M_{\text{lower}} \leq M_p \leq 2M_{\text{lower}}$. Finally, the values of the integers K_p and M_p for which the ratio K_p/M_p is closest to ρ are selected. The corresponding values of K, L, and M are then $K = K_p$, $L = M_p - K_p$, and $M = M_p - 1$, respectively. With these selections of K and L, the given value of β can be achieved accurately.

Example 4.3. Consider the specifications $\beta = 0.4\pi$ and $\gamma = 0.2\pi$. The above procedure results in $K = 17$ and $L = 9$. The order of the filter is thus $2(K + L - 1) = 50$. The amplitude response of this filter is depicted in Figure 4-20.

The transfer function having the zero-phase frequency response as given by Eq. (4.67a) or (4.67b) can be implemented using the conventional direct-form structure shown in Figure 4-7.[8] Alternatively, the transfer function can be written in the forms

$$H(z) = \left(\frac{1 + z^{-1}}{2}\right)^{2K} \sum_{n=0}^{L-1} (-1)^n d[n] z^{-(L-1-n)} \left(\frac{1 - z^{-1}}{2}\right)^{2n} \tag{4.72a}$$

and

$$H(z) = z^{-M} - (-1)^L \left(\frac{1 - z^{-1}}{2}\right)^{2L} \sum_{n=0}^{K-1} \bar{d}[n] z^{-(K-1-n)} \left(\frac{1 + z^{-1}}{2}\right)^{2n}. \tag{4.72b}$$

The advantage of realizing the transfer function in the above forms lies in the fact that these implementation forms have significantly fewer multipliers than the direct-form structure [VA84].

4-7 SOME SIMPLE FIR FILTER DESIGNS

There are two special cases where the optimum solution in the minimax sense can be obtained analytically [HE73]. The first analytically solvable case is the one where the zero-phase frequency response is monotonically decaying in the passband region and exhibits an equiripple behavior in the stopband region $[\omega_s, \pi]$. An

---

[8]In this case, it is preferred to calculate the filter coefficients by evaluating $H(\omega)$ at 2^l ($> 2M + 1$) equally spaced frequencies and using the inverse discrete Fourier transform (see Chapter 8). This guarantees that the resulting coefficient values are accurate enough. This is the procedure used by Kaiser [KA79].

equiripple behavior on $[\omega_s, \pi]$ can be achieved by mapping the Mth degree Chebyshev polynomial $T_M(x)$ to the ω-plane such that the region $[-1, 1]$, where $T_M(x)$ oscillates within the limits ± 1, is mapped to the region $[\omega_s, \pi]$ (see Figure 4-21). The desired transformation mapping $x = 1$ to $\omega = \omega_s$ and $x = -1$ to $\omega = \pi$ is

$$x = \gamma \cos \omega + (\gamma - 1), \qquad \gamma = \frac{2}{1 + \cos \omega_s}, \qquad (4.73)$$

resulting in the following zero-phase frequency response of a Type I filter of order $2M$:

$$\overline{H}(\omega) = T_M[(2 \cos \omega + 1 - \cos \omega_s)/(1 + \cos \omega_s)]. \qquad (4.74)$$

The response taking the value $1 + \delta_p$ at $\omega = 0$ is then

$$H(\omega) = (1 + \delta_p)\overline{H}(\omega)/\overline{H}(0). \qquad (4.75)$$

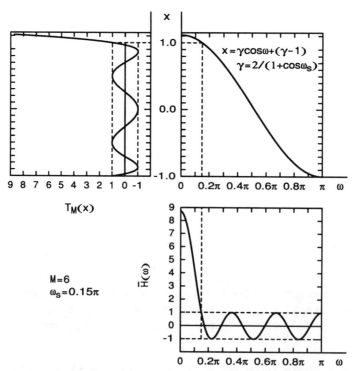

FIGURE 4-21   Generation of a zero-phase frequency response oscillating within the limits ± 1 in the stopband $[\omega_s, \pi]$ based on mapping the Mth degree Chebyshev polynomial $T_M(x)$ to the ω-plane.

This function oscillates on $[\omega_s, \pi]$ within the limits $\pm\delta_s$, where

$$\delta_s = (1 + \delta_p)/\overline{H}(0). \tag{4.76}$$

Based on the properties of Chebyshev polynomials, it can be shown that the value of M (half the filter order) giving the specified stopband ripple δ_s is [HE73]

$$M = \frac{\cosh^{-1}[(1 + \delta_p)/\delta_s]}{\cosh^{-1}[(3 - \cos \omega_s)/(1 + \cos \omega_s)]}. \tag{4.77}$$

Example 4.4. Figure 4-22 gives responses with $\omega_s = 0.1\pi$ and $\delta_p = 0.1$ for $M = 15$ and $M = 30$. The corresponding filter orders are 30 and 60, respectively. The disadvantage of these designs is that all their zeros lie on the unit circle and the passband region, where the response decays from $1 + \delta_p$ to $1 - \delta_p$, is narrow and cannot be controlled.

The response that is equiripple in the passband $[0, \omega_p]$ oscillating within the limits $1 \pm \delta_p$ and monotonically decaying in the region $[\omega_p, \pi]$ can be derived in the same manner. This solution is given by [HE73]

$$H(\omega) = 1 - \delta_p T_M[(-2 \cos \omega + 1 + \cos \omega_p)/(1 - \cos \omega_p)]. \tag{4.78}$$

If it is desired that $H(\pi) = -\delta_s$, then δ_p can be determined from

$$\delta_p = (1 + \delta_s)/T_M[(3 + \cos \omega_p)/(1 - \cos \omega_p)]. \tag{4.79}$$

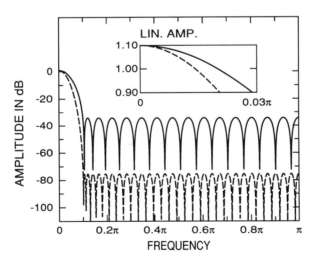

FIGURE 4-22 Responses for filters having an equiripple stopband behavior and a monotonically decaying passband response for $M = 15$ (solid line) and $M = 30$ (dashed line). $\omega_s = 0.1\pi$ and $\delta_p = 0.1$.

The minimum value of M required to meet the given ripple requirements can be determined from Eq. (4.77) by interchanging δ_p and δ_s and by replacing ω_s by $\pi - \omega_p$.

4-8 DESIGN OF FIR FILTERS IN THE MINIMAX SENSE

One of the main advantages of FIR filters over their IIR counterparts is that there exists an efficient algorithm for optimizing in the minimax sense arbitrary-magnitude FIR filters. For IIR filters, the design of arbitrary-magnitude filters is usually time-consuming and the convergence to the best solution is not always guaranteed. This section introduces this algorithm for designing linear-phase FIR filters and shows its flexibility by means of several examples. The resulting filters are optimal in the sense that they meet the given arbitrary specifications with the minimum filter order. This section considers also some properties of these optimum linear-phase FIR filters.

4-8-1 Remez Multiple Exchange Algorithm

The most efficient algorithm for designing optimum magnitude FIR filters with arbitrary specifications is the Remez multiple exchange algorithm [CH66; RI64]. The most frequently used method for implementing this algorithm is the one originally advanced by Parks and McClellan [PA72a]. Further improvements to the implementation of the Remez algorithm have been proposed by McClellan, Parks, and Rabiner [MC73a, MC73b, MC79; PA72b; RA75b]. As a result of this work, a program for designing arbitrary-magnitude FIR filters has been reported in McClellan et al. [MC73b, MC79]. This program is directly applicable to obtaining optimal designs for most types of FIR filters like lowpass, highpass, bandpass, and bandstop filters, Hilbert transformers, and digital differentiators. Also, filters having several passbands and stopbands can be designed directly. The amount of computation required for designing optimum filters can be significantly reduced by using techniques proposed by Antoniou [AN82, AN83] and Bonzanigo [BO82]. This section concentrates on the original FIR filter design program of McClellan, Parks, and Rabiner [MC73b, MC79]. This method is referred to later as the MPR algorithm.

4-8-1-1 Characterization of the Optimum Solution. The Remez multiple exchange algorithm is the most powerful algorithm for finding the coefficients $a[n]$ of the function

$$G(\omega) = \sum_{n=0}^{M} a[n] \cos n\omega \qquad (4.80)$$

to minimize on a closed subset X of $[0, \pi]$ the peak absolute value of the following weighted error function:

$$E(\omega) = \overline{W}(\omega)[G(\omega) - \overline{D}(\omega)], \qquad (4.81)$$

that is, the quantity

$$\epsilon = \max_{\omega \in X} |E(\omega)|. \tag{4.82}$$

It is required that $\overline{D}(\omega)$ be continuous on X and $\overline{W}(\omega) > 0$. This algorithm can be used in all four linear-phase cases based on the fact that the zero-phase frequency response $H(\omega)$ of a filter of order N can be expressed, according to the discussion of Section 4-3-1, in the form

$$H(\omega) = F(\omega) G(\omega), \tag{4.83}$$

where $G(\omega)$ is as given by Eq. (4.80) and

$$F(\omega) = \begin{cases} 1 & \text{for Type I} \\ \cos(\omega/2) & \text{for Type II} \\ \sin \omega & \text{for Type III} \\ \sin(\omega/2) & \text{for Type IV,} \end{cases} \qquad M = \begin{cases} N/2 & \text{for Type I} \\ (N-1)/2 & \text{for Type II} \\ (N-2)/2 & \text{for Type III} \\ (N-1)/2 & \text{for Type IV.} \end{cases} \tag{4.84}$$

If the desired function for $H(\omega)$ on X is $D(\omega)$ and the weighting function is $W(\omega)$, then the error function can be written in the form of Eq. (4.81) as follows:

$$\begin{aligned} E(\omega) &= W(\omega)[H(\omega) - D(\omega)] = W(\omega)[F(\omega)G(\omega) - D(\omega)] \\ &= W(\omega)F(\omega)[G(\omega) - D(\omega)/F(\omega)] = \overline{W}(\omega)[G(\omega) - \overline{D}(\omega)], \end{aligned} \tag{4.85a}$$

where

$$\overline{W}(\omega) = F(\omega)W(\omega), \qquad \overline{D}(\omega) = D(\omega)/F(\omega). \tag{4.85b}$$

The Remez multiple exchange algorithm can be constructed on the basis of the following characterization theorem [CH66; RI64].

Characterization Theorem. Let $G(\omega)$ be of the form of Eq. (4.80). Then $G(\omega)$ is the best unique solution minimizing ϵ as given by Eq. (4.82) if and only if there exist at least $M + 2$ points $\omega_1, \omega_2, \ldots, \omega_{M+2}$ in X such that

$$\omega_1 < \omega_2 < \cdots < \omega_{M+1} < \omega_{M+2}$$

$$E(\omega_{i+1}) = -E(\omega_i), \qquad i = 1, 2, \ldots, M + 1$$

$$|E(\omega_i)| = \epsilon, \qquad i = 1, 2, \ldots, M + 2.$$

In other words, the optimum solution is characterized by the fact that the weighted error function $E(\omega)$ alternatingly achieves the values $\pm\epsilon$, with ϵ being the peak absolute value of the weighted error, at least at $M + 2$ consecutive points in X. Figure 4-23 gives the response of a typical optimum Type I lowpass filter of order $N = 12$ and the corresponding error function. In this case, $H(\omega) \equiv G(\omega)$, $X = [0, \omega_p] \cup [\omega_s, \pi]$, $\overline{D}(\omega) = D(\omega) = 1$ and $\overline{W}(\omega) = W(\omega) = 1$ for $\omega \in [0, \omega_p]$, whereas $D(\omega) = 0$ and $W(\omega) = 2$ for $\omega \in [\omega_s, \pi]$. Note that the above weighting function makes the stopband ripple of the optimum filter to be half of that of the passband ripple. In this case, $M = N/2 = 6$ so that $G(\omega)$ contains seven unknowns $a[0], a[1], \ldots, a[6]$. The number of extremal points is $M + 2 = 8$ so that there is one more extremal frequency than there are unknowns, as required by the characterization theorem. According to the theorem, it is thus easy to check whether a given solution is the optimum one. If the relative weighting between the stopband and passband errors is k and there exists a solution $H(\omega)$ that alternatingly goes through the values $1 \pm \epsilon$ in the passband and through the values $\pm\epsilon/k$ in the stopband, and the overall number of these extrema is at least $M + 2$, then this solution is, according to the characterization theorem, the best unique solution. In the lowpass case, both ω_p and ω_s are always extremal points, and $H(\omega_p) = 1 - \epsilon$ and $H(\omega_s) = \epsilon/k$ so that $E(\omega_p) = -\epsilon$ and $E(\omega_s) = \epsilon$.

4-8-1-2   The McClellan-Parks-Rabiner (MPR) Algorithm. Given a set of $M + 2$ points on X, denoted by $\Omega = \{\omega_1, \omega_2, \ldots, \omega_{M+2}\}$, the unknown coefficients $a[0], a[1], \ldots, a[M]$ and ϵ can be determined such that $E(\omega)$ satisfies

$$E(\omega_k) = \overline{W}(\omega_k)[G(\omega_k) - \overline{D}(\omega_k)] = (-1)^k\epsilon, \qquad k = 1, 2, \ldots, M + 2.$$

$$(4.86)$$

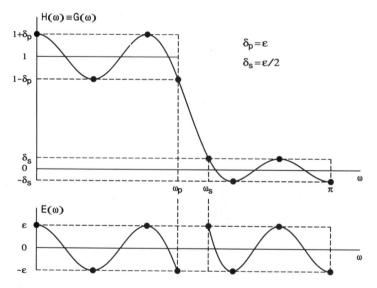

FIGURE 4-23 Zero-phase frequency response and error function for an optimum Type I linear-phase lowpass filter of order $2M = 12$. The stopband weighting is two times that of the passband.

This can be achieved by solving for the unknowns the following system of $M + 2$ linear equations:

$$\sum_{n=0}^{M} a[n] \cos n\omega_k - (-1)^k \epsilon / \overline{W}(\omega_k) = \overline{D}(\omega_k), \qquad k = 1, 2, \ldots, M + 2.$$

(4.87)

The resulting $E(\omega)$ goes alternatingly through the values $\pm \epsilon$ at the points ω_k. If X consists of the above set of $M + 2$ points, that is, $X = \Omega$, then $|\epsilon|$ is the peak absolute value of $E(\omega)$ on X and the conditions of the above characterization theorem are satisfied.[9] The Remez exchange algorithm makes use of this fact. The problem is simply to find a set Ω on X in such a way that the optimum solution on Ω is simultaneously the optimum solution on the overall set X. This is achieved if the value of $|\epsilon|$ is simultaneously the peak absolute value of $E(\omega)$ on the overall set X. The Remez algorithm iteratively finds the desired set of $M + 2$ extremal points using the following steps:

1. Select an initial set of $M + 2$ extremal points $\Omega = \{\omega_1, \omega_2, \ldots, \omega_{M+2}\}$ in X.
2. Solve the system of $M + 2$ linear equations given by Eq. (4.87) for the unknowns $a[0], a[1], \ldots, a[M]$ and ϵ.
3. Find on X, $M + 2$ extremal points of the resulting $E(\omega)$, where $|E(\omega)| \geq |\epsilon|$. If there are more than $M + 2$ extremal points, retain $M + 2$ extrema such that the largest absolute values are included with the condition that the sign of the error function $E(\omega)$ alternates at the selected points. Store the abscissas of the extrema into $\overline{\Omega} = \{\overline{\omega}_1, \overline{\omega}_2, \ldots, \overline{\omega}_{M+2}\}$.
4. If $|\omega_k - \overline{\omega}_k| \leq \alpha$ for $k = 1, 2, \ldots, M + 2$ (α is a small number), then go to the next step. Otherwise, set $\Omega = \overline{\Omega}$ and go to Step 2.
5. Calculate the filter coefficients and plot the frequency response.

The above algorithm starts by selecting $M + 2$ initial extremal points ω_k. These points can be selected, for example, to lie equidistantly on X. Then the coefficients of $G(\omega)$ and ϵ are solved for at Step 2 such that $E(\omega)$ satisfies

$$E(\omega_k) = (-1)^k \epsilon, \qquad k = 1, 2, \ldots, M + 2; \qquad (4.88)$$

that is, it alternately achieves the values $\pm \epsilon$ at the points $\omega_1, \omega_2, \ldots, \omega_{M+2}$, as required by the characterization theorem. However, all the points ω_k are not the true external points of the resulting $E(\omega)$. This is illustrated in Figure 4-24, which gives $E(\omega)$ after performing Step 2 for the first time. As seen from this figure, most of the initial extremal points are not the true extremal points and the maximum absolute value of $E(\omega)$ is higher than $|\epsilon|$ around these points. Thus the $M + 2$ new extremal points $\overline{\omega}_k$ of $E(\omega)$ are located next and these extremal points replace the ω_k's. After that, the coefficients of $G(\omega)$ and ϵ are redetermined such that Eq.

[9]When solving Eq. (4.87) for the unknowns, the resulting ϵ is either positive or negative.

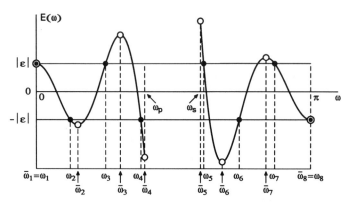

FIGURE 4-24 Error function obtained by forcing it to alternatingly go through the values $\pm\epsilon$ at the selected extremal points ω_k, $k = 1, 2, \ldots, M + 2$. $M = 6$ and $X = [0, \omega_p] \cup [\omega_s, \pi]$. The $\bar{\omega}_k$'s are the true extremal points of the error function.

(4.88) is satisfied at the new extremal points. The process is repeated until the ω_k's become the true extremal points of $E(\omega)$. In the above process, the absolute value of ϵ increases in each iteration loop.

The set of linear equations at Step 2 can be solved conveniently by first calculating ϵ analytically as

$$\epsilon = \frac{\sum\limits_{k=1}^{M+2} b_k \bar{D}(\omega_k)}{\sum\limits_{k=1}^{M+2} b_k (-1)^{k+1} / \overline{W}(\omega_k)}, \tag{4.89a}$$

where

$$b_k = \prod\limits_{\substack{i=1 \\ i \neq k}}^{M+2} \frac{1}{(\cos \omega_k - \cos \omega_i)}. \tag{4.89b}$$

After calculating ϵ, it is known that $G(\omega)$ achieves the value

$$C_k = \bar{D}(\omega_k) + (-1)^k \epsilon / \overline{W}(\omega_k) \tag{4.90}$$

at the kth extremal point. To get around the numerical sensitivity problems, the Lagrange interpolation formula in the barycentric form is used to express $G(\omega)$ as

$$G(\omega) = \frac{\sum\limits_{k=1}^{M+1} \left(\dfrac{\beta_k}{\cos \omega - \cos \omega_k} \right) C_k}{\sum\limits_{k=1}^{M+1} \left(\dfrac{\beta_k}{\cos \omega - \cos \omega_k} \right)}, \tag{4.91a}$$

where

$$\beta_k = \prod_{\substack{i=1 \\ i \neq k}}^{M+1} \frac{1}{(\cos \omega_k - \cos \omega_i)}.$$

(4.91b)

Note that after solving ϵ, $M + 1$ points, instead of $M + 2$ points, are required to uniquely determine $G(\omega)$. In the MPR algorithm [MC73b, MC79] $G(\omega)$ is expressed in the above form. This is because the actual coefficient values $a[n]$ are not needed in intermediate calculations. After the convergence of the algorithm, the $a[n]$'s are determined by evaluating $G(\omega)$ at $2M + 1$ equally spaced frequency points and then applying the inverse discrete Fourier transform (see Chapter 8). From the $a[n]$'s, the filter coefficients $h[n]$ can then be determined according to the discussion of Section 4-3-1.

In the practical implementation of the MPR algorithm, the extrema of $E(\omega)$ at Step 3 are located by evaluating $E(\omega)$ over a dense set of frequencies spanning the approximation region X. As a rule of thumb, a good selection of the number of grid points is $16M$. Typically, four to eight iterations of the above algorithm are required to arrive at the optimum solution in lowpass cases. In designing filters having several passband and stopband regions, the number of iterations is typically two or three times that required for designing lowpass filters.

4-8-2 Properties of the Optimum Filters

Before illustrating the use of the above algorithm in practical filter design problems, some properties of optimum filters are reviewed. In the lowpass case, the filter design parameters are the passband edge ω_p, the stopband edge ω_s, the passband ripple δ_p, and the stopband ripple δ_s. The remaining parameter to be determined is the minimum filter order N required to meet the given criteria. If N is prescribed, then the ripple ratio

$$k = \delta_p / \delta_s,$$

(4.92)

instead of δ_p and δ_s, is usually specified. In the latter case, the optimum solution is obtained by using the following desired response and weighting function in the MPR algorithm:

$$D(\omega) = \begin{cases} 1 & \text{for } \omega \in [0, \omega_p] \\ 0 & \text{for } \omega \in [\omega_s, \pi], \end{cases}$$

(4.93a)

$$W(\omega) = \begin{cases} 1 & \text{for } \omega \in [0, \omega_p] \\ k & \text{for } \omega \in [\omega_s, \pi]. \end{cases}$$

(4.93b)

In this case, $X = [0, \omega_p] \cup [\omega_s, \pi]$.

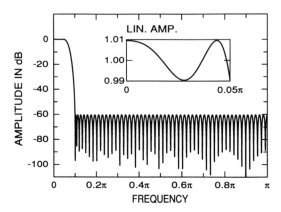

FIGURE 4-25 Amplitude response for an optimum Type I filter of order $N = 2M = 108$.

Example 4.5. Figure 4-25 gives the optimized response of a Type I filter of order $N = 108$ ($M = 54$) for $\omega_p = 0.05\pi$, $\omega_s = 0.1\pi$, and $k = 10$. The resulting ripples are given by $\delta_p = 0.00955$ and $\delta_s = 0.000955$. In this case, $N = 108$ is the minimum filter order to meet the ripple requirements $\delta_p \leq 0.01$ and $\delta_s \leq 0.001$.

Except for the case of the Chebyshev solutions considered in Section 4-7, there exist no analytic relations between the lowpass filter parameters N, ω_p, ω_s, δ_p, and δ_s. However, rather accurate estimates, based on empirical data, have been reported by Herrmann et al. [HE73], Kaiser [KA74], and Rabiner [RA73c] for the minimum filter order N. Kaiser has proposed the particularly simple formula [KA74]

$$N \approx \frac{-20 \log_{10}(\sqrt{\delta_p \delta_s}) - 13}{14.6[(\omega_s - \omega_p)/(2\pi)]} \tag{4.94}$$

for predicting the filter order N. A somewhat more accurate formula due to Herrmann et al. [HE73] is

$$N \approx \frac{D_\infty(\delta_p, \delta_s) - F(\delta_p, \delta_s)[(\omega_s - \omega_p)/(2\pi)]^2}{(\omega_s - \omega_p)/(2\pi)}, \tag{4.95a}$$

where

$$D_\infty(\delta_p, \delta_s) = [a_1(\log_{10}\delta_p)^2 + a_2\log_{10}\delta_p + a_3]\log_{10}\delta_s$$
$$- [a_4(\log_{10}\delta_p)^2 + a_5\log_{10}\delta_p + a_6] \tag{4.95b}$$

and

$$F(\delta_p, \delta_s) = b_1 + b_2[\log_{10}\delta_p - \log_{10}\delta_s] \tag{4.95c}$$

with

$$a_1 = 0.005309, \qquad a_2 = 0.07114, \qquad a_3 = -0.4761, \qquad (4.95\text{d})$$

$$a_4 = 0.00266, \qquad a_5 = 0.5941, \qquad a_6 = 0.4278, \qquad (4.95\text{e})$$

$$b_1 = 11.01217, \qquad b_2 = 0.51244. \qquad (4.95\text{f})$$

This formula has been developed for $\delta_s < \delta_p$. If $\delta_s > \delta_p$, then the estimate is obtained by interchanging δ_p and δ_s in the formula.

If the ripples of the filter are rather small, then both formulas give approximately the same result. However, when the ripple values are large, the latter formula gives a better estimate. For this formula, the estimation error is typically less than 2%. From the above formulas, it is seen that the required filter order is roughly inversely proportional to the transition bandwidth.

In the case of the specifications of Example 4.5, both of the above formulas give $N = 101$. For the optimized filter of order 101, the ripples are given by $\delta_p = 0.0157$ and $\delta_s = 0.00157$, showing that the filter order has to be increased. When determining the actual minimum filter order, it must be taken into consideration that sometimes a filter of order $N - 1$ has lower ripple values than a filter of order N. For instance, for the case $\omega_p = 0.6856\pi$, $\omega_s = 0.83246\pi$, and $k = 1$, the Type I filter of order $N = 10$ achieves $\delta_p = \delta_s = 0.1282$, whereas the Type II filter of order $N = 9$ achieves $\delta_p = \delta_s = 0.1$ [RA73b]. Based on this, it is advantageous to determine separately the minimum orders for both Type I filters (N is even) and Type II filters (N is odd), and then to select the lower order. For the specifications of Example 4.5, the minimum orders of Type I and Type II filters to meet the ripple requirements of $\delta_p = 0.01$ and $\delta_s = 0.001$ are 108 and 109, respectively, so that $N = 108$ is the minimum order.

An informative way to study the various types of optimum lowpass filter solutions is to plot the transition bandwidth

$$\Delta\omega = \omega_s - \omega_p \qquad (4.96)$$

of the filter versus ω_p for fixed values of N, δ_p, and δ_s [PA73; RA73a, RA73b]. Figure 4-26 gives such plots for Type I optimum filters with $N = 14$ ($M = 7$), $N = 16$ ($M = 8$), and $N = 18$ ($M = 9$) for $\delta_p = \delta_s = 0.1$.[10] As seen from this figure, all three curves alternate between sharp minima and flat-topped maxima. We consider in greater detail the filters corresponding to the six points, denoted by the letters A, B, C, D, E, and F, in the curve for $N = 16$ ($M = 8$). The responses of these filters are given in Figure 4-27. Filters C and F correspond to the points where the local minimum of $\Delta\omega$ occurs with respect to ω_p. These are special *extraripple* or *maximal ripple* solutions whose error function exhibits $M + 3 = 11$

---

[10]In constructing these curves, $k = 1$ has been used in the MPR algorithm. For each value of ω_p, the minimum transition bandwidth to meet the given ripple requirements has been determined by decreasing the stopband edge ω_s until the filter just meets the ripple requirements.

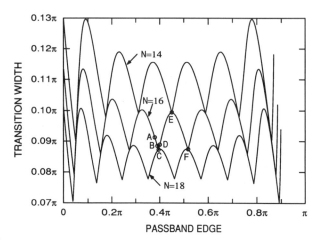

FIGURE 4-26 Transition bandwidth ($\omega_s - \omega_p$) as a function of the passband edge ω_p for $\delta_p = \delta_s = 0.1$ for filters with $N = 14$ ($M = 7$), $N = 16$ ($M = 8$), and $N = 18$ ($M = 9$).

extrema with equal amplitude. This is one more than that required by the characterization theorem. Furthermore, it follows from this theorem that these extraripple solutions are also the optimum solutions for $\overline{M} = M + 1$, or equivalently for $N = 18$. This is because the number of extrema is $\overline{M} + 2$ for these filters. The explanation to this is that the first and last impulse-response coefficients $h[n]$ of the filter with higher order become exactly zero when the filter with lower order has the extraripple solution. This gives $a[\overline{M}] = a[9] = 0$ for the filter with $N = 18$ so that the responses of the two filters coincide.

When ω_p is made smaller, the resulting filter has $M + 2$ equal amplitude extrema, as well as one smaller amplitude extremum at $\omega = 0$ (Filter B). When ω_p is further decreased, the extra extremum disappears (Filter A). On the other hand, if ω_p is made larger, the resulting filter (Filter D) has one smaller ripple at $\omega = \pi$. Also, this ripple disappears when ω_p is further increased. Filter E in Figure 4-27 corresponds to the case where the filter with $N = 14$ has the same solution (extraripple solution for the filter with $N = 14$).

Hence for Type I filters, there are three kinds of optimum solutions: solutions having $M + 2$ equal amplitude extrema, special solutions having $M + 3$ equal amplitude extrema, and solutions having, in addition to $M + 2$ equal amplitude extrema, one smaller extremum. For Type II filters, the properties are quite similar [RA73b]. The basic difference is that the Type II filters have an odd order ($N = 2M + 1$) and they have a fixed zero at $z = -1$ ($\omega = \pi$).

4-8-3 Some Useful Properties of Optimum Type I Filters

Consider a Type I transfer function of the form

$$H(z) = \sum_{n=0}^{2M} h[n]z^{-n}, \quad h[2M - n] = h[n]. \tag{4.97}$$

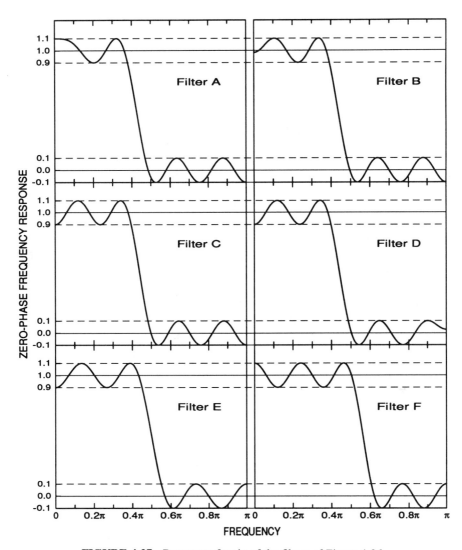

FIGURE 4-27   Responses for six of the filters of Figure 4-26.

The corresponding zero-phase frequency response is given by

$$H(\omega) = h[M] + \sum_{n=1}^{M} 2h[M - n] \cos n\omega. \qquad (4.98)$$

On the basis of $H(z)$, we can construct three Type I filters having the following transfer functions:

$$G(z) = \begin{cases} z^{-M} - H(z) & \text{for Case A} \\ (-1)^{M} H(-z) & \text{for Case B} \\ z^{-M} - (-1)^{M} H(-z) & \text{for Case C.} \end{cases} \qquad (4.99)$$

The zero-phase frequency responses of these three filters can be written as

$$G(\omega) = \begin{cases} 1 - H(\omega) & \text{for Case A} \\ H(\pi - \omega) & \text{for Case B} \\ 1 - H(\pi - \omega) & \text{for Case C.} \end{cases} \tag{4.100}$$

In Case A, the impulse-response coefficients of $G(z)$ are related to the coefficients of $H(z)$ via $g[M] = 1 - h[M]$ and $g[n] = -h[n]$ for $n = 0, 1, \ldots, M - 1$. By substituting these values into

$$G(\omega) = g[M] + \sum_{n=1}^{M} 2g[M - n] \cos n\omega, \tag{4.101}$$

we end up with $G(\omega)$ shown in Eq. (4.100). In Case B, the coefficients $g[n]$ are related to the $h[n]$'s via $g[M - n] = h[M - n]$ for n even and $g[M - n] = -h[M - n]$ for n odd. Using the facts that for n even $\cos n\omega = \cos n(\pi - \omega)$ and for n odd $-\cos n\omega = \cos n(\pi - \omega)$, we can write $G(\omega)$ in the above form. The fact that $G(\omega)$ is expressible in Case C as shown in Eq. (4.100) follows directly from the properties of the Case A and Case B filters.

In Case A, the filter pair $H(z)$ and $G(z)$ is called a *complementary filter pair* since the sum of their zero-phase frequency responses is unity; that is,

$$H(\omega) + G(\omega) = 1. \tag{4.102}$$

This means that if $H(z)$ is a lowpass design with $H(\omega)$ oscillating within the limits $1 \pm \delta_p$ on $[0, \omega_p]$ and within the limits $\pm \delta_s$ on $[\omega_s, \pi]$, then $G(z)$ is a highpass filter with $G(\omega)$ oscillating within $\pm \delta_p$ on $[0, \omega_p]$ and within $1 \pm \delta_s$ on $[\omega_s, \pi]$ (see Figures 4-28(a) and 4-28(b)). An implementation of $G(z)$ is shown in Figure 4-29. The delay term z^{-M} can be shared with $H(z)$ in this implementation. Hence at the expense of one additional adder, a complementary filter pair can be implemented.

If $H(\omega)$ is as shown in Figure 4-28(a), then the Case B filter is a highpass design with $G(\omega)$ oscillating within $\pm \delta_s$ on $[0, \pi - \omega_s]$ and within $1 \pm \delta_p$ on $[\pi - \omega_p, \pi]$ (see Figure 4-28(c)). $G(\omega)$ for the Case C filter, in turn, varies within $1 \pm \delta_s$ on $[0, \pi - \omega_s]$ and within $\pm \delta_p$ on $[\pi - \omega_p, \pi]$ (see Figure 4-28(d)). An implementation of the Case C filter is depicted in Figure 4-29. This implementation is very important in many cases as it allows us to implement a wideband filter $G(z)$ using a delay term and a transfer function that is obtained from a narrowband filter $H(z)$ by simply changing the sign of every second coefficient value. This is because there are computationally efficient implementations for narrowband filters, as will be seen in Section 4-11.

4-8-4 The Use of the MPR Algorithm

The MPR algorithm is very flexible for solving many kinds of approximation problems in the minimax sense. The user specifies first the filter type and the order of

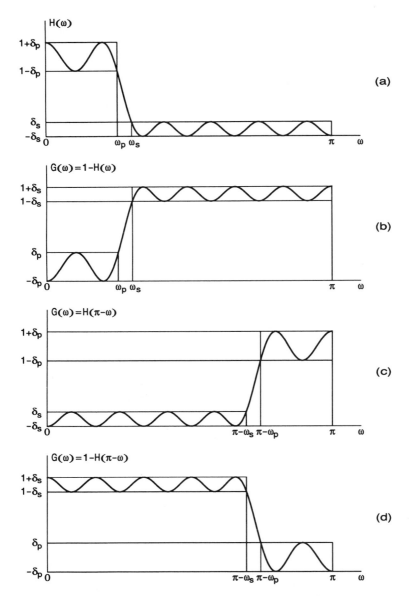

FIGURE 4-28 Responses for the prototype filter and three filters developed on the basis of this prototype filter. (a) Prototype lowpass filter. (b) Case A. (c) Case B. (d) Case C.

the filter. The different filter types are conventional frequency-selective filters having multiple passbands and stopbands, differentiators, and Hilbert transformers. The conventional filters are Type I and Type II filters, whereas in the last two cases, the filters are Type III and Type IV filters considered in greater detail in Chapter 13. A FORTRAN program implementing the MPR algorithm can be found in McClellan et al. [MC73b, MC79]. In this program, instead of the filter order N, the length of the impulse response, $N + 1$, is used. After giving the filter type

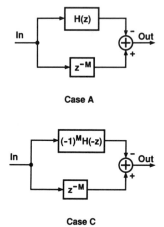

FIGURE 4-29   Implementations for Case A and Case C filters.

and the length, the program automatically checks whether the filter is a Type I, II, III, or IV design. It also changes the desired function and the weighting function correspondingly and concentrates on determining the common adjustable filter part.

The user also supplies an integer and the program selects the number of grid points to be $M + 1$ times this integer. A good selection for this integer is 16. When designing multiband filters, the user specifies the edges of the bands as well as the desired value and the weighting for each band. If other than piecewise-constant desired functions and weighting functions are desired to be used, the program has subroutines EFF and WATE, which can be used in these cases. It should be noted that the basic frequency variable f of the program is related to the angular frequency ω via

$$f = \omega/(2\pi). \tag{4.103}$$

For instance, if a desired edge angle is 0.4π, then the edge for the program is 0.2. Some examples are now given to illustrate the flexibility of the MPR algorithm.

Example 4.6. It is desired to design a lowpass filter with passband and stopband regions $[0, 0.3\pi]$ and $[0.4\pi, \pi]$, respectively. The passband ripple δ_p is restricted to be at most 0.002 on $[0, 0.15\pi]$ and at most 0.01 in the remaining region $[0.15\pi, 0.3\pi]$. The stopband ripple δ_s is at most 0.0001 (80-dB attenuation) on $[0.4\pi, 0.6\pi]$ and at most 0.001 (60-dB attenuation) on $[0.6\pi, \pi]$. Furthermore, the overall filter is implemented in the form

$$H_{\text{ove}}(z) = H_{\text{fix}}(z)H(z), \tag{4.104}$$

where the fixed term $H_{\text{fix}}(z)$ has zero pairs on the unit circle at the angular frequencies $\omega = \pm0.4\pi, \pm0.45\pi, \pm0.5\pi, \pm0.55\pi, \pm0.6\pi, \pm0.65\pi$. The desired overall filter can be obtained by designing $H(z)$ using the following desired and weighting functions:[11]

$$D(\omega) = \begin{cases} 1/|H_{\text{fix}}(\omega)|, & \omega \in [0, 0.3\pi] \\ 0, & \omega \in [0.4\pi, \pi], \end{cases} \quad (4.105a)$$

$$W(\omega) = \begin{cases} 5|H_{\text{fix}}(\omega)|, & \omega \in [0, 0.15\pi] \\ |H_{\text{fix}}(\omega)|, & \omega \in [0.15\pi, 0.3\pi] \\ 100|H_{\text{fix}}(\omega)|, & \omega \in [0.4\pi, 0.6\pi] \\ 10|H_{\text{fix}}(\omega)|, & \omega \in [0.6\pi, \pi]. \end{cases} \quad (4.105b)$$

The given criteria are met when the peak absolute value of the corresponding error function becomes smaller than or equal to 0.01. The minimum order of $H(z)$ to meet the criteria is 54. The amplitude response of the resulting overall design is depicted in Figure 4-30.

Example 4.7. This example illustrates the use of the MPR algorithm for designing FIR filters with a very flat passband and equiripple stopband. These filters have been proposed by Vaidyanathan [VA85] and their transfer function is of the form

$$H(z) = z^{-M} - (-1)^L[(1 - z^{-1})/2]^{2L}\overline{H}(z), \quad (4.106)$$

[11]In general, if the overall filter is of the form $H_{\text{ove}}(z) = H_{\text{fix}}(z)H(z)$ and the desired and weighting functions for $H_{\text{ove}}(z)$ are $D(\omega)$ and $W(\omega)$, respectively, then the given criteria are met by designing $H(z)$ using the desired function $D(\omega)/H_{\text{fix}}(\omega)$ and the weighting function $H_{\text{fix}}(\omega)W(\omega)$. This follows from the fact that the weighted error function can be written as $E(\omega) = W(\omega)[H_{\text{ove}}(\omega) - D(\omega)] = W(\omega)H_{\text{fix}}(\omega)[H(\omega) - D(\omega)/H_{\text{fix}}(\omega)]$. If $H_{\text{fix}}(\omega)$ is zero at some points in the approximation interval, then the new weighting function becomes zero at these points. This problem can be avoided by disregarding these grid points when using the MPR algorithm. The absolute values of $H_{\text{fix}}(\omega)$ are used in Eq. (4.105) to make the weighting function positive.

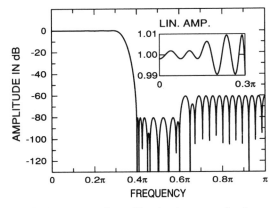

FIGURE 4-30 Amplitude response for a filter having some fixed zero pairs on the unit circle and unequal passband and stopband weightings.

where $M = L + K$ with K being half the order of $\overline{H}(z)$. Here, $\overline{H}(z)$ of even order $2K$ is designed using the following desired and weighting functions:

$$D(\omega) = \begin{cases} 0, & \omega \in [0, \omega_p] \\ 1/[\sin (\omega/2)]^{2L}, & \omega \in [\omega_s, \pi], \end{cases} \tag{4.107a}$$

$$W(\omega) = \begin{cases} [\sin (\omega_p/2)]^{2L}, & \omega \in [0, \omega_p] \\ (\delta_p/\delta_s) [\sin (\omega/2)]^{2L}, & \omega \in [\omega_s, \pi], \end{cases} \tag{4.107b}$$

where $[\sin (\omega/2)]^{2L}$ is the zero-phase frequency response of $(-1)^L [(1 - z^{-1})/2]^{2L}$. Figure 4-31 gives the resulting overall response

$$H(\omega) = 1 - [\sin (\omega/2)]^{2L} \overline{H}(\omega) \tag{4.108}$$

for the case $\omega_p = 0.6\pi$, $\omega_s = 0.7\pi$, $\delta_p \leq 0.016$, $\delta_p/\delta_s = 5$, and $2L = 16$ [VA85]. The minimum even order of $\overline{H}(z)$ to meet the criteria is $2K = 46$. In the above, the desired and weighting functions have been selected such that $[\sin (\omega/2)]^{2L} \overline{H}(\omega)$ achieves the peak absolute value of the corresponding error function ($\epsilon = 0.0144$) at $\omega = \omega_p$ and oscillates within the limits $1 \pm \epsilon/5$ on $[\omega_s, \pi]$. Correspondingly, $H(\omega)$ achieves the value $1 - \epsilon$ at $\omega = \omega_p$ and oscillates within the limits $\pm\epsilon/5$ in the stopband. On $[0, \omega_p]$, $[\sin (\omega_p/2)]^{2L} \overline{H}(\omega)$ varies within the limits $\pm\epsilon$. Because of the term $[\sin (\omega/2)]^{2L}$, $[\sin (\omega/2)]^{2L} \overline{H}(\omega)$ approximates very accurately zero and $H(\omega)$ approximates very accurately unity in the beginning of the interval $[0, \omega_p]$. Note that the fixed term has $2L$ zeros at $z = 1$ (at $\omega = 0$).

Example 4.8. Let the bandpass filter specifications be $\omega_{s1} = 0.2\pi$, $\omega_{p1} = 0.25\pi$, $\omega_{p2} = 0.6\pi$, $\omega_{s2} = 0.7\pi$, $\delta_{s1} = 0.001$, and $\delta_p = \delta_{s2} = 0.01$. In this case, the desired function is unity in the passband and zero in the stopbands, whereas the

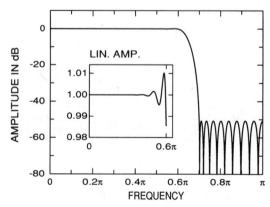

FIGURE 4-31   Amplitude response for a filter with a very flat passband and equiripple stopband.

weighting function is 10 in the first stopband and 1 in both the passband and the second stopband. The minimum order to meet the above specifications is 102. Figure 4-32(a) gives the response of the optimum filter. This response is optimal according to the characterization theorem even though it has an unacceptable transition band peak of 15 dB. This is possible because the approximation is restricted to the passband and stopband regions only and the transition bands are considered as don't-care bands. For designs with a single transition band there are no unacceptable transition band ripples. However, for filters having more than one transition band, this phenomenon of large transition band peaks occurs when the widths of the transition bands are different [RA74]; the larger the difference, the greater the problem.

The transition band peak can easily be attenuated by including the transition bands in the overall approximation interval and requiring that the response stays within the limits $-\delta_{s1}$ and $1 + \delta_p$ in the first transition band and within the limits $-\delta_{s2}$ and $1 + \delta_p$ in the second transition band. This can be done by selecting the desired function to be $\frac{1}{2}(1 + \delta_p - \delta_{s1})$ and $\frac{1}{2}(1 + \delta_p - \delta_{s2})$ in the first and second

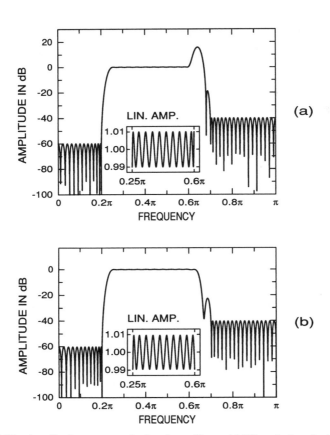

FIGURE 4-32 Amplitude responses for bandpass filters. (a) Filter designed without transition band constraints. (b) Filter designed with transition band constraints.

transition bands, respectively. If the weighting in the passband is unity, then the weighting functions in the transition bands are selected to be $\delta_p/[\frac{1}{2}(1 + \delta_p + \delta_{s1})]$ and $\delta_p/[\frac{1}{2}(1 + \delta_p + \delta_{s2})]$, respectively. These selections guarantee that if the passband ripple of the resulting filter is less than or equal to the specified δ_p, then the response stays within the desired limits in the transition bands.[12] When including the transition bands in the approximation problem, the filter order has to be increased only by one (to 103) to meet the resulting specifications. Figure 4-32(b) gives the response of this filter. For other techniques for attenuating undesired transition band ripples, see Rabiner et al. [RA74].

4-9 DESIGN OF MINIMUM-PHASE FIR FILTERS

The attractive property of Type I and Type II linear-phase FIR filters is that their delay is a constant and thus they cause no phase distortion to the signal. The delay is equal to half the filter order. This means that the delay becomes very long for high-order filters required in cases demanding a narrow transition band. In some applications, such a long delay is not tolerable. In those cases, a smaller group delay can be achieved in the passband region by using minimum-phase FIR filters. There exist also applications where linear phase is not required and the symmetry in the coefficients of linear-phase FIR filters cannot be exploited. In those cases, nonlinear-phase filters meet the same amplitude criteria with a reduced number of multipliers and delay elements. If the passband of the filter is very wide, then a saving by almost a factor of 2 can be achieved in the filter order [GO81; LE75]. This section outlines the design of nonlinear-phase filters based on the design scheme of Herrmann and Schüssler [HE70], which can be used for synthesizing filters with unweighted stopband response. For more general techniques, see Boite and Leich [BO81], Kamp and Wellekens [KA83b], and Grenez [GR83].

Consider a nonlinear-phase FIR filter with transfer function

$$H(z) = \sum_{n=0}^{M} h[n]z^{-n}. \tag{4.109}$$

The zeros of the transfer function

$$\hat{H}(z) = z^{-M}H(z^{-1}) = \sum_{n=0}^{M} h[M - n]z^{-n} \tag{4.110}$$

are reciprocal to those of $H(z)$. This implies that the function

$$G(z) = H(z)\hat{H}(z) = z^{-M}H(z)H(z^{-1}) \tag{4.111}$$

---

[12]The lower and upper edges for the first transition band are selected as $\omega_{s1} + \alpha$ and $\omega_{p1} - \alpha$, respectively, where α is a small number. Similarly, for the second transition band, the edges are $\omega_{p2} + \alpha$ and $\omega_{s2} - \alpha$. These selections prevent the desired function from becoming discontinuous at the edges.

is the transfer function of a Type I linear-phase filter of order $2M$. Since $G(z)$ must be factorizable into the terms $H(z)$ and $\hat{H}(z)$, its zeros on the unit circle have to be double. From the above equation, it follows also that the magnitude-squared function of $H(z)$ can be expressed as

$$|H(e^{j\omega})|^2 = G(\omega). \qquad (4.112)$$

Since $G(z)$ possesses double zeros on the unit circle, $G(\omega)$ has double zeros on $[0, \pi]$, making it nonnegative on $[0, \pi]$. These facts show that the design of a nonlinear-phase FIR filter of order M can be accomplished in terms of a Type I linear-phase filter of order $2M$ having double zeros on the unit circle.

Based on this, Herrmann and Schüssler [HE70] have proposed the following simple design procedure:

1. Design a Type I linear-phase FIR filter transfer function $\overline{G}(z)$ of order $2M$ using the MPR algorithm such that $\overline{G}(\omega)$ oscillates within the limits $1 \pm \overline{\delta}_p$ in the passband $[0, \omega_p]$ and within the limits $\pm\overline{\delta}_s$ in the stopband $[\omega_s, \pi]$ (see Figure 4-33(a)). This $\overline{G}(z)$ has single zeros on the unit circle.

2. Form $G(z) = \overline{\delta}_s z^{-M} + \overline{G}(z)$. The resulting $G(\omega) = \overline{\delta}_s + \overline{G}(\omega)$ is nonnegative on $[\omega_s, \pi]$, oscillating within zero and $2\overline{\delta}_s$ (see Figure 4-33(b)). On $[0, \omega_p]$, $G(\omega)$ oscillates within the limits $1 + \overline{\delta}_s \pm \overline{\delta}_p$. $G(\omega)$ has double zeros at those frequency points where $\overline{G}(\omega)$ takes the stopband minimum value of $-\overline{\delta}_s$. Correspondingly, $G(z)$ has double zeros on the unit circle at those frequencies.

3. Perform the factorization of $G(z) = H(z)z^{-M}H(z^{-1})$ such that $H(z)$ contains the zeros inside the unit circle and one each of the double zeros on the unit circle. Scale $H(z)$ such that the passband average of the resulting filter $AH(z)$ is equal to unity (see Figure 4-33(c)).

The desired scaling constant at Step 3 is

$$A = \frac{2}{\sqrt{1 + \overline{\delta}_p + \overline{\delta}_s} + \sqrt{1 - \overline{\delta}_p + \overline{\delta}_s}}. \qquad (4.113)$$

If it is required that the magnitude response of the scaled filter $AH(z)$ approximates unity in the passband with tolerance δ_p and zero in the stopband with tolerance δ_s, then the passband and stopband ripples of the linear-phase filter at Step 1 must satisfy

$$\overline{\delta}_p \leq \frac{2\delta_p}{1 + (\delta_p)^2 - (\delta_s)^2/2}, \qquad \overline{\delta}_s \leq \frac{(\delta_s)^2/2}{1 + (\delta_p)^2 - (\delta_s)^2/2}. \qquad (4.114)$$

The most difficult part in the above procedure is the factorization of $G(z)$ into the terms $H(z)$ and $z^{-M}H(z^{-1})$. The direct approach is simply to pick up the zeros of $G(z)$. However, if the order of $G(z)$ is high, conventional root-finding proce-

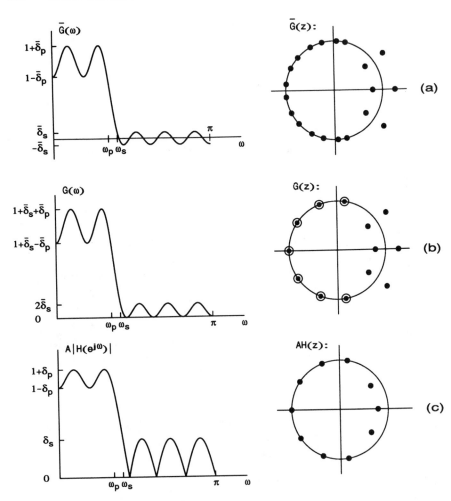

FIGURE 4-33 Steps for designing an equiripple minimum-phase FIR filter of order 10 with the aid of a linear-phase FIR filter of order 20.

dures cannot be used for locating the zeros. Another approach is to perform the factorization without finding the roots of $G(z)$. Such techniques have been proposed independently by Boite and Leich [BO81] and Mian and Naider [MI82a] (see also Boite and Leich [BO84]). For a review of different techniques for performing the factorization, see Schüssler and Steffen [SC88].

The filter obtained by selecting the zeros to lie on or inside the unit circle is called a *minimum-phase* filter. If the zeros outside the unit circle are selected, then the resulting filter is called a *maximum-phase* design.

Example 4.9. Let the specifications be $\omega_p = 0.5\pi$, $\omega_s = 0.6\pi$, $\delta_p = 0.01$, and $\delta_s = 0.00316$ (50-dB attenuation). Using Eq. (4.114), the ripples of $\overline{G}(z)$ at Step 1 become $\bar{\delta}_p \approx 0.02$ and $\bar{\delta}_s \approx 5 \times 10^{-6}$. The minimum even order to meet these criteria is 74 so that the order of the corresponding minimum-phase filter is 37. The common amplitude response of the minimum-phase and maximum-phase filters as well as their group delay responses, zero locations, and impulse responses are given in Figure 4-34. The minimum order of a linear-phase filter to meet the same criteria is 46 (see Figure 4-8) so that the saving in the filter order provided by the minimum-phase filter is 20%. For the minimum- and maximum-phase fil-

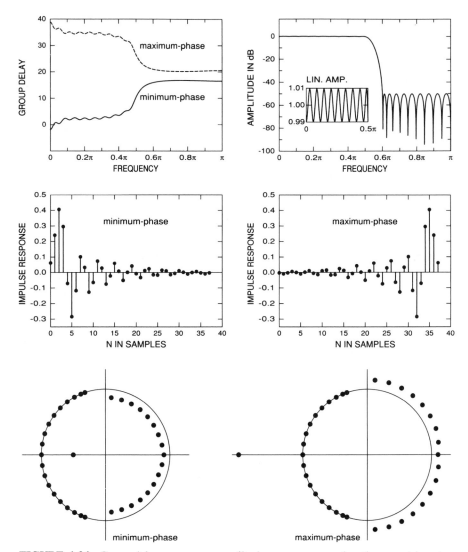

FIGURE 4-34 Group delay responses, amplitude response, zero locations, and impulse responses for minimum- and maximum-phase FIR filters having the same amplitude response.

ters, the phase and the group delay responses have the smallest and largest values, respectively, among the filters having the same amplitude response. It is also interesting to observe from Figure 4-34 that most of the energy of the impulse response of the minimum-phase (maximum-phase) term is concentrated in the beginning (end) of the response.

4-10 DESIGN OF FIR FILTERS WITH CONSTRAINTS IN THE TIME OR FREQUENCY DOMAIN

The previous sections concentrated only on designing FIR filters to meet the given amplitude criteria in some sense. However, there exist applications where there are constraints in the time domain or in the frequency domain. For example, in some applications the transient part of the step response must be constrained to vary within given limits [LI83b; RA72a, RA72b]. Another example is the design of Nyquist filters or Lth band filters with every Lth impulse-response value being zero except for the central value [LI85; MI82b; SA87b, SA88c; VA87]. Furthermore, in some cases, there are flatness constraints in the passband response of the filter [KA83a; ST79].

This section considers techniques for solving the above-mentioned approximation problems. In some cases, the desired solution can be obtained by properly modifying the design methods proposed in the previous sections. In the remaining cases, new techniques are required. Perhaps the most flexible design method for finding the optimum solution to various constrained approximation problems is linear programming [HE71a; KA83a; LI83b; RA72a, RA72b; ST79]. The advantage of this technique is that the convergence to the optimum solution is guaranteed. With linear programming, it is also possible to find the optimum solution to the unconstrained minimax approximation problems considered previously. The disadvantage, however, is that the required computation to arrive at the optimum solution is rather large. Therefore it is preferred to use linear programming only in those cases that cannot be handled with other faster design techniques.

4-10-1 Linear Programming Approach for FIR Filter Design

Linear programming is a very flexible approach for solving many constrained approximation problems in the minimax sense. Mathematically, the linear programming problem [DA63; HA63; LA73] can be stated in the form of the following primal problem: find the unknowns x_k, $k = 1, 2, \ldots, N$, subject to the constraints

$$x_k \geq 0, \quad k = 1, 2, \ldots, N, \tag{4.115a}$$

$$\sum_{k=1}^{N} \gamma_{lk} x_k = \beta_l, \quad l = 1, 2, \ldots, M \ (M < N), \tag{4.115b}$$

such that

$$\sigma = \sum_{k=1}^{N} \alpha_k x_k \qquad (4.115c)$$

is minimized. In this problem, γ_{lk}, α_k, and β_l are constants. The above problem is mathematically equivalent to the following dual problem: find the unknowns y_l, $l = 1, 2, \ldots, M$, subject to the constraints

$$\sum_{l=1}^{M} \gamma_{lk} y_l \le \alpha_k, \qquad k = 1, 2, \ldots, N, \qquad (4.116a)$$

such that

$$\rho = \sum_{l=1}^{M} \beta_l y_l \qquad (4.116b)$$

is maximized.

For digital filter design problems, the dual problem is the most natural form. There exist several well-defined procedures [DA63; HA63; LA73] for arriving at the desired solution within $M + N$ iterations. Lim has introduced an efficient special purpose algorithm for designing FIR filters [LI83b]. This is faster than general purpose algorithms.

Linear programming can be applied in a straightforward manner to those problems where the approximating function is linear; that is, it can be expressed in the form

$$H(\omega) = \sum_{n=0}^{R} b[n] \Phi(\omega, n), \qquad (4.117)$$

where the $b[n]$'s are unknowns. According to the discussion of Section 4-3-3, the zero-phase frequency response of a linear-phase FIR filter can be expressed in all four cases in the above form (see Eq. (4.32)). Also, in many other cases, the approximating function can be written in this form. For instance, in the conventional frequency-sampling methods, the filter response is expressible in the above form [RA72a, RA72b].

A general constrained frequency-domain approximation problem, which can be solved using linear programming, can be stated in the following form: find the unknowns $b[n]$ to minimize

$$\delta_1 = \max_{\omega \in X_1} |E(\omega)|, \qquad (4.118a)$$

where

$$E(\omega) = W(\omega)[H(\omega) - D(\omega)] \qquad (4.118b)$$

subject to

$$\max_{\omega \in X_2} |E(\omega)| \le \delta_2. \tag{4.118c}$$

Here, X_1 contains a part of the passband and stopband regions and X_2 contains the remaining part. For instance, by selecting X_1 and X_2 to be the stopband and passband regions of the filter, respectively, the stopband variation can be minimized for the given maximum allowable passband variation. Problems of this kind cannot be solved directly using the MPR algorithm considered in Section 4-8.

By sampling $W(\omega)$ and $D(\omega)$ along a dense grid of frequencies $\omega_1^{(1)}, \omega_2^{(1)}, \ldots,$ $\omega_{K_1}^{(1)}$ on X_1, and along a grid of frequencies $\omega_1^{(2)}, \omega_2^{(2)}, \ldots, \omega_{K_2}^{(2)}$ on X_2, the problem can be stated in the form of the dual problem as follows: find $b[0], b[1], \ldots,$ $b[R]$, and δ_1 subject to the constraints

$$\sum_{n=0}^{R} b[n]\Phi(\omega_k^{(1)}, n) - \delta_1/W(\omega_k^{(1)}) \le D(\omega_k^{(1)}), \qquad k = 1, 2, \cdots, K_1,$$

$$\tag{4.119a}$$

$$-\sum_{n=0}^{R} b[n]\Phi(\omega_k^{(1)}, n) - \delta_1/W(\omega_k^{(1)}) \le -D(\omega_k^{(1)}), \qquad k = 1, 2, \cdots, K_1,$$

$$\tag{4.119b}$$

$$\sum_{n=0}^{R} b[n]\Phi(\omega_k^{(2)}, n) \le D(\omega_k^{(2)}) + \delta_2/W(\omega_k^{(2)}), \qquad k = 1, 2, \cdots, K_2,$$

$$\tag{4.119c}$$

$$-\sum_{n=0}^{R} b[n]\Phi(\omega_k^{(2)}, n) \le -D(\omega_k^{(2)}) + \delta_2/W(\omega_k^{(2)}), \qquad k = 1, 2, \cdots, K_2,$$

$$\tag{4.119d}$$

such that

$$\rho = -\delta_1 \tag{4.119e}$$

is maximized.

Note that in the dual problem the constraints are formed in such a way that a linear combination of the unknowns is less than or equal to a constant. In the above problem, δ_2 is a constant and δ_1 is an unknown. This explains the difference between Eqs. (4.119a) and (4.119b) and Eqs. (4.119c) and (4.119d). The above equations have been constructed such that, after finding the optimum solution, $-\delta_1 \le E(\omega_k^{(1)}) \le \delta_1$ and $-\delta_2 \le E(\omega_k^{(2)}) \le \delta_2$ at the selected grid points. Note also that in the dual problem a linear combination of unknowns is maximized and maximizing $-\delta_1$ implies minimizing δ_1.

It is easy to include in the dual problem various constraints that are expressible in the form of Eq. (4.116a). For instance, it is straightforward to add constraints of the form

$$\frac{d^l H(\omega)}{d^l \omega}\bigg|_{\omega = \omega_k} = \sum_{n=0}^{R} b[n] \frac{d^l \Phi(\omega, n)}{d^l \omega}\bigg|_{\omega = \omega_k} \leq 0 \qquad (4.120a)$$

or

$$\frac{d^l H(\omega)}{d^l \omega}\bigg|_{\omega = \omega_k} = \sum_{n=0}^{R} b[n] \frac{d^l \Phi(\omega, n)}{d^l \omega}\bigg|_{\omega = \omega_k} \geq 0, \qquad (4.120b)$$

where l is an integer and ω_k is a grid point. Here, the constraint expressed by Eq. (4.120a) is directly in the desired form. The constraint of Eq. (4.120b) can be written in the form of Eq. (4.116a) by multiplying the left-hand side by -1 and replacing $\geq$ by $\leq$. By adding a constraint of the form of Eq. (4.120a) with $l = 1$ at each grid point in the passband region, the passband response of the filter can be forced to be monotonically decreasing. Steiglitz has presented a FORTRAN code for designing filters of this kind [ST79]. Furthermore, the first L derivatives of $H(\omega)$ can be forced to be zero at $\omega = \omega_k$ by simultaneously using the constraints of Eqs. (4.120a) and (4.120b) for $l = 1, 2, \ldots, L$.

In addition, if it is desired that $H(\omega)$ achieve exactly the value A at $\omega = \omega_k$, this condition can be included by using the following two constraints:

$$\sum_{n=0}^{R} b[n]\Phi(\omega_k, n) \leq A, \qquad -\sum_{n=0}^{R} b[n]\Phi(\omega_k, n) \leq -A. \qquad (4.121)$$

Example 4.10. Consider the design of a Type I linear-phase filter of order 70 having $[0, 0.3\pi]$ and $[0.4\pi, \pi]$ as the passband and stopband regions, respectively. To illustrate the flexibility of linear programming, several constraints are included. First, the filter has fixed zero pairs at the angular frequencies $\pm 0.4\pi$, $\pm 0.45\pi$, $\pm 0.5\pi$, $\pm 0.55\pi$, $\pm 0.6\pi$, and $\pm 0.65\pi$, and $H(\omega)$ achieves the value of unity at $\omega = 0$ with its first four derivatives being zero at this point. Second, the maximum deviation from unity on $[0, 0.15\pi]$ is 0.002 and the maximum deviation from zero on $[0.4\pi, 0.6\pi]$ is 0.0001, whereas the response is desired to be optimized in the remaining regions with weighting of unity on $[0.15\pi, 0.3\pi]$ and 10 on $[0.6\pi, \pi]$. The last part of this problem can be expressed in the form of Eq. (4.119) using $X_1 = [0.15\pi, 0.3\pi] \cup [0.6\pi, \pi]$ and $X_2 = [0, 0.15\pi] \cup [0.4\pi, 0.6\pi]$. $D(\omega)$ is 1 on $[0, 0.3\pi]$ and 0 on $[0.4\pi, \pi]$. $W(\omega)$ is 1 on $[0, 0.3\pi]$, 20 on $[0.4\pi, 0.6\pi]$, and 10 on $[0.6\pi, \pi]$, whereas $\delta_2 = 0.002$. To include the first part, Eq. (4.121) is used with $A = 0$ at the frequency points where the filter has fixed zeros and with $A = 1$ at the zero frequency. Equations (4.120a) and (4.120b) are used with $l = 1, 2, 3, 4$ at the zero frequency. The optimized filter response is shown in Figure 4-35. The resulting ripple values on $[0.15\pi, 0.3\pi]$ and $[0.6\pi, \pi]$ are 0.00637 and 0.000637, respectively.

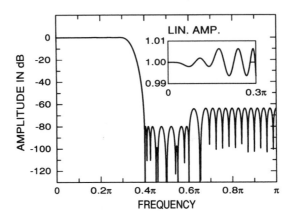

FIGURE 4-35 Amplitude response for an optimized filter of order 70 having several constraints in the frequency domain.

It is also straightforward to include time-domain constraints in the approximation problem. For instance, some of the unknowns $b[n]$, $n \in S$ can be fixed and the remaining ones can be optimized. In this case, the desired solution can be found by using the following approximating function,

$$H(\omega) = \sum_{\substack{n=0 \\ n \notin S}}^{R} b[n]\Phi(\omega, n), \tag{4.122}$$

and by including the effect of the fixed terms in the desired function by changing it to be

$$\overline{D}(\omega) = D(\omega) - \sum_{n \in S} b[n]\Phi(\omega, n). \tag{4.123}$$

Example 4.11. This example shows how linear programming can be used for designing filters with constraints on the step response, which is related to the impulse-response coefficients $h[n]$ through

$$g[n] = \sum_{m=0}^{n} h[m]. \tag{4.124}$$

As an example, Figures 4-36(a) and 4-36(b) give the amplitude and step responses for a filter of order 46 optimized without any constraints in the time domain. For this filter, the passband and stopband ripples are related via $\delta_p = \sqrt{10}\delta_s$, $\omega_p = 0.5\pi$, and $\omega_s = 0.6\pi$. The maximum undershoot of the step response occurring at $n = 21$ is -0.0921 and $g[n] = 0.9903$ for $n \geq 46$. It is desired that $g[n] = 1$ for $n \geq 46$ and

$$-\delta_{\text{step}} \leq g[n] \leq \delta_{\text{step}} \quad \text{for } 0 \leq n \leq K, \tag{4.125}$$

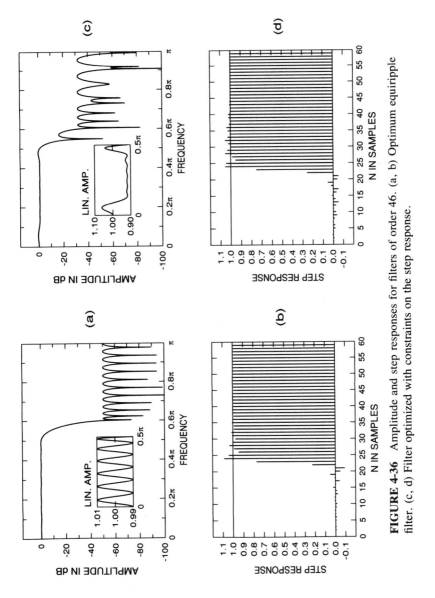

FIGURE 4-36 Amplitude and step responses for filters of order 46. (a, b) Optimum equiripple filter. (c, d) Filter optimized with constraints on the step response.

223

where $K = 21$ and $\delta_{step} = 0.05$. The first condition can be satisfied by requiring that $H(0) = 1$. The second constraint is linear in the $h[n]$'s and can thus easily be included in the dual problem. Because of this condition, it is advantageous to express $H(\omega)$ directly in terms of the $h[n]$'s as ($M = N/2$ for Type I designs)

$$H(\omega) = h[M] + \sum_{n=1}^{M} h[M - n](2 \cos n\omega), \qquad (4.126)$$

so that $\Phi(\omega, M) = 1$ and $\Phi(\omega, M - n) = 2 \cos n\omega$ for $n > 0$. The amplitude and step responses for the filter optimized with the above constraints are shown in Figures 4-36(c) and 4-36(d), respectively. It is seen that these time-domain conditions increase significantly the ripples of the amplitude response. For other examples for designing filters having constraints on the step response, see Lim [LI83b].

4-10-2 Design of Lth Band (Nyquist) Filters

Consider again a Type I linear-phase FIR filter with transfer function

$$H(z) = \sum_{n=0}^{2M} h[n]z^{-n}, \qquad h[2M - n] = h[n]. \qquad (4.127)$$

This filter is defined to be an Lth band filter if its coefficients satisfy[13] (see Figure 4-37)

$$h[M] = 1/L, \qquad (4.128a)$$

$$h[M + rL] = 0 \quad \text{for } r = \pm 1, \pm 2, \cdots, \pm \lfloor M/L \rfloor. \qquad (4.128b)$$

These filters, also called Nyquist filters, play an important role in designing digital transmission systems [LI85; MI82b; SA87b, SA88c; VA87] and filter banks [VA89]. They can also be used as efficient decimators and interpolators since every Lth impulse-response coefficient is zero except for the central coefficient. An important subclass of these filters are half-band filters, which are considered in greater detail in Section 4-10-3.

It can be shown [MI82b] that the time-domain conditions of Eq. (4.128) imply some limitations on the frequency response of the filter. First, the passband edge (in the lowpass case) is restricted to be less than π/L and the stopband edge to be larger than π/L. Usually, the edges are given in terms of an excess bandwidth factor ρ as follows (see Figure 4-37):

$$\omega_p = (1 - \rho)\pi/L, \qquad \omega_s = (1 + \rho)\pi/L. \qquad (4.129)$$

Second, if the maximum deviation of $H(\omega)$ from zero on $[\omega_s, \pi]$ is δ_s, then the maximum deviation of $H(\omega)$ from unity on $[0, \omega_p]$ is in the worst case $\delta_p =$

---

[13] $\lfloor x \rfloor$ stands for integer part of x.

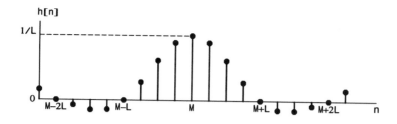

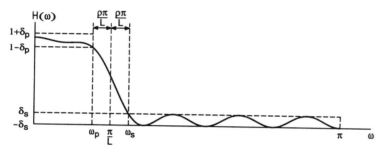

FIGURE 4-37 Typical impulse response and zero-phase frequency response for an FIR Nyquist filter.

$(L - 1)\delta_s$. Usually, δ_p is much smaller than this upper limit. Since δ_p is guaranteed to be relatively small for a small value of δ_s, the filter synthesis can concentrate on shaping the stopband response.

The stopband response can be optimized either in the minimax sense or in the least-mean-square sense. In the case of the minimax criterion, the problem is to find the coefficients of $H(z)$ such that the time-domain conditions of Eq. (4.128) are satisfied and

$$\delta_s = \max_{\omega \in [\omega_s, \pi]} |W(\omega)H(\omega)| \qquad (4.130)$$

is minimized, where $W(\omega)$ is a positive weighting function. In the case of the least-mean-square criterion, the quantity to be minimized is

$$E_2 = \int_{\omega_s}^{\pi} [W(\omega)H(\omega)]^2 \, d\omega. \qquad (4.131)$$

In some applications, it is desired to factorize $H(z)$ into the minimum-phase and maximum-phase terms. In this case, an additional constraint that $H(\omega)$ be non-negative is required. This subsection concentrates on minimax designs. For least-squared-error filters, see Vaidyanathan and Nguyen [VA87] and Nguyen et al. [NG88].

In order to find $H(z)$ minimizing δ_s as given by Eq. (4.130) and simultaneously meeting the time-domain conditions of Eq. (4.128), it is split into two parts [SA87b] as follows:

$$H(z) = H_p(z)H_s(z) = \sum_{n=0}^{2K} h_p[n]z^{-n} \sum_{n=0}^{2(M-K)} h_s[n]z^{-n}, \qquad (4.132a)$$

where

$$K = \lfloor M/L \rfloor . \qquad (4.132b)$$

Here, both $H_p(z)$ and $H_s(z)$ are Type I linear-phase filters. $H_p(z)$ has its zeros off the unit circle and is determined such that the time-domain conditions of Eq. (4.128) are satisfied, whereas $H_s(z)$ has its zeros on the unit circle and is used for providing the desired stopband response. For any $H_s(z)$, $H_p(z)$ can be determined such that the overall filter $H(z) = H_p(z)H_s(z)$ satisfies the time-domain conditions. This leads to a system of $2\lfloor M/L \rfloor + 1$ linear equations in the $2\lfloor M/L \rfloor + 1$ coefficients $h_p[n]$ of $H_p(z)$. Utilizing the fact that the coefficients of $H_p(z)$ as well as the time-domain conditions are symmetric, a system of $\lfloor M/L \rfloor + 1$ equations needs to be solved. The remaining problem is to find $H_s(z)$ to give the minimum value of δ_s. The algorithm for iteratively determining the desired $H_s(z)$ consists of the following steps [SA87b]:

1. Set $H_p(\omega) \equiv 1$ and $\Omega = \{\omega_1, \omega_2, \ldots, \omega_{M-K+1}\} = \{0, 0, \ldots, 0\}$.
2. Find $H_s(\omega)$ such that $H_s(0) = 1$ and $W(\omega)H_p(\omega)H_s(\omega)$ alternatingly achieves at least at $M - K + 1$ consecutive points on $[\omega_s, \pi]$ the extremum values $\pm\delta_s$. Store the extremal points into $\overline{\Omega} = \{\overline{\omega}_1, \overline{\omega}_2, \ldots, \overline{\omega}_{M-K+1}\}$.
3. Determine $H_p(z)$ such that the time-domain conditions of Eq. (4.128) are satisfied.
4. If $|\omega_k - \overline{\omega}_k| \le \alpha$ for $k = 1, 2, \ldots, M - K + 1$ (α is a small number), then stop. Otherwise set $\Omega = \overline{\Omega}$ and go to Step 2.

The desired $H_s(\omega)$ at Step 2 can be found using the MPR algorithm. The desired function is zero on $[\omega_s, \pi]$ and the weighting function is $W(\omega)H_p(\omega)$. $H_s(\omega)$ can be forced to take the value unity at $\omega = 0$ by selecting a very narrow passband region $[0, \epsilon]$, setting $D(\omega) \equiv 1$, and by using a large weighting function in this region. If a very narrow passband region is used, then the MPR algorithm selects automatically only one grid point ($\omega = 0$) in this region. Typically, three to five iterations of the above algorithm are needed to arrive at the desired solution. Another approach for designing Lth band filters is to use linear programming [LI85; SA88c]. However, linear programming requires significantly more computation time than the above algorithm.

Example 4.12. The specifications are $L = 4$ and $\rho = 0.2$ ($\omega_p = 0.2\pi$ and $\omega_s = 0.3\pi$), and $\delta_s = 0.01$ (40-dB attenuation). The amplitude and impulse responses for an optimized filter of order 38 are shown in Figures 4-38(a) and 4-38(b), respectively.

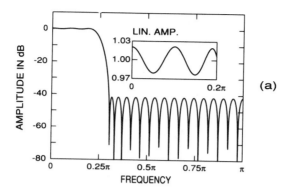

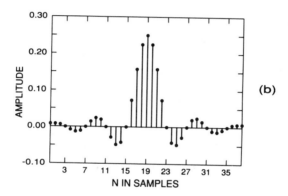

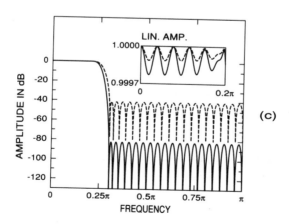

FIGURE 4-38 Fourth-band filters with $\rho = 0.2$ ($\omega_p = 0.2\pi$ and $\omega_s = 0.3\pi$). (a, b) Amplitude and impulse responses for a minimax linear-phase design of order 38. (c) Amplitude responses for the overall factorizable minimax design (solid line) of order 102 and for the minimum-phase term (dashed line) of order 51.

227

If it is desired that $H(z)$ be factorizable into the minimum- and maximum-phase terms, then the subfilter $H_s(z)$ is rewritten in the form

$$H_s(z) = [\overline{H}_s(z)]^2, \qquad \overline{H}_s(z) = \sum_{n=0}^{M-K} \overline{h}_s[n]z^{-n}. \qquad (4.133)$$

Here, $\overline{H}_s(z)$ is either a Type I linear-phase filter ($M - K$ is even) or a Type II filter ($M - K$ is odd). The resulting overall zero-phase frequency response is given by

$$H(\omega) = H_p(\omega)[\overline{H}_s(\omega)]^2. \qquad (4.134)$$

Since the zeros of $H_p(z)$ are off the unit circle, $H(\omega)$ is nonnegative, as is desired. In this case, the minimization of the stopband ripple can be performed by slightly modifying the above algorithm. The basic difference is that now $\overline{H}_s(\omega)$ is determined at Step 2 such that $\overline{H}_s(0) = 1$ and $(\sqrt{W(\omega)H_p(\omega)})\overline{H}_s(\omega)$ oscillates within the limits $\pm \overline{\delta}_s$ on $[\omega_s, \pi]$ with $\lfloor (M - K)/2 \rfloor + 1$ extremal frequencies. Correspondingly, $W(\omega)H_p(\omega)[\overline{H}_s(\omega)]^2$ oscillates within the limits 0 and $\delta_s = (\overline{\delta}_s)^2$ on $[\omega_s, \pi]$. The advantage of this approach is that both the minimum- and maximum-phase terms of $H(z)$ contain $\overline{H}_s(z)$ and only $H_p(z)$ must be factored in order to get the overall minimum-phase and maximum-phase designs.

Example 4.13. The specifications for the minimum-phase and maximum-phase filters are those of Example 4.12. The required stopband ripple for $H(\omega)$ is $(\overline{\delta}_s)^2 = 0.0001$. Figure 4-38(c) gives the amplitude responses for an optimized overall filter of order 102 (solid line) and for the minimum-phase (or maximum-phase) term of order 51 (dashed line).

4-10-3   Design of Half-Band Filters

A very important subclass of Lth band filters in many applications are half-band filters ($L = 2$). For these filters,

$$h[M] = \tfrac{1}{2}, \qquad (4.135a)$$

$$h[M + 2r] = 0 \quad \text{for } r = \pm 1, \pm 2, \ldots, \pm \lfloor M/2 \rfloor. \qquad (4.135b)$$

A filter satisfying these conditions can be generated in two steps by starting with a Type II (M is odd) transfer function

$$G(z) = \sum_{n=0}^{M} g[n]z^{-n}, \qquad g[M - n] = g[n]. \qquad (4.136)$$

In the first step, zero-valued impulse-response samples are inserted between the $g[n]$'s (see Figures 4-39(a) and 4-39(b)), giving the following Type I transfer func-

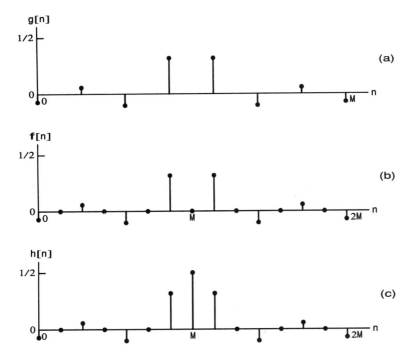

FIGURE 4-39 Generation of the impulse response of a half-band filter of order $2M$ with M odd. (a) Impulse response $g[n]$ of a Type II filter of order M. (b) Impulse response $f[n]$ resulting when inserting zero-valued samples between the $g[n]$'s. (c) Desired impulse response $h[n]$ of a half-band filter obtained by replacing $f[M] = 0$ by $h[M] = \frac{1}{2}$.

tion of order $2M$:

$$F(z) = \sum_{n=0}^{2M} f[n]z^{-n} = G(z^2) = \sum_{n=0}^{M} g[n]z^{-2n}. \tag{4.137}$$

The second step is then to replace the zero-valued impulse-response sample at $n = M$ by $\frac{1}{2}$ (see Figure 4-39(c)), resulting in the desired transfer function

$$H(z) = \sum_{n=0}^{2M} h[n]z^{-n} = \tfrac{1}{2}z^{-M} + F(z) = \tfrac{1}{2}z^{-M} + \sum_{n=0}^{M} g[n]z^{-2n}. \tag{4.138}$$

This gives $h[M] = \frac{1}{2}$, $h[n] = g[n/2]$ for n even, and $h[n] = 0$ for n odd and $n \neq M$, as is desired.

The zero-phase frequency responses of $H(z)$, $F(z)$, and $G(z)$ are related through

$$H(\omega) = \tfrac{1}{2} + F(\omega) = \tfrac{1}{2} + G(2\omega). \tag{4.139}$$

Based on these relations, the design of a lowpass half-band filter with passband edge at ω_p and passband ripple of δ can be accomplished by determining $G(z)$ such that $G(\omega)$ oscillates within $\frac{1}{2} \pm \delta$ on $[0, 2\omega_p]$ (see Figure 4-40(a)). Since $G(z)$ is a Type II transfer function, it has one fixed zero at $z = -1$ ($\omega = \pi$). $G(z)$ can be designed directly with the aid of the MPR algorithm using only one band $[0, 2\omega_p]$, $D(\omega) = \frac{1}{2}$, and $W(\omega) = 1$. Since $G(z)$ has a single zero at $z = -1$, $G(\omega)$ is odd about $\omega = \pi$. Hence $G(2\pi - \omega) = -G(\omega)$ and $G(\omega)$ oscillates within $-\frac{1}{2} \pm \delta$ on $[2\pi - 2\omega_p, 2\pi]$. The corresponding $F(\omega) = G(2\omega)$ stays within $\frac{1}{2} \pm \delta$ on $[0, \omega_p]$ and within $-\frac{1}{2} \pm \delta$ on $[\pi - \omega_p, \pi]$ (see Figure 4-40(b)). Finally, $H(\omega)$ approximates unity on $[0, \omega_p]$ with tolerance δ and zero on $[\pi - \omega_p, \pi]$ with the same tolerance δ (see Figure 4-40(c)).

For the resulting $H(\omega)$, the passband and stopband ripples are thus the same and the passband and stopband edges are related through $\omega_s = \pi - \omega_p$. In general,

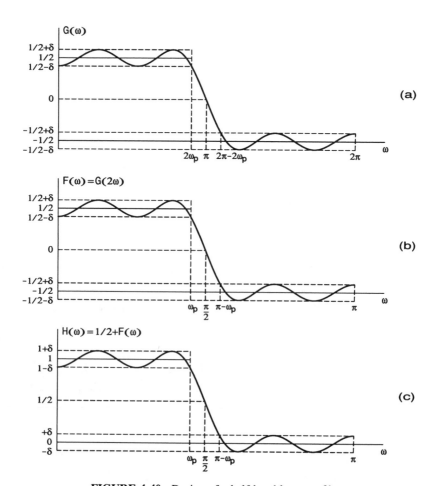

FIGURE 4-40   Design of a half-band lowpass filter.

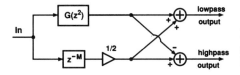

FIGURE 4-41 Implementation of a complementary lowpass–highpass half-band filter pair.

$H(\omega)$ satisfies

$$H(\omega) + H(\pi - \omega) = 1. \tag{4.140}$$

This makes $H(\omega)$ symmetric about the point $\omega = \pi/2$ such that the sum of the values of $H(\omega)$ at $\omega = \overline{\omega} < \pi/2$ and at $\omega = \pi - \overline{\omega} > \pi/2$ is equal to unity (see Figure 4-40(c)).

Figure 4-41 gives an implementation for the half-band filter as a parallel connection of $G(z^2)$ and $\frac{1}{2}z^{-M}$. This implementation is very attractive because in this case the complementary highpass output having the zero-phase frequency response $1 - H(\omega)$ is obtained directly by subtracting $G(z^2)$ from $\frac{1}{2}z^{-M}$. Note that the delay term z^{-M} can be shared with $G(z^2)$. The number of nonzero coefficient values in $G(z^2)$ is $M + 1$. By exploiting the symmetry in these coefficients, only $(M + 1)/2$ multipliers (M is odd) are needed to implement a lowpass–highpass filter pair of order $2M$. Figure 4-42 gives responses for a complementary half-band filter pair.

4-11 DESIGN OF FIR FILTERS USING PERIODIC SUBFILTERS AS BASIC BUILDING BLOCKS

One approach to reduce the cost of implementation of an FIR filter is to construct the overall filter using subfilters whose transfer function is of the form $F(z^L)$. There exist several design techniques [FA81; JI84; LI86; NE84, NE87; RA88; SA88a, SA88b, SA90], some of which are reviewed in this section.

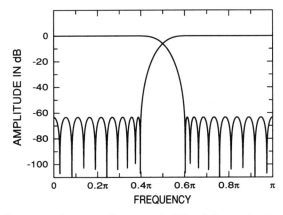

FIGURE 4-42 Responses for a complementary half-band filter pair of order 34 for $\omega_p = 0.4\pi$. The implementation of this filter pair requires only nine multipliers.

4-11-1 Periodic Filters

Consider a linear-phase transfer function of the form

$$A(z) = F(z^L) = \sum_{n=0}^{N_F} f[n]z^{-nL}, \quad f[N_F - n] = f[n]. \tag{4.141}$$

This transfer function is obtainable from a conventional transfer function

$$F(z) = \sum_{n=0}^{N_F} f[n]z^{-n} \tag{4.142}$$

by replacing z^{-1} by z^{-L}, that is, by substituting for each unit delay L unit delays. Figure 4-43 gives for $A(z)$ an implementation that exploits the coefficient symmetry. Note that there is a multiplier only after every Lth delay term. The order of $A(z)$ is LN_F and its zero-phase frequency response is

$$A(\omega) = F(L\omega). \tag{4.143}$$

$A(\omega)$ is thus a frequency-axis compressed version of $F(\omega)$ such that the interval $[0, L\pi]$ is shrunk onto $[0, \pi]$. Figure 4-44 gives the resulting $A(\omega)$ in the case where $F(\omega)$ is a lowpass design with passband and stopband edges at θ and ϕ. Since the periodicity of $F(\omega)$ is 2π, the periodicity of $A(\omega)$ is $2\pi/L$ and it contains several passband and stopband regions in the interval $[0, \pi]$. It should be noted that this applies when N_F is even.[14]

---

[14]If N_F is odd, then $A(\omega)$ changes sign at $\omega = \pi/L, 3\pi/L, 5\pi/L, \ldots$ so that $A(\omega)$ approximates alternatingly unity and minus unity in the consecutive passband regions. This is because $F(z)$ is in this case a Type II design and $F(\omega)$ is odd about the points $\omega = \pi, 3\pi, 5\pi, \ldots$ (see Figure 4-6).

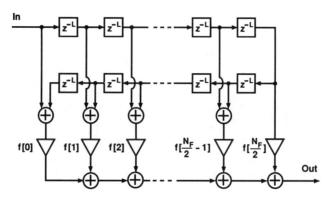

FIGURE 4-43 An implementation for a linear-phase transfer function $F(z^L)$ when N_F is even.

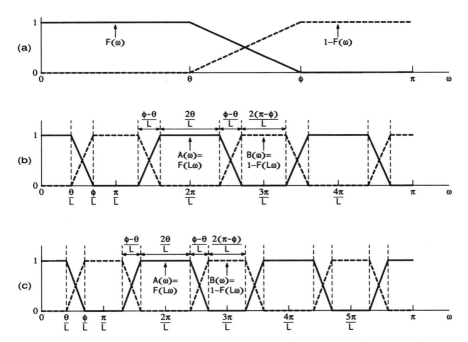

FIGURE 4-44 Periodic responses $A(\omega) = F(L\omega)$ and $B(\omega) = 1 - F(L\omega)$ when $F(\omega)$ is a lowpass design. (a) Responses $F(\omega)$ and $1 - F(\omega)$. (b) Responses $F(L\omega)$ and $1 - F(L\omega)$ for $L = 5$. (c) Responses $F(L\omega)$ and $1 - F(L\omega)$ for $L = 6$.

When N_F is even, the complementary transfer function of $A(z)$ is (cf. Section 4-8-3)

$$B(z) = z^{-LN_F/2} - A(z) = z^{-LN_F/2} - F(z^L). \tag{4.144}$$

The zero-phase frequency response of the corresponding filter is

$$B(\omega) = 1 - A(\omega) = 1 - F(L\omega) \tag{4.145}$$

and its passband regions are the stopband regions of $A(\omega)$ and vice versa (see Figure 4-44).

As seen from Figure 4-44, the periodic transfer functions $F(z^L)$ and $z^{-LN_F/2} - F(z^L)$ provide several transition bands of width $(\phi - \theta)/L$, which can be used as a transition band for a lowpass filter. The attractive property of these filters is that the number of nonzero impulse-response values to provide one of these transition bands is only $(1/L)$th of that of a conventional nonperiodic filter. This follows from the facts that the required FIR filter order is roughly inversely proportional to the transition bandwidth (cf. Section 4-8-2) and the transition bandwidth of the prototype filter $F(z)$, which determines the number of nonzero impulse-response coefficients, is $\phi - \theta$. This is L times wider. Note that the orders of the periodic filters and that of the nonperiodic filter are approximately the same, but the non-periodic filter has no zero-valued impulse-response samples.

Because of periodic responses, $F(z^L)$ or $z^{-LN_F/2} - F(z^L)$ cannot be used alone for synthesizing a lowpass filter. The desired result can be achieved by properly combining these filters with conventional nonperiodic filters.

4-11-2 Frequency-Response Masking Approach

A very elegant approach to exploiting the attractive properties of $A(z)$ and $B(z)$ has been proposed by Lim [LI86]. In this approach, the overall transfer function is constructed as

$$H(z) = F(z^L)G_1(z) + [z^{-LN_F/2} - F(z^L)]G_2(z). \qquad (4.146)$$

An implementation[15] of this transfer function is shown in Figure 4-45. The zero-phase frequency response of this filter can be written as

$$H(\omega) = F(L\omega)G_1(\omega) + [1 - F(L\omega)]G_2(\omega) \qquad (4.147)$$

provided that the delays of $G_1(z)$ and $G_2(z)$ are equal.[16]

For a lowpass design, the transition band of $H(\omega)$ can be selected to be one of the transition bands provided by $F(z^L)$ or $z^{-LN_F/2} - F(z^L)$. In the first case, referred to as Case A, the edges of $H(\omega)$ are selected to be (see Figure 4-46)

$$\omega_p = (2l\pi + \theta)/L, \qquad \omega_s = (2l\pi + \phi)/L, \qquad (4.148)$$

where l is a fixed integer, and in the second case, referred to as Case B,

$$\omega_p = (2l\pi - \phi)/L, \qquad \omega_s = (2l\pi - \theta)/L. \qquad (4.149)$$

---

[15]Note that the delay term $z^{-LN_F/2}$ can be shared with $F(z^L)$. Also, $G_1(z)$ and $G_2(z)$ can share the same delay elements if they are implemented using the transposed direct-form structure (exploiting the coefficient symmetry).

[16]This means that the orders of both $G_1(z)$ and $G_2(z)$, denoted by N_{G1} and N_{G2}, must be either even or odd and if N_{G1} and N_{G2} are not equal, then, in order to equalize the delays, the delay term $z^{-(N_{G1} - N_{G2})/2}$ $(z^{-(N_{G2} - N_{G1})/2})$ must be added to $G_2(z)$ $(G_1(z))$ if $N_{G1} - N_{G2}$ is positive (negative).

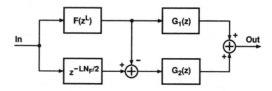

FIGURE 4-45 The structure of a filter synthesized using the frequency-response masking technique.

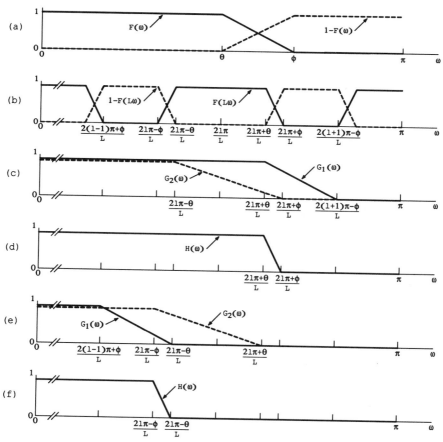

FIGURE 4-46   Design of lowpass filters using the frequency-response masking technique.

For Case A designs, $H(\omega)$ approximates unity on $[0, \omega_p]$ and zero on $[\omega_s, \pi]$ if the two masking filters $G_1(z)$ and $G_2(z)$ are lowpass filters with the following passband and stopband edges (see Figures 4-46(c) and 4-46(d)):

$$\omega_p^{(G_1)} = \omega_p = \{2l\pi + \theta\}/L, \qquad \omega_s^{(G_1)} = \{2(l+1)\pi - \phi\}/L, \qquad (4.150a)$$

$$\omega_p^{(G_2)} = \{2l\pi - \theta\}/L, \qquad \omega_s^{(G_2)} = \omega_s = \{2l\pi + \phi\}/L. \qquad (4.150b)$$

On $[0, \omega_p^{(G_2)}]$, $G_1(\omega) \approx 1$ and $G_2(\omega) \approx 1$ so that $H(\omega) \approx F(L\omega) + [1 - F(L\omega)] = 1$, as is desired. On $[\omega_p^{(G_2)}, \omega_p]$, $G_1(\omega) \approx 1$, $F(L\omega) \approx 1$, and $1 - F(L\omega) \approx 0$, giving $H(\omega) \approx F(L\omega)G_1(\omega) \approx 1$ regardless of the behavior of $G_2(\omega)$ in this region. On $[\omega_s^{(G_1)}, \pi]$, $G_1(\omega) \approx 0$ and $G_2(\omega) \approx 0$ so that $H(\omega) \approx 0$. Since $F(L\omega) \approx 0$ on $[\omega_s, \omega_s^{(G_1)}]$, the stopband region of $G_1(z)$ can start at $\omega = \omega_s^{(G_1)}$, instead of $\omega = \omega_s$.

For Case B designs, the required edges of the two masking filters $G_1(z)$ and

$G_2(z)$ are (see Figures 4-46(e) and 4-46(f))

$$\omega_p^{(G_1)} = \{2(l-1)\pi + \phi\}/L, \qquad \omega_s^{(G_1)} = \omega_s = \{2l\pi - \theta\}/L, \qquad (4.151a)$$

$$\omega_p^{(G_2)} = \omega_p = \{2l\pi - \phi\}/L, \qquad \omega_s^{(G_2)} = \{2l\pi + \theta\}/L. \qquad (4.151b)$$

In Lim [LI86], the effects of the ripples of the subresponses $G_1(\omega)$, $G_2(\omega)$, and $F(L\omega)$ on the ripples of the overall response $H(\omega)$ have been studied carefully. Based on these observations, the design of the overall filter with passband and stopband ripples of δ_p and δ_s can be accomplished for both Case A and Case B in the following two steps:

1. Design $G_k(z)$ for $k = 1, 2$ using the MPR algorithm such that $G_k(\omega)$ approximates unity on $[0, \omega_p^{(G_k)}]$ with tolerance $0.85\delta_p \ldots 0.9\delta_p$ and zero on $[\omega_s^{(G_k)}, \pi]$ with tolerance $0.85\delta_s \ldots 0.9\delta_s$.[17]

2. Design $F(L\omega)$ such that the overall response $H(\omega)$ as given by Eq. (4.147) approximates unity on

$$\Omega_p^{(F)} = \begin{cases} [\omega_p^{(G_2)}, \omega_p] = [\{2l\pi - \theta\}/L, \{2l\pi + \theta\}/L] \\ \qquad\qquad \text{for Case A} \\ [\omega_p^{(G_1)}, \omega_p] = [\{2(l-1)\pi + \phi\}/L, \{2l\pi - \phi\}/L] \\ \qquad\qquad \text{for Case B} \end{cases} \qquad (4.152a)$$

with tolerance δ_p and zero on

$$\Omega_s^{(F)} = \begin{cases} [\omega_s, \omega_s^{(G_1)}] = [\{2l\pi + \phi\}/L, \{2(l+1)\pi - \phi\}/L] \\ \qquad\qquad \text{for Case A} \\ [\omega_s, \omega_s^{(G_2)}] = [\{2l\pi - \theta\}/L, \{2l\pi + \theta\}/L] \\ \qquad\qquad \text{for Case B} \end{cases} \qquad (4.152b)$$

with tolerance δ_s.

The design of $F(L\omega)$ can be performed using linear programming [LI86]. Another, computationally more efficient, alternative is to use the MPR algorithm. It can be shown that $F(L\omega)$ meets the given criteria if the maximum absolute value of the error function given in Table 4-6 is on $[0, \theta] \cup [\phi, \pi]$ less than or equal to unity. Even though this error function looks very complicated, it is straightforward to use the subroutines EFF and WATE in the MPR algorithm for optimally designing $F(z)$.

---

[17]To reduce the order of $G_1(z)$, a smaller weighting can be used in the MPR algorithm on those regions of $G_1(z)$ where $F(L\omega)$ has one of its stopbands. As a rule of thumb, for the regions in the passband (stopband) of $G_1(z)$, the weighting can be selected to be one-tenth of the original passband (stopband) weighting. Similarly, the order of $G_2(z)$ can be reduced by using a smaller weighting on those regions where $F(L\omega)$ has one of its passbands.

TABLE 4-6   Error Function for Designing $F(z)$ in the Frequency-Response Masking Approach

$$E(\omega) = W_F(\omega)[F(\omega) - D_F(\omega)]$$

where

$$D_F(\omega) = [u(\omega) + l(\omega)]/2, \qquad W_F(\omega) = 2/[u(\omega) - l(\omega)]$$

with

$$u(\omega) = \min \left(\frac{D(\omega) - G_2[h_1(\omega)]}{G_1[h_1(\omega)] - G_2[h_1(\omega)]} + e_1(\omega), \frac{D(\omega) - G_2[h_2(\omega)]}{G_1[h_2(\omega)] - G_2[h_2(\omega)]} + e_2(\omega) \right)$$

$$l(\omega) = \max \left(\frac{D(\omega) - G_2[h_1(\omega)]}{G_1[h_1(\omega)] - G_2[h_1(\omega)]} - e_1(\omega), \frac{D(\omega) - G_2[h_2(\omega)]}{G_1[h_2(\omega)] - G_2[h_2(\omega)]} - e_2(\omega) \right)$$

$$e_1(\omega) = \frac{\delta(\omega)}{|G_1[h_1(\omega)] - G_2[h_1(\omega)]|}, \qquad e_2(\omega) = \frac{\delta(\omega)}{|G_1[h_2(\omega)] - G_2[h_2(\omega)]|}$$

and

$$D(\omega) = \begin{cases} 1 & \text{for } \omega \in [0, \theta] \\ 0 & \text{for } \omega \in [\phi, \pi], \end{cases} \qquad \delta(\omega) = \begin{cases} \delta_p & \text{for } \omega \in [0, \theta] \\ \delta_s & \text{for } \omega \in [\phi, \pi] \end{cases}$$

$$h_1(\omega) = (2l\pi + \omega)/L, \qquad h_2(\omega) = \begin{cases} (2l\pi - \omega)/L & \text{for } \omega \in [0, \theta] \\ (2(l + 1)\pi - \omega)/L & \text{for } \omega \in [\phi, \pi] \end{cases}$$

for Case A and

$$D(\omega) = \begin{cases} 0 & \text{for } \omega \in [0, \theta] \\ 1 & \text{for } \omega \in [\phi, \pi], \end{cases} \qquad \delta(\omega) = \begin{cases} \delta_s & \text{for } \omega \in [0, \theta] \\ \delta_p & \text{for } \omega \in [\phi, \pi] \end{cases}$$

$$h_1(\omega) = (2l\pi - \omega)/L, \qquad h_2(\omega) = \begin{cases} (2l\pi + \omega)/L & \text{for } \omega \in [0, \theta] \\ (2(l - 1)\pi + \omega)/L & \text{for } \omega \in [\phi, \pi] \end{cases}$$

for Case B

In practice, ω_p and ω_s are given and l, L, θ, and ϕ must be determined. To ensure that Eq. (4.148) yields a desired solution with $0 \leq \theta < \phi \leq \pi$, it is required that (see Figures 4-44 and 4-46)

$$\frac{2l\pi}{L} \leq \omega_p, \qquad \omega_s \leq \frac{(2l + 1)\pi}{L} \qquad (4.153)$$

for some positive integer l. In this case,

$$l = \lfloor L\omega_p/(2\pi) \rfloor, \qquad \theta = L\omega_p - 2l\pi, \qquad \phi = L\omega_s - 2l\pi. \quad (4.154)$$

Similarly, to ensure that Eq. (4.149) yields a desired solution with $0 \leq \theta < \phi \leq \pi$, it is required that

$$\frac{(2l - 1)\pi}{L} \leq \omega_p, \qquad \omega_s \leq \frac{2l\pi}{L} \tag{4.155}$$

for some positive integer l. In this case,[18]

$$l = \lceil L\omega_s/(2\pi) \rceil, \qquad \theta = 2l\pi - L\omega_s, \qquad \phi = 2l\pi - L\omega_p. \tag{4.156}$$

For any set of ω_p, ω_s, and L, either Eq. (4.154) or (4.156) (not both) will yield the desired θ and ϕ, provided that L is not too large. If $\theta = 0$ or $\phi = \pi$, then the resulting specifications for $F(\omega)$ are meaningless and the corresponding value of L cannot be used.

The remaining problem is to determine L to minimize the filter complexity. The following example illustrates how this problem can be solved.

Example 4.14. The filter specifications are [JI84; SA88b] $\omega_p = 0.4\pi$, $\omega_s = 0.402\pi$, $\delta_p = 0.01$, and $\delta_s = 0.0001$. The minimum order of an optimum conventional direct-form design to meet the given criteria, estimated by Eq. (4.95), is 3138, requiring 1570 multipliers. Table 4-7 gives, for the admissible values of L in the range $3 \leq L \leq 30$, l, θ, and ϕ as well as whether the overall filter is a Case A or Case B design. It shows also estimated orders for $F(z)$, $G_1(z)$, and $G_2(z)$, denoted by N_F, N_{G1}, and N_{G2}, as well as the overall number of multipliers,[19] $N_F/2 + 1 + \lfloor (N_{G1} + 2)/2 \rfloor + \lfloor (N_{G2} + 2)/2 \rfloor$. These orders have been estimated using Eq. (4.95) with the passband and stopband ripples being the specified ones.[20] N_F is approximately only $(1/L)$th of the order of an equivalent direct-form design and decreases with increasing L. The widths of the transition bands of $G_1(\omega)$ and $G_2(\omega)$ are $[2\pi - \theta - \phi]/L$ and $[\theta + \phi]/L$, respectively, so that their sum is $2\pi/L$ and decreases with increasing L. The overall number of multipliers is usually the smallest at those values of L for which these widths and, correspondingly, N_{G1} and N_{G2} are of the same order. This happens when $\theta + \phi$ is approximately equal to π. If this is true for several values of L, then the best result is obtained by increasing L until the decrease in the number of multipliers in $F(z)$ is less than the increase in the overall number of multipliers in $G_1(z)$ and $G_2(z)$. The estimated filter orders are usually so close to the actual minimum orders

---

[18] $\lceil x \rceil$ stands for the smallest integer which is larger than or equal to x.

[19] When the symmetry in the filter coefficients is exploited, a Type I linear-phase filter of order N (even) requires $N/2 + 1$ multipliers, whereas a Type II filter (N odd) requires $(N + 1)/2$ multipliers. N_F is forced, according to the discussion of Section 4-11-1, to be even in order to get the desired solution.

[20] For N_F, the nearest even value greater than or equal to the estimated value has been used. For Case A designs, the ripples in the passband $[0, \theta]$ and in the stopband $[\phi, \pi]$ are δ_p and δ_s, respectively, whereas for Case B designs they are interchanged. Note that for the estimation formula of Eq. (4.95), δ_p and δ_s are the larger and smaller ripple values, respectively. If the estimated value of N_{G1} is even (odd), then N_{G2} has also been selected to be even (odd).

TABLE 4-7 Estimation of L Minimizing the Number of Multipliers in the Frequency-Response Masking Approach

L	Case	l	θ	ϕ	N_F	N_{G1}	N_{G2}	Number of Multipliers
3	B	1	0.794π	0.8π	1046	46	10	554
4	B	1	0.392π	0.4π	786	19	31	420
6	A	1	0.4π	0.412π	524	31	47	303
7	A	1	0.8π	0.814π	448	113	27	296
8	B	2	0.784π	0.8π	392	120	32	275
9	B	2	0.382π	0.4π	350	46	72	237
11	A	2	0.4π	0.422π	286	58	84	217
12	A	2	0.8π	0.824π	262	200	46	257
13	B	3	0.774π	0.8π	242	191	51	244
14	B	3	0.372π	0.4π	224	71	113	206
16	A	3	0.4π	0.432π	196	86	120	204
17	A	3	0.8π	0.834π	184	291	65	272
18	B	4	0.764π	0.8π	174	259	73	255
19	B	4	0.362π	0.4π	166	96	156	212
21	A	4	0.4π	0.442π	150	113	157	212
22	A	4	0.8π	0.844π	142	388	84	310
23	B	5	0.754π	0.8π	136	324	92	279
24	B	5	0.352π	0.4π	130	120	200	228
26	A	5	0.4π	0.452π	120	142	192	230
27	A	5	0.8π	0.854π	116	490	102	357
28	B	6	0.744π	0.8π	112	385	113	307
29	B	6	0.342π	0.4π	108	144	246	252

that they can be used for determining the value of L minimizing the filter complexity.

With the estimated filter orders, $L = 16$ gives the best result. The actual filter orders are $N_F = 198$, $N_{G1} = 83$, and $N_{G2} = 123$. The responses of the subfilters as well as that of the overall design are given in Figure 4-47. The overall number of multipliers and adders for this design are 204 and 406, respectively, which are 13% of those required by an equivalent conventional direct-form design (1570 and 3138). The overall filter order is 3291, which is only 5% higher than that of the direct-form design (3138).

The complexity of the filter can be reduced further by using a two-level frequency-response masking. For examples, see Lim [LI86] and Saramäki and Fam [SA88b].

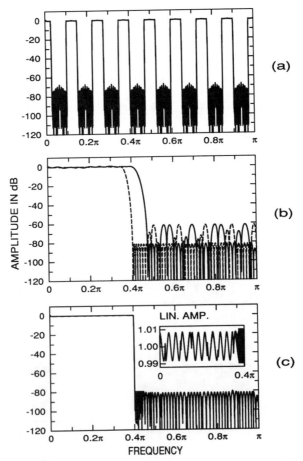

FIGURE 4-47 Amplitude responses for a lowpass filter synthesized using the frequency-response masking approach. (a) Periodic response $F(L\omega)$. (b) Subresponses $G_1(\omega)$ (solid line and $G_2(\omega)$ (dashed line) (c) Overall filter.

4-11-3 Design of Narrowband Lowpass Filters

When ω_s is less than $\pi/2$, then the overall filter can be synthesized in the following simplified form [JI84; NE84; SA88a]:

$$H(z) = F(z^L)G(z). \tag{4.157}$$

The zero-phase frequency response of this filter is given by

$$H(\omega) = F(L\omega)G(\omega). \tag{4.158}$$

If the passband and stopband edges of $F(z)$ are θ and ϕ, then the edges of the first transition band of $F(L\omega)$ are (see Figure 4-48)

$$\omega_p = \theta/L, \qquad \omega_s = \phi/L. \tag{4.159}$$

$F(L\omega)$ does not provide the desired attenuation in the regions where it has extra unwanted passbands and transition bands, that is, in the region

$$\Omega_s(L, \omega_s) = \bigcup_{k=1}^{\lfloor L/2 \rfloor} \left[k\frac{2\pi}{L} - \omega_s, \min\left(k\frac{2\pi}{L} + \omega_s, \pi \right) \right]. \tag{4.160}$$

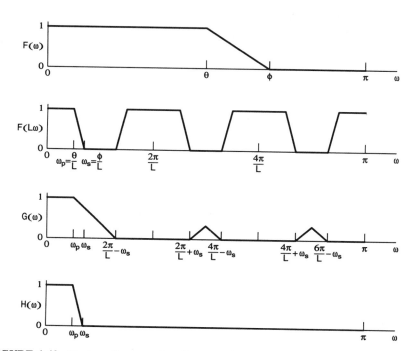

FIGURE 4-48 Design of a narrowband lowpass filter using a cascade of a periodic and a nonperiodic filter.

Therefore the role of the nonperiodic filter $G(z)$ is to provide enough attenuation in this region.

The simplest way to determine $F(z)$ and $G(z)$ such that $H(z)$ is a lowpass design with edges at ω_p and ω_s and ripples of δ_p and δ_s is to design these subfilters using the MPR algorithm to satisfy

$$1 - \delta_p^{(F)} \le F(\omega) \le 1 + \delta_p^{(F)} \quad \text{for } \omega \in [0, L\omega_p], \qquad (4.161a)$$

$$-\delta_s \le F(\omega) \le \delta_s \qquad \text{for } \omega \in [L\omega_s, \pi], \qquad (4.161b)$$

$$1 - \delta_p^{(G)} \le G(\omega) \le 1 + \delta_p^{(G)} \quad \text{for } \omega \in [0, \omega_p], \qquad (4.161c)$$

$$-\delta_s \le G(\omega) \le \delta_s \qquad \text{for } \omega \in \Omega_s(L, \omega_s), \qquad (4.161d)$$

where

$$\delta_p^{(G)} + \delta_p^{(F)} = \delta_p. \qquad (4.161e)$$

The ripples $\delta_p^{(F)}$ and $\delta_p^{(G)}$ can be selected, for example, to be half the overall ripple δ_p. In the above specifications, both $F(z^L)$ and $G(z)$ have $[0, \omega_p]$ as a passband region.

Another alternative, resulting in a considerably reduced order of $G(z)$, is to design simultaneously $F(\omega)$ to meet

$$1 - \delta_p \le F(\omega)G(\omega/L) \le 1 + \delta_p \quad \text{for } \omega \in [0, L\omega_p], \qquad (4.162a)$$

$$-\delta_s \le F(\omega)G(\omega/L) \le \delta_s \qquad \text{for } \omega \in [L\omega_s, \pi], \qquad (4.162b)$$

and $G(\omega)$ to meet

$$G(0) = 1, \qquad (4.163c)$$

$$-\delta_s \le F(L\omega)G(\omega) \le \delta_s \qquad \text{for } \omega \in \Omega_s(L, \omega_s). \qquad (4.163d)$$

In this case, $G(z)$ has all its zeros on the unit circle and concentrates on providing for the overall filter the desired attenuation on $\Omega_s(L, \omega_s)$ (see Figure 4-49). $F(L\omega)$ equalizes the passband distortion caused by $G(\omega)$ and provides the required attenuation in its stopband regions.

The desired overall response can be found by designing iteratively $F(z)$ to meet Eq. (4.162) and $G(z)$ to meet Eq. (4.163) until the difference between successive overall solutions is within the given tolerance limits. The algorithm consists of the following steps [SA88a]:

1. Set $F(\omega) \equiv 1$.

2. Determine $G(\omega)$ using the MPR algorithm to minimize on $[0, \epsilon] \cup \Omega_s(L, \omega_s)$ the peak absolute value of $E_G(\omega) = W_G(\omega)[G(\omega) - D_G(\omega)]$, where

$$D_G(\omega) = \begin{cases} 1 & \text{for } \omega \in [0, \epsilon] \\ 0 & \text{for } \omega \in \Omega_s(L, \omega_s), \end{cases} \qquad W_G(\omega) = \begin{cases} \alpha & \text{for } \omega \in [0, \epsilon] \\ F(L\omega) & \text{for } \omega \in \Omega_s(L, \omega_s). \end{cases}$$

$$(4.164)$$

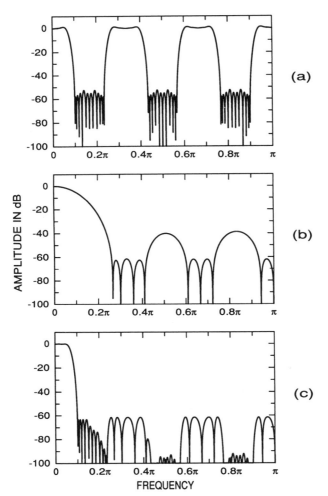

FIGURE 4-49 Amplitude responses for an optimized filter of the form $H(z) = F(z^L)G(z)$ with $L = 6$. The specifications are $\omega_p = 0.05\pi$, $\omega_s = 0.1\pi$, $\delta_p = 0.01$, and $\delta_s = 0.001$. (a) $F(z^L)$ of order 17 in z^L. (b) $G(z)$ of order 17. (c) Overall filter.

By selecting ϵ to be a very small number and α to be a very large number, the MPR algorithm uses only one grid point ($\omega = 0$) on $[0, \epsilon]$ and forces $G(\omega)$ to take the value of unity at $\omega = 0$.

3. Determine $F(\omega)$ using the MPR algorithm to minimize on $[0, L\omega_p] \cup [L\omega_s, \pi]$ the peak absolute value of $E_F(\omega) = W_F(\omega)[F(\omega) - D_F(\omega)]$, where

$$D_F(\omega) = \begin{cases} 1/G(\omega/L) & \text{for } \omega \in [0, L\omega_p] \\ 0 & \text{for } \omega \in [L\omega_s, \pi], \end{cases} \qquad (4.165a)$$

$$W_F(\omega) = \begin{cases} G(\omega/L) & \text{for } \omega \in [0, L\omega_p] \\ \delta_p G(\omega/L)/\delta_s & \text{for } \omega \in [L\omega_s, \pi]. \end{cases} \qquad (4.165b)$$

4. Repeat Steps 2 and 3 until the difference between successive solutions is within the given tolerance limits.

Typically, three to five iterations of the above algorithm are required to arrive at the desired solution.

Given the filter specifications, the remaining problem is to optimize L and the orders of $G(z)$ and $F(z^L)$ to minimize the overall number of multipliers. For the order of $F(z^L)$ in z^L, a good estimate is

$$N_F = N/L, \tag{4.166}$$

where N is the minimum order of a conventional nonperiodic FIR filter to meet the given overall criteria. For the order of $G(z)$, a good estimate has been found to be

$$N_G = \cosh^{-1}\left(\frac{1}{\delta_s}\right)\left[\frac{1}{\cosh^{-1} X\left(\omega_p, \dfrac{2\pi}{L} - \dfrac{\omega_p + 2\omega_s}{3}\right)} + \frac{L/2}{\cosh^{-1} X\left(\dfrac{L\omega_p}{2}, \pi - \dfrac{L(\omega_p + 2\omega_s)}{6}\right)}\right], \tag{4.167a}$$

where

$$X(\omega_1, \omega_2) = (2 \cos \omega_1 - \cos \omega_2 + 1)/(1 + \cos \omega_2). \tag{4.167b}$$

L has to be selected such that the stopband edge of $F(z)$, $\phi = L\omega_s$, is less than π. This means that L must be less than π/ω_s. After estimating the required orders for $G(z)$ and $F(z)$, the remaining problem is to decrease or increase the orders to find the actual minimum orders. Since the frequency-response-shaping responsibilities are very well shared with the subfilters, the minimum orders can be found rather independently. First, the minimum order of $F(z)$ can be determined for the estimated order of $G(z)$, and then the minimum order of $G(z)$ is determined. Again, an example is used to illustrate how the best value of L can be found.

Example 4.15. The specifications are $\omega_p = 0.05\pi$, $\omega_s = 0.1\pi$, $\delta_p = 0.01$, and $\delta_s = 0.001$. Table 4-8 gives for the admissible values of L ($2 \leq L \leq 9$) both the estimated and actual minimum orders of $F(z)$ and $G(z)$ as well as the edges of $F(z)$. As seen from the table, the estimated orders are very close to the actual ones, showing that the best value of L can easily be determined based on the above

TABLE 4-8 Estimated and Actual Minimum Filter Orders for $2 \leq L \leq 9$ for a Narrowband Filter Synthesized as a Cascade of a Periodic and Nonperiodic Filter

L	θ	ϕ	Estimated N_F	Estimated N_G	Actual N_F	Actual N_G	Number of Multipliers
2	0.1π	0.2π	54	3	54	3	30
3	0.15π	0.3π	36	6	35	6	22
4	0.2π	0.4π	27	9	26	9	19
5	0.25π	0.5π	22	13	21	14	19
6	0.3π	0.6π	18	17	17	17	18
7	0.35π	0.7π	15	22	14	22	20
8	0.4π	0.8π	13	27	12	28	22
9	0.45π	0.9π	12	34	10	36	25

estimation formulas. The minimum of the total number of multipliers is obtained by increasing L until the decrease in the number of multipliers of $F(z^L)$ becomes smaller than the increase in the number of multipliers of $G(z)$. The amplitude responses for the subfilters and the overall design are shown in Figure 4-49 for the best value of L, $L = 6$. The orders of both $F(z^L)$ (in z^L) and $G(z)$ are 17. This design requires 18 multipliers and 34 adders. The minimum order of a conventional direct-form design is 108, requiring 55 multipliers and 108 adders. The price paid for these reductions in the number of arithmetic operations is a slight increase in the overall filter order (from 108 to 119).

Example 4.16. Further savings in the number of arithmetic operations can be achieved by implementing $G(z)$ using special structures [KI88; SA88a, SA89b]. A particularly efficient implementation is provided by a transfer function of the form [SA89b]

$$G(z) = \prod_{r=1}^{M} T_r(z), \qquad (4.168a)$$

where

$$T_r(z) = 2^{-Pr} \sum_{k=0}^{K_r - 1} z^{-k} = 2^{-Pr} \frac{1 - z^{-Kr}}{1 - z^{-1}} \qquad (4.168b)$$

and 2^{-Pr}, with P_r integer-valued, is a scaling constant. An efficient implementation of $T_r(z)$ is depicted in Figure 4-50.[21] The implementation of the above $G(z)$ re-

---

[21]If modulo arithmetic (e.g., 1's or 2's complement arithmetic) and the worst-case scaling (corresponds to peak scaling in this case) are used, the output of $T_r(z)$ implemented as shown in Figure 4-50 is correct even though internal overflows may occur. For details, see Saramäki et al. [SA88a]. This implementation is very attractive because, in this case, the system does not need initial resetting and the effect of temporary miscalculations vanishes automatically from the output in a finite time.

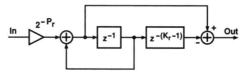

FIGURE 4-50 Efficient implementation for $T_r(z) = 2^{-Pr}(1 - z^{-Kr})/(1 - z^{-1})$.

quires no general multipliers. The zero-phase frequency response of $T_r(z)$ is

$$T_r(\omega) = 2^{-Pr} \sin (K_r\omega/2)/\sin (\omega/2) \qquad (4.169)$$

and it provides $K_r - 1$ zeros on the unit circle, located at $\omega = \pm k2\pi/K_r$ for $k = 1, 2, \ldots, \lfloor (K_r - 1)/2 \rfloor$ and at $\omega = \pi$ for K_r even. The design of $G(z)$ involves determining M, the number of $T_r(z)$'s, and the K_r's in such a way that the resulting $G(\omega)$ provides enough attenuation on $\Omega_s(L, \omega_s)$ as given by Eq. (4.160). The criteria of the previous example are met by $L = 8$, $F(z^L)$ of order 9 in z^L, and $G(z)$ consisting of four $T_r(z)$'s with $K_1 = 18$, $K_2 = 16$, $K_3 = 12$, and $K_4 = 11$. This filter requires only 5 general multipliers, 17 adders, and 129 delay elements. The amplitude responses of the subfilters $G(z)$ and $F(z^L)$ are given in Figure 4-51(a), whereas Figure 4-51(b) gives the overall response.[22]

4-11-4 Design of Wideband Lowpass Filters

The design of a wideband filter can be accomplished based on the fact (see Section 4-8-3) that if $\overline{H}(z)$ of even order $2M$ is a lowpass design with the following edges and ripples,

$$\overline{\omega}_p = \pi - \omega_s, \qquad \overline{\omega}_s = \pi - \omega_p, \qquad \overline{\delta}_p = \delta_s, \qquad \overline{\delta}_s = \delta_p, \qquad (4.170)$$

then

$$H(z) = z^{-M} - (-1)^M \overline{H}(-z) \qquad (4.171)$$

is a lowpass filter having the passband and stopband edges of ω_p and ω_s and the passband and stopband ripples of δ_p and δ_s. Hence, if ω_p and ω_s of the desired filter are larger than $\pi/2$, then $\overline{\omega}_p$ and $\overline{\omega}_s$ of $\overline{H}(z)$ are less than $\pi/2$. This enables us to design $\overline{H}(z)$ in the form

$$\overline{H}(z) = F(z^L)G(z) \qquad (4.172)$$

---

[22]The K_r's are larger than necessary to meet the criteria for $G(z)$ and they provide some zeros on $[\omega_s, 2\pi/L - \omega_s] = [0.1\pi, 0.15\pi]$, making the requirements of $F(z^L)$ less stringent in this region (see Figure 4-51). It can be shown that in order to guarantee that the overall filter meets the given criteria in this case, $F(z)$ has to be designed such that $G(\omega/L)$ is replaced by max $[|G(\omega/L)|, |G((2\pi - \omega)/L)|]$ in Eq. (4.162b) [SA89b].

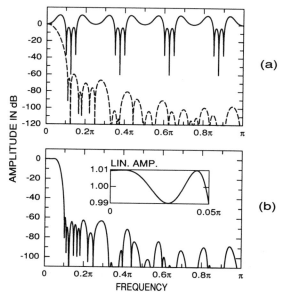

FIGURE 4-51 Amplitude responses for an optimized filter of the form $H(z) = F(z^L)G(z)$, where $G(z)$ is a cascade of four $T_r(z)$'s and $L = 8$. The filter specifications are those of Figure 4-49. (a) $F(z^L)$ (solid line) and $G(z)$ (dashed line). (b) Overall filter.

using the techniques of the previous subsection. The resulting overall transfer function is then

$$H(z) = z^{-M} - (-1)^M F((-z)^L)G(-z), \qquad (4.173)$$

where M is half the order of $F(z^L)G(z)$. An implementation of this transfer function is shown in Figure 4-52.[23] To avoid half-sample delays, the order has to be even.

Example 4.17. Let the wideband filter specifications be $\omega_p = 0.9\pi$, $\omega_s = 0.95\pi$, $\delta_p = 0.001$, and $\delta_s = 0.01$. From Eq. (4.170), the specifications of $\overline{H}(z)$ become $\overline{\omega}_p = 0.05\pi$, $\overline{\omega}_s = 0.1\pi$, $\overline{\delta}_p = 0.01$, and $\overline{\delta}_s = 0.001$. These are the narrowband

---

[23]The delay term z^{-M} can be shared with $\overline{F}(z^L)$.

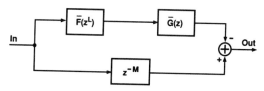

FIGURE 4-52 Implemetation for a wideband filter of the form $H(z) = z^{-M} - (-1)^M F((-z)^L)G(-z)$. $\overline{F}(z^L) \equiv (-1)^M F((-z)^L)$ and $\overline{G}(z) \equiv G(-z)$.

specifications of Example 4.15. Hence the desired wideband design is obtained by using the subfilters $F(z^L)$ and $G(z)$ of Figure 4-49. However, the overall order of the filter of Figure 4-49 is odd (119) and the resulting delay contains a half-sample delay. Therefore, in order to achieve the desired solution with even order, the order of $G(z)$ has to be increased by one. Figure 4-53 gives the amplitude response of the resulting filter. This design requires 19 multipliers, 36 adders, and 120 delay elements, whereas the corresponding numbers for a conventional direct-form equivalent of order 108 are 55, 108, and 108, respectively.

4-11-5 Generalized Designs

The Jing–Fam approach [JI84] is based on iteratively using the facts that a narrowband filter can be implemented effectively as $H(z) = F(z^L)G(z)$ and a wideband filter in the form of Eq. (4.173). If $\omega_s < \pi/2$, then the overall transfer function is first expressed as

$$H(z) \equiv \overline{H}_1(z) = G_1(z)F_1(z^{L_1}) \tag{4.174}$$

and the simultaneous criteria for $G_1(z)$ and $F_1(z)$ are stated according to Eq. (4.161). By denoting by $\delta_p^{(1)}$ the passband ripple of $G_1(z)$, the passband and stopband ripples of $F_1(z)$ are $\delta_p - \delta_p^{(1)}$ and δ_s, respectively. If L_1 is selected such that the passband and stopband edges of $F_1(z)$, $L_1\omega_p$ and $L_1\omega_s$, become larger than $\pi/2$, then $F_1(z)$ can be implemented in terms of a narrowband filter

$$\overline{H}_2(z) = G_2(z)F_2(z^{L_2}) \tag{4.175}$$

in the form

$$F_1(z) = z^{-M_1} - (-1)^{M_1}\overline{H}_2(-z). \tag{4.176}$$

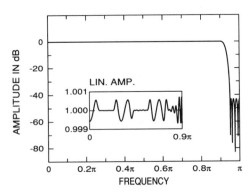

FIGURE 4-53 Amplitude response for a wideband filter implemented as shown in Figure 4-52.

The required passband and stopband ripples of $\overline{H}_2(z)$ are δ_s and $\delta_p - \delta_p^{(1)}$, respectively, whereas the passband and stopband edges are less than $\pi/2$ and given by

$$\omega_p^{(2)} = \pi - L_1 \omega_s, \qquad \omega_s^{(2)} = \pi - L_1 \omega_p. \qquad (4.177)$$

The specifications for $\overline{H}_2(z)$ are thus similar to those of $\overline{H}_1(z)$ and the process can be repeated. Continuing in this manner and designing the last stage in the form $\overline{H}_R(z) = G_R(z)$ results in the following overall transfer function [JI84; SA88b]:

$$H(z) \equiv \overline{H}_1(z) = G_1(z)F_1(z^{L_1}), \qquad (4.178a)$$

where

$$F_1(z) = z^{-M_1} - (-1)^{M_1}\overline{H}_2(-z), \qquad \overline{H}_2(z) = G_2(z)F_2(z^{L_2}), \qquad (4.178b)$$

$$F_2(z) = z^{-M_2} - (-1)^{M_2}\overline{H}_3(-z), \qquad \overline{H}_3(z) = G_3(z)F_3(z^{L_3}), \qquad (4.178c)$$

$$\vdots \qquad \qquad \vdots \qquad \qquad \vdots$$

$$F_{R-2}(z) = z^{-M_{R-2}} - (-1)^{M_{R-2}}\overline{H}_{R-1}(-z), \qquad \overline{H}_{R-1}(z) = G_{R-1}(z)F_{R-1}(z^{L_{R-1}}),$$

$$\qquad (4.178d)$$

$$F_{R-1}(z) = z^{-M_{R-1}} - (-1)^{M_{R-1}}\overline{H}_R(-z), \qquad \overline{H}_R(z) = G_R(z), \qquad (4.178e)$$

with M_r for $r = 1, 2, \ldots, R - 1$ being half the order of $\overline{H}_{r+1}(z)$.

The above equations give $H(z)$ in an implicit form. The corresponding explicit form is given later. Here, the $G_r(z)$'s for $r = 1, 2, \ldots, R$ are the filters to be designed. The overall criteria are met if the $G_r(z)$'s satisfy (cf. Eq. (4.161))

$$1 - \delta_p^{(r)} \leq G_r(\omega) \leq 1 + \delta_p^{(r)} \quad \text{for } \omega \in [0, \omega_p^{(r)}], \qquad (4.179a)$$

$$-\delta_s^{(r)} \leq G_r(\omega) \leq \delta_s^{(r)} \quad \text{for } \omega \in \Omega_s^{(r)}, \qquad (4.179b)$$

where

$$\Omega_s^{(r)} = \begin{cases} \displaystyle\bigcup_{k=1}^{\lfloor L_r/2 \rfloor} \left[k\frac{2\pi}{L_r} - \omega_s^{(r)}, \min\left(k\frac{2\pi}{L_r} + \omega_s^{(r)}, \pi \right) \right] & \text{for } r < R \\[20pt] \left[\omega_s^{(R)}, \pi \right] & \text{for } r = R. \end{cases} \qquad (4.179c)$$

The $\omega_p^{(r)}$'s and $\omega_s^{(r)}$'s for $r = 1, 2, \ldots, R$ are the edges of the $\overline{H}_r(z)$'s and can be determined iteratively as

$$\omega_p^{(r)} = \pi - L_{r-1}\omega_s^{(r-1)}, \qquad \omega_s^{(r)} = \pi - L_{r-1}\omega_p^{(r-1)}, \qquad (4.179d)$$

where $\omega_p^{(1)} = \omega_p$ and $\omega_s^{(1)} = \omega_s$, and the $\delta_s^{(r)}$'s as

$$
\delta_s^{(r)} = \begin{cases} \delta_p - \displaystyle\sum_{\substack{k=1 \\ k \text{ odd}}}^{r-1} \delta_p^{(k)} & \text{for } r \text{ even} \\[2em] \delta_s - \displaystyle\sum_{\substack{k=2 \\ k \text{ even}}}^{r-1} \delta_p^{(k)} & \text{for } r \text{ odd.} \end{cases}
\tag{4.179e}
$$

In order for the overall filter to meet the given ripple requirements, $\delta_s^{(R)}$ and the $\delta_p^{(r)}$'s have to satisfy for R even

$$
\sum_{\substack{k=2 \\ k \text{ even}}}^{R} \delta_p^{(k)} = \delta_s, \qquad \delta_s^{(R)} + \sum_{\substack{k=1 \\ k \text{ odd}}}^{R-1} \delta_p^{(k)} = \delta_p
\tag{4.180a}
$$

or for R odd

$$
\sum_{\substack{k=1 \\ k \text{ odd}}}^{R} \delta_p^{(k)} = \delta_p, \qquad \delta_s^{(R)} + \sum_{\substack{k=2 \\ k \text{ even}}}^{R-1} \delta_p^{(k)} = \delta_s.
\tag{4.180b}
$$

If $2\pi/L_r - \omega_s^{(r)} < \pi/2$ for $r < R$ or $\omega_s^{(R)} < \pi/2$, then the number of multipliers in $G_r(z)$ can be reduced by designing it, using the techniques of Section 4-11-3, in the form

$$
G_r(z) = G_r^{(1)}(z^{K_r})G_r^{(2)}(z).
\tag{4.181}
$$

After some manipulations, $H(z)$ as given by Eqs. (4.178) and (4.181) can be rewritten in the explicit form shown in Table 4-9. If $G_r(z)$ is a single-stage design, then $G_r^{(1)}(z^{K_r}) \equiv 1$ and $H_r^{(1)}(z) \equiv 1$. It can be shown that in order to obtain the desired overall solution, the order of $G_r(z)$ for $r \geq 2$, denoted by N_r in Table 4-9, has to be even. The realization of $H(z)$ is given Figure 4-54,[24] where

$$
d_r = \overline{M}_r - \overline{M}_{r+1}, \quad r = 2, 3, \ldots, R - 1, \quad d_R = \overline{M}_R.
\tag{4.182}
$$

If the edges ω_p and ω_s of $H(z)$ are larger than $\pi/2$, then we set $H(z) \equiv F_1(z)$. In this case, $\delta_p^{(1)} \equiv 0$, $L_1 \equiv 1$, and $G_1(z)$, $\omega_p^{(1)}$, and $\omega_s^{(1)}$ are absent. Furthermore, $\omega_p^{(2)} = \pi - \omega_s$ and $\omega_s^{(2)} = \pi - \omega_p$, and $H_1(z)$ is absent in Figure 4-54 and in Table 4-9.

The remaining problem is to select R, the L_r's, the K_r's, and the ripple values such that the filter complexity is minimized. The following example illustrates this.

---

[24]The use of the extra delay terms can be avoided by using the transposed structure. In this case, the delay terms can be shared with $H_R^{(1)}(z^{K_R L_R})$ or, if this filter stage is not present, with $H_R^{(2)}(z^{L_R})$. This is because the overall order of this filter stage is usually larger than the sum of the d_r's.

TABLE 4-9 Explicit Form for the Transfer Function in the Jing–Fam Approach

$$H(z) = H_1(z^{\overline{L}_1})[I_2 z^{-\overline{M}_2} + H_2(z^{\overline{L}_2})[I_3 z^{-\overline{M}_3} + H_3(z^{\overline{L}_3})[\cdots$$

$$[I_{R-1}z^{-\overline{M}_{R-1}} + H_{R-1}(z^{\overline{L}_{R-1}})[I_R z^{-\overline{M}_R} + H_R(z^{\overline{L}_R})]] \cdots]]],$$

where

$$H_r(z^{\overline{L}_r}) = H_r^{(1)}(z^{K_r \overline{L}_r})H_r^{(2)}(z^{\overline{L}_r})$$

$$H_r^{(1)}(z) = G_r^{(1)}(J_r^{(1)}z), \qquad H_r^{(2)}(z) = S_r G_r^{(2)}(J_r^{(2)}z)$$

$$S_1 = 1, \qquad S_r = -(-1)^{\overline{M}_r/\overline{L}_r}, \qquad r = 2, 3, \cdots, R$$

$$J_1^{(2)} = 1, \qquad J_2^{(2)} = -1, \qquad J_r^{(2)} = -[J_{r-1}^{(2)}]^{L_{r-1}}, \qquad r = 3, 4, \cdots, R$$

$$J_r^{(1)} = [J_r^{(2)}]^{K_r}$$

$$\overline{L}_1 = 1, \qquad \overline{L}_r = \prod_{k=1}^{r-1} L_k, \qquad r = 2, 3, \cdots, R$$

$$\overline{M}_R = \tfrac{1}{2}\overline{L}_R N_R, \qquad \overline{M}_{R-r} = \overline{M}_{R-r+1} + \tfrac{1}{2}\overline{L}_{R-r}N_{R-r}, \qquad r = 1, 2, \cdots, R - 2$$

$$I_2 = 1, \qquad I_r = [J_{r-1}^{(2)}]^{\overline{M}_r/\overline{L}_r - 1}, \qquad r = 3, 4, \cdots, R$$

$$N_r = K_r N_r^{(1)} + N_r^{(2)}$$

$N_r^{(1)}$ and $N_r^{(2)}$ are the orders of $G_r^{(1)}(z)$ and $G_r^{(2)}(z)$, respectively.

Example 4.18. Consider the specifications of Example 4.14, that is, $\omega_p = 0.4\pi$, $\omega_s = 0.402\pi$, $\delta_p = 0.01$, and $\delta_s = 0.0001$. In this case, the only alternative is to select $L_1 = 2$. The resulting passband and stopband regions for $G_1(z)$ are

$$\Omega_p^{(1)} = [0, 0.4\pi], \qquad \Omega_s^{(1)} = [0.598\pi, \pi].$$

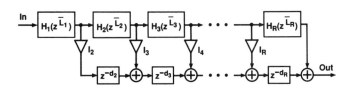

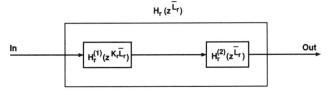

FIGURE 4-54 The structure of a filter synthesized using the Jing–Fam approach.

For $\overline{H}_2(z)$, the edges become $\omega_p^{(2)} = \pi - L_1\omega_s = 0.196\pi$ and $\omega_s^{(2)} = \pi - L_1\omega_p$ = 0.2π. For L_2, there are two alternatives to make the edges of $\overline{H}_3(z)$, $\omega_p^{(3)} = \pi - L_2\omega_s^{(2)}$ and $\omega_s^{(3)} = \pi - L_2\omega_p^{(2)}$, less than $\pi/2$. These are $L_2 = 3$ and $L_2 = 4$. We select $L_2 = 4$, giving for $G_2(z)$ the following passband and stopband regions:

$$\Omega_p^{(2)} = [0, 0.196\pi], \qquad \Omega_s^{(2)} = [0.3\pi, 0.7\pi] \cup [0.8\pi, \pi].$$

The edges of $\overline{H}_3(z)$ take the values shown in Table 4-10. By selecting $R = 5$, $L_3 = 3$, and $L_4 = 2$, the edges of $\overline{H}_4(z)$ and $\overline{H}_5(z) \equiv G_5(z)$ become as shown in Table 4-10. The passband and stopband regions for $G_3(z)$, $G_4(z)$, $G_5(z)$ are

$$\Omega_p^{(3)} = [0, 0.2\pi], \qquad \Omega_s^{(3)} = [0.4507\pi, 0.8827\pi],$$

$$\Omega_p^{(4)} = [0, 0.352\pi], \quad \Omega_s^{(4)} = [0.6\pi, \pi],$$

$$\Omega_p^{(5)} = [0, 0.2\pi], \qquad \Omega_s^{(5)} = [0.296\pi, \pi].$$

For $R = 5$, it is required that (see Eq. (4.180)) $\delta_p^{(1)} + \delta_p^{(3)} + \delta_p^{(5)} = \delta_p$ and $\delta_p^{(2)} + \delta_p^{(4)} + \delta_s^{(5)} = \delta_s$. The simplest way is to select the ripple values in these summations to be equal. In this case, the required ripples for the $G_r(z)$'s become as shown in Table 4-10. Since the stopband edges of the first and fourth subfilter are larger than $\pi/2$, they are single-stage filters and can be designed using the MPR algorithm. The remaining subfilters can be synthesized, using the techniques of Section 4-11-3, to be two-stage filters. Table 4-10 gives the parameters describing the overall filter, whereas Figure 4-55 gives the resulting overall response. This design

TABLE 4-10   Data for a Filter Designed Using the Jing–Fam Approach

	$r = 1$	$r = 2$	$r = 3$	$r = 4$	$r = 5$
$\omega_p^{(r)}$	0.4π	0.196π	0.2π	0.352π	0.2π
$\omega_s^{(r)}$	0.402π	0.2π	0.216π	0.4π	0.296π
$\delta_p^{(r)}$	$\frac{1}{3} \times 10^{-2}$	$\frac{1}{3} \times 10^{-4}$	$\frac{1}{3} \times 10^{-2}$	$\frac{1}{3} \times 10^{-4}$	$\frac{1}{3} \times 10^{-2}$
$\delta_s^{(r)}$	10^{-4}	$\frac{2}{3} \times 10^{-2}$	$\frac{2}{3} \times 10^{-4}$	$\frac{1}{3} \times 10^{-2}$	$\frac{1}{3} \times 10^{-4}$
L_r	2	4	3	2	—
K_r	—	3	2	—	3
$N_r^{(1)}$	—	26	13	—	25
$N_r^{(2)}$	38	10	10	32	19
N_r	38	88	36	32	94
L_r	1	2	8	24	48
$J_r^{(1)}$	—	-1	1	—	-1
$J_r^{(2)}$	1	-1	-1	1	-1
M_r	—	2872	2784	2640	2256
I_r	—	1	1	1	1
S_r	1	-1	-1	-1	1
d_r	—	88	144	384	2256

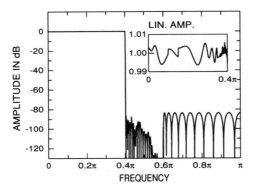

FIGURE 4-55 Amplitude response for a filter synthesized using the Jing–Fam approach.

requires only 93 multipliers and 177 adders, which are less than 6% of those required by an optimum conventional filter (1570 and 3138). The price paid for these reductions is an increased overall filter order (from 3138 to 5782). The overall filter order as well as the number of multipliers can be decreased by selecting the ripple values in a more optimum manner [JI84; SA88b]. Also, in order to arrive at the best solution, different choices of R as well as all alternatives to select the L_r's such that the $\omega_s^{(r)}$'s become smaller than $\pi/2$ are worth going through.

The above Jing–Fam synthesis technique can be used directly for almost all lowpass filter specifications. The only exceptions are filters with $\omega_p < \pi/2$ and $\omega_s > \pi/2$. One alternative to design these filters is to shift the passband and stopband edges of the filter by a factor of $\frac{3}{2}$ using a decimation by a factor of $\frac{3}{2}$ at the input of the filter and an interpolation by the same factor at the output of the filter [JI84].

When comparing the Jing–Fam approach and the frequency-response masking approach of Lim with each other, the Jing–Fam approach normally gives filters with fewer multipliers at the expense of a larger overall filter order [SA88b]. One attractive feature of the Jing–Fam approach is that it can be combined with multirate filtering to reduce the filter complexity even further [RA90] (see Chapter 14).

4-11-6 Design of Other Types of Filters

This section has concentrated on designing lowpass filters. However, the reviewed synthesis techniques can also be applied in a straightforward manner to designing other types of filters.

The design of a highpass filter $H(z)$ with edges at ω_s and ω_p and ripples of δ_s and δ_p can be performed, according to the discussion of Section 4-8-3, by first designing a lowpass filter $\overline{H}(z)$ with edges at $\overline{\omega}_p = \pi - \omega_p$ and $\overline{\omega}_s = \pi - \omega_s$ and ripples of $\overline{\delta}_p = \delta_p$ and $\overline{\delta}_s = \delta_s$. The desired highpass filter is then $H(z) = (-1)^M \overline{H}(-z)$ with M being half the filter order.

Figure 4-56 illustrates how a narrowband bandpass filter can be designed in the form $H(z) = F(z^L)G(z)$. In this case, $L = 5$, the passband and stopband edges are ω_{p1}, $\omega_{p2} = 0.7\pi \pm 0.04\pi$ and ω_{s1}, $\omega_{s2} = 0.7\pi \pm 0.06\pi$, respectively, and $\delta_p = \delta_s = 0.001$. $F(\omega)$ has been determined to be a bandpass design in such a way that $F(L\omega)$ takes care of the overall response in the interval $[0.6\pi, 0.8\pi]$, whereas $G(\omega)$ provides the desired attenuation on the extra passbands and transition bands of $F(L\omega)$. The details for designing bandpass filters of this type can be found in Saramäki et al. [SA88a]. $F(z^L)$ is of order 68 in z^L and $G(z)$ is of order 32. The overall design thus requires 52 multipliers and 100 adders, whereas the minimum order of an equivalent conventional direct-form design is 336, requiring 169 multipliers and 336 adders. The complementary bandstop filter can be implemented

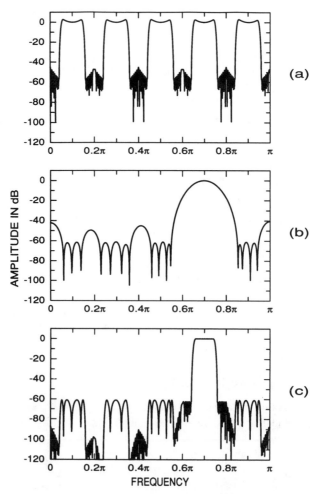

FIGURE 4-56 Amplitude responses for an optimized bandpass filter of the form $H(z) = F(z^L)G(z)$ with $L = 5$. (a) $F(z^L)$ of order 68 in z^L. (b) $G(z)$ of order 32. (c) Overall filter.

directly in the form $z^{-M} - H(z) = z^{-M} - F(z^L)G(z)$, where the delay z^{-M} can be shared with $F(z^L)$.

Another straightforward technique for designing bandpass filters has been proposed by Neuvo, Rajan, and Mitra [NE87; RA88]. This technique is based on the fact that if $\overline{H}(z)$ is a Type I or Type II linear-phase transfer function of order N, then

$$H(z) = e^{jN\omega_0/2}\overline{H}(ze^{j\omega_0}) + e^{-jN\omega_0/2}\overline{H}(ze^{-j\omega_0}) \tag{4.183}$$

is also a linear-phase transfer function of the same type and the same order. The impulse responses of these two filters are related via $h[n] = 2\overline{h}[n] \cos[(n - N/2)\omega_0]$ and the zero-phase frequency responses via (see Figure 4-57)

$$H(\omega) = \overline{H}(\omega + \omega_0) + \overline{H}(\omega - \omega_0). \tag{4.184}$$

Therefore, if $\overline{H}(z)$ is a lowpass design with edges at $\overline{\omega}_p$ and $\overline{\omega}_s$, then $H(z)$ is a bandpass design (see Figure 4-57) with passband edges at $\omega_0 \pm \overline{\omega}_p$ and stopband edges at $\omega_0 \pm \overline{\omega}_s$. The ripples of $H(z)$ are in the worst case $\delta_s = 2\overline{\delta}_s$ and $\delta_p = \overline{\delta}_p$

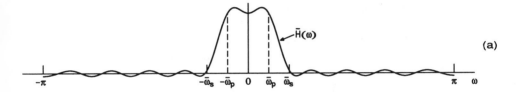

(a)

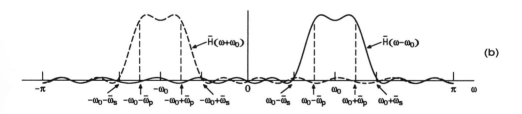

(b)

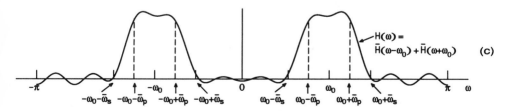

(c)

FIGURE 4-57 Design of a linear-phase bandpass filter from a lowpass prototype by modulation. (a) Response $\overline{H}(\omega)$ of the prototype with transfer function $\overline{H}(z)$. (b) Responses $\overline{H}(\omega - \omega_0)$ and $\overline{H}(\omega + \omega_0)$. (c) Response of $H(z) = e^{jN\omega_0/2}\overline{H}(ze^{j\omega_0}) + e^{-jN\omega_0/2}\overline{H}(ze^{-j\omega_0})$.

$+ \bar{\delta}_s$, where $\bar{\delta}_p$ and $\bar{\delta}_s$ are the ripples of $\bar{H}(z)$. In Neuvo et al. [NE87], an efficient structure has been derived for the overall filter in the case where the prototype lowpass filter is of the form $\bar{H}(z) = F(z^L)G(z)$, whereas in Rajan et al. [RA88] a corresponding structure has been given for the case where $\bar{H}(z)$ is designed using the frequency-response masking approach described in Section 4-11-2.

The third approach to exploit the techniques of this section is to design the bandpass filter as a cascade of a lowpass filter $H_{LP}(z)$ and a highpass filter $H_{HP}(z)$, that is, in the form

$$H(z) = H_{LP}(z)H_{HP}(z). \qquad (4.185)$$

Here, $H_{LP}(z)$ is designed to provide the second transition band $[\omega_{p2}, \omega_{s2}]$ and $H_{HP}(z)$ to provide the first transition band $[\omega_{s1}, \omega_{p1}]$. The resulting ripple in the first (second) stopband is approximately equal to the stopband ripple of the highpass (lowpass) filter, whereas the passband ripple is in the worst case equal to the sum of the passband ripples of the subfilters.

4-12   DESIGN OF FIR FILTERS USING IDENTICAL SUBFILTERS AS BASIC BUILDING BLOCKS

Another approach to reduce the cost of implementation of an FIR filter is to design it by interconnecting a number of identical subfilters with the aid of a few additional adders and multipliers. Such an approach has been suggested originally by Kaiser and Hamming [KA77a] and improved by Nakamura and Mitra [NA82]. This section concentrates on the most general approach proposed by Saramäki [SA87a]. The main advantage of using identical copies of the same filter lies in the fact that with this approach it is relatively easy to synthesize selective FIR filters without general multipliers [SA91b].

4-12-1   Filter Structures and Conditions for the Subfilter and Tap Coefficients

Figure 4-58 gives two general structures for implementing a linear-phase FIR filter as a tapped cascaded interconnection of N identical subfilters (for other alternatives, see Saramäki [SA87a]). The subfilter has a Type I linear-phase transfer function

$$F_M(z) = \sum_{n=0}^{2M} f[n]z^{-n}, \quad f[2M - n] = f[n]. \qquad (4.186)$$

The subscript M is used to emphasize that the delay of the subfilter is M. The frequency response of the structure of Figure 4-58(a) is [SA87a]

$$H(e^{j\omega}) = e^{-jNM\omega} H(\omega), \qquad (4.187a)$$

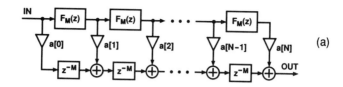

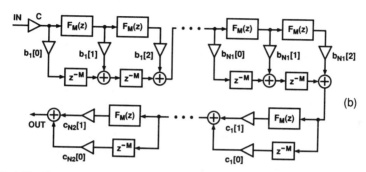

FIGURE 4-58 Two general structures for implementing a linear-phase FIR filter as a tapped cascaded interconnection of N identical subfilters of even order $2M$.

where

$$H(\omega) = \sum_{n=0}^{N} a[n] [F_M(\omega)]^n \tag{4.187b}$$

with

$$F_M(\omega) = f[M] + 2 \sum_{n=1}^{M} f[M - n] \cos n\omega. \tag{4.187c}$$

The additional tap coefficients $a[n]$ and the subfilter $F_M(z)$ can be determined such that $H(\omega)$ meets

$$1 - \delta_p \leq H(\omega) \leq 1 + \delta_p \quad \text{for } \omega \in X_p, \tag{4.188a}$$

$$- \delta_s \leq H(\omega) \leq \delta_s \quad \text{for } \omega \in X_s, \tag{4.188b}$$

where the passband and stopband regions, X_p and X_s, respectively, may consist of several bands. Based on the fact that $H(\omega)$ can be obtained from the polynomial

$$P(x) = \sum_{n=0}^{N} a[n]x^n \tag{4.189}$$

using the substitution (see Eq. (4.187b))

$$x = F_M(\omega), \tag{4.190}$$

the general simultaneous conditions for the $a[n]$'s and $F_M(z)$ can be stated as

$$1 - \delta_p \leq P(x) \leq 1 + \delta_p \quad \text{for } x_{p1} \leq x \leq x_{p2}, \qquad (4.191a)$$

$$-\delta_s \leq P(x) \leq \delta_s \quad \text{for } x_{s1} \leq x \leq x_{s2}, \qquad (4.191b)$$

$$x_{p1} \leq F_M(\omega) \leq x_{p2} \quad \text{for } \omega \in X_p, \qquad (4.192a)$$

$$x_{s1} \leq F_M(\omega) \leq x_{s2} \quad \text{for } \omega \in X_s. \qquad (4.192b)$$

Figures 4-59, 4-60, and 4-61 exemplify these relations in three different cases to be considered in more details in this section. As seen from these figures, the substitution $x = F_M(\omega)$ can be regarded as a transformation that maps the passband region $x_{p1} \leq x \leq x_{p2}$ of $P(x)$ (the stopband region $x_{s1} \leq x \leq x_{s2}$) onto the passband region X_p (stopband region X_s) of $H(\omega)$. Hence the amplitude values are preserved and only the argument axis is changed. Alternatively, $P(x)$ can be interpreted as an *amplitude change function* [KA77a], which tells that if the subfilter response $F_M(\omega)$ achieves the value x_0, then the overall response $H(\omega)$ achieves the value $P(x_0)$ without regard of the frequency. The passband and stopband regions

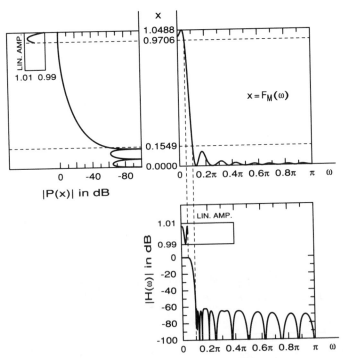

FIGURE 4-59 Design of a composite filter using four prescribed subfilters to meet the lowpass criteria: $\omega_p = 0.05\pi$, $\omega_s = 0.1\pi$, $\delta_p = 0.01$, and $\delta_s = 0.001$. In this case, $x_{s1} = 0$, $x_{s2} = 0.1549$, $x_{p1} = 0.9706$, and $x_{p2} = 1.0488$ are determined by the subfilter response $F_M(\omega)$.

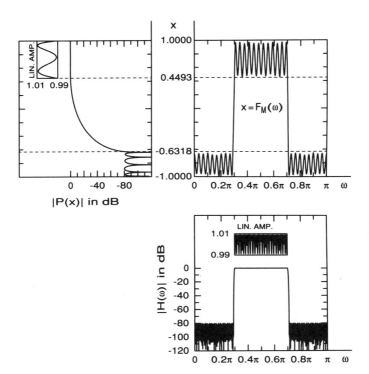

FIGURE 4-60 Design of a composite filter using eight subfilters to meet the bandpass criteria: ω_{p1}, $\omega_{p2} = 0.5\pi \pm 0.2\pi$, ω_{s1}, $\omega_{s2} = 0.5\pi \pm 0.21\pi$, $\delta_p = 0.01$, and $\delta_s = 0.0001$. Case A simultaneous specifications are used for $F_M(\omega)$ and $P(x)$ with $x_{s1} = -1$, $x_{p2} = 1$, $x_{s2} = -0.6318$, and $x_{p1} = 0.4493$.

of $F_M(\omega)$ and $H(\omega)$ are thus the same and all that happens is that the multiple use of the same subfilter reduces the large passband and stopband variations in $F_M(\omega)$ to small variations in $H(\omega)$.

The following two problems[25] can be solved in a straightforward manner:

Problem I Given N, the number of subfilters, optimize the $a[n]$'s (or, equivalently, $P(x)$) and $F_M(z)$ to meet the given criteria with the minimum subfilter order $2M$.

Problem II Given $F_M(z)$, optimize the $a[n]$'s to meet the given criteria with the minimum value of N.

For Problem II, the parameters x_{p1}, x_{p2}, x_{s1}, and x_{s2} are fixed and determined by $F_M(z)$ (see Figure 4-59). For Problem I, these parameters are adjustable and their

---

[25]In addition to these problems, $F_M(z)$ and $P(x)$ can be optimized to minimize N, the number of subfilters, for the given subfilters order $2M$. Also, the subfilter order can be minimized for the given N and the given values of the $a[n]$'s, like in the Kaiser–Hamming approach [KA77a].

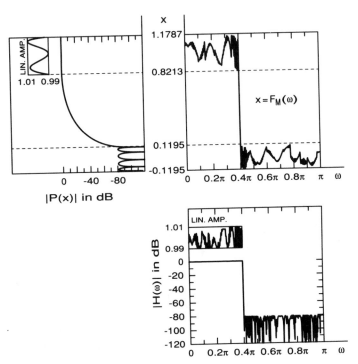

FIGURE 4-61 Design of a composite filter using eight subfilters to meet the lowpass criteria: $\omega_p = 0.4\pi$, $\omega_s = 0.402\pi$, $\delta_p = 0.01$, and $\delta_s = 0.0001$. Case B simultaneous specifications are used for $F_M(\omega)$ and $P(x)$ with x_{p1}, $x_{p2} = 1 \pm 0.1787$ and x_{s1}, $x_{s2} = \pm 0.1195$.

number can be reduced, without loss of generality, from four to two in the following two useful ways:

Case A: $x_{s1} = -1$, $x_{p2} = 1$; x_{s2} and x_{p1} are adjustable.
Case B: $x_{s1} = -\bar{\delta}_s$, $x_{s2} = \bar{\delta}_s$, $x_{p1} = 1 - \bar{\delta}_p$, $x_{p2} = 1 + \bar{\delta}_p$; $\bar{\delta}_p$ and $\bar{\delta}_s$ are adjustable.

Case A is beneficial when the subfilter is a conventional direct-form design, as in this case the subfilter is automatically peak scaled with the maximum and minimum values of $F_M(\omega)$ being $+1$ and -1, respectively (see Figure 4-60). In Case B, the subfilter criteria are conventional with the maximum passband deviation from unity (maximum stopband deviation from zero) being $\bar{\delta}_p (\bar{\delta}_s)$ (see Figure 4-61).

The additional tap coefficients in the structure of Figure 4-58(b) can be obtained by factoring $P(x)$ as given by Eq. (4.189) into the second-order and first-order terms as

$$P(x) = C \prod_{k=1}^{N_1} [b_k[2]x^2 + b_k[1]x + b_k[0]] \prod_{k=1}^{N_2} [c_k[1]x + c_k[0]], \quad (4.193)$$

where $2N_1 + N_2 = N$. The advantages of this structure compared to that of Figure 4-58(a) are that the extra delays z^{-M} can be shared with the subfilters $F_M(z)$ and its sensitivity to variations in the tap coefficients is lower [SA87a].

4-12-2 Filter Optimization

For the above problems, the design of $P(x)$ can be accomplished conveniently with the aid of an FIR filter using the substitution

$$x = \alpha \cos \Omega + \beta \tag{4.194}$$

in $P(x)$, yielding

$$G(\Omega) = P(\alpha \cos \Omega + \beta) = \sum_{n=0}^{N} g[n] \cos^n \Omega, \tag{4.195a}$$

where

$$g[n] = \sum_{r=n}^{N} a[r] \binom{r}{n} \alpha^n \beta^{r-n}. \tag{4.195b}$$

Being expressible as an Nth degree polynomial in $\cos \Omega$, $G(\Omega)$ is the zero-phase frequency response of a Type I linear-phase FIR filter of order $2N$ (see Section 4-3-3 for details) and can be designed using standard FIR filter design algorithms.[26] By selecting

$$\alpha = (x_{p2} - x_{s1})/2, \qquad \beta = (x_{p2} + x_{s1})/2, \tag{4.196}$$

the x-plane regions $[x_{p1}, x_{p2}]$ and $[x_{s1}, x_{s2}]$ are mapped, respectively, onto the Ω-plane regions $[0, \Omega_p]$ and $[\Omega_s, \pi]$, where (see Figure 4-62)

$$\Omega_p = \cos^{-1}\left[\frac{2x_{p1} - x_{p2} - x_{s1}}{x_{p2} - x_{s1}}\right], \qquad \Omega_s = \cos^{-1}\left[\frac{2x_{s2} - x_{p2} - x_{s1}}{x_{p2} - x_{s1}}\right], \tag{4.197}$$

and the conditions for $P(x)$ can be expressed in terms of $G(\Omega)$ as

$$1 - \delta_p \le G(\Omega) \le 1 + \delta_p \quad \text{for } 0 \le \Omega \le \Omega_p, \tag{4.198a}$$

$$-\delta_s \le G(\Omega) \le \delta_s \qquad \text{for } \Omega_s \le \Omega \le \pi. \tag{4.198b}$$

---

[26]These algorithms give the impulse-response coefficients $\bar{g}[n]$ of the corresponding filter. $G(\Omega)$ can be expressed as $G(\Omega) = \bar{g}[N] + 2\sum_{n=1}^{N} \bar{g}[N-n] \cos n\Omega$, which can be rewritten in the form of Eq. (4.195a) using the identity $\cos n\Omega = T_n(\cos \Omega)$, where $T_n(x)$ is the nth degree Chebyshev polynomial.

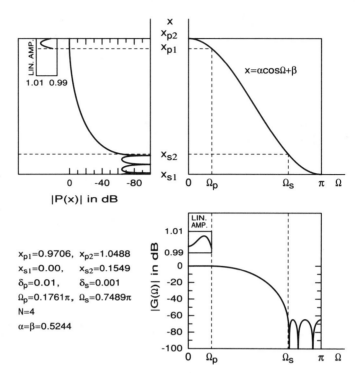

FIGURE 4-62 Design of the polynomial $P(x)$ with the aid of an FIR filter response $G(\Omega)$ for the given x_{s1}, x_{s2}, x_{p1}, and x_{p2} and the given δ_p and δ_s.

$G(\Omega)$ meeting these conventional lowpass specifications can then be converted back into the polynomial $P(x)$ using the substitution

$$\cos \Omega = [x - \beta]/\alpha. \qquad (4.199)$$

The resulting tap coefficients $a[n]$ can be determined from the $g[n]$'s according to Eq. (4.195b).

Example 4.19. This example illustrates how multiplier-free filters can be designed by first determining a computationally efficient subfilter with higher ripple values than the required ones and then using the additional tap coefficients to reduce these ripples to the desired level (Problem II). Consider again the specifications: $\omega_p = 0.05\pi$, $\omega_s = 0.1\pi$, $\delta_p = 0.01$, and $\delta_s = 0.001$. For narrowband cases of this kind, a particularly efficient subfilter transfer function is of the form [SA87a]

$$F_M(z) = \left[2^{-P} \frac{1 - z^{-K}}{1 - z^{-1}} \right]^2 [cz^{-K/2} + d(1 + z^{-K})], \qquad (4.200)$$

where 2^{-P}, with P integer-valued, is a scaling multiplier and $M = 3K/2 - 1$. An efficient implementation of this transfer function is depicted in Figure 4-63. By

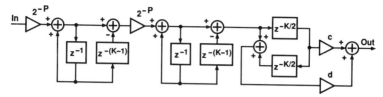

FIGURE 4-63 An implementation of the proposed subfilter.

selecting $K = 16$, $P = 4$, $c = 2$, and $d = -2^{-1}$, the resulting subfilter requires no general multipliers and $F_M(\omega)$ varies within $x_{p1} = 0.9706$ and $x_{p2} = 1.0488$ on $[0, \omega_p]$ and within $x_{s1} = 0$ and $x_{s2} = 0.1549$ on $[\omega_s, \pi]$ (see Figure 4-59). Using Eq. (4.197), the edges of $G(\Omega)$ become $\Omega_p = 0.1761\pi$ and $\Omega_s = 0.7489\pi$ (see Figure 4-62). The minimum even order $2N$ to meet the resulting criteria is 8 so that the required number of subfilters is $N = 4$. When the corresponding polynomial $P(x)$ is factored in the form of Eq. (4.193), only first-order sections are present. By fixing $c_k[1] = 1$ for $k = 1, 2, 3, 4$, the remaining coefficients take the infinite-precision values shown in Table 4-11. The given criteria are still met when these coefficient values are quantized to the easily implementable values shown also in Table 4-11. The resulting composite filter requires no general multiplications, making it very useful for hardware or VLSI implementation. The responses of Figure 4-59 are for this overall design.

Problem I can be solved by finding Ω_p and Ω_s for $G(\Omega)$ of the given even order $2N$ in such a way that it meets the criteria of Eq. (4.198) and the corresponding subfilter criteria become as mild as possible so that they can be met by the minimum even order $2M$. For any $G(\Omega)$, the corresponding polynomial $P(x)$ is obtained using the substitution of Eq. (4.199), where

$$\alpha = 1, \qquad \beta = 0 \tag{4.201}$$

TABLE 4-11 Tap Coefficients for the Filter of Example 4.19

Infinite-Precision	Coefficients
$c_1[1] = 1$	$c_1[0] = -0.009995$
$c_2[1] = 1$	$c_2[0] = -0.075844$
$c_3[1] = 1$	$c_3[0] = -0.144123$
$c_4[1] = 1$	$c_4[0] = -1.323373$
$C = -3.967595$	

Quantized	Coefficients
$c_1[1] = 1$	$c_1[0] = 0$
$c_2[1] = 1$	$c_2[0] = -2^{-4}$
$c_3[1] = 1$	$c_3[0] = -2^{-3} - 2^{-6}$
$c_4[1] = 1$	$c_4[0] = -2^{0} - 2^{-2} - 2^{-4}$
$C = -2^2$	

for Case A and

$$\alpha = \frac{2}{2 + \cos \Omega_p - \cos \Omega_s}, \qquad \beta = \frac{1 - \cos \Omega_s}{2 + \cos \Omega_p - \cos \Omega_s} \qquad (4.202)$$

for Case B. In Case A, the resulting passband and stopband regions of $P(x)$ are $[x_{p1}, 1]$ and $[-1, x_{s2}]$, where

$$x_{p1} = \cos \Omega_p, \qquad x_{s2} = \cos \Omega_s, \qquad (4.203)$$

whereas, in Case B, the corresponding regions are $[1 - \bar{\delta}_p, 1 + \bar{\delta}_p]$ and $[-\bar{\delta}_s, \bar{\delta}_s]$, where

$$\bar{\delta}_p = \frac{1 - \cos \Omega_p}{2 + \cos \Omega_p - \cos \Omega_s}, \qquad \bar{\delta}_s = \frac{1 + \cos \Omega_s}{2 + \cos \Omega_p - \cos \Omega_s}. \qquad (4.204)$$

Figure 4-64 exemplifies these relations. Note that $P(x)$ for Case B can be obtained from the Case A polynomial by simply replacing x by $[x - \beta]/\alpha$, where α and β are given by Eq. (4.202).

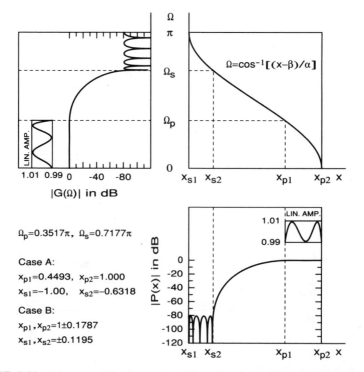

FIGURE 4-64 Relations of the Case A and Case B polynomials $P(x)$ to the best extra-ripple solution $G(\Omega)$ for the given values of δ_p and δ_s. $N = 8$, $\delta_p = 0.009$, and $\delta_s = 0.00009$.

It has turned out that the mildest subfilter criteria are typically obtained at those values of Ω_p and Ω_s for which $G(\Omega)$ has an extraripple solution for the specified values of δ_p and δ_s (see Section 4-8-2). The best extraripple solution is the one for which Ω_p and $\pi - \Omega_s$ are the most equal. As an example, Figure 4-64 gives the best extraripple solution for $N = 8$, $\delta_p = 0.009$, and $\delta_s = 0.00009$ along with the corresponding polynomials $P(x)$ in Cases A and B. Note that the allowable pass-band and stopband variations for the subfilter are in both cases huge compared to those of the overall design. In Section 4-12-3, the solutions of Figure 4-64 are used as a starting point for synthesizing multiplier-free filters for $\delta_p = 0.01$ and $\delta_s = 0.0001$.

The desired extraripple solutions can be found directly using the algorithm of Hofstetter et al. [HO71]. This algorithm can also be implemented by slightly modifying the MPR algorithm.[27]

Example 4.20. Let the specifications be $\omega_p = 0.4\pi$, $\omega_s = 0.402\pi$, $\delta_p = 0.01$, and $\delta_s = 0.0001$. Table 4-12 gives the minimum subfilter orders $2M$ for various values of N, the number of subfilters, along with the subfilter specifications in

---

[27]The modified program is available by writing to the author of this chapter.

TABLE 4-12 Data for Filters Synthesized Using Identical Subfilters

Number of Subfilters	Subfilter Order	$\cos \Omega_p$ $\cos \Omega_s$	$\bar{\delta}_p$ $\bar{\delta}_s$	Number of Distinct Coefficients	Overall Filter Order
$N = 1$	3138			1570	3138
$N = 2$	2056	0.98038	0.005000	1032	4112
		−0.94450	0.014142		
$N = 4$	1046	0.79100	0.057687	529	4184
		−0.83208	0.046348		
$N = 6$	692	0.58445	0.125781	354	4152
		−0.71930	0.084963		
$N = 8$	514	0.43774	0.183502	267	4112
		−0.62629	0.121968		
$N = 10$	408	0.33818	0.228955	216	4080
		−0.55241	0.154843		
$N = 15$	268	0.29826	0.266590	151	4020
		−0.33403	0.252999		
$N = 20$	200	0.28324	0.288587	122	4000
		−0.20044	0.321925		
$N = 30$	132	0.17262	0.355611	98	3960
		−0.15404	0.363594		
$N = 40$	98	0.11975	0.391903	91	3920
		−0.12635	0.388965		
$N = 50$	78	0.08953	0.414360	91	3900
		−0.10775	0.406070		

Cases A and B. Also, the number of distinct coefficients,[28] $N + M + 2$, and the overall filter order, $2MN$, are given in the table. The case $N = 1$ corresponds to the conventional direct-form design. It is interesting to observe that the overall filter order for all the cases is approximately 1.3 times (within 1.24–1.33) that of the direct-form design. If all the identical subfilters are implemented separately, then the overall multiplication rate per sample, $N(M + 1) + N + 1$, is higher than that of the direct-form equivalent. However, the structures of Figure 4-58 become advantageous if all the subfilters are implemented using a single subfilter by applying multiplexing. Since the subfilter order can be reduced to any value by increasing the number of subfilters, it has the potential of being realized by a fast short convolution algorithm (see Chapter 8) or implemented using an integrated FIR filter chip.

4-12-3 Design of FIR Filters Without General Multipliers

Using the structure of Figure 4-58(b), it is relatively easy to design high-order filters without general multipliers. Filters of this kind are very attractive in VLSI implementation where a general multiplier is very costly. These filters can be designed in two steps [SA91b]. In the first step, the additional tap coefficients in the structure of Figure 4-58(b) are quantized to values that are simple combinations of powers-of-two. The second step then involves designing the subfilter in such a way that there are no general multipliers. It is relatively easy to get such a subfilter without time-consuming optimization since the ripple values of the subfilter are very large (cf. Table 4-12) and, consequently, large coefficient quantization errors are allowed.[29]

To allow some quantization error for the tap coefficients in the structure of Figure 4-58(b), $G(\Omega)$ of the given order $2N$ is first designed to be the best extra-ripple solution for the passband and stopband ripples of $0.8\delta_p \cdot \cdot \cdot 0.9\delta_p$ and $0.8\delta_s \cdot \cdot \cdot 0.9\delta_s$. This $G(\Omega)$ is then converted to the Case A or Case B polynomial $P(x)$ according to the discussion of Section 4-12-2 and the passband and stopband regions of $P(x)$, $[x_{p1}, x_{p2}]$ and $[x_{s1}, x_{s2}]$, are located. In both Case A and Case B, the resulting $P(x)$ can be factored in the form

$$P(x) = D \prod_{k=1}^{N_1} (x^2 + \alpha_k x + \beta_k) \prod_{k=1}^{N_2} (x + \gamma_k). \qquad (4.205)$$

---

[28]For the structure of Figure 4-58(b), the number of distinct coefficients is $N + M + 2$ if the blocks in the structure are scaled in such a way that the coefficients $b_k[2]$ in the second-order blocks and the coefficients $c_k[1]$ in the first-order blocks become unity. This can be done without loss of generality.

[29]The rule of thumb for direct rounding of FIR filter coefficients is that if the allowed quantization error is made double, one bit is saved. Also, the order of the subfilter is significantly reduced compared to the order of the overall filter. Another rule of thumb for direct rounding is that if there are two filters with the same allowable quantization error and the order of the first filter is one-fourth that of the second filter, then the first filter requires one bit less.

A very straightforward technique to arrive at simple tap coefficients is based on expressing the coefficients of the second- and first-order terms as

$$\alpha_k = b_k[1]/b_k[2], \qquad \beta_k = b_k[0]/b_k[2], \qquad (4.206)$$

$$\gamma_k = c_k[0]/c_k[1]. \qquad (4.207)$$

The resulting $P(x)$ can be written in the following form corresponding to the structure of Figure 4-58(b):

$$P(x) = C \prod_{k=1}^{N_1} (b_k[2]x^2 + b_k[1]x + b_k[0]) \prod_{k=1}^{N_2} (c_k[1]x + c_k[0]). \qquad (4.208)$$

If the coefficients of a second-order term are desired to be quantized to two powers-of-two values, that is, values of the form $\pm 2^{-P_1} \pm 2^{-P_2}$, then a simple technique is to first set $b_k[2]$ to take all possible two powers-of-two values within $\frac{1}{2}$ and 1. Then, for each value of $b_k[2]$, the remaining coefficient values $b_k[1]$ and $b_k[0]$ are determined from Eq. (4.206) and quantized to the closest two powers-of-two values. Finally, those values that provide the closest approximations to α_k and β_k are selected. Quantized values for the $c_k[1]$'s and $c_k[0]$'s can be found in the same manner. The next step is to select C such that the average of $P(x)$ in the passband region $[x_{p1}, x_{p2}]$ is unity and it is checked whether $P(x)$ is within the limits $1 \pm \delta_p$ is the passband region $[x_{p1}, x_{p2}]$ and within the limits $\pm\delta_s$ in the stopband region $[x_{s1}, x_{s2}]$. If not, some of the coefficients require three powers-of-two representations.[30]

What remains is to design a multiplier-free subfilter such that its zero-phase frequency response $F_M(\omega)$ stays within the limits x_{p1} and x_{p2} in the passband region and within the limits x_{s1} and x_{s2} in the stopband region.

Example 4.21. Consider the ripples $\delta_p = 0.01$ and $\delta_s = 0.0001$ and the case where eight subfilters ($N = 8$) are used. Figure 4-64 gives the best extraripple solution of $G(\Omega)$ for the ripple values $0.9\delta_p$ and $0.9\delta_s$ as well as the corresponding polynomial $P(x)$ in both Case A and Case B together with its passband and stopband regions. In both cases, $P(x)$ still meets the given criteria when the additional tap coefficients are quantized, using the above procedure, to the values shown in Table 4-13. In Case A, it is required that $F_M(\omega)$ stays within the limits 0.4493 and 1 in the passband(s) and within the limits -1 and -0.6318 in the stopband(s). In Case B, the required passband and stopband ripples are $\bar{\delta}_p = 0.1787$ and $\bar{\delta}_s = 0.1195$, respectively. In both cases, the allowable passband and stopband variations are thus huge, making the quantization of the subfilter coefficients rather

---

[30]It has turned out that the second-order terms of $P(x)$ and those first-order terms of $P(x)$ that do not possess zeros in the stopband interval $[x_{s1}, x_{s2}]$ are the most sensitive and require sometimes a higher accuracy.

TABLE 4-13   Quantized Tap Coefficients for the Structure of Figure 4-58(b)

Case A		
$b_1[2] = 2^{-1} + 2^{-6}$	$b_1[1] = -2^0 + 2^{-3} - 2^{-7}$	$b_1[0] = 2^{-1} - 2^{-6} - 2^-$
$c_1[1] = 2^{-1} + 2^{-2}$	$c_1[0] = -2^0 + 2^{-4} + 2^{-5}$	
$c_2[1] = 2^0$	$c_2[0] = 2^0 - 2^{-7}$	
$c_3[1] = 2^0 - 2^{-4}$	$c_3[0] = 2^0 - 2^{-3}$	
$c_4[1] = 2^{-1} + 2^{-3}$	$c_4[0] = 2^{-1} + 2^{-6}$	
$c_5[1] = 2^0 - 2^{-3}$	$c_5[0] = 2^{-1} + 2^{-3}$	
$c_6[1] = 2^0 - 2^{-3}$	$c_6[0] = 2^{-1} + 2^{-4}$	
$C = -2^2 - 2^1 - 2^{-1} - 2^{-5}$		

Case B		
$b_1[2] = 2^0 - 2^{-3} + 2^{-6}$	$b_1[1] = -2^1 + 2^{-4} + 2^{-8}$	$b_1[0] = 2^0 + 2^{-3} - 2^{-8}$
$c_1[1] = 2^{-1} + 2^{-4} + 2^{-7}$	$c_1[0] = -2^{-1} - 2^{-2}$	
$c_2[1] = 2^{-1} + 2^{-6}$	$c_2[0] = 2^{-4} - 2^{-8}$	
$c_3[1] = 2^{-1} + 2^{-3}$	$c_3[0] = 2^{-5} + 2^{-6}$	
$c_4[1] = 2^0$	$c_4[0] = 2^{-7}$	
$c_5[1] = 2^0 - 2^{-5}$	$c_5[0] = -2^{-4}$	
$c_6[1] = 2^0 - 2^{-5}$	$c_6[0] = -2^{-3} + 2^{-6}$	
$C = -2^8 + 2^4 + 2^3 - 2^0$		

trivial. To illustrate this, the design of a bandpass filter for the passband edges of ω_{p1}, $\omega_{p2} = 0.5\pi \pm 0.2\pi$ and stopband edges of ω_{s1}, $\omega_{s2} = 0.5\pi \pm 0.21\pi$ is considered. The minimum even subfilter order to meet the Case A criteria is 112. If the subfilter order is increased to 120, then the given criteria are still met when direct rounding is used to quantize the coefficient values to the closest two powers-of-two values in 8-bit representations. These values are shown in Table 4-14.[31] Note that because of the symmetry of the filter specifications, every second impulse-response value is equal to zero. The responses of Figure 4-60 are for the resulting composite design. The overall filter order is 960, whereas the minimum order of an equivalent conventional direct-form design is 636. The price paid for getting a multiplier-free design is thus a 50% increase in the filter order. If the subfilter order is increased to 136, then with direct rounding we end up with the very simple 6-bit coefficient values of Table 4-15.[32]

Using the frequency-response masking approach described in Section 4-11-2, the lowpass filter criteria for Case B with $\omega_p = 0.4\pi$ and $\omega_s = 0.402\pi$ are met by a subfilter of the form $F_M(z) = F(z^L)G_1(z) + [z^{-N_{FL}/2} - F(z^L)]G_2(z)$, where L

---

[31]The infinite-precision filter has been designed such that the desired function is 0.7247 in the passband and -0.8159 in the stopbands and the weighting function is 1 in the passband and 1.55 in the stopbands. With this weighting, the allowable quantization errors in the passband and in the stopbands are about the same.

[32]The infinite-precision filter has been designed using a stopband weighting of 1.7.

TABLE 4-14 Quantized Coefficients[a] for a Bandpass Subfilter of Order 120

$f[0] = 8 \times 2^{-8}$	$f[2] = 20 \times 2^{-8}$	$f[4] = -15 \times 2^{-8}$	$f[6] = 6 \times 2^{-8}$	$f[8] = 3 \times 2^{-8}$
$f[10] = -4 \times 2^{-8}$	$f[12] = -2 \times 2^{-8}$	$f[14] = 5 \times 2^{-8}$	$f[16] = -1 \times 2^{-8}$	$f[18] = -5 \times 2^{-8}$
$f[20] = 4 \times 2^{-8}$	$f[22] = 4 \times 2^{-8}$	$f[24] = -6 \times 2^{-8}$	$f[26] = -1 \times 2^{-8}$	$f[28] = 7 \times 2^{-8}$
$f[30] = -4 \times 2^{-8}$	$f[32] = -6 \times 2^{-8}$	$f[34] = 8 \times 2^{-8}$	$f[36] = 3 \times 2^{-8}$	$f[38] = -12 \times 2^{-8}$
$f[40] = 4 \times 2^{-8}$	$f[42] = 12 \times 2^{-8}$	$f[44] = -12 \times 2^{-8}$	$f[46] = -7 \times 2^{-8}$	$f[48] = 20 \times 2^{-8}$
$f[50] = -4 \times 2^{-8}$	$f[52] = -28 \times 2^{-8}$	$f[54] = 28 \times 2^{-8}$	$f[56] = 34 \times 2^{-8}$	$f[58] = -120 \times 2^{-8}$
$f[60] = -48 \times 2^{-8}$				

[a] $f[n]$ is zero for n odd.

TABLE 4-15 Quantized Coefficients[a] for a Bandpass Subfilter of Order 136

$f[0] = -2 \times 2^{-6}$	$f[2] = 5 \times 2^{-6}$	$f[4] = 0$	$f[6] = 0$	$f[8] = 1 \times 2^{-6}$
$f[10] = 0$	$f[12] = -1 \times 2^{-6}$	$f[14] = 0$	$f[16] = 1 \times 2^{-6}$	$f[18] = -1 \times 2^{-6}$
$f[20] = -1 \times 2^{-6}$	$f[22] = 1 \times 2^{-6}$	$f[24] = 0$	$f[26] = -1 \times 2^{-6}$	$f[28] = 1 \times 2^{-6}$
$f[30] = 1 \times 2^{-6}$	$f[32] = -2 \times 2^{-6}$	$f[34] = 0$	$f[36] = 2 \times 2^{-6}$	$f[38] = -1 \times 2^{-6}$
$f[40] = -2 \times 2^{-6}$	$f[42] = 2 \times 2^{-6}$	$f[44] = 1 \times 2^{-6}$	$f[46] = -3 \times 2^{-6}$	$f[48] = 1 \times 2^{-6}$
$f[50] = 3 \times 2^{-6}$	$f[52] = -3 \times 2^{-6}$	$f[54] = -2 \times 2^{-6}$	$f[56] = 5 \times 2^{-6}$	$f[58] = -1 \times 2^{-6}$
$f[60] = -7 \times 2^{-6}$	$f[62] = 7 \times 2^{-6}$	$f[64] = 8 \times 2^{-6}$	$f[66] = -30 \times 2^{-6}$	$f[68] = -12 \times 2^{-6}$

[a] $f[n]$ is zero for n odd.

TABLE 4-16 Quantized Coefficients for a Subfilter Designed Using the Frequency-Response Masking Approach

$f[0] = 2 \times 2^{-6}$	$f[1] = -4 \times 2^{-6}$	$f[2] = -3 \times 2^{-6}$	$f[3] = -2 \times 2^{-6}$	$f[4] = 1 \times 2^{-6}$
$f[5] = 0$	$f[6] = -1 \times 2^{-6}$	$f[7] = -2 \times 2^{-6}$	$f[8] = 0$	$f[9] = 2 \times 2^{-6}$
$f[10] = 1 \times 2^{-6}$	$f[11] = -2 \times 2^{-6}$	$f[12] = -3 \times 2^{-6}$	$f[13] = 1 \times 2^{-6}$	$f[14] = 4 \times 2^{-6}$
$f[15] = 1 \times 2^{-6}$	$f[16] = -5 \times 2^{-6}$	$f[17] = -6 \times 2^{-6}$	$f[18] = 5 \times 2^{-6}$	$f[19] = 17 \times 2^{-6}$
$f[20] = 28 \times 2^{-6}$				

$g_1[0] = 2 \times 2^{-6}$	$g_1[1] = 3 \times 2^{-6}$	$g_1[2] = -1 \times 2^{-6}$	$g_1[3] = -2 \times 2^{-6}$	$g_1[4] = -1 \times 2^{-6}$
$g_1[5] = 3 \times 2^{-6}$	$g_1[6] = 2 \times 2^{-6}$	$g_1[7] = -4 \times 2^{-6}$	$g_1[8] = -5 \times 2^{-6}$	$g_1[9] = 4 \times 2^{-6}$
$g_1[10] = 20 \times 2^{-6}$	$g_1[11] = 28 \times 2^{-6}$			

$g_2[0] = -3 \times 2^{-6}$	$g_2[1] = -1 \times 2^{-6}$	$g_2[2] = 1 \times 2^{-6}$	$g_2[3] = 2 \times 2^{-6}$	$g_2[4] = 1 \times 2^{-6}$
$g_2[5] = -1 \times 2^{-6}$	$g_2[6] = -2 \times 2^{-6}$	$g_2[7] = 0$	$g_2[8] = 3 \times 2^{-6}$	$g_2[9] = 3 \times 2^{-6}$
$g_2[10] = -1 \times 2^{-6}$	$g_2[11] = -5 \times 2^{-6}$	$g_2[12] = -2 \times 2^{-6}$	$g_2[13] = 7 \times 2^{-6}$	$g_2[14] = 19 \times 2^{-6}$
$g_2[15] = 24 \times 2^{-6}$				

= 16, the order of $F(z)$ is 40, and the orders of $G_1(z)$ and $G_2(z)$ are 22 and 30, respectively. This filter has been slightly overdesigned[33] such that direct rounding can be used to quantize the filter coefficients to the 6-bit values shown in Table 4-16. Only one coefficient ($g_2[14] = 19 \times 2^{-6}$) requires a three powers-of-two representation. Note that no optimization has been used in finding these coefficient values. The overall order is 70% higher than that of a direct-form equivalent (5360 compared to 3138). The responses of Figure 4-61 are for the resulting overall design.

In the above, direct rounding has been used for quantizing the subfilter coefficients. Another technique, leading to better results, is to use mixed integer linear programming [LI83a, LI90].

4-13 SUMMARY

Several techniques have been reviewed for synthesizing linear-phase FIR filters along with efficient realization methods. This chapter started with very fast design methods: designs based on windowing (Section 4-4), least-squared-error designs (Section 4-5), maximally flat designs (Section 4-6), and simple analytic designs (Section 4-7). All these techniques suffer from some drawbacks. For filters designed using windowing, the passband and stopband ripples are approximately equal. Maximally flat filters are useful in applications where the signal should be preserved very accurately near the zero frequency, and least-squared-error filters in applications where white Gaussian noise in the filter stopband is desired to be attenuated as much as possible. However, if the maximum deviation from the given response is of main interest, then filters designed in the minimax sense meet the criteria with a significantly reduced filter order. These filters have been designed in Section 4-8 using the Remez multiple exchange algorithm. This algorithm is the most powerful method for designing arbitrary-magnitude FIR filters in the minimax sense. It can also be used in a straightforward manner for synthesizing nonlinear-phase FIR filters (Section 4-9).

In some applications there are additional constraints in the time domain or in the frequency domain, such as in the case of Nyquist filters or in the case where the transient part of the step response is restricted to vary within the given limits. Filters meeting these additional constraints can be designed in some cases by properly using the Remez algorithm as a subroutine (Section 4-10). However, the most flexible design method for constrained approximation problems of various kinds is linear programming, which has been considered in detail in Section 4-10.

The last two sections of this chapter have been devoted to the design of computationally efficient linear-phase FIR filters. Section 4-11 concentrated on syn-

---

[33]When designing $G_1(z)$ and $G_2(z)$, a uniform weighting has been used both in the passband and in the stopband. It has turned out that for filters with large passband and stopband ripple values, it is not worth reducing weightings in some parts of the passband and stopband like for filters with small ripple values (see Section 4-11-2). The stopband weighting of 1.6 has been used.

thesizing filters that use as basic building blocks the transfer functions obtained by replacing each unit delay element in a conventional transfer function by multiple delays. By properly combining these transfer functions together, filters with significantly fewer multipliers and adders can be designed. Section 4-12 introduced another approach to reduce the cost of implementation of FIR filters, based on the use of identical subfilters. With this approach, multiplier-free highly selective filters can be designed in a systematic manner. In addition to these design methods, filters with a reduced number of arithmetic operations can be synthesized using structures that generate piecewise-polynomial impulse responses [CH84a, CH84b] or FIR filters that mimic the performance of IIR filters [FA81; SA90]. In these two approaches, very effective implementations are achieved by using feedback loops. Furthermore, multirate filtering (see Chapter 14) and fast convolution algorithms (see Chapter 8) provide efficient implementations for FIR filters.

REFERENCES

[AN82] A. Antoniou, Accelerated procedure for the design of equiripple nonrecursive digital filters. *IEE Proc., Part G: Electron. Circuits Syst.* **129,** 1–10 (1982); and *Erratum: ibid.* p. 107 (1982).

[AN83] A. Antoniou, New improved method for the design of weighted-Chebyshev, nonrecursive, digital filters. *IEEE Trans. Circuits Syst.* **CAS-30,** 740–750 (1983).

[BL58] R. B. Blackman and J. W. Tukey, *The Measurement of Power Spectra.* Dover, New York, 1958.

[BO81] R. Boite and H. Leich, A new procedure for the design of high order minimum phase FIR digital or CCD filters. *Signal Process.* **3,** 101–108 (1981).

[BO82] F. Bonzanigo, Some improvements to the design programs for equiripple FIR filters. *Proc. IEEE Int. Conf. Acoust. Speech, Signal Process., Paris, France, 1982,* pp. 274–277 (1982).

[BO84] R. Boite and H. Leich, Comments on 'A fast procedure to design equiripple minimum-phase FIR filters.' *IEEE Trans. Circuits Syst.* **CAS-31,** 503–504 (1984).

[CH66] E. W. Cheney, *Introduction to Approximation Theory.* McGraw-Hill, New York, 1966.

[CH84a] S. Chu and C. S. Burrus, Efficient recursive realizations of FIR filters. Part I: The filter structures. *Circuits, Syst. Signal Process.* **3**(1), 3–20 (1984).

[CH84b] S. Chu and C. S. Burrus, Efficient recursive realizations of FIR filters. Part II: Design and applications. *Circuits, Syst. Signal Process.* **3**(1), 21–57 (1984).

[DA63] G. Dantzig, *Linear Programming and Extensions.* Princeton University Press, Princeton, NJ, 1963.

[DO79] J. J. Dongarra, J. R. Bunch, C. B. Moler, and G. W. Stewart, *LINPACK Users' Guide.* SIAM, Philadelphia, PA, 1979.

[FA74] D. C. Farden and L. L. Scharf, Statistical design of nonrecursive digital filters. *IEEE Trans. Acoust., Speech,, Signal Process.* **ASSP-22,** 188–196 (1974).

[FA81] A. T. Fam, MFIR filters: Properties and applications. *IEEE Trans. Acoust., Speech, Signal Process.* **ASSP-29,** 1128–1136 (1981).

[GO81] E. Goldberg, R. Kurshan, and D. Malah, Design of finite impulse response digital filters with nonlinear phase response. *IEEE-Trans. Acoust., Speech, Signal Process.* **CAS-29**, 1003–1010 (1981).

[GR83] F. Grenez, Design of linear or minimum-phase FIR filters by constrained Chebyshev approximation. *Signal Process.* **5**, 325–332 (1983).

[HA63] G. Hadley, *Linear Programming*. Addison-Wesley, Reading, MA, 1963.

[HA77] R. W. Hamming, *Digital Filters*. Prentice-Hall, Englewood Cliffs, NJ, 1977.

[HA78] F. J. Harris, On the use of windows for harmonic analysis with the discrete Fourier transform. *Proc. IEEE* **66**, 51–83 (1978).

[HA87] F. J. Harris, Multirate FIR filters for interpolating and desampling. In *Handbook of Digital Signal Processing, Engineering Applications* (D. F. Elliott, ed.), Chapter 3 (see Appendix). Academic Press, San Diego, CA, 1987.

[HE68] H. D. Helms, Nonrecursive digital filters: Design methods for achieving specifications on frequency response. *IEEE Trans. Audio Electroacoust.* **AU-16**, 336–342 (1968).

[HE70] O. Herrmann and H. W. Schüssler, Design of nonrecursive digital filters with minimum phase. *Electron. Lett.* **6**, 329–330 (1970); also reprinted in L. R. Rabiner and C. M. Rader, eds., *Digital Signal Processing*, pp. 185–186. IEEE Press, New York, 1972.

[HE71a] H. D. Helms, Digital filters with equiripple or minimax responses. *IEEE Trans. Audio Electroacoust.* **AU-19**, 87–93 (1971); also reprinted in L. R. Rabiner and C. M. Rader, eds., *Digital Signal Processing*, pp. 131–137. IEEE Press, New York, 1972.

[HE71b] O. Herrmann, On the approximation problem in nonrecursive digital filter design. *IEEE Trans. Circuit Theory* **CT-18**, 411–413 (1971); also reprinted in L. R. Rabiner and C. M. Rader, eds., *Digital Signal Processing*, pp. 202–203. IEEE Press, New York, 1972.

[HE73] O. Herrmann, L. R. Rabiner, and D. S. K. Chan, Practical design rules for optimum finite impulse response lowpass digital filters. *Bell Syst. Tech. J.* **52**, 769–799 (1973).

[HO71] E. Hofstetter, A. Oppenheim, and J. Siegel, A new technique for the design of non-recursive digital filters. *Proc. 5th Annu. Princeton Conf. Inf. Sci. Syst.*, pp. 64–72 (1971); also reprinted in L. R. Rabiner and C. M. Rader, eds., *Digital Signal Processing*, pp. 187–194. IEEE Press, New York, 1972.

[JI84] Z. Jing and A. T. Fam, A new structure for narrow transition band, lowpass digital filter design. *IEEE Trans. Acoust., Speech, Signal Process.* **ASSP-32**, 362–370 (1984).

[KA63] J. F. Kaiser, Design methods for sampled data filters. *Proc. Allerton Conf. Circuit Syst. Theory, 1st, Monticello, IL, 1963*, pp. 221–236 (1963); also reprinted in L. R. Rabiner and C. M. Rader, eds., *Digital Signal Processing*, pp. 20–34. IEEE Press, New York, 1972.

[KA66] J. F. Kaiser, Digital filters. In *System Analysis by Digital Computer* (F. F. Kuo and J. F. Kaiser, eds.), Chapter 7. Wiley, New York, 1966.

[KA74] J. F. Kaiser, Nonrecursive digital filter design using the I_0-sinh window function. *Proc. IEEE Int. Symp. Circuits Syst., 1974, San Francisco*, pp. 20–23 (1974); also reprinted in Digital Signal Processing Committee and IEEE ASSP, eds., *Selected Papers in Digital Signal Processing, II*, pp. 123–126. IEEE Press, New York, 1975.

[KA77a] J. F. Kaiser and R. W. Hamming, Sharpening the response of a symmetric non-recursive filter by multiple use of the same filter. *IEEE Trans. Acoust., Speech, Signal Process.* **ASSP-25**, 415–422 (1977).

[KA77b] J. F. Kaiser and W. A. Reed, Data smoothing using low-pass digital filters. *Rev. Sci. Instrum.* **48**, 1447–1457 (1977).

[KA79] J. F. Kaiser, Design subroutine (MXFLAT) for symmetric FIR low pass digital filters with maximally-flat pass and stop bands. In *Programs for Digital Signal Processing* (Digital Signal Processing Committee and IEEE ASSP, eds.), pp. 5.3-1–5.3-6. IEEE Press, New York, 1979.

[KA83a] J. F. Kaiser and K. Steiglitz, Design of FIR filters with flatness constraints. *Proc. IEEE Int. Conf. Acoust., Speech, Signal Process., Boston, 1983*, pp. 197–200 (1983).

[KA83b] Y. Kamp and C. J. Wellekens, Optimal design of minimum-phase FIR filters. *IEEE Trans. Acoust., Speech, Signal Process.* **ASSP-31**, 922–926 (1983).

[KE72] W. C. Kellogg, Time domain design of nonrecursive least mean-square digital filters. *IEEE Trans. Audio Electrosacoust.* **AU-20**, 155–158 (1972).

[KI88] H. Kikuchi, H. Watanabe, and T. Yanagisawa, Interpolated FIR filters using cyclotomic polynomials. *Proc. IEEE Int. Symp. Circuits Syst., Espoo, Finland, 1988*, pp. 2009–2012 (1988).

[LA73] A. Land and S. Powell, *Fortran Codes for Mathematical Programming*. Wiley, New York, 1973.

[LE75] P. Leistner and T. W. Parks, On the design of FIR digital filters with optimum magnitude and minimum phase. *Arch. Elektron. Uebertragungstech.* **29**, 270–274 (1975).

[LI83a] Y. C. Lim and S. R. Parker, FIR filter design over a discrete powers-of-two coefficient space. *IEEE Trans. Acoust. Speech, Signal Process.* **ASSP-31**, 583–591 (1983).

[LI83b] Y. C. Lim, Efficient special purpose linear programming for FIR filter design. *IEEE Trans. Acoust., Speech, Signal Process.* **ASSP-31**, 963–968 (1983).

[LI83c] Y. C. Lim and S. R. Parker, Discrete coefficient FIR digital filter design based upon an LMS criteria. *IEEE Trans. Circuits Syst.* **CAS-30**, 723–739 (1983).

[LI85] J. K. Liang, R. J. P. de Figueiredo, and F. C. Lu, Design of optimal Nyquist, partial response, Nth band, and nonuniform tap spacing FIR digital filters using linear programming techniques. *IEEE Trans. Circuits System.* **CAS-32**, 386–392 (1985).

[LI86] Y. C. Lim, Frequency-response masking approach for the synthesis of sharp linear phase digital filters. *IEEE Trans. Circuits System.* **CAS-33**, 357–364 (1986).

[LI90] Y. C. Lim, Design of discrete-coefficient-value linear phase FIR filters with optimum normalized peak ripple magnitude. *IEEE Trans. Circuits Syst.* **CAS-37**, 1480–1486, (1990).

[MC73a] J. H. McClellan and T. W. Parks, A unified approach to the design of optimum FIR linear phase digital filters. *IEEE Trans. Circuit Theory* **CT-20**, 697–701 (1973).

[MC73b] J. H. McClellan, T. W. Parks, and L. R. Rabiner, A computer program for designing optimum FIR linear phase digital filters. *IEEE Trans. Audio Electroacoust.* **AU-21**, 506–526 (1973); also reprinted in Digital Signal Processing Committee and IEEE ASSP, eds., *Selected Papers in Digital Signal Processing, II*, pp. 97–117. IEEE Press, New York, 1975.

[MC79] J. H. McClellan, T. W. Parks, and L. R. Rabiner, FIR linear phase filter design program. In *Programs for Digital Signal Processing*, (Digital Signal Processing committee and IEEE ASSP, eds.), pp. 5.1-1–5.1-13. IEEE Press, New York, 1979.

[MI82a] G. A. Mian and A. P. Naider, A fast procedure to design equiripple minimum-phase FIR filters. *IEEE Trans. Circuits Syst.* **CAS-29,** 327–331 (1982).

[MI82b] F. Mintzer, On half-band, third-band, and *N*th-band FIR filters and their design. *IEEE Trans. Acoust., Speech, Signal Process.* **ASSP-30,** 734–738 (1982).

[NA82] S. Nakamura and S. K. Mitra, Design of FIR digital filters using tapped cascaded FIR subfilters. *Circuits, Syst. Signal Process.* **1**(1), 43–56 (1982).

[NE84] Y. Neuvo, C.-Y. Dong, and S. K. Mitra, Interpolated finite impulse response filters. *IEEE Trans. Acoust., Speech, Signal Process.* **ASSP-32,** 563–570 (1984).

[NE87] Y. Neuvo, G. Rajan, and S. K. Mitra, Design of narrow-band FIR bandpass digital filters with reduced arithmetic complexity. *IEEE Trans. Circuits Syst.* **CAS-34,** 409–419 (1987).

[NG88] T. Q. Nguyen, T. Saramäki, and P. P. Vaidyanathan, Eigenfilters for the design of special transfer functions with applications in multirate signal processing. *Proc. IEEE Int. Conf. Acoust., Speech, Signal Process., New York, 1988*, pp. 1467–1470 (1988).

[OP89] A. V. Oppenheim and R. W. Schafer, *Discrete-Time Signal Processing*, Chapters 5 and 7. Prentice-Hall, Englewood Cliffs, NJ, 1989.

[PA72a] T. W. Parks and J. H. McClellan, Chebyshev approximation for nonrecursive digital filters with linear phase. *IEEE Trans. Circuit Theory* **CT-19,** 189–194 (1972).

[PA72b] T. W. Parks and J. H. McClellan, A program for the design of linear phase finite impulse response digital filters. *IEEE Trans. Audio Electroacoust.* **AU-20,** 195–199 (1972).

[PA73] T. W. Parks, L. R. Rabiner, and J. H. McClellan, On the transition width of finite impulse response digital filters. *IEEE Trans. Audio Electroacoust.* **AU-21,** 1–4 (1973).

[PA87] T. W. Parks and C. S. Burrus, *Digital Filter Design*, Chapters 2 and 3. Wiley, New York, 1987.

[RA72a] L. R. Rabiner, The design of finite impulse response digital filters using linear programming techniques. *Bell Syst. Tech. J.* **51,** 1177–1198 (1972).

[RA72b] L. R. Rabiner, Linear program design of finite impulse response (FIR) digital filters. *IEEE Trans. Audio Electroacoust.* **AU-20,** 280–288 (1972).

[RA73a] L. R. Rabiner and O. Herrmann, The predictability of certain optimum finite impulse response digital filters. *IEEE Trans. Circuit Theory* **CT-20,** 401–408 (1973).

[RA73b] L. R. Rabiner and O. Herrmann, On the design of optimum FIR low-pass filters with even impulse response duration. *IEEE Trans. Audio Electroacoustic* **AU-21,** 329–336 (1973).

[RA73c] L. R. Rabiner, Approximate design relationships for low-pass FIR digital filters. *IEEE Trans. Audio Electroacoust.* **AU-21,** 456–460 (1973); also reprinted in Digital Signal Processing Committee and IEEE ASSP, eds., *Selected Papers in Digital Signal Processing, II*, pp. 118–122. IEEE Press, New York, 1975.

[RA74] L. R. Rabiner, J. F. Kaiser, and R. W. Schafer, Some considerations in the design of multiband finite impulse-response digital filters. *IEEE Trans. Acoust., Speech, Signal Process.* **ASSP-22,** 462–472 (1974).

[RA75a] L. R. Rabiner and B. Gold, *Theory and Application of Digital Signal Processing*, Chapter 3. Prentice-Hall, Englewood Cliffs, NJ, 1975.

[RA75b] L. R. Rabiner, J. H. McClellan, and T. W. Parks, FIR digital filter design techniques using weighted Chebyshev approximation. *Proc. IEEE* **63**, 595–610 (1975); also reprinted in Digital Signal Processing Committee and IEEE ASSP, eds., *Selected Papers in Digital Signal Processing, II*, pp. 81–96. IEEE Press, New York, 1975.

[RA88] G. Rajan, Y. Neuvo, and S. K. Mitra, On the design of sharp cutoff wideband FIR filters with reduced arithmetic complexity. *IEEE Trans. Circuits Syst.* **CAS-35**, 1447–1454 (1988).

[RA90] T. Ramstad and T. Saramäki, Multistage, multirate FIR filter structures for narrow transition-band filters. *Proc. IEEE Int. Symp. Circuits Syst., New Orleans, Louisiana, 1990*, pp. 2017–2021 (1990).

[RI64] J. R. Rice, *The Approximation of Functions*, Vol. 1. Addison-Wesley, Reading, MA, 1964.

[SA87a] T. Saramäki, Design of FIR filters as a tapped cascaded interconnection of identical subfilters. *IEEE Trans. Circuits Syst.* **CAS-34**, 1011–1029 (1987).

[SA87b] T. Saramäki and Y. Neuvo, A Class of FIR Nyquist (*N*th-Band) filters with zero intersymbol interference. *IEEE Trans. Circuits Syst.* **CAS-34**, 1182–1190 (1987).

[SA88a] T. Saramäki, Y. Neuvo, and S. K. Mitra, Design of computationally efficient interpolated FIR filters. *IEEE Trans. Circuits Syst.* **CAS-35**, 70–88 (1988).

[SA88b] T. Saramäki and A. T. Fam, Subfilter approach for designing efficient FIR filters. *Proc. IEEE Int. Symp. Circuits Syst., Espoo, Finland, 1988*, pp. 2903–2915 (1988).

[SA88c] H. Samueli, On the design of optimal equiripple FIR digital filters for data transmission applications. *IEEE Trans. Circuits Syst.* **CAS-35**, 1542–1546 (1988).

[SA89a] T. Saramäki, A class of window functions with nearly minimum sidelobe energy for designing FIR filters. *Proc. IEEE Int. Symp. Circuits Syst., Portland, Oregon, 1989*, pp. 359–362 (1989).

[SA89b] T. Saramäki and S. K. Mitra, Design of efficient interpolated FIR filters using recursive running sum filters. *Proc. Int. Symp. Networks, Syst. Signal Process., 6th, Zagreb, Yugoslavia*, pp. 20–23 (1989).

[SA90] T. Saramäki and A. T. Fam, Properties and structures of linear-phase FIR filters based on switching and resetting of IIR filters. *Proc. IEEE Int. Symp. Circuits Syst., New Orleans, Louisiana, 1990*, pp. 3271–3274 (1990).

[SA91a] T. Saramäki, Adjustable windows for the design of FIR filters—A tutorial (invited paper). *Proc. Mediter. Electrotech. Conf., 6th, Ljubljana, Yugoslavia, 1991*, pp. 28–33 (1991).

[SA91b] T. Saramäki, A systematic technique for designing highly selective multiplier-free FIR filters. *Proc. IEEE Int. Conf. Circuits Syst., Singapore, 1991*, pp. 484–487 (1991).

[SC88] H. W. Schüssler and P. Steffen, Some advanced topics in filter design. In *Advanced Topics in Signal Processing*. (J. S. Lim and A. V. Oppenheim, eds.), Chapter 8. Prentice-Hall, Englewood Cliffs, NJ, 1988.

[ST79] K. Steiglitz, Optimal design of FIR digital filters with monotone passband response. *IEEE Trans. Acoust., Speech, Signal Process.* **ASSP-27**, 643–649 (1979).

[TU70] D. W. Tufts and J. T. Francis, Designing digital lowpass filters: Comparison of some methods and criteria. *IEEE Trans. Audio Electroacoust.* **AU-18,** 487–494 (1970).

[VA84] P. P. Vaidyanathan, On maximally-flat linear-phase FIR filters. *IEEE Trans. Circuits Syst.* **CAS-31,** 830–832 (1984).

[VA85] P. P. Vaidyanathan, Optimal design of linear-phase FIR digital filters with very flat passbands and equiripple stopbands. *IEEE Trans. Circuits Syst.* **CAS-32,** 904–917 (1985).

[VA87] P. P. Vaidyanathan and T. Q. Nguyen, Eigenfilters: A new approach to least-squares FIR filter design and applications including Nyquist filters. *IEEE Trans. Circuits Syst.* **CAS-34,** 11–23 (1987).

[VA89] P. P. Vaidyanathan, T. Q. Nguyen, Z. Doganata, and T. Saramäki, Improved technique for design of perfect reconstruction FIR QMF bands with lossless polyphase matrices. *IEEE Trans. Acoust., Speech, Signal Process.* **ASSP-37,** 1042–1056 (1989).

5 Infinite Impulse Response Digital Filter Design

WILLIAM E. HIGGINS

Department of Electrical and Computer Engineering
The Pennsylvania State University, University Park

DAVID C. MUNSON, JR.

Department of Electrical and Computer Engineering
University of Illinois, Urbana-Champaign

5-1 INTRODUCTION

Infinite impulse response (IIR) digital filters have transfer functions of the form

$$H(z) = \frac{a_0 + a_1 z^{-1} + \cdots + a_M z^{-M}}{1 + b_1 z^{-1} + \cdots + b_N z^{-N}}.$$

IIR filters can generally approximate a frequency response magnitude specification using a lower order filter than that required by a FIR design. This is especially true for frequency responses having short transitions from passband to stopband, since IIR filters have poles as well as zeros. A pole can be placed near the unit circle to sharply peak up the frequency response and a zero can be placed nearby on the unit circle to rapidly bring the frequency response to zero. IIR filters, however, unlike FIR filters, cannot be designed to have exact linear phase over the passband.

The most common technique used for designing IIR digital filters involves first designing an analog prototype filter and then transforming the prototype to a digital filter, as described in Sections 5-2 through 5-4. This approach is popular, because a vast literature on analog filter design exists and many approximation theories for analog filter design lead to closed-form solutions for the transfer function. Among the transformation methods available, the bilinear transformation is usually the method of choice for the design of filters having piecewise-constant frequency response (e.g., lowpass, bandpass, highpass, and bandstop filters). Section 5-5

Handbook for Digital Signal Processing, Edited by Sanjit K. Mitra and James F. Kaiser.
ISBN 0-471-61995-7 © 1993 John Wiley & Sons, Inc.

provides several examples of lowpass filters designed using the bilinear transformation method. Section 5-6 describes how to transform lowpass filters into highpass, bandpass, or bandstop filters, and Section 5-7 discusses the design of allpass phase equalizers for the purpose of linearizing the phase response of IIR filters.

For arbitrary frequency response characteristics, it is usually necessary to resort to direct numerical optimization using a computer. Section 5-8 briefly describes several such techniques. It should be emphasized that the design of an IIR filter, approximating arbitrary phase and magnitude specifications, is an extraordinarily difficult problem and that no completely satisfactory solution exists. Therefore the reader facing this problem may wish to design filters via several of the techniques of Section 5-8 and use the best resulting design. Some of the numerical methods described in Section 5-8 are also useful for designing phase equalizers to help linearize the phase of the designs produced using the transformation methods described in Sections 5-2 through 5-4.

Section 5-9 summarizes this chapter and also points the way to computer software (commercial and otherwise) that is available for the automatic design of IIR digital filters. For all but the simplest filters, the use of such software is highly recommended.

5-2 IIR DIGITAL FILTER DESIGN BASED ON TRANSFORMATION OF AN ANALOG FILTER

As indicated earlier, one popular strategy for designing an IIR digital filter is to transform an analog filter into an equivalent digital filter. The theory of analog filter design is well established, and many closed-form design formulas exist (e.g., see [HU80; SE78; TE77; WE75] and older classical texts [GU57; ST57].) Given the desired specifications of a digital filter, the derivation of the digital filter transfer function requires three steps:

1. Map the desired digital filter specifications into those for an equivalent analog filter.
2. Derive a corresponding analog filter transfer function for the analog prototype.
3. Transform the transfer function of the analog prototype into an equivalent digital filter transfer function.

Figure 5-1 depicts the specifications for the squared magnitude response of an analog lowpass filter. The various parameters in the figure have the following meanings:

Ω_p	Passband cutoff frequency (rad/sec)
Ω_s	Stopband cutoff frequency (rad/sec)
Ω_c	3-dB Cutoff frequency (rad/sec)
ϵ	Parameter specifying allowable passband error
$1/A$	Maximum allowed magnitude in stopband

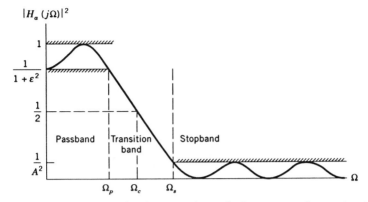

FIGURE 5-1 Specifications for the squared magnitude response of an analog filter.

This figure can be modified to apply to digital lowpass filters, as shown in Figure 5-2. Here, the critical analog frequencies Ω_p, Ω_s, and Ω_c are replaced by the normalized digital frequencies ω_p, ω_s, and ω_c. Note that ω_p, ω_s, and ω_c have units in radians and are restricted to values between 0 and π. As an additional parameter, denote N as the filter order (number of poles).

Often, a different set of parameters is used for specifying the magnitude of a digital lowpass filter's frequency response. These parameters are depicted in Figure 5-3, where δ_p represents the passband error tolerance and δ_s represents the maximum allowable magnitude in the stopband. Parameters N, ω_p, ω_s, and ω_c are defined as before. The parameters in Figures 5-2 and 5-3 can easily be related to each other. If we assume that the maximum passband magnitude of $H(e^{j\omega})$ in

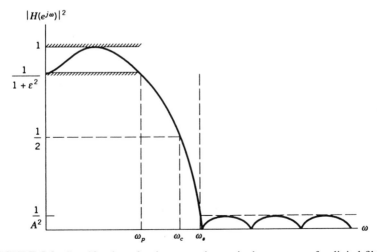

FIGURE 5-2 Specifications for the squared magnitude response of a digital filter.

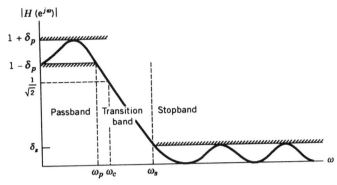

FIGURE 5-3   Alternate specifications for the magnitude response of a digital filter.

Figure 5-3 is normalized to equal 1, then

$$\epsilon = \frac{2\sqrt{\delta_p}}{1 - \delta_p},$$

$$A = \frac{1 + \delta_p}{\delta_s}.$$

It is most common to have the passband and stopband magnitude specifications given in decibels (dB). The design examples of Section 5-5 illustrate this. Throughout this chapter, we make use of the parameters defined in Figures 5-1 and 5-2.

While the approach of transforming an analog filter is popular and straightforward to apply, it is not applicable to all design problems and has the following drawbacks. First, it cannot be used to approximate all desired magnitude characteristics. For example, the transformation approach would be unsuitable for designing a differentiator [RA75]. On the other hand, filters having piecewise-constant magnitude characteristics, such as lowpass, bandpass, highpass, and bandstop, can nearly always be designed adequately with this approach. Second, the phase response is generally not preserved by the transformation, so that an analog filter having nearly linear phase may be transformed into a digital filter whose phase response is considerably nonlinear. For narrowband filters, however, the bilinear transformation (Section 5-4-1) can nearly preserve the phase in the passband.

The next several sections of this chapter describe how to apply this design procedure. Most of this material concentrates on the design of a lowpass digital filter having the specifications of Figure 5-2. Digital filters with other types of piecewise-constant magnitude responses can be designed via spectral transformations of a lowpass digital filter, as discussed in Section 5-6. It should be noted that the mapping to an equivalent set of analog specifications (Step 1) generally depends on the specific transformation chosen to convert the analog prototype filter (Step 2) into the resulting digital filter (Step 3). Section 5-3 reviews the relevant details of standard analog filters. Section 5-4 discusses how Steps 1 and 2 are performed. It also describes how to decompose a filter design into a combination of second-order sections to decrease roundoff error.

5-3 ANALOG LOWPASS FILTER DESIGNS

A general Nth order analog filter is described by a transfer function

$$H_a(s) = \frac{C(s)}{D(s)} = \frac{\displaystyle\sum_{i=0}^{M} c_i s^i}{1 + \displaystyle\sum_{i=1}^{N} d_i s^i}, \tag{5.1}$$

where $H_a(s)$ is the Laplace transform of the impulse response $h_a(t)$; that is,

$$H_a(s) = \int_{-\infty}^{\infty} h_a(t) e^{-st} \, dt$$

and $N \geq M$ must be satisfied. $H_a(s)$ represents a stable transfer function if the poles of $H_a(s)$ lie in the left half of the s-plane; that is, if the roots

$$s_k = \sigma_k + j\Omega_k, \qquad k = 0, 1, \ldots, N - 1,$$

of the denominator polynomial $D(s) = 1 + \Sigma_{i=1}^{N} d_i s^i$ satisfy $\sigma_k < 0$ for $k = 0$, $1, \ldots, N - 1$.

In this section, we review three common designs for the analog filter transfer function $H_a(s)$: Butterworth, Chebyshev, and elliptic.[1] All the relations needed for computing the poles and zeros of $H_a(s)$ for a given filter type and for a given set of specifications are provided. For many applications, the tedium of implementing these relations can be avoided by consulting the design tables of Christian and Eisenmann [CH66], Hansell [HA69], or Zverev [ZV67]. As another option, one of the software packages described in Section 5-9 can be used. More complete discussions on analog filter design can be found in Weinberg [WE75], Temes and LaPatra [TE77], Huelsman and Allen [HU80], Sedra and Brackett [SE78], Guillemin [GU57], or Storer [ST57].

---

[1]Bessel analog filters, which give maximally flat group delay, have also been used as analog prototypes for designing digital filters. But, since the group-delay property is generally not preserved after the analog-to-digital transformation, other design techniques are usually preferable for designing digital filters with approximately flat group delay [FE72; JA89; RA75].

5-3-1   Butterworth Filters

The frequency response of an Nth order Butterworth lowpass filter is defined by the squared-magnitude function

$$|H_a(j\Omega)|^2 = H_a(j\Omega)H_a(-j\Omega) = \frac{1}{1 + (\Omega/\Omega_c)^{2N}}. \tag{5.2}$$

A Butterworth filter is maximally flat at $\Omega = 0$; that is, the first $2N - 1$ derivatives of $|H_a(j\Omega)|^2$ equal zero at $\Omega = 0$. Note that the magnitude is down by 3 dB at $\Omega = \Omega_c$, regardless of the order N. Also, the magnitude monotonically decreases as $|\Omega|$ increases. As is apparent from Eq. (5.2), the filter order N completely specifies the Butterworth filter (assuming Ω_c is normalized to 1 rad/s). Figure 5-4 depicts plots of the squared magnitude responses for a Butterworth filter for various values of N. As can be seen from Figure 5-4, as N increases, the following changes

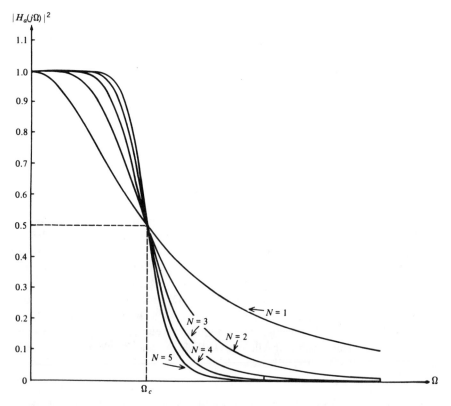

FIGURE 5-4   Squared magnitude responses for analog Butterworth lowpass filters for various N. Reprinted with permission from [PR88].

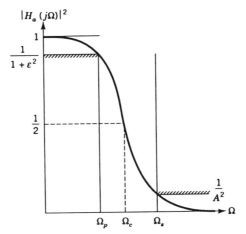

FIGURE 5-5 Specifications for the squared magnitude response of an analog Butterworth lowpass filter.

occur in the Butterworth filter's magnitude characteristic:

1. The transition from passband to stopband becomes more rapid.
2. $|H_a(j\Omega)| \approx 1$ for more of the passband.
3. $|H_a(j\Omega)| \approx 0$ for more of the stopband.

The pertinent design constraints, which can be gleaned from Figure 5-5, are given by

$$\frac{1}{1 + (\Omega_p/\Omega_c)^{2N}} \geq \frac{1}{1 + \epsilon^2} \quad \text{(passband)} \tag{5.3}$$

and

$$\frac{1}{1 + (\Omega_s/\Omega_c)^{2N}} \leq \frac{1}{A^2} \quad \text{(stopband).} \tag{5.4}$$

Assuming $\Omega_p < \Omega_c < \Omega_s$ as shown in Figure 5-5, Eqs. (5.3) and (5.4) are satisfied if

$$N \geq \frac{\log \epsilon}{\log (\Omega_p/\Omega_c)} \tag{5.5}$$

and

$$N \geq \frac{\log (A^2 - 1)}{2 \log (\Omega_s/\Omega_c)}. \tag{5.6}$$

By analytic continuation, Eq. (5.2) can be extended to the complex s-domain, giving

$$H_a(s)H_a(-s) = \frac{1}{1 + (-s^2)^N}.$$

A solution for the roots of the denominator gives the $2N$ poles of the Butterworth filter transfer function $H_a(s)$ and its mirror image $H_a(-s)$:

$$s_k = \Omega_c e^{j\pi[1/2 + (2k+1)/2N]}, \qquad k = 0, 1, \ldots, 2N - 1. \tag{5.7}$$

These poles lie equally spaced on a circle of radius Ω_c centered about $s = 0$ in the s-plane. A stable transfer function with frequency response $H_a(j\Omega)$ is obtained by choosing the left-half plane poles, corresponding to $k = 0, 1, \ldots, N - 1$. So the desired poles $s_k = \sigma_k + j\Omega_k$, $k = 0, \ldots, N - 1$, are specified by

$$\sigma_k = \Omega_c \cos\left(\frac{\pi}{2} + \frac{(2k+1)\pi}{2N}\right),$$

$$\Omega_k = \Omega_c \sin\left(\frac{\pi}{2} + \frac{(2k+1)\pi}{2N}\right). \tag{5.8}$$

Figure 5-6 illustrates the pole locations for a Butterworth filter with $N = 5$.

5-3-2 Chebyshev Filters

Chebyshev filters relax the constraint of monotonicity over either the passband or the stopband. For a given filter order N, the Chebyshev filter gives (1) minimum equiripple deviation of the magnitude characteristic over one prescribed band of frequencies and (2) monotonic behavior over the remaining band of frequencies. Depending on the design choice, the minimum equiripple error property can be associated with either the passband (Type-1 filters) or stopband (Type-2 filters). For a given set of filter specifications, this concession of allowing ripple in a band of frequencies leads to lower-order filters than Butterworth designs.

5-3-2-1 Type-1 Filters. A Type-1 Nth order lowpass Chebyshev filter, depicted in Figure 5-7(a), exhibits equiripple error in the passband and a monotonically decreasing response in the stopband. It is specified by the squared-magnitude frequency response function

$$|H_a(j\Omega)|^2 = \frac{1}{1 + \epsilon^2 T_N^2(\Omega/\Omega_p)}, \tag{5.9}$$

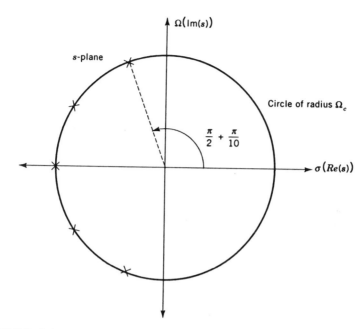

FIGURE 5-6 Pole locations for an analog Butterworth lowpass filter for $N = 5$. All poles lie on a circle of radius Ω_c centered at the origin of the s-plane. Pole s_o appears at angle $\pi/2 + \pi/10$ relative to the positive σ-axis. The other poles appear at angular increments of $\pi/5$.

where $T_N(x)$ is the Nth order Chebyshev polynomial. For nonnegative integers l, the lth order Chebyshev polynomial is given by

$$T_l(x) = \begin{cases} \cos (l \cos^{-1} x), & |x| \leq 1 \\ \cosh (l \cosh^{-1} x), & |x| \geq 1. \end{cases}$$

This polynomial can be derived from the recurrence relation

$$T_{l+1}(x) = 2xT_l(x) - T_{l-1}(x), \qquad n = 1, 2, \ldots ,$$

where $T_0(x) = 1$ and $T_1(x) = x$.

The constraint on the stopband is given by

$$\frac{1}{1 + \epsilon^2 T_N^2(\Omega_s/\Omega_p)} \leq \frac{1}{A^2}. \tag{5.10}$$

The passband constraint is satisfied by the definition of Eq. (5.9). For all Ω in the passband $[0, \Omega_p]$, the Nth order Chebyshev polynomial $T_N(\Omega/\Omega_p)$ gives the small-

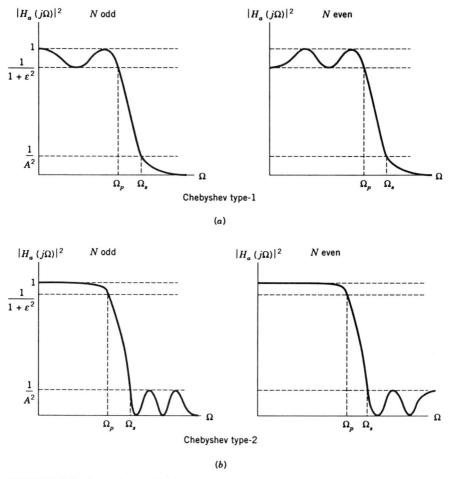

FIGURE 5-7 Squared magnitude responses for analog Chebyshev lowpass filters: (a) Type-1 and (b) Type 2.

est maximum value of any Nth order polynomial [HA89]. In particular, $|H_a(j\Omega)|^2$ oscillates between 1 and $1/(1 + \epsilon^2)$ within the passband, where it will have a total of N local maxima and local minima. ($T_N^2(\Omega/\Omega_p)$ oscillates between 0 and 1 on $[0, \Omega_p]$.) The filter order necessary to satisfy the design constraints is given by the relation

$$N \geq \frac{\cosh^{-1}(\sqrt{A^2 - 1}/\epsilon)}{\cosh^{-1}(\Omega_s/\Omega_p)}, \tag{5.11}$$

where $\cosh^{-1}(x) = \ln(x + \sqrt{x^2 - 1})$. The $2N$ poles $s_k = \sigma_k + j\Omega_k$, $k = 0, 1$ ... $2N - 1$, lie equally spaced on an ellipse centered about the origin in the s-plane. The ellipse has minor-axis length $a\Omega_p$ and major-axis length $b\Omega_p$ and is given by the equation

$$\frac{\sigma_k^2}{(a\Omega_p)^2} + \frac{\Omega_k^2}{(b\Omega_p)^2} = 1,$$

where

$$a = \frac{\gamma - \gamma^{-1}}{2}, \qquad (5.12)$$

$$b = \frac{\gamma + \gamma^{-1}}{2}, \qquad (5.13)$$

$$\gamma = \left(\frac{1 + \sqrt{1 + \epsilon^2}}{\epsilon}\right)^{1/N}, \qquad (5.14)$$

and the pole locations are specified by

$$\sigma_k = -a\Omega_p \sin\left[\frac{(2k + 1)\pi}{2N}\right], \qquad (5.15)$$

$$\Omega_k = b\Omega_p \cos\left[\frac{(2k + 1)\pi}{2N}\right]. \qquad (5.16)$$

The left-half-plane poles are selected for $H_a(s)$ to yield a stable transfer function and correspond to $k = 0, \ldots, N - 1$.

5-3-2-2 Type-2 Filters. A Type-2 Chebyshev filter, depicted in Figure 5-7(b), decays monotonically in the passband and exhibits equiripple error in the stopband. The squared-magnitude of the frequency response of a Type-2 Nth order Chebyshev filter is given by

$$|H_a(j\Omega)|^2 = \frac{1}{1 + \epsilon^2[T_N(\Omega_s/\Omega_p)/T_N(\Omega_s/\Omega)]^2}, \qquad (5.17)$$

where $|H_a(j\Omega)|^2$ has a total of N local maxima and local minima within the stopband. The constraint on the stopband is given by

$$\frac{1}{1 + \epsilon^2[T_N(\Omega_s/\Omega_p)]^2} \le \frac{1}{A^2}. \qquad (5.18)$$

Again, the passband constraint is automatically satisfied by the definition of the Chebyshev polynomial. The filter order N needed to meet these constraints is given by Eq. (5.11), the same as for Type-1 Chebyshev filters.

Unlike Type-1 Chebyshev filters, Type-2 filters have poles and zeros (since $T_N^2(\Omega_s/\Omega)$ can be zero for some Ω). The zeros are located on the imaginary axis and are specified by

$$s_k = \frac{j\Omega_s}{\cos\left[(2k+1)\pi/2N\right]}, \qquad k = 0, 1, \ldots, N-1. \qquad (5.19)$$

(If N is odd, then the zero for $k = (N-1)/2$ lies at infinity.) The poles $s_k = \sigma_k + j\Omega_k$, $k = 0, 1, \ldots, 2N-1$, are specified by

$$\sigma_k = \frac{\Omega_s \alpha_k}{\alpha_k^2 + \beta_k^2},$$

$$\Omega_k = \frac{-\Omega_s \beta_k}{\alpha_k^2 + \beta_k^2},$$

$$\alpha_k = -a\Omega_p \sin\left[\frac{(2k+1)\pi}{2N}\right],$$

$$\beta_k = b\Omega_p \cos\left[\frac{(2k+1)\pi}{2N}\right], \qquad\qquad (5.20)$$

$$a = \frac{\gamma - \gamma^{-1}}{2},$$

$$b = \frac{\gamma + \gamma^{-1}}{2},$$

$$\gamma = (A + \sqrt{A^2 - 1})^{1/N}.$$

The left-half-plane poles correspond to $k = 0, \ldots, N-1$ and do not lie on a simple geometrical figure as they do for Butterworth and Type-1 Chebyshev filters.

5-3-3 Elliptic Filters

Elliptic filters exhibit an equiripple magnitude response in both the passband and stopband. Figure 5-8 gives a plot of $|H_a(j\Omega)|^2$ for a typical lowpass elliptic filter. For a given filter order N and a given set of ripple specifications ϵ and A, no other filter provides a faster transition from passband to stopband. The theory of elliptic filters, first developed by Cauer, is involved. Only the salient details for constructing a filter are provided here. For more extensive treatments, refer to the texts by Temes and LaPatra [TE77], Sedra and Brackett [SE78], Huelsman and Allen [HU80], Gold and Rader [GO69], Parks and Burrus [PA87], Guillemin [GU57], or Storer [ST57].

The magnitude characteristic of an Nth order lowpass elliptic filter takes the general form

$$|H_a(j\Omega)|^2 = \frac{1}{1 + \epsilon^2 F_N^2(\Omega/\Omega_p)} \qquad (5.21)$$

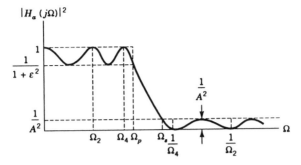

FIGURE 5-8 Squared magnitude response for an analog elliptic lowpass filter.

where $F_N(\Omega)$ is the rational function

$$
F_N(\Omega) =
\begin{cases}
\gamma^2 \dfrac{(\Omega_1^2 - \Omega^2)(\Omega_3^2 - \Omega^2) \cdots (\Omega_{2N-1}^2 - \Omega^2)}{(1 - \Omega_1^2\Omega^2)(1 - \Omega_3^2\Omega^2) \cdots (1 - \Omega_{2N-1}^2\Omega^2)}, & N \text{ even} \\[4mm]
\gamma^2 \dfrac{\Omega(\Omega_2^2 - \Omega^2)(\Omega_4^2 - \Omega^2) \cdots (\Omega_{2N}^2 - \Omega^2)}{(1 - \Omega_2^2\Omega^2)(1 - \Omega_4^2\Omega^2) \cdots (1 - \Omega_{2N}^2\Omega^2)}, & N \text{ odd}
\end{cases}
$$

and the Ω_i are constants, and γ is a parameter [GU57]. The function $F_N(\Omega)$ has the property $F_N(1/\Omega) = 1/F_N(\Omega)$. The numerator roots of $F_N(\Omega)$ lie within the interval $0 < \Omega < 1$, while the denominator roots of $F_N(\Omega)$ lie within the interval $1 < \Omega < \infty$. Note that the denominator roots are reciprocals of the numerator roots. The function $|F_N(\Omega)|^2$ needed to give the magnitude response of Figure 5-8 is shown in Figure 5-9. Within either the passband or stopband, $|H_a(j\Omega)|^2$ has a total of N local maxima and local minima.

To derive an elliptic lowpass filter $H_a(s)$ satisfying the specifications of Figure 5-1, one can use the relations below. The filter order necessary for satisfying the

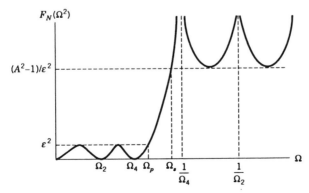

FIGURE 5-9 Function $F_N(\Omega)$ that provides the elliptic filter response in Figure 5-8.

specifications can be found using the following relations:

$$k = \frac{\Omega_p}{\Omega_s},$$

$$k_1 = \frac{\epsilon}{\sqrt{A^2 - 1}},$$

$$N \geq \frac{K(k)K(\sqrt{1 - k_1^2})}{K(k_1)K(\sqrt{1 - k^2})}, \qquad (5.22)$$

where k is known as the transition ratio and $K(\cdot)$ is the complete elliptic integral of the first kind [AB65]:

$$K(k) = \int_0^{\pi/2} \frac{d\phi}{(1 + k^2 \sin^2 \phi)^{1/2}}$$

$$= \frac{\pi}{2}\left(1 + \left(\frac{1}{2}\right)^2 k^2 + \left(\frac{1 \cdot 3}{2 \cdot 4}\right)^2 k^4 + \left(\frac{1 \cdot 3 \cdot 5}{2 \cdot 4 \cdot 6}\right)^2 k^6 + \cdots\right)$$

$$= \text{sn}^{-1}(1, k),$$

where if we let sn (u, k) be the Jacobian elliptic function and $y = \text{sn}(u, k)$, then

$$\text{sn}^{-1}(y, k) = u = \int_0^y \frac{dt}{(1 - t^2)^{1/2}(1 - k^2 t^2)^{1/2}}.$$

In lieu of Eq. (5.22), the approximation

$$N \simeq \frac{2}{\pi^2} \ln\left(\frac{4A}{\epsilon}\right) \ln [8\Omega_p/(\Omega_s - \Omega_p)] \qquad (5.23)$$

is useful in estimating the filter order [ST57; GR76].

The poles $s_i = \sigma_i + j\Omega_i$, $i = 0, 1, \ldots, N - 1$, of $H_a(s)$ are specified by the relations [GO69; ST57]

$$\sigma_i = \Omega_p \frac{\text{sn}(q, k') \text{cn}(q, k') \text{cn}(r_i, k) \text{dn}(r_i, k)}{1 - \text{sn}^2(q, k') \text{dn}^2(r_i, k)},$$

$$\Omega_i = \Omega_p \frac{\text{sn}(r_i, k) \text{dn}(q, k')}{1 - \text{sn}^2(q, k') \text{dn}^2(r_i, k)},$$

$$\text{cn}^2(u, k) = 1 - \text{sn}^2(u, k),$$

$$\text{dn}^2(u, k) = 1 - k^2 \text{sn}^2(u, k),$$

$$k' = \sqrt{1 - k^2},$$

$$q = -j \frac{K(k)}{NK(k_1)} \operatorname{sn}^{-1}\left(\frac{j}{\epsilon}, k_1\right),$$

$$r_i = -K(k) + \frac{K(k)}{N}(1 + 2i).$$

For $k \approx 1$ and $k_1 \ll 1$ (as is typical for a lowpass filter),

$$q \approx \frac{-K'(k)}{K'(k_1)} \ln\left(\frac{\sqrt{1+\epsilon^2}+1}{\epsilon}\right).$$

The Jacobian elliptic function sn (u, k) can be computed using the following relations [AB65]:

$$\operatorname{sn}(u, k) = \frac{1}{\sqrt{k}} \frac{\theta_1\left(\dfrac{u}{2K(k)}, t\right)}{\theta_4\left(\dfrac{u}{2K(k)}, t\right)},$$

$$t = e^{-\pi K(\sqrt{1-k^2})/K(k)},$$

where θ_1 and θ_4 are theta functions given by

$$\theta_1\left(\frac{u}{2K(k)}, t\right) = 2 \sum_{n=0}^{\infty} (-1)^n t^{(n+1/2)^2} \sin\left(\frac{(2n+1)\pi u}{2K(k)}\right),$$

$$\theta_4\left(\frac{u}{2K(k)}, t\right) = 1 + 2 \sum_{n=1}^{\infty} (-1)^n t^{n^2} \cos\left(\frac{2n\pi u}{2K(k)}\right).$$

The zeros, $s_i = \sigma_i + j\Omega_i$, $i = 0, 1, \ldots, N-1$, are given by the same relations as above except

$$q = -K(\sqrt{1-k^2}).$$

Note that the design constraints implied by Figure 5-8 are implicitly integrated into the above relations. The equations above can easily be implemented on a digital computer. Refer to Gold and Rader [GO69] and Gray and Markel [GR76] for more detailed discussions, or to Section 5-9 for a description of computer software for designing elliptic filters.

5-4 ANALOG-TO-DIGITAL TRANSFORMATIONS

Once the appropriate analog-filter prototype is derived, it must then be transformed using an analog-to-digital mapping. This generally involves a transformation between the s-plane and the z-plane.

The ideal transformation should have the following properties:

1. Transform a stable, causal analog filter into a stable, causal digital filter.
2. Preserve the magnitude and phase characteristics of the analog filter.

To accomplish Property 1, the transformation must (1) map the left-half s-plane into the interior of the unit circle of the z-plane and (2) map the right-half s-plane into the exterior of the unit circle in the z-plane. To achieve Property 2, the $j\Omega$-axis of the s-plane must map linearly onto the unit circle ($z = e^{j\omega}$) of the z-plane—unfortunately, no transformation can satisfy this requirement.

Several transformation approaches do exist, however, that give satisfactory results for many circumstances. This section outlines the bilinear transformation, the impulse invariance technique, and the technique of approximating differentials. (The matched z-transformation is another approach mentioned by many authors, but the aforementioned techniques generally give better filters [JA89; RA75].)

5-4-1 Bilinear Transformation Method

The bilinear transformation, the most popular analog-to-digital transformation, transforms $H_a(s)$ into $H(z)$ via the relation

$$H(z) = H_a(s)\big|_{s = \alpha(1 - z^{-1})/(1 + z^{-1})}. \tag{5.24}$$

Some authors choose $\alpha = 2/T$ or $\alpha = 1/T$ for historical reasons, but α can in fact be chosen arbitrarily (equal to 1 for instance). The analog prototype $H_a(s)$ is then designed so that the digital transfer function $H(z)$, derived using the bilinear transformation technique, is independent of the value of α. $H_a(s)$ can be recovered from the derived $H(z)$, using the inverse transformation

$$z = \frac{1 + s/\alpha}{1 - s/\alpha}. \tag{5.25}$$

The transformation in Eq. (5.24) maps the s-plane into the z-plane as follows (see Figure 5-10):

1. Left-half s-plane $\rightarrow$ interior of z-plane unit circle.
2. Right-half s-plane $\rightarrow$ exterior of z-plane unit circle.
3. Imaginary axis $s = j\Omega \rightarrow z$-plane unit circle ($z = e^{j\omega}$).

Thus the bilinear transformation maps a stable analog filter into a stable digital filter. Also, the analog frequency domain (imaginary axis) maps onto the digital frequency domain (unit circle), albeit, as we shall see, nonlinearly.

For transfer functions $H_a(s)$ of the form of Eq. (5.1), the denominator of the derived $H(z)$ will have the same order as the denominator of $H_a(s)$. The numerator

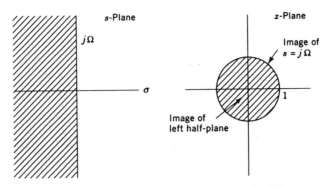

FIGURE 5-10 Mapping of the s-plane to the z-plane using the bilinear transformation.

of $H(z)$ can differ in order, however, from the numerator of $H_a(s)$. Since $H_a(s)$ has $(N - M)$ zeros at $s = \infty$, $H(z)$ will have $(N - M)$ additional zeros at $z = -1$ (consider $s = \infty$ in Eq. (5.25)).

Substituting $z = e^{j\omega}$ into Eq. (5.24) results in, after some manipulation,

$$H(e^{j\omega}) = H_a(j\alpha \tan(\omega/2)). \qquad (5.26)$$

Equation (5.26) is extremely important and gives the correspondence between the frequency responses of the analog prototype and the designed digital filter; that is, Ω maps to $\alpha \tan(\omega/2)$.

Figure 5-11 gives a plot of the inverse relation $\omega = 2 \tan^{-1}(\Omega/\alpha)$. As the figure shows, the mapping is one-to-one, but the entire analog frequency axis $(-\infty, \infty)$ is compressed nonlinearly onto the digital frequency interval $(-\pi, \pi)$. For piecewise-constant filters, such as lowpass, bandpass, highpass, and bandstop filters[2] this distortion, which is called frequency warping, can easily be compensated for. For these filters, the effect of the bilinear transformation on the magnitude of the frequency response is simply to shift the critical frequencies as shown in Figure 5-12, where $\omega_i = 2 \tan^{-1}(\Omega_i/\alpha)$. Using $\Omega_i = \alpha \tan(\omega_i/2)$, the forward relation, the critical frequencies Ω_i of the analog prototype filter must be selected so that the bilinear transformation will produce a digital filter having the desired critical frequencies ω_i. This selection of the Ω_i is commonly referred to as frequency prewarping, and the overall bilinear transformation method of design is as follows [OP89; RA75]:

1. For each of the desired critical digital frequencies (e.g., ω_p and ω_s for lowpass filters), determine the corresponding analog frequencies, using (see Eq. (5.26))

$$\Omega = \alpha \tan(\omega/2). \qquad (5.27)$$

---

[2]Ideal forms of these filters have magnitudes equal to 1 in the passbands and 0 in the stopbands.

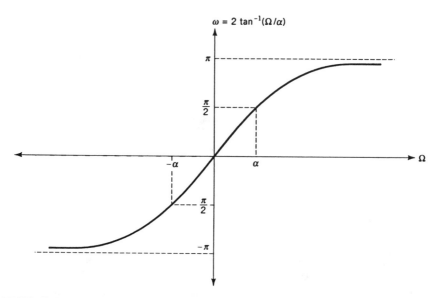

FIGURE 5-11 Correspondence between analog and digital frequency under the bilinear transformation.

2. Given the prewarped analog frequencies, derive the appropriate analog prototype transfer function $H_a(s)$.
3. Apply the bilinear transformation, Eq. (5.24), to $H_a(s)$ to obtain the desired digital filter transfer function $H(z)$.

Section 5-5 gives design examples using this technique.

5-4-2 Impulse-Invariant Method

The impulse-invariant technique derives its name from the property that the unit-pulse response of the designed digital filter is proportional to a set of uniformly spaced samples of the impulse response of the analog prototype filter. Given the analog prototype transfer function $H_a(s)$, the impulse-invariant technique gives a digital filter transfer function $H(z)$, using the following steps:

1. Express $H_a(s)$ as a partial-fraction expansion of first-order terms,

$$H_a(s) = \sum_{i=1}^{N} \frac{A_i}{s - s_i}. \tag{5.28}$$

Section 5-4-4 describes how to derive this expansion.

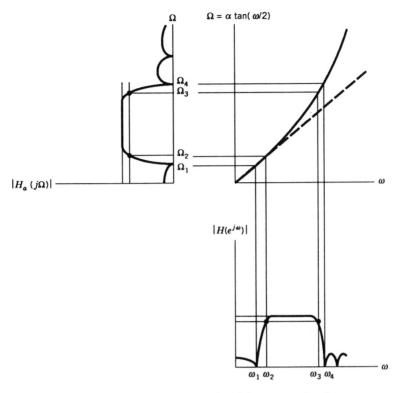

FIGURE 5-12 Digital frequency response produced from an analog frequency response via the bilinear transformation.

2. The digital filter transfer function then is given by

$$H(z) = \sum_{i=1}^{N} \frac{TA_i}{1 - e^{s_i T} z^{-1}} ; \qquad (5.29)$$

that is, the pole s_i maps to a pole z_i in the z-plane via the relation

$$z_i = e^{s_i T}.$$

Figure 5-13 illustrates how the s-plane is mapped into the z-plane when using the impulse-invariant technique. The attributes of this mapping are listed below:

1. The right-half s-plane ($\sigma > 0$) maps into the region outside the unit circle in the z-plane.

2. Each frequency interval on the $j\Omega$ axis,

$$(k - 1) \frac{\pi}{T} \leq \Omega < (k + 1) \frac{\pi}{T}, \qquad -\infty < k < \infty, \qquad (5.30)$$

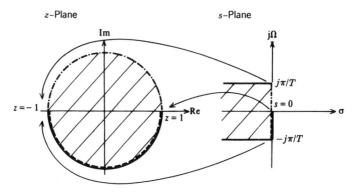

FIGURE 5-13 Mapping of the s-plane to the z-plane using impulse-invariant design. Reprinted with permission from R. Kuc, *Introduction to Digital Signal Processing*, McGraw-Hill, 1988.

maps onto the perimeter of the unit circle in the z-plane (i.e., the digital frequencies $-\pi \leq \omega < \pi$).

3. Each left-half-plane strip of the s-plane ($\sigma < 0$) bounded by the frequency intervals in Eq. (5.30) map into the interior of the unit circle in the z-plane.

Note that the functions $H_a(s)$ and $H(z)$ of Eqs. (5.28) and (5.29) correspond to the time-domain responses

$$h_a(t) = \sum_{i=1}^{N} A_i e^{s_i t} \mu(t) \tag{5.31}$$

and

$$h[n] = \sum_{i=1}^{N} A_i T e^{s_i Tn} \mu[n], \tag{5.32}$$

where $\mu(t)$ is the unit-step function (assuming the region of convergence of $H_a(s)$ is defined so that $h_a(t)$ is causal). A comparison of Eqs. (5.31) and (5.32) reveals that

$$h[n] = T \cdot h_a(nT).$$

Thus, to within a multiplicative constant, the values of the digital unit-pulse response $h[n]$ correspond to the values of the analog impulse response $h_a(t)$ at equally spaced intervals $t = nT$, $n = 0, 1, 2, \ldots$.

Since $h[n]$ is a sampled (and scaled) version of $h_a(nT)$, the digital frequency response $H(e^{j\omega})$ of Eq. (5.29) is related to the analog frequency response $H_a(j\Omega)$

by

$$H(e^{j\omega}) = \sum_{k=-\infty}^{\infty} H_a\left(j\,\frac{\omega + 2\pi k}{T}\right). \tag{5.33}$$

If

$$H_a(j\Omega) = 0, \qquad |\Omega| \geq \frac{\pi}{T}, \tag{5.34}$$

then the terms in the summation of Eq. (5.33) would not overlap, so that

$$H(e^{j\omega}) = H_a\left(\frac{j\omega}{T}\right), \qquad |\omega| \leq \pi. \tag{5.35}$$

That is, if Eq. (5.34) were to hold, then the digital filter $H(z)$ derived using the impulse-invariant technique would have a frequency response with exactly the same shape as that of the analog prototype. Although this would be desirable, it can be only an approximation in reality. Since no finite-order analog filter with transfer function of the form of Eq. (5.1) satisfies Eq. (5.34) exactly, the terms in Eq. (5.33) will overlap and $H(e^{j\omega})$ will be an aliased version of $H_a(j\Omega)$. Therefore the impulse-invariant technique cannot exactly preserve the shape of the desired frequency response, and it does not give a one-to-one mapping of the analog frequency domain onto the digital frequency domain (i.e., the $s = j\Omega$ axis onto the z-plane unit circle $z = e^{j\omega}$).

This aliasing phenomenon can be particularly troubling when designing a highpass filter. To design a highpass filter using impulse invariance, a bandpass filter should be designed instead. This prevents serious overlap of the terms in Eq. (5.33), which would invalidate Eq. (5.35). A bandpass analog prototype can be designed having an upper cutoff frequency of approximately π/T, so that Eqs. (5.34) and (5.35) will be nearly satisfied. The effect of designing a bandpass filter rather than the desired highpass filter can be minimized by decreasing T, so that the bandpass filter approximates the highpass filter across a broader range of analog frequencies (up to π/T). Alternatively, a lowpass digital filter can be designed and then transformed into a highpass filter, using the appropriate transformation described in Section 5-6. However, because the frequency response of an elliptic or Type-2 Chebyshev analog prototype may not approach zero, the effects of aliasing can be significant, even for lowpass filters or high sampling rates. Sometimes this aliasing can be profitably reduced through the use of a step-invariant (rather than impulse-invariant) design procedure. For this design method, the A_i in Eq. (5.29) are replaced by $\overline{A}_i = A_i(e^{s_i T} - 1)/s_i$ [AN79].

Assuming that the aliasing is negligible (i.e., Eq. (5.34) is approximately satisfied), it follows from Eq. (5.35) that the relationship between the digital and analog frequency domains is

$$\Omega = \frac{\omega}{T}. \tag{5.36}$$

This relationship is useful for mapping filter specifications during the design procedure. Section 5-5 gives a design example using the impulse-invariant method.

5-4-3 Approximation of Differentials

The concept of approximating continuous-time derivatives by forward or backward differences has a long history in digital system design. This concept also has been applied to digital filter design.

The transfer function $H_a(s)$, Eq. (5.1), corresponds to the following differential equation in the time domain:

$$y_a(t) + \sum_{i=1}^{N} d_i \frac{d^i y_a(t)}{dt^i} = \sum_{i=1}^{M} c_i \frac{d^i x_a(t)}{dt^i}. \tag{5.37}$$

This equation can be approximated with a digital difference equation by making the substitutions

$$y_a(nT) \leftrightarrow y[n],$$

$$\left. \frac{d^i y_a(t)}{dt^i} \right|_{t=nT} \leftrightarrow \nabla^{(i)}[y[n]],$$

where

$$\nabla^{(1)}[y[n]] = \frac{y[n] - y[n-1]}{T}, \tag{5.38}$$

$$\nabla^{(i+1)}[y[n]] = \frac{\nabla^{(i)}[y[n]] - \nabla^{(i)}[y[n-1]]}{T}. \tag{5.39}$$

Equations (5.38) and (5.39) represent a backward difference approximation to the derivatives [RA75] (analogous definitions apply for $d^i x_a(t)/dt^i$). Making the backward difference substitutions in Eq. (5.37) and taking z-transforms of both sides of the equation, we obtain the digital transfer function

$$H(z) = \frac{\displaystyle\sum_{i=0}^{M} c_i \left(\frac{1 - z^{-1}}{T}\right)^i}{1 + \displaystyle\sum_{i=1}^{N} d_i \left(\frac{1 - z^{-1}}{T}\right)^i}. \tag{5.40}$$

Equation (5.40) implies that the s-plane is mapped into the z-plane via the relation

$$s = \frac{1 - z^{-1}}{T}$$

(cf., Eqs. (5.1) and (5.40)). This mapping has the following characteristics, as depicted in Figure 5-14:

1. The left-half s-plane maps inside a circle of radius $\frac{1}{2}$ centered at $z = \frac{1}{2}$ in the z-plane,
2. The right-half s-plane maps into the region outside the circle of radius $\frac{1}{2}$ in the z-plane,
3. The $j\Omega$-axis maps onto the perimeter of the circle of radius $\frac{1}{2}$ in the z-plane.

Thus this analog-to-digital transformation technique does map a stable analog filter into a stable digital filter, but it does not preserve the shape of the analog frequency characteristic (the $j\Omega$-axis does not even map onto the $z = e^{j\omega}$ circle). If T is decreased, more of the frequency response will be concentrated near $z = 1$. For a lowpass filter, this improves the matching between the analog and digital filter frequency responses, but significant distortion will still exist and T may have to be inordinately small. Furthermore, if the desired filter is not lowpass, this procedure typically cannot be applied. As a result, the backward-difference transformation is seldom used.

Forward differences can be used in place of backward differences, but this can lead to unstable filters [RA75], which makes the forward-difference method impractical.

A more accurate approximation to the differential equation, Eq. (5.37), is obtained by integrating both sides and using a trapezoidal approximation of the in-

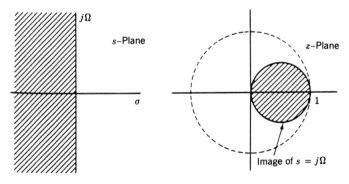

FIGURE 5-14 Mapping of the s-plane to the z-plane using a backward-difference approximation.

tegrals. Doing this and taking z-transforms, we obtain the digital transfer function [WI86]

$$H(z) = \frac{\sum_{i=0}^{M} c_i \left(\frac{1}{T} \frac{1 - z^{-1}}{1 + z^{-1}} \right)^i}{1 + \sum_{i=1}^{N} d_i \left(\frac{1}{T} \frac{1 - z^{-1}}{1 + z^{-1}} \right)^i} \tag{5.41}$$

Thus this method of approximating the differential equation describing the analog prototype leads to the bilinear transformation design method of Section 5-4-1 (cf. Eqs. (5.24) and (5.41)).

5-4-4 Decomposition into Second-Order Sections

To minimize finite-wordlength effects, it is often preferable to realize a digital filter as a cascade or parallel connection of second-order sections (see Chapters 6 and 7). Since all the analog-to-digital transformations considered here preserve the filter order, the prototype analog transfer function $H_a(s)$ can be expressed as a product or sum of second-order functions before transformation. The cascade arrangement arises by pairing up each complex-conjugate pair of poles and zeros into one function: this can always be done for $H_a(s)$ corresponding to a real $h_a(t)$. (How the sections are ordered has a significant impact on the resulting filter's roundoff noise performance. We refer the reader to Jackson [JA70] and Higgins and Munson [HI84] for a discussion of this problem.)

The equivalent cascade-form representation of $H_a(s)$ is obtained by factoring Eq. (5.1) into the form

$$H_a(s) = \prod_{i=1}^{\lfloor (N+1)/2 \rfloor} \frac{\gamma_{0i}^c + \gamma_{1i}^c s + \gamma_{2i}^c s^2}{1 + \chi_{1i}^c s + \chi_{2i}^c s^2}, \tag{5.42}$$

where $\lfloor x \rfloor$ denotes the integer part of x, and complex-conjugate poles and zeroes are paired up so that the coefficients γ_{ji}^c and χ_{ji}^c are real-valued. For N odd, one of the terms of Eq. (5.42) will be a first-order section with a real pole and real zero.

The parallel-form representation of $H_a(s)$ is given by

$$H_a(s) = a + \sum_{i=1}^{\lfloor (N+1)/2 \rfloor} \frac{\gamma_{0i}^p + \gamma_{1i}^p s}{1 + \chi_{1i}^p s + \chi_{2i}^p s^2}. \tag{5.43}$$

The coefficients γ_{ji}^p and χ_{ji}^p are all real quantities. Each second-order term in Eq. (5.43) is derived by writing Eq. (5.1) in a partial-fraction expansion and combining terms with complex-conjugate pole pairs. That is, for a complex-conjugate pole pair s_k and s_k^*, the kth second-order section in Eq. (5.43) is given by rewriting the partial-fraction expansion terms

$$\frac{A_k}{s - s_k} + \frac{A_k^*}{s - s_k^*}$$

over the common denominator $(s - s_k)(s - s_k^*)$, where the residue A_k of a simple (unique) pole s_k is given by

$$A_k = (s - s_k)H_a(s)|_{s = s_k}. \tag{5.44}$$

For the analog filter types considered here, an Nth order lowpass filter will have N unique poles. Thus Eq. (5.44) is all that one needs for computing the residues. (Refer to any standard text on linear systems or complex analysis to find how to evaluate the residues for multiple-order poles). For N odd, one term in Eq. (5.43) will be first order.

For the bilinear transformation, a simple set of relations exists for deriving the corresponding digital filter from a second-order analog filter section. Consider an analog second-order section of the form

$$H_a(s) = \frac{\gamma_0 + \gamma_1 s + \gamma_2 s^2}{1 + \chi_1 s + \chi_2 s^2}. \tag{5.45}$$

The corresponding digital second-order section, given by

$$H(z) = \frac{a_0 + a_1 z^{-1} + a_2 z^{-2}}{1 + b_1 z^{-1} + b_2 z^{-2}}, \tag{5.46}$$

can be found using the relations

$$a_0 = \left(\frac{4}{c}\right)\left(\frac{\gamma_0}{\alpha^2} + \frac{\gamma_1}{\alpha} + \gamma_2\right),$$

$$a_1 = \left(\frac{8}{c}\right)\left(\frac{\gamma_0}{\alpha^2} - \gamma_2\right),$$

$$a_2 = \left(\frac{4}{c}\right)\left(\frac{\gamma_0}{\alpha^2} - \frac{\gamma_1}{\alpha} + \gamma_2\right),$$

$$b_1 = \left(\frac{8}{c}\right)\left(\frac{1}{\alpha^2} - \chi_2\right), \tag{5.47}$$

$$b_2 = \left(\frac{4}{c}\right)\left(\frac{1}{\alpha^2} - \frac{\chi_1}{\alpha} + \chi_2\right),$$

$$c = \frac{4}{\alpha^2} + \frac{4\chi_1}{\alpha} + 4\chi_2.$$

For Butterworth and Type-1 Chebyshev filters, $\gamma_1 = \gamma_2 = 0$ and the numerator of Eq. (5.46) reduces to $4\gamma_0(1 + z^{-1})^2/\alpha^2 c$. Furthermore, for Butterworth filters,

$$\chi_1 = 2 \sin\left(\frac{(2k + 1)\pi}{2N}\right) \quad \text{and} \quad \chi_2 = 1 \tag{5.48}$$

for the kth second-order filter section. This further simplifies the design of these filters. For elliptic and Type-2 Chebyshev filters, the general form of Eq. (5.46) must be used to derive the digital transfer function.

5-5 DESIGN EXAMPLES FOR LOWPASS DIGITAL FILTERS

In this section, we give four examples of lowpass filter designs using the three-step design procedure discussed in Section 5-2. In Example 5.1, we illustrate designs derived using the bilinear transformation and the impulse-invariant techniques, described in Section 5-4. The latter three examples use only the bilinear transformation.

To facilitate understanding of the design methodology, we proceed as if each filter is being designed "by hand" using the relations presented earlier in this chapter to first design an analog prototype filter and to then transform the analog filter to a digital filter. In actuality, however, all filters in this chapter were designed using the computer software MATLAB [MA], which greatly automates this design procedure. See Section 5-9 for information on MATLAB and other software that can automatically compute IIR designs.

Example 5.1. Design a digital Butterworth lowpass filter that satisfies the following specifications:

1. Passband magnitude is constant to within 1 dB for ω less than $\omega_p = 0.2\pi$.
2. Stopband attenuation is greater than 20 dB for frequencies between $\omega_s = 0.3\pi$ and π.

Bilinear Transformation Method. Given the digital filter specifications, we map the critical frequencies $\omega_p = 0.2\pi$ and $\omega_s = 0.3\pi$ to the appropriate analog frequencies. This is done using the prewarping relationship, Eq. (5.27):

$$\omega_p \rightarrow \Omega_p = \tan\left(\frac{\omega_p}{2}\right) = 0.32492,$$

$$\omega_s \rightarrow \Omega_s = \tan\left(\frac{\omega_s}{2}\right) = 0.50953.$$

Here we have chosen $\alpha = 1$. As mentioned earlier, this choice is arbitrary. Thus the desired analog filter specifications are

$$20 \log_{10} |H_a(j\Omega)| \geq -1, \qquad 0 \leq \Omega \leq \Omega_p \qquad (5.49)$$

and

$$20 \log_{10} |H_a(j\Omega)| \leq -20, \qquad \Omega \geq \Omega_s \qquad (5.50)$$

with $\Omega_p = 0.32492$ and $\Omega_s = 0.50953$. Design of the analog prototype Butterworth filter requires that Ω_c and N be chosen to satisfy Eqs. (5.49) and (5.50). Since Ω_c is unknown, we cannot find N via Eq. (5.5) and proceed directly. Equations (5.49) and (5.50) state, however, that

$$|H_a(j\Omega)| \geq 0.89125, \qquad 0 \leq \Omega \leq \Omega_p, \qquad (5.51)$$

$$|H_a(j\Omega)| \leq 0.1, \qquad \Omega \geq \Omega_s. \qquad (5.52)$$

Since the Butterworth filter's magnitude characteristic, Eq. (5.2), monotonically decreases with Ω, the constraints of Eqs. (5.51) and (5.52) will be met if

$$|H_a(j\Omega_p)| = 0.89125,$$

$$|H_a(j\Omega_s)| = 0.1,$$

or, equivalently,

$$\frac{1}{1 + \left(\dfrac{\Omega_p}{\Omega_c}\right)^{2N}} = (0.89125)^2, \qquad (5.53)$$

$$\frac{1}{1 + \left(\dfrac{\Omega_s}{\Omega_c}\right)^{2N}} = (0.1)^2. \qquad (5.54)$$

Solving these equations for Ω_c and N gives $\Omega_c = 0.35989$ and $N = 6.60834$.

Since N must be an integer, we round the filter order up to the nearest integer and select $N = 7$. Substituting for N in the passband condition, Eq. (5.53), gives $\Omega_c = 0.35784$. With these values for Ω_c and N, the passband constraint is met exactly and the stopband constraint is exceeded. The poles of the filter are found using Eq. (5.8):

Pole pair 1	$-0.07963 \pm j0.34887$
Pole pair 2	$-0.22311 \pm j0.27977$
Pole pair 3	$-0.32240 \pm j0.15526$
Real pole	-0.39784

As explained in Section 5-3-1, these poles lie equally spaced in the s-plane on a circle of radius Ω_c. Multiplying the complex-conjugate pole terms together gives $H_a(s)$ as a cascade of second-order sections. Then Eqs. (5.45)–(5.48) can be applied to find $H(z)$ in cascade form via the bilinear transformation. Alternatively, the pole terms of the analog prototype filter can all be multiplied together, giving[3]

$$H_a(s) = \frac{1}{\begin{array}{l} 1.000 + 12.559s + 78.859s^2 + 318.45s^3 \\ + 889.92s^4 + 1721.0s^5 + 2140.4s^6 + 1331.0s^7 \end{array}},$$

where the transfer function is normalized to have unit gain at dc. Application of the bilinear transformation, Eq. (5.24), with $\alpha = 1$, then gives the digital transfer function

$$H(z) = \frac{\begin{array}{l} 0.0002 + 0.0011z^{-1} + 0.0032z^{-2} + 0.0054z^{-3} \\ + 0.0054z^{-4} + 0.0032z^{-5} + 0.0011z^{-6} + 0.0002z^{-7} \end{array}}{\begin{array}{l} 1.0000 - 3.9190z^{-1} + 7.0109z^{-2} - 7.2790z^{-3} \\ + 4.6943z^{-4} - 1.8690z^{-5} + 0.4236z^{-6} - 0.0420z^{-7} \end{array}}.$$

The magnitude and phase of the frequency response of the resulting digital filter are shown in Figure 5-15. Note the monotonically decaying magnitude response

[3]Due to width limitations, two lines are used for some denominators and numerators in this chapter.

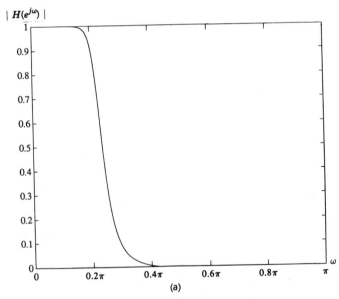

(a)

FIGURE 5-15 Frequency response of a 7th-order lowpass Butterworth filter with $\omega_p = 0.2\pi$ and $\omega_s = 0.3\pi$, designed using the bilinear transformation: (a) magnitude response, (b) magnitude response in dB, and (c) phase response.

20 log| $H(e^{j\omega})$ |

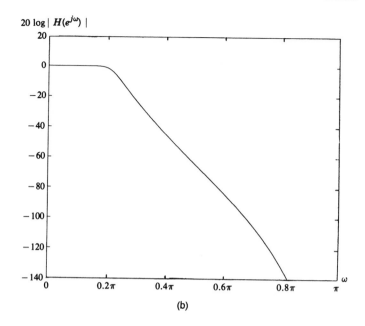

(b)

arg [$H(e^{j\omega})$]

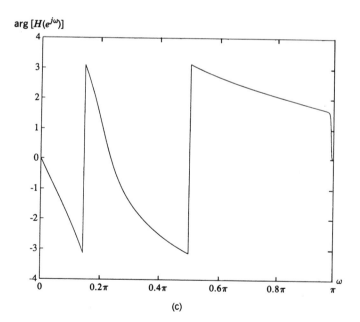

(c)

FIGURE 5-15 (*Continued*)

across both the passband and stopband, which is characteristic of a Butterworth filter. Also, the phase response is nearly linear across the passband, but it then becomes noticeably nonlinear in the transition band, $0.2\pi < \omega < 0.3\pi$.

Impulse-Invariant Method. We choose the value of T arbitrarily at the outset of the design procedure and then design the analog prototype filter accordingly. Letting $T = 1$ in Eq. (5.36) gives the critical analog frequencies $\Omega_p = 0.2\pi$ and $\Omega_s = 0.3\pi$. The analog filter specifications then become

$$20 \log_{10} |H_a(j\Omega)| \geq -1, \qquad |\Omega| \leq 0.2\pi,$$

$$20 \log_{10} |H_a(j\Omega)| \leq -20, \qquad |\Omega| \geq 0.3\pi.$$

Again, the filter constraints are stated in Eqs. (5.51) and (5.52) and equivalently in Eqs. (5.53) and (5.54), but now with $\Omega_p = 0.2\pi$ and $\Omega_s = 0.3\pi$. Solving Eqs. (5.53) and (5.54) for Ω_c and N gives $\Omega_c = 0.68896$ and $N = 7.33263$. Rounding N up to the nearest integer gives $N = 8$. Note that for this example the analog prototype filter has a slightly higher order than the prototype used for the above bilinear transformation example (8 versus 7). Solving Eq. (5.53) for Ω_c gives $\Omega_c = 0.68368$. The poles of the analog prototype filter are thus obtained from Eq. (5.8) and are given by:

Pole pair 1	$-0.13338 \pm j0.67054$
Pole pair 2	$-0.37893 \pm j0.56846$
Pole pair 3	$-0.56856 \pm j0.37893$
Pole pair 4	$-0.67054 \pm j0.13338$

and the numerator of $H_a(s)$ is the constant, 0.0477, to provide unit gain at dc. Expanding $H_a(s)$ in a partial-fraction expansion gives

$$H_a(s) = \sum_{i=1}^{8} \frac{A_i}{s - s_i},$$

where s_1 and s_2 are the first pole pair above, s_3 and s_4 are the second pole pair, and so on. The values of the residues are

$$A_1 = 0.2010 - j0.1343,$$
$$A_2 = A_1^*,$$
$$A_3 = 0.2371 + j1.1918,$$
$$A_4 = A_3^*,$$
$$A_5 = -2.8774 - j0.5723,$$
$$A_6 = A_5^*,$$
$$A_7 = 2.4393 - j3.6507,$$
$$A_8 = A_7^*.$$

From Eq. (5.29), the poles of the digital filter are given by e^{s_i} (we have chosen $T = 1$), which have the values

$$p_1 = 0.6857 + j0.5438,$$
$$p_2 = p_1^*,$$
$$p_3 = 0.5764 + j0.3682,$$
$$p_4 = p_3^*,$$
$$p_5 = 0.5260 + j0.2100,$$
$$p_6 = p_5^*,$$
$$p_7 = 0.5069 + j0.0680,$$
$$p_8 = p_7^*.$$

Thus, from Eq. (5.29), the transfer function of the digital filter is

$$H(z) = \sum_{i=1}^{8} \frac{A_i z}{z - p_i},$$

where the A_i and p_i are given above. Putting the terms in the sum over a common denominator and normalizing the numerator to provide unit gain at dc yields

$$H(z) = \frac{\begin{array}{l} 0.0455 - 0.2089z^{-1} + 0.04405z^{-2} - 0.5504z^{-3} \\ + 0.4463z^{-4} - 0.2311z^{-5} + 0.0805z^{-6} - 0.0155z^{-7} + 0.0013z^{-8} \end{array}}{\begin{array}{l} 1.0000 - 4.5900z^{-1} + 9.6779z^{-2} - 12.1013z^{-3} \\ + 9.7478z^{-4} - 5.1555z^{-5} + 1.7422z^{-6} - 0.3430z^{-7} + 0.0301z^{-8} \end{array}}.$$

Typically, this transfer function would be realized in either cascade or parallel form using second-order sections.

The magnitude and phase of the frequency response of $H(z)$ are given in Figure 5-16. Note that because of aliasing that is inherent in the impulse-invariant design procedure, the frequency response exhibits some ripple in the passband, and the response does not decay across the stopband (even though the analog prototype was a Butterworth filter, which has no ripple in the passband and a response that decays monotonically in the stopband). Since the analog filter, however, exceeded the original design specifications, the resulting digital filter still meets the specifications. Nevertheless, aliasing artifact, arising from the nonzero stopband, degrades the digital filter's response. Comparing with Figure 5-15, we see that the eighth-order lowpass filter designed via impulse invariance exhibits a poorer magnitude response than the seventh-order filter designed via the bilinear transformation. Because of the aliasing problem with impulse invariance, the bilinear transformation method generally gives better filter designs.

Example 5.2. Design Type-1 and Type-2 digital Chebyshev lowpass filters that satisfy the design specifications of Example 5.1. Use the bilinear transformation design technique.

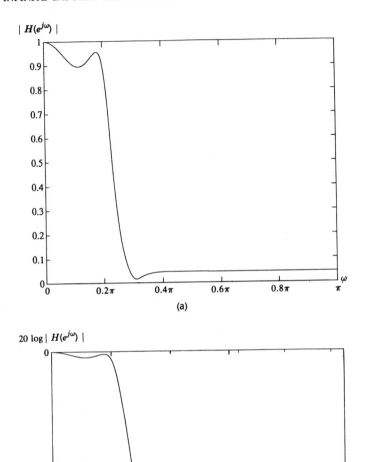

FIGURE 5-16 Frequency response of an 8th-order lowpass filter with $\omega_p = 0.2\pi$ and ω_s $= 0.3\pi$, designed using the impulse-invariant method with a Butterworth analog prototype: (a) magnitude response, (b) magnitude response in dB, and (c) phase response.

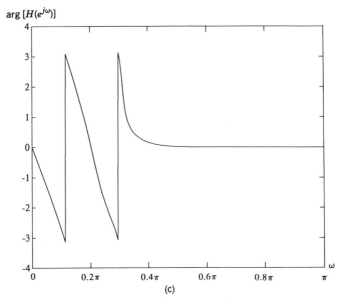

FIGURE 5-16 *(Continued)*

Type-1 Chebyshev Design. As in the Butterworth design using the bilinear transformation in Example 5.1, the prewarped analog critical frequencies are given by $\Omega_p = 0.32492$ and $\Omega_s = 0.50953$ ($\alpha = 1$ in the prewarping relationship, Eq. (5.27)), and Eqs. (5.49) and (5.50) apply, which state

$$20 \log_{10} |H_a(j\Omega)| \geq -1, \qquad 0 \leq \Omega \leq \Omega_p, \tag{5.55}$$

$$20 \log_{10} |H_a(j\Omega)| \leq -20, \qquad \Omega \geq \Omega_s. \tag{5.56}$$

Within the passband (cf., Eq. (5.9)), the maximum value of $T_N^2(\Omega/\Omega_p) = 1$, which implies that the minimum value of $|H_a(j\Omega)| = \sqrt{1/(1 + \epsilon^2)}$. Thus, from Eq. (5.55), $\sqrt{1/(1 + \epsilon^2)} = 10^{-0.05}$, giving $\epsilon = 0.50885$. From Eqs. (5.10) and (5.56), $A = 10$. Equation (5.11) gives the filter order $N = 3.59005$. Rounding this up to the nearest integer gives $N = 4$. Using Eqs. (5.12)–(5.14) gives $\gamma = 1.42903$, $a = 0.36462$, and $b = 1.06440$. Hence, from Eqs. (5.15) and (5.16), the poles of the filter are:

> Pole pair 1 $-0.04534 \pm j0.31952$
> Pole pair 2 $-0.10945 \pm j0.13235$

Multiplying the complex-conjugate pole terms together gives $H_a(s)$ as a cascade of two second-order sections. Then Eqs. (5.45)–(5.47) can be applied to find $H(z)$ in cascade form via the bilinear transformation. Alternatively, all pole terms can be multiplied together, giving the analog prototype

$$H_a(s) = \frac{0.89125}{1.0000 + 8.2922s + 49.965s^2 + 100.78s^3 + 325.52s^4}.$$

Since this Type-1 Chebyshev filter has an even order ($N = 4$), we have normalized the numerator to provide a gain of $\sqrt{1/(1 + \epsilon^2)}$ at dc, so that the maximum gain in the passband will be one (see Figure 5-7(a)). Application of the bilinear transformation, Eq. (5.24) with $\alpha = 1$, gives the digital transfer function

$$H(z) = \frac{0.0018 + 0.0073z^{-1} + 0.0110z^{-2} + 0.0073z^{-3} + 0.0018z^{-4}}{1.0000 - 3.0543z^{-1} + 3.8290z^{-2} - 2.2925z^{-3} + 0.5507z^{-4}}.$$

The magnitude and phase of the frequency response of the resulting digital filter are shown in Figure 5-17. Note the equiripple magnitude response in the passband and the monotonically decaying response in the stopband, which are characteristics of a Type-1 Chebyshev filter. For the magnitude response, the number of local maxima and local minima within the passband equals 4, the same number as the filter order. The response exactly meets the specification in the passband, and since the necessary filter order was rounded up to $N = 4$, the response is better than the specification in the stopband. The phase response is somewhat more nonlinear

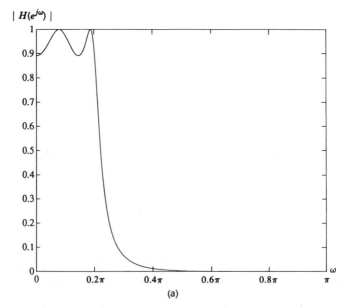

FIGURE 5-17   Frequency response of a 4th-order lowpass Type-1 Chebyshev filter with $\omega_p = 0.2\pi$ and $\omega_s = 0.3\pi$, designed using the bilinear transformation: (a) magnitude response, (b) magnitude response in dB, and (c) phase response.

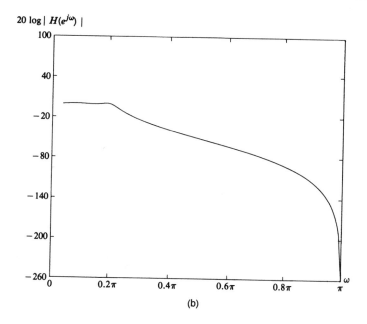

(b)

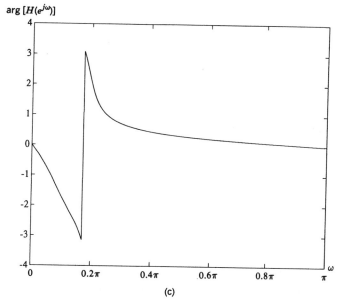

(c)

FIGURE 5-17 (*Continued*)

across the passband and transition band than the phase response for the Butterworth filter in Figure 5-15(c).

Type-2 Chebyshev Design. The prewarped analog critical frequencies are given by $\Omega_p = 0.32492$ and $\Omega_s = 0.50953$, as before. The filter should follow the requirements given in Eqs. (5.55) and (5.56) (repeated here):

$$20 \log_{10} |H_a(j\Omega)| \geq -1, \qquad 0 \leq \Omega \leq \Omega_p, \qquad (5.57)$$

$$20 \log_{10} |H_a(j\Omega)| \leq -20, \qquad \Omega \geq \Omega_s. \qquad (5.58)$$

Within the passband, $|H_a(j\Omega)|$, as given by Eq. (5.17), has a minimum value of $\sqrt{1/(1 + \epsilon^2)}$ at $\Omega = \Omega_p$. Thus, from Eq. (5.57), $\sqrt{1/(1 + \epsilon^2)} = 10^{-0.05}$, giving $\epsilon = 0.50885$. From Eqs. (5.18) and (5.58), $A = 10$. These values are the same as for the Type-1 example. Also, from Eq. (5.11), the filter order is $N = 3.59005$. (Equation (5.11) holds for both Type-1 and Type-2 Chebyshev filters.) Rounding up gives a filter order $N = 4$. Using Eq. (5.20), with $\gamma = 2.11342$, $a = 0.82012$, and $b = 1.29329$, gives the pole locations:

$$
\begin{array}{ll}
\text{Pole pair 1} & -0.32247 \pm j1.22775 \\
\text{Pole pair 2} & -1.45070 \pm j0.94759
\end{array}
$$

The Type-2 Chebyshev filter also has zeros, given by Eq. (5.19), at locations

$$\pm j0.55151, \qquad \pm j1.33147.$$

As discussed in the Type-1 Chebyshev example, the design can now proceed based on second-order filter sections. Alternatively, all pole terms and zero terms can be multiplied together, giving the analog prototype

$$H_a(s) = \frac{1.0000 + 3.8518s^2 + 1.8545s^4}{1.0000 + 4.2060s + 12.697s^2 + 21.369s^3 + 18.545s^4}.$$

Here the gain has been normalized to one at dc. Applying the bilinear transformation, Eq. (5.24), with $\alpha = 1$, gives

$$H(z) = \frac{0.1160 - 0.0591z^{-1} + 0.1630z^{-2} - 0.0591z^{-3} + 0.1160z^{-4}}{1.0000 - 1.8076z^{-1} + 1.5891z^{-2} - 0.6201z^{-3} + 0.1153z^{-4}}.$$

The magnitude and phase of the frequency response of this filter are shown in Figure 5-18. Note the monotonically decaying magnitude response in the passband and the equiripple response in the stopband, which are characteristics of a Type-2 Chebyshev filter. For the magnitude response, the number of local maxima and local minima within the stopband is the same as the filter order $N = 4$. The response exactly meets the specification in the stopband, and, since the necessary

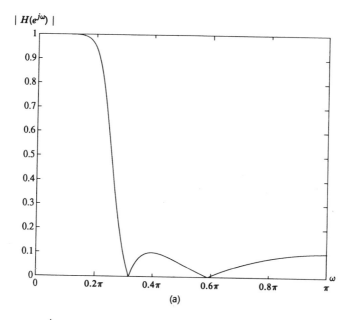

(a)

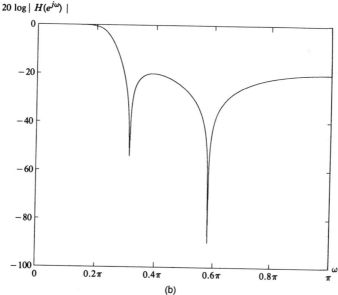

(b)

FIGURE 5-18 Frequency response of a 4th-order lowpass Type-2 Chebyshev filter with $\omega_p = 0.2\pi$ and $\omega_s = 0.3\pi$, designed using the bilinear transformation: (a) magnitude response, (b) magnitude response in dB, and (c) phase response.

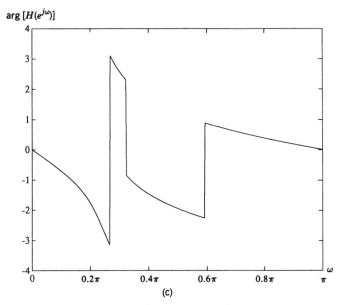

FIGURE 5-18   (*Continued*)

filter order was rounded up to $N = 4$, the response is better than the specification in the passband. The phase response is comparable to that of the Type-1 Chebyshev filter in Figure 5-17(c) and is more nonlinear than that of the Butterworth filter in Figure 5-15(c).

Example 5.3.   Design a digital elliptic lowpass filter that satisfies the design constraints of Examples 5.1 and 5.2, using the bilinear transformation.

The critical frequencies of the digital filter are as before: $\omega_p = 0.2\pi$ and $\omega_s = 0.3\pi$. Also, the filter constraints of Eqs. (5.55)–(5.56) apply, giving $\epsilon = 0.50885$ and $A = 10$. That is, these values of ϵ and A imply that the minimum value of $|H_a(j\Omega)| = \sqrt{1/(1 + \epsilon^2)}$ is -1 dB within the passband and that the maximum value of $|H_a(j\Omega)| = 1/A^2$ is -20 dB in the stopband. Thus we have the same results as for the Chebyshev designs of Example 5.2. From Eq. (5.22), $N = 2.685$. Rounding up to the nearest integer gives $N = 3$. Using the prewarped analog frequencies $\Omega_p = 0.32492$ and $\Omega_s = 0.50953$, the pole and zero locations can now be found as explained in Section 5-3-3; but this is extremely tedious. Instead, it is recommended that filter design software be used to find $H_a(s)$ and to apply the bilinear transformation. Given the specifications ω_p, ω_s, ϵ, and A, the computer software MATLAB [MA] determines the necessary filter order $N = 3$ and produces the digital filter transfer function

$$H(z) = \frac{0.09542 - 0.0152z^{-1} - 0.0152z^{-2} + 0.9542z^{-3}}{1.0000 - 2.0165z^{-1} + 1.6850z^{-2} - 0.5098z^{-3}}.$$

The magnitude and phase of the frequency response of this filter are shown in Figure 5-19. Note the equiripple nature of the magnitude response in both the passband and stopband, which is characteristic of elliptic filters. Also, note that the number of local maxima and local minima of the magnitude characteristic within the passband or stopband equals the filter order $N = 3$. The elliptic design gives the lowest-order filter of all the designs, but the phase response is more nonlinear across the passband and transition band than that for the Butterworth and Chebyshev filters.

Example 5.4. Design a digital elliptic lowpass filter that satisfies the following constraints:

1. Passband magnitude is constant to within 0.5 dB for $\omega < 0.1\pi$.
2. Stopband attenuation is greater than 40 dB for frequencies between 0.15π and π.

Use the bilinear transformation technique.

We will use the software MATLAB [MA], which accepts the digital specifications as input and then, with a single user command, performs the design of the analog prototype filter and the bilinear transformation.

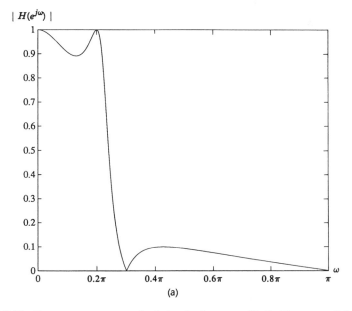

(a)

FIGURE 5-19 Frequency response of a 3rd-order lowpass elliptic filter $\omega_p = 0.2\pi$, $\omega_s = 0.3\pi$, designed using the bilinear transformation: (a) magnitude response, (b) magnitude response in dB, and (c) phase response.

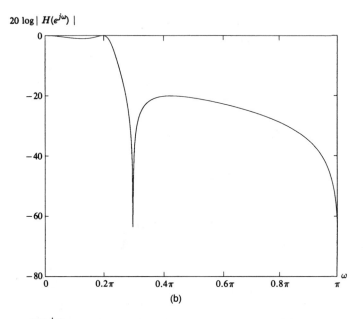

(b)

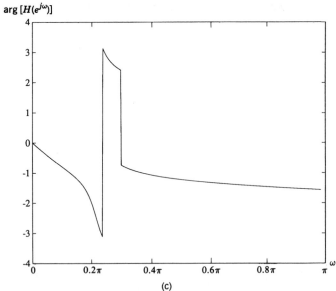

(c)

FIGURE 5-19   (*Continued*)

The digital filter specifications can be written as

$$20 \log_{10} |H(e^{j\omega})| \geq -0.5, \qquad 0 \leq \omega \leq \omega_p, \tag{5.59}$$

$$20 \log_{10} |H(e^{j\omega})| \leq -40, \qquad \omega_s \leq \omega \leq \pi, \tag{5.60}$$

where $\omega_p = 0.1\pi$ and $\omega_s = 0.15\pi$. The passband constraint states $|H(e^{j\omega})|^2 \geq 1/(1 + \epsilon^2)$ for $0 \leq \omega \leq \omega_p$. Thus, from Eq. (5.59), $1/(1 + \epsilon^2) = 10^{-0.05}$, giving $\epsilon = 0.34931$. The stopband constraint is $|H(e^{j\omega})|^2 \leq 1/A^2$ for $\omega_s \leq \omega \leq \pi$. Equating this with Eq. (5.60) gives $A = 100$. From Eq. (5.22), $N = 4.397$. Rounding up to the nearest integer gives $N = 5$. Given the digital filter specifications, the software MATLAB [MA] confirms that the necessary filter order is $N = 5$ and produces the digital filter transfer function

$$H(z) = \frac{0.0073 - 0.0184z^{-1} + 0.0115z^{-2} + 0.0115z^{-3} - 0.0184z^{-4} + 0.0073z^{-5}}{1.0000 - 4.5064z^{-1} + 8.2615z^{-2} - 7.6908z^{-3} + 3.6326z^{-4} - 0.6961z^{-5}}.$$

The magnitude and phase of the frequency response of this filter are shown in Figure 5-20. The filter has an equiripple magnitude characteristic in both the passband and stopband. The number of local maxima and local minima of the magnitude characteristic within the passband or stopband equals the filter order $N = 5$.

5-6 DIGITAL FREQUENCY TRANSFORMATIONS

Sections 5-2 through 5-5 described how to design a lowpass IIR digital filter in a three-step procedure by transforming an analog prototype filter. By adding one more step to this process, the design of highpass, bandpass, and bandstop filters

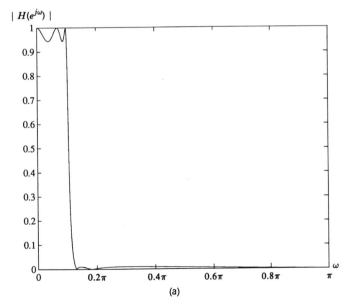

(a)

FIGURE 5-20 Similar to Figure 5-19, but for an elliptic filter with $\omega_p = 0.1\pi$, $\omega_s = 0.15\pi$, and tighter ripple specifications.

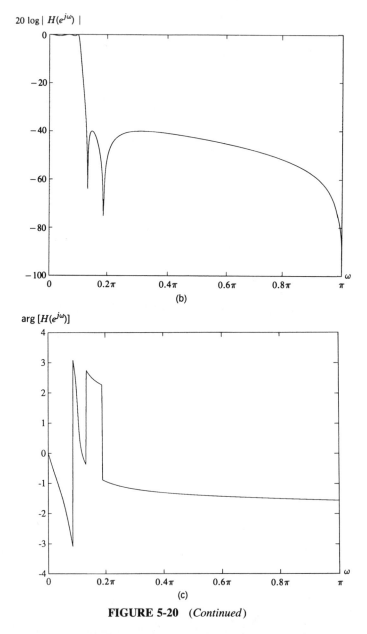

FIGURE 5-20 (*Continued*)

can be accomplished. This additional step can be performed in one of two ways:

1. Analog frequency transformation—the analog prototype lowpass filter is first transformed to the type of filter desired prior to applying the analog-to-digital transformations of Section 5-4.

2. Digital frequency transformation—the designed digital lowpass filter is directly transformed to the type of digital filter desired.

Although these two procedures in general yield somewhat different results, when using the bilinear transformation of Section 5-4-1, the results will be identical. Thus we consider here only the second approach, digital frequency transformation. For information on analog frequency transformations, the reader is referred to Proakis and Manolakis [PR88].

A lowpass digital filter with transfer function $H(z)$ is transformed to a highpass, bandpass, or bandstop digital filter by replacing the variable z^{-1} with a rational function of the form

$$f(z^{-1}) = \pm \prod_{i=1}^{K} \frac{\beta_i - z^{-1}}{1 - \beta_i z^{-1}}.$$

Here $|f(e^{j\omega})| = 1$ for all ω; that is, the transformation is allpass, which ensures that the unit circle is mapped onto itself. Table 5-1 lists digital frequency transformations for converting a given lowpass digital filter into several other types of filters: a lowpass filter having a different cutoff frequency, or highpass, bandpass, or bandstop filters. Derivations of these relations can be found in Constantinides [CO70] and Antoniou [AN79].

Performing the substitutions given in Table 5-1 is tedious, except for filters of very low order. Thus it is recommended that highpass, bandpass, and bandstop

TABLE 5-1 Digital Frequency Transformations

Type of Transformation	Transformation	Parameters
Lowpass	$z^{-1} \rightarrow \dfrac{z^{-1} - \beta}{1 - \beta z^{-1}}$	$\omega_c' =$ cutoff frequency of new filter $\beta = \dfrac{\sin\ [(\omega_c - \omega_c')/2]}{\sin\ [(\omega_c + \omega_c')/2]}$
Highpass	$z^{-1} \rightarrow -\dfrac{z^{-1} - \beta}{1 - \beta z^{-1}}$	$\omega_c' =$ cutoff frequency of new filter $\beta = \dfrac{\cos\ [(\omega_c + \omega_c')/2]}{\cos\ [(\omega_c - \omega_c')/2]}$
Bandpass	$z^{-1} \rightarrow -\dfrac{z^{-2} - \beta_1 z^{-1} + \beta_2}{\beta_2 z^{-2} - \beta_1 z^{-1} + 1}$	$\omega_l =$ lower cutoff frequency $\omega_u =$ upper cutoff frequency $\beta_1 = 2\gamma K/(K + 1)$ $\beta_2 = (K - 1)/(K + 1)$ $\gamma = \dfrac{\cos\ [(\omega_u + \omega_l)/2]}{\cos\ [(\omega_u - \omega_l)/2]}$ $K = \cot \dfrac{\omega_u - \omega_l}{2} \tan \dfrac{\omega_c}{2}$
Bandstop	$z^{-1} \rightarrow \dfrac{z^{-2} - \beta_1 z^{-1} + \beta_2}{\beta_2 z^{-2} - \beta_1 z^{-1} + 1}$	$\omega_l =$ lower cutoff frequency $\omega_u =$ upper cutoff frequency $\beta_1 = 2\gamma/(K + 1)$ $\beta_2 = (1 - K)/(1 + K)$ $\gamma = \dfrac{\cos\ [(\omega_u + \omega_l)/2]}{\cos\ [(\omega_u - \omega_l)/2]}$ $K = \tan \dfrac{\omega_u - \omega_l}{2} \tan \dfrac{\omega_c}{2}$

filters be designed using one of the software packages described in Section 5-9, which automatically implement these relations. Alternatively, one can easily solve for the transformed poles and zeros [JA89]. Figures 5-21, 5-22, and 5-23 show example frequency responses of highpass, bandpass, and bandstop filters designed using MATLAB [MA], which utilizes state-space versions of the transformations in Table 5-1. Figure 5-21 shows the frequency response for a highpass Type-2 Chebyshev filter with parameters $N = 7$, $\omega_s = 0.5\pi$, and a response that is 0.5 dB down at the edge of the passband and 35 dB down in the stopband. Figure 5-22 gives the response for a bandpass elliptic filter with specifications $N = 8$, cutoff frequencies 0.3π and 0.5π, 0.1-dB ripple in the passband, and a minimum of 60-dB attenuation in the stopband. Figure 5-23 shows the response for a Butterworth bandstop filter with parameters $N = 5$ and cutoff frequencies 0.2π and 0.3π.

5-7 PHASE EQUALIZATION

The phase responses for the Butterworth and Chebyshev filters designed in Section 5-5 are nearly linear over the passband, as shown earlier in Figures 5-15 through 5-18. Thus the group delay, defined as

$$\tau_H(\omega) = \frac{-dB(\omega)}{d\omega},$$

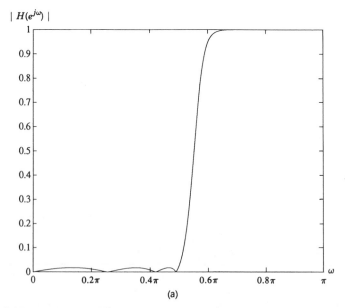

FIGURE 5-21 Frequency response of a 7th-order highpass Type-2 Chebyshev filter designed using the bilinear transformation and a lowpass-to-highpass mapping: (a) magnitude response, (b) magnitude response in dB, and (c) phase response.

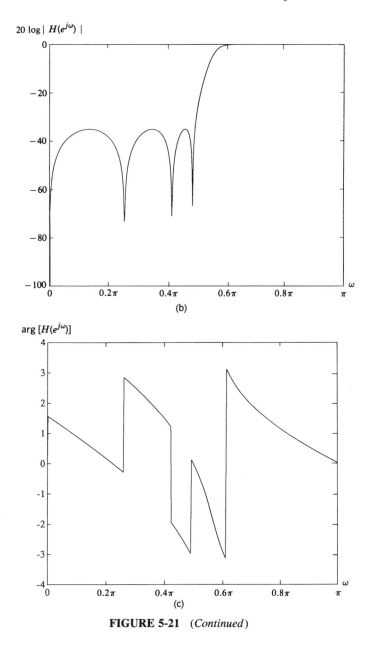

FIGURE 5-21 (*Continued*)

where $B(\omega) = \arg H(e^{j\omega})$ (phase in radians), is nearly constant for input components at all frequencies in the passband. Sometimes, however, the phase response can be quite nonlinear within the passband, especially for elliptic filters. Such nonlinearity causes frequency dispersion, which distorts signal components lying in the passband of the filter. (There is little concern about signal components in

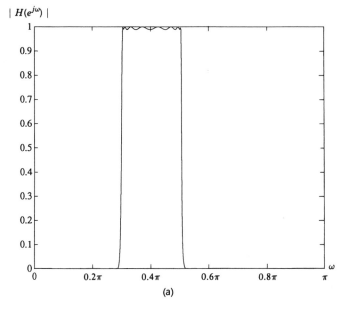

(a)

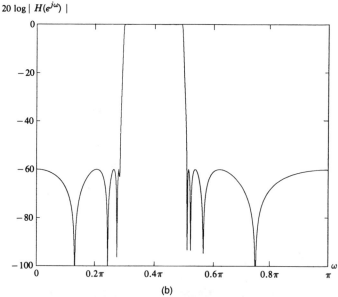

(b)

FIGURE 5-22 Frequency response of an 8th-order bandpass elliptic filter designed using the bilinear transformation and a lowpass-to-bandpass mapping: (a) magnitude response, (b) magnitude response in dB, and (c) phase response.

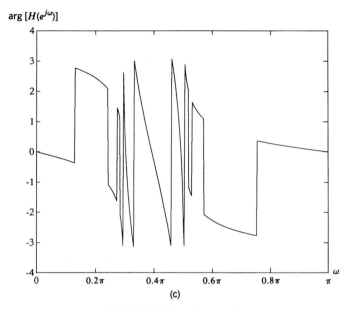

(c)

FIGURE 5-22   (*Continued*)

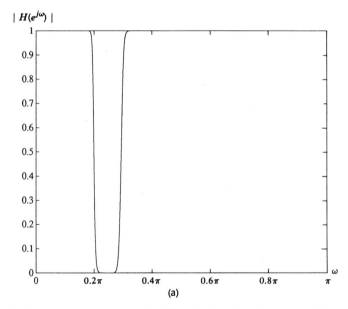

(a)

FIGURE 5-23 Frequency response of a 5th-order bandstop Butterworth filter designed using the bilinear transformation and a lowpass-to-bandstop mapping: (a) magnitude response, (b) magnitude response in dB, and (c) phase response.

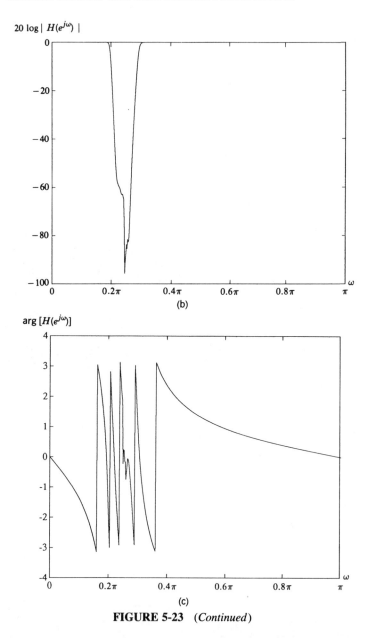

FIGURE 5-23 (*Continued*)

the stopband since they are greatly attenuated in amplitude by the filter.) The amount of nonlinearity in the phase tends to increase as one moves from a Butterworth to a Chebyshev, and then to an elliptic design (with a concomitant improvement in the magnitude response).

For cases where the phase response is too nonlinear, the situation can be improved by cascading an allpass phase equalizer with the original IIR filter. The overall frequency response of the filter–equalizer combination is

$$H_T(e^{j\omega}) = H(e^{j\omega})H_E(e^{j\omega}), \tag{5.61}$$

where $H(z)$ is the transfer function of the original filter and $H_E(z)$ is the transfer function of the equalizer. $H_E(z)$ is restricted to have the form

$$H_E(z) = \prod_{i=1}^{J} \frac{\beta_{2i} + \beta_{1i}z^{-1} + z^{-2}}{1 + \beta_{1i}z^{-1} + \beta_{2i}z^{-2}}, \tag{5.62}$$

where each second-order section has real-valued coefficients and is a product of two first-order allpass transfer functions. It can be shown that $|H_E(e^{j\omega})| = 1$ so that, from Eq. (5.61), $|H_T(e^{j\omega})| = |H(e^{j\omega})|$. That is, the shape of the magnitude of the frequency response of H is unaffected by the equalization operation. However, the phase relationship corresponding to Eq. (5.61) is

$$\arg [H_T(e^{j\omega})] = \arg [H(e^{j\omega})] + \arg [H_E(e^{j\omega})]. \tag{5.63}$$

Thus the design of $H_E(z)$ involves selecting the coefficients in Eq. (5.62) so that the right-hand side of Eq. (5.63) is nearly linear in ω. This problem is difficult and is generally solved using numerical optimization techniques, such as those described in Section 5-8. Section 5-8-1 gives an example showing the improvement attainable in the phase response through the use of a phase equalizer.

5-8 COMPUTER-AIDED DESIGN OF IIR FILTERS

The IIR digital filter design approach of Sections 5-2 through 5-5 works well when one can take advantage of analog designs based on closed-form or easily computed formulas. Such is the case for lowpass, highpass, bandpass, and bandstop filters. In general, however, simple procedures do not exist for the design of either analog or digital filters approximating an arbitrary frequency response. Instead, for the general case, it is often necessary to rely on numerical optimization techniques implemented on a computer. Such techniques apply equally well to both analog and digital filter design. So when numerical optimization is required, the digital design problem is attacked directly.

Optimization-based approaches for designing IIR filters involve the following steps:

1. $H(z)$ is assumed to be a rational function with fixed-order numerator and denominator polynomials.

2. An approximation-error criterion is defined that is suitable for the application in mind. This error criterion may involve the magnitude of the frequency response, the phase of the frequency response, or both, or it may be a function of the time-domain response of the filter.

3. Using an appropriate algorithm, the coefficients of $H(z)$ are chosen (sometimes iteratively) in an attempt to minimize the approximation error between the desired frequency response and $H(e^{j\omega})$ or between the time-domain response and $h[n]$.

Section 5-8-1 discusses Deczky's IIR design method that attempts to minimize an ℓ_p measure of error in the frequency domain. Section 5-8-2 describes a time-domain design procedure using Prony's method that produces a digital filter having an impulse response whose first L terms match the desired impulse response. Section 5-8-3 contains a brief description of four other frequency-domain IIR design methods, three of which are intended for the general design problem. The fourth utilizes the Remez exchange algorithm to design lowpass, highpass, bandpass, and bandstop filters.

5-8-1   A Frequency-Domain Design Procedure (Deczky)

Let $G(e^{j\omega})$ be the desired frequency response, and $H(e^{j\omega})$ be the actual frequency response obtained by choosing the parameters in

$$H(z) = A \prod_{k=1}^{K} \frac{1 + a_{k1}z^{-1} + a_{k2}z^{-2}}{1 + b_{k1}z^{-1} + b_{k2}z^{-2}}. \tag{5.64}$$

Then Deczkys' algorithm [DE72] seeks to minimize

$$E = (1 - \lambda) \sum_{i=1}^{Q} W_i[|H(e^{j\omega_i})| - |G(e^{j\omega_i})|]^p$$

$$+ \lambda \sum_{i=1}^{Q} V_i[\tau_H(\omega_i) - \tau(\omega_o) - \tau_G(\omega_i)]^p, \tag{5.65}$$

where the ω_i represent a grid of Q frequencies, $\tau_H(\omega)$ is the delay associated with the filter being designed $(\tau_H(\omega) = -d/d\omega\{\arg[H(e^{j\omega})]\})$, $\tau_G(\omega)$ is the desired group delay, and $\tau(\omega_o)$, called the "nominal delay," is a constant to be chosen in the optimization. The choice of λ $(0 \leq \lambda \leq 1)$ in Eq. (5.65) determines whether the error measure emphasizes error in the magnitude response, error in the delay, or some combination. The $\{W_i\}$ and $\{V_i\}$ sequences are chosen by the filter designer to adjust the goodness of the approximation as a function of frequency, and p is an even integer that can be selected to provide any ℓ_p error norm from ℓ_2 up to nearly minimax ($p = 40$ will provide a nearly minimax design [DI79]).

The Deczky approach to minimizing Eq. (5.65) involves setting equal to zero the derivatives of E with respect to each of the $4K + 1$ unknowns in Eq. (5.64). The resulting equations are nonlinear in the unknowns and are solved using the iterative Fletcher–Powell technique [FL63]. It is guaranteed that a local minimum of E will be achieved through this procedure; however, the minimum found may not be the global minimum. Therefore it is best to try this design procedure several times with different sets of initial values for the unknowns in Eq. (5.64), and to keep the resulting design that is best. Suitable initial guesses for the unknowns can be obtained by using noniterative algorithms described later in this section. A computer program that implements the Deczky algorithm is available [DI79].

As an example of an application of the Deczky procedure, we cite the design of a phase equalizer as described in Deczky [DE72]. A fourth-order lowpass elliptic filter was designed with the attenuation characteristic $(-20 \log_{10}|H(e^{j\omega})|)$ shown in Figure 5-24(a) and group delay shown by the lower curve in Figure 5-24(b). Since the elliptic filter has a sharp magnitude cutoff but a rather nonlinear phase (nonconstant delay), the Deczky procedure was used to design an allpass equalizer such that when cascaded with the elliptic filter it would improve the phase char-

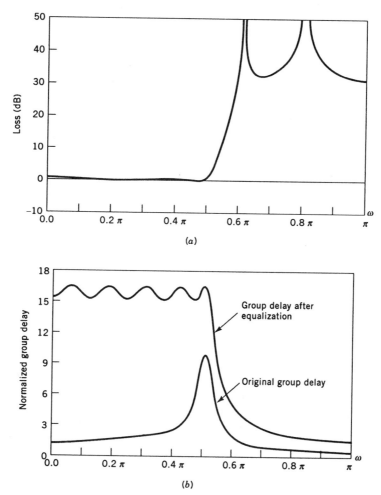

FIGURE 5-24 Example of phase equalization: (a) attenuation characteristic of a fourth order lowpass elliptic filter and (b) group delay (lower curve) and equalized group delay (upper curve). After Deczky [DE72]. Reprinted with permission.

acteristic, that is flatten the group delay over the passband. Designing a fourth-order phase equalizer using a value $p = 10$ in the error norm, the vastly improved group delay shown as the upper curve in Figure 5-24(b) resulted for the cascaded combination. It can be seen that the group delay is nearly equiripple over the passband, which occurs because p was chosen to be fairly large.

5-8-2   A Time-Domain Design Procedure (Prony's Method)

This section shows how to design an IIR filter with transfer function

$$H(z) = \frac{A(z)}{B(z)} = \frac{a_0 + a_1 z^{-1} + \cdots + a_M z^{-M}}{1 + b_1 z^{-1} + \cdots + b_N z^{-N}} \qquad (5.66)$$

so that the resulting unit-pulse response will exactly match the desired unit-pulse response for the first $N + M + 1$ samples. Our development follows Parks and Burrus [PA87] and utilizes a matrix description of Prony's method [BU70].

From Eq. (5.66) we have $A(z) = H(z)B(z)$ so that $\{a_n\}$ is given by the convolution $a_n = h[n] \circledast b_n$. The first $N + M + 1$ "outputs" of this convolution can be written in matrix form as

$$
\begin{bmatrix} a_0 \\ a_1 \\ a_2 \\ \vdots \\ a_M \\ 0 \\ \vdots \\ \vdots \\ 0 \end{bmatrix}
=
\begin{bmatrix}
h[0] & 0 & 0 & \cdots & 0 \\
h[1] & h[0] & 0 & & \\
h[2] & h[1] & h[0] & & \\
\vdots & & & & \\
h[M] & & & & \\
h[M+1] & & & & \\
\vdots & & & & \\
\vdots & & & & \\
h[M+N] & & & & h[M]
\end{bmatrix}
\begin{bmatrix} 1 \\ b_1 \\ b_2 \\ \vdots \\ \vdots \\ b_N \end{bmatrix}
. \qquad (5.67)
$$

Substituting the first $M + N + 1$ samples of the desired unit-pulse response for the $h[n]$ in Eq. (5.67), and solving, will give coefficients $\{a_n\}$ and $\{b_n\}$ that will produce these same $M + N + 1$ unit-pulse response samples. The problems of finding the a_n and b_n can be decoupled by partitioning Eq. (5.67) as

$$
\begin{bmatrix} \mathbf{a} \\ \cdots \\ \mathbf{0} \end{bmatrix}
=
\begin{bmatrix} \mathbf{H_1} \\ \cdots \quad \cdots \\ \mathbf{h_1} \quad \vdots \quad \mathbf{H_2} \end{bmatrix}
\begin{bmatrix} 1 \\ \cdots \\ \mathbf{b'} \end{bmatrix}
, \qquad (5.68)
$$

where $\mathbf{a}$ is the vector of the $M + 1$ numerator coefficients, $\mathbf{b'}$ is the vector of the N unknown denominator coefficients $b_1, b_2 \ldots, b_N$, $(b_0 = 1)$, $\mathbf{h_1}$ is the vector of the last N terms of the unit pulse response, $\mathbf{H}_1$ is the $(M + 1) \times (N + 1)$ upper part of the matrix in Eq. (5.67), and $\mathbf{H}_2$ is the $N \times N$ remaining part. The lower N equations are

$$0 = \mathbf{h}_1 + \mathbf{H}_2 \mathbf{b}'$$

or

$$\mathbf{H}_2 \mathbf{b}' = -\mathbf{h}_1, \tag{5.69}$$

which can be solved for $\mathbf{b}'$, the denominator coefficients in Eq. (5.66), assuming that $\mathbf{H}_2$ is nonsingular. The upper $M + 1$ equations of Eq. (5.68) are

$$\mathbf{a} = \mathbf{H}_1 \mathbf{b},$$

which gives $\mathbf{a}$, the numerator coefficients of the transfer function. If $\mathbf{H}_2$ is singular, then Eq. (5.69) may have many solutions, which indicates that $\{h[n]\}_{n=0}^{N+M}$ can be generated by a lower order system.

The limitation of this method is that the first $N + M + 1$ unit-pulse response samples are exactly matched, but the quality of the approximation is not controlled after that. Other techniques exist, though, for approximating the desired unit-pulse response $\{h[n]\}$ over a broader range of n. The reader is referred to Parks and Burrus [PA87] for an extension of the basic Prony method.

5-8-3 Alternative IIR Design Techniques

In this section we briefly mention four frequency-domain IIR filter design techniques, which are alternatives to those described in Sections 5-2 through 5-5, and 5-8-1 and 5-8-2. The first three are intended for the design of filters having a frequency response of arbitrary shape. These are:

1. An iterative nonlinear programming method by Dolan and Kaiser [DI79] that minimizes the maximum deviation from the desired response (Chebyshev norm).
2. A linear programming approach that minimizes the Chebyshev norm [RA74].
3. A frequency sampling design technique [PA87].

Method 1 is an alternative to the Deczky algorithm described in Section 5-8-1. Neither of these filter design algorithms can guarantee convergence to a global optimum and each of these methods has different numerical properties. Therefore it may be wise to try both of these approaches (each with several sets of initial conditions) and to keep the resulting design that is best. A computer program that implements Method 1 is available [DI79]. Method 2 is theoretically elegant, but the resulting linear program is quite large and difficult to solve for many problems of practical interest. Method 3 designs the filter so that the resulting frequency response agrees either exactly or approximately with the desired response on a discrete grid of frequencies. The disadvantage of this method, as with FIR design using frequency sampling, is that the response of the designed filter can sometimes depart significantly from the desired response at frequencies between those on the design grid.

The fourth alternative design approach, applicable to lowpass, highpass, bandpass, and bandstop filters, permits adjustment of the shapes of the passband and stopband responses of a filter. For these types of filters, the zeros of the numerator of the transfer function primarily control the stopband characteristics of the filter, whereas the zeros of the denominator mainly control the passband, so that the effects of the two are somewhat uncoupled. Because of this, the Remez exchange algorithm can be applied, alternating between the passband and stopband, to provide an effective method for designing IIR filters with a Chebyshev error criterion [DE74; JA90; MA78; SA83]. For numerator and denominator polynomials of the same order and a desired filter of an ideal lowpass shape, this approach should produce an elliptic filter. However, the Remez exchange method permits numerators and denominators of different orders. For narrowband applications, choosing the denominator order to be higher than the numerator order can provide advantages over elliptic designs in terms of reduced number of filter coefficients or increased stopband attenuation. In wideband cases, filters with a higher order numerator than denominator can be superior to elliptic filters.

5-9 SUMMARY AND DISCUSSION

This chapter developed the most commonly used methods for IIR filter design. Two distinct approaches were considered. In the first, the digital filter was designed by transformation of an analog prototype filter. Design relations were presented for Butterworth, Chebyshev (Type-1 and Type-2), and elliptic analog lowpass prototypes. The transformation methods discussed were the bilinear transformation, impulse invariance, and the approximation of differentials. Frequency transformations were given to convert the designed lowpass filter into a lowpass filter having a different cutoff frequency, or to a highpass, bandpass, or bandstop filter. Among the transformation methods, the bilinear transformation has proved to be the most popular. The impulse-invariant method suffers from aliasing, but in some cases it may better preserve the phase response of the analog prototype [AN79]. All three transformation methods rely on the availability of an analog prototype or the means for easily designing one. Closed-form design formulas are generally available, though only for lowpass, highpass, bandpass, and bandstop analog filters.

A second design approach, based on numerical optimization, was described for the more general case. Depending on the specific error criterion used, this approach attempts to minimize either a frequency-domain or time-domain measure of the error between the actual and desired response. For particular frequency-domain error measures, the general IIR design problem can be attacked directly using computer programs developed by Deczky [DE72; DI79] or Dolan and Kaiser [DI79]. These algorithms are iterative and require an initial set of filter coefficients as a starting point. These initial (suboptimal) values can be obtained by applying a simple version of Prony's method or by using a frequency-sampling design. Prony's method and frequency sampling can also be used as full-fledged design procedures for matching up specified samples of the desired and actual time-domain or frequency-domain responses. The IIR design problem also can be formulated as a linear program, but the resulting optimization problem is generally large and difficult to solve. Finally, we noted that the design of lowpass, highpass, bandpass, and bandstop filters can be solved using the Remez exchange method, even for the case with varying shapes in the passbands and stopbands. This approach permits numerators and denominators of unequal orders, which sometimes produces designs that are superior to elliptic filters.

Computer software is available for conveniently implementing the IIR filter design techniques that we have described. For all but the simplest filter designs, the use of such software is highly recommended. We conclude this chapter by discussing a representative set of such software.

In the commercial realm, Signal Technology, Inc. was one of the first companies to produce signal processing software. Their ILS package [IL] includes software for IIR filter design as well as other signal processing operations. Another software package by Atlanta Signal Processors, Inc. [AT84] is capable of bilinear transformation design of all the standard Butterworth, Chebyshev, and elliptic filters and can produce machine-language code for the Texas Instruments TSM320 digital signal processing chip. We make special mention of MATLAB [MA] by The Math Works, Inc., which was used to generate all the filter design examples in this chapter. This software can design Butterworth, Chebyshev, and elliptic filters, and it includes a host of other useful signal processing commands, some of which could facilitate one's own development of IIR filter design procedures.

Some years ago the IEEE published a book [DI79] and computer tape containing a large selection of programs useful for digital signal processing. Several IIR filter design routines are listed including the general design programs of Deczky and of Dolan and Kaiser. The IEEE computer tape also includes a program for designing lowpass, highpass, bandpass, and bandstop filters of the Butterworth, Chebyshev, and elliptic types using the bilinear transformation. As an additional feature, this program can optimize the noise performance of the designed filter by properly pairing and ordering the poles and zeros of the filter as a cascade of second-order sections. Another set of general signal processing software is the SIG package from Lawrence Livermore National Laboratory [SI85]. A part of this package designs IIR filters. In addition, books by Embree and Kimble [EM91], Parks and Burrus [PA87], Bose [BO85], Antoniou [AN79], and many others contain extensive software listings for different IIR filter design methods.

REFERENCES

[AB65] M. Abramowitz and I. A. Stegun, eds., *Handbook of Mathematical Functions*. Dover, New York, 1965.

[AN79] A. Antoniou, *Digital Filters: Analysis and Design*. McGraw-Hill, New York, 1979.

[AT84] Atlanta Signal Processors, Inc., *Digital Filter Design Package, DFDP*. Atlanta Signal Processors, Atlanta, GA, 1984.

[BO85] N. K. Bose, *Digital Filters: Theory and Applications*. Elsevier, New York, 1985.

[BU70] C. S. Burrus and T. W. Parks, Time domain design of recursive digital filters. *IEEE Trans. Audio Electroacoust.* **AU-18**, 137–141 (1970).

[CH66] E. Christian and E. Eisenmann, *Filter Design Tables and Graphs*. Wiley, New York, 1966.

[CO70] A. G. Constantinides, Spectral transformation for digital filters. *Proc. Inst. Electr. Eng.* **117**, 1585–1590 (1970).

[DE72] A. G. Deczky, Synthesis of recursive digital filters using the minimum P error criterion. *IEEE Trans. Audio Electroacoust.* **AU-20**, 257–263 (1972).

[DE74] A. G. Deczky, Equiripple and minimum (Chebyshev) approximations for recursive digital filters. *IEEE Trans. Acoust., Speech, Signal Process.* **ASSP-22**, 98–111 (1974).

[DI79] Digital Signal Processing Committee, IEEE Acoustics, Speech, and Signal Processing Society, ed., *Programs for Digital Signal Processing*. IEEE Press, New York, 1979.

[EM91] P. M. Embree and B. Kimble, *C Language Algorithms for Digital Signal Processing*. Prentice-Hall, Englewood Cliffs, NJ, 1991.

[FE72] A. Fettweis, A simple design of maximally flat delay digital filters. *IEEE Trans. Audio Electroacoust.* **AU-20**, 112–114 (1972).

[FL63] R. Fletcher and M. J. D. Powell, A rapidly convergent descent method for minimization. *Comput. J.* **6**, 163–168 (1963).

[GO69] B. Gold and C. M. Rader, *Digital Processing of Signals*. McGraw-Hill, New York, 1969.

[GR76] A. H. Gray, Jr. and J. D. Markel, A computer program for designing digital elliptic filters. *IEEE Trans. Acoust., Speech, Signal Process.* **ASSP-24**(6), 529–538 (1976).

[GU57] E. A. Guillemin, *Synthesis of Passive Networks*. Wiley, New York, 1957.

[HA69] G. E. Hansell, *Filter Design and Evaluation*. Van Nostrand-Reinhold, New York, 1969.

[HA89] R. W. Hamming, *Digital Filters*, 3rd ed. Prentice-Hall, Englewood Cliffs, NJ, 1989.

[HI84] W. E. Higgins and D. C. Munson, Jr., Optimal and suboptimal error spectrum shaping for cascade-form digital filters. *IEEE Trans. Circuits Syst.* **CAS-31**, 429–437 (1984).

[HU80] L. P. Huelsman and P. E. Allen, *Introduction to the Theory and Design of Active Filters*. McGraw-Hill, New York, 1980.

[IL] *ILS: Interactive Signal Processing Software*. Signal Technology, Inc., Goleta, CA.

[JA70] L. B. Jackson, Roundoff-noise analysis for fixed-point digital filters realized in cascade on parallel form. *IEEE Trans. Audio Electroacoust.* **AU-18**, 107–122 (1970).

[JA89] L. B. Jackson, *Digital Filters and Signal Processing*, 2nd ed., Kluwer Academic Publishers, Boston, MA, 1989.

[JA90] L. B. Jackson and G. J. Lemary, A simple Remez exchange algorithm to design IIR filters with zeros on the unit circle. *Proc. IEEE Int. Conf. Acoust., Speech, Signal Process. Albuquerque, New Mexico, 1990*, pp. 1333–1336 (1990).

[MA] *MATLAB*. The Math Works, Inc., South Natick, MA.

[MA78] H. G. Martinez and T. W. Parks, Design of recursive digital filters with optimum magnitude and attenuation poles on the unit circle. *IEEE Trans. Acoust., Speech, Signal Process.* **ASSP-26,** 150–156 (1978).

[OP89] A. V. Oppenheim and R. W. Schafer, *Discrete-Time Signal Processing.* Prentice-Hall, Englewood Cliffs, NJ, 1989.

[PA87] T. W. Parks and C. S. Burrus, *Digital Filter Design.* Wiley, New York, 1987.

[PR88] J. G. Proakis and D. G. Manolakis, *Digital Signal Processing: Principles, Algorithms, and Applications*, 2nd ed., Macmillan, New York, 1992.

[RA74] L. R. Rabiner, N. Y. Graham, and H. D. Helms, Linear programming design of IIR digital filters with arbitrary magnitude function. *IEEE Trans. Acoust., Speech, Signal Process.* **ASSP-22,** 117–123 (1974).

[RA75] L. R. Rabiner and B. Gold, *Theory and Application of Digital Signal Processing*, Prentice-Hall, Englewood Cliffs, NJ, 1975.

[SA83] T. Saramaki, Design of optimum recursive digital filters with zeros on the unit circle. *IEEE Trans. Acoust., Speech, Signal Process.* **ASSP-31,** 450–458 (1983).

[SE78] A. S. Sedra and P. O. Brackett, *Filter Theory and Design: Active and Passive.* Matrix, Forest Grove, OR, 1978.

[SI85] *SIG: A General Purpose Signal Processing, Analysis, and Display Program.* Lawrence Livermore National Laboratory, Livermore, CA, 1985.

[ST57] J. E. Storer, *Passive Network Synthesis.* McGraw-Hill, New York, 1957.

[TE77] G. C. Temes and J. W. LaPatra, *Circuit Synthesis and Design.* McGraw-Hill, New York, 1977.

[WE75] L. Weinberg, *Network Analysis and Synthesis.* R. E. Krieger Publishing Co., Huntington, NY, 1975.

[WI86] C. S. Williams, *Designing Digital Filters.* Prentice-Hall, Englewood Cliffs, NJ, 1986.

[ZV67] A. I. Zverev, *Handbook of Filter Synthesis.* Wiley, New York, 1967.

6 Digital Filter Implementation Considerations

YRJÖ NEUVO

Nokia Corporation
Helsinki, Finland

6-1 INTRODUCTION

Digital filter design and implementation consist of several strongly interacting steps. At each step there is a significant amount of freedom for the designer to select different solutions. This means that for best performance the design cycle may have to be iterated several times.

First, there are typically infinitely many filter transfer functions that meet the same frequency- or time-domain specifications. Similarly, the number of different and, in some sense, good filter structures is large. Many of the filter design methods produce transfer functions that are ''structure specific'' and the structures, on the other hand, to a large extent dictate many of the properties of the final implementation like speed, required arithmetic accuracy, and suitable hardware architectures. These interactions are schematically shown in Figure 6-1.

In order to reduce the number of iterations in the design phase it is desirable to be able to predict the performance of the filter hardware before actually implementing the filter. In this chapter we introduce the tools for analyzing the effects resulting from finite accuracy arithmetic used in the implementation. Also, methods to achieve high-speed multiprocessor implementations are discussed. The actual hardware and software implementations are discussed in Chapters 11 and 12.

The detailed properties of the arithmetic used in the implementation play a central role in this chapter. In Section 6-2 binary fixed- and floating-point arithmetic schemes are discussed. A general view to implementation-related issues is given in Section 6-3. These include complexity, finite wordlength effects, design time, and flexibility of the design to changing specifications.

In Section 6-4 transformations for filter structure modifications are discussed. They can be used to change the frequency-domain, finite wordlength, or hardware

Handbook for Digital Signal Processing, Edited by Sanjit K. Mitra and James F. Kaiser.
ISBN 0-471-61995-7 © 1993 John Wiley & Sons, Inc.

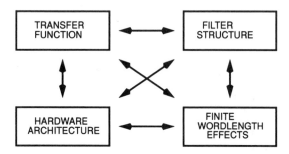

FIGURE 6-1   Digital filter design aspects and their interactions.

implementation aspects of the filter. The block implementation as an efficient way to obtain structures suitable for parallel processing is introduced in Section 6-5.

Section 6-6 deals with those aspects of the filter structure that are related to the hardware and/or software organization of the implementation and also strongly influence the achievable speed of the implementation. A method to partition the filtering task to processors operating in parallel is introduced here.

The remaining five sections deal with the finite wordlength properties of filter structures. Analysis of the effects of quantization and overflow are of great importance in carefully optimized implementations. Statistical and deterministic models for the roundoff operations are given in Section 6-7. It also includes a discussion of overflow operations.

Methods to analyze the required coefficient accuracy are introduced in Section 6-8. In the following section, the signal-to-noise ratio (SNR) resulting from quantizing the analog input signal to the finite accuracy of the digital implementation is discussed. In Section 6-10, statistical models for the roundoff error are used to derive the signal-to-noise ratio concept for digital filters. Also, the method of scaling is introduced to improve the signal-to-noise ratio. As examples, the roundoff noise properties of first- and second-order sections are analyzed. Motivated by the increasing use of floating-point signal processors, an approach to roundoff noise analysis of floating-point implementations is introduced in Section 6-11. Concluding remarks are presented in Section 6-12.

6-2   NUMBER REPRESENTATION AND ARITHMETIC SCHEMES

The operations in a digital signal processor are basically operations on numbers stored in binary form. These include binary addition, multiplication, negation, and quantization. In this section we introduce commonly used schemes for the binary representations of numbers and outline some of the arithmetic rules [FL63; KW79].

6-2-1   Fixed-Point, Floating-Point, and Block Floating-Point Representations

Binary representations of numbers consist of two symbols, usually denoted by 1 and 0, called the *bi*nary digi*ts* (or bits), which are associated with weights of appropriate powers of 2. For example, the representation $1001_\Delta 01$ stands for 1×2^3 $+ 0 \times 2^2 + 0 \times 2^1 + 1 \times 2^0 + 0 \times 2^{-1} + 1 \times 2^{-2}$, which is equal to 9.25 in the decimal number system. The meaning of the binary point$_\Delta$ is also clear from here: the bits to the right of the binary point constitute the fractional part and those to the left determine the integer part of the number. In a b-bit fixed-point representation of a positive number, a prescribed number of bits, say, b_1, is associated with the integer part, the remaining $b_2 = b - b_1$ bits representing the fractional part. In the above example, $b_1 = 4$ and $b_2 = 2$. If $b_1 = 0$, the number is a fixed-point fraction, and if $b_2 = 0$ it is a fixed-point integer. Integers are usually represented without showing explicitly the binary point. For example, 1001 would just represent 9 in decimal form and is an abbreviation for $1001_\Delta 0$.

Because of the finite wordlength of a register, the digital representation of a number in a digital machine can have a finite range of values. For example, the set of all 8-bit positive integers is restricted to be in the range $0 \le x \le 255$. Such a range is called the *dynamic range* of the number representation. If the operands of an arithmetic operation are within this range, the result cannot be faithfully represented because it exceeds the range. Such a phenomenon is called an *overflow*.

If two fixed-point numbers of length b bits each are added, the result may require more than b bits. For example, the addition of two 3-bit fractions, $_\Delta 110$ and $_\Delta 111$, results in $1_\Delta 001$, which is 4 bits long. Thus, in general, addition of fixed-point numbers causes an overflow.

When two fixed-point numbers having b bits are multiplied, the result in general has $2b$ bits. If both numbers have an integer part, the result might overflow. For example, the product of $10_\Delta 11$ and $11_\Delta 01$ yields $1000_\Delta 1111$, indicating an overflow in the integer part. If both the numbers are fixed-point fractions, the result remains a fraction and clearly cannot overflow. Accordingly, it is convenient to assume all numbers in fixed-point schemes to be scaled such that they are fractions.

Multiplication of two fixed-point fractions, even though free of overflow, does give rise to a larger number of bits. For example, if we multiply the 3-bit numbers $_\Delta 101$ and $_\Delta 110$, the result is $_\Delta 011110$, which requires 6 bits in the fractional part. In order to prevent repeated accumulation of bits in this manner, the result is usually approximated so as to have b_2 bits in the fractional part. In the present example, we can truncate the 6-bit answer to obtain a 3-bit representation given by $_\Delta 011$. Alternatively, the result can be rounded to 3 bits, giving $_\Delta 100$. Both operations are often loosely referred to as "rounding."

A floating-point representation of a decimal number x is of the form $x = 2^c M$, where c is called the *exponent* and M the *mantissa*. The mantissa is a b-bit fixed-point number in the range $0.5 \le M < 1$. Both M and c are stored in binary form. These could be of either sign depending on x (representation of negative quantities will be considered soon). If we multiply two numbers $x_1 = 2^{c_1} M_1$ and $x_2 = 2^{c_2} M_2$ in floating-point form, the result is $x = 2^{c_1 + c_2} M_1 M_2$, which means that to obtain the product we add the exponents and multiply the mantissa parts. In practice, the resulting exponent c is readjusted so that the mantissa of the product $M = M_1 M_2$ is brought back to the range $0.5 \le M < 1$. If we wish to add x_1 and x_2, the steps are a little more involved; the first step is to adjust the exponent and mantissa of the smaller number until the exponents of x_1 and x_2 match. The mantissas are then added. Finally, the resulting representation is rescaled so that its mantissa in turn is in the range [0.5, 1).

In summary, fixed-point fractional implementations have the advantages of simplicity; addition does not cause roundoff error, but multiplication does. However, addition *can* cause overflow error but multiplication cannot. Floating point implementations, on the other hand, permit a very large dynamic range, and there is virtually no possibility of an overflow. For example, if the exponent c has 10 bits, then the range of decimal numbers that can be represented is approximately 2^{511} to 2^{-512}. However, in a floating-point implementation both addition and multiplication can cause roundoff errors. Finally, the bookkeeping and additional overhead involved makes this arithmetic much slower than fixed-point arithmetic.

A compromise between fixed- and floating-point schemes is the block floating-point arithmetic [OP70]. Here, the set of signals to be handled is divided into subclasses. Numbers belonging to a particular subclass have the same value for the exponent. Accordingly, arithmetic within a subclass is almost like fixed-point arithmetic, and, moreover, only one exponent per subclass needs to be stored. This scheme is attractive when the set of signals can be classified such that each subset is most likely to occupy a particular dynamic range for most of the time, as in certain FFT flow graphs and in digital audio applications.

A special type of fixed-point arithmetic is integer arithmetic, where each number is an integer (i.e., $b_2 = 0$). Such arithmetic is entirely free from roundoff errors, but both addition and multiplication operations can cause overflow errors. Note that roundoff errors are "small" errors, whereas overflow errors are large and should be avoided. Accordingly, integer arithmetic is not commonly used, except in residue number systems [JE77]. These are discussed more in Chapter 9.

6-2-2 Representation of Negative Numbers

A fixed-point number can have either sign. Likewise, the exponent and mantissa in a floating-point number can be either positive or negative. Since the exponent and mantissa are fixed-point numbers themselves, it suffices to describe here how to represent negative fixed-point numbers.

For simplicity of notation and discussion, we consider only fixed-point *fractions* here. There are three standard conventions for representing such signed numbers. These are the *sign–magnitude convention, one's complement convention*, and the *two's complement convention*. A b-bit fixed-point fraction is shown in Figure 6-2. Such a number has a sign bit denoted by s, followed by the b binary bits α_1, α_2,

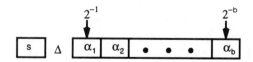

FIGURE 6-2   A b-bit fixed-point fraction with a sign bit.

..., α_b. Regardless of the convention adopted, $s = 0$ for a positive number and for a negative number $s = 1$. For a positive number the magnitude is always represented by $\sum_{k=1}^{b} \alpha_k 2^{-k}$. The three conventions differ only in the way in which the bits α_k represent the magnitude of a negative number (i.e., when $s = 1$).

In the sign-magnitude representation, the quantity $\sum_{k=1}^{b} \alpha_k 2^{-k}$ represents the magnitude, regardless of the sign bit s. In the two's complement system, the *value* of the number is always given by $-s + \sum_{k=1}^{b} \alpha_k 2^{-k}$. It can be shown [FL63; KW79] that, for such a representation, the magnitude of a negative number is obtained by complementing each bit (including s) and adding 2^{-b}. A two's complement number can be negated simply by complementing each bit and adding 2^{-b}. For example, let $b = 4$ and let $1_{\Delta}1001$ be the two's complement representation of $-x$. Then its magnitude is represented by $0_{\Delta}0111$, whose decimal equivalent is $\frac{7}{16}$. Addition of x_1 and x_2 in this representation is performed by treating the sign bit as just another bit. (If both numbers have the same sign, this might cause overflow.) For example, let $x_1 = -\frac{5}{8}$ and $x_2 = \frac{6}{8}$ which are represented by $1_{\Delta}011$ and $0_{\Delta}110$, respectively. Then a straight addition of x_1 and x_2 is given by $1_{\Delta}011$ $+ 0_{\Delta}110 = 10_{\Delta}001$. Discarding the leftmost bit corresponding to 2^1 we get the result $0_{\Delta}001$, which correctly represents the answer $\frac{1}{8}$. The new bit (which we had to discard) should not be confused with an overflow. Since x_1 and x_2 in this example have opposite signs, there cannot indeed be an overflow. (Detection of true overflow can be done based on simple rules; see Flores [FL63].) The justification for discarding the bit corresponding to 2^1 is that two's complement arithmetic rules are compatible with modulo 2 arithmetic, and hence 2^1 is equivalent to zero.

When a two's complement number is shifted to the right (which corresponds to dividing by a power of 2), it is necessary to shift the sign bit as well; the vacant space created by the sign bit must then be *padded* with the correct sign bit. For example, let $x_1 = -\frac{5}{8}$ with the representation $1_{\Delta}0110$. Then $x_1/2 = -\frac{5}{16}$ is represented by $1_{\Delta}1011$. Next, a left shift of a number corresponds to multiplication by a power of 2. This can cause overflow for obvious reasons. Further rules for two's complement arithmetic can be found in Oppenheim and Schafer [OP75], Rabiner and Gold [RA75], Flores [FL63], and Kwang [KW79].

In one's complement representations, the magnitude of a negative number is obtained simply by complementing each bit. The value of any number is given in terms of the representation by $-s(1 - 2^{-b}) + \sum_{k=1}^{b} \alpha_k 2^{-k}$. This type of representation is not very commonly used in DSP implementations. Two's complement arithmetic is simple to implement, since the rules are the same regardless of the sign of the operands. The three different conventions for negative numbers are summarized in Table 6-1 using $b = 2$. Note that in the sign-magnitude and one's complement representations zero can be presented in two different ways.

Example 6.1. A commonly used FIR filter is the recursive running sum or sliding average, which implements the transfer function

$$H(z) = \frac{1}{N}(1 + z^{-1} + z^{-2} + \cdots + z^{-N+1}) = \frac{1}{N}\frac{1 - z^{-N}}{1 - z^{-1}}.$$

TABLE 6-1 Binary Number Representations

	−4/4	−3/4	−2/4	−1/4	0	1/4	2/4	3/4
Sign Magnitude	NA	$1_\Delta11$	$1_\Delta10$	$1_\Delta01$	$0_\Delta00$ $1_\Delta00$	$0_\Delta01$	$0_\Delta10$	$0_\Delta11$
Two's Complement	$1_\Delta00$	$1_\Delta01$	$1_\Delta10$	$1_\Delta11$	$0_\Delta00$	$0_\Delta01$	$0_\Delta10$	$0_\Delta11$
One's Complement	NA	$1_\Delta00$	$1_\Delta01$	$1_\Delta10$	$0_\Delta00$ $1_\Delta11$	$0_\Delta01$	$0_\Delta10$	$0_\Delta11$

This is often implemented in the recursive form as shown in Figure 6-3. This implementation is based on the overflow properties of two's complement arithmetic. Consider a constant nonzero input signal $u[n]$. Obviously, the signal $x_1[n]$ overflows at regular intervals. However, the effect of these overflows is compensated at the second adder as long as the signal $x_2[n]$ is within the permissible range $[-1, 1 - 2^{-b}]$. In order to see more clearly how this filter operates, let us take $N = 2$ and $b = 2$. These selections make the filter to be of little practical use but let us observe the essential properties of this structure. We assume that initially (at time $n = 0$) there is an arbitrary number in the feedback delay element; let us select $x_1[-1] = \frac{1}{2}$. Clearly, after one iteration (at $n = 1$) with zero input signal, $u[n] = 0$ for $n \geq 0$, the output $y[n]$ also becomes zero. We now apply a constant input signal $u[n] = \frac{1}{4}$. Table 6-2 shows the evolution of the associated signals. At $n = 1$, $x_2[n] = \frac{1}{4}$ and the output also becomes $\frac{1}{4}$ if rounding is applied after the multiplication. For $n \geq 2$ the output stays at $\frac{1}{4}$ even though internal overflows occur. Note that the initial values stored in the delay elements only have a finite duration effect on the output as determined by the length of the impulse response. Also note that the implementation requires only two additions and one multiplication irrespective of the filter length. If the length is selected to be $N = 2^k$ with k a positive integer, the multiplication reduces to an arithmetic right shift.

Canonic Sign Digit Code. Sign digit code (SD code) is a useful representation of numbers, with three possible digits ($1, \bar{1}, 0$) instead of two digits ($1, 0$) as in

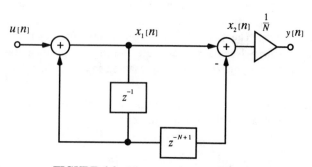

FIGURE 6-3 The recursive running sum.

TABLE 6-2   Evolution of the Signals in the Recursive Running Sum

n	$u[n]$	$x_1[n]$	$x_1[n-1]$	$x_1[n-2]$	$x_2[n]$	$y[n]$
-1	$0_\Delta 00$	$0_\Delta 10$	$0_\Delta 10$	undef	undef	undef
0	$0_\Delta 00$	$0_\Delta 10$	$0_\Delta 10$	$0_\Delta 10$	$0_\Delta 00$	$0_\Delta 00$
1	$0_\Delta 01$	$0_\Delta 11$	$0_\Delta 10$	$0_\Delta 10$	$0_\Delta 01$	$0_\Delta 00$
2	$0_\Delta 01$	$1_\Delta 00$	$0_\Delta 11$	$0_\Delta 10$	$0_\Delta 10$	$0_\Delta 01$
3	$0_\Delta 01$	$1_\Delta 01$	$1_\Delta 00$	$0_\Delta 11$	$0_\Delta 10$	$0_\Delta 01$
4	$0_\Delta 01$	$1_\Delta 10$	$1_\Delta 01$	$1_\Delta 00$	$0_\Delta 10$	$0_\Delta 01$

the binary system (here $\overline{1}$ is used to represent -1). For example, the binary number $0_\Delta 011111111$ can be rewritten as $0_\Delta 10000000\overline{1}$ in the SD code. In the decimal system this corresponds to $2^{-1} - 2^{-9}$. The SD representation often requires fewer nonzero digits resulting in a lower complexity for implementing multiplications. In the above example, if a signal x is multiplied with $0_\Delta 011111111$, this is equivalent to seven addition operations in standard binary format, whereas multiplying with the SD form $0_\Delta 10000000\overline{1}$ corresponds to only one subtraction operation. In fact, given any binary fixed-point number, there exist algorithms to write it in the SD form with the smallest number of nonzero digits. Such a representation is called the *canonic sign digit code*.

6-2-3   IEEE Standard for Binary Floating-Point Arithmetic

With the advent of floating-point processors, more and more complex DSP systems, and high-level DSP system design tools, definite advantages would be obtained if the code derived for one type of floating-point processor is portable to later versions of the processor as well as to other floating-point processors. This would facilitate and encourage the exchange of existing algorithms with tested performance and shorten the development time of DSP systems.

A complete floating-point number system is quite complicated when all the fine details of exceptions handling and so on are taken into account. ANSI (American National Standards Institute) and IEEE have published a standard that gives certain recommended and well-defined binary floating-point formats [IN85] for 32- and 64-bit arithmetic units. The recommended 32-bit format or its subset is used in certain commercial floating-point signal processors.

We now briefly describe the most important features of the 32-bit format. The format presents numbers in the form

$$v = (-1)^s 2^E (\alpha_{0\Delta} \alpha_1 \alpha_2 \cdots \alpha_{23}), \tag{6.1}$$

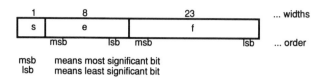

msb means most significant bit
lsb means least significant bit

FIGURE 6-4 The 32-bit floating-point format: msb, most significant bit; lsb, least significant bit.

where s is the sign bit, and E is the exponent within the number range $[-126, 127]$, limits inclusive. Thus the mantissa is represented in sign–magnitude form. The mantissa without the sign bit is described by bits $\alpha_{0\Delta}\alpha_1 \cdots \alpha_{23}$ and is called the *significand*. Normalization is such that α_0 is 1, implying that the value of the significand is between one and two. The format of the presentation is shown in Figure 6-4. Note that only the fractional part f (bits $\alpha_1 \ldots \alpha_{23}$) of the significand is given because for normalized numbers α_0 is always 1. The exponent E is coded in biased form $e = E + 127$ using 8 bits.

Special meanings are reserved for representations with $e = 0$ and $e = 255$ in order to maintain integrity of operations under all circumstances:

1. If $e = 255$ and $f \neq 0$, then v is a symbolic entity encoded in the floating-point format. These entities are called NaN's (Not a Number). They can inform about invalid operations and can be used to propagate useful diagnostic and other information to the user.
2. If $e = 255$ and $f = 0$, then $v = (-1)^s \infty$.
3. If $e = 0$ and $f \neq 0$, then $v = (-1)^s 2^{-126}(0_\Delta f)$. These are denormalized numbers.
4. If $e = 0$ and $f = 0$, then $v = (-1)^s 0$. Thus we have a positive and negative zero in the representation.

The standard specifies four rounding operations: to the nearest, toward $+\infty$, toward $-\infty$, and toward zero. To the nearest rounding gives the number with smaller magnitude if two numbers are equally apart. The standard also specifies the regular arithmetic operations: add, subtract, multiply, divide, remainder, and square root, as well as conversions between different floating-point and integer formats.

6-3 CHARACTERISTICS OF FILTER STRUCTURES

6-3-1 Computational Complexity

By computational complexity of a filter structure we mean the number of certain types of operations in the filter implementation that consume resources, such as time or chip area. A digital filter structure is composed of adders, multipliers, and delays or storage elements. These are generally considered to be good descriptors of the complexity of a filter. Traditionally, research on filter structures has concentrated on minimizing the number of multipliers without much concern about the number of adders and delays. This is a reasonable goal only if the multipliers are much more complicated to implement than the adders and delays.

In modern signal processors, with hardware multipliers, additions and multiplications tend to take the same time. Sometimes there is a multiply–add operation making this combined operation as time consuming as a single addition. In addition, depending on the instruction set, data moves, looping, and branching instructions consume time that may not be an insignificant part of the time consumed by the arithmetic instructions. As a consequence, for high-speed applications the filter structure and the signal processor's arithmetic and instruction set should be matched with each other. Even for a fixed filter structure the filtering program can be written in a variety of ways, the goal being, for example, *maximum execution speed* or *minimum program size*. The available program and data memory sizes are often critical design issues when high-speed signal processors are used.

In general purpose computers and microprocessors without a hardware multiply unit, the multiplication is performed in a shift and add type fashion. Then the multiplications for b-bit multipliers are roughly b times more time consuming than the additions. However, there are schemes to speed up the multiplications like the canonic sign digit code described earlier in this chapter. In this kind of environment the available memory size is often not a limiting factor.

In VLSI implementations there are several types of design trade-offs between the speed and the chip area. A major design issue is the type of arithmetic used. This largely affects the relative cost of additions and multiplications. In bit serial implementations using serial/parallel multipliers, we need b one-bit adders, making the surface area of a multiplier roughly b times the area taken by a serial adder. If parallel arithmetic is used with combinatorial multipliers, the multiplier is also approximately b times as complex as the parallel adder. The space taken by memories, control unit, wiring, and I/O pads also needs to be taken into account.

For a certain type of resource and a certain type of implementation, a reasonable figure of merit F_M that can be used to aid in the selection of a suitable filter structure is the weighted sum of the number of additions, A, number of multipliers, M, and the number of delays, D:

$$F_M = a_A A + a_M M + a_D D. \qquad (6.2)$$

For example, in a VLSI implementation with serial arithmetic, one may have for area minimization $10a_A = a_M = 2a_D$. If a signal processor is used it may be reasonable to take $a_A = a_M = 1$ and $a_D = 0$ if processing time is to be minimized and if the delays can be implemented in connection with the arithmetic operations.

A study about the suitability of several allpass structures for signal processor implementation is reported in Renfors and Zigouris [RE88b]. The results indicated that with representative processors the direct form 1 structure was the fastest in implementing first- and second-order sections even though specific allpass sections require only one-half of the multipliers of the direct form structure. This is primarily due to the good match between the direct form structure and the processor instruction sets.

6-3-2   Finite Wordlength Effects

Digital filters are implemented with finite accuracy using typically fixed- or floating-point arithmetic. The available accuracy has continuously increased, most notably in general purpose microprocessors and signal processors [LE90]. With the 16- or 24-bit fixed-point and 32-bit floating-point processors, one can implement many signal processing applications without much concern of the finite wordlength effects. It is sufficient to use an algorithm that is known to be reasonably good for the particular type of application.

On the other hand, the advent of high-accuracy and high-speed processors brings up new, more demanding applications, which again require a thorough analysis and optimization of the finite wordlength properties.

In custom VLSI implementations the situation is often different because there one can directly convert the reduction of accuracy to savings in chip area and/or to higher speed. However, even in custom circuits issues like modularity, generality, and design time can be more important aspects than optimization of the circuit area to an extent that makes the design effort applicable only in the particular application.

In signal processing, finite arithmetic manifests in three different ways: coefficient quantization, rounding or truncation of intermediate results, and arithmetic overflows [LI71; OP72].

Filter coefficients are quantized (rounded or truncated) from the ideal values derived from the filter approximation stage. This makes the transfer function differ from the ideal one. This error is commonly characterized by the *coefficient sensitivity* of the filter. For a fixed filter this is a deterministic type of error that can be judged ahead of time. However, general evaluation of a certain filter class calls for statistical measures of coefficient sensitivity.

Different filter structures have different sensitivity properties. For example, cascaded second-order sections tend to have very good stopband sensitivity properties because the zeros remain on the unit circle at approximately the same locations even after coefficient quantization, whereas parallel allpass structures have good passband sensitivity properties provided the allpass structures remain allpass under coefficient quantization (see Chapter 7).

In fixed-point arithmetic, multiplication of data samples with filter coefficients produces results that are longer than the available accuracy permits. For example, the product of two 16-bit numbers occupies 32 bits. If in a recursive filter no rounding or truncation of intermediate results is performed, the required accuracy increases with time. Thus it is necessary to quantize the results of multiplications to the available accuracy, causing errors that can be seen at the output of the filter as small fluctuations of the output signal around the ideal response, which would otherwise be obtained with unrestricted arithmetic accuracy.

In fixed-point addition, the required wordlength can increase by one bit per addition. In digital filtering this increase of wordlength is normally taken care of by properly limiting the signal levels to be sufficiently small so that the sum of two numbers remains in the proper number range. In floating-point arithmetic both multiplications and additions cause the mantissa length to increase, requiring the intermediate results to be quantized.

The effect of roundoff operations can be studied using the setup shown in Figure 6-5. Here the filter H_I is implemented with accurate arithmetic. In simulations double precision can be used. The filter H_Q is the same as the filter H_I except that it uses the actual finite wordlength. By analyzing the error sequence $e[n]$, valuable information about the level and the statistical properties of the roundoff errors can be obtained.

The roundoff errors can be categorized into two classes. One type of error is a random type variation, which at the point of incidence can be modeled as white noise called *roundoff noise*. Roundoff noise level at filter output depends strongly on the filter structure. For a given filter the *signal to roundoff noise ratio* can be maximized by *scaling* of the internal signals. In floating-point arithmetic scaling is not normally required as the normalization of floating-point numbers performs essentially the same operation.

The other error arising from quantization of intermediate results can be seen as a highly correlated component in the output signal. This is often in the form of deterministic oscillations called *limit cycles* of *granular type* [ER85; PA71]. Limit cycles arise in looped calculations and can thus exist only in recursive digital filters.

Limit cycles can be analyzed treating the filter as a nonlinear system [ER85]. However, for higher than second-order IIR structures this is very difficult. The granular type limit cycles can be eliminated from certain IIR filter structures by using *magnitude truncation* as it has the property of reducing the magnitude of the signal, thus gradually eliminating the oscillation.

Typically granular oscillations, if they exist, are of the same order of magnitude as the roundoff noise level. For a second-order section the amplitude of limit cycles with zero input signal is typically a few quantization steps.

Roundoff noise and limit cycles are two extreme descriptors of the quantization operation. The former applies best when the signal changes relatively fast and occupies a significant portion of the available number range. The latter applies best when the signal is essentially constant or has a simple deterministic oscillation whose period is an integer multiple of the sampling period.

Arithmetic overflows in recursive filters can produce oscillations of large am-

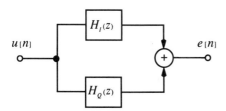

FIGURE 6-5 Model for roundoff error analysis.

plitude [EB69]. Especially dangerous in this respect is the two's complement fixed-point number system because in the case of overflow the error caused is of the size of the full number range. *Saturation arithmetic* is commonly used to eliminate *overflow oscillations*. In saturation arithmetic the overflow is detected and the output is set to the largest positive or smallest negative number depending on if the overflow was from the positive or negative side, respectively.

Limit cycles and overflow oscillations are discussed further in Section 6-7. The effect of both the coefficient quantization and the roundoff are seen at the output through the same mechanism, namely, through transfer functions from the point of incidence to the output. This suggests that there could be some kind of a relationship between these two errors. As a general observation from research of filter structures, one can state that if the structure is good in one of these respects it is not very bad in the other respect, either. For properly scaled cascaded and parallel second-order sections a relationship between the coefficient sensitivity and a lower bound of roundoff noise level has been derived in Jackson [JA76]. For floating-point filters a certain similarity between the roundoff noise model and coefficient sensitivity model exists, as is shown in Section 6-11.

6-3-3 Some Design and Implementation Issues

The way the delays, multipliers, and adders are connected together in a given filter structure dictates, in addition to the finite arithmetic properties, the suitability of the structure for implementation of a specific transfer function on a certain hardware or software organization.

In signal processing algorithm design, a critical constraint is the execution time of the algorithm. In *serial processing*, there is only one processing unit and it implements the algorithm in a certain order. In this case the speed of the implementation depends on the speed and the instruction set of the processing unit. For example, many audio and speech processing algorithms can be implemented with one modern signal processor at one of the standard audio sampling rates around 40 kHz.

However, by adding more filtering tasks or by making the algorithms more sophisticated one easily comes to a situation that cannot be handled by only one processor. The only way to implement these algorithms is to use several processing units operating simultaneously and sharing the filtering tasks in a *parallel processing mode*. As an example of an application requiring parallel processing one could mention a complete digital HIFI audio processing unit implementing equalization, reverberation, and crossover filters in stereo. This application requires approximately four modern signal processors depending on the algorithms used. Similarly, the need for parallel processing arises in digital video where the speed demands are much higher than in audio, but the algorithms are much simpler.

Parallel processing opens up the possibility of obtaining very high processing speeds. With proper hardware organizations sampling intervals shorter than the multiplication time of the processing unit can be achieved, the limit being the order of a gate delay.

With the advent of low cost signal processors supporting communications tasks the implementation of parallel processing systems has become relatively easy. In addition, some of the new signal processors themselves contain multiple arithmetic units. In VLSI implementations, especially if serial arithmetic is used, parallel processing is the natural choice of implementation.

Parallel processing is well suited for signal processing implementations as the algorithms are typically decided on ahead of time and can thus be optimized with the hardware. In addition, signal processing algorithms do not typically contain data-dependent branching instructions where the next operation depends on the outcome of the current instruction. This enables the design of a good fixed job partitioning for the parallel operating units.

There are several ways to classify parallel processing architectures. The traditional way is based on the type of control and data flow in the architecture [BL88; FL66].

Single Instruction Single Data Stream (SISD) machines are conventional processors, where operations are performed sequentially using one processing unit. Most microprocessor and signal processor chips belong to this class. A general block diagram is shown in Figure 6-6(a). In this figure the processing element (PE) and the instruction memory (IM) are shown explicitly. The processing element contains the necessary memory space for storing the data.

The *Multiple Instruction Single Data Stream* (MISD) architecture is also called the *pipeline* machine because multiple processing stages are cascaded to perform many operations on a single data stream (see Figure 6-6(b)). Each processing element may execute different program codes. Particularly suitable for image processing applications is the *Single Instruction Multiple Data* (SIMD) architecture, where the same operations are performed simultaneously using several processors on several data streams as shown in Figure 6-6(c). The most general is the *Multiple Instruction Multiple Data* (MIMD) architecture, shown in Figure 6-6(d) for a single-input–single-output filter. The connection matrix (CM) provides the necessary communication paths between the processors. In this architecture the processors apply different instructions on different data sequences. The aim is to distribute the processing tasks among the processors in such a way that maximal performance is achieved. This will be discussed in more detail in Section 6-6.

There are several methods to find the maximal parallelism of a signal processing algorithm, for example, the precedence graph method [CR75b], the precedence matrix method [BA78], the critical path method [BR78], and the use of Petri nets [NO80]. These approaches allow one to derive a MIMD-type architecture for the implementation.

For digital filters realized using parallel or cascaded second-order sections a rather natural solution is to have each second-order section implemented in one processing unit. In the parallel structure the final addition can be divided into the second-order sections. The multiprocessor organization implementing the parallel filter structure can be called the SIMD architecture as each processor executes the same second-order section program but the architecture contains multiple parallel data streams. Naturally, the SIMD architecture allows the input data streams to the second-order sections to be different too.

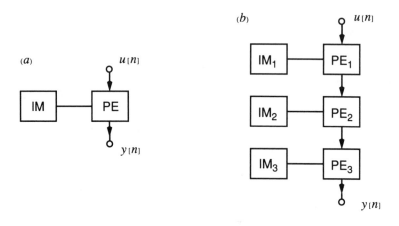

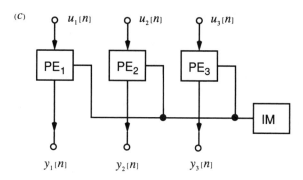

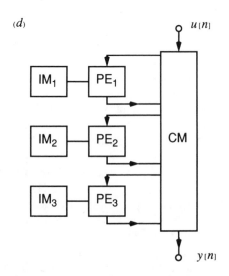

FIGURE 6-6 Parallel processing architectures: (a) SISD, (b) MISD, (c) SIMD, and (d) MIMD.

The multiprocessor implementation of the cascaded structure is the MISD architecture because only one data stream goes through the chain of processors. However, the MISD architecture would allow the programs in different processors to be different, implying that the cascaded filter structures could be different.

The fundamental difference between the parallel and cascade structure is in the *latency* of the implementation. In a digital filter, latency is the time it takes the first nonzero value of the impulse response to appear at the output of the filter. In the parallel structure the latency is normally one sampling period.

In cascaded second-order sections a filter module implementing one second-order section can finish its operation only after the output signal from the previous section has been calculated and transferred to the module. If each section is executed in one sampling period, different sections operating in series operate on different input data samples. This kind of operation is called pipelining. For M second-order sections the latency is now M sampling periods. This arithmetic delay added up with the algorithmic delay, which is often taken to be the group delay of the algorithm, gives the total delay of the implementation. This total delay can be a critical design issue in some applications such as communications systems.

In *block processing* [AN80; BU72; MI78; MO76] the input data are first vectorized to a certain block size and these blocks are processed with highly parallel algorithms. By increasing the block size the *throughput*, meaning the number of samples processed in a given unit of time (u.t.), can be increased with the penalty of increased latency and some increase in the number of arithmetic operations. The latency comes from the vectorizing operations. For the block length of L one has to collect first L samples for processing. This produces a latency of L samples. If block processing is implemented in one sampling period the output vector has been calculated in $L + 1$ sampling periods, implying a latency of $L + 1$. Block processing methods represent new filter structures as they change the internal computations of the filter and thus also the finite wordlength properties. Block processing is discussed in more detail in Section 6-5.

Parallelism can also be implemented at a lower level, for example, by having a separate hardware multiplier for each individual coefficient. The *precedence graph* or *timing diagram* of the algorithm indicates what has to be calculated before a certain operation can be started.

An example of parallel FIR filter structures is the transposed direct form structure shown in Figure 6-7 for implementation with M processors. As shown in the figure this structure can be partitioned for parallel processing after each addition. The input sequence $x[n]$ is broadcasted to all processors simultaneously. Data transfers between the processors contain a unit delay. This means that during each sampling interval each processor can perform its tasks using the input sample and the data stored in the delay elements. Thus there is no need to obtain any data corresponding to the current sampling instant from the neighboring processors. For highest possible speed one processor would implement only one multiplication and one addition. The Cascadable Signal Processor made by Inmos [IN87] implements a 32-tap FIR filter on one chip using this structure. Chips can be cascaded for implementation of higher order filters without any speed penalty. The coefficients can be updated conveniently in adaptive applications.

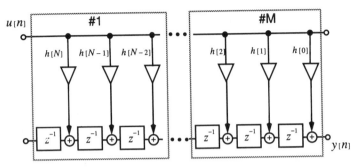

FIGURE 6-7 Block diagram of the transposed form FIR structure implemented using M filter units operating in parallel.

The implementations based on second-order sections as well as the FIR structure described above are highly *modular* as the same basic structure can be repeatedly used to implement more complex filters. Modularity helps in the design, implementation, and testing of both hardware and software signal processing systems. An aspect closely related to modularity is *data transfer complexity* between modules. The amount of data transferred between modules should be small and from each module concentrated to only certain neighboring modules. This simplifies hardware implementations and may also increase the speed of the implementation. Data transfer should also be as regular as possible, enabling simple control of the operations.

Irrespective of the type of implementation, the overall design time, including testing of the system and design of later modifications, is becoming a more and more important aspect in product design. This calls for good *design support* software for digital filter design [MA90]. Starting from high-level algorithmic descriptions these tools automatically generate code for signal processors or provide an easy connection to a high-level integrated circuit design environment like a *silicon compiler*. As an intermediate result the tools enable thorough emulation of the signal processing system before actually building any hardware.

This trend will call for filter structures that are straightforward to obtain from the algorithmic description of the filter and are general enough to implement practically any algorithm with reasonable efficiency. An example of current design support systems is the Digital Filter Design Package [MA88] from Atlanta Signal Processors Inc. The software, running on a personal computer, generates the signal processor code starting from the frequency-domain specifications of the filter. It uses the direct form FIR and cascaded second-order sections IIR structures in a very straightforward but acceptable manner. The code can be downloaded to the signal processor board and tested in real time or using test signals stored on disk.

Some signal processing software packages for workstations already provide an integrated environment for signal analysis, system design, simulation, and code generation. Good examples are the Signal Processing Worksystem (SPW) from Comdisco Systems, and Gabriel from University of California, Berkeley [BI90].

6-4 STRUCTURAL TRANSFORMATIONS

We now describe several transformations that can be applied to a digital filter structure to change the properties of a filter. The first class of transformations are *structural frequency transformations*, which modify the frequency response of the filter and can be used in many applications requiring tunable filters. The other class of transformations, called *similarity transformations*, retains the overall transfer function but can change the complexity and the finite wordlength properties of the filter. The *equivalence transformations* retain the complexity and finite wordlength properties and are used to manipulate the internal timing relationships in order to obtain more efficient parallel implementations. The equivalence transformations can change the overall arithmetic delay of the filter.

6-4-1 Structural Frequency Transformations

6-4-1-1 IIR Filters. For IIR filters the frequency transformations of Chapter 5 provide the general tools for changing the corner frequencies and filter type without redesigning the transfer function. In these transformations a delay element is in principle replaced by an allpass section that warps the frequency axis in the desired manner. This replacement, if applied directly, produces in general delay-free loops.

However, in the case of a lowpass to bandpass transformation, which does not change the bandwidth, this replacement can be implemented directly:

$$z^{-1} \rightarrow -z^{-1} \frac{z^{-1} - \alpha}{1 - \alpha z^{-1}} \tag{6.3}$$

with $\alpha = -\cos \omega_0$, where ω_0 is the desired center frequency. This replacement is shown in Figure 6-8 and it provides single parameter tuning of the center frequency at the expense of the N additional multipliers for an Nth order prototype lowpass filter.

For tuning of the bandwidth of the parallel allpass structure one can use an approximate method that is based on the Taylor series expansion of the lowpass to lowpass frequency transformation [MI90]. The parallel allpass structure [RE88a;

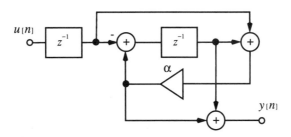

FIGURE 6-8 Structural lowpass to bandpass transformation.

VA86] is characterized by the equation

$$H(z) = \tfrac{1}{2}[A_0(z) + A_1(z)]. \tag{6.4}$$

The allpass sections $A_0(z)$ and $A_1(z)$ are implemented in cascade forms using first-
and second-order allpass sections. The sections are implemented using filter structures where multiplier values are the same as the coefficients in the corresponding
transfer function. There are several such structures with the additional benefit of
requiring only one and two multipliers in the first- and second-order sections, respectively [MI74; SZ88].

The lowpass to lowpass transformation is of the form

$$z^{-1} \rightarrow \beta(z^{-1}) = \frac{z^{-1} - \alpha_1}{1 - \alpha_1 z^{-1}} \tag{6.5a}$$

with

$$\alpha_1 = \frac{\sin\left[(\omega_1 - \omega_2)/2\right]}{\sin\left[(\omega_1 + \omega_2)/2\right]}, \tag{6.5b}$$

where ω_1 and ω_2 are the old and new cutoff frequencies, respectively. For a small
change of the cutoff frequency Eq. (6.5b) can be approximated at $\omega_2 \approx \omega_1 -
2\alpha_1 \sin \omega_1$, indicating linear tuning of the cutoff frequency as α_1 is varied. In the
case of a narrowband filter a further approximation can be made resulting in $\omega_2 \approx
\omega_1(1 - 2\alpha_1)$.

By performing the substitution of Eq. (6.5a) for a first-order allpass section

$$A_1(z) = \frac{z^{-1} - d_1}{1 - d_1 z^{-1}}, \tag{6.6}$$

we find

$$A_1(\beta(z^{-1})) = \frac{z^{-1} - (d_1 + \alpha_1)/(1 + \alpha_1 d_1)}{1 - (d_1 + \alpha_1)/(1 + \alpha_1 d_1)z^{-1}}. \tag{6.7}$$

By expanding the coefficient $(d_1 + \alpha_1)/(1 + \alpha_1 d_1)$ in a Taylor series in α_1 and
truncating after the linear term, we obtain the approximation

$$A_1(\beta(z^{-1})) \approx \frac{z^{-1} - [d_1 + (1 - d_1^2)\alpha_1]}{1 - [d_1 + (1 - d_1^2)\alpha_1]z^{-1}}. \tag{6.8}$$

This can be implemented by adding to the multiplier branch of nominal value d_1
the parallel tuning branch of value $\beta_0 \alpha_1$, where $\beta_0 = 1 - d_1^2$. Likewise, if $A_2(z)$

is a second-order allpass function

$$A_2(z) = \frac{d_2 - d_1 z^{-1} + z^{-2}}{1 - d_1 z^{-1} + d_2 z^{-2}},\qquad(6.9)$$

a similar procedure results in tuning branches of the type $\beta_i \alpha_1$, $i = 1, 2$, in parallel with the nominal multiplier values d_1 and d_2 with

$$\beta_1 = 2 + 2d_2 - d_1^2,\qquad(6.10a)$$

$$\beta_2 = d_1(1 - d_2).\qquad(6.10b)$$

Although the tuning method is approximate, tuning ranges of several octaves have been obtained in the case of narrowband elliptic lowpass filters [MI90]. The number of additional arithmetic operations needed depends on the desired tuning strategy. One can implement the parallel tuning branches in the filtering hardware resulting in an increased number of multipliers and adders. Alternatively, one can recalculate the values for the coefficients in Eqs. (6.6) and (6.9) after each tuning event. If tuning takes place relatively seldom, the latter method results in practically no additional complexity in the actual filtering task.

In Figure 6-9 we show some frequency responses of a tunable fifth-order elliptic lowpass filter obtained by performing the approximate bandwidth tuning for the lowpass elliptic prototype filter with a nominal bandwidth of 0.1π and a 1-dB passband ripple. The transition width is 0.03π and the resulting stopband attenuation is 45 dB.

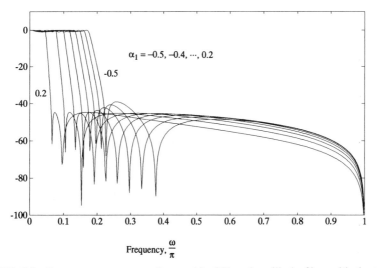

FIGURE 6-9 Frequency responses of a tunable fifth-order elliptic filter with the tuning parameter α_1 varying between 0.2 and -0.5.

By adding the structural lowpass to bandpass transformation of Eq. (6.3) to this lowpass structure, we obtain a tenth-order elliptic bandpass filter with single parameter tuning for both the center frequency and the bandwidth.

6-4-1-2 FIR Filters.

For tuning of FIR filters there is no simple method that would retain the optimality of the filter. However, the transformations of Section 5-6 retain the sizes of the passband and stopband ripples. The transformations alter the transition width. The lowpass to lowpass transformation increases the transition width as it essentially stretches part of the prototype filter frequency response to cover the whole frequency axis [CR76; OP76].

Let us consider a causal linear-phase filter $h[n]$ of length $2N + 1$, $n = 0, 1, \ldots, 2N + 1$. The corresponding zero-phase (noncausal) impulse response is $h_0[n] = h[n - N]$. The two transfer functions are related as $H(z) = z^{-N} H_0(z)$. The zero-phase transfer function can now be expressed as

$$H_0(z) = h_0(n) + \sum_{n=1}^{N} h_0[n][z^n + z^{-n}].\tag{6.11}$$

$H_0(z)$ can be rewritten as

$$H_0(z) = \sum_{n=0}^{N} a[n] \left[\frac{z + z^{-1}}{2} \right]^n \tag{6.12}$$

by using

$$z^n + z^{-n} = 2T_n \left[\frac{z + z^{-1}}{2} \right],$$

where $T_n(x)$ is a Chebyshev polynomial of nth order. The coefficients $a[n]$ are related to $h_0[n]$ through the coefficients of the Chebyshev polynomials. The zero-phase frequency response corresponding to Eq. (6.12) is

$$H_0(e^{j\omega}) = \sum_{n=0}^{N} a[n][\cos \omega]^n.\tag{6.13}$$

A direct implementation of Eq. (6.12) is shown in Figure 6-10(a). This structure is called the Taylor structure for linear-phase filters.

A structural lowpass to lowpass transformation is now obtained by substituting in Eq. (6.13)

$$\cos \omega = A_0 + A_1 \cos \hat{\omega}.\tag{6.14}$$

It is convenient to require that the prototype frequency response value $H(e^{j\omega})$ at $\omega = 0$ maps to the transformed frequency response value $\hat{H}(e^{j\hat{\omega}})$ at $\hat{\omega} = 0$. This requires that $A_1 = 1 - A_0$ with $0 \le A_0 < 1$. If the prototype filter is a lowpass

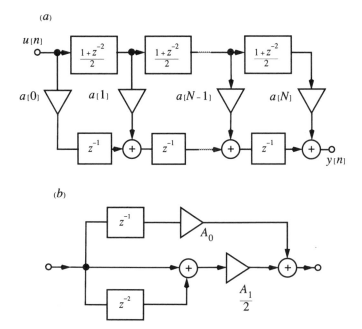

FIGURE 6-10 Tunable linear-phase FIR filter: (a) the Taylor structure and (b) subnetwork that implements the frequency transformation.

filter with cutoff frequency ω_c, then the transformed filter will have a cutoff frequency $\hat{\omega}_c$, where

$$\hat{\omega}_c = \cos^{-1}\left(\frac{\cos \omega_c - A_0}{1 - A_0}\right). \tag{6.15}$$

This transformation reduces the cutoff frequency. When the cutoff frequency of the transformed filter is required to be greater than or equal to that of the prototype filter, we select $\omega = \pi$ to be the frequency of invariance in the transformation resulting in $A_1 = 1 + A_0$ with $-1 < A_0 \leq 0$. In this case the transformation is given by

$$\cos \omega = A_0 + (1 + A_0) \cos \hat{\omega}. \tag{6.16}$$

A subnetwork that replaces the $(1 + z^{-2})/2$ blocks in Figure 6-10(a) and implements the transformation of Eq. (6.14) is shown in Figure 6-10(b). Note that the Taylor structure is not in general canonic in delays. However, there is a fourth-order Taylor structure that implements the transformation of Eq. (6.14) in delay canonic manner [OP76]. By cascading these fourth-order sections higher-order tunable filters can be obtained.

Variable FIR bandpass structures based on the Taylor structure have been studied in Ahuja and Dutta Roy [AH79] and Hazra [HA84]. Alternatively, one can build the bandpass filter as a parallel or cascade connection of tunable lowpass or highpass filters.

For direct form FIR filters there are also approximate frequency transformations that only change the coefficients of the filter [JA88]. These transformations are discussed here as limiting cases of structural frequency transformations because they can be used in filter implementations in a very similar manner as the other transformations discussed in this Section.

As these transformations retain the number of adders, delays, and multipliers of the direct form FIR structure, the methods are quite efficient even though the prototype filters may need to be overdesigned to compensate for the nonideal behavior of the tuning process.

A symmetric bandpass filter is obtained from a lowpass prototype filter with frequency response $H(e^{j\omega})$ using the complex modulation scheme

$$H_{BP}(e^{j\omega}) = H(e^{j(\omega - \omega_0)}) + H(e^{j(\omega + \omega_0)}). \tag{6.17}$$

The impulse response of the bandpass filter can be expressed as

$$h_{BP}[n] = (e^{j\omega_0 n} + e^{-j\omega_0 n})h[n] = 2\cos[\omega_0 n]h[n], \tag{6.18}$$

where ω_0 is the desired center frequency of the bandpass filter. The resulting bandpass filter has the worst case maximum error $\delta_p + \delta_s$ in the passband and $2\delta_s$ in the stopband, where δ_p and δ_s are the peak errors of the lowpass filter in the passband and stopband, respectively.

For linear-phase filters, coefficient symmetry is maintained when the zero-phase impulse response $h_0[n]$ is modulated as indicated in Eq. (6.18). Tunable bandstop filters are obtained from a highpass prototype filter, which in turn can be designed directly, or one can take the zero-phase complementary filter of a lowpass prototype filter.

For changing the bandwidth of a lowpass linear-phase filter, a simple approximate method can be derived by expressing the coefficients as simple trigonometric functions of the cutoff frequency ω_c. The ideal zero-phase frequency response of a lowpass filter is

$$H_{0ID}(e^{j\omega}) = \begin{cases} 1 & \text{for } 0 \le \omega \le \omega_c \\ 0 & \text{for } \omega_c < \omega \le \pi. \end{cases} \tag{6.19}$$

The ideal impulse response is obtained as the inverse discrete Fourier transform of Eq. (6.19)

$$h_{0ID}[n] = \frac{\sin[\omega_c n]}{\pi n} = \frac{\omega_c}{\pi}\text{sinc}[\omega_c n/\pi]. \tag{6.20}$$

Here the center coefficient, $n = 0$, is a linear function of ω_c and the other coefficients are of the form $h_{0ID}[n] = c[n]\sin[\omega_c n]$, $|n| \ge 1$, where $c[n]$ is a constant for each n and $c[-n] = -c[n]$. As an extension of Eq. (6.20) one can now define a filter class where the coefficients of the filter are of the same form as in Eq. (6.20); that is,

$$h_0[n] = \begin{cases} c[n]\omega_c & \text{for } n = 0 \\ c[n] \sin[\omega_c n] & \text{for } 1 \le |n| \le N \\ 0 & \text{otherwise,} \end{cases} \qquad (6.21)$$

where ω_c is the 6-dB cutoff frequency. Any given fixed lowpass filter of odd length can be expressed in the form of Eq. (6.21). However, filters with approximately equal passband and stopband ripples can best be approximated with Eq. (6.21). The prototype filter can be designed by the window method or using an optimal filter design method, like the Parks–McClellan method, with equal stopband and passband weights. Equating the filter coefficients with Eq. (6.21) the constants $c[n]$ can be solved. In order to avoid numerical problems, the prototype filter should be such that each coefficient is sufficiently different from zero.

As an example we show in Figure 6-11 variable cutoff frequency lowpass filters of length 51 designed with two different transition bandwidths and equal weights in both bands of the prototype filter.

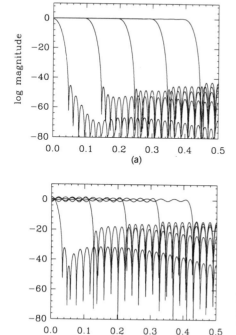

FIGURE 6-11 Frequency responses of tunable linear-phase FIR filters of length 51. The prototype filter has equal weights in the passband and stopband and transition bandwidth of (a) 0.10π and (b) 0.02π.

Variable bandwidth bandpass filters can be obtained by combining Eq. (6.18) and Eq. (6.21). The tuning can be based on the bandwidth and center frequency or on the two cutoff frequencies ω_{c1} and ω_{c2} via the equations $\omega_0 = (\omega_{c1} + \omega_{c2})/2$ and $\omega_c = |\omega_{c2} - \omega_{c1}|/2$.

For unequal ripples in the passband and stopband a more complicated approximation is required that is based on two lowpass prototype designs. For odd-length filters a good approximation for the impulse response is

$$h_0[n] = \begin{cases} c[n]\omega_c + d[n] & \text{for } n = 0 \\ c[n]\sin[\omega_c n] + d[n]\cos[\omega_c n] & \text{for } 1 \le |n| \le N. \end{cases} \quad (6.22)$$

The constants $c[n]$ and $d[n]$ can be solved by designing two optimal prototype filters with nonzero coefficients and equating them with Eq. (6.22). The method does not give good results for very narrowband or wideband filters. In these cases multirate filtering or the IFIR approach can be used to bring the tunable filter to a more accurate domain; see Chapters 4 and 14. Variable cutoff frequency lowpass designs of length 51 and transition bandwidth of 0.1π are shown in Figure 6-12 with different passband/stopband weights. The even-length frequency transformations are obtained through similar derivations [JA88].

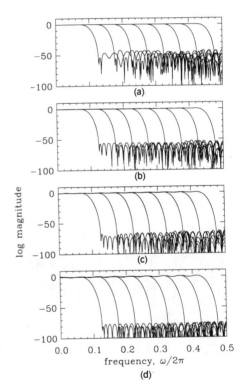

FIGURE 6-12 Variable cutoff lowpass filters of length 51, transition bandwidth of 0.1π, and passband/stopband weights of (a) 1/1, (b) 1/10, (c) 1/100, and (d) 1/1000.

6-4-2 Similarity Transformations

There are several alternative ways to modify a filter structure in order to obtain a new structure implementing the same transfer function with modified structural properties and possibly with better suitability for a certain type of implementation.

6-4-2-1 Flow-Graph Methods. Consider a graphical representation of a digital filter network. Draw over it a simple closed curve that does not pass through any node and does not contain the input or output node of the filter. This is exemplified in Figure 6-13. Then multiply by an arbitrary constant c the transmittance of each branch crossing the curve and coming out, and divide the transmittance of each branch crossing the curve and going in by the same constant c. Due to linearity, only the signal levels in the subnetwork N'' were changed, multiplied by $1/c$. Transfer functions between any two nodes outside the curve remained the same.

Flow-graph *transposition* is another useful method to obtain another structure implementing the same transfer function as described in Chapter 3.

6-4-2-2 Matrix Methods. The state-space representation of an Nth order filter is described by the equations

$$\mathbf{x}[n + 1] = \mathbf{A}\mathbf{x}[n] + \mathbf{B}u[n], \tag{6.23a}$$

$$y[n] = \mathbf{C}\mathbf{x}[n] + Du[n]. \tag{6.23b}$$

The state vector $\mathbf{x}[n]$ is a column vector of dimensionality N. The state matrix $\mathbf{A}$ is of dimensionality $N \times N$. For single-input–single-output filter $\mathbf{B}$ is a $N \times 1$ column vector, $\mathbf{C}$ is a $1 \times N$ row vector, and D is a scalar quantity.

This structure can be modified by applying a linear transformation on the vector $\mathbf{x}[n]$, $\mathbf{x}'[n] = \mathbf{P}\mathbf{x}[n]$, where $\mathbf{P}$ is any nonsingular matrix of dimensionality $N \times N$. By substituting $\mathbf{x}[n] = \mathbf{P}^{-1}\mathbf{x}'[n]$ into Eqs. (6.23a) and (6.23b) we obtain

$$\mathbf{x}'[n + 1] = \mathbf{P}\mathbf{A}\mathbf{P}^{-1}\mathbf{x}'[n] + \mathbf{P}\mathbf{B}u[n], \tag{6.24a}$$

$$y[n] = \mathbf{C}\mathbf{P}^{-1}\mathbf{x}'[n] + Du[n]. \tag{6.24b}$$

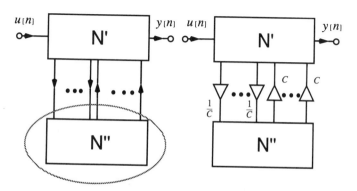

FIGURE 6-13 Digital filter structural transformation based on flow-graph manipulation.

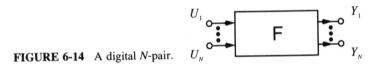

FIGURE 6-14 A digital N-pair.

For modifying other than state-space structures one can apply a similar linear transformation to the matrix representation of the filter (see Chapters 3 and 7).

6-4-2-3 Use of Constrained N-Pairs.
Digital filters can be described as constrained N-pairs. An N-pair has N inputs and outputs connected via a linear network F as shown in Figure 6-14. The transfer matrix $\mathbf{T}$ of the N-pair has as elements the internal transfer functions t_{ij} from input signal U_j to output signal Y_i. In matrix form this can be expressed as

$$\mathbf{Y}(z) = \mathbf{T}\mathbf{U}(z). \tag{6.25}$$

For single-input–single-output filters the use of N-pairs with $N > 1$ makes it possible to take the multipliers or the delays or both of them outside the network and use them as constraints for the N-pair. By modifying the internal structure of F in a suitable way one can modify the structure without affecting certain characteristics of the structure such as the number of multipliers or delays [SZ75, SZ88].

As a simple example let us consider a realization F of a transfer function $T_1(z) = Y_1(z)/U_1(z)$ including a multiplier α_1. This realization can be partitioned into two parts with multiplier α_1 taken out as shown in Figure 6-15. The two-pair can be characterized by the transfer matrix $\mathbf{T}_2$:

$$\begin{bmatrix} Y_1 \\ Y_2 \end{bmatrix} = \begin{bmatrix} t_{11} & t_{12} \\ t_{21} & t_{22} \end{bmatrix} \begin{bmatrix} U_1 \\ U_2 \end{bmatrix} = \mathbf{T}_2 \begin{bmatrix} U_1 \\ U_2 \end{bmatrix}. \tag{6.26}$$

Using Eq. (6.26) and the constraining equation $U_2(z) = \alpha_1 Y_2(z)$, we obtain the transfer function $H_1(z) = Y_1(z)/U_1(z)$ as a function of the multiplier α_1:

$$H_1(z) = t_{11} + t_{12}t_{21} \frac{\alpha_1}{1 - t_{22}\alpha_1}. \tag{6.27}$$

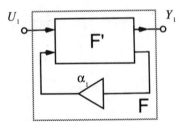

FIGURE 6-15 A digital two-pair constrained by one multiplier.

One can now impose restrictions on the transfer parameters t_{ij}. In order to avoid delay-free loops there cannot be a direct path in t_{22}. It is also feasible to require the implementation be such that there are no multipliers in mere cascade connection because this would lead to nonminimal realizations. With this kind of reasoning the search for different feasible structures can be simplified.

6-4-3 Equivalence Transformations

Equivalence transformations do not change the transfer function or the internal computations of the filter. Since all computations in the filter remain numerically the same, the finite wordlength effects are not affected by equivalence transformations. The only change that can be seen in the input/output relationship is the possible change of the transfer function by a constant delay factor [FE76; RE81]. Two digital filter networks are said to be *essentially equivalent* if they can be transformed from one into another by a sequence of equivalence transformations.

The equivalence transformation is a straightforward extension of the graphical similarity transformation described in Section 6-4-2 and illustrated by Figure 6-13. The delays are modified without affecting the computations if the transmittances of each branch crossing the closed curve and going in are multiplied by a complex constant $c = z^{-\tau_0/T}$ and correspondingly the outgoing transmittances are multiplied by $c = z^{\tau_0/T}$. Here T is the sampling period. This transformation adds the delay of τ_0 to the incoming branches and of $-\tau_0$ to the outgoing branches.

A sequence of equivalence transformations can be used to introduce fractional delays in the filter and to move them to places where they can be used advantageously such as to take care of the computation times of arithmetic operations. For realizability, the transformations have to be applied in such a way that the resulting network does not contain negative delays (i.e., advance operation). Figure 6-16 shows how the delays in a second-order section can be partly moved to the multiplier branches to accommodate for the time it takes to perform the multiplications.

6-5 BLOCK IMPLEMENTATION

In block processing the input data $u[n]$ is first converted to vector form using a serial input, parallel output (SIPO) shift register of length L. These L-dimensional vectors $\mathbf{u}[k]$ are then processed in the block processing unit, giving the output vectors $\mathbf{y}[k]$ of dimension L, which are again converted to serial form by the parallel input, serial output (PISO) register as shown in Figure 6-17. The output data sequence $y[n]$ is obtained after the latency time arising from the time it takes to collect the input vector $\mathbf{u}[k]$, to perform the block processing operations, and to load the output register.

The main objective in block processing is to obtain a signal processing scheme with high inherent parallelism. This enables us to map the algorithm on several processing units and results in a high throughput.

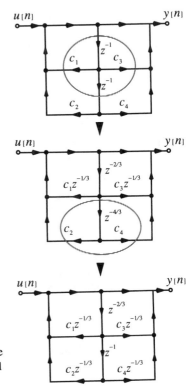

FIGURE 6-16 Example of the use of equivalence transformations to fraction the delays © IEEE, 1981 [RE81].

Block processing was first suggested by Gold and Jordan [GO68]. Burrus has derived block structures of digital filters making use of the matrix representation of the convolution [BU71, BU72]. Several block state-space structures have been presented and analyzed in Mitra and Gnanasekaran [MI78], Ananthakrishna and Mitra [AN80], Barnes and Shinnaka [BA80], and Zeman and Lindgren [ZE81]. Mitra and Gnanasekaran [MI78] noticed that in block processing the poles of the filter are closer to the origin than in conventional filter implementations, resulting in improved finite wordlength properties.

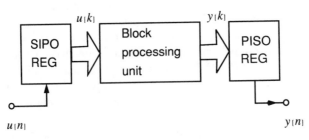

FIGURE 6-17 Principle of block processing.

The *state decimating block structure* [ZE81], also known as *block state struc-ture* [AN80; BA80], is obtained from the state-space representation of Eq. (6.23) by calculating the state vector at only every p time instants. This can be done as the state and the output at any future time instant can be calculated from the current state and the input signal.

Starting with the state-space filter structure

$$\mathbf{x}[n+1] = \mathbf{A}\mathbf{x}[n] + \mathbf{B}u[n],$$
$$y[n] = \mathbf{C}\mathbf{x}[n] + Du[n],$$

we obtain for time instant $n+1$

$$\mathbf{x}[n+2] = \mathbf{A}^2\mathbf{x}[n] + \mathbf{A}\mathbf{B}u[n] + \mathbf{B}u[n+1],$$
$$y[n+1] = \mathbf{C}\mathbf{A}\mathbf{x}[n] + \mathbf{C}\mathbf{B}u[n] + Du[n+1].$$

In a similar manner we obtain

$$\mathbf{x}[n+3] = \mathbf{A}^3\mathbf{x}[n] + \mathbf{A}^2\mathbf{B}u[n] + \mathbf{A}\mathbf{B}u[n+1] + \mathbf{B}u[n+2],$$
$$y[n+2] = \mathbf{C}\mathbf{A}^2\mathbf{x}[n] + \mathbf{C}\mathbf{A}\mathbf{B}u[n] + \mathbf{C}\mathbf{B}u[n+1] + Du[n+2].$$

This can be expressed in matrix form for a block size of p as

$$\mathbf{x}[k+L] = \mathbf{A}'\mathbf{x}[k] + \mathbf{B}'\mathbf{u}[k], \tag{6.28}$$
$$\mathbf{y}[k] = \mathbf{C}'\mathbf{x}[k] + \mathbf{D}'\mathbf{u}[k]. \tag{6.29}$$

Equations (6.28) and (6.29) are of the state-space representation form with

$$\mathbf{y}[k] = [y[kL], y[kL+1], \dots, y[kL+L-1]]', \tag{6.30}$$
$$\mathbf{u}[k] = [u[kL], u[kL+1], \dots, u[kL+L-1]]', \tag{6.31}$$

and

$$\mathbf{A}' = \mathbf{A}^L, \tag{6.32}$$
$$\mathbf{B}' = [\mathbf{A}^{L-1}\mathbf{B} \quad \mathbf{A}^{L-2}\mathbf{B} \quad \cdots \quad \mathbf{A}\mathbf{B} \quad \mathbf{B}], \tag{6.33}$$

$$\mathbf{C}' = \begin{bmatrix} \mathbf{C} \\ \mathbf{C}\mathbf{A} \\ \mathbf{C}\mathbf{A}^2 \\ \vdots \\ \mathbf{C}\mathbf{A}^{L-1} \end{bmatrix}, \tag{6.34}$$

and

$$\mathbf{D'} = \begin{bmatrix} D & 0 & \cdots & 0 & 0 \\ \mathbf{CB} & D & \cdots & 0 & 0 \\ \mathbf{CAB} & \mathbf{CB} & \cdots & 0 & 0 \\ \vdots & \vdots & & & \\ \mathbf{CA}^{L-2}\mathbf{B} & \mathbf{CA}^{L-3}\mathbf{B} & \cdots & \mathbf{CB} & D \end{bmatrix}. \tag{6.35}$$

Note that all the matrices in Eqs. (6.28) and (6.29) can be calculated in advance as indicated by Eqs. (6.32), (6.33), (6.34), and (6.35).

In block implementation of an Nth order state-space digital filter, the number of multiplications (and additions) per calculated output value is

$$m = [N^2 + NL + NL + L(L + 1)/2]/L. \tag{6.36}$$

The multiplication rate is minimized when $L = \sqrt{2}N$ (or nearest integer value), resulting in about $(2 + \sqrt{2})N + \frac{1}{2}$ multiplications per sample. This is only about 1.4 times more than the number of multiplications required in a cascaded second-order section implementation (five multiplications for one second-order section).

Equations (6.28) and (6.29) are highly parallel. In fact, all the multiplications of one computational period (producing L output samples) can be done concurrently. However, in that case the number of multiplier units becomes large as indicated by Eq. (6.36). Pipelining can be used to speed up the additions.

The number of distinct multipliers in state decimation filters can be quite large as indicated roughly by the numerator of Eq. (6.36). This can be reduced by implementing a high-order filter in cascade or parallel connection of low-order state decimation filters.

The eigenvalues (poles of the filter) of $\mathbf{A'} = \mathbf{A}^L$ remain inside the unit circle but move closer to the origin. This has been shown to reduce the finite wordlength effects: coefficient sensitivity and roundoff noise level. Occasionally, limit cycles are eliminated or at least their level is reduced.

The need for higher and higher speeds in digital signal processing together with the advent of low-cost processing units is likely to make block processing a realistic implementation alternative for a number of different applications.

6-6 MAXIMUM SAMPLING RATE AND MULTIPROCESSOR IMPLEMENTATIONS

In Section 6-3 we discussed the complexity of a filter structure in terms of the number of additions, multiplications, and delays in the structure. In high-speed applications another issue arises: how effectively the filter structure can be mapped on several arithmetic units or processors operating in parallel.

For FIR filters the situation is rather straightforward. Let us assume for example that the hardware is capable of performing an addition in time τ_a and multiplication in time τ_m with $\tau_m \geq \tau_a$. The direct form FIR filter can be implemented up to sampling rates of $1/\tau_m$ by reserving one separate arithmetic unit for each multiplier and by performing the additions in a pipeline fashion simultaneously with the multiplications. In this pipelined implementation the latency depends on the relative speeds of additions and multiplications but is necessarily more than one sampling period as the multipliers and adders operate on data belonging to different sampling instants.

For even higher speeds the multipliers can be implemented in a pipeline fashion too. This makes it possible to perform the multiplications in a few gate delays and results in implementations operating at sampling periods of the order of a gate delay. See Chapter 11 for a more detailed description on signal processing hardware.

6-6-1 Maximum Sampling Rate of IIR Structures

For IIR structures the situation is somewhat more complicated due to the feedback loops. If we assume that only additions and multiplications are time-consuming operations in the implementation, we can find the maximal sampling rate quite easily by inspecting all directed loops in the filter's signal flow diagram. We assume that these operations are executed in constant times τ_a and τ_m, respectively, irrespective of the operands. Later we discuss how one can take into account delays whose position in the implementation depends on other factors such as on the job partitioning between the processing units.

Let us denote by τ_{ai} the execution time of the ith arithmetic operation (addition or multiplication). We call these execution times *arithmetic delays*. For adders this delay is considered to be at the output of the adder. Clearly, the sampling period T must be so large that the following equation holds in every directed loop l:

$$D_l \leq n_l T, \tag{6.37}$$

where

$$D_l = \sum_{i \in l} \tau_{ai} \tag{6.38}$$

is the total arithmetic delay in loop l and n_l is the number of delays in loop l. Motivated by Eq. (6.37) we calculate

$$T_0 = \max_l \{D_l/n_l\}, \tag{6.39}$$

where the maximum is over all directed loops. The loop in which the maximum is reached is called a *critical loop*. It can be shown that the filter can be implemented with a minimum sampling period T_0. That is, the unit delays can be replaced with fractional delays to account for arithmetic delays through a sequence of equivalence transformations in such a way that T_0 is reached.

In addition to the arithmetic delays, the implementation will contain *shimming delays* that take care of the time slack between the execution of a certain arithmetic operation and the time the result is needed for further computations [FE76].

The length of the critical loop depends on the structure. The direct form is in this respect excellent as the critical loop is the one that contains only the innermost feedback loop with one delay, one multiplier, and one adder (assuming that the multiplication takes a much longer time than the addition).

A lower bound n_p for the number of processors needed, with addition and multiplication times of τ_a and τ_m, respectively, can be obtained from the sampling period T and the sum D of all arithmetic delays in the network:

$$n_p \geq \lceil D/T \rceil, \tag{6.40}$$

where $\lceil x \rceil$ is equal to the smallest integer greater than or equal to x. For many filter structures there seem to be a wide variety of sampling rates at which they can be realized using the minimal number of processors given by Eq. (6.40) [RE81].

6-6-2 Maximum Sampling Rate Implementation

Given the arithmetic delays τ_{ai} of the implementation, our objective is to find the shimming delays required in the structure for the maximal sampling rate implementation. After that we show how the job partitioning for multiple processing units can be done.

As an example we use a fourth-order Jaumann wave digital filter [NO74] network shown in Figure 6-18. This filter is essentially a parallel connection of first- and third-order allpass sections and implements bandpass or bandstop characteristics. Assuming that the multiplications take 5 u.t. (units of time) and additions 1 u.t., the critical loop is easily found as shown in Figure 6-18(a) with minimal sampling period, according to Eq. (6.39), of $T_0 = 16$ u.t.

The shimming delays are found using simple graph-theoretical concepts. We assume that the digital filter network G is computable and *proper*, meaning that there is a direct path from the input to every node and from every node to the output. A connected subgraph of G that does not contain any loops is called a *spanning tree* of G. The branches that are not in the spanning tree are called *link branches*. When a link branch is inserted a loop is formed. The set of loops determined by the link branches is called a *complete set of fundamental loops* [FE74].

It is relatively easy to show that two proper digital filter networks with the same topology and the same coefficients are essentially equivalent, if and only if the total delays are equal in each loop of some complete set of fundamental loops. Here each delay enters the sum as τ_i or $-\tau_i$, depending on whether or not the direction of branch i coincides with the orientation of the link branch [RE81].

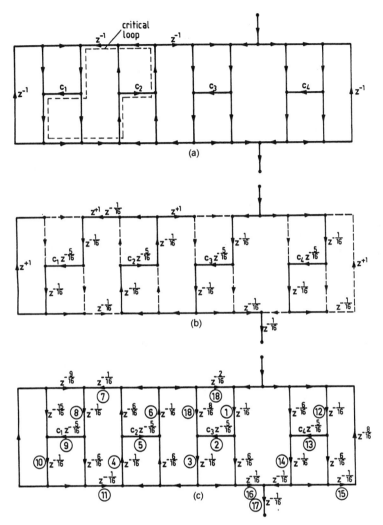

FIGURE 6-18 (a) Fourth-order example structure showing the critical loop. (b) The network N'' and the maximal distance spanning tree. (c) The essentially equivalent form of N containing the arithmetic delays © IEEE, 1981 [RE81].

Example 6.2. In Figure 6-19 we have a second-order section with one spanning tree shown. The link branches indicated by $l_1, \ldots, l_4$ define a complete set of fundamental loops. Three second-order sections with the same topology and coefficients are shown in Figure 6-16. By adding the delays in each fundamental loop in Figure 6-19 and in any of the structures shown in Figure 6-16, it is easy to see that all these structures are essentially equivalent. In fact, the transfer function $H_1(z)$ corresponding to Figure 6-19 is related to the transfer function $H_2(z)$ corresponding to the structures shown in Figure 6-16 as $H_1(z) = z^{-1/3}H_2(z)$.

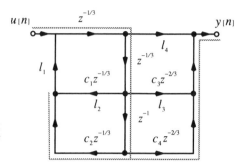

FIGURE 6-19 Essentially equivalent second-order section with structures of Figure 6-16.

Let us now consider a network G' obtained from G by removing the unit delays and inserting the arithmetic delays τ_{ai}. Using equivalence transformations we can now move all the arithmetic delays of G' and all the unit delays of G from the corresponding spanning trees of G' and G to the link branches and make the two networks essentially equivalent by adding to the link branches of G' shimming delays that make the total delays in the fundamental loops of G' and G equal. The length of the unit delays has to be at least T_0 as given by Eq. (6.39).

The above procedure does not, however, guarantee that the shimming delays corresponding to nondirected fundamental loops will all be nonnegative. The values of the shimming delays will depend on the spanning tree selected.

In order to find an implementation with positive shimming delays, we take a slightly different approach. Let G'' be the network containing the arithmetic delays and also the unit delays of G with value $-T_0$. Now we make the total delays of the fundamental loops of G'' equal to zero by adding suitable shimming delays. When the delays $-T_0$ are removed from the resulting network, we obtain a network essentially equivalent to G. For the spanning tree we select the *maximal distance spanning tree* of G'', which has the property that the sum of delays along the directed paths from the input to every other node in the network is maximal. Note that the spanning tree used in Example 6.2 and shown in Figure 6-19 is the maximal distance spanning tree. Due to this selection of the spanning tree, the situation in all nondirected fundamental loops is as shown in Figure 6-20, where the sum of delays along path P_1 is longer than along path P_2. This means that the shimming delay to be inserted to the link branch will be positive. The network G'' and the maximal distance spanning tree for our example structure are shown in Figure 6-18(b) and the final filter structure containing arithmetic and shimming delays in Figure 6-18(c).

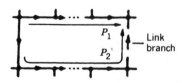

FIGURE 6-20 A nondirected fundamental loop in the maximal distance spanning tree © IEEE, 1981 [RE81].

6-6-3   Job Partitioning for Multiprocessor Implementations

Let us first assume that the number of processing elements is not limited. In this case it is easy to find a partitioning that gives the maximal sampling rate by using the method of the previous section. After this has been found the number of processors can easily be decreased at the expense of slower operation.

The timing relationships can be studied conveniently with the aid of a *timing diagram*. This can be constructed easily from the essentially equivalent filter network containing the arithmetic delays; see Figure 6-18. The time instant at which each operation begins can be found by calculating the sum of delays from the input node to the node corresponding to this operation. This time is counted modulo T_0, indicating that different parts of the network operate on data belonging to different sampling instants. Thus if the sum of delays is $mT_0 + k$, where m is an integer, the operation begins at time instant k in the timing diagram. The operations are assigned to different processors in such a way that interprocessor communication is minimized. The number of processors required will be equal to the number of concurrent operations in the timing diagram.

Continuing the example of the previous section, a timing diagram corresponding to Figure 6-18 is developed as shown in Figure 6-21(a). Note that three processors are required for the maximal sampling rate implementation. The bold lines represent the time consumed by the operations numbered in Figure 6-18(c) and the

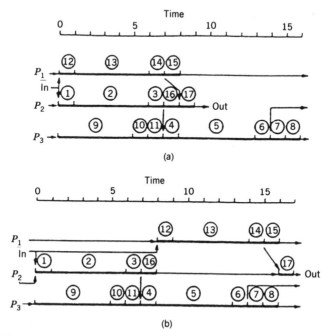

FIGURE 6-21   Timing diagram of the example filter: (a) the maximal rate realization requiring three processors and (b) two-processor realization © IEEE, 1981 [RE81].

arrows indicate which operations must be performed before the execution of a certain operation can begin. In the figure only critical precedence relations are shown. Many of the precedence relations are in practice implied by some other precedence relations. The arrows extending to the right of $t = 16$ continue at $t = 0$ on the same line, indicating precedence relations between consecutive computational periods.

The number of processors can sometimes be decreased by shifting some operations in time so that the tasks of two or more processors can be combined. By increasing the sampling period the number of processors can be further decreased. Naturally, the timing diagram has to be modified so that the precedence relations are not violated, meaning that the arrows point to the right. In Figure 6-21(b) the sampling period has been increased to $T = 17$ u.t. and operations 12, . . . , 15 and 17 have been delayed by 8 u.t. Since the tasks of processors P_1 and P_2 can now be combined, this timing diagram can be implemented using two processors.

We have assumed in the above discussion that arithmetic operations are the only time-consuming operations in the implementation and that they are always executed in constant time. In practice, the situation is more complicated. Arithmetic operations may require different times to execute. Sometimes the execution time depends on whether the operand is in the accumulator or in the memory. Data transfer between processors often requires execution of WRITE and READ type instructions. These instructions send and receive, respectively, a data word to and from the port or memory location providing the interface between the processors. For synchronization purposes, looping instructions may also be needed to facilitate communication between processors. These delays can be divided into *partition independent delays* and *partition dependent delays*. The additional partition independent delays, like a looping instruction at the end of the program of a processor, can be added, maintaining the precedence relations, to the timing diagram after the number of processors has been fixed. The partitioning dependent delays can be added after the exact job partitioning between the processors has been decided. A good starting point for these timing diagram manipulations is always the ideal timing diagram obtained using only the arithmetic delays.

6-7 QUANTIZATION AND OVERFLOW OPERATIONS

Consider Figure 6-22, which represents a first-order IIR filter with transfer function

$$H(z) = 1/(1 - az^{-1}). \tag{6.41}$$

Let $w[n]$ be represented by a b-bit fixed-point fraction as in Figure 6-2. Assuming for simplicity that the multiplier a is also a b-bit fraction, the signal $v[n + 1] = aw[n]$ requires $2b$ bits for faithful representation. Thus $w[n + 1]$ requires at least $2b$ bits. After one more recursion, we see that $w[n + 2]$ requires at least $3b$ bits and so on. In order to prevent such indefinite accumulation of bits (which is caused

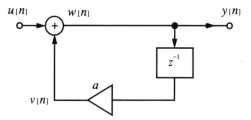

FIGURE 6-22 Implementation of a first-order transfer function.

by the multiplier a in the feedback loop), we have to introduce a *quantizer*, which is denoted by Q in Figure 6-23. The purpose of this quantizer is to convert the $2b$-bit number into a b-bit number. This introduces a quantization error into the feedback loop, usually modeled as in Figure 6-24. The error sequence $e[n]$ can in most cases be modeled as a random process with certain statistical properties. The exact nature of this model depends on the number representation and the quantization rules. In this section we first outline these properties.

6-7-1 Modeling the Quantization Error

Let x represent a b_1-bit fixed-point fraction and let $y = Q(x)$ be the b-bit quantized version as shown in Figure 6-25. The quantization error is defined as

$$e = y - x = Q(x) - x. \tag{6.42}$$

(a)

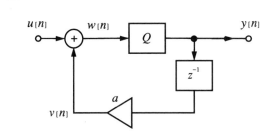

(b)

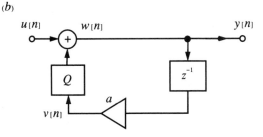

FIGURE 6-23 Two possible quantization schemes.

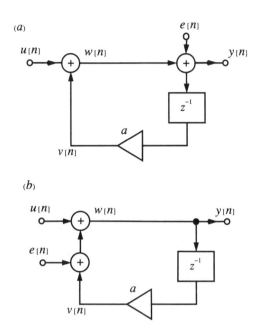

FIGURE 6-24   Modeling the quantizer noise.

The properties of the error e depend not only on the number representation (such as sign–magnitude or two's complement) but on the quantization rules as well (i.e., whether we perform truncation or rounding, etc.).

Truncation arithmetic (or truncation type of quantization), by definition, is a process whereby the rightmost $b_1 - b$ bits in x are simply discarded, regardless of what the sign bit s is. Let us denote $\Delta = 2^{-b} - 2^{-b_1}$. For a positive number we see that $-\Delta \le e \le 0$. For a negative number, the range of e depends on the representation [OP75]. Thus for the case of sign–magnitude and one's complement representations, we have $0 \le e \le \Delta$, whereas for two's complement representation, $-\Delta \le e \le 0$.

As a consequence, the sign of e is opposite to that of x in the case of sign–magnitude and one's complement systems. This means that the quantization error correlates with the signal. However, this correlation can normally be neglected in standard roundoff noise analysis. For the two's complement case, we have $e \le 0$ regardless of the sign of x, which means that the sign of e is uncorrelated to x. Truncation arithmetic therefore results in $|Q(x)| \le |x|$ for sign–magnitude and

FIGURE 6-25   The $(b + b_1)$-bit fixed-point number and the quantized version.

one's complement schemes, whereas for the two's complement case, the magnitude *increases* when a negative number is quantized. In other words, truncation arithmetic implies magnitude truncation in the case of sign–magnitude and one's complement representations, whereas it represents *value truncation* (i.e., $Q(x) \leq x$) for the two's complement case.

This observation is crucial when we attempt to suppress limit cycles, as we shall see in Section 7-10. Briefly, in order to suppress limit cycles, it is necessary to ensure that the quantizer does not increase the energy of the sequence to be quantized. For this reason magnitude-truncation quantizers are preferred. Two's complement truncation can be converted into this type by adding the quantity 2^{-b} to $Q(x)$ whenever $x < 0$ and $e \neq 0$. Examples of limit cycles are given in Section 6-7-3. The *roundoff* quantizer is a device that quantizes x to the nearest available b-bit fraction. As a result, the error is in the range $-\Delta/2 \leq e \leq \Delta/2$, regardless of the sign of x. Clearly the sign of e is uncorrelated to that of x; but the magnitude of $Q(x)$ can be greater than that of x.

In practice we usually have $1 \gg 2^{-b} \gg 2^{-b_1}$. Under this condition, regardless of the statistics of x, e can be regarded as a uniform random variable, with probability density functions $p(e)$. These are shown in Figure 6-26 along with the statistical mean and variance values. For the rest of the chapter assume

$$2^{-b} \gg 2^{-b_1}$$

so that

$$\Delta = 2^{-b}, \tag{6.43}$$

which is referred to as the *quantization step size*.

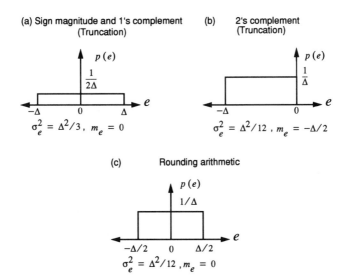

(a) Sign magnitude and 1's complement (Truncation)

$$\sigma_e^2 = \Delta^2/3, \; m_e = 0$$

(b) 2's complement (Truncation)

$$\sigma_e^2 = \Delta^2/12, \; m_e = -\Delta/2$$

(c) Rounding arithmetic

$$\sigma_e^2 = \Delta^2/12, \; m_e = 0$$

FIGURE 6-26 Probability functions for fixed-point quantizer error.

Note that the variance of e is lower for rounding arithmetic (regardless of representation of negative numbers), and the statistical mean of e is zero. For two's complement truncation the variance of e is again low, but e has a nonzero mean value.

6-7-2 Overflow Arithmetic Operations

In fixed-point arithmetic with numbers represented as in Figure 6-2, it is possible for the results of an arithmetic operation to exceed the magnitude of unity. Since this cannot be represented within the dynamic range permitted by Figure 6-2, an overflow has occurred. It is possible to reduce the probability of overflow, but if we wish to *completely avoid* overflow, we can do so only by permitting an enormous sacrifice of the signal-to-noise ratio (see Section 6-10). In practical structures it is therefore common to reduce the probability of overflow rather than eliminate it completely. Accordingly, it is necessary to set up some rules of action to be taken after an overflow has occurred. Specifically, if the result x of some arithmetic operation has exceeded the dynamic range, then x should be replaced with a number y that belongs to the permitted range. Figure 6-27 depicts two possible rules. In Figure 6-27(a), whenever $|x| > 1$ we replace it with 1 or -1, as appropriate. This is called *saturation arithmetic*. In Figure 6-27(b), if x does not belong to the range $[-1, 1)$, we take $y = x \bmod 2$. This is called the *two's complement overflow* feature for the following reason: in two's complement arithmetic, if a number exceeds the permitted dynamic range (which is $[-1, 1)$), and if we simply ignore the overflow bits (i.e., bits to the left of the sign bit in Figure 6-2), this is precisely the same as performing the mod 2 operation. Accordingly, the scheme of Figure 6-27(b) is natural and easy to implement in two's complement arithmetic. The recursive running sum discussed in Example 6.1 is based on the properties of two's complement overflow arithmetic. However, in general, the saturating characteristics of Figure 6-27(a) should be applied in practice.

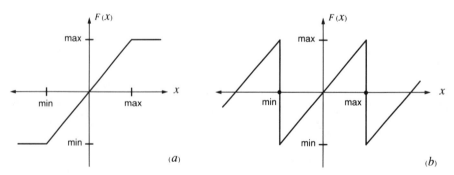

FIGURE 6-27 Two schemes to handle a number when it exceeds the dynamic range: (a) saturation arithmetic and (b) two's complement type of overflow.

6-7-3 Limit Cycles and Overflow Oscillations

If the input signal to a digital filter is either a constant, or varies very slowly, or exhibits certain types of deterministic oscillations, the statistical analysis methods for finite wordlength effects cannot be justified, as already indicated in Section 6-3-2. In this section we take a closer look at the oscillations caused by rounding errors and arithmetic overflows. Section 7-10 gives a more detailed treatment on how to avoid these oscillations.

Let us consider the second-order direct form filter with the quantizer $Q(\ )$ as shown in Figure 6-28. The corresponding difference equation is

$$y[n] = Q(x[n] - \beta_1 y[n - 1] - \beta_2 y[n - 2]). \qquad (6.44)$$

In order to make our examples more illustrative, we use a decimal fractional number system with the quantization step size $\Delta = 0.1$. The largest number in the presentation is taken as 0.9 and the smallest as -1.0. This selection is analogous with the two's complement binary representation.

Example 6.3. Let us first consider limit cycles arising from roundoff errors. We assume that the input signal $x[n] = 0.0$ for $n \geq 0$. We use the rounding arithmetic of Figure 6-29 and set in Eq. (6.44) $\beta_1 = 0.0$ and $\beta_2 = 0.9$. The poles in this case are on the imaginary axis at points $\pm \sqrt{0.9} = \pm 0.95$. Proper selection of the initial conditions now gives rise to limit cycle oscillations of the type $v, 0, -v, 0, v, 0, \ldots$, where v is a positive number. The amplitude v of the limit cycles depends on the initial conditions. If we set $y[-1] = 0$ and $y[-2] = 0.5$, $v = 0.5$ as $Q(0.9 \times 0.5) = 0.5$. It is easy to see that similar but smaller amplitude oscillations arise if $y[-2]$ is initially set to $0.1, 0.2, 0.3,$ or 0.4.

Example 6.4. If we set $\beta_2 = 0.0$ and $\beta_1 = 0.9$ corresponding to a first-order recursive filter, the same amplitude limit cycle oscillations as in the second-order case can occur but now the cycle of the oscillations is two. By setting $\beta_1 = -0.9$ we obtain limit cycles with period one, meaning that a zero input signal produces a nonzero constant output signal.

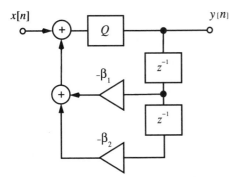

FIGURE 6-28 Second-order filter with quantizer.

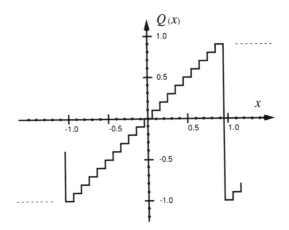

FIGURE 6-29 Rounding arithmetic used in the examples. The solid line and the dashed line correspond to standard overflow and saturation characteristics, respectively.

The maximum amplitude of the limit cycles when only one of the multipliers is nonzero can easily be calculated. Let us set $\beta_1 = 0.0$ in Eq. (6.44). Limit cycles are sustained in the second-order filter if $|Q(\beta_2 v)| = v$. In the case of rounding arithmetic, $|\beta_2 v| \geq v - \Delta/2$. By noting that the limit cycle amplitude is an integer multiple of the quantization step, the above inequality gives the maximum limit cycle amplitude as

$$v_{max} = \Delta \, \frac{1}{2(1 - |\beta_2|)} \,. \tag{6.45}$$

It has been shown that Eq. (6.45) can be used to get an estimate of the maximum limit cycle magnitude also in the more general case of both multipliers in Eq. (6.44) being nonzero [JA86]. In this case the oscillations can take more complicated forms than our simple examples indicate. According to Eq. (6.45) the level of limit cycles can be reduced by moving the poles away from the unit circle and by reducing the quantization step size.

Example 6.5. Let us now assume that the quantization characteristic being used is the magnitude truncation as shown in Figure 6-30. With the same pole positions and initial conditions $y[-1] = 0$, $y[-2] = 0.5$, as in Example 6.3 and with zero-input signal we obtain the output sequence corresponding to $n = 0, 1, \ldots$ as

$$-0.4, 0, 0.3, 0, -0.2, 0, 0.1, 0, 0, \ldots$$

It is easy to see that limit cycles of the type described in Example 6.3 cannot be sustained.

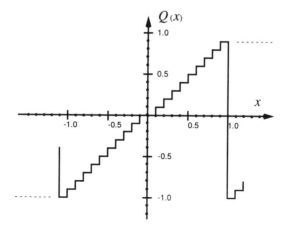

FIGURE 6-30 Magnitude truncation arithmetic used in the examples. The solid line and the dashed line correspond to standard overflow and saturation characteristics, respectively.

In general, magnitude truncation is an effective way to eliminate limit cycles. There are several filter structures that have been shown to be free of limit cycles if magnitude truncation is used but not if rounding arithmetic is used. One example is the coupled form second-order structure of Figure 7-9.

Another way to eliminate limit cycles is to introduce randomness in the roundoff operation. In the technique called *random rounding*, a standard two's complement rounding quantizer is used for quantizing the numbers to b-bit accuracy. However, the numbers being quantized are modified by replacing the $(b + 1)$th bit randomly by 0 or 1 independent of the actual signal being quantized (see Section 7-10-3).

Our example filter of Eq. (6.44) also supports overflow oscillations. Let us assume a constant input signal. An overflow oscillation with period of two is maintained if the following conditions are satisfied: $y[n] = y[n - 2]$, $y[n + 1] = y[n - 1]$ with $y[n] \neq y[n - 1]$, and overflows occur in additions.

Example 6.6. Let us select $\beta_1 = -1$ and $\beta_2 = 0.5$ in Eq. (6.44). This corresponds to poles at stable positions of $0.5 \pm j0.5$. With zero input signal, overflow oscillations occur if

$$y[0] = Q(y[1] - 0.5y[0]) \tag{6.46}$$

and

$$y[1] = Q(y[0] - 0.5y[1]). \tag{6.47}$$

Using the rounding arithmetic of Figure 6-29 with the standard overflow characteristics and setting $y[-2] = y[0] = -0.8$ and $y[-1] = y[1] = 0.8$ give $y[0] = Q(1.2) = -0.8$ and $y[1] = Q(-1.2) = 0.8$, indicating overflow oscillation.

However, if the saturation arithmetic of Figure 6-29 is used in the same situation, we obtain from Eq. (6.44) $y[0] = Q(1.2) = 0.9$, $y[1] = Q(0.9 - 0.5 \times 0.8) = 0.5$, $y[2] = Q(0.5 - 0.5 \times 0.9) = 0.0$ followed by $-0.3, -0.3, -0.1, 0.1,$ $0.2, 0.1, 0.0, -0.1, -0.1, 0.1, 0.1, 0.1, 0.0, -0.1, -0.1, \ldots$. Thus the final response is a limit cycle with cycle length of 6. If, instead, the saturation arithmetic with magnitude truncation of Figure 6-30 is used, the output converges to zero.

In practice, overflow oscillations may be triggered by a suitable input signal segment or by an uncontrolled loading of the delay elements. This can happen, for instance, if the filter circuitry temporarily loses electric power or an electromagnetic interference makes the circuits malfunction for a short time. The filter recovers from the overflow oscillation when the input signal varies over a sufficiently large dynamic range, or after the delay elements have been reset to zero. In practice, one should use saturation arithmetic whenever possible. However, even with the saturation operation certain types of nonlinear large-amplitude limit cycles can exist.

Example 6.7. Let us consider the saturation arithmetic of Figure 6.29 in the filter

$$y[n] = Q(0.9y[n - 1] + 0.1u[n]) \qquad (6.48)$$

with constant input signal $u[n] = v$. For constant input signals and without the quantization operation, the gain of the filter is one. This means that the output $y[n]$ should also approach v after the input has been applied. However, if for some reason the initial state $y[-1] = 0.9$ the output remains at 0.9 if the input signal is given a value $v \geq 0.4$. This is a limit cycle type oscillation with period of one arising from the rounding operation. A similar situation occurs naturally with negative input signals.

In real signal processing applications one can sometimes notice that the output signal gets "stuck" to some level for a longer time than one would expect from the properties of the input signal. In other words, the output signal does not follow the input signal variation. This is called the *dead band effect* and it results from the roundoff operation.

Example 6.8. Let us apply the input signal of Figure 6-31(a) to the filter of Example 6.7. The response is shown in the same figure. First, the output signal does not respond at all to the input signal variation. In addition, the maximum level of the output is limited to 0.6. Note that the output does not reach zero even though the input signal returns to zero. Let us now take the input signal as in Figure 6-31(b) and as initial condition set 0.9 in the delay register. The output remains at 0.9 in spite of the input signal decreasing to 0.4. Clearly, this rough quantization of the numbers makes the filter completely useless. The output signal corresponding to a constant input signal of magnitude 0.4 can as well be 0.0 or 0.9. It is easy to see that one more decimal digit in the internal number presentation would eliminate the nonlinear effects from this particular example.

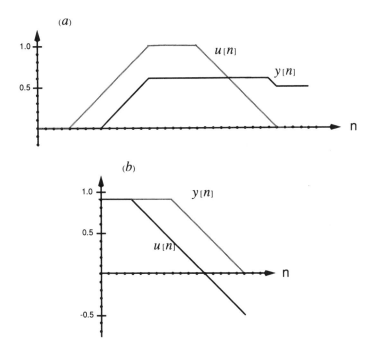

FIGURE 6-31 Responses of the filter of Eq. (6.48) to test signals: (a) starting with zero initial value and (b) with initial value set of 0.9.

The common property with magnitude truncation and saturation arithmetic is that

$$|Q(x)| \leq |x|, \tag{6.49}$$

where $Q(\ )$ denotes the truncation or saturation operation. Equation (6.49) is the reason why these operations often make the existence of nonlinear oscillations impossible. An analysis of the existence of limit cycles in several filter structures with different types of nonlinearities is given in Erickson and Michel [ER85].

6-7-4 Floating-Point Quantization Error Model

From the quantization point of view, floating-point arithmetic is beneficial. Due to mantissa normalization, all mantissa bits are significant, meaning that over a very wide range of numbers the full accuracy of the mantissa can be utilized. On the other hand, extra bits are needed to represent the exponent.

With floating-point arithmetic, the results of both multiplications and additions need to be quantized because both can otherwise exceed the length of the mantissa. Due to normalization of numbers to full mantissa accuracy, the quantization errors are relative to the signal magnitude:

$$Q(x[n]) = x[n](1 + \epsilon[n]). \tag{6.50}$$

Here $\epsilon[n]$ is the relative quantization error. It is easy to show that for a $(b + 1)$-bit mantissa, where one bit is reserved for the sign bit, the relative error satisfies

$$-2^{-b} < \epsilon[n] \leq 2^{-b} \tag{6.51}$$

if rounding arithmetic is used [FE74]. For one's complement and sign-and-magnitude truncation of the mantissa,

$$-2 \cdot 2^{-b} < \epsilon[n] \leq 0, \tag{6.52}$$

and for two's complement truncation,

$$-2 \cdot 2^{-b} < \epsilon[n] \leq 0, \qquad x[n] \geq 0$$
$$0 \leq \epsilon[n] < 2 \cdot 2^{-b}, \qquad x[n] < 0. \tag{6.53}$$

With rounding arithmetic and generally used mantissa lengths, it is feasible to assume that the (relative) quantization error sequences in a floating-point digital filter are white noise sequences that do not correlate with each other or any node variables in the filter. However, $\epsilon[n]$ is not uniformly distributed in its range [FE74]. With rounding arithmetic a good approximation to the variance of the relative quantization noise is

$$\sigma^2 = \frac{2^{-2b}}{6}. \tag{6.54}$$

As mentioned in Section 6-2-3 one bit is saved in the mantissa in the normalized representation of floating-point numbers. This is not taken into account in Eq. (6.54). Due to the different mechanism of roundoff noise generation, floating-point roundoff noise analysis is quite different from the fixed-point case, as will be shown in Section 6-11.

6-8 COEFFICIENT SENSITIVITY

A fundamental difference between an analog and digital filter is that in the latter all copies of the same filter always have exactly the same coefficient values. This makes it easy to check that the design with quantized coefficients meets the specifications. It also gives the opportunity to improve the properties of a digital filter using integer type optimization methods over the quantized coefficients.

The number of bits required to represent a filter coefficient can be calculated in several different ways. Sometimes the sign bit is included; sometimes it is not. In this book we therefore keep on mentioning how the sign bit is taken into account. Assuming binary fractional arithmetic, we normally include in the number of bits also the leading zeros and ones (in the case of two's complement negative numbers). In some hardware structures it is possible to use only the significant bits to represent the coefficients. The leading zeros (and ones) are taken care of by the arithmetic shift unit of the hardware. Also, floating-point implementations use only the significant bits in the bit count. Traditionally, the coefficient sensitivity analysis methods also count the leading bits in the bit count. This gives a somewhat pessimistic measure in cases where each coefficient is scaled to have only significant bits.

As already mentioned in Section 6-3-2, different filter structures can exhibit quite different amounts of degradation under coefficient quantization.

Example 6.9. Let us consider a third-order elliptic filter with passband edge at ω_p = 0.3π, passband ripple of 0.92 dB, and stopband attenuation of approximately 20 dB. In linear scale the passband ripple is between 1 and 0.9. The transfer function satisfying these specifications is

$$H(z) = \frac{0.1336 + 0.0563z^{-1} + 0.0563z^{-2} + 0.1336z^{-3}}{1 - 1.5055z^{-1} + 1.2630z^{-2} - 0.3778z^{-3}}. \tag{6.55a}$$

$H(z)$ can also be expressed as the sum of two allpass transfer functions,

$$H(z) = \frac{-0.4954 + z^{-1}}{1 - 0.4954z^{-1}} + \frac{0.7626 - 1.0101z^{-1} + z^{-2}}{1 - 1.0101z^{-1} + 0.7626z^{-2}}. \tag{6.55b}$$

It is easy to check that the gain of $\omega = 0$ and $\omega = \pi$ is 1 and 0, respectively, by setting $z = 1$ and $z = -1$ in Eq. (6.55). Based on the above expressions, the filter has been implemented using the direct form and the parallel allpass structure. The magnitude responses obtained with 6- and 7-bit coefficient accuracies are shown in Figure 6-32 along with the ideal response. In the passband, the parallel allpass structure is clearly much more tolerant to coefficient quantization than the direct form.

The coefficient sensitivities of a filter structure can be used for general characterization of the filter's susceptibility to coefficient quantization errors. Let us consider the single-input–single-output filter structure of Figure 6-33, which realizes the transfer function $H(z)$ between the nodes Ⓘ and Ⓞ. The sensitivity of $H(z)$ with respect to the coefficient k between nodes Ⓝ and Ⓜ of the structure can be expressed as [OP75]

$$\frac{\partial H(z)}{\partial k} = T_{in}(z)T_{mo}(z). \tag{6.56}$$

Here $T_{in}(z)$ is the transfer function from the input to node Ⓝ and $T_{mo}(z)$ is the transfer function from node Ⓜ to the output. The transfer functions are calculated with the coefficient k present. Equation (6.56) allows us to calculate all coefficient sensitivities conveniently using the matrix representation of the filter structure.

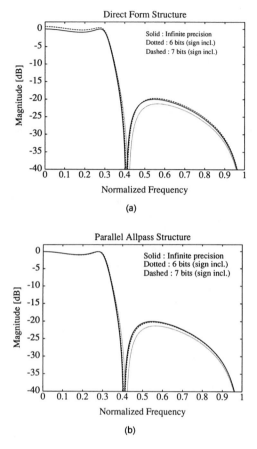

FIGURE 6-32 Frequency response of the third-order filter with quantized coefficients: (a) direct form implementation and (b) parallel allpass structure.

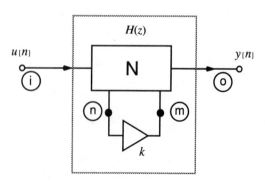

FIGURE 6-33 A general filter structure with multiplier k shown explicitly.

The sensitivity of the magnitude of the transfer function with respect to the coefficient k is given as [CR75b]

$$\frac{\partial |H(z)|}{\partial k} = \text{Re}\left[\frac{|H(z)|}{H(z)} T_{in}(z) T_{mo}(z) \right]. \qquad (6.57)$$

The expressions in Eqs. (6.56) and (6.57) can be used for calculating the changes in the transfer function due to a small change Δk in k.

For large Δk there is a convenient expression to calculate the corresponding change $\Delta H(z)$ in the transfer function:

$$\Delta H(z) = \frac{T_{in}(z) T_{mo}(z) \Delta k}{1 - T_{mn}(z) \Delta k}, \qquad (6.58)$$

requiring the calculation of the additional transfer function $T_{mn}(z)$ from the output of the multiplier k to its input. In nonrecursive structures there is no signal flow backward. Thus $T_{mn}(z) = 0$ and the change $\Delta H(z)$ depends linearly on Δk.

6-8-1 Statistical Measures of Coefficient Quantization Errors

For a filter with m multipliers of values k_i ($i = 1, \ldots, m$), the overall magnitude error $\Delta |H(e^{j\omega})|$ due to small changes Δk_i in the multipliers is

$$\Delta |H(e^{j\omega})| \approx \sum_{i=1}^{m} \frac{\partial |H(z)|}{\partial k_i} \Delta k_i. \qquad (6.59)$$

Assuming that the coefficient quantization errors Δk_i are independent with equal variance σ_k^2, the variance of $\Delta |H(e^{j\omega})|$ can be expressed as

$$\sigma_{\Delta H}^2 \approx \sum_{i=1}^{m} \left[\frac{\partial |H(z)|}{\partial k_i} \right]^2 \sigma_k^2 = S^2(e^{j\omega}) \sigma_k^2. \qquad (6.60)$$

The variance σ_k^2 is $q^2/12$ for uniform quantization error distribution over the quantization step size q. Equation (6.60) gives a frequency-dependent measure $S(e^{j\omega})$,

$$S^2(e^{j\omega}) = \sum_{i=1}^{m} \left[\frac{\partial |H(z)|}{\partial k_i} \right]^2,$$

that can be used to characterize filter structures [CR75a, CR75b]. By the Central Limit Theorem the distribution of $\Delta |H(e^{j\omega})|$ is approximately Gaussian. This allows us to select the coefficient quantization step size q such that with the given probability y the error $\Delta |H(e^{j\omega})|$ will be less than $x\sigma_{\Delta H}$ for an arbitrary set of coefficients. That is,

$$P[\|\Delta|H(e^{j\omega})\| \leq x\sigma_{\Delta H}] \approx y. \tag{6.61}$$

Let us assume that we want to design a filter to meet the passband (or stopband) specified gain with a tolerance $\pm\delta_{tot}$. This tolerance is composed of the infinite precision tolerance $\delta_i = \|H_0(e^{j\omega})\| - |H_I(e^{j\omega})\|$ and the quantization error $\Delta|H(e^{j\omega})|$. Here, $|H_0(e^{j\omega})|$ denotes the magnitude response obtained with infinite coefficient accuracy and $H_I(e^{j\omega})$ the specified ideal magnitude response. Now

$$\delta_{tot} = \delta_i + \Delta|H(e^{j\omega})|. \tag{6.62}$$

The coefficient wordlength of the filter W can be defined as

$$W = 1 + i_M - i_L, \tag{6.63}$$

where 2^{iM} is the power of the two represented by the most significant bit and 2^{iL} is the power of two represented by the least significant bit. Equation (6.63) does not include the sign bit.

Using the above equations, we can derive an expression for the quantization step size q, which can then be used to find

$$i_L = \log_2 q \tag{6.64}$$

and consequently the statistical wordlength as a function of frequency as

$$w(\omega) = 1 + i_M - \log_2 \left[\frac{\sqrt{12}(\delta_{tot} - \delta_i)}{xS(e^{j\omega})} \right]. \tag{6.65}$$

Note that the only frequency-dependent term in Eq. (6.65) is $S(e^{j\omega})$. This makes it easy to check the effect of different error combinations to achieve a certain δ_{tot} in the passband and stopband of the filter.

Example 6.10. Let us calculate the statistical wordlengths as given by Eq. (6.65) for the two filters of Example 6.9. The infinite precision tolerance in the passband and stopband is $\delta_i = 0.1$. We assume that fractional arithmetic is used making $i_M = -1$ in Eq. (6.65). This implies that the numerator and denominator coefficients in Eq. (6.55a) have been scaled down by a factor of two in order to make all coefficients less than one in magnitude in the direct form implementation. Corresponding scaling of the coefficients has been performed for the parallel allpass structure. We select $\delta_{tot} = 0.12$ and $x = 1$. This means that the quantization step size is selected such that one standard deviation, $\sigma_{\Delta H}$, of the statistical overall magnitude error corresponds to $\delta_{tot} - \delta_i = 0.02$. The statistical wordlengths are shown in Figure 6-34. Note that in the passband and especially at the two fre-

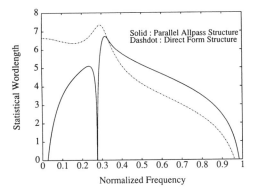

FIGURE 6-34 Statistical wordlength of the third-order direct form and parallel allpass structures.

quencies where the gain of the filter is 1, the parallel allpass structure has very good sensitivity properties.

Now W can be written as

$$W = \max_{\omega} w(\omega). \qquad (6.66)$$

Typically $w(\omega)$ tends to have maxima around the edges of the passband(s) and stopband(s). As can be expected, the statistical wordlength often has local maxima near the extremum points of the frequency response. For a given tolerance δ_{tot} there is a continuum of candidate transfer functions. By carefully selecting the transfer function, the required coefficient accuracy can be somewhat reduced [CR75a].

Another approach to analyzing the sensitivity is by designing a number of filters and measuring the error in the frequency response with a certain coefficient accuracy. For the passband sensitivity characterization one can use the relative error measure [CR72]

$$RE = \begin{cases} \dfrac{H_{\text{MAX}} - H_{\text{MIN}} - A_M}{A_M}, & H_{\text{MAX}} - H_{MIN} > A_M \\ 0, & H_{\text{MAX}} - H_{\text{MIN}} \le A_M, \end{cases} \qquad (6.67)$$

where H_{MAX} and H_{MIN} are the maximum and minimum magnitudes in decibels in the passband and A_M is the infinite precision passband variation in decibels. By calculating RE for various coefficient accuracies and representative filter specifications, general conclusions of the sensitivity and required accuracy can be drawn [CR72].

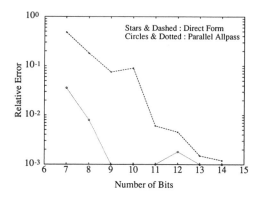

FIGURE 6-35   Relative passband error as function of coefficient accuracy.

Example 6.11. The relative error measure for the filters of Example 6.9 are shown in Figure 6-35. For the parallel allpass filter the relative error becomes less than 10^{-3} for more than 8-bit coefficient accuracies. The horizontal separation between the two error measures suggests that the direct form requires from 3 to 5 more bits for the same accuracy in the passband than the parallel allpass structure.

Coefficient quantization naturally affects the pole and zero positions. An idea of the characteristic sensitivities can in certain cases be obtained by calculating the sensitivities of pole and zero positions to coefficient quantization. If the poles and zeros move very little when coefficients are quantized, one concludes that the coefficient sensitivities are good. For second-order sections the *pole (zero) grid* offers a way to visualize the pole (zero) sensitivities. It displays on the z-plane the admissible pole (zero) positions with a certain coefficient accuracy [OP75]. The denser the grid is in the vicinity of the desired pole (zero) position the better.

6-8-2   Coefficient Sensitivities of FIR Filters

For FIR filters statistical wordlengths have been derived by Chan and Rabiner [CH73]. For simplicity, let us consider a linear-phase odd-length N FIR filter with the ideal zero-phase response

$$\hat{H}(\omega) = h[(N-1)/2] + \sum_{i=0}^{(N-3)/2} 2h[i] \cos\left[\left(\frac{N-1}{2} - i\right)\omega\right]. \quad (6.68)$$

The quantized coefficients are $h_q[n] = h[n] + e[n]$. The error sequence $e[n]$ is the impulse response of a linear-phase filter of length N in parallel with the original filter $H(z)$. The error filter has the zero-phase response

$$\hat{E}(\omega) = e[(N-1)/2] + \sum_{i=0}^{(N-3)/2} 2e[i] \cos\left[\left(\frac{N-1}{2} - i\right)\omega\right]. \quad (6.69)$$

Since $|e[n]| \leq q/2$, a bound on $\hat{E}(\omega)$ can readily be obtained as

$$|\hat{E}(\omega)| \leq N\frac{q}{2} \tag{6.70}$$

by approximating the cosines by one. This limit is quite conservative in practice. More realistic bounds are obtained by using statistical analysis of the quantization error.

From Eq. (6.69) it is seen that for any ω, $\hat{E}(\omega)$ is the sum of independent random variables which by the Central Limit Theorem makes $\hat{E}(\omega)$ to be essentially Gaussian for normally used N. From Eq. (6.69) the standard deviation of $\hat{E}(\omega)$ can be obtained as

$$\sigma_{EL}(\omega) = \frac{q}{2}\sqrt{\frac{2N-1}{3}}\left[\frac{1}{2} + \frac{1}{2N-1}\left(-\frac{1}{2} + \frac{\sin N\omega}{\sin \omega}\right)\right]^{1/2}. \tag{6.71}$$

As the term in brackets is less than or equal to 1 we can obtain a frequency-independent upper limit,

$$\sigma_{EL}(\omega) \leq \frac{q}{2}\sqrt{\frac{2N-1}{3}}. \tag{6.72}$$

For arbitrary-phase direct form a similar analysis can be performed, giving for the standard deviation of the error

$$\sigma_{EA}(\omega) \leq \frac{q}{2}\sqrt{\frac{N}{3}}. \tag{6.73}$$

Example 6.12. Let us consider a length $N = 121$ linear-phase FIR lowpass filter designed using the Parks–McClellan algorithm with $\omega_p = 0.3\pi$ and $\omega_s = 0.35\pi$. The passband and stopband relative weights are both one. The resulting frequency response is shown in Figure 6-36(a) with infinite precision and 10-bit coefficients. The error frequency response of Eq. (6.69) is shown in Figure 6-36(b) along with the upper bound and the frequency-independent upper limit for the standard deviation of Eqs. (6.70) and (6.72), respectively.

It is worth noticing that the error frequency response of Eq. (6.69) depends only on the filter length. Note also that the bounds calculated in Example 6.12 are valid for all length 121 linear-phase filters. In the case of IIR filters, the shape of the frequency response has a strong effect on the sensitivities.

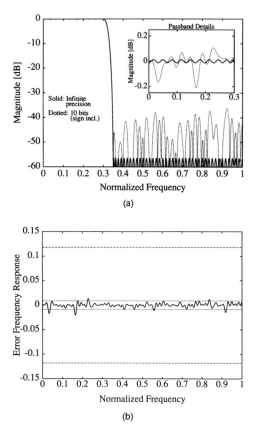

FIGURE 6-36   Coefficient quantization of a FIR filter: (a) frequency responses; (b) solid line shows the zero-phase response of the error, dotted lines correspond to the standard deviation based error measure of Eq. (6.72), and the dash-dot lines are the upper bounds of Eq. (6.70).

6-9   INPUT QUANTIZATION ERROR

Digital signal processing algorithms are often employed in the processing of analog signals. The interface between the analog and digital worlds is via analog-to-digital and digital-to-analog converters. For the designer, the goal is to achieve a good enough signal quality in the system by taking into account the imperfections of both the analog and digital parts of the system. The analog parts generate thermal noise and also may pick up some disturbing signals from, for example, the digital part [DI88]. The nonlinearities in the analog circuits cause signal clipping and harmonic distortion. If modern 16- to 32-bit signal processors are used in the digital part of the system, the analog circuitry tends to be more critical for obtaining high-fidelity signals. The analog-to-digital and digital-to-analog converters are discussed in Chapter 10.

We now take a closer look at the quantization error arising from analog-to-digital conversion and study the effect of signal level scaling on the signal-to-noise ratio in the digital signal processing part of the overall system. The issues discussed here are closely related to roundoff noise analysis discussed in Sections 6-7, 6-10, and 6-11.

In the analog-to-digital converter an analog discrete-time sample $u_a[n]$ is converted to a digital form $u[n]$ using either a uniform or nonuniform converter. In the uniform converter, commonly used in signal processing, the quantization step size Δ is constant, independent of the signal level (within the dynamic range of the converter) [GR90]. The quantization error $e[n] = u[n] - u_a[n]$ is bounded by $-\Delta/2 \leq e \leq \Delta/2$. The difference between the quantization errors resulting from analog and digital signal quantization (Section 6-7) is that in the former the error e is a continuous variable within its range, whereas roundoff operations result in discrete error distributions.

In most cases we can assume that the analog-to-digital conversion error $e[n]$ has the following properties: (1) the error sequence $e[n]$ is a sample sequence of a wide-sense stationary (WSS) random process; (2) the error sequence is uncorrelated with $u[n]$ and other signals in the system; and (3) the error is a white noise process with uniform amplitude probability distribution over the range of the quantization error.

Example 6.13. In order to see the validity of the above assumptions, 2000 samples of music have been converted to 8-bit accuracy. The normalized autocorrelation function of the quantization error $R_{ee}[k] = E\{e[n]e[n + k]\}/E\{(e[n])^2\}$ and the normalized cross-correlation function $R_{ue}[k] = E\{u[n]e[n + k]\}/E\{(e[n])^2\}E\{(u[n])^2\}^{1/2}$ calculated from this music sample are shown in Figure 6-37. Clearly, these graphs support the assumptions that the error is white noise and uncorrelated with the actual signal.

The variance of the error sequence is obtained as

$$\sigma_\epsilon^2 = \frac{1}{\Delta} \int_{-\Delta/2}^{\Delta/2} e^2 \, de = \frac{\Delta^2}{12}. \tag{6.74}$$

With good accuracy also the roundoff error variance can be approximated by Eq. (6.74), as has already been done in Section 6-7.

The signal-to-noise ratio after the analog-to-digital conversion depends, in addition to the conversion accuracy, also on how large the signal level is. Let us assume that the number range is $[-1, 1]$ and that we have $b + 1$ bits representing a digital sample. For a sine wave input of amplitude 1 and power $\sigma_u^2 = \frac{1}{2}$ the signal-to-noise ratio (SNR) becomes

$$\text{SNR} = 10 \log_{10}\left(\frac{\sigma_u^2}{\sigma_\epsilon^2}\right) = 10 \log_{10}\left(\frac{1/2}{\Delta^2/12}\right) = 6.02b + 7.8. \tag{6.75}$$

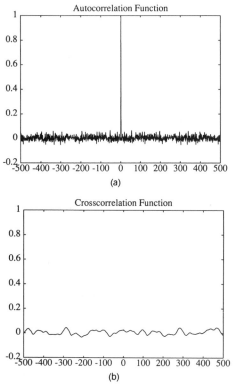

FIGURE 6-37   (a) Normalized autocorrelation function of a music sample. (b) Normalized cross-correlation between the quantized signal and quantization error.

Adding one bit to the representation increases the SNR by about 6 dB. In general, the input signal should be scaled such that the probability of reaching the overflow range of the converter is small enough. The acceptable frequency of overflows depends on the application. This implies that a certain number of bits have to be reserved in the number representation above the rms level of the signal [KR90]. The signal range reserved above the rms level is sometimes called the *headroom*. In the case of a sinusoidal input signal the headroom can be as little as half a bit as assumed by Eq. (6.75).

In practice, the required headroom depends on the signal amplitude distribution (on the peakedness of the signal), on the uncertainty of the exact amplitude statistics, and on how harmful an overflow is considered to be.

Example 6.14. It is instructive to compare the amplitude histograms of different types of musical signals played on a CD player. The music is played from a CD player. The analog signal is sampled at 43,000 samples/s. We have selected as examples three different pieces of music: orchestral classical [CI84], modern [YE86], and vocal [VE90]. Normalized amplitude histograms (density functions) and the corresponding complements of the distribution functions have been calculated from these signals and are shown in Figure 6-38. In each graph the signal levels have been normalized by the rms value of the signal sample measured over the whole music sample. The density functions clearly show the very different amplitude statistics of the samples. The frequency of the signal exceeding a certain level can directly be seen from the complemented distribution functions in percent. As one might expect, the nonstationary nature of the orchestral sample results in a wide amplitude histogram. For this sample the required headroom is about 4 bits (the histogram suggests that the large volume part, the gunfire, of the sample may have been compressed during the recording process). The other two samples of music also seem to have some quite large peak values in comparison to the rms value.

Let the signal be Gaussian and we want overflows to occur with a probability of 10^{-3}. This implies that the 3.3σ level of the signal has to be scaled to correspond to the maximum level of one. The SNR ratio now becomes

$$\text{SNR} \approx 6.02b + 0.4 \approx 6b. \tag{6.76}$$

Example 6.15. Music on CD records is stored with $(15 + 1)$-bit accuracy. If the music satisfies the assumptions behind Eq. (6.76) we obtain the SNR of approximately 90 dB. However, if the maximum signal level is 15σ as suggested by our previous example, the SNR is approximately $-12.7 + 6b$, giving 78 dB.

The effect of signal level scaling on the SNR is shown in Figure 6-39. The SNRs for fixed-point number representations are calculated for sinusoidal signals using Eq. (6.75). In fixed-point representations the SNR decreases as the signal level decreases, as expected. The SNRs for floating-point representations have been calculated using Eq. (6.54). Note that within the accuracy of the assumptions made about the floating-point roundoff noise, the SNR is independent of the signal amplitude characteristics. The maximum SNR of fixed- and floating-point representations are about the same if the mantissa length equals the fixed-point wordlength. In the figure we have not taken into account that one saves one bit in the normalized representation of floating-point mantissas (see Section 6-8-3).

For the output SNR of a digital filter one has to add up the various noise powers at the output of the filter arising from analog-to-digital conversion and from roundoff errors. This is discussed in more detail in Section 6-10 and Chapter 7.

6-10 ROUNDOFF NOISE AND DYNAMIC RANGE CONSIDERATIONS

Consider again the first-order IIR filter structure of Figure 6-22. Assuming it represents a stable filter, we have $|a| < 1$. As outlined earlier, it is necessary to introduce a quantizer in the feedback loop. Two ways of doing this are shown in Figure 6-23, and for the rest of our discussion, we consider the scheme of Figure

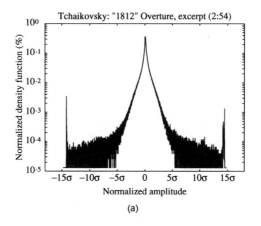

(a)

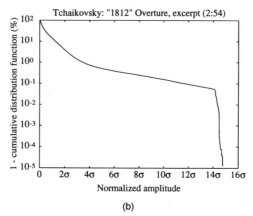

(b)

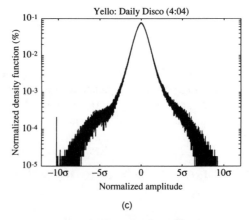

(c)

FIGURE 6-38 Amplitude statistics of different types of music: parts (a), (c), and (e) show the normalized density functions, while parts (b), (d), and (f) show the corresponding frequencies in percent when the amplitude exceeds a certain multiple of its rms value.

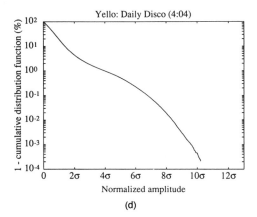

(d)

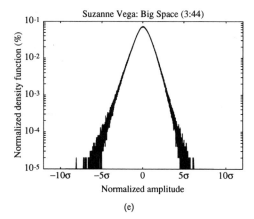

(e)

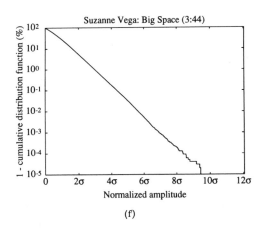

(f)

FIGURE 6-38 (*Continued*)

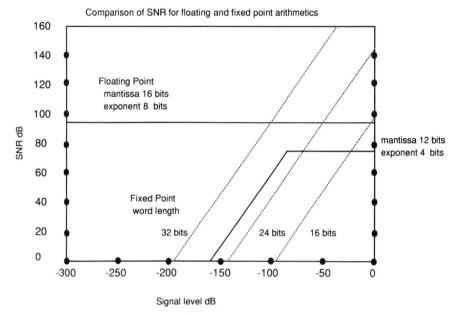

FIGURE 6-39 The SNR resulting from different number representations using sinusoidal input signal. The solid lines correspond to floating-point representations and dotted lines to fixed-point representations.

6-23(a), p. 373. This scheme matches with the architecture of current signal processors that quantize the results after the multiply–accumulate operation. In Figure 6-24, the effect of the quantizer is shown by incorporating a noise source $e[n]$.

Throughout this section simplifying assumptions are made about the roundoff noise $e[n]$ generated by a quantizer. These assumptions are valid when the signals being quantized are much stronger than the quantization error itself [BA85; SR77]. This is usually the case when the number of bits b (see Figure 6-25) at the quantizer output is large (e.g., exceeding 8). These assumptions cause the roundoff errors to satisfy the same properties as the input quantization error discussed in Section 6-9.

It is assumed that $e[n]$ represents a wide-sense stationary (WSS) random process. Accordingly, the statistical mean $m_e = E(e[n])$ is independent of n, and the covariance sequence $C_e[m] = E([e[n] - m_e][e[n + m] - m_e])$ depends only on the lag m. It is furthermore assumed that $e[n]$ is actually white, that is, $C_e[m] = \sigma_\epsilon^2 \delta[m]$, and in addition that it is uncorrelated to other signals in the system, such as $u[n]$, $y[n]$, $w[n]$, and so on. The statistical distribution of $e[n]$ is assumed to be uniform (any one of the types in Figure 6-26). Moreover, if there are several quantizers in a structure giving rise to noise sources $e_k[n]$, then each source is assumed uncorrelated to the others.

6-10-1   Noise Transfer Function and Noise Gain

In Figure 6-24, p. 374, the transfer function from the noise source to the filter output is seen to be

$$G(z) = \frac{1}{1 - az^{-1}}, \tag{6.77}$$

which turns out to be the same as the system transfer function given by Eq. (6.41). $G(z)$ is known as the noise transfer function. The component of the noise at the filter output, denoted by $y_e[n]$, is the convolution of $e[n]$ with the impulse response of Eq. (6.77); namely,

$$g[n] = a^n \mu[n]. \tag{6.78}$$

Accordingly, the mean and variance of the output noise sequence are given by

$$m_f = E[y_e[n]] = m_e \sum_{n=0}^{\infty} g[n], \tag{6.79}$$

$$\sigma_f^2 = E[y_e[n] - m_f]^2 = \sigma_e^2 \sum_{n=0}^{\infty} |g^2[n]|. \tag{6.80}$$

Recall that m_e and σ_e^2 are characteristics of the quantizer noise. As discussed before, their values are as in Figure 6-26. Based on the expression (6.78) for $g[n]$, we obtain

$$m_f = \frac{m_e}{1 - a}, \qquad \sigma_f^2 = \frac{\sigma_e^2}{1 - a^2}. \tag{6.81}$$

The significant point is that the noise variance of the quantizer appears amplified at the filter output. The ratio σ_f^2 / σ_e^2 is called the *noise gain*. If in Eq. (6.81) a^2 is close to unity, there is a large noise gain. Accordingly, if the pole a of the first-order section is close to the unit circle, the noise gain is large.

If a digital filter has several quantizers, there is a noise gain associated with each of these. We return to this issue in a later section. For the time being, note that if the noise transfer function from a particular noise source to the filter output is $G(z)$ then the noise gain for this noise source is given by

$$\sum_{n=0}^{\infty} g^2[n] = \frac{1}{2\pi} \int_0^{2\pi} |G(e^{j\omega})|^2 \, d\omega, \tag{6.82}$$

which is the energy of the impulse response sequence corresponding to $G(z)$.

One of the most popular structures for digital filter realization is the cascade form, which is a cascade of first- and second-order sections. First-order sections have real poles whereas second-order sections have complex-conjugate pole pairs. If the zeros of the transfer function are ignored for a moment, the first-order sections are of the form of Eq. (6.41), whereas the second-order sections have transfer functions of the form

$$H(z) = \frac{1}{1 - 2r \cos \theta z^{-1} + r^2 z^{-2}},$$ (6.83)

whose poles are $z = re^{j\theta}$ and $z = re^{-j\theta}$. For stability, the pole radius r should be less than unity. Section 7-3 contains an analysis of roundoff noise in cascade form structures with each section having both poles and zeros. However, analysis of the all-pole sections is helpful in drawing important conclusions about the noise gain as a function of the pole location. Figure 6-40 shows the direct form structure for Eq. (6.83) with a quantizer in the feedback loop. There are other places in the loop where the quantizer could have been inserted; in fact, it is possible to use two quantizers [OP75] rather than one. Such minor variations do not significantly affect our results on the output noise variance, so the simplest scheme of Figure 6-40 is considered here.

It can be seen by analogy with the first-order case that the quantizer noise source in Figure 6-40 sees a noise transfer function $G(z)$, which turns out to be the same as $H(z)$. The impulse response corresponding to Eq. (6.83) is

$$h[n] = \frac{r^n \sin (n + 1)\theta}{\sin \theta} \mu[n].$$ (6.84)

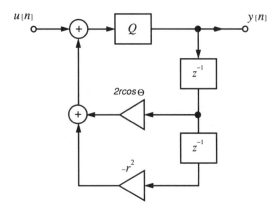

FIGURE 6-40   The second-order direct form structure with one quantizer.

Accordingly, the noise gain is equal to $\Sigma_{n=0}^{\infty} h^2[n]$, which can be shown to be given by

$$\sum_{n=0}^{\infty} h^2[n] = \left[\frac{1 + r^2}{1 - r^2} \right] \left[\frac{1}{r^4 - 2r^2 \cos 2\theta + 1} \right]. \tag{6.85}$$

Once again, if the pole is close to the unit circle, that is, if r is close to unity, the noise gain is very large. As a general rule, direct form implementations always have a large noise gain when the poles are close to the unit circle.

The noise level at the filter output does not tell the whole story about finite wordlength effects. The output signal level that can be obtained without causing *internal signal overflow* is a crucial parameter that determines the signal-to-noise ratio at the output. The topic of signal scaling is related to the overflow problem, which in turn governs the SNR. This is the subject of the next subsection.

6-10-2 Dynamic Range Considerations and Scaling

Consider Figure 6-22 again. For a moment, let us ignore the quantizer and concentrate on the scaling problem. The signal $w[n]$ is related to the input $u[n]$ by

$$w[n] = \sum_{m=0}^{\infty} h[m]u[n - m], \tag{6.86}$$

where $h[n]$ is the impulse response (same as $g[n]$ in Eq. (6.78)). If $u[n]$ is represented by fixed-point fractions, then we know $|u[n]|$ cannot exceed unity. Accordingly, $w[n]$ is bounded as

$$|w[n]| \leq \sum_{m=0}^{\infty} |h[m]|. \tag{6.87}$$

The right-hand side in Eq. (6.87) may not be bounded by unity. For example, with $a = 0.99$ in Figure 6-22, we have $\Sigma_{m=0}^{\infty} |h[m]| = 1/(1 - |a|) = 100$. In other words, all we can say is that $|w[n]|$ is surely less than or equal to 100. This implies a great likelihood of an overflow at the adder output. Note that if an overflow occurs it sets up a transient, which takes time to decay. Accordingly, the probability of an overflow should be kept minimal.

One way to completely avoid overflow (i.e., to reduce overflow probability to zero) in the above example is to restrict the input to be such that $|u[n]| < \frac{1}{100}$. This is equivalent to the introduction of a scale factor $1/L$ as shown in Figure 6-41, with $L = 100$.

The side effect of scaling is to reduce the signal level at the filter output. As an example, assume that $u[n]$ is itself a wide-sense stationary (WSS) white process with variance σ_u^2. Then the output signal $y[n]$ is a WSS process, with power spec-

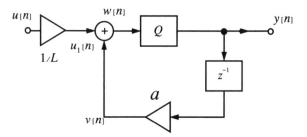

FIGURE 6-41   Insertion of the scale factor $1/L$.

tral density proportional to $|H|^2$, and has variance equal to

$$\sigma_y^2 = \sigma_u^2 \frac{1/L^2}{1 - a^2}. \tag{6.88}$$

The roundoff noise variance σ_f^2 at the output is still given by Eq. (6.81); hence the signal-to-noise ratio is

$$\text{SNR} = \frac{\sigma_y^2}{\sigma_f^2} = \frac{\sigma_u^2}{L^2 \sigma_e^2}. \tag{6.89}$$

If $L = 1$ we have an unscaled structure. The signal-to-noise ratio is then large and is independent of the noise gain! However, the probability of overflow is large. At the other extreme, if

$$L = \sum_{n=0}^{\infty} |h[n]|, \tag{6.90}$$

then we have $L = 1/(1 - |a|)$, and there is complete freedom from overflow but the SNR decreases to $(1 - |a|)^2$ times the unscaled value. For example, with $a = 0.99$, we have $(1 - a)^2 = 0.0001$ and this is the factor by which we lose the SNR and is the price paid for avoiding overflow. This scaling is called sum scaling, since the sum of absolute values of the impulse response is bounded by unity.

In practice, scaling is never done as conservatively as this. The factor L is chosen such that a compromise between the SNR and overflow probability is accomplished. The knowledge on the input signal also plays a crucial role in this regard, as elaborated soon.

Figure 6-42 helps us to visualize the dynamic range of various signals in the structure. The input $u[n]$ is a b-bit fixed-point number in the range $[-1, 1)$ (assuming two's complement arithmetic as an illustration), and the scaled signal $u_1[n]$ has certain leading zeros (unshaded region on the left). With $u[n]$ so scaled, the signal $w[n]$ does not overflow and is a fixed-point fraction in the range $[-1, 1)$. The quantized signal $y[n]$ is also a b-bit fixed-point fraction in the range $[-1, 1)$. The cross-hatched area in $w[n]$ is representative of the error introduced by the

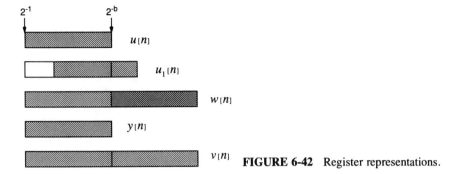

FIGURE 6-42 Register representations.

quantizer Q. Assuming that a is a b-bit number (restricted to be in the range $-1 < a < 1$ for stability), the signal $v[n]$ is a $2b$-bit fixed-point fraction in the range $[-1, 1)$.

6-10-3 Less Conservative Scaling Rules

In Figure 6-41, if we know nothing about the input signal except the obvious fact that it is in the range $[-1, 1)$, then the *only way* to avoid overflow at the adder output is to choose L as in Eq. (6.90). If, however, we have additional knowledge about the input, it might be possible to choose a smaller L that would guarantee freedom from overflow. The SNR would then be better. For example, suppose we know *ahead of time* that the energy of the input is bounded as follows:

$$\sum_{n=0}^{\infty} |u[n]|^2 < 1. \tag{6.91}$$

From Figure 6-41 it is clear that $w[n]$ is given by

$$w[n] = \sum_{m=0}^{\infty} f[m]u[n - m], \tag{6.92}$$

where

$$f[n] = \frac{1}{L} h[n]. \tag{6.93}$$

From here it is possible to deduce [JA70] that

$$|w[n]| \leq \sqrt{\sum_{n=0}^{\infty} |f[n]|^2 \sum_{n=0}^{\infty} |u[n]|^2}. \tag{6.94}$$

Since the input is bounded as in Eq. (6.91), it is clear that, as long as

$$\sum_{n=0}^{\infty} |f[n]|^2 \leq 1, \tag{6.95}$$

there will be no overflow. This is in turn guaranteed simply by choosing

$$L = \sqrt{\sum_{n=0}^{\infty} |h[n]|^2}. \tag{6.96}$$

If we assume $a = 0.99$ as before, then the value of L resulting from Eq. (6.96) is about 50. The SNR of the scaled system is therefore four times better than for conservative scaling.

The scaling policy just described is called $\mathcal{L}_2$ scaling. The reason can be seen by noting that Eq. (6.94) can be written, using Parseval's relation, as

$$|w[n]| \leq \sqrt{\left(\frac{1}{2\pi}\int_0^{2\pi} |F(e^{j\omega})|^2 \, d\omega\right)\left(\frac{1}{2\pi}\int_0^{2\pi} |U(e^{j\omega})|^2 \, d\omega\right)}. \tag{6.97}$$

Now the $\mathcal{L}_p$ norm of a sequence $s[n]$ is defined as

$$\|S\|_p = \left[\frac{1}{2\pi}\int_0^{2\pi} |S(e^{j\omega})|^p \, d\omega\right]^{1/p}. \tag{6.98}$$

Accordingly, Eq. (6.97) can be written as

$$|w[n]| \leq \|F\|_2\|U\|_2. \tag{6.99}$$

Thus, for an input satisfying Eq. (6.91), if the $\mathcal{L}_2$ norm of $f[n]$ is scaled to be equal to unity, there is no overflow. More generally, it can be shown [JA70] that given any three sequences $w[n]$, $f[n]$, $u[n]$ related by Eq. (6.92), the inequality

$$|w[n]| \leq \|F\|_p\|U\|_q \tag{6.100}$$

always holds, where p and q are integers such that

$$\frac{1}{p} + \frac{1}{q} = 1. \tag{6.101}$$

Setting $p = q = 2$ results in Eqs. (6.94), (6.97), and (6.99). Equation (6.100) tells us that if an input has its q-norm less than unity, and if we scale the filter so that $f[n]$ has p-norm less than or equal to unity, then there is no overflow. This type of scaling is called $\mathcal{L}_p$ scaling.

So far, we have been considering only Figure 6-41. However, all the basic concepts required for noise/dynamic range analysis are included in this example. The transfer function $F(z)$ from the filter input to the node $w[n]$ (which is the node being scaled) is called the *scaling transfer function*, and $G(z)$ is the noise transfer function. In more general digital filter structure, there are several internal signals $w_k[n]$ that must be prevented from overflowing. Accordingly, there are several scaling transfer functions involved. It is important to identify those nodes that require scaling, as against those that need not be scaled. If saturation arithmetic is used and the filter coefficients are all less than one in magnitude, then only the adder outputs need to be scaled. With two's complement arithmetic, it is only necessary to scale those internal signals that are input to multipliers. As long as these signals are within the desired dynamic range, it is permissible to have overflow at other internal nodes. This strange property can be proved based on the fact that two's complement arithmetic corresponds to modulo 2 arithmetic [JA70; OP75].

In general, let nodes 1, 2, . . . , M be the nodes where overflow should be avoided. Let the transfer function to the kth node from the filter input be denoted $F_k(z)$, with associated impulse response $f_k[n]$. Moreover, let $G_l(z)$ represent the noise transfer function from the lth quantizer output with $l = 1, 2, . . . , L$. This is shown schematically for the kth node and lth quantizer in Figure 6-43.

In order to derive a scaled structure for this system, an appropriate transformation should be performed without changing the shape of the overall transfer function $H(z)$. Let $F_k'(z)$ and $G_l'(z)$ denote the scaling and noise transfer functions of the transformed structure. This structure is said to be scaled in the $\mathcal{L}_p$ sense if $\|F_k\|_p \le 1$ for $k = 1, 2, . . . , M$. $\mathcal{L}_p$ scaling ensures that if the input signal has a $\mathcal{L}_q$ norm less than unity, then there is no overflow at any node, at any time. On the other hand, if complete freedom from overflow is required for all possible input signals, then the scaled structure should satisfy $\Sigma_{n=0}^{\infty} |f_k'[n]| \le 1$ for all k. This type of scaling, again, is called sum scaling.

The most commonly used scaling policies are: sum scaling, $\mathcal{L}_\infty$ scaling, $\mathcal{L}_2$ scaling, and $\mathcal{L}_1$ scaling. It can be shown that the $\mathcal{L}_\infty$ norm of a sequence is equal to the peak value of the Fourier transform, and hence $\mathcal{L}_\infty$ scaling is also called

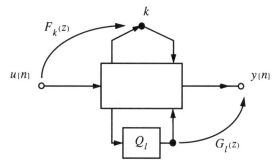

FIGURE 6-43 Block diagram explicitly showing the node k to be scaled and the lth quantizer along with the pertinent transfer functions for scaling the structure.

peak scaling. It turns out that, for any sequence $f[n]$, we have

$$\|F\|_\infty \geq \|F\|_p \quad \text{for all } p. \tag{6.102}$$

Accordingly, $\mathcal{L}_\infty$ scaling reduces the signal levels to a greater extent than other types of $\mathcal{L}_p$ scaling policies. The result is a decreased probability of overflow and a decreased SNR at the filter output. As a result, $\mathcal{L}_\infty$ scaling is considered to be the most stringent among all $\mathcal{L}_p$ scaling methods. At the other extreme, $\mathcal{L}_1$ scaling corresponds to $p = 1$, which means there will be freedom from overflow only if the input signal $u[n]$ satisfies $\|U\|_\infty < 1$. This is a very stringent condition on the input signal. Finally, $\mathcal{L}_2$ scaling, which is midway between the other types, has the advantage of mathematical simplicity; it is directly related to the energy of the impulse response sequence and has a nice symmetry property, namely, when $p = 2$ we have $q = 2$ as well.

It helps to remember the following intuitive notions: a *stringent* scaling policy implies *less* probability of overflow and results in *smaller* SNR at the filter output. This phenomenon is the celebrated *interaction between dynamic range and noise* [JA70]. A *stringent* scaling policy must be employed if sufficient information is not available about the input. In summary, the roundoff noise at the filter output is a meaningful measure of performance only if the output signal level is also specified. However, for a given output noise, the signal level can be made arbitrarily large (e.g., by decreasing L in Figure 6-41) and, accordingly, the SNR itself is not yet meaningful. The only meaningful measure of performance is the SNR under the condition that the internal signals are *scaled* so as to be within the permissible dynamic range (with a certain specified probability).

Example 6.16. Let us consider the first-order filter shown in Figure 6-44. If the scale factors s_1 and s_2 are both set to unity, the transfer function is

$$H(z) = c + \frac{q}{1 - az^{-1}}. \tag{6.103}$$

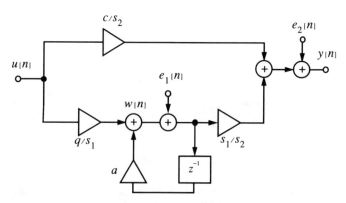

FIGURE 6-44   Example filter for scaling.

By selecting $a = 0.99$, $q = -0.01005$, and $c = 1.00505$, this filter has a zero at $z = 1$ and the gain at $z = -1$ is $H(-1) = 1$. It is easy to see that the adder output $w[n]$ and the filter output $y[n]$ are the critical signals to be scaled. We assume that quantization is performed after the additions, implying that there are two quantization error sources, $e_1[n]$ and $e_2[n]$, in the filter. We further assume that the input signal $u[n]$ is white noise. Let us apply $\mathcal{L}_2$ scaling. From the impulse response of the filter, $h[n] = c + q, qa, qa^2, \ldots$ it is easy to calculate $\|H\|_2^2$,

$$\|H\|_2^2 = (c + q)^2 + \frac{q^2 a^2}{1 - a^2}, \qquad (6.104a)$$

and the $\mathcal{L}_2$ norm to the adder output signal $w[n]$,

$$\|F\|_2^2 = q^2 + \frac{q^2 a^2}{1 - a^2}. \qquad (6.104b)$$

The above equations give $\|H\|_2 = 0.9975$ and $\|F\|_2 = 0.0713$. Clearly, the output node is much more likely to overflow. In order to avoid overflow, the input signal rms value can be at most $\sigma_u = 1/0.9975 = 1.0025$. The output SNR now becomes

$$\text{SNR} = 10 \log_{10} \left(\frac{\sigma_y^2}{\sigma_n^2} \right) = 10 \log_{10} \left(2^{2b} 12 \frac{\sigma_u^2 \sum\limits_{n=0}^{\infty} (h[n])^2}{1 + 1/(1 - a^2)} \right) \text{dB}. \qquad (6.105a)$$

With $(5 + 1)$-bit accuracy $(b = 5)$ and the maximal input signal level, we obtain $\text{SNR} = 23.88$ dB. In this case, only a small fraction of the input signal power goes through the recursive path to the output summation. However, quite a lot of noise is generated in this path because the noise gain for $e_1[n]$ is $1/(1 - a^2) = 50.25$. Let us now introduce the scale factor s_1 as shown in Figure 6-44. This enables us to make $\|F\|_2 = \|H\|_2 = 0.9975$, implying that signal levels at both adder outputs will be the same. Note that in this case scaling does not produce additional noise sources as quantization is performed after the summations. The scale factor s_1 becomes $s_1 = 0.0713/0.9975 = 0.0714$. The scaled structure SNR is

$$\text{SNR}_{\text{SC1}} = 10 \log_{10} \left(\frac{\sigma_y^2}{\sigma_n^2} \right) = 10 \log_{10} \left(2^{2b} 12 \frac{\sigma_u^2 \sum\limits_{n=0}^{\infty} (h[n])^2}{1 + s_1^2/(1 - a^2)} \right) \text{dB}. \qquad (6.105b)$$

With the numerical values used above we obtain $\text{SNR}_{\text{SC1}} = 39.89$ dB. If we want the filter to tolerate white noise input signals with variance $\sigma_u^2 = 2$ without overflows, we need to introduce the scale factor s_2. For the output not to overflow we need $s_2 = 2(0.9975) = 1.995$ and correspondingly $s_1' = 0.0714 s_2 = 0.1425$. In order to avoid overflows the signal levels in the filter have now been decreased, resulting in a decreased signal-to-noise ratio of $\text{SNR}_{\text{SC2}} = 36.89$ dB.

Normalized Structure Versus Scaled Structure. If a particular scaling scheme, say the $\mathcal{L}_p$ scheme, is implemented in such a way that the $\mathcal{L}_p$ norms of $F'_k(z)$ are exactly equal to unity for all k, then we have a *normalized structure*. In practice, in order to avoid an excessive number of multipliers, normalization is often not done. Exact normalization typically requires scaling multipliers, whereas scaling without normalization (i.e., the property $\|F'_k\| \leq 1$ without strict equality) can be accomplished by replacing extra scaling multipliers with approximate powers of two.

Example 6.17. Let us consider the filter of Example 6.16 scaled for $\sigma_u^2 = 2$ with $s'_1 = 0.1425$ and $s_2 = 1.995$. Using power of two scale factors, we obtain $s'_1 = 2^{-2} = 0.25$ and $s_2 = 2$. This slightly decreases the signal-to-noise ratio, which now is 35.35 dB.

Roundoff Noise Gain of a General Digital Filter Structure. The noise transfer functions of the unscaled and scaled structures are, respectively, $G_k(z)$ and $G'_k(z)$, for $1 \leq k \leq L$, where L is the number of quantizers. If we assume that each noise source has variance σ_e^2, then based on the standard assumption of uncorrelatedness stated earlier, we can show that the total noise variance at the filter output is

$$\sigma_f^2 = \begin{cases} \sigma_e^2 \sum_{k=0}^{L} \|G_k\|_2^2 & \text{(unscaled)} \\ \sigma_e^2 \sum_{k=0}^{L} \|G'_k\|_2^2 & \text{(scaled).} \end{cases} \tag{6.106}$$

Note that regardless of the scaling policy the noise variance always involves the $\mathcal{L}_2$ norm of the noise transfer functions.

6-10-4 Roundoff Noise/Dynamic Range Interaction in First- and Second-Order Filters

One of the commonly used structures for digital filtering is a cascade of first-and second-order stages. Scaling of the cascade structure is discussed in detail in Section 7.3. The purpose of this section is to give an understanding of the roundoff noise properties of first- and second-order sections. For the first-order section of Figure 6-22 characterized by the transfer function of Eq. (6.41), the roundoff noise variance at the filter output is given by σ_f^2 in Eq. (6.81) because the noise gain is $1/(1 - a^2)$. The signal power at the output depends on two quantities, namely, the scaling multiplier L in Figure 6-41 and the nature of the input. Let us assume that two's complement overflow arithmetic is used. As a result, it suffices to scale the signals that are inputs to multipliers. In Figure 6-40, it is therefore sufficient to scale the adder output $w[n]$. For the unscaled structure, the scaling transfer function is $F_k(z) = 1/(1 - az^{-1})$, from which we find that

$$\sum_{n=0}^{\infty} |f_k[n]| = \frac{1}{1 - |a|}, \quad \|F_k\|_2 = \frac{1}{\sqrt{1 - a^2}}, \quad \|F_k\|_\infty = \frac{1}{1 - |a|}, \tag{6.107}$$

which gives us the values of L required to perform sum scaling, $\mathcal{L}_2$ scaling, or $\mathcal{L}_\infty$ scaling, respectively. In this case the sum scaling and $\mathcal{L}_\infty$ scaling give the same value for L.

The statistics of the input signal also enters into the picture while computing the output SNR. For a moment, assume that we have chosen to perform sum scaling. If the input signal is white and WSS with uniform distribution in the range $(-1, 1)$, then $\sigma_u^2 = \frac{1}{3}$, and the output signal variance is

$$\sigma_y^2 = \frac{1}{3L^2} \times \frac{1}{1 - a^2}. \tag{6.108}$$

Hence the output SNR is

$$\text{SNR} = \frac{1}{3L^2 \sigma_e^2} = \frac{(1 - |a|)^2}{3\sigma_e^2}, \tag{6.109}$$

which shows that the SNR is small (i.e., the structure is *noisy*) when the pole a is close to the unit circle.

If the input distribution is Gaussian rather than uniform, the SNR is worse. Assume that the fixed-point number 1 and -1 correspond to $3\sigma_u$ and $-3\sigma_u$ so that for about 99.72% of the time the input signal faithfully "fits" into the b-bit register. With this, we have $\sigma_u = \frac{1}{3}$ and the output signal-to-noise ratio is three times smaller than given by Eq. (6.109).

Let us now assume that the input is a sinusoid of known frequency. This assumption is nearly true when we have a narrowband signal to be filtered from background noise of low variance. The scaling rule for this situation is completely different from anything we have stated so far in this chapter: we wish to choose L such that the output signal (which is sinusoidal as well) has an amplitude less than unity. With L so chosen, the output signal "power" is almost equal to $\frac{1}{2}$, whence the SNR is

$$\text{SNR} = \frac{S}{\sigma_f^2} = \frac{1 - a^2}{2\sigma_e^2}. \tag{6.110}$$

In order to get a feeling for these formulas, let $a = 0.99$, then $(1 - |a|)^2 = 0.0001$, whereas $(1 - a^2) = 0.0199$, which is about 200 times larger than 0.0001. For a^2 close to *unity*, the SNR in Eq. (6.110) is thus much better than in Eq. (6.109). Basically, in deriving Eq. (6.110), we had a specific knowledge about the input, which we *exploited* when choosing the scaling policy.

Example 6.18. In order to see the effect of different scaling policies let us calculate the signal-to-noise ratios in decibels for the first-order filter of Eq. (6.41). We select $a = 0.99$ and $a = 0.90$. The signals are represented with 16-bit accuracy

**TABLE 6-3 SNR (in dB) for
a First-Order Filter**

	SUM	L_2
$a = 0.99$	56.33	72.39
$a = 0.90$	76.33	89.12

($b = 15$). Let the input signal to the filter be again white and WSS with uniform distribution in the range $(-1, 1)$. The input signal power is thus $\sigma_u^2 = \frac{1}{3}$. The scaling strategies used are sum scaling and $\mathcal{L}_2$ scaling with scaling coefficients obtained from Eq. (6.107). The results are shown in Table 6-3. Note that the SNR can be improved by about 6 dB for every additional bit of internal accuracy. Thus if a digital filter with internal wordlengths equal to 16 bits gives an output SNR of 40 dB, then the same filter with 20 bits per internal signal would have output SNR $= 64$ dB. In this example the use of $\mathcal{L}_2$ scaling instead of sum scaling saves almost 4 bits in the internal accuracy if $a = 0.99$.

Next, consider second-order transfer functions of Eq. (6.83). Figure 6-40 shows the direct form structure with a quantizer. The noise model along with the scaling multiplier L is shown in Figure 6-45. The noise gain is given by Eq. (6.85). In order to perform sum scaling the quantity L must be taken as $L = \sum_{n=0}^{\infty} |h[n]|$, where $h[n]$ is as in Eq. (6.84). Even though an exact closed form expression for

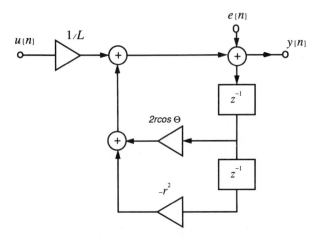

FIGURE 6-45 The second-order direct form.

this sum is not available, it can be shown [UN75] that L is bounded as

$$\frac{1}{(1 - r)^2(1 - 2r \cos 2\theta + r^2)} \le L^2 \le \frac{16}{\pi^2(1 - r^2)^2 \sin^2\theta}. \qquad (6.111)$$

As an example, if the input signal $u[n]$ is a white WSS process uniformly distributed in $(-1, 1)$, then its variance is $\sigma_u^2 = \frac{1}{3}$ and the output signal variance σ_y^2 is $\frac{1}{3}L^2$ times the quantity given by Eq. (6.85). With a roundoff quantizer, the output noise variance σ_f^2 is the product of the quantizer noise variance with the noise gain of Eq. (6.85). Accordingly, the SNR ratio is $\frac{1}{3}L^2 \sigma_e^2$. The quantity L depends on the scaling policy. For complete freedom from overflow L_2 is in the range of Eq. (6.111). As a result

$$\frac{2^{-2b}}{4(1 - r)^2(1 - 2r \cos 2\theta + r^2)} \le \frac{N}{S} \le \frac{4}{\pi^2} \frac{2^{-2b}}{(1 - r^2)^2 \sin^2\theta}. \qquad (6.112)$$

Let us introduce the notation $\delta = 1 - r$, which represents the radial distance of the pole from the unit circle. It is helpful to simplify the expression in Eq. (6.112) for the special case when δ and θ are very small. These conditions are met when implementing narrowband lowpass filters. In this case, Eq. (6.112) simplifies to

$$\frac{2^{-2b}}{4\delta^2(\delta^2 + 4\theta^2)} \le \frac{N}{S} \le \frac{1}{\pi^2} \frac{2^{-2b}}{\delta^2\theta^2}. \qquad (6.113)$$

The main conclusion from here is that, if δ is very small (i.e., $r \to 1$), and/or if θ is very small, then the SNR ratio is very small. In other words, for poles close to the unit circle, and/or close to the real-axis, direct form structures display poor SNR.

6-11 ROUNDOFF NOISE ANALYSIS OF FLOATING-POINT FILTERS

The roundoff error analysis of floating-point filters is complicated by the fact that in the roundoff error model the additive error term depends on the signal level as indicated by Eq. (6.50). Tools for analyzing floating-point filter structures have been developed in Sandberg [SA67], Liu and Kanek [LI69], and Kan and Aggarwal [KA71]. However due to the complicated nature of the problem, relatively little is known about the floating-point roundoff noise properties of common filter structures.

Here our aim is to show how an approximate analysis of floating-point roundoff errors can be performed. Even though the method is approximate, taking into account only the additive error terms but not their effects on each other, it produces reasonably good results in practice.

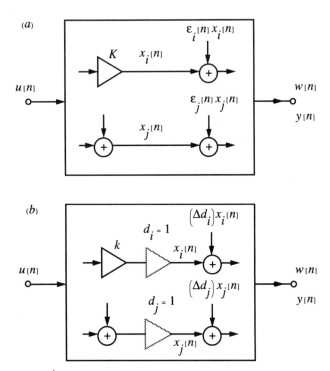

FIGURE 6-46 (a) Floating-point roundoff noise models. (b) Corresponding models utilizing the concept of pseudosensitivity.

Floating-point quantization is characterized by Eq. (6.50) and is shown in Figure 6-46(a) explicitly for one adder and multiplier in a digital filter. Here $w[n]$ and $y[n]$ denote, respectively, the infinite precision output and the output with the rounding operations performed. Thus the roundoff error $e[n]$ at the output is

$$e[n] = y[n] - w[n]. \tag{6.114}$$

We assume that the relative roundoff errors $\epsilon_i[n]$ and $\epsilon_j[n]$ are random variables with variances given by Eq. (6.54). We further assume that the roundoff errors are independent of each other and of the associated signals. It is obvious that errors due to multiple roundings are comparatively small so that little error is made by neglecting them.

We assume that the input is a WSS stochastic signal. Let us denote by $\Phi_u(z)$ the power spectral density (PSD) of the input signal $u[n]$. Since the relative roundoff errors are assumed to be white, the additive roundoff noises are also white. The variance of the ith source $\epsilon_i[n]x_i[n]$ is

$$\sigma_i^2 = \sigma_e^2 E\{x_i^2\} = \sigma_e^2 \frac{1}{2\pi j} \oint |H_{u,x_i}(z)|^2 \Phi_u(z) \frac{dz}{z}. \tag{6.115}$$

Here $H_{u, x_i}(z)$ is the transfer functions from the input $u[n]$ to the variable $x_i[n]$. The variance of the rounding error σ_ϵ^2 is given by Eq. (6.54). If the input signal is white noise with variance σ_u^2 Eq. (6.115) reduces to

$$\sigma_i^2 = \sigma_\epsilon^2 \sigma_u^2 \frac{1}{2\pi j} \oint |H_{u, x_i}(z)|^2 \frac{dz}{z}. \tag{6.116}$$

The noise power contribution $\sigma_{e_i}^2$ of the ith noise source to the output roundoff noise is

$$\sigma_{e_i}^2 = \sigma_i^2 \frac{1}{2\pi j} \oint |H_{x_i, w}(z)|^2 \frac{dz}{z}. \tag{6.117}$$

where $H_{x_i, w}(z)$ denotes the transfer function form $x_i[n]$ to the output $w[n]$.

There exists a loose connection between the coefficient sensitivity and roundoff noise analysis of floating-point filters [ZE91]. In Figure 6-46(b) we have added dummy multipliers d_i and d_j after the real multiplier and adder, respectively. These dummy multipliers have nominal values of one and thus do not change the transfer function. However, it is obvious that if we let the dummy multipliers change their values in an appropriate manner we can produce exactly the same error sequence at each rounding location as the roundoff error model of Figure 6-46(a) produces. In Figure 6-46(b) the dummy multipliers have been perturbed by Δ_i and Δ_j. It is easy to see that the internal transfer functions needed in the coefficient sensitivity analysis (see Section 6-8) of the dummy multipliers are the same as used in floating-point roundoff error analysis.

The total roundoff noise variance at the output is obtained by adding up the contributions from all rounding operations:

$$\sigma_e^2 = \sum_i \sigma_{e_i}^2. \tag{6.118}$$

When calculating Eq. (6.118) one can add up separately the noise due to additions and due to multiplications in order to get an understanding of the individual noise contributions. Note that standard flow-graph or matrix methods can be used to calculate the pertinent transfer functions in Eqs. (6.115), (6.116), and (6.117). The output signal power $E\{w^2[n]\}$ is

$$E\{w^2[n]\} = \frac{1}{2\pi j} \oint |H(z)|^2 \Phi_u(z) \frac{dz}{z}. \tag{6.119}$$

Using Eqs. (6.118) and (6.119), one can now calculate the output SNR.

It is worth noting that the SNR is independent of the input signal level assuming that there are no exponent underflows. This can be seen from Eqs. (6.117) and (6.119). By multiplying the input signal $u[n]$ by a constant K, both expressions become K^2 times larger. This property is one of the reasons why floating-point format has gained popularity in recently proposed audio standards. In floating-point format one can also make a clear distinction between the SNR and dynamic range. The former is determined by the mantissa accuracy and the latter by the exponent range of the representation.

Example 6.19. Let us study the floating-point roundoff noise characteristics of the filter used in Example 6.16. We use $5 + 1$ bits to represent the mantissa and assume that the exponent range is large enough to prevent exponent underflow. In Example 6.16 we used $5 + 1$ bits in the fixed-point representation. We further assume that quantization is performed after the additions in order to make this example comparable with Example 6.16. By using Eq. (6.117), the roundoff noise variances at the output of the filter can be calculated as

$$\sigma_{e_i}^2 = \sigma_u^2 \frac{2^{-2b}}{6} \frac{q^2}{(1 - a^2)^2} \tag{6.120}$$

and

$$\sigma_{e_2}^2 = \sigma_u^2 \frac{2^{-2b}}{6} \|H\|_2^2 = \sigma_u^2 \frac{2^{-2b}}{6} \left((c + q)^2 + \frac{q^2 a^2}{1 - a^2} \right). \tag{6.121}$$

The SNR at the output is obtained using Eqs. (6.118) and (6.119):

$$\text{SNR} = 10 \log_{10} \left(\frac{\sigma_y^2}{\sigma_n^2} \right) = 10 \log_{10} \left(\frac{6 \cdot 2^{2b} \|H\|_2^2}{q^2/(1 - a^2)^2 + \|H\|_2^2} \right) \text{dB}. \tag{6.122}$$

Substituting the numerical values for the parameters, we obtain SNR = 35.33 dB. Note that this result is independent of the input signal level. It is also easy to see that scaling does not affect the output SNR. In Example 6.16 the fixed-point case with optimal scaling and $\sigma_u^2 = 1$ gave 39.89 dB. However, it is easy to see that for input signals with $\sigma_u^2 > 2.87$ the floating-point implementation will give a better SNR.

Example 6.20. Let us study the effect of different section orderings in the cascade form and different orderings of additions in the parallel form. The example filter is an eighth order elliptic filter with $\omega = 0.085\pi$ and $\omega_s = 0.11\pi$. The passband and stopband ripples are 0.1 dB and 72.5 dB, respectively. Totally, there are 24 different orderings of the second-order sections for both the cascade and parallel forms. With WSS white noise as the input signal and 24 bits in the mantissa of the floating-point number representation, the best and worst orderings were searched using computer simulations. For the cascade realization the best SNR that can be achieved is 62.72 dB and the worst is 61.69 dB. Of the parallel structures, the best ordering gives a SNR of 63.33 dB and the worst gave 63.17 dB. In both structures the ordering has only a small effect on the performance. For comparison purposes, direct form 1 gives 43.92 dB and direct form 2 results in 44.10 dB. Clearly, the direct forms are much worse. It is worth noting that the two direct form structures give approximately the same SNR.

Despite the simplicity of the approach used, the results presented in this section give reasonable guidelines for implementation. In Rao [RA92] a systematic approach to floating-point roundoff noise analysis is given. The method is detailed enough to allow evaluation of different orderings of additions in IIR and FIR filters.

6-12 CONCLUDING REMARKS

In this chapter the aim was to bridge the gap between digital filter structures and their hardware or software implementations using finite precision arithmetic. The properties of different arithmetic schemes form the basis that allow one to derive models describing the effect of coefficient quantization and roundoff errors.

A significant part of this chapter is devoted to the signal-to-noise ratio and coefficient sensitivity analysis methods of FIR and IIR filters. Scaling of the internal variables of fixed-point filters is shown to have a significant effect on the SNR. In addition to the traditionally used fixed-point representations, special emphasis is given to floating-point representations. In addition to the large dynamic range, floating-point arithmetic simplifies the design phase by eliminating the need for scaling and, for example, the need to study the effect of section ordering in cascade form implementations. However, in low-cost or very-high-speed applications, the simplicity of fixed-point hardware is likely to overrule the benefits of floating-point arithmetic.

With the cost of computation decreasing and the applications range of digital signal processing widening, the use of parallel processing schemes is becoming common practice. Two approaches to arrive at parallel implementations were described: the graph theoretic, giving rise to a MIMD structure, and the block processing method, resulting in highly parallel matrix formulations.

The time it takes to design and implement a digital filter is an issue of increasing importance and calls for efficient design and simulation software tools. In many applications the properties of the filter have to be changed in real-time. To facilitate this, structural frequency transformations were introduced, allowing easy and continuous tuning of the corner frequencies over a certain range of frequencies.

REFERENCES

[AH79] S. S. Ahuja and S. C. Dutta Roy, Linear phase variable digital bandpass filters. *Proc. IEEE* **67,** 173–174 (1979).

[AN80] P. Ananthakrishna and S. K. Mitra, Block-state recursive digital filters with minimum round-off noise. *Proc. IEEE Int. Conf. Acoust. Speech, Signal Process., Denver, Colorado, 1980*, pp. 81–84 (1980).

[BA78] T. P. Barnwell, S. Gaglio, and R. M. Price, A multi-microprocessor architecture for digital signal processing. *Proc. Int. Conf. Parallel Process.*, *Shanty Creek, Michigan, 1978*, pp. 115-121 (1978).

[BA80] C. W. Barnes and S. Shinnaka, Finite word effects in block-state realizations of fixed-point digital filters. *IEEE Trans. Circuits Syst.* **CAS-27**, 345-350 (1980).

[BA85] C. W. Barnes, B. N. Tran, and S. H. Leung, On the statistics of fixed point round-off error. *IEEE Trans. Acoust. Speech, Signal Process.* **ASSP-33**, 595-606 (1985).

[BI90] J. C. Bier, E. E. Goei, W. H. Ho, P. D. Lapsley, M. P. O'Reilly, G. C. Sih, and E. A. Lee, Gabriel: A design environment for DSP. *IEEE Micro.* **10**, 28-45 (1990).

[BL88] W.-E. Blanz, D. Petković, and J. L. Sanz, Algorithms and architectures for machine vision. *Signal Processing Handbook* (C. H. Chen, ed.), Chapter 10. Dekker, New York, 1988.

[BR78] J. P. Brafman, J. Szczupak, and S. K. Mitra, An approach to the implementation of digital filters using microprocessors. *IEEE Trans. Acoust. Speech, Signal Process.* **ASSP-26**, 442-446 (1978).

[BU71] C. S. Burrus, Block implementation of digital filters. *IEEE Trans. Circuit Theory.* **CT-18**, 697-701 (1971).

[BU72] C. S. Burrus, Block realization of digital filters. *IEEE Trans. Audio Electroacoust.* **AU-20**, 230-235 (1972).

[CH73] D. S. K. Chan and L. R. Rabiner, Analysis of quantization errors in the direct form for finite impulse response filters. *IEEE Trans. Audio Electroacoust.* **AU-21**, 354-366 (1973).

[CI84] Cincinnati Symphony Orchestra, *1812 Overture, Excerpt: Tchaikovsky*, Telarc Digital Compact Discs, Sampler Vol. 2, CD-80102. Telarc Records, Cleveland, OH, 1984.

[CR72] R. E. Crochiere, Digital ladder structures and coefficient sensitivity. *IEEE Trans. Audio Electroacoust.* **AU-20**, 240-246 (1972).

[CR75a] R. E. Crochiere, A new statistical approach to the coefficient word length problem for digital filters. *IEEE Trans. Circuits Syst.* **CAS-22**, 190-196 (1975).

[CR75b] R. E. Crochiere and A. V. Oppenheim, Analysis of linear digital networks. *Proc. IEEE* **63**, 581-595 (1975).

[CR76] R. E. Crochiere and L. R. Rabiner, On the properties of frequency transformations for variable cutoff linear phase digital filters. *IEEE Trans. Circuits Syst.* **CAS-23**, 684-686 (1976).

[DI88] D. F. Dinn, Analog signal acquisitions, conditioning, and conversion to digital format. In *Signal Processing Handbook* (C. H. Chen, ed.), Chapter 2. Dekker, New York, 1988.

[EB69] P. M. Ebert, J. E. Mazo, and M. C. Taylor, Overflow oscillations in digital filters. *Bell Syst. Tech. J.* **48**, 2999-3020 (1969).

[ER85] K. T. Erickson and A. N. Michel, Stability analysis of fixed-point digital filters using generated Lyapunov functions. Part I and II. *IEEE Trans. Circuits Syst.* **CAS-32**, 113-142 (1985).

[FE74] A. Fettweis, On properties of floating point roundoff noise. *IEEE Trans. Acoust. Speech, Signal Process.* **ASSP-22**, 149-151 (1974).

[FE76] A. Fettweis, Realizability of digital filter networks. *Arch. Elektr. Uebertragungstech.* **30**, 90-96 (1976).

[FL63] I. Flores, *The Logic of Computer Arithmetic*. Prentice-Hall, Englewood Cliffs, NJ, 1963.

[FL66] M. Flynn, Very high computing systems. *Proc. IEEE* **54**, 1901–1909 (1966).

[GO68] B. Gold and K. L. Jordan, A note on digital filter synthesis. *Proc. IEEE* **56**, 1717–1718 (1968).

[GR90] R. M. Gray, *Source Coding Theory*. Kluwer Academic Publishers, Boston, MA, 1990.

[HA84] S. N. Hazra, Linear phase bandpass digital filters with variable cutoff frequencies. *IEEE Trans. Circuits Syst.* **CAS-31**, 661–663 (1984).

[IN85] Institute of Electrical and Electronics Engineers, *IEEE Standard for Binary Floating-Point Arithmetic*, ANSI/IEEE Standard 754-1985, IEEE, New York, 1985.

[IN87] INMOS Limited, *IMS A100 Cascadable Signal Processor*, IMS A100 Preliminary Data Sheet. INMOS, Ltd., Bristol, UK, 1987.

[JA70] L. B. Jackson, On the interaction of roundoff noise and dynamic range in digital filters. *Bell Syst. Tech. J.* **49**, 159–184 (1970).

[JA76] L. B. Jackson, Roundoff noise bounds derived from coefficient sensitivities for digital filters. *IEEE Trans. Circuits Syst.* **CAS-23**, 481–485 (1976).

[JA86] L. B. Jackson, *Digital Filters and Signal Processing*. Kluwer Academic Publishers, Boston, MA, 1986.

[JA88] P. Jarske, Y. Neuvo, and S. K. Mitra, A simple approach to the design of linear phase FIR digital filters with variable characteristics. *Signal Process.* **14**, 313–326 (1988).

[JE77] W. K. Jenkins and B. J. Leon, The use of residue number systems in the design of FIR digital filters. *IEEE Trans. Circuits Syst.* **24**, 191–201 (1977).

[KA71] E. F. P. Kan and J. K. Aggarwal, Error analysis of digital filters employing floating-point arithmetic. *IEEE Trans. Circuit Theory* **CT-18**, 678–686 (1971).

[KR90] M. F. Krause and H. Petersen, How can the headroom of digital recordings be used optimally? *J. Audio Eng. Soc.* **38**, 857–863 (1990).

[KW79] K. Kwang, *Computer Arithmetic*. Wiley, New York, 1979.

[LE90] E. A. Lee, Programmable DSP: A brief overview. *IEEE Micro.* **10**, 14–16 (1990).

[LI69] B. Liu and T. Kaneko, Error analysis of digital filters realized with floating-point arithmetic. *Proc. IEEE* **57**, 1735–1747 (1969).

[LI71] B. Liu, Effects of finite wordlength on the accuracy of digital filters—A review. *IEEE Trans. Circuit Theory* **CT-18**, 670–677 (1971).

[MA88] B. C. Mather, Software review. Digital filter design package (DFDP2). Version 2.12. *IEEE Spectrum* **25**, 16 (1988).

[MA90] B. C. Mather, Needed DSP software emerges. *IEEE Spectrum* **27**, 52 (1990).

[MI74] S. K. Mitra and K. Hirano, Digital allpass network. *IEEE Trans. Circuits Syst.* **CAS-21**, 688–700 (1974).

[MI78] S. K. Mitra and R. Gnanasekaran, Block implementation of recursive digital filters—New structures and properties. *IEEE Trans. Circuits Syst.* **CAS-25**, 200–207 (1978).

[MI90] S. K. Mitra, Y. Neuvo, and H. Roivainen, Design and implementation of recursive digital filters with variable characteristics. *Int. J. Circuit Theory Appl.* **18**, 107–119 (1990).

[MO76] A. L. Moyer, An efficient parallel algorithm for digital IIR filters. *Proc. IEEE Int. Conf. Acoust. Speech, Signal Process.*, *Philadelphia, 1976*, pp. 525–528 (1976).

[NO74] R. Nouta, The Jaumann structure in wave digital filters. *Int. J. Circuit Theory Appl.* **2**, 163–174 (1974).

[NO80] R. Nouta and O. Simula, On multiprocessor implementations of digital signal processing algorithms. *Proc. IEEE Int. Symp. Circuits Syst. Houston, Texas, 1980*, pp. 1133–1137 (1980).

[OP70] A. V. Oppenheim, Realization of digital filters using block-floating-point-arithmetic. *IEEE Trans. Audio Electroacoust.* **AU-18**, 130–136 (1970).

[OP72] A. V. Oppenheim and C. J. Weinstein, Effects of finite register length in digital filtering and the fast Fourier transform. *Proc. IEEE* **60**, 8, 957–976 (1972).

[OP75] A. V. Oppenheim and R. W. Schafer, *Digital Signal Processing.* Prentice-Hall, Englewood Cliffs, NJ, 1975.

[OP76] A. V. Oppenheim, W. F. G. Mecklenbräuker, and R. M. Mersereau, Variable cutoff linear phase digital filters. *IEEE Trans. Circuits Syst.* **CAS-23**, 199–203 (1976).

[PA71] S. R. Parker and S. F. Hess, Limit-cycle oscillations in digital filters. *IEEE Trans. Circuit Theory* **CT-8**, 687–697 (1971).

[RA75] L. R. Rabiner and B. Gold, *Theory and Application of Digital Signal Processing.* Prentice-Hall, Englewood Cliffs, NJ, 1975.

[RA92] B. D. Rao, Floating point arithmetic and digital filters. *IEEE Trans. Signal Process.* **40**, 85–95 (1992).

[RE81] M. Renfors and Y. Neuvo, The maximum sampling rate of digital filters under hardware speed constraints. *IEEE Trans. Circuits Syst.* **CAS-28**, 196–202 (1981).

[RE88a] P. A. Regalia, S. K. Mitra, and P. P. Vaidyanathan, The digital all-pass filter: A versatile signal processing building block. *Proc. IEEE* **76**, 19–36 (1988).

[RE88b] M. Renfors and E. Zigouris, Signal processor implementation of digital all-pass filters. *IEEE Trans. Acoust. Speech, Signal Process.* **ASSP-36**, 714–729 (1988).

[SA67] I. W. Sandberg, Floating-point-roundoff accumulation in digital filter realizations. *Bell Syst. Tech. J.* **46**, 1775–1791 (1967).

[SR77] A. S. Sripad and D. L. Snyder, A necessary and sufficient condition for quantization errors to be uniform and white. *IEEE Trans. Acoust. Speech, Signal Process.* **25**, 442–448 (1977).

[SZ75] J. Szczupak and S. K. Mitra, Digital filter realization using successive multiplier-extraction approach. *IEEE Trans. Acoust. Speech,, Signal Process.* **ASSP-23**, 235–239 (1975).

[SZ88] J. Szczupak, S. K. Mitra, and J. Fadavi-Ardekani, A computer-based synthesis method of structurally LBR digital allpass networks. *IEEE Trans. Circuits Syst.* **CAS-35**, 755–760 (1988).

[UN75] Z. Unver and K. Abdullah, A tighter practical bound on quantization errors in second-order digital filters with complex conjugate poles. *IEEE Trans. Circuits Syst.* **CAS-22**, 632–633 (1975).

[VA86] P. P. Vaidyanathan, S. K. Mitra, and Y. Neuvo, A new approach to the realization of low sensitivity IIR digital filters. *IEEE Trans. Acoust. Speech, Signal Process.* **ASSP-34**, 350–361 (1986).

[VE90] S. Vega, Big space. *Days of Open Hand*, 395293-2 A&M Records. Inc., Los Angeles, 1990.

[YE86] Yello, Daily disco. *Yello 1980–1985*, 826 773-2Q. The New Mix in One Go, Phonogram Gmbh, Hamburg, Germany, 1986.

[ZE81] J. Zeman and A. G. Lindgren, Fast digital filters with low round-off noise. *IEEE Trans. Circuits Syst.* **CAS-28,** 716–723 (1981).

[ZE91] B. Zeng and Y. Neuvo, Analysis of floating point roundoff errors using dummy multiplier coefficient sensitivities. *IEEE Trans. Circuits Syst.* **38,** 590–601 (1991).

7 Robust Digital Filter Structures

P. P. VAIDYANATHAN

Department of Electrical Engineering
California Institute of Technology, Pasadena

7-1 INTRODUCTION

Digital filters, and more generally digital signal processing algorithms, can be implemented in several ways [KA66]. The implementation could be a simulation on a general purpose computer, or it could be a program running on a commercial digital signal processing (DSP) chip, or it could be a dedicated piece of hardware, for example, a VLSI chip. In any case, the resources such as computational units, time, memory, and chip area are finite. One consequence of this fact is that the external and internal signals involved and the filter coefficients are represented by binary words of finite length. As explained and demonstrated in Section 6-3-2, this causes three kinds of errors in the output of the system [LI71; OP72, OP75; RA75; VA87f]. First, there is coefficient quantization, the effect of which is to cause errors in the transfer function being realized. This is a deterministic type of error, and its effect can be evaluated ahead of time. The second type of error is due to quantization of signals, particularly the internal state variables (and often nonstate variables). One component of this is a random type of error (called roundoff noise) and should accordingly be characterized by stochastic models (Section 6-10). The other component is a highly correlated type of error, occurring in the form of periodic oscillations called limit cycles. These oscillations can in turn be classified as either *granular type* or *overflow type*; the former are usually of small amplitude and are significant only when the signal level is low, whereas overflow oscillations are very large disturbances and make the filter output entirely meaningless. These effects were demonstrated in Section 6-7-3.

Recall (Section 6-4-2) that there exist an infinite number of structures to realize a digital filter. The multipliers that appear in the structure are said to be the coefficients of the structure. For example, Figure 7-1 is the direct form structure (Sec-

Handbook for Digital Signal Processing, Edited by Sanjit K. Mitra and James F. Kaiser.
ISBN 0-471-61995-7 © 1993 John Wiley & Sons, Inc.

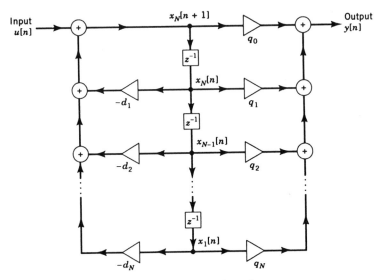

FIGURE 7-1   The Nth order direct form structure.

tion 3-5-1) of an IIR transfer function of order N given by

$$H(z) = \frac{\sum_{i=0}^{N} q_i z^{-i}}{1 + \sum_{i=1}^{N} d_i z^{-i}}. \tag{7.1}$$

The poles of the transfer function are typically crowded around the bandedge [AN79; OP75; RA75], which causes the pole locations to be sensitive functions of the multipliers d_i. (See page 336 of [OP89].) Accordingly, the actual frequency response $H(e^{j\omega})$ of the direct form structure is very sensitive to digitization of the coefficients, particularly for large N and sharp-cutoff filters. These structures are therefore not used in practice except when $N \leq 2$. As we shall see in this chapter, there are other structures that are less sensitive (i.e., more *robust* to coefficient quantization) and are generally called *low-sensitivity structures*.

The direct form structure is also known to have large roundoff noise, particularly for poles close to the unit circle (and close to the real axis), as seen in Section 6-10-1.

The importance of finite wordlength effects should not be overlooked. For example, with a commercial 16-bit fixed-point DSP chip, roundoff noise effects can be quite significant for sharp-cutoff IIR filters as demonstrated in this chapter. Moreover, in any fixed-point implementation, it is possible to have overflow oscillations unless the structure and the arithmetic are carefully chosen.

In this chapter we study structures having low sensitivity and low roundoff noise. We also present structures that are free from self-sustained limit cycle oscillations.

It is important to note in this context that, in implementations using VLSI techniques, several other considerations should be taken into account, in addition to finite wordlength effects. These considerations include pipelineability, modularity, and localizability of data transfers (Section 6-3-3). Some of the low-sensitivity/low-noise structures we include here satisfy these additional properties as well.

The error feedback system is introduced in Section 7-2. Noise and dynamic range considerations in cascade form structures are discussed in Section 7-3. Section 7-4 considers low-noise structures based on a state-space approach. This approach allows us to formulate the noise minimization problem in an elegant manner. Section 7-5 provides a quantitative analysis of the coefficient sensitivity properties for second-order sections. In Section 7-6 we introduce the concept of structural passivity and thereby develop a simple structure based on two allpass functions, which enjoys low passband sensitivity. This structure also has many other robust properties as elaborated in later sections. This section also discusses orthogonal digital filters. In Section 7-7 we introduce wave digital filters. Section 7-8 discusses low-sensitivity implementation of FIR filters and filter banks in the form of (''lossless'') lattice structures. Section 7-9 includes a discussion of roundoff noise and dynamic range properties in structurally lossless systems. It is proved, in particular, that the allpass lattice structure not only enjoys the minimum noise property but is also scaled in the $\mathcal{L}_2$ sense automatically. Section 7-10 is a study of limit cycles in digital filters. We develop conditions for suppression of these and show that many structurally lossless systems including the lattice structures are free from limit cycles if the quantizers are chosen to be of magnitude truncation type.

7-2 ROUNDOFF NOISE REDUCTION USING ERROR FEEDBACK

In Section 6-10-1 it was shown that direct form structures have poor SNRs for poles close to the unit circle and/or close to the real axis. For first- and second-order sections, there is a very simple way to overcome this problem, via a technique called *error feedback*, abbreviated as EFB [TH77; CH81; HI84; VA85a]. Figure 7-2 describes the idea for the first-order direct form section. The error se-

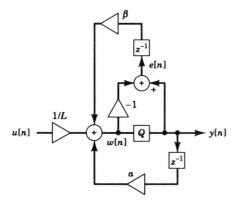

FIGURE 7-2 Feeding back the quantizer error.

quence $e[n]$ produced by the quantizer is circulated through a feedback path, instead of being discarded. The multiplier β is typically an integer, so that there is no additional roundoff error generated by the EFB path. If the structure is analyzed we find that the transfer function (in the absence of quantization) is still given by $1/L(1 - az^{-1})$, but in the presence of the quantizer, we have

$$w[n] = aw[n - 1] + \frac{u[n]}{L} + (a + \beta)e[n - 1], \tag{7.2}$$

$$y[n] = ay[n - 1] + \frac{u[n]}{L} + e[n] + \beta e[n - 1], \tag{7.3}$$

which tells us that the noise transfer function is given by

$$G_1(z) = \frac{(a + \beta)z^{-1}}{1 - az^{-1}} \quad \text{(if } w[n] \text{ is the output)}, \tag{7.4}$$

$$G_1(z) = \frac{1 + \beta z^{-1}}{1 - az^{-1}} \quad \text{(if } y[n] \text{ is the output)}, \tag{7.5}$$

which reduces to Eq. (6.77) for $\beta = 0$. This then is a simple way to change the noise transfer function, without changing either the scaling transfer function or the input/output transfer function. In practice, since $y[n] = w[n] + e[n]$, it really does not matter which one of the signals $y[n]$ or $w[n]$ is defined as the output. However, the above expressions for noise gain help us to choose β judiciously. If we pick $\beta = -a$, Eq. (7.4) tells us that the output noise variance is reduced to zero, but this is a deceptive conclusion for the following reason: since $|a| < 1$ (for stability reasons), β is a fraction whenever $\beta = -a$. Such a choice of β would generate additional roundoff noise (in the EFB path), which we ignored. In fact, it has been shown [MU82] that the choice $\beta = -a$ corresponds to a structure without EFB, but whose quantizer would permit $2b$ bits at its output (i.e., this corresponds to a double precision implementation). Accordingly, with $\beta = -a$ the quantization error is at the 2^{-2b} level rather than at the 2^{-b} level.

As a more meaningful choice, we can take β to be the integer closest to $-a$ (i.e., $\beta = 0$, 1, or -1 as the case may be). With this choice, there is no secondary roundoff error generated by the feedback path. For $|a| < 0.5$ the EFB structure is then the same as a structure without EFB. When $|a| \geq 0.5$ we have the output noise variance

$$\sigma_f^2 = \frac{(1 - |a|)^2}{1 - a^2} \sigma_e^2 = \frac{1 - |a|}{1 + |a|} \sigma_e^2, \tag{7.6}$$

assuming that $w[n]$ is the system output. This implies a noise reduction by a factor of $(1 - |a|)^2$, which is substantial for $|a| \to 1$. The extra hardware required for mechanization of EFB includes the two new adders and an additional storage element (to prevent the delay in the EFB path). Basically, the EFB technique is a clever compromise between single and double precision. The hardware complexity is much less than for double precision, but the improvement is almost equivalent to that obtained by use of double precision, when the poles are very close to the unit circle.

The noise variance in the absence of EFB (given by σ_f^2 in Eq. (6.81)) can get arbitrarily large when $a^2 \rightarrow 1$, whereas with EFB, the noise variance at the adder output, given by Eq. (7.6), cannot exceed $\frac{1}{3}$ for any a in the range $|a| \geq 0.5$. And, in fact, as $|a| \rightarrow 1$, Eq. (7.6) gets arbitrarily small! Under this condition, if the signal $y[n]$ is taken as the output, then the difference $e[n]$ between $w[n]$ and $y[n]$ provides a lower limit on the output noise variance.

In practice, the noise variance at the node representing $w(n)$ cannot really get "arbitrarily" small because, if $|a|$ is very close to unity, it requires many more than b bits to represent a in binary. Accordingly, the signal $w[n]$, which is yet unquantized, has to have multiple rather than just double precision. In other words, for a given permissible precision at the quantizer input, $|a|$ cannot get arbitrarily close to unity.

For example, let $b = 8$, and assume that at the quantizer input, the permissible precision is $2b = 16$. Then a should be represented by no more than $b = 8$ bits (without counting the sign-bit). The largest positive number representable in binary format with 8 bits is $a = 0_\Delta 11111111$, which in decimal translates to $a = 0.9960937$. From Eq. (7.6) we then see that the noise gain from the quantizer noise source to the adder output is nearly 0.002. Thus the adder output is virtually noise-free. If $y[n]$ rather than $w[n]$ is taken as the output, then the difference $e[n]$, which has variance $\sigma_e^2 = 2^{-2b}/12$, is the only significant noise. In other words, the effective noise gain has been reduced to *unity*! This can also be understood by looking at the noise transfer function from $e[n]$ to $y[n]$ given in Eq. (7.5), which approximates unity when $\beta = -1$ and $a \rightarrow 1$.

From Eq. (7.5) we see that the noise transfer function with EFB is equal to that without EFB, multiplied by $(1 + \beta z^{-1})$. This shows how the error spectrum is being shaped by the EFB circuit. This is demonstrated in Figure 7-3, for the case $\beta = -1$ (which is the best integer value of β when $a > 0.5$). The unshaped quantizer error is white, with variance σ_e^2. The shaped error has a spectrum $4 \sin^2 (\omega/2) \sigma_e^2$, which is plotted in the figure. This shaped error now passes through the transfer function $1/(1 - az^{-1})$ (which is lowpass since $a > 0.5$) to produce the output noise. As the noise energy has essentially been "pushed" into the stopband of the transfer function $1/(1 - az^{-1})$, the output noise power is reduced. Because

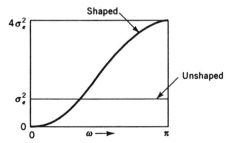

FIGURE 7-3 The new and conventional error spectra.

of this interpretation, EFB techniques have also been called error spectrum shaping (ESS) techniques [HI84].

The EFB idea can easily be extended to second-order sections as shown in Figure 7-4. It can be verified that, if $y[n]$ is taken as the filter output, the noise transfer function (i.e., the transfer function from the quantizer output $e[n]$ to the filter output $y[n]$) is given by

$$G(z) = \frac{1 + \beta_1 z^{-1} + \beta_2 z^{-2}}{1 + d_1 z^{-1} + d_2 z^{-2}}. \tag{7.7}$$

Once again, the strategy would be to choose β_1 and β_2 to be integers (so that there is no additional quantization error in the EFB path), nearest to d_1 and d_2. For example, assume that the poles of the transfer function are at $re^{\pm j\theta}$ so that $d_1 = -2r \cos \theta$ and $d_2 = r^2$. As indicated in Section 6-10-1, the noise gain of the structure without EFB is very large, whenever $r \to 1$ and $\theta \to 0$. Under this condition, we have $d_1 \to -2$ and $d_2 \to 1$; hence we choose $\beta_1 = -2$ and $\beta_2 = 1$. With this choice of parameters, the noise gain is

$$\|G(e^{j\omega})\|_2^2 = 1 + \frac{\theta^4}{4\delta(\theta^2 + \delta^2)}, \tag{7.8}$$

where $\|\cdot\|_2$ denotes the $\mathcal{L}_2$ norm, and δ stands for $1 - r$.

Table 7-1 shows the SNR with and without EFB, for the second-order structure. The input is assumed to be white and uniformly distributed. For small θ and δ, the

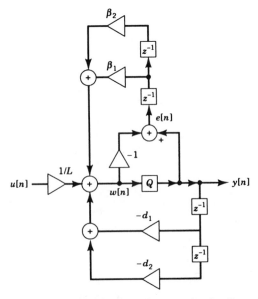

FIGURE 7-4   Error feedback on the second order direct-form.

TABLE 7-1 SNR for Second-Order Section with Poles $re^{\pm j\theta}$ With and Without Error Spectrum Shaping (ESS)[a].

		Example SNR in dB $r = 0.995,\ \theta = 0.07\pi$ $b = 16,\ \beta_1 = -2,\ \beta_2 = 1$	
SNR			
Without ESS	With ESS	Without ESS	With ESS
$16\delta^2\theta^2 2^{2b}$	$16\dfrac{\delta^2}{\theta^2}2^{2b}$	49	75

[a]We assume $\delta,\ \theta \to 0$ and $\delta \ll \theta$, where $\delta = 1 - r$. Input $u(n)$ is white and uniformly distributed.

improvement is self-evident, as supported by the numerical example shown in the table.

7-3 CASCADE FORM DIGITAL FILTER STRUCTURES

First- and second-order IIR digital filter sections are often used as building blocks in cascade form and parallel form implementations of higher order transfer functions. These realizations are preferred to a direct form implementation, as they help to reduce the pole-crowding effect, resulting in reduced coefficient sensitivity. Such implementations also have better *signal to roundoff noise* ratio.

Figure 7-5 shows a cascade form implementation, whose overall transfer function is given by

$$H(z) = q_0 \prod_{k=1}^{M} H_k(z). \tag{7.9}$$

The transfer function of the kth building block is

$$H_k(z) = \frac{1 + q_{1k}z^{-1} + q_{2k}z^{-2}}{1 + d_{1k}z^{-1} + d_{2k}z^{-2}} = \frac{A_k(z)}{B_k(z)}. \tag{7.10}$$

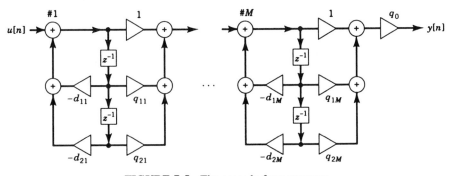

FIGURE 7-5 The cascade form structure.

It is assumed for simplicity that all sections are of second order. If there exist first-order sections, they can be accommodated by setting $q_{2k} = d_{2k} = 0$. Even though Figure 7-5 implicitly assumes that each second-order section is in direct form, this is not the only possibility. As we shall see in Section 7-5, there exist other types of second-order sections such as the coupled form, normal form, modified coupled form, and so on, which have better sensitivity properties.

In any fixed-point implementation, it is necessary to scale the structure of Figure 7-5, in order to reduce the probability of arithmetic overflow (Section 6-10-2). Accordingly, scale factors are inserted as shown in Figure 7-6(a). (For the time being, ignore the quantizers shown in the figure.) Note that the overall transfer function $H(z)$ is unaltered by the scaling operation. In order to find the values of the scale factors $s_1, s_2, \ldots, s_M$, let us refer to the unscaled structure of Figure 7-5 and define the transfer functions

$$F_k(z) = \frac{1}{1 + d_{1k}z^{-1} + d_{2k}z^{-2}} \prod_{l=1}^{k-1} H_l(z), \quad k = 1, 2, \ldots, M. \quad (7.11)$$

These are precisely the transfer functions from the input to the internal nodes that need to be scaled. If we wish Figure 7-6 to represent a structure scaled in the $\mathcal{L}_p$

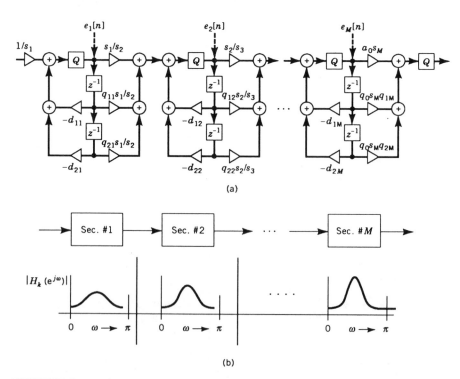

(a)

(b)

FIGURE 7-6 (a) The cascade form with quantizers and scaling multipliers. (b) Suggested ordering of sections for $\mathcal{L}_\infty$ scaling.

sense, we define

$$s_k = \|F_k\|_p. \tag{7.12}$$

Having scaled the structure, let us consider the noise transfer functions from the outputs of the quantizers to the filter output. For the noise source $e_k[n]$ this is

$$G'_k(z) = q_0 s_k \prod_{l=k}^{M} H_l(z) = s_k G_k(z) = \|F_k\|_p G_k(z). \tag{7.13}$$

If we make the usual assumptions about the noise sources (Section 6-10-1), we obtain, for the output noise variance of the scaled structure,

$$\sigma_f^2 = \sigma_e^2 \left(1 + \sum_{k=1}^{M} \|G'_k\|_2^2 \right), \tag{7.14}$$

which can be rewritten as

$$\sigma_f^2 = \sigma_e^2 (1 + \|F_1\|_p^2 \|G_1\|_2^2 + \|F_2\|_p^2 \|G_2\|_2^2 + \cdots). \tag{7.15}$$

The above scaling operation leaves the output signal power unchanged, since $H(z)$ is unchanged. As a result of scaling, there is reduced probability of overflow; this advantage is gained at the cost of increased *output noise variance*, as seen by the appearance of the factors $\|F_k\|_p^2$ in the above equation.

Section Ordering. The scale factors and the SNR of the cascade form structure depend on the ordering of the sections and also on pole–zero pairing. If the number of sections is small, an exhaustive search can be performed to find the ordering and pairing that maximize the SNR. If this is not feasible, a more efficient approach, proposed by Liu and Peled [LI75], can be employed. This is a useful heuristic approach for optimizing the section ordering and pole–zero pairing and leads to satisfactory results.

A second approach for determining the ordering and pole–zero pairing is based on Jackson's concept of "peakedness" [JA70a]. For a given second-order section, the peakedness is defined to be the quantity

$$\rho_k = \frac{\|H_k\|_\infty}{\|H_k\|_2}. \tag{7.16}$$

If a pole is paired with the nearest zero, the peakedness tends to be smallest. This helps to keep the internal signals within the dynamic range without too much scaling effort (which therefore improves the SNR). It has been suggested [JA70b] that when $\mathcal{L}_\infty$ scaling is being employed, the section with the smallest peakedness should be on the left; that is, the sections should be ordered according to increasing ρ_k. Figure 7-6(b) demonstrates this ordering.

7-4 STATE-SPACE APPROACH FOR LOW-NOISE DESIGNS

A powerful tool in the understanding and minimization of finite wordlength effects in digital filters is based on the state-space formalism (Section 3-6). This approach can be used to design filters with minimum noise, filters free from limit cycles, and often low-sensitivity filters.

7-4-1 State-Space Descriptions and State-Space Implementations

Consider an IIR transfer function as in Eq. (7.1). Assuming that there are no common factors between the numerator and denominator polynomials, the quantity N denotes the order of the system. We can implement the filter with only N delays (such implementations are said to be *minimal in delays* or just *minimal*). Let $x_k[n]$ denote the output signal of the kth delay element, and let

$$\mathbf{x}[n] = [x_1[n] \quad x_2[n] \quad \ldots \quad x_N[n]]'.$$

The state-space equations describing the structure, which were given in Eq. (6.23), are reproduced below:

$$\mathbf{x}[n + 1] = \mathbf{A}\mathbf{x}[n] + \mathbf{B}u[n], \tag{6.23a}$$

$$y[n] = \mathbf{C}\mathbf{x}[n] + Du[n]. \tag{6.23b}$$

Here $\mathbf{A} = [A_{kl}]$ is an $N \times N$ matrix, $\mathbf{B} = [B_k]$ is an $N \times 1$ matrix (i.e., a column-vector), $\mathbf{C} = [C_l]$ is a $1 \times N$ matrix (i.e., a row-vector) and D is a scalar quantity. Note that, among all implementations of a transfer function, a minimal implementation has the smallest size for the $\mathbf{A}$ matrix.

As an example, for the direct form structure of Figure 7-1, if the state variables are numbered starting from the bottom up, we obtain the above description with

$$\mathbf{A} = \begin{bmatrix} 0 & 1 & 0 & \cdots & 0 \\ 0 & 0 & 1 & \cdots & 0 \\ \vdots & \vdots & \vdots & & \vdots \\ 0 & 0 & 0 & \cdots & 1 \\ -d_N & -d_{N-1} & -d_{N-2} & \cdots & -d_1 \end{bmatrix}, \tag{7.17}$$

$$\mathbf{B} = [0 \quad 0 \quad \cdots \quad 0 \quad 1]',$$

$$\mathbf{C} = [q_N - q_0 d_N \quad q_{N-1} - q_0 d_{N-1} \quad \cdots \quad q_1 - q_0 d_1], \quad D = q_0. \tag{7.18}$$

If a digital filter structure is such that the elements A_{kl}, B_k, C_l, D of the state-space description are also the multiplier coefficients in the structure, then it is called a *state-space implementation*, *state-space realization*, or a *state-space struc-*

ture. Note that the direct form is not a state-space implementation, since the elements C_l are not the multipliers in Figure 7-1.

The transfer function $H(z)$ is related to the state-space parameters by

$$H(z) = D + \mathbf{C}(z\mathbf{I} - \mathbf{A})^{-1}\mathbf{B},$$

whereas the impulse response corresponding to $H(z)$ is given by

$$h[n] = \begin{cases} D & \text{for } n = 0 \\ \mathbf{CA}^{n-1}\mathbf{B} & \text{for } n > 0. \end{cases} \tag{7.19}$$

Since the system is causal, we have $h[n] = 0$, $n < 0$.

We can show [AN79] that the eignevalues of $\mathbf{A}$ are the poles of the transfer function $H(z)$. Accordingly, a minimal system is stable if and only if $\mathbf{A}$ has all eigenvalues λ_k satisfying $|\lambda_k| < 1$.

Given a state-space description $\mathbf{A}$, $\mathbf{B}$, $\mathbf{C}$, D, suppose we replace these matrices with $\mathbf{A}_1$, $\mathbf{B}_1$, $\mathbf{C}_1$, D_1, where

$$\mathbf{A}_1 = \mathbf{P}^{-1}\mathbf{AP}, \qquad \mathbf{B}_1 = \mathbf{P}^{-1}\mathbf{B}, \qquad \mathbf{C}_1 = \mathbf{CP}, \qquad D_1 = D, \tag{7.20}$$

and where $\mathbf{P}$ is any $N \times N$ nonsingular matrix. This leaves the transfer function unchanged. Accordingly, we can derive an equivalent description by using the transformation Eq. (7.20), which is called the *similarity transformation*. For a given transfer function, there exist an infinite number of state-space realizations, since $\mathbf{P}$ is arbitrary. However, some of these realizations have better SNR than others. Accordingly, as we show later, it is possible to find the best state-space structure by minimizing the roundoff noise variance at the output (for a given signal level) under scaled conditions. This is the main advantage of the state-space approach: it offers a convenient mathematical framework to minimize a performance measure [MU76].

It should be noted that a state-space realization (i.e., a realization of $H(z)$ with multiplier coefficients $[A_{kl}]$, $[B_k]$, $[C_l]$, D) requires a total of $(N + 1)^2$ multipliers, which, for large N, is far in excess of the number of multipliers in a direct form implementation. Accordingly, minimum noise state-space realizations are usually restricted to second-order sections, which are then used in a cascade to produce higher order filters.

7-4-2 Propagation of Roundoff Noise

Figure 7-7 shows the signal flow-graph representation of the state equations. Quantizers are introduced in the feedback loop to prevent infinite bit accumulation. In terms of Eq. (6.23a), this means that the right-hand side is first computed with no error, and then the N components of the results are quantized. Even though it is typical to quantize the right-hand side of Eq. (6.23a) as well, this quantization noise does not go through a feedback loop, and its effect on the filter output can

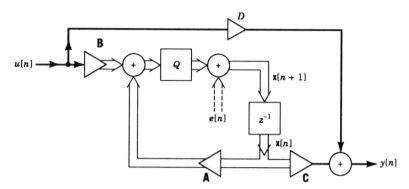

FIGURE 7-7   The signal flow-graph representing the state equations.

be studied separately. Accordingly, for our discussions, the noise model is as in Figure 7-7.

In an implementation with fixed-point arithmetic, some scaling policy should be used to reduce the probability of overflow. It is sufficient to scale the signal $x_k[n]$, which are inputs to **A** and **C**. Thus the scaling transfer functions $F_k(z)$ of interest are the quantities $X_k(z)/U(z)$ (where $X_k(z)$ is the z-transform of $x_k[n]$). These depend only on the quantities **A** and **B**. In this section, we shall consider $\mathcal{L}_2$ scaling policy, because this leads to a mathematically tractable noise minimization problem as we show later. The noise transfer functions $G_k(z)$ are the transfer functions from the state-variable nodes to the filter output $y[n]$. The impulse responses $f_k[n]$ and $g_k[n]$ corresponding to $F_k(z)$ and $G_k(z)$ are given by

$$f_k[n] = [\mathbf{A}^{n-1}\mathbf{B}]_k\mu[n-1], \qquad g_k[n] = [\mathbf{CA}^n]_k\mu[n], \tag{7.21}$$

where the notation $[\cdot]_k$ stands for the kth component. Assuming that the outputs of the quantizers are fixed-point fractions, the noise variance at the filter output under usual assumptions is

$$\sigma_f^2 = \sigma_e^2 \sum_{k=1}^{N} \|G_k\|_2^2 = \sigma_e^2 \sum_{k=1}^{N} \sum_{n=0}^{\infty} |g_k[n]|^2. \tag{7.22}$$

This variance depends on $g_k[n]$, which in turn depends on the realization **A**, **B**, **C**, **D**, which is not unique for a given $H(z)$. Given an arbitrary state-space realization for $H(z)$, the aim here is to find a similarity transformation **P** such that the output noise variance is minimized under scaled conditions. For this, it is first necessary to know how the norms $\|F_k\|_2$ and $\|G_k\|_2$ change when a similarity transformation is applied. This is facilitated [MU76] by defining the matrices

$$\mathbf{K} = \sum_{n=0}^{\infty} \mathbf{A}^n\mathbf{B}(\mathbf{A}^n\mathbf{B})^t, \qquad \mathbf{W} = \sum_{n=0}^{\infty} (\mathbf{CA}^n)^t\mathbf{CA}^n. \tag{7.23}$$

By comparing Eq. (7.21) with Eq. (7.23) it is clear that the diagonal elements K_{kk} and W_{kk} of $\mathbf{K}$ and $\mathbf{W}$ have a very simple interpretation:

$$K_{kk} = \|F_k\|_2^2, \qquad W_{kk} = \|G_k\|_2^2. \tag{7.24}$$

In other words, W_{kk} is the noise gain for the quantizer error associated with the kth state variable $x_k[n]$, and $\sqrt{K_{kk}}$ is the scaling factor associated with the same state variable, for $\mathcal{L}_2$ scaling. When a similarity transformation is applied, the matrices $\mathbf{K}$ and $\mathbf{W}$ are replaced with

$$\mathbf{K}_1 = \mathbf{P}^{-1}\mathbf{K}\mathbf{P}^{-t}, \qquad \mathbf{W}_1 = \mathbf{P}^t\mathbf{W}\mathbf{P}. \tag{7.25}$$

The diagonal entries of $\mathbf{K}_1$ and $\mathbf{W}_1$ therefore tell us about the scaling factors and noise gains of the transformed system.

We say that the structure of Figure 7-7 is scaled in the $\mathcal{L}_2$ sense (see Section 6-10-3) if $\|F_k\|_2 = 1$ (i.e., if $K_{kk} = 1$) for $1 \le k \le N$. Given an unscaled realization $\mathbf{A}$, $\mathbf{B}$, $\mathbf{C}$, and D, it is possible to obtain a scaled realization $\mathbf{A}_1$, $\mathbf{B}_1$, $\mathbf{C}_1$, and D by using Eq. (7.20) provided $\mathbf{P}$ is chosen to be a diagonal matrix such that $P_{kk} = \sqrt{K_{kk}}$. Having done this, the scaled realization has a $\mathbf{W}_1$ matrix given as in Eq. (7.25), and we have $(W_1)_{kk} = K_{kk}W_{kk}$. Accordingly, the noise gain of the scaled structure in terms of the parameters of the unscaled structure is

$$\mathfrak{N} = \sum_{k=1}^{N} K_{kk}W_{kk}. \tag{7.26}$$

The advantage of the above figure of merit is that it helps us to compare scaled structures and unscaled structures on a common ground. Since $H(z)$ is unchanged by the transformation $\mathbf{P}$, the signal power at the filter output is the same for all equivalent state-space realizations. Accordingly, if $\mathfrak{N}$ increases, the SNR decreases and vice versa. For a given transfer function $H(z)$, a state-space realization with smallest $\mathfrak{N}$ is therefore said to have "minimum noise." The main result of interest here is due to Mullis and Roberts [MU76], which gives a set of necessary and sufficient conditions so that a realization has minimum noise.

Minimum Noise Condition.   For a minimal structure with state-space description $\mathbf{A}$, $\mathbf{B}$, $\mathbf{C}$, D and noise model as above, the quantity $\mathfrak{N}$ defined above is minimum [MU76] if and only if $\mathbf{K}$ and $\mathbf{W}$ are such that:

1. $\mathbf{K} = \mathbf{D}_0\mathbf{W}\mathbf{D}_0$ for some diagonal matrix $\mathbf{D}_0$ having positive elements.
2. $K_{kk}W_{kk} = $ constant independent of k.

For a $\mathcal{L}_2$-scaled realization (i.e., when $K_{kk} = 1$ for all k), we can restate the above result as follows: the roundoff noise is minimum if and only if $\mathbf{W} = c\mathbf{K}$, where $c > 0$ is a scalar constant. Under this condition, we clearly have $W_{kk} = c$

for all k; that is, the noise gains from all the noise sources to the output are equalized.

As pointed out earlier, state-space structures require $(N + 1)^2$ multipliers and are therefore not economical for large N [MI81a]. Accordingly, the above results are usually applied for the derivation of second-order sections having minimum noise. Jackson et al. [JA79] have shown that, for a second-order system, if the state-space structure is such that

$$A_{11} = A_{22}, \tag{7.27}$$

$$B_1 C_1 = B_2 C_2, \tag{7.28}$$

then the conditions for minimum noise are satisfied. Accordingly, the above two equations constitute a set of sufficient conditions for a minimum noise structure. In fact, closed form expressions for the matrix elements A_{kl}, B_k, C_l, D of the minimum noise second-order sections have been derived by some authors [BA84; BO85] based on this.

7-5 SECOND-ORDER IIR STRUCTURES WITH LOW SENSITIVITY

Since second-order IIR sections are the building blocks in cascade and parallel form implementations, the coefficient-sensitivity properties of these sections should be understood. Careful choice of the structure of a second-order section (depending on pole locations) often leads to reduced coefficient sensitivity for the pole locations.

The direct form structure of Figure 6-45 has poles at $re^{\pm j\theta}$. The multipliers in the structure are

$$m_1 = 2r \cos \theta, \qquad m_2 = -r^2, \tag{7.29}$$

and we can obtain an estimate of the sensitivities of r and θ with respect to these multipliers, by evaluating their first partial derivatives. These are given by

$$\frac{\partial r}{\partial m_1} = 0, \qquad \frac{\partial r}{\partial m_2} = \frac{-1}{2r}, \qquad \frac{\partial \theta}{\partial m_1} = \frac{-1}{2r \sin \theta}, \qquad \frac{\partial \theta}{\partial m_2} = \frac{-1}{2r^2 \tan \theta}. \tag{7.30a}$$

In practice, it is unlikely for r to get very close to 0, but for narrowband lowpass and highpass filters, the pole angle θ can be very close to 0 or π. This leads to large sensitivity, as seen from the above equation. However, if θ is close to $\pi/2$ (which can happen for narrowband bandpass filters), the direct form structure actually has *low* sensitivity.

This conclusion can also be obtained from the pole-grid pattern of Figure 7-8, which shows the permissible pole locations for a particular level of quantization of the multipliers m_1, m_2 (4 bits in the figure). The poor sensitivity for $\theta \to 0$ arises

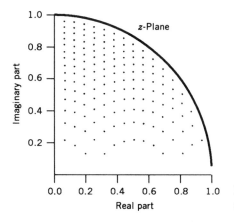

FIGURE 7-8 The pole-grid pattern for second-order direct form.

because of the sparseness of permissible pole locations around $\theta = 0$. It is clear that a structure with better sensitivity properties can be obtained by constraining the multipliers to be $m_1 = r \cos \theta$ and $m_2 = r \sin \theta$, so as to obtain a uniform pole-grid pattern. Such a structure is called the coupled form [GO69; OP75] and is shown in Figure 7-9 along with the pole-grid pattern in Figure 7-10. The sensitivity expressions for this structure are given by

$$\frac{\partial r}{\partial m_1} = \cos \theta, \qquad \frac{\partial r}{\partial m_2} = \sin \theta, \qquad \frac{\partial \theta}{\partial m_1} = \frac{-\sin \theta}{r}, \qquad \frac{\partial \theta}{\partial m_2} = \frac{\cos \theta}{r}. \qquad (7.30b)$$

Clearly, unless r is very small (which is unlikely), these quantities are small for all θ.

Two other types of coupled form structure, called *modified coupled forms*, have been reported in Yan and Mitra [YA82]. These have low sensitivity for certain particular pole locations. All these results are summarized in Vaidyanathan [VA87f, Table X].

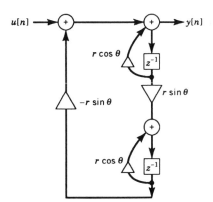

FIGURE 7-9 The coupled form circuit.

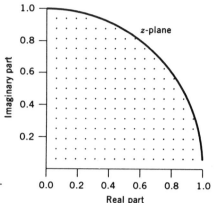

FIGURE 7-10   Pole-grid pattern for the coupled form.

7-6   LOW-SENSITIVITY IIR DESIGNS BASED ON STRUCTURAL PASSIVITY

Perhaps the earliest type of filters known to engineers are *passive filters*, made of analog electrical components such as inductors, capacitors, and resistors [BA69; GU57; TE77]. These are continuous-time filters, and their realization in terms of electrical network elements is well developed, based on rich theoretical frameworks such as the two-port extraction approaches [GU57], scattering matrix factorizations [BE68], and state-space formulations [AN73]. There is a subclass of these networks, called the doubly terminated lossless networks [TE77] which, if appropriately designed, lead to filters with low passband sensitivity. The reason for this, based on what is called the *maximum power transfer principle*, can be found in Orchard [OR66], Temes and LaPatra [TE77], and Antoniou [AN79].

The overall appearance of a doubly terminated lossless two-port is shown in Figure 7-11, where the rectangular box is an interconnection of lossless elements (inductors and capacitors). The transfer function is $H(s) = V_o(s)/V_i(s)$. Let Ω_k denote the frequencies where $|H(j\Omega)|$ attains its maximum value. If the network elements are designed such that the source $v_i(t)$ transfers maximum power at frequencies Ω_k, then it can be argued [OR66] that the passband sensitivity of $|H(j\Omega)|$ is low.

Now it is possible to convert the network of Figure 7-11 into a digital filter, by employing appropriate mappings from the s-domain to the z-domain. Such a mapping preserves the low-sensitivity properties but at the same time leads to delay-

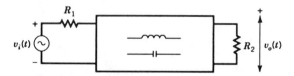

FIGURE 7-11   The lossless two-port, terminated at both ports with resistors.

free loops unless the voltages and currents are converted into variables of the form $v + R\,i$ and $v - R\,i$ called wave variables. This observation is the origin of wave digital filters [FE71], which inherit the low-sensitivity properties of the LC network (see Section 7-7).

A second approach for low-sensitivity digital filter design would be to work directly in the z-domain and synthesize the structure [DE80; VA85b,c, and VA86c]. Such an independent approach has the advantage that it can be comprehended by designers not familiar with classical network theory. In addition, it avoids the mapping of digital specifications into analog domain, and then the reverse mapping of analog filters back into the digital domain. Moreover, the direct z-domain approach gives rise to a number of new digital filter structures that are not derivable from standard, documented analog structures. Finally, new applications in the area of multirate filter banks are opened up by formulating the passivity concepts in the z-domain [NG88; VA87b, VA87e, VA88].

In this section we first introduce the notions of structural passivity and losslessness in digital filters, as a basis for low-sensitivity designs. Some forms of orthogonal filters [RA84], which also belong to the passive class, are discussed at the end of this section. Wave digital filters are discussed in Section 7-7. Passive FIR lattice structures based on cascaded lossless building blocks are introduced in Section 7-8.

Results on noise propagation in these structures are included in Section 7-9. In Section 7-10 we present conditions for suppression of limit cycles, along with examples of passive structures free from limit cycles.

7-6-1 Basic Requirements for Low Sensitivity

Consider a digital filter structure with multiplier coefficients m_i, implementing a transfer function $G(z)$ with a typical magnitude response as in Figure 7-12(a). Assume that m_i are restricted to be in a well-defined range $\Re$ (e.g., such as $-1 < m_k < 1$). The magnitude response in Figure 7-12(a) corresponds to the ideal situation where the m_i are not quantized. When the m_i are quantized, the response changes in a way depending on the structure. Suppose the structure is such that, as long as the m_i belong to the permissible range $\Re$, $|G(e^{j\omega})|$ never exceeds unity. At the frequency ω_k where $|G(e^{j\omega})| = 1$ under ideal conditions, the response can therefore only decrease, no matter how the multiplier coefficient m_i changes. Fig-

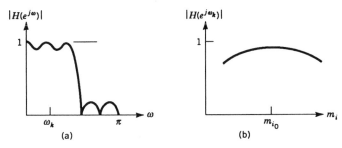

FIGURE 7-12 Explanation of low sensitivity induced by structural passivity.

ure 7-12(b) shows this situation. The first-order sensitivity of $|G(e^{j\omega})|$ with respect to the m_i is therefore zero at $\omega = \omega_k$. Since this holds for all m_i and all ω_k, the structure is expected to have low sensitivity with respect to quantization of m_i, in the neighborhood of each extremal frequency ω_k.

An implementation satisfying the property $|G(e^{j\omega})| \leq 1$ for all ω regardless of the *values* of m_i in a range $\mathfrak{R}$ is said to be *structurally bounded* or *structurally passive* in the range $\mathfrak{R}$ (the reasons for these names are explained below). The special case where an implementation is such that $|G(e^{j\omega})| = 1$ for all ω (regardless of values of m_i) is useful in implementing allpass filters, which remain allpass despite quantization of m_i. Such an implementation is said to be *structurally lossless*.

Some Terminology. A stable transfer function $G(z)$ satisfying $|G(e^{j\omega})| \leq 1$ for all ω is said to be *bounded* or *passive*. The term "passive" is motivated by the fact that for such systems, the output energy is less than or equal to the energy of the input sequence. If, in addition, the impulse response of $G(z)$ is real (so that $G(z)$ is real for all real z) we say that $G(z)$ is *bounded real* (BR for short). Stable allpass transfer functions (which satisfy $|G(e^{j\omega})| = 1$) are also called *lossless* transfer functions. (For such transfer functions, the energy of the output sequence is equal to that of the input sequence, for every finite energy input.) If a lossless function has real impulse response it is said to be *lossless bounded real* (abbreviated LBR). Any stable transfer function can be scaled so that it becomes bounded.

From the above discussion we conclude this: if a bounded transfer function is such that $|G(e^{j\omega})|$ is exactly equal to unity for some passband frequencies, and if it is implemented in a structurally passive manner, the resulting system has low passband sensitivity.

We shall find it useful (see Section 7-9) to extend these definitions to the case of transfer matrices: a stable transfer matrix $\mathbf{T}(z)$ is said to be *passive* or *bounded* if $\mathbf{T}^\dagger(e^{j\omega})\mathbf{T}(e^{j\omega}) \leq \mathbf{I}$ for all ω, and *lossless* if equality holds for all ω. These concepts are essentially discrete-time versions of the properties of scattering functions (and matrices) of continuous-time passive and lossless multiports. Details on these classical concepts are available in network theory literature [AN73; BE68] but will not be required here.

Throughout this section, we shall assume that there are no pole–zero cancellations in any of the transfer functions involved (unless mentioned otherwise). Such cancellations would give rise to trivial counterexamples to some of the results presented in the section.

7-6-2 Structures Based on Two Allpass Functions

Suppose $G(z)$ is a transfer function of the form

$$G(z) = \tfrac{1}{2}[A_0(z) + A_1(z)], \tag{7.31}$$

where $A_0(z)$ and $A_1(z)$ are stable allpass functions with frequency responses

$$A_0(e^{j\omega}) = e^{j\phi_0(\omega)}, \qquad A_1(e^{j\omega}) = e^{j\phi_1(\omega)}, \qquad (7.32)$$

with $\phi_0(\omega)$ and $\phi_1(\omega)$ denoting the phase responses. Clearly, $|G(e^{j\omega})| \leq 1$, with equality when $\phi_0(\omega) = \phi_1(\omega) + 2\pi i$ for any integer i. If the allpass functions $A_k(z)$ are implemented such that they remain (stable and) allpass despite coefficient quantization, then $G(z)$ remains bounded despite quantization. This leads to low passband sensitivity. At the passband extema ω_k, where $|G(e^{j\omega})| = 1$, the phases of $A_0(e^{j\omega})$ and $A_1(e^{j\omega})$ are aligned, whereas at the transmission zeros (in the stopband of $G(z)$), the phases differ by (odd integral multiples of) π.

This opens up two questions: first, how to implement an allpass function such that the allpass property is preserved despite quantization and second, what kind of transfer functions can be written as a sum of two allpass functions. In this subsection we answer the second question. The next subsection deals with the first question.

It can be shown that classical Butterworth, Chebyshev, and elliptic digital filters (and in fact a much wider class of filters) can be represented in the form of Eq. (7.31). For lowpass and highpass filters, if the order N is odd, the orders of $A_1(z)$ and $A_0(z)$ are r and $N - r$, respectively, for an appropriate integer r, and these allpass functions have real coefficients. If N is even, then $A_0(z)$ has order $N/2$ and has complex coefficients, and $A_1(z)$ is obtained from $A_0(z)$ by conjugation of coefficients. Accordingly, for even N, the output of $G(z)$ in response to a real input sequence is equal to the real part of the output of $A_0(z)$ in response to the same input. The proofs of these statements can be found in Vaidyanathan et al. [VA86a, VA87c]. The purpose of this section is to outline the main results, including a procedure for finding the coefficients of the functions $A_0(z)$ and $A_1(z)$.

Example 7.1. As an example, consider the third-order elliptic transfer function

$$G(z) = \frac{0.23179 + 0.36021z^{-1} + 0.36021z^{-2} + 0.23179z^{-3}}{1 - 0.38409z^{-1} + 0.70390z^{-2} - 0.13581z^{-3}},$$

whose magnitude response is shown in Figure 7-13. It can be verified that this transfer function is decomposable as

$$G(z) = 0.5 \left[\frac{-0.20356 + z^{-1}}{1 - 0.20356z^{-1}} + \frac{0.66715 - 0.18053z^{-1} + z^{-2}}{1 - 0.18053z^{-1} + 0.66715z^{-2}} \right],$$

which is a sum of two allpass functions.

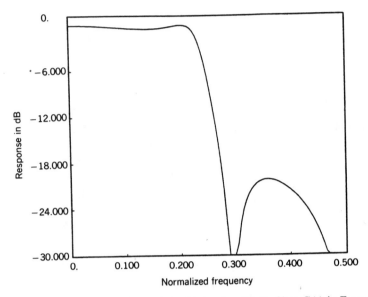

FIGURE 7-13   Magnitude response of the third-order elliptic filter $G(z)$ in Example 7.1.

Conditions for Allpass Decomposition. The most general Lth order allpass transfer function has the form

$$A(z) = c \frac{d_L^* + d_{L-1}^* z^{-1} + \cdots + z^{-L}}{1 + d_1 z^{-1} + \cdots + d_L z^{-L}}, \tag{7.33}$$

where c is a nonzero constant. In this section we assume $c = 1$ so that $A(z) = z^{-L} \tilde{B}(z)/B(z)$.[1]

Let us first assume that $G(z)$ is a real-coefficient transfer function that can be expressed in the form of Eq. (7.31), where $A_0(z)$ and $A_1(z)$ are real-coefficient stable allpass functions of the form

$$A_0(z) = \frac{z^{-(N-r)} \tilde{D}_0(z)}{D_0(z)} \quad \text{and} \quad A_1(z) = \frac{z^{-r} \tilde{D}_1(z)}{D_1(z)}, \tag{7.34}$$

with no uncanceled factors between $D_k(z)$ and $\tilde{D}_k(z)$. Assume further that there are no common factors between $D_0(z)$ and $D_1(z)$ (because such a common factor $(1 + dz^{-1})$ would manifest itself as a factor $(d^* + z^{-1})/(1 + dz^{-1})$ in Eq. (7.31), which does not affect the magnitude response of $G(z)$). Under these fair assumptions the following facts can be verified [VA86a].

---

[1]For a review of notations such as $\tilde{B}(z)$, $\hat{B}(z)$, and so on, see Section 1-5.

Fact 7.1. If $G(z)$ is expressible as in Eq. (7.31), then its numerator $P(z)$ is a symmetric polynomial; that is, with $P(z) = \sum_{n=0}^{N} p_n z^{-n}$, we have

$$p_n = p_{N-n} \qquad (7.35)$$

Fact 7.2. If $G(z)$ is expressible as in Eq. (7.31), then the new transfer function $H(z)$ defined according to

$$H(z) \triangleq \tfrac{1}{2} [A_0(z) - A_1(z)] \qquad (7.36)$$

is such that

(a) $G(z)$ and $H(z)$ are power complementary, that is,

$$|G(e^{j\omega})|^2 + |H(e^{j\omega})|^2 = 1, \qquad (7.37)$$

and

(b) the numerator $Q(z)$ of $H(z)$ is antisymmetric. That is, with $Q(z) = \sum_{n=0}^{N} q_n z^{-n}$ we have

$$q_n = -q_{N-n}. \qquad (7.38)$$

Note that $G(z)$ and $H(z)$ share a common denominator $D(z) = D_0(z)D_1(z)$. The essence of the above discussions is that if $G(z)$ is expressible as in Eq. (7.31), then we can derive two conclusions, namely, Facts 7.1 and 7.2. The converse of this observation is summarized in the following result [SA85; VA86a].

The Allpass Decomposition Result. Let $G(z) = P(z)/D(z)$ be an Nth order BR function and let $P(z)$ be a symmetric polynomial (i.e., satisfying Eq. (7.35)). Furthermore, let there exist a BR rational function $H(z) = Q(z)/D(z)$ such that $Q(z)$ is an antisymmetric polynomial (i.e., satisfying Eq. (7.38)) and such that the power complementary property Eq. (7.37) holds. Then $G(z)$ and $H(z)$ can be decomposed as in Eqs. (7.31) and (7.36), where $A_0(z)$ and $A_1(z)$ are stable allpass functions of orders $N - r$ and r for some integer r.

The power complementary property Eq. (7.37) can be expressed equivalently as

$$\tilde{P}(z)P(z) + \tilde{Q}(z)Q(z) = \tilde{D}(z)D(z). \qquad (7.39)$$

Given a BR function $G(z) = P(z)/D(z)$ with symmetric $P(z)$, one often wishes to test whether it can be expressed as in Eq. (7.31). From the preceding result we know that if there exists an antisymmetric $Q(z)$ satisfying Eq. (7.39), the $G(z)$ can be decomposed as in Eq. (7.31), and $H(z)$ automatically takes the form of Eq.

(7.36). Such a $Q(z)$ is a spectral factor of the function

$$|D(e^{j\omega})|^2 - |P(e^{j\omega})|^2, \tag{7.40}$$

which is guaranteed to be nonnegative because $G(z)$ is BR.[2] If no such antisymmetric $Q(z)$ exists, then $G(z)$ is *not* expressible as in Eq. (7.31).

Identifying the Allpass Functions. Assuming that $G(z)$ is expressible as in Eq. (7.31), it remains to find the factors $D_0(z)$ and $D_1(z)$ of its denominator $D(z)$ so that $A_0(z)$ and $A_1(z)$ can be identified using Eq. (7.34). To this end note that Eq. (7.39) implies, along with Eqs. (7.35) and (7.38),

$$[P(z) + Q(z)][P(z) - Q(z)] = z^{-N}D(z^{-1})D(z). \tag{7.41}$$

Let z_k, $1 \le k \le N$, denote the zeros of the Nth order polynomial $P(z) + Q(z)$. Some of these are zeros of $D(z)$ and some are those of $D(z^{-1})$. None of these can therefore be on the unit circle. Let z_k be inside the unit circle for $|k| \le r$ and let the rest be outside. Then the zeros of $D(z)$ (which are strictly inside the unit circle) are uniquely identified as

$$z_k \quad \text{for } 1 \le k \le r,$$

$$\frac{1}{z_k} \quad \text{for } r + 1 \le k \le N.$$

By writing $P(z) + Q(z)$, $P(z) - Q(z)$, and $D(z)$ in terms of their zeros, and simplifying the resulting expressions for $[P(z) \pm Q(z)]/D(z)$, it can be verified that $G(z)$ and $H(z)$ take the forms given by Eqs. (7.31) and (7.36) with

$$A_0(z) = \prod_{k=r+1}^{N} \frac{z^{-1} - (z_k^*)^{-1}}{1 - z^{-1}z_k^{-1}}, \qquad A_1(z) = \prod_{k=1}^{r} \frac{z^{-1} - z_k^*}{1 - z^{-1}z_k}, \tag{7.42}$$

which completely defines the allpass functions.

Computation of the Spectral Factor $Q(z)$. Apparently, the computation of a spectral factor of Eq. (7.40) involves finding the zeros of the polynomial $\tilde{D}(z)D(z) - \tilde{P}(z)P(z)$, and assigning an appropriate subset to $Q(z)$. However, the antisymmetric nature of $Q(z)$ enables us to find its coefficients in a more efficient manner. For this let

$$R(z) \triangleq z^{-N} [\tilde{P}(z)P(z) - \tilde{D}(z)D(z)], \tag{7.43}$$

---

[2]The spectral factor of a nonnegative valued function $F(e^{j\omega})$ is defined to be any function $S(z)$ such that $F(e^{j\omega}) = |S(e^{j\omega})|^2$. Since $F(e^{j\omega})$ is real, we can show that if $F(\alpha) = 0$ then $F(1/\alpha^*) = 0$. Moreover, if α is on the unit circle, then there is a double zero at α (since $F(e^{j\omega}) \ge 0$). Let the zeros of $F(z)$ be α_k and $1/\alpha_k^*$ for $0 \le k \le L - 1$. Then a spectral factor of $F(z)$ can be obtained as $S(z) = c\prod_{k=0}^{L-1}(1 - \alpha_k z^{-1})$, where c is an appropriate constant.

which is a $2N$th degree polynomial $\sum_{n=0}^{2N} r_n z^{-n}$. Compute the coefficients q_n according to

$$q_n = -q_{N-n} = \frac{r_n - \sum_{k=1}^{n-1} q_k q_{n-k}}{2q_0}, \qquad n \geq 2. \qquad (7.44)$$

The recursion is initialized with $q_0 = \sqrt{r_0}$ and $q_1 = r_1/2q_0$. If $Q(z)$ is not antisymmetric, the above procedure does not give the correct values of q_n, and this can be used as a test for $Q(z)$. Thus assume $Q(z)$ to be antisymmetric and compute it according to Eq. (7.44). Then compute the coefficients of $P(z^{-1})P(z) + Q(z^{-1})Q(z)$ and those of $D(z^{-1})D(z)$. These two sets of coefficients will turn out to be equal according to Eq. (7.39), if $Q(z)$ is indeed antisymmetric. If this equality is not true, then the assumption that $Q(z)$ is antisymmetric is incorrect and there does not exist a decomposition of the form Eq. (7.31).

The symmetric and antisymmetric restrictions on $P(z)$ and $Q(z)$ are not at all severe. Consider, for example, a fifth-order lowpass elliptic filter $G(z)$ with response as in Figure 7-14. All zeros of $P(z)$ are on the unit circle with none at $z = 1$; hence $P(z)$ is symmetric. Next, by definition of $H(z)$, $|H(e^{j\omega})|$ satisfies Eq. (7.37), so the response $|H(e^{j\omega})|$ is as shown by the broken lines (and can be verified to be elliptic as well). All zeros of $Q(z)$ are then on the unit circle, including a zero at $z = 1$. This ensures that $Q(z)$ is antisymmetric. If, on the other hand, the order N is *even*, then $|G(e^{j\omega})| < 1$ for $\omega = 0$. This means that $H(z)$, defined according to Eq. (7.37), does not have a zero at $z = 1$. As a result, $Q(z)$ is symmetric rather than antisymmetric. The decompositions, Eqs. (7.31) and (7.36), are then not possible with real coefficients for $A_k(z)$.

Note that for even N, the allpass decomposition with real allpass filters is possible if $H(z)$ still has only one zero at $z = 1$ (as in bandpass filters). As a simple example, in the above fifth-order case, if we replace z with z^2, then we do obtain even-order filters $G(z)$ and $H(z)$ of the form Eqs. (7.31) and (7.36), with real-coefficient allpass functions.

In order to handle situations where $P(z)$ and $Q(z)$ are both symmetric, it becomes necessary to allow allpass functions with complex coefficients. Assuming

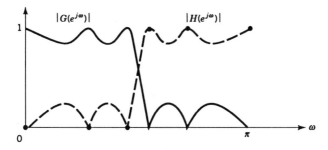

FIGURE 7-14 Explanation of how the symmetric and antisymmetric numerators arise.

N is even (which is the most frequent situation requiring symmetric $P(z)$ and $Q(z)$), the results can be stated as follows [VA87c].

A Second Allpass-Decomposition Result.   Let $G(z) = P(z)/D(z)$ and $H(z) = Q(z)/D(z)$ be BR functions satisfying Eq. (7.37), and let $P(z)$ and $Q(z)$ be symmetric. $G(z)$ and $H(z)$ are then expressible in the form

$$G(z) = \frac{\mathcal{C}_0(z) + \mathcal{C}_1(z)}{2}, \qquad H(z) = \frac{\mathcal{C}_0(z) - \mathcal{C}_1(z)}{2j}, \qquad (7.45)$$

where

$$\mathcal{C}_0(z) = \beta \prod_{k=1}^{M} \frac{z^{-1} - z_k}{1 - z^{-1}z_k^*},$$

$$\mathcal{C}_1(z) = \beta* \prod_{k=1}^{M} \frac{z^{-1} - z_k^*}{1 - z^{-1}z_k}, \qquad M = \frac{N}{2}, \qquad (7.46)$$

and $|\beta| = 1$.

Here $\mathcal{C}_k(z)$ are allpass functions with complex coefficients, and the coefficients of $\mathcal{C}_1(z)$ are complex conjugates of those of $\mathcal{C}_0(z)$. Given a BR function $G(z) = P(z)/D(z)$ with symmetric $P(z)$, we have to find the symmetric spectral factor $Q(z)$ of Eq. (7.40), so as to identify the power complementary function $H(z)$. The symmetry of $Q(z)$ enables us to compute it by a recursion similar to Eq. (7.44). For this, compute the coefficients of $R(z)$ as in Eq. (7.43) and then execute the modified recursion

$$q_n = q_{N-n} = \frac{-r_n - \sum_{k=1}^{n-1} q_k q_{n-k}}{2q_0}, \qquad 2 \leq n \leq N/2, \qquad (7.47)$$

with $q_0 = \sqrt{-r_0}$ and $q_1 = -r_1/2q_0$. (The definition of $R(z)$ guarantees that r_0 is negative, so q_0 is real.) Once again, if there does not exist a symmetric $Q(z)$ satisfying Eq. (7.39), the polynomial identified according to Eq. (7.47) does not satisfy Eq. (7.39), and the representation Eq. (7.45) does not exist.

If a symmetric $Q(z)$ is successfully identified, it remains to find the poles z_k of $\mathcal{C}_0(z)$. The procedure for this is as follows. Compute the coefficients of the polynomial $P(z) + jQ(z)$ and then find its roots z_k, $1 \leq k \leq N$. Out of these $M = N/2$ are guaranteed to be inside the unit circle, and the rest will be outside (and none on the unit circle [VA87c]). Let z_k, $1 \leq k \leq M$, be the ones inside. In terms of these, define $\mathcal{C}_0(z)$ and $\mathcal{C}_1(z)$ as in Eq. (7.46). Then the representation Eq. (7.45) holds. The constant β can be identified by making use of some a priori knowledge about $G(z)$ and $H(z)$ (such as the value at $\omega = 0$).

The decompositions of the form in Eqs. (7.31) and (7.36), where $A_0(z)$ and $A_1(z)$ are real-coefficient allpass functions, can be represented as shown in Figure 7-15(a). In the next subsection we show how the allpass functions $A_0(z)$ and $A_1(z)$

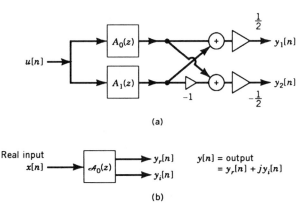

FIGURE 7-15 (a) Realization of a power-complementary pair using allpass functions. Here $G(z) = Y_1(z)/U(z)$ and $H(z) = Y_2(z)/U(z)$. (b) Realization of a power-complementary pair using a single complex allpass filter. Here $G(z) = Y_r(z)/U(z)$ and $H(z) = Y_i(z)/U(z)$.

can be implemented with $N - r$ and r real multipliers, respectively. A total of N multipliers is thus required in Figure 7-15(a) and gives rise to two transfer functions $G(z)$ and $H(z)$. In comparison, a direct form implementation would require a total of about $2N$ multipliers (even after sharing the common denominator $D(z)$ and taking into account the symmetry properties of numerators).

For the decomposition of the type in Eq. (7.45), the representation is as in Figure 7-15(b), where the input is assumed to be real. The real part of the output corresponds to the output of $G(z)$ and the imaginary part to that of $H(z)$. A total of $N/2$ complex multiplications are required in the implementation of Figure 7-15(b) and corresponds to $2N$ real multiplications, which is the same as for a direct form implementation.

Design examples for the decomposition Eqs. (7.31) and (7.36) can be found in Vaidyanathan et al. [VA86a]. For the decomposition of Eq. (7.45) with complex allpass functions, we reproduce below an example from Vaidyanathan et al. [VA87c].

Example 7.2. Let $G(z) = P(z)/D(z)$ be an eighth-order elliptic lowpass BR function with coefficients p_n and d_n as in Table 7-2. The response $|G(e^{j\omega})|$ has the nature shown in Figure 7-16. A qualitative plot of $|H(e^{j\omega})|$ satisfying Eq. (7.37) is also shown in the figure. Clearly, $H(z)$ is of eighth order, with all zeros on the unit circle, and none at $z = 1$. So $Q(z)$ is guaranteed to be symmetric, and its coefficients q_n computed according to Eq. (7.47) are shown in Table 7-2. The zeros of $P(z) + jQ(z)$ are then computed and $\mathcal{A}_0(z)$ in Eq. (7.46) identified. The relevant coefficients β, z_1, z_2, z_3, and z_4 are included in Table 7-2. Figure 7-17 shows the responses of $G(z)$ and $H(z)$ with unquantized coefficients. With the allpass function $\mathcal{A}_0(z)$ implemented in direct form with 4 bits real multiplier (in canonic SD code), the response of $G(z)$ is as shown in Figure 7-18. In comparison if we im-

TABLE 7-2   Coefficients of $G(z)$ and $H(z)$ in Example 7.2

n	p_n	q_n	d_n
0	0.025913507	0.321023225	1.000000000
1	−0.028763702	−1.901491193	−3.982269600
2	0.074164524	5.455303983	8.498003759
3	−0.041691969	−9.724285248	−11.486223632
4	0.084194289	11.705785106	10.707331654
5	−0.041691969	−9.724285248	−6.906407027
6	0.074164524	5.455303983	3.018707576
7	−0.028763702	−1.901491193	−0.809265853
8	0.025913507	0.321023225	0.103727421

$\beta = 0.4698 + j0.8828$; $z_1 = 0.4344 - j0.2253$; $z_2 = 0.4831 + j0.5675$; $z_3 = 0.5244 - j0.7367$; and $z_4 = 0.5492 + j0.8075$.

plement $G(z)$ itself in direct form, the passband response is as in Figure 7-19(b) even with 9 bits per coefficient in canonic SD code. The low passband sensitivity behavior of the allpass-based structure is thus demonstrated.

Adjustable Multilevel Filters. Still more general classes of power complementary functions can be constructed. An example of this can be seen by referring to Figure 7-15(a), where $A_0(z)$ and $A_1(z)$ have real coefficients. If we replace the 2×2 crisscross in the figure by a more general orthogonal matrix, we obtain Figure 7-20 (which reduces to Figure 7-15(a), except for a scale factor, when $k = \hat{k} = 1/\sqrt{2}$). We can now verify that

$$|G(e^{j\omega})|^2 + |H(e^{j\omega})|^2 = 2. \tag{7.48}$$

When the phases of $A_0(z)$ and $A_1(z)$ are aligned (in the passband of $G(z)$), we have

$$|G(e^{j\omega})| = |k + \hat{k}|, \qquad |H(e^{j\omega})| = |k - \hat{k}|. \tag{7.49}$$

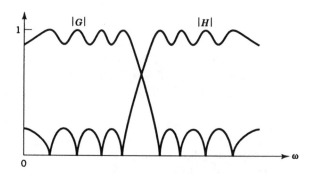

FIGURE 7-16   Qualitative plots of $|G(e^{j\omega})|$ and $|H(e^{j\omega})|$ (eighth-order elliptic filters).

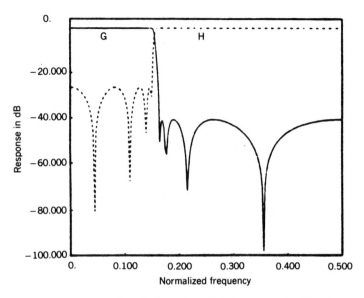

FIGURE 7-17 The responses $|G(e^{j\omega})|$ and $|H(e^{j\omega})|$ for unquantized implementation of $\mathcal{Q}_0(z)$ © IEEE, 1987 [VA87c].

When the phases of $A_0(z)$ and $A_1(z)$ differ by π (in the stopband of $G(z)$), then Eq. (7.49) holds with G and H interchanged. Thus the structure of Figure 7-20 can be used to approximate a two-band filter $G(z)$ with nominal passband and stopband gain $|k + \hat{k}|$ and $|k - \hat{k}|$, respectively (rather than the conventional "1" and "0"). Design examples can be found in Ansari [AN86], Koilpillai et al. [KO90], and Renfors et al. [RE91].

7-6-3 Structurally Lossless Implementations of Allpass Filters

There exist many ways of implementing stable allpass functions with minimum number of multipliers, such that they remain stable and allpass despite multiplier quantization. Such implementations are said to be structurally lossless.

First consider allpass functions of the form

$$A(z) = \frac{d_N + d_{N-1}z^{-1} + \cdots + d_1 z^{-(N-1)} + z^{-N}}{1 + d_1 z^{-1} + \cdots + d_{N-1} z^{-(N-1)} + d_N z^{-N}} \tag{7.50}$$

with real d_k. These can be written as the product of first- and second-order sections (as we did in Section 7-3 to obtain cascade form structures). As long as these sections are implemented in a structurally lossless manner, the cascade (which realizes Eq. (7.50)) is clearly structurally lossless.

A first-order allpass section of the form shown in Figure 7-21 implements the

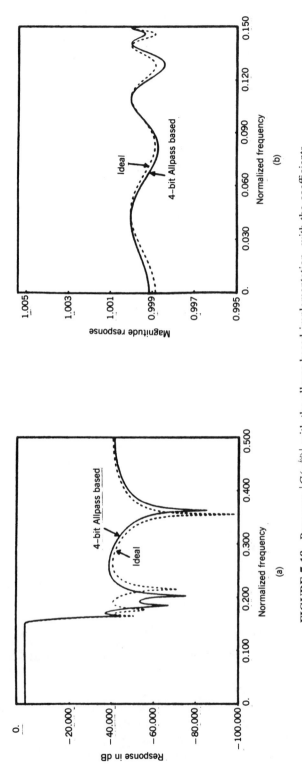

FIGURE 7-18 Response $|G(e^{j\omega})|$ with the allpass-based implementation, with the coefficients of $\alpha_0(z)$ quantized to 4 bits in canonic SD code © IEEE, 1987 [VA87c].

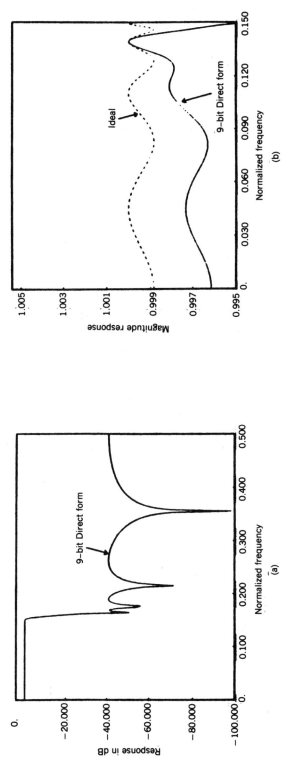

FIGURE 7-19 Response $|G(e^{j\omega})|$ with $G(z)$ implemented in direct form with coefficients quantized to 9 bits in canonic SD code © IEEE, 1987 [VA87c].

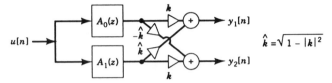

FIGURE 7-20 A generalization of the realization of power-complementary pairs.

transfer function

$$A(z) = \frac{d_1 + z^{-1}}{1 + d_1 z^{-1}}. \qquad (7.51)$$

The input–output relation remains allpass even if d_1 is quantized. The only disadvantage of this structure is the use of two delays, which is not the minimum number. A systematic procedure to obtain first-and second-order allpass sections with the smallest number of delays and multipliers has been advanced in Mitra and Hirano [MI74]. The method is based on the principle of multiplier extraction. To explain this assume that we wish to obtain second-order allpass sections with only two multipliers, to implement

$$A(z) = \frac{d_2 + d_1 z^{-1} + z^{-2}}{1 + d_1 z^{-1} + d_2 z^{-2}}, \qquad (7.52)$$

where the d_k are real. We would like d_1 and d_2 to be the multiplier coefficients. So we "extract" these multipliers from the proposed structure. The result is a three-input–three-output system (Figure 7-22) characterized by a 3×3 transfer matrix $\mathbf{T}(z)$. Two out of these three inputs are constrained externally as

$$x_2[n] = d_1 y_2[n], \quad x_3[n] = d_2 y_3[n]. \qquad (7.53)$$

The elements T_{ij} of $\mathbf{T}(z)$ are independent of d_1 and d_2, so that they can be realized without multipliers. The overall transfer function $A(z) = Y_1(z)/X_1(z)$ can be expressed in terms of $T_{ij}(z)$, d_1, and d_2. If the resulting expression is equated to Eq.

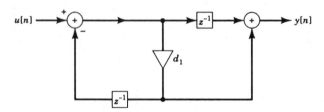

FIGURE 7-21 The first-order allpass function, realized with one multiplier and two delays.

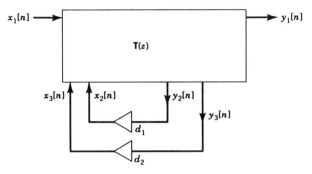

FIGURE 7-22 The multiplier extraction approach.

(7.52), it is possible to identify the elements $T_{ij}(z)$. The solutions are by no means unique and give rise to several equivalent structures, which are tabulated in Mitra and Hirano [MI74]. Moreover, instead of using d_1 and d_2 as multipliers, other combinations (such as d_1 and d_2/d_1) can be used. Based on variations of these, 24 two-multiplier structures for second-order allpass functions (and four one-multiplier structures for first-order functions) are derived in Mitra and Hirano [MI74]. Of these some have the smallest number of delays (two delays for second-order and one for first-order). Figure 7-23 shows examples of such minimum-multiplier, minimum-delay sections. All of these structures retain the allpass property despite quantization. For the second-order sections, stability can be maintained under quantization by ensuring that the coefficients d_1 and d_2 belong to the stability triangle (see Figure 7-61 later). However, we cannot quantize the multipliers *independently of each other* to arbitrarily few bits and still expect the filters to be stable (since the shaded region in Figure 7-61 is not a *rectangle* with edges parallel to the axes).

It is possible to obtain further generalizations of the element-extraction approach. For example, we can extract two delays and two multipliers to obtain all realizations having only two multipliers and two delays. The structure then takes the form of a 5×5 matrix, four of whose inputs are constrained in terms of the

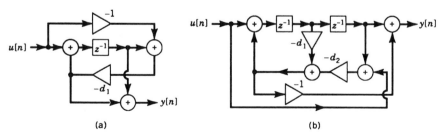

FIGURE 7-23 Allpass filters realized with the smallest number of delays and multipliers: (a) first-order filter and (b) second-order filter.

corresponding outputs. This approach results in a very large number of equivalent structures indeed. Details can be found in Szczupak et al. [SZ87].

A different class of structures for allpass functions, called the cascaded lattice structures, naturally arises from the mathematics of linear prediction [GR73; MA76]. These structures are also related to the Schur-Cohn/Jury theory for testing the stability of transfer functions [JU64; VA87d].[3] The basic appearance of the lattice structure is a cascade of two-input–two-output sections as shown in Figure 7-24, with consecutive sections separated by a delay. Each section has the internal details as shown in Figure 7-25. The transfer function $Y(z)/U(z)$ is the desired allpass function of order N. The *lattice coefficients* k_m are real (since the d_i in Eq. (7.50) are assumed real). The main features of the structure are the following:

1. With $k_m^2 < 1$ for all m, the transfer function $A(z)$ is stable *and* allpass. Accordingly, k_m can be *independently* quantized without affecting the stable allpass property, as long as we ensure that the quantized k_m satisfies $k_m^2 < 1$.
2. *Any* stable allpass function of the form Eq. (7.50) can be realized as in Figure 7-24 by appropriate choice of k_m.

We now outline the steps required in order to compute the values of k_m from the given allpass transfer function $A(z)$ of order N. For convenience, let $G_N(z)$ denote the transfer function $A(z)$. By denoting the denominator polynomial as $D_N(z)$, we an write $G_N(z) = \hat{D}_N(z)/D_N(z)$, where the hat denotes reversal of coefficients of the polynomial (Section 1-5). Given the allpass function $G_N(z)$, our procedure is to compute a sequence of lower order allpass functions $G_m(z) = \hat{D}_m(z)/D_m(z)$ by using the recursion

$$z^{-1}G_{m-1}(z) = \frac{G_m(z) - k_m}{1 - k_m G_m(z)} \qquad (7.54)$$

Or in terms of the polynomials,

$$z^{-1}N_{m-1}(z) = N_m(z) - k_m D_m(z), \qquad (7.55)$$

$$D_{m-1}(z) = D_m(z) - k_m N_m(z), \qquad (7.56)$$

[3]For a review of the Schur-Cohn stability test, see Section 3-4-3.

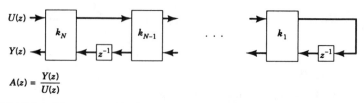

$$A(z) = \frac{Y(z)}{U(z)}$$

FIGURE 7-24 The cascaded lattice structure for allpass functions. Here $A(z)$ is synonymous to $G_N(z)$.

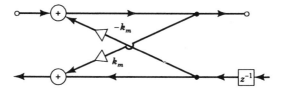

FIGURE 7-25   The two-multiplier lattice section.

where the general form of $D_m(z)$ is

$$D_m(z) = \sum_{n=0}^{m} d_{m,n} z^{-n}, \tag{7.57}$$

and $N_m(z) = \hat{D}_m(z)$. By appropriate choice of k_m, it is possible to ensure that the coefficient of z^{-m} on the right-hand side (RHS) of Eq. (7.56) is zero, so that $D_{m-1}(z)$ is of order $m - 1$. This choice, which is

$$k_m = G_m(\infty) = d_{m,m}/d_{m,0}, \tag{7.58}$$

simultaneously ensures that the constant term on the RHS of Eq. (7.55) is also equal to zero. As a result, $G_{m-1}(z)$ is indeed an $(m - 1)$th order allpass function of the form Eq. (7.50). By repeating this order reduction process starting from $m = N$, we eventually arrive at $G_0(z) = 1$. We have thus obtained all the coefficients k_m.

Example 7.3.   To demonstrate the basic idea, consider the allpass function

$$G_4(z) = \frac{N_4(z)}{D_4(z)} = \frac{1 + z^{-1} + 2z^{-2} + 3z^{-3} + 4z^{-4}}{4 + 3z^{-1} + 2z^{-2} + z^{-3} + z^{-4}}.$$

The quantity k_4, evaluated according to Eq. (7.58), is

$$k_4 = \tfrac{1}{4}.$$

The reduced order polynomial $D_3(z)$ computed according to Eq. (7.56) is

$$D_3(z) = 15 + 11z^{-1} + 6z^{-2} + z^{-3}.$$

This defines the allpass function

$$G_3(z) = \frac{N_3(z)}{D_3(z)} = \frac{1 + 6z^{-1} + 11z^{-2} + 15z^{-3}}{15 + 11z^{-1} + 6z^{-2} + z^{-3}}.$$

By repeating this process, we obtain in a similar way

$$k_3 = \tfrac{1}{15}, \qquad D_2(z) = 224 + 159z^{-1} + 79z^{-2},$$

$$k_2 = \tfrac{79}{224}, \qquad D_1(z) = 43935 + 23055z^{-1},$$

$$k_1 = \tfrac{23055}{43935}, \qquad D_0(z) = \text{constant}.$$

With all the k_m thus computed, the lattice structure for $G_4(z)$ is completely determined.

Interpretation in Terms of Digital Two-Pair Extraction. The procedure for generating the allpass function $G_{m-1}(z)$ in terms of $G_m(z)$ can be interpreted using the notion of a digital two-pair [MI74], [MI77]. A digital two-pair is a two-input–two-output system as shown in Figure 7-26 and is characterized by a 2×2 transfer matrix

$$\mathbf{T}(z) = \begin{bmatrix} T_{11}(z) & T_{12}(z) \\ T_{21}(z) & T_{22}(z) \end{bmatrix}, \tag{7.59}$$

so that the inputs and outputs are related by

$$\begin{bmatrix} Y_1(z) \\ Y_2(z) \end{bmatrix} = \mathbf{T}(z) \begin{bmatrix} X_1(z) \\ X_2(z) \end{bmatrix}. \tag{7.60}$$

Now consider the arrangement of Figure 7-27, which is called a *constrained two-pair* [MI74]. Here the input $X_2(z)$ is constrained in terms of $Y_2(z)$ according to the relation

$$X_2(z) = G_{m-1}(z)Y_2(z). \tag{7.61}$$

The *input transfer function* $G_m(z)$, which is defined to be $Y_1(z)/X_1(z)$, is given in terms of the two-pair parameters by

$$G_m(z) = T_{11}(z) + \frac{T_{12}(z)T_{21}(z)G_{m-1}(z)}{1 - T_{22}(z)G_{m-1}(z)}. \tag{7.62}$$

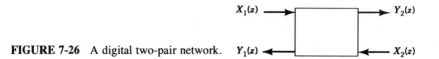

FIGURE 7-26   A digital two-pair network.

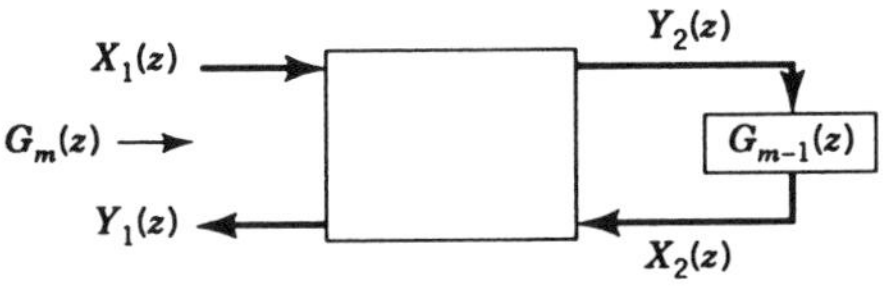

$X_1(z)$ ⟶

$Y_2(z)$

$G_m(z)$ ⟶ $\boxed{G_{m-1}(z)}$

$Y_1(z)$ ⟵

$X_2(z)$

FIGURE 7-27 The constrained two-pair. Here $X_2(z) = G_{m-1}(z)Y_2(z)$ and $G_m(z) = Y_1(z)/X_1(z)$.

Returning now to Eq. (7.54), we see that it can be inverted to obtain

$$G_m(z) = k_m + \frac{(1 - k_m^2)z^{-1}G_{m-1}(z)}{1 + k_m z^{-1}G_{m-1}(z)}. \tag{7.63}$$

Comparing Eq. (7.62) with Eq. (7.63), we see that the allpass function $G_m(z)$ in Eq. (7.63) is related to $G_{m-1}(z)$ precisely as in Figure 7-27 with two-pair parameters such that

$$T_{11}(z) = k_m, \quad T_{12}(z)T_{21}(z) = z^{-1}(1 - k_m^2), \quad T_{22}(z) = -k_m z^{-1}. \tag{7.64}$$

One choice of the two-pair parameters satisfying these conditions is

$$T_{11}(z) = k_m, \quad T_{12}(z) = z^{-1}(1 - k_m^2), \quad T_{21}(z) = 1, \quad T_{22}(z) = -k_m z^{-1}. \tag{7.65}$$

The process of obtaining the lower order function $G_{m-1}(z)$ from $G_m(z)$ such that Eq. (7.63) is satisfied is called *two-pair extraction*. In other words, a two-pair with a certain transfer matrix has been extracted from $G_m(z)$ to obtain a *remainder* function $G_{m-1}(z)$. In our specific example, we have extracted a two-pair with transfer matrix having the elements given in Eq. (7.65) from an allpass function $G_m(z)$ in such a way that the remainder $G_{m-1}(z)$ is a lower order allpass function. It should be noticed that the two-pair is not unique because only the product $T_{12}(z)T_{21}(z)$ matters in Eq. (7.62). For example, two-pairs with any of the following transfer matrices

$$T_{11}(z) = k_m, \quad T_{12}(z) = z^{-1}(1 + k_m), \quad T_{21}(z) = 1 - k_m, \quad T_{22}(z) = -k_m z^{-1} \tag{7.66}$$

and

$$T_{11}(z) = k_m, \quad T_{12}(z) = \hat{k}_m z^{-1}, \quad T_{21}(z) = \hat{k}_m, \quad T_{22}(z) = -k_m z^{-1} \tag{7.67}$$

can be used in place of Eq. (7.65) resulting in the same remainder $G_{m-1}(z)$. Once again, the notation $\hat{k}$ stands for $\hat{k} = \sqrt{1 - |k|^2}$.

The Lattice Structure. Figure 7-28 is a cascaded connection of the extracted two-pairs (ignore the quantity $H_N(z)$ for the present). If the rightmost two-pair is con-

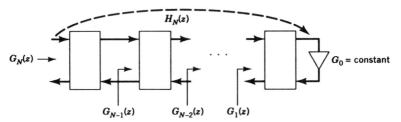

FIGURE 7-28 The result of repeated two pair extraction.

strained (or terminated) by the zero order remainder G_0, then the intermediate transfer functions $G_m(z)$ appear as physical transfer functions, as shown in Figure 7-28. As long as $G_N(z)$ is stable to begin with, we can be assured that the $G_m(z)$ are stable and allpass and that $k_m^2 < 1$ for all m. The rectangular boxes in the figure (the two-pairs) have transfer matrices given by Eq. (7.65) (or one of the equivalent forms, such as Eq. (7.66) or (7.67)). The appearance of these two-pairs when the transfer matrices have the forms Eqs. (7.65), (7.66), and (7.67) are shown in Figures 7-25, 7-29, and 7-30, respectively. The cascaded structure with any one of the three two-pair building blocks is called a *lattice structure*.

The cascaded lattice structure of Figure 7-28 with the two-pairs as in Figure 7-25 or 7-29 or 7-30 is therefore an implementation of $G_N(z)$. If the two-pairs are as in Figure 7-29, then the structure has only N multipliers (and N delays). On the other hand, if each two-pair is as in Figure 7-30, then the structure has $4N$ multipliers. This appears to be inefficient at first but is sometimes preferred for reasons that will be clear in Section 7-9-2.

The transfer matrix of the two-pair of Figure 7-30 has the form

$$\mathbf{T}(z) = \begin{bmatrix} \cos\theta & z^{-1}\sin\theta \\ \sin\theta & -z^{-1}\cos\theta \end{bmatrix}, \tag{7.68}$$

where $\cos\theta = k_m$ and $\sin\theta = \sqrt{1 - k_m^2} = \hat{k}_m$. This represents an LBR matrix and the two-pair is said to be *normalized*. We can therefore interpret the order reduction as a process of extracting a lossless two-pair from a lossless function $G_m(z)$ to obtain a lower order lossless function $z^{-1}G_{m-1}(z)$. The corresponding lattice

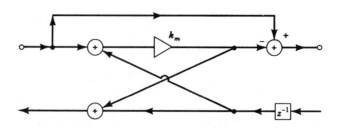

FIGURE 7-29 Denormalized one-multiplier version.

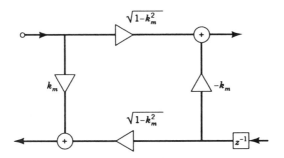

FIGURE 7-30 The normalized (four-multiplier) lattice section.

structure is called the normalized lattice. The reason is, in the normalized lattice, the $\mathcal{L}_2$ norm of the transfer function from the input node to any multiplier input is equal to unity [GR75], and further scaling effort is unnecessary (Section 7-9-2).

With two-pairs as in Figures 7-25 and 7-29, the lattice structure is said to be *denormalized*. The denormalized two-pairs can be represented in terms of the normalized versions as in Figure 7-31, for appropriate choice of α_m.

The lattice structure presented here has the advantage that the transfer function remains stable and allpass when the k_m are independently quantized in the range $(-1, 1)$. In addition, there are other advantages. First, it can be made free from all zero-input limit cycles (Section 7-10). Second, it provides the smallest possible roundoff noise among all allpass structures in a certain class (Section 7-9).

The two-pair in Eq. (7.68) is lossless in the sense that $\mathbf{T}(z)$ is stable and $\mathbf{T}(e^{j\omega})$ is unitary. Other types of digital lossless two pairs have also been introduced in the literature [SW75; VA84], with a view to obtaining low-sensitivity structures. An example is the Type 1A two-pair [VA84], which has transfer matrix

$$\mathbf{T}(z) = \frac{1}{1 + \sigma z^{-1}} \begin{bmatrix} 1 - \sigma & \sqrt{\sigma}(1 + z^{-1}) \\ \sqrt{\sigma}(1 + z^{-1}) & -(1 - \sigma)z^{-1} \end{bmatrix}, \quad 0 < \sigma < 1.$$

If this type of two-pair is used for every building block in Figure 7-28 (and if $G_0 = 1$), then $G_m(z)$ is allpass for all z. If we replace the elements $T_{12}(z)$ and $T_{21}(z)$ with

$$T_{12}(z) = \sigma \frac{1 + z^{-1}}{1 + \sigma z^{-1}}, \qquad T_{21}(z) = \frac{1 + z^{-1}}{1 + \sigma z^{-1}},$$

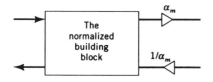

FIGURE 7-31 Representation of denormalized lattice sections in terms of the normalized section.

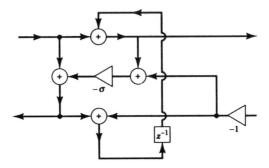

FIGURE 7-32 Implementing the Type 1A two-pair with only one multiplier.

then the transfer functions $G_m(z)$ are unaffected. This modified two-pair can be implemented with only one multiplier, as shown in Figure 7-32. It turns out that this building block, independently derived in the z-domain using the lossless concept, coincides with a fundamental building block in wave digital filter theory (Section 7-7).

7-6-4 Orthogonal Digital Filter Structures

The normalized Gray and Markel lattice structure, which can be implemented as in Figure 7-28 with building blocks as in Figure 7-30, is said to be an orthogonal filter, because the matrix

$$\begin{bmatrix} k_m & \sqrt{1 - k_m^2} \\ \sqrt{1 - k_m^2} & -k_m \end{bmatrix}$$

is orthogonal (with $-1 < k_m < 1$). The primary application of this lattice is in the implementation of allpass filters (and also in the implementation of the allpole transfer function $H_N(z)$ (Figure 7-28), which finds application in speech synthesis [MA76]).

More general classes of orthogonal digital filters have been developed by several authors [DE80; HE83; RA84], to implement arbitrary (not necessarily allpass or allpole) transfer functions. These are low-sensitivity structures (again because of structural passivity) and, in fact, share all the robustness properties of the structurally passive implementations described earlier in this chapter. One main feature of these structures is that they can be implemented with *planar rotation operators*

$$\begin{bmatrix} \cos \theta_m & \sin \theta_m \\ \sin \theta_m & -\cos \theta_m \end{bmatrix}$$

as the only type of computational building blocks. The other feature is that the overall transfer matrix being synthesized is lossless (the *components* of the matrix being *bounded*). Essentially, given a bounded transfer function $G(z)$, one can construct a lossless transfer matrix (e.g., 2×2), one of whose elements say, $T_{11}(z)$, is equal to $G(z)$. The lossless matrix is then synthesized as an interconnection of first- and second-order lossless building blocks.

The purpose of this section is to introduce one type of orthogonal digital filter, based on an extension of the lattice structure of Figure 7-24 (with building blocks as in Figure 7-30) for allpass synthesis. The results stated here were first introduced in Henrot and Mullis [HE83], and Rao and Kailath [RA84] and later reviewed in the context of structurally passive implementations in Vaidyanathan [VA85d]. Let $G(z) = P(z)/D(z)$ be a BR transfer function. We can always construct the new transfer function $H(z) = Q(z)/D(z)$ as in Section 7-6-2, such that the power complementary property Eq. (7.37) holds. Now construct the LBR transfer matrix $\mathbf{G}_N(z) = [G(z)\ H(z)]'$. We now have the vector analog of the situation in Section 7-6-3, where we synthesized an Nth order scalar LBR (allpass) function $G_N(z)$ by extracting the LBR two-pair described in Eq. (7.67). Each step in the synthesis procedure in Section 7-6-3 is equivalent to extracting the two-pair Eq. (7.67) from $G_m(z)$ with k_m as in Eq. (7.58), so as to obtain a remainder LBR $z^{-1}G_m(z)$.

Not surprisingly, a similar trick works in the vector case. In order to describe this, let us write $\mathbf{G}_m(z) = \mathbf{N}_m(z)/D_m(z)$, where $\mathbf{N}_m(z) = [P_m(z)\ Q_m(z)]'$. Here $P_m(z)$, $Q_m(z)$, and $D_m(z)$ are mth order polynomials in z^{-1}. Assuming that $\mathbf{G}_m(z)$ is 2×1 LBR, we can form the lower order 2×1 LBR $\mathbf{G}_{m-1}(z) = \mathbf{N}_{m-1}(z)/D_{m-1}(z)$ in a manner analogous to Eqs. (7.55) and (7.56):

$$z^{-1}\mathbf{N}_{m-1}(z) = (\mathbf{I} - \mathcal{K}_m\mathcal{K}_m')^{-1/2}(\mathbf{N}_m(z) - \mathcal{K}_m D_m(z)), \qquad (7.69)$$

$$D_{m-1}(z) = (1 - \mathcal{K}_m'\mathcal{K}_m)^{-1/2}(D_m(z) - \mathcal{K}_m'\mathbf{N}_m(z)), \qquad (7.70)$$

where $\mathcal{K}_m$ is a 2×1 matrix given by a relation analogous to Eq. (7.58):

$$\mathcal{K}_m = \mathbf{G}_m(\infty). \qquad (7.71)$$

If this process is repeated starting with $m = N$, we eventually end up with the constant matrix

$$\mathbf{G}_0 = [\cos\theta\ \ \sin\theta]', \qquad (7.72)$$

resulting in a cascaded structure of the form shown in Figure 7-33. Each building block in this cascade (shown separately in Figure 7-34) is an extended two-pair with transfer matrix $\mathbf{T}_m$ given by[4]

$$\mathbf{T}_m = \begin{bmatrix} \mathcal{K}_m & (\mathbf{I} - \mathcal{K}_m\mathcal{K}_m')^{1/2} \\ (1 - \mathcal{K}_m'\mathcal{K}_m)^{1/2} & -\mathcal{K}_m'\left(\dfrac{\mathbf{I} - \mathcal{K}_m\mathcal{K}_m'}{1 - \mathcal{K}_m'\mathcal{K}_m}\right)^{-t/2} \end{bmatrix}. \qquad (7.73)$$

---

[4]Review Section 1-5 for superscript notations such as $t/2$, $-t/2$, and so on.

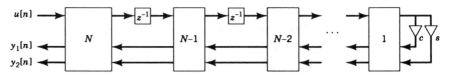

FIGURE 7-33 The vector version of the cascaded lattice structure. Here $G(z) = Y_1(z)/U(z)$ and $H(z) = Y_2(z)/U(z)$; the terminations are $c = \cos\theta$ and $s = \sin\theta$.

The matrix Eq. (7.73) is orthogonal and can be implemented in the form

$$
\mathbf{T}_m = \begin{bmatrix} 1 & 0 & 0 \\ 0 & \cos\theta_2 & \sin\theta_2 \\ 0 & \sin\theta_2 & -\cos\theta_2 \end{bmatrix} \begin{bmatrix} -\cos\theta_1 & \sin\theta_1 & 0 \\ \sin\theta_1 & \cos\theta_1 & 0 \\ 0 & 0 & 1 \end{bmatrix}, \qquad (7.74)
$$

which is a sequence of two planar rotations. The orthogonal filter structure of Figure 7-33 can be used to implement any scalar BR transfer function $G(z)$. A natural by-product here is the complementary transfer function $H(z)$, which is available in Figure 7-33. Because of the delays separating adjacent sections, the structures of Figures 7-24 and 7-33 are highly pipelineable and are suitable for VLSI implementation [DE85]. Unfortunately, in Figure 7-33, there is a total of $2N + 1$ planar rotation angles involved, whereas in the allpass-based structure of Figure 7-15(a), we have only N rotations. For the most commonly used type of transfer functions, the structure of Figure 7-33 is therefore very expensive.

We conclude this section with a synthesis example.

Example 7.4. Let the bounded real transfer function $G(z)$ be

$$
G(z) = \frac{0.5(1 + 0.5z^{-1})}{1 + 0.75z^{-1} + 0.5z^{-2}}.
$$

First, we construct the lossless vector $\mathbf{G}_2(z)$ as

$$
\mathbf{G}_2(z) = \frac{1}{1 + 0.75z^{-1} + 0.5z^{-2}} \begin{bmatrix} 0.5(1 + 0.5z^{-1}) \\ \frac{1}{\sqrt{2}}(1 + z^{-1} + z^{-2}) \end{bmatrix}.
$$

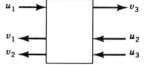

FIGURE 7-34 The extended LBR two-pair building block.

The coefficient $\mathcal{K}_2$ for the first extraction is given by

$$\mathcal{K}_2 = \mathbf{G}_2(\infty) = \begin{bmatrix} 0.5 \\ 1/\sqrt{2} \end{bmatrix}.$$

After extracting the lossless two-pair, the remainder is

$$\mathbf{G}_1(z) = \frac{0.5 + z^{-1}}{\sqrt{3}(1 + 0.5z^{-1})} \begin{bmatrix} -1 \\ \sqrt{2} \end{bmatrix}.$$

The next stage of extraction requires

$$\mathcal{K}_1 = \mathbf{G}_1(\infty) = \frac{1}{2\sqrt{3}} \begin{bmatrix} -1 \\ \sqrt{2} \end{bmatrix}.$$

After this extraction, the remainder is a constant, given by

$$\mathbf{G}_0 = \frac{1}{\sqrt{3}} \begin{bmatrix} -1 \\ \sqrt{2} \end{bmatrix}.$$

This completes the synthesis procedure. It can be verified that $\mathbf{G}_0' \mathbf{G}_0 = 1$.

7-7 WAVE DIGITAL FILTERS

In a classic work in 1971, Fettweis introduced the concept of wave digital filters [FE71], as a means of obtaining low-sensitivity digital filter structures. Since then, several papers have appeared on this subject [FE71, FE73, FE74, FE75, FR68; GA85; SE73], leading to a well established family of techniques toward this end. These structures again belong to the structurally passive family and display good features such as low sensitivity, freedom from limit cycles, and low roundoff noise. Historically, wave filters are the earliest class of digital filter structures with passivity properties.

Given a set of specifications for a digital filter (e.g., lowpass as in Figure 5-2), the first step in the design of wave filters is to map the specifications into the continuous time domain using the bilinear transform (Section 5-4-1),

$$s = (1 - z^{-1})/(1 + z^{-1}). \tag{7.75}$$

The mapped specifications are shown in Figure 7-35, where $\Omega_p = \tan(\omega_p/2)$ and $\Omega_s = \tan(\omega_s/2)$. The second step is to design a LCR filter of the form shown in Figure 7-11 (the doubly terminated lossless two-port). This is done in such a way that, at the frequencies Ω_k corresponding to passband maxima, the source $V_i(j\Omega)$

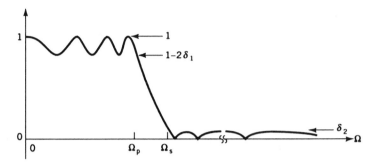

FIGURE 7-35 The lowpass specifications, translated into continuous-time domain.

transfers maximum power to the load R_2. When the passive elements are perturbed due to a variety of practical reasons, the power transferred at these frequencies Ω_k can only decrease. This results in low passband sensitivity for the same reason as given in Section 7-6-1.

The final step is to translate the passive electrical network into a digital filter. This involves the mapping of the variable s back into z^{-1}, and also certain judicious mappings of voltages and currents into what are called the wave variables. (The need for the use of wave variables arises in an attempt to avoid delay-free loops [FE71].) Our aim in this section is to give some elaboration on this third step. The literature in this area is extensive and very well documented, so the exposition here is brief. Details of the second step above can be found in any one of the standard references such as Temes and LaPatra [TE77] and Sedra and Brackett [SE78].

Consider an inductor L (Figure 7-36), which is electrically described by $V(s) = sLI(s)$. Let us define the *wave variables* $A_1(s)$ and $B_1(s)$ as

$$A_1(s) = V(s) + RI(s), \qquad B_1(s) = V(s) - RI(s), \tag{7.76}$$

with $R > 0$. The quantity R is called the port resistance of the "one-port" in Figure 7-36. Substituting $V(s) = sLI(s)$ in the above equation, we obtain the relation

$$B(z) = \frac{(1 - z^{-1})L - (1 + z^{-1})R}{(1 - z^{-1})L + (1 + z^{-1})R} A(z), \tag{7.77}$$

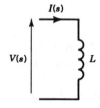

FIGURE 7-36 The inductor as a one-port.

where $B(z)$ and $A(z)$ are the quantities obtained by replacing s in $B_1(s)$ and $A_1(s)$ with Eq. (7.75). The choice of R, which so far has been arbitrary, can be made so as to simplify Eq. (7.77). Letting $R = L$, Eq. (7.77) reduces to $B(z) = -z^{-1}A(z)$. In other words, an inductor can be simulated in the wave digital domain by means of a delay element (with a minus sign). See Figure 7-37. In a similar manner, one can obtain wave digital equivalents of other standard electrical elements. In Table 7-3 we tabulate these for common passive network elements, along with the suggested choice of R. The quantities $A(z)$ and $B(z)$ are called the incident and reflected wave variables, respectively. Note that the resistor R_1 translates into a *wave sink* (which is an element for which the reflected wave $B(z)$ is identically zero, regardless of $A(z)$).

Suppose we want to obtain the wave digital equivalent of the complete electrical network of Figure 7-38. It is easy to draw the equivalent of various elements by referring to Table 7-3. However, before we attempt to *interconnect* these digital elements, some thinking is necessary: the various wave variables appearing at the nodes of the equivalent elements are not "compatible" in the sense that the values of R used in their definitions are not the same. The proper interconnection is actually done by creating wave digital equivalents of the "interconnections" in the electrical circuit. These equivalents are called *wave adaptors*. Consider, for example, the parallel connection of three "one-ports" in Figure 7-39, and let R_1, R_2, R_3 be the port resistances. As the port voltages and currents are constrained by

$$V_1 = V_2 = V_3, \qquad I_1 + I_2 + I_3 = 0, \tag{7.78}$$

the wave variables $A_k = V_k + R_k I_k$ and $B_k = V_k - R_k I_k$ are accordingly constrained by

$$B_k = 2A_3 - A_k + \alpha_1(A_1 - A_3) + \alpha_2(A_2 - A_3), \qquad k = 1, 2, 3, \tag{7.79}$$

where

$$\alpha_n = \frac{2G_n}{G_1 + G_2 + G_3}, \tag{7.80}$$

with $G_n = 1/R_n$. The structure that generates B_k's in response to A_k's according to Eq. (7.79) is schematically indicated in Figure 7-40 and is called the *parallel*

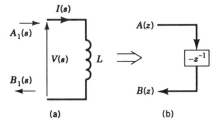

(a) (b)

FIGURE 7-37 Converting the inductor into the wave digital one-port.

TABLE 7-3   Wave Digital Equivalents of Commonly Used One-Ports

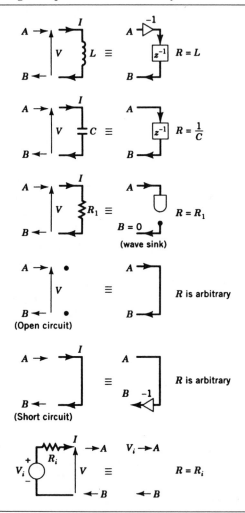

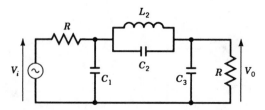

FIGURE 7-38   Example of a doubly terminated lossless two-port.

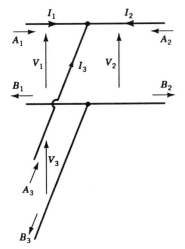

FIGURE 7-39 Three one-ports connected in parallel.

adaptor. If we wish to interconnect any three one-ports (with port resistances R_1, R_2, and R_3) in parallel, the procedure would be to design the structure of Figure 7-40 (with corresponding values of R_1, R_2, R_3) and connect the one-ports to its terminals.

The parallel adaptor described by Eq. (7.79) requires seven adders and two multipliers. A simplification of this structure is possible if the designer has the freedom to constrain the port resistances such that

$$G_2 = G_1 + G_3, \tag{7.81}$$

with G_n defined to be $1/R_n$. This freedom happens to be available in practice (see Example 7.5). Substituting Eq. (7.81) in Eq. (7.79) results in

$$B_1 = (-1 + \alpha)A_1 + A_2 + (1 - \alpha)A_3,$$

$$B_2 = \alpha A_1 + (1 - \alpha)A_3, \qquad B_3 = \alpha A_1 + A_2 - \alpha A_3, \tag{7.82}$$

where α is equal to α_1 evaluated from Eq. (7.80). Equation (7.82) can be implemented with one multiplier and four adders (Figure 7-41). Since B_2 is not connected directly to A_2 port #2 is called "reflection free" and is indicated by the

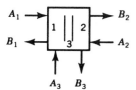

FIGURE 7-40 The three-port parallel adaptor.

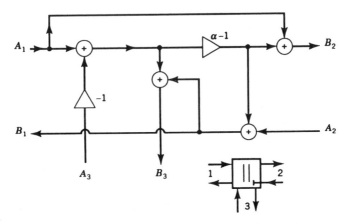

FIGURE 7-41   The three-port parallel adaptor with a reflection-free port (port 2).

notation shown in Figure 7-41. Perhaps the most important result of the reflection-free feature is that it enables us to interconnect the adaptors to adjacent elements without giving rise to delay-free loops.

A series interconnection (Figure 7-42) can be simulated in a similar manner by a *series adaptor* (Figure 7-43). With port resistances R_1, R_2, and R_3, and with A_k and B_k defined in the usual manner, the series constraint (i.e., $V_1 + V_2 + V_3 = 0$ and $I_1 = I_2 = I_3$) can be written as $B_k = A_k - \beta_k(A_1 + A_2 + A_3)$, with $k = 1, 2, 3$, where

$$\beta_k = 2R_k/(R_1 + R_2 + R_3). \tag{7.83}$$

While interconnecting series adaptors with other elements, it is necessary to avoid delay-free loops. This is often ensured by constraining R_k according to

$$R_2 = R_1 + R_3, \tag{7.84}$$

which results in a reflection-free port #2. Figure 7-44, which implements this version of the series adaptor, again requires only one multiplier and four adders.

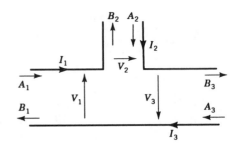

FIGURE 7-42   Three one-ports, connected in series.

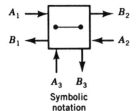

Symbolic
notation

FIGURE 7-43 The notation for a three-port serial adaptor.

Example 7.5. For a complete design example, consider Figure 7-38. First identify the interconnections as shown in Figure 7-45. Interconnection #2 is a series connection, while the others are parallel connections. Once we assign numbers to the three individual ports in each interconnection, it remains only to replace the electrical elements using Table 7-3 and to replace the interconnections using adaptors. During each such replacement, the only judgment required is the choice of the appropriate port resistances. Let R_{ik} denote the port resistances of the kth interconnection in Figure 7-45, with $k = 1, 2, 3, 4$ and $i = 1, 2, 3$. These are chosen as follows: $R_{11} = R$, which is the source resistance. From Table 7-3,

$$R_{31} = 1/C_1, \qquad R_{14} = L_2, \qquad R_{34} = 1/C_2, \qquad R_{33} = 1/C_3, \qquad R_{23} = R.$$
(7.85)

The resistances R_{21}, R_{24}, and R_{22} are chosen to ensure that appropriate ports of the adaptors are reflection-free:

$$G_{21} = G_{11} + G_{31}, \qquad G_{24} = G_{14} + G_{34}, \qquad R_{22} = R_{12} + R_{32}, \quad (7.86)$$

with $G_{ik} \triangleq 1/R_{ik}$. The resistances R_{12}, R_{32}, and R_{13} are chosen so that two interconnected ports have the same resistance:

$$R_{12} = R_{21}, \qquad R_{32} = R_{24}, \qquad R_{13} = R_{22}. \qquad (7.87)$$

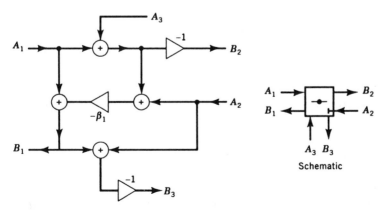

Schematic

FIGURE 7-44 The simplified series adaptor.

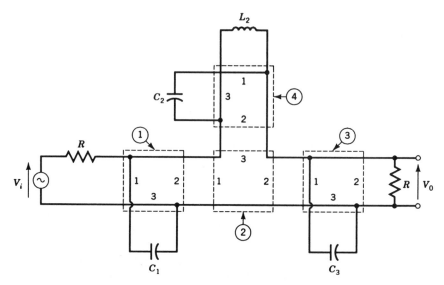

FIGURE 7-45   Pertaining to Example 7.5.

This results in a unique choice for all the port resistances. The values of α's and β's required in the adaptors can now be found from relations of the form in Eqs. (7.80) and (7.83).

It is interesting to note that if the series adaptor in Figure 7-44 is terminated on port #3 with a delay, the result is identical to the Type 1A two-pair shown in Figure 7-32. In this way, the z-domain synthesis approach based on two-pair extraction encompasses the wave filters [VA84, VA85d]. In an article by Swamy and Thyagarajan [SW75] and an article by Constantinides [CO76], certain new approaches to wave filter design are proposed based on the viewpoint that electrical elements should be simulated as two-ports rather than one-ports. The resulting structures do not make explicit use of adaptors. Instead, they are based on a simple set of first- and second-order digital two-pairs. The Type 1A LBR two-pair belongs to this class of two-pairs. Further discussions on the relations between wave filters, orthogonal filters, and the structurally passive approach (or the LBR approach) can be found in Vaidyanathan [VA85d].

7-8 PASSIVE FIR LATTICE STRUCTURES BASED ON LBR BUILDING BLOCKS

A FIR transfer function $G(z)$ with real coefficients, satisfying $|G(e^{j\omega})| \leq 1$, is called a FIR BR function. Such functions can always be implemented [VA86b] by a cascaded interconnection of lossless building blocks. Figure 7-46 shows the overall appearance of these structures. The auxiliary transfer function $H(z)$ is the complementary function introduced in Section 7-6-2 and satisfies Eq. (7.37). The

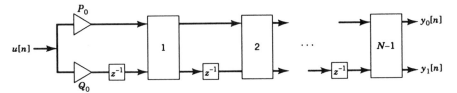

FIGURE 7-46 Overall appearance of the proposed FIR lattice structures. Here $G(z) = Y_0(z)/U(z)$ and $H(z) = Y_1(z)/U(z)$.

vector

$$\mathbf{G}_{N-1}(z) = \begin{bmatrix} G(z) \\ H(z) \end{bmatrix} \tag{7.88}$$

is FIR LBR and satisfies

$$\mathbf{G}_{N-1}^{\dagger}(e^{j\omega})\mathbf{G}_{N-1}(e^{j\omega}) = 1. \tag{7.89}$$

The subscript $N - 1$ above is equal to the order of the filters $G(z)$ and $H(z)$.

7-8-1 The Synthesis Procedure

Given an arbitrary FIR BR function $G(z)$, the procedure to synthesize the structure of Figure 7-46 will be described in this section. The first step is to construct the complementary FIR function $H(z)$. This is done by identifying $H(z)$ as a spectral factor of $1 - |G(e^{j\omega})|^2$. Once the FIR LBR vector $\mathbf{G}_N(z)$ is constructed, we construct lower order FIR LBR vectors by repeated application of an order reduction process.

This reduction process can be explained by referring to Figure 7-47. Here, $\mathbf{G}_m(z) = [P_m(z)\ Q_m(z)]^t$ is a FIR LBR vector of order m, where

$$P_m(z) = \sum_{n=0}^{m} p_{m,n}z^{-n}, \qquad Q_m(z) = \sum_{n=0}^{m} q_{m,n}z^{-n}. \tag{7.90}$$

We assume the extreme coefficients $p_{m,m}$ and $p_{m,0}$ to be nonzero to avoid trivialities. Given $\mathbf{G}_m(z)$, we wish to create a lower order FIR LBR vector $\mathbf{G}_{m-1}(z) =$

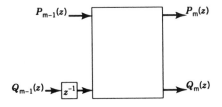

FIGURE 7-47 The order-reduction problem.

$[P_{m-1}(z)\ Q_{m-1}(z)]^t$, where

$$P_{m-1}(z) = \sum_{n=0}^{m-1} p_{m-1,n} z^{-n}, \qquad Q_{m-1}(z) = \sum_{n=0}^{m-1} q_{m-1,n} z^{-n}. \qquad (7.91)$$

It can be shown that this is achieved if we define $P_{m-1}(z)$ and $Q_{m-1}(z)$ to be

$$P_{m-1}(z) = k_m P_m(z) + \hat{k}_m Q_m(z), \qquad (7.92)$$

$$z^{-1} Q_{m-1}(z) = -\hat{k}_m P_m(z) + k_m Q_m(z), \qquad (7.93)$$

with k_m and $\hat{k}_m$ chosen as

$$k_m = \frac{-q_{m,m}}{\sqrt{p_{m,m}^2 + q_{m,m}^2}}, \qquad \hat{k}_m = \frac{p_{m,m}}{\sqrt{p_{m,m}^2 + q_{m,m}^2}}. \qquad (7.94)$$

The choices in Eqs. (7.94) ensure that the coefficient of z^{-m} on the RHS of Eq. (7.92) is zero, as verified by direct substitution. Because of the LBR property of $\mathbf{G}_m(z)$, this choice of k_m *simultaneously* ensures that the coefficient of z^0 on the RHS of Eq. (7.93) is also zero [VA86b]. Note that $k_m^2 + \hat{k}_m^2 = 1$, so that $0 \le k_m^2 \le 1$.

We can interpret this order reduction process by saying that a 2×2 transfer matrix $\mathbf{T}_m(z)$ has been extracted from the FIR LBR vector $\mathbf{G}_m(z)$ to obtain the remainder FIR LBR vector $\mathbf{G}_{m-1}(z)$. The extracted transfer matrix

$$\mathbf{T}_m(z) = \begin{bmatrix} k_m & -\hat{k}_m z^{-1} \\ \hat{k}_m & k_m z^{-1} \end{bmatrix} \qquad (7.95)$$

can be verified to be LBR too. The recursion Eqs. (7.92) and (7.93) can be initiated with $P_{N-1}(z) = G(z)$ and $Q_{N-1}(z) = H(z)$. Repeated application of the order reduction step results in the lattice structure of Figure 7-46, which is a sequence of planar rotation operators (as in Figure 7-48) separated by delays. We can interpret the FIR filtering operation of Figure 7-46 as a sequence of complex multiplications, with a delay inserted on the imaginary channel at each stage. *Any* FIR (BR) function $G(z)$ can be implemented like this.

In Figure 7-46, it can be verified that the transfer functions from the input of

FIGURE 7-48 The operation of the orthogonal matrix on a real vector is equivalent to a complex multiplication $y_1 + jy_2 = (x_1 + jx_2)e^{j\theta}$.

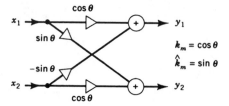

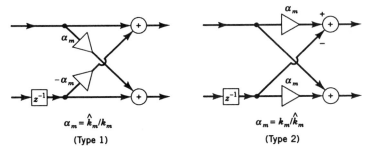

(Type 1)   (Type 2)

FIGURE 7-49   The denormalized FIR lattice sections.

the structure to any multiplier input is BR, so that the corresponding $\mathcal{L}_2$ norm is bounded by unity. The price paid for this automatic scaling property is the existence of four multipliers per section. Denormalized versions of the lattice structure (with two multipliers per section) can be obtained by scaling all four entries of Eq. (7.95) by a constant, say, k_m or $\hat{k}_m$. Clearly, this affects $G(z)$ and $H(z)$ only by a common scale factor. This results in two types of denormalized sections, shown in Figure 7-49. With such sections, the total number of multipliers in the structure is $2N - 1$ (approximately $2N$). If $G(z)$ and $H(z)$ are implemented in direct form, a total of about $2N$ multipliers would in general be required, unless one of them has linear phase. If both $G(z)$ and $H(z)$ are nonlinear-phase functions (as required in certain filter bank applications [VA88]), then the complexity of the lattice structure is about the same as the direct form, but the lattice structure offers better passband sensitivity. It is worth noting in this context that if two FIR transfer functions satisfy the power complementary property, then they cannot *both* be linear-phase functions [VA85f] (unless they have trivial frequency responses like $\cos K\omega$ and $\sin K\omega$). If one of the transfer functions, say, $P_{N-1}(z)$, has linear phase, it can be shown [VA86b] that the lattice coefficients satisfy $k_m = k_{N-1-m}$ and $\hat{k}_m = \Delta \hat{k}_{N-1-m}$ for $0 \le m \le N - 1$. Here, $\Delta = 1$ if the coefficients of $G(z)$ are symmetric and $\Delta = -1$ if they are antisymmetric.

It is worth noting the similarity and differences of this lattice structure with the FIR linear prediction lattice (or LPC lattice [MA76, MA78]), reproduced in Figure 7-50. The similarity is in the overall appearance. The difference is in the details of each section. The sections in Figures 7-48 and 7-49 are orthogonal, whereas those in Figure 7-50 are not. Because of this, with k_m^2 bounded by unity, the LPC lattice can realize only minimum phase $G(z)$, whereas the lattice section of Figure 7-48 can be used to realize arbitrary FIR BR functions.

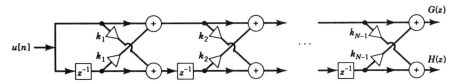

FIGURE 7-50   The FIR lattice structure associated with linear prediction.

Example 7.6. Let $G(z)$ be a linear-phase, equiripple, FIR lowpass filter with specifications $\omega_p = 0.196\pi$, $\omega_s = 0.27\pi$, peak-to-peak passband ripple $2\delta_1 \leq 0.0026$, and stopband attenuation $A_s \geq 32$ dB. Using McClellan–Parks algorithm (Section 4-8), the order of $G(z)$ turns out to be $N - 1 = 60$. Assuming $|G(e^{j\omega})| \leq 1$, the complementary function $H(z)$ can be computed. Since $G(z)$ has linear phase, $H(z)$ cannot have linear phase. In any case, $H(z)$ can be computed as a spectral factor of $1 - |G(e^{j\omega})|^2$ in an efficient manner (without solving for polynomial factors) using a standard approach such as the one in Mian and Nainer [MI82]. The method outlined in Mian and Nainer [MI82] works even if $|H(e^{j\omega})|^2$ has zeros on the unit circle. Once we have $G(z)$ and $H(z)$, we let $P_{N-1}(z) = G(z)$ and $Q_{N-1}(z) = H(z)$ and initiate the recursion Eqs. (7.92) and (7.93). The resulting k_m and $\hat{k}_m$ are shown in Table 7-4. Because of the symmetry of the coeffi-

TABLE 7-4   The Coefficients of the FIR Lossless Lattice in Example 7.6

Section Number	k_m	$\hat{k}_m$
0	0.002331	0.999997
1	0.999871	0.016038
2	0.999939	0.011081
3	0.999974	0.007208
4	−1.000000	0.
5	−0.999956	0.009332
6	−0.999894	0.014533
7	−0.999927	0.012089
8	−0.999999	0.001686
9	0.999926	0.012183
10	0.999762	0.021803
11	0.999793	0.020337
12	0.999981	0.006167
13	−0.999889	0.014912
14	−0.999505	0.031474
15	−0.999481	0.032228
16	−0.999915	0.013047
17	0.999824	0.018739
18	0.998941	0.046000
19	0.998748	0.050015
20	0.999760	0.021926
21	−0.999587	0.028726
22	−0.997249	0.074126
23	−0.996791	0.080049
24	−0.999647	0.026561
25	0.997416	0.071841
26	0.987856	0.155375
27	0.991294	0.131669
28	−0.995787	0.091698
29	−0.867826	0.496868
30	−0.713467	0.700689

cients of $G(z)$, we have $k_m = k_{60-m}$ and $\hat{k}_m = \hat{k}_{60-m}$. Accordingly, the coefficients are listed in Table 7-4 only for $0 \leq m \leq 30$. Figure 7-51 shows the simulated responses of the lattice transfer functions. The effect of coefficient quantization can be seen in Figure 7-52 and 7-53. Figure 7-52 is the response of $G(z)$ for the denormalized lattice with coefficients quantized to 3 bits in canonic SD code. Figure 7-53 is the direct form response with coefficients quantized to the same extent. These results demonstrate the low passband sensitivity property of the lattice. The price paid for this is in terms of the number of multipliers: there are a total of 121 multipliers in the denormalized lattice structure. In comparison, the direct form implementation of $[G(z), H(z)]$ requires about 90 multiplications (because $G(z)$ has linear phase). In addition to low sensitivity, the denormalized lattice has the advantage that $G(z)$ and $H(z)$ remain power complementary (except for a fixed scale factor independent of frequency) despite quantization! This feature is impossible to achieve with the direct form.

7-8-2 Extension to Arbitrary Number of Filters

There are applications in filter bank theory where it is necessary to have a set of M FIR BR functions $G_k(z)$, $0 \leq k \leq M-1$, which satisfy the property

$$\sum_{k=0}^{M-1} |G_k(e^{j\omega})|^2 = 1, \tag{7.96}$$

which is an extension of the power complementary property. The set of M filters basically split the input signal into M frequency bands for subband processing.

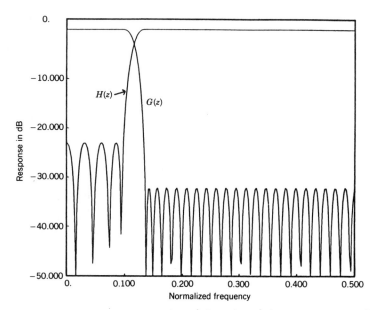

FIGURE 7-51 The simulated responses $|G(e^{j\omega})|$ and $|H(e^{j\omega})|$ for the FIR lattice © IEEE, 1986 [VA86b].

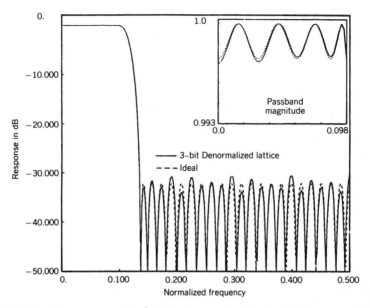

FIGURE 7-52   The response $|G(e^{j\omega})|$ for the quantized lattice (3 bits per coefficient in SD code) © IEEE, 1986 [VA86b].

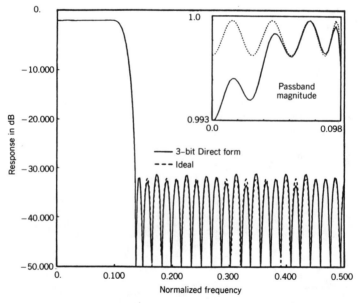

FIGURE 7-53   The response $|G(e^{j\omega})|$ for the 3-bit direct form structure © IEEE, 1986 [VA86b].

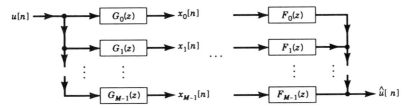

FIGURE 7-54 An analysis/synthesis system.

These signals $x_k[n]$ are eventually recombined at the "receiver end." Figure 7-54 shows an analysis/synthesis system based on this idea. The "analysis filters" are $G_k(z)$, $0 \leq k \leq M - 1$, and the M synthesis filters (or the recombination filters) are defined by

$$F_k(z) = z^{-(N-1)}G_k(z^{-1}), \qquad (7.97)$$

with $N - 1$ denoting the order of $G_k(z)$. The overall system transfer function $T(z)$ $= \hat{U}(z)/U(z)$ is therefore equal to

$$T(z) = z^{-(N-1)} \sum_{k=0}^{M-1} G_k(z^{-1})G_k(z), \qquad (7.98)$$

which reduces to $T(z) = z^{-(N-1)}$ because of Eq. (7.96).

Given a set of M FIR functions satisfying Eq. (7.96), how can we synthesize them in a structurally lossless form? One way is to define a FIR LBR vector

$$\mathbf{G}_{N-1}(z) = [G_0(z) \quad G_1(z) \quad \cdots \quad G_{M-1}(z)]^t \qquad (7.99)$$

and then initiate an order reduction process, whereby successively lower order FIR LBR vectors (with M components) are generated by extraction of $M \times M$ LBR building blocks. It is shown in Vaidyanathan [VA86b] that this procedure works for any FIR LBR vector $\mathbf{G}_{N-1}(z)$. The resulting structure is shown in Figure 7-55. Each $M \times M$ building block $\mathcal{K}_n$ here is a sequence of $M - 1$ planar rotations.

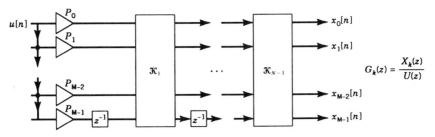

FIGURE 7-55 Implementation of a set of M power-complementary FIR filters as a cascade of lossless building blocks. Here $\sum_{k=0}^{M-1} P_k^2 = 1$ and $\mathcal{K}_m$ are orthogonal matrices.

The structure has $N - 1$ delays (which is minimal). Figure 7-55 is an automatically $\mathcal{L}_2$-scaled realization for the same reason explained with reference to Figure 7-46.

A more common application of the filter bank system of Figure 7-54 is in situations where the subband signals $x_k[n]$ (outputs of $G_k(z)$) are *decimated* before transmission. The reconstruction filters in this case should be chosen to cancel aliasing [SM87] in addition to the power-complementary requirement. The solution to this problem and lattice structures that *structurally* cancel aliasing and simultaneously force Eq. (7.96) can be found in Vaidyanathan [VA87b, VA87e] and Doganata et al. [DO88]. These are called *structurally perfect* reconstruction multirate systems.

7-9 ROUNDOFF NOISE IN STRUCTURALLY PASSIVE AND LOSSLESS SYSTEMS

In the last few sections we found that passive implementations based on lossless building blocks have low-sensitivity properties. This is true with wave filters, orthogonal filters, and other structures described in Sections 7-6, 7-7, and 7-8. A natural question in this context is: How do these implementations behave when the internal signals (and not just the multipliers) go through the quantization process? As mentioned in Section 6-3-2, two types of effects due to internal signal quantization can be distinguished, namely, roundoff noise and limit cycles. In this section we outline some results on roundoff noise in passive and lossless implementations. The next section deals with limit cycles in digital filters, including passive/lossless systems.

7-9-1 State-Space Manifestation of Losslessness

In Section 7-4, we dealt with state-space structures and descriptions and presented results on roundoff noise propagation, scaling, and the attainment of minimum noise. Because of the rich theory available for state-space systems, it is natural to ask how the allpass and lossless properties manifest in state-space form.

In response to this question, we begin by looking at one of the most popular structures for lossless functions, namely, the normalized allpass lattice (i.e., Figure 7-24 with building blocks as in Figure 7-30). If we denote the outputs of the delay elements as state variables (Figure 7-56) the state-space equations can be

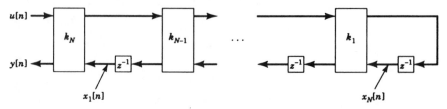

FIGURE 7-56 The state variables, identified in the IIR allpass lattice. Here $G_N(z) = Y(z)/U(z)$ is IIR allpass.

written out explicitly. For example, with $N = 2$, we find

$$\mathbf{A} = \begin{bmatrix} -k_2 k_1 & \hat{k}_1 \\ -k_2 \hat{k}_1 & -k_1 \end{bmatrix}, \quad \mathbf{B} = \begin{bmatrix} \hat{k}_2 k_1 \\ \hat{k}_2 \hat{k}_1 \end{bmatrix}, \quad \mathbf{C} = [\hat{k}_2 \ \ 0], \quad D = k_2. \quad (7.100)$$

If we put together the matrices $\mathbf{A}$, $\mathbf{B}$, $\mathbf{C}$, D into the form

$$\mathbf{R} = \begin{bmatrix} \mathbf{A} & \mathbf{B} \\ \mathbf{C} & D \end{bmatrix}, \quad (7.101)$$

we can verify that $\mathbf{R}$ is an *orthogonal* matrix! More generally it can be shown [VA85e] that for arbitrary N, the state-space description for Figure 7-56 is such that $\mathbf{R}$ is orthogonal. Now we know that given any stable allpass function $G_n(z)$, we can synthesize the structure of Figure 7-56 with $k_m^2 < 1$. We also know that the number of sections (hence the number of delays) is the smallest possible for the given order N, so that this is a minimal realization. Putting these together, we conclude that given any stable allpass function, we can always find a minimal realization whose state-space description has orthogonal $\mathbf{R}$. The converse of this result, which is not really obvious, also turns out to be true. These can be compactly stated as follows [VA85e].

State-Space Manifestation of Allpass Property. A transfer function $G(z)$ with real coefficients is stable allpass if and only if there exists a minimal realization whose state-space description is such that $\mathbf{R}$ in Eq. (7.101) is orthogonal.

7-9-2 Roundoff Noise in Allpass Lattice Structures

Since Figure 7-56 represents a recursive structure, it is necessary to insert quantizers. It is *sufficient* to insert these quantizers prior to the delay elements as indicated in Figure 7-57(a), so as to avoid infinite bit accumulation. With the quantizer scheme shown in this figure, the internal wordlength grows as we move from section to section toward the right. A better quantization scheme would be to introduce *two* quantizers between adjacent sections as shown in Figure 7-57(b).

We first consider the arrangement of Figure 7-57(a). Since the state variables are the only ones that are being quantized, the noise model of Figure 7-7 holds; hence we can analyze noise and dynamic range properties simply by looking at

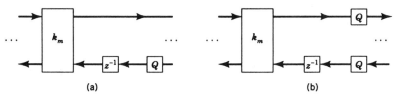

(a)   (b)

FIGURE 7-57 Two possible quantizer arrangements for allpass lattice: (a) one quantizer per section and (b) two quantizers per section.

the **K** and **W** matrices defined in Eq. (7.23). These matrices are easily computed, because of the impressive way in which the allpass property of the structure reflects itself into the state space, as discussed next.

We know that for the realization in Figure 7-56 with the normalized building blocks of Figure 7-30, the matrix **R** is orthogonal (i.e., $\mathbf{R}'\mathbf{R} = \mathbf{I}$). One of the consequences is the equation

$$\mathbf{A}'\mathbf{A} + \mathbf{C}'\mathbf{C} = \mathbf{I}. \tag{7.102}$$

The orthogonality of **R** also implies $\mathbf{RR}' = \mathbf{I}$, from which we obtain the condition

$$\mathbf{AA}' + \mathbf{BB}' = \mathbf{I}. \tag{7.103}$$

By using the above two equations in Eq. (7.23) (and by using the fact that $\mathbf{A}^n \to \mathbf{0}$ as $n \to \infty$ for stable **A**), it can be shown that $\mathbf{K} = \mathbf{W} = \mathbf{I}$. This important conclusion can be stated as follows.

Minimum Noise Property of the Lattice Structure. The Mullis–Roberts conditions for minimum noise gain (presented in Section 7-4-2) are clearly satisfied by the normalized lattice structure, simply because it has $\mathbf{K} = \mathbf{W} = \mathbf{I}$. In other words, *it represents a scaled, minimum noise structure!*

Furthermore, the total roundoff noise at the output terminal $y[n]$ of the structure is

$$\sigma_f^2 = \sum_{n=1}^{N} W_{nn}\sigma_e^2 = N\sigma_e^2, \tag{7.104}$$

where the quantizer noise variance is $\sigma_e^2 = 2^{-2b}/12$. The total noise gain, N, is therefore independent of the pole location.

Automatic Scaling Property of the Lattice. Since $\mathbf{K} = \mathbf{I}$, we have $K_{nn} = 1$ for all n, so the state variables are automatically scaled in the $\mathcal{L}_2$ sense. It can in fact be shown, based on this, that the impulse response from the filter input to any multiplier input has $\mathcal{L}_2$ norm equal to unity. In summary, the structure is scaled in the $\mathcal{L}_2$ sense.

In Section 7-6-3 we saw that the normalized structure and the denormalized structures (based on building blocks of Figures 7-25 and 7-29, respectively) are equivalent as far as the input–output relation is concerned. It can be verified [VA87a] that the state-space descriptions of the three types of lattice structures are related by similarity transformations, which are *diagonal* matrices. As a result, the noise gain parameter $\sum_{n=1}^{N} K_{nn}W_{nn}$ (Eq. (7.26)) is the same for these three structures. The conclusion is that the structure of Figure 7-57 with any of these building blocks has the same noise gain parameter. Its value is equal to the minimum possible value N *regardless of the pole locations.* The choice of a specific lattice building block affects only the scaling properties.

An immediate application of this result is in the parallel allpass-based implementation of Figure 7-15(a), where each allpass function is implemented as a
lattice. The total noise gain from the input $u[n]$ to the output $y_1[n]$ is equal to
$(N - r) + r = N$, where N is the order of the transfer function $G(z)$. (This assumes
that the noise sequences due to the two allpass functions are uncorrelated to each
other.) This result is true *regardless* of the pole locations of the filter!

Let us now turn attention to the more practical scheme of Figure 7-57(b), where
two quantizers are used per section. Assume again that the noise generated by each
quantizer is uncorrelated with others. Now the transfer function from the location
of a quantizer Q_1 on the top row to the location of the corresponding quantizer Q_2
at the bottom row is an allpass function (since the $G_m(z)$ in Figure 7-28 are allpass
whenever the rectangular building blocks take on any one of the forms in Figure
7-25 or 7-29 or 7-30). The noise gain for Q_2 is therefore the same as for Q_1,
namely, W_{nn}, which is equal to unity. In conclusion, the total noise variance at the
output terminal is simply twice the amount in Figure 7-57(a).

7-9-3   A Noise Gain Result for LBR Transfer Matrices

Consider Figure 7-58, which represents a $p \times r$ transfer matrix $\mathbf{T}(z)$. Assume
that the r inputs $e_k[n]$ are random sequences (such as the quantization error). Then
the output sequences $f_k[n]$ are also random. If $e_k[n]$ represent roundoff noise
sources, they are typically white and uncorrelated with each other. In other words,
assuming zero mean and equal variance,

$$E[e_k[n]e_l[n']] = \sigma_e^2\delta[k - l]\delta[n - n'], \qquad (7.105)$$

where $\delta[n]$ is the unit-pulse sequence defined in Chapter 2.

If the transfer matrix $\mathbf{T}'(z)$ is LBR, then $f_k[n]$ are also white and uncorrelated.
More specifically, we have[5] the following noise gain.

Noise Gain Due to Lossless Systems. Let $\mathbf{T}(z)$ be a $p \times r$ transfer matrix whose
inputs $e_k[n]$, $0 \le k \le r - 1$, are uncorrelated, zero mean, white random processes

---

[5]We assume all coefficients and sequences to be real for simplicity; all results here extend easily to the
complex case.

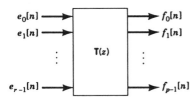

FIGURE 7-58 Propagation of a random process through a linear system.

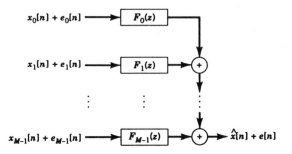

FIGURE 7-59 A power-complementary synthesis bank.

with common variance σ_e^2. If $\mathbf{T}'(z)$ is LBR, then the output sequences $f_k[n]$ are also uncorrelated white random process, with zero mean and equal variance σ_e^2.

This result finds applications in the analysis of noise propagation in several filter bank systems and subsystems. As an example, consider Figure 7-59, where the M filters $F_k(z)$ are power complementary; that is,

$$\sum_{k=0}^{M-1} |F_k(e^{j\omega})|^2 = 1.$$

If we define

$$\mathbf{T}(z) = [F_0(z) \quad F_1(z) \quad \cdots \quad F_{M-1}(z)], \tag{7.106}$$

then $\mathbf{T}'(z)$ is clearly LBR. If $e_k[n]$ are noise sources generated from an earlier stage (such as an analysis bank as in Figure 7-54) and if they satisfy the standard assumptions stated earlier, then the noise affecting $\hat{x}[n]$ is white, with variance σ_e^2.

As a second application, recall the FIR analysis bank of Figure 7-55, where the M filters $G_k(z)$ are again power complementary. Here, the roundoff noise is generated by the orthogonal matrices $\mathcal{K}_m$. Each of these matrices generates a set of noise sources $e_{k,m}[n]$, $0 \le k \le M - 1$. These are typically zero mean, uncorrelated, and white, with variance, say, σ_e^2. The transfer matrix from the location of these M sources of the mth stage to the M output terminals of the system is a cascade of orthogonal matrices separated by delays and is therefore LBR. As a result, the contribution $f_{k,m}[n]$, $0 \le k \le M - 1$, at these output terminals, due to the noise sources generated by the mth orthogonal matrix $\mathcal{K}_m$, can be analyzed using the above methods. Thus, for a given m, the sequences $f_{k,m}[n]$ form a set of M white uncorrelated noise sequences with zero mean and variance σ_e^2. If the noise sources due to different orthogonal matrices (i.e., for different m) are uncorrelated, then the total noise $f_k[n]$ at the kth output terminal in Figure 7-55 is white with variance $N\sigma_e^2$. Moreover, $f_k[n]$ and $f_l[n']$ are uncorrelated for $k \ne l$.

7-10   IIR FILTER STRUCTURES FREE FROM LIMIT CYCLES[6]

When the input $u[n]$ to a digital filter is turned off, the signals stored in the delay elements (the state variables) are still nonzero and result in a nonzero output sequence. Under ideal conditions where we do not have quantizers (i.e., an implementation with infinite precision and dynamic range), the state variables and the output eventually approach zero (assuming that the poles of the ideal system are inside the unit circle). When we have quantizers, the story is different [PA71] because quantization is a nonlinear operation. The quantizers form closed loops with linear elements (Figure 7-60) and this causes instability problems. It is a nontrivial matter to force the internal energy in the state variables to go to zero, unless the nonlinear feedback loops satisfy certain conditions for stability. These behaviors were discussed in Section 6-7-3 along with examples.

Thus, under certain conditions, a nonzero amount of energy gets "trapped" into the system, resulting in sustained oscillations called *granular* or *roundoff limit cycles*. It is also sometimes possible for the internal energy to grow with time and result in an overflow of internal register(s); the overflow phenomenon can be sustained in the form of steady oscillations called *overflow oscillations* or overflow limit cycles.

For a second-order direct form structure (Figure 7-1 with $N = 2$), it is well known [EB69; OP75] that in the absence of quantizers, the system is stable if and only if d_1 and d_2 belong to the shaded region in Figure 7-61 (i.e., anywhere inside the big triangle). However, with two's complement arithmetic, it can be shown that under zero input it is necessary (and sufficient) for d_1 and d_2 to belong to the cross-hatched subregion of Figure 7-61, in order to avoid overflow oscillations with arbitrary initial states.

Stability properties of nonlinear loops have been well understood in system theory [CL75; VI78] and there exist "sufficient conditions" under which oscillations can be suppressed. For example, if the quantizer is the magnitude truncation type (so that it does not increase the energy) and if $T(z)$ is strictly BR (i.e., $|T(e^{j\omega})| < 1$ for all ω), granular limit cycles are absent. In practice, however, there are several quantizers, and the complete feedback loop is in the form of a transfer matrix $\mathbf{T}(z)$ looking back into a set of quantizers. It is in general a nontrivial task to force $\mathbf{T}(z)$ to be strictly BR. A more convenient tool is required to handle this problem.

---

[6]In this section, "limit cycles" always stand for "zero-input limit cycles." Limit cycles with nonzero inputs are not considered here.

FIGURE 7-60   Closed loops formed by quantizers.

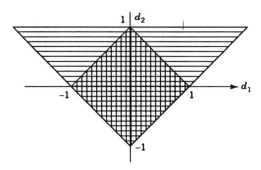

FIGURE 7-61   Stability regions for a second-order system.

For first-order sections (Figure 6-22), if we use magnitude truncation arithmetic (and as long as $|a| < 1$), limit cycles are absent. For second-order sections, Classen et al. [CL73] have shown that with magnitude truncation arithmetic, the probability of granular limit cycles is very small. Since transfer functions of arbitrary order can be implemented in cascade form with first-and second-order sections, we can greatly reduce the possibility of granular limit cycles simply by using the cascade form with magnitude truncation arithmetic.

It is sometimes desirable to suppress (both types of) limit cycles *completely*. It is also often necessary to use low-sensitivity structures that are more advanced than the cascade form. For these reasons, several techniques have been independently reported in the literature [BA77; FE75; GR80; MI78, VA87a] for complete suppression of limit cycles of both types. A vast majority of them can be understood based on a single result [VA87a], which uses passivity and state-space description of the structures. The purpose of this section is to indicate several instances (wave filters, orthogonal filters, lattice structures, minimum norm implementations, and second-order minimum *noise* implementations) where this result helps one to see how limit cycles (of both types) are suppressed. Unless mentioned otherwise, b-bit fixed-point arithmetic is assumed.

In contrast to the above instances, there are certain other approaches where the limit cycle is suppressed by very careful design of the quantizer alone. For example, it is shown in Mitra and Lawrence [MI81b] how the second-order direct-form structure can be designed to be free from limit cycles, by a complicated choice of quantization rules. The details of this are outside the scope of this section.

7-10-1   A Useful Set of Sufficient Conditions for Absence of Limit Cycles

A quantizer takes a number x and converts it into a b-bit fixed-point fraction $Q(x)$. The quantizer is said to be *passive* if it satisfies

$$|Q(x)| \leq |x| \quad \text{for all } x. \tag{7.107}$$

If x belongs to the permitted dynamic range[7] $\Re$, then quantization creates the error e defined as $e = Q(x) - x$. If we adopt a *magnitude truncation* scheme, then this quantization is clearly passive. Next, if x happens to belong to the region outside $\Re$ (i.e., x has overflowed; e.g., $x \geq 1$ in two's complement arithmetic), the quantizer has to adopt a certain convention so as to map x back into the permitted range $\Re$. As elaborated in Section 6-7-2 (and Figure 6-31), two conventions are commonly employed, namely, *saturation arithmetic* and *two's complement overflow arithmetic*. In saturation arithmetic, we define

$$Q(x) = \begin{cases} 1 - 2^{-b} & \text{for } x \geq 1 - 2^{-b} \\ -1 & \text{for } x \leq -1. \end{cases} \tag{7.108}$$

In two's complement overflow arithmetic, we simply add or subtract an integer multiple of 2 to x, and bring $Q(x)$ to the desired range $\Re$. In either case, it is clear that the quantization operation is passive. Thus a magnitude truncation quantizer, with such a well-defined scheme to handle overflow, is passive.

Recall now that we can write a state-space description and obtain models with quantizers preceding the delay (as in Figure 7-7) for a number of structures. Such a description holds for the lattice structures of Figure 7-57(a) and for wave filters, even though these are not state-space *structures* (see Section 7-4). With the model of Figure 7-7, we can suppress limit cycles of both types with passive quantizers, as long as **A** is strictly passive; that is,

$$v'A'Av < v'v, \qquad v \neq 0. \tag{7.109}$$

This means that the energy in the vector Av is strictly less than the energy of the nonzero vector v. The model of Figure 7-7 redrawn for zero input as Figure 7-62, is described by the equations

$$w[n + 1] = Ax[n], \tag{7.110}$$

$$x[n + 1] = Q(w[n + 1]), \tag{7.111}$$

where the notation $Q(v)$ means that each component of the vector v is quantized independently of the others. Because of Eq. (7.109), the energy in $w[n + 1]$ is strictly less than that in $x[n]$ (unless $x[n] = 0$). Also, the passivity of the quan-

---

[7]For example, the permitted range $\Re$ is $-1 \leq x < 1$ in two's complement arithmetic.

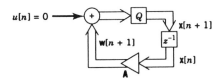

$u[n] = 0$
$w[n + 1]$
$x[n + 1]$
$x[n]$

FIGURE 7-62 The closed loop with zero input.

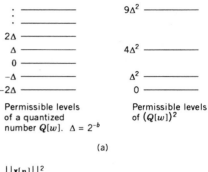

Permissible levels of a quantized number $Q[w]$. $\Delta = 2^{-b}$

Permissible levels of $(Q[w])^2$

(a)

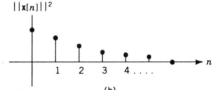

FIGURE 7-63 Pertaining to limit cycle suppression by incorporation of passivity.

(b)

tizers ensures that the energy of $x[n + 1]$ is no larger than that of $w[n + 1]$. As a result, the energy of $x[k]$ is a strictly decreasing function of k. Since the quantizer permits only a quantized set of energy levels (Figure 7-63(a)), the energy decreases at least by an amount Δ^2 each unit of time, where $\Delta = 2^{-b}$. This is demonstrated in Figure 7-63(b). After a finite number of recursions, the energy clearly becomes zero, and there can be no self-sustained limit cycles.

The condition Eq. (7.109) is not satisfied by an arbitrary state transition matrix A, even if all its eigenvalues are strictly inside the unit circle. For example, let

$$A = \begin{bmatrix} 0.9 & 0 \\ \eta & 0.5 \end{bmatrix}.$$

The eigenvalues of this matrix are 0.9 and 0.5 and do not depend on η. Now

$$v = \begin{bmatrix} 1 \\ 0 \end{bmatrix} \Rightarrow Av = \begin{bmatrix} 0.9 \\ \eta \end{bmatrix}.$$

So the energy of Av can be made arbitrarily large by increasing η. So the A matrix is stable but not passive for arbitrary η.

Let

$$\alpha_M = \max_k |\lambda_k|,$$

where λ_k, $1 \le k \le N$, are the N eigenvalues of A. For a stable system $\alpha_M < 1$. It can be shown that

$$\max_{v \neq 0} \frac{v'A'Av}{v'v} \ge \alpha_M^2. \tag{7.112}$$

The quantity on the left-hand side is the square of the $\mathcal{L}_2$ norm of $\mathbf{A}$ (denoted $\|\mathbf{A}\|$). The above inequality says that $\|\mathbf{A}\|$ is at least as large as α_M. In particular, even if $\alpha_M < 1$, it is possible for $\|\mathbf{A}\|$ to exceed unity.

If $\mathbf{A}$ happens to be such that $\|\mathbf{A}\|$ is *equal to* the smallest possible value α_M, the structure is called a minimum norm structure. For such structures Eq. (7.109) holds. For this reason, minimum norm realizations are free from limit cycles when passive quantizers are used. A family of minimum norm structures is the class for which $\mathbf{A}$ satisfies the property $\mathbf{A}'\mathbf{A} = \mathbf{A}\mathbf{A}'$. These are called *normal form* structures [BA77] (and $\mathbf{A}$ is said to be a normal matrix). These are precisely the set of matrices for which $\mathbf{A}$ can be diagonalized by a unitary matrix [FR68].

Example 7.7. Consider the second-order *coupled form structure* of Figure 7-9. The $\mathbf{A}$ matrix here has the form

$$\mathbf{A} = r \begin{bmatrix} \cos\theta & -\sin\theta \\ \sin\theta & \cos\theta \end{bmatrix}, \qquad (7.113)$$

so that $\mathbf{A}'\mathbf{A} = r^2\mathbf{I}$. This implies that $\|\mathbf{A}\|$ is equal to $r < 1$. Higher order normal form structures can be obtained by using a parallel interconnection of the coupled form sections (and possible first-order sections). In these systems, the matrix $\mathbf{A}$ is block-diagonal, with diagonal blocks of the form Eq. (7.113) (and possible 1×1 diagonal blocks representing first-order sections). Such block-diagonal normal forms can always be found as long as the poles of the system are distinct [BA77].

Further Examples of Limit-Cycle-Free Structures. There exist other kinds of normal form realizations [LI88] with *circulant* and *skew circulant* $\mathbf{A}$ matrices. These have the advantage that the state recursion is essentially a circular convolution. Because of the minimum norm property, these structures are also free from limit cycles.

7-10-2 An Improved Set of Sufficient Conditions

There exist many structures, such as the lattice structures of Section 7-6-3, which do not satisfy the minimum norm condition, even though they are free from limit cycles. This behavior is explained by invoking an improved set of sufficient conditions [VA87a] for absence of limit cycles. This improvement comes in two respects. First, the condition Eq. (7.109) need not be a strict inequality. Second, a diagonal similarity transformation on $\mathbf{A}$ does not affect the limit-cycle-free behavior. The following result covers these generalizations.

More General Conditions for Limit Cycle Suppression. Let $\mathbf{A}$ be a stable matrix (i.e., all eigenvalues inside the unit circle), which in addition satisfies

$$\mathbf{A}'\mathbf{D}\mathbf{A} \leq \mathbf{D}, \qquad (7.114)$$

for some diagonal matrix $\mathbf{D}$ with positive diagonal entries, and let the quantizers (Figure 7-62) be passive. Then there can be no zero-input limit cycles of either type. With zero input, the internal energy $\mathbf{x}'[n]\mathbf{x}[n]$ decreases to zero within a finite amount of time.[8]

As a special case, it is clear that with $\mathbf{D} = \mathbf{I}$ and with strict inequality in Eq. (7.114), we obtain the condition Eq. (7.109). If a state-space structure satisfies Eq. (7.114), and if we apply the similarity transformation Eq. (7.20) with diagonal $\mathbf{T}$, the resulting $\mathbf{A}_1$ satisfies $\mathbf{A}_1' \mathbf{D}_1 \mathbf{A}_1 \leq \mathbf{D}_1$ for $\mathbf{D}_1 = \mathbf{TDT}$ and hence continues to be free from limit cycles (because $\mathbf{D}_1$ is also diagonal with positive diagonal elements).

Example 7.8. As a first application of this, consider the *normalized allpass lattice structure* (i.e., Figure 7-24 with the normalized building blocks of Figure 7-30). We know that the state-space description satisfies Eq. (7.102), so that

$$v'\mathbf{A}'\mathbf{A}v = v'v - v'\mathbf{C}'\mathbf{C}v, \tag{7.115}$$

which implies $\mathbf{A}'\mathbf{A} \leq \mathbf{I}$. Thus Eq. (7.114) holds with $\mathbf{D} = \mathbf{I}$. This explains why this structure is free from limit cycles when passive quantizers are used. However, the state energy $\mathbf{x}'[n]\mathbf{x}[n]$ does not necessarily decrease for every increment of n (unlike in Section 7-10-1, where we had $\mathbf{A}'\mathbf{A} < \mathbf{I}$ rather than $\mathbf{A}'\mathbf{A} \leq \mathbf{I}$). It can be shown [VA87a], however, that this energy decreases (at least by Δ^2) at least once in N units of time. As a result, $\mathbf{x}(n)$ does become zero in a finite amount of time.

The same result holds if the denormalized building blocks are used, because these are related to the normalized structure via a *diagonal* similarity transformation (as mentioned in Section 7-9-2). By writing out the state-space descriptions explicitly, it can be verified for the same reasons that wave digital filters and orthogonal filters are also free from limit cycles if passive quantizers are used.

Use of Two Quantizers per Lattice Section. All these results are based on the model of Figure 7-62. This model assumes that quantizers are not introduced anywhere expect prior to the delay elements. Even though this is a sufficient condition to avoid infinite bit accumulation, this is not always a convenient scheme. For example, as pointed out in Section 7-9-2, it is often desirable to have two quantizers per section in the lattice structure (Figure 7-57(b)). With this scheme, the preceding results cannot be applied directly, but an indirect energy-balance argument [VA87g] shows that this scheme is also free from limit cycles, when the quantizers are passive.

7-10-3 Random Rounding: Another Technique to Suppress Limit Cycles

Recall that any real-coefficient transfer function can be implemented as a cascade of first- and second-order *direct form* sections. The first-order sections do not sup-

---

[8]Please review Section 1-5 for notations such as $\mathbf{P} \leq \mathbf{Q}$.

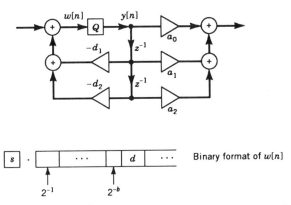

FIGURE 7-64 Pertaining to the "random rounding" method of suppressing granular limit cycles.

port limit cycles as long as quantizers are passive. The second-order sections can, in principle, exhibit limit cycle oscillations. Several authors have, however, proposed a scheme called random rounding [BA77; KI76; LA78], which has been found to suppress granular limit cycles in second-order sections.

To describe the idea consider Figure 7-64, which shows a second-order section with one quantizer. The quantizer converts its input $w[n]$ to a b-bit number $y[n]$, by *rounding*. The binary format of $w[n]$ is shown in the figure. Consider the binary digit corresponding to the $(b + 1)$th bit (indicated as d). Normally, the value of this digit governs whether the quantized number $y[n]$ is larger or smaller than the unquantized number $w[n]$.

In the technique called *random rounding*, this digit d is replaced with a binary random variable (a variable whose value is 0 or 1 with equal probability for each). This random replacement is done for each time instant n. (This can be done by generating an uncorrelated binary random sequence $r[n]$, and replacing d with $r[n]$ for each n.) It has been found [BU77] that this technique suppresses zero-input granular limit cycles. Heuristically speaking, random rounding introduces randomness into the quantizer error, thereby preventing the possibility of a periodic behavior (which would cause a limit cycle). The price paid is an increased amount of *roundoff noise*. For further details, see [BU77].

7-11 CONCLUDING REMARKS

In this chapter several techniques have been presented for robust implementation of digital filters. This includes cascade form structures, structures with error feedback, wave digital filters, and, finally, a parallel connection of two allpass filters.

The parallel connection of two allpass filters (Section 7-6-2) can be used to realize a wide variety of filters, including Butterworth, Chebyshev, and elliptic digital filters. This structure has several distinct advantages. First, the number of multipliers is very small. For example, an Nth order (odd N) elliptic lowpass filter $G(z)$ can be implemented with only N multipliers (less than that for direct form, which requires $1.5N$ multipliers). Moreover, an additional complementary transfer function $H(z)$ is available with no extra cost (except a two-input adder). Next, if each allpass filter is implemented in a structurally lossless manner (Section 7-6-3), the filter $G(z)$ remains stable (and satisfies $|G(e^{j\omega})| \leq 1$). This results in low passband sensitivity. If each allpass function is implemented using the lattice structure with passive quantizers, the structure is free from zero-input limit cycle oscillations. Finally, the lattice structure offers a *minimum noise* implementation of each of the two allpass functions.

The reader should also note that most of the well-known structures that enjoy robustness properties (such as low sensitivity, low noise, and freedom from limit cycles) have some kind of passivity incorporated into them.

REFERENCES

[AN73] B. D. O. Anderson and S. Vongpanitlerd, *Network Analysis and Synthesis*. Prentice-Hall, Englewood Cliffs, NJ, 1973.

[AN79] A. Antoniou, *Digital Filters: Analysis and Synthesis*. McGraw-Hill, New York, 1979.

[AN86] R. Ansari, Multilevel IIR digital filters. *IEEE Trans. Circuits Syst.* **CAS-33**, 337–341 (1986).

[BA69] N. Balabanian and T. Bickart, *Electrical Network Theory*. Wiley, New York, 1969.

[BA77] C. W. Barnes and A. T. Fam, Minimum norm recursive digital filters that are free of overflow limit cycles. *IEEE Trans. Circuits Syst.* **CAS-24**, 569–574 (1977).

[BA84] C. W. Barnes, On the design of optimal state-space realizations of second order digital filters. *IEEE Trans. Circuits Syst.* **CAS-31**, 602–608 (1984).

[BE68] V. Belevitch, *Classical Network Synthesis*. Holden-Day, San Francisco, CA, 1968.

[BO85] B. W. Bomar, New second order state space structures for realizing low roundoff noise digital filters. *IEEE Trans. Acoust. Speech, Signal Process.* **ASSP-33**, 106–110 (1985).

[BU77] M. Buttner, Elimination of limit cycles in digital filters with very low increase in the quantization noise, *IEEE Trans. Circuits and Syst.* **CAS-24**, 300–304 (1977).

[CH81] T. L. Chang and S. A. White, An error cancellation digital filter structure and its distributed arithmetic implementation. *IEEE Trans. Circuits Syst.* **CAS-28**, 339–342 (1981).

[CL73] T. A. C. M. Classen, W. F. G. Mecklenbräuker, and J. B. H. Peek, Second order digital filter with only one magnitude truncation quantizer and having practically no limit cycles. *Electron. Lett.* **9**, 531–532 (1973).

[CL75] T. A. C. M. Classen, W. F. G. Mecklenbräuker, and J. B. H. Peek, Frequency domain criteria for the absence of zero-input limit cycles in nonlinear discrete-time systems, with applications to digital filters. *IEEE Trans. Circuits Syst.* **CAS-22**, 232–239 (1975).

[CO76] A. G. Constantinides, Design of digital filters from LC ladder networks. *Proc. Inst. Electr. Eng.* **123,** 1307–1312 (1976).

[DE80] E. Deprettere and P. DeWilde, Orthogonal cascade realization of real multiport digital filters. *Int. J. Circuit Theory Appl.* **8,** 245–277 (1980).

[DE85] P. Dewilde, E. Deprettere, and R. Nouta, Parallel and pipelined VLSI implementation of signal processing algorithms. In *VLSI and Modern Signal Processing* (S. Y. Kung, H. J. Whitehouse, and T. Kailath, eds.). Prentice-Hall, Englewood Cliffs, NJ, 257–276 1985.

[DO88] Z. Doganata, P. P. Vaidyanathan, and T. Q. Nguyen, General synthesis procedures for FIR lossless transfer matrices for perfect-reconstruction multirate filter bank applications. *IEEE Trans. Acoust., Speech, Signal Process.* **ASSP-36,** 1561–1574 (1988).

[EB69] P. M. Ebert, J. E. Mazo, and M. G. Taylor, Overflow oscillations in digital filters. *Bell Syst. Tech. J.* **48,** 2999–3020 (1969).

[FE71] A. Fettweis, Digital filter structures related to classical filter networks. *AEU* **25,** 79–81 (1971).

[FE73] A. Fettweis, Pseudopassivity, sensitivity, and stability of wave digital filter. *IEEE Trans. Circuit Theory* **CT-19,** 668–673 (1973).

[FE74] A. Fettweis, Wave digital lattice filters. *Int. J. Circuit Theory Appl.* **2,** 203–211 (1974).

[FE75] A. Fettweis and K. Meerkötter, Suppression of parasitic oscillations in wave filters. *IEEE Trans. Circuits Syst.* **22,** 239–246 (1975).

[FR68] J. Franklin, *Matrix Theory.* Prentice-Hall, Englewood Cliffs, NJ, 1968.

[GA85] L. Gaszi, Explicit formulas for lattice wave digital filters. *IEEE Trans. Circuits Syst.* **CAS-32,** 68–88 (1985).

[GO69] B. Gold and C. M. Rader, *Digital Processing of Signals.* McGraw-Hill, New York, 1969.

[GR73] A. H. Gray, Jr. and J. D. Markel, Digital lattice and ladder filter synthesis. *IEEE Trans. Audio Electroacoust.* **AU-21,** 491–500 (1973).

[GR75] A. H. Gray, Jr. and J. D. Markel, A normalized digital filter structure. *IEEE Trans. Acoust., Speech, Signal Process.* **ASSP-23,** 268–277 (1975).

[GR80] A. H. Gray, Jr., Passive cascaded lattice digital filters. *IEEE Trans. Circuits Syst.* **CAS-27,** 337–344 (1980).

[GU57] E. A. Guillemin, *Synthesis of Passive Networks.* Wiley, New York, 1957.

[HE83] D. Henrot and C. T. Mullis, A modular and orthogonal digital filter structure for parallel processing. *IEEE Int. Conf. Acoust. Speech Signal Process., 1983* pp. 623–626 (1983).

[HI84] W. E. Higgins and D. C. Munson, Jr., Optimal and suboptimal error spectrum shaping for cascade-form digital filters. *IEEE Trans. Circuits System.* **CAS-31,** 429–437 (1984).

[JA70a] L. B. Jackson, On the interaction of roundoff noise and dynamic range in digital filters. *Bell Syst. Tech. J.* **49,** 159–184 (1970).

[JA70b] L. B. Jackson, Roundoff noise analysis for fixed point digital filters realized in cascade or parallel form. *IEEE Trans. Audio Electroacoust.* **AU-18,** 107–122 (1970).

[JA79] L. B. Jackson, A. G. Lindgren, and Y. Kim, Optimal synthesis of second-order state-space structures for digital filters. *IEEE Trans. Circuits Syst.* **CAS-26,** 149–153 (1979).

[JU64] E. I. Jury, *Theory and Application of the Z-transform Method.* Wiley, New York, 1964.

[KA66] J. F. Kaiser, Digital filters. In *System Analysis by Digital Computer,* (F. F. Kuo and J. F. Kaiser, eds.), Chapter 7. Wiley, New York, 1966.

[KI76] R. B. Kieburtz, V. B. Lawrence, and K. V. Mina, Control of limit cycles in recursive digital filters by randomized quantization, *Proc. 9th IEEE Int. Symp. Cir. Syst.* Munich, Germany, 624–627, (1976).

[KO90] R. D. Koilpillai, P. P. Vaidyanathan, and S. K. Mitra, On arbitrary-level IIR and FIR filters. *IEEE Trans. Circuits Syst.* **CAS-37,** 280–284 (1990).

[LA78] V. B. Lawrence and K. V. Mina, Control of limit cycle oscillations in second-order recursive digital filters using constrained random quantization, *IEEE Trans. Acoustics, Speech and Signal Processing,* **ASSP-26,** (2) 127–134 (1978).

[LI71] B. Liu, Effects of finite wordlength on the accuracy of digital filters—A review. *IEEE Trans. Circuit Theory* **CT-18,** 670–677 (1971).

[LI75] B. Liu and A. Peled, Heuristic optimization of the cascade realization of fixed-point digital filters. *IEEE Trans. Acoust., Speech, Signal Process.* **ASSP-23,** 467–473 (1975).

[LI88] V. C. Liu and P. P. Vaidyanathan, Circulant and skew-circulant matrices as new normal-form realization of IIR digital filters. *IEEE Trans. Circuits Syst.* **CAS-35,** 625–635 (1988).

[MA76] J. D. Markel and A. H. Gray, Jr., *Linear Prediction of Speech.* Springer-Verlag, New York, 1976.

[MA78] J. Makhoul, A class of all-zero lattice digital filters: properties and applications. *IEEE Trans. Acoust., Speech, Signal Process.* **ASSP-26,** 304–314 (1978).

[MI74] S. K. Mitra and K. Hirano, Digital allpass networks. *IEEE Trans. Circuits Syst.* **CAS-21,** 688–700 (1974).

[MI77] S. K. Mitra, P. S. Kamat, and D. C. Huey, Cascaded lattice realization of digital filters. *Int. J. Circuit Theory Appl.* **5,** 3–11 (1977).

[MI78] W. L. Mills, C. T. Mullis, and R. A. Roberts, Digital filter realizations without overflow oscillations. *IEEE Trans. Acoust., Speech, Signal Processing.* **ASSP-26,** 334–338 (1978).

[MI81a] W. L. Mills, C. T. Mullis, and R. A. Roberts, Low roundoff noise and normal realizations of fixed point IIR digital filters. *IEEE Trans. Acoust., Speech, Signal Process.* **ASSP-29,** 893–903 (1981).

[MI81b] D. Mitra and V. B. Lawrence, Controlled rounding arithmetic, for second-order direct form digital filters, that eliminate all self-sustained oscillations. *IEEE Trans. Circuits Syst.* **CAS-28,** 894–905 (1981).

[MI82] G. A. Mian and A. P. Nainer, A fast procedure to design equiripple minimum-phase FIR filters. *IEEE Trans. Circuits Syst.* **CAS-29,** 327–331 (1982).

[MU76] C. T. Mullis and R. A. Roberts, Synthesis of minimum roundoff noise fixed point digital filters. *IEEE Trans. Circuit Syst.* **CAS-23,** 551–562 (1976).

[MU82] C. T. Mullis and R. A. Roberts, An interpretation of error spectrum shaping digital filter structures. *IEEE Trans. Acoust., Speech, Signal Process.* **ASSP-30,** 1013–1015 (1982).

[NG88] T. Q. Nguyen and P. P. Vaidyanathan, Maximally decimated perfect-reconstruction FIR filter banks with pairwise mirror-image analysis (and synthesis) frequency responses. *IEEE Trans. Acoust., Speech, Signal Process.* **ASSP-36**, 693–706 (1988).

[OP72] A. V. Oppenheim and C. J. Weinstein, Effects of finite register length in digital filtering and the fast fourier transform. *Proc. IEEE* **60**, 957–976 (1972).

[OP75] A. V. Oppenheim and R. W. Schafer, *Digital Signal Processing*. Prentice-Hall, Englewood Cliffs, NJ, 1975.

[OP89] A. V. Oppenheim and R. W. Schafer, *Discrete-Time Signal Processing*. Prentice-Hall, Englewood Cliffs, NJ, 1989.

[OR66] H. J. Orchard, Inductorless filters. *Electron. Lett.* **2**, 224–225 (1966).

[PA71] S. R. Parker and S. F. Hess, Limit cycle oscillations in digital filters. *IEEE Trans. Circuit Theory* **CT-8**, 687–697 (1971).

[RA75] L. R. Rabiner and B. Gold, *Theory and Application of Digital Signal Processing*. Prentice-Hall, Englewood Cliffs, NJ, 1975.

[RA84] S. K. Rao and T. Kailath, Orthogonal digital filters for VLSI implementation. *IEEE Trans. Circuits Syst.* **CAS-31**, 933–945 (1984).

[RE91] M. Renfors, S. K. Mitra, P. A. Regalia, and Y. Neuvo, Weighted complementary IIR digital filters. *Proc. IEEE Int. Symp. Circuits Syst., Singapore, 1991*, pp. 136–139 (1991).

[SA85] T. Saramaki, On the design of digital filters as a sum of two allpass filters. *IEEE Trans. Circuits Syst.* **CAS-32**, 1191–1193 (1985).

[SE73] A. Sedlmeyer and A. Fettweis, Digital filters with true ladder configuration. *Int. J. Circuit Theory Appl.* **1**, 5–10 (1973).

[SE78] A. S. Sedra and P. O. Brackett, *Filter Theory and Design: Active and Passive*. Matrix Publishers, Inc. Beaverton, OR, 1978.

[SM87] M. J. T. Smith and T. P. Barnwell, III, A new filter bank theory for time-frequency representation. *IEEE Trans. Acoust., Speech, Signal Process.* **ASSP-35**, 314–327 (1987).

[SW75] M. N. S. Swamy and K. Thyagarajan, A new type of wave digital filters. *J. Franklin Inst.* **300**, 41–58 (1975).

[SZ87] J. Szczupak, S. K. Mitra, and J. Fadavi-Ardekani, A computer-based method of realization of structurally LBR digital allpass networks. *IEEE Trans. Circuits Syst.* **CAS-35**, 755–760 (1988).

[TE77] G. C. Temes and J. W. LaPatra, *Circuit Synthesis and Design*. McGraw-Hill, New York, 1977.

[TH77] T. Thong and B. Liu, Error spectrum shaping in narrowband recursive digital filters. *IEEE Trans. Acoust., Speech, Signal Process.* **ASSP-25**, 200–203 (1977).

[VA84] P. P. Vaidyanathan and S. K. Mitra, Low passband sensitivity digital filters: A generalized viewpoint and synthesis procedures. *Proc. IEEE* **72**, 404–423 (1984).

[VA85a] P. P. Vaidyanathan, On error spectrum shaping in state space digital filters. *IEEE Trans. Circuits Syst.* **CAS-32**, 88–92 (1985).

[VA85b] P. P. Vaidyanathan, The doubly terminated lossless digital two-pair in digital filtering. *IEEE Trans. Circuits Syst.* **CAS-32**, 197–200 (1985).

[VA85c] P. P. Vaidyanathan, A general theorem for degree-reduction of a digital BR function. *IEEE Trans. Circuits Syst.* **CAS-32**, 414–415 (1985).

[VA85d] P. P. Vaidyanathan, A unified approach to orthogonal digital filters and wave digital filters, based on LBR two-pair extraction. *IEEE Trans. Circuits Syst.* **CAS-32**, 673–686 (1985).

[VA85e] P. P. Vaidyanathan, The discrete-time bounded real lemma in digital filtering. *IEEE Trans. Circuits Syst.* **CAS-32**, 918–924 (1985).

[VA85f] P. P. Vaidyanathan, On power complementary FIR filters. *IEEE Trans. Circuits Syst.* **CAS-32**, 1308–1310 (1985).

[VA86a] P. P. Vaidyanathan, S. K. Mitra, and Y. Neuvo, A new approach to the realization of low sensitivity IIR digital filters. *IEEE Trans. Acoust., Speech, Signal Process.* **ASSP-34**, 350–361 (1986).

[VA86b] P. P. Vaidyanathan, Passive cascaded lattice structures for low sensitivity FIR filter design, with applications to filter banks. *IEEE Trans. Circuits Syst.* **CAS-33**, 1045–1064 (1986).

[VA86c] P. P. Vaidyanathan, Impact of classical network-theoretic concepts in modern digital signal processing. *Proc. 20th Annu. Asilomar Conf. Signals, Syst. Comput. Pacific Grove, California*, (1986).

[VA87a] P. P. Vaidyanathan, and V. Liu, An improved sufficient condition for absence of limit cycles in digital filters. *IEEE Trans. Circuits Syst.* **CAS-34**, 319–322 (1987).

[VA87b] P. P. Vaidyanathan, Theory and design of M-channel maximally decimated quadrature mirror filters with arbitrary M, having perfect reconstruction property. *IEEE Trans. Acoust., Speech, Signal Process.* **ASSP-35**, 476–492 (1987).

[VA87c] P. P. Vaidyanathan, P. A. Regalia, and S. K. Mitra, Design of doubly complementary IIR digital filters using a single complex allpass filter, with multirate applications. *IEEE Trans. Circuits Syst.* **CAS-34**, 378–389 (1987).

[VA87d] P. P. Vaidyanathan, and S. K. Mitra, A unified structural interpretation of some well-known stability test procedures for linear systems. *Proc. IEEE* **75**, 478–497 (1987).

[VA87e] P. P. Vaidyanathan, Quadrature mirror filter banks, M-band extensions and perfect-reconstruction techniques. *IEEE ASSP Mag.* **4**, 4–20 (1987).

[VA87f] P. P. Vaidyanathan, Low-noise and low-sensitivity digital filters. In *Handbook of Digital Signal Processing*, (D. F. Elliot, ed.). Academic Press, Orlando, FL, 359–479 1987.

[VA87g] P. P. Vaidyanathan, *Quantization Effects in Multirate Filter Banks*, Tech. Rep. California Institute of Technology, Pasadena, 1987.

[VA88] P. P. Vaidyanathan, P.-Q. Hoang, Lattice structures for optimal design and robust implementation of two-channel perfect reconstruction QMF banks. *IEEE Trans. Acoust., Speech, Signal Process.* **ASSP-36**, 81–94 (1988).

[VI78] M. Vidyasagar, *Nonlinear System Analysis*. Prentice-Hall, Englewood Cliffs, NJ, 1979.

[YA82] G. T. Yan and S. K. Mitra, Modified coupled-form digital filter structures. *Proc. IEEE* **70**, 762–763 (1982).

8 Fast DFT and Convolution Algorithms

HENRIK V. SORENSEN
Department of Electrical Engineering
University of Pennsylvania, Philadelphia

C. SIDNEY BURRUS
Department of Electrical and Computer Engineering
Rice University, Houston, Texas

The concepts of discrete convolution and of the discrete Fourier transform (DFT) are central to most of digital signal processing. Unfortunately, directly computing the DFT of a sequence and direct convolution of two sequences both have a high complexity; that is, they require many operations (additions, multiplications or data manipulations/indexing) even for moderate length sequences. Hence much effort has gone into developing alternative and more efficient ways of implementing discrete convolutions and DFTs. This chapter discusses the most common algorithms and points out the trade-offs that are involved.

There are several methods for describing both convolution and DFT algorithms. This chapter derives each algorithm using the description that seems most appropriate rather than trying to put everything into a common framework that does not always fit.

The noncyclic (also called aperiodic or linear) convolution $y[n]$ of the two sequences $x[n]$ and $h[n]$ of length N and M, respectively, is defined as

$$y[n] = \sum_{m = \max(n + 1 - M, 0)}^{\min(n, N - 1)} x[m]h[n - m] \equiv x[n] \circledast h[n],$$

$$n = 0, 1, \ldots, N + M - 2, \tag{8.1}$$

where max( ) and min( ) are functions that return the maximum and minimum of their arguments, and the symbol $\circledast$ denotes noncyclic convolution. If it is assumed

Handbook for Digital Signal Processing, Edited by Sanjit K. Mitra and James F. Kaiser.
ISBN 0-471-61995-7 © 1993 John Wiley & Sons, Inc.

that $h[n]$ is defined as zero for $n < 0$ and for $n \geq M$, then Eq. (8.1) can be written

$$y[n] = \sum_{m=0}^{N-1} x[m]h[n - m], \tag{8.2}$$

which can also be written

$$y[n] = \sum_{m=n-N+1}^{n} h[m]x[n - m]. \tag{8.3}$$

Both of these expressions are more convenient to program than Eq. (8.1).

The cyclic convolution $y[n]$ of the sequences $x[n]$ and $h[n]$, both of length N, is defined as

$$y[n] = \sum_{m=0}^{N-1} x[m]h[\langle n - m \rangle_N] \equiv x[n] \,\textcircled{N}\, h[n], \qquad n = 0, 1, \ldots, N - 1, \tag{8.4}$$

where $\langle n - m \rangle_N$ is the residue of $n - m$ evaluated modulo N, and $\textcircled{N}$ denotes the length N cyclic convolution operation.

The DFT of $x[n]$ is defined by

$$X[k] = \sum_{n=0}^{N-1} x[n] W_N^{nk}, \qquad k = 0, 1, \ldots, N - 1, \tag{8.5}$$

and the inverse DFT is defined as

$$x[n] = \frac{1}{N} \sum_{k=0}^{N-1} X[k] W_N^{-nk}, \qquad n = 0, 1, \ldots, N - 1, \tag{8.6}$$

where $W_N = e^{-j2\pi/N}$. If $H[k]$ is the DFT of $h[n]$, the convolution operation in Eq. (8.4) can be calculated indirectly by multiplying the DFTs of $x[n]$ and $h[n]$. It can be proved [BU85; MC79b] that the sequence $y[n]$ from Eq. (8.4) can be computed as

$$y[n] = \frac{1}{N} \sum_{k=0}^{N-1} \{X[k]H[k]\} W_N^{-nk}, \qquad n = 0, 1, \ldots, N - 1, \tag{8.7}$$

which is the inverse DFT of $Y[k] = X[k]H[k]$.

Hence there is a strong relationship between DFTs and cyclic convolutions, and one way to view the DFT is as a method to transform a cyclic convolution into the Fourier (or frequency) domain, where the convolution operation is converted into the simpler multiplication operation (exactly as the log (x) function is a transform into the log domain, which converts a product into a simpler addition). As will be shown later, the majority of the operations in performing a convolution in the Fourier domain are in calculating the DFT and IDFT.

Section 8-1 of this chapter describes methods to compute digital convolution in the time domain (as opposed to Fourier domain) in detail and discusses which way is more efficient to compute both cyclic and noncyclic convolutions. Section 8-2 describes the various fast Fourier transforms (FFTs), which are efficient algorithms for computing the DFT. Aside from using the DFT to compute convolutions, it is also used to compute the frequency (Fourier) spectrum of a sequence. Other domains exist that have a similar convolution property and hence can be used for computing convolutions. Most of these transforms do not have a clean physical interpretation as does the DFT, and since they also normally require more operations than the DFT, few of them are used in practice. Section 8-3 describes some of the more well-known methods: the Hartley transform, the sine/cosine transforms, number theoretic transforms, and the Walsh–Hadamard transform.

Five programs to compute the DFT and IDFT are supplied in Appendix 8-A.

8-1 COMPUTATION OF CONVOLUTION AND FILTERING

This section describes algorithms for computing the discrete convolution without transforming the problem into the frequency domain. The algorithms will be derived using three descriptions of convolution. First, the summation notation in Eqs. (8.1) and (8.4) will be used. Second, a polynomial description [NU81] will be used. This may seem a little tedious, but it is essential for understanding some of the details of discrete convolutions. Third, a matrix notation that sometimes gives insight into FFT algorithms is reviewed [BL84; BU85].

Given a length N sequence $x[n]$, the polynomial $X(z)$ of degree $N - 1$ can be defined as

$$X(z) = \sum_{n=0}^{N-1} x[n]z^n; \tag{8.8}$$

that is, the coefficient to the term z^i equals the $(i + 1)$th element, $x[i]$, of the sequence. This is the same as the definition of the z-transform of $x[n]$ except for the sign of the exponent of z. If $H(z)$ is similarly defined as the polynomial of degree $M - 1$ of the sequence $h[n]$, the noncyclic convolution from Eq. (8.1) can be written [BL84; MC79b; NU81].

$$Y(z) = X(z)H(z). \tag{8.9}$$

The $M + N - 1$ elements of the output sequence $y[n]$ can be found as the coefficients of the polynomial $Y(z)$. In fact, the actual operations in computing convolution, polynomial multiplication, and integer multiplication (except for carrying) are the same.

To compute the cyclic convolution $y[n] = x[n] \, \circledN \, h[n]$, the sequences $x[n]$

$3 + 4z^3 + 5z^7 + 2z^{12} + 7z^{15} \bmod (z^6 - 1)$

$= 3 + 4z^3 + 5z^6 z + 2z^6 z^6 + 7z^6 z^6 z^3 \bmod (z^6 - 1) = 3 + 4z^3 + 5z + 2 + 7z^3$

$= 5 + 5z + 11z^3$

FIGURE 8-1 Example of polynomial reduction.

and $h[n]$ must have the same length, N. These sequences are converted to polynomials as above, and now the convolution is performed by

$$Y(z) = X(z)H(z) \quad \bmod(z^N - 1); \tag{8.10}$$

that is, the polynomial multiplication is reduced modulo the polynomial $(z^N - 1)$ [BL84; MC79b; NU81]. Evaluating a polynomial $X(z)H(z)$ modulo $(z^N - 1)$ is a well-known operation from algebra, and it means that all factors in $X(z)H(z)$ of z^N are equal to 1, since $z^N - 1 \cong 0 \Leftrightarrow z^N \cong 1$. Figure 8-1 shows an example of polynomial reduction modulo $(z^6 - 1)$. In matrix notation the noncyclic convolution in Eq. (8.1) can be written

$$
\begin{bmatrix}
y[0] \\
y[1] \\
\vdots \\
y[M-1] \\
y[M] \\
\vdots \\
y[N+M-1]
\end{bmatrix}
=
\begin{bmatrix}
h[0] & 0 & 0 & \cdots & 0 \\
h[1] & h[0] & 0 & \cdots & 0 \\
\vdots & \vdots & \vdots & & \vdots \\
h[M-1] & h[M-2] & h[M-3] & \cdots & 0 \\
0 & h[M-1] & h[M-2] & \cdots & 0 \\
\vdots & \vdots & \vdots & & \vdots \\
0 & 0 & 0 & \cdots & h[0]
\end{bmatrix}
$$

$$
\cdot
\begin{bmatrix}
x[0] \\
x[1] \\
\vdots \\
x[N-1]
\end{bmatrix}
\tag{8.11}
$$

and the cyclic convolution in Eq. (8.4) can be written

$$
\begin{bmatrix}
y[0] \\
y[1] \\
\vdots \\
y[N-1]
\end{bmatrix}
=
\begin{bmatrix}
h[0] & h[N-1] & \cdots & h[1] \\
h[1] & h[0] & \cdots & h[2] \\
\vdots & \vdots & & \vdots \\
h[N-1] & h[N-2] & \cdots & h[0]
\end{bmatrix}
\begin{bmatrix}
x[0] \\
x[1] \\
\vdots \\
x[N-1]
\end{bmatrix}
.
$$

$$\tag{8.12}$$

Example 8.1 shows a noncyclic and a cyclic convolution of the same two sequences $\{a_0, a_1, a_2\}$ and $\{b_0, b_1, b_2\}$ in both the polynomial and matrix notation.

Example 8.1. The noncyclic convolution

$$y[n] = \{a_0, a_1, a_2\} \circledast \{b_0, b_1, b_2\}$$

can be transformed into the polynomial product

$$
\begin{aligned}
Y(z) &= (a_0 + a_1 z^1 + a_2 z^2)(b_0 + b_1 z^1 + b_2 z^2) \\
&= (a_0 b_0) + (a_0 b_1 + a_1 b_0) z^1 + (a_0 b_2 + a_1 b_1 + a_2 b_0) z^2 \\
&\quad + (a_1 b_2 + a_2 b_1) z^3 + (a_2 b_2) z^4.
\end{aligned}
$$

So the noncyclic convolution of the two sequences is

$$y[n] = \{a_0 b_0,\ a_0 b_1 + a_1 b_0,\ a_0 b_2 + a_1 b_1 + a_2 b_0,\ a_1 b_2 + a_2 b_1,\ a_2 b_2\}.$$

It can also be written as a matrix product

$$
Y = \begin{bmatrix} a_0 & 0 & 0 \\ a_1 & a_0 & 0 \\ a_2 & a_1 & a_0 \\ 0 & a_2 & a_1 \\ 0 & 0 & a_2 \end{bmatrix} \begin{bmatrix} b_0 \\ b_1 \\ b_2 \end{bmatrix}.
$$

The cyclic convolution

$$y[n] = \{a_0, a_1, a_2\} \, ③ \, \{b_0, b_1, b_2\}$$

can be transformed into

$$
\begin{aligned}
Y(z) &= (a_0 + a_1 z^1 + a_2 z^2)(b_0 + b_1 z^1 + b_2 z^2) \bmod(z^3 - 1) \\
&= (a_0 b_0 + a_1 b_2 + a_2 b_1) + (a_0 b_1 + a_1 b_0 + a_2 b_2) z^1 \\
&\quad + (a_0 b_2 + a_1 b_1 + a_2 b_0) z^2.
\end{aligned}
$$

So the cyclic convolution is

$$y[n] = \{a_0 b_0 + a_1 b_2 + a_2 b_1,\ a_0 b_1 + a_1 b_0 + a_2 b_2,\ a_0 b_2 + a_1 b_1 + a_2 b_0\},$$

which is equivalent to the matrix product

$$
\mathbf{Y} = \begin{bmatrix} a_0 & a_2 & a_1 \\ a_1 & a_0 & a_2 \\ a_2 & a_1 & a_0 \end{bmatrix} \begin{bmatrix} b_0 \\ b_1 \\ b_2 \end{bmatrix}.
$$

By comparing the two results, it is obvious how the cyclic convolution is obtained by wrapping around the result of the noncyclic convolution.

Most fast convolution algorithms (including those based on the DFT) compute a cyclic convolution, while most applications call for a noncyclic convolution, and hence it will be discussed how to compute the noncyclic convolution using a cyclic convolution. Also, in most applications one of the sequences (e.g., the filter impulse response) is much shorter than the other (the input), so methods for sectioning the input into smaller blocks to minimize the delay on the output and to save operations will be described.

The noncyclic convolution, Eq. (8.1), can be computed directly using MN multiplications and $(M - 1)(N - 1)$ additions, while the cyclic convolution, Eq. (8.4), can be computed in N^2 multiplications and $N(N - 1)$ additions. These numbers grow very fast as N and M increase, and an alternative way to compute the convolutions is by converting the one-dimensional convolution into a multidimensional convolution. This method, which is presented in detail later, was first described for noncyclic convolution by Agarwal and Burrus [AG74a] and later refined to cyclic convolution, which is more computationally efficient, by Agarwal and Cooley [AG77] and Pollard [PO79]. Each of the one-dimensional convolutions in this multidimensional convolution are then performed using short-length, very efficient algorithms.

8-1-1   Implementation of Noncyclic Convolution as Cyclic Convolutions

When computing a cyclic convolution, the two sequences to be convolved are periodic; that is, the indices are evaluated modulo N as indicated in Eq. (8.4). Comparing the results of a cyclic and a noncyclic convolution of the same two length N sequences, one finds that the result of the cyclic convolution is obtained from the noncyclic by wrapping around the result of the noncyclic convolution, as illustrated in Example 8.1. This is equivalent to the aliasing that occurs when undersampling a time function, as described in Chapter 2. To compute a noncyclic convolution of a length N and a length M sequence using a cyclic convolution algorithm, the problem has to be converted to a form such that the wrapping around of the sequence does not alter the result [BL84; MC79b; NU81]. This is done by extending both sequences to length $N + M - 1$ by appending zeros and then computing a length $N + M - 1$ cyclic convolution. Because of the zeros in the sequences, the two polynomials will have degree $N - 1$ and degree $M - 1$ (the higher order coefficients are zero). Hence the highest order present in the polynomial product is $N + M - 2$, but since the convolution was computed modulo $(z^{N+M-1} - 1)$, there is no wraparound. Note that any number $L \geq M + N - 1$ will work equally well. Because some lengths have more efficient cycle convolution algorithms than others, it is sometimes advantageous to choose L greater than the minimum $M + N - 1$.

To compute the noncyclic convolution in Example 8.1 using a cyclic convolution, the sequences are extended to length $N = 3 + 3 - 1 = 5$, and the length 5 cyclic convolution $\{a_0, a_1, a_2, 0, 0\}$ ⑤ $\{b_0, b_1, b_2, 0, 0\}$ is then computed. This produces the five elements of the original length 3 noncyclic convolution.

8-1-2 Computation of Convolutions and Correlations by Sectioning

In most filtering applications, N, the length of the signal $x[n]$, is much greater than M, the length of the filter impulse response $h[n]$. One way of computing the noncyclic convolution in Eq. (8.1) is to convert the noncyclic convolution to a cyclic convolution as described in the previous section, and to implement this cyclic convolution using FFTs. However, acquiring the whole signal sequence $x[n]$ before the output $y[n]$ in Eq. (8.1) is computed requires a large amount of memory and implies a large delay in the output. To prevent this, the signal sequence is sectioned into smaller blocks, which are convolved with the filter impulse response $h[n]$ to obtain sections of the output. There are two well-known methods for doing this: the overlap-add and the overlap-save methods [BU85; RA75], which are both special cases of block processing techniques described in Burrus [BU72].

In the overlap-add method the input $x[n]$ is broken up into consecutive nonoverlapping blocks $x_i[n]$. The output $y_i[n]$ for each input $x_i[n]$ is computed separately by convolving (noncyclic) $x_i[n]$ with $h[n]$. The output blocks $y_i[n]$ are longer than the corresponding input blocks $x_i[n]$, and hence the output sections $y_i[n]$ will overlap each other in time. Convolution is a linear operation, and therefore the output $y[n]$ is obtained by adding the individual outputs $y_i[n]$, as shown in Figure 8-2.

In the overlap-save method, the input $x[n]$ is also broken up into blocks $x_i[n]$. If the blocks $x_i[n]$ are longer than the filter impulse response $h[n]$, each convolution $h[n]$ ⊛ $x_i[n]$ produces some values that are the actual values of the output $y[n]$, while some that are not. The overlap-save method keeps only the correct values and discards the rest. So to obtain the total output sequence $y[n]$, each block $x_i[n]$ of the input $x[n]$ has to contain overlapping input elements, as shown in Figure 8-3. The number of correct values per convolution equals the difference between the length of the input blocks $x_i[n]$ and the filter impulse response $h[n]$.

8-1-3 Short-Length Convolution and Polynomial Product Algorithms

This section describes short-length, very efficient cyclic and noncyclic convolution algorithms. These algorithms are used as building blocks for longer algorithms.

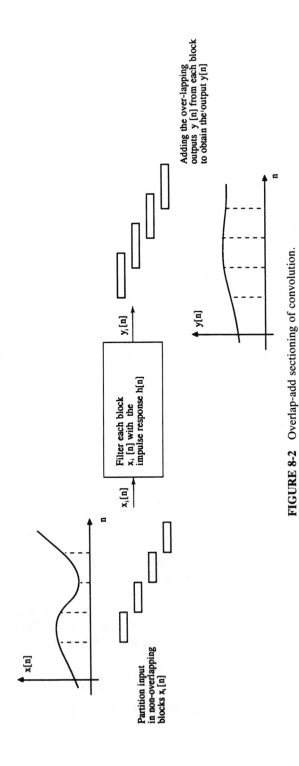

FIGURE 8-2 Overlap-add sectioning of convolution.

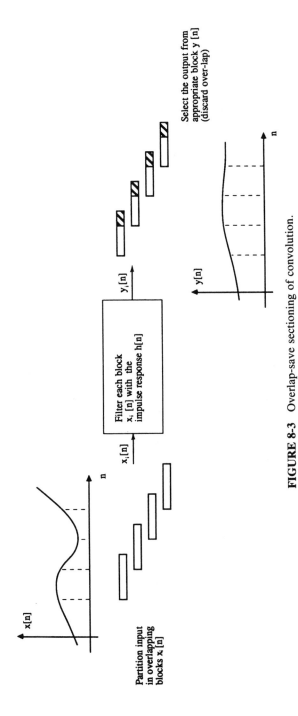

FIGURE 8-3 Overlap-save sectioning of convolution.

Toom–Cook Algorithm. The Toom–Cook algorithm computes short-length non-cyclic convolutions [BL84; MC79b; WI78]. Let the noncyclic convolution be written as a polynomial product (Eq. (8.9)),

$$Y(z) = X(z)H(z), \tag{8.13}$$

where $X(z)$, $H(z)$, and $Y(z)$ have degrees $N - 1$, $M - 1$, and $N + M - 2$, respectively. If $\alpha_0, \alpha_1, \ldots, \alpha_{N+M-2}$ are $N + M - 1$ distinct real numbers, then $Y(z)$ can be interpolated from $Y(\alpha_0), Y(\alpha_1), \ldots, Y(\alpha_{N+M-2})$ using Lagrange's interpolation formula [BL84; MC79b]

$$Y(z) = \sum_{i=0}^{N+M-2} Y(\alpha_i) \frac{\prod\limits_{j \neq i} (z - \alpha_j)}{\prod\limits_{j \neq i} (\alpha_i - \alpha_j)} \tag{8.14}$$

The Toom–Cook algorithm calculates $X(\alpha_i)$ and $H(\alpha_i)$, computes

$$Y(\alpha_i) = X(\alpha_i)H(\alpha_i), \quad i = 0, 1, \ldots, N + M - 1, \tag{8.15}$$

and then uses Lagrange's interpolation formula of Eq. (8.14) to calculate $Y(z)$. $X(\alpha_i)$ and $H(\alpha_i)$ can be computed without multiplications if α_i is a small integer (like $0, \pm 1, \pm 2$, etc.) or infinity (∞). Equation (8.15) takes $N + M - 1$ multiplications. The Lagrange reconstruction, Eq. (8.14), can also be computed using only additions, if it is assumed that one of the sequences (e.g., $H(z)$ – the filter) is fixed. The divisions that occur in Eq. (8.14) are factored into the fixed polynomials and hence do not influence the operation count for the convolution. Example 8.2 shows an example of a 2×2 noncyclic convolution.

Example 8.2. The noncyclic convolution

$$\{y_0, y_1, y_2\} = \{x_0, x_1\} \circledast \{h_0, h_1\}$$

can be transformed into the polynomial product

$$Y(z) = X(z)H(z) = (x_0 + x_1 z)(h_0 + h_1 z).$$

Now choose $\alpha_0 = 0$, $\alpha_1 = 1$, $\alpha_2 = \infty$, and calculate

$$Y(0) = X(0)H(0) = (x_0)(h_0),$$

$$Y(1) = X(1)H(1) = (x_0 + x_1)(h_0 + h_1),$$

$$Y(\infty) = X(\infty)H(\infty) = (x_1)(h_1).$$

$X(\infty)$ is evaluated using the fact that

$$X(z)|_{z=\infty} = X\left(\frac{1}{z}\right)\Bigg|_{1/z=0} = \left(x_0 \frac{1}{z} + x_1\right)\Bigg|_{1/z=0} = x_1;$$

and similar for $H(\infty)$. Use the Lagrange interpolation

$$Y(z) = y_0 + y_1 z + y_2 z^2 = Y(0)(1-z) + Y(1)(z) + Y(\infty)(z^2 - z),$$

from which the sequence $y[n]$ can be found:

$$y_0 = Y(0) = x_0 h_0,$$

$$y_1 = [Y(1) - Y(0) - Y(\infty)] = x_0 h_1 + x_1 h_0,$$

$$y_2 = Y(\infty) = x_1 h_1.$$

The algorithm requires three multiplications ($X(0) \times H(0)$, $X(1) \times H(1)$, and $X(\infty) \times H(\infty)$) and three additions/subtractions ($x_0 + x_1$ and $Y(1) - Y(0) - Y(\infty)$); note it is assumed that $(h_0 + h_1)$ is precomputed. Direct computation takes four multiplications and one addition, so the new algorithm has less multiplications but more total operations. It is, however, sometimes advantageous to minimize or lower the number of multiplications at the expense of total operations.

In Appendix 8-B a 2 × 2 and a 3 × 3 noncyclic convolution algorithm can be found (see also Nussbaumer [NU81] and Blahut [BL84]). Table 8-1 from Blahut [BL84] shows the operation counts for some noncyclic convolutions. Note that variations in the development allow trade-offs between the number of multiplications and additions.

TABLE 8-1 Number of Operations to Compute Short-Length Noncyclic Convolutions Using the Toom–Cook Algorithm

Operation Count for Noncyclic Convolution Algorithms		
Length	#Multiplications	#Additions
2 × 2	3	3
2 × 2	4	1
3 × 3	5	20
3 × 3	6	10
3 × 3	9	4
4 × 4	7	41
4 × 4	9	15

Winograd's Short-Length Cyclic Convolution Algorithms. For an efficient computation of short-length cyclic convolutions, the problem is again transformed into a polynomial product (Eq. (8.10)) [MC79b; SC84; WI77; WI78]:

$$Y(z) = X(z)H(z) \mod(z^N - 1). \qquad (8.16)$$

The problem is broken down using the Chinese remainder theorem (CRT) for polynomials [MC79b; SC84; WI77; WI78]. The polynomial $(z^N - 1)$ is well known from number theory, and it factors into [MC79b; NI80]

$$(z^N - 1) = \prod_{d|N} C_d(z), \qquad (8.17)$$

where C_d is called the dth cyclotomic polynomial, and $d|N$ means that d divides N. There are as many divisors of $(z^N - 1)$ as there are divisors of N (1 and N included), and this number is sometimes written as $\phi(N)$ in number theory. The cyclotomic polynomials can be found as

$$C_d(z) = \prod_{\substack{(i,d)=1 \\ i \le d}} (z - W_d^i), \qquad (8.18)$$

where (i, d) is the greatest common divisor of i and d, and $W_d = e^{-j2\pi/d}$. The cyclotomic polynomials are irreducible (equivalent to being prime for integers); that is, they cannot be factored into two polynomials of lower degree with rational coefficients. They are particularly interesting, because the cyclotomic polynomials up to the 105th degree have only coefficients that are ± 1 or 0. Example 8.3 shows how $(z^6 - 1)$ factors.

Example 8.3. The polynomial $(z^6 - 1)$ factors into

$$(z^6 - 1) = C_1(z)C_2(z)C_3(z)C_6(z),$$

where

$$C_1(z) = (z - W_1^1) = (z - 1),$$
$$C_2(z) = (z - W_2^1) = (z + 1),$$
$$C_3(z) = (z - W_3^1)(z - W_3^2) = (z^2 + z + 1),$$
$$C_6(z) = (z - W_6^1)(z - W_6^5) = (z^2 - z + 1).$$

In Winograd's approach, both $X(z)$ and $H(z)$ are reduced modulo the cyclotomic polynomials $C_i(z)$, yielding the residual polynomials $X_i(z)$ and $H_i(z)$. The products of these residual polynomials are then recombined using the polynomial CRT to give $Y(z)$ using fewer multiplications than Eq. (8.16). Specifically, $X(z)$ and $H(z)$ are first reduced by

$$X_i(z) \equiv X(z) \mod C_i(z), \quad \forall \, i \,|N, \qquad (8.19)$$
$$H_i(z) \equiv H(z) \mod C_i(z), \quad \forall \, i \,|N.$$

Next the products

$$Y_i(z) = X_i(z)H_i(z) \mod C_i(z), \quad \forall \, i \,|N, \qquad (8.20)$$

are formed from which $Y(z)$ can then be recovered, if its degree is less than N, by

$$Y(z) = \sum_{i|N} S_i(z) Y_i(z) \mod(z^N - 1), \qquad (8.21)$$

where $S_i(z)$ is a unique polynomial such that

$$S_i(z) \equiv \delta_{ij} \mod C_j(z), \quad \forall \, i, j \,|N \qquad (8.22)$$

and

$$\delta_{ij} = \begin{cases} 1, & \text{for } i = j \\ 0, & \text{otherwise.} \end{cases} \qquad (8.23)$$

$S_i(z)$ is chosen of the form [MC79b; NU81]

$$S_i(z) = P(z) \prod_{\substack{j|N \\ j \neq i}} C_j(z), \qquad (8.24)$$

such that

$$S_i(z) \equiv 0 \mod C_j(z), \quad \forall \, j \neq i, \, j \,|N. \qquad (8.25)$$

$P(z)$ is then determined such that

$$P(z) \prod_{\substack{j|N \\ j \neq i}} C_j(z) \equiv 1 \mod C_i(z), \qquad (8.26)$$

which can be restated as

$$P(z) \prod_{\substack{j|N \\ j \neq i}} C_j(z) + T(z)C_i(z) = 1 \qquad (8.27)$$

for some $T(z)$. This Diophantine equation can be solved with Euclid's algorithm for polynomials by a continued fraction expansion

$$\frac{C_i(z)}{\prod_{\substack{j|N \\ j \neq i}} C_j(z)} = R_0(z) + \cfrac{1}{R_1(z) + \cfrac{1}{R_2(z) + \cdots + \cfrac{1}{R_k(z)}}}. \tag{8.28}$$

From the approximations

$$\frac{\Psi_0(z)}{\Theta_0(z)} = R_0(z),$$

$$\frac{\Psi_1(z)}{\Theta_1(z)} = R_0(z) + \frac{1}{R_1(z)},$$

$$\vdots$$

$$\frac{\Psi_k(z)}{\Theta_k(z)} = \frac{C_i(z)}{P(z) \prod_{\substack{j|N \\ j \neq i}} C_j(z)}, \tag{8.29}$$

the two solutions are found as

$$P(z) = \beta(-1)^k \Psi_{k-1}(z),$$

$$T(z) = \beta(-1)^k \Theta_{k-1}(z), \tag{8.30}$$

where β is the leading coefficient of $\Psi_k(z)$, or it can easily be determined from Eq. (8.26). More details can be found in McClellan and Rader [MC79b] and Schroeder [SC84].

Winograd used this method to show that the minimum number of multiplications required to compute a length N cyclic convolution is

$$2N - K, \tag{8.31}$$

where $K = \phi(N)$ is the number of factors of $z^N - 1$ and also the number of divisors (including 1 and N) of N. Winograd proved [WI78] that the minimum number of multiplications always is achieved by the CRT approach. The method, however, only considers multiplications from computing Eq. (8.20) and does not count the rational coefficients that appear in the reconstruction polynomials $S_i(z)$ as illustrated in the next example. For small N these coefficients are fairly simple and can easily be computed by a set of shifts and additions, while for large N they become very difficult to apply without multiplications.

Example 8.4. The length 4 cyclic convolution $\{x_0, x_1, x_2, x_3\}$ ④ $\{h_0, h_1, h_2, h_3\}$ can be written

$$Y(z) = X(z)H(z) \quad \text{mod}(z^4 - 1),$$

where $X(z) = \sum_{i=0}^{3} x_i z^i$ and $H(z) = \sum_{i=0}^{3} h_i z^i$. Now factor $(z^4 - 1)$ into its cyclotomic polynomials

$$(z^4 - 1) = C_1(z) C_2(z) C_4(z) = (z - 1)(z + 1)(z^2 + 1).$$

Compute the polynomial reductions

$$X_1(z) \equiv X(z) \quad \text{mod } C_1(z) = (x_0 + x_1 + x_2 + x_3),$$

$$X_2(z) \equiv X(z) \quad \text{mod } C_2(z) = (x_0 - x_1 + x_2 - x_3),$$

$$X_4(z) \equiv X(z) \quad \text{mod } C_4(z) = (x_0 - x_2) + (x_1 - x_3)z,$$

and similarly for $H_1(z)$, $H_2(z)$, and $H_4(z)$. Compute next the $2N - K = 2 \times N - \phi(N) = 8 - 3 = 5$ multiplications:

$$m_1 = (x_0 + x_1 + x_2 + x_3)(h_0 + h_1 + h_2 + h_3),$$

$$m_2 = (x_0 - x_1 + x_2 - x_3)(h_0 - h_1 + h_2 - h_3),$$

$$m_3 = (x_0 - x_2)(h_0 - h_2),$$

$$m_4 = (x_1 - x_3)(h_1 - h_3),$$

$$m_5 = [(x_0 - x_2) + (x_1 - x_3)][(h_0 - h_2) + (h_1 - h_3)],$$

from which $Y_1(z)$, $Y_2(z)$, and $Y_4(z)$ are computed

$$Y_1(z) \equiv X_1(z) H_1(z) \quad \text{mod } C_1(z) = m_1,$$

$$Y_2(z) \equiv X_2(z) H_2(z) \quad \text{mod } C_2(z) = m_2,$$

$$Y_4(z) \equiv X_4(z) H_4(z) \quad \text{mod } C_4(z)$$

$$= (m_3 - m_4) + (m_5 - m_3 - m_4)z.$$

It can be found that $S_1(z) = \frac{1}{4}(z^3 + z^2 + z + 1)$, $S_2(z) = \frac{1}{4}(-z^3 + z^2 - z + 1)$, and $S_4(z) = -\frac{1}{2}(z^2 - 1)$, from which $Y(z)$ is calculated as

$$Y(z) \equiv y_0 + y_1 z + y_2 z^2 + y_3 z^3 = S_1(z) Y_1(z) + S_2(z) Y_2(z) + S_4(z) Y_4(z).$$

By evaluating the above expression and comparing coefficients, it is found that

$$y_0 = \frac{m_1 + m_2}{4} + \frac{m_3 - m_4}{2},$$

$$y_1 = \frac{m_1 - m_2}{4} + \frac{m_5 - m_3 - m_4}{2},$$

$$y_2 = \frac{m_1 + m_2}{4} - \frac{m_3 - m_4}{2},$$

$$y_3 = \frac{m_1 - m_2}{4} - \frac{m_5 - m_3 - m_4}{2},$$

and the length 4 cyclic convolution is computed.

Example 8.4 shows how a length 4 cyclic convolution algorithm using the CRT approach is obtained. The algorithm requires 5 multiplications and 23 additions, if the same term is not counted more than once. This can be reduced to 15 additions, if it is assumed that $H_1(z)$, $H_2(z)$, and $H_4(z)$ are precomputed, which is a reasonable assumption, since the filter is usually known ahead of time. This assumption also makes it possible to get rid of the 4 in the denominator in the expressions for y_0, y_1, y_2, and y_3. The divisors of 2 and 4 can be factored into the m_i's and hence into the $H_i(z)$'s, which are precomputed.

The cyclic convolution is often described in terms of matrix multiplication and vector products in the form

$$\begin{bmatrix} y[0] \\ y[1] \\ \vdots \\ y[N-1] \end{bmatrix} = \mathbf{A} \left(\mathbf{B} \begin{bmatrix} x[0] \\ x[1] \\ \vdots \\ x[N-1] \end{bmatrix} \diamond \mathbf{C} \begin{bmatrix} h[0] \\ h[1] \\ \vdots \\ h[N-1] \end{bmatrix} \right),$$

$$(8.32)$$

where $\diamond$ denotes element-by-element multiplication of the two vectors. The matrices $\mathbf{B}$ and $\mathbf{C}$ compute the reductions of the $h[n]$ and $x[n]$ sequences, respectively, in Eq. (8.19). The vector product computes the polynomial multiplication in Eq. (8.20), and the matrix $\mathbf{A}$ computes the reconstruction in Eq. (8.21).

Example 8.5. The $\diamond$ operator works like this

$$\begin{bmatrix} 1 \\ 2 \\ 3 \\ 4 \end{bmatrix} \diamond \begin{bmatrix} 8 \\ 7 \\ 6 \\ 5 \end{bmatrix} = \begin{bmatrix} 1 \times 8 \\ 2 \times 7 \\ 3 \times 6 \\ 4 \times 5 \end{bmatrix} = \begin{bmatrix} 8 \\ 14 \\ 18 \\ 20 \end{bmatrix}.$$

Since the reconstruction (**A** matrix) can be viewed as an inverse operation of the reductions (the **B** and **C** matrices), the **A** matrix represents roughly twice the computational complexity of the **B** or **C** matrices individually. However, it can be proved [JO85] that Eq. (8.32) can also be written

$$
\begin{bmatrix} y[0] \\ y[1] \\ \vdots \\ y[N-1] \end{bmatrix} = \mathbf{R}\mathbf{C}^T \left(\mathbf{B} \begin{bmatrix} x[0] \\ x[1] \\ \vdots \\ x[N-1] \end{bmatrix} \;\odot\; \mathbf{A}^T\mathbf{R} \begin{bmatrix} h[0] \\ h[1] \\ \vdots \\ h[N-1] \end{bmatrix} \right),
$$

(8.33)

where **R** is any permutation matrix such that for any Toeplitz matrix **P**, $\mathbf{R}\mathbf{P}^T\mathbf{R} =$ **P**. In other words, by permuting the **A** and **C** matrices (which does not affect their computational complexity), the **A** and **C** matrices can be switched. Often one of the input sequences (the filter $h[n]$) is fixed, and then this trick makes it possible to move the **C** matrix, which has the highest computational complexity, into the precomputed part of the algorithm.

Example 8.6. The length 4 cyclic convolution from Example 8.4 can be written

$$
\begin{bmatrix} y_0 \\ y_1 \\ y_2 \\ y_3 \end{bmatrix} = \frac{1}{4} \begin{bmatrix} 1 & 1 & 2 & -2 & 0 \\ 1 & -1 & -2 & -2 & 2 \\ 1 & 1 & -2 & 2 & 0 \\ 1 & -1 & 2 & 2 & -2 \end{bmatrix} \left(\begin{bmatrix} 1 & 1 & 1 & 1 \\ 1 & -1 & 1 & -1 \\ 1 & 0 & -1 & 0 \\ 0 & 1 & 0 & -1 \\ 1 & 1 & -1 & -1 \end{bmatrix} \begin{bmatrix} x_0 \\ x_1 \\ x_2 \\ x_3 \end{bmatrix} \right.
$$

$$
\left. \;\odot\; \begin{bmatrix} 1 & 1 & 1 & 1 \\ 1 & -1 & 1 & -1 \\ 1 & 0 & -1 & 0 \\ 0 & 1 & 0 & -1 \\ 1 & 1 & -1 & -1 \end{bmatrix} \begin{bmatrix} h_0 \\ h_1 \\ h_2 \\ h_3 \end{bmatrix} \right).
$$

This is the regular form, and the reconstruction matrix obviously has higher computational complexity than the reduction matrices. If the **A** and **C** matrices are switched, then

$$
\begin{bmatrix} y_0 \\ y_1 \\ y_2 \\ y_3 \end{bmatrix} = \begin{bmatrix} 1 & -1 & 0 & -1 & -1 \\ 1 & 1 & -1 & 0 & -1 \\ 1 & -1 & 0 & 1 & 1 \\ 1 & 1 & 1 & 0 & 1 \end{bmatrix} \left(\begin{bmatrix} 1 & 1 & 1 & 1 \\ 1 & -1 & 1 & -1 \\ 1 & 0 & -1 & 0 \\ 0 & 1 & 0 & -1 \\ 1 & 1 & -1 & -1 \end{bmatrix} \begin{bmatrix} x_0 \\ x_1 \\ x_2 \\ x_3 \end{bmatrix} \right.
$$

$$
\left. \odot \tfrac{1}{4} \begin{bmatrix} 1 & 1 & 1 & 1 \\ -1 & 1 & -1 & 1 \\ 2 & -2 & -2 & 2 \\ 2 & 2 & -2 & -2 \\ -2 & 0 & 2 & 0 \end{bmatrix} \begin{bmatrix} h_0 \\ h_1 \\ h_2 \\ h_3 \end{bmatrix} \right),
$$

where the most complex matrix can be precomputed. The **R** matrix used here is

$$
\mathbf{R} = \begin{bmatrix} 0 & 0 & \cdots & 0 & 1 \\ 0 & 0 & \cdots & 1 & 0 \\ \vdots & \vdots & & \vdots & \vdots \\ 0 & 1 & \cdots & 0 & 0 \\ 1 & 0 & \cdots & 0 & 0 \end{bmatrix} .
$$

The CRT for polynomials can be considered a generalization of Lagrange's interpolation used in the Toom–Cook algorithms. If Eq. (8.13) is evaluated modulo any $P(z)$, where the degree of $P(z)$ is greater than $N + M - 1$, then the result will not change. Hence it can be evaluated modulo

$$
P(z) \equiv \prod_{i=0}^{N+M-2} P_i(z) = \prod_{i=0}^{N+M-2} (z - \alpha_i), \tag{8.34}
$$

which means that

$$
X_i(z) \equiv X(z) \mod P_i(z) \equiv X(z) \mod(z - \alpha_i) \equiv X(\alpha_i), \tag{8.35}
$$

which is exactly the way the Toom–Cook algorithm computed the convolution (see Eq. (8.15)).

The algorithms developed here are very efficient for short lengths, but the number of additions becomes very large as N increases, and thus even though these algorithms achieve the minimum number of multiplications, suboptimal (requiring

**TABLE 8-2   Number of Operations to Compute
Winograd's Short-Length Cyclic Convolution
Algorithms**[a]

Operation Count for Cyclic Convolution Algorithms			
N	#Multiplications	#Additions	$2N - K$
2	2	4	2
3	4	11	4
4	5	15	5
5	10	31	8
6	8	34	8
7	16	70	12
8	14	46	12
9	19	74	15

[a]Some have been modified slightly to give better balance between
multiplications and additions.

more than the minimum number of multiplications) algorithms are preferred for
longer lengths. Table 8-2 shows the operation counts for some short-length cyclic
convolution algorithms, which can be found in Appendix 8-C and in Nussbaumer
[NU81] and Winograd [WI78]. These algorithms do not all achieve the minimum
number of multiplications from Eq. (8.31). Some of them (length 5, 7, 8, and 9
modules) are chosen as suboptimal to reduce the number of additions (e.g., the
length 5 cyclic convolution can be computed in 8 multiplications and 62 additions,
but 10 multiplications and 31 additions are normally preferable).

8-1-4   Computation of Noncyclic Convolution by Short-Length Algorithms

The Toom–Cook algorithms require too many additions when N gets large. Instead, the convolution is mapped into a multidimensional convolution [AG74a],
where each dimension is small and can be computed efficiently using the Toom–
Cook algorithms.

Let N have two factors N_1 and N_2 (everything else can easily be generalized
from this case). Assuming the sequence $h[n]$ is defined to be zero for $n < 0$ and
for $n \geq M$, the $N \times N$ noncyclic convolution (see Eq. (8.1)) can be rewritten

$$y[n] = \sum_{m=0}^{N-1} x[m]h[n-m], \qquad n = 0, 1, \ldots, 2N - 2. \qquad (8.36)$$

By substituting

$$
\begin{aligned}
n &= n_1 + N_1 n_2, & n_1, m_1 &= 0, 1, \ldots, N_1 - 1, \\
m &= m_1 + N_1 m_2, & n_2, m_2 &= 0, 1, \ldots, N_2 - 1,
\end{aligned}
\qquad (8.37)
$$

Eq. (8.36) becomes

$$\tilde{y}[n_1 + N_1 n_2] = \tilde{y}[n_1, n_2]$$

$$= \sum_{m_1 = 0}^{N_1 - 1} \sum_{m_2 = 0}^{N_2 - 1} x[m_1 + N_1 m_2] h[n_1 - m_1 + N_1(n_2 - m_2)] \quad (8.38)$$

$$= \sum_{m_1 = 0}^{N_1 - 1} \sum_{m_2 = 0}^{N_2 - 1} x[m_1, m_2] h[n_1 - m_1, n_2 - m_2]$$

and the convolution has been converted into a two-dimensional convolution. Because of the assumption that $h[n]$ is zero outside $0 \le n \le N - 1$, the relationship between $\tilde{y}[n]$ and $y[n]$ is not as straightforward as Eq. (8.37) might indicate. As an example, a 6×6 noncyclic convolution gives a length $2 \times 6 - 1 = 11$ result, but if the length 6 sequences are converted into a 2 by 3 sequence, when they convolve they produce a length $(2 \times 2 - 1)$ by $(2 \times 3 - 1) = 3$ by 5 result with 15 variables in the result instead of 11. It is shown in Agarwal and Burrus [AG74a] that the relationship between $\tilde{y}[n]$ and $y[n]$ is

$$y[n_1 + N_1 n_2] = \tilde{y}[n_1, n_2] + \tilde{y}[n_1 + N_1, n_2 - 1],$$
$$n_1 = 0, 1, \cdots, N_1 - 1, \quad n_2 = 0, 1, \ldots, 2N_2 - 1, \quad (8.39)$$

and hence the method is a generalization of the overlap-add method.

To see how it is efficiently computed, the polynomial formulation in Eq. (8.9) is used [NU81]. If z^{N_1} is defined as z_1 then Eq. (8.8), when substituting Eq. (8.37), becomes

$$X(z, z_1) = \sum_{n_1 = 0}^{N_1 - 1} \sum_{n_2 = 0}^{N_2 - 1} x[n_1 + N_1 n_2] z^{n_1} z_1^{n_2} = \sum_{n_2 = 0}^{N_2 - 1} X_{n_2}(z) z_1^{n_2}, \quad (8.40)$$

where

$$X_{n_2}(z) = \sum_{n_1 = 0}^{N_1 - 1} x[n_1 + N_1 n_2] z^{n_1}, \quad n_2 = 0, 1, \ldots, N_2 - 1, \quad (8.41)$$

and similarly for $H(z, z_1)$ and $Y(z, z_1)$. The convolution can now be written

$$Y(z, z_1) = X(z, z_1) H(z, z_1) \quad (8.42)$$

or

$$Y_{n_2}(z) = \sum_{m_2 = 0}^{N_2 - 1} X_{m_2}(z) H_{n_2 - m_2}(z). \quad (8.43)$$

Hence the two-dimensional convolution can be viewed as a length N_2 one-dimensional convolution of length N_1 polynomials instead of individual points. This N_2 by N_2 convolution can now be computed using, for example, the Toom–Cook algorithm, but all the multiplications are polynomial multiplications of two length N_1 polynomials (which is exactly a noncyclic convolution as shown in Eq. (8.9)), and the additions are polynomial additions.

Assuming that the length N_1 noncyclic convolution can be done in M_1 multiplications and A_1 additions, and that the length N_2 noncyclic convolution can be computed in M_2 multiplications and A_2 additions, then the length $N = N_1 N_2$ convolutions can be computed in

$$M = M_1 M_2 \qquad (8.44)$$

multiplications and

$$A = M_2 A_1 + A_{1,2} N_1 + A_{2,2}(2N_1 - 1) + (N_1 - 1)(2N_2 - 2) \qquad (8.45)$$

additions [NU81], where $A_{1,2}$ and $A_{2,2}$ are the number of input and output additions required to compute a length $2N_2 - 1$ convolution. Table 8-3 from Nussbaumer [NU81] shows the number of operations to compute noncyclic convolution using length 2 and 3 short-length algorithms.

Example 8.7. To compute the length $N = 6$ noncyclic convolution

$$\{x_0, x_1, x_2, x_3, x_4, x_5\} \circledast \{h_0, h_1, h_2, h_3, h_4, h_5\},$$

TABLE 8-3 Number of Real Operations to Compute the Noncyclic Convolution of Two Real Sequences, or Number of Complex Operations to Compute Convolution of Two Complex Sequences

Operation Count for Noncyclic Convolution Algorithms		
Length	#Multiplications	#Additions
2	3	3
3	5	20
4	9	19
9	25	194
16	81	295
32	243	993
81	625	7412
128	2187	10041
243	3125	40040
256	6561	31015

let $N_1 = 3$ and $N_2 = 2$ and use the mapping

$$n = n_1 + N_1 n_2 = n_1 + 3n_2, \qquad n_1 = 0, 1, 2, \quad n_2 = 0, 1.$$

Now compute Eq. (8.41)

$$X_0(z) = x_0 + x_1 z + x_2 z^2,$$

$$X_1(z) = x_3 + x_4 z + x_5 z^2,$$

and similarly for $H_0(z)$ and $H_1(z)$ (which may be precomputed). Equation (8.43) now says that the convolution can be written as

$$\{X_0(z), X_1(z)\} \circledast \{H_0(z), H_1(z)\},$$

which is a length 2 noncyclic convolution of polynomials. This length 2 convolution is computed by

$$m_0(z) = X_0(z) H_0(z),$$

$$m_1(z) = (X_0(z) + X_1(z))(H_0(z) + H_1(z)),$$

$$m_2(z) = X_1(z) H_1(z),$$

where each multiplication of two polynomials is a length 3 noncyclic convolution and each addition is a polynomial addition. The output is obtained by

$$Y_0(z) = m_0(z) = X_0(z) H_0(z),$$

$$Y_1(z) = m_1(z) - m_0(z) - m_2(z) = X_0(z) H_1(z) + X_1(z) H_0(z),$$

$$Y_2(z) = m_2(z) = X_1(z) H_1(z).$$

Using an inverse mapping, $Y(z)$ is found

$$Y(z) = Y_0(z) z_1^0 + Y_1(z) z_1^1 + Y_2(z) z_1^2$$

$$= Y_0(z) + Y_1(z) z^3 + Y_2(z) z^6,$$

since $z_1 = z^{N_1} = z^3$. The coefficients of $Y(z)$ are the output sequence $y[n]$.

8-1-5 Computation of Cyclic Convolution by Short-Length Algorithms

It was shown by Agarwal and Cooley [AG77] that the cyclic convolution

$$y[n] = \sum_{m=0}^{N-1} x[m] h[\langle n - m \rangle_N], \qquad n = 0, 1, \ldots, N-1, \qquad (8.46)$$

can be computed efficiently by mapping it into a multidimensional convolution. Assuming that N has two relatively prime factors N_1 and N_2, this is done by the substitution

$$n = N_2 n_1 + N_1 n_2 \quad \text{mod } N, \quad n_1, m_1 = 0, 1, \ldots, N_1 - 1, \tag{8.47}$$

$$m = N_2 m_1 + N_1 m_2 \quad \text{mod } N, \quad n_2, m_2 = 0, 1, \ldots, N_2 - 1,$$

which gives

$$
\begin{aligned}
y[n_1, n_2] &\equiv y[\langle N_2 n_1 + N_1 n_2 \rangle_N] \\
&= \sum_{m_1=0}^{N_1-1} \sum_{m_2=0}^{N_2-1} x[m_1, m_2] h[\langle n_1 - m_1 \rangle_{N_1}, \langle n_2 - m_2 \rangle_{N_2}] \\
&\equiv \sum_{m_1=0}^{N_1-1} \sum_{m_2=0}^{N_2-1} x[\langle N_2 m_1 + N_1 m_2 \rangle_N] \\
&\quad \cdot h[\langle N_2(n_1 - m_1) + N_1(n_2 - m_2) \rangle_N]
\end{aligned}
\tag{8.48}
$$

and a two-dimensional convolution is obtained. Note here the important requirement is that N_1 and N_2 are relatively prime. Expressing the convolution in terms of polynomials by defining

$$X_{n_2}(z) = \sum_{n_1=0}^{N_1-1} x[\langle N_2 n_1 + N_1 n_2 \rangle_N] z^{n_1} \tag{8.49}$$

and similarly for $H_{n_2}(z)$ and $Y_{n_2}(z)$, the cyclic convolution can be written

$$Y_{n_2}(z) = \sum_{m_2=0}^{N_2-1} X_{m_2}(z) H_{n_2-m_2}(z) \quad \text{mod}(z^{N_2} - 1),$$

$$n_2 = 0, 1, \ldots, N_2 - 1. \tag{8.50}$$

The length N cyclic convolution is now expressed as a length N_2 cyclic convolution of polynomials, which can be computed efficiently by a length N_2 Winograd convolution algorithm. This method for computing convolution is also called the rectangular transform method [AG77].

The above technique is easily extended for N with more factors, and if the length N_1 short-length algorithm takes M_1 multiplications and A_1 additions, and so on, the length $N = N_1 N_2 N_3 \cdots N_j$ cyclic convolution can be computed in

$$M = M_1 M_2 \cdots M_j \tag{8.51}$$

multiplications and

$$
\begin{aligned}
A = {}& A_1 N_2 N_3 \ldots N_{j-1} N_j + M_1 A_2 N_3 \ldots N_{j-1} N_j \\
& + M_1 M_2 A_3 \cdots N_{j-1} N_j + \ldots + M_1 M_2 M_3 \ldots M_{j-1} A_j \quad (8.52)
\end{aligned}
$$

additions. Note here that the order of the factors N_1, N_2, . . . influences the number of operations, and for two factors they should be selected such that [NU81]

$$\frac{M_1 - N_1}{A_1} \leq \frac{M_2 - N_2}{A_2}. \tag{8.53}$$

For more factors the rule is a little more complicated and can be found in Agarwal and Cooley [AG77]. Table 8-4 shows the number of operations required to compute cyclic convolutions. The last column shows an improvement described by Nussbaumer [NU78; NU81], which nests the computation in the two dimensions N_1 and N_2 by factorization of the polynomial $z^{N_2} - 1$.

A program to compute the rectangular transform can be obtained by modifying the WFTA program in McClellan and Rader [MC79b] by replacing the DFT modules with short-length convolution modules.

Example 8.8. To compute the length $N = 6$ cyclic convolution

$$\{x_0, x_1, x_2, x_3, x_4, x_5\} ⑥ \{h_0, h_1, h_2, h_3, h_4, h_5\},$$

TABLE 8-4 Number of Real Operations to Compute Cyclic Convolution of Two Real-Valued Sequences, or Number of Complex Operations to Compute Convolution of Two Complex Sequences[a]

	Operation Count for Cyclic Convolution Algorithms		
Length	#Multiplications	#Additions Agarwal–Cooley	#Additions Nussbaumer
18	38	184	184
20	50	230	215
24	56	272	244
30	80	418	392
36	95	505	461
48	132	900	840
60	200	1120	964
72	266	1450	1186
84	320	2100	1784
120	560	3096	2468
180	950	5470	4382
210	1280	7958	6458
240	1056	10176	9696
360	2280	14748	11840
420	3200	20420	15256
504	3648	26304	21844
840	7680	52788	39884
1008	10032	71265	56360
1260	12160	95744	72268
2520	29184	241680	190148

[a]From Nussbaumer [NU81].

let $N_1 = 3$ and $N_2 = 2$ and use the mapping

$$n = N_2 n_1 + N_1 n_2 = 2n_1 + 3n_2, \qquad n_1 = 0, 1, 2, \quad n_2 = 0, 1$$

Now compute Eq. (8.49)

$$X_0(z) = x_0 + x_2 z + x_4 z^2,$$
$$X_1(z) = x_3 + x_5 z + x_1 z^2,$$

and similarly for $H_0(z)$ and $H_1(z)$ (could be precomputed). Equation (8.50) now says that the convolution can be written as

$$\{X_0(z), X_1(z)\} \; ② \; \{H_0(z), H_1(z)\},$$

which is a length 2 cyclic convolution of polynomials. This length 2 convolution is computed as

$$m_0(z) = (X_0(z) + X_1(z)) \frac{H_0(z) + H_1(z)}{2},$$

$$m_1(z) = (X_0(z) - X_1(z)) \frac{H_0(z) - H_1(z)}{2},$$

where each multiplication of two polynomials is a length 3 cyclic convolution and each addition is a polynomial addition. The output is obtained from

$$Y_0(z) = m_0(z) + m_1(z) = X_0(z) H_0(z) + X_1(z) H_1(z),$$

$$Y_1(z) = m_1(z) - m_0(z) = X_0(z) H_1(z) + X_1(z) H_0(z).$$

Using an inverse mapping, $Y(z)$ is found:

$$Y(z) = Y_0(z) z^0 + Y_1(z) z^1.$$

The coefficients of $Y(z)$ are the output sequence $y[n]$.

8-1-6 Comparison of Convolution Algorithms

As explained in the introduction, convolutions can also be computed using FFTs. Table 8-5 shows the number of operations for computing convolutions using FFTs. The table is based on operation counts for either the split-radix FFT if N is a power-of-2 or the prime factor FFT if N has several different factors. These two algorithms will be discussed in Sections 8-2-5 and 8-2-7, respectively.

It is apparent that the Agarwal–Cooley algorithm in Table 8-4 is only competitive for short lengths, while for large N, the FFT approach requires less operations. Since the Agarwal–Cooley algorithm furthermore requires more overhead

TABLE 8-5   Real Operation to Compute Cyclic
Convolution of Two Real Sequences Using FFTs[a]

Operation Count for Cyclic Convolution Using FFTs		
N	#Multiplications	#Additions
4	5(5)	35(51)
8	15(17)	49(69)
9	33(43)	80(114)
16	43(53)	141(201)
18	66(86)	196(282)
20	69(89)	207(297)
24	79(101)	241(345)
30	144(194)	370(534)
32	115(149)	373(537)
36	133(173)	463(669)
48	195(257)	613(885)
60	289(389)	859(1245)
64	291(389)	933(1353)
72	303(401)	1105(1605)
84	429(581)	1495(2181)
120	639(869)	2017(2937)
128	707(965)	2245(3273)
180	1029(1409)	3535(5169)
210	1494(2084)	4744(6960)
240	1459(2009)	4693(6861)
256	1667(2309)	5253(7689)
360	2239(3089)	7969(11685)
420	2989(4169)	10327(15177)
504	3279(4541)	12913(18993)
840	6399(8969)	22753(33501)
1008	7315(10217)	28597(42141)
1260	10089(14189)	37699(55605)
2560	21439(30269)	173473(258321)

[a]The first set of numbers assume that one of the two sequences are
known a priori and hence one of the forward FFTs can be precomputed, while the second set of numbers does not. The numbers do
not include scaling by–in the inverse FFT.

(indexing, etc.) than the FFT approach, it is seldom used. For short-length convolutions a direct implementation often turns out to be fastest.

The noncyclic convolution complexity in Table 8-3 cannot be compared directly
to that in Table 8-5, since the latter contain operation counts for cyclic convolutions. However, as described in a previous section, a length $2N - 1$ (or longer)
cyclic convolution algorithm can be used for computing noncyclic convolutions.
So, for example, the operation counts for the length 128 noncyclic convolution in
Table 8-3 should be compared to the operation counts for the length 256 cyclic
convolution in Table 8-5. Again, it is apparent that the FFT approach is superior

for longer sequences and is thus normally used. Also, for noncyclic convolutions a direct implementation is often used for short-length convolutions.

Since computers have finite wordlength, quantization errors are introduced with convolution algorithms. The error depend on whether a floating-point or a fixed-point machine is used and on the algorithm computed. A detailed discussion can be found in Oppenheim and Weinstein [OP72].

8-2 FAST COMPUTATION OF THE DFT

The Discrete Fourier Transform (DFT) $X[k]$ of the sequence $x[n]$ defined in Eq. (8.5) is repeated below for convenience:

$$X[k] = \sum_{n=0}^{N-1} x[n] W_N^{nk}, \qquad k = 0, 1, \ldots, N - 1, \qquad (8.54)$$

where $W_N = e^{-j2\pi/N}$. Direct computation of this length N DFT takes N^2 complex multiplications and $N(N - 1)$ complex additions, but methods exist that take far fewer operations. The DFT can also be written as a matrix product:

$$
\begin{bmatrix} X[0] \\ X[1] \\ \vdots \\ X[N-1] \end{bmatrix} =
\begin{bmatrix}
W_N^0 & W_N^0 & \cdots & W_N^0 \\
W_N^0 & W_N^1 & \cdots & W_N^{N-1} \\
\vdots & \vdots & & \vdots \\
W_N^0 & W_N^{N-1} & \cdots & W_N^{(N-1)(N-1)}
\end{bmatrix}
\begin{bmatrix} x[0] \\ x[1] \\ \vdots \\ x[N-1] \end{bmatrix}
$$

$$
= \mathbf{W}
\begin{bmatrix} x[0] \\ x[1] \\ \vdots \\ x[N-1] \end{bmatrix}. \qquad (8.55)
$$

Algorithms for computing the DFT efficiently are called fast Fourier transform (FFT) algorithms, and one method for deriving them is a factorization of the DFT matrix operator. The factorization depends strongly on the nature of the transform length, N. The basic Cooley–Tukey algorithm [CO65] assumes that N is a power of 2 and hence is called a radix-2 algorithm. Depending on how the factorization is done, either a decimation-in-time (DIT) or a decimation-in-frequency (DIF) algorithm is obtained. The technique used by Cooley and Tukey can also be applied to DFTs where N is a power of R. The resulting algorithms are named radix-R algorithms, and it turns out that radix-4, radix-8, and radix-16 DFTs are especially interesting. A new algorithm called the split-radix FFT [DU84; SO86a], like the radix-2 FFTs, also assumes N is a power of 2, but it factors the even-indexed elements as a radix-2 and the odd-indexed elements as a radix-4, and a very efficient algorithm is obtained. All the FFTs above have a similar theory, and only the decimation-in-time (DIT) FFT will be described in detail.

If N is composed of several different components, $N = N_1 N_2 N_3 \ldots N_j$, a mixed-radix algorithm can be used. If it is furthermore assumed that the factors are relatively prime, the more efficient prime factor algorithm (PFA) can and should be used. For example, if 450 is factored as $2 \cdot 3 \cdot 5 \cdot 3 \cdot 5$, a mixed-radix algorithm must be used, but if it is factored as $2 \cdot 9 \cdot 25$, a prime factor algorithm can be used.

The Winograd FFT algorithm minimizes the number of multiplications (usually at the expense of an increased number of additions). It is based on a reorganization of the PFA, which reduces the required number of multiplications.

Most algorithms compute the DFT in-place, which means that the same memory locations are used for the input sequence and its DFT on the output. Since the DFT of a real-valued sequence is complex, most algorithms assume that the input sequence is complex as well. However, in most applications the input sequence is real-valued, so special algorithms exist that reduce memory and computational requirements by about half for real-valued input sequences.

A special algorithm, Goertzel's algorithm [BL84; BU85], does not improve the operation count significantly when computing the whole transform, but if only a few spectral components are wanted, it may be a viable alternative.

8-2-1   Goertzel's Algorithm

Goertzel's algorithm [GO58] is not a true FFT, since it requires on the order of N^2 operations and therefore does not lower the operation count significantly for computing the DFT of a sequence. However, if only a few values of $X[k]$ are needed, it can compete successfully with the FFT [BL84; BU85]. It computes each value $X[k]$ individually and is derived by rewriting Eq. (8.5) as a polynomial:

$$X[k] = \sum_{n=0}^{N-1} x[n] z^n \big|_{z = W_N^k} \equiv P(z)\big|_{z = W_N^k}. \qquad (8.56)$$

$X[k]$ is obtained by evaluating the polynomial $P(z)$ at $z = W_N^k$. By factoring the polynomial $P(z)$ into

$$P(z) = (z - W_N^k)(z - W_N^{-k})Q(z) + R(z), \qquad (8.57)$$

where $Q(z)$ is the quotient and $R(z)$ is the remainder (of degree 1), $P(z)$ can be evaluated as

$$P(W_N^k) = (W_N^k - W_N^k)(W_N^k - W_N^{-k})Q(W_N^k) + R(W_N^k) = R(W_N^k). \quad (8.58)$$

Since $R(z)$ has degree 1, it can be evaluated with only one multiplication and one addition, and the main effort in computing $X[k]$ is the factorization in Eq. (8.58).

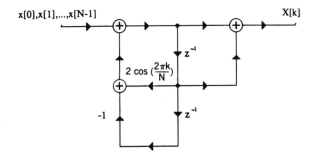

FIGURE 8-4 Flow-graph of Goertzel's algorithm.

In Figure 8-4 the factorization is done by Horner's method in the feedback loop, while $X[k]$ is only computed once after all the input has been processed [BL84; BU85]. Figure 8-5 shows a section of FORTRAN code that implements the flow-graph and computes the kth element ($= $ REAL $+ j$ IMAG) of the DFT of the sequence $x[n]$ (real part in $X(I)$ and imaginary part in $Y(I)$). The DO 10 loop computes the factorization, while the evaluation of $R(z)$ is the last two lines.

Each element $X[k]$ takes $2N + 2$ real multiplications and $4N$ real additions, and since a FFT computes the total DFT on the order of $N \log_2 (N)$ real multiplications and additions (as will be shown in the next section), the Goertzel algorithm is only efficient if less than $\log_2 (N)$ elements are wanted or if the algorithm fits the computer architecture better. A recent paper [BO86] provides a different approach to computing single DFT points.

```
C = COS(2*3.1415*K/N)
S = SIN(2*3.1415*K/N)
CC = C+C
A2 = 0
A1 = X(1)
B1 = 0
B2 = Y(1)
DO 10 I=2,N
    TEMP = A1
    A1 = CC*A1 − A2 + X(I)
    A2 = TEMP
    TEMP = B1
    B1 = CC*B1 − B2+Y(I)
    B2 = TEMP
10  CONTINUE
    REAL = C*A1 − A2 − S*B1
    IMAG = S*A1 + C*B1 − B2
```

FIGURE 8-5 FORTRAN code for computing $X[k]$ of the length N sequence $x[n]$ (from Burrus and Parks [BU85]).

8-2-2 Decimation-in-Time Radix-2 Algorithm

There exist several ways to derive the different FFT algorithms. It can be done using polynomial factorization [MA84b; NU81], by matrix factorization, or by index mapping [BU77; BU85]. This chapter for the most part uses the concept of index mapping, which is a method of mapping a one-dimensional sequence of length $N = N_1 N_2$ into a two-dimensional sequence of size $N_1 \times N_2$. This approach allows easy implementation of practical programs.

The mapping is done by substituting

$$
\begin{aligned}
n &\equiv K_1 n_1 + K_2 n_2 \mod N, \\
k &\equiv K_3 k_1 + K_4 k_2 \mod N,
\end{aligned}
\tag{8.59}
$$

in the definition for the DFT, Eq. (8.54). If N_1 and N_2 are relatively prime, a prime factor map (PFM) can be derived, while if N_1 and N_2 are not relatively prime, a common factor map (CFM) will result. The factors K_1 to K_4 depend on the type of map used, and the specific requirement can be found in Burrus [BU77]. The basic Cooley–Tukey decimation-in-time FFT algorithm to compute DFTs of length $N = R^M$, where R is called the radix, is obtained by using the CFM

$$
n \equiv R n_1 + n_2, \quad n_1 = 0, 1, \ldots, \frac{N}{R} - 1, \quad n_2 = 0, 1, \ldots, R - 1;
$$

$$
k \equiv k_1 + \frac{N}{R} k_2, \quad k_1 = 0, 1, \ldots, \frac{N}{R} - 1, \quad k_2 = 0, 1, \ldots, R - 1.
$$

$$
\tag{8.60}
$$

Let $N = 2^M$ and therefore $R = 2$ in the index mapping in Eq. (8.60). Substituting the index map in Eq. (8.54), it becomes

$$
X\left[k_1 + \frac{N}{2} k_2\right] = \sum_{n_1=0}^{N/2-1} \sum_{n_2=0}^{1} x[2n_1 + n_2] W_N^{2n_1 k_1 + N n_1 k_2 + n_2 k_1 + (N/2) n_2 k_2}
\tag{8.61}
$$

for $k_1 = 0, 1, \ldots, N/2 - 1$ and $k_2 = 0, 1$. Evaluating the sum over n_2, we obtain

$$
X\left[k_1 + \frac{N}{2} k_2\right] = \sum_{n_1=0}^{N/2-1} x[2n_1] W_{N/2}^{n_1 k_1} + W_N^{k_1 + (N/2)k_2} \sum_{n_1=0}^{N/2-1} x[2n_1 + 1] W_{N/2}^{n_1 k_1},
\tag{8.62}
$$

since $W_N^{2n} = W_{N/2}^n$ and $W_N^N = 1$. Hence the length N DFT has been broken up into two length $(N/2)$ DFTs, one of which is multiplied by $W_N^{k_1 + (N/2)k_2}$, which is called a twiddle factor. By realizing that $W_N^{k_1 + (N/2)k_2} = (-1)^{k_2} W_N^{k_1}$, a length N DFT can be computed using two length $(N/2)$ DFTs and some extra operations, as shown for $N = 8$ in Figure 8-6. These operations are called "butterfly operations," and one of them is shown in Figure 8-7. Each of these "butterflies" is a length 2 DFT,

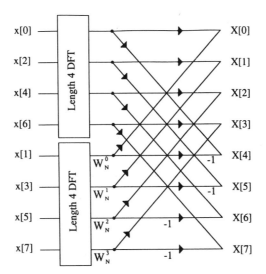

FIGURE 8-6 Last stage of radix-2 DIT FFT.

so one way to view the index mapping is that it breaks a length N DFT up into a two-dimensional length $(2 \times N/2)$ DFT. The penalty for doing this is the twiddle factors that appear between the row and column DFTs.

Direct implementation of a length N DFT requires N^2 complex multiplications and $N(N - 1)$ complex additions, so two length $(N/2)$ DFTs require $2(N/2)^2 = (N^2/2)$ multiplications and $2(N/2)(N/2 - 1) = N(N/2 - 1)$ additions. Including $N/2$ multiplications to account for the twiddle factors and N additions to compute the sum of the two parts, the length N DFT now requires $(N^2 + N)/2$ multiplications and $N^2/2$ additions. Hence a reduction in arithmetic has been achieved by breaking up the DFT. This is possible because one of the W_N terms disappeared in Eq. (8.61). The same reduction does not occur in breaking up convolutions, since no orthogonalities exist to reduce the number of terms.

The scheme above can now be applied to the two length $(N/2)$ DFTs, each producing two length $(N/4)$ DFTs in the next stage. This approach is continued until length 2 DFTs are obtained, which can be computed in a "butterfly operation" without any twiddle factors, and a radix-2 fast Fourier transform (FFT) is obtained. The flow-graph of a FFT to compute a length 8 DFT is shown in Figure 8-8.

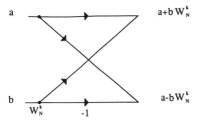

FIGURE 8-7 Butterfly of radix-2 DIT FFT.

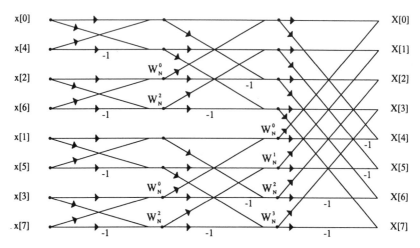

FIGURE 8-8 Length 8 radix-2 decimation-in-time FFT.

As can be seen from Figure 8-8, the input is required to be in a scrambled order. To compute the DFT in-place, which means that the input, output, and intermediate data occupy the same memory locations (so the memory requirement for the DFT is of order N), the input data to the FFT must be prescrambled. The scrambling used is bit-reversing; that is, the element $x[n]$ in location n is mapped to location m, where m is obtained by reversing the binary representation of n. Because the input represents the time domain and the input is scrambled, this algorithm is called the decimation-in-time (DIT) FFT. In Appendix 8-A program 2 computes a radix-2 DIT FFT using the above scheme. The program consists of two major parts, the bit-reverser and the butterfly computations. The butterfly computations consist of 3 loops, where the DO 10 loop steps through the different stages and the DO 20 steps through all the different butterflies (i.e., they have different twiddle factors). Finally, the DO 30 loops selects the blocks, that is, selects all the butterflies in the stage which have the same twiddle factors.

Each factor of two in N requires a separate stage, so there are $\log_2 (N) = M$ stages in the algorithm. The algorithm described above requires $N/2$ complex multiplications and N complex additions per stage. Each complex multiplication requires four real multiplications and two real additions, while a complex addition requires two real additions; hence the total algorithm requires $2N \log_2 (N)$ real multiplications and $3N \log_2 (N)$ real additions. Some of the multiplications are by unity, since $W_N^0 = 1$, so the multiplication count can be reduced by deselecting these operations in the program code. However, this requires a special section of code to compute the butterflies with these twiddle factors. Also, multiplications by $W_N^{N/4} = j$ and by $W_N^{N/8} = (1/\sqrt{2})(1 - j)$ can be computed by a special butterfly to save operations. Again, each of them requires a special butterfly. Deselecting as much as possible, the DIT radix-2 FFT requires $2N \log_2 (N) - 7N + 12$ real multiplications and $3N \log_2 (N) - 3N + 4$ real additions.

Finally, a complex multiplication can be implemented in three real multiplications and three real additions [BU85], and using this strategy results in $\frac{3}{2} N \log_2 N$ $- 5N + 8$ real multiplications and $\frac{7}{2} N \log_2 (N) - 5N + 8$ real additions.

8-2-3 Decimation-in Frequency Radix-2 Algorithm

The radix-2 decimation-in-frequency (DIF) FFT is closely related to the radix-2 DIT FFT, as it is derived from the CFM

$$n \equiv n_1 + \frac{N}{R} n_2, \quad n_1 = 0, 1, \cdots, \frac{N}{R} - 1, \quad n_2 = 0, 1, \cdots, R - 1,$$

$$k \equiv Rk_1 + k_2, \quad k_1 = 0, 1, \cdots, \frac{N}{R} - 1, \quad k_2 = 0, 1, \cdots, R - 1,$$

$$(8.63)$$

with $R = 2$. Substituting these expressions in Eq. (8.54), it is found that

$$X[2k_1] = \sum_{n_1 = 0}^{N/2 - 1} \left\{ x[n_1] + x\left[n_1 + \frac{N}{2} \right] \right\} W_{N/2}^{n_1 k_1},$$

$$X[2k_1 + 1] = \sum_{n_1 = 0}^{N/2 - 1} \left\{ x[n_1] - x\left[n_1 + \frac{N}{2} \right] \right\} W_N^{n_1 k_2} W_{N/2}^{n_1 k_1},$$

$$(8.64)$$

since $W_2^{n_2 k_2} = (-1)^{n_2 k_2}$ and $W_N^N = 1$. Hence a length N DFT can be computed using two length $(N/2)$ DFTs, as shown in Figure 8-9. Iterating this approach by

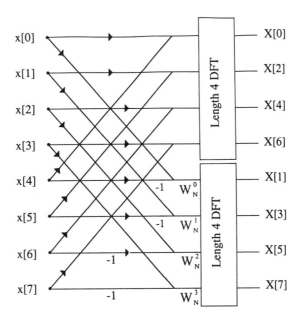

FIGURE 8-9 First stage of DIF radix-2 FFT.

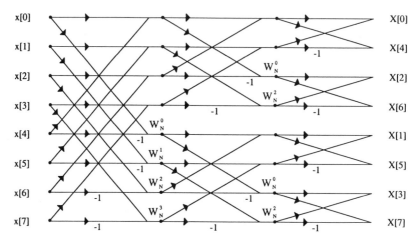

FIGURE 8-10   Length 8 radix-2 decimation-in-time FFT.

computing the length $(N/2)$ DFTs using the scheme above, a DIF FFT is obtained. Figure 8-10 shows the flow-graph of a length 8 radix-2 DIF FFT.

By comparing Figure 8-10 to Figure 8-8, it is apparent that there are strong ties between the DIF and the DIT algorithms, and that one can be obtained from the other by a flow-graph reversal. The complexity for the DIF FFT is $2N \log_2 (N)$ real multiplications and $3N \log_2 (N)$ real additions, which is the same as the radix-2 DIT FFT. Also, as for the radix-2 DIT FFT, several butterflies can be deselected and computed in special butterflies to achieve an operation count of $\frac{3}{2}N \log_2 (N)$ $- 5N + 8$ real multiplications and $\frac{7}{2}N \log_2 (N) - 5N + 8$ real additions. Appendix 8-A program 1 is a radix-2 DIF FFT program.

8-2-4   Higher-Radix Algorithms

If $N = 4^M$ then $R = 4$, and the index map, Eq. (8.60), when substituted in Eq. (8.54), produces

$$
\begin{aligned}
X\left[k_1 + \frac{N}{4} k_2\right] = & \sum_{n_1=0}^{N/4-1} x[4n_1] W_{N/4}^{n_1 k_1} \\
& + (-j)^{k_1} W_N^{k_1} \sum_{n_1=0}^{N/4-1} x[4n_1 + 1] W_{N/4}^{n_1 k_1} \qquad (8.65) \\
& + (-1)^{k_1} W_N^{2k_1} \sum_{n_1=0}^{N/4-1} x[4n_1 + 2] W_{N/4}^{n_1 k_1} \\
& + (j)^{k_1} W_N^{3k_1} \sum_{n_1=0}^{N/4-1} x[4n_1 + 3] W_{N/4}^{n_1 k_1},
\end{aligned}
$$

which is the first stage of a DIT radix-4 FFT. Figure 8-11 shows the flow-graph for a length $16 = 4^2$ FFT. Note that the input is scrambled, but that the order is

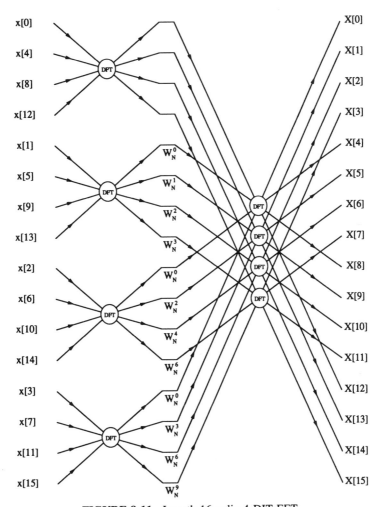

FIGURE 8-11 Length 16 radix-4 DIT FFT.

different from that of the radix-2 DIT FFT. The order here is obtained by reversing the base-4 representation of the element index; for example, $6 = 12_4$ (base-4 representation), which reverses to $21_4 = 9$, so $x[6]$ maps into $x[9]$. It has been shown that any radix-R algorithm can be modified to produce the output in bit-reversed order by a simple rearrangement of the butterfly output [BU88]. The complexity for a radix-4 FFT with all special butterflies deselected is $\frac{9}{4} N \log_2 (N) - \frac{43}{12}N + \frac{16}{3}$ real multiplications and $\frac{25}{4} N \log_2 (N) - \frac{43}{12}N + \frac{16}{3}$ real additions, assuming that the complex multiplications are done using three real multiplications and three additions.

Similarly, by choosing $R = 4$ in the index map Eq. (8.63), a radix-4 DIF FFT can be obtained. It has the same complexity as the DIT radix-4 FFT but has the output scrambled instead of the input as in the case of the DIT algorithm. The approach described here generalizes to other radices such as 8 and 16 as well, but these are not described here because of their strong similarities to the radix-2 and -4 algorithms. An efficient program is described in Morris [MO83] and IEEE Press book of programs for digital signal processing [IE79].

8-2-5 Split-Radix Algorithm

By applying a radix-2 index map to the even-indexed terms and a radix-4 map to the odd-indexed terms, it is found that

$$X[2k_1] = \sum_{n_1=0}^{N/2-1} \left\{ x[n_1] + x\left[n_1 + \frac{N}{2}\right] \right\} W_{N/2}^{n_1 k_1} \tag{8.66}$$

for the even-indexed terms, and

$$X[4k_1 + 1] = \sum_{n_1=0}^{N/4-1} \left\{ \left(x[n_1] - x\left[n_1 + \frac{N}{2}\right]\right) \right.$$
$$\left. - j\left(x\left[n_1 + \frac{N}{4}\right] - x\left[n_1 + \frac{3N}{4}\right]\right) \right\} W_N^{n_1} W_{N/4}^{n_1 k_1},$$

$$X[4k_1 + 3] = \sum_{n_1=0}^{N/4-1} \left\{ \left(x[n_1] - x\left[n_1 + \frac{N}{2}\right]\right) \right.$$
$$\left. + j\left(x\left[n_1 + \frac{N}{4}\right] - x\left[n_1 + \frac{3N}{4}\right]\right) \right\} W_N^{3n_1} W_{N/4}^{n_1 k_1},$$

for the odd-indexed terms. This decomposition yields the very efficient split-radix algorithm [DU84; SO86a]. This results in a L-shaped "butterfly" as shown in Figure 8-12, which relates a length N DFT to one length $(N/2)$ DFT and two length $(N/4)$ DFTs with twiddle factors. This process can be repeated for the

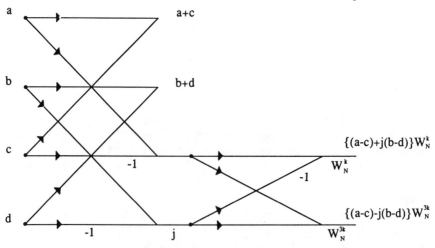

FIGURE 8-12 Split-radix butterfly.

length $(N/2)$ and the two length $(N/4)$ DFTs, until only length 2 DFTs remain. At this point length 2 DFTs are applied to complete the transform, and a DIF split-radix FFT has been obtained. Figure 8-13 shows the flow-graph of a length 16 split-radix FFT. The flow-graph strongly resembles the radix-2 DIF FFT except that the twiddle factors are in different locations. The split-radix FFT does not, like nearly all other algorithms, progress stage by stage. The even part (top half) progresses one stage as a radix-2, while the odd part (bottom half) progresses two stages as a radix-4; hence its name, "split-radix." This complicates the indexing somewhat, but the savings obtained may be significant enough to account for this.

The operation count for the split-radix FFT is $N \log_2 (N) - \frac{2}{3}N + \frac{2}{3}(-1)^{\log_2(N)}$ real multiplications and $3N \log_2 (N) - \frac{2}{3}N + \frac{2}{3}(-1)^{\log_2(N)}$ real additions, assuming that the complex multiplications are done using three real multiplications and three real additions [BU85]. A simple program that implements these operation counts can be found in Sorensen et al. [SO86a], which also has a DIT split-radix FFT. As for the previously discussed algorithms, certain butterflies can be deselected and computed separately. By doing this, a length N split-radix FFT can be computed in $N \log_2 (N) - 3N + 4$ real multiplications and $3N \log_2 (N) - 3N + 4$ real additions, which is lower than any other known constant-radix

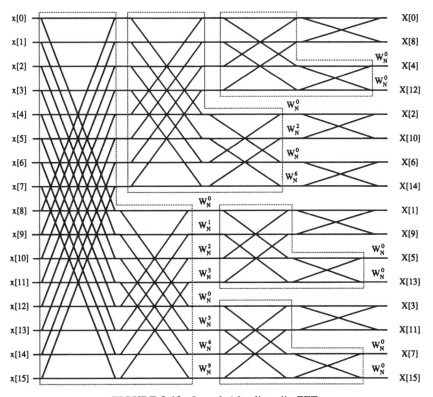

FIGURE 8-13 Length 16 split-radix FFT.

FFT. The split-radix FFT is known to be optimum for length up to and including 16 and is conjectured to be optimum for all lengths that are powers of two if both multiplications and additions are considered. Appendix 8-A program 3 is a split-radix FFT with three different butterflies that achieves these counts. The program uses a recursive computation of the sine and cosine values needed and timings indicate it is a very efficient program. If many DFTs are needed, it is more efficient to compute these trigonometric functions once and let the program use a table lookup as in program 4 in Appendix 8-A. Assuming the initialization is done, this program is faster than program 3.

8-2-6 Mixed-Radix Algorithm

If N is not a power of 2, the common factor map results in a mixed-radix FFT described by Singleton [SI69]. The mixed-radix FFT does not make any assumption on the factors of N, and the idea is described here by analyzing a length 36 DFT. Assuming that $N = 36 = 3 \times 2 \times 3 \times 2$, and using the index map

$$
\begin{aligned}
n &\equiv 3n_1 + n_2, & n_1 &= 0, 1, \ldots, 11, & n_2 &= 0, 1, 2, \\
k &\equiv k_1 + 12k_2, & k_1 &= 0, 1, \ldots, 11, & k_2 &= 0, 1, 2,
\end{aligned}
\tag{8.67}
$$

the DFT can be written

$$
X[k_1 + 12k_2] = \sum_{n_1=0}^{11} x[3n_1] W_{12}^{n_1 k_1} + W_{36}^{k_1} W_3^{k_2} \sum_{n_1=0}^{11} x[3n_1 + 1] W_{12}^{n_1 k_1}
$$

$$
+ W_{36}^{2k_1} W_3^{k_2} \sum_{n_1=0}^{11} x[3n_1 + 2] W_{12}^{n_1 k_1}.
\tag{8.68}
$$

Hence the length 36 DFT can be computed as three length 12 DFTs and twelve length 3 DFTs with some twiddle factors. Figure 8-14 shows one of the twelve length 3 DFTs in the last stage of this algorithm. Note that there are two different kinds of multiplications; there are twiddle factors (W_{36}) and there are multiplications in the length 3 DFTs (W_3). The length 12 DFTs are now computed as two length 6 DFTs and six length 2 DFTs with twiddle factors. Continuing this approach until only scalars remain, a mixed-radix FFT is obtained. A length N DFT, where N is an odd prime, always has some nontrivial multiplications, and since the number of multiplications increase with N, short lengths are most efficient. Singleton [SI69] shows that radix-2, -4, -8, and -16 algorithms have a higher relative efficiency (smaller number of multiplications per output point) than any odd-length algorithm. Hence the mixed-radix FFT is generally not very attractive; however, it allows a greater variety of lengths than other algorithms. A program that implements the mixed-radix FFT can be found in Singleton [SI69, SI79], which allows factors up to 23 but is "efficient" only for factors of 2, 3, and 5. For these lengths, a special module is included, while for any prime factor greater than 5 the DFT is computed using a nearly direct implementation with $(N - 1)^2$ real multiplications.

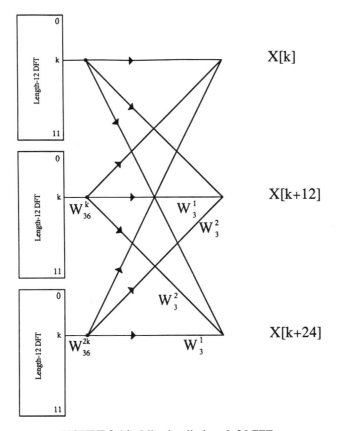

FIGURE 8-14   Mixed-radix length 36 FFT.

8-2-7   Prime Factor Algorithm

By restricting the factors in N to be relatively prime, it is possible to gain some of the flexibility of the mixed-radix FFT, while obtaining operation counts that are very close to those of the split-radix FFT [BU81; KO77]. Assume that N has Q relatively prime factors, $N = N_1 N_2 \cdots N_Q$. Define the factors

$$M_i = \left\langle \left(\frac{N}{N_i}\right)^{-1} \right\rangle_{N_i}, \qquad i = 1, 2, \ldots, Q, \tag{8.69}$$

or in other words $\langle M_i N / N_i \rangle_{N_i} = 1$, so M_i is the inverse of (N/N_i) in the finite Galois field $\mathrm{GF}(N_i)$. The prime factor map (PFM) then is

$$n \equiv \left\langle \frac{N}{N_1} M_1 n_1 + \frac{N}{N_2} M_2 n_2 + \cdots + \frac{N}{N_Q} M_Q n_Q \right\rangle_N,$$

$$k \equiv \left\langle \frac{N}{N_1} k_1 + \frac{N}{N_2} k_2 + \cdots + \frac{N}{N_Q} k_Q \right\rangle_N. \tag{8.70}$$

Inserting this map into Eq. (8.54) gives

$$
X\left[\left\langle \frac{N}{N_1}k_1 + \frac{N}{N_2}k_2 + \cdots + \frac{N}{N_Q}k_Q \right\rangle_N\right]
$$

$$
= \sum_{n_1=0}^{N_1-1}\sum_{n_2=0}^{N_2-1}\cdots\sum_{n_Q=0}^{N_Q-1} x\left[\left\langle \frac{N}{N_1}M_1n_1 + \frac{N}{N_2}M_2n_2 + \cdots + \frac{N}{N_Q}M_Qn_Q \right\rangle_N\right]
$$

$$
\cdot W_{N_1}^{n_1k_1} W_{N_2}^{n_2k_2} \cdots W_{N_Q}^{n_Qk_Q} \qquad (8.71)
$$

which can be rewritten as

$$
X[k_1, k_2, \cdots, k_Q] = \sum_{n_1=0}^{N_1-1}\sum_{n_2=0}^{N_2-1}\cdots\sum_{n_Q=0}^{N_Q-1} x[n_1, n_2, \cdots n_Q]
$$

$$
\cdot W_{N_1}^{n_1k_1} W_{N_2}^{n_2k_2} \cdots W_{N_Q}^{n_Qk_Q}. \qquad (8.72)
$$

This is the prime factor FFT algorithm (PFA), which converts the one-dimensional DFT into a multidimensional DFT where each dimension is transformed independently. There are no twiddle factors to be multiplied between each dimension as was the case for the FFTs discussed so far. Each dimension is transformed using very efficient short-length modules that are usually the Winograd modules described in the next section.

Example 8.9. To compute a length 15 DFT, let $N_1 = 5$ and $N_2 = 3$. Hence $M_1 = 2$ and $M_2 = 2$, since $\langle 2 \times 3 \rangle_5 \equiv 1$ and $\langle 2 \times 5 \rangle_3 \equiv 1$. The index map then is

$$
n \equiv \langle 6n_1 + 10n_2 \rangle_N,
$$

$$
k \equiv \langle 3k_1 + 5k_2 \rangle_N,
$$

which gives

$$
X[\langle 3k_1 + 5k_2 \rangle_N] = \sum_{n_1=0}^{N_1-1}\sum_{n_2=0}^{N_2-1} x[\langle 6n_1 + 10n_2 \rangle_N] W_{N_1}^{n_1k_1} W_{N_2}^{n_2k_2},
$$

or written as a two-dimensional DFT

$$
X[k_1, k_2] = \sum_{n_1=0}^{N_1-1}\sum_{n_2=0}^{N_2-1} x[n_1, n_2] W_{N_1}^{n_1k_1} W_{N_2}^{n_2k_2}.
$$

Figure 8-15 shows how the length 15 DFT is broken up into five length 3 DFTs and three length 5 DFTs. Note there are no twiddle factors between the two blocks.

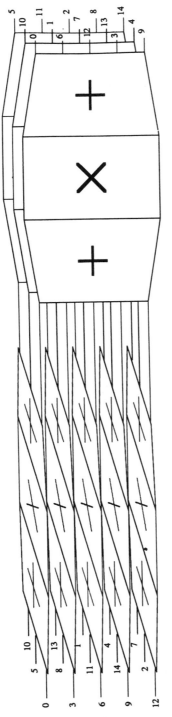

FIGURE 8-15 Length 15 prime factor FFT with length 3 and length 5 modules.

531

The modules are length 3 and 5 Winograd modules and are described in the next section.

It can be proved that the factorization into shorter relatively prime length DFTs can also be written as matrix Kronecker products of the DFT matrices of all the factors in N [BL84]. This turns out to be an important feature when dealing with the Winograd–Fourier transform algorithm. First, we consider an example.

Example 8.10. The length 15 DFT discussed above can be written as

$$
\begin{bmatrix}
X[0] \\ X[3] \\ X[6] \\ X[9] \\ X[12] \\ X[5] \\ X[8] \\ X[11] \\ X[14] \\ X[2] \\ X[10] \\ X[13] \\ X[1] \\ X[4] \\ X[7]
\end{bmatrix}
=
\begin{bmatrix}
W_{15}^0 & W_{15}^0 & W_{15}^0 \\
W_{15}^0 & W_{15}^5 & W_{15}^{10} \\
W_{15}^0 & W_{15}^{10} & W_{15}^5
\end{bmatrix}
\otimes
\begin{bmatrix}
W_{15}^0 & W_{15}^0 & W_{15}^0 & W_{15}^0 & W_{15}^0 \\
W_{15}^0 & W_{15}^3 & W_{15}^6 & W_{15}^9 & W_{15}^{12} \\
W_{15}^0 & W_{15}^6 & W_{15}^{12} & W_{15}^3 & W_{15}^9 \\
W_{15}^0 & W_{15}^9 & W_{15}^3 & W_{15}^{12} & W_{15}^6 \\
W_{15}^0 & W_{15}^{12} & W_{15}^9 & W_{15}^6 & W_{15}^3
\end{bmatrix}
\begin{bmatrix}
x[0] \\ x[6] \\ x[12] \\ x[3] \\ x[9] \\ x[10] \\ x[1] \\ x[7] \\ x[13] \\ x[4] \\ x[5] \\ x[11] \\ x[2] \\ x[8] \\ x[14]
\end{bmatrix}
$$

where $\otimes$ is the Kronecker product. Note that the two matrices are the length 3 and length 5 DFT matrices.

This concept generalizes, such that the DFT vector **X** of the sequence vector **x** can be written

$$\mathbf{X} = \mathbf{W}(N_1) \otimes \mathbf{W}(N_2) \otimes \cdots \otimes \mathbf{W}(N_Q)\mathbf{x}, \tag{8.73}$$

where $\mathbf{W}(N_i)$ is the length N_i DFT matrix and N_i are the factors of N. The two vectors **x** and **X** are permuted according to the index map used, as indicated in the example.

Because some of the twiddle factors are trivial (0 or 1), length 2 and 4 FFTs do not require any multiplications, and because of symmetries of the unit circle, length 8 and 16 FFTs require very few multiplications. Short odd-length FFTs are much less efficient compared to their length (a length 3 DFT requires a minimum of two nontrivial multiplications and a length 5 DFT requires a minimum of five nontrivial multiplications). Because of the relatively prime constraint only one factor in the PFA can be even (2, 4, 8, or 16); all others are odd, and one might erroneously assume that the PFA is less efficient than, say, the radix-4 FFT. This is not necessarily true, because the radix-4 has twiddle factors between each factor of 4 (i.e., between each dimension) as shown in Eq. (8.65), whereas the PFA has no twiddle factors. As can be seen in the tables in Section 8-2-10, the PFA has a very low operation count compared to other FFT algorithms while retaining a flexibility that allows a wide variety of lengths rather than just powers of 2. The actual operation counts depend on which factors are used, since not all factors are equally efficient. Most PFA algorithms use the factors shown in Table 8-6, which also show the operation counts for the Winograd modules (see next section). From here it is apparent that even factors are much more efficient than odd factors.

In Appendix 8-A, program 6 is a PFA FFT with modules of length 2, 3, 4, 5, and 7. Other modules can be found in Burrus and Parks [BU85]. The DO 10 loop steps through the factors N_i of N; the DO 20 loop steps through the transforms, and the DO 30 loop sets up the indices in the I array, such that when the modules are called in the computed GOTO statement, the elements are addressed indirectly through I.

The order of the output from the PFA is scrambled, and there are basically four ways to deal with that [BU88]. First, it might not be necessary to unscramble the

TABLE 8-6 Computional Complexity of Short-Length PFA Modules [BU85]

| | Number of Real Multiplications and Additions to Compute a Length N DFT | | | |
| | Real Input | | Complex Input | |
N	Multiplications	Additions	Multiplications	Additions
2	0	2	0	4
3	2	4	4	12
4	0	6	0	16
5	5	13	10	34
7	8	30	16	72
8	2	20	4	52
9	10	34	20	84
11	20	74	40	168
13	20	82	40	188
16	10	60	20	148
17	35	141	70	314
19	38	168	76	372
25	66	186	132	420
32	34	164	68	388

output at all; this is often the case when doing convolution. Second, an unscrambler can be added, as was done to the Cooley–Tukey algorithms. Third, unscrambling can be done within the modules. This, however, is unique to a specific length, and if it changes, the modules have to be changed as well [BU81]. Finally, the program can use separate input and output pointers in each module, as described in Burrus and Parks [BU85].

8-2-8 Winograd's Algorithm

Like all the FFT algorithms described so far, the Winograd FFT also factors the DFT into a multidimensional DFT. Each dimension is computed using Winograd short-length modules, which consist of three parts: the preweave additions, the multiplications, and the postweave additions. These modules are based on Rader's permutation, which transforms a DFT into a cyclic convolution and then uses Winograd's efficient short-length cyclic convolution algorithms.

The Winograd–Fourier transform algorithm (WFTA) does not, like the FFTs presented so far, complete one dimension at a time, but rather computes all dimensions simultaneously. The WFTA first computes the preweave additions for all the dimensions, then computes the multiplications for all the dimensions, and finally computes the postweave additions for all dimensions. In this way it is possible to nest and hence reduce the number of multiplications.

Rader's Permutation. Assume that N is a prime, and rewrite Eq. (8.54):

$$X[0] = \sum_{n=0}^{N-1} x[n],$$

$$X[k] = x[0] + \sum_{n=1}^{N-1} x[n] W_N^{nk}, \qquad k = 1, 2, \ldots, N - 1. \qquad (8.74)$$

It is now a well-known fact from number theory [NI80] that if N is a prime, there exists a primitive root g of N. A primitive root g of N is a number such that $\langle g^i \rangle_N \neq \langle g^j \rangle_N$ for $i \neq j$, $0 \leq i, j \leq N - 2$, and $\langle g^i \rangle_N \neq 0$ $\forall i$, where $\langle \ \rangle_N$ denotes modulo N. In other words, $\langle g^i \rangle_N$, $i = 0, 1, \ldots, N - 2$, is a permutation of the numbers $1, 2, \ldots, N - 1$. Substituting $k \rightarrow \langle g^k \rangle_N$ and $n \rightarrow \langle g^{-n} \rangle_N$ into Eq. (8.74) gives

$$X[0] = \sum_{n=0}^{N-1} x[n],$$

$$X[\langle g^k \rangle_N] = x[0] + \sum_{n=0}^{N-2} x[\langle g^{-n} \rangle_N] W_N^{g^{k-n}}, \qquad k = 0, 1, \ldots, N - 2,$$

$$(8.75)$$

where the last summation is recognized as a cyclic convolution in n and k. The quantity $\langle g^{-n} \rangle_N$ is defined as $\langle (g^{-1})^n \rangle_N$, where g^{-1} is the inverse of g; that is,

$\langle gg^{-1}\rangle_N = 1$. This method, which converts a length N DFT into a length $N - 1$ cyclic convolution, is called Rader's permutation [RA68].

Example 8.11. The length 5 DFT can be rewritten

$$X[0] = \sum_{n=0}^{4} x[n],$$

$$X[k] = x[0] + \sum_{n=1}^{4} x[n] W_5^{nk}, \qquad k = 1, 2, 3, 4.$$

Now $g = 2$ is a primitive root of 5, since

$$\langle 2^0\rangle_5 = 1, \qquad \langle 2^1\rangle_5 = 2, \qquad \langle 2^2\rangle_5 = 4, \qquad \langle 2^3\rangle_5 = 3,$$

and $g^{-1} = 3$, since $\langle 2 \times 3\rangle_5 = 1$. Substituting $n \to 2^{-n}$ and $k \to 2^k$ in the DFT, it becomes

$$X[0] = \sum_{n=0}^{4} x[n],$$

$$X[\langle 2^k\rangle_5] = x[0] + \sum_{n=0}^{3} x[\langle 2^{-n}\rangle_5] W_5^{2^{k-n}}, \qquad k = 0, 1, 2, 3,$$

or written in matrix form

$$X[0] = [1\ 1\ 1\ 1\ 1] \begin{bmatrix} x[0] \\ x[1] \\ x[2] \\ x[3] \\ x[4] \end{bmatrix}$$

$$\begin{bmatrix} X[\langle 2^0\rangle_5] \\ X[\langle 2^1\rangle_5] \\ X[\langle 2^2\rangle_5] \\ X[\langle 2^3\rangle_5] \end{bmatrix} = \begin{bmatrix} W_5^{2^0} & W_5^{2^{-1}} & W_5^{2^{-2}} & W_5^{2^{-3}} \\ W_5^{2^1} & W_5^{2^0} & W_5^{2^{-1}} & W_5^{2^{-2}} \\ W_5^{2^2} & W_5^{2^1} & W_5^{2^0} & W_5^{2^{-1}} \\ W_5^{2^3} & W_5^{2^2} & W_5^{2^1} & W_5^{2^0} \end{bmatrix} \begin{bmatrix} x[\langle 2^{-0}\rangle_5] \\ x[\langle 2^{-1}\rangle_5] \\ x[\langle 2^{-2}\rangle_5] \\ x[\langle 2^{-3}\rangle_5] \end{bmatrix} + \begin{bmatrix} x[0] \\ x[0] \\ x[0] \\ x[0] \end{bmatrix} \Leftrightarrow$$

$$\begin{bmatrix} X[1] \\ X[2] \\ X[4] \\ X[3] \end{bmatrix} = \begin{bmatrix} W_5^1 & W_5^3 & W_5^4 & W_5^2 \\ W_5^2 & W_5^1 & W_5^3 & W_5^4 \\ W_5^4 & W_5^2 & W_5^1 & W_5^3 \\ W_5^3 & W_5^4 & W_5^2 & W_5^1 \end{bmatrix} \begin{bmatrix} x[1] \\ x[3] \\ x[4] \\ x[2] \end{bmatrix} + \begin{bmatrix} x[0] \\ x[0] \\ x[0] \\ x[0] \end{bmatrix},$$

where the W matrix is a cyclic convolution matrix. Combining the two results again, the length 5 DFT becomes

$$
\begin{bmatrix} X[0] \\ X[1] \\ X[2] \\ X[4] \\ X[3] \end{bmatrix} = \begin{bmatrix} 1 & 1 & 1 & 1 & 1 \\ 1 & W_5^1 & W_5^3 & W_5^4 & W_5^2 \\ 1 & W_5^2 & W_5^1 & W_5^3 & W_5^4 \\ 1 & W_5^4 & W_5^2 & W_5^1 & W_5^3 \\ 1 & W_5^3 & W_5^4 & W_5^2 & W_5^1 \end{bmatrix} \begin{bmatrix} x[0] \\ x[1] \\ x[3] \\ x[4] \\ x[2] \end{bmatrix}.
$$

If N is a power of a prime, say $N = p^r$, then the length N DFT can be converted into a length $p^{r-1}(p-1)$ cyclic convolution using a variant of Rader's permutation. This case is more complicated than the above, because no primitive root exists that will generate all the numbers 1 to p^r. All numbers that contain p as a factor cannot be generated. The actual scheme is too cumbersome to describe here but can be found in McClellan and Rader [MC79b], Blahut [BL84], and Nussbaumer [NU81]; also, an example for $N = 3^2$ is shown below.

Example 8.12. Permuting the length 9 DFT with $g = 2$ ($g^{-1} = 5$), it becomes

$$
\begin{bmatrix} X[0] \\ X[3] \\ X[6] \\ X[1] \\ X[5] \\ X[7] \\ X[8] \\ X[4] \\ X[2] \end{bmatrix} = \begin{bmatrix} 0 & 0 & 0 & 0 & 0 & 0 & 0 & 0 & 0 \\ 0 & 0 & 0 & 3 & 6 & 3 & 6 & 3 & 6 \\ 0 & 0 & 0 & 6 & 3 & 6 & 3 & 6 & 3 \\ 0 & 3 & 6 & 1 & 2 & 4 & 8 & 7 & 5 \\ 0 & 6 & 3 & 5 & 1 & 2 & 4 & 8 & 7 \\ 0 & 3 & 6 & 7 & 5 & 1 & 2 & 4 & 8 \\ 0 & 6 & 3 & 8 & 7 & 5 & 1 & 2 & 4 \\ 0 & 3 & 6 & 4 & 8 & 7 & 5 & 1 & 2 \\ 0 & 6 & 3 & 2 & 4 & 8 & 7 & 5 & 1 \end{bmatrix} \begin{bmatrix} x[0] \\ x[3] \\ x[6] \\ x[1] \\ x[2] \\ x[4] \\ x[8] \\ x[7] \\ x[5] \end{bmatrix},
$$

where the entries of the matrix are exponents of W_9. The lower right-hand block is a $3^{2-1}(3-1) = 6$ cyclic convolution. Besides that, there are several length 2 convolutions as well. This shows that the length 9 DFT can be computed as in Figure 8-16.

Short-Length DFT Modules. Winograd [WI78] introduced in 1976 a new approach for computing a short-length DFT that is very efficient. It converts the DFT to a cyclic convolution using Rader's permutation and then computes the cyclic con-

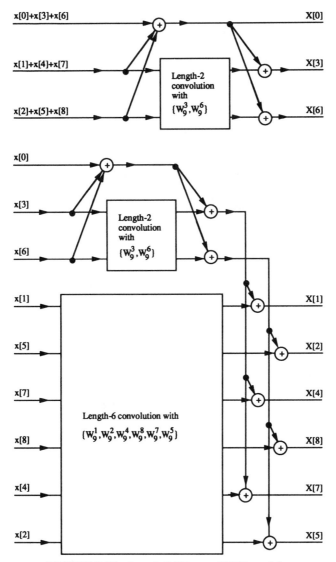

FIGURE 8-16 Length 9 Winograd DFT module.

volution using Winograd's short-length convolution algorithms described in Section 8-1-3.

To save additions in the resulting algorithm, it turns out that a slightly different form of Eq. (8.74) is useful. By realizing that

$$\sum_{n=0}^{N-1} W_N^{nk} = \begin{cases} N, & k = 0 \\ 0, & k \neq 0 \end{cases} \tag{8.76}$$

it immediately follows that Eq. (8.74) can also be written

$$X[0] = \sum_{n=0}^{N-1} x[n],$$

$$X[k] = X[0] + \sum_{n=1}^{N-1} x[n](W_N^{nk} - 1), \qquad k = 1, 2, \ldots, N - 1. \quad (8.77)$$

The last summation can now be converted to a convolution by Rader's permutation, and computed efficiently using Winograd's convolution modules.

To see how the short-length DFT algorithm is developed, an example is used.

Example 8.13. To compute a length 5 DFT, convert it to a length 4 convolution as in Example 8.11,

$$\{x[1], x[3], x[4], x[2]\} \textcircled{4} \{W_5^1, W_5^2, W_5^4, W_5^3\}.$$

However, by using Eq. (8.77), it can easily be found that the following convolution results instead:

$$\{x[1], x[3], x[4], x[2]\} \textcircled{4} \{(W_5^1 - 1), (W_5^2 - 1), (W_5^4 - 1), (W_5^3 - 1)\}.$$

First define

$$
\begin{bmatrix} B_1 \\ B_2 \\ jB_3 \\ jB_4 \\ jB_5 \end{bmatrix}
= \tfrac{1}{4}
\begin{bmatrix}
1 & 1 & 1 & 1 \\
-1 & 1 & -1 & 1 \\
2 & -2 & -2 & 2 \\
2 & 2 & -2 & -2 \\
-2 & 0 & 2 & 0
\end{bmatrix}
\begin{bmatrix}
(W_5^1 - 1) \\
(W_5^2 - 1) \\
(W_5^4 - 1) \\
(W_5^3 - 1)
\end{bmatrix}
$$

$$
=
\begin{bmatrix}
\tfrac{1}{2}(\cos(2\theta) + \cos(\theta)) - 1 \\
\tfrac{1}{2}(\cos(2\theta) - \cos(\theta)) \\
j(\sin(2\theta) - \sin(\theta)) \\
-j(\sin(2\theta) + \sin(\theta)) \\
j\sin(\theta)
\end{bmatrix},
$$

where $\theta = 2\pi/5$. Now use the length 4 cyclic convolution algorithm in Examples 8.4 and 8.6:

$$
\begin{bmatrix} X[1] \\ X[2] \\ X[3] \\ X[4] \end{bmatrix}
=
\begin{bmatrix}
1 & -1 & 0 & -1 & -1 \\
1 & 1 & -1 & 0 & -1 \\
1 & -1 & 0 & 1 & 1 \\
1 & 1 & 1 & 0 & 1
\end{bmatrix}
\left(
\begin{bmatrix}
1 & 1 & 1 & 1 \\
1 & -1 & 1 & -1 \\
1 & 0 & -1 & 0 \\
0 & 1 & 0 & -1 \\
1 & 1 & -1 & -1
\end{bmatrix}
\begin{bmatrix} x[1] \\ x[2] \\ x[3] \\ x[4] \end{bmatrix}
\; \diamond \;
\begin{bmatrix} B_1 \\ B_2 \\ jB_3 \\ jB_4 \\ jB_5 \end{bmatrix}
\right),
$$

where the last matrix is obtained by multiplying and evaluating the last matrix product in Example 8.6. The $\diamond$ operation is commutative, and term-by-term multiplication can also be written as a matrix multiplication by expanding the B vector to a diagonal matrix, so the algorithm can be written

$$
\begin{bmatrix} X[1] \\ X[2] \\ X[3] \\ X[4] \end{bmatrix} = \begin{bmatrix} 1 & -1 & 0 & -1 & -1 \\ 1 & 1 & -1 & 0 & -1 \\ 1 & -1 & 0 & 1 & 1 \\ 1 & 1 & 1 & 0 & 1 \end{bmatrix} \begin{bmatrix} B_1 & 0 & 0 & 0 & 0 \\ 0 & B_2 & 0 & 0 & 0 \\ 0 & 0 & jB_3 & 0 & 0 \\ 0 & 0 & 0 & jB_4 & 0 \\ 0 & 0 & 0 & 0 & jB_5 \end{bmatrix}
$$

$$
\cdot \begin{bmatrix} 1 & 1 & 1 & 1 \\ 1 & -1 & 1 & -1 \\ 1 & 0 & -1 & 0 \\ 0 & 1 & 0 & -1 \\ 1 & 1 & -1 & -1 \end{bmatrix} \begin{bmatrix} x[1] \\ x[2] \\ x[3] \\ x[4] \end{bmatrix} .
$$

Finally, expand the matrices, so they also compute $X[0]$, and factor the j's into the first matrix

$$
\begin{bmatrix} X[0] \\ X[1] \\ X[2] \\ X[3] \\ X[4] \end{bmatrix} = \begin{bmatrix} 1 & 0 & 0 & 0 & 0 & 0 \\ 1 & 1 & -1 & 0 & -j & -j \\ 1 & 1 & 1 & -j & 0 & -j \\ 1 & 1 & -1 & 0 & j & j \\ 1 & 1 & 1 & j & 0 & j \end{bmatrix}
$$

$$
\cdot \begin{bmatrix} B_0 & 0 & 0 & 0 & 0 & 0 \\ 0 & B_1 & 0 & 0 & 0 & 0 \\ 0 & 0 & B_2 & 0 & 0 & 0 \\ 0 & 0 & 0 & B_3 & 0 & 0 \\ 0 & 0 & 0 & 0 & B_4 & 0 \\ 0 & 0 & 0 & 0 & 0 & B_5 \end{bmatrix} \begin{bmatrix} 1 & 1 & 1 & 1 & 1 \\ 0 & 1 & 1 & 1 & 1 \\ 0 & 1 & -1 & 1 & -1 \\ 0 & 1 & 0 & -1 & 0 \\ 0 & 0 & 1 & 0 & -1 \\ 0 & 1 & 1 & -1 & -1 \end{bmatrix}
$$

$$
\cdot \begin{bmatrix} x[0] \\ x[1] \\ x[2] \\ x[3] \\ x[4] \end{bmatrix} ,
$$

where $B_0 = 1$. If the same addition is only counted once, this length 5 DFT algorithm takes six real multiplications (of which one is trivial since $B_0 = 1$) and 17 real additions (of which four are trivial, since adding purely real and imaginary is not an actual addition), assuming that the input sequence is real. Figure 8-17 shows the resulting module.

Example 8.13 shows how a length 5 convolution is converted into a length 4 cyclic convolution, which is then computed using Winograd's short-length cyclic convolution algorithms. Note that the algorithm has the form

$$
\begin{bmatrix} X[0] \\ \vdots \\ X[N-1] \end{bmatrix} = \mathbf{CBA} \begin{bmatrix} x[0] \\ \vdots \\ x[N-1] \end{bmatrix}, \tag{8.78}
$$

where $\mathbf{A}$ is a matrix that computes the input additions (preweave additions), $\mathbf{B}$ is a diagonal matrix that represents the multiplications, and $\mathbf{C}$ computes the output additions (postweave additions). Since the DFT can also be written as Eq. (8.55), the product $\mathbf{CBA}$ can be viewed as a factorization of the matrix $\mathbf{W}$. Note in the example, which is true in general, that the matrices $\mathbf{A}$ and $\mathbf{C}$ are not square. This means that the preweave additions expand the input vector to L elements, where L is the number of multiplications required by the DFT, and the postweave additions reduce the L elements from the multiplications to N elements again. For this reason the DFT modules are often shown as for the length 5 DFT in Figure 8-18.

The same scheme described for the length 5 DFT extends to all other DFTs of odd-prime lengths. For powers of odd primes a variant of the above scheme is used. This gets somewhat cumbersome, because there are several convolutions in this case. For powers of 2, however, a different scheme has to be used. The Cooley–Tukey structure can be used for some applications (e.g., as a module in the prime factor FFT), but for the Winograd–Fourier transform algorithm (WFTA) it is important that the module has the form of Eq. (8.78), so a different technique is used. It can be shown [BL84] that the length 2^M DFT can be converted into a two-dimensional ($2 \times N/2$) cyclic convolution using the generalized Rader's permutation. This two-dimensional cyclic convolution is computed using Winograd's

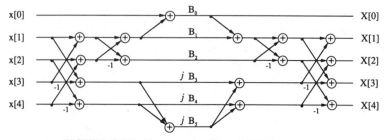

FIGURE 8-17   Flow-graph of length 5 DFT module.

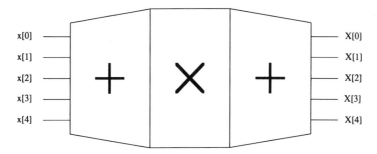

FIGURE 8-18   Length 5 DFT module.

length 2 cyclic convolution algorithm in a recursive manner. For details see Section 10.4 in Blahut [BL84].

Code for modules of different lengths can be found in Blahut [BL84], Nussbaumer [NU81], Burrus and Parks [BU85], and McClellan and Rader [MC79b]. The number of operations required for the different lengths are shown in Table 8-7.

The Winograd–Fourier Transform Algorithm (WFTA). As was indicated in Eq. (8.73), a prime factor DFT can be written as a Kronecker product of short-length DFT matrices. Only the case of two factors will be discussed, since the extension to more factors is straightforward. If N has two factors N_1 and N_2, then $\mathbf{W}$, the length N DFT matrix, factors like

$$\mathbf{W} = \mathbf{W}_1 \otimes \mathbf{W}_2, \qquad (8.79)$$

TABLE 8-7   Complexity of Short-Length Winograd DFT Modules[a]

	Number of Real Multiplications and Additions to Compute a Length N DFT			
	Real Input		Complex Input	
N	Multiplications	Additions	Multiplications	Additions
2	2(2)	2	4(4)	4
3	3(1)	4	6(2)	12
4	4(4)	6	8(8)	16
5	6(1)	13	12(2)	34
7	9(1)	30	18(2)	72
8	8(6)	20	16(12)	52
9	11(1)	34	22(2)	84
11	21(1)	74	42(2)	168
13	21(1)	82	42(2)	188
16	18(8)	60	36(16)	148
17	36(4)	141	72(2)	157
19	39(1)	168	78(2)	372

[a]The numbers in parentheses are the number of trivial multiplications (by -1).

where $\mathbf{W}_1$ and $\mathbf{W}_2$ are the length N_1 and length N_2 DFT matrices, respectively, and $\otimes$ is the Kronecker product operator. From the previous section, it was shown that a short-length DFT factors like **CBA**, so

$$\mathbf{W}_1 = \mathbf{C}_1\mathbf{B}_1\mathbf{A}_1 \quad \text{and} \quad \mathbf{W}_2 = \mathbf{C}_2\mathbf{B}_2\mathbf{A}_2. \tag{8.80}$$

Combining these two equations, it can be shown that the computation of the modules can be nested like

$$\mathbf{W} = (\mathbf{C}_1\mathbf{B}_1\mathbf{A}_1) \otimes (\mathbf{C}_2\mathbf{B}_2\mathbf{A}_2) = (\mathbf{C}_1 \otimes \mathbf{C}_2)(\mathbf{B}_1 \otimes \mathbf{B}_2)(\mathbf{A}_1 \otimes \mathbf{A}_2). \tag{8.81}$$

This algorithm for the length N DFT has the same form as the short-length DFTs, Eq. (8.78) ($\mathbf{A} = \mathbf{A}_1 \otimes \mathbf{A}_2$, etc.); in other words, by nesting, all the multiplications are merged into the center of the algorithm. The way the algorithm is derived, the input and output are permuted, but it is easy to rearrange rows and columns of $\mathbf{A}$ and $\mathbf{C}$ so that they are both in natural order. As was discussed in the previous section, the matrices $\mathbf{A}$ and $\mathbf{C}$ are not square, but rectangular. This causes a data expansion in the middle of the modules, and because of the nesting, it causes an expansion of the whole algorithm as seen in Figure 8-19. After the two addition blocks $\mathbf{A}_1$ and $\mathbf{A}_2$, there are 18 intermediate variables instead of 15 in the length 15 DFT. This means that the WFTA cannot be computed in-place (i.e., an extra array of size L is needed), which is a major drawback of the WFTA.

The WFTA requires modules of the type described in Eq. (8.78), while the prime factor FFT does not care how the modules are structured and hence can use any FFT as a module. This is because the prime-factor FFT computes each module (dimension) individually and hence their internal structure does not matter, while the WFTA merges all the modules. The merging is done by first computing all the preweave additions for all modules, then performing all the multiplications, and finally computing all the postweave additions. Figure 8-19 shows a length 15 WFTA, which should be compared to the length 15 PFA in Figure 8-15. A program that computes the WFTA can be found in McClellan and Rader [MC79b] and McClellan and Nawab [MC79a]. The program has modules of length 2, 3, 4, 5, 7, 8, 9, and 16.

Since all the multiplications are nested in the middle of the algorithm, the Kronecker products $\mathbf{B}_1 \otimes \mathbf{B}_2 \otimes \cdots \otimes \mathbf{B}_Q$ can be precomputed, and the total number of multiplications required in the WFTA are

$$M(N) = M(N_1)M(N_2) \cdots M(N_Q), \tag{8.82}$$

where $M(N_i)$ is the number of multiplications for the length N_i module. The number of additions are somewhat difficult to compute, because they depend on the order of the preweave and postweave additions. For two factors in N it is

$$A(N) = N_1A(N_2) + M(N_2)A(N_1), \tag{8.83}$$

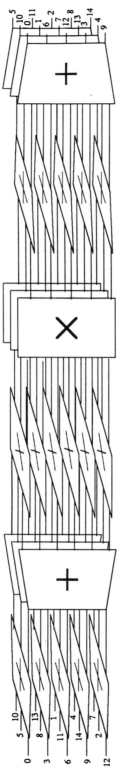

FIGURE 8-19 Length 15 Winograd–Fourier transform algorithm.

where $A(N_i)$ is the number of additions in the length N_i module. The formula for more factors can be found in Nussbaumer [NU81]. The tables in Section 8-2-10 shows the operation count compared to several other FFTs.

8-2-9 Real-Valued FFTs

All the algorithms derived so far have dealt with complex data. In most applications, however, the sequence to be transformed is real-valued, so algorithms that utilize this fact to reduce the required arithmetic are preferred. One approach is to use the complex FFTs (CFFTs) already developed and to modify the input/output. The other approach is to modify the algorithm to suit the properties of the input data. The first approach is more frequently used but is less efficient than the second. Specialized real-valued FFTs (RFFTs) are all derived from their complex counterparts in nearly the same manner, so only the radix-2 case will be discussed in detail. See also Sorensen et al. [SO87]. For other redundancies in the input sequence, see Rabiner [RA79], Markel [MA71], and Nagai [NA86].

Real-Valued FFTs Using Complex FFTs. There are three techniques using this approach. One very simplistic technique assigns the real-valued input sequence $x[n]$ to the complex sequence $z[n]$ with zero imaginary part and then computes the DFT of $z[n]$. This turns out to be wasteful, because it about doubles the arithmetic needed.

A second technique uses the symmetries of the DFT to transform two real-valued sequences simultaneously by computing one complex FFT (CFFT) [CO70]. If $x[n]$ and $y[n]$ are real-valued, then their transforms $X[k]$ and $Y[k]$ have an even real part and an odd imaginary part. The DFT is a linear transform, so the DFT of $z[n] = x[n] + jy[n]$ is

$$\text{DFT } \{z[n]\} = Z[k] = Z_r[k] + jZ_i[k], \tag{8.84}$$

where subscripts r and i denote real and imaginary part, respectively. The transforms $X[k]$ and $Y[k]$ can then be obtained as [SO87]

$$X[k] = \tfrac{1}{2}(Z_r[k] + Z_r[N - k]) + j\tfrac{1}{2}(Z_i[k] - Z_i[N - k]), \qquad k = 0, 1, \ldots, \frac{N}{2},$$

$$Y[k] = \tfrac{1}{2}(Z_i[N - k] + Z_i[k]) + j\tfrac{1}{2}(Z_r[N - k] - Z_r[k]), \qquad k = 0, 1, \ldots, \frac{N}{2}.$$

$$\tag{8.85}$$

The above equations reduce for $k = 0$ and $k = N/2$:

$$X[0] = Z_r[0] \quad \text{and} \quad Y[0] = Z_i[0],$$

$$X\left[\frac{N}{2}\right] = Z_r\left[\frac{N}{2}\right] \quad \text{and} \quad Y\left[\frac{N}{2}\right] = Z_i\left[\frac{N}{2}\right], \tag{8.86}$$

and hence require $2N - 4$ real additions. This means that two length N real-valued transforms require one length N complex FFT and $2N - 4$ real additions. If two transforms are needed, as would be the case when convolving two arbitrary sequences or doing block processing, the operations per real transform are half of the above.

The third technique is useful when only one transform is needed. It uses the observation [CO70] that the final stage of a decimation-in-time radix-2 FFT, Eq. (8.62), combines two independent length $(N/2)$ transforms. If the input is real-valued, the two half-length DFTs are also transforms of real-valued data, and they can be computed as in Eq. (8.85) by computing the length $(N/2)$ complex DFT of $z[n] = x[2n] + jx[2n + 1]$. The separation of the two half-length transforms and the computation of the last stage require $N - 6$ real multiplications and $\frac{5}{2} N - 6$ real additions (assuming four real multiplications and two real additions per complex multiplication), which must be added to the operation counts for a half-length complex FFT to find the total operation count.

Radix-2 Decimation-in-Time RFFT. The fundamental idea is that symmetries due to the real-valued nature of the data can be used at every stage of the CFFT algorithm to remove redundancies. If $x[n]$ is real it is easily verified that

$$X[0] \quad \text{and} \quad X\left[\frac{N}{2}\right] \text{ are real,}$$

$$X[k] = X^*[N - k], \qquad 1 \le k \le \frac{N}{2} - 1. \tag{8.87}$$

Only half of the complex values need be computed, since the others are then known by the complex conjugate symmetry. One possibility is to compute $X[0]$ through $X[N/2]$, but a better choice is to compute $X[0]$ through $X[N/4]$ and $X[N/2]$ through $X[3N/4]$. The latter choice is preferred because these are the values computed by the first $N/4 + 1$ complex butterflies in the standard complex-valued algorithm.

To compute the RFFT in-place, the redundancies must be utilized to halve the storage requirement compared to the corresponding CFFT. This is possible if the real part of the kth complex-valued coefficient is placed in the kth location in the real array, and the imaginary part is stored in the redundant $(N - k)$th location. Since the imaginary part of the kth complex output $(1 \le k \le N/2 - 1)$ is stored in the $(N - k)$th memory location, the value in that location from the previous stage must be used before it is overwritten. This is really not a problem, since the current content of this memory location is needed only by the current butterfly. The butterfly operations for the RFFT and the CFFT are identical; only the memory locations from which the data are fetched are different [SO87]. The resulting butterfly closely resembles the butterfly for the fast Hartley transform presented in Sorensen et al. [SO85].

A special butterfly in the final stage computes $X[0]$ and $X[N/2]$ separately because they are real, and since it requires no multiplications this also reduces the computational complexity. A further savings of four additions and a complex multiply in the last stage is achieved by noting that the butterfly computing $X[N/4]$ and $X[3N/4]$ requires no multiplications or additions (only a sign change) for real-valued data. These four values are computed separately in a "special butterfly." This scheme exactly halves the number of multiplications and storage locations and reduces the additions to two less than half that required to compute one stage of the complex-valued transform.

This approach can be applied at every stage by recognizing that the half-length transforms of the even- and odd-indexed data samples are again DFTs of real-valued data, so the same approach can be applied recursively to shorter DFTs to complete the algorithm. The result is a radix-2 decimation-in-time FFT algorithm for real-valued data that requires exactly half the multiplications and storage, and $N - 2$ less than half the additions of the algorithm for complex data. In other words, if the algorithm for complex data requires A_c additions, the corresponding algorithm for real data requires

$$A_r = A_c - N + 2 \tag{8.88}$$

additions. Figure 8-20 illustrates the algorithm for a length 16 transform. At every stage, the values that are real are indicated by an X, and the complex values by an O. Locations that are the complex conjugate of another location (and where the imaginary part of the complex number is stored) are connected by a dashed arc. Only the butterflies drawn in solid lines are actually computed.

It is important to note that the decimation-in-time algorithm is natural for the fast calculation of the spectrum of a real-valued series because the shorter sequences are also real-valued, and the symmetry of their transforms can be ex-

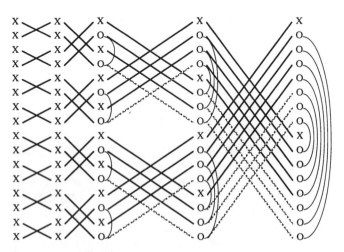

FIGURE 8-20   Length 16 radix-2 RFFT. The symbol X indicates a real value and O a complex value. Arcs show complex values that are conjugates of each other. Only the butterflies in solid lines need to be computed.

ploited at every stage to reduce operations and storage. Although a decimation-in-frequency algorithm (for complex inputs) at the end yields a spectrum with the proper symmetry, the redundancies are not apparent within the algorithm and cannot easily be exploited to reduce the operation count and storage from that required for complex data.

For the inverse discrete Fourier transform (IDFT) that transforms the complex conjugate symmetric spectrum to a real-valued time series, the decimation-in-frequency algorithm is the natural choice. If the even-indexed and odd-indexed frequency samples are separated into two half-length series, symmetry of both series is maintained. At every stage of the decimation-in-frequency IDFT algorithm, the redundancies are explicit and easily removed.

Higher-Radix Algorithms. The most well-known efficient FFT algorithm for real-valued sequences is probably Bergland's radix-8 RFFT [BE69; IE79]. Like the radix-4 and higher radix algorithms, it is developed in the same manner as the radix-2 RFFT. The symmetry of the transform implies that only half of the butterflies need be computed and also results in simplification of the butterflies with simple coefficients. This reduction saves half of the multiplications and storage, and $N - 2$ more than half of the additions over the algorithm for complex data.

The Split-Radix RFFT. The decimation-in-time split-radix FFT described in [SO86a] is modified exactly the same way as was the radix-2 FFT. It leads to a very efficient algorithm requiring only $\frac{1}{2} N \log_2 (N) - \frac{3}{2}N + 2$ real multiplications and $\frac{3}{2} N \log_2 (N) - \frac{5}{2}N + 4$ real additions to compute a length N split-radix RFFT, which is half the multiplications and $N - 2$ less than half the additions required by the split-radix CFFT (see Eq. (8.88)) [SO86b]. An interesting feature is that, while it is not possible to compute a real-valued radix-2 DIF FFT, a split-radix DIF algorithm has been developed in Duhamel [DU86a], and Martens [MA84a]. The indexing for this scheme may be slightly more efficient than the standard split-radix indexing, but the output data are in a strange order that requires a complicated unscrambler. If, however, unscrambling is not needed, the DIF split-radix FFT may be a good choice.

In Appendix 8-A, programs 7 and 8 are very efficient forward and inverse split-radix RFFTs.

Prime Factor RFFTs. Only the case of two factors in N will be discussed, since the ideas generalize easily. The redundancies due to real-valued data in the prime factor algorithm can be removed as follows. The first set of short (length N_1) transforms (over the columns) are of real-valued data, so half of the multiplications and $N_1 - 1$ ($N_1 - 2$ if N_1 is even) more than half of the additions in the short-length DFTs can be saved by specializing DFT modules for real-valued data. When computing the second dimension, the row corresponding to the zero frequency index of the first stage of transforms is real-valued (as is the $[N_1/2]$th row if N_1 is even) and requires half the multiplications and $N_2 - 1$ ($N_2 - 2$ if N_2 is even) less than half the additions (see Eq. (8.88)). A second-stage real-valued module computes the DFT of this (or these) row(s). The other $N_1 - 1$ or $N_1 - 2$ rows are complex, but since the input data are real-valued, the $(N_1 - k_1)$th row is the complex conjugate of the k_1th row after the DFT over the first dimension. Only half of the length N_2 complex transforms need be computed or stored, since the DFT of the complex conjugate of a sequence is the time-reversed complex conjugate of the DFT of the sequence. The real part can be stored in the k_1th row, and the imaginary part can be stored in the $(N_1 - k_1)$th row. If the imaginary parts of the output are stored in reverse $(N_2 - k_2)$ order, the output order of the transform is given by the standard PFA mapping, where the real part is in $X(k_1, k_2)$ and the corresponding imaginary part is in $X(N_1 - k_1, N_2 - k_2)$. Hence using the symmetries of the complex PFA algorithm reduces the storage and multiplications to half and reduces the additions to $N - 1$ ($N - 2$ if N is even) less than half that of a complex-valued prime factor algorithm (see Eq. (8.88)) [HE84; SO87].

The scheme described above requires both a real-valued and a complex-valued short DFT module for each stage. As an alternative, the complex transforms can be computed by applying a real-valued module to the real and imaginary parts of the data and adding the results appropriately to get a complex transform. This requires exactly the same number of operations as a complex module. Another alternative is to commute the combining additions to the end of the algorithm [HE84]. The floating-point operation counts are the same, but recalculation of the indices at the final combination state makes this technique somewhat slower.

Real-Valued Winograd Fourier Transform Algorithm. The short-length DFT modules in Eq. (8.78) have complex entries only in the **C** matrix, which means that if the input is purely real (or imaginary) it stays real (imaginary) right until the postweave additions. A real-valued module can hence easily be derived by removing half the multiplications and $N - 1$ ($N - 2$ if N is even) more than half the additions in the length N complex module (see Eq. (8.88)). In nesting these real-valued modules, the real and imaginary parts remain separate until the postweave additions, and the real-valued WFTA is obtained. The algorithm is discussed further in Parsons [PA79]. A program can be derived by modifying the complex WFTA program in McClellan and Nawab [MC79a] and McClellan and Rader [MC79b].

8-2-10   Comparison and Computational Aspects of FFT Algorithms

Operation counts for the different algorithms discussed in this chapter are shown in Table 8-8. The table shows the absolute minimal number of operations that it is possible to achieve with that algorithm. For the WFTA, the number of multiplications in parentheses is the number achieved by the program by McClellan and Nawab [MC79a]. For PFA and WFTA, many more lengths are possible; for more detailed tables, see Burrus and Parks [BU85]. One might argue that operation counts may not be a realistic measure for the performance of FFTs (and rightfully so!), but it is the measure that gives the most unbiased comparison independent of compilers and machines. A better measure would be CPU time to execute certain

TABLE 8-8 Operation Counts for Algorithms FFT with Complex Input Data[a]

Multiplicative/Additive Complexity for Computing Complex FFTs with 3/3 Multiplication Algorithm

Length N	Radix-2 1 BF		Radix-2 5 BF		Radix-4 5 BF		Radix-8 5 BF		Split-Radix 3 BF		PFA		WFTA	
	Multiplies	Adds	Multiplies	Adds	Multiplies	Adds	Multiplies	Adds	Multiplies	Adds	Multiplies	Adds	Multiplies	Adds
8	40	80	4	52			4	52	4	52	4	52	4(16)	52
12											16	96	16(24)	96
15											50	162	34(36)	162
16	104	216	24	152	20	148			20	148	20	148	20(36)	148
30											100	384	68(72)	384
32	256	544	88	408					68	388				
35											150	598	106(108)	666
60											200	888	136(144)	888
64	608	1312	264	1032	208	976	204	972	196	964				
72											196	1140	164(176)	1156
112											396	2188	308(324)	2332
120											460	2076	276(288)	2076
128	1408	3072	712	2504					516	2308				
140											600	2952	424(432)	3224
252											1136	5952	784(792)	6584
256	3200	7040	1800	5896	1392	5488			1284	5380				
280											1340	6604	852(864)	7148
504											2524	13164	1572(1584)	14428
512	7168	15872	4360	13576			3204	12420	3076	12292				
560											3100	14748	1928(1944)	17168
1008											5804	29100	3548(3564)	34416
1024	15872	35328	10248	30728	7856	28336			7172	27652				
2048	34816	77824	23560	68616					16388	61444				
2520											17660	82956	9492(9504)	99068
4096	75776	169984	53256	151560	40624	138928	38404	136708	36868	135172				
5040											39100	179772	21368(21384)	232668

[a]It is assumed that the complex multiplications are done using three multiplications and three additions.

length FFTs. Unfortunately, the numbers resulting from such a comparison say as much about the compiler and CPU used as about the actual algorithm [MO83]. Here are some examples of questions that arise when comparing run-times. (1) How many internal registers does the CPU have? (2) How efficient are the compiler's trigonometric functions? (3) How well does the compiler optimize the code? (4) How fast is the disk/external memory? (5) Does the computer have pipelining or a cache? All these and several more issues influence the performance of the different algorithms differently, so it is impossible to make machine-independent statements about the performance of the algorithms in terms of run-time. A much better approach is to use the operation counts as a guideline and test some of the better algorithms on the target system.

Table 8-8 shows that, as the radix increases, the number of both multiplications and additions decreases. However, as the radix increases, so does the length of program code implementing the FFT. For that reason the radix-2 and radix-4 FFTs are both used frequently, while the radix-8 is used less frequently. The split-radix FFT, which was discovered recently, has a lower operation count than any regular radix FFTs, while the program code (see Appendix 8-A) is about the same length as for a radix-4, and hence it often turns out to be an excellent trade-off. The drawback of the split-radix is a more complicated indexing scheme. The prime factor FFT is more flexible than both the split-radix and regular radix FFTs, since it allows much more different lengths. It also requires much less coefficient storage than a WFTA or a table look-up FFT. The price paid for this flexibility is a longer program, since a different module is needed for each factor in the data length N. Especially for fixed-length FFTs, the PFA turns out to be a good choice. The Winograd Fourier transform algorithm has the lowest number of multiplications, but it has achieved this by increasing the number of additions and the indexing. The WFTA also does not allow in-place calculations and requires a large coefficient storage. Hence the WFTA is seldom used except for special cases where multiplications are very expensive compared to additions and overhead.

TABLE 8-9   Operation Counts for FFT Algorithms with Complex Input Data[a]

	Multiplicative/Additive Complexity for Computing Complex FFTs with 4/2 Multiplication Algorithm							
	Radix-2 1 BF		Radix-2 2 BF		Radix-2 3 BF		Radix-2 5-BF	
Length N	Multiplications	Additions	Multiplications	Additions	Multiplications	Additions	Multiplications	Additions
8	48	72	20	58	8	52	4	52
16	128	192	68	162	40	148	28	148
32	320	480	196	418	136	388	108	388
64	768	1152	516	1026	392	964	332	964
128	1792	2688	1284	2434	1032	2308	908	2308
256	4096	6144	3076	5634	2568	5380	2316	5380
512	9216	13824	7172	12802	6152	12292	5644	12292
1024	20480	30720	16388	28674	14344	27652	13324	27652
2048	45056	67584	36868	63490	32776	61444	30732	61444
4096	98304	147456	81924	139266	73736	135172	69644	135172

[a]It is assumed that the complex multiplications are done using four multiplications and two additions.

Special butterflies and the required deselection operations are introduced in all the algorithms above to achieve the absolute minimum operation count. However, it is seldom necessary or even desirable to deselect as many butterflies as possible. Only the second and perhaps the third butterfly brings any significant reduction in the operation counts or execution times; the rest just lengthen the program code. How many should be introduced is very application dependent and has to be decided in each case. To give some idea of the savings possible, check the radix-2 FFT in Table 8-9. Note that going from three to five butterflies saves no additions and only a few multiplications. One problem is introducing more butterflies, besides the increased program length, is that it often becomes necessary to introduce IF statements in the loop structure of the FFT for butterfly deselection, which decreases the performance of the algorithm, or several different loops must be used, which increases the program size.

Trigonometric functions are used in all FFTs, and there are several ways to compute them. First, they can be computed as they are needed. Assuming that the standard FFT has a three-loop structure as did the radix-2 FFT, it is desirable to arrange the computations such that the trigonometric function evaluation takes place in the second loop, not the innermost loop, to save evaluations. Second, the trigonometric functions can be computed in a recursive manner. At each stage, an initial value and a step-size are computed, and these values are used to compute the needed trigonometric function. Again, to save operations, the trigonometric function should be calculated in the second loop. Third, a table can be used, in which all the required trigonometric functions are precomputed. This is the fastest way, although the recursive method is almost as fast on most machines; however, the sin/cos table requires a fair amount of memory. In this case it does not matter where the trigonometric functions are evaluated.

Complex multiplications by twiddle factors can be computed directly;

$$(x + jy)(\cos \theta + j \cos \theta) = (x \cos \theta - y \sin \theta) + j(x \sin \theta + y \cos \theta),$$

TABLE 8-9 (*Continued*)

Multiplicative/Additive Complexity for Computing Complex FFTs with 4/2 Multiplication Algorithm							
Radix-4 2 BF		Radix-4 3 BF		Split-Radix 2 BF		Split-Radix 2+ BF	
Multiplications	Additions	Multiplications	Additions	Multiplications	Additions	Multiplications	Additions
				8	52	4	52
36	146	28	144	32	144	28	144
				104	372	92	372
324	930	284	920	288	912	268	912
				744	2164	700	2164
2052	5122	1884	5080	1824	5008	1740	5008
				4328	11380	4156	11380
11268	26114	10588	25944	10016	25488	9676	25488
				22760	56436	22076	56436
57348	126978	54620	126296	50976	123792	49612	123792

which involves four real multiplications and two real additions. An alternative way is to compute $D = (\cos \theta)(x - y)$, $E = (\cos \theta - \sin \theta)y$, and $F = (\cos \theta + \sin \theta)x$, so the real part of the complex multiplication is given by $E + D$ and the imaginary part of the complex multiplication by $E - F$. Now assuming that $(\cos \theta + \sin \theta)$ and $(\cos \theta - \sin \theta)$ are precomputed, this scheme takes three real multiplications and three real additions. For this scheme to work, the trigonometric functions have to be precomputed; otherwise, there are five and not three additions. However, now three tables are needed (cos, cos + sin, and cos − sin), where in the previous case only two (cos and sin) were needed, and hence more memory is occupied. On computers where multiplications are slower than additions, it might be worthwhile. By comparing the radix-2 FFT in Table 8-8 to the one in Table 8-9, it is possible to get an idea of the "savings" involved.

Real-valued FFTs can always be derived which require exactly half the multiplications and $N - 1 (N - 2$ if N is even) less than half the additions of the corresponding complex FFT (see Eq. (8-88)) [SO87]. For convenience, Table 8-10 shows operation counts for some real-valued FFTs, and as reference the operation counts for Bergland's two programs [BE69] are shown as well. The indexing is also roughly halved, and the memory requirements are halved, so a RFFT should always be used when dealing with real-valued data instead of using the packing or doubling techniques.

Program optimization is an important part of all programming, and that also holds for FFTs. The programs supplied with this book have not been optimized

TABLE 8-10 Operation Counts for FFT Algorithms with Real Input Data[a]

	Multiplicative/Additive Complexity for Computing Real-Valued FFTs with 3/3 Multiplication Algorithm							
	CFFT Direct		CFFT Double		CFFT Packing		Radix-2 RFFT	
Length	Multiplications	Additions	Multiplications	Additions	Multiplications	Additions	Multiplications	Additions
8	4	52	2	32	2	30	2	20
12								
15								
16	20	148	10	88	12	88	12	62
30								
32	68	388	34	224	40	228	44	174
35								
60								
64	196	964	98	544	112	556	132	454
72								
112								
120								
128	516	2308	258	1280	288	1308	356	1126
140								
252								
256	1284	5380	642	2944	704	3004	900	2694
280								
504								
512	3076	12292	1538	6656	1664	6780	2180	6278
560								
1008								
1024	7172	27652	3586	14848	3840	15100	5124	14342

[a]It is assumed that the complex multiplications are done using three multiplications and three additions. The programs "FAST" and "FFA" have been changed accordingly.

for any specific compiler but have been kept simple to enhance readability. Bergland [BE69] has two highly optimized programs (radix-8 and radix-4) in IEEE Press book [IE79] and Morris [MO83] has done a lot of work in compiler-optimized FFTs and has a very fast radix-4 FFT.

Since computers have finite wordlength, the W_N terms of the FFT cannot be represented exactly and there will always, except in very special cases, be errors in the DFT coefficients due to quantization. The nature and amount of quantization errors depend on the type of arithmetic performed. For floating-point implementations the error is normally not a problem, while for fixed-point implementations, great care has to be taken to avoid overflow. The error also depends on the transform length and the specific algorithm, since the output noise basically is a function of the number of arithmetic operations performed. These issues are discussed in detail in Oppenheim and Weinstein [OP72].

8-3   OTHER TRANSFORMS

The DFT is not the only transform that possesses a "convolution property" (i.e., cyclic convolution turns into multiplication when transformed). Other transforms with a similar property to be discussed here are the number theoretic transforms, Hartley transform, cosine/sine transforms, and the Walsh–Hadamard transform.

TABLE 8-10   (*Continued*)

Multiplicative/Additive Complexity for Computing Real-Valued FFTs with 3/3 Multiplication Algorithm									
FAST [IE79] Radix-4		FFA [IE79] Radix-8		Real Split-Radix		Real PFA		Real WFTA	
MULT	ADD	MULT	ADD	MULT	ADD	MULT	ADD	MULT	ADD
2	20	2	20	2	20	2	20	2	20
						8	38	8	38
						25	67	17	67
11	62	12	62	10	60	10	60	10	60
						50	164	34	164
37	170	37	167	34	164				
						75	265	53	299
						100	386	68	386
109	442	103	425	98	420				
						98	500	82	508
						198	984	154	1056
						230	920	138	920
285	1082	283	1053	258	1028				
						300	1338	212	1474
						568	2726	392	3042
717	2586	683	2477	642	2436				
						670	3024	426	3296
						1262	6080	786	6712
1709	5978	1611	5709	1538	5636				
						1550	6816	964	8026
						2902	13544	1774	17208
4013	13658	3851	13069	3586	12804				

Number theoretic transforms (NTTs) turn out to be useful for special purpose hardware implementations [MC79b; PO71]. NTTs are normally computed over finite fields (GF(p^n)—Galois fields) and use integers as the kernel instead of W_N as used by the DFT. The Fermat number transforms (FNTs), which are computed in the field GF$(2^{2^n} + 1)$, are especially useful because they are often the easiest to implement [AG74b].

The discrete Hartley transform (DHT) [BR83] resembles the DFT very much. As a kernel it uses $\cos \theta + \sin \theta$, whereas the DFT uses $\cos \theta - j \sin \theta$. Hence the DHT is a real-valued transform, but with properties very similar to the DFT. The DHT has the same forward and inverse transforms, but it has been shown that the real-valued FFT always requires slightly less operations and almost exactly the same overhead as the DHT, so the DHT should only be used in special cases.

The discrete cosine/sine transforms (DCTs/DSTs) [JA79] are a whole family of transforms with properties very similar to those of the DFT as well. They are primarily used in image processing because of the symmetries of their spectrum.

The Walsh–Hadamard transforms (WHTs) [AH71] are also widely used in image processing and special hardware systems. Their kernel consists of a square wave rather than a sine/cosine as for the trigonometric transforms.

8-3-1 Number Theoretic Transforms to Compute Cyclic Convolution

As was proved in the introduction to this chapter, the DFT has a simple convolution property; that is, convolution in the time domain of two sequences converts into simple point-by-point multiplication of their DFTs. It is proved in Agarwal and Burrus [AG74b] that a transform with kernel α, which is defined as

$$X(k) = \sum_{n=0}^{N-1} x(n)\alpha^{nk}, \qquad k = 0, 1, \ldots, N-1, \tag{8.89}$$

has a simple convolution property if and only if α is a Nth root of unity; that is,

$$\alpha^N = 1 \ \wedge \ \alpha^k \neq 1, \qquad \forall k \ni 1 \leq k < N. \tag{8.90}$$

Since $e^{j2\pi k/N}$, where $(k, N) = 1$, are the only roots of unity in the complex field, the DFT is the only transform that has a simple convolution property in the fields of reals. The proof in Agarwal and Burrus [AG74b] assumes that there is no data expansion, so the transform is of the same length as the sequence itself (simple convolution property). Relaxing this requirement, other transforms exist in the field of reals with a "convolution property," such as the Hartley transform.

Assuming that the sequences to be convolved consist of only integers, which can always be obtained by proper scaling of quantized data, the convolution can be performed in a finite field. Such fields are called Galois fields and are denoted GF(p^n) in number theory, where p is prime and n is an integer. Doing the arithmetic in these fields allow a simple point-by-point convolution property as does the DFT [PO71; RE75]. We will mainly look at GF(p), which always has p number of elements, since there exist ''nice'' transforms in these fields. The arithmetic is done modulo p and hence only numbers between 0 and $p - 1$ or between $(1 - p)/2$ and $(p - 1)/2$ can be represented in $GF(p)$.

Fermat Number Transforms. Fermat numbers are of the form $F_n = 2^{2^n} + 1$ and are primes for $n = 1, 2, 3, 4$, while composite for $n = 5, 6$. The two cases, depending on F_n being prime or not, are very different, so they are treated separately. If F_n is prime the Fermat number transform (FNT) is computed in the field $GF(2^{2^n} + 1)$, and is defined as Eq. (8.89). All arithmetic is done modulo the Fermat number F_n and the transform is only valid if α is an Nth root of unity in the field. That is, α has to have order[1] N in the field, where N is the desired transform length. From number theory it is known that there exist numbers of order $2, 2^2, \ldots, 2^{2^n}$ in the field $GF(2^{2^n} + 1)$, and hence there exist transforms of these lengths.

The transform length thus depends on α and one α that gives the maximum length $N_{max} = 2^{2^n}$ is 3. In hardware setups multiplications by 3 are somewhat difficult, and a much better transform algorithm can be obtained choosing $\alpha = 2$, since multiplication by 2 can be implemented by shifts in hardware [MC76]. However, 2 only has the order 2^n in the field $GF(F_n)$, which limits the transform length. One alternative is to use $\sqrt{2} = 2^{2^{n-2}}(2^{2^{n-1}} - 1)$. Every power of $\sqrt{2}$ is of the form $2^a \pm 2^b$ and hence is still easy to implement in hardware. The order of $\sqrt{2}$ is twice that of 2 and hence the transforms can be twice as long. Table 8-11 shows the maximum transform length for α chosen as 2, $\sqrt{2}$, or 3.

If the lengths that can be obtained this way are not enough, there are two ways to extend them. First, using numbers other than 2 as kernel allows much longer lengths, but that introduces more complicated multiplications than by 2. However, since the purpose of the NTT in the first place is to perform convolution, multiplications are required to multiply the two NTTs together, so some sort of multiplier is needed anyway. By choosing α appropriately, it is possible to break the transform into several smaller transforms with a kernel of 2 by index mapping. These transforms can then be computed efficiently without multiplications, while the first stage requires multiplications [BL84].

[1]The order of a number α in the field $GF(p)$ is the smallest integer Ψ such that $\alpha^{\Psi} \equiv 1$.

TABLE 8-11 Parameters for Several Fermat Number Transforms (FNTs)

		Parameters for Several Possible FNTs			
n	F_n	N $\alpha = 2$	N $\alpha = \sqrt{2}$	N_{max}	α for N_{max}
3	$2^8 + 1$	16	32	256	3
4	$2^{16} + 1$	32	64	65,536	3
5	$2^{32} + 1$	64	128	128	$\sqrt{2}$
6	$2^{64} + 1$	128	256	256	$\sqrt{2}$

A second way to solve the problem of too short transform lengths is to use F_5 or F_6. Even though these two Fermat numbers are not prime, it turns out that for $\alpha = 2$ longer lengths are possible [AG74b]. Table 8-11 shows the possible lengths for $\alpha = 2$ and $\alpha = \sqrt{2}$ and shows $N_{\max}$.

One drawback of the FNT is that it does not fit a binary representation very well. The Fermat numbers are always one greater than a power of 2, which means that you need one extra bit just to represent one number (b bits can represent 2^b values, not $2^b + 1$). Another alternative is to ignore the "extra" value and deal with the error that introduces [AG74b].

Mersenne Number Transforms.   The Mersenne number transforms (MNTs) [RA72] are computed in the fields GF(p), where p is a Mersenne prime of the form $2^n - 1$, $n = 2, 3, 5, 7, 13, 17, 19, 31, 61$. One of the advantages of the MNT is that arithmetic modulo a Mersenne prime is easy to implement in hardware. Since $2^n = 1$, any overflow gets added to the low-order bits of the number, which is exactly the same way overflow in ones-complement arithmetic is handled [NU81].

A MNT can be defined for all lengths N such that N divides $p - 1 = 2^n - 2$, since only elements of these orders exist. However, $2^n - 2$ is not a power of 2, so the Cooley–Tukey approach cannot be used. Instead, a mixed-radix, a prime factor, or a Winograd algorithm can be used. Blahut [BL84] discusses how Winograd's approach can be used to obtain efficient short-length MNT modules, which can then be nested to obtain longer transforms.

Several other NTTs have been presented in the literature, and the most interesting extension is to deal with complex data. Common for all NTTs are that they perform residue arithmetic, which limits the dynamic input range (the range of inputs that the MNT can represent uniquely) to avoid overflow. However, since all arithmetic is done on integers and they stay integers, there are no quantization errors as in the trigonometric transforms.

8-3-2 Hartley Transform

The discrete Hartley transform (DHT) [BR83] and its inverse are defined as

$$
\begin{aligned}
H[k] &= \sum_{n=0}^{N-1} x[n] \operatorname{cas}\left(\frac{2\pi}{N} kn\right), \qquad 0 \le k \le N - 1, \\
x[n] &= \frac{1}{N} \sum_{k=0}^{N-1} H[k] \operatorname{cas}\left(\frac{2\pi}{N} kn\right), \qquad 0 \le n \le N - 1,
\end{aligned}
\tag{8.91}
$$

where $\operatorname{cas} x \triangleq \cos x + \sin x$. Note the only difference from the DFT is the lack of $-j$ multiplying the sine term. Another important factor is that the DHT is a real-valued transform. The discrete Hartley transform was discovered independently by Wang, who named it the W transform [WA84].

The symmetry of the transform pair is a valuable feature of the DHT, which means that one program can do both forward and inverse transforms. For real-valued FFTs the differences between the forward and the inverse transform are so great that two different programs are needed. The DHT possesses a convolution property, but it is not a point-by-point multiplication. It can easily be verified that

$$\text{DHT} \{x_1[n] \, \text{Ⓝ} \, x_2[n]\} [k] = \tfrac{1}{2} \{H_1[k] H_2[k]$$

$$+ \, H_1[k] H_2[N - k] + H_1[N - k] H_2[k]$$

$$- \, H_1[N - k] H_2[N - k]\}, \tag{8.92}$$

which strongly resembles a complex multiplication in which the real and imaginary parts are added.

The existence of the fast Hartley transform (FHT) was first demonstrated by Bracewell [BR84], who derived a decimation-in-time radix-2 FHT. It was later extended to decimation-in-frequency algorithms, higher radix algorithms, the split-radix algorithm, the prime factor, and the Winograd–Hartley transform algorithm (WHTA) in Sorensen et al. [SO85]. The methods used in Sorensen et al. [SO85] for deriving the FHTs are very similar to those used in this chapter to derive the corresponding FFTs. It is concluded in Sorensen et al. [SO85] that the FHT always requires more arithmetic (about N more additions) than its corresponding FFT, and since everything else (overhead memory, etc.) is the same, the FHT should only be used in special cases. Recently, a hybrid FHT/FFT program has been presented [DU86b] that computes the FHT in only two more additions than the FFT. However, implementation of this algorithm either has more overhead than the same length FFT/FHT or, by using in-line coding, requires longer code.

8-3-3 Cosine/Sine Transforms

The discrete sine transform (DST) of the sequence $x[n]$ is defined to be

$$S[k] = \sqrt{\frac{2}{N}} \sum_{n=1}^{N-1} x[n] \sin \left(\frac{\pi}{N} nk \right), \qquad k = 1, 2, \ldots, N - 1, \tag{8.93}$$

which is its own inverse. This definition has a normalization constant, $\sqrt{2/N}$, on both the forward and inverse transforms (to make them equal) as opposed to the definition of the DFT, where only the inverse transform has a normalization constant.

There exist several variations for the kernel of the DST, and Table 8-12 shows some of them. It also shows the kernels for different cosine transforms (DCT) [CH77; YI84]. Each of the eight transforms have properties that are very similar, and the transforms have computational complexities for similar algorithms that also are very much alike. Hence the main decision, when choosing one transform over the other, is if shifted samples are wanted and/or the desired type of symmetries of the time and frequency samples.

The development of fast algorithms for the DCT/DST is done using the same principles as are used to develop the FFT and FHT algorithms [MA80]. However,

TABLE 8-12   Kernels for Sine/Cosine Transforms

	Kernels for DSTs and DCTs	
Name	Kernel	Inverse Kernel
S_{N-1}^I	$\sqrt{\dfrac{2}{N}}\,\sin\left((k)(n)\dfrac{\pi}{N}\right),\qquad n,k=1,2,\ldots,N-1$	S_{N-1}^I
S_N^{II}	$t_k\sqrt{\dfrac{2}{N}}\,\sin\left((k)(n-\tfrac{1}{2})\dfrac{\pi}{N}\right),\qquad n,k=1,2,\ldots,N$	S_N^{III}
S_N^{III}	$t_n\sqrt{\dfrac{2}{N}}\,\sin\left((k+\tfrac{1}{2})(n)\dfrac{\pi}{N}\right),\qquad n,k=1,2,\ldots,N$	S_N^{II}
S_N^{IV}	$\sqrt{\dfrac{2}{N}}\,\sin\left((k+\tfrac{1}{2})(n+\tfrac{1}{2})\dfrac{\pi}{N}\right),\qquad n,k=0,1,\ldots,N-1$	S_N^{IV}
C_{N+1}^I	$t_n t_k\sqrt{\dfrac{2}{N}}\,\cos\left((k)(n)\dfrac{\pi}{N}\right),\qquad n,k=0,1,\ldots,N$	C_{N+1}^I
C_N^{II}	$t_k\sqrt{\dfrac{2}{N}}\,\cos\left((k)(n+\tfrac{1}{2})\dfrac{\pi}{N}\right),\qquad n,k=0,1,\ldots,N-1$	C_{N+1}^{III}
C_N^{III}	$t_n\sqrt{\dfrac{2}{N}}\,\cos\left((k+\tfrac{1}{2})(n)\dfrac{\pi}{N}\right),\qquad n,k=0,1,\ldots,N-1$	C_{N+1}^{II}
C_N^{IV}	$\sqrt{\dfrac{2}{N}}\,\cos\left((k+\tfrac{1}{2})(n+\tfrac{1}{2})\dfrac{\pi}{N}\right),\qquad n,k=0,1,\ldots,N-1$	C_{N+1}^{IV}

$$t_i = \begin{cases} \dfrac{1}{\sqrt{2}} & \text{if } i = 0 \text{ or } N \\[2mm] 1 & \text{if } i \neq 0 \text{ or } N \end{cases}$$

the factorization for the DST/DCT is not as regular as for the DFT, and hence it is harder to distinguish between the different algorithms. As was the case for the DFT and the DHT, there exist split-radix, radix-2, radix-4, and so on, algorithms; each of them has both a decimation-in-time and decimation-in-frequency version. Table 8-13 gives some of the best operation counts for different DCT/DST algorithms. They are all more or less split-radix algorithms. Very little work has gone into developing prime factor and Winograd algorithms, since it is not clear that they would have many advantages over the power-of-2 algorithms.

The sine-cosine transforms always require more operations than the real-valued FFT counterparts, so they should only be used in special cases. They are often used for filtering purposes in image processing, because the DCT/DST contains symmetries not present in the DFT. If the image to be transformed has a white edge and a dark edge, the repetitions inherent to the DFT will produce a significant

TABLE 8-13 Number of Operations to Compute Different Power-of-2 DCTs/DSTs

	Operation Counts for DST/DCT Algorithms		
Kernel Type	Multiplications	Additions	Reference
C_{N+1}^{I}	$\dfrac{N}{2}\log_2(N) - \dfrac{3N}{4}$	$\dfrac{3N}{2}\log_2(N) - \dfrac{3N}{4} + 1$	[YI84]
C_N^{II}	$\dfrac{N}{2}\log_2(N)$	$\dfrac{3N}{2}\log_2(N) - N + 1$	[LE84; VE84]
C_N^{III}	$\dfrac{N}{2}\log_2(N) + 1$	$\dfrac{3N}{2}\log_2(N) - N + 1$	[WA85]
C_N^{IV}	$\dfrac{N}{2}\log_2(N) + 1$	$\dfrac{3N}{2}\log_2(N)$	[SU86]
S_{N-1}^{I}	$\dfrac{N}{2}\log_2(N) - N + 1$	$2N\log_2(N) - 4N + 4$	[YI84]
S_N^{II}	$\dfrac{N}{2}\log_2(N)$	$\dfrac{3N}{2}\log_2(N) - N + 1$	[YI84]
S_N^{III}	$\dfrac{N}{2}\log_2(N)$	$\dfrac{3N}{2}\log_2(N) - N + 1$	[YI85]
S_N^{IV}	$\dfrac{N}{2}\log_2(N) + N + 1$	$\dfrac{3N}{2}\log_2(N) - 1$	[YI84]

error term in the image spatial spectrum, since the white and black edges misalign and hence produce a sharp transition in the repeated image. The DCT, instead of a repetition symmetry, has a mirroring symmetry, which prevents any sharp transitions due to edge phenomena and hence gives a sharper spectrum.

The cosine transforms are also used in image coding since it has been shown that their performance resemble the Karhunen–Loève (KL) transform on first-order Markov processes [FL82; JA79; RO82]. The KL transform of a random vector $\bar{\nu}$, uses the eigenvectors of the covariance matrix $[\bar{\nu}\bar{\nu}']$ as basis functions, and hence it optimizes the decorrelation of the transform coefficients and thus minimizes the energy distribution.

8-3-4 Hadamard Transform

The Hadamard (or Walsh–Hadamard or BIFORE) transform of the length N sequence $x[n]$ is defined as [AH71]

$$X[k] = \mathbf{H}_N \begin{bmatrix} x[0] \\ x[1] \\ \vdots \\ x[N-1] \end{bmatrix}, \tag{8.94}$$

where $\mathbf{H}_N$ is the $N \times N$ Hadamard matrix, which is defined recursively as

$$\mathbf{H}_{2j} = \begin{bmatrix} [\mathbf{H}_j] & [\mathbf{H}_j] \\ [\mathbf{H}_j] & -[\mathbf{H}_j] \end{bmatrix}, \tag{8.95}$$

and $\mathbf{H}_1 = [1]$.

Example 8.14. The 8×8 Hadamard matrix is given by

$$\mathbf{H}_8 = \begin{bmatrix} 1 & 1 & 1 & 1 & 1 & 1 & 1 & 1 \\ 1 & -1 & 1 & -1 & 1 & -1 & 1 & -1 \\ 1 & 1 & -1 & -1 & 1 & 1 & -1 & -1 \\ 1 & -1 & -1 & 1 & 1 & -1 & -1 & 1 \\ 1 & 1 & 1 & 1 & -1 & -1 & -1 & -1 \\ 1 & -1 & 1 & -1 & -1 & 1 & -1 & 1 \\ 1 & 1 & -1 & -1 & -1 & -1 & 1 & 1 \\ 1 & -1 & -1 & 1 & -1 & 1 & 1 & -1 \end{bmatrix}.$$

The eight rows of the Hadamard matrix $\mathbf{H}_8$ are the eight Walsh functions of length 8 [AH71], and hence the name Walsh–Hadamard transform. The Walsh functions can be interpreted as binary (sampled) versions of the sin and cos, which are the basis functions of the DFT. This interpretation led to the name BInary FOurier REpresentation (BIFORE). The inverse Hadamard transform is given by the transpose of the Hadamard matrix scaled by the factor $1/N$.

From the definition of $\mathbf{H}_N$, it can be seen that N has to be a power of 2. This turns outs to be very useful when programming the Hadamard transform. Because of the recursive nature of $\mathbf{H}_8$ the computation of the Hadamard transform can be done using the flow-graph in Figure 8-21. This flow-graph resembles the radix-2

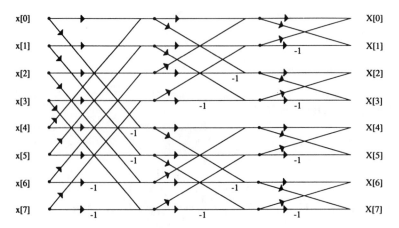

FIGURE 8-21 Length 8 Hadamard transform.

decimation-in-frequency FFT flow-graph, but the Hadamard transform has no twiddle factors since its basis functions are square waves of either 1 or -1. Hence the Hadamard transform requires no multiplications and only $N \log (N)$ additions.

Since there are no twiddle factors, the complexity of the Hadamard transform, unlike that of the FFT, cannot be improved by using "higher" radix or split-radix algorithms. Also unlike the FFT, the Hadamard transform does not produce the data in bit-reversed order. The Hadamard transform does, however, like the FFT, have frequency information about the input sequence, and things like a power spectrum can be defined [AH71]. The most common applications of the Hadamard transform are in image coding [PR78].

8-4 CONCLUSION

This chapter describes several convolution and FFT algorithms and some miscellaneous other transforms. It shows that both noncyclic (linear) and cyclic convolution algorithms require more operations than the comparable FFT method except for small lengths (small N). Even for small lengths it will often be advantageous to use the FFT, since these methods have quite a bit less overhead (indexing). However, if the target machine is not a main-frame/micro computer, issues other than the number of operations might advocate one of the convolution algorithms or NTTs.

The chapter also shows that, of the FFT algorithms, the WFTA has the lowest number of multiplications, but as N grows the number of additions and memory requirements get very large. The PFA allows the same great variety of lengths as the WFTA but has a far better trade-off between multiplications and additions for most computers and requires less memory. The split-radix FFT is shown to have the flexibility of a radix-2 algorithm, the complexity (number of operations) better than a radix-8, and the length of code of a radix-4. Hence the split-radix is often the best choice for power-of-2 algorithms, although it requires more indexing. If a more regular algorithm is wanted, a radix-4 FFT is often a good choice. The split-radix and PFA require about the same number of operations. The choice here should depend on the flexibility of the transform length, N, required. The PFA obviously allows far more lengths than the split-radix but is slightly more complicated to program. It is further shown that for transforms for real data, a special real-valued transform should always be used. Such a transform is easily obtained from the complex transform and requires far fewer operations than any other method.

The chapter also discusses the Hartley, cosine, and sine transforms. It is shown that they always require more operations than the comparable FFT for real data and hence should only be used if some of the special properties of the kernel are required.

The algorithms mentioned are only discussed with respect to programming of sequential computers and do not necessarily prove useful on parallel computers or for VLSI implementation.

ACKNOWLEDGMENTS

The authors would like to thank the National Science Foundation and Texas Instruments for supporting some of the research for the results given here. They also would like to thank Douglas L. Jones for his contributions and advice.

REFERENCES

[AG74a] R. C. Agarwal and C. S. Burrus, Fast one-dimensional digital convolution by multidimensional techniques. *IEEE Trans. Acoust., Speech, Signal Process.* **ASSP-22**(1), 1–10 (1974) Reprinted in "Selected Pagers in Digital Signal Processing II" ed. Digital Signal Processing Committee, pp. 18–27, New York, IEEE Press, 1975.

[AG74b] R. C. Agarwal and C. S. Burrus, Fast convolution using Fermat number transforms with applications to digital filtering, *IEEE Trans. Acoust., Speech, Signal Process.* **ASSP-22**(2), 87–97 (1974); reprinted in *Number Theory in Digital Signal Processing* (J. H. McClellan and C. M. Rader, eds.), pp. 168–178. Prentice-Hall, Englewood Cliffs, NJ, 1979.

[AG77] R. C. Agarwal and J. W. Cooley, New algorithms for digital convolution. *IEEE Trans. Acoust., Speech, Signal Process.* **ASSP-25**(5), 392–410 (1977); reprinted in *Number Theory in Digital Signal Processing* (J. H. McClellan and C. M. Rader, eds.), pp. 106–124. Prentice-Hall, Englewood Cliffs, NJ, 1979.

[AH71] N. Ahmed, K. R. Rao, and A. L. Abdussattar, BIFORE or Hadamard transform. *IEEE Trans. Audio Electroacoust.* **AU-19**(4), 225–234 (1971); reprinted in *Discrete Transforms and their Applications* (K. R. Rao, ed.), pp. 13–18. Van Nostrand-Reinhold, New York, 1985.

[BE69] G. D. Bergland, A Radix-eight fast Fourier transform subroutine for real-valued series. *IEEE Trans. Audio Electroacoust.* **AU-17**(2), 138–144 (1969).

[BL84] R. E. Blahut, *Fast Algorithms for Digital Signal Processing.* Addison-Wesley, Reading, MA, 1984.

[BO86] C. G. Boncelet, Jr., A rearranged DFT algorithm requiring $N^2/6$ multiplications. *IEEE Trans. Acoust., Speech, Signal Process.* **ASSP-34**(6), 1658–1659 (1986).

[BR83] R. N. Bracewell, Discrete Hartley transform. *J. Opt. Soc. Am.* **73**(12), 1832–1835 (1983).

[BR84] R. N. Bracewell, The fast Hartley transform. *Proc. IEEE* **72**(8), 1010–1018 (1984).

[BU72] C. S. Burrus, Block realization of digital filters. *IEEE Trans. Audio Electroacoust.* **AU-20**(4), 230–235 (1972).

[BU77] C. S. Burrus, Index mappings for multidimensional formulation of the DFT and convolution. *IEEE Trans. Acoust., Speech, Signal Process.* **ASSP-25**(3), 239–242 (1977).

[BU81] C. S. Burrus and P. W. Eschenbacher, An in-place, in-order prime factor FFT algorithm. *IEEE Trans. Acoust., Speech, Signal Processing.* **ASSP-29**(4), 806–817 (1981).

[BU85] C. S. Burrus and T. W. Parks, *DFT/FFT and Convolution Algorithms.* Wiley, New York, 1985.

[BU88] C. S. Burrus, Unscrambling for fast DFT algorithms. *IEEE Trans. Acoust., Speech, Signal Process.* **ASSP-36**(7), 1086–1087 (1988).

[CH77] W. H. Chen, C. H. Smith, and S. C. Fralick, A fast computational algorithm for the discrete cosine transform. *IEEE Trans. Commun.*, **COM-25**(9), 1004–1009 (1977); reprinted in *Discrete Transforms and their Applications* (K. R. Rao, ed.), pp. 13–18. Van Nostrand-Reinhold, New York, 1985.

[CO65] J. W. Cooley and J. W. Tukey, An algorithm for the machine calculation of complex Fourier series. *Math. Comput.* **19**(90), 297–301 (1965); reprinted in *Digital Signal Processing* (L. R. Rabiner and C. M. Rader, eds.), pp. 223–227. IEEE Press, New York, 1972; also reprinted in *Papers on Digital Signal Processing* (A. V. Oppenheim, ed.), pp. 146–150. MIT Press, Cambridge, MA, 1969.

[CO70] J. W. Cooley, P. A. W. Lewis, and P. D. Welch, The fast Fourier transform algorithm: Programming considerations in the calculation of sine, cosine, and Laplace transforms. *J. Sound Vibr.* **12**(3), 315–337 (1970); reprinted in *Digital Signal Processing* (L. R. Rabiner and C. M. Rader, eds.), pp. 271–293. IEEE Press, New York, 1972.

[DU84] P. Duhamel and H. Hollmann, 'Split radix' FFT algorithm. *Electron. Lett.* **20**(1), 14–16 (1984).

[DU86a] P. Duhamel, Implementation of 'split radix' FFT algorithms for complex, real and real-symmetric data. *IEEE Trans. Acoust., Speech, Signal Process.* **ASSP-34**(2), 285–295 (1986).

[DU86b] P. Duhamel and M. Vetterli, Cyclic convolution of real sequences: Hartley versus Fourier and new schemes. *Proc. IEEE Int. Conf. Acoust., Speech, Signal Process. Tokyo, 1986* pp. 229–232, (1986).

[FL82] M. D. Flickner and N. Ahmed, A derivation of the discrete cosine transform. *Proc. IEEE* **70**(9), 1132–1134 (1982); reprinted in *Discrete Transforms and their Applications* (K. R. Rao, ed.), pp. 33–35. Van Nostrand-Reinhold, New York, 1985.

[GO58] G. Goertzel, An algorithm for the evaluation of finite trigonometric series. *Am. Math. Mon.* **65**(1), 34–35 (1958).

[HE84] M. T. Heideman, C. S. Burrus, and H. W. Johnson, Prime factor FFT algorithms for real-valued series. *Proc. IEEE Int. Conf. Acoust., Speech, Signal Process. San Diego, California, 1984*, pp. 28A.7.1–28A.7.4, 1984.

[IE79] IEEE Acoustic, Speech and Signal Processing Society (IEEE ASSP Soc.), *Programs for Digital Signal Processing.* IEEE Press, New York, 1979.

[JA79] A. K. Jain, A sinusoidal family of unitary transforms. *IEEE Trans. Pattern Anal. Mach. Intell.* **PAMI-1**(4), 356–365 (1979); reprinted in *Discrete Transforms and their Applications* (K. R. Rao, ed.), pp. 69–78. Van Nostrand-Reinhold, New York, 1985.

[JO85] H. W. Johnson and C. S. Burrus, On the structure of efficient DFT algorithms. *IEEE Trans. Acoust., Speech, Signal Process.* **ASSP-33**(1), 248–254 (1985).

[KO77] D. P. Kolba and T. W. Parks, A prime factor FFT algorithm using high-speed convolution. *IEEE Trans. Acoust., Speech, Signal Process.* **ASSP-25**(4), 281–294 (1977); reprinted in *Number Theory in Digital Signal Processing* (J. H. McClellan and C. M. Rader, eds.), pp. 90–103. Prentice-Hall, Englewood Cliffs, NJ, 1979.

[LE84] B. Lee, A new algorithm to compute the discrete cosine transform. *IEEE Trans. Acoust., Speech, Signal Process.* **ASSP-32**(6), 1243–1245 (1984).

[MA71] J. Markel, FFT pruning. *IEEE Trans. Audio Electroacoust.* **AU-19**(4), 305–311 (1971).

[MA80] J. Makhoul, A fast cosine transform in one and two dimensions. *IEEE Trans. Acoust., Speech, Signal Process.* **ASSP-28**(1), 27–34 (1980).

[MA84a] J. B. Martens, Discrete Fourier transform algorithms for real valued sequences. *IEEE Trans. Acoust., Speech, Signal Process.* **ASSP-32**(2), 390–396 (1984).

[MA84b] J. B. Martens, Recursive cyclotomic factorization—a new algorithm for calculating the discrete Fourier transform. *IEEE Trans. Acoust., Speech, Signal Process.* **ASSP-32**(4), 750–761 (1984).

[MC76] J. H. McClellan, Hardware realization of a Fermat number transform. *IEEE Trans. Acoust., Speech, Signal Process.*, **ASSP-24**(3), 216–225 (1976); reprinted in *Number Theory in Digital Signal Processing* (J. H. McClellan and C. M. Rader, eds.), pp. 189–198. Prentice-Hall, Englewood Cliffs, NJ, 1979.

[MC79a] J. H. McClellan and H. Nawab, Complex general-N Winograd Fourier Transform Algorithm (WFTA). In *Programs for Digital Signal Processing* (Digital Signal Processing Committee, eds.), pp. 1.7-1–1.7-22. IEEE Press, New York, 1979.

[MC79b] J. H. McClellan and C. M. Rader, eds., *Number Theory in Digital Signal Processing.* Prentice-Hall, Englewood Cliffs, NJ, 1979.

[MO83] L. R. Morris, *Digital Signal Processing Software.* DSPS Inc., Ottawa, Canada, 1983.

[NA86] K. Nagai, Pruning the decimation-in-time FFT algorithm with frequency shift, *IEEE Trans. Acoust., Speech, Signal Process.* **ASSP-34**(4), 1008–1010 (1986).

[NI80] I. Niven and H. S. Zuckerman, *An Introduction to the Theory of Numbers.* Wiley, New York, 1980.

[NU78] H. J. Nussbaumer, New algorithms for convolution and DFT based on polynomial transforms. *Conf. Rec. IEEE Int. Conf. Acoust., Speech, Signal Process. Tulsa, Oklahoma, 1978*, pp. 638–641 (1978).

[NU81] H. J. Nussbaumer, *Fast Fourier Transform and Convolution Algorithms.* Springer-Verlag, Berlin, Heidelberg, and New York, 1981.

[OP72] A. V. Oppenheim and C. J. Weinstein, Effects of finite register length in digital filtering and the fast Fourier transform. *Proc. IEEE* **60**(8), 957–976 (1972); reprinted in *Selected Papers in Digital Signal Processing, II* (Digital Signal Processing Committee, eds.), pp. 335–354. IEEE Press, New York, 1975.

[PA79] T. W. Parsons, A Winograd-Fourier transform algorithm for real-valued data. *IEEE Trans. Acoust., Speech, Signal Process.* **ASSP-27**(4), 398–402 (1979).

[PO71] J. M. Pollard, The fast Fourier transform in a finite field. *Math. Comput.* **25**(114), 365–374 (1971); reprinted in *Number Theory in Digital Signal Processing* (J. H. McClellan and C. M. Rader, eds.), pp. 158–167. Prentice-Hall, Englewood Cliffs, NJ, 1979.

[PO79] J. M. Pollard, Remarks on the convolution algorithm of Agarwal and Cooley. *Electron. Lett.* **15**(19), 593–594 (1979).

[PR78] W. K. Pratt, *Digital Image Processing.* Wiley, New York, 1978.

[RA68] C. M. Rader, Discrete Fourier transforms when the number of data samples is prime. *Proc. IEEE* **56**(6), 1107–1108 (1968); reprinted in *Digital Signal Processing* (L. R. Rabiner and C. M. Rader, eds.), p. 329. IEEE Press, New York, 1972; also reprinted in *Number Theory in Digital Signal Processing* (J. H. McClellan and C. M. Rader, eds.), pp. 104–105. Prentice-Hall, Englewood Cliffs, NJ, 1979.

[RA72] C. M. Rader, Discrete convolution via Mersenne transforms. *IEEE Trans. Comput.* **C-21**(12), 1269–1273 (1972); reprinted in *Selected Papers in Digital Signal Processing, II* (Digital Signal Processing Committee, eds.), pp. 2–6. IEEE Press, New York, 1975.

[RA75] L. R. Rabiner and B. Gold, *Theory and Application of Digital Signal Processing.* Prentice-Hall, Englewood Cliffs, NJ, 1975.

[RA79] L. R. Rabiner, On the use of symmetry in FFT computation, *IEEE Trans. Acoust., Speech, Signal Process.* **ASSP-27**(3), 233–239 (1979).

[RE75] I. S. Reed and T. K. Truong, The use of finite fields to compute convolutions. *IEEE Trans. Inf. Theory* **IT-21**(2), 208–213 (1975); reprinted in *Number Theory in Digital Signal Processing* (J. H. McClellan and C. M. Rader, eds.), pp. 203–208. Prentice-Hall, Englewood Cliffs, NJ, 1979.

[RO82] A. Rosenfeld and A. C. Kak, *Digital Picture Processing*, Vol. 1. Academic Press, New York, 1982.

[SC84] M. R. Schroeder, *Number Theory in Science and Communication.* Springer-Verlag, Berlin, Heidelberg, New York, and Tokyo, 1984.

[SI69] R. C. Singleton, An algorithm for computing the mixed radix fast Fourier transform. *IEEE Trans. Audio Electroacoust.* **AU-17**(2), 93–103 (1969); reprinted in *Digital Signal Processing* (L. R. Rabiner and C. M. Rader, eds.), pp. 294–304. IEEE Press, New York, 1972; also reprinted in *Papers on Digital Signal Processing* (A. V. Oppenheim, ed.), pp. 163–170. MIT Press, Cambridge, MA, 1969.

[SI79] R. C. Singleton, Mixed radix fast Fourier transforms. In *Programs for Digital Signal Processing* (Digital Signal Processing Committee, eds.), pp. 1.4-1–1.4-18. IEEE Press, New York, 1979.

[SO85] H. V. Sorensen, D. L. Jones, C. S. Burrus, and M. T. Heideman, "On computing the discrete Hartley transform. *IEEE Trans. Acoust., Speech, Signal Process.* **ASSP-33**(5), 1231–1238 (1985).

[SO86a] H. V. Sorensen, M. T. Heideman, and C. S. Burrus, "On computing the split-radix FFT. *IEEE Trans. Acoust., Speech, Signal Process.* **ASSP-34**(1), 152–156 (1986).

[SO86b] H. V. Sorensen, D. L. Jones, M. T. Heideman, and C. S. Burrus, "A split-radix real-valued fast Fourier transform," *Proc. Euro. Signal Process. Conf., 3rd, 1986, The Hague, Netherlands*, pp. 287–290 (1986).

[SO87] H. V. Sorensen, D. L. Jones, M. T. Heideman, and C. S. Burrus, Real-valued fast Fourier transform algorithms. *IEEE Trans. Acoust., Speech, Signal Process.* **ASSP-35**(6), 849–863 (1987).

[SU86] N. Suehiro and M. Hatori, Fast algorithms for the DFT and other sinusoidal transforms. *IEEE Trans. Acoust., Speech, Signal Process.* **ASSP-34**(3), pp. 642–644 (1986).

[VE84] M. Vetterli and H. J. Nussbaumer, Simple FFT and DCT algorithms with reduced number of operations. *Signal Process.* **6**(4), 267–278 (1984).

[WA84] Z. Wang, Fast algorithms for the discrete W transform and for the discrete Fourier transform. *IEEE Trans. Acoust., Speech, Signal Process.* **ASSP-32**(4), 803–816 (1984).

[WA85] Z. Wang, On computing the discrete Fourier and cosine transforms. *IEEE Trans. Acoust., Speech, Signal Process.* **ASSP-33**(4), 1341–1344 (1985).

[WI77] S. Winograd, Some bilinear forms whose multiplicative complexity depends on the field of constants. *Math. Syst. Theory* **10**(2), 169–180 (1977); reprinted in *Number The-*

ory in Digital Signal Processing (J. H. McClellan and C. M. Rader, eds.), pp. 217–228. Prentice-Hall, Englewood Cliffs, NJ, 1979.

[WI78] S. Winograd, "On computing the discrete Fourier transform. *Math. Comput.* **32**(141), 175–199 (1978); reprinted in *Number Theory in Digital Signal Processing* (J. H. McClellan and C. M. Rader, eds.), pp. 125–149. Prentice-Hall, Englewood Cliffs, NJ, 1979.

[YI84] P. Yip and K. R. Rao, "Fast decimation-in-time algorithms for a family of discrete sine and cosine transforms," *Circuits, Syst., Signal Process.* **3**(4), 387–408 (1984).

[YI85] P. C. Yip and K. R. Rao, DIF algorithms for DCT and DST. *Proc. IEEE Int. Conf. Acoust., Speech, Signal Process. Tampa, Florida, 1985*, pp. 776–779 (1985).

APPENDIX TO CHAPTER 8

8-A   FFT Programs

Program 1: Radix-2 DIF FFT

```
CC======================================================================CC
CC                                                                      CC
CC   Subroutine CFFT2DF(X,Y,M):                                         CC
CC       A Cooley-Tukey in-place, radix-2 complex FFT program           CC
CC       Decimation-in-frequency, cos/sin in second loop               CC
CC                                                                      CC
CC   Input/output:                                                      CC
CC       X     Array of real part of input/output (length >= N)        CC
CC       Y     Array of imaginary part of input/output (length >= N)   CC
CC       M     Transform length is N=2**M                              CC
CC                                                                      CC
CC   Author:                                                            CC
CC       H. V. Sorensen,     University of Pennsylvania,   Dec. 1984    CC
CC                           Internet address: hvs@ee.upenn.edu         CC
CC                                                                      CC
CC       This program may be used and distributed freely provided      CC
CC       this header is included and left intact                       CC
CC                                                                      CC
CC======================================================================CC
        SUBROUTINE CFFT2DF (X,Y,M)
        REAL X(1), Y(1)
C-------Main FFT loops----------------------------------------------------CC
        N = 2**M
        N2 = N
        DO 10 K = 1, M
            N1 = N2
            N2 = N2/2
            E  = 6.283185307179586/N1
            A  = 0
            DO 20 J = 1, N2
                C = COS (A)
```

```
                S = SIN (A)
                A = J*E
                DO   30 I = J, N, N1
                     L = I + N2
                     XT   = X(I) - X(L)
                     X(I) = X(I) + X(L)
                     YT   = Y(I) - Y(L)
                     Y(I) = Y(I) + Y(L)
                     X(L) = C*XT + S*YT
                     Y(L) = C*YT - S*XT
30                CONTINUE
20             CONTINUE
10         CONTINUE
C-------Digit reverse counter-------------------------------------------C
100   J = 1
      DO   104 I=1, N-1
          IF (I.GE.J) GOTO 101
              XT = X(J)
              X(J) = X(I)
              X(I) = XT
              XT   = Y(J)
              Y(J) = Y(I)
              Y(I) = XT
101       K = N/2
102       IF  (K.GE.J) GOTO 103
              J = J - K
              K = K/2
              GOTO 102
103       J = J + K
104   CONTINUE
      RETURN
      END
```

Program 2: Radix-2 DIT FFT

```
CC======================================================================CC
CC                                                                      CC
CC   Subroutine CFFT2DT(X,Y,M):                                         CC
CC       A Cooley-Tukey in-place, radix-2 complex FFT program           CC
CC       Decimation-in-time, cos/sin in second loop                     CC
CC                                                                      CC
CC   Input/output:                                                      CC
CC       X    Array of real part of input/output (length >= N)          CC
CC       Y    Array of imaginary part of input/output (length >= N)     CC
CC       M    Transform length is N=2**M                                CC
CC                                                                      CC
CC   Author:                                                            CC
CC       H.V. Sorensen,     University of Pennsylvania,    Dec. 1984    CC
CC                          Internet address: hvs@ee.upenn.edu          CC
```

```
CC                                                                        CC
CC        This program may be used and distributed freely provided       CC
CC        this header is included and left intact                        CC
CC==================================================================CC
        SUBROUTINE CFFT2DT (X,Y,M)
        REAL X(1), Y(1)
        N = 2**M
C-------Digit reverse counter------------------------------------------C
  100   J = 1
        DO  104 I = 1,N-1
            IF (I.GE.J) GOTO 101
                XT   = X(J)
                X(J) = X(I)
                X(I) = XT
                XT   = Y(J)
                Y(J) = Y(I)
                Y(I) = XT
  101       K = N/2
  102       IF (K.GE.J) GOTO 103
                J = J - K
                K = K/2
                GOTO 102
  103       J = J + K
  104   CONTINUE
C-------Main FFT loops-------------------------------------------------C
        N1 = 1
        DO  10 K = 1,M
            N2 = N1
            N1 = N2*2
            E  = 6.283185307179586/N1
            A  = 0
            DO  20 J = 1,N2
                C = COS (A)
                S = SIN (A)
                A = J*E
                DO  30 I = J,N,N1
                    L = I + N2
                    XT   = C*X(L) + S*Y(L)
                    YT   = C*Y(L) - S*X(L)
                    X(L) = X(I)      - XT
                    X(I) = X(I)      + XT
                    Y(L) = Y(I)      - YT
                    Y(I) = Y(I)      + YT
  30            CONTINUE
  20        CONTINUE
  10    CONTINUE
        RETURN
        END
```

Program 3: Split-Radix FFT Without Table Look-up

```
CC=====================================================================CC
CC                                                                     CC
CC   Subroutine CFFTSR(X,Y,M):                                         CC
CC       An in-place, split-radix complex FFT program                 CC
CC       Decimation-in-frequency, cos/sin in second loop              CC
CC       and is computed recursively                                  CC
CC                                                                     CC
CC   Input/output:                                                     CC
CC       X    Array of real part of input/output (length >= N)        CC
CC       Y    Array of imaginary part of input/output (length >= N)   CC
CC       M    Transform length is N=2**M                              CC
CC                                                                     CC
CC   Calls:                                                            CC
CC        CSTAGE,CBITREV                                               CC
CC                                                                     CC
CC   Author:                                                           CC
CC       H.V. Sorensen,      University of Pennsylvania,  Dec. 1984    CC
CC                           Arpa address: hvs@ee.upenn.edu.           CC
CC   Modified:                                                         CC
CC       H.V. Sorensen,      University of Pennsylvania,  Jul. 1987    CC
CC                                                                     CC
CC   Reference:                                                        CC
CC       Sorensen, Heideman, Burrus: "On computing the split-radix    CC
CC       FFT", IEEE Tran. ASSP, Vol. ASSP-34, No. 1, pp. 152-156      CC
CC       Feb. 1986                                                     CC
CC                                                                     CC
CC       This program may be used and distributed freely provided     CC
CC       this header is included and left intact                       CC
CC                                                                     CC
CC=====================================================================CC
        SUBROUTINE CFFTSR(X,Y,M)
        REAL X(1), Y(1)
        N = 2**M
C-----L shaped butterflies------------------------------------------C
        N2 = 2*N
        DO  10 K = 1, M-1
            N2 = N2/2
            N4 = N2/4
            CALL CSTAGE(N,N2,N4,X(1),X(N4+1),X(2*N4+1),X(3*N4+1),
      $                         Y(1),Y(N4+1),Y(2*N4+1),Y(3*N4+1))
   10   CONTINUE
C-----Length two butterflies----------------------------------------C
        IS = 1
        ID = 4
   20       DO 30 I1   = IS,N,ID
                T1     = X(I1)
                X(I1)  = T1 + X(I1+1)
                X(I1+1) = T1 - X(I1+1)
```

```
                T1        = Y(I1)
                Y(I1)     = T1 + Y(I1+1)
                Y(I1+1)   = T1 - Y(I1+1)
30              CONTINUE
                IS = 2*ID - 1
                ID = 4*ID
         IF (IS.LT.N) GOTO 20
C-------Digit reverse counter----------------------------------------C
        CALL CBITREV(X,Y,M)
        RETURN
        END
CC================================================================CC
CC                                                                CC
CC   Subroutine CSTAGE - the workhorse of the CFFTSR              CC
CC      Computes one stage of a complex split-radix length N      CC
CC      transform.                                                CC
CC                                                                CC
CC   Author:                                                      CC
CC      H.V. Sorensen,    University of Pennsylvania,   Jul. 1987 CC
CC                                                                CC
CC      This program may be used and distributed freely provided  CC
CC      this header is included and left intact                   CC
CC                                                                CC
CC================================================================CC
        SUBROUTINE CSTAGE(N,N2,N4,X1,X2,X3,X4,Y1,Y2,Y3,Y4)
        REAL X1(1),X2(1),X3(1),X4(1),Y1(1),Y2(1),Y3(1),Y4(1)
        N8 = N4/2
C-------Zero butterfly-----------------------------------------------C
        IS = 0
        ID = 2*N2
10          DO 20  I1 = IS+1,N,ID
                T1        = X1(I1) - X3(I1)
                X1(I1)    = X1(I1) + X3(I1)
                T2        = Y2(I1) - Y4(I1)
                Y2(I1)    = Y2(I1) + Y4(I1)
                X3(I1)    = T1     + T2
                T2        = T1     - T2
                T1        = X2(I1) - X4(I1)
                X2(I1)    = X2(I1) + X4(I1)
                X4(I1)    = T2
                T2        = Y1(I1) - Y3(I1)
                Y1(I1)    = Y1(I1) + Y3(I1)
                Y3(I1)    = T2     - T1
                Y4(I1)    = T2     + T1
20          CONTINUE
            IS = 2*ID - N2
            ID = 4*ID
        IF (IS .LT. N) GOTO 10
C
        IF (N4-1) 100,100,30
```

```
C-------N/8 butterfly----------------------------------------------C
   30     IS = 0
          ID = 2*N2
   40        DO  50 I1 = IS+1+N8,N,ID
                T1     = X1(I1) - X3(I1)
                X1(I1) = X1(I1) + X3(I1)
                T2     = X2(I1) - X4(I1)
                X2(I1) = X2(I1) + X4(I1)
                T3     = Y1(I1) - Y3(I1)
                Y1(I1) = Y1(I1) + Y3(I1)
                T4     = Y2(I1) - Y4(I1)
                Y2(I1) = Y2(I1) + Y4(I1)
                T5     = (T4 - T1)*0.707106778
                T1     = (T4 + T1)*0.707106778
                T4     = (T3 - T2)*0.707106778
                T2     = (T3 + T2)*0.707106778
                X3(I1) = T4 + T1
                Y3(I1) = T4 - T1
                X4(I1) = T5 + T2
                Y4(I1) = T5 - T2
   50        CONTINUE
          IS = 2*ID - N2
          ID = 4*ID
       IF (IS .LT. N-1) GOTO 40
C
       IF (N8-1) 100,100,60
C-------General butterfly. Two at a time---------------------------C
   60     E  = 6.283185307179586/N2
          SS1 = SIN(E)
          SD1 = SS1
          SD3 = 3.*SD1-4.*SD1**3
          SS3 = SD3
          CC1 = COS(E)
          CD1 = CC1
          CD3 = 4.CD1**3-3.*CD1
          CC3 = CD3
          DO  90  J = 2,N8
             IS = 0
             ID = 2*N2
             JN = N4 - 2*J + 2
   70           DO  80 I1=IS+J,N+J,ID
                   T1     = X1(I1) - X3(I1)
                   X1(I1) = X1(I1) + X3(I1)
                   T2     = X2(I1) - X4(I1)
                   X2(I1) = X2(I1) + X4(I1)
                   T3     = Y1(I1) - Y3(I1)
                   Y1(I1) = Y1(I1) + Y3(I1)
                   T4     = Y2(I1) - Y4(I1)
                   Y2(I1) = Y2(I1) + Y4(I1)
                   T5 = T1 - T4
```

```
                        T1 = T1 + T4
                        T4 = T2 - T3
                        T2 = T2 + T3
                        X3(I1) =   T1*CC1 - T4*SS1
                        Y3(I1) = -T4*CC1 - T1*SS1
                        X4(I1) =   T5*CC3 + T2*SS3
                        Y4(I1) =   T2*CC3 - T5*SS3
                        I2 = I1 + JN
                        T1      = X1(I2) - X3(I2)
                        X1(I2) = X1(I2) + X3(I2)
                        T2      = X2(I2) - X4(I2)
                        X2(I2) = X2(I2) + X4(I2)
                        T3      = Y1(I2) - Y3(I2)
                        Y1(I2) = Y1(I2) + Y3(I2)
                        T4      = Y2(I2) - Y4(I2)
                        Y2(I2) = Y2(I2) + Y4(I2)
                        T5 = T1 - T4
                        T1 = T1 + T4
                        T4 = T2 - T3
                        T2 = T2 + T3
                        X3(I2) =   T1*SS1 - T4*CC1
                        Y3(I2) = -T4*SS1 - T1*CC1
                        X4(I2) = -T5*SS3 - T2*CC3
                        Y4(I2) = -T2*SS3 + T5*CC3
80                 CONTINUE
                   IS = 2*ID - N2
                   ID = 4*ID
           IF (IS .LT. N) GOTO 70
C
           T1  = CC1*CD1 - SS1*SD1
           SS1 = CC1*SD1 + SS1*CD1
           CC1 = T1
           T3  = CC3*CD3 - SS3*SD3
           SS3 = CC3*SD3 + SS3*CD3
           CC3 = T3
90      CONTINUE
100     RETURN
        END
```

```
CC===================================================================CC
CC                                                                   CC
CC  Subroutine CBITREV(X,Y,M):                                       CC
CC      Bit reverses the array X of length 2**M. It generates a      CC
CC      table ITAB (minimum length is SQRT(2**M) if M is even        CC
CC      or SQRT(2*2**M) if M is odd). ITAB need only be generated    CC
CC      once for a given transform length.                           CC
CC                                                                   CC
CC  Author:                                                          CC
CC      H.V. Sorensen,   University of Pennsylvania,   Aug. 1987     CC
CC                       Arpa address: hvs@ee.upenn.edu              CC
```

```
CC                                                                    CC
CC   Reference:                                                       CC
CC       D. Evans, Tran. ASSP, Vol. ASSP-35, No. 8, pp 1120-1125,     CC
CC       Aug. 1987                                                    CC
CC                                                                    CC
CC       This program may be used and distributed freely provided     CC
CC       this header is included and left intact                      CC
CC                                                                    CC
CC================================================================CC
      SUBROUTINE CBITREV(X,Y,M)
      DIMENSION X(1),Y(1),ITAB(256)
C--------Initialization of ITAB array--------------------------------C
      M2 = M/2
      NBIT = 2**M2
      IF (2*M2.NE.M) M2 = M2 + 1
      ITAB(1) = 0
      ITAB(2) = 1
      IMAX = 1
      DO   10 LBSS = 2, M2
          IMAX = 2 * IMAX
      DO   10 I = 1, IMAX
          ITAB(I)      = 2 * ITAB(I)
          ITAB(I+IMAX) = 1 + ITAB(I)
  10    CONTINUE
C-----The actual bit reversal---------------------------------------C
      DO   20 K = 2, NBIT
          J0 = NBIT * ITAB(K) + 1
          I = K
          J + J0
          DO   20 L = 2, ITAB(K)+1
              T1   = X(I)
              X(I) = X(J)
              X(J) = T1
              T1   = Y(I)
              Y(I) = Y(J)
              Y(J) = T1
              I = I + NBIT
              J = J0 + ITAB(L)
  20    CONTINUE
        RETURN
        END
```

Program 4: Split-Radix with Table Look-up

```
CC====================================================================CC
CC                                                                    CC
CC   Subroutine CTFFTSR(X,Y,M,CT1,CT3,ST1,ST3,ITAB):                  CC
CC        An in-place, split-radix complex FFT program                CC
CC        Decimation-in-frequency, cos/sin in third loop              CC
CC        and is looked-up in table. Tables CT1,CT3,ST1,ST3           CC
CC        have to have length >= N/8-1. The bit reverser uses partly  CC
CC        table lookup.                                               CC
CC                                                                    CC
CC   Input/output:                                                    CC
CC        X     Array of real part of input/output (length >= N)      CC
CC        Y     Array of imaginary part of input/output (length >= N) CC
CC        M     Transform length is N=2**M                            CC
CC        CT1   Array of cos() table (length >= N/8-1)                CC
CC        CT3   Array of cos() table (length >= N/8-1)                CC
CC        ST1   Array of sin() table (length >= N/8-1)                CC
CC        ST3   Array of sin() table (length >= N/8-1)                CC
CC        ITAB  Array of bit reversal indices (length >= sqrt(2*N))   CC
CC                                                                    CC
CC   Calls:                                                           CC
CC        CTSTAG                                                      CC
CC                                                                    CC
CC   Note:                                                            CC
CC        TINIT must be called before this program!!                 CC
CC                                                                    CC
CC   Author:                                                          CC
CC        H.V. Sorensen,     University of Pennsylvania,    Dec. 1984 CC
CC                                Internet address: hvs@ee.upenn.edu  CC
CC   Modified:                                                        CC
CC        Henrik Sorensen,   University of Pennsylvania,    Jul. 1987 CC
CC                                                                    CC
CC        This program may be used and distributed freely provided    CC
CC        this header is included and left intact                    CC
CC                                                                    CC
CC====================================================================CC
      SUBROUTINE CTFFTSR(X,Y,M,CT1,CT3,ST1,ST3,ITAB)
      REAL X(1),Y(1),CT1(1),CT3(1),ST1(1),ST3(1)
      INTEGER ITAB(1)
      N = 2**M
C-------L shaped butterflies-------------------------------------------C
      ITS = 1
      N2 = 2*N
      DO  10 K = 1, M-1
          N2 = N2/2
          N4 = N2/4
          CALL CTSTAG(N,N2,N4,ITS,X(1),X(N4+1),X(2*N4+1),X(3*N4+1),
     $                            Y(1),Y(N4+1),Y(2*N4+1),Y(3*N4+1),
     $                            CT1,CT3,ST1,ST3)
          ITS = 2 * ITS
   10 CONTINUE
```

```
C-------Length two butterflies-----------------------------------------C
      IS = 1
      ID = 4
 20       DO  30 I1 = IS,N,ID
              T1       = X(I1)
              X(I1)    = T1 + X(I1+1)
              X(I1+1) = T1 - X(I1+1)
              T1       = Y(I1)
              Y(I1)    = T1 + Y(I1+1)
              Y(I1+1) = T1 - Y(I1+1)
 30       CONTINUE
          IS = 2*ID - 1
          ID = 4*ID
      IF (IS.LT.N) GOTO 20
C-------Digit reverse counter------------------------------------------C
      M2 = M/2
      NBIT = 2**M2
      DO  50 K = 2, NBIT
          J0 = NBIT * ITAB(K) + 1
          I = K
          J = J0
          DO  40 L = 2, ITAB(K)+1
              T1    = X(I)
              X(I) = X(J)
              X(J) = T1
              T1    = Y(I)
              Y(I) = Y(J)
              Y(J) = T1
              I = I + NBIT
              J = J0 + ITAB(L)
 40       CONTINUE
 50   CONTINUE
      RETURN
      END
CC=================================================================CC
CC                                                               CC
CC  Subroutine CTSTAG - the workhorse of CTFFTSR                 CC
CC      Computes one stage of a length N split-radix transform   CC
CC                                                               CC
CC  Author:                                                      CC
CC      Henrik Sorensen,   University of Pennsylvania,   Jul. 1987 CC
CC                                                               CC
CC      This program may be used and distributed freely provided  CC
CC      this header is included and left intact                  CC
CC                                                               CC
CC=================================================================CC
      SUBROUTINE CTSTAG(N,N2,N4,ITS,X1,X2,X3,X4,Y1,Y2,Y3,Y4,
     $                  CT1,CT3,ST1,ST3)
      REAL X1(1),X2(1),X3(1),X4(1),Y1(1),Y2(1),Y3(1),Y4(1)
      REAL CT1(1),CT3(1),ST1(1),ST3(1)
      N8 = N4/2
```

```
C-------Zero butterfly------------------------------------------C
      IS = 0
      ID = 2*N2
 10        DO  20  I1 = IS+1,N,ID
              T1     = X1(I1) - X3(I1)
              X1(I1) = X1(I1) + X3(I1)
              T2     = Y2(I1) - Y4(I1)
              Y2(I1) = Y2(I1) + Y4(I1)
              X3(I1) = T1     + T2
              T2     = T1     - T2
              T1     = X2(I1) - X4(I1)
              X2(I1) = X2(I1) + X4(I1)
              X4(I1) = T2
              T2     = Y1(I1) - Y3(I1)
              Y1(I1) = Y1(I1) + Y3(I1)
              Y3(I1) = T2     - T1
              Y4(I1) = T2     + T1
 20        CONTINUE
           IS = 2*ID - N2
           ID = 4*ID
      IF (IS .LT. N) GOTO 10
C
      IF (N4-1) 100,100,30
C-------N/8 butterfly------------------------------------------C
 30   IS = 0
      ID = 2*N2
 40        DO  50 I1 = IS+1+N8,N,ID
              T1     = X1(I1) - X3(I1)
              X1(I1) = X1(I1) + X3(I1)
              T2     = X2(I1) - X4(I1)
              X2(I1) = X2(I1) + X4(I1)
              T3     = Y1(I1) - Y3(I1)
              Y1(I1) = Y1(I1) + Y3(I1)
              T4     = Y2(I1) - Y4(I1)
              Y2(I1) = Y2(I1) + Y4(I1)
              T5     = (T4 - T1)*0.707106778
              T1     = (T4 + T1)*0.707106778
              T4     = (T3 - T2)*0.707106778
              T2     = (T3 + T2)*0.707106778
              X3(I1) = T4 + T1
              Y3(I1) = T4 - T1
              X4(I1) = T5 + T2
              Y4(I1) = T5 - T2
 50        CONTINUE
           IS = 2*ID - N2
           ID = 4*ID
      IF (IS .LT. N-1) GOTO 40
C
      IF (N8-1) 100,100,60
```

```
C-------General butterfly. Two at a time---------------------------C
  60      IS = 1
          ID = N2*2
  70          DO  90  I = IS, N, ID
                  IT = 0
                  JN = I + N4
                  DO  80  J=1, N8-1
                      IT = IT+ITS
                      I1 = I+J
                      T1      = X1(I1)  -  X3(I1)
                      X1(I1) = X1(I1)  +  X3(I1)
                      T2      = X2(I1)  -  X4(I1)
                      X2(I1) = X2(I1)  +  X4(I1)
                      T3      = Y1(T1)  -  Y3(I1)
                      Y1(I1) = Y1(I1)  +  Y3(I1)
                      T4      = Y2(I1)  -  Y4(I1)
                      Y2(I1) = Y2(I1)  +  Y4(I1)
                      T5 = T1  -  T4
                      T1 = T1  +  T4
                      T4 = T2  -  T3
                      T2 = T2  +  T3
                      X3(I1) =   T1*CT1(IT)  -  T4*ST1(IT)
                      Y3(I1) =  -T4*CT1(IT)  -  T1*ST1(IT)
                      X4(I1) =   T5*CT3(IT)  +  T2*ST3(IT)
                      Y4(I1) =   T2*CT3(IT)  -  T5*ST3(IT)
                      I2 = JN  -  J
                      T1      = X1(I2)  -  X3(I2)
                      X1(I2) = X1(I2)  +  X3(I2)
                      T2      = X2(I2)  -  X4(I2)
                      X2(I2) = X2(I2)  +  X4(I2)
                      T3      = Y1(I2)  -  Y3(I2)
                      Y1(I2) = Y1(I2)  +  Y3(I2)
                      T4      = Y2(I2)  -  Y4(I2)
                      Y2(I2) = Y2(I2)  +  Y4(I2)
                      T5 = T1  -  T4
                      T1 = T1  +  T4
                      T4 = T2  -  T3
                      T2 = T2  +  T3
                      X3(I2) =   T1*ST1(IT)  -  T4*CT1(IT)
                      Y3(I2) =  -T4*ST1(IT)  -  T1*CT1(IT)
                      X4(I2) =  -T5*ST3(IT)  -  T2*CT3(IT)
                      Y4(I2) =  -T2*ST3(IT)  +  T5*CT3(IT)
  80              CONTINUE
  90          CONTINUE
          IS = 2*ID - N2 + 1
          ID = 4*ID
      IF (IS .LT. N) GOTO 70
 100  RETURN
      END
```

```
CC=====================================================================CC
CC                                                                     CC
CC  Subroutine TINIT:                                                  CC
CC       Initialize sin/cos and bit reversal tables                    CC
CC                                                                     CC
CC  Author:                                                            CC
CC       Henrik Sorensen,    University of Pennsylvania,   Jul. 1987   CC
CC                                                                     CC
CC       This program may be used and distributed freely provided      CC
CC       this header is included and left intact                       CC
CC                                                                     CC
CC=====================================================================CC
        SUBROUTINE TINIT(M,CT1,CT3,ST1,ST3,ITAB)
        REAL CT(1),CT3(1),ST1(1),ST3(1)
        INTEGER ITAB(1)
C-------Sin/cos table-------------------------------------------------C
        N = 2**M
        ANG = 6.283185307179586/N
        DO   10 I=1,N/8-1
            CT1(I) = COS(ANG*I)
            CT3(I) = COS(ANG*I*3)
            ST1(I) = SIN(ANG*I)
            ST3(I) = SIN(ANG*I*3)
  10    CONTINUE
C-------Bit reversal table--------------------------------------------C
        M2 = M/2
        NBIT = 2**M2
        IF (2*M2.NE.M) M2 = M2 + 1
        ITAB(1) = 0
        ITAB(2) = 1
        IMAX = 1
        DO   30 LBSS = 2, M2
            IMAX = 2 * IMAX
            DO   20 I = 1, IMAX
                ITAB(I)      = 2 * ITAB(I)
                ITAB(I+IMAX) = 1 + ITAB(I)
  20        CONTINUE
  30    CONTINUE
        RETURN
        END
```

Program 5: Inverse Split-Radix FFT

```
CC==================================================================CC
CC                                                                CC
CC   Subroutine ICSRFFT(X,Y,M):                                   CC
CC       An in-place, inverse split-radix complex FFT program     CC
CC       Decimation-in-frequency, cos/sin in second loop          CC
CC       and is computed recursively                              CC
CC                                                                CC
CC   Input/output:                                                CC
CC       X     Array of real part of input/output (length >= N)   CC
CC       Y     Array of imaginary part of input/output (length >= N) CC
CC       M     Transform length is N=2**M                         CC
CC                                                                CC
CC   Calls:                                                       CC
CC       ICSTAGE,CBITREV                                          CC
CC                                                                CC
CC   Author:                                                      CC
CC       H.V. Sorensen,    University of Pennsylvania,  Dec. 1984 CC
CC                         Arpa address: hvs@ee.upenn.edu         CC
CC   Modified:                                                    CC
CC       H.V. Sorensen,    University of Pennsylvania,  Jul. 1987 CC
CC                                                                CC
CC   Reference:                                                   CC
CC       Sorensen, Heideman, Burrus: "On computing the split-radix CC
CC       FFT", IEEE Tran. ASSP, Vol. ASSP-34, No.1, pp. 152-156   CC
CC       Feb. 1986                                                CC
CC                                                                CC
CC       This program may be used and distributed freely provided CC
CC       this header is included and left intact                  CC
CC                                                                CC
CC==================================================================CC
       SUBROUTINE ICSRFFT(X,Y,M)
       REAL X(1),Y(1)
       N = 2**M
C-----L shaped butterflies ---------------------------------------C
       N2 = 2*N
       DO  10  K = 1, M-1
          N2 = N2/2
          N4 = N2/4
          CALL ICSTAGE(N,N2,N4,X(1),X(N4+1),X(2*N4+1),X(3*N4+1),
      $                       Y(1),Y(N4+1),Y(2*N4+1),Y(3*4N+1))
  10   CONTINUE
C-----Length two butterflies--------------------------------------C
       IS = 1
       ID = 4
  20      DO 30 I1 = IS,N,ID
             R1      = X(I1)
             X(I1)   = R1 + X(I1+1)
             X(I1+1) = R1 - X(I1+1)
```

```
                R1       = Y(I1)
                Y(I1)    = R1 + Y(T1+1)
                Y(I1+1)  = R1 - Y(I1+1)
 30         CONTINUE
            IS = 2*ID - 1
            ID = 4*ID
        IF (IS.LT.N) GOTO 20
C-----Digit reverse counter--------------------------------------------C
        CALL CBITREV(X,Y,M)
C-----Divide by N------------------------------------------------------C
        DO  40 I=1,N
            X(I)=X(I)/N
            Y(I)=Y(I)/N
 40      CONTINUE
         RETURN
         END
CC====================================================================CC
CC                                                                    CC
CC  Subroutine ICSTAGE - the workhorse of the ICFFTSR                 CC
CC      Computes one stage of an inverse complex split-radix          CC
CC      length N transform.                                           CC
CC                                                                    CC
CC  Author:                                                           CC
CC      H.V. Sorensen,   University of Pennsylvania,   Jul. 1987      CC
CC                                                                    CC
CC      This program may be used and distributed freely provided      CC
CC      this header is included and left intact                       CC
CC                                                                    CC
CC====================================================================CC
        SUBROUTINE  ICSTAGE(N,N2,N4,X1,X2,X3,X4,Y1,Y2,Y3,Y4)
        REAL X1(1),X2(1),X3(1),X4(1),Y1(1),Y2(1),Y3(1),Y4(1)
        N8 = N4/2
C-----Zero butterfly---------------------------------------------------C
        IS = 0
        ID = 2*N2
 10         DO  20  I0 = IS+1,N,ID
                R1       = X1(I0) - X3(I0)
                X1(I0) = X1(I0) + X3(I0)
                R2       = X2(I0) - X4(I0)
                X2(I0) = X2(I0) + X4(I0)
                S1       = Y1(I0) - Y3(I0)
                Y1(I0) = Y1(I0) + Y3(I0)
                S2       = Y2(I0) - Y4(I0)
                Y2(I0) = Y2(I0) + Y4(I0)
                X3(I0) = R1 - S2
                X4(I0) = R1 + S2
                Y4(I0) = S1 - R2
                Y3(I0) = R2 + S1
 20         CONTINUE
            IS = 2*ID - N2
            ID = 4*ID
```

```
      IF (IS.LT.N) GOTO 10
C
      IF (N4-1) 100,100,30
C-----N/8 butterfly-------------------------------------------------------C
  30  IS = 0
      ID = 2*N2
  40     DO  50   IO = IS+1+N8,N,ID
            R1      = X1(IO) - X3(IO)
            X1(IO) = X1(IO) + X3(IO)
            R2      = X2(IO) - X4(IO)
            X2(IO) = X2(IO) + X4(IO)
            S1      = Y1(IO) - Y3(IO)
            Y1(IO) = Y1(IO) + Y3(IO)
            S2      = Y2(IO) - Y4(IO)
            Y2(IO) = Y2(IO) + Y4(IO)
            S3     = R1 - S2
            R1     = R1 + S2
            S2     = R2 - S1
            R2     = R2 + S1
            X3(IO) = (S3 - R2)*0.707106778
            Y3(IO) = (R2 + S3)*0.707106778
            X4(IO) = (S2 - R1)*0.707160778
            Y4(IO) = (S2 + R1)*0.707160778
  50     CONTINUE
         IS = 2*ID - N2
         ID = 4*ID
    IF (IS.LT.N) GOTO 40
C
    IF (N8-1) 100,100,60
C-----General butterfly. Two at a time---------------------------------C
  60  E = 6.283185307179586/N2
      SS1 = SIN(E)
      SD1 = SS1
      SD3 = 3.*SD1-4.*SD1**3
      SS3 = SD3
      CC1 = COS(E)
      CD1 = CC1
      CD3 = 4.*CD1**3-3.*CD1
      CC3 = CD3
      DO  90   J = 2, N8
          IS  = 0
          ID  = 2*N2
          JN = N4 - 2*J + 2
  70          DO  80   IO = IS +J,N+J,ID
                R1      = X1(IO) - X3(IO)
                X1(IO) = X1(IO) + X3(IO)
                R2      = X2(IO) - X4(IO)
                X2(IO) = X2(IO) + X4(IO)
                S1      = Y1(IO) - Y3(IO)
                Y1(IO) = Y1(IO) + Y3(IO)
```

```
                 S2       = Y2(I0) - Y4(I0)
                 Y2(I0) = Y2(I0) + Y4(I0)
                 S3     = R1 - S2
                 R1     = R1 + S2
                 S2     = R2 - S1
                 R2     = R2 + S1
                 X3(I0) = S3*CC1 - R2*SS1
                 Y3(I0) = R2*CC1 + S3*SS1
                 X4(I0) = R1*CC3 + S2*SS3
                 Y4(I0) =-S2*CC3 + R1*SS3
                 I1 = I0 + JN
                 R1       = X1(I1) - X3(I1)
                 X1(I1) = X1(I1) + X3(I1)
                 R2       = X2(I1) - X4(I1)
                 X2(I1) = X2(I1) + X4(I1)
                 S1       = Y1(I1) - Y3(I1)
                 Y1(I1) = Y1(I1) + Y3(I1)
                 S2       = Y2(I1) - Y4(I1)
                 Y2(I1) = Y2(I1) + Y4(I1)
                 S3     = R1 - S2
                 R1     = R1 + S2
                 S2     = R2 - S1
                 R2     = R2 + S1
                 X3(I1) = S3*SS1 - R2*CC1
                 Y3(I1) = R2*SS1 + S3*CC1
                 X4(I1) =-R1*SS3 - S2*CC3
                 Y4(I1) = S2*SS3 - R1*CC3
80           CONTINUE
             IS = 2*ID - N2
             ID = 4*ID
      IF (IS.LT.N) GOTO 70
C
      T1   = CC1*CD1 - SS1*SD1
      SS1 = CC1*SD1 + SS1*CD1
      CC1 = T1
      T3   = CC3*CD3 - SS3*SD3
      SS3 = CC3*SD3 + SS3*CD3
      CC3 = T3
90    CONTINUE
100   RETURN
      END
CC================================================================CC
CC                                                                CC
CC   Subroutine CBITREV(X,Y,M):                                   CC
CC       Bit reverses the array X of length 2**M. It generates a  CC
CC       table ITAB (minimum length is SQRT(2**M) if M is even    CC
CC       or SQRT(2*2**M) if M is odd). ITAB need only be generated CC
CC       once for a given transform length.                       CC
CC                                                                CC
CC                                                                CC
```

```
CC   Author:                                                       CC
CC       H.V. Sorensen,    University of Pennsylvania,  Aug. 1987  CC
CC                         Arpa address: hvs@ee.upenn.edu          CC
CC                                                                 CC
CC   Reference:                                                    CC
CC       D. Evans, Tran. ASSP, Vol. ASSP-35, No. 8, pp.1120-1125,  CC
CC       Aug. 1987                                                 CC
CC                                                                 CC
CC       This program may be used and distributed freely provided  CC
CC       this header is included and left intact                   CC
CC                                                                  CC
CC================================================================CC
      SUBROUTINE CBITREV(X,Y,M)
      DIMENSION X(1),Y(1),ITAB(1)
C-----Initialization of ITAB array-------------------------------------C
      M2 = M/2
      NBIT = 2**M2
      IF (2*M2.NE.M) M2 = M2 + 1
      ITAB(1) = 0
      ITAB(2) = 1
      IMAX = 1
      DO  10 LBSS = 2, M2
          IMAX = 2 * IMAX
          DO  10 I = 1, IMAX
              ITAB(I)      = 2 * ITAB(I)
              ITAB(I+IMAX) = 1 + ITAB(I)
  10   CONTINUE
C-----The actual bit reversal-----------------------------------------C
      DO  20 K = 2, NBIT
          J0 = NBIT * ITAB(K) + 1
          I = K
          J = J0
          DO  20 L = 2, ITAB(K)+1
              T1     = X(I)
              X(I) = X(J)
              X(J) = T1
              T1     = Y(I)
              Y(I) = Y(J)
              Y(J) = T1
              I = I + NBIT
              J = J0 + ITAB(L)
  20   CONTINUE
      RETURN
      END
```

Program 6: Prime Factor FFT

```
CC=====================================================================CC
CC                                                                   CC
CC   Subroutine PFA(X,Y,N,M,NI):                                     CC
CC        A prime factor FFT program. In-place and in-order.         CC
CC        Length N transform with M factors in array NI              CC
CC             N = NI(1)*NI(2)*...*NI(M)                              CC
CC        Factors are implemented for NI = 2,3,4,5,7                 CC
CC                                                                   CC
CC   Input/output:                                                   CC
CC        X    Array with real part of input/output (length >= N)    CC
CC        Y    Array with imaginary part of input/output (length >= N)CC
CC        N    Transform length                                      CC
CC        M    Number of factors in NI                               CC
CC        NI   Array with factors of N (length >= M)                 CC
CC                                                                   CC
CC   Author:                                                         CC
CC        C.S. Burrus,       Rice University,        Aug. 1987       CC
CC                                                                   CC
CC        This program may be used and distributed freely provided   CC
CC        this header is included and left intact                    CC
CC                                                                   CC
CC=====================================================================CC
      SUBROUTINE PFA(X,Y,N,M,NI)
      INTEGER  NI(4),  I(16),  IP(16),  LP(16)
      REAL X(1),  Y(1)
      DATA  C31,  C32  /  -0.86602540,-1.50000000  /
      DATA  C51,  C52  /   0.95105652,-1.53884180  /
      DATA  C53,  C54  /  -0.36327126,  0.55901699  /
      DATA  C55        /  -1.25  /
      DATA  C71,  C72  /  -1.16666667,-0.79015647  /
      DATA  C73,  C74  /   0.055854267,  0.7343022  /
      DATA  C75,  C76  /   0.44095855,-0.34087293  /
      DATA  C77,  C78  /   0.53396936,  0.87484229  /
C-------Nested loops--------------------------------------------------C
      DO  10 K = 1, M
          N1 = NI(K)
          N2 = N/N1
          L = 1
          N3 = N2 - N1*(N2/N1)
          DO  15 J = 2, N1
              L = L + N3
              IF (L.GT.N1) L = L - N1
              LP(J) = L
15        CONTINUE
C
          DO  20 J = 1, N, N1
              IT   = J
              I(1) = J
```

```
              IP(1) = J
              DO   30 L = 2, N1
                  IT = IT + N2
                  IF (IT.GT.N)   IT = IT - N
                  I(L) = IT
                  IP(LP(L)) = IT
30            CONTINUE
              GOTO (20,102,103,104,105,20,107), N1
C-------WFTA N = 2-----------------------------------------------------C
  102         R1        = X(I(1))
              X(I(1))   = R1 + X(I(2))
              X(I(2))   = R1 - X(I(2))
              R1        = Y(I(1))
              Y(IP(1))  = R1 + Y(I(2))
              Y(IP(2))  = R1 - Y(I(2))
              GOTO 20
C-------WFTA N = 3-----------------------------------------------------C
  103         R2 = (X(I(2)) - X(I(3))) * C31
              R1 =  X(I(2)) + X(I(3))
              X(I(1)) = X(I(1)) + R1
              R1        = X(I(1)) + R1 * C32
              S2 = (Y(I(2)) - Y(I(3))) * C31
              S1 =  Y(I(2)) + Y(I(3))
              Y(I(1)) = Y(I(1)) + S1
              S1        = Y(I(1)) + S1 * C32
              X(IP(2)) = R1 - S2
              X(IP(3)) = R1 + S2
              Y(IP(2)) = S1 + R2
              Y(IP(3)) = S1 - R2
              GOTO 20
C-------WFTA N = 4-----------------------------------------------------C
  104         R1 = X(I(1)) + X(I(3))
              T1 = X(I(1)) - X(I(3))
              R2 = X(I(2)) + X(I(4))
              X(IP(1)) = R1 + R2
              X(IP(3)) = R1 - R2
              R1 = Y(I(1)) + Y(I(3))
              T2 = Y(I(1)) - Y(I(3))
              R2 = Y(I(2)) + Y(I(4))
              Y(IP(1)) = R1 + R2
              Y(IP(3)) = R1 - R2
              R1 = X(I(2)) - X(I(4))
              R2 = Y(I(2)) - Y(I(4))
              X(IP(2)) = T1 + R2
              X(IP(4)) = T1 - R2
              Y(IP(2)) = T2 - R1
              Y(IP(4)) = T2 + R1
              GOTO 20
```

```
C-------WFTA N = 5------------------------------------------------C
 105          R1 = X(I(2)) + X(I(5))
              R4 = X(I(2)) - X(I(5))
              R3 = X(I(3)) + X(I(4))
              R2 = X(I(3)) - X(I(4))
              T = (R1 - R3) * C54
              R1 = R1 + R3
              X(I(1)) = X(I(1)) + R1
              R1       = X(I(1)) + R1 * C55
              R3 = R1 - T
              R1 = R1 + T
              T = (R4 + R2) * C51
              R4 = T + R4 * C52
              R2 = T + R2 * C53
              S1 = Y(I(2)) + Y(I(5))
              S4 = Y(I(2)) - Y(I(5))
              S3 = Y(I(3)) + Y(I(4))
              S2 = Y(I(3)) - Y(I(4))
              T = (S1 - S3) * C54
              S1 = S1 + S3
              Y(I(1)) = Y(I(1)) + S1
              S1       = Y(I(1)) + S1 * C55
              S3 = S1 - T
              S1 = S1 + T
              T = (S4 + S2) * C51
              S4 = T + S4 * C52
              S2 = T + S2 * C53
              X(IP(2)) = R1 + S2
              X(IP(5)) = R1 - S2
              X(IP(3)) = R3 - S4
              X(IP(4)) = R3 + S4
              Y(IP(2)) = S1 - R2
              Y(IP(5)) = S1 + R2
              Y(IP(3)) = S3 + R4
              Y(IP(4)) = S3 - R4
              GOTO 20
C-------WFTA N = 7------------------------------------------------C
 107          R1 = X(I(2)) + X(I(7))
              R6 = X(I(2)) - X(I(7))
              S1 = Y(I(2)) + Y(I(7))
              S6 = Y(I(2)) - Y(I(7))
              R2 = X(I(3)) + X(I(6))
              R5 = X(I(3)) - X(I(6))
              S2 = Y(I(3)) + Y(I(6))
              S5 = Y(I(3)) - Y(I(6))
              R3 = X(I(4)) + X(I(5))
              R4 = X(I(4)) - X(I(5))
              S3 = Y(I(4)) + Y(I(5))
              S4 = Y(I(4)) - Y(I(5))
              T3 = (R1 - R2) * C74
```

```
          T  = (R1 - R3) * C72
          R1 = R1 + R2 + R3
          X( I (1)) = X( I (1)) + R1
          R1        = X( I (1)) + R1 * C71
          R2 = (R3 - R2) * C73
          R3 = R1 - T + R2
          R2 = R1 - R2 - T3
          R1 = R1 + T + T3
          T  =(R6 - R5) * C78
          T3 =(R6 + R4) * C76
          R6 =(R6 + R5 - R4) * C75
          R5 =(R5 + R4) * C77
          R4 = R6 - T3 + R5
          R5 = R6 - R5 - T
          R6 = R6 + T3 + T
          T3 = (S1 - S2) * C74
          T  = (S1 - S3) * C72
          S1 = S1 + S2 + S3
          Y( I (1)) = Y( I (1)) + S1
          S1        = Y( I (1)) + S1 * C71
          S2 =(S3 - S2) * C73
          S3 = S1 - T + S2
          S2 = S1 - S2 - T3
          S1 = S1 + T + T3
          T  = (S6 - S5) * C78
          T3 = (S6 + S4) * C76
          S6 = (S6 + S5 - S4) * C75
          S5 = (S5 + S4) * C77
          S4 = S6 - T3 + S5
          S5 = S6 - S5 - T
          S6 = S6 + T3 + T
          X( IP(2)) = R3 + S4
          X( IP(7)) = R3 - S4
          X( IP(3)) = R1 + S6
          X( IP(6)) = R1 - S6
          X( IP(4)) = R2 - S5
          X( IP(5)) = R2 + S5
          Y( IP(4)) = S2 + R5
          Y( IP(5)) = S2 - R5
          Y( IP(2)) = S3 - R4
          Y( IP(7)) = S3 + R4
          Y( IP(3)) = S1 - R6
          Y( IP(6)) = S1 + R6
20     CONT I NUE
10   CONT I NUE
     RETURN
     END
```

Program 7: Real-Valued Split-Radix FFT

```
CC=======================================================================CC
CC                                                                       CC
CC  Subroutine RSRFFT(X,M):                                              CC
CC       A real-valued, in-place, split-radix FFT program                CC
CC       Decimation-in-time, cos/sin in second loop                     CC
CC       and computed recursively                                        CC
CC       Output in order:                                                CC
CC               [ Re(0),Re(1),.....,Re(N/2),Im(N/2-1),...Im(1)]         CC
CC                                                                       CC
CC  Input/output:                                                        CC
CC       X     Array of input/output (length >= N)                       CC
CC       M     Transform length is N=2**M                                CC
CC                                                                       CC
CC  Calls:                                                               CC
CC       RSTAGE,RBITREV                                                  CC
CC                                                                       CC
CC  Author:                                                              CC
CC       H.V. Sorensen,   University of Pennsylvania,   Oct. 1985        CC
CC                            Arpa address: hvs@ee.upenn.edu             CC
CC  Modified:                                                            CC
CC       F. Bonzanigo,     ETH-Zurich,                  Sep. 1986        CC
CC       H.V. Sorensen,    University of Pennsylvania,  Mar. 1987        CC
CC       H.V. Sorensen,    University of Pennsylvania,  Oct. 1987        CC
CC                                                                       CC
CC  Reference:                                                           CC
CC       H.V. Sorensen, Jones, Heideman, Burrus, "Real-valued fast       CC
CC       Fourier transform algorithms", IEEE Tran. ASSP,                 CC
CC       Vol. ASSP-35, No. 6, pp. 849-864, June 1987,                    CC
CC                                                                       CC
CC       This program may be used and distributed freely provided        CC
CC       this header is included and left intact                         CC
CC                                                                       CC
CC=======================================================================CC
        SUBROUTINE   RSRFFT(X,M)
        REAL X(2)
        N = 2**M
C-------Digit reverse counter-------------------------------------------C
        CALL RBITREV(X,M)
C-----Length two butterflies-------------------------------------------C
        IS = 1
        ID = 4
50      DO  60  I0  = IS, N, ID
                T1       = X(I0)
                X(I0)    = T1 + X(I0+1)
                X(I0+1)  = T1 - X(I0+1)
60      CONTINUE
        IS = 2*ID - 1
        ID = 4*ID
      IF (IS .LT. N) GOTO 50
```

```
C-------L shaped butterflies-------------------------------------------C
      N2 = 2
      DO  70  K = 2, M
         N2 = N2*2
         N4 = N2/4
         CALL RSTAGE (N,N2,N4,X(1),X(N4+1),X(2*N4+1),X(3*N4+1))
70    CONTINUE
      RETURN
      END
```

```
CC================================================================CC
CC                                                                CC
CC  Subroutine RSTAGE - the workhorse of the RFFT                 CC
CC      Computes one stage of a real-valued split-radix length N  CC
CC      transform.                                                CC
CC                                                                CC
CC  Author:                                                       CC
CC      H.V. Sorensen,   University of Pennsylvania,   Mar. 1987  CC
CC                                                                CC
CC      This program may be used and distributed freely provided  CC
CC      this header is included and left intact                   CC
CC                                                                CC
CC================================================================CC
      SUBROUTINE   RSTAGE(N,N2,N4,X1,X2,X3,X4)
      DIMENSION   X1(1), X2(1), X3(1), X4(1)
      N8 = N2/8
      IS = 0
      ID = N2*2
10       DO  20  I1 = IS+1, N, ID
            T1     = X4(I1) + X3(I1)
            X4(I1) = X4(I1) - X3(I1)
            X3(I1) = X1(I1) - T1
            X1(I1) = X1(I1) + T1
20       CONTINUE
         IS = 2*ID - N2
         ID = 4*ID
      IF (IS .LT. N) GOTO 10
C
      IF (N4-1) 100,100,30
30    IS = 0
      ID = N2*2
40       DO  50  I2 = IS+1+N8, N, ID
            T1     = (X3(I2) + X4(I2)) * .7071067811865475
            T2     = (X3(I2) - X4(I2)) * .7071067811865475
            X4(I2) =  X2(I2) - T1
            X3(I2) = -X2(I2) - T1
            X2(I2) =  X1(I2) - T2
            X1(I2) =  X1(I2) + T2
50       CONTINUE
         IS = 2*ID - N2
         ID = 4*ID
      IF (IS .LT. N) GOTO 40
```

```
C
      IF (N8-1) 100,100,60
60    E = 2.* 3.14159265358979323/N2
      SS1 = SIN(E)
      SD1 = SS1
      SD3 = 3.*SD1 - 4.*SD1**3
      SS3 = SD3
      CC1 = COS(E)
      CD1 = CC1
      CD3 = 4.*CD1**3 - 3.*CD1
      CC3 = CD3
      DO  90 J=2,N8
          IS = 0
          ID = 2*N2
          JN = N4 - 2*J + 2
70            DO  80 I1 = IS+J, N, ID
                  I2 = I1 + JN
                  T1 = X3(I1)*CC1 + X3(I2)*SS1
                  T2 = X3(I2)*CC1 - X3(I1)*SS1
                  T3 = X4(I1)*CC3 + X4(I2)*SS3
                  T4 = X4(I2)*CC3 - X4(I1)*SS3
                  T5 = T1 + T3
                  T3 = T1 - T3
                  T1 = T2 + T4
                  T4 = T2 - T4
                  X3(I1) = T1 - X2(I2)
                  X4(I2) = T1 + X2(I2)
                  X3(I2) = -X2(I1) - T3
                  X4(I1) =  X2(I1) - T3
                  X2(I2) = X1(I1) - T5
                  X1(I1) = X1(I1) + T5
                  X2(I1) = X1(I2) + T4
                  X1(I2) = X1(I2) - T4
80            CONTINUE
              IS = 2*ID - N2
              ID = 4*ID
          IF (IS .LT. N) GOTO 70
C
          T1  = CC1*CD1 - SS1*SD1
          SS1 = CC1*SD1 + SS1*CD1
          CC1 = T1
          T3  = CC3*CD3 - SS3*SD3
          SS3 = CC3*SD3 + SS3*CD3
          CC3 = T3
90    CONTINUE
C
100   RETURN
      END
```

```
CC==============================================================CC
CC                                                              CC
CC   Subroutine RBITREV(X,M):                                   CC
CC        Bit reverses the array X of length 2**M. It generates a   CC
CC        table ITAB (minimum length is SQRT(2**M) if M is even     CC
CC        or SQRT(2*2**M) if M is odd). ITAB need only be generated CC
CC        once for a given transform length.                   CC
CC                                                              CC
CC   Author:                                                    CC
CC        H.V. Sorensen,    University of Pennsylvania,  Aug. 1987   CC
CC                          Apra address: hv@ee.upenn.edu       CC
CC                                                              CC
CC   Reference:                                                 CC
CC        D. Evans, Tran. Assp, Vol. ASSP-35, No. 8. pp. 1120-1125,   CC
CC        Aug. 1987                                             CC
CC                                                              CC
CC        This program may be used and distributed freely provided   CC
CC        this header is included and left intact              CC
CC                                                              CC
CC==============================================================CC
      SUBROUTINE RBITREV(X,M)
      DIMENSION X(1),ITAB(256)
C-------Initialization of ITAB array------------------------------C
      M2 = M/2
      NBIT = 2**M2
      IF (2*M2.NE.M) M2 = M2+1
      ITAB(1) = 0
      ITAB(2) = 1
      IMAX = 1
      DO   10 LBSS = 2, M2
              IMAX = 2 * IMAX
              DO   10 I = 1, IMAX
                  ITAB(I)        = 2 * ITAB(I)
                  ITAB(I+IMAX) = 1 + ITAB(I)
10      CONTINUE
C-------The actual bit reversal-----------------------------------C
      DO   20   K = 2, NBIT
              J0 = NBIT * ITAB(K) + 1
              I = K
              J = J0
              DO   20 L = 2, ITAB(K)+1
                  T1    = X(I)
                  X(I) = X(J)
                  X(J) = T1
                  I = I + NBIT
                  J = J0 + ITAB(L)
20      CONTINUE
      RETURN
      END
```

Program 8: Inverse Real-Valued Split-Radix FFT

```
CC======================================================================CC
CC                                                                      CC
CC   Subroutine IRSRFFT(X,M):                                           CC
CC       A inverse real-valued, in-place, split-radix FFT program       CC
CC       Decimation-in-frequency, cos/sin in second loop                CC
CC       and computed recursively                                       CC
CC       Symmetric input in order:                                      CC
CC               [ Re(0),Re(1),....,Re(N/2),Im(N/2-1),...Im(1)]         CC
CC       The output is real-valued                                      CC
CC                                                                      CC
CC   Input/output                                                       CC
CC       X     Array of input/output (length >= N)                      CC
CC       M     Transform length is N = 2**M                             CC
CC                                                                      CC
CC   Calls:                                                             CC
CC       IRSTAGE, RBITREV                                               CC
CC                                                                      CC
CC   Author:                                                            CC
CC       H.V. Sorensen,   University of Pennsylvania,   Oct. 1985       CC
CC                        Arpa address: hvs@ee.upenn.edu                CC
CC   Modified:                                                          CC
CC       F. Bonzanigo,    ETH-Zurich,                   Sep. 1986       CC
CC       H.V. Sorensen,   University of Pennsylvania,   Mar. 1987       CC
CC                                                                      CC
CC   Reference:                                                         CC
CC       Sorensen, Jones, Heideman, Burrus: "Real-valued fast          CC
CC       Fourier transform algorithms, IEEE Tran. ASSP,                CC
CC       Vol. ASSP-35, No. 6, pp. 849-864, June 1987                   CC
CC                                                                      CC
CC       This program may be used and distributed freely provided      CC
CC       this header is included and left intact                        CC
CC                                                                      CC
CC======================================================================CC
        SUBROUTINE IRSRFFT(X,M)
        REAL X(1)
C-------L shaped butterflies------------------------------------------C
        N = 2**M
        N2 = 2*N
        DO  10 K = 1, M-1
            N2 = N2/2
            N4 = N2/4
            CALL IRSTAGE(N,N2,N4,X(1),X(N4+1),X(2*N4+1),X(3*N4+1))
  10    CONTINUE
C-----Length two butterflies------------------------------------------C
        IS = 1
        ID = 4
  70        DO  60 I1 = IS,N,ID
                T1        = X(I1)
```

```
                X(I1)    = T1 + X(I1+1)
                X(I1+1)  = T1 - X(I1+1)
60          CONTINUE
            IS = 2*ID - 1
            ID = 4*ID
        IF (IS.LT.N) GOTO 70
C-----Digit reverse counter-------------------------------------------------C
        CALL RBITREV(X,M)
C-----Divide by N-----------------------------------------------------------C
        DO  99 I=1,N
            X(I) = X(I)/N
99      CONTINUE
        RETURN
        END
CC==================================================================CC
CC                                                                  CC
CC  Subroutine IRSTAGE - the work-horse of the IRFFT                CC
CC      Computes a stage of an inverse real-valued split-radix      CC
CC      length N transform.                                         CC
CC                                                                  CC
CC  Author:                                                         CC
CC      H.V. Sorensen,   University of Pennsylvania,   Mar. 1987    CC
CC                                                                  CC
CC      This program may be used and distributed freely provided    CC
CC      this header is included and left intact                     CC
CC                                                                  CC
CC==================================================================CC
        SUBROUTINE  IRSTAGE(N,N2,N4,X1,X2,X3,X4)
        DIMENSION X1(1),X2(1),X3(1),X4(1)
        N8 = N4/2
        IS = 0
        ID = 2*N2
10          DO  20  I1 = IS+1,N,ID
                T1    = X1(I1) - X3(I1)
                X1(I1) = X1(I1) + X3(I1)
                X2(I1) = 2*X2(I1)
                T2    = 2*X4(I1)
                X4(I1) = T1 + T2
                X3(I1) = T1 - T2
20          CONTINUE
            IS = 2*ID - N2
            ID = 4*ID
        IF (IS .LT. N) GOTO 10
C
        IF (N4-1) 100,100,30
30      IS = 0
        ID = 2*N2
40          DO  50 I1 = IS+1+N8,N,ID
                T1    = (X2(I1) - X1(I1))*1.4142135623730950488
                T2    = (X4(I1) + X3(I1))*1.4142135623730950488
```

```
                    X1(I1) = X1(I1) + X2(I1)
                    X2(I1) = X4(I1) - X3(I1)
                    X3(I1) = -T2-T1
                    X4(I1) = -T2+T1
50          CONTINUE
            IS = 2*ID - N2
            ID = 4*ID
      IF (IS .LT. N-1) GOTO 40
C
      IF  (N8-1) 100,100,60
60    E   = 6.283185307179586/N2
      SS1 = SIN(E)
      SD1 = SS1
      SD3 = 3.*SD1-4.*SD1**3
      SS3 = SD3
      CC1 = COS(E)
      CD1 = CC1
      CD3 = 4.*CD1**3-3.*CD1
      CC3 = CD3
      DO  90  J = 2,N8
            IS = 0
            ID = 2*N2
            JN = N4 - 2*J + 2
70                DO  80  I1=IS+J,N,ID
                    I2 = I1 + JN
                    T1    = X1(I1) - X2(I2)
                    X1(I1) = X1(I1) + X2(I2)
                    T2    = X1(I2) - X2(I1)
                    X1(I2) = X2(I1) + X1(I2)
                    T3    = X4(I2) + X3(I1)
                    X2(I2) = X4(I2) - X3(I1)
                    T4    = X4(I1) + X3(I2)
                    X2(I1) = X4(I1) - X3(I2)
                    T5 = T1 - T4
                    T1 = T1 + T4
                    T4 = T2 - T3
                    T2 = T2 + T3
                    X3(I1) =  T5*CC1 + T4*SS1
                    X3(I2) = -T4*CC1 + T5*SS1
                    X4(I1) =  T1*CC3 - T2*SS3
                    X4(I2) =  T2*CC3 + T1*SS3
80                CONTINUE
            IS = 2*ID - N2
            ID = 4*ID
      IF (IS .LT. N) GOTO 70
C
      T1  = CC1*CD1 - SS1*SD1
      SS1 = CC1*SD1 + SS1*CD1
      CC1 = T1
      T3  = CC3*CD3 - SS3*SD3
```

```
            SS3 = CC3*SD3 + SS3*CD3
            CC3 = T3
90    CONTINUE
C
100   RETURN
      END
```

```
CC==============================================================CC
CC                                                              CC
CC   Subroutine RBITREV(X,M):                                   CC
CC        Bit reverses the array X of length 2**M. It generates a   CC
CC        table ITAB (minimum length is SQRT(2**M) if M is even     CC
CC        or SQRT(2*2**M) if M is odd). ITAB need only be generated CC
CC        once for a given transform length.                        CC
CC                                                              CC
CC   Author:                                                    CC
CC        H.V. Sorensen,    University of Pennsylvania, Aug. 1987   CC
CC                          Arpa address: hvs@ee.upenn.edu          CC
CC                                                              CC
CC   Reference:                                                 CC
CC        D. Evans, Tran. Assp, Vol. ASSP-35, No. 8. pp. 1120-1125,    CC
CC        Aug. 1987                                             CC
CC                                                              CC
CC        This program may be used and distributed freely provided   CC
CC        this header is included and left intact                    CC
CC                                                              CC
CC==============================================================CC
      SUBROUTINE RBITREV(X,M)
      DIMENSION X(1),ITAB(256)
C-------Initialization of ITAB array--------------------------------C
      M2 = M/2
      NBIT = 2**M2
      IF (2*M2.NE.M) M2 = M2+1
      ITAB(1) = 0
      ITAB(2) = 1
      IMAX = 1
      DO  10 LBSS = 2, M2
            IMAX = 2 * IMAX
            DO  10 I = 1, IMAX
                ITAB(I)      = 2 * ITAB(I)
                ITAB(I+IMAX) = 1 + ITAB(I)
10    CONTINUE
C-------The actual bit reversal-------------------------------------C
      DO  20  K = 2, NBIT
            J0 = NBIT * ITAB(K) + 1
            I = K
            J = J0
            DO  20  L = 2, ITAB(K)+1
                T1   = X(I)
                X(I) = X(J)
```

```
            X(J)  =  T1
            I  =  I  +  NBIT
            J  =  JO  +  ITAB(L)
20          CONTINUE
      RETURN
      END
```

8-B   Noncyclic Short-Length Convolution Modules

This section describes some noncyclic convolution modules using the ideas and notation from Section 8-1-3.

2×2 Noncyclic Convolution ($\alpha_0 = 0$, $\alpha_1 = 1$, $\alpha_2 = \infty$)

$$
\begin{bmatrix} y_0 \\ y_1 \\ y_2 \end{bmatrix}
=
\begin{bmatrix} 1 & 0 & 0 \\ -1 & 1 & -1 \\ 0 & 0 & 1 \end{bmatrix}
\begin{bmatrix} G_0 & 0 & 0 \\ 0 & G_1 & 0 \\ 0 & 0 & G_2 \end{bmatrix}
\begin{bmatrix} 1 & 0 \\ 1 & 1 \\ 0 & 1 \end{bmatrix}
\begin{bmatrix} x_0 \\ x_1 \end{bmatrix},
$$

where

$$
\begin{bmatrix} G_0 \\ G_1 \\ G_2 \end{bmatrix}
=
\begin{bmatrix} 1 & 0 \\ 1 & 1 \\ 0 & 1 \end{bmatrix}
\begin{bmatrix} h_0 \\ h_1 \end{bmatrix}.
$$

Algorithm (three multiplications and three additions):

$$t_0 = x_0 + x_1,$$

$$m_0 = x_0 G_0,$$

$$m_1 = t_0 G_1,$$

$$m_2 = x_1 G_2,$$

$$y_0 = m_0,$$

$$y_1 = m_1 - m_0 - m_2,$$

$$y_2 = m_2.$$

3 × 3 Noncyclic Convolution ($a_0 = 0$, $a_1 = 1$, $a_2 = -1$, $a_3 = 2$, $a_4 = \infty$)

$$
\begin{bmatrix} y_0 \\ y_1 \\ y_2 \\ y_3 \\ y_4 \end{bmatrix}
=
\begin{bmatrix}
2 & 0 & 0 & 0 & 0 \\
-1 & 2 & -2 & -1 & 2 \\
-1 & 1 & 3 & 0 & -1 \\
1 & -1 & -1 & 1 & -2 \\
0 & 0 & 0 & 0 & 1
\end{bmatrix}
\begin{bmatrix}
G_0 & 0 & 0 & 0 & 0 \\
0 & G_1 & 0 & 0 & 0 \\
0 & 0 & G_2 & 0 & 0 \\
0 & 0 & 0 & G_3 & 0 \\
0 & 0 & 0 & 0 & G_4
\end{bmatrix}
$$

$$
\cdot
\begin{bmatrix}
1 & 0 & 0 \\
1 & 1 & 1 \\
1 & -1 & 1 \\
1 & 2 & 4 \\
0 & 0 & 1
\end{bmatrix}
\begin{bmatrix} x_0 \\ x_1 \\ x_2 \end{bmatrix},
$$

where

$$
\begin{bmatrix} G_0 \\ G_1 \\ G_2 \\ G_3 \\ G_4 \end{bmatrix}
=
\begin{bmatrix}
\frac{1}{2} & 0 & 0 & 0 & 0 \\
0 & \frac{1}{2} & 0 & 0 & 0 \\
0 & 0 & \frac{1}{6} & 0 & 0 \\
0 & 0 & 0 & \frac{1}{6} & 0 \\
0 & 0 & 0 & 0 & 1
\end{bmatrix}
\begin{bmatrix}
1 & 0 & 0 \\
1 & 1 & 1 \\
1 & -1 & 1 \\
1 & 2 & 4 \\
0 & 0 & 1
\end{bmatrix}
\begin{bmatrix} h_0 \\ h_1 \\ h_2 \end{bmatrix}.
$$

Algorithm (five multiplications and 20 additions):

$$t_0 = x_1 + x_2, \qquad\qquad t_5 = m_1 + m_1,$$

$$t_1 = x_2 - x_1, \qquad\qquad t_6 = m_2 + m_2,$$

$$t_2 = x_0 + t_0, \qquad\qquad t_7 = m_4 + m_4,$$

$$t_3 = x_0 + t_1, \qquad\qquad t_8 = t_7 - m_0 - m_3,$$

$$t_4 = t_0 + t_0 + t_1 + t_2, \qquad t_9 = m_1 + m_2,$$

$$m_0 = x_0 G_0, \qquad\qquad y_0 = m_0 + m_0,$$

$$m_1 = t_2 G_1, \qquad\qquad y_1 = t_5 - t_6 + t_8,$$

$$m_2 = t_3 G_2, \qquad\qquad y_2 = t_2 + t_9 - y_0 - m_4,$$

$$m_3 = t_4 G_3, \qquad\qquad y_3 = -t_8 - t_9,$$

$$m_4 = x_2 G_4, \qquad\qquad y_4 = m_4.$$

8-C   Cyclic Short-Length Convolution Modules

This section describes several cyclic convolution modules using the ideas and notation from Section 8-1-3.

Length 2 Cyclic Convolution

$$\begin{bmatrix} y_0 \\ y_1 \end{bmatrix} = \begin{bmatrix} 1 & 1 \\ 1 & -1 \end{bmatrix} \begin{bmatrix} G_0 & 0 \\ 0 & G_1 \end{bmatrix} \begin{bmatrix} 1 & 1 \\ 1 & -1 \end{bmatrix} \begin{bmatrix} x_0 \\ x_1 \end{bmatrix},$$

where

$$\begin{bmatrix} G_0 \\ G_1 \end{bmatrix} = \begin{bmatrix} \frac{1}{2} & 0 \\ 0 & \frac{1}{2} \end{bmatrix} \begin{bmatrix} 1 & 1 \\ 1 & -1 \end{bmatrix} \begin{bmatrix} h_0 \\ h_1 \end{bmatrix}.$$

Algorithm (two multiplications and four additions):

$$t_0 = x_0 + x_1, \qquad m_1 = t_1 G_1,$$
$$t_1 = x_0 - x_1, \qquad y_0 = m_0 + m_1,$$
$$m_0 = t_0 G_0, \qquad y_1 = m_0 - m_1.$$

Length 3 Cyclic Convolution

$$\begin{bmatrix} y_0 \\ y_1 \\ y_2 \end{bmatrix} = \begin{bmatrix} 1 & 1 & 0 & -1 \\ 1 & -1 & -1 & 2 \\ 1 & 0 & 1 & -1 \end{bmatrix} \begin{bmatrix} G_0 & 0 & 0 & 0 \\ 0 & G_1 & 0 & 0 \\ 0 & 0 & G_2 & 0 \\ 0 & 0 & 0 & G_3 \end{bmatrix} \begin{bmatrix} 1 & 1 & 1 \\ 1 & 0 & -1 \\ 0 & 1 & -1 \\ 1 & 1 & -2 \end{bmatrix} \begin{bmatrix} x_0 \\ x_1 \\ x_2 \end{bmatrix},$$

where

$$\begin{bmatrix} G_0 \\ G_1 \\ G_2 \\ G_3 \end{bmatrix} = \begin{bmatrix} \frac{1}{3} & 0 & 0 & 0 \\ 0 & 1 & 0 & 0 \\ 0 & 0 & 1 & 0 \\ 0 & 0 & 0 & \frac{1}{3} \end{bmatrix} \begin{bmatrix} 1 & 1 & 1 \\ 1 & 0 & -1 \\ 0 & 1 & -1 \\ 1 & 1 & -2 \end{bmatrix} \begin{bmatrix} h_0 \\ h_1 \\ h_2 \end{bmatrix}.$$

Algorithm (four multiplications and 11 additions):

$$t_0 = x_0 + x_1, \qquad m_2 = t_3 G_2,$$

$$t_1 = x_0 - x_2, \qquad m_3 = t_4 G_3,$$

$$t_2 = t_0 + x_2, \qquad t_5 = m_1 - m_3,$$

$$t_3 = x_1 - x_2, \qquad t_6 = m_2 - m_3,$$

$$t_4 = t_1 + t_3, \qquad y_0 = m_0 + t_5,$$

$$m_0 = t_2 G_0, \qquad y_1 = m_0 - t_5 - t_6,$$

$$m_1 = t_1 G_1, \qquad y_2 = m_0 + t_6.$$

Length 4 Cyclic Convolution

$$
\begin{bmatrix} y_0 \\ y_1 \\ y_2 \\ y_3 \end{bmatrix}
=
\begin{bmatrix}
1 & 1 & 1 & 0 & -1 \\
1 & -1 & 1 & 1 & 0 \\
1 & 1 & -1 & 0 & 1 \\
1 & -1 & -1 & -1 & 0
\end{bmatrix}
\begin{bmatrix}
G_0 & 0 & 0 & 0 & 0 \\
0 & G_1 & 0 & 0 & 0 \\
0 & 0 & G_2 & 0 & 0 \\
0 & 0 & 0 & G_3 & 0 \\
0 & 0 & 0 & 0 & G_4
\end{bmatrix}
$$

$$
\cdot
\begin{bmatrix}
1 & 1 & 1 & 1 \\
1 & -1 & 1 & -1 \\
1 & 1 & -1 & -1 \\
1 & 0 & -1 & 0 \\
0 & 1 & 0 & -1
\end{bmatrix}
\begin{bmatrix} x_0 \\ x_1 \\ x_2 \\ x_3 \end{bmatrix},
$$

where

$$
\begin{bmatrix} G_0 \\ G_1 \\ G_2 \\ G_3 \\ G_4 \end{bmatrix}
=
\begin{bmatrix}
\frac{1}{4} & 0 & 0 & 0 & 0 \\
0 & \frac{1}{4} & 0 & 0 & 0 \\
0 & 0 & \frac{1}{2} & 0 & 0 \\
0 & 0 & 0 & \frac{1}{2} & 0 \\
0 & 0 & 0 & 0 & \frac{1}{2}
\end{bmatrix}
\begin{bmatrix}
1 & 1 & 1 & 1 \\
1 & -1 & 1 & -1 \\
1 & 1 & -1 & -1 \\
1 & 0 & -1 & 0 \\
0 & 1 & 0 & -1
\end{bmatrix}
\begin{bmatrix} h_0 \\ h_1 \\ h_2 \\ h_3 \end{bmatrix}.
$$

Algorithm (five multiplications and 15 additions):

$$t_0 = x_0 + x_2, \qquad m_3 = t_5 G_3,$$
$$t_1 = x_1 + x_3, \qquad m_4 = t_6 G_4,$$
$$t_2 = t_0 + t_1, \qquad t_7 = m_0 + m_1,$$
$$t_3 = t_0 - t_1, \qquad t_8 = m_0 - m_1,$$
$$t_4 = x_0 - x_2, \qquad t_9 = m_4 - m_3,$$
$$t_5 = x_1 - x_3, \qquad t_{10} = m_4 - m_2,$$
$$t_6 = t_4 + t_5, \qquad y_0 = t_7 + t_9,$$
$$m_0 = t_2 G_0, \qquad y_1 = t_8 + t_{10},$$
$$m_1 = t_3 G_1, \qquad y_2 = t_7 - t_9,$$
$$m_2 = t_4 G_2, \qquad y_3 = m_8 - t_{10}.$$

Length 5 Cyclic Convolution

Algorithm (10 multiplications and 31 additions):

$$t_0 = x_0 - x_4, \qquad\qquad m_6 = t_6 G_6,$$
$$t_1 = x_1 + x_4, \qquad\qquad m_7 = t_7 G_7,$$
$$t_2 = t_0 + t_1, \qquad\qquad m_8 = t_8 G_8,$$
$$t_3 = x_2 - x_4, \qquad\qquad m_9 = t_9 G_9,$$
$$t_4 = x_3 - x_4, \qquad\qquad t_{10} = m_0 + m_2,$$
$$t_5 = t_3 + x_4, \qquad\qquad t_{11} = m_1 + m_2,$$
$$t_6 = t_0 - t_3, \qquad\qquad t_{12} = m_3 + m_5,$$
$$t_7 = t_1 - t_4, \qquad\qquad t_{13} = m_4 + m_5,$$
$$t_8 = t_2 - t_5, \qquad\qquad t_{14} = m_6 + m_8,$$
$$t_9 = x_0 + x_1 + x_2 + x_3 + x_4, \qquad t_{15} = m_7 + m_8,$$
$$m_0 = t_0 G_0, \qquad\qquad y_0 = m_9 + t_{10} - t_{14},$$
$$m_1 = t_1 G_1, \qquad\qquad y_1 = m_9 - t_{10} - t_{11} - t_{12} - t_{13},$$
$$m_2 = t_2 G_2, \qquad\qquad y_2 = m_9 + t_{13} + t_{15},$$
$$m_3 = t_3 G_3, \qquad\qquad y_3 = m_9 + t_{12} + t_{14},$$
$$m_4 = t_4 G_4, \qquad\qquad y_5 = m_9 + t_{11} - t_{15},$$
$$m_5 = t_5 G_5,$$

where

$$G_0 = h_0 - h_2 + h_3 - h_4, \qquad G_1 = h_1 - h_2 + h_3 - h_4,$$

$$G_2 = \tfrac{1}{5}(3h_2 + 3h_4 - 2h_0 - 2h_1 - 2h_3), \qquad G_3 = h_1 + h_3 - h_0 - h_2,$$

$$G_4 = h_1 + h_4 - h_0 - h_2, \qquad G_5 = \tfrac{1}{5}(3h_0 - 2h_1 + 3h_2 - 2h_3 - 2h_4),$$

$$G_6 = h_3 - h_2, \qquad G_7 = h_1 - h_2, \qquad G_8 = \tfrac{1}{5}(4h_2 - h_0 - h_1 - h_3 - h_4),$$

$$G_9 = \tfrac{1}{5}(h_0 + h_1 + h_2 + h_3 + h_4).$$

Length 7 Cyclic Convolution
Algorithm (16 multiplications and 70 additions):

$$t_0 = x_0 - x_6,$$
$$t_1 = x_1 - x_6,$$
$$t_2 = x_2 - x_6,$$
$$t_3 = x_3 - x_6,$$
$$t_4 = x_4 - x_6,$$
$$t_5 = x_5 - x_6,$$
$$t_6 = t_2 + t_1,$$
$$t_7 = t_6 + t_0,$$
$$t_8 = x_2 - x_1,$$
$$t_9 = t_0 + t_8,$$
$$t_{10} = t_7 + t_6 + t_6 + t_8,$$
$$t_{11} = t_4 + t_5,$$
$$t_{12} = t_3 + t_{11},$$
$$t_{13} = x_5 - x_4,$$
$$t_{14} = t_3 + t_{13},$$
$$t_{15} = t_{12} + t_{11} + t_{11} + t_{13},$$
$$t_{16} = x_3 - x_6,$$
$$t_{17} = t_{12} - t_7,$$

$$t_{18} = t_{14} - t_9,$$
$$t_{19} = t_{15} - t_{10},$$
$$t_{20} = x_5 - x_2,$$
$$t_{21} = x_0 + x_1 + x_2,$$
$$t_{22} = t_{17} + t_{21} + t_{21} + x_6,$$
$$m_0 = t_0 G_0,$$
$$m_1 = t_7 G_1,$$
$$m_2 = t_9 G_2,$$
$$m_3 = t_{10} G_3,$$
$$m_4 = t_2 G_4,$$
$$m_5 = t_3 G_5,$$
$$m_6 = t_{12} G_6,$$
$$m_7 = t_{14} G_7,$$
$$m_8 = t_{15} G_8,$$
$$m_9 = t_5 G_9,$$
$$m_{10} = t_{16} G_{10},$$
$$m_{11} = t_{17} G_{11},$$
$$m_{12} = t_{18} G_{12},$$

$$m_{13} = t_{19} G_{13},$$

$$m_{14} = t_{20} G_{14},$$

$$m_{15} = t_{22} G_{15},$$

$$t_{23} = m_0 + m_{10},$$

$$t_{24} = m_1 + m_{11},$$

$$t_{25} = m_2 + m_{12},$$

$$t_{26} = m_3 + m_{13},$$

$$t_{27} = m_4 + m_{14},$$

$$t_{28} = m_5 - m_{10},$$

$$t_{29} = m_6 - m_{11},$$

$$t_{30} = m_7 - m_{12},$$

$$t_{31} = m_8 - m_{13},$$

$$t_{32} = m_9 - m_{14},$$

$$t_{33} = t_{24} + t_{26},$$

$$t_{34} = t_{25} + t_{33},$$

$$t_{35} = t_{23} + t_{34},$$

$$t_{36} = t_{26} + t_{33},$$

$$t_{37} = t_{36} - t_{25},$$

$$t_{38} = t_{36} + t_{26} + t_{26} + t_{27},$$

$$t_{39} = t_{38} + t_{25},$$

$$t_{40} = t_{38} + t_{35} + t_{36},$$

$$t_{41} = t_{29} + t_{31},$$

$$t_{42} = t_{41} + t_{30},$$

$$t_{43} = t_{42} + t_{28},$$

$$t_{44} = t_{41} + t_{31},$$

$$t_{45} = t_{44} - t_{30},$$

$$t_{46} = t_{44} + t_{31} + t_{31} + t_{32},$$

$$t_{47} = t_{46} + t_{30},$$

$$t_{48} = t_{46} + t_{43} + t_{44},$$

$$y_0 = m_{15} + t_{35},$$

$$y_1 = m_{15} - t_{40} - t_{48},$$

$$y_2 = m_{15} + t_{47},$$

$$y_3 = m_{15} + t_{45},$$

$$y_4 = m_{15} + t_{43},$$

$$y_5 = m_{15} - t_{39},$$

$$y_6 = m_{15} + t_{37},$$

where

$$G_0 = \tfrac{1}{2}(2h_0 + h_1 - 2h_2 - h_3 - 3h_4 - 2h_5 - h_6),$$

$$G_1 = \tfrac{1}{14}(3h_2 - h_1 - 4h_0 + 10h_3 - 11h_4 + 3h_5 + 10h_6),$$

$$G_2 = \tfrac{1}{6}(3h_2 - h_1 - 2h_3 - h_4 + 3h_5 - 2h_6),$$

$$G_3 = \tfrac{1}{6}(h_1 + h_4 - h_3 - h_6),$$

$$G_4 = h_0 - 2h_1 - h_2 + 3h_3 - h_5 + 2h_6,$$

$$G_5 = \tfrac{1}{2}(3h_1 - h_0 - 2h_2 - h_3 + 2h_4 + h_5 - 2h_6),$$

$$G_6 = \tfrac{1}{14}(10h_0 - h_{11} + 3h_2 + 10h_3 - 4h_4 - 11h_5 + 3h_6),$$

$$G_7 = \tfrac{1}{6}(3h_2 - 2h_0 - h_1 - 2h_3 - h_5 + 3h_6),$$

$$G_8 = \tfrac{1}{6}(h_1 - h_0 - h_3 + h_5),$$

$$G_9 = 3h_0 - 2h_1 - h_2 + 2h_3 + h_4 - 2h_5 - h_6,$$

$$G_{10} = \tfrac{1}{2}(h_1 - 2h_2 - h_3 + 2h_4),$$

$$G_{11} = \tfrac{1}{14}(5h_2 - 2h_0 - 9h_1 + 12h_3 - 2h_4 - 2h_5 - 2h_6),$$

$$G_{12} = \tfrac{1}{6}(3h_2 - h_1 - 2h_3),$$

$$G_{13} = \tfrac{1}{6}(h_1 - h_3),$$

$$G_{14} = h_0 - 2h_1 - h_2 + 2h_3,$$

$$G_{15} = \tfrac{1}{7}(h_0 + h_1 + h_2 + h_3 + h_4 + h_5 + h_6).$$

Length 8 Cyclic Convolution
Algorithm (14 multiplications and 46 additions):

$$
\begin{aligned}
t_0 &= x_0 + x_4, & t_{14} &= t_3 + t_7, \\
t_1 &= x_0 - x_4, & t_{15} &= t_1 + t_5, \\
t_2 &= x_1 + x_5, & t_{16} &= t_{15} - t_{14}, \\
t_3 &= x_1 - x_5, & t_{17} &= t_5 - t_7, \\
t_4 &= x_2 + x_6, & t_{18} &= t_1 - t_{13}, \\
t_5 &= x_2 - x_6, & t_{19} &= t_9 + t_{11}, \\
t_6 &= x_3 + x_7, & m_0 &= t_1 G_0, \\
t_7 &= x_3 - x_7, & m_1 &= t_3 G_1, \\
t_8 &= t_0 + t_4, & m_2 &= t_5 G_2, \\
t_9 &= t_0 - t_4, & m_3 &= t_7 G_3, \\
t_{10} &= t_2 + t_6, & m_4 &= t_9 G_4, \\
t_{11} &= t_2 - t_6, & m_5 &= t_{11} G_5, \\
t_{12} &= t_8 + t_{10}, & m_6 &= t_{12} G_6, \\
t_{13} &= t_8 - t_{10}, & m_7 &= t_{13} G_7,
\end{aligned}
$$

$$m_8 = t_{14} G_8, \qquad t_{30} = t_{27} + t_{20},$$

$$m_9 = t_{15} G_9, \qquad t_{31} = t_{28} - t_{22},$$

$$m_{10} = t_{16} G_{10}, \qquad t_{32} = t_{26} + t_{21},$$

$$m_{11} = t_{17} G_{11}, \qquad t_{33} = t_{29} + t_{23},$$

$$m_{12} = t_{18} G_{12}, \qquad t_{34} = t_{20} - t_{27},$$

$$m_{13} = t_{19} G_{13}, \qquad t_{35} = t_{28} + t_{24},$$

$$t_{20} = m_6 + m_7, \qquad t_{36} = t_{21} - t_{26},$$

$$t_{21} = m_6 - m_7, \qquad t_{37} = t_{37} + t_{25},$$

$$t_{22} = m_{11} + m_3, \qquad y_0 = t_{30} + t_{31},$$

$$t_{23} = m_{11} - m_2, \qquad y_1 = t_{32} + t_{33},$$

$$t_{24} = m_1 + m_{12}, \qquad y_2 = t_{34} + t_{35},$$

$$t_{25} = m_0 - m_{12}, \qquad y_3 = t_{36} + t_{37},$$

$$t_{26} = m_{13} + m_4, \qquad y_4 = t_{30} - t_{31},$$

$$t_{27} = m_{13} - m_5, \qquad y_5 = t_{32} - t_{33},$$

$$t_{28} = m_8 + m_{10}, \qquad y_6 = t_{34} - t_{35},$$

$$t_{29} = m_9 - m_{10}, \qquad y_7 = t_{36} - t_{37}.$$

$$G_0 = \tfrac{1}{2} (-h_0 - h_1 + h_2 + h_3 + h_4 + h_5 - h_6 - h_7)$$

$$G_1 = \tfrac{1}{2} (-h_0 + h_1 + h_2 + h_3 + h_4 - h_5 - h_6 - h_7)$$

$$G_2 = \tfrac{1}{2} (h_0 + h_1 + h_2 + h_3 - h_4 - h_5 - h_6 - h_7)$$

$$G_3 = \tfrac{1}{2} (h_0 + h_1 + h_2 - h_3 - h_4 - h_5 - h_6 + h_7)$$

$$G_4 = \tfrac{1}{4} (-h_0 + h_1 + h_2 - h_3 - h_4 + h_5 + h_6 - h_7)$$

$$G_5 = \tfrac{1}{4} (h_0 + h_1 - h_2 - h_3 + h_4 + h_5 - h_6 - h_7)$$

$$G_6 = \tfrac{1}{8} (h_0 + h_1 + h_2 + h_3 + h_4 + h_5 + h_6 + h_7)$$

$$G_7 = \tfrac{1}{8} (h_0 - h_1 + h_2 - h_3 + h_4 - h_5 + h_6 - h_7)$$

$$G_8 = \tfrac{1}{2} (h_0 - h_3 - h_4 + h_7)$$

$$G_9 = \tfrac{1}{2}(h_0 + h_1 - h_4 - h_5)$$

$$G_{10} = \tfrac{1}{2}(h_0 - h_4)$$

$$G_{11} = \tfrac{1}{2}(h_0 + h_2 - h_4 - h_6)$$

$$G_{12} = \tfrac{1}{2}(-h_0 + h_2 + h_4 - h_6)$$

$$G_{13} = \tfrac{1}{2}(h_0 - h_2 + h_4 - h_6)$$

8-D Short-Length DFT Modules

This section describes some short-length DFT modules using the ideas and notation from Section 8-2-8. Several of the multiplications in each module will be by one and hence trivial, but when nested, these must be counted also. Both the total number of multiplications and the number of nontrivial multiplications are given for each module.

Length 2 DFT Module

$$\begin{bmatrix} X_0 \\ X_1 \end{bmatrix} = \begin{bmatrix} 1 & 0 \\ 0 & 1 \end{bmatrix} \begin{bmatrix} G_0 & 0 \\ 0 & G_1 \end{bmatrix} \begin{bmatrix} 1 & 1 \\ 1 & -1 \end{bmatrix} \begin{bmatrix} x_0 \\ x_1 \end{bmatrix},$$

where

$$G_0 = G_1 = 1.$$

Algorithm (two (zero nontrivial) multiplications and two additions):

$$X_0 = G_0(x_0 + x_1),$$
$$X_1 = G_1(x_0 - x_1).$$

Length 3 DFT Module

$$\begin{bmatrix} X_0 \\ X_1 \\ X_2 \end{bmatrix} = \begin{bmatrix} 1 & 0 & 0 \\ 1 & 1 & -j \\ 1 & 1 & j \end{bmatrix} \begin{bmatrix} G_0 & 0 & 0 \\ 0 & G_1 & 0 \\ 0 & 0 & G_2 \end{bmatrix} \begin{bmatrix} 1 & 1 & 1 \\ 0 & 1 & 1 \\ 0 & 1 & -1 \end{bmatrix} \begin{bmatrix} x_0 \\ x_1 \\ x_2 \end{bmatrix},$$

where

$$G_0 = 1, \qquad G_1 = \cos\left(\frac{2\pi}{3}\right) - 1, \qquad G_2 = \sin\left(\frac{2\pi}{3}\right).$$

Algorithm (three (two nontrivial) multiplications and six additions):

$$t_0 = x_1 + x_2, \qquad m_2 = t_2 G_2,$$
$$t_1 = x_0 + t_0, \qquad t_3 = m_0 + m_1,$$
$$t_2 = x_1 - x_2, \qquad X_0 = m_0,$$
$$m_0 = t_1 G_0, \qquad X_1 = t_3 - jm_2,$$
$$m_1 = t_0 G_1, \qquad X_2 = t_3 + jm_2.$$

Length 4 DFT Module

$$
\begin{bmatrix} X_0 \\ X_1 \\ X_2 \\ X_3 \end{bmatrix}
=
\begin{bmatrix}
1 & 0 & 0 & 0 \\
0 & 0 & 1 & -j \\
0 & 1 & 0 & 0 \\
0 & 0 & 1 & j
\end{bmatrix}
\begin{bmatrix}
G_0 & 0 & 0 & 0 \\
0 & G_1 & 0 & 0 \\
0 & 0 & G_2 & 0 \\
0 & 0 & 0 & G_3
\end{bmatrix}
$$

$$
\cdot
\begin{bmatrix}
1 & 1 & 1 & 1 \\
1 & -1 & 1 & -1 \\
1 & 0 & -1 & 0 \\
0 & 1 & 0 & -1
\end{bmatrix}
\begin{bmatrix} x_0 \\ x_1 \\ x_2 \\ x_3 \end{bmatrix},
$$

where

$$G_0 = G_1 = G_2 = G_3 = 1.$$

Algorithm (four (zero nontrivial) multiplications and eight additions):

$$t_0 = x_0 + x_2, \qquad m_1 = t_3 G_1,$$
$$t_1 = x_1 + x_3, \qquad m_2 = t_4 G_2,$$
$$t_2 = t_0 + t_1, \qquad m_3 = t_5 G_3,$$
$$t_3 = t_0 - t_1, \qquad X_0 = m_0,$$
$$t_4 = x_0 - x_2, \qquad X_1 = m_2 - jm_3,$$
$$t_5 = x_1 + x_3, \qquad X_2 = m_1,$$
$$m_0 = t_2 G_0, \qquad X_3 = m_2 + jm_3.$$

Length 5 DFT Module

$$
\begin{bmatrix} X_0 \\ X_1 \\ X_2 \\ X_3 \\ X_4 \end{bmatrix}
=
\begin{bmatrix}
1 & 0 & 0 & 0 & 0 & 0 \\
1 & 1 & 1 & -j & j & 0 \\
1 & 1 & -1 & -j & 0 & -j \\
1 & 1 & -1 & j & 0 & j \\
1 & 1 & 1 & j & -j & 0
\end{bmatrix}
$$

$$
\begin{bmatrix}
G_0 & 0 & 0 & 0 & 0 & 0 \\
0 & G_1 & 0 & 0 & 0 & 0 \\
0 & 0 & G_2 & 0 & 0 & 0 \\
0 & 0 & 0 & G_3 & 0 & 0 \\
0 & 0 & 0 & 0 & G_4 & 0 \\
0 & 0 & 0 & 0 & 0 & G_5
\end{bmatrix}
\begin{bmatrix}
1 & 1 & 1 & 1 & 1 \\
0 & 1 & 1 & 1 & 1 \\
0 & 1 & -1 & -1 & 1 \\
0 & 1 & -1 & 1 & -1 \\
0 & 0 & -1 & 1 & 0 \\
0 & 1 & 0 & 0 & -1
\end{bmatrix}
$$

$$
\begin{bmatrix} x_0 \\ x_1 \\ x_2 \\ x_3 \\ x_4 \end{bmatrix}
$$

where

$$
G_0 = 1, \qquad G_1 = \frac{1}{2}\left\{ \cos\left(\frac{2\pi}{5}\right) + \cos\left(\frac{4\pi}{5}\right) \right\} - 1,
$$

$$
G_2 = \frac{1}{2}\left\{ \cos\left(\frac{2\pi}{5}\right) - \cos\left(\frac{4\pi}{5}\right) \right\},
$$

$$
G_3 = \sin\left(\frac{2\pi}{5}\right),
$$

$$
G_4 = \sin\left(\frac{2\pi}{5}\right) + \sin\left(\frac{4\pi}{5}\right),
$$

$$
G_5 = \sin\left(\frac{4\pi}{5}\right) - \sin\left(\frac{2\pi}{5}\right).
$$

Algorithm (six (five nontrivial) multiplications and 17 additions):

$$t_0 = x_1 + x_4, \qquad m_4 = t_3 G_4,$$

$$t_1 = x_1 - x_4, \qquad m_5 = t_1 G_5,$$

$$t_2 = x_3 + x_2, \qquad t_8 = m_0 + m_1,$$

$$t_3 = x_3 - x_2, \qquad t_9 = t_8 + m_2,$$

$$t_4 = t_0 + t_2, \qquad t_{10} = t_8 - m_2,$$

$$t_5 = t_0 - t_2, \qquad t_{11} = m_3 - m_4,$$

$$t_6 = x_0 + t_4, \qquad t_{12} = m_3 + m_5,$$

$$t_7 = t_3 + t_1, \qquad X_0 = m_0,$$

$$m_0 = t_6 G_0, \qquad X_1 = t_9 - jt_{11},$$

$$m_1 = t_4 G_1, \qquad X_2 = t_{10} - jt_{12},$$

$$m_2 = t_5 G_2, \qquad X_3 = t_{10} + jt_{12},$$

$$m_3 = t_7 G_3, \qquad X_4 = t_9 + jt_{11}.$$

Length 7 DFT Module

Algorithm (nine (eight nontrivial) multiplications and 36 additions):

$$t_0 = x_1 + x_6, \qquad t_{12} = t_1 - t_5,$$

$$t_1 = x_1 - x_6, \qquad t_{13} = t_5 - t_3,$$

$$t_2 = x_2 + x_5, \qquad t_{14} = t_3 - t_1,$$

$$t_3 = x_2 - x_5, \qquad m_0 = t_7 G_0,$$

$$t_4 = x_4 + x_3, \qquad m_1 = t_6 G_1,$$

$$t_5 = x_4 - x_3, \qquad m_2 = t_8 G_2,$$

$$t_6 = t_0 + t_2 + t_4, \qquad m_3 = t_9 G_3,$$

$$t_7 = t_6 + x_0, \qquad m_4 = t_{10} G_4,$$

$$t_8 = t_6 - t_4, \qquad m_5 = t_{11} G_5,$$

$$t_9 = t_4 - t_2, \qquad m_6 = t_{12} G_6,$$

$$t_{10} = t_2 - t_0, \qquad m_7 = t_{13} G_7,$$

$$t_{11} = t_1 + t_3 + t_5, \qquad m_8 = t_{14} G_8,$$

$$t_{15} = m_0 + m_1,$$

$$t_{16} = m_2 + m_3,$$

$$t_{17} = m_2 + m_4,$$

$$t_{18} = m_4 - m_3,$$

$$t_{19} = m_5 + m_6 + m_7,$$

$$t_{20} = m_5 - m_6 - m_8,$$

$$t_{21} = m_7 - m_5 - m_8,$$

$$t_{22} = t_{15} + t_{16},$$

$$t_{23} = t_{15} - t_{17},$$

$$t_{24} = t_{15} + t_{18},$$

$$X_0 = m_0,$$

$$X_1 = t_{22} - jt_{19},$$

$$X_2 = t_{23} - jt_{20},$$

$$X_3 = t_{24} + jt_{21},$$

$$X_4 = t_{24} + jt_{21},$$

$$X_5 = t_{23} + jt_{20},$$

$$X_6 = t_{22} + jt_{19},$$

where

$$G_0 = 1,$$

$$G_1 = \frac{1}{3}\left\{\cos\left(\frac{2\pi}{7}\right) + \cos\left(\frac{4\pi}{7}\right) + \cos\left(\frac{6\pi}{7}\right)\right\} - 1,$$

$$G_2 = \frac{1}{3}\left\{2\cos\left(\frac{2\pi}{7}\right) - \cos\left(\frac{4\pi}{7}\right) - \cos\left(\frac{6\pi}{7}\right)\right\},$$

$$G_3 = \frac{1}{3}\left\{\cos\left(\frac{2\pi}{7}\right) - 2\cos\left(\frac{4\pi}{7}\right) + \cos\left(\frac{6\pi}{7}\right)\right\},$$

$$G_4 = \frac{1}{3}\left\{\cos\left(\frac{2\pi}{7}\right) + \cos\left(\frac{4\pi}{7}\right) - 2\cos\left(\frac{6\pi}{7}\right)\right\},$$

$$G_5 = \frac{1}{3}\left\{\sin\left(\frac{2\pi}{7}\right) + \sin\left(\frac{4\pi}{7}\right) - \sin\left(\frac{6\pi}{7}\right)\right\},$$

$$G_6 = \frac{1}{3}\left\{2\sin\left(\frac{2\pi}{7}\right) - \sin\left(\frac{4\pi}{7}\right) + \sin\left(\frac{6\pi}{7}\right)\right\},$$

$$G_7 = \frac{1}{3}\left\{\sin\left(\frac{2\pi}{7}\right) - 2\sin\left(\frac{4\pi}{7}\right) - \sin\left(\frac{6\pi}{7}\right)\right\},$$

$$G_8 = \frac{1}{3}\left\{\sin\left(\frac{2\pi}{7}\right) + \sin\left(\frac{4\pi}{7}\right) + \sin\left(\frac{6\pi}{7}\right)\right\}.$$

Length 8 DFT Module

Algorithm (eight (two nontrivial) multiplications and 26 additions):

$$t_0 = x_0 + x_4, \qquad t_{18} = m_6 + m_7,$$

$$t_1 = x_0 - x_4, \qquad t_{19} = m_6 - m_7,$$

$$t_2 = x_1 + x_5, \qquad m_0 = t_{12} G_0,$$

$$t_3 = x_1 - x_5, \qquad m_1 = t_{13} G_1,$$

$$t_4 = x_2 + x_6, \qquad m_2 = t_9 G_2,$$

$$t_5 = x_2 - x_6, \qquad m_3 = t_1 G_3,$$

$$t_6 = x_3 + x_7, \qquad m_4 = t_{15} G_4,$$

$$t_7 = x_3 - x_7, \qquad m_5 = t_{11} G_5,$$

$$t_8 = t_0 + t_4, \qquad m_6 = t_5 G_6,$$

$$t_9 = t_0 - t_4, \qquad m_7 = t_{14} G_7,$$

$$t_{10} = t_2 + t_6, \qquad X_0 = m_0,$$

$$t_{11} = t_2 - t_6, \qquad X_1 = t_{16} - jt_{18},$$

$$t_{12} = t_8 + t_{10}, \qquad X_2 = m_2 - jm_5,$$

$$t_{13} = t_8 - t_{10}, \qquad X_3 = t_{17} - jt_{19},$$

$$t_{14} = t_3 + t_7, \qquad X_4 = m_1,$$

$$t_{15} = t_3 - t_7, \qquad X_5 = t_{17} + jt_{19},$$

$$t_{16} = m_3 + m_4, \qquad X_6 = m_2 + jm_5,$$

$$t_{17} = m_3 - m_4, \qquad X_7 = t_{16} + jt_{18},$$

where

$$G_0 = G_1 = G_2 = G_3 = G_5 = G_6 = 1,$$

$$G_4 = \cos\left(\frac{2\pi}{8}\right), \qquad G_7 = \sin\left(\frac{2\pi}{8}\right).$$

9 Finite Arithmetic Concepts

W. KENNETH JENKINS
Department of Electrical and Computer Engineering
University of Illinois, Urbana

9-1 INTRODUCTION

In the first century A.D. the Chinese scholar Sun-Tsu wrote an obscure verse that described a rule to determine a number having the remainders 2, 3, and 2 when divided by the numbers 3, 5, and 7, respectively [GR66; KN69; TA85a]. This was perhaps the first time in history that the notion of multiple-moduli residue number systems (RNS) arithmetic was documented in a published work. Beginning with the rapid evolution of the electronic digital computer in the 1950s, RNS arithmetic has become the object of intense interest by modern engineers and scientists interested in both general computing and digital signal processing.

RNS arithmetic has a modular structure that leads naturally to modularity and parallelism in digital hardware. The carry-free structure of RNS arithmetic has attracted considerable attention over the years for high-speed computation, especially when high precision (large wordlength) is required. The carry-free property holds only for addition, subtraction, and multiplication, that is, for arithmetic operations that can be performed naturally with integers. Because general division and division by a constant (scaling) do not directly produce integer results, they both turn out to be more complicated operations that involve interactions among all residue digits in the operands. Due to the fact that there is no ordered significance among residue digits, it is not as simple to determine relative size and polarity as in weighted number systems. In general, RNS operations fall into two categories: (1) simple operations (addition, subtraction, multiplication), where the carry-free property holds, and (2) complicated operations (division, scaling, magnitude comparison), where it does not hold. In recent years RNS arithmetic has been used more for digital signal processing, where the simple operations domi-

Handbook for Digital Signal Processing, Edited by Sanjit K. Mitra and James F. Kaiser.
ISBN 0-471-61995-7 © 1993 John Wiley & Sons, Inc.

nate, than for general computing, where all operations tend to be equally important.

The first published report on modern applications of RNS arithmetic in electronic computers was that of Svoboda and Valach in Czechoslovakia, first appearing in 1955 [SV55, SV57, SV58]. They conducted experiments on a small-moduli RNS computer constructed from the vacuum tube technology of the day. Vacuum tube computers were slow and unreliable, largely due to excessive heat buildup that resulted from large banks of vacuum tubes that served as elemental switching devices. Svoboda and Valach applied RNS theory in an effort to increase speed and control errors through RNS error correction codes. At approximately the same time Garner (GA59) published work describing the fundamentals of RNS arithmetic and suggested that it could be used to advantage in high-speed electronic computers. During the late 1950s and early 1960s extensive research on the subject was conducted at the Harvard Computation Laboratory [AI59], RCA [BA61], Westinghouse [SL63], and Lockheed Missiles and Space Company [TA62] in an effort to further develop the theory and establish its utility in general purpose computers. Much of the interest behind research at this time was oriented toward airborne or spaceborne computers, where power is at a premium and reliability is of crucial importance. In 1967 Szabo and Tanaka [SZ67] published the first definitive book on the subject. Until it went out of print in the mid-1970s this book was the only complete reference that was available on the subject. Unfortunately, the difficulties with general division, sign detection, and magnitude comparison tended to nullify the advantages of high-speed multiplication, addition, and subtraction in general purpose computers where test-and-branch operations are frequently required. When RNS techniques did not provide clear advantages for general computers, researchers began to investigate their use in more specialized computational problems, which are dominated by large numbers of repetitive operations.

In the mid-1960s digital signal processing began to emerge as a discipline distinct from digital computing. Digital hardware was becoming more integrated, and low-cost memories and arithmetic-logic units (ALUs) became available on a single chip. The digital filter, a simple assemblage of shift registers, adders, multipliers, and A/D and D/A converters began to make its way out of the telephone communications industry where it was developed and was quickly assimilated into avionics systems and consumer products. The microprocessor arrived in 1970, stimulating a burst of interest in new digital signal processing applications. It was during this time that a number of researchers recognized that RNS principles may offer important advantages for the specialized needs of signal processing, where typical algorithms are dominated by highly repetitive sequences of multiply and add operations. For example, digital filtering, digital correlation, FFT processing, and matrix manipulations are dominated by large numbers of additions, subtractions, and multiplications. One of the earliest publications describing the use of RNS arithmetic in a signal processing application was that of Cheney [CH61], in which he presented the design of a digital correlator based entirely on residue arithmetic. Soderstrand et al. [SO86] provide a modern survey on the theory and applications of RNS arithmetic to problems in signal processing.

9-2 FUNDAMENTALS OF MODULAR ARITHMETIC

This section introduces some important notations and then reviews the basics of simple finite fields and finite rings. Throughout the discussion, the notation $\langle x \rangle_m$ is used to denote the least positive remainder of x modulo m; that is, x is represented by r, where $x = km + r$ and where k is the largest integer such that $0 \leq r \leq m - 1$. A simplified notation is used when a set of moduli $\{m_1, m_2, \ldots, m_L\}$ is used to define a multiple-moduli RNS and a number x is represented in RNS form. In this case, $\langle x \rangle_{m_i}$ is abbreviated x_i. These conventions are used exclusively throughout the discussions in this chapter.

9-2-1 Finite Field Arithmetic

The set $S_m = \{0, 1, \ldots, m - 1\}$ together with modulo m addition and multiplication forms a finite algebra, denoted $\{S_m, +, \cdot\}$. Two distinctly different structures result, depending on whether m is a prime or a composite integer. If $m = p$ is prime, $GF(p) = \{S_p, +, \cdot\}$ is a finite (Galois) field. A field is, relatively speaking, a very sophisticated algebraic structure. For every element of the field, $a \in S_p$, there exists a multiplicative inverse $a^{-1} \in S_p$ such that $\langle a \cdot a^{-1} \rangle_p = 1$. It is important to note that, in a finite field, multiplication by a multiplicative inverse does not correspond to division as it does in the conventional real number field with which we are familiar. For example, in the field $GF(5)$, the multiplicative inverse of 2 is $2^{-1} = 3$. But $\langle 2^{-1} \cdot 3 \rangle_5 = 4$, which is not equal to $\frac{3}{2}$.

It is well known from number theory that every finite field contains a generator $g \in S_p$ that multiplicatively generates all nonzero field elements, that is, the first $p - 1$ powers of g, together with 0, constitute the set S_p. For example, in $GF(7)$, $g = 3$ is found to be a generator. Hence a one-to-one correspondence can be established as follows:

$$S_7 = \{0, 3^1, 3^2, 3^3, 3^4, 3^5, 3^6\} = \{0, 3, 2, 6, 4, 5, 1\}.$$

Note that in the first representation of S_7 the exponents of the generator, referred to in some literature as indices, form a set of finite field logarithms that can be arranged into a one-to-one correspondence with the nonzero elements of S_7. This correspondence can be used to construct a pair of log–antilog tables that are used to replace general multiplication of nonzero elements by an index addition. If $x, y \in GF(p)$, and if $x = g^\alpha$ and $y = g^\beta$, then

$$\langle x \cdot y \rangle_p = g^{\langle \alpha + \beta \rangle_{(p-1)}}. \tag{9.1}$$

This mechanism provides a trade-off between memory and computational complexity that is considerably less memory intensive than using completely stored multiplication tables. Since the 0 element does not have a logarithm, multiplication by 0 must always be handled as a special case. This is usually done by simply storing a "flag" in the log table for the 0 element, and this flag is subsequently used to load a 0 into the final product register. Note that in the case of a prime modulus, there will always be $2^n - p$ states at the top of the address space in the stored tables that are unused. One of these states can be used to store the flag, so that the special handling of the 0 does not increase the overall hardware complexity.

9-2-2   Finite Ring Arithmetic

If m is a composite integer rather than a prime, then $R(m) = \{S_m, +, \cdot\}$ becomes a finite ring in which certain elements do not have multiplicative inverses. The algebraic structure of a finite ring is considerably weaker than that of a field in that a generator does not exist, implying that a complete set of indices does not exist for a ring. Any element in the ring that has a nonunity factor in common with the ring modulus will have neither a logarithm nor a multiplicative inverse.

The modulus $m = 2^n$ is particularly important because for this case the elements of S_m are exactly represented in an n-bit wordlength, and because modulo 2^n addition and multiplication are simple to mechanize by simply retaining only the n last significant bits after a conventional binary operation. However, since 2^n is always composite, the algebraic structure resulting from this choice is always a ring. Other important moduli are those of the form $m = 2^n - 1$ and $m = 2^n + 1$, often called the "diminished radix-2" and "augmented radix-2" forms. Modular arithmetic for a diminished radix-2 modulus is the same as the carry-add arithmetic of 1's-complement binary codes. Augmented radix-2 moduli lead to carry-subtract arithmetic, which is similar in complexity to the carry-add type.

Modular addition for an arbitrary modulus m is rather simple to implement with a standard 2's-complement adder. Define 2^n to be the smallest power-of-2 that is larger than or equal to m; that is, $2^n = \min \{2^k \ni : 2^k \geq m\}$. The first step in forming $Z = \langle x + y \rangle_m$ is to form $Z' = x + y$ in a standard 2's-complement adder. If $Z \in [0, m - 1]$, then overflow has not occurred and $Z = Z'$. If $Z \in [m, 2^n - 1]$, then Z' must be "corrected" by subtracting m; that is, $Z = Z' - m$. The overflow correction can be implemented by adding the 2's-complement of m in the same adder in which the Z' was formed, so that $\langle Z \rangle_m = \langle Z + 2^n - m \rangle_{2^n}$. The constant $C = 2^n - m$ is a fixed constant that has been called a *generalized end-around carry*. Note that for $m = 2^n$, $C = 0$. Similarly, $m = 2^n - 1$ implies $C = 1$ and $m = 2^n + 1$ implies $C = -1$. When characterized in this way, it is clear that the popular binary codes used in commercially produced computers result in slightly different forms of modular arithmetic in which the end-around carry has a particularly simple form.

Since semiconductor read-only memories (ROMs) have become readily available, for modest sized wordlengths it may be simpler to use Z' as an address to a ROM that contains a stored correction table and produces Z by a simple memory fetch, as illustrated in Figure 9-1. In many applications a buffer can be placed between the adder and ROM and the structure can be pipelined so that the throughput rate is limited by just one add cycle. Another circuit for modular addition based on an end-around carry addition is shown in Figure 9-2. The comparator deter-

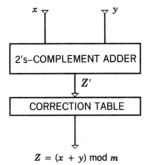

$$Z = (x + y) \bmod m$$

FIGURE 9-1 A modular adder configured with a 2's-complement adder and a correction table stored in ROM.

mines whether the 2's-complement sum exceeds $m - 1$, while at the same time the overflow adder adds the end-around carry in a second 2's-complement adder. The output of the comparator then controls the multiplexer to produce either the original sum or the corrected sum. Although this approach seems to perform an unneeded addition every time the original sum is correct, it has been observed that in high-speed VLSI designs the combination of silicon area required for the comparator, overflow adder, and multiplexer is significantly less than that required for the correction. However, note that other operations that may be required, such as index encoding or decoding, can be combined into the ROM without increasing the hardware requirements. Therefore it has been found that both approaches to modular addition are quite useful, depending on the particular tasks that are required by the specific application.

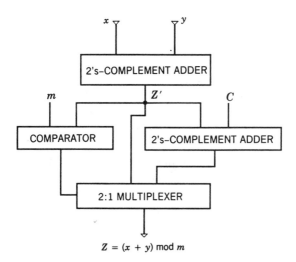

$$Z = (x + y) \bmod m$$

FIGURE 9-2 A modular adder configured with a 2's-complement adder and an overflow correction adder.

9-2-3   Multiple-Moduli Residue Number Systems

A more general class of modular systems is constructed as a direct sum of several simple modular structures (either fields or rings) that have moduli that are pairwise relatively prime integers (no two have a nonunity common factor). If $M = m_1 m_2 \cdots m_L$, where $\{m_1, m_2, \ldots, m_L\}$ is the set of moduli, then

$$R(M) \approx R_1(m_1) + \cdots + R_L(m_L), \tag{9.2}$$

where "$\approx$" is used to denote that $R(M)$ is an integer ring that is isomorphic to the direct sum of the L subrings $R(m_i)$, $i = 1, \ldots, L$. The direct sum representation of $R(M)$, as denoted by the right side of Eq. (9.2), is typically called a *nonredundant residue number system*. The interval $[0, M - 1]$ is called the *legitimate range* of the RNS because it represents the useful computational range of the number system. Usually the legitimate range is partitioned into positive and negative regions so that signed arithmetic can be handled in complement notations. If M is odd, the dynamic range of a RNS becomes $[-(M - 1)/2, (M - 1)/2]$; if M is even, it is $[-M/2, (M/2 - 1)]$. Each natural integer X in the dynamic range is mapped onto the legitimate range and represented as an L-tuple of residue digits $(x_1, x_2, \ldots, x_L)$, where $x_i = \langle X \rangle_{m_i}$ for X in the positive portion of the dynamic range, and $x_i = m_i - \langle X \rangle_{m_i}$ for X in the negative portion of the dynamic range. Note that if M is odd $[-M/2, -1]$ maps onto $[(M + 1)/2, M - 1]$; if M is even $[-M/2, -1]$ maps onto $[M/2, M - 1]$. Hence negative numbers map onto the upper half of the legitimate range through RNS-complement coding.

Residue arithmetic is defined by

$$(x_1, x_2, \ldots, x_L) \circ (y_1, y_2, \ldots, y_L) = (z_1, z_2, \ldots, z_L), \tag{9.3}$$

where $z_i = \langle x_i \circ y_i \rangle_{m_i}$, $i = 1, \ldots, L$ and $\circ$ denotes one of addition, subtraction, or multiplication. Since z_i is determined entirely from x_i and y_i, RNS arithmetic is carry-free in the sense that there is no propagation of information from the ith channel to the qth channel, $i \neq q$.

As an example, consider the RNS defined by $m_1 = 7$, $m_2 = 9$, $m_3 = 11$, and $m_4 = 13$. For this case, $M = 9009$. Thus the legitimate range is $[0, 9008]$, and the dynamic range is $[-4504, 4504]$. A positive number $X = 300$ is encoded as $(6, 3, 3, 1)$, whereas $Y = -2$ is encoded as $(5, 7, 9, 11)$. Then $X + Y \approx (4, 1, 1, 12) \approx 298$, and $X \cdot Y \approx (2, 3, 5, 11) \approx -600$. Note that the signed numbers are easily manipulated by exactly the same rules as positive numbers after the initial complement encoding is done. Note also that it is rather difficult to see that $(5, 7, 9, 11)$ represents a negative number because there is no explicit sign bit in the code. Therefore it is easy to calculate with signed numbers in the RNS code but difficult to do sign testing often needed to control data-dependent decisions.

9-2-4   Residue-to-Binary and Binary-to-Residue Conversion

The one-to-one relationship between $R(M)$ and $R(m_1) + \cdots + R(m_L)$ is known as the *Chinese Remainder Theorem* (CRT) in honor of its ancient Chinese origins.

A mathematical representation of the CRT is given by Eq. (9.4a), or alternatively by Eq. (9.4b):

$$\langle X \rangle_M = \left\langle \sum_{i=1}^{L} \hat{m}_i \langle \hat{m}_i^{-1} x_i \rangle_{m_i} \right\rangle_M, \tag{9.4a}$$

$$\langle X \rangle_M = \left\langle \sum_{i=1}^{L} \hat{m}_i \langle \hat{m}_i^{-1} \rangle_{m_i} x_i \right\rangle_M, \tag{9.4b}$$

where $\hat{m}_i = M/m_i$ and $\hat{m}_i^{-1}$ is the multiplicative inverse of $\hat{m}_i$ mod m_i; that is, $\langle \hat{m}_i \hat{m}_i^{-1} \rangle_{m_i} = 1$. Both forms of the CRT are encountered in the literature, although the form given by Eq. (9.4a) is probably the most frequently used in the number theory literature. As far as implementation is concerned, the form in Eq. (9.4a) requires an extra $\langle \cdot \rangle_{m_i}$ operation to be computed in the inner loop. However, the constants $\langle \hat{m}_i \langle \hat{m}_i^{-1} \rangle_{m_i} \rangle_M$ in Eq. (9.4b) are large integers, and hence the arithmetic in the second form requires multiplication by rather large numbers, a feature that is undesirable from the hardware point of view.

The CRT can be implemented without performing any explicit multiplication by applying the ROM-ACCUMULATOR algorithm of Peled and Liu [PE74] that was proposed for the implementation of digital filters. Suppose that $r_i = \langle \hat{m}_i^{-1} x_i \rangle_{m_i}$ in Eq. (9.4a), and suppose each r_i is encoded as a binary integer of the form

$$r_i = \sum_{l=0}^{B-1} 2^l b_{il}, \tag{9.5}$$

where the b_{il} terms are the bits representing r_i, and where B is chosen large enough to accommodate the largest r_i (smaller r_i's will simply contain leading zeros). The desired algorithm is obtained by substituting Eq. (9.5) into Eq. (9.4a), interchanging the order of summations (note both summations are finite), and precomputing the resulting inner summation to form a function to be stored in read-only memory (ROM). The result is

$$\langle X \rangle_M = \left\langle \sum_{l=0}^{B-1} 2^l F(A_l) \right\rangle_M, \tag{9.6a}$$

where

$$F(A_l) = \left\langle \sum_{i=1}^{L} \hat{m}_i b_{il} \right\rangle_M, \tag{9.6b}$$

and $A_l = b_{1l}, b_{2l}, \ldots, b_{nl}$ is an address formed by "bit slicing" the r_i terms, that is, by taking the lth bit from each r_i and arranging this set of n bits as the address A_l. Note that $\langle X \rangle_M$ can be computed by successively addressing a ROM that contains $F(A_l)$, adding the result in an accumulator, shifting one bit to the left, and then repeating the cycle B times. The total computation includes B memory fetches, $B - 1$ shifts, and $(B - 1) \bmod M$ additions. The function $F(A_l)$ requires 2^L words of storage, with each word requiring $\log_2 M$ bits. Since B is typically a small number, and since modular addition is simple to implement for an arbitrary modulus, the ROM-ACCUMULATOR algorithm provides an attractive implementation of the CRT. A modified form of the ROM-ACCUMULATOR algorithm results if the second form of the CRT algorithm (Eq. (9.4b)) is taken as the starting point. Let $\hat{m}_i' = \hat{m}_i \cdot \langle \hat{m}_i^{-1} \rangle_{m_i}$ and let

$$F'(A_l') = \left\langle \sum_{i=1}^{L} \hat{m}_i' b_{i,l} \right\rangle_M, \tag{9.7}$$

where the address A_l' is formed by bit slicing the x_i terms in Eq. (9.4b). In this form the multiplication of the $\hat{m}_i^{-1}$ terms in the inner brackets is eliminated, thereby reducing the computational complexity. However, a larger memory is needed because the wordlength required for $F'(A_l')$ is approximately $\log_2 M$, rather than $\log_2 \hat{m}_i$ as required by the first form. The add-shift algorithm is identical in both cases. Regardless of the form selected, the ROM-ACCUMULATOR algorithm provides an effective implementation for the CRT in either of the above forms.

9-2-5 The Associated Mixed-Radix Number System

It is possible to translate numbers from a residue representation to a mixed-radix representation, which is a weighted representation that facilitates sign detection and magnitude comparisons. If the moduli of the RNS are chosen to be the weights in the mixed-radix representation, the mixed-radix system is said to be associated with the RNS, and the translation operation is greatly facilitated. More specifically, a number X that falls into the total range of RNS with moduli $\{m_1, \ldots, m_L\}$ can be represented by L mixed-radix digits $(a_L, \ldots, a_1)$ defined by

$$X = a_L (m_{L-1} m_{L-2} \cdots m_2 m_1) + \cdots + a_2 (m_1) + a_1. \tag{9.8}$$

A more notationally compact, although somewhat more cryptic, form for Eq. (9.8) is

$$X = \sum_{k=1}^{L} a_k \prod_{i=1}^{k-1} m_i, \tag{9.9}$$

where $\Pi_{i=1}^{0}$ is defined to be equal to 1. The associated mixed-radix representation is quite important in RNS theory because the mixed-radix digits $(a_L, \ldots, a_2, a_1)$ provide a weighted representation of the residue number X, which is quite easy to generate. From Eq. (9.8) it can be seen that $a_1 = x_1$ directly. The higher order digits can be generated sequentially by $a_2 = \langle (X - a_1) m_1^{-1} \rangle_{m_2}$, $a_3 = \langle ((X - a_2) m_1^{-1}) - a_2) m_2^{-1} \rangle_{m_3}$, and so on. Figure 9-3 shows a ROM array that generates the mixed-radix digits from the residue input for a 4-moduli RNS. This mixed-

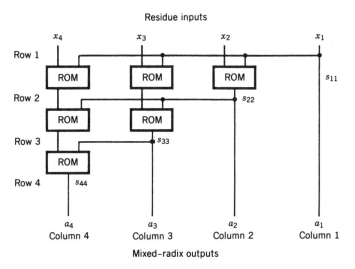

FIGURE 9-3 A mixed-radix converter realized as a ROM array for a RNS with four moduli.

radix converter is a highly regular structure that consists primarily of ROM and hence is ideally suited for VLSI realization. In general, the circuit of Figure 9-3 requires $L - 1$ memory access cycles to produce the full set of mixed-radix digits. If the circuit is pipelined, then a complete set of mixed-radix digits can be produced during every memory access cycle, although there will be a latency of $L - 1$ cycles relative to the application of the residue inputs. Once the a_i terms are computed, the binary representation of X can be obtained to the desired degree of accuracy by selecting the appropriate set of most significant a_i terms and computing Eq. (9.8) with either a Peled–Liu algorithm or a ROM-ADDER tree (note that the "coefficients" in Eq. (9.8) are fixed RNS parameters, so that general multiplication is not required). In applications requiring an analog representation of X, each a_i can be converted to an analog voltage with a short wordlength D/A converter and Eq. (9.8) can be implemented in the analog domain with a simple summing circuit [JE78].

9-2-6 Overflow Detection, Scaling, and Base Extension

Overflow detection, scaling, and base extension are RNS operations that, although not as difficult as general division, are considerably more difficult to implement than addition, subtraction, and multiplication. In all three cases the mixed-radix converter of Figure 9-3 forms the basis of the operation, since a mixed-radix representation is required as an intermediate step in the procedure. In order to determine if overflow has occurred, it is necessary to provide additional dynamic range in the RNS, and to then test the result of a computation to see if it has overflowed into the "extra" range. In general, only the final results of a computation must be tested in this way, since overflow has no meaning within the residue algebra itself, but rather only when the ring is mapped onto an interval of real numbers when decoding. The extra range needed for this purpose is provided by adding a redundant modulus whose only purpose is overflow detection. It is well known that a necessary and sufficient condition to check for overflow with one redundant modulus is that the redundant modulus be the largest modulus [BA73]. The occurrence of overflow is then detected if $a_{L+1} \neq 0$, where a_{L+1} is the highest order mixed-radix digit of the redundant RNS representation of X. This assumes that the quantity being tested, which has possibly overflowed the original RNS range, is not so large as to overflow the augmented range of the redundant system. This also illustrates that overflow detection requires a mixed-radix converter, similar to that of Figure 9-3 although designed to accommodate the augmented residue representation needed for redundancy. It can be seen from this discussion that overflow detection and mixed-radix conversion are similar in complexity and are both considerably more complicated than RNS addition and multiplication. It is fortunate that overflow detection is a relatively infrequent operation in many signal processing problems, in contrast to much more frequently required addition, subtraction, and multiplication.

In signal processing, scaling is a special form of multiplication in which one of the operands I_s is a fixed scale factor that allows the implementation to be greatly simplified in comparison to a general multiply. In weighted binary systems, scaling is often accomplished by simple right or left shifts, corresponding to scaling by positive (right shift) or negative (left shift) powers-of-2. In residue systems, scaling by a factor that is composed of one or more inverse moduli is particularly easy. Since this limitation on scale factors is usually acceptable in most practical situations, the following discussion concentrates on implementing the operation $X_s = Q[I_s X]$, where $I_s = 1/m_i$ and $Q[\cdot]$ denotes the quantization required to produce an integer result.

First note that $X_s = \langle m_i^{-1} X \rangle_{\hat{m}_i}$ if X is an exact multiple of m_i and m_i^{-1} is the multiplicative inverse of $m_i \bmod \hat{m}_i$. If the mixed-radix representation of X as given in Eq. (9.8) is implemented with the circuit of Figure 9-3, it can be seen that after the subtraction of x_1 and multiplication of m_1^{-1} in the first row of the ROM array, the residue representation of $Q[m_1^{-1} X]$ is automatically available at the outputs of row 1, although one digit is missing, namely, the first one, $x_{s1} = \langle X_s \rangle_{m_1}$. The next step is to generate x_{s1} from the digits $x_{s2}, \ldots, x_{sL}$. This can be done by noting that

$$X_s = (m_L \cdots m_2)a_L + \cdots + (m_2)\, a_3 + a_2$$

so that

$$x_{s1} = \langle C_{L-1} a_L + \cdots + C_2 a_3 + C_1 a_2 \rangle_{m_1},$$

where

$$C_1 = 1 \quad \text{and} \quad C_i = \prod_{k=2}^{i} m_k$$

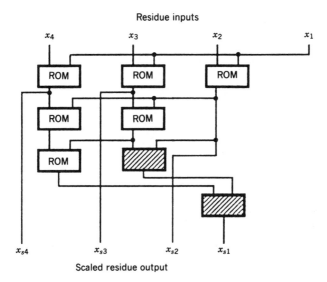

FIGURE 9-4   A mixed-radix converter modified to scale by the factor m_1.

for $i = 2, \ldots, L - 1$. As the mixed-radix digits are generated in the mixed-radix converter, the above expression for x_{s1} can be implemented with the modified mixed-radix converter shown in Figure 9-4. The general procedure of extending the residue representation of a quantity from one set of residues to a larger set is called base extension. It should now be apparent why the ROM-ARRAY mixed-radix converter circuit plays a central role in RNS hardware designs; its speed and complexity determine how efficiently the designer can implement mixed-radix conversion, scaling, and base extension, all of which are important operations for signal processing and which fall into the category of difficult RNS operations.

There are many other approaches to RNS overflow detection, scaling, and base extension that can be found in the literature [SO86], but they are beyond the scope of this discussion. Many of these alternate techniques are based on the modifications of the Chinese Remainder Theorem, which first obtain a weighted number representation (analogous to the mixed-radix conversion process) and then perform various tasks to produce the desired result. Another type of approach is based on the concept called "autoscale," which was first introduced by Taylor et al. for residue-to-decimal conversion [TA81] and controlled-precision multiplication [TA82] for modular systems that are characterized with relatively few large moduli (typically three 8-bit moduli). However, when the designer wishes to keep the moduli small in number and to use quite a few of them, the mixed-radix conversion techniques described above are best suited for VLSI implementation. Also, since all the computations required in the circuits of Figures 9-3 and 9-4 are operations with RNS arithmetic, it is possible to detect and correct errors that occur within the circuits themselves, a feature that is not so directly provided in other approaches. The significance of this property will become clearer in Section 9-4, where RNS fault tolerance is treated in considerable detail.

9-2-7   Advantages and Disadvantages of RNS Arithmetic in DSP Applications

An obvious question that should be discussed at this point concerns the usefulness of modular arithmetic in signal processing applications, in light of the fact that RNS techniques have not previously achieved widespread use in general purpose computers. Operations within the RNS can be classified into those that are particularly fast and easy (addition, subtraction, multiplication, negation, etc.) and those that are rather difficult (division, scaling, sign determination, magnitude comparison, etc.). Common signal processing tasks such as digital filtering, correlation, interpolation, prediction, and spectral analysis are characterized as requiring large numbers of easy operations from the first category, as well as an infrequent scaling operation from the second. Since wordlengths are fixed and the dynamic range of the input data is often known, it is usually possible to prescale the calculations so that overflow does not occur. Also, image processing tasks, such as noise reduction, edge detection, and feature extraction, require large numbers of simple repetitive operations that mostly fall into the simple class of RNS operations.

In contrast, a general purpose programmable computer must be designed for a much broader set of tasks that require sign detection, magnitude comparison, and general division, that is, operations that fall into the more difficult category. Often the range of the input data is not known a priori so that floating-point representation is the best choice of data representation. In general, the flexibility required of a general purpose processor does not allow the overall system to take advantage of the inherent efficiencies of RNS coding. For all these reasons RNS concepts have not yet found any significant application in general computer design.

On the other hand, the inherent modularity of RNS algebra translates into a natural partitioning of the hardware that is particularly attractive for VLSI designs. Since much of the current activity in digital signal processing involves finding algorithms that are suitable for integrated circuit implementation, the RNS approach is also attractive in this regard. Identical hardware modules are designed at the chip level and then are simply programmed during the last stages of metalization in order to customize each module for different mod m_i operations. This modularity results in simpler designs, simpler testing, and higher operating speeds than the traditional weighted binary arithmetic, which has to deal with the problems of carry propagation. Modular redundancy, error detection and correction, and the design of fault-tolerant systems all become more feasible when the signal processing task is well enough specified to permit a RNS design.

9-2-8   Relationships Between RNS Arithmetic and Number Theoretic Transforms

During the 1970s a considerable amount of work was reported on number theoretic transforms (NTTs). Since these NTTs are defined within a real or complex finite ring (field) and strive to reduce computational complexity and simplify hardware, they are closely related to the concepts of RNS arithmetic. The relationship between the NTT and RNS arithmetic is analogous to the relationship between the

FFT and conventional complex arithmetic. The NTT can be thought of as a special fast transform that has a FFT structure and a convolutional property and that is typically defined within a single modulus residue system. Note that a NTT can be defined within a multiple-modulus RNS, although the block length for which a NTT can be defined is limited by the smallest m_i, and hence combining the NTT with RNS arithmetic results in somewhat conflicting algebraic requirements [JE75]. The range in which this hybrid form is clearly useful is for moduli in the vicinity of 256. These moduli can be processed in an 8-bit microprocessor and are large enough to accommodate long enough block lengths for signal processing. In particular, block lengths on the order of 256 are useful for many types of image processing applications, which are computation intensive due to the two-dimensional format of the data. NTTs are not discussed further in this chapter, since the subject is already more than adequately covered in McClellan and Rader [MC79] and Blahut [BL85].

9-3 THE DESIGN OF VLSI DIGITAL FILTERS USING RNS ARITHMETIC

RNS design techniques have been most successful to date for finite impulse response (FIR) filters, which can take full advantage of fast RNS operations while avoiding the problems associated with scaling. They have also been studied extensively for infinite impulse response (IIR) digital filters but have attracted less interest for this class because IIR filters require scaling due to their recursive nature. A FIR filter is characterized by

$$y[n] = \sum_{k=0}^{Q-1} h[k]s[n - k], \qquad (9.10)$$

where $y[n]$ is the filter output at integer time n, the $h[k]$ terms are the filter coefficients, the $s[n]$ terms are samples of the input signal, and Q denotes the length of the FIR filter. When Eq. (9.10) is encoded in RNS arithmetic, the equations to be computed are described by

$$y_i[n] = \sum_{k=0}^{Q-1} h_i[k]s_i[n - k], \qquad i = 1, \ldots, L, \qquad (9.11)$$

where the subscript i denotes mod m_i residue digits, and all additions and multiplications in Eq. (9.11) are defined to be modulo m_i operations. In a digit-parallel design each circuit in Eq. (9.11) for $i = 1, \ldots, L$ is implemented independently and the ensemble is operated in parallel for maximum operating speed.

Two experimental integrated circuits that were designed with RNS arithmetic are described in this section, illustrating the application of RNS theory to FIR filter design. Both of these VLSI circuits were designed and fabricated for one module

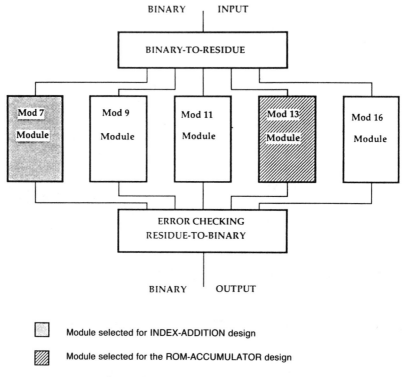

FIGURE 9-5   RNS digital filter architecture.

of a RNS finite impulse response digital filter, the architecture of which is illustrated in Figure 9-5. Both of these designs were developed for fully parallel RNS arithmetic. The first example is a custom-designed circuit for the mod 13 module of a digital filter implemented with a ROM-ACCUMULATOR architecture, which eliminates general multipliers through the application of distributed arithmetic [PA84]. (This is the same algorithm that was discussed but not described in detail in Section 9-2 for Chinese Remainder Theorem residue-to-binary conversion.) The second example is a semicustom integrated circuit designed with the IBM MVISA CAD system [JE87b]. This design system, based on a standard cell approach, was developed at the IBM Manasses facility and was made available by IBM on the campus of the University of Illinois as part of a university–industry cooperative program. The architecture of this second chip, designed for the same FIR digital filter as the first chip, uses finite field logarithms (indices) to eliminate general multiplication. This structure is referred to here as an INDEX-ADDITION architecture. Finite field logarithms provide for general multiplication, so that the filter characteristic can be programmed at the time of application or even used in an adaptive filtering application, a flexibility that is otherwise lost through the selection of distributed arithmetic in the first design.

9-3-1   FIR RNS Digital Filters Designed with Distributed Arithmetic

General multipliers are eliminated in the ROM-ACCUMULATOR design through the use of distributed arithmetic, resulting in a filter whose frequency characteristic is fixed in the ROMs at the time of circuit fabrication. In this design the input residues $s_i[n]$ in Eq. (9.11) are represented as binary integers:

$$s_i[n] = \sum_{k=0}^{b-1} 2^k s_{ik}[n] \qquad (9.12)$$

where $s_{ik}[n]$ denotes the kth bit in the binary representation of $s_i[n]$. If expression (9.12) is substituted into Eq. (9.11) and the order of summations interchanged, the filter equation becomes

$$y_i[n] = \left\langle \sum_{i=0}^{b-1} 2^i F_i(S_i[n]) \right\rangle_{m_i}, \qquad (9.13)$$

where

$$F(S_i[n]) = \left\langle \sum_{k=0}^{N-1} s_{ik}[n - k]h_i[k] \right\rangle_{m_i} \qquad (9.14)$$

and $S_i[n] = [s_{ik}[n], \ldots, s_{ik}[n - N]]$ is an N-bit address formed from ''bit slices'' of the input residues. Using this algorithm and the RNS architecture shown in Figure 9-5, the FIR linear phase lowpass filter function illustrated in Figure 9-6 was realized in a VLSI design for an RNS with moduli $\{16, 15, 13, 11, 7\}$. Each b-bit ($b = 4$) module is designed on a separate chip and is customized for a particular modulus by the programming of the ROM. The filter function of Figure 9-6 corresponds to a length $N = 8$ impulse response function with coefficients $h[0] = 1.107 = h[7]$, $h[1] = 3.265 = h[6]$, $h[2] = 5.332 = h[5]$, and $h[3] = 6.963 = h[4]$. It was assumed in this design that the input samples are available with 6 bits of accuracy, and that the largest two moduli $\{16, 15\}$ are used for error detection and correction (to be discussed in more detail in Section 9-4). This leaves a computational range of the RNS of $[0, 1001]$, or a dynamic range for internal computation of $[-500, 500]$, that is, approximately 10 bits of internal accuracy. In this design Eq. (9.13) is implemented recursively according to

$$G_i[n, q] = \langle F_i(S_i[n]) + 2G_i[n, q - 1] \rangle_{m_i}, \qquad (9.15)$$

where $G_i(n, 0] = 0$ and $y_i(n] = G_i(n, b - 1]$. A block diagram of the chip architecture is shown in Figure 9-7, where the ROM at the bottom of the circuit performs the residue encoding on the input samples, the ROM at the upper left stores the function $F_i(\cdot)$, the ROM at the upper right stores the function $G_i[n, q]$, and the shift registers at the lower left form the delay line for the $N = 8$ stored

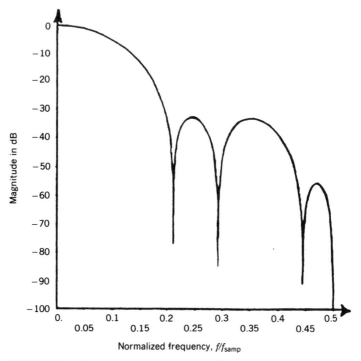

FIGURE 9-6   Frequency response of the FIR design example ($N = 8$).

data samples. The input encoding ROM is a 128-bit ROM organized as 32 words by 4 bits/word, whereas the ROMs that implement the convolution are 1K-bit ROMs organized as 256 words by 4 bits/word.

The IC modules were designed with 4-micron custom nMOS technology. Due to cost limitations, only the mod 13 module was actually fabricated. Five packaged circuits were processed through MOSIS and were later tested at the University of Illinois. A microphotograph of one of these ICs is shown in Figure 9-8. All modules were determined to be fully functional. Clock frequencies as high as 6.7 MHz were recorded, which corresponds to a sampling frequency of 1.7 MHz. Power dissipation was measured to be 80 mW per module. In general, this custom design represents a high-performance chip that was very costly in terms of design time and effort. The next approach to be described uses an automated design system to obtain a semicustom chip. This approach lowers the cost in terms of design time but also results in less efficient use of silicon area due to the loss of flexibility.

9-3-2   FIR RNS Digital Filters Designed with Finite Field Arithmetic

In a finite field RNS design the moduli p_i, $i = 1, \ldots, L$, are chosen to be distinct prime integers so that modular multiplication can be implemented by finite loga-

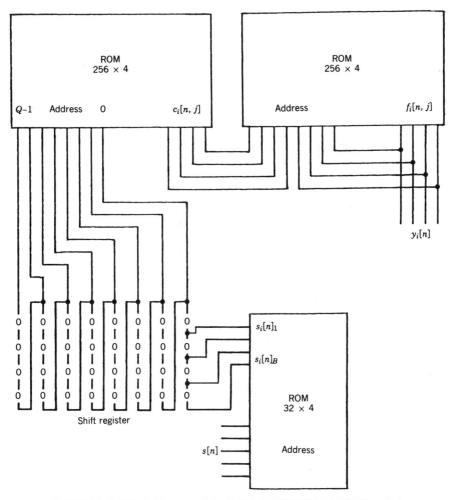

FIGURE 9-7 Block diagram of the ROM-ACCUMULATOR IC module.

rithm (index) addition, as illustrated in Eqs. (9.16):

$$S[k] = \{\log(h_i[k]) + \log(s_i[n - k])\}, \tag{9.16a}$$

$$y_i[n] = \sum_{k=0}^{Q-1} \text{antilog}\{\langle S[k]\rangle_{(p_i - 1)}\}. \tag{9.16b}$$

With this type of design the index encoding of the input residues $s_i[n]$ is done in the same look-up table used to convert the binary input samples into residues, and the filter coefficients $h_i[n]$ are stored in residue index form. A block diagram for the FIR INDEX-ADDITION design is shown in Figure 9-9, where it is seen that

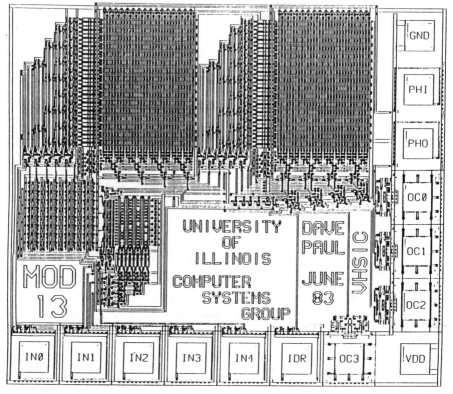

FIGURE 9-8   Microphotograph of the ROM-ACCUMULATOR IC module.

the core of the filtering operation is computed with two modular adders, one for the index addition (multiplication) and a second for the accumulation of products.

An experimental integrated circuit was designed and fabricated for a modulo 7 (3-bit) module using a semicustom design package called MVISA [JE87b]. The designer describes the circuit functionality at a high level, and the software tools translate these specifications into a standard cell layout that automatically adheres to IBM's design rules. Although the final designs are less efficient in terms of silicon area than a fully custom design, the entire design–fabrication cycle can be completed very rapidly with a high probability of success on the first design cycle.

The basic element in the MVISA circuit for implementing the filtering function is shown in Figure 9-10 for this INDEX-ADDITION architecture. The adder at the top of Figure 9-10 implements Eq. (9.16a) (assuming the filter coefficients $h[n]$ and the signal samples $s[n]$ have already been residue encoded and converted to logarithm form at the filter input), and the RAM and multiplexers implement the logarithm decoding and accumulation described by Eq. (9.16b). Thus the index adder adds the 3-bit index residue data samples and the index residue filter coefficients, resulting in 4-bit sums. The RAM decodes the index sums and accumulates the partial sums according to Eq. (9.16b). The MUX REG FILE buffers the

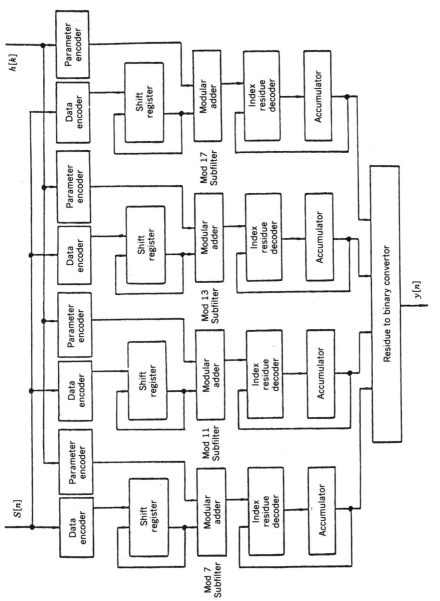

FIGURE 9-9 Block diagram of the INDEX-ADDITION FIR digital filter.

629

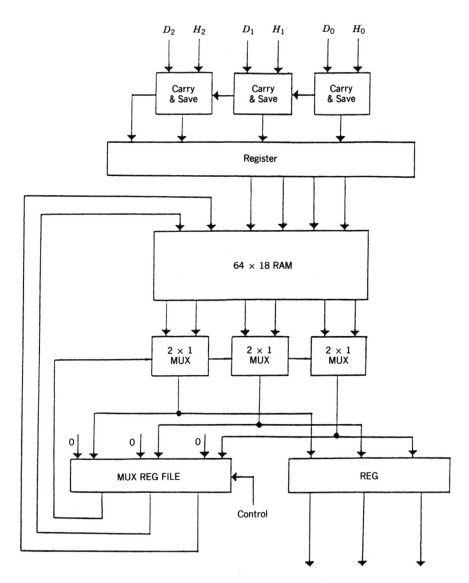

FIGURE 9-10   Block diagram of the IC module for the INDEX-ADDITION filter module designed with the MVISA CAD system.

partial sums to be fed back to the RAM, with the register storing the residue outputs of the filter.

MVISA predicts that power dissipation of the 4.2-mm chip should be approximately 89 mW. This design used a total of 529 of the available 560 standard cells allowed by MVISA. During computer simulation with MVISA, the chips were simulated at a system–cycle frequency of 10 MHz, which corresponds to a data–

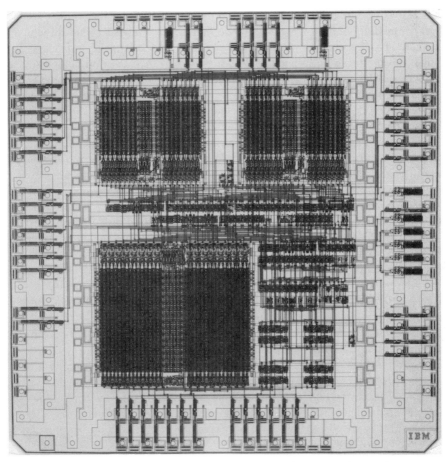

FIGURE 9-11   Final layout of the INDEX-ADDITION IC module.

cycle frequency of 1.2 MHz. The final layout of the INDEX-ADDITION chip is shown in Figure 9-11.

In comparing these two architectures it was found that the ROM-ACCUMU-LATOR architecture is faster and cheaper for low-order filtering, provided that the application permits fixed filter coefficients. However, the memory requirement for the ROM-ACCUMULATOR architecture increases rapidly with higher filter orders. For higher order filters that require constantly changing coefficients (such as adaptive filters), the INDEX-ADDITION architecture becomes more attractive.

9-3-3 FIR RNS Digital Filters Designed with Digit Serial Arithmetic

An interesting trade-off between speed and hardware complexity can be achieved with a RNS digit-serial scheme called a *serial-by-modulus* (SBM) architecture. With this scheme the residue digits $x_1, x_2, \ldots, x_L$ are transmitted through the processor in a serial fashion; that is, x_1 arrives first, then x_2, and so on. Assume that the moduli are b-bit integers, so that there are b parallel bit lines throughout the entire processor. In rough terms, the serial-by-modulus architecture operates L times slower than a fully parallel design, although the hardware complexity is substantially reduced. Serial-by-modulus arithmetic is similar in principle to bit serial binary arithmetic that was popular in the early days of digital filter design. However, it represents a compromise between slow speed of bit serial binary arithmetic and the high cost of bit parallel binary arithmetic.

A SBM architecture requires an efficient realization for both modulo $(p_i - 1)$ index addition (multiplication) and modulo p_i addition (accumulation). The design uses a standard 2's-complement adder whose output is used to address a correction table, which produces the correct modular sum from the binary sum. When the adder forms an index addition, the inverse logarithmic mapping is encoded into the same correction table so translation from index form to residue form does not require any additional hardware. Consider an RNS defined by the four 5-bit moduli 31, 29, 23, and 19. When two residues are added in a 2's-complement adder, the result may require as many as 6 bits, that is, one additional "buffer" bit is needed to represent all possible results. Furthermore, since there are four moduli, it will take 2 bits in order to uniquely label each residue so that it can be identified unambiguously with the correct timing interval. In total, each residue, together with its timing label, can be encoded with 8 bits, as illustrated in the block diagram of Figure 9-12. The ROM in Figure 9-12 consists of four independent correction tables for the four different m_i, $i = 1, 2, 3, 4$. The 2-bit timing label, which is appended to the corresponding residue by two extra bits in the word-length, serves as the two highest order bits of the address, so the correct look-up table is automatically addressed. Note that the adder shown in Figure 9-12 is really only a 6-bit adder. The label bits do not take part in the addition but are simply passed on through the adder for subsequent processing.

Since the label bits have been attached firmly to each residue (as opposed to being generated locally by a synchronized clock), it is possible to place simple comparators at strategic locations on the chip to check that no timing skew has occurred and to guarantee that digit synchronization is properly maintained throughout the processor. The design presented here realizes more than 18 bits of

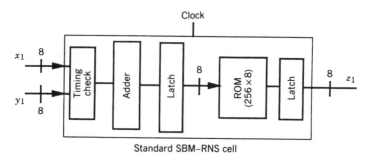

Standard SBM–RNS cell

FIGURE 9-12   Block diagram of a serial-by-modulus computational element.

effective wordlength with an integrated module that has only an 8-bit wordlength. Not only has the silicon area (and corresponding power requirement) been significantly reduced as compared to the fully digit parallel form, but the module can be more easily tested for faults due to its short wordlength. Of course, the effective speed of the processor has been reduced by an approximate factor of 4 due to the digit serial arithmetic. Note that two modules similar to the one shown in Figure 9-12, one for multiplication and one for addition, constitute the essential structure of a FIR digital filter. This is illustrated in Figure 9-13, where the standard IA cell is programmed for mod($p_i - 1$) addition and the standard A cell is programmed for mod p_i addition.

An important element in any RNS design is the mixed-radix converter, which is used in output conversion, error checking, scaling, and so on. Figure 9-14 shows the architecture for a fully parallel mixed-radix converter for a RNS with $L = 4$ moduli. The function F_{ki} that must be stored in the ROM located in the kth row and ith column of the array (as reverse labeled in Figure 9-14) is given by

$$s_{k+1,i} = F_{ki}(s_{k,i}, s_{k,k}) = \langle(s_{k,i} - s_{k,k})m_k^{-1}\rangle_{m_i}, \tag{9.17}$$

for $k = 1, \ldots, L - 1$ and $i = k + 1, \ldots, L$ and with $s_{1i} = x_i$ and $a_q = s_{q,q}$, $i, q = 1, \ldots, L$. Figure 9-15 shows a modified mixed-radix converter that operates with a SBM-RNS system. The digits are buffered so that the latches at the top of Figure 9-14 can be loaded in parallel. The structure is then clocked four times, after which the mixed-radix digits a_1, a_2, a_3, and a_4 appear in place of the original residue digits. Figures 9-15(a)–(c) show the states of the converter after the first three cycles. While the converter is cycling, the RNS digits for the next sample are loaded into the buffer (not shown) in preparation for the next conversion cycle. It can be shown [JE87c] that the order of complexity of the hardware for the parallel mixed-radix converter is $(L^2 - L)/2$, whereas that of the SBM-RNS mixed-radix converter is L. Also, the parallel structure operates L times faster. This demonstrates that for large L the savings in hardware within the mixed-radix converter are proportional to the reduction in speed when the SBM architecture is selected over a fully parallel RNS structure. Serial-by-modulus RNS arithmetic is attractive for reducing the hardware complexity of VLSI digital processors. The scheme described above sacrifices speed to achieve simpler hardware designs, while

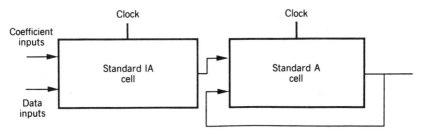

FIGURE 9-13 Two SBM computational elements forming a FIR filter module.

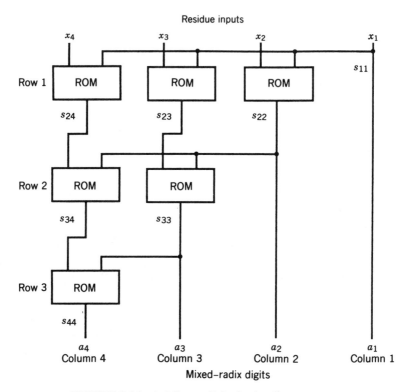

Residue inputs

FIGURE 9-14   A fully parallel mixed-radix converter.

maintaining a modular architecture that is desirable for partitioning, fault tolerance, and testability in VLSI designs.

9-4 FAULT-TOLERANT SYSTEMS DESIGNED WITH RNS ARITHMETIC

The lack of communication in RNS arithmetic suggests that if an error occurs in
one digit it cannot be propagated into other digit positions during subsequent operations involving addition, subtraction, or multiplication. This property provides
a basis for fault tolerance that is inherent in the basic algebraic structure and that
can be used to obtain fault-tolerant hardware architectures. During some of the
more difficult RNS operations, such as scaling, division, or magnitude comparison, there is interaction between residue digits and this error isolation property is
not preserved. Therefore the fault-tolerant properties of RNS arithmetic are particularly useful for signal processing applications where most of the computation
consists of addition, subtraction, and multiplication.

The nonweighted structure of the RNS code is another basic property that makes
residue arithmetic useful in the design of fault-tolerant processor structures. If a

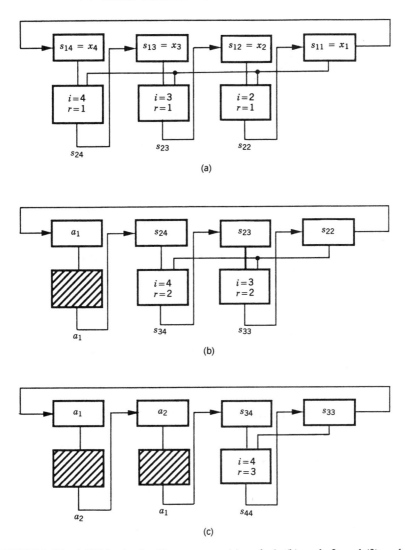

FIGURE 9-15 A SBM mixed-radix converter: (a) cycle 1, (b) cycle 2, and (3) cycle 3.

particular residue digit is found to be consistently erroneous, the corresponding faulty module can be identified by RNS error checking techniques and disconnected without affecting other modules. If the original RNS contains enough dynamic range, the reduced processor can continue to function with a reduced dynamic range. This concept is called *soft failure* because the processor does not catastrophically fail when a hardware failure occurs, but rather the faulty module is disabled, and the remaining modules continue functioning in a useful although restricted manner. If desirable, error correction can be used to replace the function of the faulty module provided enough redundancy is designed into the code.

9-4-1 Redundant Residue Number Systems

A redundant RNS is formed by adding extra moduli that are not allocated for increasing the computational range, but rather are used to provide enough degrees of freedom for error detection and correction. Suppose that r redundant moduli are added to create $L + r$ moduli in total. All the $L + r$ moduli must be pairwise relatively prime to ensure a unique representation for each state in the system. A redundant RNS has all the properties described previously for a nonredundant RNS. The moduli $\{m_1, m_2, \ldots, m_L\}$ are called the *fundamental moduli*. The additional r moduli $\{m_{L+1}, \ldots, m_{L+r}\}$ are called *redundant moduli*. In a redundant RNS a number is encoded by $L + r$ residue digits $x_1, x_2, \ldots, x_{L+r}$, where x_1, $x_2 \ldots, x_L$ are the fundamental residue digits and $x_{L+1}, \ldots, x_{L+r}$ are the redundant digits. The total range $[0, M_T - 1]$, where $M_T = m_1 m_2 \ldots m_{L+r}$, is the complete set of states representable by $L + r$ digits. The interval $[0, M - 1]$, where $M = m_1 m_2 \cdots m_L$, is the *legitimate range*, and $[M, M_T - 1]$ is called the *illegitimate range*. Any integer contained in the legitimate range is called legitimate and any number contained in the illegitimate range is illegitimate.

In order that the redundancy is used properly for error detection and correction (and not simply to increase the dynamic range), initial operands and results of all arithmetic computations must be constrained to the legitimate range. Hence, as in the nonredundant case, the dynamic range in a redundant RNS is $[-(M - 1)/2, (M-1)/2]$ if M is odd and $[-M/2, M/2 - 1]$ if M is even. It was described earlier how a RNS-complement code establishes a one-to-one correspondence between signed integers in the dynamic range and the states in the legitimate range. In a redundant RNS, the usual RNS-complement code maps negative numbers from the negative part of the dynamic range, $[-(M - 1)/2, (M-1)/2]$ assuming M is odd, to the interval $[M_T - (M - 1)/2, M_T - 1]$. This shows that legitimate negative numbers are inadvertently mapped to the top of the illegitimate range and will be confused with illegitimate numbers caused by errors unless some further action is taken. This problem is easily corrected by simply rotating the modular ring by adding the constant $(M-1)/2$ to any state using mod M_T addition, as illustrated in Figure 9-16. This operation, called a *polarity shift*, must be executed before error detection is applied to determine if a residue operand falls into the legitimate range.

9-4-2 RNS Error Detection and Correction

A simple example will be used to illustrate the principles of RNS error detection and correction. Consider a redundant RNS defined by $m_1 = 3$, $m_2 = 4$, $m_3 = 5$, and $m_4 = 7$, where the legitimate (computational) range is provided by m_1 and m_2, and where m_3 and m_4 provide the redundancy necessary for error checking. All numerical quantities must be scaled so that the results of all calculations always remain within the legitimate range $[0, 11]$. The illegitimate range $[12, 419]$ is used only when a single digit error occurs. Note that a legitimate state can be represented correctly by any three of the four residues (as well as by any two of the residues). For example, $x = 5 \approx (2, 1, 0, 2)$ is correctly represented by $(2, 1, 0)$

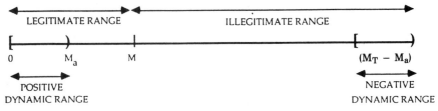

ORIGINAL RNS COMPLEMENT ENCODING

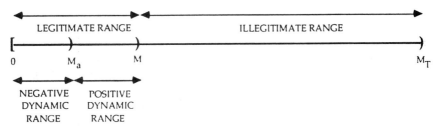

ENCODING AFTER POLARITY SHIFT

FIGURE 9-16 Polarity shifting for error detection in redundant RNS systems.

in the reduced RNS defined by $\{m_1, m_2, m_3\}$, by (2, 1, 2) in $\{m_1, m_2, m_4\}$, by (2, 0, 2) in $\{m_1, m_3, m_4\}$, or by (1, 0, 2) in $\{m_2, m_3, m_4\}$. Therefore if one of the original residue digits is erroneous, and if one digit is discarded from the original set of four residues, a correct representation results if and only if the erroneous digit is discarded. This is the mechanism used for error location. Once an erroneous digit is located, the correct value can be found by base extension using the subset of correct digits.

The mechanism by which a single digit error transforms a legitimate residue number into an illegitimate one is easily explained using the Chinese Remainder Theorem. Whenever an error e_i occurs in the ith digit, the resulting illegitimate number x' can be expressed as the correct state plus an error term as follows:

$$x' = \langle x + \hat{m}_i \langle \hat{m}_i^{-1} e_i \rangle_{m_i} \rangle_{M_T}. \tag{9.18}$$

The error term is bounded according to min $\{\hat{m}_i\} \leq \hat{m}_i \langle \hat{m}_i^{-1} e_i \rangle_{m_i} \leq M_T -$ min $\{m_i\}$. This guarantees that the minimum error moves x' out of the legitimate range since $\hat{m}_i > M$ for all i and $r \geq 2$. The maximum possible error is small enough so that large errors cannot wrap around into the legitimate range. (Note that the mod M_T operator can be omitted in Eq. (9.18).) For the example used above, the Chinese Remainder Theorem takes the form

$$\langle x \rangle_{420} = \langle 140 \langle 2x_1 \rangle_3 + 105x_2 + 84 \langle 4x_3 \rangle_5 + 60 \langle 2x_4 \rangle_7 \rangle_{420},$$

from which it is seen that the minimum possible error is 60, while the maximum is 360. Therefore a single error always transforms x into an x' that falls into the illegitimate range [12, 419].

The error detection/correction procedure requires three basic steps: (1) it is first determined if the residue number being checked is legitimate; (2) the digits are discarded one at a time until a legitimate reduced representation is found; and (3) the correct digit can be derived by a base extension using the reduced set of digits found to be legitimate in step 2. There are many variations on this procedure, and many different approaches to carrying out the details of each step, which are described in the literature. Note that overflow also causes illegitimate states to appear if the computation is not scaled properly. Overflow detection and error detection/correction are closely related issues. Proper care must be given to the design when the RNS code is required to handle simultaneous overflow and single digit errors.

The reduced representations that result from discarding one digit at a time are called projections. The qth projection that results from discarding the qth residue digit is equivalent to $X_q = \langle x \rangle_{\hat{m}_q}$, where $[0, \hat{m}_q - 1]$ is the total range of the reduced RNS. The simplest way to determine if X_q is legitimate is to form its mixed-radix representation associated with the reduced RNS. It is well known that X_q will be legitimate if and only if the highest order mixed-radix digit is zero. Consider a redundant RNS defined by $m_1 = 7$, $m_2 = 9$, $m_3 = 11$, $m_4 = 13$, $m_5 = 16$, and $m_6 = 17$. If m_5 and m_6 are defined to be redundant ($L = 4$ and $r = 2$), then the condition $m_R \geq \max\{m_i m_q\}$, $i \neq q$, guarantees single error correction capability. Suppose a legitimate residue number $x = 500 \in (3, 5, 5, 6, 4, 7)$ is mapped to an illegitimate state $x' \approx (5, 5, 4, 6, 4, 7)$ by an error $e_3 = 10$ in x_3. Table 9-1 shows the representations of the X_q terms in a decimal code (which is not calculated in an actual implementation), a mixed-radix code, and a residue code. The only legitimate projection is X_3, since it falls into the legitimate range $[0, 9008]$. This implies that x'_3 is in error, and the correct value of x_3 can be formed from $x_3 = \langle a_4 \cdot 63 + a_2 \cdot 7 + a_1 \rangle_{11}$, where a_4, a_2, and a_1 are the nonzero mixed-radix digits of X_3 in the reduced RNS $\{7, 9, 13, 16, 17\}$. Note that the legitimate projection is uniquely identified by the fact that the highest order mixed-radix digit a_6 is zero. The efficiency of this error checking procedure depends entirely on the ability to efficiently generate the mixed-radix representations of all the projections.

9-4-3 An Error Checker Based on a Modified Mixed-Radix Converter

From the above discussion it is clear that error location and error correction require that a quantity in question be ''projected'' onto various subspaces within the RNS

TABLE 9-1   An Example of Error Detection and Correction

Projection		Decimal	Mixed-Radix	Residue
X_1	=	127,796	(6, 3, 3, 9, 5, —)	(—, 5, 4, 6, 4, 7)
X_2	=	50,004	(3, 1, 12, 4, —, 3)	(3, —, 4, 6, 4, 7)
X_3	=	500	(0, 0, 7, —, 8, 3)	(3, 5, —, 6, 4, 7)
X_4	=	154,724	(13, 15, —, 2, 8, 3)	(3, 5, 4, —, 4, 7)
X_5	=	84,038	(9, —, 4, 2, 8, 3)	(3, 5, 4, 6, —, 7)
X_6	=	66,020	(—, 9, 4, 2, 8, 3)	(3, 5, 4, 6, 4, —)

algebra. A projection $X_q = \langle X \rangle_{\hat{m}_q}$ is a representation of X in the subspace defined by $\{m_1, \ldots, m_{q-1}, m_{q+1}, \ldots, m_{L+r}\}$, and hence the RNS representation of X_q is given by $(x_1, \ldots, x_{q-1}, x_{q+1}, \ldots, x_{M+r})$. Recall that the mixed-radix representation of X is given by

$$X = \sum_{k=1}^{L+r} a_k \prod_{i=1}^{k-1} m_i. \tag{9.19}$$

Equation (9.19) can be modified to express the mixed-radix representation of X_q as follows:

$$X_q = \sum_{\substack{k=1 \\ k \neq q}}^{L+r} a_k^q \prod_{\substack{i=1 \\ i \neq q}}^{k-1} m_i. \tag{9.20}$$

Note that this expression for X_q (Eq. (9.20)) is identical to that for X (Eq. (9.19)) if $a_q = 0$ and $m_q = 1$ in Eq. (9.19). This observation reveals a mathematical principle that permits the calculation of the a_k^q terms for all projections by a simple modification of the general mixed-radix converter.

Recall that the a_k terms in Eq. (9.19) can be computed by a recursive procedure that is characterized by

$$s_{k+1,i} = \langle (s_{k,i} - a_k) m_k^{-1} \rangle_{m_i}, \tag{9.21}$$

for $k = 1, \ldots, L + r - 1$ and $i = r + 1, \ldots, L + r$ with $s_{1,i} = x_i$ (initial conditions) and $a_q = s_{q,q}$ (final conditions) for $i, q = 1, \ldots, L + r$. If it is desired to calculate the qth mixed-radix digit for an arbitrary residue number $X \approx x_1, \ldots, x_{L+r}$, the rescursive equation is initialized with $s_{1,q}$, where x_q is the qth residue digit of X, and the recursion is iterated $q - 1$ times. Since the lower order a_i terms are used in the recursive relation, it is always assumed when calculating a_q that $a_1, a_2, \ldots, a_{q-1}$ have previously been computed. Figure 9-17 shows a detailed block diagram for the architecture of a ROM array that will produce all $L + r$ mixed-radix digits for a RNS with $L = 3$ and $r = 2$. Each of the blocks represents a particular iteration of the recursive Eq. (9.21). The shaded blocks are not actually present in the physical realization but have been included so that the structure can be interpreted as a square array with rows and columns referenced according to the labels shown in Figure 9-17. Therefore the physical structure is the upper triangular portion of the square array.

By characterizing the mixed-radix converter as a square array, it is easy to determine the modifications that are needed to use the same structure to obtain the mixed-radix digits for an arbitrary projection X_q, $q = 1, \ldots, L + r$, as required for error location. For the qth projection, the $L + r - 1$ mixed-radix digits a_1^q, $\ldots, a_{q-1}^q, a_{q+1}^q, \ldots, a_{L+r}^q$ are computed, and the highest order digit a_{L+r}^q is tested for zero. If $a_{L+r}^q \neq 0$, then X_q is illegitimate and the error is not in x_q. If $a_{L+r}^q = 0$, then the error is in x_q. Furthermore, the correct digit can be obtained

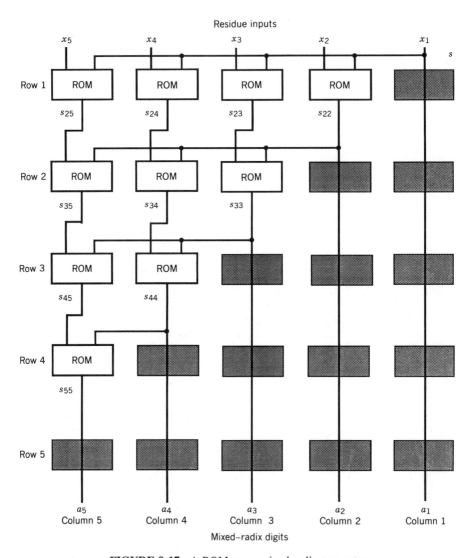

FIGURE 9-17 A ROM array mixed-radix converter.

as $x_q = \langle X_q \rangle_{m_q}$. An examination of the mixed radix representation given earlier in Eq. (9.20) reveals that X_q differs from X in that the qth term is excluded from the summation and the modulus m_q is excluded from each of the product terms. In fact, Eq. (9.19) represents X_q if a_q is set equal to zero (to exclude the qth term in the summation) and if m_q is set to 1 (to exclude m_q in each product term). Since the mixed-radix conversions for X and X_q are directly dependent in Eqs. (9.19) and (9.20), it follows that a similar change of parameters in the recursive relation of Eq. (9.21) will allow the square matrix structure of Figure 9-17 to generate the mixed-radix digits for X_q.

The variable $s_{k+1,i}$ in Eq. (9.21) is the output of the block at the intersection of the kth row and ith column (with rows and columns reverse labeled as in Figure 9-17). Setting $a_q = 0$ is equivalent to setting the output of block (q, q) to zero, which suggests that the qth column should be removed from operation. Setting $m_q^{-1} = 1$ with $s_{q,q} = 0$ is equivalent to removing the qth row from the array and simply transferring the inputs $s_{q,i}$, $i = 1, \ldots, L + r$, directly through the blocks in the qth row unaltered. These observations are summarized as follows:

A general mixed-radix converter, characterized as a triangular subarray within an $(L + r) \times (L + r)$ square ROM array with rows and columns labeled as in Figure 9-17, will generate the mixed-radix digits for an arbitrary projection X_q if the blocks in the qth column are disabled, the variable $s_{q,q}$ (physical output of the qth column) is set to zero, and the blocks in the qth row are "shorted" so as to transmit their input variables unchanged. The mixed-radix digits a_1^q, $\ldots$, a_{q-1}^q, a_{q+1}^q, $\ldots$, a_{L+r}^q constitute a valid mixed-radix representation for X_q and hence can be used for error detection/correction.

Figure 9-18 shows the physical ROM array as it would be modified for calculating the mixed-radix digits for X_1. The shaded blocks in row 1 are "shorted" so the corresponding x_i is passed through the block unchanged. The rest of the array remains as it was in the general mixed-radix converter. Figure 9-19 shows a similar case for X_3, where the third row is "shorted" (cross-hatch shading) and the third column is disabled (solid shading) with its output set to zero (denoted by ground).

The mechanism described above for calculating X_q can also be used to modify the general mixed-radix converter if a permanent failure occurs in the qth module, and it is subsequently removed from operation. In this case the qth column is permanently shorted, and the variable $s_{q,q}$ is permanently set to zero. Further reduction proceeds in the same manner; that is, rows and columns of the ROM array are removed pairwise as the system gradually degrades due to hardware failures.

To optimize the data throughput rates, the ROM array structure can be pipelined as shown in Figure 9-20. Latch registers are inserted between each level of ROM, so that at any given time, there are $L + r - 1$ projections progressing through the array simultaneously, with each in a different stage of computation. It is convenient to design the array structure so the projections are produced at the output in the order $X_5, X_4, \ldots, X_1$. In this case the complete representation of X itself is produced simultaneously with X_5, since the mixed-radix representation of X is available for use in output translation, or for other types of additional processing.

The circuit of Figure 9-20 is designed so that at each iteration $q = 1, \ldots, 5$ the full set of residue digits $x_1, \ldots, x_5$ enters the top of the array. The pipeline is shown in Figure 9-20 after $L + r - 1$ iterations, where the shaded latches denote invalid numerical values due to the fact that a complete set of residues is used, rather than the reduced set that properly characterizes a projection. Up to this point in the iteration cycle none of the incorrect data values have propagated out of their original columns, and hence they do not invalidate the other calculations up to this stage. During the $(L + r)$th iteration, each block in the array structure "shorts,"

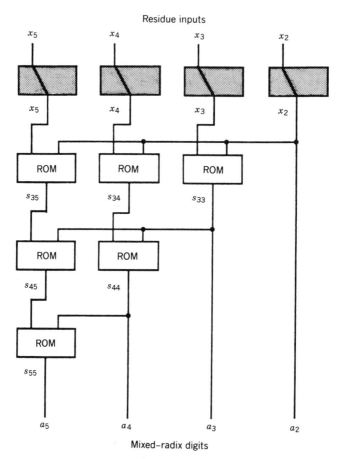

Residue inputs

FIGURE 9-18 Modification of the ROM array for calculating the mixed-radix digits of X_1.

thereby passing the contents of its input latch to the output latch unaltered. The incorrect data are therefore not propagated to the left but rather are sent ahead one level in the pipeline and will eventually occupy the position of the undefined mixed-radix digit for that particular projection. Since the undefined mixed-radix digit is never used, the erroneous value latched into that position is simply ignored. Note that by "shorting" all blocks at the $(L + r)$th iteration, each X_q is spared the multiplication of m_q^{-1}, which is equivalent to setting $m_q = 1$ during that iteration, as required by Eq. (9.18).

Preventing the propagation $s_{q,q}$ at the $(L + r)$th iteration is equivalent to setting $a_q = 0$ in Eq. (9.21). By means of this design, the pipelined mixed-radix converter sequentially produces a stream of ordered projections in mixed-radix form, with the $(L + r)$th projection being produced simultaneously with the mixed-radix representation of the true X. It then becomes a simple matter to construct a logic circuit, or to use a microprocessor, to monitor the a_k^q terms as they come streaming

Residue inputs

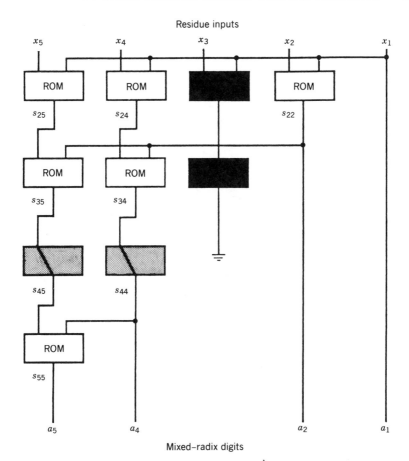

Mixed–radix digits

FIGURE 9-19 Modification of the ROM array for calculating the mixed-radix digits of X_3.

through the pipeline and to flag the location of any error that is detected. Error correction requires additional circuitry, although once the a_k^q terms are available, the correct value of x_q can be obtained from the following expression:

$$x_q = \langle X_q \rangle_{m_q} = \left\langle \sum_{\substack{k=1 \\ k \neq q}}^{L+r} a_k^q \prod_{\substack{i=1 \\ i \neq q}}^{k-1} m_i \right\rangle_{m_q} \qquad (9.22)$$

Since Eq. (9.22) is calculated in mod m_q arithmetic, the correct digit can be calculated in a modular circuit that uses ROM to store the partial sums $S_k(\cdot)$ and a modular adder to accumulate the partial sums according to Eq. (9.23):

$$x_q = \left\langle \sum_{\substack{k=1 \\ k \neq q}}^{L+r} S_k(a_k^q) \right\rangle_{m_q}, \qquad (9.23)$$

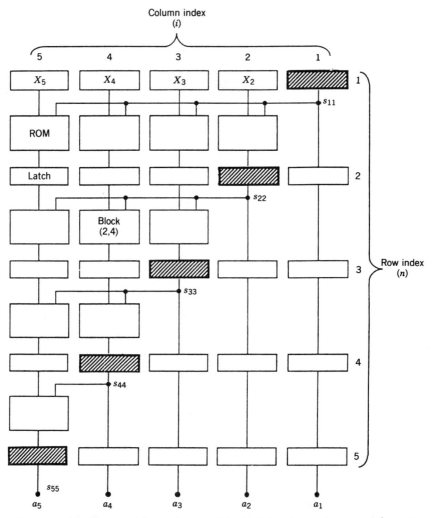

FIGURE 9-20 Mixed-radix error checker with pipelining for maximum data rate (shown after $(L + r - 1)$th iterations).

where

$$S_k(a_k^q) = \left\langle a_k^q \prod_{\substack{i=1 \\ i \neq q}}^{k-1} m_i \right\rangle_{m_q}$$

A hardware prototype of the pipelined error checker in Figure 9-20 was constructed from MSI circuit packages and tested to verify the theory. A block diagram for each of the array elements is shown in Figure 9-21; photographs of the hardware module are shown in Figure 9-22. Note that each array element is an

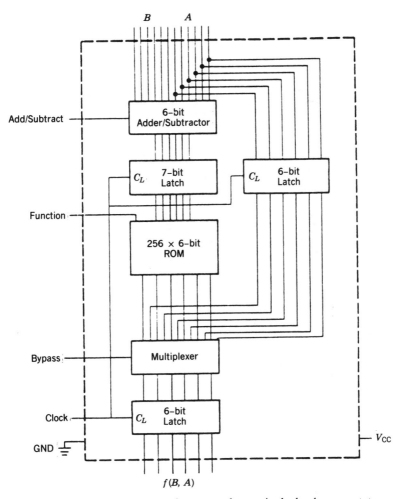

FIGURE 9-21   Block diagram of an array element in the hardware prototype.

adder-ROM structure in which the look-up table implements the multiplication by the appropriate inverse modulus. A multiplexer is used to transfer the A input directly to the output latch when this function is required by the algorithm. This experimental module was interfaced to the S100 bus of a Motorola 68000 microprocessor, which generated the residue test sequences and performed the monitoring function on the mixed-radix digits. Experiments confirmed that the error checker performed correctly according to the theory that was described above. In a realistic problem a VLSI circuit would be developed for the entire hardware module, thereby resulting in an "error checker on a chip."

A question that invariably arises in fault-tolerant processor design concerns the reliability of the error checker itself. Although this issue is beyond the scope of this discussion, it is interesting to observe that the hardware array that constitutes

FIGURE 9-22 Hardware prototype mixed-radix error checker: (a) front and (b) top.

the error checker is a self-checking structure. This is due to the fact that the mixed-radix digits for all the projections are calculated with RNS arithmetic. It is immaterial whether an error occurs in one of the original residues entering at the top of the error checker, or whether an error occurs in the array itself, say, in the qth column. In either case the result produced in the qth column, $s_{q,q}$, contains an error. Since the qth column is electronically disabled during the mixed-radix calculations for X_q, such an error can be detected and located by the same procedure used for error in the original residues. Jenkins and Altman [JE88] present a complete analysis of these self-checking properties of a mixed-radix error checker.

9-5 COMPLEX RNS ARITHMETIC

In recent years the efficient implementation of complex digital arithmetic has become increasingly important because many signal processing applications require processing of complex signals with complex digital filters, that is, digital filters that have complex coefficients. Examples of this can be found in baseband processing for narrowband radiofrequency signals in homomorphic speech processing, in spectral analysis, and in matched filtering for coherent radars. Conventional complex signal processing requires the formulation of real and imaginary channels and special handling of cross-product terms during complex multiplication. In general, currently available integrated microprocessors and specialized DSP devices do not have dedicated hardware for complex arithmetic, but rather they depend on a software (or firmware) implementation of complex arithmetic. This results in limited processing rates for applications that require large numbers of complex arithmetic operations.

During the last few years many researchers have worked with certain types of modular number systems that have an unusual representation of complex data, which leads to a decoupling of the real and imaginary channels [JE87a; JU87; TA85b]. This decoupling simplifies complex multiplication and provides error isolation between the real and imaginary channels. The following section develops the theory of quadratic residue number system arithmetic (QRNS) and shows how it can be used in the design of special-purpose digital signal processors that can switch between *ordinary complex arithmetic* and *quadratic complex arithmetic*.

9-5-1 Complex Modular Arithmetic

The set of integers $M = \{0, \ldots, m - 1\}$, together with mod m addition and multiplication, forms a finite integer ring $R(m)$ for any positive integer m. $R(m)$ is a real ring in the sense that all elements of M are real integers. Recall that the ordinary complex number system results because the equation $x^2 = -1$ does not have a solution among the real numbers. Similarly, to form a complex modular structure it is necessary to determine if Eq. (9.24) has a solution in $R(m)$:

$$x^2 = \langle -1 \rangle_m. \tag{9.24}$$

If there exists a solution to Eq. (9.24), $\hat{j} \in R(m)$, then Eq. (9.24) is *solvable* and -1 is a quadratic residue modulo m. Otherwise Eq. (9.24) is *nonsolvable* and -1 is a quadratic nonresidue. The first step in constructing a complex modular system for a given m is to determine if Eq. (9.24) is solvable or nonsolvable. A well-known result from number theory states that the number -1 is a quadratic residue mod p of all primes of the form $p = 4k + 1$ and a quadratic nonresidue modulo p of all primes of the form $p = 4k + 3$. Hence for an arbitrary m, if m is prime, Eq. (9.24) is solvable if and only if m has the form $4k + 1$. (Note that the expressions above include all odd primes. For the special case $p = 2$, Eq. (9.24) is solvable modulo 2, but it can be shown that a quadratic representation does not exist.) If m is not prime, it is necessary to analyze its prime factorization, $m = p_1^{r_1}, \ldots, p_L^{r_L}$. It is known from number theory that an integer $\hat{j}$ is a solution to Eq. (9.24) if and only if -1 is also a quadratic residue of all primes that divide m. Hence it is sufficient to consider solutions to Eq. (9.25):

$$x^2 = \langle -1 \rangle_{p_i^{r_i}}, \qquad i = 1, \ldots, L, \tag{9.25}$$

since the solutions to Eq. (9.24) can be obtained from the L solutions to Eq. (9.25) and the Chinese Remainder Theorem. For a prime of the form $p_i = 4k_i + 1$, Eq.

(9.26) always has two solutions, $\hat{j}_{i1}$ and $\hat{j}_{i2}$, that are both additive and multiplicative inverses of each other.

$$x^2 = \langle -1 \rangle_{p_i}, \qquad i = 1, \ldots, L. \tag{9.26}$$

If $\hat{j}_{i_1}$ is a solution to Eq. (9.26), then a solution to Eq. (9.25), $\hat{j}_i$, always exists and is congruent to $\hat{j}_{i_1}$ modulo p_i. Therefore it can be determined if Eq. (9.24) is solvable for a given modulus m by examining each prime factor of m. A consequence of this theory is that Eq. (9.24) is solvable if and only if all the prime factors of m are of the form $p_i = 4k_i + 1, i = 1, \ldots, L$.

9-5-1-1   Complex Extension Fields. First consider the case when the modulus is a prime of the form $p = 4k + 3$, so that Eq. (9.24) is nonsolvable and $\hat{j} = \langle \sqrt{-1} \rangle_p$ cannot be found in $F(p)$. A complex modular structure, represented by the second degree extension field $F(p^2)$, can be formed by taking ordered pairs $(x_r, x_i) \approx x_r + \hat{j}x_i$, with $x_r \in F(p)$ and $x_i \in F(p)$, and defining addition "$+$" and multiplication "$\bullet$" in $F(p^2)$ by:

$$
\begin{aligned}
(x_r, x_i) + (y_r, y_i) &= (u_r, u_i), \\
(x_r, x_i) \bullet (y_r, y_i) &= (z_r, z_i),
\end{aligned}
\tag{9.27a}
$$

where

$$
\begin{aligned}
u_r &= \langle x_r + y_r \rangle_p, \\
u_i &= \langle x_i + y_i \rangle_p, \\
z_r &= \langle x_r y_r - x_i y_i \rangle_p, \\
z_i &= \langle x_i y_r + x_r y_i \rangle_p.
\end{aligned}
\tag{9.27b}
$$

Note that the arithmetic defined by Eqs. (9.27) is similar to ordinary complex arithmetic except that the real and imaginary components are computed with mod p arithmetic. Formally, $F(p^2)$ is a second degree finite extension field containing p^2 elements and, as such, possesses all the algebraic properties of a complex modular field (multiplicative inverses, generators, log tables, etc.). Such complex finite fields have interesting properties that are useful in signal processing, although they will not be considered further in this discussion.

9-5-1-2   Complex Extension Rings. Suppose now that the requirements are relaxed so that the modulus m is not prime, but that $M = \{(x_r, x_i)/x_r, x_i \in R(m)\}$ and ordered pairs are treated by the same rules defining complex arithmetic given by Eqs. (9.27). This structure, referred to here as the *extended structure* of $R(m)$, is itself a finite ring $R(m^2)$. Note that this extended structure $R(m^2)$ is a complex ring whether or not Eq. (9.24) is solvable, as long as the rules of arithmetic given by Eqs. (9.27) are used to add and multiply the elements of $R(m^2)$.

Now consider the same situation described in the previous paragraph, but with the additional constraint that all the prime factors of m are of the form $p_i = 4k_i + 1$, so that Eq. (9.24) has a solution $\hat{j} \in R(m)$, and -1 is a quadratic residue modulo m. Under these conditions, the ring $R(m^2)$ can be mapped onto a *quadratic ring* $QR(m^2)$, which is isomorphic to $R(m^2)$; that is,

$$(x_r, x_i) \in R(m^2) \leftrightarrow (X, \hat{X}) \in QR(m^2),$$

where $X = \langle x_r + \hat{j}x_i \rangle_m$, $\hat{X} = \langle x_r - \hat{j}x_i \rangle_m$, $x_r = \langle 2^{-1}(X + \hat{X}) \rangle_m$, and $x_i = \langle 2^{-1}\hat{j}^{-1}(X - \hat{X}) \rangle_m$. It is essential that $\hat{j} \in R$ so that both X and $\hat{X}$ are real numbers that form an ordered pair $(X, \hat{X})$ to represent (x_r, x_i) in the quadratic ring $QR(m^2)$. Addition and multiplication within $QR(m^2)$ are defined by:

$$(X, \hat{X}) \circ (Y, \hat{Y}) = (Z, \hat{Z}), \tag{9.28a}$$

where

$$Z = \langle X \circ Y \rangle_m, \tag{9.28b}$$
$$\hat{Z} = \langle \hat{X} \circ \hat{Y} \rangle_m,$$

and where "∘" denotes QRNS addition, subtraction, or multiplication. In the QRNS system, addition, subtraction, and multiplication have the same simple form as addition in conventional complex arithmetic; that is, both are computed by independent operations in the uncoupled channels. The elimination of cross-product terms in the complex multiply translates to improved speed, reduced hardware complexity, error isolation between channels, better testability for VLSI realization, and a regularity in structure that is beneficial for hardware multiplexing.

There are two distinctly different types of quadratic rings that can be obtained, depending on whether the modulus m is a prime of the form $4k + 1$, or whether it is a composite integer whose prime factors all have this particular form. In the first case the underlying real structure is a finite field, whereas in the second it is a finite ring. However, in either case the extended structures are complex rings; that is, $R(m^2)$ and $QR(m^2)$ are isomorphic rings regardless of whether the modulus m is a prime or not. However, if m is prime the component arithmetic becomes finite field arithmetic, which offers the possibility of using finite field logarithms to implement the real mod p multiplication.

The particular form selected for the modulus m leads to four possible complex structures, depending on whether m is prime or whether Eq. (9.24) is solvable. These four cases are summarized in Table 9-2. The quadratic ring listed in Class 4 is the most promising structure in the sense that the choice of m is flexible. The next question is how to best choose m to obtain the desired properties while minimizing the complexity of the hardware realization.

9-5-2 Dual-Mode Machines Designed with Quadratic RNS Arithmetic

In order to design a processor that can operate efficiently with both normal complex arithmetic and quadratic complex arithmetic, that is, with both $R(m^2)$ and $QR(m^2)$,

TABLE 9-2 Summary of Complex Modular Structures

Class	Modulus Type	Prime Factors	Eq. (9.24) Solvable?	Real Structure	Extended Structure	Quadratic Structure
1	p Prime	$4n+3$	No	$GF(p)$	$GF(p^2)$	None
2	p Prime	$4n+1$	Yes	$GF(p)$	$R(p^2)$	$QR(p^2)$
3	m Composite	At least one prime factor of the form $p_i = 4n_i + 3$	No	$R(m)$	$R(m^2)$	None
4	m Composite	All prime factors of the form $p_i = 4n_i + 1$	Yes	$R(m)$	$R(m^2)$	$\boxed{QR(m^2)}$

respectively, it is important to select a modulus m that leads to a simple implementation of the ring arithmetic, as well as to simple translation algorithms between the two modes of operation. It is well known that moduli of the form $2^n - 1$, 2^n, and $2^n + 1$ result in ring arithmetic that can be implemented with carry-add, 2's-complement, and carry-subtract binary arithmetic, respectively. It is clear that simple power-of-2 moduli do not admit a quadratic representation, since the only prime factor is 2, for which it is known that the quadratic representation does not exist. The diminished power-of-2 moduli also do not admit the quadratic structure, since $2^n - 1 = 4(2^{n-2} - 1) + 3$, which has the form $4k + 3$ with $k = 2^{n-2} - 1$, $n > 1$. (The cases of $n = 0$ and $n = 1$ are degenerate and of no practical interest.)

This leaves the augmented power-of-2 moduli as the most promising possibility. Table 9-3 shows all the augmented power-of-2 integers for $0 \le n \le 16$, their prime factorizations, and whether or not the modulus admits the quadratic representation. Table 9-3 shows explicitly that the quadratic representation exists for all even n up to 16. Also, a reasonable number of these are prime, so that both Class 2 and Class 4 quadratic rings are possible within this range. The pattern observed in Table 9-3 is true in general; that is, for all moduli of the form $m = 2^n + 1$, the number -1 is a quadratic residue mod m if and only if $n > 0$ is an even integer. This means that the quadratic representation exists for all augmented power-of-2 moduli with even n.

The above discussion establishes that a carry-subtract binary processor with an odd number of bits can implement complex arithmetic in the isomorphic quadratic ring. Next, it is important to examine the complexity of the algorithms needed to translate between $R(m^2)$ and $QR(m^2)$, assuming it will be desirable to execute some operations in $R(m^2)$, although most of the complex multiplications will be done in the quadratic ring. Note that for augmented power-of-2 moduli with $m = 2^{2r} + 1$, the square root of -1 is a simple power-of-2; that is, $\hat{j} = \sqrt{-1} = 2^r$. It can also be shown that $2^{-1} = -2^{2r-1}$, $\hat{j}^{-1} = -2^r$, and $(2^{-1}\hat{j}^{-1}) = -2^{r-1}$. Therefore the mapping between $R(m^2)$ and $QR(m^2)$ is described by:

$$X = f_1(x_r, x_i), \qquad x_r = g_1(X, \hat{X}),$$
$$\hat{X} = f_2(x_r, x_i), \qquad x_i = g_2(X, \hat{X}),$$

$$(9.29a)$$

TABLE 9-3   Augmented Power-of-2 Moduli

n	$m = 2^n + 1$	Prime Factorization	Eq. (9.24) Solvable?
0	1	1 (prime)	Yes
1	3	3 (prime)	No
2	5	5 (prime)	Yes
3	9	3^2	No
4	17	17 (prime)	Yes
5	33	(3)(11)	No
6	65	(5)(13)	Yes
7	129	(3)(43)	No
8	257	257 (prime)	Yes
9	513	$(3^2)(57)$	No
10	1,025	$(5^2)(41)$	Yes
11	2,049	(3)(683)	No
12	4,097	4097(prime)	Yes
13	8,193	(3)(2731)	No
14	16,385	(5),(11),(331)	Yes
15	36,769	$(3^2)(11)(331)$	No
16	65,637	65537(prime)	Yes

where

$$f_1(x_r, x_i) = \langle x_r + 2^r x_i \rangle_m,$$

$$f_2(x_r, x_i) = \langle x_r - 2^r x_i \rangle_m,$$

$$g_1(X, \hat{X}) = \langle -2^{2r-1}(X + \hat{X}) \rangle_m,$$ (9.29b)

$$g_2(X, \hat{X}) = \langle 2^{r-1}(\hat{X} - X) \rangle_m.$$

It is significant that all the constants in Eqs. (9.29b) are simple powers-of-2, so that the translations between $R(m^2)$ and $QR(m^2)$ can be implemented by shift, add, and complement operations.

In their work on Fermat number transforms with moduli of the form $2^b + 1$, where $b = 2^t$, Agarwal and Burrus [AG74] showed that modular operations for augmented power-of-2 moduli (the Fermat numbers are a subset of the entire set of augmented power-of-2 integers) can be realized with carry-subtract binary arithmetic. The complexity of carry-subtract arithmetic is similar to that of carry-add arithmetic, which is the type implemented in computers using 1's-complement binary codes. Numerous computer manufacturers (e.g., Control Data Corporation and Sperry Univac) have used 1's-complement codes in general purpose scientific computers and have successfully competed with IBM's choice of 2's-complement codes. One of the undesirable features of augmented power-of-2 codes is the extra bit required to represent the state 2^{2r}, since this extra bit does not provide a significant increase in dynamic range.

9-5-2-1   Realization with Diminished-1 Binary Coding.   A diminished-1 binary code results from performing a linear code translation on a standard 2's-comple-

ment code, of the form $x_D = x_{2C} - 1$. The idea behind this translation is to let the state $10 \cdot \cdot \cdot 0$ represent zero, so that all nonzero elements are represented by $2r$ bits; that is, this one troublesome state no longer represents the largest state in the ring, 2^{2r}, but now represents the zero state. Since arithmetic operations involving zero are trivial, this state is detected and handled as a special case. Arithmetic operations involving all other nonzero states can be implemented using only $2r$ bits of the operands, with rules similar to those of 1's-complement arithmetic. Although a $(2r + 1)$-bit wordlength is still required in the memories and data paths of such a processor, the arithmetic elements (adders, multipliers, etc.) are required to handle only $2r$ bits and hence are kept as simple as possible. Table 9-4 shows a comparison of diminished-1 $(2r = 4)$, 2's-complement $(n = 4)$, and 1's-complement $(n = 4)$ codes for both integer and fractional number systems. Note that in the diminished-1 code the $(2r + 1)$th bit serves to identify zero, whereas the $2r$th bit serves as a sign bit, much the same as the nth bit serves as a sign for both the 1's- and 2's-complement codes.

In the following discussion the rules for implementing mod $(2^{2r} + 1)$ arithmetic in a diminished-1 binary code are presented and examples are presented using $n = 2r = 4$ (5-bit wordlength) to clarify the basic principles. Let x be represented by $x_n, \ldots, x_0$, where each x_i, $i = 0, \ldots, n$, is a binary bit. Also let x' denote the n lower bits of x $(x_{n-1}, \ldots, x_0)$, and let x_c denote the bit-by-bit complement of x. Also, the special sign $\boxplus$ will be used to denote carry-complement addition, which is normal 1's-complement addition except that the end-around carry is complemented before it is added back to the least significant bit position. (Note that carry-complement addition has essentially the same complexity as carry-add addition characteristic of 1's-complement systems.) With the aid of this notation it

TABLE 9-4 Comparison of Three Binary Codes ($n=4$)

Binary	Diminished-1		2's-Complement		1's-Complement	
00000	1	0.125	0	0.000	0	0.000
00001	2	0.250	1	0.125	1	0.125
00010	3	0.375	2	0.250	2	0.250
00011	4	0.500	3	0.375	3	0.375
00100	5	0.625	4	0.500	4	0.500
00101	6	0.750	5	0.625	5	0.625
00110	7	0.875	6	0.750	6	0.750
00111	8	1.000	7	0.875	7	0.875
01000	9 (−8)	−1.000	8 (−8)	−1.000	8 (−7)	−0.875
01001	10 (−7)	−0.875	9 (−7)	−0.875	9 (−6)	−0.750
01010	11 (−6)	−0.750	10 (−6)	−0.750	10 (−5)	−0.625
01011	12 (−5)	−0.625	11 (−5)	−0.625	11 (−4)	−0.500
01100	13 (−4)	−0.500	12 (−4)	−0.500	12 (−3)	−0.375
01101	14 (−3)	−0.375	13 (−3)	−0.375	13 (−2)	−0.250
01110	15 (−2)	−0.250	14 (−2)	−0.250	14 (−1)	−0.125
01111	16 (−1)	−0.125	15 (−1)	−0.125	15 (−0)	−0.000
10000	17 (0)	0.000	xxxxxx	xxxxxx	xxxxxx	xxxxxx

is now possible to give concise definitions of all the basic diminished-1 operations that are important in digital signal processing. This basic theory was originally worked out by Leibowitz [LE76] for the mechanization of Fermat number transforms.

1. *Negation.* If $x_n = 1$, then $x \equiv 0$ and $-x = x$. If $x_n = 0$, then $-x = (0, x'_c)$. For example, if $x = 00010 = 3$, then $-x = 01101$. (See Table 9-3.) Therefore changing the sign of a diminished-1 number is very simple.

2. *Addition.* Suppose it is desired to form the sum $s = x + y$. If $x_n = 1$, then $x \equiv 0$ and $s = y$. If $y_n = 1$, then $y \equiv 0$ and $s = x$. If $x_n = 1$ and $y_n = 1$, then $x \equiv 0$, $y \equiv 0$, and $s \equiv 0$. If $x_n = 0$ and $y_n = 0$, then $s' = x' \boxplus y'$, and $s = (0, s')$. For example, consider the addition $s = (-5) + (7) = 2$:

$$
\begin{array}{r}
-5 \approx 0\ 1011 \\
7 \approx 0\ 0110 \\
\hline
1\ 0001 \\
0 \quad \text{(complement and add carry)} \\
\hline
2 \approx 0\ 0001 \quad \text{(final result)}
\end{array}
$$

Note that addition of nonzero quantities is essentially the same as 1's-complement addition.

3. *Multiplication by 2^k.* If $x_n = 1$, then $x \equiv 0$ and the multiplication is inhibited so that $y = 2^k x \equiv 0$. If $x_n = 0$, then the bits of x' are left circularly rotated by k positions while each bit that is shifted out of the $(n - 1)$th bit position is complemented before it is shifted into the least significant bit position. For example, consider the operation $y = 2^3(-2)$:

$$
\begin{array}{r}
-2 \approx 0\ 1110 \\
2^3(-8) \approx 0\ 0000 \Rightarrow 1
\end{array}
$$

Since $\langle 8(-2) \rangle_{17} = 1$, it can be seen that this algorithm produces the correct result.

4. *Translation Between Diminished-1 and 2's-Complement Codes.* Before discussing general multiplication, it is important to establish simple algorithms for converting between diminished-1 and 2's-complement representation, and vice versa. Let x_D denote the diminished-1 representation of x, and x_{2C} the 2's-complement representation of x. First recall that in the diminished-1 representation, x_n serves to identify zero, and x_{n-1} serves as a sign bit; that is, a number x is negative if and only if $x_{n-1} = 1$. Then x_{2C} can be derived as follows: (a) if $x_n = 1$, then $x_{2C} = x'$; (b) if $x_n = 0$ and $x_{n-1} = 1$, then $x_{2C} = x'$; (c) if $x_n = 0$ and $x_{n-1} = 0$, then $x_{2C} = x' + 1$. For case (c), x_{2C} can be formed by complementing each bit of x', starting from the least significant position, until the first zero is encountered. The first zero is complemented, but all bits further to the left remain unchanged. For example, if $x_D = 00100$, then $x_{2C} = 0101$; if $x_D = 01001$, then $x_{2C} = 1001$. Using this rule, the conversion from diminished-1 to 2's-complement can be implemented very quickly in a simple logic circuit.

The rules for translating for a 2's-complement to diminished-1 representation are essentially the reverse of the above procedure: (a) if x is zero, then x_D is formed by adding an extra 1 bit on the left end of x_{2C}; (b) if x is negative, then x_D is formed by adding an extra 0 bit on the left end of x_{2C}; (c) if x is positive, then $x_D = x_{2C} - 1$. For case (c), x_D can be derived from x_{2C} by complementing each bit from the right end of the word until the first 1 is encountered. This first 1 is complemented, but all bits further to the left remain unchanged, and an extra zero is added in the nth bit position. For example, if $x_{2C} = 0110$, then $x_D = 00101$; if $x_{2C} = 1000$, then $x_D = 01000$.

5. *General Multiplication.* Note first that due to the simple rules described in the previous paragraph for diminished-1 to 2's-complement translation, it is quite easy to translate from one representation to the other as necessary. The simplest general multiplication algorithm first converts the operands to 2's-complement representation, performs a normal 2's-complement multiply, and then completes the operation with a modular reduction, which is accomplished by a diminished-1 subtraction of the n most significant bits from the n least significant bits. For example, suppose it is desired to multiply $x = 5$ by $y = -7$ to produce $p = \langle -35 \rangle_{17} = 16$. Then $x_D = 00100$ and $y_D = 01001$, $x_{2C} = 0101$ and $y_{2C} = 1001$. The intermediate 8-bit product is $p_{2C} = 11011101$ and $p_D = 11011100$, and the final corrected product is $p' = (1100 \boxplus 0010) = 1111$, and $p_D = 01111 \approx 16$. Similarly, for an example when both operands are positive, consider $x = 5$ and $y = 6$ multiplied to produce $p = \langle 30 \rangle_{17} = 13$. Then $x_D = 0100$, $y_D = 0101$, $x_{2C} = 0101$, $y_{2C} = 0110$, $p_{2C} = 00011110$, $p_D = 00011101$, $p' = (1101 \boxplus 1111) = 1100$, and $p_D = 01100 \approx 13$.

Note that a real modular multiplication is more difficult within a diminished-1 code than in a 2's-complement code in that it requires the translations into and out of the intermediate 2's-complement representation. However, when implemented in the quadratic ring $QR(m^2)$, a complex multiplication has been decomposed into two noninteracting real multiplications, which represents a considerable net savings.

9-5-2-2 Scaling and Quantization in Diminished-1 Arithmetic. In most signal processing applications it is important to be able to scale by a power-of-2 and quantize the result so the data remain within the dynamic range of the fixed-point number code. In complex arithmetic, it is often necessary to scale the real and imaginary parts of complex quantities. In many circumstances scaling should be done in $R(m^2)$ rather than in $QR(m^2)$, since quantization errors in the quadratic ring can result in very large errors in the real and imaginary parts when translated back into $R(m^2)$. This is one reason the dual-mode nature of the design is important, and why the translation algorithms between the two modes should be as simple as possible.

To scale $(x_r, x_i) \in R(m^2)$ by 2^{-k}, both the real and imaginary parts (encoded in the diminished-1 binary code) are shifted right while extending the sign bit (x_{n-1}) by k bits and quantizing to an integer. If the quantization is done by simply truncating the result, a strange form of quantization occurs in which positive numbers are quantized to the right nearest integer (away from zero) and negative numbers are quantized to the left nearest integer (also away from zero). This form of quantization has been called *sign-magnitude* (S-M) *augmentation* to denote its similar but opposite nature to sign-magnitude truncation.

To illustrate S-M augmentation, consider scaling the number $7 \approx 00110$ by 2^{-2} in the diminished-1 code. Note that $2^{-2}(7) = 1\frac{3}{4}$, so that S-M truncation would result in a quantized value of 1. However, in the diminished-1 code $[2^{-2}(7)]_t = 00001 \approx 2$, where $[\cdot]_t$ denotes truncation. This illustrates that the exact quantity $1\frac{3}{4}$ has been quantized away from zero. Similarly, consider $2^{-2}(-7) = -(1\frac{3}{4})$. In the diminished-1 code the result is $[2^{-2}(-7)]_t = 01110 \approx -2$. Again the quantization has moved the result away from zero. At the present time, very little is known about the effects of S-M augmentation in special purpose processors such as digital filters, where the details of the quantization scheme have a significant impact on limit cycle behavior and quantization error accumulation.

Since scaling generally requires a translation from $QR(m^2)$ to $R(m^2)$, it appears that quadratic modular arithmetic is best suited for applications that require many complex adds and multiplies, but relatively few scaling operations. Figures 9-23 and 9-24 show block diagrams of the converters required to translate back and forth between the operating modes of the modular arithmetic. Figure 9-25 shows one channel of a binary adder (diminished-1), and Figure 9-26 shows one channel of a binary (diminished-1) multiplier. The disconnected structure of the two "channels" in the quadratic mode of operation provides error isolation between channels, as well as modularity in the hardware, both of which are features that are desirable for highly integrated VLSI systems. A block diagram for the scalar is not shown explicitly, since it consists of the combined elements in Figures 9-23 and 9-24, with a truncation operation performed between the stages.

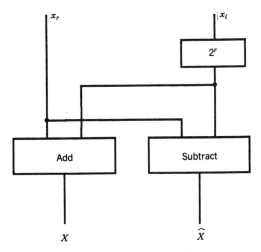

FIGURE 9-23   An $R(m^2)$-to-$QR(m^2)$ converter.

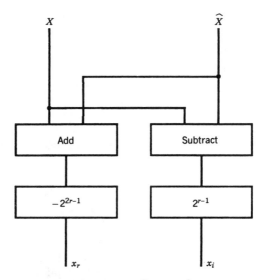

FIGURE 9-24   A $QR(m^2)$-to-$R(m^2)$ converter.

9-5-3   A Complex Recursive Digital Filter Design

Suppose it is desired to realize a complex baseband recursive filter described by

$$y[n] = \sum_{k=0}^{N} a_k x[n - k] + \sum_{k=1}^{N} b_k y[n - k], \tag{9.30}$$

where $x[n]$ and $y[n]$ are the complex input and output signals, respectively. This type of complex filter might arise in an application if a lowpass digit filter with a real impulse response $h[n]$ were converted into a tunable bandpass filter by modulating the impulse response; that is, the modified filter with passband center at ω_0 is characterized by $h'[n] = e^{j\omega_0 n} h[n]$. The direct form recursive filter described by

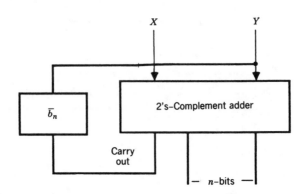

FIGURE 9-25   A diminished-1 binary adder (one channel).

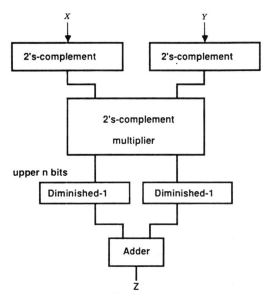

FIGURE 9-26   A diminished-1 multiplier (one channel).

Eq. (9.30) will be realized with a ROM-ACCUMULATOR architecture. Let $m = 2^{2r} + 1$ be the modulus of a $(2r + 1)$-bit quadratic modular code that will be used to design this filter. Assume that $x[n] = x_R[n] + \hat{j}x_I[n]$ and $y[n] = y_R[n] + \hat{j}y_I[n]$, each with $(2r + 1)$-bit integer real and imaginary parts, are encoded in the QRNS by

$$X[n] = \langle x_R[n] + 2^r x_I[n]\rangle_m,$$
$$\hat{X}[n] = \langle x_R[n] - 2^r x_I[n]\rangle_m \qquad (9.31)$$

($Y[n]$ and $\hat{Y}[n]$ are similarly encoded), that each is then expressed as a binary integer according to

$$X[n] = \sum_{i=0}^{2r} 2^i X_i[n] \quad \text{and} \quad \hat{X}[n] = \sum_{i=0}^{2r} 2^i \hat{X}_i[n], \qquad (9.32)$$

and that $Y[n]$ and $\hat{Y}[n]$ are represented similarly. The complex coefficients a_k and b_k are encoded as $(A_k, \hat{A}_k)$ and $(B_k, \hat{B}_k)$, and the filter is decomposed into two parallel subfilters described by

$$Y[n] = \left\langle \sum_{i=0}^{2r} 2^i F(S_i[n]) \right\rangle_m,$$
$$\hat{Y}[n] = \left\langle \sum_{i=0}^{2r} 2^i \hat{F}(S_i[n]) \right\rangle_m, \qquad (9.33)$$

where

$$F(S_i[n]) = \left\langle \sum_{k=0}^{N} X_i[n - k]A_k + \sum_{k=1}^{N} Y_i[n - k]B_k \right\rangle_m,$$

$$\hat{F}(\hat{S}_i[n]) = \left\langle \sum_{k=0}^{N} \hat{X}_i[n - k]\hat{A}_k + \sum_{k=1}^{N} \hat{Y}_i[n - k]\hat{B}_k \right\rangle_m,$$

and

$$S_i[n] = [X_i[n], \ldots, X_i[n - N], Y_i[n - 1], \ldots, Y_i[n - N]],$$

$$\hat{S}_i[n] = [\hat{X}_i[n], \ldots, \hat{X}_i[n - N], \hat{Y}_i[n - 1], \ldots, \hat{Y}_i[n - N]],$$

are $(2N - 1)$-bit addresses that are formed from *bit slices*.

The filter output must be scaled (down) and quantized before the subsequent iteration to prevent overflow. This is accomplished by first forming

$$y_R[n] = \langle 2^{-1}(Y[n] + \hat{Y}[n]) \rangle_m \qquad (9.34a)$$

and

$$y_I[n] = \langle 2^{-1}\hat{j}(Y[n] - \hat{Y}[n]) \rangle_m, \qquad (9.34b)$$

and then scaling and truncating these quantities. (Note that during the above operation the outputs from the separate channels interact.) If $S_k^R(\cdot)$ denotes a right arithmetic shift of the argument by k bits, and $[\cdot]_Q$ denotes quantization (rounding or truncating), the scaling is accomplished by forming

$$y_R'[n] = [S_k^R(y_R[n])]_Q \quad \text{and} \quad y_I'[n] = [S_k^R(y_I[n])]_Q, \qquad (9.35)$$

and then reencoding to obtain

$$\begin{aligned} Y'[n] &= \langle y_R'[n] + 2^r y_I'[n] \rangle_m, \\ \hat{Y}'[n] &= \langle y_R'[n] - 2^r y_I'[n] \rangle_m. \end{aligned} \qquad (9.36)$$

The scaled quantities $(Y'[n], \hat{Y}'[n]$ are then used in the next iteration. Although the scaling operation appears rather complicated, note that it is essentially equivalent to four additions and four circular rotations, since the multipliers in this system are all powers-of-2. Scaling by this algorithm is a good example of a *dual-mode design*, in which data are translated between $R(m^2)$ and $QR(m^2)$ in order to take advantage of special properties of both structures. If the quantization operation denoted by $[\cdot]_Q$ is simply chopping of the noninteger part (rightmost k bits after right shifting), then the quantization scheme implemented by this algorithm is the sign-magnitude augmentation quantization discussed previously.

For example, consider a digital filter designed from a fourth-order analog

Chebyshev prototype. The bilinear transformation is employed with analog cutoff frequency prewarped to give a digital cutoff of 0.3π radians. Then the frequency response of the filter is shifted $\pi/2$ radians, which corresponds to modulating the complex carrier $e^{(j\pi/2)n}$ with the original impulse response $h[n]$. The impulse response of the complex filter is then $h[n]e^{(j\pi/2)n}$. The filter can be realized in the QRNS using a modulus of $2^{16} + 1$ and a scaling factor of 2^8. Thus the filter is operating with a scaled dynamic range of roughly $(-128, 128)$ and the accuracy of arithmetic operations and coefficient representation is about 8 bits.

The input signal $x[n] = 50\cos(0.1\pi n) + 50\cos(0.05\pi n)$ shown in Figure 9-27 consisting of two real tones was applied to the filter for 150 iterations. Figure 9-28 shows the real part of the output realized with the full 32-bit precision of the computer. This is used as a baseline for comparisons with various realizations based on the QRNS. Figure 9-29 shows the real part of the output from a direct form implementation and Figure 9-30 is the magnitude of the associated error. Similarly, Figures 9-31, 9-32, 9-33, and 9-34 show the real part of the output and the error magnitude for parallel structures consisting of second-order sections and (complex) first-order sections, respectively. In all cases the error is a significant component of the output, which is expected due to the 8-bit accuracy of this implementation. All three configurations exhibit roughly the same error power while the excitation is applied. Also, it is interesting to note that the size of the deadband limit cycles increases with increasing parallelism. This example demonstrates the feasibility of designing digital filter structures using quadratic modular concepts,

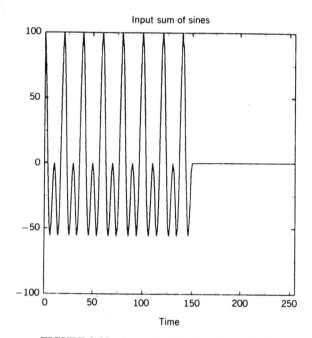

FIGURE 9-27 Sum of cosines input test signal.

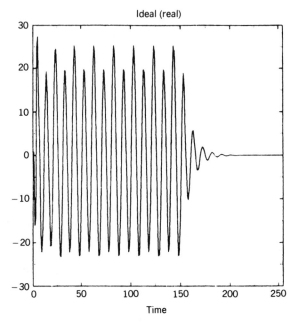

FIGURE 9-28 Real part of the filtered output signal (full precision).

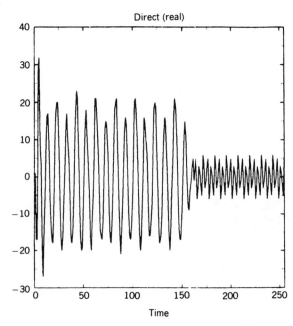

FIGURE 9-29 Real part of the filtered output signal (QRNS direct form).

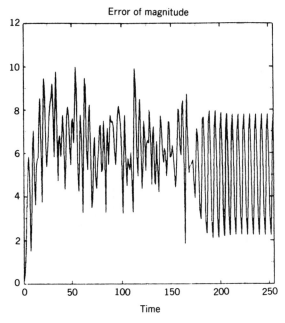

FIGURE 9-30 Magnitude of output error (QRNS direct form).

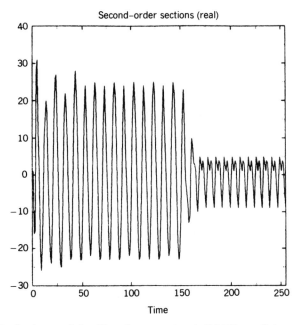

FIGURE 9-31 Real part of the filtered output signal (QRNS parallel second-order sections).

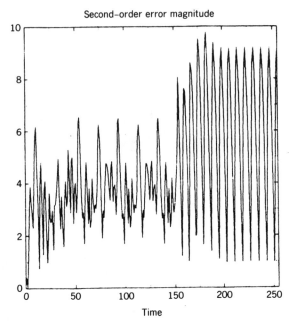

FIGURE 9-32   Magnitude of output error (QRNS parallel second-order sections).

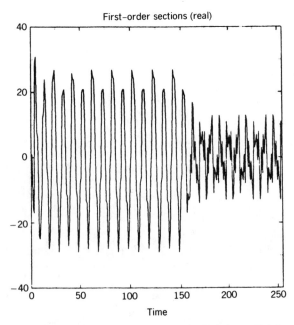

FIGURE 9-33   Real part of the filtered output signal (QRNS parallel first-order complex sections).

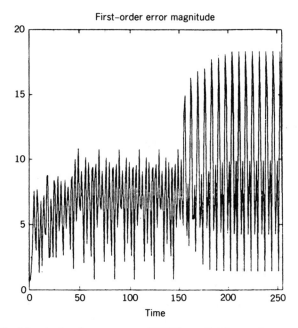

First-order error magnitude

FIGURE 9-34 Magnitude of output error (QRNS parallel first-order complex sections).

and it also demonstrates the occurrence of "complex limit cycles" due to sign-magnitude augmentation.

QRNS architectures are attractive for digital processors that are required to implement large numbers of high-speed complex arithmetic operations. These are based on a dual-mode concept in which the processor can operate in a normal complex modular ring for certain operations, or it can operate in an isomorphic quadratic complex ring where complex multiplication is greatly simplified through the elimination of cross-product terms. The best efficiencies are achieved when a large percentage of the total computational burden can be handled in the quadratic mode of operation, when the channels remain completely decoupled. Architectures based on the dual-mode concept may become attractive for VLSI realizations in which modularity, error isolation, testability, and efficient computational capabilities are all important features.

9-6 DESIGN EXAMPLE: A RNS DIGITAL CORRELATOR

This section discusses a high-speed digital correlator that was designed and constructed to implement the real-time correlation function required in an ultrasonic blood flow meter. In order to achieve the necessary real-time processing rates, a RNS architecture was selected for the hardware. The RNS architecture not only allows high-speed real-time processing rates to be achieved, but it also facilitates the design, construction, and testing of the hardware. The hardware correlator is controlled by a PC and is designed so that eventually it can be implemented in VLSI circuits for operation in a clinical environment.

9-6-1 Background

Recently, a new ultrasonic flow measurement method employing time-domain correlation of consecutive pairs of echoes was reported [EM86; FO84, HE90]. The time shift between a pair of range gated echoes is determined by searching for the shift that results in the maximum correlation. This shift indicates the distance a group of scatters has moved between pulses. A firm theoretical foundation of the scheme has been developed and the technique has been verified experimentally in controlled laboratory experiments. In order to implement this technique in a clinical blood flow measuring instrument, high-speed digital correlation must be executed in real-time. A design analysis revealed that highly repetitive correlations of length 41 with data samples that have 8 bits of accuracy must be implemented in a specially designed hardware correlator in order to achieve the data rates required for real-time operation.

It is now well known that high-speed digital correlation is one signal processing function that is ideally suited for RNS design techniques [SO86]. This is because the RNS is extremely effective in executing high-speed multiplication and addition. Furthermore, the modular structure of the RNS leads to modularity in the hardware that facilitates design, construction, and testing. The following discussion first describes the time-domain correlation flow measurement techniques. It then presents the design of a special purpose RNS digital correlator for use in a clinical prototype of an ultrasonic blood flow meter that operates with the time-domain correlation technique.

9-6-2 Flow Measurement Technique

Currently, all available flow meters utilize a Doppler technique to estimate volumetric blood flow. However, due to the relative inaccuracy of that method, a new method using time-domain correlation is being developed. This method relies on shift in the time domain rather than the frequency domain. In this method, ultrasonic pulses are transmitted at regular intervals. Adjacent echoes collected from these pulses are compared and used to calculate the change in position of the scattering media, in this case, blood cells. It has been discovered that a particular set of scatters will emit a characteristic echo when an ultrasonic pulse is transmitted. Thus if a set of scatters is in the cone of a transmitted pulse and remains in the cone of a subsequent pulse, the characteristic of that set of scatters will be present in the echoes from both pulses, as shown in Figure 9-35. By comparing these echoes, a time shift can be determined from which the scatterer velocity can be calculated.

In order to determine the scatterer displacement between two echoes, one echo is shifted until the scatterer characteristics present in each echo are directly superimposed. Since the correlation function is maximized when two identical wave-

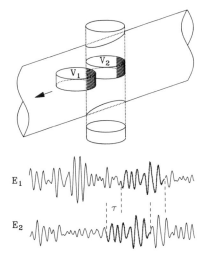

E_1

E_2

FIGURE 9-35 Two sample echoes each containing a shaded scatterer characteristic. The variable τ represents the time shift between the two echoes.

forms are exactly superimposed, correlation is used to determine when the correct shift has been located, as depicted in Figure 9-36. Because the spacing in time between consecutive ultrasonic pulses is fixed, the velocity of the scatterer can be quickly calculated once the shift is determined.

Since blood flow in a vessel is not of uniform velocity (it is faster in the center, slower at the edge), it is necessary to divide the echoes into distinct ranges to improve the accuracy. Each range is handled separately and correlated with the same range from other echoes. Thus a one-dimensional flow profile takes on a parabolic shape, as illustrated in Figure 9-37. In addition, a two-dimensional flow

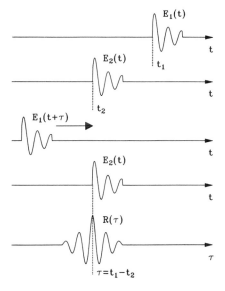

$E_1(t)$

t

$E_2(t)$

t_1

t

t_2

$E_1(t+\tau)$

t

$E_2(t)$

t

$R(\tau)$

τ

$\tau=t_1-t_2$

FIGURE 9-36 E_1 and E_2 represent two sample scatterer characteristics. E_1 is shifted with respect to E_2 and correlated, resulting in function R, which is maximized when E_1 and E_2 are exactly superimposed.

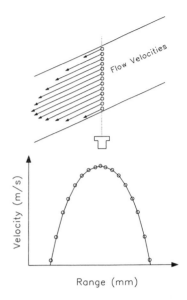

FIGURE 9-37 An example of a result from a one-dimensional scan.

profile of a blood vessel is produced by making several one-dimensional scans at different angles through the blood vessel. In this manner, a complete image of the blood flow can be determined as demonstrated in Figure 9-38.

9-6-3 RNS Correlator Architecture

The purpose of the hardware is to perform a correlation function on the samples of two echoes. Each echo contains 1024 8-bit samples divided into 25 ranges of

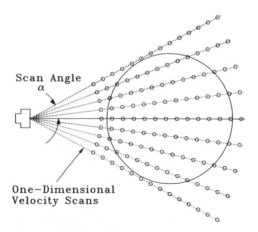

FIGURE 9-38 Several one-dimensional scans taken together to form a two-dimensional scan of a blood vessel.

41 samples each. In a given range, the hardware performs the function

$$f(a, b) = \sum_{i=0}^{40} x[a + i]y[b + i]. \qquad (9.37)$$

In Eq. (9.37), x and y each represent echoes, while a and b represent the initial offsets into the respective echoes. Each echo is composed of 1024 8-bit samples divided into 24 forty-sample ranges, which represent different depths across the blood vessel. The samples from a particular range are correlated across different echoes to derive a flow rate for that range.

The maximum wordlength required to calculate a value from Eq. (9.37) is 21 bits, determined as follows:

$$8\text{-bits (signed)} \times 8\text{-bits (signed)} = 15 \text{ bits}$$
$$+ \; 40 \text{ partial products} = \underline{\;\; 6 \text{ bits}}$$
$$\text{Total} = 21 \text{ bits}$$

To prevent overflow during calculation, four 6-bit moduli were selected: 64, 63, 61, and 59. (Note that 64 is admissible since the largest residue from a mod 64 number is 63.) This gives a RNS dynamic range of 24 bits, which easily accommodates the largest possible result from Eq. (9.37). The 6-bit moduli were selected because the largest high-speed ROMs that are commonly available are 8192 × 8 bits, with a 13-bit address space. This allows two 6-bit residues to be multiplied or added using a high-speed ROM look-up table instead of a more time-consuming and complicated hardware multiplier. Note that the use of smaller moduli would not improve the speed but would increase the hardware required.

9-6-4 Correlator Hardware

The RNS correlator prototype, shown in Figure 9-39, is composed of several functional units: four RNS correlator channels, a residue reconstruction unit, echo memory, and an Intel 80286 microprocessor. In addition, a Compaq-386 personal computer is used to graphically display the blood flow profile and to program and control the correlator. An ultrasound scanner is used to identify the desired blood vessel and to generate the echo signal for the digitizer.

A TRW 75-MHz A/D converter is used to quantize the signal from the transducer. Currently, the transducer operational frequency is 5 MHz and the A/D is clocked at 50 MHz. The signal is windowed so that only the part containing the energy reflected from the blood vessel is sent to the A/D, where it is converted into 1024 samples. This ensures that the samples will span only the blood vessel and include as little of the surrounding tissue as possible. A total of 256 echoes are stored in the memory from which a one-dimensional flow profile is computed.

The residue correlator channel, shown in Figure 9-40, consists of a pipelined architecture that performs an add and a multiply simultaneously for each of the 40

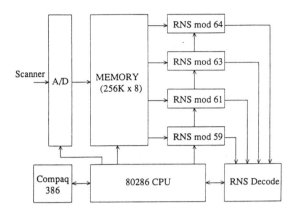

FIGURE 9-39   A block diagram of the correlator hardware.

data samples. The samples are encoded into residue form as they enter the pipeline. The arithmetic operations in the correlator channel are performed using a ROM look-up table. Since the product or sum can be at most 6 bits in wordlength, a ROM of size $2^{12} \times 6$ (4096 × 6) is required. Each operation can then be performed in the time required to look up the result, which in this case is approximately 40 ns. The system includes four correlator channels, each of which is architecturally identical and differs only in the contents of the residue encoding and arithmetic ROMs.

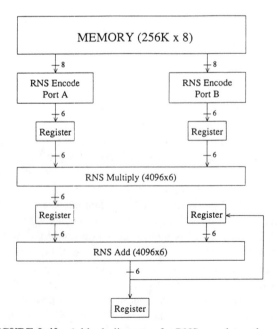

FIGURE 9-40   A block diagram of a RNS correlator channel.

The result from the four residue correlator channels must be translated into a binary value before it can be evaluated. This unit gathers the resultant residues from the four correlator channels and applies the Chinese Remainder Theorem to obtain the binary result. The normal form of the Chinese Remainder Theorem for the chosen set of parameters is

$$
\begin{aligned}
X &\approx f(x_{64}, x_{63}, x_{61}, x_{59}) \\
&= \langle x_{64}\hat{m}_{64}^{-1}\hat{m}_{64} + x_{63}\hat{m}_{63}^{-1}\hat{m}_{63} + x_{61}\hat{m}_{61}^{-1}\hat{m}_{61} + x_{59}\hat{m}_{59}^{-1}\hat{m}_{59} \rangle_M,
\end{aligned} \tag{9.38}
$$

where $M = (m_{64})(m_{63})(m_{61})(m_{59})$, $\hat{m}_{64} = M/m_{64}$, and $\hat{m}_{64}^{-1}$ is the multiplicative inverse of $\hat{m}_{64}$ mod m_{64}, and so on. Although 21 bits are required to handle the full dynamic range of the binary output, the full 21 bits of accuracy are not needed for subsequent processing. Therefore a scaled output, denoted $f_s(x_{64}, x_{63}, x_{61}, x_{59})$, is produced by applying the scaled Chinese Remainder Theorem as follows:

$$
X_s \approx f_s(x_{64}, x_{63}, x_{61}, x_{59}) = \langle f_{s1}(x_{64}, x_{63}) + f_{s2}(x_{61}, x_{59}) \rangle_{\hat{m}_{63}}, \tag{9.39}
$$

where

$$
f_{s1}(x_{64}, x_{63}) = \langle x_{64}\hat{m}_{64}^{-1}\hat{m}_{64s} + x_{63}\hat{m}_{63}^{-1}\hat{m}_{63s} \rangle_{\hat{m}_{63}},
$$

$$
f_{s2}(x_{61}, x_{59}) = \langle x_{61}\hat{m}_{61}^{-1}\hat{m}_{61s} + x_{59}\hat{m}_{59}^{-1}\hat{m}_{59s} \rangle_{\hat{m}_{63}},
$$

In Eq. (9.39), $\hat{m}_{63}$ denotes the scaled dynamic range, which in this design is 18 bits. The parameters $\hat{m}_{64s} = \hat{m}_{64}/m_{63}$, $\hat{m}_{61s} = \hat{m}_{61}/m_{63}$, and $\hat{m}_{59s} = \hat{m}_{59}/m_{63}$ are all integers. The one exception, $\hat{m}_{63s} = [\hat{m}_{63}/m_{63}]_Q$, is made into an integer by rounding $\hat{m}_{63}/m_{63}$ to the nearest integer. After the result is translated into 2's-complement form, only the most significant 16 bits are retained for further processing. This 16-bit result provides adequate precision for further processing and is consistent with a 16-bit integer format used in the Intel 80286 processor. The scaled Chinese Remainder Theorem is implemented by precalculating two subtotals, denoted by $f_{s1}(\cdot)$ and $f_{s2}(\cdot)$ in Eq. (9.39), and storing them in ROM. The two subtotals are then added together and checked for overflow. If an overflow occurs, the result is taken modulo $\hat{m}_{63}$. In order to save time and improve the conversion rate, both the normal sum and the sum modulo $\hat{m}_{63}$ are calculated and the correct one is selected based on the carry out of the adder [JE78], as illustrated in Figure 9-41.

The echo memory consists of a 256K × 8 dual-port memory with 25-ns access time, during which two 8-bit samples can be read out simultaneously. This feature is necessary so that the correlator can perform an operation on every clock cycle. In addition, when writing into the memory, the write port accepts four 8-bit samples per write cycle. Since the samples are written by a 50-MHz A/D converter and the memory cannot sustain that write frequency, the write port of the memory was made 32 bits wide so that samples could be multiplexed to make the write frequency more manageable.

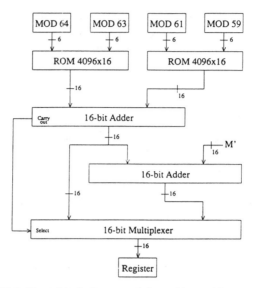

FIGURE 9-41 A block diagram of the residue-to-binary converter.

An Intel 80286 microprocessor is used to control the correlator and run an averaging algorithm on the various correlations performed by the residue correlator channels. The processor initializes the system, calculates and sets the addresses at which to correlate the echo samples, and generates flow results that are transmitted to the personal computer for graphical display. A microprocessor is included so that the system can be more configurable and flexible as the blood flow measurement technique is being verified.

9-6-5   Personal Computer

A Compaq 386 PC is used to determine which correlations should be performed and to calculate the flow profile. Since the blood velocity is determined by averaging the time shifts obtained from various pairs of echoes, it is advantageous to control the operation in software. In addition, the PC calculates the starting point (initial time shift) of each correlation based on the results of the previous correlation. This is both difficult and inflexible to implement solely in hardware. The PC interface allows the PC to control which samples are sent to the RNS correlator channels. The PC loads the starting addresses of the two 41 sample data ranges to be correlated. The interface then proceeds to fetch those samples and the 40 succeeding samples following each of those two for presentation to the correlator channels. Upon completion of the correlation, the interface reconstructs the result from the four residues produced by the correlator channels, scales the result to 16 bits, and returns that value to the PC. The interface provides all the control and synchronization signals for the RNS channels, the memory, and the A/D.

9-6-6 Experimental Results

In the earliest blood flow phantom experiments, the correlations were performed using a small Z80 based machine, which performed arithmetic operations at 500 ns per operation. Due to the high overhead associated with the equipment supporting this correlator, a single 1-D scan required 15 min to compute. With the initiation of the first animal experiments, the system was changed to run completely in software on an Intel 80386 based machine with a Weitek floating-point accelerator. This system required approximately 15 sec to calculate a single 1-D scan. While this represented a vast improvement over the previous system, it was still too slow to gather significant usable data, since it was difficult to keep a handheld transducer aligned for that amount of time. The RNS correlator prototype being used for the second round of animal experiments is able to perform a 1-D scan in approximately 2 sec.

During the initial testing phase for the correlator, a blood flow phantom was used to generate echo data for the correlator. A sample of a blood flow velocity profile produced by this hardware system operating on a laboratory phantom is shown in Figure 9-42. The blood vessel phantom system consists of a temperature controlled water bath, a peristaltic pump, and a fluid flow regulation system. Dialysis tubing with an inside diameter of approximately 6.3 mm was used to mimic the blood vessel, and a commercially available substance called Sephadex® was used to simulate a flowing fluid with the same reflective properties as blood. The improvement in processing speeds achieved with this new system now allows the

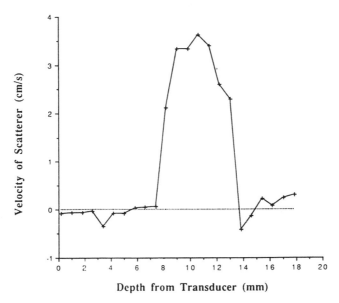

FIGURE 9-42 A sample blood flow velocity profile produced by the ultrasound hardware system operating on an experimental phantom.

sonographer to get far more immediate feedback regarding the proper alignment and placement of the transducer. With the RNS correlator, the blood flow technique can be verified experimentally with noninvasive animal studies and the resulting velocity profiles can be generated in near real-time.

9-6-7 VLSI Circuit Fabrication

Once the correlator prototype is thoroughly verified through blood flow phantom and animal studies, further improvements in performance will be achieved prior to the initiation of clinical verification. To achieve this higher speed and improved reliability, the RNS correlator will be fabricated as a monolithic VLSI circuit. This circuit will include the four residue channels and the residue reconstruction unit on a single chip. However, in the monolithic VLSI design, it will no longer be possible to directly utilize high-speed ROM look-up tables for arithmetic operations because the ROM sizes required for this are too large.

In order to maintain high throughput while avoiding the inclusion of a hardware multiplier, the design will be altered to use a logarithmic approach for performing multiplication, similar to the approach described earlier for the INDEX-ADDITION FIR digital filter design presented in Section 9-3-2. In the modified design, a new set of moduli will be selected consisting of smaller prime numbers, of which more will be used. This will allow the multiplication operations to be implemented by addition of finite field logarithms and log–antilog tables, which require considerably smaller ROMs than the complete stored-table look-up operations used in the prototype. Finite field logarithms can be used as long as the moduli are prime integers. Figure 9-43 shows the new components that will replace the large (8192

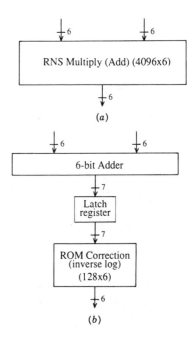

FIGURE 9-43 Modified designs for RNS adders/multipliers suitable for custom chip designs and fabrication.

$\times$ 8)-bit commercial ROMs used for the multiply and add operations in the prototype. The structure shown in Figure 9-43(b) is suitable for both the adder and the multiplier. In the case of the multiplier, the look-up tables used to encode the 8-bit samples into residue form will be modified to produce the logarithms of the residues directly. Note that this will require only a change in the ROM contents, and no additional hardware. The adder then sums the logarithms and a ROM correction table does both a modular correction and an inverse logarithmic mapping to produce the correct residue product. For the adder, the hardware structure remains the same and the ROM is programmed only for the modular correction.

Two interesting and useful features result from the modified design. First, each large ROM in the original design is replaced with an adder, a latch register, and a much smaller ROM. Second, the latch between the adder and the ROM allows a higher degree of pipelining. It is anticipated that the new design will be much faster than the commercial components used in the prototype. It is also expected that the resulting VLSI chip set will greatly reduce power consumption and increase reliability in the complete instrument.

9-7 SUMMARY

This chapter has reviewed the theory and discussed many applications of finite arithmetic concepts for the design of special purpose VLSI digital signal processors. In particular, the basic principles of residue number system (RNS) arithmetic were explained in considerable detail and illustrated by practical examples where possible. Although the mathematical theory of RNS arithmetic is a very old subject that became the object of intense study over 30 years ago by the newly emerging digital computer industry, applications of the concepts have never been successful in general purpose digital computers where the deficiencies of RNS techniques tend to counterbalance the advantages. However, the modern requirements of real-time digital signal processing and the capabilities of VLSI implementation form an ideal setting for RNS techniques. This chapter demonstrated the capabilities of RNS processor architectures for achieving high computational speeds, modularity, error isolation, fault tolerance, and simplified complex arithmetic in special purpose VLSI digital signal processors. Today, these finite arithmetic concepts are being used primarily by engineers in research and development, and only occasionally find their way into practical solutions to engineering problems. It remains to be seen if finite arithmetic concepts will provide important solutions for signal processing problems of the future.

ACKNOWLEDGMENTS

Much of the research summarized in this chapter was supported from 1978 to the present by the National Science Foundation under grants ENG-79-01686, ESC-84-05987 and MIP 91-00212, the Joint Services Electronics Program under contracts N00014-79-C-0424 and N00014-84-C-0149, the National Institutes of Health under grant HL 39704, and the Army Research Office under contract DAAL 03-86-K-0111. The author acknowledges the contributions of Prof. W. D. O'Brien and graduate research assistants J. T. Chen and I. A. Hein to the high-speed correlator design example discussed in Section 9-6.

REFERENCES

[AG74] R. C. Agarwal and C. S. Burrus, Fast convolution using Fermat number transforms with applications to digital filtering. *IEEE Trans. Acoust., Speech, Signal Process.* **ASSP-22,** (2), 87–97, (1974).

[AI59] H. H. Aiken and W. Semon, *Advanced Digital Computer Logic*, Tech. Rep. WADC TR 59–472. Cambridge, MA, 1959.

[BA61] R. A. Baugh and E. C. Day, *Electronic Sign Evaluation for Residue Number Systems*, Tech. Rep. No. TR-60-597-32. RCA, Camden, NJ, and Burlington, MA, 1961.

[BA73] F. Barsi and P. Maestrini, Error correcting properties or redundant residue number systems. *IEEE Trans. Comput.* **C-18,** 307–316, (1973).

[BL85] R. E. Blahut, *Fast Algorithms for Digital Signal Processing*. Addison-Wesley, Reading, MA, 1985.

[CH61] P. W. Cheney, A digital correlator based on the residue number system. *IRE Trans. Electron. Comput.* **EC-11,** 63–70, (1961).

[EM86] P. M. Embree, The accurate ultrasonic measurement of the volume flow of blood by time domain correlation. Ph.D. dissertation, Department of Electrical and Computer Engineering, University of Illinois at Urbana-Champaign, Urbana, 1986.

[FO84] S. G. Foster, A pulsed ultrasonic flowmeter employing time domain methods. Ph.D. dissertation, Department of Electrical and Computer Engineering, University of Illinois at Urbana-Champaign, Urbana, 1984.

[GA59] H. L. Garner, The residue number system. *IRE Trans. Electron. Comput.* **EC-8,** 140–147, (1959).

[GR66] E. Grosswald, *Topics From the Theory of Numbers*. Macmillan, New York, 1966.

[HE90] I. A. Hein, An accurate and precise measurement of blood flow using ultrasound and time domain correlation. Ph.D. dissertation, Department of Electrical and Computer Engineering, University of Illinois at Urbana-Champaign, Urbana, 1990.

[JE75] W. K. Jenkins, Composite number theoretic transforms for digital filtering. *Proc. Asilomar Conf. Circuits, Syst. Comput., 9th, Pacific Grove, California, 1975.* pp. 458–462, (1975).

[JE78] W. K. Jenkins, Techniques for residue to analog conversion for residue encoded digital filters. *IEEE Trans. Circuits Syst.* **CAS-25,** (7), 555–562, (1978).

[JE87a] W. K. Jenkins and J. V. Krogmeier, The design of dual-mode complex signal processors based on quadratic modular number codes. *IEEE Trans. Circuits Syst.* **CAS-34,** (4), 354–364, (1987).

[JE87b] W. K. Jenkins and S. F. Lao, The design of an RNS digital filter module using the IBM MVISA design system. *Proc. IEEE Int. Symp. Circuits Syst., Philadelphia, 1987,* pp. 122–125, (1987).

[JE87c] W. K. Jenkins and J. T. Chen, New architectures for VLSI digital signal processors using serial-by-modulus RNS arithmetic. *Proc. Asilomar Conf. Signals, Syst. Comput., 21st, Pacific Grove, California, 1987.* (1987).

[JE88] W. K. Jenkins and E. J. Altman, Self-checking properties of residue number error checkers based on mixed radix conversion. *IEEE Trans. Circuits and Syst.* **CAS-35,** (2), 159–167, (1988).

[JU87] G. A. Jullien, R. Krishnan, and W. C. Miller, Complex digital signal processing over complex fields. *IEEE Trans. Circuits Syst.* **CAS-34,** (4), 365–377, (1987).

[KN69] D. E. Knuth, *The Art of Computer Programming*, Vol. 2, Addison-Wesley, Reading, MA, 1969.

[LE76] L. M. Leibowitz, A simplified binary arithmetic for the Fermat number transform. *IEEE Trans. Acoust., Speech, Signal Process.* **ASSP-24,** (5), 356–359, (1976).

[MC79] J. H. McClellan and C. M. Radar, *Number Theory in Digital Signal Processing.* Prentice-Hall, Englewood Cliffs, NJ, 1979.

[PA84] D. F. Paul, W. K. Jenkins, and E. S. Davidson, Residue arithmetic for real-time applications: High throughput and reliability using customized modules. *Proc. Conf. Circuits Comput. Port Chester, New York, 1984,* pp. 689–694, (1984).

[PE74] A. Peled and B. Liu, A new hardware realization of digital filters. *IEEE Trans. Acoust., Speech, Signal Process.* **ASSP-22,** 456–462, (1974).

[SL63] D. L. Slotnick, *Modular Arithmetic Computing Techniques*, Tech. Rep. ASD-TDR-63-280. Westinghouse Electric Corp., Air Arm Division, Baltimore, MD, 1963.

[SO86] M. A. Soderstrand, W. K. Jenkins, G. A. Jullien, and F. J. Taylor, eds., *Residue Number System Arithmetic: Modern Applications in Digital Signal Processing.* IEEE Press, New York, 1986.

[SV55] A. Svoboda and M. Valach, Operational circuits. *Stroje Na Zpracovani Informaci,* Vol. 3, Nakl. CSAV V, Prague, 1955.

[SV57] A. Svoboda, Rational numerical system of residual classes. *Stroje Na Zpracovani Informaci,* Vol. 5, pp. 9–37, Nakl. CSAV, Prague, 1957.

[SV58] A. Svoboda, The numerical system of residual classes in mathematical machines. *Proc. Congr. Int. Automa., Madrid, Spain, 1958,* also *Info. Process. (Proc. UNESCO Conf, 1959)*, pp. 419–422, (1960).

[SZ67] N. S. Szabo and R. I. Tanaka, *Residue Arithmetic and Its Applications to Computer Technology.* McGraw-Hill, New York, 1967.

[TA62] R. I. Tanaka, *Modular Arithmetic Techniques*, Tech. Rep. 2-38-62-1A, ASTDR. Lockheed Missiles and Space Co., 1962.

[TA81] F. J. Taylor and A. S. Ramnarayanan, An efficient residue-to-decimal converter. *IEEE Trans. Circuits Syst.* **CAS-28,** (12), 1164–1169, (1981).

[TA82] F. J. Taylor and C. H. Huang, An autoscale residue multiplier. *IEEE Trans. Comput.* **C-31,** (4), 321–325, (1982).

[TA85a] F. J. Taylor, The impact of residue arithmetic on digital signal processing. *Proc. IASTED Int. Symp., Paris, 1985.* (1985).

[TA85b] F. J. Taylor, Single modulus ALU for signal processing. *IEEE Trans. Acoust., Speech, Signal Process.* **ASSP-33,** (5), 1302–1315, (1985).

10 Signal Conditioning and Interface Circuits

LAWRENCE E. LARSON

Hughes Research Laboratories
Malibu, California

GABOR C. TEMES

Department of Electrical & Computer Engineering
Oregon State University, Corvallis

Signal conditioning and interface structures constitute the interface between the continuous-time phenomena, which represent data inputs to an electronic system, and digital signal processing algorithms and hardware. As a result, the quality of the data generated by the signal conditioning and interface structures is crucial for determining the accuracy of the complete signal processing system.

The interface between analog and digital signals can be separated into two realms: the analog signal conditioning stages, which consist of anti-aliasing and/ or smoothing filters, and the conversion stages, which consist of analog-to-digital or digital-to-analog converters. A block diagram of a typical digital signal processing system, including analog interface structures, is shown in Figure 10-1. The purpose of this chapter is to describe each of the most important interface techniques and to illuminate the various considerations that go into their design and implementation.

10-1 ANTI-ALIASING FILTERS

10-1-1 Sampling, Aliasing, and Anti-Aliasing Filters

As defined in Chapter 2, a signal is a function of an independent variable, which is usually time. There are two types of signals that are of concern in this chapter: continuous-time and discrete-time. A continuous-time signal has a well-characterized value at all times over a given interval, and its dependent variable is usually voltage, current, charge, or power (Figure 10-2(a)). A discrete-time signal has a

Handbook for Digital Signal Processing, Edited by Sanjit K. Mitra and James F. Kaiser.
ISBN 0-471-61995-7 © 1993 John Wiley & Sons, Inc.

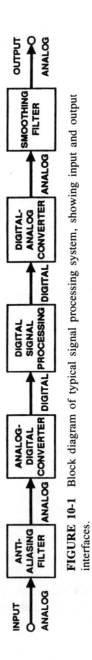

FIGURE 10-1 Block diagram of typical signal processing system, showing input and output interfaces.

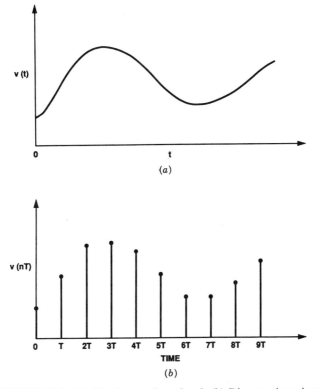

FIGURE 10-2 (a) Continuous-time signal. (b) Discrete-time signal.

well-characterized value only at discrete-time instances; at other times it is indeterminate and can be assumed to be zero (Figure 10-2(b)). Its dependent variable may be a dimensionless number. As a result, the relationship between a continuous-time signal and the corresponding discrete-time signal can be expressed in the form

$$v(nT) = v(t)_{t=nT}, \qquad n = 0, 1, 2, \ldots, \qquad (10.1)$$

where T is the sampling period.

The samples in the discrete-time signal are usually represented by a finite number of digits, or bits if the signal is binary coded. Such a signal is referred to as a *digital* signal. Therefore the digital signal can only assume discrete values, the smallest of which is determined by the smallest available digit, and the largest is determined by the maximum range of the digital word. By contrast, a signal that can attain *any* value whatsoever is called an *analog* signal.

The relationship between the frequency spectrums of the continuous-time and discrete-time signals can be derived by considering the circuit of Figure 10-3(a) [GR86]. This circuit presents a simplified representation of the *sampling process*,

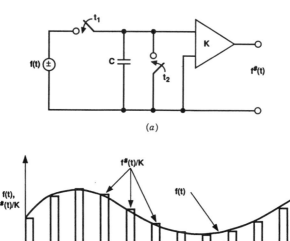

FIGURE 10-3 (a) Circuit demonstrating sampling process. (b) Continuous-time signal $f(t)$ and its sampled equivalent $f^{\#}(t)$.

where continuous-time signals are converted into discrete-time signals. In this case, the switches S_1 and S_2 close and then open instantaneously at time intervals t_1 and t_2. When S_1 closes at time t_1, capacitor C is charged to the instantaneous value of voltage $f(t_1)$. When S_2 closes at time $t_2 = t_1 + \tau$, capacitor C is completely discharged. The amplifier multiplies the voltage on capacitor C by its gain k, and the amplified voltage then appears at the output. The relationship between the continuous-time input and the resulting *sampled* output waveform appears in Figure 10-3(b). This sampled output is equivalent to the discrete-time version of the original continuous-time signal as τ approaches 0. The nth output pulse, which occurs between times nT and $nT + \tau$, is given by

$$f_n(t) = kf(nT)[\mu(t - nT) - \mu(t - nT - \tau)], \qquad (10.2)$$

where $\mu(t)$ is the *step function*

$$\mu(t) \equiv \begin{cases} 1, & t \geq 0, \\ 0, & t < 0. \end{cases}$$

As a result, the *total* sampled output is the sum of the individual pulses from $t = 0$ to $t = \infty$, or

$$f^{\#}(t) = \sum_{n=0}^{\infty} f_n(t) = k \sum_{n=0}^{\infty} f(nT)[\mu(t - nT) - \mu(t - nT - \tau)]. \qquad (10.3)$$

If we calculate the *unilateral* (or one-sided) Laplace transform, $F^{\#}(s)$, of $f^{\#}(t)$ [PA80], we obtain

$$F^{\#}(s) = k \frac{1 - e^{-s\tau}}{s} \sum_{n=0}^{\infty} f(nT) \, e^{-snT}. \tag{10.4}$$

This result can be simplified as the sampling interval τ grows smaller and smaller. Then the expression $k(1 - e^{-s\tau})/s$ at the beginning of Eq. (10.4) becomes equal to $k\tau$. If k remained constant, this quantity would approach zero as τ approached zero. But by setting $k = 1/\tau$, Eq. (10.4) becomes

$$F^{\#}(s) = \sum_{n=0}^{\infty} f(nT) \, e^{-snT}. \tag{10.5}$$

The frequency spectrum of this signal can be determined by replacing s in Eq. (10.5) with $j\Omega$. The resulting spectrum is

$$F^{\#}(j\Omega) = \sum_{n=0}^{\infty} f(nT) \, e^{-jn\Omega T}. \tag{10.6}$$

This spectrum is a periodic function of Ω, with a period of $2\pi/T$. In addition, it can be shown that the $F^{\#}(j\Omega)$ spectrum of the discrete-time signal is related to the Fourier transform $F(j\Omega)$ of the continuous-time signal $f(t)$ [OP88] by[1]

$$F^{\#}(j\Omega) = \frac{1}{T} \sum_{k=-\infty}^{\infty} F(j\Omega - j2\pi k/T). \tag{10.7}$$

Therefore the spectrum of the resulting discrete-time signal is a *periodic* function in Ω, repeating every $2\pi/T$, whereas the original continuous-time spectrum is aperiodic.

The requirement for anti-aliasing filters results from the relationship between the two spectra, which is described by Eq. (10.7). In particular, consider a continuous-time signal whose frequency spectrum $F_a(j\Omega)$ appears in Figure 10-4(a). In this case it is completely *bandlimited*, that is, contained inside the frequency range $-\Omega_A$ to Ω_A, where $\Omega_A < \pi/T$, and there is no signal power outside this range. A filter that accomplishes this bandlimiting is not physically realizable, but it can be closely approximated by using standard analog lowpass filter design techniques.

When this continuous-time signal is sampled, and converted into a discrete-time signal, the resulting spectrum repeats at every multiple of $2\pi/T$ and the replicas $F_a^{\#}(j\Omega)$ do not overlap each other, as shown in Figure 10-4(b); there is a one-to-one correspondence between the original continuous-time signal and the result-

---

[1]See Section 2-9 for the derivation.

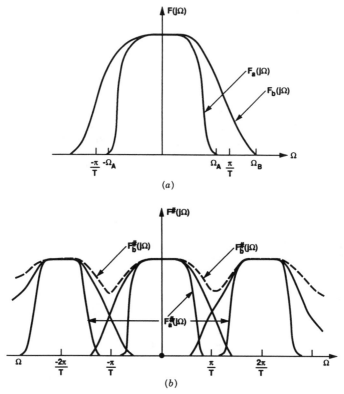

FIGURE 10-4   (a) Continuous-time signal spectra. (b) Discrete-time signal spectra.

ing discrete-time signal. This discrete-time signal can then be lowpass filtered and the original continuous-time signal recovered without error.

Next, consider the case where the original continuous-time signal, $F_b(j\Omega)$, is not strictly bandlimited within the bounds of $\pm\pi/T$, as shown in Figure 10-4(a). In this case, the resulting discrete-time replicas overlap each other (Figure 10-4(b)), and the resulting spectrum is the sum of all the overlapping frequencies. Thus there is no longer a one-to-one correspondence between the original continuous-time input signal and the resulting discrete-time output. This phenomenon is known as aliasing and is an undesirable nonlinear distortion of the original input signal. In this case, there is no possibility of recovering the spectrum of the original continuous-time signal.

The resulting condition on the sampling rate for the elimination of aliasing,

$$\frac{\pi}{T} > \Omega_A, \tag{10.8}$$

was first observed by Nyquist, and the resulting lower bound on the sampling frequency (Ω_A/π) is usually referred to as the *Nyquist rate*.

In common engineering practice, Eq. (10.8) is met by minimizing T and Ω_A. The bandlimit Ω_A is usually reduced by passing the continuous-time signal $f(t)$ through a lowpass filter prior to discrete-time sampling. This filter has an ideal transfer function given by

$$H(j\Omega) = \begin{cases} 1, & |\Omega| \leq \pi/T, \\ 0, & |\Omega| \leq \pi/T, \end{cases} \qquad (10.9)$$

and is referred to as an *anti-aliasing filter* (AAF). Using such a filter before the sampling operation is undertaken ensures the absence of aliasing in the sampled version of the continuous-time signal. As a result, the spectra of the discrete-time signal and its parent continuous-time signal are identical in the frequency range from 0 to π/T, except for a scale factor.

It also follows from Eq. (10.7) that the original spectrum $F(j\Omega)$, and the original continuous-time signal $f(t)$, can be recovered from the discrete-time signal by passing it through a filter having the frequency response of Eq. (10.9). This filter is referred to as a *smoothing filter*.

As a result, a complete discrete-time signal processing system, with continuous-time inputs and outputs, is usually constructed as shown in the block diagram of Figure 10-5. An aliasing filter on the input prevents the unwanted distortion that would result from *undersampling* the input, and the smoothing filter at the output reconstructs a continuous-time output.

10-1-2 Anti-aliasing Filter Requirements

The previous section outlined the requirements for an ideal anti-aliasing filter. In particular, a lowpass filter is desired with a response defined by Eq. (10.9). A "brick-wall" filter of this type is impossible to realize in practice [OP88], and a variety of techniques have been devised to approximate it. The next section summarizes these techniques, but they all require more and more components (resistors, capacitors, inductors, amplifiers) of increasingly precise values as the filter more closely approximates the "ideal" response.

In order to minimize this problem, the sampling interval T can be lowered, so that the continuous-time filter has less stringent stopband requirements. Ideally, the ratio of the sampling frequency $(1/T)$, to the passband frequency of the anti-aliasing should be large—between 10 and several hundred—in order to keep the

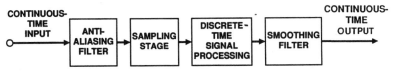

FIGURE 10-5 Block diagram of discrete-time signal processing system with continuous-time inputs and outputs.

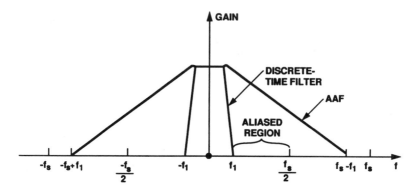

FIGURE 10-6   Anti-aliasing filter (AAF) for lowpass discrete-time signal processor.

anti-aliasing filter sufficiently simple. Unfortunately, there are practical technological limits, involving circuit speed and complexity, to how far T can be lowered, so this approach is of limited usefulness.

This dilemma can be resolved by a variety of techniques. The first follows from the observation that the discrete-time filter following the anti-aliasing filter is usually of a bandpass or lowpass nature. In this case, the strict Nyquist criterion on the bandwidth of the original continuous-time signal can be eased somewhat. Consider the frequency responses of Figure 10-6, where the discrete-time filter has a lowpass characteristic, with a cutoff frequency $f_1 \ll f_s$, and the continuous-time anti-aliasing filter has a cutoff frequency of $f_s - f_1$. As a result of the relatively high cutoff frequency of the anti-aliasing filter, that part of the input signal with frequencies between $f_s/2$ and $f_s - f_1$ is replicated (and aliased) into the region between f_1 and $f_s/2$. Fortunately, the discrete-time filter has a large loss in this region, and the aliasing creates only a small distortion in the final output. Since $f_1 \ll f_s$, the required stopband frequency of the anti-aliasing filter is effectively doubled over that required by the classic Nyquist criteria, simplifying the design of the filter. Of course, this technique is only applicable in cases where the discrete-time filter does not pass high frequencies.

The second technique for simplifying the design of the anti-aliasing filter is the use of decimation techniques[2] [MA80], where a relatively simple lowpass discrete-time *prefilter*, operating at a higher multiple of f_s, follows the anti-aliasing filter and precedes the final discrete-time filter, as shown in Figure 10-7(a). The spectra of the anti-aliasing filter, the decimator, and the final discrete-time filter all appear in Figure 10-7(b).

In this case, the cutoff frequency of the anti-aliasing filter is set to $Df_s - f_1$, where D is the ratio of the decimator's sampling frequency to that of the final discrete-time filter. Usually D is an integer power of 2 in order to simplify the implementation of the decimator. Note that the anti-aliasing filter cutoff frequency is now almost $2D$ times higher than Nyquist's criterion would dictate, simplifying

---

[2]See Chapter 14 for a discussion on the decimation process.

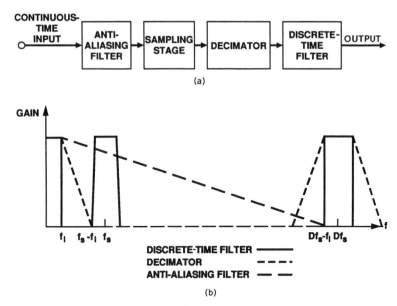

FIGURE 10-7 (a) Block diagram of discrete-time signal processor with decimator. (b) Frequency spectra of AAF, decimator, and discrete-time filter.

the design of the anti-aliasing filter even further. The decimation filter has a higher cutoff frequency $f_s - f_1$ although its passband limit is the same as that of the final discrete-time filter (f_p).

As a result of the relatively high cutoff frequency of the decimation filter and the anti-aliasing filter, there will be substantial aliasing of the input in the frequency region between f_1 and ($Df_s - f_1$). Fortunately, the decimator has a large loss at frequencies above $f_s - f_1$, and the discrete-time filter has a large loss between f_1 and $f_s - f_1$. So the resulting filter response has no gain in the regions where aliasing does occur, and it is unaltered in the regions where there is no aliasing.

Both of these techniques reduce the requirements on the anti-aliasing filter by placing the aliased components of the signal into regions where the decimation or discrete-time filter rejects them. Therefore the filter rejects the unwanted out-of-band input signals without distorting the final in-band spectrum.

10-1-3 Anti-aliasing Filter Implementation

The design of anti-aliasing filters is based on the well-known principles of continuous-time analog lowpass filter design [TE73]. These filters can be implemented in discrete form, using inductors, capacitors, resistors, and operational amplifiers, or in integrated form using resistors, capacitors, and operational amplifiers. In general, active RC filters—those that employ operational amplifiers—are preferred, because they require less area and are easier to realize in an integrated circuit form.

The realization of integrated anti-aliasing filters is especially attractive, because they make possible the realization of a complete digital signal processing system on one integrated circuit die.

The preferred implementation of low-order active RC anti-aliasing filters is typically a cascade of first- and second-order lowpass sections. Their design is relatively straightforward, and their responses can often be altered to account for the "$\sin x/x$" distortion resulting from the sample-and-hold process. By contrast, the preferred embodiment of higher-order anti-aliasing filters is in the simulated ladder form, which takes advantage of a doubly terminated matched reactance network's inherently low sensitivity to component value changes [OR68].

The general transfer function of an all-pole lowpass filter of the order p is of the form

$$A(s) = \frac{A_0}{1 + c_1 s + c_2 s^2 + \cdots + c_p s^p}, \tag{10.10}$$

where $c_1, c_2, \ldots$ are positive real numbers.

It is possible to rewrite Eq. (10.10) in a factored form. If complex poles are permitted, a product of quadratic expressions of the form

$$A(s) = \frac{A_0}{(1 + a_1 s + b_1 s^2)(1 + a_2 s + b_2 s^2) \cdots} \tag{10.11}$$

results.

Anti-aliasing filter design requires choosing the proper coefficients a_i, b_i for the realization of a filter that closely approximates the ideal transfer function for a given order. Many lowpass analog filters make use of the *Butterworth* polynomial approximation to the ideal "brick-wall" filter response. The Butterworth response is achieved by a maximally flat approximation to the ideal response at $\omega = 0$. As a result, the natural modes of a Butterworth filter are equally spaced on a circle around the origin of the complex s-plane. A plot of the frequency response of Butterworth lowpass filters is shown in Figure 10-8. It can be shown that the resulting coefficients are, for even-order p,

$$a_i = 2 \cos \frac{(2i - 1)\pi}{2p} \quad \text{and} \quad b_i = 1 \tag{10.12}$$

for $i = 1, \ldots p/2$, and for odd-order p,

$$a_i = 2 \cos \frac{(i - 1)\pi}{p} \quad \text{and} \quad b_i = 1$$

for $i = 2, \ldots, (p + 1)/2$.

Other polynomial approximations to the ideal lowpass filter can be employed, including the Chebyshev, Bessel, and elliptic responses [ZV67]. These techniques

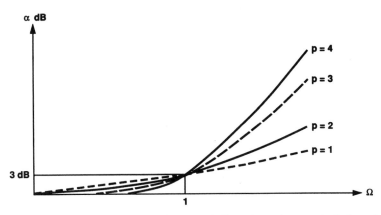

FIGURE 10-8   Loss responses of Butterworth filters for order $p = 1, 2, 3$, and 4.

require fewer elements for the realization of a desired response or exhibit improved pulse responses compared with the Butterworth filter. The resulting coefficients can be calculated, or derived from previously tabulated values.

Once the coefficients have been obtained for the desired response of the anti-aliasing filter, the next step is the actual design of the filter. In the case of low-order active RC filters, this can usually be accomplished by a cascade connection of first- and second-order active RC filters.

An example of a first-order lowpass active RC section is shown in Figure 10-9. The corresponding transfer function is given by

$$A(s) = \frac{-R_2/R_1}{1 + sR_2C_1}. \tag{10.13}$$

By choosing appropriate values for R_2, R_1, and C_1, the desired first-order transfer function can be realized.

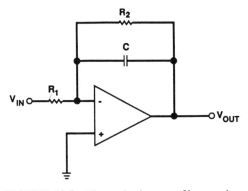

FIGURE 10-9   First-order lowpass filter section.

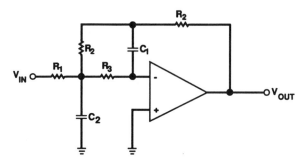

FIGURE 10-10   Second-order lowpass filter section.

A second-order lowpass section is shown in Figure 10-10. In this case, the transfer function is given by

$$A(s) = \frac{A_0}{1 + s[C_1(R_1 + R_2) + (1 - A_0)R_1C_2] + s^2R_1R_2C_1C_2}, \quad (10.14)$$

where $A_0 = -R_2/R_1$.

By setting $A_0 = 1$, $R_1 = R_2 = R$, and $C_2 = 2C_1 = C$, the transfer function of this circuit reduces to

$$H(s) = \frac{1}{1 + (1/Q\omega_0)s + s^2/\omega_0^2}, \quad (10.15)$$

where the pole frequency is $\omega_0 = 1/\sqrt{RC}$, and the pole-Q is given by $Q = 1/\sqrt{2}$, which is independent of the actual values of the resistors and capacitors as long as the matching conditions hold. This is an example of a second-order Sallen–Key active RC lowpass filter, and the resulting transfer function is that of a second-order Butterworth filter.

The preferred technique for the design of higher-order ($p > 4$) continuous-time anti-aliasing filters is the use of simulated ladder networks. This is due to the well-known lower sensitivity of the transfer function to component variations compared with a cascade realization [OR68]. An active RC version of a passive RLC ladder can be derived, which employs active RC integrators to simulate the action of inductances and capacitors. The design techniques for these circuits are described in Ghausi and Laker [GH81].

Circuit designers often find the implementation of the active RC circuits of Figures 10-9 and 10-10 impractical on an integrated circuit die. This is due to the large area required for the implementation of the high-valued resistors required to realize the large time constants that are often required. In addition, it is often difficult to realize accurate resistors and capacitors on an integrated circuit die without resorting to expensive trimming techniques. As a result, improved design techniques are often required for the implementation of monolithic continuous-time anti-aliasing filters.

One attractive alternative approach is shown in Figure 10-11 [BA83]. Figure 10-11(a) shows a typical active *RC* integrator. Its resistor could, in principle, be replaced by a MOSFET biased in the linear active region, as shown in Figure 10-11(b). Unfortunately, for large input signals, the distortion that results from the nonlinearity of the channel resistance of the MOSFET is unacceptable. However, the even-order harmonic distortion can be eliminated by employing a fully differential integrator, as shown in Figure 10-11(c). It consists of two identical capacitors *C* and two identical MOSFETs, as well as a fully differential operational amplifier. The output voltage of the operational amplifier is kept symmetric with respect to ground. With this approach, the active *RC* integrator has been replaced with a "MOSFET-C" integrator, and the resulting area required for implementation on an integrated circuit die has been greatly reduced.

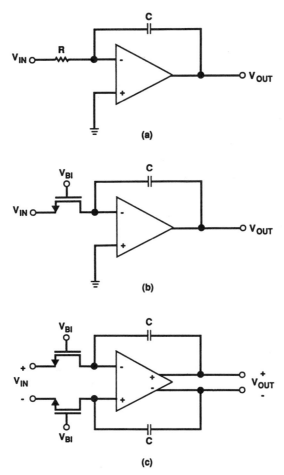

FIGURE 10-11 (a) Active *RC* integrator. (b) Active MOS-C integrator. (c) Improved MOS-C integrator.

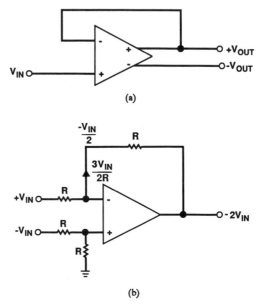

FIGURE 10-12   (a) Single-ended to differential conversion circuit. (b) Differential to single-ended conversion.

This approach for the integration of anti-aliasing filters possesses the advantages of relatively low distortion and easy integrability along with analog and digital signal processing components. Its main disadvantage is the requirement of a fully differential input signal, which can be synthesized from a single-ended input using two operational amplifiers, as shown in Figure 10-12(a). The circuit of Figure 10-12(b) performs the reverse operation, differential to single-ended conversion.

10-2 ANALOG-TO-DIGITAL CONVERTERS

Until now, we have assumed that the conversion of the continuous-time signal into its discrete-time equivalent has been carried out with perfect fidelity. In other words, we have assumed that there are an infinite number of digits available for the representation of the discrete-time signal, so that the signal-to-noise ratio of the discrete-time signal is the same as that of the original continuous-time signal.

In fact, only a finite number of digits, or bits in the case of a binary-coded signal, can be employed to represent the discrete-time signal. An analog-to-digital (A/D) converter, often abbreviated as ADC, performs the conversion between the analog continuous-time domain and the digital discrete-time domain. The ADC is a crucial building block in any discrete-time signal processing system, since it determines the achievable dynamic range and fidelity of the overall system.

This section outlines the performance limitations, characteristics, architectures, and circuit implementations of ADCs.

10-2-1 Characterization of A/D Converter Performance

There are a number of ways of assessing the performance of an ADC. For the purposes of signal processing, one of the most useful figures of merit is the *signal-to-noise ratio* (SNR). This quantity is the ratio of the maximum output signal power to the total output noise power, integrated over the full Nyquist bandwidth $(f_s/2)$. The SNR for an ideal ADC is given by the expression

$$SNR = (6N + 1.8) \text{ dB}, \tag{10.16}$$

where N is the resolution of the ADC, measured in bits.

However, the achievable SNR of a given ADC implementation is usually limited by thermal noise, aperture jitter, comparator uncertainty, component non-idealities, settling errors, and a variety of other static and dynamic errors, rather than strictly by N. For example, the thermal noise intrinsic to the environment limits the maximum practical value of N to approximately

$$N_{max} \approx \log_2 \left(\frac{V_{max}^2}{8kTR \, \Delta f} \right)^{1/2}, \tag{10.17}$$

where R is the impedance level of the input signal, V_{max} is the maximum input voltage amplitude, Δf is the desired input bandwidth, T is temperature, and k is Boltzmann's constant.

This equation is simply an expression of the well-known trade-off between dynamic range and bandwidth. Figure 10-13 compares the thermal limit on ADC

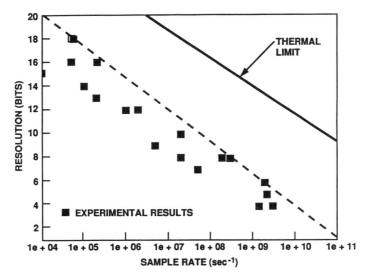

FIGURE 10-13 Reported sample rates for analog-to-digital converters and comparison to fundamental thermal limit.

performance operating at 270 K in a 50-Ω system, with a 1.0-V maximum input amplitude, to the performances of ADCs that have been reported in the literature [WA90]. It is clear that the thermal noise limit has yet to be approached, and that other technological factors must be currently limiting ADC resolution.

Another basic SNR limitation is due to the random variations in the sampling interval ($1/f_s$). These variations are known as *aperture jitter* and degrade the resulting output signal-to-noise ratio by distorting the output waveform. The allowable uncertainty in the sampling interval decreases as the resolution and bandwidth increase. This uncertainty arises from low-frequency noise modulation of the incoming input sampling clock, temperature drift, or other instabilities. The precise degradation of the SNR due to aperture jitter depends on the statistical properties of the input signal and the jitter process itself. However, if the rms value of the aperture noise is limited to less than the rms value of the quantization noise, in order to minimize the degradation in SNR, then we must satisfy [LA88]

$$2^N f_{\text{sig}} \tau_a \leq 1, \tag{10.18}$$

where f_{sig} is the maximum signal frequency and τ_a is the rms aperture jitter.

Another limitation on the ADC performance is the comparator *regeneration time*. All ADCs compare the analog input signal against some known reference and produce a digital output depending on their difference. Any uncertainty in that decision causes a degradation in the resulting SNR. A simplified model of the comparator and its time-domain response appear in Figures 10-14(a) and 10-14(b). The comparator acts like an amplifier in the "track" mode and a regenerative latch in the "latch" mode. During the latch mode, the comparator multiplies the amplified input voltage by a growing exponential, e^{t/τ_c}, until the output saturates or the latch mode ends. At the end of the latch period, a digital output signal is available for decoding.

An indecisive comparator output at the end of the latch period will result in a degradation of the SNR. The probability of such an event occurring is

$$P_i = \frac{2V_L}{A_0 q} e^{-t/\tau_c}, \tag{10.19}$$

where t is the duration of the latch period, A_0 is the gain of the comparator in the track mode, q is the quantization level, V_L is the logic level at the output of the comparator, and τ_c is the comparator regeneration time. As τ_c is reduced, the probability of an indecisive comparator output decreases. The decrease in τ_c is limited by the intrinsic speed of the transistors in the comparator. Assuming a two-transistor circuit, and using a simplified small-signal equivalent circuit model of a comparator, the resulting comparator regeneration time is

$$\tau_c \approx \frac{1}{2\pi f_T}, \tag{10.20}$$

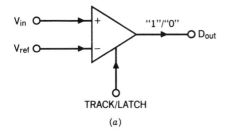

TRACK/LATCH

(a)

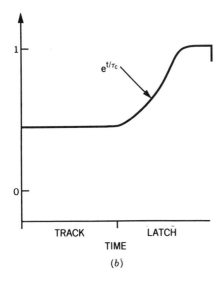

FIGURE 10-14 (a) Simplified comparator model. (b) Time domain response.

where f_T is the unity-current-gain cutoff frequency of the transistors in the comparator. Practical implementations of comparator circuits usually exhibit regeneration times longer than this minimum, because of the effects of parasitic capacitances and resistances.

The excess quantization noise that results from comparator indecision can be approximated by [AE76]

$$\sigma_{ex}^2 \approx \frac{p_i 2^{-N+1}}{12}. \qquad (10.21)$$

Figure 10-15 shows the achievable resolution as a function of input bandwidth for two different comparator regeneration times. The plot assumes that the excess quantization noise is limited to 10% of the ideal ADC quantization noise.

A variety of static (or dc) errors can also compromise the performance of ADCs. These errors are illustrated in Figure 10-16 and can be categorized as offset errors (Figure 10-16(a)), gain errors (Figure 10-16(b)), and nonlinearity (Figure 10-16(c), (d)).

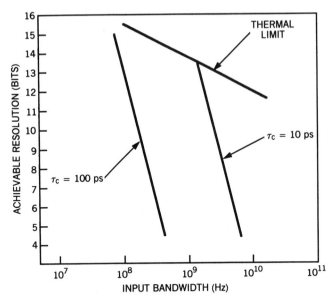

FIGURE 10-15 Achievable ADC resolution for different comparator regeneration times.

Offset and gain errors do not necessarily comprise the SNR of the converter, if the input signal is properly adjusted to account for them. However, ADC nonlinearities can severely degrade the SNR. Nonlinearities can be further classified into integral and differential nonlinearities. The integral linearity error is the maximum deviation (in LSBs) of the output of an ADC from an (ideal) straight-line transfer function. Differential linearity errors are the maximum relative deviations in the $\Delta V_{in}/\text{LSB}$. The maximum peak error in an ideal rounding ADC is $\pm Q/2$, where Q is the quantization level (Figure 10-17). The rms error of the converter is $Q/2\sqrt{3}$. A list of these and other commonly used conversion error definitions is shown in Table 10-1.

10-2-2 A/D Converter Architectures

The usefulness of an ADC structure can be (somewhat arbitrarily) measured by three parameters: accuracy, speed, and complexity. The accuracy of an ADC, depending on its application, may be expressed in terms of its SNR, as given in Eq. (10.16), or its differential and/or integral linearity errors. The speed is given in terms of the number of samples converted per second or per sampling clock period. Finally, the complexity includes the number of components and the difficulty of their realization in both the analog and digital segments of the ADC. Generally, these parameters interact in such a way that optimal performance in any two areas precludes good results in the third. Accordingly, if practical limits are set on the complexity of the ADC, then increasing accuracy generally can only be attained by slower converters.

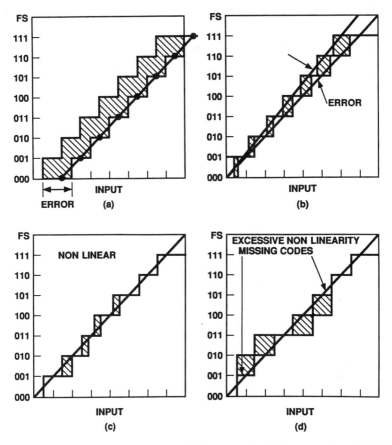

FIGURE 10-16 Static ADC errors. (a) Offset error. (b) Gain error. (c) and (d) Nonlinearity error.

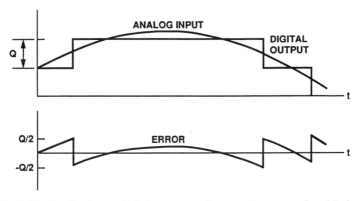

FIGURE 10-17 Analog input, digital output, and conversion error of an ideal rounding A/D converter.

TABLE 10-1   ADC Conversion Error Definitions

Differential nonlinearity (DNL)	The maximum difference between any actual step and the ideal step size (1 LSB)
Integral nonlinearity (INL)	The maximum difference between the conversion curve and a straight-line reference
Monotonicity	Larger input gives larger output
Absolute linearity	Maximum difference between ideal and actual characteristics, including offset and gain errors
Missing code in A/D converter	Some digital codes cannot be generated

Note: INL < 1 LSB usually implies DNL < 1 LSB but not vice versa. DNL ≥ 1 LSB usually implies nonmonotonicity or missing code.

The commonly used ADC structures may be classified into one of the following groups listed in order of increasing accuracy and decreasing speed:

1. Parallel (''flash'') and serial–parallel A/D converters.
2. Pipelining and multiplexing A/D converters.
3. Serial (successive-approximation) A/D converters.
4. Oversampling A/D converters.
5. Counting A/D converters.

These converter types are now briefly discussed.

Flash Converters. The block diagram of a flash converter is shown in Figure 10-18. It contains three stages. The first one is a reference generator, which consists of the dc voltage references V_R^+ and V_R^-, as well as a string of $2^N + 1$ resistors. This string supplies comparison voltages for the $2^N + 1$ comparators in the second stage. These comparators provide a binary ''thermometer'' code output, in which the location of the sign change between two adjacent comparator outputs indicates the range of V_{in} and also the presence of any overflow or underflow. Finally, the thermometer code is converted into an N-bit binary-coded decimal output by the digital logic contained in the third stage.

Ideally, the comparison and code conversion operations can be performed in a single clock cycle and the data conversion is thus nearly instantaneous. Also, the monotonicity of conversion is assured unless the offset voltages of the comparators are excessive. However, the complexity of the circuit becomes very high for $N \geq$ 8 and so the chip area and dc power dissipation also become excessive. In addition, the necessary accuracy of the comparisons needed also increases exponentially with N. For these reasons, flash converters are seldom used to achieve 9-bit or higher resolution.

Serial–Parallel Converters. To reduce the complexity required by the flash structure, the serial–parallel (also called *half-flash*) configuration may be used. Two

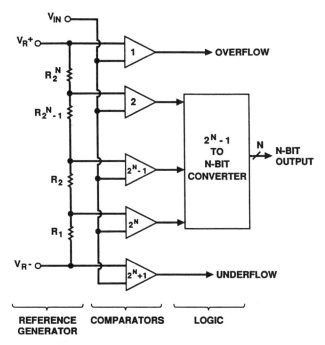

FIGURE 10-18 A flash A/D converter.

examples are shown in Figure 10-19. In the first structure, often called a *subranging* A/D converter, V_{in} is entered into an $N/2$-bit flash "coarse" converter, which gives the $N/2$ MSBs. Next, these are reconverted into analog form and subtracted from V_{in}. The result is the quantization error, which is then amplified by $2^{N/2}$ and converted by a second $N/2$-bit flash "fine" A/D converter.

An alternative scheme (the *ripple* converter) is illustrated in Figure 10-19(b). Here, the coarse converter, in addition to providing the $N/2$ MSBs, also controls a reference generator, which derives the reference voltage set V_r' for the fine A/D converter. The latter then provides the $N/2$ LSBs. In a typical implementation, the reference voltages of the coarse converter may be generated by a resistor string connected between V_R^+ and V_R^-, as shown in Figure 10-18; the MSBs thus obtained are then used to locate those two tap voltages on the coarse resistor string that bracket V_{in}. The resistor string of the fine converter is next connected between these two voltages, suitably buffered, and provides the V_R' references.

In the described sequence of operations for the converters of Figure 10-19, the fine converter is idle while the MSBs are derived by the coarse converter and, vice versa, the coarse converter is idle while the fine converter is active. Hence it is possible to time-share a single $N/2$-bit flash converter to perform both tasks at substantial saving, both in complexity and in power dissipation.

An alternative way of fully utilizing the stages of the half-flash system is *pipelining*. In this process the coarse converter derives the MSBs of the nth sample of V_{in}, while at the same time the LSBs of the previous (i.e., $(n-1)$st) sample

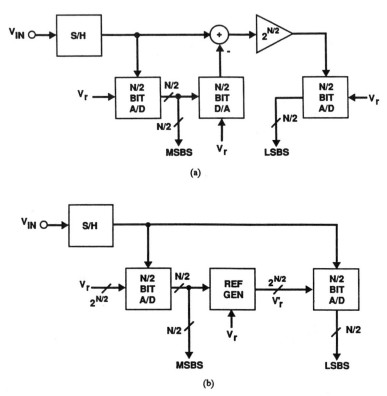

FIGURE 10-19   Serial-parallel ADCs. (a) Subranging ADC. (b) Ripple converter.

are derived by the fine converter. In order to obtain all bits of a converted sample at the output of the converter simultaneously, the MSBs of each word must be stored until the LSBs of the word become available. This can be achieved using shift registers as delay devices.

Pipelined Converters.   The pipelining principle outlined above can be generalized. Thus, instead of two subsystems, a larger number (up to N) may be used. The general scheme is illustrated in Figure 10-20. Here, the ith stage derives N_i bits, which are sent to the output via the ith shift register SR_i to delay them appropriately. The same N_i bits and/or their analog equivalent signal are also entered (possibly along with a sample of V_{in}) into the next stage, which then derives the next N_{i+1} bits of the word, and so on.

Two examples of the general scheme are shown in Figures 10-21 and 10-22. The former [LE87] is based on the subranging scheme of Figure 10-19(a). It requires M S/H stages and A/D as well as D/A converters and $M - 1$ analog gain stages. The operation of Stage 1 must be accurate to the full N-bit solution but Stage 2 requires only an $(N - n_1)$-bit accuracy, and so on.

The other scheme (shown in Figure 10-22 for single-bit stages) is based on the

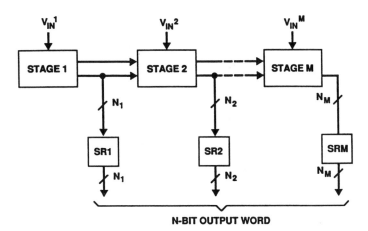

FIGURE 10-20 Pipelined converter.

ripple converter of Figure 10-19(b). It functions as follows [TE85]. When clock phase ϕ_1 is high, comparator C1 compares $V_{in}^{(1)}$ with reference voltage $V_r^{(1)} = V_r/2$, assuming a conversion range of $0 \sim V_r$. This operation give the MSB, which enters an $(N - 1)$-stage shift register. During $\phi_2 = 1$, the MSB is used to derive the reference voltage $V_r^{(2)}$ for C2: $V_r^{(2)} = 3V_r/4$ if MSB = 1 and $V_r/4$ if MSB = 0. In a similar manner, $V_r^{(3)}$, $V_r^{(4)}$, $\cdots$, $V_r^{(N)}$ can be found from the earlier bits derived from $V_{in}^{(1)}$. The switches keep all comparators busy all the time, and the output words appear at the fast clock rate of $\phi_1 + \phi_2 + \cdots \phi_N$. This scheme avoids the cumulative errors usually associated with pipelined systems.

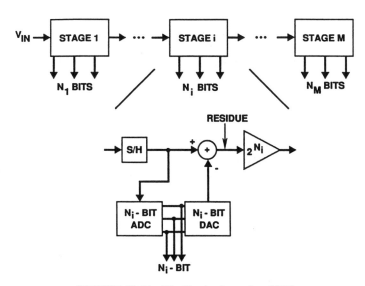

FIGURE 10-21 Pipelined subranging ADC.

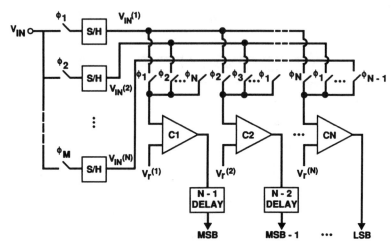

FIGURE 10-22   Pipelined ripple converter.

Both architectures can be modified to incorporate digital error correction for improved accuracy.

Multiplexing Converters. Figure 10-23 shows the block diagram of a time-multiplexing A/D converter [BL80]. In this system, each converter is allowed M clock periods to perform an *N*-bit A/D conversion. Thus added complexity is introduced to buy speed. Any skew or jitter in the timing of the ϕ_1 is transformed into noise. Also, any mismatch of the *M* paths introduces "fixed-pattern" noise into the output data stream. This converter is often called an *interleaved* A/D structure.

Serial Converters. In a serial converter, a single analog stage with associated digital logic is used to derive, one by one, all *N* bits of the output word. A possible implementation is illustrated in Figure 10-24. It requires *N* cycles to obtain all *N* bits. In the first cycle, S_1 is closed (but S_2 is open), and hence the comparator C compares $2V_{in}$ with V_r. If $2V_{in} > V_r$, the MSB is $b_1 = 1$ and V_r is subtracted from

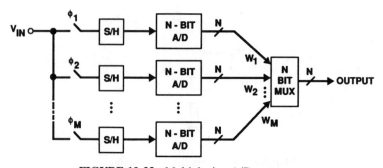

FIGURE 10-23   Multiplexing A/D converter.

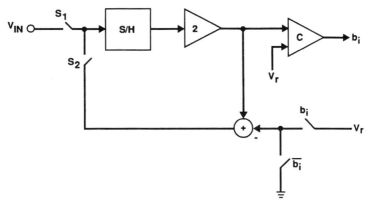

FIGURE 10-24 A serial ADC.

$2V_{in}$. Next, the S/H is reset to the remainder $V = 2V_{in} - V_r$ and $2V = 4V_{in} - 2V_r$ is compared to V_r, and so on. Thus, after N cycles, the bits $b_1, b_2, \cdots, b_N$ are found. The general scheme, which is called the *multiplied remainder* method, is illustrated in Figure 10-25. Note that the bit b_0 is the sign bit: $b_0 = 1$ if $V_{in} > 0$, otherwise it is 0.

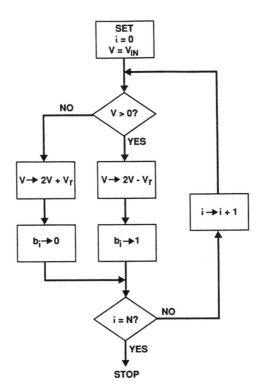

FIGURE 10-25 The flowchart of the multiplied remainder conversion algorithm.

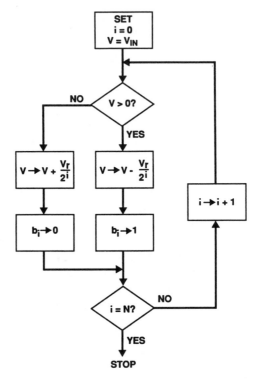

FIGURE 10-26   The flowchart of the divided reference conversion algorithm.

An alternative successive-approximation conversion process is based on the *divided reference* method. This uses steps very similar to those illustrated in Figure 10-25 except that instead of multiplying the remainder V in each step by 2, V_r is divided by 2. The flowchart of the process is shown in Figure 10-26.

As the preceding discussions illustrate, the main components of a serial converter are a S/H stage, which holds initially V_{in} and then the remainder voltage V; a reference voltage generator, a comparator C, and digital logic and storage (often called *successive-approximation register* or SAR), which control the derivation of the next remainder and reference voltages.

Oversampling A/D Converters. These converters are used to achieve high resolution using only low-accuracy analog components. Figure 10-27 illustrates the block diagram of an oversampling A/D converter, often called (for historical reasons) a *delta-sigma* converter [CA91]. The equivalent circuit, in which the integrator has been replaced by a sampled-data accumulator and the quantization error $e[n]$ specifically indicated, is also shown. As discussed in Section 10-2-1, the quantization error (for an appropriately bandlimited analog input signal) is restricted to $-Q/2 \le e[n] \le Q/2$ and has a rms error $Q/2\sqrt{3}$. For the system of Figure 10-27, it is usually assumed that the $e[n]$ are samples of a random noise

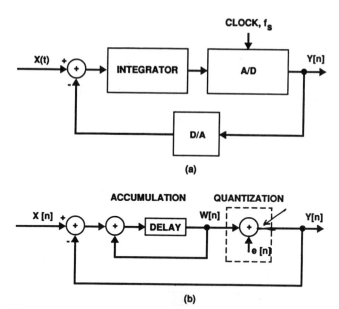

FIGURE 10-27 The block diagram of the delta-sigma quantizer and its sampled-data equivalent circuit.

sequence with a uniform (white) spectral distribution. While this assumption is not strictly correct, since $x(t)$ and the initial state fully determine the $e[n]$, it is a good approximation if $x(t)$ is "busy," that is, varies fairly rapidly and is nearly random itself. Under these conditions, ignoring the correlation between the $x[n]$ and the $e[n]$, the system of Figure 10-27(b) may be analyzed to yield the z-domain *signal transfer function*

$$H_S(z) \triangleq \left. \frac{Y(z)}{X(z)} \right|_{E(z) \equiv 0} = \frac{H(z)}{H(z) + 1} \tag{10.22}$$

and the *noise transfer function*

$$H_N(z) \triangleq \left. \frac{Y(z)}{E(z)} \right|_{X(z) \equiv 0} = \frac{1}{H(z) + 1} \tag{10.23}$$

Here, $H(z) = z^{-1}/(1 - z^{-1})$ is the transfer function of the accumulator, which plays the role of the loop filter. Substituting the expression for $H(z)$, we get the transfer functions:

$$H_S(z) = z^{-1} \quad \text{and} \quad H_N(z) = 1 - z^{-1}. \tag{10.24}$$

Thus in the time domain $y[n] = x[n-1] + (e[n] - e[n-1])$, indicating that the output equals the delayed input plus a modulation noise which is the first dif-

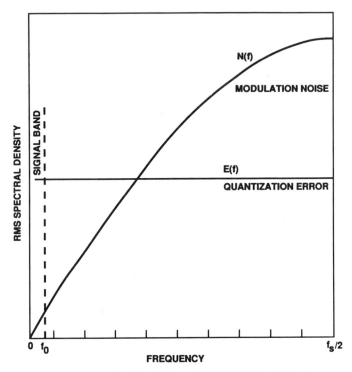

FIGURE 10-28 The spectral density of the noise $N(f)$ from delta-sigma quantization compared with the direct quantization noise $E(f)$.

ference of the quantization error $e[n]$. In the frequency domain, $H_N(f) = 1 - \exp(-j2\pi fT)$, where $T = 1/f_s$ the sampling period. $H_N(f)$ is a highpass function; Figure 10-28 compares the spectra of $E(f)$ and $N(f) = H_N(f)E(f)$. Clearly, the noise spectral density is greatly reduced at low frequencies, at the cost of increasing it at higher ones. Thus if the signal band extends from dc to a bandwidth $f_o \ll f_s$, then the in-band noise power may be much lower than that of quantization error. The enhanced high-frequency out-of-band noise can be filtered out by a digital postfilter.

The delta-sigma modulator suffers from two shortcomings. First, to achieve large signal-to-noise ratios, the oversampling ratio OSR given by $f_s/(2f_o)$ must be huge. Thus OSR ≥ 1024 is required to achieve an 84-dB SNR, equivalent to a 14-bit accuracy. This is because of the limited highpass filtering ("noise shaping") ability of the first-order loop filter. Second, due to the simple structure of the loop, the correlation between $x[n]$ and $e[n]$ remains strong and for dc or very low frequency input signal $x(t)$ resonance effects may occur. These cause in-band tones to be generated in the $e[n]$ and may make the device unusable. Both problems can be overcome by the use of a higher-order loop filter. Figure 10-29 shows a modulator with a second-order loop filter. Linear analysis shows that now $H_N(z) = (1 - z^{-1})^2$, while $H_S(z)$ remains unchanged. Thus the noise filtering corresponds to

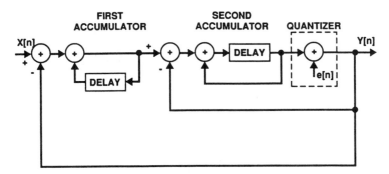

FIGURE 10-29   A second-order noise-shaping quantizer.

two cascaded first-order filters, and the in-band noise power is reduced. Figure 10-30 illustrates the dependence of the in-band noise power on the OSR and the order L of the loop filter [CA91].

The second-order delta-sigma modulator also has some potential problems. Its internal signal levels may be higher than those of a first-order system and it may become unstable due to inaccuracies or parasitic delays. The stability may be enhanced by realizing the second-order noise shaping by a cascade of two first-order delta-sigma loops (Figure 10-31). In this system, an analog signal containing the conversion error $e'[n]$ of the first stage is converted into a digital signal by a second

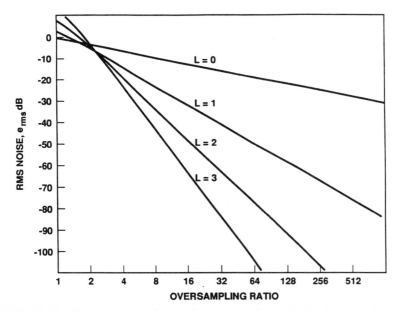

FIGURE 10-30   The rms baseband noise for oversampling ratios in the range 1 through 512, assuming busy input signals.

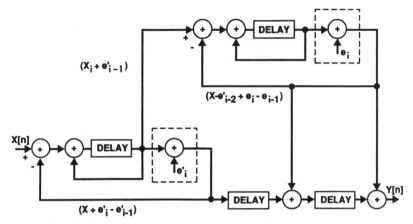

FIGURE 10-31 A cascade of two first-order delta-sigma modulators.

stage. The two digital outputs are then postfiltered and combined such that $e'[n]$ is canceled while the error $e[n]$ of the second stage is second-order filtered. The cancellation of $e'[n]$ requires that the analog accumulator in the first stage be accurately realized and may require special circuit techniques as well as careful design. Higher-order loop filters may also be used but require extra care since the resulting systems are potentially unstable.

Since the delta-sigma modulators contain a D/A converter in the feedback loop, any error caused by its nonlinearity is directly subtracted from $x(t)$ and thus from $y[n]$. Hence any such nonlinearity must be extremely small. The simplest way to achieve this is to use a single-bit A/D converter (i.e., a comparator) and a single-bit D/A converter in the loop. The latter can be realized by a toggle switch and is inherently *linear*. Other solutions, based on randomization [CA89b], frequency modulation [LE92], digital correction [CA89a], and a combination of a single-bit feedback and multibit feedforward paths [LE90] also exist. Figure 10-32 illustrates the last-named technique. The output contains only the difference of the multibit quantization error.

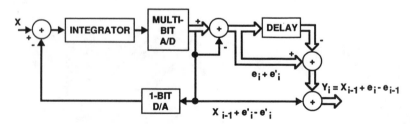

FIGURE 10-32 A first-order delta-sigma modulator with a single-bit feedback and multibit feedforward path.

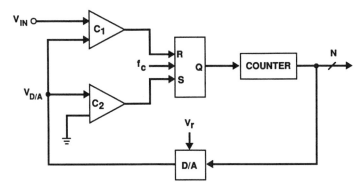

FIGURE 10-33 An A/D converter based on counting.

Counting Converters. The most accurate but slowest analog-to-digital converters are based on the counting principle. A simple implementation is illustrated in Figure 10-33. Comparator C_2 initiates the count when the D/A converter output $V_{D/A}$ reaches zero and C_1 stops it when $V_{D/A}$ reaches V_{in}. The number of steps taken between these two conditions, registered by the counter, is proportional to V_{in}/V_r and is the digital output. To achieve an N-bit accuracy, the step size should be at most $2^{-N}V_r$ and hence 2^N steps are needed for the conversion.

A more accurate version of this system, which to a good degree is independent of the accuracy of the components such as comparators and D/A converter, is the *dual-slope* converter. The concept is illustrated in Figure 10-34. In one version of this system, first V_{in} is integrated for a fixed time t_{in}, and then $-V_r$ is integrated for as long a time t_r as it takes to return the output signal V_o of the integrator (monitored by the comparator C) to zero. The ratio t_r/t_{in}, as measured by a counter in the logic block, is ideally equal to V_{in}/V_r. In another version [RO87], pulses proportional to the input are integrated until V_o. After 2^N input pulses, the number N_r of the reset pulses gives the converted version of V_{in} according the formula $V_{in} = 2^{-N}N_rV_r$.

The properties of the five converter types discussed in this section are summarized in Table 10-2.

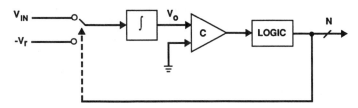

FIGURE 10-34 A dual-slope A/D converter.

TABLE 10-2   Properties of Basic ADC Architectures

Converter Type	Minimum Number of Nyquist Periods per Sample	Typical Accuracy (bits)	Minimum Number of Analog Stages	Complexity of Digital Circuitry
Flash (half-flash)	1 (2)	8 (10)	2^N ($2^{N/2}$)	Medium
Pipeline/multiplexing	1, also N periods of delay (latency)	9	N	Medium
Serial	N	14	1	High
Oversampling	OSR $\sim$ 128	16	1	Very high
Counting	2^N	18	1	Medium

10-3   DIGITAL-TO-ANALOG CONVERTERS

10-3-1   General Considerations

A digital-to-analog (D/A) converter, often abbreviated as DAC, takes a binary-weighted digital input signal and converts it into a sampled-and-held analog output signal, which can be voltage or current. The maximum signal-to-noise ratio (SNR) at the output of the DAC is limited by its accuracy and is the same as for an ADC (Eq. (10.16)). The limit on the achievable resolution N is determined by a variety of static and dynamic factors. An illustration of some of the potential static DAC errors appears in Figure 10-35 for the case of an idealized three-bit DAC [SH72]. These errors can be classified as offset errors (Figure 10-35(a)), gain errors (Figure 10-35(b)), and nonlinearity errors (Figure 10-35(c) and 10.35(d)), which in extreme cases can lead to a nonmonotonic response.

The extent to which these static errors degrade system performance is system dependent. Some analog systems can accept a certain degree of gain error and nonlinearity without suffering catastrophic failures. Linearity errors can be subdivided into integral linearity error, which is the maximum deviation of the input–output characteristic from a straight line over the full range of outputs, and differential linearity error, which is the maximum relative change of $\Delta V_{out}/$LSB over the full range of analog outputs.

Dynamic errors can also cause a degradation in the SNR of the output of the DAC. One well-known source of dynamic errors in high-speed DAC operation is the presence of "glitches," which usually occur at major code transition steps. This is illustrated in Figure 10-36, which shows a glitch occurring at the transition from "01111111" to "10000000" in the output of an 8-bit DAC. Excessive settling time can also create major problems in the dynamic response of a DAC. Ideally, the output of the DAC should settle to within $\pm\frac{1}{2}$ LSB during the sampling interval. If it does not, a nonlinear (output-dependent) error is created in the output that leads to harmonic distortion.

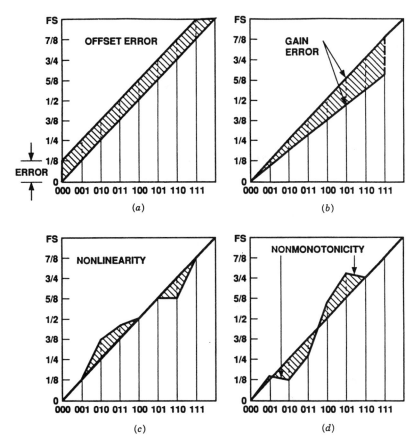

FIGURE 10-35 Static DAC errors. (a) Offset error. (b) Gain error. (c) and (d) Nonlinearity error.

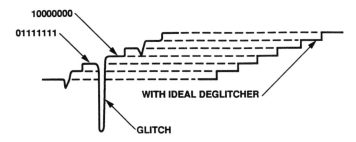

FIGURE 10-36 Illustration of DAC "glitch" at major carry transition.

10-3-2 DAC Architectures

As discussed for ADCs, the performance of DACs can also be evaluated in terms of their accuracy, speed, and complexity. The commonly used DAC architectures can be classified into the same categories as the ADC structures and are briefly discussed next in order of increasing accuracy (but decreasing speed). The categories used are listed below.

1. Parallel D/A converters.
2. Pipelining and multiplexing D/A converters.
3. Serial D/A converters.
4. Oversampling D/A converters.
5. Counting D/A converters.

In discussing these structures, we shall also briefly describe some available alternatives for their physical realizations.

Parallel Converters. In these converters all bits of the input digital word are used simultaneously to derive a physical quantity, such as voltage, current, or charge, which is proportional to the word. A typical realization for a *voltage-based conversion* is shown in Figure 10-37 for the simple case of $N = 3$. The advantages

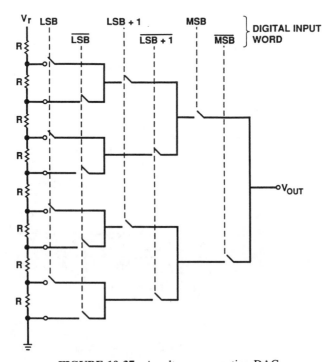

FIGURE 10-37   A voltage-segmenting DAC.

of this circuit are its monotonicity and simplicity. However, for high accuracy ($N > 10$) and fast conversion, the structure has significant drawbacks. The string resistor values must be small to make the associated RC time constants negligible, causing a large dc power dissipation. Also, the path from the resistor string to the output terminal includes N conduction switches behaving like RC transmission lines, and the parasitic capacitances of the N nonconductive switches also load the nodes of the path. These effects slow down the settling of the converter after a new input word is entered.

At the cost of increased digital complexity, the number of conducting switches in the path can be reduced to one. This is achieved by using a decoding logic that converts the N-bit binary input word into a 2^N-bit one that contains a 1 at the appropriate location and 0's everywhere else. The resistor string now only has the 2^N LSB and $\overline{\text{LSB}}$ switches connected and the 2^N-bit decoded word closes the appropriate one between the string and the output terminal.

For a *current-based converter*, it is possible simply to use $2N$ equal-valued current sources. The binary input word is then decoded into a $2N$-bit "thermometer" code in which the number of 1's is equal to the value of the input. Each "1" then results in the current I of a source being included in the current output of the converter. For more compact layout and better matched current sources, the cells realizing the sources are arranged in a square array, and a column decoder as well as a row decoder is used to steer the cell currents to the output or to a current sink. (Note that current steering is a much faster operation than turning on and off the sources.)

An alternative technique of current-based D/A conversion is illustrated in Figure 10-38. As the diagram illustrates, the currents flowing in the shunt $2R$ resistors are binary scaled by the "R–$2R$" ladder. To see this, we note that the impedances seen at each node of the ladder equal R. It follows that the node voltages are binary scaled and form the sequence $V_r, V_r/2, \ldots, V_r/2^{N-1}$. Hence the currents in the shunt branches are $I = V_r/2R, I/2, \ldots, I/2^{N-1} = V_r/2^N R$. The bits of the input word decide which of these currents contribute to the output current I_{out} and hence to the output voltage $V_{\text{out}} = -RI_{\text{out}}$. This scheme requires much less total resistance than the voltage division scheme of Figure 10-37 ($3NR + R$ versus $2^N R$).

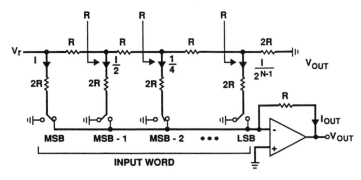

FIGURE 10-38 A current-segmenting DAC.

However, the sensitivity of the analog output to the values of the resistors associated with the MSBs is high, and hence the accuracy requirements are quite stringent; in the kth section the allowable tolerance is $(\Delta R/R)_k \leq 1/2^{N-k}$, $k = 0, 1, \ldots, N-1$.

It is also possible to generate binary-scaled currents by scaling the W/L ratios of MOSFETs or the emitter areas of bipolar devices. It is, however, hard to obtain very-high-accuracy scaling using this technique. More sophisticated circuits, based on dynamic matching or current copying, can be used to obtain high resolution for current-based DACs.

Charged-based DACs can be constructed by charging selected elements of a binary-scaled capacitor array (Figure 10-39). The conversion operation is performed in two steps. In Step 1, the *reset step*, all switches that are connected to ground close, discharging all capacitors. In Step 2 (the *conversion step*), switch S opens and the toggle switches associated with bits equal to 1 switch to V_r while those associated with 0-valued bits remain at ground. The resulting output voltage is $V_{\text{out}} = \Sigma_{i=1}^{N} b_i 2^{-1} V_r$, where the b_i are the bits and b_1 is the LSB.

Alternative charge redistribution DAC schemes may be based on C–$2C$ ladders or on gain programmable switched-capacitor (SC) amplifiers [GR83].

Pipelining and Multiplexing DACs. A pipelining DAC built from single-bit DACs (essentially toggle switches), delays, and divide-by-2 stages is illustrated in Figure 10-40. It is easy to follow the progress of a sample through the pipeline and to verify that V_{out} is now given by $\Sigma_{i=1}^{N} b_i 2^{-i+1} V_r$. Since the first stage becomes idle after it processes the LSB of a word, it can then be used to process the LSB of the next word, and so on. Thus, as discussed in connection with the pipelining ADCs, by feeding to all stages in every cycle a new bit, the conversion rate can be made 1 word/clock period. Again, however, there is an unavoidable delay of N clock periods between the appearance of the LSB of a word and that of the corresponding analog output V_{out}. The proper timing of the bits can be obtained by using an "organ-pipe" array of shift registers.

Figure 10-41 shows a simple switched-capacitor realization of a pipelining charge-redistribution DAC [WA89]. Here, the delaying and scaling-by-2 of the charges are obtained by connecting adjacent capacitors via the switches S_1 to S_N.

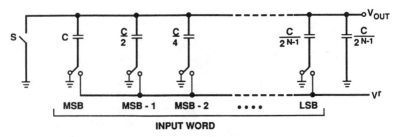

FIGURE 10-39   A charge-redistributing DAC.

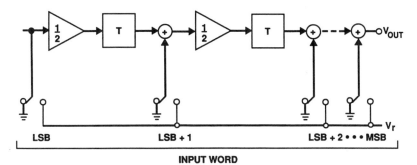

FIGURE 10-40 A pipelining voltage-based DAC.

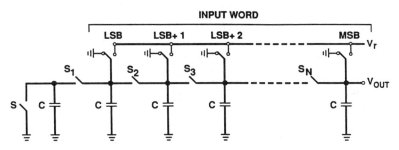

FIGURE 10-41 A pipelining charge-redistribution DAC.

Pipelining DACs can also be constructed by using SC-op amp stages to realize the divide-by-2 and delay functions in the block diagram of Figure 10-40.

Multiplexing (interleaved) D/A converters can be constructed in a scheme similar to that used for multiplexing ADCs (Figure 10-42). By using M DACs rather than only one, the available conversion time can be extended from 1 to M clock periods. Thus slower converters can be used in a given application.

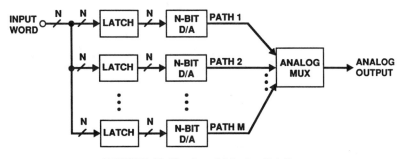

FIGURE 10-42 A multiplexing DAC.

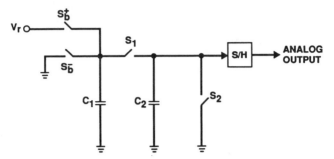

FIGURE 10-43   A serial D/A converter.

Serial D/A Converters. In a serial DAC , the N-bit digital input word is converted, bit by bit, in a time interval of N clock periods. As an example, the scheme of Figure 10-40 may be used as a serial DAC if the bits of the input word are fed serially to the toggle switches (with the LSB first) and the next word is entered only N clock periods later. The same operation, however, can also be performed using a single stage composed of SC elements (Figure 10-43). In this circuit, which is equivalent to one section of the pipelining converter of Figure 10-41, the two equal-valued capacitors are initially reset (i.e., discharged) by closing S_1 and S_2. Next, the bits b_i are used to develop the analog output voltage, as follows. S_1 and S_2 are opened, and the bit switches S_b^+ and S_b^- used to charge C_1 to V_r if the LSB b_1 is 1 or to remain discharged if $b_1 = 0$. Next, S_b^+ and S_b^- open, S_1 closes, and C_1 and C_2 share charges. Now, S_1 opens and S_b^+ and S_b^- allow C_1 to recharge to V_r or ground as controlled by b_2. After this, C_1 and C_2 share charges again, and so on. In each charge-sharing operation, the charges earlier stored in C_2 are halved, so after N operations the LSB charge will be divided by 2^N, the charge due to the next LSB by 2^{N-1}, and so on, as required. Thus the voltage detected across C_2, after all N bits have been entered, is the desired DAC output.

Oversampling D/A Converters. Like the oversampling ADCs, oversampling DACs are used to overcome the accuracy limitations of integrated analog components. Oversampling reduces the problem of accurate N-bit data conversion (which is difficult for, say, $N > 10$ bits) to a rapid succession of 1-bit D/A conversions. Since the latter operation involves only a toggle switch and reference voltage, it can be performed with a high accuracy and linearity. The basic scheme is illustrated in Figure 10-44. The low input sampling frequency f_{SL} is increased to a much higher rate $f_{SH} = \text{OSR} \times f_{SL}$ in the interpolator. The latter is a cascade of

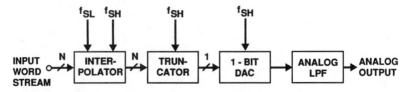

FIGURE 10-44   The block diagram of an oversampling DAC.

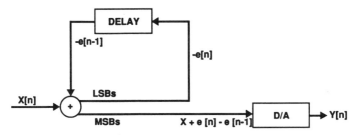

FIGURE 10-45 A digital quantizer loop.

digital lowpass filters that suppress the signal spectra around f_{SL}, $2f_{SL}$, ... The filtered high-sample-rate signal is then entered into the truncator (quantizer), which is essentially a digital delta-sigma loop. It accepts a multibit data stream and creates a 1-bit output stream such that the input and output spectra in the signal band are very similar. This requires a shaping of the spectrum of the quantization error such that the error power in the baseband is negligible. A simple quantizer is shown in Figure 10-45. By inspection, the output is $y[n] = x[n] + e[n] - e[n - 1]$, where x is the current input and $-e[n]$ is the current value of the neglected LSBs (in our case, these include all bits except the MSB). Clearly, the modulation noise is $\eta[n] = e[n] - e[i - 1]$, the first difference of the quantization error $e[n]$. Thus the signal transfer function equals 1 and the noise transfer function is $H_N(z) = 1 - z^{-1}$ as for a first-order noise-shipping ADC. Thus in the baseband the quantization noise power is suppressed, as required. The high-frequency noise is converted into an analog noise and must be suppressed by the analog lowpass filter following the DAC (Figure 10-44).

Higher-order quantization loops may also be constructed by using two delays in the quantization loop [CA91] or using cascaded quantization loop. In the latter system, the LSBs contained in the $-e[n]$ error of the first loop are entered into the second loop. The outputs of the two loops are then D/A converted and combined using appropriate analog postfilters. Alternatively, the digital outputs may be first combined and then converted [CA91]. In any case, in the system the first-stage error $-e[n]$ is canceled by the second stage and is replaced by the second-order filtered quantization error originating in the second loop.

Counting D/A Converters. These slow but accurate converters rely on adding or integrating equal-valued analog quantities (e.g., charge packets) to generate the analog output. To achieve an N-bit resolution, at least $2N$ steps are needed. An implementation is illustrated in Figure 10-46. In the system, the switch S closes when the counter starts generating a digital ramp and remains closed until the digital comparator indicates that the digital ramp reached the level of the input word. In a sampled-analog implementation, the switch S can be replaced by the SC input branch of a SC integrator, which continues to enter charges $C_{in} V_r$ into the integrator as long as the ramp is being generated.

Some key properties of the D/A converter types discussed above are summarized in Table 10-3.

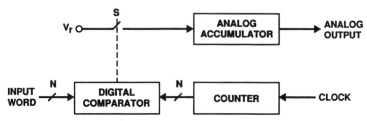

FIGURE 10-46   The block diagram of a counting DAC.

TABLE 10-3   Properties of Basic DAC Architectures

Converter Type	Minimum Number of Nyquist Periods per Sample	Typical Accuracy (bits)	Minimum Number of Analog Stages	Complexity of Digital Circuitry
Parallel	1	9	2^N	Medium
Pipeline/multiplexing	1, also an N periods of delay (latency)	8	N	Medium
Serial D/A converters	N	10	1	High
Oversampling D/A converters	OSR $\sim$ 128	16	1	Very high
Counting D/A converters	2^N	18	1	Medium

10-4   SMOOTHING FILTERS

The output signal of the D/A converter is usually a sampled-and-held voltage or current wave and has a staircase waveform. Its spectrum may be obtained from Eqs. (10.4) and (10.7); the end result is

$$F_{SH}(j\Omega) = e^{-j\Omega T/2} \frac{\sin (\Omega T/2)}{\Omega T/2} \sum_{k=-\infty}^{\infty} F\left(j\Omega - j\frac{2\pi k}{T} \right), \qquad (10.25)$$

where $F(j\Omega)$ is the spectrum of the continuous-time signal that we are attempting to reconstruct. Figure 10-47 illustrates the relation between $F(j\Omega)$ and $F_{SH}(j\Omega)$. Clearly, to regain $F(j\Omega)$, two tasks must be performed: the replicas centered at ω_S, $2\Omega_S$, . . . , must be removed and the amplitude distortion (''droop'') represented by the $\sin (\Omega T/2)/(\Omega T/2)$ factor must be equalized. The former task is very similar to that performed by the anti-aliasing filters discussed in Section 10-1 and many considerations discussed there apply here as well. Thus if the passband spectrum $F(j\Omega)$ is in the frequency range 0 to f_1, then it is sufficient if the smoothing filter is an analog lowpass filter with a passband edge at $f = f_1$ and a stopband edge at $f_S - f_1$. Since the smoothing filter must be realized in analog form, it tends to be bulky and expensive, and, as discussed in Section 10-1-3, it is preferably kept

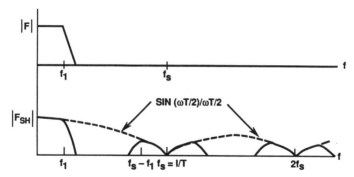

FIGURE 10-47 The spectra of a continuous-time signal and its sampled-and-held equivalent.

very simple (of order $p = 3$ or lower) especially if an on-chip realization is desired. This will usually be possible if $f_S \gg f_1$. If this is not the case, then the sampling rate at the smoothing filter input may be increased using a process called *interpolation*. This can often be performed in the digital domain using techniques discussed in Chapter 14 of this book. Interpolation (which is the inverse of the decimation process discussed in Section 10-1-2 and illustrated in Figure 10-7) raises the sampling rate by a factor R to Rf_S, and hence the smoothing filter now needs to cut off only at $Rf_S - f_1$. The digital interpolation ratio R may be limited by the available speed of the D/A converter. In this case, it is also possible to perform part or all of the interpolation procedure after the D/A converter, in the analog domain. This involves sampled-data analog filtering, which is usually carried out using switched-capacitor (SC) filters. Figure 10-48 illustrates the general scheme of an interpolation-based smoothing filter system, including both digital and analog interpolator stages. Figure 10-49 shows a SC interpolator [GR86] that performs a linear interpolation. For the timing diagram shown, $R = 4$. Its transfer function is given by [GR86]

$$|H(j\omega)| = \frac{1}{R}\left|\frac{\sin(\Omega R/2)}{\sin(\Omega/2)}\right|. \tag{10.26}$$

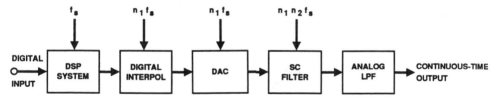

FIGURE 10-48 An interpolating smoothing filter system.

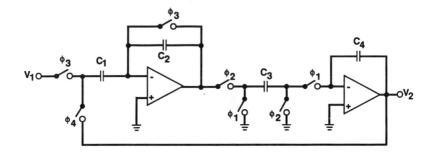

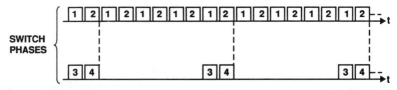

FIGURE 10-49   Switched-capacitor interpolator. The element values are $C_2 = C_4 = \sqrt{RC_1} = \sqrt{RC_3}$. The input need *not* be a sampled-and-held signal.

This function has zeros at $f = f_S, 2f_S, \ldots, (R - 1)f_S$, and thus it preserves the spectral replicas only around $f = 0$, Rf_S, $2Rf_S$, and so on, as required for the interpolation process.

The actual smoothing filter is the last stage of the system. Its requirements, and hence also its realization, are very similar to those of the analog anti-aliasing filter discussed in Section 10-1. Often, its zero and pole locations are slightly perturbed in order to introduce a gain peak at the passband edge. This can compensate for the droop due to the $\sin(\Omega T/2)/(\Omega T/2)$ factor introduced by the sample-and-hold effect.

10-5   SUMMARY

This chapter has summarized the various components of signal conditioning and interface circuits. We began with a discussion of the mathematics of the sampling process and how aliasing results from undersampling of an insufficiently bandlimited analog signal. We then introduced a variety of sampling techniques and analog filters that would minimize the undesired distortion that results from aliasing in the sampling process. Our next topic was analog-to-digital converters, where we discussed characterization and figures of merit for these circuits. A variety of analog-to-digital converter architectures were introduced including flash, pipelined, and oversampled. We next discussed digital-to-analog converters, their characterization, design, and performance. Our final section introduced the concept of the smoothing filter and how it could be used to reconstruct the original analog spectrum from a discrete-time digital output signal.

Analog interface structures constitute an important interface between the analog continuous-time signals that characterize phenomena external to electronic systems and the systems themselves. It is hoped that this chapter has given the reader some understanding of their importance, and some perspective on their design and performance.

REFERENCES

[AE76] Aerospace Group, *500-MHz Analog-to-Digital Converter*, Final report, Rep. P72–64, Hughes Aircraft Co., Los Angeles, CA, 1976.

[BA83] M. Banu and Y. Tsividis, Fully integrated active *RC* filters in MOS technology. *IEEE J. Solid-State Circuits* **SC-18,** 644–651 (1983).

[BL80] W. Black and D. Hodges, Time interleaved converter arrays. *IEEE J. Solid-State Circuits* **SC-15,** 1022–1029 (1980).

[CA89a] T. Cataltepe, G. C. Temes, and L. E. Larson, Digitally corrected and multi-bit $\Sigma - \Delta$ data converters. *Proc. IEEE Int. Symp. Circuits Syst. 1989*, pp. 647–650 (1989).

[CA89b] R. Carley, A noise-shaping coder topology for 15 + bit converters. *IEEE J. Solid-State Circuits* **SC-24,** 267–273 (1989).

[CA91] J. C. Candy and G. C. Temes, Oversampling methods for A/D and D/A conversion. In *Over-sampling Delta-Sigma Converters*. IEEE Press, New York, NY, 1991.

[GH81] M. S. Ghausi and K. R. Laker, *Modern Filter Design: Active RC and Switched-Capacitor*. Prentice-Hall, Englewood Cliffs, NJ, 1981.

[GR83] R. Gregorian, A switched-capacitor DTMF chip set, *Microelectron. J.* **12,** 10–13 (1983).

[GR86] R. Gregorian and G. C. Temes, *Analog MOS Integrated Circuits for Signal Processing*. Wiley, New York, 1986.

[LA88] L. Larson, High-speed analog-to-digital conversion with GaAs technology: Prospects, trends, and obstacles, *Proc. IEEE Inter. Symp. Circ. Sys., 1989* pp. 2871–2878 (1988).

[LE87] S. Lewis and P. R. Gray, A pipelined 5m sample/sec 9-bit analog-to-digital converter. *IEEE J. Solid-State Circuits* **SC-22,** 954–961 (1987).

[LE90] T. C. Leslie and B. Singh, An improved sigma-delta modulator architecture. *Proc. IEEE Int. Conf. Circuits Syst.* pp. 372–375 (1990).

[LE92] B. Leung and S. Sutarja, to appear.

[MA80] K. Martin and A. Sedra, Easing prefiltering requirements of SC filters. *Electron. Lett.* **16,** 613–614 (1990).

[OP88] A. V. Oppenheim and R. W. Schafer, *Discrete-Time Signal Processing*. Prentice-Hall, Englewood Cliffs, 2nd ed. NJ, 1988.

[OR68] H. J. Orchard and G. C. Temes, Filter design using transformed variables. *IEEE Trans. Circuit Theory* **CT-15,** (4), 385–408 (1968).

[PA80] A. Papoulis, *Circuit and Systems*. Holt, Rinehart, & Winston, New York, 1980.

[RO87] J. Robert, G. C. Temes, V. Valencic, et al., A 16-bit low-voltage CMOS A/D converter. *IEEE J. Solid-State Circuits* **SC-22,** 157–163 (1987).

[SH72] A. S. Sheingold and B. L. Ferrero, Understanding A/D and D/A converters. *IEEE Spectrum* **9,** 47–56 (1972).

[TE73] G. C. Temes and S. K. Mitra, eds., *Modern Filter Theory and Design*. Wiley, New York, 1973.

[TE85] G. C. Temes, High accuracy pipeline A/D converter configuration. *Electron. Lett.* **21,** 762–763 (1985).

[WA89] F. J. Wang, G. C. Temes, and S. Law, A quasi-passive CMOS pipeline D/A converter. *IEEE J. Solid-State Circuits* **SC-24,** 1752–1755 (1989).

[WA90] R. H. Walden, private communication (1990).

[ZV67] A. I. Zverev, *Handbook of Filter Synthesis*. Wiley, New York, 1967.

11 Hardware and Architecture

TRAN THONG

Tektronix Federal Systems, Inc.
Beaverton, Oregon

YIH C. JENQ

Department of Electrical Engineering
Portland State University, Oregon

11-1 WHY HARDWARE?

11-1-1 Hardware Versus Software

In many applications, digital signal processing algorithms are implemented by computer programs on a general purpose computer (see Chapter 12). For example, digital filter designs (see Chapter 4) are done almost exclusively on a general purpose digital computer. In this case the output is a set of filter coefficients, that is, a set of numbers. Simulations of physical (analog) systems are carried out on a computer also. Analyses of (real or artificial) stored data are also performed on computers. Examples of this would be the analysis of stock market trends or the statistical analysis of a database. Most of these problems are characterized by a non-real-time requirement (i.e., the processing rate does not have to keep up with the input data rate) or by a very low input data rate (e.g., stock market data analysis).

However, when dynamic signals are acquired, and it is desirable to analyze data in a real-time mode, the processor must be able to keep up with the flow of data. If the processor cannot keep up with the input data stream, then invariably data will be lost since in real-time systems the backlog of data will continue to build up and no amount of storage will be sufficient. Most small to medium size computers can keep up with data rates on the order of a few thousand samples per second. Above these rates, a general purpose computer is not an effective solution. However, the flexibility of a general purpose computer, which is also its weak point, is such that for certain applications one may still want to remain in this environment. The user will typically turn to an *array processor* or hardware accelerator to speed up the calculations. An array processor is a dedicated arithmetic

Handbook for Digital Signal Processing, Edited by Sanjit K. Mitra and James F. Kaiser.
ISBN 0-471-61995-7 © 1993 John Wiley & Sons, Inc.

processor. Since most signal processing operations are of the vector type—for example, a sum of products is the sum of the term-by-term multiplication of two vectors—the array processors are typically optimized to perform an array operation efficiently, thus their names. Available array processors span a very wide range of performance, from a few million instructions per second (Mips) to thousands of Mips. The lower performance array processors are typically just stripped down microcomputers with hardware floating-point arithmetic processors. Code optimization is the key to success here. The higher performance array processors can be full-fledged arrays of processors with an instruction set optimized for vector operations like the sum of products, for DO loops, and so on.

Typically, the user will not program the array processor directly. This involves tedious assembly language programming since compilation from a high-level language does not make full use of the processor computing capability. The array processor vendor will typically provide the user with a set of subroutines that will make the necessary system calls to highly optimized routines in the array processor [WU72].

However, when the flexibility of a general purpose computer is not needed, one should consider a hardware solution. A number of vendors are now providing digital signal processors complete with a data acquisition subsystem and output drivers for interface to a (personal) computer and/or a display device. The disadvantage here is some loss in flexibility. Should one need to change the filtering algorithm, the update cycle can be fairly complex since it is necessary to go through at least two systems, namely, the (host) computer and the digital signal processor. Typically, an assembler or compiler is used to generate the codes on the host computer and this is then down-loaded into the signal processor.

In a production environment (e.g., a digital measurement instrument like a digital spectrum analyzer), one has no choice but to build directly a hardware processor. Since these processors are then manufactured, the large design cost can be amortized over many units.

For other applications, where the highest performance is required, custom hardware must be built. Examples of this can be found in radar and sonar applications.

11-1-2 Hardware Considerations

When the decision is made to implement the digital signal processing in hardware as opposed to a software solution on a general purpose processor (either a mainframe computer or minicomputer or embedded microcomputer), the user faces a number of choices.

Fixed-Point Versus Floating-Point Arithmetic. The first choice is whether to perform the operations in fixed-point arithmetic or to use floating-point arithmetic. The latter choice is made possible by the availability of the floating-point single-chip digital signal processor (e.g., the TRW TDC1022) [EL83]. In general, fixed-point arithmetic with about 16 bits is sufficient for 90-db A/D and D/A, but not for most DSP operations because of roundoff noise. Most available digital-to-analog converters can just barely provide this level of signal-to-noise ratio.

As with any general statement, there are exceptions. In a number of areas (e.g., seismic and sonar signal processing), the dynamic range requirement can be larger. While locally 16 bits of precision is sufficient, the data can span several orders of magnitude over the length of the record. These signals are typically characterized by a decreasing amplitude. Seismic signals have been reported to decay 100 dB in the first 4 s [WO75]. Since no processing will involve both the large initial signals and the weak signals toward the end of the record, a monotonically increasing weighting function can be applied to the record in order to boost up the signals at the end of the record. Floating-point arithmetic makes this scaling unnecessary.

Another operation that requires floating-point arithmetic is the power calculation. After a Fourier transform is performed, the instantaneous power can be evaluated by squaring the real and imaginary parts of the Fourier transform. In order to preserve data amplitude, twice the dynamic range is needed after the squaring operation.

Another set of operations that require floating-point arithmetic are the matrix manipulations. While not strictly digital signal processing operations (filter, Fourier transform), they are frequently encountered in adaptive systems (see Chapter 15) and in modern spectral analysis techniques (see Chapter 16).

Single-Chip Processor Versus Custom Hardware. The user then needs to decide whether to use one of the single-chip digital signal processors discussed in Section 11-3 or to build a custom hardware. The choice here is fairly simple. The implementation of choice is the single-chip digital signal processor, provided that the computational requirements can be met. In Section 11-2 we discuss this choice.

Off-the-Shelf Hardware Versus Application-Specific IC. Next, the user will need to trade-off the development time and cost against the size of the processor and its power consumption. This is discussed in Section 11-4.

11-2 DIGITAL SIGNAL PROCESSING COMPUTATIONAL REQUIREMENTS

The computational requirements for a digital signal processor can vary significantly from one field to another. The type of hardware to be used, either a general purpose digital signal processor or custom hardware, will depend on the throughput rate required. In this section we review the requirements for a number of typical applications.

11-2-1 Telecommunication Applications

Real-time digital signal processing in telecommunication applications [BR82; DU80; FR78] is characterized by an 8 kHz sampling frequency. Signals are sampled every 125 μs. All the processing for one sample has to be completed in 125 μs. This, at first glance, appears to be a very long time. In reality, the same hardware is multiplexed over many channels. For example, a digital filter operating on a T-1 class of signal is multiplexed over the 24 time-multiplexed channels. Thus the time available to process a sample is only around 5 μs. To implement just a simple second-order recursive digital filter requiring three multiplications (neglecting additions), a cycle time of 1.6 μs is needed. If a fourth-order digital filter is used, the cycle needs to be halved.

Thus in telecommunication applications, with the sampling rate fixed at 8 kHz, the computational requirements are really dictated by the number of multiplexed channels.

11-2-2 Audio Applications

Audio frequency typically ranges from dc to about 20 kHz; hence sampling rates from 44 to 49 kHz have been considered for standardization. Compared to the telecommunication application, the audio application [BL75; BL78; ST75] of digital technique needs about five to six times more processing power for real-time operation. Since the emphasis in audio signal processing is on quality (the signals are typically digitized to 16-bit precision), 24 bit arithmetic is preferred and the operations are typically more involved. Thus, while audio processing involves only two channels (stereo), the higher sampling rate and the more involved processing make this a problem on par with that in telecommunications.

11-2-3 Test and Measurement Applications

The simplest digital signal processing operation in test and measurement applications is signal averaging. A waveform is digitized and then averaged with an earlier record. Averaging is done to enhance measurement accuracy under the following signal acquisition conditions:

1. Noisy signal environment. The signal-to-noise (power) ratio increases proportionally with the number of averages.
2. Inaccurate sampling time. This is especially important at the highest sampling rate, where a digitizer may not be able to keep up with the required sampling rate. In this case the instrument is switched to the equivalent time sampling mode, where, on each sampling cycle, only a fraction of the required samples of the *periodic signal* are taken; the others are acquired on successive cycles. Averaging reduces the effect of the sampling jitter.
3. A coarse A/D is used. For example, an 8-bit A/D converter is used and averaging can be used to achieve the equivalent of 10 bits. Integer and fractional dithers are used to overcome converter nonlinearity and to break up any correlation in the quantization error.

From 512 to 4096 samples are acquired for each record. The computational requirement here is typically not severe since averaging can be achieved by just summing the corresponding points and then dividing by the number of samples per output point, typically a power of 2. This last operation can be implemented by a shift operation. Exponential averaging of the form

$$y[n] = ax[n] + (1 - a) y[n - 1],$$

where $y[n]$ is the stored record and $x[n]$ is the new record, is also frequently used. Since the next record does not have to be acquired until the system is ready, the computational load is not a primary consideration here. However, without sufficient processing capability, the overall acquisition time suffers. In most cases it has been desirable to keep the update rate higher than 60 Hz in order to prevent any perceptible flicker. The update rate is typically kept below 100 Hz in order to reduce the cost of hardware. Thus on the order of 400K samples need to be processed each second.

Signal interpolation is an operation involving digital filtering. Linear interpolation is the simplest. Once the slope of the line between two samples has been determined, linear interpolation just requires a recursive summation of the incremental value. An undesirable aspect of linear interpolation is that it usually does not match one's expectation of a bandlimited signal, which typically has a smooth curve. Better interpolation can be achieved with digital filters. FIR filters are typically used in order not to cause phase distortion. For example, in the Tektronix 468 oscilloscope, a 2X interpolation is done with windowed $\sin x/x$ filters. For real-time operations, the refresh rate would be at least 60 Hz. With a 1024 display point, and assuming five multiplications per output point (with a windowed $\sin x/x$ filter, the filter length will be 20 times the interpolation factor since alternate coefficients are zero and the filter is symmetric) a multiplication rate of 300,000/s is needed.

Digital spectrum analyzers are the first real-time digital signal processing instruments. The analog waveform is digitized, and the frequency band of interest is heterodyned down to baseband and then digitally filtered. This is then resampled and a FFT is performed. An early example is the Hewlett–Packard 3561A [PA84], where the digital lowpass filters are implemented with custom hardware and the FFT is performed on a Texas Instrument TMS 32010. While the custom digital lowpass multirate filter can keep up with signals as high as 100 kHz, the throughput rate of the analyzer is limited by the rate of computation of the FFT, about 30 ms for a 1024-point FFT. Since the TMS 32010 must also perform other operations besides the FFT, the real-time bandwidth is quoted as being only 7.5 kHz. A more modern digital spectrum analyzer is the Tektronix 3052 with its 2-MHz real-time bandwidth and 200-μs update [TE88b]. The computational requirement just for the FFT is on the order of 100 million multiplications per second. Taking into account the decimation filters that bring the initial 25.6-MHz sampling rate, corresponding to a 10-MHz bandwidth signal, down to a frequency span of 1 kHz, the total number of multiplications required is close to a billion operations per second! Clearly, this rate can only be achieved through parallel processing.

11-2-4   Radar Applications

Radar signal processing [MC78] is characterized by requirements to process signals with large bandwidths, 10–100 MHz in real time. The most demanding operation in a digital radar environment is the matched filtering that is performed at the beginning. The received signal is convolved with a time-reversed version of the original signal. This time-compresses the received signal. Because the signal bandwidth is extremely high here, fast convolution is used. This consists of a FFT, followed by multiplication by the transfer function of the matched filter, and then an inverse FFT.

Range and velocity estimation of a target are obtained by performing FFTs on the resulting data stream. The processing requirements here are more modest since it is a function of the interpulse period and the desired resolution. Digital filters are used also in moving target indicators (MTI) to discriminate moving targets from static ground clutter.

11-2-5   Sonar Applications

While conceptually similar to radar, the environment in which sonar [BA78] has to perform makes it a much harder signal processing function. While bandwidths are much smaller, in the 10^3-Hz range, random background noise is much stronger, correlated noise sources (due to multiple paths) exist, return signals are much weaker (high attenuating medium), and the water environment is nonuniform (the speed of sound in water is a complex function of depth and temperature). One typically has to process an array of sensors instead of the single sensor more typical of the radar environment.

Because of the multisensor environment of sonar, beamforming is a commonly encountered function. A beamformer is just a finite impulse response filter where the time-delay elements have been replaced by (not necessarily uniform) time-delay spatial-difference elements reflecting the times of arrival of the wavefront at the various sensors. While the sampling rate requirements are modest, the multichannel aspect and the ability to focus the beam on a large number of directions make the sonar problem even more challenging from a computational aspect. Spectrum analysis is also done in conjunction with beamforming, for example, $k-\omega$ beamforming [ME80].

Because of its harsher environment, more extensive postprocessing is done in sonar than in radar. Adaptive filters are more extensively used to improve signal and to remove interferences. The interested reader is referred to Chapter 15.

From a hardware point of view, in sonar applications one finds that the signal processors are time multiplexed, that is, a processor will be associated with a number of tasks over time. This is in contrast to the radar environment, where the dedicated hardware can barely keep up with the data.

11-2-6   Video Applications

Two-dimensional interpolation and decimation filters (see Chapter 14) are also used in video special effects, where a television image is either enlarged, shrunk, rotated, or distorted in real time (30 or 25 frames per second). Incorrect filtering will yield artifacts, the most bothersome of which is flickering, which occurs when aliasing takes place. While horizontal and vertical interpolation and decimation artifacts can be understood relatively easily, more complex operations (e.g., rotation) [TH85] may yield unexpected artifacts.

For video applications, the hardware must be able to process at least one of the interlaced fields[1] at a 60-Hz rate. Assuming that the SMPTE-EBU 13.5 Ms/s sampling standard is used, and using a 53.5-μs active line period, about 722 samples are gathered per line. With 480 active lines, 10.4 million samples are collected per second. According to the digital standard, each sample consists of a luminance sample and a sample from one of the two chrominance (color) signals. Thus the processing rate is over 20 Ms/s.

The 13.5-Ms/s standard is a compromise between the various existing television standards. It is not well suited for sampling existing composite signals like the NTSC signals. The preferred sampling rate for the NTSC signals is four times the color subcarrier frequency, namely, 14.3 Ms/s [GO77]. At this rate the decoding of the color signals is greatly simplified. A digital filter is then needed to convert from the 14.3-Ms/s rate to the 13.5-Ms/s rate.

More recently, with the availability of video analog-to-digital converters, digital signal processing has been applied to color decoding [YA83], ghost cancellation [SH83], and color interpolation [TH83]. Prior to the development of the SMPTE-EBU standard, digital interpolation filters [BA77] have been used to convert from NTSC to either PAL or SECAM.

11-2-7 Summary

The hardware requirements for the above six applications areas vary by as much as four orders of magnitude. We can divide these areas into three groups:

1. Low computational requirements: telecom (T-1 system), audio, non-real-time test, and instrumentation (digital oscilloscope). The digital signal processing requirements for these applications can be met by single-chip digital signal processors.

2. Medium computational requirements: sonar and video. The processing needs here can be addressed through custom very large scale integrated (VLSI) circuits or through a parallel processing architecture based on single-chip digital signal processors.

3. High computational requirements: radar and real-time instrumentation (spectrum analyzer). Due to the high throughput rate needed here, only a parallel processing architecture using (custom designed) VLSI circuits will meet the processing needs.

---

[1] An image in present television systems is generated at a 60-Hz rate in the NTSC system. Only alternate lines are generated in a field. On the next field the other lines are generated. The two field sequences make up a frame. Thus the frame rate is 30 Hz. There are 525 lines in a frame, of which only 480 lines are active, the rest being lines associated with the vertical retrace time. In the PAL and SECAM systems used outside North America, in Japan, and in some South American countries, the field and frame rates are 50 and 25 Hz. There are 625 lines in a frame in these systems.

11-3 GENERAL PURPOSE DSP CHIPS AND DEVELOPMENT SYSTEMS

A DSP chip is a digital signal processor implemented in a single integrated circuit chip. In Section 11-3-1, we examine some important architecture considerations for DSP chip design. With the background presented here, we then study the architectures of representative commercially available DSP chips. We also comment on the architecture performance trade-offs. Finally, a simple and cost effective development support system for DSP chip-based applications is discussed in Section 11-3-3.

11-3-1 DSP Architectures and Performances

The key to high performance in digital signal processing hardware is to have the hardware computational elements organized in such a way that they match the structure of the computation algorithm at hand. Different algorithms require different hardware architectures to achieve the best performance. Hence the challenge to the designer of a general purpose DSP chip is to come up with one DSP architecture that can be packed into a single IC chip, which can achieve a relatively good performance for a wide variety of digital signal processing applications.

The three most important DSP applications are the FIR filter, the IIR filter, and the FFT. Both the FIR and the IIR filters have the same computational structure of the "sum of product" type:

$$X * Y + A \rightarrow A,$$

where * denotes the multiplication operation. The data accessing structure of these two applications is mainly of the sequential type. The FFT has a computational structure of the "butterfly" type:

$$w * v + u \rightarrow u,$$

$$-w * v + u \rightarrow v,$$

where $w = \cos (2\pi kn/N) - j \sin (2\pi kn/N)$ and N is the FFT size. The data accessing structure of the FFT is not purely sequential. Note that u and v are complex samples.

In this section we first discuss some important architecture design considerations and then examine their architecture implications.

DSP Architecture Consideration. Some of the basic building blocks of a DSP chip are shown in Figure 11-1. The architecture design of a DSP chip involves the specifications of these basic building blocks and the determination of the way in which they are interconnected.

The program sequencer (PS) generates the address of the instruction to be executed. The program memory (PM) stores the program instructions and the instruc-

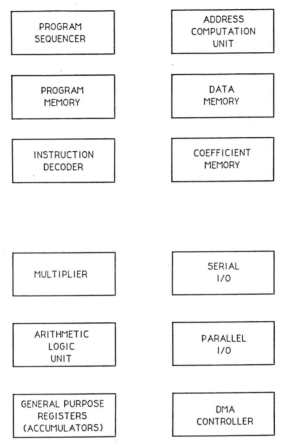

FIGURE 11-1 Basic building blocks of a DSP chip.

tion decoder (ID) decodes the instruction to generate the control signals to carry out the execution of the instruction. The data memory (DM) stores data to be processed and some intermediate results, while the coefficient memory (CM) stores the fixed coefficient constants such as the filter coefficients and the sine and cosine tables for the FFT and the sinusoidal signal generation applications. The address computation unit (ACU) computes the addresses for the operands under the direction of the control signals generated by the ID. The PS, PM, ID, and ACU are sometimes grouped together and referred to as the control unit (CU), while DM and CM are collectively referred to as the memory unit (MU).

The multiplier (MPY) in a DSP chip is in general a parallel multiplier. The arithmetic logic unit (ALU) performs addition, subtraction, shifting, and some logic functions. The general purpose registers (GPR) hold temporary data and are usually directly connected to the data bus and other basic building blocks. In the simplest case, there is just one register, which is called the accumulator. The MPY, ALU, and GPR form the arithmetic unit (AU).

The input/output interface group includes the serial I/O (S-I/O) unit, the parallel I/O (P-I/O) unit, and the direct memory access controller (DMA).

The architecture and design of the DSP chip involves the specification of the capabilities of these building blocks and the determination of their appropriate interconnections. The architecture of a DSP chip has a profound effect on the ultimate performance of the system based on the DSP chip. Two major performance measures of a DSP-based system are the speed and the precision. The achievable precision is dictated by the word size of the memory units and the arithmetic unit, while the speed of a DSP chip depends on many factors such as the instruction cycle time and the power of the instruction set supported by the architecture. If it takes N instructions to complete the execution of an algorithm, then the total time needed for this algorithm execution is $N * t$, where t is the instruction cycle time. Note that the ultimate speed performance measure is the product of N and t, so just a shorter instruction cycle time t alone is not sufficient. For example, the NEC μPD7720 has a 250-ns instruction cycle time and the TMS32010 has a shorter instruction cycle time of 200 ns. However, it takes the μPD7720 less time to execute a FIR filter algorithm because that algorithm consists of essentially all "sum of product" type operations. It takes the TMS32010 two instruction cycles to do the "multiply and accumulate" operation, whereas only one instruction cycle is needed in the μPD7720. It is interesting to note that the TMS32010 is so far the most successful DSP chip on the market. This fact implies that it takes a lot more than just performance to be a successful product. We will come back to this point later.

From the above discussion it is clear that a desirable architecture is one that can support a powerful instruction set such that most important DSP algorithms can be accomplished in a small number of instruction cycles, while having the shortest possible instruction cycle time achievable by the technology used to make the chip.

Some important architecture design considerations that are essential to achieve a desirable DSP architecture are listed below [AL75; BO81; PE76].

1. Most DSP algorithms are computation intensive; hence an AU capable of efficiently executing the "sum of product" and the "butterfly" is important.

2. Most DSP algorithms can be pipelined; hence an architecture capable of supporting a pipeline structure efficiently is important.

3. Most data accessing in DSP algorithms is highly structured (although not necessarily sequential); hence an architecture capable of supporting various structured access efficiently is important.

4. The memory-AU bandwidth and the AU speed should be carefully balanced, such that the total system performance is not seriously degraded because of this bottleneck.

5. A good programming language for supporting parallel processing is needed.

Now let us examine these considerations in more detail and derive *architecture implications* from them. Since arithmetic operations dominate the execution of most DSP algorithms, a high-speed parallel multiplier is a must. Currently, the multiplication time is still the dominant factor in the determination of the instruction cycle time of a DSP chip. In fact, for the DSP chips we examine in the next section, all but one have the instruction cycle time the same as the multiplication time. This implies that if the multiplication time can be further shortened beyond just technology improvement, it will be a real challenge to the DSP architecture designer to keep the instruction cycle time as short as the multiplication time. An ALU with preshifting and postshifting capability would be useful for scaling operands to prevent the overflow in fixed-point operations. Some nonlinear hardware operations such as arithmetic rounding and saturation, taking absolute value and format conversion without an extra instruction cycle, would be desirable for speed improvement. Having more than one general purpose register is convenient for handling temporary variables in the delay operation and the calculation of complex multiplications.

To support a highly pipelined operation, each building block must operate in parallel. Hence they should be carefully interconnected to avoid bus interferences. As mentioned in a previous discussion, the speed is determined by the product $N * t$, of the instruction cycle time and the number of instructions needed to accomplish a task; hence these basic building blocks should be so interconnected such that the instruction cycle time is minimized and, at the same time, most algorithm steps (such as $X * Y + A \rightarrow A$) can be accomplished in one instruction. Currently, for most DSP chips, an instruction cycle time is dictated by the speed of the multiplier. Therefore one should try to accomplish all other operations, such as instruction fetch, instruction decoding, operands (coefficients and data) fetch, and storing results back to data memory, in the time it takes to complete a multiplication. To achieve this, several architecture requirements are necessary. First, a long instruction word is needed in order to decode the control signals quickly (i.e., in parallel) for each unit operating in parallel. This calls for a separation of program memory from data memory so that different wordlengths can be used in different memory units. Second, it is highly desirable to have a separate program bus and data bus so that the instruction fetch and operands fetch can proceed in parallel. Furthermore, it is even more desirable to have separate data memory and coefficient memory units with the coefficient bus separated from the data bus and directly connected to one of the multiplier's registers. In this way, two operands can be accessed simultaneously. Bell's DSP-1 uses the same bus for both the program bus and the data bus; also, its coefficient memory and program memory are in the same memory unit. In order to implement the "multiply and accumulate" operation in one instruction cycle to support the pipelining structure, three sequential bus accesses are required in one instruction cycle. This results in a long (800-ns) instruction cycle time, although its multiplication time is only 400 ns [BO81; WE82].

To support a structured access, usually a separate address computation unit with various indexing capabilities is implemented in a DSP chip. The inclusion of a dedicated address computation unit also has contributed greatly to the shortening of the instruction cycle time. Some DSP chips go even further to have a separate address computation unit dedicated to each memory unit, which gives even greater flexibility for structured data and coefficients accesses.

Although the greatest emphasis of the DSP chip design has been on the arithmetic unit, it is now generally recognized that the memory-AU bandwidth is also an important consideration. In fact, the separation of PM and DM, the separation of the program bus from the data bus, and the dedicated address computation unit are all aimed to improve the memory-AU bandwidth. A good measure of the memory-AU bandwidth is the ratio of the total number of "multiply and accumulate" operations in an algorithm to the total number of instruction cycles it takes to accomplish the algorithm execution.

To ease the program development for a DSP chip-based system, a multiple field type of assembly language is more desirable than the traditional assembly language. Since, in a DSP chip, there are several basic building blocks operating concurrently and somewhat independently, a multiple field type of assembly instruction can better reveal the operation of the instruction and can give more flexibility in assembly programming. Several DSP chips (such as the Bell DSP-1 and NEC μPD7720) use the multiple field instruction format. Although the TMS32010 does support parallel processing, this is not reflected in its assembly instructions. This fact makes it difficult to remember all the micro-operations performed in the various processing units during an instruction.

11-3-2   Some Representative DSP Chips

In the previous section, we discussed the DSP architectures and performances in a general setting to provide a framework for assessing a particular DSP chip. In this section we describe the architectures of some representative DSP chips. Ten chips are discussed in order of their announcement time. In this way, the reader can easily see how the architectures of different DSP families (vendors) evolve, and where they are heading.

11-3-2-1   AMI S2811. The functional block diagram of the S2811 signal processing peripheral [AD79] is shown in Figure 11-2. This chip was introduced in 1978 [CU85] and was one of the earlier products in the single-chip DSP area. (Gould AMI no longer offers this chip; they are second-sourcing the NEC μPD7720.) The main features of the S2811 are now described.

Technology. The S2811 is fabricated in VMOS. It has a 300-ns instruction cycle time with a 20-MHz clock.

Control Unit. The S2811 has a 256 × 17 program memory with 17-bit instruction word. In addition to a traditional auto-increment program counter, it has a return address register (RAR) for one-level subroutine nesting. It also has a 5-bit loop counter to support the "repeat" instruction. The "repeat" instruction allows the processor to execute the current instruction up to 32 times without a new instruction fetch. This may not be important in terms of speed in the environment where

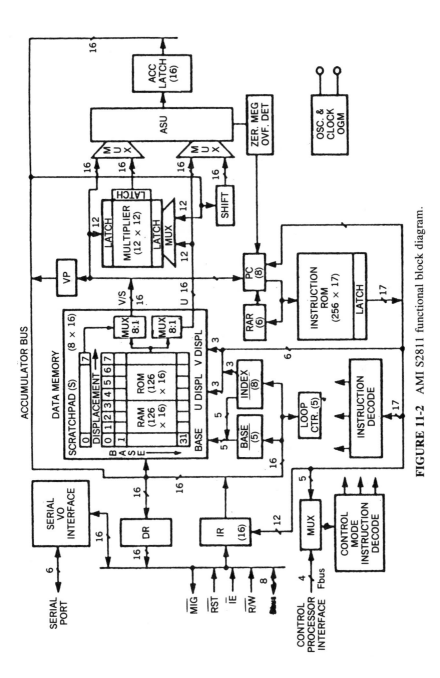

FIGURE 11-2 AMI S2811 functional block diagram.

733

the instruction fetch is usually pipelined with the instruction execution. It does have the advantage of saving program space in implementing many digital signal processing algorithms with the 256-word program memory. It also has a base register and an index register to provide fairly flexible data and coefficient addressing.

Arithmetic Unit. The S2811 has a $12 \times 12 \rightarrow 16$ parallel multiplier that can multiply two 12-bit operands to produce a 16-bit product in 300 ns. Its ALU has a 16-bit 2's complement ADD/SUBTRACT unit and a 16-bit accumulator. It also has a 1-bit right shifter for implementing the operation of "divide-by-2." A special register VP is provided to implement the digital filter z^{-1} delay. The VP register can be used to temporarily store the contents of one of the multiplier input latches. This capability proves to be very handy in implementing a second-order IIR filter section.

Memory Unit. The S2811 has a 128×16 data memory (RAM), a 128×16 coefficient memory (ROM), and an 8×16 scratch pad memory (RAM). The data memory and the coefficient memory are organized in a two-dimensional array structure, which allows a simultaneous fetch of the data and the coefficient from the memory unit. This is a very desirable feature for digital filtering applications.

I/O Unit. The S2811 has a serial port for direct interface to an A/D converter. It has separate input and output registers to exchange data with the S2811 data port. Its serial interface logic can perform the conversion between 2's complement and sign magnitude formats.

It is interesting to note that this DSP chip has almost all the desirable characteristics discussed in the previous section despite its early introduction date. It will be seen later that NEC μPD7720, Hitachi HD61810, Fujitsu MB8764, and the most recent NEC μPD77230 all have similar architectures except for the enlargement of the word size and some minor functional improvements. Although NEC's floating-point processor μPD77230 represents a quantum jump from the fixed-point processor to the floating-point processor, it still possesses a very similar overall processor architecture.

The instruction pipelining is organized around the multiplication cycle. It takes 300 ns to generate the product of the multiplier. The instruction decoding, memory address setup, and operands access (data RAM and coefficient ROM) are performed before the beginning of the multiplication cycle (which is in the middle of the instruction cycle); then the two inputs and one output of the multiplier are latched (which starts a new multiplication cycle), and finally all ALU operations and the write operation are performed in the second half of the instruction cycle. Since the program bus is separated from the main data bus, the instruction fetch is performed in parallel with other operations.

The instruction wordlength is 17 bits. It is partitioned into three fields. The first field consists of the five most significant bits and handles the load, the data transfer, the register manipulation, and the branch instructions. The second field has the next four significant bits and handles all arithmetic operations. The last field has the remaining 8 bits to store the operands' address information. These three fields can be programmed independently in assembly code with only slight restrictions.

We now give a simple programming example to illustrate the importance of the data structure of both the data RAM and the coefficient ROM in achieving the high performance. Let us consider a simple second-order IIR filter section, which can be described by the following two state equations:

$$w[n] = x[n] + a1 * w[n - 1] + a2 * w[n - 2],$$

$$y[n] = w[n] + b1 * w[n - 1] + b2 * w[n - 2],$$

where $x[n]$ is the input, $y[n]$ is the output, and $w[n - 1]$ and $w[n - 2]$ are state variables. In the beginning of the operation, the two memory units V and U are organized as follows:

$$v[0] = w[n - 1], \quad v[1] = w[n - 2], \quad u[4] = a1, \quad u[5] = a2, \quad u[6] = b1,$$

$$\text{and} \quad u[7] = b2.$$

The input $x[n]$ is in the accumulator ACC. At the end of the operation, the required data structures are $v[0] = w[n]$ and $v[1] = w[n - 1]$; U memory remains unchanged and $y[n]$ is in the accumulator ACC. The program listing with comments are as follows:

OP1	OP2	Operands	Comments
No operation	No operation	$u[4], v[0]$	$a1 * w[n - 1] \to P$ (product latch) $x[n]$ is in ACC
P + ACC → ACC	No operation	$u[5], v[1]$	$a2 * w[n - 2] \to P$ $x[n] + a1 * w[n - 1]$ is in ACC
P + ACC → ACC	ACC → v[0]	$u[6], v[0]$	$b1 * w[n - 1] \to P$ $w[n]$ is in ACC ACC $\to v[0]$ (replace $w[n - 1]$) $w[n - 1] \to$ VP (temporary store)
P + ACC → ACC	VP → v[1]	$u[7], v[1]$	$b2 * w[n - 2] \to P$ $w[n] + b1 * w[n - 1]$ is in ACC VP $\to v[1]$ (replace $w[n - 2]$)
P + ACC → ACC	Return	—	$y[n]$ is in ACC return to main program

It is easy to see that this subroutine can easily be extended to implement a higher-order IIR filter by stepping through the base register without compromising the efficiency.

11-3-2-2 Bell DSP-1. The functional block diagram of the Bell DSP-1 (or West-ern Electric F-61329A) [BO81; WE82] is shown in Figure 11-3. This chip was introduced in 1980 [CU85] and was intended for Bell's internal use in the area of voice and low-speed data modem applications. The main features of the DSP-1 are now described.

Technology. The DSP-1 is fabricated in depletion-load NMOS and packaged in a 40-pin DIP. It has an 800-ns instruction cycle time with a 5-MHz clock.

Control Unit. The DSP-1 has a 1024 × 16 ROM that can be used as a program memory or a combination of a program memory and a coefficient memory. The instruction wordlength is 16 bits. It has a program counter for storing the address of the next instruction and a program return register (PR) for one-level of subrou-tine capability. It also has a 6-bit loop counter used to provide looping within an algorithm. However, it does not support the ''repeat'' instruction as discussed in AMI S2811. The DSP-1 has a fairly sophisticated address computation unit. Fixed coefficients in ROM can be addressed by the program counter or, alternatively, by the auxiliary register RX, which can point to either ROM or RAM. It has two auxiliary registers RY and RYA for reading data from RAM and another two aux-iliary registers RD and RDA for writing data into RAM. Its address computation unit also provides a selection of possible increments for postmodification of these registers. Increments of $+1$, 0, -1 or by the contents of the 8-bit registers I, J, or K are provided.

Arithmetic Unit. The DSP-1 has a 16 × 20 → 36 parallel multiplier that can multiply a 16-bit coefficient and a 20-bit data to produce a 36-bit product in 400 ns. Compared to the S2811's 12 × 12 → 16 300-ns multiplier, the DSP-1 has a fairly fast multiplier. However, due to the limited bus architecture, the DSP-1 has a much longer instruction cycle time (800 ns compared to 300 ns for the S2811). We will come back to this point later. The DSP-1 has a 36-bit 2's complement fixed-point ALU with a 40-bit accumulator. The full capability of its arithmetic unit can be described by the following expression:

$$x * f(y \text{ or } w) + g(a) \rightarrow a \ [\rightarrow w],$$

where x is the 16-bit coefficient, y is the 20-bit data word, a is the accumulator, w is the rounded or truncated 20-bit AU output, f is a function of y or w, such as the actual value, the absolute value, or the sign function, and g is a scaling (2 or 8) function of a or a logical function of a and p. The results are available in the accumulator a and optionally in the register w. The DSP-1 also has an on-chip μ-255 to linear PCM conversion function.

Memory Unit. The DSP-1 has a 128 × 16 data RAM. If the coefficients are fixed, they are normally stored in the ROM shared with the program instruction words.

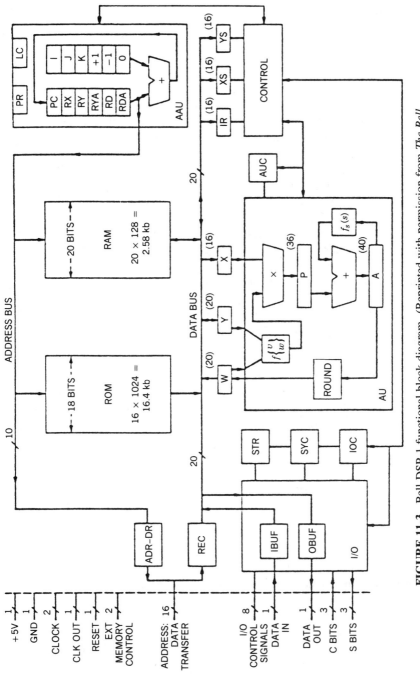

FIGURE 11-3 Bell DSP-1 functional block diagram. (Reprinted with permission from *The Bell System Technical Journal.*)

I/O Unit. The DSP-1 has two buffer registers for I/O operation. It has a programmable serial I/O that can accommodate a serial transfer of 8-, 16-, or 20-bit words.

Development Support. The DSP-1 has a very well supported hardware/software development system. It has a UNIX^{TM}-based C cross assembler, a software simulator, a hardware in circuit emulator, and a large number of debugging tools. Compared to the DSP-1, the S2811 has essentially no development support.

It is clear that the DSP-1 was designed primarily with telecommunication applications in mind. It is the only DSP chip discussed here that has a built-in hardware μ-255 law to linear PCM converter. Although the 800-ns instruction cycle time makes it one of the slowest chips on the market, it was quite satisfactory for voice and low-speed modem applications with input signals limited to 8K samples per second. As mentioned earlier, the DSP-1 has a much more powerful arithmetic unit and much more sophisticated address computation unit than the S2811. However, due to its limited bus architecture, the DSP-1 must have its program memory, its coefficient memory, and its data memory share the same bus. This factor greatly limits its potential performance. Compared to the S2811, the DSP-1 takes two more bus cycles (one for instruction fetch and one for coefficient fetch) to complete an instruction cycle. This is the major reason why the DSP-1, with only a marginally longer multiplication time than the other processors, has a substantially longer instruction cycle time. A performance gain factor of 2 can be realized through an architecture improvement that will reduce its instruction cycle time to its multiplication time.

The DSP-1 has a 16-bit instruction word and has two types of instructions: normal and auxiliary. The normal instructions deal with computational type operations while the auxiliary instructions are used to control noncomputational type operations. For a normal instruction, the assembler input line consists of up to four almost independent instruction fields (similar to that of S2811) with each field specifying the operations of the four functional units of the chip.

11-3-2-3   NEC μPD7720. The functional block diagram of the μPD7720 signal processing interface [NE82] is shown in Figure 11-4. This chip was introduced in 1980 [CU85] and was designed for a variety of signal processing applications with a multiple-bus architecture. The main features of the μPD7720 are now described.

Technology. The μPD7720 is fabricated in NMOS and packaged in a 28-pin DIP. It has a 250-ns instruction cycle time with an 8-MHz clock.

Control Unit. The μPD7720 has a 512 $\times$ 23 instruction ROM addressed by a 9-bit program counter. It also contains a four-level program stack pointer for up to four levels of subroutine capability and interrupt handling. The instruction word is 23 bits long, which allows for a powerful multiple field instruction set. The address computation unit of the μPD7720 is divided into two separate units: one is dedicated to the data RAM and the other is dedicated to the data ROM (or

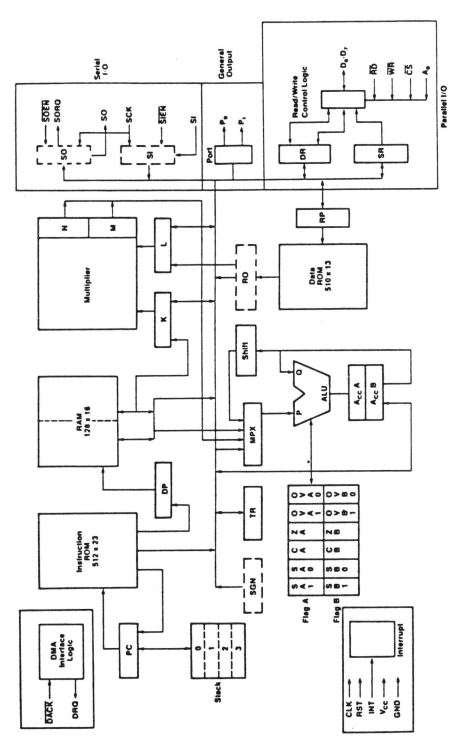

FIGURE 11-4 NEC μPD7720 functional block diagram.

739

coefficient ROM). The 9-bit ROM pointer RP can either stay unchanged or be decremented by one after each instruction execution. The 7-bit data RAM pointer DP is divided into two sections: the operations for the lower 4 bits can be no operation, increment, decrement, or clear and it can act as a modulo-16 counter, while the upper 3 bits can be Exclusive ORed with the Mask defined by the 3 bits in the instruction word to provide essentially a random access capability. This technique helps keep the instruction cycle time short, while maintaining a random paging capability.

Arithmetic Unit. The μPD7720 has a $16 \times 16 \rightarrow 31$ parallel multiplier that can multiply two 16-bit 2's complement number to produce a 31-bit product in 250 ns. The ALU is a 16-bit 2's complement unit capable of executing 16 distinct operations on virtually any of the μPD7720's internal registers, which is the benefit of its extensive multiple-bus architecture. It has a pair of accumulators that allow the separate manipulation of two values, such as in the case of complex multiplications. It also has a sign register to implement saturation arithmetic.

Memory Unit. The μPD7720 has a 510×13 data (or coefficient) ROM and a 128×16 data RAM, each with its dedicated address register directly connected to it. In other words, including the instruction ROM, it has three separate address buses for its three memory sections.

I/O Unit. The μPD7720 has three communication ports: two serial and one 8-bit parallel. The parallel port also has DMA control lines for high-speed data transfer with little processor overhead.

Development System. A development system for assembly code editing, debugging, and assemblying is available from NEC. The Evakit-7720 Evaluation System is also available for hardware testing.

The key to the good performance of the μPD7720 is its extensive multiple-bus architecture. For almost all the data transfers that have a fixed source and destination registers, a dedicated path is provided. This architecture allows concurrent operations of up to three data transfers in addition to the multiplier and ALU operations in one instruction cycle. In addition to several secondary buses, it also has a primary bus that provides a data path between all the registers (including I/O registers), a memory, and AU to provide flexibility.

The μPD7720 has a 23-bit instruction word. The instruction is also organized into several independent fields. The first field distinguishes the normal arithmetic type instructions from the nonarithmetic instructions. The remaining part of a normal arithmetic instruction has seven fields. The first field selects the ALU input (RAM, *M* register of the multiplier output, *N* register of the multiplier output, or the primary bus). The second field selects the ALU operation. The third field selects the accumulator (A or B). The fourth field determines the DP operation, and the fifth field selects the RP operation. The last two fields specify the source and destination registers (including RAM), respectively. A typical instruction reads as follows: load data RAM and coefficient ROM into multiplier's input registers, K and L; add the high 16-bit register M to the accumulator A; decrement the ROM pointer; Exclusive OR the RAM row pointer (higher 3 bits of DP); clear the lower 4 bits of DP; and return from a subroutine. All six statements are accomplished in one instruction cycle. Multiplication is not specified as an instruction statement, because it is performed automatically during each instruction cycle. (The multiplier section is considered as a combinational circuit.)

11-3-2-4 TI TMS32010. The functional block diagram of the TMS32010 digital signal processor [TE83] is shown in Figure 11-5. This chip was introduced in 1982 and was designed as a high-speed digital controller with numerical capability. The TMS32010 uses a modified Harvard architecture. In a strict Harvard architecture, the program bus and the data bus are separate, which allows a full overlap of the instruction fetch and execution. In the TMS32010, paths are provided between the two buses to provide some flexibility. The main features of the TMS32010 are now described.

Technology. The TMS32010 is fabricated in 2.7-micron NMOS and packaged in a 40-pin DIP. It has a 200-ns instruction cycle time with a 20-MHz clock.

Control Unit. The TMS32010 has two versions: the first one is equipped with a 1536×16 program ROM, while a second one has no internal program ROM. It has the capability of accessing up to 4K words of 16-bit wide program memory without losing speed if the external memory has an access time less than 100 ns. Like the μPD7720, the TMS32010 has a program counter and a four-level stack pointer that enable the user to perform subroutine calls, branches, and interrupts. Two auxiliary registers, AR0 and AR1, selected by an auxiliary register pointer (ARP) are used to perform the indirect addressing of the data RAM. The auxiliary registers can be made to auto-increment/decrement after each instruction is executed. A 1-bit data page pointer is provided to combine with the 7-bit address in the instruction word to directly address the 144×16 data RAM. The two auxiliary registers serve many different functions. As discussed earlier, the lower 8 bits of the AR can be used for indirect addressing purposes. The lower 9 bits can also be used as a 9-bit loop counter. The upper 7 bits of the AR are unaffected by any auto-increment/decrement operation. The two auxiliary registers can also be used as general purpose 16-bit registers for temporary storage use.

Arithmetic Unit. The TMS32010 has a $16 \times 16 \rightarrow 32$ parallel multiplier that can multiply two 16-bit 2's complement numbers to produce a 32-bit product in 200 ns. Three other arithmetic elements are the ALU, the accumulator, and two shifters. In a typical arithmetic operation, the data coming from the 16-bit data bus first pass through the barrel shifter, which can left-shift a word 0 to 15 bits depending on the value specified in the instruction. The data then enter a 32-bit ALU, where they are loaded into or added/subtracted from the 32-bit accumulator. Another shifter is provided to perform the scaling on the output of the accumulator

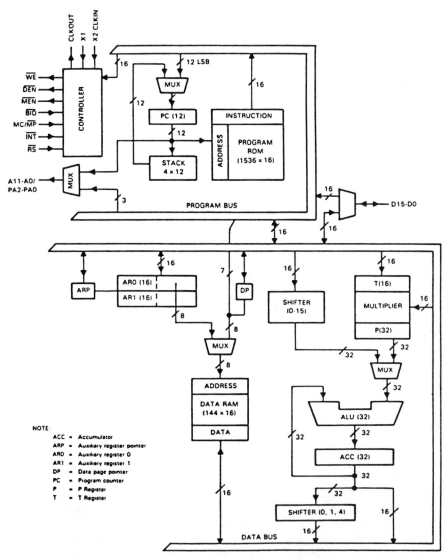

FIGURE 11-5 TI TMS32010 functional block diagram. (Reprinted with permission from Texas Instruments.)

as they are being stored back to the data RAM. Although it seems, from the block diagram, that the architecture allows for the ''multiply and accumulate'' operation to be completed in one instruction, the instruction set does not support it. Hence it takes two instructions (i.e., 400 ns) to perform the ''multiply and accumulate'' operation. It is not surprising to see that the recently announced TMS32020 does have a one-instruction cycle ''multiply and accumulate'' instruction.

Memory Unit. All nonimmediate data operands are stored in the 144 × 16 data RAM. Two instructions, table read (TBLR) and table write (TBLW), are provided to transfer values from program memory to the on-chip data RAM and from data RAM to program memory (presumably in the form of off-chip RAM), respectively.

I/O Unit. Two instructions, IN and OUT, allow users to transfer data between the peripheral and the data RAM. The three multiplexed least significant bits of the address bus are used as a port address by the IN and OUT instructions; hence up to eight I/O ports can be implemented with the TMS32010.

Development System. Ready availability of the chip and a good development system are two of the main reasons why the TMS32010 has become the most successful DSP chip. For software development, a macro assembler/linker, a software library, and a simulator are available. For hardware support, there are the TMS32010 Evaluation Module, the TMS32010 Emulator (XDS), and the TMS32010 Analog Interface Board (AIB). There is also an active third-party support program. Extensive documentation support coupled with workshop offerings and availability of public domain programs from universities makes the TMS32010 the most popular DSP chip among both industrial and academic users.

11-3-2-5   Hitachi HD61810. The functional block diagram of the HD61810 high-performance signal processor [HI84] is shown in Figure 11-6. This chip was introduced in 1984 [CU85] and was the first single-chip DSP with some limited floating-point capability. The main features of the HD61810 are now described.

Technology. The HD61810 is fabricated in CMOS and packaged in a 40-pin DIP. It has a 250-ns instruction cycle time with a 16-MHz clock.

Control Unit. The HD61810 has a 512 × 22 instruction ROM addressed by a 9-bit program counter. It has a two-level stack pointer for up to two levels of subroutine capability and interrupt handling. Its address computation unit is very similar to that of AMI S2811. A distinct feature is the dual RAM pointers A and B, which allow operations with complex numbers to be efficiently implemented. It also has a repeat counter for implementing the "repeat" instruction and a delay register to implement the delay operation in the FIR filter applications.

Arithmetic Unit. The major difference between the HD61810's arithmetic unit and other signal processors' arithmetic units is the combined fixed-floating-point arithmetic unit. In the fixed-point operation, it is basically a 16-bit machine, while in the floating-point mode, it becomes a 12-bit mantissa with a 4-bit exponent machine. This allows a flexible trade-off between precision and dynamic range. It also has two 20-bit accumulators, which is convenient for storing temporary results.

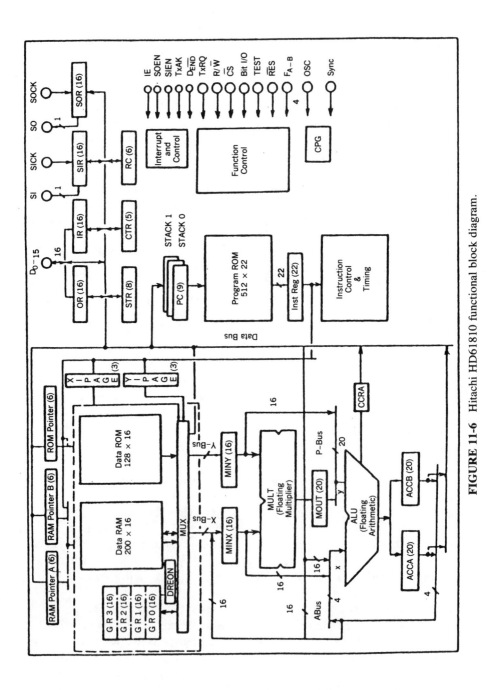

FIGURE 11-6 Hitachi HD61810 functional block diagram.

Memory Unit. The HD61810 has a 200 × 16 data RAM, a 128 × 16 data (coefficient) ROM, and four general purpose registers. Its data RAM is partitioned into four pages, each of which consists of 50 words. The RAM is of dual-port structure, which permits different data words to be read out at the same time from different pages by using two page pointers X and Y. This capability is useful in many digital signal processing applications such as auto-correlation and cross-correlation computations.

I/O Unit. The HD61810 has a standard parallel I/O interface with 8-bit and 16-bit microprocessor. It also has a DMA capability. A serial I/O interface for up to 16 bits is also provided.

11-3-2-6 Fujitsu MB8764. The functional block diagram of the MB8764 general purpose digital signal processor is shown in Figure 11-7 [FU84]. This chip was introduced in 1984 and is so far the fastest chip in terms of instruction cycle time. The main feature of MB8764 are now described.

Technology. The MB8764 is fabricated in silicon-gate CMOS technology and packaged in an 88-pin grid array. It has a 100-ns instruction cycle time with a 20-MHz clock.

Control Unit. The MB8764 has a 1024 × 24 program ROM addressed by a 10-bit program counter. Either internal or external ROM can be selected as the program memory. It has two instruction registers, IR0 and IR1. The program ROM output instruction is transferred to IR0 at the beginning of an instruction cycle and moved to IR1 at the beginning of the following cycle. The instructions fetched from the program memory and transferred to IR0 and IR1 are moved to the look-ahead decoder (LAD) and the decoder (DEC), respectively. The address computation usually takes place when the instruction is in IR0 and the execution of an instruction usually takes place while the instruction is in IR1. The program counter has a program counter stack for one level of subroutine capability. Two loop counters are provided for applications where an embedded loop is required. Two independent address computation units are provided for each of the AU operands. Each unit consists of an index register with a stack, an adder, and some other special registers. The index stack permits indexing of addresses in a program having an embedded loop structure.

Arithmetic Unit. The MB8764 has a 16 × 16 → 26 multiplier that can multiply two 16-bit 2's complement numbers to produce a 26-bit product. Although this chip has the shortest instruction cycle time, it takes two instruction cycles to obtain the multiplication result. However, because of its pipelined structure, the "multiply and add" operation can still be accomplished in one instruction cycle if the data structure in the memory unit is properly arranged. The multiplier structure of the MB8764 represents a possible way to make the instruction cycle shorter than the multiplication cycle. It is obvious that a machine with 100-ns instruction cycle

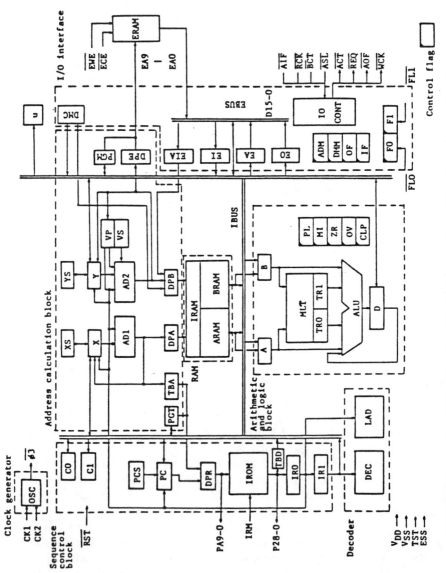

FIGURE 11-7 Fujitsu MB8764 functional block diagram. (Reprinted with permission from Fujitsu Microelectronics.)

746

time and 200-ns multiplication cycle time is preferred to the one with a 200-ns instruction cycle time. The ALU unit is a standard one with a 26-bit accumulator.

Memory Unit. The MB8764 has two 128 × 16 data RAM areas called ARAM and BRAM. These two RAMs can be used either as two independent RAMs or a single RAM with a continuous address space. An external RAM capability is also provided. Up to 1K of external RAM can be addressed.

I/O Unit. The MB8764 has a 16-bit parallel I/O port with DMA capability. It also has several different modes of I/O operations.

Development System. The MB8764 development support is built around a Fujitsu personal computer running CP/M86. It consists of a source code text editor, an assembler, and an evaluation board.

The most interesting feature of the MB8764 is its pipelined multiplier, which enables the chip to have an instruction cycle time equal to half of the complete multiplication time. All the DSP chips discussed previously have an instruction cycle time either greater or equal to the multiplication time. This chip clearly shows that a high performance can be achieved through architecture innovation.

11-3-2-7 TI TMS32020. The functional block diagram of the TMS32020 digital signal processor [TE85] is shown in Figure 11-8. This chip was introduced in early 1985 and is the direct upgrading from the TMS32010. The overall modified Harvard architecture remains unchanged. However, in almost every basic building block, performance is greatly improved. Here are some of the highlights.

Technology. The TMS32020 is fabricated in 2.4-micron NMOS and packaged in a 68-pin grid array. The instruction cycle time remains at 200 ns with a 20-MHz clock.

Control Unit. The program counter has been expanded from 12 bits to 16 bits, thus expanding the address capability from 4K to 64K. A repeat counter is added to implement the "repeat" instruction. Also, a global memory allocation register is provided to facilitate multiprocessor configuration. The address computation unit is greatly enhanced. The number of auxiliary registers has been increased from two to five, and a dedicated unsigned arithmetic unit for address manipulation has been added.

Arithmetic Unit. The major improvement of the AU is the implementation of the single-instruction "multiply and accumulate" operation. Also, a shifter is added to the output of the product register.

Memory Unit. The memory unit has also been greatly expanded. The TMS32020 provides three separate on-chip memory units: a 32 × 16 data RAM, a 256 × 16

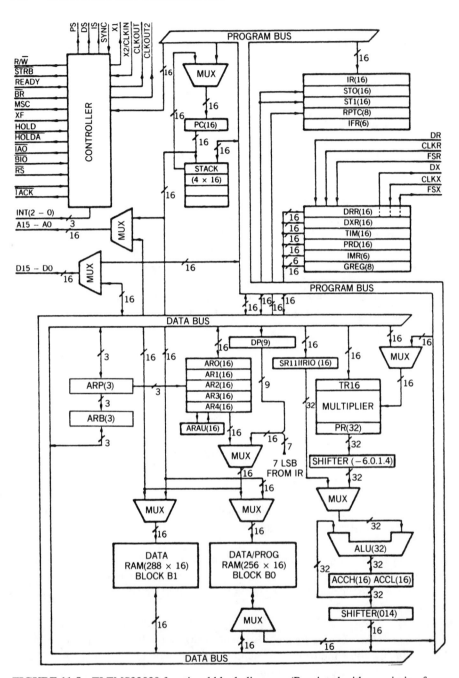

FIGURE 11-8   TI TMS32020 functional block diagram. (Reprinted with permission from Texas Instruments.)

data RAM, and a 256 × 16 data/program RAM. The data page pointer is also expanded from 1 bit to 9 bits.

I/O Unit. The TMS32020 provides up to 16 I/O ports capability, which doubles that of the TMS32010. A DMA capability is also provided. About 10 systems/ I/O control registers are included in the new chip.

Although the instruction wordlength remains at 16 bits, the instruction set is almost doubled (compared to TMS32010) due to the enhancement of the capabilities of all the basic building blocks.

11-3-2-8 NEC μPD77230. The functional block diagram of the μPD77230 advanced signal processor [NE85] is shown in Figure 11-9. This chip is very powerful and is probably close to the ultimate of what a single DSP chip can offer. Just as the TMS32020 grew out of the TMS32010 architecture, the μPD77230 still maintains the same multiple-bus architecture of the μPD7720. However, many enhancements have been made to every basic building block of the DSP chip.

Technology. The μPD77230 is fabricated in < 1.75 micron CMOS and packaged in a 64-pin shrink DIP. The instruction cycle time is around 150 ns.

Control Unit. The μPD77230 has a 2K × 32 instruction ROM; hence the instruction wordlength is extended to 32 bits. The program counter is 12 bits long; hence a 4K external program memory capability is there. The level of the stack is expanded from four to eight. Two independent address computation mechanisms are provided for two independent data RAMs. Each unit has a base and an index register with a dedicated adder. In the μPD77230, a direct path from the instruction decoder to the data ROM is provided to give additional flexibility.

Arithmetic Unit. The big enhancement is the full 32-bit floating-point arithmetic unit. The multiplier can perform either a 24 × 24 → 47 fixed-point multiplication or a 32 × 32 → 55 floating-point multiplication in a 150-ns instruction cycle time. It also has a complete 55-bit floating-point ALU with a 47-bit barrel shifter and eight 55-bit general purpose registers (accumulators).

Memory Unit. The memory unit is also greatly expanded. A 1K × 32 data (coefficient) ROM for storing fixed constant data. Two identical 512 × 32 data RAMs allow efficient FFT operation for up to 512 data points. This capability will further advance the applications of the FFT technology to much wider fields.

I/O Unit. The μPD77230 has a 32-bit parallel interface. The serial I/O port has speeded up from 2 MHz (in μPD7720) to 5 MHz and supports 8/16/24/32 bits data length.

11-3-2-9 AT&T DSP32 Floating-Point Processor. The AT&T DSP 32 is a programmable digital signal processor with a 32-bit floating-point arithmetic unit. It is shown in Figure 11-10. Some key features of the chip are now described.

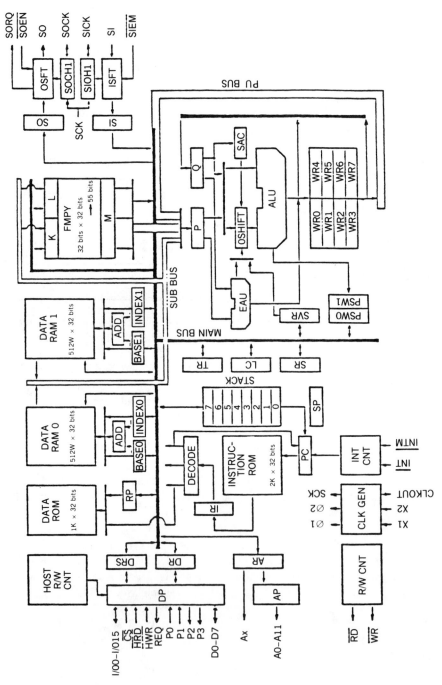

FIGURE 11-9 NEC μPD77230 functional block diagram.

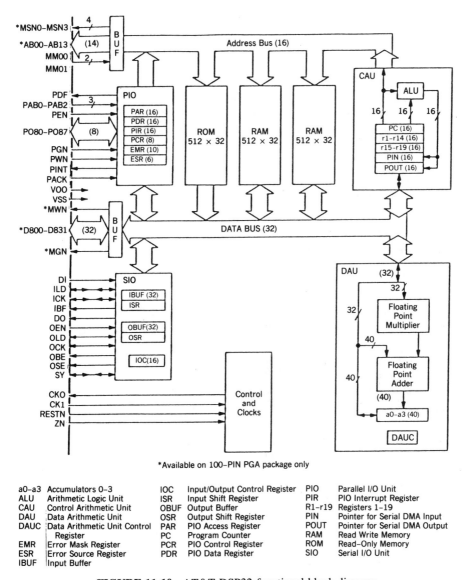

FIGURE 11-10 AT&T DSP32 functional block diagram.

a0–a3	Accumulators 0–3	IOC	Input/Output Control Register	PIO	Parallel I/O Unit
ALU	Arithmetic Logic Unit	ISR	Input Shift Register	PIR	PIO Interrupt Register
CAU	Control Arithmetic Unit	OBUF	Output Buffer	R1–r19	Registers 1–19
DAU	Data Arithmetic Unit	OSR	Output Shift Register	PIN	Pointer for Serial DMA Input
DAUC	Data Arithmetic Unit Control	PAR	PIO Access Register	POUT	Pointer for Serial DMA Output
	Register	PC	Program Counter	RAM	Read Write Memory
EMR	Error Mask Register	PCR	PIO Control Register	ROM	Read–Only Memory
ESR	Error Source Register	PDR	PIO Data Register	SIO	Serial I/O Unit
IBUF	Input Buffer				

Technology. The chip is fabricated in 1.5-micron NMOS technology and packaged in either a 40-pin DIP or a 100-pin array. The 100-pin package is chosen when external memory expansion is required. It has a 250-ns instruction cycle time and 16-MHz clock rate.

Control and Memory Units. It has a 512 × 32 ROM and two 512 × 32 RAMs. The memory can be expanded to include additional 5K bytes of external memory without degrading performance. The 32-bit instruction can be stored in either ROM or RAM. The address computation unit has twenty-one 16-bit general purpose registers and a full function ALU. These registers can also be used as pointers with auto-increment capability.

Arithmetic Unit. It has full function floating-point multiplier and adder. The multiplication operation can be accomplished in 125 ns. The 32-bit floating-point number has a 24-bit mantissa and an 8-bit exponent. It also has four accumulators for increased precision and flexibility.

I/O Unit. It has a serial I/O port and an 8-bit parallel I/O port. Both I/O ports are equipped with DMA capability.

11-3-2-10   TI TMS320C30.   The TMS320C30 is the third-generation device in the TMS320 family and has shown great improvement in both raw speed and capability from its predecessor.

Technology. The TMS320C30 is fabricated in 1-micron CMOS and is packaged in a 180-pin grid array. The instruction cycle time has been reduced to merely 60 ns.

Control Units. The program counter has been expanded from 16 bits to 24 bits, thus extending addressing capacity from 64K to 16M. The data bus has been expanded from 16 bits to 32 bits, and the number of auxiliary registers has been increased from five to seven. There are seven extended-precision registers capable of storing and supporting operation on 32-bit integer and 40-bit floating-point numbers. The "repeat" instruction of TMS32020 has been generalized to "block repeat" instruction in TMS320C30 by the addition of a repeat start address register.

Arithmetic Unit. The multiplier can perform a single-cycle multiplication on either 24-bit integer or 32-bit floating-point numbers. The ALU performs single-cycle operations on 32-bit integer, 32-bit logical, and 40-bit floating-point data including single-cycle integer and floating-point conversion.

Memory Units. The memory unit has been greatly expanded compared to that of TMS32020. The TMS320C30 provides two 1K × 32 RAM blocks, one 4K × 32 ROM block, and a 64 × 32 instruction cache memory. The data page pointer is expanded to 32 bits with eight LSBs used by the direct addressing mode as a pointer to the page of data addressed.

I/O Units. All TMS320C30 peripherals are controlled through memory mapped registers on a dedicated peripheral bus composed of a 32-bit data bus and a 24-bit address bus. This is a big improvement over the 16 I/O ports of TMS32020. The TMS320C30 peripherals include two serial ports and two timers.

11-3-3 Development Systems

As indicated in the previous section, some DSP chips have very little support for the development of a DSP chip-based system, while other chips (such as the Bell DSP-1 and the TMS320 family) have extensive development supports. In this section, instead of describing what is available from the DSP chip vendors, we examine some essential and desirable elements of a development system for DSP chip-based applications. Hopefully, this section will provide enough information so that a reader can create (either by building his/her own or by acquiring one from a vendor) a cost effective development system for DSP chip-based applications.

The basic utility of a development system is to help an engineer develop and test signal processing application code. The most direct way to check if a piece of code will perform the intended function satisfactorily is to have the code running in the hardware and check its result. In order to do this, one first has to edit the assembly code, have it assembled into the machine code, and then load the machine code into the memory (usually an external RAM in the prototype development). Next, a signal generator for generating the test signals is connected to the input and an oscilloscope (or a speaker), for observing (hearing) the output waveforms, is hooked up to the output. If the output waveform does not look (sound) right, the code has to be debugged and the whole procedure is repeated. A good development system should make this iteration loop short and painless.

The block diagram of a simple development system setup is as shown in Figure 11-11. It consists of a host, a signal generator, an oscilloscope, and a "module." The module contains a controller, a memory unit, an input subsystem (A/D), an

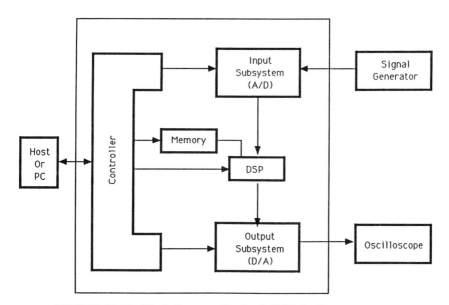

FIGURE 11-11 Block diagram of a simple DSP development system.

output subsystem (D/A), and the DSP chip itself. Usually, the program editing and assembly are done in the host. The host can be a personal computer or a terminal with a link (such as an RS-232 port) to a mainframe. Most hosts already have a good screen editor and the installation of the DSP assembler is straightforward. To make the job of the controller in the module easier, a simple code to handle the down-loading of the machine code may have to be written. But it should be a small effort. Although it is possible to implement a simple resident line editor and assembler in the module (e.g., the TMS320 evaluation board) to avoid the need for down-loading the code, it will become clear quickly that a line editor with limited capability is far from desirable. The controller in the module handles all the "setup" work. It has a simple resident monitor that accepts the commands from the host terminal to set up the operating environments such as the sampling clock rate of the input subsystem and the display mode of the output subsystem. After the setup is completed, the controller passes the control to the DSP by placing the beginning address of the DSP code in the DSP program counter and the experiment starts. To facilitate the debugging process, the controller should make it possible to examine/modify the contents of the memory unit and some important DSP registers, such as the program counter and the accumulator. Also, a single-step operation capability is a very useful feature for debugging the code and should be included in the controller. The simplest way to acquire a controller with the above-mentioned capability is to use an industrial standard single-board computer. The resident monitor, which comes with the single-board computer, usually has all the capabilities except the single-step operation, which can easily be implemented.

The input subsystem contains an A/D converter and a sampling system with programmable sampling clock. The output subsystem has a display memory and a display controller. It is desirable to have a dual-port display memory so that the output waveform will not be interrupted when the DSP is updating the display memory. Some flexibility should be built into the display controller to make observation of the output waveform easy. For example, a stable waveform should be obtained easily on the oscilloscope if so desired. One way to accomplish this is to provide a triggering pulse synchronized to the display period. Another desirable feature of the display controller is the capability to display only a few selective sections of the display memory. This capability will prove to be very useful for interactive debugging.

11-4 CUSTOM HARDWARE

The digital signal processing integrated circuits discussed in Section 11-3 can be used as part of general purpose signal processing microcomputers to implement all signal processing algorithms. For certain applications, these microcomputers will not meet the performance or cost criteria. When performance is an issue, an alternative to using these signal computers is to build a signal processor based on more fundamental building blocks, such as the multipliers, adders, or counters, and the large number of TTL and ECL parts (MSI and SSI) that are widely available. When cost and/or performance is the issue, custom integrated circuits can also be built, integrating a number of the above basic building blocks. Another alternative is to use the signal processing microprocessors described in Section 11-3 as the basic building blocks.

The disadvantage of these approaches over the microcomputer approach is that the design cycle is longer and the development costs are higher. However, the designer has more flexibility and substantial product cost reduction and/or product performance increases can be achieved. Whereas the microcomputer implementation is limited to processing rates on the order of a few million operations per second, with these latter approaches orders of magnitude increases in the processing speed can be achieved through the use of faster devices and/or greater parallelism.

In this section we first review the different semiconductor technologies available. We then review the standard digital signal processing components that are available. Custom designed components are discussed next. We then present a number of basic signal processing architectures with examples taken from the recent literature.

11-4-1 Semiconductor Technologies

Since even custom components will need to interface to some standard components, we review the semiconductor technologies by looking at the logic families available.

First is the bipolar *emitter-coupled-logic* (ECL) family of integrated circuits that is characterized by (sub-)nanosecond gate delays and negative power supplies. Power dissipation is rather high. Since power dissipation is the limiting factor in circuit integration, ECL has not fared well in the large scale integration (LSI) revolution that has taken place in this decade. An example of an ECL gate is shown in Figure 11-12(a).

ECL is typically used only for the very highest speed applications. A disadvantage of ECL is that there is not as wide a selection of supporting components as for the other logic families. Furthermore, its higher power dissipation is a source of hidden costs. Not only is a power supply with more current sourcing capability required but one also needs to consider the problem of heat removal. Bulky heat sinks and noisy fans may be needed.

Next is the bipolar *transistor–transistor-logic* (TTL) family. Since the original TTL circuits became available, namely, the 54/7400 series, there have been a large number of improvements to the original concept through the use of Schottky barrier diodes (to prevent transistors from entering saturation so that they can be switched off faster) and advanced processing. The resulting logic families are the 54/7400X series, where the suffix X is S(chottky), L(ow-power) S(chottky), A(dvanced) L(ow-power) S(chottky), and so on.

To these bipolar TTL circuits, we should also include the *TTL-compatible* logic circuits made using non-TTL circuit techniques but having compatible external

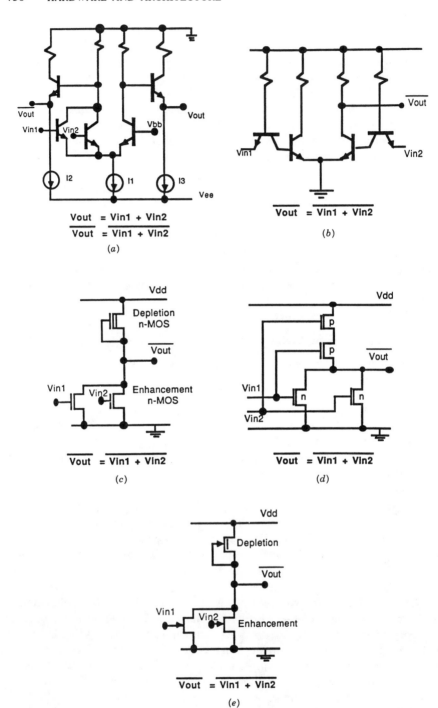

FIGURE 11-12   (a) ECL NOR gate. (b) TTL NOR gate. (c) n-MOS NOR gate. (d) CMOS NOR gate. (e) E/D GaAs NOR gate.

signal specifications. These include most of the *MOSFET* (metal oxide semiconductor field effect transistor) circuits, namely, *p-MOS*, *n-MOS*, and *C(omplementary)MOS*,[2] and even circuits that internally are ECL but have TTL buffers.[3]

The TTL-compatible family is characterized by a power supply[4] of 5 volts and from very low to fairly high power dissipation. In addition, they offer a wide range of speeds. Some examples of gate designs in TTL, n-MOS, and CMOS are illustrated in Figures 11-12(b) to 11-12(d).

The main advantage of designing circuits using the TTL signal format is the availability of a very large range of support circuits, the "glue" circuits. This will ensure that this signal format will remain dominant for the foreseeable future. However, there is a trend to replace bipolar TTL circuits with CMOS circuits in the commercial market. An advantage of CMOS circuits is that one can trade-off power consumption for speed. By increasing the power supply voltage, one can increase the switching speed of the circuit at the cost of increased power consumption. And by reducing the power supply voltage, both speed and power decrease. Another advantage of CMOS is that power dissipation is roughly proportional to the speed of operation of the circuit. Thus for low-speed operation, power dissipation is minimal.

Currently, almost all circuits are made using silicon. *Gallium Arsenide* (GaAs) is a competing base material that has a faster intrinsic switching speed. GaAs circuits are currently much more expensive than silicon circuits, but the cost differential will be narrowing down in the future as more GaAs manufacturing capacity is brought on-line. Standard circuits are currently not available, but they should start appearing in the near future. These will most likely be ECL compatible. While silicon circuits with a few hundred thousand transistors have been fabricated, GaAs circuits are currently limited to a few thousand transistors.

The current generation of GaAs components is based on depletion[5] MESFETs (metal semiconductor field effect transistors). The circuits tend to be much faster than similar ECL circuits, but they also consume considerably more power. The development of the enhancement[6] GaAs MESFETs has made possible the design of n-MOS-like circuits as shown in Figure 11-12(e). These circuits will have ECL-like speed with much lower power consumption. These E/D (enhancement/depletion) GaAs circuits hold the key to high-speed large-scale integration for GaAs.

The next generation of high-speed GaAs devices, beyond the MESFET and its derivative (like HEMT—high electron mobility transistors), will be based on the heterojunction bipolar transistor (HBT). These GaAs bipolar transistors will be able to make use of the large body of design techniques that have been developed over the past decades for silicon bipolar transistors. These will be at least an order

[2]p-MOS is considered old technology now and should be avoided. The bulk of LSI circuits, such as microprocessors, are built using n-MOS. However, CMOS, with its lower static power consumption, is rapidly becoming the technology of choice.
[3]An example is the Am 2901 bit slice microprocessor by Advanced Micro Devices (AMD).
[4]The true bipolar TTL circuits typically only require the 5-volt supply. The TTL-compatible MOS circuits sometimes require other power supplies. The trend is towards a single 5-volt supply.
[5]A depletion FET is normally ON. The switching action is to turn it OFF.
[6]An enhancement FET is normally OFF. The switching action is to turn it ON.

of magnitude faster than the current GaAs MESFET circuits. However, new circuit techniques will need to be developed for these high switching speeds since parasitic effects that are currently ignored will be a major limiting factor.

In summary, the preferred technologies are CMOS for low to medium speed, bipolar ECL, and in the future E/D GaAs, for high-speed applications. For the very highest speed applications, depletion GaAs MESFET technology is used.

While standard (merchant market) components will use only one of the above two signal formats (ECL, TTL), one is free to choose a different format for custom integrated circuits if there is no requirement for interconnecting with standard components. This is especially true with CMOS and GaAs, where speed and power optimization invariably points to a different (typically lower) supply voltage. Even when input and output must be according to one of the above signal formats, one may still want, inside the integrated circuit, to use a different format for efficiency reasons. For example, all TTL-compatible CMOS circuits internally use a logic threshold that is higher than that of TTL; namely, the CMOS threshold is around 2 volts, whereas for TTL a logic "0" is defined to be below 0.8 volt and a logic "1" is above 2.0 volts. Translation buffers are used to go between signal formats at the interfaces.

11-4-2   Standard Digital Signal Processing Components

One of the basic operations in digital signal processing is addition. The most widely used off-the-shelf *adders* are the 4-bit slices. To make longer wordlength adders, a number of these slices are operated in parallel. The speed of the resulting adder is determined by the time it takes for the carry from the least significant circuit to ripple through to the most significant circuit. Carry-look-ahead circuits, which are used to determine the carries into the various adder slices from the input data and a set of signals generated by each of the slices, are used to speed up these large adders. Examples of adder slices are the 54/74381 Arithmetic Logic Unit/Function Generator circuits, which can perform addition, subtraction, and logical operations like AND, OR, Exclusive-OR. Examples of look-ahead carry generators are the 54/74182 circuits.

The other basic operation in digital signal processing is multiplication. *Multipliers* come in various wordlengths from 8 bits up to 24 bits. In specifying multipliers, one needs to consider the required wordlength, the multiply time (60–200 ns), the power dissipation (heat sinks and high-volume air flow may be required for certain circuits), and the way the final result is made available. The latter is an issue since the wordlength of the result is typically twice that of the inputs, and most manufacturers provide either a rounded or truncated result and/or a time-multiplexed representation of the result in two parts. The type of multiplication also needs to be considered. Both unsigned multiplication, where one or both operands are considered to be positive numbers, and 2's complement multiplication, where both operands are in 2's complement format, can be performed by these off-the-shelf devices. Multiplier-accumulators are also used frequently. In these devices, the multiplication result is added to the content of an internal register, the accumulator.

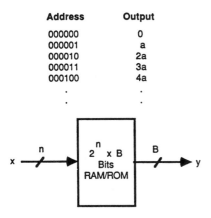

Address	Output
000000	0
000001	a
000010	2a
000011	3a
000100	4a

FIGURE 11-13 RAM multiplication.

In the past, when multiplication was considered to be an "expensive" operation either due to part count or delay, there was a lot of interest in performing the operation by table look-up. The data are used to address a RAM table where the result of the multiplication by a predefined constant has been stored, as shown in Figure 11-13. Array multiplication table look-up [PE74, PE76], as shown in Figure 11-14, is another multiplier-less operation. There has also been considerable interest in reducing the multiplication to simpler operations (see Chapter 9) and in making the coefficient some power of 2 [LI79, LI83]. With the availability of multipliers and with the multiplication–addition penalty now almost nonexistent, at least at the component level, there has been a decrease in interest in these areas, except for custom circuits where transistor count difference can be significant, especially when a large number of multipliers are used.

The other basic operation is the delay, which can be implemented either by shift registers, by random access memories (RAM), or by registers.

To the adder, multiplier, and memory one can add a long list of other basic components such as comparators and multiplexers.

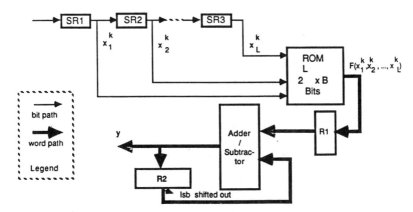

FIGURE 11-14 Inner product multiplication by table look-up.

In the above discussion we dealt mostly with fixed-point operations. Recently, floating-point components and signal processing circuits [EL83; KE85] have started to become available. The rationale for using floating-point operation is that one does not have to worry about the scaling of the input, output, and internal signal variables to ensure the dynamic range is not exceeded. This is typically not a major concern in simple signal processing operations, such as filtering and Fourier transform. However, it is a problem for adaptive systems where the parameters may cover a very wide dynamic range. Certain applications (e.g., seismic) may also deal with signals with very large dynamic ranges and thus require floating-point numbers. While we expect that floating-point operators will ultimately prevail, most of the next generation of signal processors will still use fixed-point operators. The floating-point arithmetic components that are becoming available are adders, multipliers, and monolithic signal processors.

11-4-3 Application-Specific Integrated Circuits

An alternative to designing with off-the-shelf components is to design custom components, or application-specific integrated circuits (ASIC), to just fit one's particular needs. Three approaches are currently available: gate array design, standard (library) cell design, and full custom design [MA83; ME80; NE83].

The technology of choice is currently CMOS for low to medium-speed applications. Bipolar ECL and, eventually, GaAs are used for high-speed applications.

To put things in perspective, we should first review the economics of the various alternatives.

Each of the digital signal processing integrated circuits discussed in Section 11-3, and their successors, will cost anywhere from $20 to $1000 depending on the stage of their life cycle. The initial higher price is needed to recover the engineering cost associated with their development. To the cost of the integrated circuit one needs to add the cost of the program and data memory and that of the SSI and MSI "glue" circuitry and the physical carrier (circuit board, ceramics, rack, etc.).

If the complexity of the custom circuit is low, for example, less than a few thousand gates, a gate array[7] or a transistor array is used. The turnaround time of these designs is typically good. However, the development cost is still fairly high, on the order of $40K, depending on the amount of work done by the vendor.[8] For

---

[7]In a gate array, the transistors of the integrated circuit have already been defined and may have been prefabricated. The design involves interconnecting these transistors to make up the required gates. Since the number of transistors and their locations are fixed in advance, the design of the circuit is highly constrained. Extensive computer tools are used to come up with an acceptable layout. The final one or more metallization masks are then fabricated and the circuit processing is completed. Since the circuit can be fabricated in advance, except for the final metallization masks, fabrication cost can be lower. The design cycle can be short with the use of computer tools.

[8]The design can be handed over to the gate array vendor at the block diagram level, at the logic level, or at the layout level. The cost decreases accordingly.

higher complexity, a standard cell approach[9] is used to design the custom integrated circuit. Only for the highest performance or for very high volume will a fully custom design be undertaken.

For a standard cell design or a fully custom design, the initial fabrication cost is typically around $35K. To this must be added the design and layout costs, which are much higher for the fully custom approach than for the standard cell design. Both approaches are more costly than the gate array approach and also have a long turnaround time. With automated design tools, the design of a standard cell circuit can actually be faster than that of a gate array circuit, since the designer has more predefined circuits available and has considerably more freedom with regard to how the wiring is done.

When a custom integrated circuit is required, one needs to consider the amortization of the higher development costs. Typically, the design cycle for an integrated circuit is much longer than that of a circuit board. Since an integrated circuit cannot be modified once fabricated, the designer needs to be more thorough in the design. The layout of the integrated circuit is also time consuming. There are also mask, processing, and packaging costs. Turnaround time is also typically longer than that of a circuit board design using off-the-shelf parts. However, as we shall discuss later, especially in performance-limiting cases, this may be the only viable approach.

Table 11-1 summarizes the relative advantages of the various design approaches [IC85].

11-4-4   Basic Processor Architecture

While most existing off-the-shelf digital signal processing circuits are sequential systems with some type of modified Harvard architecture, custom systems offer more flexibility.

Single-Chip Signal Processor. Most of the signal processors presented in Section 11-3 have ROM versions that can be used as single-chip custom signal processors.

For example, the AMI S2814A is a ROM preprogrammed version of the AMI S2811 (see Section 11-3-21) that is used for evaluating FFTs. It can evaluate a 32-point complex FFT in 1.3 ms with the processor operating at a 20-MHz clock rate. It functions as a hardware peripheral of a microprocessor that controls the data flow, including I/O, and initiates the execution of the FFT routines in the S2814.

---

[9]In a standard cell design, the basic building blocks are obtained from a library of predefined cells, which includes the standard gates such as NAND and NOR, but which may also include so-called macrocells, which can be up to full-fledged microprocessors. Each cell is typically of some standard height, or a multiple of the basic height, and with the widths being integer multiples of some basic unit. Working with these cells, the user will design the integrated circuit. The next stage, which is the layout stage, is typically automated. The layout of the circuit is done by computer-aided design tools such as placement and routing programs.

TABLE 11-1   Comparison of Advantages of Various Design Approaches[a]

Advantages	Full Custom	Cell Library	Gate Array	Non-LSI
Design costs	4	3	2	1
Design time	4	2	3	1
Mask costs	4	3	2	1
Redesign flexibility	4	3	2	1
Test program costs	3	3	2	1
Circuit purchase price	1	2	3	4
System power required	1	2	2	4
Reliability	1	2	2	4
PC board and costs	1	2	2	4
Production labor	1	2	2	4
Security	1	3	2	4
Added features/board	1	3	3	4

[a]A rating of one indicates the best.

A technique for speeding up these general purpose signal processors, which are now dedicated to special functions, is to use *straight-line coding* of the key routines. In straight-line coding the loops are expanded. The penalty is that the program code will be larger. This method should be compared with the conventional methods of using loops, which are more code efficient but incur a running-time penalty due to loop overheads.

However, since these are general purpose processors, the amount of overhead is still large. A more customized version of these processors will typically achieve higher throughput rates.

An example of a single-chip custom processor is the 32-point monolithic FFT processor chip by Covert [CO82], the TMC2032. Its block diagram is given in Figure 11-15. With a clock rate of 50 MHz, this circuit can evaluate a 32-point complex FFT in 47 μs. In order to simplify the design of the multiplier-accumulator (MAC), the width of the read-only memory (ROM) has been increased from the normal 16 bits to 24 bits. This enables Covert to use a direct Booth [WA82] multiplication algorithm. The MAC, the 17-bit adder, and the register stack have been optimized for the radix-2 FFT. Control of the operations is done through PLA, which will sequence the circuit through all the steps for a 32-point FFT without external intervention.

In this custom chip, Covert has also made provisions for operating a number of them in parallel in order to handle larger FFTs at high rate. Figure 11-16 is the block diagram for the implementation of a 1024-point FFT.

A direct comparison of the S2814 and the TMC2032 is not really valid since different technologies are involved and the two circuits were designed for different roles. Furthermore, the S2814 is restricted to serial communication. A comparison of the TMS32010 and the TMC2032 appears to be appropriate. The best time for a 32-point FFT, achieved by straight-line coding the FFT on the 32010, is 291 μs. Normalizing the clock rates, we find that the TMC2032 is relatively faster than

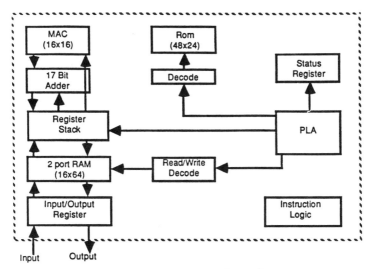

FIGURE 11-15 TMC2032, 32-point FFT processor.

the TMS32010, namely, 47 μs at 50 MHz versus 291 μs at 20 MHz. This is somewhat of a validation of our statement that custom design yields greater performance. On the other hand, the design cost of the TMC2032 is significantly higher than that of converting the general purpose TMS32010 to a special purpose processor by customizing its ROM (à la S2814)!

In general, a custom single-chip signal processor will be a reduced form of the standard signal processors discussed in Section 11-3. Typically, the first modification is the instruction path. Instead of allowing off-chip program memory, most of these chips will use internal program ROM. The instruction counter will typically have considerably less flexibility in order to reduce chip area.

The next modification is to the arithmetic unit. The logic capability of the arithmetic unit is typically deleted and the shift operations will become more restrictive. Where a parallel multiplier is not used, one can attempt to simplify the shift-and-

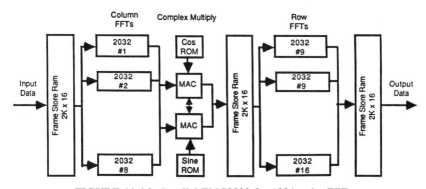

FIGURE 11-16 Parallel TMC2032 for 1024-point FFT.

```
          x0 . x1 x2 x3 x4 x5 x6 x7
       *  y0 . y1 y2 y3
       -----------------------------
(x0 x0 x0 x0 x1 x2 x3 x4 x5 x6 x7) * y3
(x0 x0 x0 x1 x2 x3 x4 x5 x6 x7   ) * y2
(x0 x0 x1 x2 x3 x4 x5 x6 x7      ) * y1
(x0 x1 x2 x3 x4 x5 x6 x7         ) * (-y0)
-------------------------------------

z0.z1 z2 z3 z4 z5 z6 z7 z8 z9 z10
```

(*a*)

```
          x0 . x1 x2 x3 x4 x5 x6 x7
       *  y0 . y1 y2 y3
       -----------------------------
(x0 x0 x0 x0 x1 x2 x3 x4         ) * y3
(x0 x0 x0 x1 x2 x3 x4 x5         ) * y2
(x0 x0 x1 x2 x3 x4 x5 x6         ) * y1
(x0 x1 x2 x3 x4 x5 x6 x7         ) * (-y0)
-------------------------------------

z0.z1 z2 z3 z4 z5 z6 z7
```

(*b*)

FIGURE 11-17   (a) Normal 2's complement multiplication. (b) Left-sided 2's complement multiplication.

add/subtract multiplication by imposing restrictions on the coefficients. For example, the coefficients are restricted to be represented by canonical signed digit[10] numbers with only a few nonzero digits, for example, two nonzero digits for the equivalent of a 16-bit coefficient. Another possibility is to use certain of the pseudomultiplication techniques discussed earlier, such as the table look-up schemes illustrated in Figures 11-14 and 11-15. When a parallel multiplier is used, one may consider using a left-sided multiplication [TH78] as opposed to the full multiplication, as illustrated in Figure 11-17. The gate saving can be significant since up to half of the terms will not have to be evaluated. For both the parallel multiplier and the case of the shift-and-add multiplier, one may also restrict the coefficient to be smaller than the usual 8 or 16 bits, or one may make it larger to 24 or 32 bits. One may also consider trade-offs between the size of the coefficient ROM and the multiplier circuit. Past this point, one needs to consider adding processing blocks to the circuit in order to improve its capability. For example, extra adders or even simplified multipliers can be added. Another function that may be added is a (roundoff) error spectrum shaping quantizers[11] [CH81; MU81; TH76; TH77].

---

[10]In the canonical signed digit (CSD) representation, which is related to the Booth recoding of multiplication, a 2's complement number is represented in a ternary system, which consists of the digits $\{-1, 0, 1\}$. The main property of the CSD representation is that of two consecutive digits; only one can be different from 0.

[11]Rounding typically introduces an error whose spectrum is essentially white. For certain narrowband filters, it is desirable to have a certain roundoff noise shape that has a zero at some appropriate frequencies. This is the concept of error spectrum shaping, which can be implemented by feeding the error back. This is discussed in detail in Section 7-6.

By shaping the error spectrum and matching it to the system, one can achieve improved signal-to-noise ratios. This can also be used to suppress limit cycle oscillations. Another function is the μ-law coder in the Bell DSP-1.

Most often a single-chip signal processor is developed to meet cost constraints. The custom chip can fill the role of the general purpose integrated circuits discussed in Section 11-3 without the need for additional peripheral circuitry. Thus for high-volume applications, a custom single-chip signal processor can be cost effective.

Custom Signal Processor. All the comments made above for the single-chip signal processor, except those about cost, also apply to a custom signal processor built with off-the-shelf components such as gates, adders, multipliers, RAM, ROM, sequencer, and microprocessors. The difference is only a matter of scale and design time and cost.

11-4-5 System Architecture

Normally, from the specifications of the problem to be solved, a system architecture is derived. As part of the system architecture, specifications for the processor(s) are derived. We have chosen to cover the system and processor issues in reverse order since a lot of the system considerations to be discussed below also apply to the architecture of the basic processor.

In the following, a processor can be either a circuit board, an integrated circuit, or a building block in an integrated circuit. No effort will be made to differentiate between these implementations since they are technology dependent. What is a board today can be just a macrocell tomorrow!

11-4-5-1 Parallel Processing. Certain signal processing problems are characterized by very high throughput rate requirements due to the need for processing signals in real time. No matter how fast a processor can operate, there will be situations where a single processor cannot maintain the required throughput rate. Storing the data for later processing is not feasible because of the rate at which data accumulates (e.g., at a modest sampling rate of 1 Ms/s, even the best storage medium to date can store at best a few hours of information). For these compute-bound situations, the solution is to move to a parallel processing architecture. The purpose of these high-speed processors is essentially one of data reduction, since only the relevant data after the transformation need be stored.

While all the parallel processing techniques of computer architecture can be used, a further increase in efficiency can be achieved by taking into account the signal flow requirement of the digital signal processing tasks in order to reduce the interprocessor communication bottleneck.

A key element in the successful design of a parallel processor is the identification of a simple unit that can be replicated many times. Another consideration is the communication network linking processors.

Efficient parallel processor design is still an art. While there are a number of general techniques for designing parallel processors, each situation is unique. Since

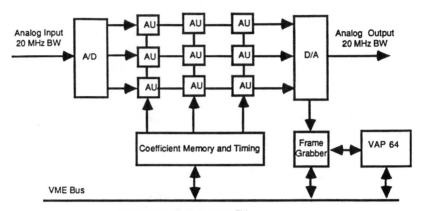

FIGURE 11-18   GOPS™ processor.

it is not possible for us to cover all possible parallel processor implementations, we comment on only a few examples from the recent literature.[12]

Schirm et al. [SC85b] have proposed a digital signal processing architecture that is claimed to be capable of achieving throughput rates in excess of 10 billion fixed-point operations[13] per second. This is achieved through massive parallelism of off-the-shelf hardware. Figure 11-18 shows the basic architecture of the GOPS™ processor. Each of the nine arithmetic units (AUs) in Figure 11-18 is a 24-tap finite impulse response filter operating at 16.7 Ms/s. Their functional block diagram is shown in Figure 11-19. (Their actual implementation is with multiplier-accumulators.) By cascading three of these AUs, a 72-tap filter is obtained. By a three-way time-interleaving,[14] a sampling rate of 50 Ms/s can be achieved for a 72-tap filter, yielding a 7.2-giga-operations-per-second processing rate. Higher rates can be achieved through increased parallelism. The operations are controlled by the VAP-64 processor. Applications are limited to adaptive filtering and cross-correlation operations. In these applications, the array of AUs will perform the signal operations, while the VAP-64 will perform the adaptation calculations.

In this type of *time-interleaving*[15] *parallelism* each path is capable of performing the full operation. The amount of hardware increases proportionally with the speed. While multiple copies of a basic processor are fairly straightforward to implement, the problem of data sharing is more challenging. The same input data may be needed in more than one processor. For example, the nth input sample is

---

[12]Disclaimer: The authors do not claim that any of the systems discussed in this section are realizable, practical, or available. The reader should refer to the appropriate references for more information.

[13]An operation is defined to be either an addition or a multiplication.

[14]For example, each row of AUs is used to compute a different output sample. At any instant in time three output samples are being processed.

[15]Time-interleaving is frequently confused with time-multiplexing, which is the reverse operation. In time-multiplexing, a single processor will perform the same set of tasks on a number of independent channels. This is most frequently done in speech systems, where the data rate is only 8 ks/s, whereas the processor is capable of performing millions of operations each second.

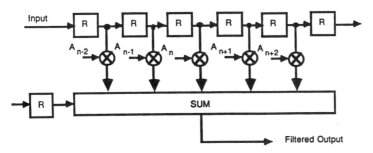

FIGURE 11-19   AU of GOPS.

needed in the evaluation of the *m*th and (*m* + 1)th output. This problem of data sharing actually causes the hardware complexity to increase faster than linearly with the number of paths. Controlling the data flow is also more complex. This is the issue of input/output bandwidth that Kung [KU82] has identified and tried to remedy with his systolic architecture. Thus, while conceptually simple, time-interleaving can be very difficult to implement in practice, especially when a large amount of hardware is involved.

For high-speed applications, *time-interleaving enables one to achieve a high throughput rate with slower processors. The delay[16] through the system is determined by the basic processor. The throughput rate[17] is equal to the product of the throughput rate of one unit and the number of parallel units.* Hardware complexity increases at a rate slightly greater than linear.

The above example uses an off-the-shelf processor. Swartzlander [SW83, SW84] has proposed a pipeline FFT processor using off-the-shelf floating-point arithmetic processors [EL83] and a standard-cell delay commutator circuit. A block diagram of the *pipeline* FFT processor is shown in Figure 11-20. In pipeline processing, which is another parallel processing technique, a number of incomplete results are present at various points in the pipeline. The data distribution problem has been solved by the sequential nature of the FFT. A computational element (CE) has been used to handle each of the FFT passes. The data throughput rate is limited to four complex points every cycle.

Pipelining is a technique for splitting up the operation into smaller sequential tasks. Each task is assigned to a processor. The throughput rate is determined by the throughput rate associated with the slowest of the tasks. The delay through the system is equal to the sum of the delays of all the task.

Let us diverge for a moment from the topic of parallel processing to look at the rationale for the development of the custom circuit mentioned above. Swartzlander found that using a standard-cell custom circuit to replace the many chips needed to implement the widely used delay commutator led to a reduction in overall system complexity by 60%. The block diagram of the delay commutator is shown in

---

[16]The delay is measured from the time a new piece of data is delivered to the processor to the time when the corresponding output appears.

[17]The throughput rate is the rate at which data can be accepted by the processor.

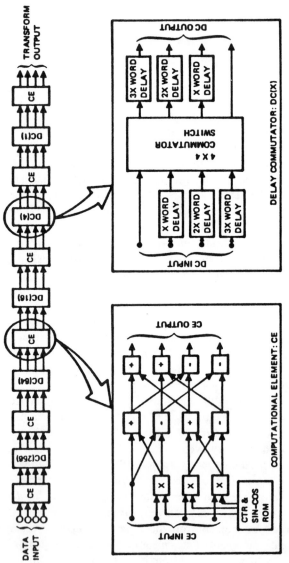

FIGURE 11-20 Fast pipeline FFT processor.

Figure 11-21. It is interesting to note the following characteristics of most successful custom circuits. The delay commutator has been overdesigned. The 256/512/768 word shift registers are needed only in the first stage of the FFT. Much smaller registers[18] are required in later stages. Rather than design one circuit for each stage of the FFT, the largest circuit was designed and the 5 : 1 MUX added to customize it for the other stages. Another advantage of the overdesign is that the design cost is now amortized over a large number of parts.

The above two architectures are based on *bit parallel* paths for the data. When a large number of processors are used in parallel, data distribution is an issue. A common data bus can be used. However, this can quickly become the bottleneck if the processing tasks are I/O intensive. Furthermore, if the bus is very long, synchronization can become a problem. For the word being transmitted in parallel to be received properly, the signals must have settled to their final values when the receiving circuit samples the bus. Now if one line of the bus is for some reason more heavily loaded than the others, the corresponding signal may settle to its final value after the bus has been sampled—thus causing an error. This becomes an even greater problem when the system clock is increased.

One can increase the number of data buses and make them local to a number of processors. This will consume a large amount of circuit board real-estate or lead to an increase in the number of circuit board layers, a source of manufacturing and reliability problems.

The advantage of a bit parallel architecture is that one word can be processed each clock cycle. An alternative to this bit parallel architecture is a bit serial architecture.

11-4-5-2 Bit Serial Architecture. A bit serial architecture was first proposed for DSP by Jackson et al. [JA68]. Recently, interest has revived in bit serial arithmetic [AL85; KA81; MO78; PO74; PO75; PO76; PO81; TE88b; TH82] and it has mostly been generated by the need to perform large-scale parallel processing.

While it takes longer to transmit a bit serial word, n clock cycles for an n-bit word, the interprocessor connection is reduced to a single wire. Synchronization is less of a problem since all the data are now time serial. While it is true that one now needs more processors to achieve the same performance, and thus consumes more circuit board real-estate, it has been found in practice that the trade-off has typically been favorable when a large number of processors are involved.

Since the bit serial operations are now simpler, the processors can be much smaller. Thus it is possible to integrate more functionality into a circuit. It is also possible to trade-off throughput rate for accuracy. Bit serial processors should be considered only for large parallel systems with a heavy interprocessor communication load.

An example [TH82] is illustrated in Figure 11-22. This bit serial circuit is referred to as a digital operational amplifier because of its similarity to the analog

[18]Each successive stage of the FFT requires only one-quarter of the storage requirement of the previous stage.

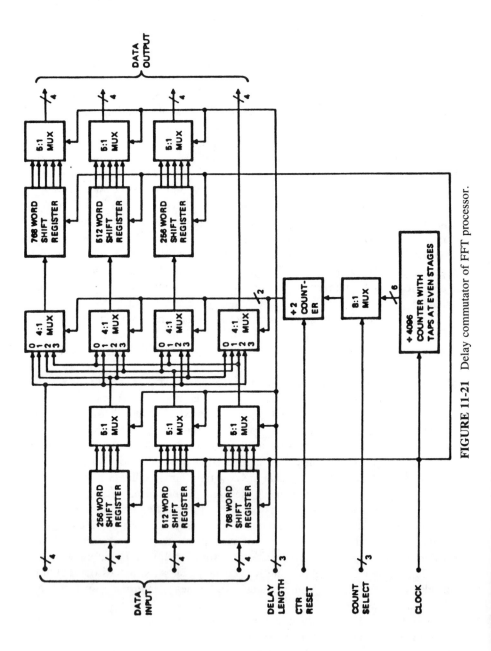

FIGURE 11-21 Delay commutator of FFT processor.

770

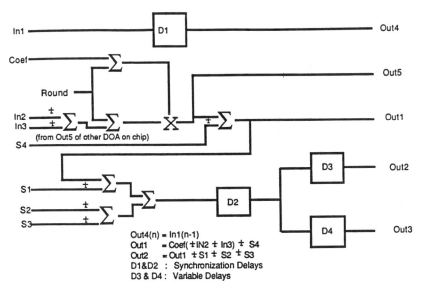

FIGURE 11-22 Bit serial processor.

op-amp. We have a summing junction before the multiplier. Since the design of this particular circuit was aimed at a radix-4 FFT, there are two summing junctions after the multiplier. The first summing junction allows the evaluation of the complex multiplication. Following the complex multiplication, four complex numbers need to be summed; thus the four-input summing junction. Since the FFT requires both addition and subtraction, programmable adder-subtractor circuits are used. By conventional count, up to six operations are performed each word time in the circuit of Figure 11-22.

Only the n most significant bits of the data are used in the multiplication with the 16-bit multiplier COEF. The resulting double precision result is $n + 16$ bits long. The additions are performed on $(n + 16)$-bit words. The word time is defined to be $(n + 16)$-bit time. In this particular case, the length n is variable. One can trade-off speed for accuracy by altering the value of n used. A difficulty in using these bit serial circuits is that the signals at each of the outputs are slightly shifted in time. Thus one should not mix a signal coming from an OUT1 output with one coming from an OUT5 output. This is the reason for the restrictive interconnection rules [TH82].

As shown in Figure 11-23, two of the above circuits were integrated in a chip together with a cross-bar structure that allows on-chip programmable connection. Some of the programmable features, such as the control signals for the adder-subtractor and for the cross-bar connections are loaded in by way of a serial data port. This was done to minimize chip pin-out.

A similar circuit is the basis for the FFT processor in the Tektronix 3052 [TE88b]. With eight complex data points being processed at each stage of the radix-4 FFT processor, 160 processors are used to achieve a 5-mega-sample per

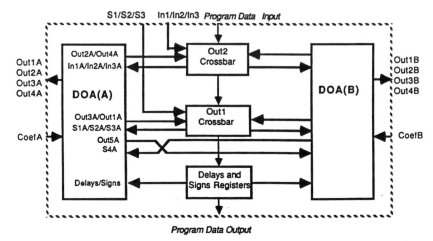

FIGURE 11-23   Bit serial circuit.

second 1024-point FFT rate.[19] Additional processors are used to implement the window function.

Powell [PO81] has used similar bit serial circuits to build a 10-mega-sample per second 1024-point FFT processor. To achieve this high throughput rate, 640 custom bit serial p-MOS processors operating at a bit clock rate of 2.5 MHz were used. This large number of custom processors were packaged in 40 identical printed wire boards.

In both of these FFT processors, data flow is controlled strictly by hard wiring. Thus in spite of the large number of processors, control is fairly simple. It consists only of memory management and input and output control.

Aliphas [AL85] has proposed the high-resolution digital filter, KSC 2408, shown in Figure 11-24. As opposed to the bit serial circuits discussed above, which have essentially hard-wired functions, this circuit is more like a microprocessor. Bit serial arithmetic was used in order to implement a number of 24-bit, bit serial multipliers with minimum silicon real-estate. Two outputs are provided. The "Muxed serial output" is tristateable and is used when time-interleaving is done with a number of chips. The computational rate is 640K multiplications per second. By time-multiplexing the arithmetic units, eight independent second-order sections, or four independent fourth-order sections, and so on, can be evaluated at a 16-kHZ rate. Control is achieved by loading a bit serial data stream, as in the digital operational amplifier [TH82]. Multiple chips can also be used with almost no "glue" logic.

11-4-5-3 Systolic Architecture.
Parallel processing until recently has been implemented on a fairly ad hoc basis. While some general rules have been identi-

---

[19]This corresponds to roughly 5000 FFTs per second. The multiplication rate is about 100 million per second. Using a ratio of 2:1 for addition versus multiplication, the overall operation rate is 300 million operations per second.

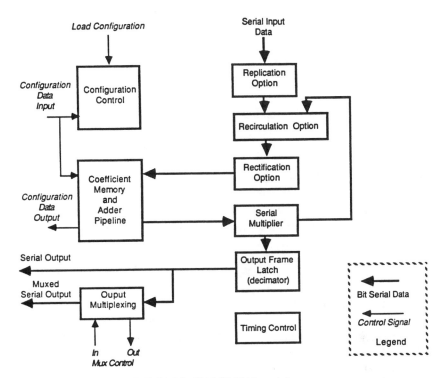

FIGURE 11-24 KSC 2048 bit serial processor.

fied—for example, time-interleaving and pipelining—each case was treated separately and there was no consistent buildup of a body of knowledge. Kung [KU82] came up with an architectural concept that provides a methodology for implementing parallel processing. One cornerstone of this architectural concept is the belief that typically I/O is the limiting factor. He has identified a number of key factors for successful parallel processing design:

1. *Simple and Regular Design.* Because of the large nonrecurring cost of design of the basic processor, it behooves the designer to make this as simple as possible. A regular design will enable the designer to integrate more function into a special purpose processor, and thus spread the nonrecurring cost over many processors instead of having to make lots of expensive exceptions.

2. *Concurrency and Communication.* This comes from the observation that processing speed at the device (gate) level has not really increased much over the past decade. Parallel processing is a more efficient means for achieving high throughput rate rather than just through raw speed increase: thus the need to develop algorithms that will maximize the number of simultaneous operations. As mentioned earlier, communication can quickly become the bottleneck of a parallel processing system. A simple, regular communication and control system is needed for efficient implementation.

3. *Balancing Computation with I/O.* As with any system, the limiting factor is the weakest link in the system. Too often we have concentrated our attention on issues associated with computation throughput rather than on the I/O bandwidth issues. For example, if the memory can be accessed every x microseconds, a processor capable of operating at twice this rate will do no better than a processor running at x microseconds per sample. Optimal performance is reached when the bandwidths of all the operators are equal.

Based on the above observations, Kung has developed an architectural concept that has been termed systolic,[20] which is based on the notion of pulsing data through the system. A large body of literature has appeared on systolic computing recently [AR85; BR81; HE83; KU80; KU81; KU82]. In this section we review only a few examples as they apply to digital filtering.

Consider the problem of finite impulse response filtering. The expression to be evaluated is

$$y[n] = a_0 x[n] + a_1 x[n - 1] + \cdots + a_N x[n - N].$$

This can be implemented by the structure shown in Figure 11-25. First note that in order for this filter to operate properly, data are sent through the system only every other clock cycle. Thus $x[n]$ is followed by a blank interval, which is then followed by $x[n + 1]$. Similarly, the output $y[n]$ is followed by a blank interval. One can turn this deficiency into a useful feature by processing two independent data streams. Another point to notice is that communication between processors is local. There is no global communication. This is Kung's W1 case [KU82], where the weights stay in place, and the input and output data move in different directions.

There is also the W2 case, where the weights stay, and all the data move in the same direction, as shown in Figure 11-26. The disadvantage of W2 over W1 is that the output will not be available for N cycles until the last of its input enters

---

[20]According to Webster, systole is the usual rhythmic contraction of the heart, especially of the ventricles, following each dilation, during which the blood is driven onward from the chambers.

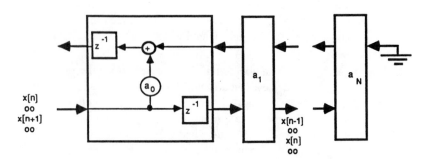

FIGURE 11-25 Systolic architecture, W1 FIR filter.

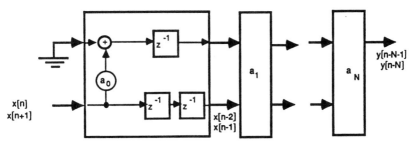

FIGURE 11-26 Systolic architecture, W2 FIR filter.

the system. In the case of W1, the output is available on the cycle after the last input enters the system. Other cases of systolic architectures are discussed in Kung [KU82].

A recursive system defined by the equation

$$y[n] = a_0 x[n] + \cdots + a_{N-1} x[n - N]$$

$$- b_1 y[n - 1] - \cdots - b_N y[n - N].$$

can be implemented with a structure similar to W1. This is shown in Figure 11-27. Again note that data are pulsed every other cycle through the system.

The examples presented are illustrations of linear systolic arrays. There are also higher dimensional array structures. An example of a two-dimensional array for sparse matrix multiplication is shown in Figure 11-28.

It should be noted that the systolic architecture is by no means universally accepted. One of the disadvantages of systolic arrays is that they do not have as high a utilization efficiency as some of the more ad hoc methods for parallel processing. For example, the systolic filter in Figure 11-27 is not as efficient as the one shown in Figure 11-29. One of the key assumptions of the systolic architecture—that

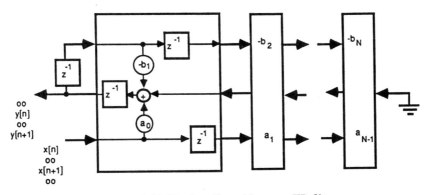

FIGURE 11-27 Systolic architecture, IIR filter.

$$
\begin{bmatrix}
a_{11} & a_{12} & & & & \\
a_{21} & a_{22} & a_{23} & & 0 & \\
a_{31} & a_{32} & a_{33} & a_{34} & & \\
& & a_{42} & & \ddots & \\
0 & & & & &
\end{bmatrix}
\begin{bmatrix}
b_{11} & b_{12} & b_{13} & & & \\
b_{21} & b_{22} & b_{23} & b_{24} & 0 & \\
& b_{32} & b_{33} & b_{34} & b_{35} & \\
& & b_{43} & & \ddots & \\
0 & & & & &
\end{bmatrix}
=
\begin{bmatrix}
c_{11} & c_{12} & c_{13} & c_{14} & & \\
c_{21} & c_{22} & c_{23} & c_{24} & 0 & \\
c_{31} & c_{32} & c_{33} & c_{34} & & \\
c_{41} & c_{42} & & & \ddots & \\
0 & & & & &
\end{bmatrix}
$$

$$A \qquad\qquad\qquad B \qquad\qquad\qquad C$$

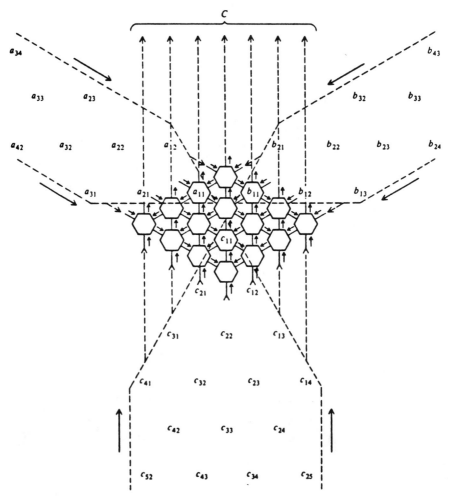

FIGURE 11-28 Two-dimensional systolic array for sparse matrix multiplier.

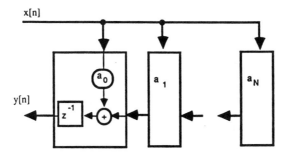

FIGURE 11-29 Nonsystolic FIR filter.

communication is costly and thus should be local—may not be valid here. Where broadcasting of information imposes no penalty, greater processor efficiency can be achieved with the architecture of Figure 11-29.

Even for an architecture with only local communication, the low processor utilization of the systolic architecture can be traced to the requirement that all the data transfers between cells be done simultaneously [SC84]. By relaxing this requirement more efficient architectures can be obtained [SC85a].

11-5 SUMMARY

Issues in the hardware implementation of digital signal processing algorithms were discussed in this chapter. The following methods for implementing digital signal processors were presented.

1. *Single-Chip Digital Signal Processor.* This is by far the simplest implementation. If a standard computer bus is used, circuit boards containing the desired single-chip digital signal processor are available from a number of manufacturers. Algorithm implementation then just becomes a firmware development issue. Even when a circuit board is not available, the manufacturer of the processor can provide almost all the information needed to build one to meet the user's special requirements. The key issues with a single-chip implementation are fixed-point and performance versus floating-point and flexibility.

2. *Off-the-Shelf Hardware.* Using basic building blocks such as adders, multipliers, ROM, RAM, and shift-registers, the user can build up a digital signal processor. This development method is rapidly falling out of favor as the performance of the single-chip digital signal processor increases. The only exception would be when very high throughput rates are needed. Even in this case, the third solution below becomes preferable.

3. *Application-Specific Integrated Circuits.* Since all the features are custom designed, this solution offers the ultimate in flexibility. The following trade-offs will have to be made:

Design type: gate array, standard cell, full custom

Technology: CMOS, silicon ECL, GaAs

Once the basic processor has been defined, the designer whose needs cannot be met by a single processor can choose from a number of parallel processing architectures:

Time-interleaving

Pipelining

Systolic

In parallel processing architecture, a key concern is the interprocessor communication. For digital signal processing tasks that require massive parallelism with minimal flexibility, instead of considering the standard word parallel arithmetic most commonly found, the designer may want to consider a bit serial arithmetic system with fixed point-to-point communication.

The area of digital signal processor hardware and architecture is rapidly changing. The key to the dynamics in this area is the progress in integrated circuit technology. Single-chip systems with 30 megaFLOPS (FLoating point Operations Per Second) are now becoming available [TE88a]. Also, 100-MHz FFT processors [MA88] have appeared.

REFERENCES

[AD79] Advanced Microsystems, Inc., *Signal Processing Peripherals*, S2811. Advanced Microsystems, Inc., 1979.

[AL75] J. Allen, Computer architectures for signal processing. *Proc. IEEE* **63**(4), 624–633 (1975).

[AL85] A. Aliphas et al., High resolution digital filter chip. *Proc. IEEE Int. Conf. Acoust., Speech, Signal Process., Tampa, Florida, 1985*, pp. 224–227 (1985).

[AR85] E. Arnould et al., A systolic array computer. *Proc. IEEE Int. Conf. Acoust., Speech, Signal Process., Tampa, Florida, 1985*, pp. 232–235 (1985).

[BA77] J. Baldwin, Digital standards conversion. In *Digital Video*, pp. 84–93. Society of Motion Picture and Television Engineers, Inc., Scarsdale, NY, 1977.

[BA78] A. B. Baggeroer, Sonar signal processing. In *Applications of Digital Signal Processing* (A. V. Oppenheim, ed.), pp. 331–437. Prentice-Hall, Englewood Cliffs, NJ, 1978.

[BL75] B. Blesser et al., A real time digital computer for simulating audio system. *J. Audio Eng. Soc.* **23**(9), pp. 698–707 (1975).

[BL78] B. Blesser and J. M. Kates, Digital processing in audio signals. In *Applications of Digital Signal Processing* (A. V. Oppenheim, ed.), pp. 29–116. Prentice-Hall, Englewood Cliffs, NJ, 1978.

[BO81] J. R. Boddie et al., DSP architecture and performance. *Bell Syst. Tech. J.* **60,** (7) 1449–1462 (1981).

[BR81] K. Bromley, J. J. Symanski, J. M. Speiser, and H. J. Whitehouse, Systolic array processor development. In *VLSI Systems and Computations* (H. T. Kung, B. Sproull, and G. Steele, eds.), pp. 273–284. Computer Science Press, Rockville, MD, 1981.

[BR82] A. B. Brown et al., Telecommunications: Point-to-point and mobile systems. In *Electronics Engineers' Handbook* (D. G. Fink and D. Christiansen, eds.) 2nd ed., Sect. 22, pp. 22-46–22-62. McGraw-Hill, New York, 1982.

[CH81] T. L. Chang, Suppression of limit cycles in digital filters designed with one magnitude-truncation quantizer. *IEEE Trans. Circuits Syst.* **CAS-28**(2), 107–111 (1981).

[CO82] G. D. Covert, A 32 point monolithic FFT processor chip. *Proc. IEEE Int. Conf. Acoust., Speech, Signal Process., Paris, 1982*, pp. 1081–1083 (1982).

[CU85] R. H. Cushman, Third generation DSPs put advanced functions on chip. *EDN,* **30,** 16, pp. 58–69 (July, 1985).

[DU80] D. L. Dutweiler and Y. S. Chen, A single chip VLSI echo canceller. *Bell Syst. Tech. J.* **59,** 149–160 (1980).

[EL83] J. Eldon, A 32-bit floating point registered arithmetic logic unit. *Proc. IEEE Int. Conf. Acoust., Speech, Signal Process., Boston, 1983*, pp. 943–946 (1983).

[FR78] S. L. Freeny et al., Some applications of digital signal processing in telecommunications. In *Applications of Digital Signal Processing* (A. V. Oppenheimer, ed.), pp. 1–28. Prentice-Hall, Englewood Cliffs, NJ, 1978.

[FU84] Fujitsu, *DSP MB8764 Hardware Manual.* Fujitsu, San Jose, CA, 1984.

[GO77] A. A. Goldberg, PCM NTSC television characteristics. In *Digital Video*, pp. 12–16. Society of Motion Picture and Television Engineers, Inc., Scarsdale, NY, 1977.

[HE83] D. E. Heller, Decomposition of recursive filters for linear systolic arrays. *Proc. SPIE—Int. Soc. Opt. Eng.* **431,** 55–59 (1983).

[HI84] Hitachi, *Digital Signal Processor HD61810B User's Manual.* Hitachi, 1984.

[IC85] IC Master, Hearst Business Communications, Inc., Garden City, NY, 1985.

[JA68] L. B. Jackson et al., An approach to the implementation of digital filters. *IEEE Trans. Audio, Electroacsout.* **AU-16,** 413–421 (1968).

[KA81] A. Kanemasa et al., An LSI chip set for DSP hardware implementation. *Proc. IEEE Int. Conf. Acoust., Speech, Signal Process., Atlanta, Georgia, 1981*, pp. 644–647 (1981).

[KE85] R. N. Kershaw et al., A programmable digital signal processor with 32b floating point arithmetic. *Proc. Int. Solid State Circuits Conf., New York, 1985*, pp. 92–93 (1985).

[KU80] H. T. Kung and C. L. Leiserson, Algorithms for VLSI processor arrays. In *Introduction to VLSI Systems* (C. Mead and L. Conway, eds.), Sect. 8.3, pp. 271–292. Addison-Wesley, Reading, MA, 1980.

[KU81] H. T. Kung, L. M. Ruane, and D. W. L. Yen, A two-level pipelined systolic array for convolutions. In *VLSI Systems and Computations* (H. T. Kung, B. Sproull, and G. Steele, eds.), pp. 255–264. Computer Science Press, Rockville, MD, 1981.

[KU82] H. T. Kung, Why systolic architectures. *Computer* **15**(1), 37–46 (1982).

[LI79] Y. C. Lim and A. G. Constantinides, Linear phase FIR digital filter without multipliers. *Proc. IEEE Int. Symp. Circuits Syst. 1979*, pp. 185–188 (1979).

[LI83] Y. C. Lim and S. R. Parker, FIR filter design over a discrete powers-of-two coefficient space. *Proc. IEEE Int. Symp. Circuits Syst. 1983*, pp. 1071–1074 (1983).

[MA83] J. Mavor, M. A. Jack, and P. B. Denyer, *Introduction to MOS LSI DESIGN*. Addison-Wesley, Reading, MA, 1983.

[MA88] S. Magar et al., An application specific DSP chip set for 100 MHz data rate. *Proc. IEEE Int. Conf. Acoust., Speech, Signal Process. 1988*, pp. 1989–1992 (1988).

[MC78] J. H. McClellan and R. J. Purdy, Applications of digital signal processing to radar. In *Applications of Digital Signal Processing* (A. V. Oppenheim, ed.), pp. 239–329. Prentice-Hall, Englewood Cliffs, NJ, 1978.

[ME80] C. Mead and L. Conway, *Introduction to VLSI Systems*. Addison-Wesley, Reading, MA, 1980.

[MO78] A. L. Moyer et al., A highly parallel processor for matrix computations. *Jt. Autom. Control Conf., 1978* (1978).

[MU81] D. C. Munson, Jr. and B. Liu, Narrowband recursive filters with error spectrum shaping. *IEEE Trans. Circuits Syst.* **CAS-28**(2), 160–163 (1981).

[NE82] NEC Electronics USA, Inc., *D7720 Digital Signal Processor*. NEC Electronics USA, Inc., Mountain View, CA, 1982.

[NE83] J. Newkirk and R. Mathews, *The VLSI Designer's Library*. Addison-Wesley, Reading, MA, 1983.

[NE85] NEC Electronics USA, Inc., *Advanced Signal Processor D77230*. NEC Electronics USA, Inc., Mountain View, CA, 1985.

[PA84] C. R. Panek and S. F. Kator, Custom digital filters for dynamic signal analysis. *Hewlett-Packard J.* **35**, 12, December, pp. 28–36 (1984).

[PE74] A Peled and B. Liu, A new hardware realization of digital filters. *IEEE Trans. Acoust., Speech, Signal Process.* **ASSP-22**(6), 456–462 (1974).

[PE76] A. Peled and B. Liu, *Digital Signal Processing: Theory, Design, and Implementation*. Wiley, New York, 1976.

[PO74] N. R. Powell and J. M. Irwin, Flexible high speed FFT with MOS monolithic chips. *Asilomar Conf. Circuits, Syst., Comput. 8th, 1974*, pp. 524–528 (1974).

[PO75] N. R. Powell and J. M. Irwin, A MOS monolithic chip for high speed flexible FFT microprocessors. *Proc. IEEE Int. Solid State Circuits Conf., 1975*, pp. 18–19 (1975).

[PO76] N. R. Powell and J. M. Irwin, Integrated functional processors in signal processing. *Proc. EASCON, 1976*, pp. 92A–92D (1976).

[PO81] N. R. Powell, Functional parallelism in VLSI systems and computations. In *VLSI Systems and Computations* (H. T. Kung, B. Sproull, and G. Steele, eds.), pp. 41–49. Computer Science Press, Rockville, MD, 1981.

[SC84] D. A. Schwartz and T. P. Barnwell, III, A graph theoretic technique for the generation of systolic implementation for shift-invariant flow graphs. *Proc. IEEE Int. Conf. Acoust., Speech, Signal Process., San Diego, California, 1984*, pp. 8.3.1–8.3.4 (1984).

[SC85a] D. A. Schwartz and T. P. Barnwell, III, Cyclo-static multiprocessor scheduling for the optimal realization of shift-invariant flow graphs. *Proc. IEEE Int. Conf. Acoust., Speech, Signal Process., Tampa, Florida, 1985*, pp. 1384–1387 (1985).

[SC85b] L. Schirm and R. de Koyer, GOPS^{TM} digital signal processor provides 20 MHz analog bandwidth. *Proc. IEEE Int. Conf. Acoust., Speech, Signal Process., Tampa, Florida, 1985*, pp. 1625–1628 (1985).

[SH83] H. Shimbo et al., Automatic ghost equalizer with digital signal processing. *Proc. IEEE. Symp. Circuits Syst., Newport Beach, California, 1983*, pp. 180–183 (1983).

[ST75] T. G. Stockham, Jr. et al., Blind deconvolution through digital signal processing. *Proc. IEEE* **63**(4), 678–692 (1975).

[SW83] E. E. Swartzlander, L. S. Lome, and G. Hallnor, Digital signal processing with VLSI technology. *Proc. IEEE Int. Conf. Acoust., Speech, Signal Process., Boston, 1983*, pp. 951–954 (1983).

[SW84] E. E. Swartzlander and G. Hallnor, Fast transform processor implementation. *Proc. IEEE Int. Conf. Acoust., Speech, Signal Process., San Diego, California, 1984*, pp. 25A.5.1–25A.5.4 (1984).

[TE83] Texas Instruments, Inc., *TMS32010 User's Guide*. Texas Instruments, Inc., Houston, TX, 1983.

[TE85] Texas Instruments, Inc., *TMS32020 User's Guide*. Texas Instruments, Inc., Houston, TX, 1985.

[TE88a] Texas Instruments, Inc., *TMS320C30 User's Guide*. Texas Instruments, Inc., Houston, TX, 1988.

[TE88b] Tektronix, Inc., *The Tek 3052 - Dc to 10 MHz Fastest Real Time Bandwidth - 2 MHz*, 3052 Digital Spectrum Analyzer Data Sheet. Tektronix, Inc., Beaverton, OR, 1988.

[TH76] T. Thong and B. Liu, A recursive digital filter using DPCM. *IEEE Trans. Commun.* **COM-24**(1), 2–11 (1976).

[TH77] T. Thong and B. Liu, Error spectrum shaping in narrow band filters. *IEEE Trans. Acoust., Speech, Signal Process.* **ASSP-25**(2), 200–203 (1977).

[TH78] T. Thong, A new sum of products implementation for digital signal processing. *IEEE Trans. Circuits Syst.* **CAS-25**(1), 27–31 (1978).

[TH82] T. Thong and R. G. Sparkes, A building block for digital signal processing: The digital operational amplifier. *Proc. IEEE Circuits Comput., New York, 1982*, pp. 360–362 (1982).

[TH83] T. Thong, Reconstruction of color images for a line sequential chroma system. *Proc. IEEE Int. Symp. Circuits Syst., Newport Beach, California, 1983*, pp. 184–185 (1983).

[TH85] T. Thong, Frequency domain analysis of two-pass rotation algorithm. *Proc. IEEE Int. Conf. Acoust., Speech, Signal Process., Tampa, Florida, 1985*, pp. 1333–1336 (1985).

[WA82] S. Waser and M. J. Flynn, *Introduction to Arithmetic for Digital Systems Designers*. Holt, Rinehart, & Winston, New York, 1982.

[WE82] Western Electric, *F-61329A IC Data Sheet*. Western Electric, 1982.

[WO75] L. C. Wood and S. Treitel, Seismic signal processing. *Proc. IEEE* **63**(4), 649–661 (1975).

[WU72] Y. S. Wu, Architectural considerations of a signal processor under microprogram control. *AFIPS Conf. Proc.* **40**, 675–683 (1972); also in *Digital Signal Computers and Processors* (A. C. Salazar ed.), pp. 84–92. IEEE Press, New York, 1977.

[YA83] Y. Yasumoto et al., Digital video NTSC and PAL signal processor for VLSI. *Proc. IEEE Int. Symp. Circuits Syst., Newport Beach, California, 1983*, pp. 172–175 (1983).

12 Software Considerations

JALIL FADAVI-ARDEKANI and KALYAN MONDAL

AT&T Bell Laboratories
Allentown, Pennsylvania

12-1 INTRODUCTION

The digital filtering algorithms described in Chapters 3, 4, 8, 9, and 13–16 can be implemented in various ways depending on the application. In the hardware approach, discussed in the previous chapter, the aim is to actually design and construct the filter using digital circuitry. The ultimate goal in this case is design and implementation of a special purpose processor that would be more or less committed to a specific signal processing application. This processor can take the form of a specially designed VLSI chip or a system constructed using commercially available IC modules, such as DSP chips or, in some cases, multipliers, adders, shift registers, and so on. In the software approach, considered in this chapter, the end product is most often a computer program for a general purpose computer or minicomputer. Other computing media include multiprocessor systems such as the supercomputers or special purpose processors (DSP chips) with programmable characteristics.

The signal processing algorithms serving as a basis for the computer programs can be classified broadly into three groups: (1) nonrecursive algorithms, (2) recursive algorithms, and (3) fast transform algorithms. This chapter provides a review of each of these issues by considering the implementation of a finite impulse response (FIR) digital filter, an infinite impulse response (IIR) digital filter, and fast Fourier transform algorithms.

12-1-1 Basic Issues for the Implementation of a Digital Filter

As described in Chapters 5 and 6, the implementation of a digital filter begins with the selection of an appropriate filter structure realizing the specified transfer function. Next, the time-domain characterization of the realized structure is determined by developing the relations between the input and output signal variables and cer-

Handbook for Digital Signal Processing, Edited by Sanjit K. Mitra and James F. Kaiser.
ISBN 0-471-61995-7 © 1993 John Wiley & Sons, Inc.

tain internal signal variables. A software (or hardware) implementation of these time-domain relations, given as difference equations, yields the desired result. As pointed out in these chapters, there are many different structures realizing a digital transfer function, and the structure chosen for the implementation of the specified transfer function is usually decided on the basis of issues such as cost, speed of operation, and effects of finite wordlengths.

For a FIR digital filter of length N, the transfer function is of the form

$$H(z) = \sum_{n=0}^{N-1} h[n] z^{-n}, \tag{12.1}$$

where $\{h[n]\}$ is the impulse response of the filter. The direct form realization of the above transfer function, shown in Figure 12-1 for $N = 3$, corresponds to the implementation of the convolution sum:

$$y[n] = \sum_{k=0}^{N-1} h[k] u[n - k], \tag{12.2}$$

where $\{u[n]\}$ and $\{y[n]\}$ denote, respectively, the input and output sequences. This structure is typically employed in the implementation of a FIR filter in most applications.

The IIR digital filter of interest in this book is characterized by a transfer function of the form

$$H(z) = \frac{\sum_{k=0}^{N} q_k z^{-k}}{1 + \sum_{k=1}^{N} d_k z^{-k}} \tag{12.3}$$

or, equivalently, by the difference equation given by

$$y[n] = \sum_{k=0}^{N} q_k u[n - k] - \sum_{k=1}^{N} d_k y[n - k], \tag{12.4}$$

where $\{u[n]\}$ and $\{y[n]\}$ denote, respectively, the input and output sequences, and N denotes the order of the filter. As indicated in Chapter 3, a realization based on Eq. (12.4) is known as the direct form I structure.

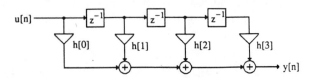

FIGURE 12-1   Direct form third-order ($N = 3$) FIR structure.

Because of the unavoidable finite wordlength effects, an IIR digital filter of order higher than two is always implemented as a cascade of second-order sections and, possibly, one first-order section. To this end, the transfer function of Eq. (12.3) is expressed in the form

$$H(z) = \prod_{i=1}^{K} \frac{\beta_{0i} + \beta_{1i}z^{-1} + \beta_{2i}z^{-2}}{1 - \alpha_{1i}z^{-1} - \alpha_{2i}z^{-2}},$$

(12.5)

where $K = \lceil N/2 \rceil$ with $\lceil x \rceil$ denoting the closest integer greater than or equal to x. If N is odd, one of the second-order sections degenerates to a first-order one. The ith factor in the above expression is given by

$$H_i(z) = \frac{Y_i(z)}{U_i z} = \frac{\beta_{0i} + \beta_{1i}z^{-1} + \beta_{2i}z^{-2}}{1 - \alpha_{1i}z^{-1} - \alpha_{2i}z^{-2}},$$

(12.6)

where $Y_K(z) = Y(z)$, $U_i(z) = Y_{i-1}(z)$, and $U_1(z) = U(z)$. The corresponding difference equation representation of the ith section is given by

$$y_i[n] = \beta_{0i}u_i[n] + \beta_{1i}u_i[n-1] + \beta_{2i}u_i[n-2]$$
$$+ \alpha_{1i}y_i[n-1] + \alpha_{2i}y_i[n-2]$$

(12.7)

A direct form II realization of the above is shown in Figure 12-2. The difference equation for direct form II can be rewritten as

$$w_i[n] = u_i[n] + \alpha_{1i}w_i[n-1] + \alpha_{2i}w_i[n-2],$$

(12.8a)

$$y_i[n] = \beta_{0i}w_i[n] + \beta_{1i}w_i[n-1] + \beta_{2i}w_i[n-2].$$

(12.8b)

The IIR digital filter can also be implemented in various other forms as discussed in Chapters 5 and 6. In this chapter we illustrate the implementation of an IIR filter based on the cascade of second-order structures described by Eq. (12.7) or (12.8).

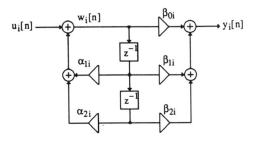

FIGURE 12-2   A second-order IIR direct form II structure.

12-1-2 Basic Issues in the FFT Implementation

There are various different FFT algorithms as discussed in Chapter 8. The two most popular ones are the decimation-in-time (DIT) and the decimation-in-frequency (DIF) FFT algorithms. In the radix-2 based FFT algorithm, the basic component is a 2-point DFT called a *butterfly*, and the computation of an N-point DFT with $N = 2^\nu$ involves the processing of the data through a total of ν stages with $N/2$ butterflies per stage leading to a total of $(N/2)\log_2 N$ butterflies. For example, in the radix-2 DIT algorithm, the butterfly computational scheme is as illustrated in Figure 12-3, where the signal variables in general are complex quantities. The input–output relations of the basic butterfly are given as

$$A[m] = A[m-1] + W_N^r B[m-1], \tag{12.9a}$$

$$B[m] = A[m-1] - W_N^r B[m-1], \tag{12.9b}$$

where $W_N^r = e^{-j(2\pi r/N)}$ is called the *twiddle factor*. The above equations can be rewritten as

$$\operatorname{Re} A[m] = \operatorname{Re} A[m-1] + \{\operatorname{Re} B[m-1]\}\cos\left(\frac{2\pi r}{N}\right)$$

$$+ \{\operatorname{Im} B[m-1]\}\sin\left(\frac{2\pi r}{N}\right), \tag{12.10a}$$

$$\operatorname{Im} A[m] = \operatorname{Im} A[m-1] - \{\operatorname{Re} B[m-1]\}\sin\left(\frac{2\pi r}{N}\right)$$

$$+ \{\operatorname{Im} B[m-1]\}\cos\left(\frac{2\pi r}{N}\right), \tag{12.10b}$$

$$\operatorname{Re} B[m] = \operatorname{Re} A[m-1] - \{\operatorname{Re} B[m-1]\}\cos\left(\frac{2\pi r}{N}\right)$$

$$- \{\operatorname{Im} B[m-1]\}\sin\left(\frac{2\pi r}{N}\right), \tag{12.10c}$$

$$\operatorname{Im} B[m] = \operatorname{Im} A[m-1] + \{\operatorname{Re} B[m-1]\}\sin\left(\frac{2\pi r}{N}\right)$$

$$- \{\operatorname{Im} B[m-1]\}\cos\left(\frac{2\pi r}{N}\right). \tag{12.10d}$$

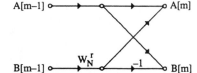

FIGURE 12-3 Flow-graph of the butterfly computation in the decimation-in-time FFT algorithm.

To avoid the possibility of overflow, each stage of the FFT is scaled down by a factor of 2. Thus the overall output gets scaled down by $N = 2^\nu$ for ν stages of an N-point FFT. Even with this scaling, overflow may still occur because of the maximum magnitude value for complex input data.

Another interesting aspect of the FFT computation, as pointed out in Chapter 8, is that the input time-domain samples $\{x[n]\}$ when processed in their natural sequential order lead to DFT samples $\{X[k]\}$ at the output in bit-reversed order. Alternately, to compute the output samples $\{X[k]\}$ in sequential order, the input samples $\{x[n]\}$ must be processed in bit-reversed order. The bit-reversal operation can be performed in-place, which means that in addition to N complex data registers, the use of one additional temporary data memory location is needed for the FFT computation.

The inverse DFT computation is almost identical to the DFT calculation except for the difference in sign of the twiddle factor and a division by a scalar N. As a result, an identical FFT software routine can be used for calculating both DFT and IDFT.

12-2 IMPLEMENTATION ON A GENERAL PURPOSE COMPUTER[1]

Until the mid-1980s, the implementation of digital signal processing algorithms was almost exclusively carried out on a general purpose computer. Since then, DSP chips are increasingly being used for implementation in most applications oriented toward specific signal processing tasks. However, there are still various situations where a general purpose computer implementation is preferable for flexibility and other reasons. Such implementations, in general, do not provide real-time signal processing capabilities and are used for simulation purposes or in applications where real-time computing is not of major concern.

An important consideration in a general purpose computer implementation is the type of programming language to be used, which is dictated mostly by the computer being employed. The three most widely used languages are BASIC, FORTRAN, and C. The objective of this section is to outline the basic ideas involved in programming a computer to simulate a digital filtering algorithm. Without any loss of generality, the language selected for illustration is FORTRAN. Moreover, attention is focused here on FIR and IIR digital filter simulations. No attempt has been made to optimize the FORTRAN programs included here for illustration purposes. Chapter 8 contains the program listings of fast Fourier transform algorithm implementations.

12-2-1 Implementation of Finite Impulse Response Filters

As indicated earlier, for most applications, the software implementation of a FIR filter involves the running of a computer program that simulates the convolution

---

[1]This section has been contributed by Dr. Ognjan V. Shentov of Pennie & Edmonds Intellectual Proprietary Law Firm, New York, NY.

sum of Eq. (12.2) or its equivalent and computing the output samples $y[n]$ for a specified input sequence $\{x[n]\}$, starting at a given value $n = n_0$ (typically $n_0 = 0$). A FORTRAN subroutine implementing the convolution sum of Eq. (12.2) is shown in Figure 12-4.

In many applications, digital filters with linear-phase responses are preferred. FIR filters with exact linear phase can easily be designed, whereas in the IIR case the linear-phase property can only be met approximately. As indicated in Section 4-3-1, linear-phase FIR filters exhibit a symmetry in their impulse response coefficients. Four types of coefficient symmetry can be defined depending on the filter length N:

$$\text{Type I} \quad h[n] = h[N - 1 - n] \quad \text{with } N \text{ odd,} \tag{12.11a}$$

$$\text{Type II} \quad h[n] = h[N - 1 - n] \quad \text{with } N \text{ even,} \tag{12.11b}$$

$$\text{Type III} \quad h[n] = -h[N - 1 - n] \quad \text{with } N \text{ odd,} \tag{12.11c}$$

$$\text{Type IV} \quad h[n] = -h[N - 1 - n] \quad \text{with } N \text{ even.} \tag{12.11d}$$

```
      subroutine firdir(x,n,h,nh,hx,err)
c-
c-This routine filters a n-point input sequence x with a FIR
c-filter h of length nh.
c-The transfer function of the filter is given by
c-    H(z) = h(0)+h(1)*z**(-1)+...+h(nh-1)*z**(-nh+1)
c-The vector hx of length nh contains the past values
c-of the input x (most current sample first). Output
c-overwrites the input and is of length n+nh-1.   err is a
c-flag indicating an error if the output exceeds 1.0e10.

      dimension x(0:n+nh-2), h(0:nh-1), hx(0:nh-1)
      err = 1
      do 30 i = 0,n+nh-2
         if(i.lt.n-1) then
         hx(0)  = x(i)
         else
         hx(0)  = 0.0
         endif
         x(i) = 0.0
         do 10 j = 0,nh-1
           x(i) = x(i) + h(j)*hx(j)
10         continue

         if (abs(x(i)).gt.1.0e10) return
            do 20 k = nh,1,-1
               hx(k)  = hx(k-1)
20             continue
30       continue

      err = 0
      return
      end
```

FIGURE 12-4   The FORTRAN program code "firdir" for the simulation of a direct form FIR structure.

The above symmetry can be exploited to reduce the number of multiplications required per output sample to about one-half of that needed in the direct implementation of the filter using Eq. (12.7).[2] Figure 12-5 shows the program code to implement a linear-phase FIR filter where the symmetry of Eq. (12.11) has been exploited to minimize the number of multiplications.

12-2-2 Implementation of an Infinite Impulse Response Filter

An implementation of an IIR digital filter on a computer involves the execution of a computer program simulating either the difference equation of Eq. (12.4) or its equivalent, and computing the output sequence $\{y[n]\}$ starting at a given value of $n = n_0$ (typically $n_0 = 0$), for a series of values of the input sequence $\{x[n]\}$ beginning at n_0 and a given set of initial conditions $\{y[n], n = n_0 - 1, n_0 - 2,$ $\ldots, n_0 - N\}$. (The input samples prior to $x[n_0 - N]$ are assumed to be zero.)

As indicated earlier, a direct form realization of the overall transfer function is seldom used in practice. Instead, the filter is implemented as a cascade of second-order IIR sections and, possibly, one first-order IIR section. A FORTRAN subroutine implementing the difference equation of Eq. (12.4) is given in Figure 12-6. A program to implement a higher-order IIR filter using a cascade of second-order sections is given in Figure 12-7.

As described in Chapters 6 and 7, there are various other structures realizing an IIR filter. Computer simulations of any of these structures can be developed following similar lines. For example, the program "biquad" of Figure 12-6 with different coefficient vectors can be used to implement a parallel form IIR structure.

12-2-3 FIR Filtering Using FFT

In Section 12-2-1, it has tacitly been assumed that the input data size M and the FIR filter length N are relatively short. In this case the convolution sum of Eq. (12.2) can be evaluated explicitly and the filtering process still has acceptable computational complexity. For filter lengths of 20 or larger, however, the explicit evaluation of the convolution sum is computationally inefficient and it is preferable to use other methods.

One approach to the evaluation of the convolution sum is via DFT, which in turn can be implemented efficiently using FFT algorithms. As pointed out in Section 2-12, the N-point inverse DFT of the product of the N-point DFTs of two sequences is of length N and is equivalent to the sequence obtained by a circular convolution. On the other hand, the linear convolution of a sequence of length N and a sequence of length M results in a sequence of length $N + M - 1$. To make use of the DFT-based approach for computing the linear convolution, the two

---

[2]The number of multiplications is $N/2$ for Types II and IV, and $(N + 1)/2$ for Types I and III FIR filters.

```
c-
      subroutine symfir(x,n,h,nh,hx,sym,err)
c-This routine filters an n-point input sequence x with a
c-symmetric FIR filter h of length nh. Parameter sym indicates
c-the type of symmetry: even symmetry - 0, odd symmetry - 1.
c-The user supplies half the coefficient values (0:nh/2) and
c-the center value in the odd length case. The vector hx of length
c-nh contains the past values of the input x. err is a flag
c-indicating an error if the response value exceeds 1.0e10.

      dimension x(0:n+nh-2), h(0:nh-1), hx(0:nh-1)
      err = 1
      do 60 i = 0,n+nh-2
        if(i.lt.n-1) then
        hx(0) = x(i)
        else
        hx(0) = 0.0
        endif

        x(i) = 0.0
        if ((nh/2)*2.eq.nh) then
          if (sym.eq.0.0) then
            do 10 j = 0,nh/2-1
            x(i) = x(i) + h(j)*(hx(j)+hx(nh-1-j))
10          continue
          else
            do 20 j = 0,nh/2-1
            x(i) = x(i) + h(j)*(hx(j)-hx(nh-1-j))
20          continue
          endif

        else.

          if (sym.eq.0.0) then
            do 30 j = 0,nh/2-1
            x(i) = x(i) + h(j)*(hx(j)+hx(nh-1-j))
30          continue
            x(i) = x(i) + h(nh/2)*hx(nh/2)
          else
            do 40 j = 0,nh/2-1
            x(i) = x(i) + h(j)*(hx(j)-hx(nh-1-j))
40          continue
            x(i) = x(i) + h(nh/2)*hx(nh/2)
          endif

        if (abs(x(i)).gt.1.0e10) return
          do 50 k = nh,1,-1
            hx(k) = hx(k-1)
50          continue

60      continue
        error = 0
        return
        end
```

FIGURE 12-5 The FORTRAN program code ''symfir'' for the simulation of a direct form
linear-phase FIR structure.

```
c-
      subroutine biquad(x,n,p,q,hx,hy,err)
c-This subroutine filters an input sequence x of length n
c-with a second order IIR filter section. The numerator and
c-the denominator coefficients are in vectors q and p,
c-respectively. The user supplies the initial conditions:
c-hx - the input past values; hy - the output past values.
c-The transfer function is of the form
c-            q(0)+q(1)*z**(-1)+q(2)*z**(-2)
c-    h(z) = ----------------------------
c-             1+p(1)*z**(-1)+p(2)*z**(-2)
c-The input vector is overwritten by the output.
c-err=0 no error. routine checks if the output exceeds 1.0e10
c-and exits if true.

      dimension x(0:n-1),q(0:2),p(2),hx(0:2),hy(2)
      err=1
      do 10 i=0,n-1
        hx(0)=x(i)
        x(i)=q(0)*hx(0)
        do 20 j=1,2
          x(i)=x(i)+q(j)*hx(j)-p(j)*hy(j)
20      continue
        if(abs(x(i)).gt.1.0e10) return
        do 40 k=2,1,-1
          hx(k)=hx(k-1)
40      continue
        hy(2)=hy(1)
        hy(1)=x(i)
10    continue
      err=0
      return
      end
```

FIGURE 12-6 The FORTRAN program code "biquad" for the simulation of an IIR direct form II second-order structure.

sequences therefore should be extended to length $L \geq N + M - 1$ each by padding with the appropriate number of zero-valued samples. An L-point inverse DFT of the product of the L-point DFTs of these two extended sequences results in a linear convolution. Note that for most FFT algorithms, the DFT length L must also be a power of 2; that is, $L = 2^v$. Hence in implementing the above approach, L must be selected to satisfy the constraint $L = 2^v \geq N + M - 1$. The program code "fastconv" implementing the DFT-based FIR filtering is provided in Figure 12-8.

A modification of the fast convolution routine "fastconv" of Figure 12-8 can also be used to filter very long or infinite length input sequences. To this end the input sequence is first split into a set of short, length-N sequences. Fast convolution is then performed on each input segment with the given FIR filter. The resulting output segments are then appropriately recombined to give the final output. To this end, two basic algorithms have been proposed [OP89]. Figure 12-9 presents a program "fftfilt" implementing one such method, called the *overlap-add approach*.

```
c-
      subroutine iircas(x,n,p,q,ns,hx,hy,err)
c-IIR filtering of an input sequence x of length n.
c-The filter consists of ns second - order filter sections
c-with numerator and denominator polynomials q and p.
c-The user supplies the initial conditions:
c-hx - the input past values;
c-hy - output past values.
c-For a k - stage transfer function
c-          ___  q(0,k)+q(1,k)*z**(-1)+q(2,k)*z**(-2)
c-      h(z) =| | -----------------------------------
c-          k        1.0+p(1,k)*z**(-1)+p(2,k)*z**(-2)
c-output vector is of same length and replaces the input
c-vector. err=0  error condition - checks if output exceeds
c-1.0e6 and returns the filter section number in which the
c-error occurred.

             dimension x(0:n1),p(2,ns),q(0:2,ns),hx(0:2,ns),hy(2,ns)
             do 10 i=1,ns
             err=1
               do 20 j=0,n-1
                 hx(0,i)=x(j)
                 x(j)=q(0,i)*hx(0,i)
                 do 30 m=1,2
                   x(j)=x(j)+q(m,i)*hx(m,i)-p(m,i)*hy(m,i)
30             continue
                 if(abs(x(j)).gt.1.e6) return
                 hx(2,i) = hx(1,i)
                 hx(1,i) = hx(0,i)
                 hy(2,i) = hx(1,i)
                 hy(1,i) = x(j)

20           continue
10           continue
             err=0
             return
             end
```

FIGURE 12-7   The FORTRAN program code "iircas" for the simulation of an IIR filter as a cascade of direct form II second-order structures.

12-3   PARALLEL COMPUTER IMPLEMENTATION[3]

The computational requirements for the processing of wideband signals often exceed the capability of current digital signal processors. While VLSI systolic arrays can provide a higher throughput, their inflexibility in the implementation of complex algorithms limits their use in many applications. Thus system implementation methods that allow wideband signal processing as well as programmability in software are often needed. The use of parallel computer architecture is the most promising approach for this purpose. Computers with highly pipelined functional units, usually called the pipelined vector processors, and systems composed of a number of identical processing units, multiprocessors, are frequently used in many com-

[3]This section has been contributed by Dr. Wonyong Sung of the Seoul National University, Seoul, Korea.

```
c-
      subroutine fastconv(x,n,h,err)
c-The subroutine implements a fast convolution of the sequences
c-x and h via FFT and IFFT computations of length n=2**k.
c-It assumes a complex input with the real and imaginary parts
c-stored in consecutive memory locations.
c-The trailing ends of both sequences should be padded with
c-zero elements to avoid wraparound error.  Flag
c-err is an error indicating improper zero-padding.
c-FFT/IFFT computation is done using the subroutine
c- "fourier" with usage fourier(data,length,sgn), where sgn=1
c- for a forward transform, sgn=-1 for an inverse transform.

      dimension x(0:2*n), h(0:2*n)
      complex tmp
      err = 1
      tst = n
c-test zero padding
      do 10 i=2*(n-1),0,-1
         if(x(i).ne.0.0) go to 5
10       continue
5        do 20 j=2*(n-1),0,-1
            if(h(j).ne.0.0) go to 15
20          continue
15          if(n-i/2+1.le.j/2-1) go to 3
c-call forward FFT transform for both sequences
         call fourier(x,n,1)
         call fourier(h,n,1)
         do 30 k = 0,n-1
         tmp = cmplx(x(2*k),x(2*k+1))*cmplx(h(2*k),h(2*k+1))
         x(2*k) = real(tmp)/n
         x(2*k+1) = aimag(tmp)/n
30       continue
c-call inverse FFT transform to obtain the result.
         call fourier(x,n,-1)
         err=0
3        return
         end
```

FIGURE 12-8 The FORTRAN program code "fastconv" for the implementation of linear convolution of two finite-length sequences via DFT.

puting applications. Pipelined vector processors reduce their cycle time by employing a pipelining technique in the design of functional units. A substantial speed improvement is possible without a large hardware overhead. Currently, the pipelined vector processor architecture is widely used in Class VI supercomputers, such as Cray-1, Cray X-MP, Cray-2, and Cyber 205. It appears that this architecture will be adopted in the near future to the design of digital signal processors. Another parallel computer architecture, the multiprocessor system, is becoming popular as device costs drop due to the advancement of VLSI technology. Single instruction multiple data (SIMD) type systems, such as the Massively Parallel Processor (MPP) system, have been used extensively for image signal processing applications [BA80]. At present, multiple instruction multiple data (MIMD) type multiprocessor systems are of interest because they can readily be built using single-chip signal processors [AG87; SU90].

Although the potential processing speed is increased by using the parallel computer systems, not all algorithms can be implemented efficiently. The algorithms

```
c-
      subroutine fftfilt(h,nh,x,nx,y,nfft,err)
c-This subroutine filters an input sequence x of length nx
c-with a filter sequence h of length nh using the overlap-add
c-method with specified FFT length nfft=2**k.
c-The output is stored in an array y of length nx+nh-1.
c-The routine uses division of the input x into nonoverlapping
c-sections of length nfft-nh+1, padded with zeros to avoid
c-wraparound error. The DFT calculations are done with a
c-routine "fourier(data,length,sgn)" where sgn=1 is used for
c-forward transform, and sgn=(-1) for the inverse. err is an
c-error flag, indicating improper input data dimensions. All
c-sequences are assumed complex, with real and imaginary parts
c-stored in consecutive memory locations.
      dimension x(0:2*nx), h(0:2*nh), y(0:2*(nx+nh-1))
      dimension hdum(0:1024), xdum(0:1024)
      complex tmp

      err=1
      if (nfft.lt.32.or.nh.gt.nfft) go to 3
      len = nfft-nh+1

      if (nh.lt.nfft) then

      do 10 i = 0, 2*nh-1
      hdum(i) = h(i)
10    continue

      do 20 j = 2*nh,2*nfft-1
      hdum(j) = 0.0
20    continue
      endif

c-computes the DFT of the filter
      call fourier(hdum,nfft,1)

      ibeg = 0
70    if (ibeg.le.2*nx) then
             if(2*nx-1-ibeg.ge.2*len) then
             do 30 i = 0,2*len-1
             xdum(i) = x(ibeg+i)
30           continue
             do 40 j = 2*len,2*nfft-1
             xdum(j) = 0.0
40           continue

             else
             short=2*nx-1-ibeg
             do 35 i = 0,short-1
             xdum(i) = x(ibeg+i)
35           continue

             do 45 j = short,2*nfft-1
             xdum(j) = 0.0
45           continue
             endif

c-computes the DFT of the input segment
      call fourier(xdum,nfft,1)

      do 50 k = 0,nfft-1
      tmp =
      cmplx(xdum(2*k),xdum(2*k+1))*cmplx(hdum(2*k),hdum(2*k+1))
      xdum(2*k) = real(tmp)/nfft
      xdum(2*k+1) = aimag(tmp)/nfft
```

FIGURE 12-9 The FORTRAN program code ''fftfilt'' for the implementation of the overlap-add method.

```
50      continue
c-use IFFT to obtain the desired intermediate output.
        call fourier(xdum,nfft,-1)

c-add the overlapping segments to obtain the desired output
c-sequence.
        yend = min(2*(nx+nh-1),ibeg+2*nfft-1)
        do 60 k = ibeg,yend
           y(k) = y(k) + xdum(k-ibeg)
60         continue
        ibeg = ibeg + 2*len
        go to 70
        endif

        err = 0
3       return
        end

        subroutine fourier(x,nn,isign)
        real*8 wr,wi,wpr,wpi,wtemp,phi
        dimension x(0:2*nn)
        n=2*nn
        j=0
        do 10 i =0,n-1,2
                if(j.gt.i) then
                        tmpr=x(j)
                        tmpi=x(j+1)
                        x(j) = x(i)
                        x(j+1) = x(i+1)
                        x(i) = tmpr
                        x(i+1) = tmpi
                endif
                m=n/2
5               if((m.ge.2).and.(j+1.gt.m)) then
                        j=j-m
                        m = m/2
                go to 5
                endif
                j = j+m
10      continue
        mmax=2

15      if(n.gt.mmax) then
                step=2*mmax
        phi=6.28318530717959d0/(isign*mmax)
                wpr= -2.d0*dsin(0.5d0*phi)**2
                wpi=dsin(phi)
                wr=1.d0
                wi=0.d0
        do 30 m=0,mmax-1,2
                do 20 i = m,n,step
                        j=i+mmax
                tmpr=sngl(wr)*x(j)-sngl(wi)*x(j+1)
                tmpi=sngl(wr)*x(j+1)+sngl(wi)*x(j)
                x(j)=x(i)-tmpr
                x(j+1)=x(i+1)-tmpi
                x(i)=x(i)+tmpr
                x(i+1)=x(i+1)+tmpi
20      continue
        wtemp=wr
        wr=wr*wpr-wi*wpi+wr
        wi=wi*wpr+wtemp*wpi+wi
30      continue
        mmax=step
        go to 15
        endif
        return
        end
```

FIGURE 12-9 (*Continued*)

should have parallel structures to utilize the advantage of these systems. The performance is very much dependent on the structure of an algorithm. Thus it is necessary to employ or develop algorithms that have parallel structures even if they may need a larger number of arithmetic operations. Another problem is the programming technique. We are accustomed to sequential programming, not concurrent programming. In addition, most of the programs available are made for single processor systems. Thus it is necessary to use parallel or vectorizing compilers to aid the development of programs. These compilers detect parallel structures of a program and execute them in parallel. However, it is still the programmer's responsibility that the program should employ parallel algorithms and follow some programming rules that can be acknowledged by the compilers.

Multiprocessor systems can provide nearly unlimited speed-up compared to single processor systems. However, systematic implementation methods are yet to be developed. The implementation methods using pipelined vector processors have been established fairly well since they are employed in most popular supercomputer systems. In this section, the architecture, algorithm, and program development methods using a pipelined vector processor are introduced. Implementation examples for FIR and recursive filters are also presented.

12-3-1 Architecture and Modeling

The performance of an algorithm in a vector processor is dependent on many factors, not just the number of arithmetic operations. As a result, the performance should be evaluated based on the architectural model of a system. The modeling for the pertinent units, which are pipelined functional units and interleaved memory, is now briefly described.

12-3-1-1 Pipelined Functional Units. The characteristics of the pipelined vector processors are determined primarily by the pipelined functional units. A pipelined unit is organized like an assembly line. During each cycle, a new pair of operands arrives at the beginning of the pipeline, and the operands already in the pipeline are moved one step forward, with the computational results being gradually built up in the pipeline. Therefore the operands should have a vector form to utilize the pipelined functional units, and the average computation rate is dependent on the vector length as well as the cycle time. A simple loop in FORTRAN program is

```
C       A SIMPLE FORTRAN LOOP
        DO 10 I = 1, M
10      C(I) = A(I) + B(I)                (12.12)
```

Let τ be the clock period and D represent the fixed start-up clock cycles, which include setting up and pipelining pass-through time. The first output, $C(1)$, comes after a start-up time $D\tau$. And then the subsequent results, $C(2), C(3), \ldots, C(M)$, are obtained at every clock. The time, T_n, to perform an operation on a vector of length n can thus be represented as [HO81]

$$T_n = (D + n)\tau. \tag{12.13}$$

The corresponding computation rate, R_n, for a vector of length n is then given by

$$R = \frac{n}{T_n} = \frac{1}{\left(\dfrac{D}{n} + 1\right)\tau} \quad \text{(operations per second).} \tag{12.14}$$

Note that the computation rate is dependent on the vector length and thus gets higher as the vector length increases. In many cases, it is necessary to reformulate programs for obtaining long vectors. This process is called the *vectorization*. The computation rate R_∞ for a vector of infinite length, called the *maximum performance rate*, is equal to $1/\tau$. The vector length $n_{1/2}$ for which the computation rate is half the maximum performance rate is called the *half-performance length*. From Eq. (12.14), it follows that $n_{1/2}$ is equal to D. The maximum performance rate and the half-performance length are the two most important parameters for defining the performance of a vector processor [HO81]. If the half-performance length is long, the computational rate for a short vector or a scalar is comparatively poor. The half-performance vector length for conventional serial computers is just zero, indicating the same performance for both vector and scalar data. The execution time T_n for processing a vector of length n can alternatively be represented using the above mentioned two parameters:

$$T_n = (n + n_{1/2})\tau = \frac{n + n_{1/2}}{R_\infty}. \tag{12.15}$$

The execution time for a vector add operation as a function of the vector length is shown in Figure 12-10. The corresponding program has been written in the FOR-TRAN language and was executed using one CPU of the Cray X-MP. The result shows that the maximum performance rate R_∞ is about 50 Mflops, and $n_{1/2}$ is about 50.

Another technique that is used for increasing the computation rate is *chaining*. Several functional units can be chained together if the output of the previous functional unit is routed to the input of the next one without waiting for the completion of the first vector instruction. Since the increase of the time due to chaining is not significant, the time for performing a chained vector multiply and add operation can also be simply modeled by Eq. (12.15).

Most of the programs need some scalar operations even after careful vectorization. Vector processors usually have scalar functional units to perform the scalar operations efficiently. In this case, the scalar unit requires several cycles for each output, though the start-up overhead is not needed.

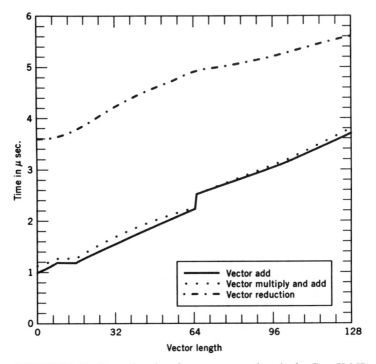

FIGURE 12-10 Execution time for vector operations in the Cray X-MP.

It should be noted that the maximum performance rate, R_∞, is mainly related to the device speed, not to the architecture. Hence the maximum performance rate can be neglected, by normalization, in deriving the computational complexity of an algorithm. The parameter that characterizes a vector processor is only the half-performance length $n_{1/2}$. The execution time for the program shown in Eq. (12.12) can be modeled as $(D + M)$, when we normalize the clock cycle time τ. Thus the average processing time for one add operation is $(D + M)/M$, which becomes 1 when the vector length M is sufficiently large. This example shows that the performance degradation due to the start-up time can be compensated by proper vectorization.

12-3-1-2 Interleaved Memory Organization. Interleaved memory systems, where multiple banks of slow memory are arranged in parallel, are frequently used for increasing the effective memory bandwidth. In the case of the Cray-1, the memory is organized into 16 banks, with a bank cycle time of 4 clock periods. The bank cycle time is the number of cycles needed to complete a memory access operation. If the same memory bank is accessed within a bank cycle time, the memory access is delayed. This is usually called a *bank conflict*.

The FORTRAN program shown below illustrates the memory bank conflicts in the Cray-1:

$$\text{DIMENSION A}(16,64), \text{ B}(16,64)$$

$$\text{DO 1 I} = 1,16$$

$$\text{DO 1 J} = 1,64$$

$$1 \quad \text{A}(I,J) = \text{A}(I,J) + \text{B}(I,J) \quad \quad (12.16)$$

In FORTRAN programs, arrays are stored by columns; hence the array $A(16,64)$ is located as shown in Figure 12-11. Consequently, the row access in the program leads to bank conflicts. However, by modifying the array dimension to $A(17,64)$ and $B(17,64)$, the memory bank conflict can be avoided. The increase in execution speed by this modification is almost 3 to 1 in the Cray-1 [SI84]. Besides the bank conflict, limited bandwidth between memory and functional units frequently degrades the performance.

12-3-2 Parallel Algorithms

Consider a parallel implementation of a finite impulse response (FIR) filtering equation given by the convolution sum of Eq. (12.2). The number of output samples, M, is assumed to be much larger than the filter length N. One implementation approach is to compute the output samples sequentially, one by one. The number of processors that can be used in this method is limited to N and, in this case, all the processors may not be fully utilized or the vector length is limited to N. This approach is called the *inner product method*. The other approach is to compute all the output samples at the same time using M processors or by vectorization of length M in a pipelined vector processor, which is called the *outer product method*.

Bank 1	Bank 2	.....	Bank 16
A(1,64)		.....	A(16,64)
.	.	.....	.
.	.	.....	.
.	.	.....	.
A(1,3)	A(2,3)	.....	A(16,3)
A(1,2)	A(2,2)	.....	A(16,2)
A(1,1)	A(2,1)	.....	A(16,1)

FIGURE 12-11 Allocation of two-dimensional FORTRAN array in an interleaved memory.

The outer product method should be considered the better parallel computation method because it results in a longer vector length. However, the outer product method is not directly applicable to all digital filtering algorithms. It is difficult to employ this approach when the computation sequences for each output sample are dependent on each other.

One frequently encountered situation is when there is a feedback path in an algorithm, such as the Nth order recursive equation

$$y[n] = \sum_{i=1}^{N} a_i y[n-1] + x[n], \qquad n = 1, 2, \ldots, M. \qquad (12.17)$$

In this case, the inner product method can be applied directly even though the vector length N is very short. However, it is not possible to compute the multiple output samples concurrently or vectorize the computation of $y[n]$ because the determination of $y[n]$ is dependent on previous output samples, such as $y[n-1]$. The *dependency relation* should be removed to apply the outer product method in this equation. One simple approach for removing the dependency relation is to expand the equation so that it can be represented only in terms of the initial condition and the input samples. In the first-order case, the following set of equations is obtained by the expansion:

$$y[1] = ay[0] + x[1],$$
$$y[2] = a^2 y[0] + ax[1] + x[2], \qquad\qquad (12.18)$$
$$\vdots$$
$$y[M] = a^M y[0] + a^{M-1} x[1] + \cdots + ax[M-1] + x[M].$$

The output samples are computed independently in the above equations, and therefore the outer product method can be used. However, the number of arithmetic operations is increased a great deal. Therefore it is difficult to increase the processing speed. This example shows that the conversion of a sequential algorithm into a parallel algorithm can be very costly and can eclipse the advantage of parallel processing. For that reason, it is important to minimize the overhead as well as to increase the degree of parallelism.

12-3-2-1 Parallel Block Processing Method. One method for overcoming the dependency relation as well as minimizing the overhead of block processing is the parallel block processing method [SU87a; SU87b]. It is well known that the solution of recursive equations consists of particular and transient solutions. Thus it is possible to compute them separately and add them later to obtain the complete solution. The particular solution is solely dependent on the input; hence it corresponds to the solution with zero initial condition. In contrast, the transient solution is only dependent on the initial condition with zero input.

The parallel block processing method consists of two stages of computation:

x[M]	x[2M]	. . . . .	$x[M^2]$
.	.	. . . . .	.
.	.	. . . . .	.
x[2]	x[M+2]	. . . . .	x[(M−1)M+2]
x[1]	x[M+1]	. . . . .	x[(M−1)M+1]
Block #1	Block #2	. . . . .	Block #M

FIGURE 12-12 Data structure for parallel block processing.

the first stage is for computing the particular solution and the second is for the transient solution. This approach processes M blocks of data, where each block consists of M samples, at a time, and the data structure is shown in Figure 12-12. This method is explained for a first-order recursive equation and then generalized to the case of an Nth order equation.

In the first stage of computation, the particular solution, $y_p[n]$, is computed in parallel between the blocks, which is performed by using Eq. (12.17) with zero initial condition for each block. Since the initial condition for each block is assumed to be zero, the particular solutions for M blocks can be computed in parallel. This means that $y_p[j]$, $y_p[M+j]$, . . . , and $y_p[(M-1)M+j]$ are computed in parallel, although the computational procedure within a particular block, that is, the computation of $y_p[1]$, $y_p[2]$, . . . , and $y_p[M]$, is sequential.

In the second stage, the effect of the nonzero initial conditions for each block is calculated as the transient solution and added to the particular solution. As the initial condition for the first block is known, the output for the first block, $y[1]$, $y[2]$, . . . , $y[M]$, can be represented as follows based on the particular solution of the first block, the weighting factors, and the initial condition $y[0]$:

$$\begin{bmatrix} y[1] \\ y[2] \\ \vdots \\ y[M] \end{bmatrix} = \begin{bmatrix} y_p[1] \\ y_p[2] \\ \vdots \\ y_p[M] \end{bmatrix} + \begin{bmatrix} W[1] \\ W[2] \\ \vdots \\ W[M] \end{bmatrix} y[0]. \qquad (12.19)$$

The weighting factors $W[i]$ can be computed a priori and are simply $[a, a^2, a^3, . . . , a^M]$ for the case of a first-order recursive equation. The main point to note here is that the procedure involves no recurrence relation. It is possible to calculate $y[1]$, $y[2]$, . . . , $y[M]$ in parallel without any dependency relations among them. After computing the output for the first block, the output for the second block can be computed because the initial condition for the second block, which is $y[M]$, has been prepared by the output of the first block. The remaining blocks can be processed in the same way.

The above procedure can easily be generalized for solving Nth order recursive equations by using vector and matrix variables. The number of arithmetic steps for the constant coefficient case is only proportional to the order of the recursive equation and is about 50% larger than that of the sequential computation method [SU87a; SU87b]. The 50% of overhead can easily be compensated by the advantage of vectorization or parallel computation.

12-3-3   Vectorizing Compilers

The Cray FORTRAN Compiler is introduced since there is no widely recognized standardized high-level language for vector processors. The Cray FORTRAN Compiler follows the principle of automatic vectorization, and therefore typical FORTRAN programs can be executed without modification. However, the technique of vectorization is essential to take advantage of vector processors. Some rules for vectorization follows.

Rule 1. Only the innermost DO loop is vectorized. However, an equivalent IF statement forming a loop is not vectorized. For example, the following program can be vectorized:

$$
\begin{aligned}
&\text{DO 1}\quad \text{I} = 1,\ \text{M} \\
&\quad \text{A(I)} = \text{B(I)} * \text{C(I)} \\
&1\quad \text{D(I)} = \text{B(I)} + \text{C(I)}
\end{aligned}
\tag{12.20}
$$

The resulting sequence of the above program in a serial computer is $A(1)$, $D(1)$, $A(2)$, $D(2)$, . . . , and it is different from that obtained in a vector processor, in which case it is $A(1)$, $A(2)$, . . . , $A(M)$, $D(1)$, $D(2)$, . . . , $D(M)$. Thus the vector length can be M. In nested DO loops, it is more efficient to make the innermost DO loop the longest loop.

Rule 2. A DO loop that has a dependency relation cannot be vectorized. The following loop has a dependency because the computation of $Y(I)$ requires the value of the previous result $Y(I - 1)$, which is still in the pipeline.

$$
\begin{aligned}
&\text{C}\quad \text{SIMPLE RECURSIVE FILTERING} \\
&\quad \text{DO 1 I} = 1,\ \text{M} \\
&1\quad \text{Y(I)} = \text{A} * \text{Y(I-1)} + \text{X(I)}
\end{aligned}
\tag{12.21}
$$

Rule 3. A DO loop containing a vector to scalar reduction operation, such as a vector summation operation, is also vectorizable. The following is an example:

$$
\begin{aligned}
&\text{C}\quad \text{AN OPERATION FOR VECTOR SUMMING UP} \\
&\quad \text{DO 1}\quad \text{I} = 1,\ \text{M} \\
&1\quad \text{Y} = \text{Y} + \text{X(I)}
\end{aligned}
\tag{12.22}
$$

However, the half-performance length for it is usually longer than the other vector operations. Thus the computational complexity for the vector reduction can be modeled as

$$t_{\text{red}} = M + n_{1/2,\text{red}}, \tag{12.23}$$

when $n_{1/2,\text{red}}$ denotes the half-performance length for the vector reduction, which is a few times higher than the $n_{1/2}$.

In addition to the above rules, it is also necessary not to use an IF, GOTO, or CALL statement in a loop to be vectorized. Vectorization for some of the programs is machine dependent. For example, the memory reference in Cray-1 should be uniform for vectorization, implying a constant increase or decrease of memory address at the loop. But vectorization of nonuniform memory addressing is possible when the machine has an indirect addressing unit. There are still many limitations in the Cray FORTRAN Compiler. However, the problem for the vectorization of the dependency relation may be regarded as an inherent problem of the parallel computation, not that of the compiler.

12-3-4 Implementation Examples

Implementation examples for outer product based FIR, inner product based recursive, and outer product recursive filterings are now illustrated.

12-3-4-1 Outer Product Based FIR Filtering. A FIR filtering program based on the outer product method is shown below:

```
C       OUTER PRODUCT FIR FILTERING
        DO 1 J = 1, N
        DO 1 I = 1, M
1       Y(I) = Y(I) + H(J) * X(I-J)
```
(12.24)

Here, the vector length M can be very large regardless of the filter length N. Also, the multiply and add operations can be chained. The computation time for one sample can be modeled as

$$t_{\text{outer, FIR}} = N\left(1 + \frac{n_{1/2}}{M}\right) \approx N. \tag{12.25}$$

The second term, $n_{1/2}/M$, can be neglected in most cases because the vector length can be very large. The outer product method can also be applied to the linear-phase FIR filter implementation, where the number of multiplications is reduced by about half. However, there is no significant saving for the linear-phase implementation because the chaining operation cannot be fully utilized.

12-3-4-2 Inner Product Based Recursive Filtering. The equation for the direct form recursive filter is

$$y[n] = \sum_{i=1}^{N} a_i y[n - i] + \sum_{k=1}^{N} b_k x[n - k]. \qquad (12.26)$$

In the equation, the feedforward term can be computed efficiently by using the outer product method used in FIR filtering, for which the number of computation steps is approximately $N + 1$ per output sample. Then $y[n]$ can be computed using the following, inner product based, program. The variable $F(I)$ represents the previously computed feedforward term given by the second term in Eq. (12.26).

```
C       DIRECT FORM RECURSIVE FILTER IN INNER PRODUCT
        DO 1 I = 1, M
        YTMP = F(I)
        DO 2 J = 1, N
2       YTMP = YTMP + A(J) * Y(I-J)
1       YOUT(I) = YTMP                                    (12.27)
```

The vector length in the above equation is limited to the filter order N, and the above operation contains a vector to scalar reduction operation. Therefore it is not efficient in the vector-processor environment unless the filter order is large. The computational complexity can be modeled as

$$t = 2N + 1 + n_{1/2, \text{red}}, \qquad (12.28)$$

which shows that the program needs a large start-up time.

12-3-4-3 Outer Product Based Recursive Filtering. As discussed previously, the outer product implementation of a recursive equation produces a dependency problem. Thus the parallel block processing algorithm should be employed to remove the dependency relation. The first stage of the computation, where the recursive filtering equation is applied separately to each block with the initial conditions for the blocks set to zero, can be programmed as follows. A first-order recursive filtering case has been assumed for simplicity.

```
C       COMPUTATION OF PARTICULAR SOLUTION, Z(I,J)
C       INITIAL CONDITION FOR EACH BLOCK, Z(I,0), IS ZERO
C       A IS THE FIRST ORDER FILTER COEFFICIENT
        DO 1 J = 1, M
        DO 1 I = 1, M
1       Z(I, J) = A * Z(I, J-1) + X(I, J)                (12.29)
```

In the preceding program, the inner loop can be vectorized because there is no dependency relation on index I, which is the block number. The vector length M is not limited by the filter order but, rather, by the amount of data stored at the memory. Hence it is usually possible to make the vector length much larger than $n_{1/2}$. In this case, the computational complexity per sample for the above program can be represented as

$$t_{\text{part, first order}} = \frac{M(M + n_{1/2})}{M^2} \approx 1. \tag{12.30}$$

In the second stage of computation, the transient solutions in a block are computed in parallel just by weighting the initial condition. The corresponding program is shown below. The term $y(M, M)$ represents the initial condition for the first block, which is the last output sample of the previous computation.

```
C   TRANSIENT SOLUTION FOR RECURSIVE FILTERING
C   W(J) IS THE WEIGHTING FACTOR WHOSE VALUE IS
C   [A,A², · · · , Aᴹ]
    DO 1 I = 1, M                                      (12.31)
    IM1 = I - 1
    IF (IM1.EQ.0) IM1 = M
    DO 1 J = 1, M
1   Y(I, J) = Z(I, J) + W(J) * Y(IM1,M)
```

The inner loop can be vectorized with a vector length of M.

The vectorization method can also be extended to the general Nth order recursive filtering equation. The program for the first stage of computation is as follows, where $F(I, J)$ is the previously computed feedforward term:

```
C     PARTICULAR SOLUTION FOR M-TH ORDER FILTER

C     Z(I, J) ARE SET TO ZERO INITIALLY

      DO 1 J = 1, M

      DO 2 K = 1, N

      DO 2 I = 1, M                                    (12.32)

2     Z(I, J) = A(K) * Z(I, J-K) + Z(I, J)

      DO 3 I = 1, M

3     Z(I, J) = Z(I, J) + F(I, J)

1     CONTINUE
```

The weighting factor W, used for the second stage of computation, is now a matrix of size $M \times N$.

The total number of computation cycles for obtaining M^2 samples of the particular solutions is $M(NM + 2Nn_{1/2} + M + n_{1/2})$ for the Nth order case. The computation of the transient solution needs similar computation steps. Also, it takes about $N + 1$ steps per sample to compute the feedforward term by using the outer product method. As a result, assuming that M is much larger than $n_{1/2}$, the computational complexity per sample is given by

$$t_{\text{parallel, REC}} = 3N + 2. \tag{12.33}$$

The beauty of the above result is that it is not a function of $n_{1/2}$. Comparing this result with that of the other implementations, it can be seen that the proposed method is much better unless the filter order is considerably higher than the half-performance length. The proposed method can easily be applied to other structures, such as the parallel or cascade forms, where a similar computational advantage can be obtained. In Figure 12-13, the performance of the outer product based method implemented on a Cray X-MP computer is compared with the direct form inner product, cascade form inner product, and scalar arithmetic based methods. The result shows that there is virtually no start-up overhead in the implementation of the outer product based method.

12-3-5 Concluding Remarks

The implementation of recursive digital filtering algorithms in a vector processor is presented to show the opportunities and difficulties in the use of parallel com-

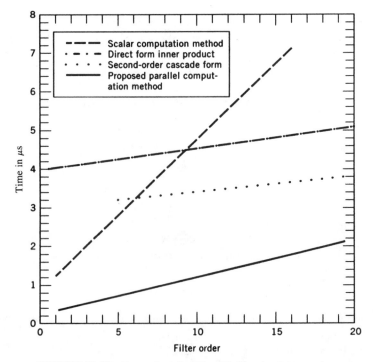

FIGURE 12-13   Execution time for IIR filter implementation.

puter systems for digital signal processing. The results show that conventional implementation methods, which are optimized for a sequential computer, do not show good computational performance in a parallel computer environment. It is only possible to take advantage of parallel computer systems when we employ efficient parallel algorithms and follow the programming rules of the compilers. Implementation examples for adaptive filtering and fast Fourier transform algorithms can be found in Sung and Mitra [SU87b] and Agarwal and Cooley [AG87].

12-4 A REVIEW OF REPRESENTATIVE DSP CHIP PROGRAMMING[4]

In this section, we investigate the structures and methods of programming of several digital signal processor (DSP) integrated circuits. For brevity we have chosen a few of the commonly used DSPs to explain the basic techniques of programming these devices. They include devices from manufacturers such as AT&T, Motorola, and Texas Instruments (TI). To illustrate the difference in programming of these DSPs, we include complete codes for implementing finite and infinite impulse response (FIR and IIR) filters and fast Fourier transform (FFT) algorithm on each of the DSPs under consideration. However, similar methods are also employed in programming DSP chips manufactured by others.

There are two distinct classes of DSPs: fixed point and floating point DSPs. In the first category, we treat TMS320C25 (from TI), WE® DSP16A (from AT&T), and DSP56001 (from Motorola). In the second category, we deal with WE® DSP32C (from AT&T), DSP96002 (from Motorola), and TMS320C30 (from TI).

There are many factors to be considered when choosing a DSP for an application. The choice of fixed-point or floating point is usually dictated by the requirements of the algorithm (such as dynamic range and accuracy). However, a compromise between various other (many times conflicting) requirements of speed, precision, program/data memory space, power consumption, cost, ease of use, and available technical support for the device often leads to the selection of a particular device.

Extensive useful metrics and data for a comparison of various DSPs have been compiled by Shear [SH88]. In this article, two important parameters, namely, the time of execution and total memory requirements for various benchmark algorithms executed on different DSPs, have been collected and contrasted. These data become invalid as soon as a new generation of products is announced by various vendors or improvements are made to the existing products. In order to make a good decision on the capabilities of various chips at any point in time, it is worthwhile to recompute the benchmark results. Also, one needs to focus on the current application and choose a chip that best fits the requirements of the application although the chip may not be the best in all benchmarks.

---

[4]The text in this section on Texas Instruments and Motorola DSPs has been contributed by Dr. Sanjit K. Mitra of the University of California, Santa Barbara.
WE is a trademark of AT&T.

12-4-1 Programming Fixed-Point TMS320C25 DSP Device

The TMS320C25 is an enhanced, pin-for-pin, and object-code upward compatible version of the older part TMS32020. Its salient features are as follows:

- Instruction cycle time: 100 ns
- Low-power CMOS technology with powerdown mode
- 64K Word data memory, 64K word program memory spaces
- 544 × 16-bit on-chip data/program RAM
- 4K words of on-chip masked ROM
- Single-cycle multiply-accumulate operations
- 32-bit ALU with several shift options
- Five address registers with dedicated arithmetic unit
- Eight auxiliary registers with a dedicated arithmetic unit
- Eight-level hardware stack
- Bit-reversed addressing modes for radix-2 FFTs
- Extended-precision arithmetic and adaptive filtering support
- Full-speed operation of MAC/MACD from external memory
- Accumulator carry bit and unsigned multiply for complex arithmetic
- On-chip hardware timer/period registers
- Fully static double-buffered 8/16-bit serial port
- Multiprocessor interface with clock synchronization
- 16-bit Parallel interface for data, program, and I/O access with wait state capability
- T1/G.711 transmission interface
- Concurrent DMA using an extended HOLD operation
- 68-pin PLCC package
- Single 5-volt supply

Architecture of TMS320C25. A block diagram of the TMS320C25 chip [TE90] is shown in Figure 12-14. The Harvard architecture of this chip is based on the use of two memory spaces (data and instruction memory spaces), with each of these spaces containing on-chip and off-chip physical memory blocks. One physical memory block consists of an on-chip RAM of 256 16-bit words (block B0) and it can be allocated to either the data or program memory spaces. There are two other physical memory blocks, namely, block B1 of 256 words and block B2 of 32 words. These two RAMs are always used as data memory. The reason for assigning two RAM blocks to the second memory space is that they have noncontiguous address spaces. There is no (program or coefficients) ROM on chip. But a total of 128K words of off-chip memory address space is provided. External memory (RAM or ROM) can be supplied for both data (total data memory—64K words) and program memory (total program memory—64K words). Since the data and address buses for the corresponding memories are multiplexed, control signals are pro-

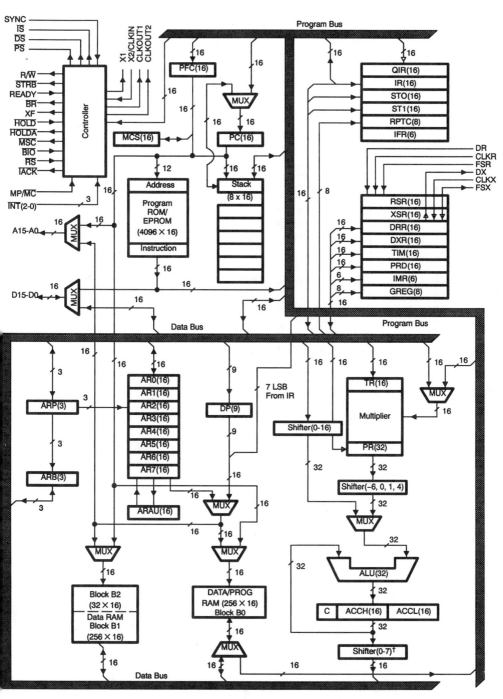

FIGURE 12-14 Functional block diagram of the TMS320C25.

vided to select correct physical memory at any given time. There are two instructions, namely, "configure block as data memory (CNFD)" and "configure block as program memory (CNFP)" for specifying the desired configuration of the memories. In Figure 12.15(a), we see the address assigned to each block of memory. In particular, notice that block B0 starts from the hexadecimal address FF00 (of

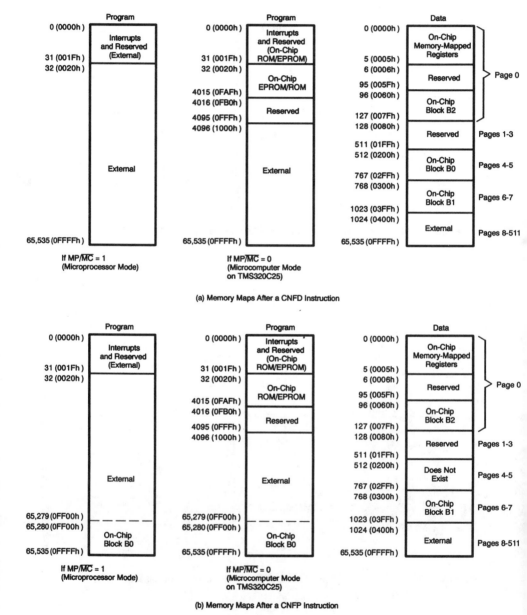

(a) Memory Maps After a CNFD Instruction

(b) Memory Maps After a CNFP Instruction

FIGURE 12-15 TMS320C25 memory maps.

program memory) when configured as program memory and from the hexadecimal address 0200 (of data memory) otherwise. (Hexadecimal addresses will be distinguished by a "greater than" sign, >, from now on.) Also note that block B1 always starts at address >0300.

There are two 16-bit wide buses inside the chip called data bus and program bus. The program bus carries the instruction code and immediate operands from the program memory. The data bus interconnects various elements, such as the central arithmetic logic unit (CALU) and the auxiliary register file to the data RAM. An important building block inside this chip is the 16 × 16 two's complement multiplier. Normally, with one internal data bus, two instructions are needed for obtaining the results of multiplication of two numbers—one for bringing in the multiplier and the other for the multiplicand. By allowing transfers from the data and program memory spaces to the multiplier inputs, two operands can be fetched simultaneously. Thus an effective single instruction cycle multiplication-accumulation becomes possible. The multiplier can have its operands simultaneously from the two on-chip data spaces, one via the data bus (block B1 or block B2) and the other (block B0) from the program bus. Therefore, when this mode of operation is desired, block B0 of RAM should be configured as program memory so that the operand residing there can be addressed as a program item and can be transferred to the multiplier via the program bus. This configuration of the memory is shown in Figure 12-15(b). Other manufacturers [KI83; NI81] have opted for two bypass buses directly between the data RAMs and the multiplier inputs. This, at the expense of more buses, simplifies the needed communication. The result of a multiplication is always captured in a register called product register (PR). The output of this register always feeds the ALU and therefore the result of the previous multiplication gets accumulated (in ALU) while a new multiplication is taking place (pipelining). The multiplication and accumulation operations take 100 ns. The instruction cycle is 100 ns as well. Because of pipelining, successive instructions of multiply-accumulate (along with proper data movements) can be executed to have an effective throughput of single instruction cycle time, that is, 100 ns. In fact, an instruction exists (RPTK) that arranges for repeated execution of its following instruction. Therefore, for FIR filter, every tap calculation (in a repeat instruction mode) can be done in 100 ns. One should be aware that for a small number of multiplication-accumulations, the overhead of setting up the repeat instruction mode may defeat the advantages mentioned above, and therefore it is not recommended [TE90]. In such a case, in-line coding (writing the multiplication-accumulation instruction as many times as desired) is advised. A high degree of parallelism in TMS320C25 also allows simultaneous operation of CALU and arithmetic/logic operations in the auxiliary register arithmetic unit (ARAU).

TMS320C25 Instruction Set Summary. Tables 12-1 through 12-6 provide a summary of all instructions for the TMS320C25 chip. Data move type instructions are listed in Tables 12-1 and 12-2. Table 12-3 summarizes the multiply-accumulate instructions. The branch, subroutine, and control flow instructions are included in Tables 12-4 and 12-5. Table 12-6 shows the I/O instructions.

TABLE 12-1 TMS320C25 Accumulator Memory Reference Instructions

INSTRUCTION	DESCRIPTION	INSTRUCTION	DESCRIPTION
ABS	Absolute value of accumulator	ADD	Add to accumulator with shift
ADDC	Add to accumulator with carry	ADDH	Add to high accumulator
ADDK	Add to accumulator short immediate	ADDS	Add to low accumulator with sign extension suppressed
ADDT	Add to accumulator with shift specified by T register	ADLK	Add to accumulator long immediate with shift
AND	Logical AND with accumulator	ANDK	Logical AND immediate with accumulator and shift
CMPL	Complement accumulator	LAC	Load accumulator with shift
LACK	Load accumulator immediate short	LACT	Load accumulator with shift specified by T register
LALK	Load accumulator long immediate with shift	NEG	Negate accumulator
NORM	Normalize contents of accumulator	OR	Logical OR with accumulator
ORK	Logical OR immediate with accumulator and shift	ROL	Rotate accumulator left
ROR	Rotate accumulator right	SAC	Store high accumulator with shift
SACL	Store low accumulator with shift	SBLK	Subtract from accumulator long immediate with shift
SFL	Shift accumulator left	SFR	Shift accumulator right
SUB	Subtract from accumulator with shift	SUBB	Subtract from accumulator with borrow
SUBC	Conditional subtract (used in division routine)	SUBH	Subtract from high accumulator
SUBK	Subtract from accumulator short immediate	SUBS	Subtract from low accumulator with sign extension suppressed
SUBT	Subtract from accumulator with shift specified by T register	XOR	Logical Exclusive OR with accumulator
XORK	Logical Exclusive OR immediate with accumulator and shift	ZAC	Zero accumulator
ZALH	Zero low accumulator and load high accumulator	ZALR	Zero low accumulator and load high accumulator with rounding
ZALS	Zero accumulator and load low accumulator with sign extension suppressed		

12-4-2 Programming Fixed-Point WE DSP16A DSP Device

The WE DSP16A Digital Signal Processor is a 16-bit, high-performance, CMOS integrated circuit. The WE DSP16A Digital Signal Processor features:

- 25- or 33-ns Instruction cycle
- 16 × 16-bit Multiply-add in one instruction cycle

TABLE 12-2 TMS320C25 Auxiliary Registers and Data Page Pointer Instructions

INSTRUCTION	DESCRIPTION	INSTRUCTION	DESCRIPTION
ADRK	Add to auxiliary register short immediate	CMPR	Compare auxiliary register with auxiliary register AR0
LAR	Load auxiliary register	LARK	Load auxiliary register immediate short
LARP	Load auxiliary register pointer	LDP	Load data memory page pointer
LDPK	Load data memory page pointer immediate	LRLK	Load auxiliary register long immediate
MAR	Modify auxiliary register	SAR	Store auxiliary register
SBRK	Subtract from auxiliary register short immediate value		

- Two 36-bit accumulators
- 4096 Words of ROM and 2048 words of RAM on-chip
- Complete set of ALU operations
- Instruction cache for high-speed, ROM-efficient vector operations
- Serial and parallel I/O ports with multiprocessor capability

Architecture of WE DSP16A. The major elements of the DSP16/DSP16A architecture are the memories, ROM and RAM; the cache; the address arithmetic units; the data arithmetic unit; the I/O, serial and parallel; and the control unit that con-

TABLE 12-3 TMS320C25 T, P Registers and Multiply Instructions

INSTRUCTION	DESCRIPTION	INSTRUCTION	DESCRIPTION
APAC	Add P register to accumulator	LPH	Load high P register
LT	Load T register	LTA	Load T register and accumulate previous product
LTD	Load T register, accumulate previous product and move data	LTP	Load T register and store P register in accumulator
LTS	Load T register and subtract previous product	MAC	Multiply and accumulate
MACD	Multiply and accumulate with data move	MPY	Multiply (with T register, store product in P register)
MPYA	Multiply and accumulate previous product	MPYK	Multiply immediate operand
MPYS	Multiply and subtract previous product	MPYU	Multiply unsigned
PAC	Load accumulator with P register	SPAC	Subtract P register from accumulator
SPH	Store high P register	SPL	Store low P register
SPM	Set P register output shift mode	SQRA	Square and accumulate
SQRS	Square and subtract previous product		

TABLE 12-4 TMS320C25 Branch and Call Instructions

INSTRUCTION	DESCRIPTION	INSTRUCTION	DESCRIPTION
B	Branch unconditionally	BACC	Branch to address specified by accumulator
BANZ	Branch on auxiliary register not zero	BBNZ	Branch if test/control status (TC) bit is not zero
BBZ	Branch if test/control status (TC) bit is zero	BC	Branch on carry being true
BGEZ	Branch if accumulator is greater than or equal to zero	BGZ	Branch if accumulator is greater than zero
BIOZ	Branch on I/O status = 0	BLEZ	Branch if accumulator is less than or equal to zero
BLZ	Branch if accumulator is less than zero	BNC	Branch on no carry
BNV	Branch if no overflow	BNZ	Branch if accumulator is not zero
BV	Branch on overflow	BZ	Branch if accumulator is zero
CALA	Call subroutine indirect	CALL	Call subroutine
RET	Return from subroutine	TRAP	Software interrupt

TABLE 12-5 TMS320C25 Control Instructions

INSTRUCTION	DESCRIPTION	INSTRUCTION	DESCRIPTION
BIT	Test bit of the specified data memory	BITT	Test bit specified by T register
CNFD	Configure block as data memory	CNFP	Configure block as program memory
DINT	Disable interrupt	EINT	Enable interrupt
IDLE	Idle until interrupt	LST	Load status register ST0
LST1	Load status register ST1	NOP	No operation
POP	Pop top of stack to low accumulator	POPD	Pop top of stack to data memory
PSHD	Push data memory value onto stack	PUSH	Push low accumulator onto stack
RC	Reset carry bit	RHM	Reset hold mode
ROVM	Reset overflow mode	RPT	Repeat instruction as specified by data memory value
RPTK	Repeat instruction as specified by immediate value	RSXM	Reset sign extension mode
RTC	Reset test/control flag	SC	Set carry bit
SHM	Set hold mode	SOVM	Set overflow mode
SST	Store status register ST0	SST1	Store status register ST1
SSXM	Set sign extension mode	STC	Set test/control flag

TABLE 12-6 TMS320C25 I/O and Data Memory Instructions

INSTRUCTION	DESCRIPTION	INSTRUCTION	DESCRIPTION
BLKD	Block move from data memory to data memory	BLKP	Block move from program memory to data memory
DMOV	Data move in data memory	PORT	Format serial port registers
IN	Input data from port	OUT	Output data to port
RFSM	Reset serial port frame synchronization mode	RTXM	Reset serial port transmit mode
RXF	Reset external flag	SFSM	Set serial port frame synchronization mode
STXM	Set serial port transmit mode	SXF	Set external flag
TBLR	Table read	TBLW	Table write

nects these elements in a pipeline. Figure 12-16 [AT89a] is a block diagram of the DSP16/DSP16A.

The arithmetic unit contains a 16×16-bit parallel 2's complement multiplier that generates a full 32-bit product in one instruction cycle. The product can be accumulated with one of two 36-bit accumulators. The data in these accumulators can directly be loaded from or stored into memory in 16-bit words with automatic saturation on overflow. The ALU supports a full set of arithmetic and logic operations on either 16- or 32-bit data. A standard set of ALU conditions can be tested for conditional branches and subroutine calls. This procedure allows the processor to function as a powerful 16- or 32-bit microprocessor for logical and control applications.

Two addressing units support high-speed, register-indirect memory addressing with postmodification of the register. Four address registers can be used for either read or write addresses to the RAM without restrictions. One address register is dedicated to the ROM for table look-up. Direct and immediate addressing is supported at a cost of only one additional instruction cycle and one ROM location.

WE DSP16/DSP16A Instruction Set Summary. All DSP16/DSP16A instructions are 16 bits wide and have a C-like syntax. Pipelining of instructions is necessary to achieve high real-time performance but latency effects have been eliminated. Five different categories of instructions are supported: Multiply/ALU, Special Function, Control, Data Move, and Cache.

Table 12-7 lists the mnemonics (CON) that are used in conditional instructions. One example of such an instruction is

```
if c1lt goto loop
```

Table 12-8 introduces the DSP16/DSP16A Multiply/ALU instructions. These instructions consist of two parts: Multiply/ALU Functions (F1) explained in Table 12-9 and the Data Move operations.

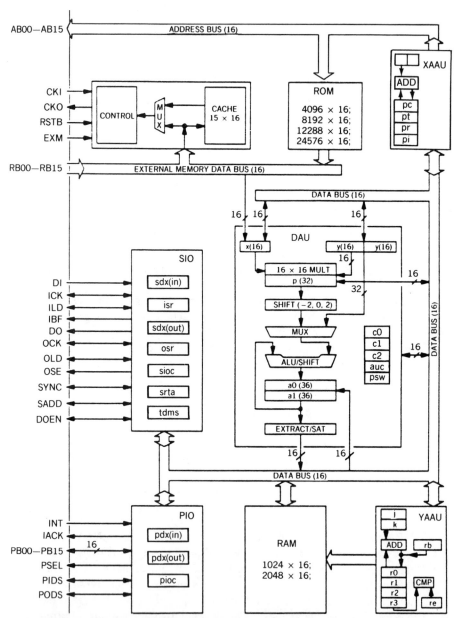

FIGURE 12-16 DSP16/DSP16A digital signal processor block diagram.

Note that in Table 12-8, aD, aS, and aT represent one of the two accumulators: a0, a1. The pointers X, Y, and Z allow for various addressing schemes available in DSP16/DSP16A. X takes the general form *pt+ +i that refers to a ROM location pointed to by pt, which gets postmodified by i. Y takes the values *rM, *rM+ +, *rM − −, and *rM+ +j, where M = 0, 1, 2, 3. Y represents a RAM location pointed to by the register rM, which gets postmodified by 0, 1, −1, and

TABLE 12-7   WE DSP16/DSP16A Conditional Mnemonics

CONDITION	DESCRIPTION	CONDITION	DESCRIPTION
pl	Result is nonnegative (checking sign bit 35)	mi	Result is negative
eq	Result is equal to zero	ne	Result is not equal to zero
gt	Result is greater than zero	le	Result is less than or equal to zero
lvs	Logical overflow set (36-bit overflow)	lvc	Logical overflow clear
mvs	Mathematical overflow set (32-bit overflow)	mvc	Mathematical overflow clear
c0ge	Counter 0 (c0) greater than or equal to zero and auto-increment c0	c0lt	Counter 0 (c0) less than zero and auto-increment c0
c1ge	Counter 1 (c1) greater than or equal to zero and auto-increment c1	c1lt	Counter 1 (c1) less than zero and auto-increment c1
heads	Pseudorandom sequence bit set	tails	Pseudorandom sequence bit clear
true	The condition is always true in an if instruction	false	The condition is never satisfied in an if instruction

j, respectively. Z:R represents read/write compound addressing with values *rMzp:R, *rMpz:R, *rMm2:R, *rMjk:R, where M = 0, 1, 2, 3. the registers rM and R cannot be identical. Register R is loaded with the contents of memory pointed to by rM, rM gets postmodified by 0, 1, -1, and j, respectively, rM is loaded with the contents of R, and rM gets postmodified by 1, 0, 2, and k, respectively.

The accumulators denoted by aD or aS in Table 12-10 represents a0 or a1. All the special function instructions in Table 12-10 can be conditionally executed as "if CON" and with an event counter as "ifc CON" with CON listed in Table 12-7.

TABLE 12-8   WE DSP16/DSP16A Multiply/ALU Instructions

INSTRUCTION	DESCRIPTION	INSTRUCTION	DESCRIPTION
F1 Y	multiply/ALU operation with postmodification of pointer register	F1 Y = aT[l]	multiply/ALU operation with parallel accumulator store
F1 x = Y	multiply/ALU operation with parallel load of x register	F1 y[l] = Y	multiply/ALU operation with parallel load of y register
F1 y = Y x = X	multiply/ALU operation with parallel load of x and y registers	F1 y = aT x = X	multiply/ALU operation with parallel load of x and y (from accumulator) registers
F1 aT[l] = Y	multiply/ALU operation with parallel load of accumulator register	F1 Y = y[l]	multiply/ALU operation with parallel store of y register
F1 Z:y[l]	multiply/ALU operation with compound data move	F1 Z:aT[l]	multiply/ALU operation with parallel compound accumulator move
F1 Z:y x = X	multiply/ALU operation with compound data move and parallel load of x register		

TABLE 12-9   WE DSP16/DSP16A Multiply/ALU Function Statements (F1)

FUNCTION	DESCRIPTION	FUNCTION	DESCRIPTION
p = x * y	multiply contents of x and y and leave result in p register	aD = p p = x * y	copy contents of p register into accumulator aD, multiply contents of x and y registers, and leave result in p register
aD = aS + p p = x * y	add the contents of accumulator aS and p register, leave result in accumulator aD, multiply contents of x and y registers, and leave result in p register	aD = aS - p p = x * y	subtract the contents of p register from accumulator aS, leave result in accumulator aD, multiply contents of x and y registers, and leave result in p register
aD = p	copy contents of p register into accumulator aD	aD = aS + p	add the contents of accumulator aS and p register and leave result in accumulator aD
aD = aS - p	subtract the contents of p register from accumulator aS and leave result in accumulator aD	aD = y	copy contents of y register into accumulator aD
aD = aS + y	add the contents of accumulator aS and y register and leave result in accumulator aD	aD = aS - y	subtract the contents of y register from accumulator aS and leave result in accumulator aD
aD = aS & y	Logical AND the contents of accumulator aS and y register and leave result in accumulator aD	aD = aS | y	Logical OR the contents of accumulator aS and y register and leave result in accumulator aD
aD = aS ^ y	Logical XOR the contents of accumulator aS and y register and leave result in accumulator aD	aS - y	subtract contents of y register from accumulator aS and set ALU flags
aS & y	Logical AND contents of y register from accumulator aS and set ALU flags		

TABLE 12-10   WE DSP16/DSP16A Special Function Instructions

FUNCTION	DESCRIPTION
aD = aS >> N	Sign preserved arithmetic right shift by N = 1, 4, 8, 16 of a 36 bit accumulator
aD = aS, aD = -aS	Transfer one accumulator content to another
aD = rnd(aS)	Round upper 20 bits of accumulator
aDh = aSh + 1	Increment high half of accumulator (lower half cleared)
aD = aS + 1	Increment accumulator
aD = y, aD = p	Contents of y or p registers are written to an accumulator
aD = aS << N	Sign extended logical left shift by N = 1, 4, 8, 16 of the least significant 32-bits of the 36-bit accumulators

TABLE 12-11 WE DSP16/DSP16A Control Instructions

FUNCTION	DESCRIPTION
goto JA	Move the immediate value JA into the lower 12 bits of the program counter (pc) register
goto pt	Move the contents of pt register into the program counter (pc) register
call JA	Move the contents of the program counter (pc) register into the program return (pr) register and the immediate data JA into the lower 12 bits of the pc register
call pt	Move the contents of the program counter (pc) register into the program return (pr) register and the data in pt into the pc register
icall	Move the contents of the program counter (pc) register into the program interrupt (pi) register and the address 2 into the pc register
return / goto pr	Move the contents of the program return (pr) register into the program counter (pc) register
ireturn / goto pi	Move the contents of the program interrupt (pi) register into the program counter (pc) register

In Table 12-11, JA represents a 12-bit value and all the control instructions (except icall and ireturn) can be conditionally executed similar to Special Function instructions.

In Table 12-12, aS and aT represent the high half of either accumulator a0 or a1, N represents a 16-bit value, M represents a 9-bit value, and Y and Z represent the same pointer addresses as used in the Multiply/ALU instructions. The register R stands for DAU registers—x, y, yl, auc, c0, c1, and c2; YAAU registers—r0, r1, r2, r3, rb, rc, j, k; XAAU registers—pt, pr, pi, and i; Processor status word—psw; SIO registers—sioc, sdx, tdms, and srta; and PIO registers—pioc, pdx0, and pdx1.

The Cache instructions (see Table 12-13) treat the specified N1 ($1 \leq N1 \leq 15$) instructions as a low overhead loop to be executed K ($2 \leq K \leq 127$) times to conserve program memory.

TABLE 12-12 WE DSP16/DSP16A Data Move Instructions

FUNCTION	DESCRIPTION
R = N	Load immediate data value, N, into specified destination register R
R = M	Load a 9-bit immediate data value, M, into one of the YAAU registers
R = Y	Load the contents of Y source into the specified destination register R
R = aS	Load the most significant 16-bit data from specified accumulator into destination register R
aT = R	Load the content of register R into most significant 16 bits of the specified accumulator. The guard bits are loaded with the value of bit 31
Y = R	Load the content of source register R into the specified destination register Y
Z:R	Write data from the specified Z source to the destination register R and write the old data in register R to the Z destination

TABLE 12-13   WE DSP16/DSP16A Cache Instructions

FUNCTION	DESCRIPTION
do K { instruction_1 ... instruction_N1 }	This cache instruction uses one ROM location and executes in one instruction cycle
redo K	Execute the N1 instructions currently in the cache's memory K times

12-4-3   Programming Fixed-Point DSP56000/1 DSP Device

The DSP56001 is a 24-bit fixed-point, general purpose DSP fabricated in low-power HCMOS technology. The DSP56001 features 512 words of full-speed, on-chip, program RAM; two preprogrammed data ROMs; and special on-chip bootstrap hardware to permit convenient loading of user programs into the program RAM. The DSP56001 is an off-the-shelf part since there are no user-programmable, on-chip ROMs. The DSP56000 part features 3.75K words of full-speed, on-chip, program ROM instead of 512 words of program RAM.

The heart of the processor consists of three execution units operating in parallel: the data arithmetic logic unit (ALU), the address generation unit (AGU), and the program controller. The DSP56001 has on-chip peripherals, program memory, data memory, and a memory expansion port.

The Motorola DSP56001 features:

- 75 or 100-ns Instruction cycle
- 24 × 24-bit Multiply/add in one instruction cycle
- Two 56-bit accumulators
- 544 Words of ROM and 1024 words of RAM on-chip
- Complete set of ALU operations
- Three types of I/O ports

Architecture of DSP56001. The DSP56001 architecture has been designed to maximize throughput in data-intensive DSP applications. This objective has resulted in a dual-natured, expandable architecture with sophisticated on-chip peripherals and general purpose I/O. The architecture is double Harvard since besides the program memory, there are two independent, expandable data memory spaces, two address generation units (AGUs), and a data arithmetic logic unit (ALU) having two accumulators and two shifter/limiter circuits. Such partitioning of data into two segments can be useful in graphics and image processing for working on X and Y spaces, in filtering applications for storing data and coefficients, and in complex arithmetic for separating the real and imaginary parts. Figure 12-17 [MO90] shows a block diagram of the DSP56001.

The DSP56001 is organized around the registers of a central processor composed of three independent execution units. The buses move data and instructions while instructions are being executed inside the execution units. Data movement

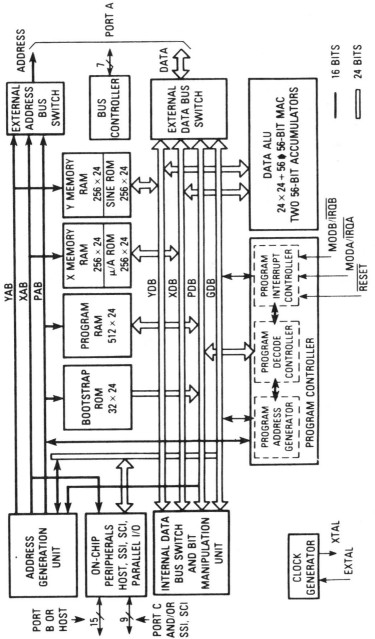

FIGURE 12-17 DSP56001 block diagram.

821

on the chip occurs over four bidirectional 24-bit buses: the X data bus (XDB), the Y data bus (YDB), the program data bus (PDB), and the global data bus (GDB). The X and Y data buses may also be treated by certain instructions as one 48-bit data bus.

Addresses are specified for internal X data memory and Y data memory on two, unidirectional 16-bit buses: X address bus (XAB) and Y address bus (YAB). Program memory addresses are specified on the bidirectional program address bus (PAB). External memory spaces are addressed via a single 16-bit, unidirectional address bus driven by a three-input multiplexer that can select the XAB, the YAB, or the PAB. Only one external memory access can be made in an instruction cycle. There is no speed penalty if only one external memory space is accessed in an instruction cycle. If two or three external memory spaces are accessed in a single instruction, there will be a one- or two-instruction cycle execution delay, respectively.

The data ALU performs all the arithmetic and logical operations on data operands. It consists of four 24-bit input registers, two 48-bit accumulator registers, two 8-bit accumulator extension registers, an accumulator shifter, two data bus shifter/limiter circuits, and a parallel, single-cycle nonpipelined multiply-accumulator (MAC) unit.

The I/O structure consists of a 47-pin expansion port (port A) and 24 additional I/O pins. These pins may be used as general purpose I/O pins, called port B and port C, or allocated to on-chip peripherals under software control. Three on-chip peripherals are provided on the DSP56001: an 8-bit parallel host microprocessor unit/direct memory access (DMA) interface, a serial communication interface (SCI), and a synchronous serial interface.

DSP56000 Instruction Set Summary. The DSP56000/56001 instruction set is divided into six groups: Arithmetic, Logical, Bit Manipulation, Loop, Move, and Program Control. The general syntax of instructions includes four fields: Opcode, Operands, XDB_data_move, and YDB_data_move.

The arithmetic instructions (see Table 12-14) execute in one instruction cycle within the data ALU. The XDB, YDB, and global data bus (GDB) are free to perform parallel data movement during arithmetic instruction execution. The conditions "cc" used in some instructions in Table 12-14 are: CC(HS) (carry clear, higher or same), CS(LO) (carry set, lower), EC (extension clear), EQ (equal), ES (extension set), GE (greater than or equal), GT (greater than), LC (limit clear), LE (less than or equal), LS (limit set), LT (less than), MI (minus), NE (not equal), NR (normalized), PL (plus), NN (not normalized). An example of an arithmetic instruction is

MAC X0,Y0,A   X:(R0)+,X0   Y:(R4)+,Y0

The logical instructions (see Table 12-15), which execute in one instruction cycle, perform all the logical operations within the data ALU (except ANDI or ORI).

TABLE 12-14 DSP56000 Arithmetic Instructions

INSTRUCTION	DESCRIPTION	INSTRUCTION	DESCRIPTION
ABS D (parallel move)	Store the absolute value of the destination operand D in the destination accumulator	ADC S,D (parallel move)	Add the source operand S and the carry bit C of the condition code register to the destination operand D and store the result in the destination accumulator
ADD S,D (parallel move)	Add the source operand S to the destination operand D and store the result in the destination accumulator	ADDL S,D (parallel move)	Add the source operand S to two times the destination operand D and store the result in the destination accumulator
ADDR S,D (parallel move)	Add the source operand S to one-half the destination operand D and store the result in the destination accumulator	ASL D (parallel move)	Arithmetically shift the destination operand D one bit to the left and store the result in the destination accumulator
ASR D (parallel move)	Arithmetically shift the destination operand D one bit to the right and store the result in the destination accumulator	CLR D (parallel move)	Clear the 56-bit destination accumulator
CMP S1,S2 (parallel move)	Subtract the source one operand, S1, from the source two accumulator, S2, and update the condition code register	CMPM S1,S2 (parallel move)	Subtract the magnitude of the source one operand, S1, from the magnitude of the source two accumulator, S2, and update the condition code register
DIV S,D	Divide the destination operand D (48-bit positive fraction dividend sign extended to 56-bits) by the source operand S (24-bit signed fraction divisor and store the partial remainder and the formed quotient (one new bit) in the destination accumulator D	MAC {+,-}S1,S2,D (parallel move)	Multiply the two signed 24-bit source operands S1 and S2 and add/subtract the product to/from the specified 56-bit destination accumulator D
MACR {+,-}S1,S2,D (parallel move)	Multiply the two signed 24-bit source operands S1 and S2 and add/subtract the product to/from the specified 56-bit destination accumulator D, and then round the result using convergent rounding	MPY {+,-}S1,S2,D (parallel move)	Multiply the two signed 24-bit source operands S1 and S2 and store the resulting product (with optional negation) in the specified 56-bit destination accumulator
MPYR {+,-}S1,S2,D (parallel move)	Multiply the two signed 24-bit source operands S1 and S2, round the result using convergent rounding, and store the resulting product (with optional negation) in the specified 56-bit destination accumulator	NEG D (parallel move)	Negate (56-bit twos-complement) the destination operand D and store the result in the destination accumulator
NORM Rn,D	Based upon the result of one 56-bit normalization iteration on the specified destination operand D, update the specified address register Rn and store the result back in the destination accumulator	RND D (parallel move)	Round the 56-bit value in the specified destination operand D by convergent rounding and store the result in the most significant portion of the destination accumulator (A1 to B1)

TABLE 12-14   DSP56000 Arithmetic Instructions (*Continued*)

INSTRUCTION	DESCRIPTION	INSTRUCTION	DESCRIPTION
SBC S,D (parallel move)	Subtract the source operand S and the carry bit C of the condition code register from the destination operand D and store the result in the destination accumulator	SUB S,D (parallel move)	Subtract the source operand S from the destination operand D and store the result in the destination operand D
SUBL S,D (parallel move)	Subtract the source operand S from two times the destination operand D and store the result in the destination accumulator	SUBR S,D (parallel move)	Subtract the source operand S from one-half the destination operand D and store the result in the destination accumulator
Tcc S1,D1	Transfer data from the specified source register S1 to the specified destination accumulator D1 if the specified condition "cc" is true	TFR S,D (parallel move)	Transfer data from the specified source data ALU register S to the specified destination data ALU accumulator D
TST S (parallel move)	Compare the specified source accumulator S with zero and set the condition codes accordingly		

TABLE 12-15   DSP56000 Logical Instructions

INSTRUCTION	DESCRIPTION	INSTRUCTION	DESCRIPTION
AND S,D (parallel move)	Logically AND the source operand S with bits 47-24 of the destination operand D and store the result in bits 47-24 of the destination accumulator	AND(I) #xx,D	Logically AND the 8-bit immediate operand (#xx) with the contents of the destination control register D and store the result in D
EOR S,D (parallel move)	Logically Exclusive OR the source operand S with bits 47-24 of the destination operand D and store the result in bits 47-24 of the destination accumulator	LSL D (parallel move)	Logically shift bits 47-24 of the destination operand D one bit to the left and store the result in the destination accumulator
LSR D (parallel move)	Logically shift bits 47-24 of the destination operand D one bit to the right and store the result in the destination accumulator	NOT D (parallel move)	Store the result of ones complement of bits 47-24 of the destination operand D back in bits 47-24 of the destination accumulator
OR S,D (parallel move)	Logically OR the source operand S with bits 47-24 of the destination operand D and store the result in bits 47-24 of the destination accumulator	OR(I) #xx,D	Logically OR the 8-bit immediate operand (#xx) with the contents of the destination control register D and store the result in D
ROL D (parallel move)	Rotate bits 47-24 of the destination operand D one bit to the left and store the result in the destination accumulator	ROR D (parallel move)	Rotate bits 47-24 of the destination operand D one bit to the right and store the result in the destination accumulator

TABLE 12-16 DSP56000 Bit Manipulation Instructions

INSTRUCTION	DESCRIPTION	INSTRUCTION	DESCRIPTION
BCLR #n,D	Test the nth bit of the destination operand D, clear it and store the result in the destination location	BSET #n,D	Test the nth bit of the destination operand D, set it, and store the result in the destination location
BCHG #n,D	Test the nth bit of the destination operand D, complement it and store the result in the destination location	BTST #n,D	Test the nth bit of the destination operand D and store its state in the carry bit C of the condition code register

The bit manipulation instructions (see Table 12-16) test the state of any single bit in a memory location and then optionally set, clear, or invert the bit and store the result in the carry bit of the CCR.

The loop instructions in Table 12-17 allow the execution of a DO loop in hardware.

The move instructions (see Table 12-18) perform data movement over the XDB and YDB or over the GDB.

TABLE 12-17 DSP56000 Loop Instructions

INSTRUCTION	DESCRIPTION	INSTRUCTION	DESCRIPTION
DO S,expr	Begin a no overhead hardware DO loop that is to be repeated the number of times specified in the source operand and whose range of execution is terminated by "expr"	ENDDO	Terminate the current hardware DO loop before the current loop counter (LC) equals one

TABLE 12-18 DSP56000 Move Instructions

INSTRUCTION	DESCRIPTION	INSTRUCTION	DESCRIPTION
LUA ea,D	Load the updated address (specified by the effective addresss "ea") into the destination address register D	MOVE S,D	Move the contents of the specified data source S to the specified destination. (Data ALU is inactive during this instruction's execution)
MOVE(C) S1,D2	Move the contents of the specified source control register S1 to the specified destination or move the specified source to the specified destination control register D2	MOVE(M) S,D	Move the specified operand to/from the specified program memory location (specified by S or D being of the form P:ea)
MOVE(P) S,D	Move the specified operand from/to the specified X or Y I/O peripheral (specified by S or D being of the form X:pp or Y:pp)		

TABLE 12-19 DSP56000 Program Control Instructions

INSTRUCTION	DESCRIPTION	INSTRUCTION	DESCRIPTION
ILLEGAL	Illegal instruction interrupt (NOP)	Jcc ea	If the specified condition "cc" is true, jump to the location in program memory given by "ea"
JMP ea	Jump to the location in program memory given by the effective address "ea"	JCLR #n,S,xxxx	Jump to the 16-bit absolute address in program memory specified in the instruction's 24-bit extension word if the nth bit of the source operand S is clear
JScc ea	Jump to the subroutine whose location in program memory is given by the effective address "ea" if the specified condition "cc" is true	JSET #n,S,xxxx	Jump to the 16-bit absolute address in program memory specified in the instruction's 24-bit extension word if the nth bit of the source operand S is set
JSR ea	Jump to the subroutine whose location in program memory is given by the effective address "ea" unconditionally	JSCLR #n,S,xxxx	Jump to the subroutine at the 16-bit absolute address in program memory specified in the instruction's 24-bit extension word if the nth bit of the source operand S is clear
JSSET #n,S,xxxx	Jump to the subroutine at the 16-bit absolute address in program memory specified in the instruction's 24-bit extension word if the nth bit of the source operand S is set	NOP	Just increment the program counter (PC)
REP S	Repeat the single-word instruction immediately following the REP instruction the specified number of times (specified by S)	RESET	Reset the interrupt priority register and all on-chip peripherals
RTI	Return from interrupt by pulling the PC and the status register (SR) from the system stack	RTS	Return from subroutine by pulling the PC from the system stack
STOP	Stop the instruction processing (low power standby)	SWI	Begin software interrupt processing
WAIT	Wait for interrupt (low power standby)		

The program control instructions (see Table 12-19) include conditional and unconditional branches and other instructions that affect the program counter (PC) or the system stack (SS). The condition codes denoted by ''cc'' are listed in Table 12-14 under the instruction ''Tcc.''

12-4-4 Programming Floating-Point WE DSP32C DSP Device

The WE DSP32C Digital Signal Processor is a member of the AT&T family of 32-bit, *floating-point* DSPs with 32-bit instructions.

A block diagram of WE DSP32C is shown in Figure 12-18. Two on-chip memory options are available.

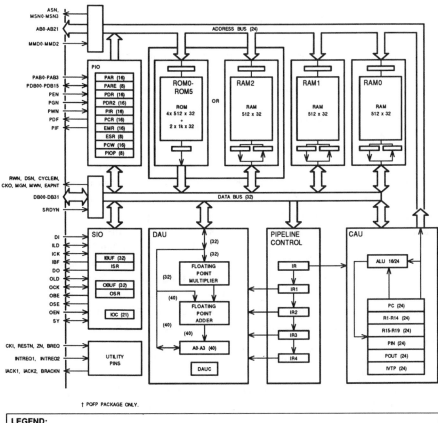

FIGURE 12-18 Block diagram of the DSP32C.

1. ROM-less chip with 1536 32-bit words of RAM.
2. ROM version with 1024 32-bit words of RAM and 4096 32-bit words of ROM.

Additionally, up to 16 Mbytes of external memory can be addressed directly by the external memory interface that supports wait states and bus arbitration.

The WE DSP32C has two execution units: the control arithmetic unit (CAU) and the data arithmetic unit (DAU). The CAU is a 24-bit, fixed-point integer unit that supports control, addressing, arithmetic, and logic functions for 16- or 24-bit data. The DAU is a 32-bit floating-point unit for signal processing computations.

This chip has several on-chip I/O units. The serial I/O unit interfaces with most codecs and time division multiplex (TDM) lines with few (if any) additional components. The parallel I/O interfaces the WE DSP32C device with a microprocessor. Data transfers can be made directly with the WE DSP32C memory (DMA) without program intervention.

Architecture of WE DSP32C. The DAU is the main execution unit for filtering and other signal processing algorithms. It contains a floating-point multiplier and an adder, four static 40-bit accumulators, and the data arithmetic unit control register (DAUC). The DAU performs multiply-accumulate operations on signal processing data and does data type conversions. The form of instruction executed in the DAU is $a = b + c \times d$. The DAU multiplier inputs can be from memory, I/O registers, or from one of four accumulators. These inputs are 32-bit floating-point numbers with a 24-bit mantissa and an 8-bit exponent. The adder inputs can be from memory, I/O registers, or from an accumulator and are 8, 16, or 40 bits wide. An accumulator input of 40 bits (nominal 32 bits plus 8 mantissa guard bits) can only come from one of the four accumulators, a0 through a3, or from the multiplier.

The control arithmetic unit (CAU) is used for generating addresses to memory and executing instructions that operate on 16- or 24-bit integers. It is capable of executing a full set of such instructions. The CAU contains twenty-two 24-bit general purpose registers (r1–r19, pin, pout, ivtp), a 24-bit program counter (PC), and an arithmetic logic unit (ALU).

Although in this chip there are two memory banks, the architecture is not truly Harvard. Here, the two memory banks share the same bus. Therefore transfer of the operands to the DAU and transfer of data to the CAU are time multiplexed on the internal bus. Every processor cycle (of 80 ns) is composed of four clock cycles. In three of the four clock cycles, read-from-the-memories operations are scheduled, while one clock cycle is reserved for writing back into the memory. Usually, the three read operations in every processor cycle are devoted to instruction fetch and two operand fetches, respectively. The instructions of this processor are more complex than other DSPs in that a single instruction can specify multiple microoperations. For example, a single instruction can request two operands from the memory, send them to the floating-point DAU to get them multiplied, the result to be added to one of the four available accumulators, then the new result to be registered in another accumulator, and finally a copy of the result to be written back in the memory at the address pointed to by one of the 21 registers of the CAU (the pointers themselves can also be postmodified inside the CAU). Interestingly enough, the processor is capable of executing other instructions while the data are being worked on in the DAU. Since the internal sections of the DAU itself are pipelined (two-stage multiply-accumulate), the next instruction can also send its operands to the DAU, without affecting the previous calculations. In this manner, at every given processor cycle, up to six instructions can be in different stages of fetch, decode, and execute. This also means that the result of an instruction may not be ready for use in the following instruction. The amount of *latency* for obtaining the result of each instruction varies and we discuss them as we encounter them.

There is only one internal data and address bus. Still the time multiplexing of the bus has created a virtual Harvard architecture. At the other extreme, one could have provided a separate bus for each memory bank. Then the bottleneck would have been the speed of floating-point multiplication. In that case, knowing that for a given technology, floating-point multiplication takes a certain amount of time, either the CAU had to sit idle or more pipelining would have been necessary in the DAU section. The present scheme of time multiplexing the bus is a good compromise between the two extremes. Also, this level of pipelining makes WE DSP32C a very efficient machine [KO81]. Another important point here is that the floating-point multiplier and accumulator of the WE DSP32C produce normalized results. This is very convenient and allows correct computations of the signal processing algorithms without the need for any special measures for normalization.

The following shows some salient features of WE DSP32C device:

- On-chip 1024 words of RAM and 4096 words of ROM
- Up to 16 Mbytes of external memory (24-bit address space)
- Memory can be addressed as 8, 16, or 32 bits
- Bit-reverse addressing
- 50 MHz operation/(12.5 MIPS, or 25 MFLOPS)
- 32-bit Floating-point (normalized) multiplication and addition
- Single-precision IEEE floating-point conversion capability
- 16- and 24-bit Integer operations
- Four 40-bit accumulators
- 16 Mbits/sec Serial I/O port with DMA option
- 16-bit Parallel I/O port with DMA option
- 133-pin Square PGA package
- Single 5-volt supply
- Low-power CMOS technology
- Zero-overhead looping

WE DSP32C Instruction Set Summary. The WE DSP32C supports five major types of instructions: data arithmetic (DA), 32-bit floating-point multiply/accumulate and data conversion special function instructions executed in the DAU and control arithmetic (CA) control, arithmetic/logic, and data move instructions.

Table 12-20 lists the data arithmetic multiply/accumulate instructions. In these instructions aN and aM denote one of the four DAU accumulators a0–a3.

TABLE 12-20 WE DSP32C DA Multiply/Accumulate Instructions

INSTRUCTION	DESCRIPTION	INSTRUCTION	DESCRIPTION
[Z =] aN = [-]aM {+,-} Y * X	The product of the X and Y data is added/subtracted to/from accumulator aM and the result is stored in accumulator aN. The result can also be sent to destination Z.	aN = [-]aM {+,-} (Z = Y) * X	The Y field operand is output according to the Z field. The product of the X and Y fields is added to accumulator aM and the sum is stored in accumulator aN.
[Z =] aN = [-]Y {+,-} aM * X	The product of the X field and accumulator aM is added/subtracted to/from the Y field. The result is placed in accumulator aM and can also be output according to the Z field.	[Z =] aN = [-]Y * X	The product of the X and Y fields is added/subtracted to/from zero. The result is stored in accumulator aN and can also be output according to the Z field.
aN = [-](Z = Y) * X	The value of the Y field is output according to the Z field. The product of the Y and X fields is stored in accumulator aN.	[Z =] aN = [-]Y {+,-} X	The sum or difference of the Y and X fields is stored in accumulator aN, and the result can also be output according to the Z field.
[Z =] aN = [-]Y	The value of the Y field or its negation is placed in accumulator aN and can also be output according to the Z field.	aN = [-](Z = Y) {+,-} X	The sum or difference of the X and Y fields is stored in accumulator aN, and Y is output according to the Z field.

X and Y denote the data sources: 32-bit memory location pointed to by registers r1–r14 (denoted by rP) and optionally incremented by registers r15–r19 (denoted by r1) in the form *rP, *rP++, *rP−−, *rP++r1; one of the four accumulators a0–a3, SIO input buffer ibuf.

Z denotes the data destination: 32-bit memory location pointed in the form *rP, *rP++, *rP−−, *rP++r1; *rP++r1r (r1r indicates carry reverse add); SIO output buffer obuf; PIO data register pdr.

In the special function instructions included in Table 12-21, aN represents one of four accumulators: a0–a3; Y represents memory locations pointed to by *rP, *rP++, *rP−−, *rP++r1, one of the four accumulators: a0–a3, SIO input buffer ibuf, and PIO data register pdr; Y denotes memory locations pointed to by *rP, *rP++, *rP−−, *rP++r1, SIO output buffer obuf, and PIO data register pdr.

CAU control instructions include unconditional branch instructions, conditional branch instructions, and subroutine call and return instructions. These are included in Table 12-22. The notation **rH** represents any register **r1–r22** and program counter **pc**. The register notation **rM** represents one of the 22 CAU registers: **r1–r22**. The notation **N** represents a 16-bit 2's complement integer and **M** represents a 24-bit 2's complement integer.

CA_COND includes pl (plus), mi (minus), ne (not equal to zero), eq (equal to zero), vc (no overflow), vs (overflowed), cc (no carry), cs (carry set), ge (greater

TABLE 12-21   WE DSP32C DA Special Function Instructions

INSTRUCTION	DESCRIPTION	INSTRUCTION	DESCRIPTION
[Z =] aN = ic(Y)	Convert data Y from either μ-law, A-law, or 8-bit linear format (as specified in the DAUC register) to the float format, storing the result in accumulator aN and optionally output according to the Z field.	[Z =] aN = oc(Y)	Convert data in Y in the float format to either μ-law, A-law, or 8-bit linear format (as specified in the DAUC register), storing the result in accumulator aN and optionally output according to the Z field.
[Z =] aN = float(Y)	Convert 16-bit 2's complement integer in Y to float, storing the result in accumulator aN and optionally output according to the Z field.	[Z =] aN = float24(Y)	Convert 24-bit 2's complement integer in Y to float, storing the result in accumulator aN and optionally output according to the Z field.
[Z =] aN = int(Y)	Convert a float in Y to 16-bit 2's complement integer (using rounding or truncation as specified by bit 4 of DAUC register), storing the result in accumulator aN and optionally output according to the Z field.	[Z =] aN = int24(Y)	Convert a float in Y to 24-bit 2's complement integer (using rounding or truncation as specified by bit 4 of DAUC register), storing the result in accumulator aN and optionally output according to the Z field.
[Z =] aN = round(Y)	Convert a 40 bit float data in Y to a 32 bit float, storing the result in accumulator aN and optionally output according to the Z field.	[Z =] aN = ifalt(Y)	If aN is negative, then [Z =] aN = Y, else [Z =] aN
[Z =] aN = ifaeq(Y)	If aN is equal to zero, then [Z =] aN = Y, else [Z =] aN	[Z =] aN = ifagt(Y)	If aN is positive, then [Z =] aN = Y, else [Z =] aN
[Z =] aN = dsp(Y)	Convert an IEEE floating point format number in Y to the DSP32C floating point format, storing the result in accumulator aN and optionally output according to the Z field.	[Z =] aN = ieee(Y)	Convert an DSP32C floating point format number in Y to the IEEE floating point format number, storing the result in accumulator aN and optionally output according to the Z field.
[Z =] aN = seed(Y)	Create a seed (reciprocal of Y) with mantissa good to 3 bits of accuracy by inverting all bits of Y, except the sign bit, storing the result in accumulator aN and optionally output according to the Z field.		

than or equal to), lt (less than), gt (greater than), le (less than or equal to), hi (greater than for unsigned number), and ls (less than or equal to for unsigned number).

DA_COND corresponds to conditions based on accumulator values: ane (not equal to zero), aeq (equal to zero), age (greater than or equal to zero), alt (less than zero), avc (no overflow), avs (overflowed), auc (no underflow), aus (underflowed), agt (greater than zero), and ale (less than or equal to zero).

IO_COND represents I/O conditions of DSP32C: ibe (input buffer empty), ibf (input buffer full), obe (output buffer empty), obf (output buffer full), pde (parallel

TABLE 12-22   WE DSP32C CA Control Instructions

INSTRUCTION	DESCRIPTION	INSTRUCTION	DESCRIPTION
if (CA_COND) goto {rH,N,rH+N}	Depending on the values of the flags from the result of the previous CA instruction, perform a conditional branch by loading program counter (pc) with one of the values specified in the braces { }.	if (DA_COND) goto {rH,N,rH+N}	Depending on the values of the flags from the result of the previous DA instruction, perform a conditional branch by loading program counter (pc) with one of the values specified in the braces { }.
if (IO_COND) goto {rH,N,rH+N}	Depending on the values of the flags from the result of the previous I/O instruction, perform a conditional branch by loading program counter (pc) with one of the values specified in the braces { }.	if (rM-- >= 0) goto {rH,N,rH+N}	Loop counter instruction where the value of rM is treated as a 16-bit number that is decremented each time. If the value of rM is greater than or equal to zero, the value of the pc changes to one of the locations specified in the braces { }, else the next instruction is executed.
call {rH,N,rH+N,M} (rM)	Store the value of the program counter (pc) in the specified rM register (return address) and move the value in the braces { } to the pc to call a subroutine.	return (rM)	Move the value of rM into the program counter (pc) to return from a subroutine call.
ireturn	Return from interrupt.	do M,{K,rH}	Do without interrupt the next M + 1 instructions K + 1 (or rH + 1) times, where K = rH = 0, 1, ..., 2047.
goto {rH,N,rH+N,M}	Unconditional branch to the instruction pointed to by the value specified in the braces { }.	nop	No operation

data register empty), pdf (parallel data register full), pie (parallel interrupt register empty), pif (parallel interrupt register full), syc (sync signal low), sys (sync signal high), fbc (serial frame boundary clear), fbs (serial frame boundary set), ireq1–hi (INTREQ1 pin is negated 1), ireq1_lo (INTREQ1 pin is asserted 0), ireq2_hi (INTREQ2 pin is negated 1), and ireq2_lo (INTREQ2 pin is asserted 0).

In Table 12-23, rH represents pc or one of the 22 CAU registers r1_r22; rD and rS represent one of the 22 CAU registers r1_r22; and N represents a 16-bit integer.

TABLE 12-23   WE DSP32C CA Arithmetic/Logic Instructions

INSTRUCTION	DESCRIPTION	INSTRUCTION	DESCRIPTION
rD[e] = rH + N	Three-operand add with 16-bit sign-extended immedite operand N	rD[e] = rS1 + rS2	Triadic add (two source registers summed and result in the destination register)
rD[e] = rD + rS	Diadic add (one source register value added to destination register value)	rD[e] = rS1 - rS2	Triadic left 2's complement subtract
rD[e] = rS2 - rS1	Triadic right 2's complement subtract	rD[e] = rD - {N,rS}	Right subtract (with immediate or source register data)
rD[e] - {N,rS}	Compare (with immediate or source register data)	rD[e] = {N,rS} - rD	Left subtract (with immediate or source register data)
rD[e] = rD & {N,rS}	Logical dyadic AND (with immediate or source register data)	rD[e] = rS1 & rS2	Logical triadic AND
rD[e] & {N,rS}	Bit test (with immediate or source register data)	rD[e] = rD | {N,rS}	Logical dyadic OR (with immediate or source register data)
rD[e] = rS1 | rS2	Logical triadic OR	rD[e] = rD ^ {N,rS}	Logical dyadic XOR (with immediate or source register data)
rD[e] = rS1 ^ rS2	Logical triadic XOR	rD[e] = rS/2	Arithmetic right shift
rD[e] = rS >> 1	Logical right shift	rD[e] = rS >>> 1	Rotate right through carry
rD[e] = rS <<< 1	Rotate left through carry	rD[e] = -rS	Negate source register value and store in destination register
rD[e] = rS * 2	Arithmetic left shift	rD[e] = rD # {N,rS}	Dyadic carry reverse add
rD[e] = rS1 # rS2	Triadic carry reverse add	rD[e] = rS1 & rS2	Logical triadic AND with complement
rD[e] = rD & {N,rS}	Logical dyadic AND with complement	rD[e] = rS	Assignment
rD[e] = rS {+,-} 1	Increment/decrement		

The optional e suffix in the instructions is for 24-bit (extended) operations.

Most of the CA arithmetic/logic instructions listed in Table 12-23 that do not use immediate operands may also be conditionally executed on the basis of CA_COND as

`if (CA_COND) instruction`

The CA Data Move instruction in Table 12-24 uses the following notations. The notation **rH** represents one of the 22 general purpose CAU registers **r1–r22** or the program counter **pc**. The register symbols **rM, rS,** and **rD** represent one of the 22 general purpose CAU registers **r1–r22**. The notations **rDh** and **rSh** represent high-order bits 8–15 of one of the 22 general purpose CAU registers **r1–r22** for data move (where the low-order bits 0–7 are cleared for **rD** and remain unchanged for **rS**). Similarly **rDl** and **rSl** represent low-order bits 0–7 of one of the 22 general purpose CAU registers **r1–r22** for data move (where the high-order bits 8–15 are cleared for **rD** and remain unchanged for **rS**). The notation **MEM** rep-

TABLE 12-24 WE DSP32C CA Data Move Instructions

INSTRUCTION	DESCRIPTION	INSTRUCTION	DESCRIPTION
rD = N	16-bit immediate load	rDe = M	24-bit immediate load
{ioc,dauc} = VALUE	5- or 21-bit immediate load	{MEM,*N,obuf,piop} = {rSh,rSl}	Move 8-bit data from the high- or low-order byte of the register rS to one of the locations specified on the left-hand side in the braces { }.
{MEM,*N,obuf,pdr,pdr2,pcw,pir} = {rS,pcsh}	Move the 16-bit data from the register rS or pcsh to one of the locations specified on the left-hand side in the braces { }.	{MEM,*N,obuf,pdr} = {rSe,pcshe}	Move the 24-bit data from the register rS or pcsh to one of the locations specified on the left-hand side in the braces { }.
{rDh,rDl} = {MEM,*N,ibuf,piop}	Move 8-bit data from the memory, the serial input buffer, or the parallel I/O port into the high- or low-order byte of the register rD.	rD = {MEM,*N,ibuf,pdr,pdr2,pir,pcw}	Move the 16-bit data to the register rD from one of the locations specified on the right-hand side in the braces { }.
rDe = {MEM,*N,ibuf,pdr}	Move the 24-bit data to the register rD from one of the locations specified on the right-hand side in the braces { }.	MEM = {ibufl,piop}	Move 8-bit data from the 8 LSBs of the serial input buffer or the parallel I/O port into the memory.

resents a pointer to a 32-, 16-, or 8-bit location with **rP** being the pointer register and **rl** the increment register that can have one of the following forms: ***rP**, ***rP + +**, ***rP − −**, ***rP + +rl**; ***rP + +rlr**, where P, 1 = 1,. . . , 22. The symbol N represents a 16-bit 2's complement number and **M** represents a 24-bit 2's complement number. The notation **VALUE** is a 21-bit integer for the **ioc** word or a 5-bit integer for the **dauc** word. Additionally, **piop** represents the parallel I/O port register, **pcw** represents the processor control word register, and **pcsh** represents the program counter shadow register.

12-4-5 Programming Floating-Point DSP96002 DSP Device

The DSP96002 is the first member of a family of dual-port IEEE floating-point programmable CMOS digital signal processors. The family concept defines a core

TABLE 12-24 (*Continued*)

INSTRUCTION	DESCRIPTION	INSTRUCTION	DESCRIPTION
MEM = {ibuf,pdr,pdr2,pir,pcw}	Move the 16-bit data from the 16 LSBs of one of the locations specified on the right-hand side in the braces { } into the memory.	MEM = {ibufe,pdre}	Move 32-bit data from the serial input buffer or the PIO data register into the memory.
{obufl,piop} = MEM	Move 8-bit data from an 8-bit memory location to the lower 8-bits of the serial output buffer or the parallel I/O port.	{obuf,pdr,pdr2,pir,pcw} = MEM	Move the 16-bit data from a 16-bit memory location to the 16 LSBs of one of the locations specified on the left-hand side in the braces { }.
{obufe,pdre} = MEM	Move 32-bit data from the memory into the serial output buffer or the PIO data register.		

as the data arithmetic logic unit (ALU), address generation unit (AGU), program controller, and associated instruction set.

The main characteristics of the DSP96002 are full support of IEEE 754 Single Precision (SP, 8-bit exponent and 24-bit mantissa) and Single Extended Precision (SEP, 11-bit exponent and 32-bit mantissa) floating-point and 32-bit fixed-point arithmetic, coupled with two identical external memory expansion ports. Additional features of DSP96002 are:

- 16.5 MIPS with 33-MHz clock
- 49.5 Million floating-point operations per second (MFLOPS) peak with a 33-MHz clock
- Highly parallel instruction set with special DSP addressing modes
- Nested hardware DO loops
- Fast auto-return interrupts
- Two independent on-chip 512 × 32-bit data RAMs and 1024 × 32-bit program RAM
- Two independent on-chip 1024 × 32-bit data ROMs and 64 × 32-bit bootstrap ROM
- Off-chip expansion to 2×2^{32} 32-bit words of data memory and 2^{32} 32-bit words of program memory
- Two 32-bit parallel microprocessor interfaces with DMA
- On-chip two-channel DMA controller and emulator interface

Architecture of DSP96002. A block diagram of the DSP96002 chip is shown in Figure 12-19. Data movement on the chip occurs over five bidirectional 32-bit

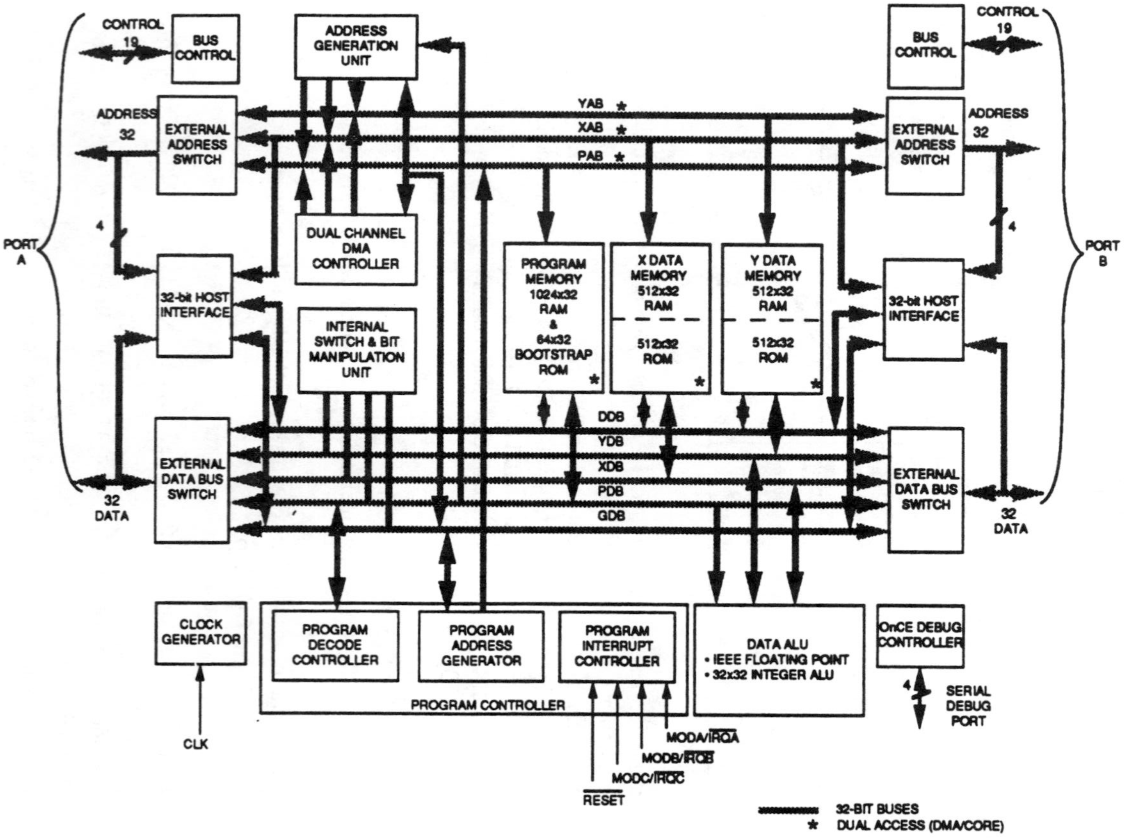

FIGURE 12-19 DSP96002 block diagram.

buses: X data bus (XDB), Y data bus (YDB), the DMA data bus (DDB), program data bus (PDB), and the global data bus (GDB). Data transfer between the data ALU and the X data memory and Y data memory occur over the XDB and YDB. Program memory data and DMA data transfers occur over dedicated buses. All other data transfers occur over the GDB.

Addresses are specified for internal X data memory and Y data memory on two unidirectional 32-bit buses: X address bus (XAB) and Y address bus (YAB). The program memory addressing is done over the program address bus (PAB).

The data ALU consists of ten 96-bit general purpose registers (D0–D9), a 32-bit barrel shifter, a 32-bit adder, and a 32 × 32-bit parallel multiplier (with 64-bit result). It is capable of multiplication, addition, subtraction, format conversion, shifting, and logical operations in one instruction cycle. Data ALU registers may be read or written over the XDB and YDB as 32- or 64-bit operands. Each 96-bit data ALU register is divided into three subregisters: high, middle, and low (e.g., D0.H, D1.M, D2,L) Floating-point data ALU operations always have a 96-bit result, whereas fixed-point operations have a 32- or 64-bit result. All the floating-point computations are performed using the SEP format and the results are automatically rounded to SP or SEP numbers as programmed. In addition to supporting all IEEE arithmetic exception handling modes, a "Flush to Zero" mode is also provided that forces all floating-point result underflows to zero without an additional instruction cycle.

The AGU performs (in parallel with other chip resources) all the address storage and effective address calculations necessary to address data operands in memory and it is used by both the core and the on-chip DMA controller. It contains eight 32-bit address registers (R0–R7), eight 32-bit offset registers (N0–N7), and eight 32-bit modifier registers (M0–M7). An offset register will be accessed for an address register update calculation involving an address register of the same number (i.e., N0 is accessed when R0 is updated, N1 for R1, etc.). Each address register may be accessed for output to the XAB, YAB, and PAB. AGU registers may be read or written over the GDB as 32-bit operands. The AGU can generate two 32-bit addresses every instruction cycle—one for any two of the XAB, YAB, or PAB.

The program control logic performs instruction prefetch, instruction decoding, and exception processing. A 32-bit program counter (PC) register is used to address the program memory space. The system stack is a separate internal RAM (64 bits wide and 15 locations "deep" and divided into two banks: High, SSH and Low, SSL), which stores the PC (in SSH), the status register (SR, in SSL) for subroutine calls and long interrupts, the loop counter (LC, in SSL), and the loop address (LA, in SSH) register for program looping. The SR consists of an 8-bit mode register (MR, affected by processor reset, exception processing, and looping operations), an 8-bit IEEE exception register (IER, affected by processor reset and data ALU IEEE floating-point operations), an 8-bit exception register (ER, affected by processor reset and data ALU floating-point operations), and an 8-bit condition code register (CCR, affected by the data ALU operations). When a subroutine call or long interrupt occurs, the contents of the PC and SR are pushed on the "top" location in the system stack. On a return from subroutine, the PC is restored. Both the PC and the SR are restored on return from an interrupt. Upon interrupt, the current instruction in decode will execute normally, unless it is the first word of a two-word instruction, in which case it will be aborted and refetched at the completion of exception processing. The next two fetch addresses are supplied by the interrupt controller. If one of the words fetched by the interrupt controller is a jump to subroutine, a long interrupt routine is formed, and a context switch is performed using the stack. If neither interrupt instruction word causes a change of control flow, then the two instructions fetched are executed as a fast interrupt routine with minimum overhead and the system stack is not used. This mechanism is commonly used to move data between memory and an I/O device. The DSP96002 provides a hardware loop instruction called DO (similar in concept to the DO loop in a FORTRAN or other high-level programming language) that allows repeated execution of a given set of instructions. When a no-overhead hardware program loop is initiated with the execution of a DO instruction, the following events occur:

- The current LC and LA register contents are pushed onto the system stack to allow nested loops.
- The LC and LA registers are initialized with values specified in the DO instruction.
- The address of the first instruction in the program loop and the current SR contents are transferred onto the system stack.
- The loop flag bit in the SR is set.

A program loop begins execution after the DO instruction and continues until the program address fetched equals the LA register contents (last address of program loop). The contents of the LC are then tested for 1. If the LC is not 1, it is decremented and the top location in the stack RAM is read into the PC to return to the start of the loop. If the LC is 1, the program loop is terminated by incrementing the PC, reading the previous loop flag bit from the top location in the stack into the SR, purging the stack, and restoring the LA, LC registers, and the system stack pointer.

The DSP96002 instruction set contains a full set of operand addressing modes that can be grouped into three categories: register direct, address register indirect (Mn; Nn; Rn with n being the register number), and special (to specify operand or the address of the operand directly).

DSP96002 Instruction Set Summary. DSP96002 supports 38 floating-point arithmetic, 30 fixed-point arithmetic, 13 logical, 4-bit manipulation, 4 loop, 9 data move, and 35 program control instructions.

Table 12-25 lists all floating-point arithmetic instructions that operate on the 96-bit data ALU registers with register direct addressing modes used for operands. This means that the X data bus, Y data bus, and global data bus are free for optional parallel move operations. These instructions always execute in a single instruction cycle in the Flush-to-Zero mode (set by bit 27 of the status register).

TABLE 12-25 DSP96002 Floating-Point Arithmetic Instructions

INSTRUCTION	DESCRIPTION	INSTRUCTION	DESCRIPTION
FABS.S D	Take the absolute value of the destination operand, round to single precision and store the result in the destination operand D	FABS.X D	Take the absolute value of the destination operand and store the result in the destination operand D.
FADD.S S,D	Add (D + S) the two specified operands, round to single precision and store the result in the destination operand D.	FADD.X S,D	Add (D + S) the two specified operands, round to single extended precision and store the result in the destination operand D.
FADDSUB.S D1,D2	Add (D1 + D2) and subtract (D1 - D2) the two specified operands and round to single precision. Store the rounded result of the addition in D2 and of the subtraction in D1.	FADDSUB.X D1,D2	Add (D1 + D2) and subtract (D1 - D2) the two specified operands and round to single extended precision. Store the rounded result of the addition in D2 and of the subtraction in D1.
FCLR D	All 96 bits of the destination operand are cleared to zero.	FCMP S1,S2	Subtract the two specified operands (S2 - S1) and set proper CCR condition codes Z, N, I, LR, RB, A
FCMPG S1,S2	Subtract the two specified operands (S2 - S1) and set proper CCR condition codes for graphics compare with trivial accept/reject flags	FCMPM S1,S2	Subtract the absolute value (magnitude) of the two specified operands (S2 - S1) and set proper CCR condition codes Z, N, I
FCOPYS.S S,D	Copy the sign of the floating-point operand S to the floating-point operand D, round the resulting operand D to single precision and store the result in the specified destination D	FCOPYS.X S,D	Copy the sign of the floating-point operand S to the floating-point operand D.
FGETMAN S,D	Extract the mantissa and sign of the floating-point operand S, normalize the mantissa (so the result is in the range 1-2), and store the result as a floating-point value in the specified destination D.	FINT S,D	Round the floating-point source operand S to an integer value using the current rounding mode specified by bits R1-R0 in the IER register, and store the result as a floating-point number in the specified destination D.
FLOAT.S D	Convert the 2's complement 32-bit integer located in the low portion of the operand D into a floating-point operand, round to single precision and store the result in the operand D.	FLOAT.X D	Convert the 2's complement 32-bit integer located in the low portion of the operand D into a floating-point operand, round to single extended precision and store the result in the operand D.

Also, in the IEEE mode they execute in a single instruction cycle if denormalized numbers are not detected; otherwise additional instruction cycles will be required.

The fixed-point arithmetic instructions (see Table 12-26) perform all operations within the data ALU and are register-based (register direct addressing modes used for operands). This means that the X data bus, Y data bus, and global data bus are free for optional parallel move operations. These instructions execute in one instruction cycle.

The logical instructions (see Table 12-27), except ANDI and ORI, perform all

TABLE 12-25   DSP96002 Floating-Point Arithmetic Instructions (*Continued*)

INSTRUCTION	DESCRIPTION	INSTRUCTION	DESCRIPTION
FMPY S1,S2,D1 FSUB.X S3,D2	Multiply the two operands S1 and S2, round to single extended precision and store the result in the specified destination register D1. Simultaneously, subtract S3 from D2, round to single extended precision and store the result in the desination operand D2.	FMPY.S S1,S2,D	Multiply the operands S1 and S2, round to single precision and store the result in the destination operand D.
FMPY.X S1,S2,D	Multiply the operands S1 and S2, round to single extended precision and store the result in the destination operand D.	FNEG.S D	Subtract the destination operand D from zero, round to single precision, and store the result in the destination operand D.
FNEG.X D	Subtract the destination operand D from zero and store the result in the destination operand D.	FSCALE.S S,D	Scale the destination operand D according to the scale factor contained in the 11 LSBs of the high portion of the source register S, round to single precision and store the result in the destination operand D. An 8-bit immediate Short scaling factor (in place of S), sign-extended to 11 bits may also be used.
FSCALE.X S,D	Scale the destination operand D according to the scale factor contained in the 11 LSBs of the high portion of the source register S, round to single extended precision and store the result in the destination operand D. An 8-bit immediate Short scaling factor (in place of S), sign-extended to 11 bits may also be used.	FSEEDD S,D	Determine an approximation to 1.0/S, and store the result in the destination operand D. Used for initiating floating-point division.
FSEEDR S,D	Determine an approximation to sqrt(1.0/S), and store the result in the destination operand D. Used for initiating floating-point square root calculation.	FSUB.S S,D	Subtract the two specified operands (D - S), round to single precision and store the result in the destination operand D.

the logical operations within the data ALU. This means that the X data bus, Y data bus, and global data bus are free for optional parallel move operations. These instructions execute in one instruction cycle.

The bit manipulation instructions (see Table 12-28) test the state of any single bit in a data memory location or register and then optionally set, clear, or invert the bit. The Carry bit in the CCR register will contain the result of the bit test. Parallel moves are not allowed with any of these instructions. The destination D (or source S in the case of BTST instruction) could be either a register or a memory location denoted as X:ea, X:aa, X:pp, Y:ea, Y:aa, or Y:pp (ea stands for effective address, aa stands for absolute short address, and pp stands for I/O short address).

TABLE 12-25 (*Continued*)

INSTRUCTION	DESCRIPTION	INSTRUCTION	DESCRIPTION
FLOATU.S D	Convert the unsigned 32-bit integer located in the low portion of the operand D into a floating-point operand, round to single precision and store the result in the operand D.	FLOATU.X D	Convert the unsigned 32-bit integer located in the low portion of the operand D into a floating-point operand, round to single extended precision and store the result in the operand D.
FLOOR S,D	Round the floating-point source operand S to an integer value using the round to minus infinity mode and storee the result as a floating-point number in the specified destination D.	FMPY S1,S2,D1 FADD.S S3,D2	Multiply the two operands S1 and S2, round to the precision indicated by the MP mode bit and store the result in the specified destination register D1. Simultaneously, add the two operands S3 and D2, round to single precision and store the result in the destination operand D2.
FMPY S1,S2,D1 FADD.X S3,D2	Multiply the two operands S1 and S2, round to single extended precision and store the result in the specified destination register D1. Simultaneously, add the two operands S3 and D2, round to single extended precision and store the result in the destination operand D2.	FMPY S1,S2,D1 FADDSUB.S D3,D2	Multiply the two operands S1 and S2, round to the precision indicated by the MP mode bit and store the result in the specified destination register D1. Simultaneously, add the two operands D2 & D3, subtract D2 from D3, round both results in single precision and store the result of the addition in register D2 and of the subtraction in register D3.
FMPY S1,S2,D1 FADDSUB.X D3,D2	Multiply the two operands S1 and S2, round to single extended precision and store the result in the specified destination register D1. Simultaneously, add the two operands D2 & D3, subtract D2 from D3, round both results in single extended precision and store the result of the addition in register D2 and of the subtraction in register D3.	FMPY S1,S2,D1 FSUB.S S3,D2	Multiply the two operands S1 and S2, round to the precision indicated by the MP mode bit and store the result in the specified destination register D1. Simultaneously, subtract S3 from D2, round to single precision and store the result in the desination operand D2.
FSUB.X S,D	Subtract the two specified operands (D - S), round to single extended precision and store the result in the destination operand D.	FTFR.S S,D	Round source operand S to single precision and store the result in the destination operand D.
FTFR.X S,D	Store source operand S in the destination operand D.	FTST S	Compare the specified operand with zero and set CCR condition codes Z, N, I.

TABLE 12-26   DSP96002 Fixed-Point Arithmetic Instructions

INSTRUCTION	DESCRIPTION	INSTRUCTION	DESCRIPTION
ABS D	Take the absolute value of the destination operand low portion and store the result in the low portion of D.	ADD S,D	Add the low portion of the two specified operands and store the result in the low portion of the destination operand D.
ADDC S,D	Add the low portion of the two specified operands along with the C bit of the condition code register and store the result in the low portion of the destination operand D.	ASL S,D	Arithmetically shift the low portion of the specified operand D left by one bit (when S is not specified) or by 11-bit unsigned integer located in the 11 LSBs of the high portion of S or by a 6-bit immediate field in the instruction (in place of S). The result is stored in the low portion of D.
ASR S,D	Arithmetically shift the low portion of the specified operand D right by one bit (when S is not specified) or by 11-bit unsigned integer located in the 11 LSBs of the high portion of S or by a 6-bit immediate field in the instruction (in place of S). The result is stored in the low portion of D.	CLR D	Clear the low portion of the destination operand D to zero.
CMP S1,S2	Subtract the low portion of the operand S1 from that of S2 and set CCR condition codes C, V, Z, N, LR appropriately.	CMPG S1,S2	Subtract the low portion of the operand S1 from that of S2 and set CCR condition codes C, V, Z, N, LR, RB, A appropriately. This is graphics compare with trivial accept/reject flags.
DEC D	Decrement by 1 the low portion of the specified operand D and store the result in the low portion of D.	EXT D	Sign extend the lower 16 bits of D.L into the upper 16 bits of D.L.
EXTB D	Sign extend the lower byte of D.L into the upper 24 bits of D.L.	GETEXP S,D	Extract the exponent of the single extended precision floating-point operand S and store it as an unbiased, 2's complement, 32-bit integer in the low portion of D.
INC D	Increment by one the low portion of the specified operand D and store the result in the low portion of D.	INT D	Convert the specified floating-point operand to 32-bit, 2's complement integer and store the result in the low portion of D.

The loop instructions (see Table 12-29) control no-overhead hardware looping by saving registers used by a program loop (LA and LC) on the system stack. The source operand S can either be from a register or an immediate value #count, or from memory location denoted by X:ea, Y:ea (ea stands for effective address).

The data move instructions (see Table 12-30) perform data movement over the X, Y, global, and program data buses.

The program control instructions (see Table 12-31) include jumps, conditional jumps, branches, conditional branches, and other instructions that affect the PC

TABLE 12-26 (*Continued*)

INSTRUCTION	DESCRIPTION	INSTRUCTION	DESCRIPTION
INTRZ D	Convert the specified floating-point operand to 32-bit, 2's complement integer rounding toward zero and store the result in the low portion of D.	INTU D	Convert the specified floating-point operand to 32-bit, unsigned integer and store the result in the low portion of D.
INTURZ D	Convert the specified floating-point operand to 32-bit, unsigned integer rounding toward zero and store the result in the low portion of D.	JOIN S,D	Transfer the 16 LSBs of the lower portion of source operand S into the 16 MSBs of the lower portion of destination D.
JOINB S,D	Transfer the 8 LSBs of the lower portion of source operand S into bits 15-8 of the lower portion of destination D. The 16 MSBs of the lower portion of D are zeroed.	MPYS S1,S2,D	Multiply two signed 32-bit operands from the low portions of S1 and S2 and store the 64-bit signed integer product in the middle and low portions of D.
MPYU S1,S2,D	Multiply two unsigned 32-bit operands from the low portions of S1 and S2 and store the 64-bit unsigned integer product in the middle and low portions of D.	NEG D	Subtract from zero the low portion of the destination operand and store the result in the low portion of D.
NEGC D	Subtract from zero the low portion of the destination operand along with the C bit of the condition code register and store the result in the low portion of D.	SETW D	Set to all ones the low portion (long word) of the destination operand.
SPLIT S,D	Transfer the 16 MSBs of the lower portion of source operand S into the 16 LSBs of the lower portion of destination D and sign-extend to 32 bits.	SPLITB S,D	Transfer bits 15-8 of the lower portion of source operand S into the 8 LSBs of the lower portion of destination D and sign-extend to 32 bits.
SUB S,D	Subtract the low portion of the source operand S from that of D and store the result in the low portion of D.	SUBC S,D	Subtract the low portion of the source operand S from that of D along with the C bit of the condition code register and store the result in the low portion of D.
TFR S,D	Transfer data from the low portion of S to the low portion of D without affecting the condition code bits.	TST S	Compare the low portion of the specified operand with zero and appropriately set CCR condition codes V, Z, and N.

and system stack. In Table 12-31, ''cc'' represents one of the following mnemonics: CC(HS) (carry clear), CS(LO) (carry set), EQ (equal), GE (greater or equal), GT (greater than), HI (higher), LE (less or equal), LS (lower or same), LT (less than), MI (minus), NE(Q) (not equal), PL (plus), VC (overflow clear), and VS (overflow set).

12-4-6 Programming Floating-Point TMS320C30 DSP Device

The TMS320C30 Digital Signal Processor (DSP) is a third-generation high-performance CMOS 32-bit device in the TMS320 family of single-chip DSPs. Some

TABLE 12-27   DSP96002 Logical Instructions

INSTRUCTION	DESCRIPTION	INSTRUCTION	DESCRIPTION
AND S,D	Logically AND the low portion of the two specified operands and store the result in the low portion of D.	ANDC S,D	Logically AND the low portion of D with the logical complement of the low portion of S, and store the result in the low portion of D.
AND(I) #byte,D	Logically AND the contents of the control register D with an 8-bit immediate operand and store the result back into D.	BFIND S,D	Return the position of the source operand S leading one, considered from left to right, as a 2's complement integer in the high portion of destination operand D.
EOR S,D	Logically exclusive OR the low portion of the two specified operands and store the result in the low portion of D.	LSL S,D	Logically left shift the low portion of the specified operand D either by 1 bit (when S is not specified) or by the 11-bit unsigned integer located in the 11 LSBs of the high portion of S, or by a 6-bit immediate field in the instruction (in place of S) and store the result in the low portion of D.
LSR S,D	Logically right shift the low portion of the specified operand D either by 1 bit (when S is not specified) or by the 11-bit unsigned integer located in the 11 LSBs of the high portion of S, or by a 6-bit immediate field in the instruction (in place of S) and store the result in the low portion of D.	NOT D	Take the one's complement of the low portion of the destination operand D and store the result in D.
OR S,D	Logically OR the low portion of the two specified operands and store the result in the low portion of D.	ORC S,D	Logically OR the low portion of D with the logical complement of the low portion of S, and store the result in the low portion of D.
OR(I) #mask,D	Logically OR the contents of the control register D with an 8-bit immediate operand and store the result back into D.	ROL D	Rotate the low portion of operand D one bit to the left and store the result in the low portion of D.
ROR D	Rotate the low portion of operand D one bit to the right and store the result in the low portion of D.		

key features of the TMS320C30 are listed below:

- 60-ns Single-cycle instruction execution time resulting in 33.3 MFLOPS or 16.7 MIPS
- On-chip 4K × 32-bit dual-access ROM
- On-chip two 1K × 32-bit dual access RAM blocks
- 64 × 32-bit Instruction cache
- 32-bit Instruction and data words, 24-bit addresses
- 32-bit Floating-point and integer multiplier, ALU, and barrel shifter

TABLE 12-28   DSP96002 Bit Manipulation Instructions

INSTRUCTION	DESCRIPTION	INSTRUCTION	DESCRIPTION
BCHG #bit,D	Test the nth bit of the destination operand D and store its state in the C condition code bit. After the test, change the state of the nth bit in the destination.	BCLR #bit,D	Test the nth bit of the destination operand D and store its state in the C condition code bit. After the test, clear the nth bit in the destination.
BSET #bit,D	Test the nth bit of the destination operand D and store its state in the C condition code bit. After the test, set the nth bit in the destination.	BTST #bit,S	Test the nth bit of the source operand D and store its state in the C condition code bit.

- Eight extended precision accumulators
- On-chip DMA controller for concurrent I/O
- Two- and three-operand instructions
- Zero overhead loops with single-cycle branches
- Two serial ports to support 8/16/32-bit transfers
- Two 32-bit timers
- 180-pin PGA package

Architecture of TMS320C30. The TMS320C30 architecture (a block diagram of which is shown in Figure 12-20) is characterized by the accuracy and precision of the floating-point units, large on-chip memory, a high degree of parallelism, and the DMA controller.

The central processing unit (CPU) of the TMS320C30 includes:

- A floating-fixed-point multiplier (floating-point: $32 \times 32 \to 40$; fixed-point: $24 \times 24 \to 32$)
- A 32-bit barrel shifter

TABLE 12-29   DSP96002 Loop Instructions

INSTRUCTION	DESCRIPTION	INSTRUCTION	DESCRIPTION
DO S,label	Begin a hardware DO loop that is to be repeated the number of times specified in the instruction source operand S and whose range of execution is terminated by the destination operand "label".	DOR S,label	Begin a PC relative hardware DO loop that is to be repeated the number of times specified in the instruction source operand S and whose range of execution is terminated by the destination operand "label".
ENDDO	Terminate the current hardware DO loop before the current loop counter (LC) equals 1.	REP S	Execute the single word instruction following the REP instruction LC times repetitively, where LC is the value of the loop counter obtained from the source operand S.

TABLE 12-30 DSP96002 Data Move Instructions

INSTRUCTION	DESCRIPTION	INSTRUCTION	DESCRIPTION
LEA address,D	Calculate effective address as specified in "address" operand and store in the destination register D without affecting the source address registers.	LRA displacement,D	Add the PC to the specified displacement and store the result in destination D.
MOVE	Move the contents of the specified source to the specified destination. Usually the register moves are done parallel with some Data ALU instructions and the instructions are written without the mnemonic MOVE.	MOVETA	Move the contents of the specified source to the specified destination and update the C, V, N, and Z flags in the CCR according to the result of the address calculation.
MOVE(C) S2,D1	Move the contents of the specified control register to the specified destination or move the specified source to the specified control register	MOVE(I) #data,D	The 16-bit immediate short operand is sign extended to a word operand and is stored in the destination register D.
MOVE(M) S,D	Move the specified program memory word (S = P:ea) to the specified destination register or from the specified source register to the specified program memory location (D = P:ea).	MOVE(P) S,D	Move the word operand to or from the X and Y I/O peripherals.
MOVE(S) S,D	Move the word operand to or from the lower 128 memory locations in X and Y data memories.		

- A 32-bit arithmetic logic unit (ALU)
- Two auxiliary register arithmetic units (ARAUs) to generate two operand addresses in a single cycle
- Eight 40-bit extended precision registers (R0–R7) for accumulation
- Eight 32-bit auxiliary registers (AR0–AR7) for the two ARAUs
- A 32-bit data page pointer (DP) register for pointing to 256 64K words long data pages
- Two 32-bit index registers (IR0–IR1) for indexed addressing
- A 32-bit block size register (BK) used by ARAU in circular addressing
- A 32-bit system stack pointer (SP)
- A 32-bit status register (ST) that contains global information on the state of the CPU
- A 32-bit CPU/DMA interrupt enable register (IE)
- A 32-bit CPU interrupt flag register (IF)
- A 32-bit I/O flags register (IOF) to control the function of the dedicated external pins
- XF0 and XF1 for chip I/O
- A 32-bit repeat counter (RC) for looping operation

TABLE 12-31 DSP96002 Program Control Instructions

INSTRUCTION	DESCRIPTION	INSTRUCTION	DESCRIPTION
Bcc label	If the specified condition is true, program execution continues at location PC + displacement.	BRA label	Branch always to location PC + displacement
BRCLR #bit,S,label	Branch to location PC + displacement if the nth bit in the source operand is clear	BRSET #bit,S,label	Branch to location PC + displacement if the nth bit in the source operand is set
BScc label	Branch to subroutine if the condition is true	BSCLR #bit,S,label	Branch to subroutine if the nth bit of source operand is clear
BSR label	Branch to subroutine	BSSET #bit,S,label	Branch to subroutine if the nth bit in the source operand is set
DEBUGcc	If the specified condition is true, enter debug mode	FBcc label	Branch if the specified floating point condition is true
FBScc label	Branch to subroutine for floating-point condition	FDEBUGcc	If the specified floating-point condition is true, enter debug mode
FFcc (FFcc.U)	Conditional Data ALU operation without (with) CCR update	FJcc label	Jump if the floating-point condition is true
FJScc	Jump to subroutine if the floating-point condition is true	FTRAPcc	If the specified floating-point condition is true, initiate software exception processing
IFcc	Conditional Data ALU operation without CCR update	IFcc.U	Conditional Data ALU operation with CCR update
ILLEGAL	Illegal instruction interrupt	Jcc	Jump if the condition is true
JCLR #bit,S,label	Jump if the nth bit in the source operand is clear	JMP label	Jump
JScc label	Jump to subroutine if the specified condition is true	JSCLR #bit,S,label	Jump to subroutine if the nth bit in the source operand is clear
JSET #bit,S,label	Jump if the nth bit in the source operand is set	JSR label	Jump to subroutine
JSSET #bit,S,label	Jump to subroutine if the nth bit in the source operand is set	NOP	No operation
RESET	Reset peripheral devices	RTI	Return from interrupt
RTR	Return from subroutine and restore status register	RTS	Return from subroutine
STOP	Stop processing for low power stand-by	TRAPcc	If the specified integer condition is true, normal instruction execution is suspended and software exception processing is initiated.
WAIT	Wait for interrupt for low power stand-by		

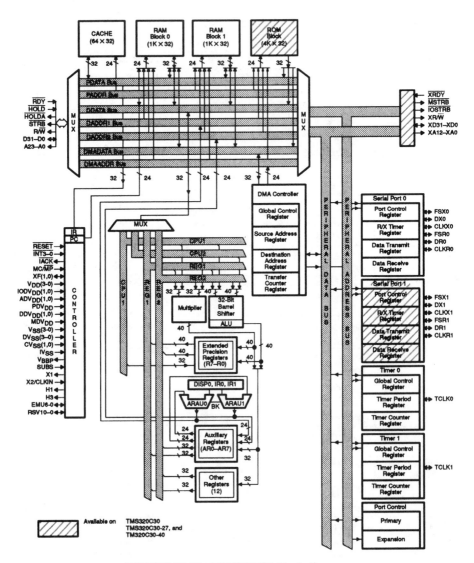

FIGURE 12-20 TMS320C30 block diagram.

- A 32-bit repeat start address register (RS)
- A 32-bit repeat end address register (RE)
- A 32-bit program counter (PC) register

A 64 × 32-bit instruction cache stores sections of the code that can be fetched when repeatedly accessing time-critical code. This greatly reduces the number of off-chip accesses necessary for code stored off-chip in slower, lower cost memories and external buses can be used by the DMA or other system elements. The cache can operate in a completely automatic fashion without the need for user intervention. Three cache control bits are located in the CPU SR: the cache clear (CC) bit, cache enable (CE) bit, and cache freeze (CF) bit. By writing proper data into these control bits the operation of cache can be controlled.

The total memory space of the TMS320C30 is 16 million 32-bit words. The memory map (shown in Figure 12-21) is dependent on whether the processor is running in the microprocessor mode (where the 4K on-chip ROM is not mapped into the memory map) or the microcomputer mode (where the 4K on-chip ROM is mapped into locations 0x0 through 0x0fff.

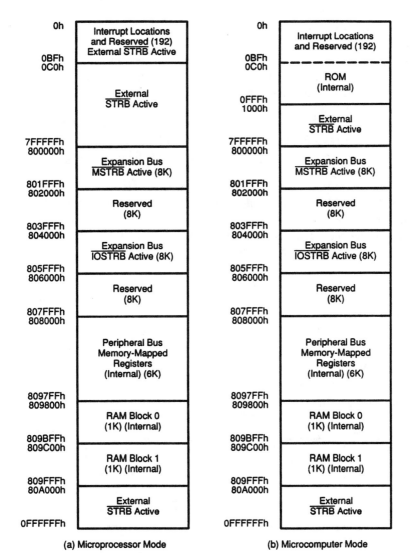

(a) Microprocessor Mode (b) Microcomputer Mode

FIGURE 12-21 TMS320C30 memory maps.

Five groups of addressing modes are provided on the TMS320C30:

- General addressing—register (any CPU register), short immediate (16-bit operand), direct (24-bit address), and indirect
- Three-operand addressing—register and indirect
- Parallel addressing—register (R0–R7) and indirect
- Long immediate (24-bit operand)
- Conditional branch addressing—register and PC relative (a signed 16-bit displacement is added to the PC)

TMS320C30 operation is controlled by five major functional units: fetch, decode, read, execute, and DMA. To provide for maximum processor throughput, these units can operate in parallel, with each unit operating on a different instruction creating a pipeline. The breaking of a pipeline in the midst of processing results in pipeline conflicts. The first class of pipeline conflicts occurs when a branching instruction is executed that requires reading and/or modifying the PC. Flushing the pipeline is necessary in these cases to guarantee that portions of succeeding instructions do not inadvertently get partially executed. The second class of pipeline conflicts are the register and memory conflicts. *Register conflicts* occur when reading or writing of some specific registers (not ready for such a read or write operation) is attempted. Extra delays get inserted to avoid such a conflict. Similarly, possible *memory conflicts* occur when the memory bandwidth of a physical memory space is exceeded. It is resolved by delaying program fetch until data access for the preceding instruction is completed. Careful programming avoids the use of "nop" instructions so as not to lose CPU cycles unnecessarily.

TMS320C30 Instruction Set Summary. The instruction set contains 114 instructions organized into six functional groups, which are included in Tables 12-32 through 12-38.

Programming of TMS320C30 can either be done in the assembly language or C. Use of the C compiler increases the transportability of applications that have been tested on large, general purpose computers and decreases their porting time.

TABLE 12-32 TMS320C30 Load and Store Instructions

INSTRUCTION	DESCRIPTION	INSTRUCTION	DESCRIPTION
LDE	Load floating point exponent	POP	Pop integer from top of stack
LDF	Load floating point value	POPF	Pop floating-point value from stack
LDF*cond*	Load floating-point value conditionally	PUSH	Push integer on stack
LDI	Load integer	PUSHF	Push floating-point value on stack
LDI*cond*	Load integer conditionally	STF	Store floating-point value
LDM	Load floating-point mantissa	STI	Store integer

TABLE 12-33   TMS320C30 Two-Operand Instructions

INSTRUCTION	DESCRIPTION	INSTRUCTION	DESCRIPTION
ABSF	Absolute value of a floating-point number	NORM	Normalize floating-point value
ABSI	Absolute value of an integer	NOT	Bitwise logical complement
ADDC	Add integers with carry	OR	Bitwise logical OR
ADDF	Add floating point values	RND	Round floating-point value
ADDI	Add integers	ROL	Rotate left
AND	Bitwise logical-AND	ROLC	Rotate left through carry
ANDN	Bitwise logical-AND with complement	ROR	Rotate right
ASH	Arithmetic shift	RORC	Rotate right through carry
CMPF	Compare floating-point values	SUBB	Subtract integers with borrow
CMPI	Compare integers	SUBC	Subtract integers conditionally
FIX	Convert floating-point value to integer	SUBF	Subtract floating-point values
FLOAT	Convert integer to floating-point value	SUBI	Subtract integer
LSH	Logical shift	SUBRB	Subtract reverse (source value - destination value) integer with borrow
MPYF	Multiply floating-point values	SUBRF	Subtract reverse (source value - destination value) floating-point value
MPYI	Multiply integers	SUBRI	Subtract reverse (source value - destination value) integer
NEGB	Negate integer with borrow	TSTB	Test bit fields
NEGF	Negate floating-point value	XOR	Bitwise exclusive OR
NEGI	Negate integer		

TABLE 12-34   TMS320C30 Three-Operand Instructions

INSTRUCTION	DESCRIPTION	INSTRUCTION	DESCRIPTION
ADDC3	Add operands with carry	MPYF3	Multiply floating-point values
ADDF3	Add floating-point values	MPYI3	Multiply integers
ADDI3	Add integers	OR3	Bitwise logical-OR
AND3	Bitwise logical-AND	SUBB3	Subtract integers with borrow
ANDN3	Bitwise logical-AND with complement	SUBF3	Subtract floating-point values
ASH3	Arithmetic shift for a given count	SUBI3	Subtract integers
CMPF3	Compare floating-point values	TSTB3	Test bit fields
CMPI3	Compare integers	XOR3	Bitwise exclusive-OR
LSH3	Logical shift for a given count		

TABLE 12-35 TMS320C30 Program Control Instructions

INSTRUCTION	DESCRIPTION	INSTRUCTION	DESCRIPTION
B*cond*	Branch conditionally	IDLE	Idle until interrupt
B*cond*D	Delayed branch conditionally	NOP	No operation
BR	Branch unconditionally	RETI*cond*	Return from interrupt conditionally
BRD	Delayed branch unconditionally	RETS*cond*	Return from subroutine unconditionally
CALL	Call subroutine	RPTB	Repeat block of instructions
CALL*cond*	Call subroutine conditionally	RPTS	Repeat single instruction
DB*cond*	Decrement specified auxiliary register and branch conditionally	SWI	Software interrupt
DB*cond*	Delayed decrement specified auxiliary register and branch conditionally	TRAP*cond*	Trap interrupt conditionally

For best use of the C compiler it is recommended that the application in C is run in real time. If it does not run in the desired time, one has to identify places where most of the time is spent and optimize those areas by writing equivalent assembly language routines. When writing a C program, a simple way to increase the execution speed is to maximize the use of register variables.

In order to write efficient assembly language code the following hints are useful:

- Use delayed branches (B*cond*D, BRD) that take a single cycle (as opposed to four for regular branches, B*cond*, BR) to execute. The three instructions following the branch instruction are always executed.
- Apply the repeat single/block construct (RPTS, RPTB) to achieve looping with no overhead.
- Use parallel instructions (such as MPYF3 || ADDF3) to increase the number of operations executed in a single cycle. For example, it is possible to have a multiply in parallel with an add and store the results.
- Maximize the use of registers by using them in parallel instructions and as scratch pad memory.
- Use the cache specially in conjunction with external slow memory.

TABLE 12-36 TMS320C30 Interlocked Op Instructions for Multiprocessor Communication

INSTRUCTION	DESCRIPTION	INSTRUCTION	DESCRIPTION
LDFI	Load floating-point value, interlocked	STFI	Store floating-point value, interlocked
LDII	Load integer, interlocked	STII	Store integer, interlocked
SIGI	Signal interlocked operation		

TABLE 12-37   TMS320C30 Parallel Operations Instructions

INSTRUCTION	DESCRIPTION	INSTRUCTION	DESCRIPTION
ABSF || STF	Absolute value of a floating-point number in parallel with store	MPYF3 || STF	Multiply floating-point values in parallel with store
ABSI || STI	Absolute value of an integer in parallel with store	MPYF3 || SUBF3	Multiply and subtract floating-point values
ADDF3 || STF	Add floating-point values in parallel with store	MPYI3 || ADDI3	Multiply and add integer values
ADDI3 || STI	Add integer values in parallel with store	MPYI3 || STI	Multiply integer values in parallel with store
AND3 || STI	Bitwise logical-AND in parallel with store	MPYI3 || SUBI3	Multiply and subtract integer values
ASH3 || STI	Arithmetic shift in parallel with store	NEGF || STF	Negate floating-point value in parallel with store
FIX || STI	Convert floating-point value to integer in parallel with store	NEGI || STI	Negate integer value in parallel with store
FLOAT || STF	Convert integer value to floating-point in parallel with store	NOT3 || STI	Complement integer in parallel with store
LDF || LDF	Parallel load floating-point data	OR3 || STI	Bitwise logical-OR in parallel with store
LDF || STF	Load floating-point data in parallel with store	STF || STF	Parallel store floating-point data
LDI || LDI	Parallel load integers	STI || STI	Parallel store integers
LDI || STI	Load integer data in parallel with store	SUBF3 || STF	Subtract floating-point values in parallel with store
LSH3 || STI	Logical shift in parallel with store	SUBI3 || STI	Subtract integer data in parallel with store
MPYF3 || ADDF3	Multiply and add floating-point numbers	XOR3 || STI	Bitwise exclusive-OR in parallel with store

TABLE 12-38   TMS320C30 Condition Codes

CONDITION	DESCRIPTION	CONDITION	DESCRIPTION
U	Unconditional	GE (NN)	Greater than or equal (Nonnegative)
LO (C)	Lower than (Carry)	NV	No overflow
LS	Lower or same	V	Overflow
HI	Higher than	NUF	No underflow
HS (NC)	Higher or same (No carry)	UF	Underflow
EQ (Z)	Equal (Zero)	NLV	No latched overflow
NE (NZ)	Not equal (Not zero)	LV	Latched overflow
LT (N)	Less than (Negative)	NLUF	No latched floating-point underflow
LE	Less than or equal	LUF	Latched floating-point underflow
GT (P)	Greater than (Positive)	ZUF	Zero or floating-point underflow

- Use the faster on-chip memory instead of external memory. Use the DMA in parallel with the CPU to transfer data to internal memory before operating on them.
- Avoid pipeline conflicts as discussed earlier.

In Table 12-33 two-operand instructions are listed. The two operands are the source and destination. The source operand may be a memory word, a register, or a part of the instruction word. The destination operand is always a register.

Tables 12-34, 12-35, and 12-36 list three-operand, program control, and multiprocessor communication instructions for TMS320C30.

The parallel instructions (in Table 12-37) allow:

- Parallel loading of registers
- Parallel arithmetic operations (add, subtract, multiply) or
- Arithmetic/logical operations in parallel with a store instruction

Each instruction in a pair is entered as a separate source statement with the second instruction always preceded by two vertical bars (‖).

The 20 condition codes in Table 12-38 can be used with any of the conditional instructions (indicated by *cond*). The conditions include signed and unsigned comparisons, comparisons to zero, and comparisons based on the status of seven individual condition flags:

- N—negative condition flag
- Z—zero condition flag
- V—overflow condition flag
- C—carry flag
- UF—floating-point underflow condition flag
- LV—latched overflow condition flag (set whenever V is set)
- LUF—latched underflow condition flag (set whenever UF is set)

All conditional instructions can accept the suffix "U" to indicate unconditional operation.

12-5 EXAMPLES OF DSP CHIP IMPLEMENTATION OF FIR FILTERS[5]

In this section we include the source-file listings for FIR programs. These programs are intended to demonstrate the use of various programming techniques and may easily be modified for use in different applications. Comments within the

---

[5]The text in this section on Texas Instruments and Motorola DSPs has been contributed by Dr. Sanjit K. Mitra of the University of California, Santa Barbara.

source files provide all the information necessary to understand the program's function and to make minor modifications.

12-5-1 FIR Filtering on TMS320C25

In this program [TE90] the delay elements are implemented by consequent data RAM locations. One multiplication and accumulation with the previous results will be done at every instruction cycle. We employ all the parallelism available in TMS320C25 to ensure maximum throughput. In order to benefit from the single-instruction cycle multiply-accumulate operation (MACD), we have to configure the memory block B0 as program memory. We also have to arrange the signal samples and the coefficients in the memory in the manner shown in Figure 12-22.

The instruction MACD stands for "multiply and accumulate with data move." It has several modes of operation. Among them, the following case is of interest. **MACD $\langle$pma$\rangle$, $\{*-\}$** will multiply the content of the program memory (block B0), addressed by the value of program memory address (i.e., $\langle$pma$\rangle$), by the content of the data memory (block B1), addressed by the current pointer (desired pointer register should be selected before this instruction). In the same instruction, after the operands have been fetched, the operand in the data memory will be copied to the next higher on-chip RAM location. This in effect realizes the z^{-1} delay action. Also, the current data memory pointer will be decremented (as requested by the $*-$convention in the MACD instruction above). This pointer decrement will prepare the machine to address the next coefficient. The program memory address $\langle$pma$\rangle$, which in this instruction gets stored in the program counter (PC), is also automatically incremented. It can be recalled that the multiply-accumulate block is pipelined. Thus when the MACD instruction is completed (after one instruction cycle), the result of the latest multiplication did not get accumu-

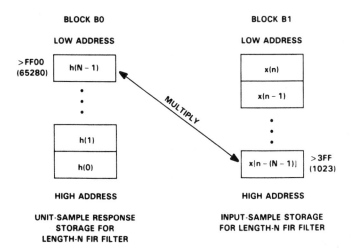

FIGURE 12-22 TMS320C25 memory storage scheme for FIR filtering.

```
*************************************************************************
*
*       FIR Filtering Code for TMS320C25.
*
*************************************************************************
*
        CNFP                    * Use block B0 as program area.
*
* This section of code polls the input port to bring in the next
* sample.
*
WAIT BIOZ NXT                   * "BIO" pin goes low when a new sample
                                * is available. This instruction looks
                                * to find out if "BIO" pin is zero. If
                                * so, a branch to NXT takes place. Else,
                                * the next instruction loops the control
                                * back to WAIT.
*
        B    WAIT               * Branch to WAIT.
*
NXT  IN    XN,PA0               * XN is a constant showing the desired place
                                * for storing the incoming data. This
                                * address, i.e. XN is specified as an offset to
                                * the  beginning page of selected memory block.
                                * Its value depends on the order of the filter.
                                * PA0 refers to the I/O port address 0.
                                * IN brings the data (signal sample) from
                                * peripheral on port address 0.
                                * NXT is the label of this instruction. It
                                * will be used whenever we like to jump to this
                                * instruction for bringing the next sample.
*
        LRLK AR1,>3FF           * The auxiliary register AR1 is loaded with the
                                * address of the highest point in block B1.
*
        LARP AR1                * By loading the ARP register with 1, we are
                                * selecting AR1, as our pointer register.
*
        MPYK 0                  * Multiplication by constant 0, will clear the
                                * Product Register.
*
        ZAC                     * Zero accumulator clears the accumulator.
*
* This section of code implements the sum expression from Eqn. (12.1.1)
*
        RPTK NM1                * NM1 is a constant equal to N - 1.
```

FIGURE 12-23 FIR filtering code for TMS320C25. (The code fragment is reproduced by permission of Texas Instruments, Inc., with additional annotations by the author.)

```
                        * RPTK as explained above, will repeat the
                        * following instruction N times.
*
    MACD >FF00,*-       * This instruction was explained in detail
                        * above. It multiplies and accumulates according
                        * to the above FIR expression.
*
    APAC                * Add Product Register to accumulator.
*
    SACH YN,1           * YN is a constant indicating the desired
                        * location in the data memory (B2)
                        * for storage of the output samples.
                        * We have assumed that all the input samples
                        * and the coefficients have values ≤1.
                        * Therefore, the result of all the multiply-
                        * accumulations is 32 bits, with the decimal
                        * point between the 30th and 31st bits.
                        * SACH copies the entire accumulator into a
                        * shifter. It then shifts this entire 32-bit
                        * number 1 bit (as we have asked) to the left,
                        * and stores the upper 16 bits of the shifted
                        * value into data memory at location YN.
*
    OUT  YN,PA1         * Outputs the filter response now stored at
                        * location YN, to peripheral on port address 1.
*
    B    WAIT           * Branch to get next input sample.
```

FIGURE 12-23 (*Continued*)

lated to the previous result yet. This will be taken care of by an extra APAC ("add p register to accumulator") instruction.

There is another instruction that we like to take advantage of. That instruction is **RPTK** ⟨**constant**⟩ ("repeat following instruction as specified by immediate value"). In this instruction, the 8-bit immediate value (constant) is loaded into a dedicated counter, called RPTC. Several other instructions (such as MACD) condition their operations on the content of this counter. That is, if the content of RPTC is greater than zero, they will continue their normal operation and arrange for their reexecution while at the same time decrementing RPTC after each execution. However, if the content of RPTC is zero, these instructions will finish the execution of current instruction and arrange for the normal execution of the next instruction in the sequence. Note that in order to repeat an instruction N times, one has to set the initial value of RPTC to $N - 1$.

With the memory arrangement of the signal samples, and the coefficients as depicted in Figure 12-22, and the two instructions described above, we can now write the program as shown in Figure 12-23 for an Nth order FIR filter ($N \leq 256$, since we use the RPTK instruction to repeat tap multiply-accumulate and RPTK will repeat no more than 256 times).

```
/*
        DSP16/DSP16A FIR filter design example:

        The following code represents a 20th order FIR filter.
        Input samples are provided via the serial input.
        The output samples are sent to the serial output.  */


  .ram
X20:          19*int
X1:           int
ibuf:         int
  .endram

/*    20th order FIR filter

        This FIR example uses a modulo-addressed delay line (circular
        buffering).
        Outputs are calculated from X[n-20] toward X[n].
        The inner loop of the filter uses 21 multiplies and requires
        2.1 microseconds.

              a0:    scratch calculations
              r0:    input buffer locations
              r1:    delay line pointer
              pt:    coefficient pointer

        Also modifies rb, re, i, x, y, p, sdx  */

fir20:          auc = 0x02       /* Use fractional notation */
                pt = H20         /* Initialize coefficient pointer  */
                r0 = ibuf
                i = -20
                rb = X20         /* The first location of the state
                                    variable buffer in RAM */
                re = X1          /* The last location of the state
                                    variable buffer in RAM  */
                r1 = X20
loop:           y = 0x0010       /* Main loop */

wait:           a0 = pioc        /* Wait for valid input  */
                a0&y
       if eq    goto wait
                *r0 = sdx        /* Get input sample from serial input  */

              /* Perform Convolution  */

                a0 = p                          y = *r1++       x = *pt++
       do 19 {
                a0 = a0 - p      p = x*y      y = *r1++        x = *pt++
       }
```

FIGURE 12-24 FIR filtering code for DSP16A. (The code fragment is reproduced by permission of AT&T with additional annotations by the author.)

```
          a0 = a0 - p    p = x*y    y = *r0              x = *pt++i
          a0 = a0 - p    p = x*y    *r1++ = y
          a0 = a0 - p
          sdx = a0              /* Output samples to serial output */
endl:     goto loop

/*   Coefficients from h[n-20] to h[n]   */

H20: int   -0.001021
          .
          .
          .
```

FIGURE 12-24   *(Continued)*

For an FIR filter with 80 coefficients, it takes 93 cycles to execute the above program. Therefore with a TMS320C25 that has 100 ns instruction cycle time, we can filter signals at a rate as fast as $93 \times 100 = 9300$ ns intervals, or at 108 kHz sampling rate.

12-5-2   FIR Filtering on DSP16A

A FIR filtering code fragment implemented on DSP16A [AT89a] is shown in Figure 12-24.

12-5-3   FIR Filtering on DSP56000

Figure 12-25 shows the code fragment for a 20-tap FIR filter implemented on Motorola 56000.

Note that during the execution of the FIR filter algorithm, the address register r4 is post-incremented a total of *ntaps* times. Thus the address value in r4 wraps around and points to the coefficient buffer's lower boundary location. The first coefficient b_0 resides in this buffer location.

Note also that during the filter program execution, the address register m 0 is postincremented a total of *ntaps* times. Thus the address value in r0 wraps around and points to the state variable buffer's x-memory location M (that contains the current input sample u[n]). In conjunction with the *macr* instruction, r0 is post-decremented, resulting in r0 pointing to the state variable buffer's x-memory location $M - 1$. The next time the program is executed, a new (next) input sample $u(n + 1)$ will then overwrite the value in x-memory location $M - 1$.

12-5-4   FIR Filtering on WE DSP32C

We now examine programming the WE DSP32C [AT90] for an Nth order FIR filter (Figure 12-26). A direct form implementation of the filter is available as "fir" routine in WE DSP32C Library Routine. A single WE DSP32C instruction can perform all the operations needed to perform one tap of FIR filter: fetch two operands, multiply, accumulate, and shift data samples.

```
;
;       20-tap FIR filter implemented on Motorola 56000/1
;
;       Maximum sample rate at 27 MHz chip clock frequency:  500 kHz
;       Memory size:  Prog - 4 + 6 words;  Data - 2 x 20 words
;       Number of clock cycles:  54
;
;       This FIR filter reads the input sample from the memory location
;       Y:input and writes the output sample to the memory location
;       Y:output
;
;       The samples are stored in the X memory
;       The coefficients are stored in the Y memory
;
;       The filter equation is:
;
```
$$y[n] \;=\; \sum_{k=0}^{N-1} b_k u[n-k]$$
```
;
;       initialization
;
ntaps       equ         20
start       equ         $40          ; start address in HEX
states      equ         $0
coef        equ         $0
input       equ         $ffe0
output      equ         $ffe1
;
            org         p:start
            move        #states,r0   ; r0 pointing to samples
            move        #coef,r4     ; r4 pointing to coefficients
            move        #ntaps - 1,m0 ; set modulo arithmetic
            move        m0,m4        ; for the 2 circular buffer
;
            opt         cc
;
;       filter loop:  8 + (ntaps - 1) cycles
;
            movep       y:input,x:(r0) ; input sample in memory
;
            clr         a       x:(r0)+,x0        y:(r4)+,y0
;
;       the clr instruction above clears the a-accumulator and simultaneously
;       a.  moves the input sample from the x-memory location pointed
;           to by address register r0 to the data ALU's input register x0 and
;       b.  moves the coefficient from the y-memory location pointed to
;           by address register r4 to the data ALU's input register y0,
;       and also post-increments both address registers r0 and r4
;
```

FIGURE 12-25 FIR filtering code for DSP56000. (The code fragment is reproduced by permission of Motorola, Inc., with additional annotations by the author.)

```
        rep         #ntaps - 1              ; repeat next instruction
;                                           ntaps - 1 times
;

        mac         x0,y0,a   x:(r0)+,x0              y:(r4)+,y0
;
;       the mac instruction above multiplies the filter state variable
;       in x0 by the coefficient in y0, adds the product to the
;       a-accumulator, and simultaneously
;       a.  moves the next state variable from the x-memory location
;           pointed to by the address register r0 to input register x0 and
;       b.  moves the next coefficient from the y-memory location
;           pointed to by address register r4 to input register y0,
;       and also post-increments both address registers r0 and r4
;

        macr        x0,y0,a   (r0)-
;
;       at the last step of final tap computation for a given sample,
;       the preceding macr instruction performs signed
;       multiply-accumulate with rounding and post-decrements the
;       address register r0.
;

        movep       a,y:output       ; output filtered sample
;
        end
```

FIGURE 12-25   (*Continued*)

12-5-5   FIR Filtering on DSP96002

Figure 12-27 illustrates the code for FIR filtering on DSP96002 [MO89].

12-5-6   FIR Filtering on TMS320C30

Figure 12-28 shows the actual code fragment for the implementation of FIR filtering on TMS320C30 [TE91].

12-6   EXAMPLES OF DSP CHIP IMPLEMENTATION OF IIR FILTERS[6]

In this section we include the source-file listings for some representative IIR filter implementation programs. These programs are intended to demonstrate the use of various programming techniques and may easily be modified for use in different applications. Comments within the source files provide all the information necessary to understand the program's function and to make minor modifications.

---

[6]The text in this section on Texas Instruments and Motorola DSPs has been contributed by Dr. Sanjit K. Mitra of the University of California, Santa Barbara.

```
/*
This section of code implements the FIR filtering expression
```

$$\sum_{i=0}^{N-1} h[i]\, u[n-i]$$

```
on WE DSP32C processor.
```

Arguments:

N	Length of the FIR filter (N > 2).
&in	Memory location storing the input sample
&out	Memory location for the output sample
&h(N-1)	Memory location of the (N - 1)th filter coefficient where the coefficients are stored in consecutive memory locations and arranged in reversed order
&firS	Starting memory location for the state variables

Execution Time:

   14+N instructions for a filter of length N

Synopsis:

```
main:
             .
             .
             .
             call fir (r14)
             nop
int24        N
int24        &in, &out
int24        &h(N-1), &firS
 .align      4
             .
             .
             . */
```

```
/*  Initialize CAU registers  */
```

```
fir:        r1e = *r14++         /* r1 = 24-bit integer length of the
                                    FIR filter, N, done by immediately
                                    loading the contents of the location
                                    pointed to by r14 into r1 and
                                    post-incrementing r14   */

            r5e = *r14++         /* r5 points to 24-bit input sample &in
                                    */

            r6e = *r14++         /* r6 points to 24-bit output sample
                                    &out   */
```

FIGURE 12-26 FIR filtering code for WE DSP32C. (The code fragment is reproduced by permission of AT&T with additional annotations by the author.)

```
r2e = *r14++          /*  r2 points to 24-bit coefficients
                      &h(N-1) from table  */
r3e = *r14++          /*  r3 points to 24-bit state variable
                      &firS  */
r1e = r1 + (-3)       /*  r1 = N - 3 is used as 24-bit do loop
                      counter  */
r4e = r3              /*  r4 points to 24-bit state variable
                      */
a1 = *r4++ * *r2++    /*  a1 = u[n-N+1]h[N-1]  */

do 0, r1              /*  repeat N - 2 times the following  */

a1 = a1 + (*r3++ = *r4++) * *r2++
                      /*  sum of h[N-i]u[n-N+i] is
                      calculated for i from 2 to N - 1.  At
                      every step u[j] is replacing u[j-1]. */

a0 = a1 + (*r3 = *r5) * *r2
                      /*  u[n]h[0] is added to the
                      total sum, i.e., the output.
                      u[n] is replaced by u[n+1]. */

*r6 = a0 = a0         /*  store the convolution sum to the
                      memory location pointed to by r6  */

nop                   /*  guarantees the last store
                      instruction completion before
                      subroutine ends  */
return   (r14)
nop                   /*  end of fir routine  */
```

FIGURE 12-26   (*Continued*)

12-6-1 IIR Filtering on TMS320C25

Now we consider the implementation of IIR filters [TE90] in the form of cascaded second-order direct form II sections. One should emphasize that because of the finite precision and fixed-point multiplier of TI TMS320C25, one has to pair appropriately the poles and the zeros. Also, one should insert interstage multipliers for gain distribution [JA86]. However, in the present implementation these points are not explicitly shown.

We assume that there are S second-order sections in the code shown in Figure 12-30. Also, we store all the coefficients of the S sections as shown in Figure 12-29.

It takes 24 cycles to compute each second-order section. Therefore signals sampled up to 139 kHz can conveniently be filtered with IIR filters composed of three second-order sections.

```
;
;       N-tap FIR Filter with Data Shift on DSP96002.
;
;       Uses 10 program words and N+12 instruction cycles
;
N       equ  20

        org  x:0
data ds    N

        org  y:0
coef ds    N

        move #data,r0            ; point to states
        move #coef,r4            ; point to coefficients
        move #N - 1,m0           ; modulo addressing

        fclr d1                           m0,m4

;       clear floating-point register d1
;       d1 = 0
;       and set modulo addressing

        movep                     x:input,x:(r0)

;       get input sample from X I/O peripheral "input" and load into X
;       memory location pointed to by r0

        fclr d0                   x:(r0)-,d4.s   y:(r4)+,d6.s

;       clear floating-point register d0,
;       d0 = 0
;       move floating-point state u[n] from X memory location pointed
;       to by r0 into register d4 (and postdecrement r0),
;       d4 = u[n]
;       and move floating-point coefficient h[0] from Y memory location
;       pointed to by r4 into register d6 (and postincrement r4)
;       d6 = h[0]

        rep #N                            ; repeat next instruction N times

        fmpy d4,d6,d1  fadd.s d1,d0   x:(r0)-,d4.s   y:(r4)+,d6.s

;       multiply in floating-point the state (in d4) and the coefficient
;       (in d6), round the result and store in register d1
;       d1 = h[i-1]u[n-i+1]
;       simultaneously, add in floating-point the two operands in d1 and d0,
;       round to single precision and store the accumulated result in d0
;       move floating-point previous state u[n-i] from X memory location
```

FIGURE 12-27 FIR filtering code for DSP96002. (The code fragment is reproduced by permission of Motorola, Inc. with additional annotations by the author.)

```
;    pointed to by r0 into register d4 (and postdecrement r0),
;    d4 = u[n-i]
;    and move corresponding floating-point coefficient h[i] from Y
;    memory location pointed to by r4 into register d6 (and
;    postincrement r4)
;    d6 = h[i]
```

```
                fadd.s d1,d0
```

```
;    add the final tap value h[N-1]u[n-N+1] in d1 into accumulated
;    sum in d0 and store the result in d0
```

```
    movep                        d0.s,x:output
```

```
;    output sample value into X I/O peripheral "output" from d0
```

FIGURE 12-27 (*Continued*)

12-6-2 IIR Filtering on DSP16A

In Figure 12-31, we include the code for IIR filtering on DSP16A [AT89b]. Here "do" looping is used to evaluate biquad sections.

12-6-3 IIR Filtering on DSP56000

Figure 12-32 illustrates the code for implementing cascaded biquad sections on DSP56000.

12-6-4 IIR Filtering on WE DSP32C

The WE DSP32 and DSP32C software support includes an optimizing C compiler. The use of a high-level language like C language in developing DSP application programs makes the code modular, easy to understand, and portable [AT88]. Additionally, the DSP application code can be embedded in a behavioral level description of the overall system. The application code development time reduces considerably when the high-level language is used.

There are three libraries of functions distributed with the WE DSP32 and DSP32C C language compiler:

- A subset of the UNIX system standard C library, **libc.** (It includes only those functions needed to support error handling in the math library and the "printf" function for use with the WE DSP32/DSP32C software simulator.)
- The UNIX system standard math library, **libm.**
- The WE DSP32-AL Application Software Library, **libap.** (It includes functions for arithmetic, matrix arithmetic, filtering, adaptive filtering, fast Fourier transforms, and graphics/imaging.)

```
********************************************************************************
*
*       N tap FIR filter routine on TMS320C30
*
********************************************************************************
*
*       Implements:
*
*              N-1
*       y[n] = Σ h[i]u[n−i]
*              i=0
*
*       Synopsis:
*
*       LOAD    AR0
*       LOAD    AR1
*       LOAD    RC
*       LOAD    BK
*       CALL    FIR
*
*       Arguments:
*
*       AR0     Address of impulse response coefficients {h[N-1]}
*       AR1     Address of state variables {u[n-N+1]}
*       RC      Length of the filter - 2 (= N - 2)
*       BK      Length of the filter (= N)
*
*       Registers modified:  R0, R2, AR0, AR1, RC
*
*       Register containing result:  R0
*
*       Uses 6 words of program memory and 10+N instruction cycles
*
        .global    FIR
*                             ;  initialize R0:
FIR     MPYF3      *AR0++(1),*AR1++(1)%,R0
*                             ;  compute h[N−1]u[n−N+1] in floating-point   and
*                             ;  store result in R0, also postincrement AR0 & AR1
*
        LDF        0.0,R2     ;  initialize R2 to 0
*
*       Filter loop (1 ≤ i < N)
*
        RPTS       RC         ;  set up the repeat cycle of N - 2 for
*                             ;  the following parallel instruction
*
        MPYF3      *AR0++(1),*AR1++(1)%,R0
*                             ;  compute h[N−1−i]u[n−N+1+i] in floating-point and
*                             ;  store result in R0, also postincrement AR0 & AR1
```

FIGURE 12-28 FIR filtering code for TMS320C30. (The code fragment is reproduced by permission of Texas Instruments, Inc., with additional annotations by the author.)

```
*
|  |    ADDF3        R0,R2,R2    ;   in parallel accumulate in floating-point the
*                                ;   content of R0 into R2 and store result in R2
*
       ADDF3        R0,R2,R0    ;   accumulate last product h[0]u[n] in
*                                ;   floating-point and store result in R0
*       Return sequence
*
       RETS                     ;   loop return
*
*       End of the FIR routine
*
    .end
```

FIGURE 12-28   (*Continued*)

The program in Figure 12-33 calls the **iir** function from *libap* to calculate the unit pulse response of a two biquad section IIR lowpass filter. The input values are read from array **in** and the corresponding output values are stored in array **out.** Ten sample points are used.

12-6-5   IIR Filtering on FSP96002

Next, in Figure 12-34, we include the code fragment for implementing cascaded biquad sections on DSP96002 [MO89].

12-6-6   IIR Filtering on TMS320C30

Figure 12-35 shows the code for implementing cascaded biquad sections on TMS320C30 [TE91].

α_{11}
α_{21}
β_{21}
β_{11}
β_{01}
α_{12}
α_{22}
β_{22}
β_{12}
β_{02}
α_{13}
α_{23}
β_{23}
.
.
.

etc. **FIGURE 12-29**   TMS320C25 coefficient storage scheme for IIR filtering.

```
***********************************************************************
*
*      TMS320C25 Code for IIR Filtering
*
***********************************************************************

* This section of code polls the input port to bring in the next
* sample.
*
WAIT BIOZ NXT         * The function of these two lines was
     B    WAIT        * explained in the FIR program.
*
NXT  IN   XN,PA2      * Bring in the new sample u[n].
*
* This section of code sets up the counters.
*
     LARK AR3,SM1     * Load the auxiliary register AR3 with S - 1,
                      * where S is the number of desired sections.
*
     LAR  AR0,COEFP   * Load the auxiliary register AR0 with the
                      * first address of the coefficients.
*
     LAR  AR1,DNP     * Load the auxiliary register AR1 with the
                      * address of the d_i[n].
*
* This section of code implements the following equations for Direct Form II
*
```

$$* \; d_i[n] \; = \; y_{i-1}[n] \; + \; \alpha_{1i} d_i[n-1] \; + \; \alpha_{2i} d_i[n-2]$$

$$* \; y_i[n] \; = \; \beta_{0i} d_i[n] \; + \; \beta_{1i} d_i[n-1] \; + \; \beta_{2i} d_i[n-2]$$

$$* \; \text{where} \; i = 1, \; \ldots, \; \left\lceil \frac{N}{2} \right\rceil$$

```
*
LOOP LAC  XN,15       * Load (with correct bit alignment) the
                      * accumulator with the input to the filter.
*
     LARP AR1         * Choose AR1 as current pointer.
     MAR  *+          * Point at d_i[n-1].
     LT   *+,AR0      * Load T register (one input to the multiplier)
                      * with the data from d_i[n-1].
                      * Also point at the address of d_i[n-2].
                      * Choose AR0 as next pointer
                      * (for addressing the coefficients.)
*
     MPY  *+,AR1      * Calculate α_{1i}d_i[n-1]. Point at α_{2i}.
                      * Choose AR1 as next pointer.
*
     LTA  *-,AR0      * Load T register with d_i[n-2].
                      * Calculate u[n] + α_{1i}d_i[n-1]. Point at d_i[n-1].
                      * Choose AR0 as next pointer.
```

FIGURE 12-30 IIR filtering code on TMS320C25. (The code fragment is reproduced by permission of Texas Instruments, Inc., with additional annotations by the author.)

```
*
        MPY   *+,AR1      * Calculate α₂ᵢdᵢ[n-2].  Point at β₂ᵢ.
                          * Choose AR1 as next pointer.
*
        APAC              * By adding the product register to the accumulator,
                          * calculate u(n) + α₁ᵢdᵢ[n-1] + α₂ᵢdᵢ[n-2].
*
        MAR   *-          * Point at dᵢ[n].
        SACH  *+,1        * Store dᵢ[n] (with proper bit alignment).
                          * Point at dᵢ[n-1].
*
        ZAC               * Clear accumulator. Also, notice that the
                          * multiplier T register is still loaded with dᵢ[n-2].
*
        LARP  AR0         * Choose AR0 (for coefficients) pointer.
        MPY   *+,AR1      * Calculate β₂ᵢdᵢ[n-2].  Point at β₁ᵢ.
                          * Choose AR1 (for data) pointer.
*
        LTD   *-,AR0      * Load multiplier T register with dᵢ[n-1].
                          * Move β₂ᵢdᵢ[n-2] to the accumulator.
                          * Also, dᵢ[n-2] = dᵢ[n-1].  Point at dᵢ[n].
                          * Choose AR0 (for coefficients) pointer.
*
        MPY   *+,AR1      * Calculate β₁ᵢdᵢ[n-1].  Point at β₀ᵢ.
                          * Choose AR1 (for data) pointer.
*
        LTD   *+,AR0      * Load multiplier T register with dᵢ[n].
                          * Calculate β₂ᵢdᵢ[n-2] + β₁ᵢdᵢ[n-1].
                          * Also, dᵢ[n-1] = dᵢ[n].  Point at dᵢ[n-1].
                          * Choose AR0 (for coefficients) pointer.
*
        MPY   *+,AR1      * Calculate β₀ᵢdᵢ[n].
                          * Point at the α₁ᵢ of the next section.
                          * Choose AR1 (for data) pointer.
*
        MAR   *+          * Point at dᵢ[n-2].
        MAR   *+          * Point at dᵢ[n] of the next section.
*
        APAC              * By adding the product register to the accumulator,
                          * Calculate yᵢ[n] = β₀ᵢdᵢ[n] + β₁ᵢdᵢ[n-1] + β₂ᵢdᵢ[n-2].
*
        SACH  XN,1        * Store the output of the ith section in the
                          * same place where the input of the (i + 1)th section
                          * is expected.
*
        LARP  AR3         * Use third auxiliary register.
        BANZ  LOOP        * If all of the sections have not been
                          * calculated, go to process the next section.
*
OTPT    OUT   XN,PA2      * Output the filter response y[n] (now in the
                          * same location as previous XN).
*
        B     WAIT        * Branch to process next input.
```
$$\text{MPY} \quad *+,\text{AR1} \quad * \text{ Calculate } \alpha_{2i}d_i[n-2].$$
FIGURE 12-30 (*Continued*)

```
/*

        DSP16/DSP16A IIR filter design example:

        The following code represents a 5th order IIR filter.
        It is implemented as 3 five-multiply biquad sections with
        numerator coefficients:  a₁ᵢ and a₂ᵢ and
        denominator coefficients:  b₀ᵢ, b₁ᵢ, and b₂ᵢ.
        Input samples are provided via the SIO and prescaled with no
        loss of precision and the output samples are sent to the SIO.

        The filter was designed by performing a bilinear transformation
        on an analog elliptic filter designed with the following
        parameters.

                3-dB Cutoff Frequency = 1    kHz
                Stopband Edge         = 1.5  kHz
                Stopband Attenuation  = 50   dB
                Passband Ripple       = 0.2  dB
                Sampling Frequency    = 10   kHz   */
```

```
  .ram
statev:           6*int            /* Six state variables */
  .endram
/*    This IIR coding example uses the following DSP16 resources:

        a0:   input/output for each section
        r1:   state variable pointer
        pt:   pointer to filter coefficients
        j, k: for compound addressing mode

      Here compound addressing is used to maintain the state
      variable delay line.  */
```

```
iir5:             auc = 0x02       /* Use fractional notation  */
                  j = -2           /* Initialize registers  */
                  k = 3            /* for compound addressing  */

loop:             pt = coef        /* Main program loop  */
                  r1 = statev
                  y = 0x0010       /* SIO input buffer full mask in HEX  */

wait:             a0 = pioc        /* Wait for valid input  */
                  a0&y
          if eq   goto wait
                  a0 = sdx         /* Read from SIO input buffer  */
                  a0 = a0>>1       /* Prescale input value (/2)  */
                  a0 = a0>>1       /* Divide by 2  */
```

FIGURE 12-31 IIR filtering code on DSP16A. (The code fragment is reproduced by permission of AT&T with additional annotations by the author.)

```
/*          Execute three second-order sections   */

                                     y = *r1++      x = *pt++ /* a₁₁ */
                          p = x*y    y = *r1--      x = *pt++ /* a₂₁ */
    do 3 {
            a0 = a0 - p   p = x*y    y = *r1++      x = *pt++ /* b₁ᵢ */
            a0 = a0 - p   p = x*y    *r1zp:y        x = *pt++ /* b₂ᵢ */
```

```
/*    Here compound addressing mnemonic *r1zp:y implies three steps:
      temp = y;  y = *r1;  *r1++ = temp;  */
```

```
            a0 = p        p = x*y    y = a0         x = *pt++ /* b₀ᵢ */
            a0 = a0 + p   p = x*y    *r1jk:y        x = *pt++ /* a₁ᵢ₊₁ */
```

```
/*    Here compound addressing mnemonic *r1jk:y implies three steps:
      temp = y;  y = *r1++j;  *r1++k = temp;  */
```

```
            a0 = a0 + p   p = x*y    y = *r1--      x = *pt++ /* a₂ᵢ₊₁ */
    }
            sdx = a0                 /* Write to SIO output buffer  */
```

```
/*    First order section coefficients   */
coef:       int   -0.759982      /* a₁₁ */
            int   0.0             /* a₂₁ */
            int   0.060045        /* b₁₁ */
            int   0.0             /* b₂₁ */
            int   0.060045        /* b₀₁ */

            int   -1.508821       /* a₁₂ */
            int   0.70049         /* a₂₂ */
            int   -0.108068       /* b₁₂ */
            int   0.245454        /* b₂₂ */
            int   0.245952        /* b₀₂ */

            int   -1.530470       /* a₁₃ */
            int   0.905134        /* a₂₃ */
            int   -1.976028       /* b₁₃ */
            int   1.758388        /* b₂₃ */
            int   1.759004        /* b₀₃ */
```

FIGURE 12-31 (*Continued*)

12-7 EXAMPLES OF DSP CHIP IMPLEMENTATION OF FFT ALGORITHMS[7]

As discussed in Chapter 8, there are several FFT algorithms that have been proposed in the literature and coded on commercial DSP chips. Due to the architectural and technological limitations, there is some upper bound on the number of points that can be accommodated in a specific chip. In this section, we only include code fragments that are readily available for the various chips and show some of

---

[7]The text in this section on Texas Instruments and Motorola DSPs has been contributed by Dr. Sanjit K. Mitra of the University of California, Santa Barbara.

```
;
;       DSP56000/1 Cascaded Canonic Biquad IIR filter
;
;       The IIR filter reads the input sample from the memory location
;       Y:input
;       and writes the filtered output sample to the memory location
;       Y:output
;
;       The samples are stored in the X memory
;       The coefficients are stored in the Y memory
;
;       The equations of the filter implemented here are:
```

$$w[n]/2 = u[n]/2 - a_{1i}w[n-1]/2 - a_{2i}w[n-2]/2$$
$$y[n]/2 = w[n]/2 + b_{1i}w[n-1]/2 + b_{2i}w[n-2]/2$$

```
;
;       Initialization
;
nsec        equ     3
start       equ     $40
data        equ     0
coef        equ     0
input       equ     $ffe0
output      equ     $ffe1
igain       equ     0.5
            ori     #$08,mr         ;  set scaling mode
            move    #data,r0        ;  point to filter states
            move    #coef,r4        ;  point to filter coefficients
            move    #2*nsec - 1,m0
            move    #4*nsec - 1,m4
            move    #igain,y1       ;  y1 = initial gain

            opt     cc
;
;       filter loop: 4*nsec+9
;
            movep   y:input,y0      ;  get input sample
;
;       y0 = input sample, y1 = 0.5
;
            mpy     y0,y1,a         x:(r0)+,x0          y:(r4)+,y0
;
```

; x0 = 1st section state variable $w[n-2]$, y0 = $a_{21}/2$

```
;
            do      #nsec,endl      ;  do each section
            mac     -x0,y0,a        x:(r0)-,x1          y:(r4)+,y0
;
```

; x1 = state variable $w[n-1]$, y0 = $a_{1i}/2$

```
;
            macr    -x1,y0,a        x1,x:(r0)+          y:(r4)+,y0
;
```

; push $w[n-1]$ to $w[n-2]$ in x0, y0 = $b_{2i}/2$

```
;
            mac     x0,y0,a         a,x:(r0)+           y:(r4)+,y0
```

FIGURE 12-32 IIR filtering code on DSP56000. (The code fragment is reproduced by permission of Motorola, Inc., with additional annotations by the author.)

```
;
;        push w[n] to w[n-1] in x1, y0 = b₁ᵢ/2
;
         mac          x1,y0,a          x:(r0)+,x0              y:(r4)+,y0
;
;        for next iteration: x0 = w[n-2], y0 = a₂ᵢ/2
;
         endl
;
         rnd          a                ;  round result
         movep        a,y:output       ;  output sample
;
         end
```

FIGURE 12-32 (*Continued*)

the complexities encountered in coding FFT algorithms. We have included brief explanations on the working of the programs in the form of comments in each code fragment.

12-7-1 FFT Implementation on TMS320C25

Figure 12-36 provides the TMS320C25 code for an 8-point DIT FFT implementation [TE90] using "special" butterflies. Although any N-point FFT (where N is a power of 2) can be directly implemented with the general butterfly only, special butterflies are used to speed up the execution.

Special butterflies can take advantage of the symmetry of the sine and cosine values of certain twiddle factors (e.g., 45 degrees). These special butterflies can reduce the size of the code.

Also, the first two stages of an N-point radix-2 FFT can be performed simultaneously with a special radix-4 butterfly. It includes four separate radix-2 butterflies and is equivalent to a 4-point DFT. This can speed up the execution of the FFT code.

12-7-2 FFT Implementation of DSP16A [AT89b]

The WE DSP16 and DSP16A Application Software Library provides subroutines written in assembly language suitable for many signal processing applications. Here we show the use of the 8-point FFT computation routine –fft8. Routines for computing FFTs up to 1024 points are included in the library.

In the program in Figure 12-37, an 8-point complex FFT is computed on real data. It implements the in-place, decimation-in-time, radix-2 FFT algorithm. The data consist of a 1-kHz and a 2-kHz sinusoid (sampled at 8-kHz rate). The two tones have equal amplitude and the phase of the 1-kHz tone is 45 degrees ahead of the 2-kHz tone.

```
        Cascaded 2nd order IIR filter sections with four multiplications
        per section for DSP32C.

        The transfer function of the IIR filter is given by:
```

$$H(z) \;=\; K \prod_{i=1}^{N} \frac{1+\beta_{1i}z^{-1}+\beta_{2i}z^{-2}}{1+\alpha_{1i}z^{-1}+\alpha_{2i}z^{-2}}$$

```
        where K is the overall constant multiplier and N is the number of
        biquad sections.
*/

#include <libap.h>

#define    N    2                        /*  4th order filter  */

float in[] = {                           /*  unit pulse input samples  */
     1.0,  0.,  0.,  0.,  0.,  0.,  0.,  0.,  0.,  0.
};

float iircoe[] = {                       /*  filter coefficients  */
     2.637256e-2,                        /*  K  */
     -1.022403,  0.3256819,              /*  α11 ,α21   */
     1.452737,  1.,                      /*  β11 ,β21   */
     -1.072848,  0.7197602,              /*  α12 ,α22   */
     0.1447465,  1.                      /*  β12 ,β22   */
};

float iirsv[2*N];                        /*  state variables  */
float out[10];

main()
{
     register int i;
     register float *p = in;
     register float *q = out;
     register float *coef = iircoe;
     register float *sv = iirsv;

     for (i = 9; i-- >= 0; )

/*    call iir() function  */

          *q++ = iir(N, *p++, coef, sv);

}

/*    After the program is run, the results stored in array out are:

          0.02637256    0.09738685    0.2058420    0.2934080
          0.2996999     0.1998714     0.04697882   -0.07330985
          -0.1075733    -0.06420625
*/
```

FIGURE 12-33 IIR filtering code on WE DSP32C. (The code fragment is reproduced by permission of AT&T with additional annotations by the authors.)

```
;
;        N Cascaded Real Biquad IIR Filters on DSP96002
;
;        The ith biquad section is characterized by:
;
;        w_i[n]  =  u_i[n]  -  a_{1i}w_i[n-1]  -  a_{2i}w_i[n-2]
;        y_i[n]  =  w_i[n]  +  b_{1i}w_i[n-1]  +  b_{2i}w_i[n-2]
;
;        Uses 19 program words and 4N+18 instruction cycles
;
;        X Memory Organization              Y Memory Organization
;
;                                           b_{1N}          coef + 4N - 1
;                                           b_{2N}
;                                           a_{1N}
;                                           a_{2N}
;                                             .
;                                             .
;        w_N[n-1]         data + 2N - 1       .
;        w_N[n-2]                             .
;          .                                  .
;          .                                  .
;          .                                b_{11}
;          .                                b_{21}
;        w_1[n-1]                           a_{11}
;        w_1[n-2]         data,r0,r1        a_{21}          coef,r4
;
N        equ   2

         org   x:0
data     ds    2N

         org   y:0
coef     ds    4N

         move  #$ffffffff,m0          ; initialize for modulo addressing
         move  m0,m4
         move  m0,m1

         move  #data,r0               ; point to states
         move  r0,r1

         move  #coef,r4               ; point to coefficients

         movep                                 x:input,d0.s

; get floating-point input sample from X I/O peripheral "input"
; and load into d0
; d0 = u[n]

         fclr  d1                              x:(r0)+,d4.s   y:(r4-,d6.s

; clear floating-point register d1
; d1 = 0
```

FIGURE 12-34 IIR filtering code on DSP96002. (The code fragment is reproduced by permission of Motorola, Inc., with additional annotations by the author.)

```
;    move floating-point state w₁[n-2] from X memory location
;    pointed to by r0 into register d4 (and postincrement r0),
;    d4 = w₁[n-2]
;    and move floating-point coefficient a₂₁ from Y memory
;    location pointed to by r4 into register d6 (and postincrement r4)
;    d6 = a₂₁

     do    #N,end

;    repeat the following instructions until end for N times

     fmpy d4,d6,d1  fadd.s   d1,d0    x:(r0)+,d5.s   y:(r4)+,d6.s

;    multiply in floating-point the state (in d4) and the coefficient
;    (in d6), round the result and store in register d1
;    d1 = wᵢ[n-2] a₂ᵢ
;    simultaneously, add in floating-point the two operands in d1 and d0,
;    round to single precision and store the accumulated result in d0
;    move floating-point state wᵢ[n-1] from X memory location
;    pointed to by r0 into register d5 (and postincrement r0),
;    d5 = wᵢ[n-1]
;    and move floating-point coefficient a₁ᵢ from Y memory
;    location pointed to by r4 into register d6 (and postincrement r4)
;    d6 = a₁ᵢ

     fmpy d5,d6,d1  fsub.s   d1,d0    d5.s,x:(r1)+   y:(r4)+,d6.s

;    multiply in floating-point the state (in d5) and the coefficient
;    (in d6), round the result and store in register d1
;    d1 = wᵢ[n-1] a₁ᵢ
;    simultaneously, subtract in floating-point the operand in d1 from d0,
;    round to single precision and store the accumulated result in d0
;    move floating-point state wᵢ[n-1] from register d5 into X
;    memory location pointed to by r1 (and postincrement r1),
;    and move floating-point coefficient b₂ᵢ from Y
;    memory location pointed to by r4 into register d6 (and
;    postincrement r4)
;    d6 = b₂ᵢ

     fmpy d4,d6,d1  fsub.s   d1,d0    x:(r0)+,d4.s   y:(r4)+,d6.s

;    multiply in floating-point the state (in d4) and the coefficient
;    (in d6), round the result and store in register d1
;    d1 = wᵢ[n-2] b₂ᵢ
;    simultaneously, subtract in floating-point the operand in d1 from d0,
;    round to single precision and store the accumulated result in d0
;    move floating-point state wᵢ₊₁[n-1] from X memory location
;    pointed to by r0 into register d4 (and postincrement r0),
;    d4 = wᵢ₊₁[n-1]
;    and move floating-point coefficient b₁ᵢ from Y
;    memory location pointed to by r4 into register d6 (and
;    postincrement r4)
;    d6 = b₁ᵢ
```

FIGURE 12-34   IIR filtering code on DSP96002. (*Continued*)

```
     fmpy d5,d6,d1   fadd.s   d1,d0        d4.s,x:(r1)+    y:(r4)+,d6.s
```

```
;         multiply in floating-point the state (in d5) and the coefficient
;         (in d6), round the result and store in register d1
;         d1 = wᵢ[n-1] b₁ᵢ
;         simultaneously, add in floating-point the two operands in d1 and d0,
;         round to single precision and store the accumulated result in d0
;         move floating-point state w_{i+1}[n-1] from register d4 into X
;         memory location pointed to by r1 (and postincrement r1),
;         and move floating-point coefficient a_{2i+1} from Y memory
;         location pointed to by r4 into register d6 (and postincrement r4)
;         d6 = a_{2i+1}
```

```
end
```

```
                    fadd.s     d1,d0        ; final addition

     movep                                  d0.s,x:output
```

```
;         output sample value into X I/O peripheral "output" from d0
```

FIGURE 12-34 (*Continued*)

```
*
*
*         N Biquad section IIR filter (N > 1) routine on TMS320C30
*
*         Implements the ith section with 0 ≤ i < N:
*
*         wᵢ[n] = y_{i-1}[n] − a_{1i}wᵢ[n−1] − a_{2i}wᵢ[n−2]
*
*         yᵢ[n] = b_{0i}wᵢ[n] + b_{1i}wᵢ[n−1] + b_{2i}wᵢ[n−2]
*
*         Data memory organization:
*
*         Filter Coefficients (AR0)          State Variables (AR1)
*
*         a₂₀                                w₀[n]
*         b₂₀                                w₀[n−1]
*         a₁₀                                w₀[n−2]
*         b₁₀                                empty
*         b₀₀                                .
*         .                                  .
*         .                                  .
*         .                                  w_{N−1}[n]
*         a_{2N−1}                           w_{N−1}[n−1]
*         b_{2N−1}                           w_{N−1}[n−2]
*         a_{1N−1}                           empty
*         b_{1N−1}
*         b_{0N−1}
*
```

FIGURE 12-35 IIR filtering code on TMS320C30. (The code fragment is reproduced by permission of Texas Instruments, Inc., with additional annotations by the author.)

```
*       Synopsis:
*
*       LOAD    R2
*       LOAD    AR0
*       LOAD    AR1
*       LOAD    IR0
*       LOAD    IR1
*       LOAD    BK
*       LOAD    RC
*       CALL    IIR2
*
*       Arguments:
*
*       R2      Input sample {u[n]}
*       AR0     Address of filter coefficients {a₂₀}
*       AR1     Address of state variables {w₀[n-2]}
*       BK      Block register initialized to 3
*       IR0     Index register 0 initialized to 4
*       IR1     Index register 1 initialized to 4N - 4
*       RC      Repeat counter initialized to N - 2
*
*       Registers modified: R0, R1, R2, AR0, AR1, RC
*
*       Register containing the result:  R0
*
*       Uses 17 words of program memory and 17 + 6N instruction cycles
*
        .global    IIR2
*
IIR2 MPYF3      *AR0,*AR1,R0
*                        ;  compute a₂₀w₀[n-2]
*                        ;  in floating-point and store in register R0
*
     MPYF3      *++AR0(1),*AR1--(1)%,R1
*                        ;  preincrement AR0 to point to the coefficient b₂₀,
*                        ;  compute b₂₀w₀[n-2] in floating-point and
*                        ;  store result in register R1 postdecrement AR1 to
*                        ;  point to the state w₀[n-1]
*
     MPYF3      *++AR0(1),*AR1,R0
*                        ;  preincrement AR0 to point to the coefficient a₁₀,
*                        ;  compute a₁₀w₀[n-1] in floating-point and
*                        ;  store result in register R0
*
||   ADDF3      R0,R2,R2
*                        ;  in parallel add the contents of registers R0 & R2 in
*                        ;  floating-point and store the result in R2
*                        ;  R2 = a₂₀w₀[n-2]  +  y₋₁[n]
*
     MPYF3      *++AR0(1),*AR1--(1)%,R0
*                        ;  preincrement AR0 to point to the coefficient b₁₀,
```

FIGURE 12-35 IIR filtering code on TMS320C30. (*Continued*)

```
*                            ;   compute b_10 w_0[n-1] in floating-point and
*                            ;   store result in register R0, postdecrement AR1 to
*                            ;   point to the state w_0[n]
*
||   ADDF3      R0,R2,R2
*                            ;   in parallel add the contents of registers R0 & R2 in
*                            ;   floating-point and store the result in R2
*                            ;   R2 = w_0[n]
*
     MPYF3      *++AR0(1),R2,R2
*                            ;   preincrement AR0 to point to the coefficient b_00,
*                            ;   compute b_00 w_0[n] in floating-point and
*                            ;   store result in register R2
*
||   STF  R2,*AR1--(1)%
*                            ;   store w_0[n] from register R2 into the location
*                            ;   pointed to by AR1, postdecrement AR1 to
*                            ;   point to the state w_0[n-2]
*
     RPTB LOOP
*                            ;   block repeat for 1 <= i < N
*
     MPYF3      *++AR0(1),*++AR1(IR0),R0
*                            ;   preincrement AR0 to point to the coefficient a_{2i},
*                            ;   preincrement AR1 modified by 4 to point to the state
*                            ;   w_i[n-2], compute a_{2i} w_i[n-2]
*                            ;   in floating-point and store the result in register R0
*
||   ADDF3      R0,R2,R2
*                            ;   in parallel add the contents of registers R0 & R2 in
*                            ;   floating-point and store the result in R2
*                            ;   R2 = b_{0i-1} w_{i-1}[n] + b_{1i-1} w_{i-1}[n-1]
*
     MPYF3      *++AR0(1),*AR1--(1)%,R1
*                            ;   preincrement AR0 to point to the coefficient b_{2i},
*                            ;   compute b_{2i} w_i[n-2] in floating-point and store
*                            ;   the result in register R1, postdecrement AR1 to
*                            ;   point to the state w_i[n-1]
*
||   ADDF3      R1,R2,R2
*                            ;   in parallel add the contents of registers R1 & R2 in
*                            ;   floating-point and store the result in R2
*                            ;   R2 = y_{i-1}[n]
*
     MPYF3      *++AR0(1),*AR1,R0
*                            ;   preincrement AR0 to point to the coefficient a_{1i},
*                            ;   compute a_{1i} w_i[n-1] in floating-point and store
*                            ;   the result in register R0
*
||   ADDF3      R0,R2,R2
*                            ;   in parallel add the contents of registers R0 & R2 in
```

FIGURE 12-35 (*Continued*)

```
*                                  ;  floating-point and store the result in R2
*                                  ;  R2 = a_2i w_i[n-2]  +  y_{i-1}[n]
*
     MPYF3      *++AR0(1),*AR1--(1)%,R0
*                                  ;  preincrement AR0 to point to the coefficient b_1i,
*                                  ;  compute b_1i w_i[n-1] in floating-point and store
*                                  ;  the result in register R0, postdecrement AR1 to
*                                  ;  point to the state w_i[n]
*
||   ADDF3      R0,R2,R2
*                                  ;  in parallel add the contents of registers R0 & R2
*                                  ;  in floating-point and store the result in R2
*                                  ;  R2 = w_i[n]
*
     STF   R2,*AR1--(1)%
*                                  ;  store w_i[n] from register R2 into the location
*                                  ;  pointed to by AR1, postdecrement AR1 to
*                                  ;  point to the state w_i[n-2]
*
LOOP MPYF3      *++AR0(1),R2,R2
*                                  ;  preincrement AR0 to point to the coefficient b_0i,
*                                  ;  compute b_0i w_i[n], restart loop until RC is 0
*
*    Final summation
*
     ADDF  R0,R2
*                                  ;  R2 = b_{0N-1} w_{N-1}[n]  +  b_{1N-1} w_{N-1}[n-1]
*
     ADDF3      R1,R2,R0
*                                  ;  R2 = y_{N-1}[n]
*
     NOP   *AR1--(IR1)
*                                  ;  return to first biquad
*
     NOP   *AR1--(1)%
*                                  ;  point to w_0[n-1]
*
*    Return from routine
*
     RETS
*
*    End of the routine
*
     .end
```

FIGURE 12-35   IIR filtering code on TMS320C30. (*Continued*)

```
*
*         An 8-point DIT FFT Code on TMS320C25
*
* Store complex input sample values consecutively in memory
*
X0R     EQU     00              * X0R = Re x[0]
X0I     EQU     01              * X0I = Im x[0]
X1R     EQU     02              * X1R = Re x[1]
X1I     EQU     03              * X1I = Im x[1]
X2R     EQU     04              * X2R = Re x[2]
X2I     EQU     05              * X2I = Im x[2]
X3R     EQU     06              * X3R = Re x[3]
X3I     EQU     07              * X3I = Im x[3]
X4R     EQU     08              * X4R = Re x[4]
X4I     EQU     09              * X4I = Im x[4]
X5R     EQU     10              * X5R = Re x[5]
X5I     EQU     11              * X5I = Im x[5]
X6R     EQU     12              * X6R = Re x[6]
X6I     EQU     13              * X6I = Im x[6]
X7R     EQU     14              * X7R = Re x[7]
X7I     EQU     15              * X7I = Im x[7]
W       EQU     16
        AORG    >20
WTABLE      DATA    >5A82       * value for sin(45) or cos(45)
*
* Initialize
*
INIT SPM  0                     * No shift at outputs of AR (Re A)
     SSXM                       * Select sign-extension mode
     ROVM                       * Reset overflow mode
     LDPK 4                     * Choose data page 4
     LALK WTABLE                * Get twiddle factor address
     TBLR W                     * Store sin(45) or cos(45) in W
*
* Macro for input bit reversal
*
* AR = Re A, AI = Im A, BR = Re B, BI = Im B
*
BITREV      $MACRO      AR,AI,BR,BI
        ZALH  :AR:
        ADDS  :BR:
        SACL  :AR:
        SACH  :BR:
        ZALH  :AI:
        ADDS  :BI:
        SACL  :AI:
```

FIGURE 12-36 Eight-point DIT FFT code on TMS320C25. (The code fragment is reproduced by permission of Texas Instruments, Inc., with additional annotations by the author.)

```
           SACH :BI:
           $END
*
*       Macro for special radix-4 combination butterfly
*
* The first two stages of a radix-2 N-point DIT FFT can be implemented
* with a special radix-4 which has a unity twiddle factor.
*
* Total number of instructions is 37.
* Execution time is equivalent to 39 machine cycles.
*
* The inputs are:
*
* A = R1 + jI1, B = R2 + jI2, C = R3 + jI3, and D = R4 + jI4
*
* The intermediate outputs (from first stage) are:
*
* A' = (R1 + R2) + j(I1 + I2), B' = (R1 - R2) + j(I1 - I2)
* C' = (R3 + R4) + j(I3 + I4), D' = (R3 - R4) + j(I3 - I4)
*
* The final outputs (from the second stage) are:
*
* B1 = A' + C', B2 = B' - jD', B3 = A' - C', B4 = B' + jD'
*
* This macro computes the real and imaginary parts of the complex quantities
* B1, B2, B3, and B4 and scales them down by 2
*
COMBO        $MACRO     R1,I1,R2,I2,R3,I3,R4,I4
*
*       calculate partial terms for R3, R4, I3, and I4
*
           LAC  :R3:,14      * ACC := (1/4)(R3)
           ADD  :R4:,14      * ACC := (1/4)(R3 + R4)
           SACH :R3:,1       * ACC := (1/2)(R3 + R4)
           SUB  :R4:,15      * ACC := (1/4)(R3 + R4) - (1/2)(R4)
           SACH :R4:,1       * R4 := (1/2)(R3 - R4)
           LAC  :I3:,14      * ACC := (1/4)(I3)
           ADD  :I4:,14      * ACC := (1/4)(I3 + I4)
           SACH :I3:,1       * I3 := (1/2)(I3 + I4)
           SUB  :I4:,15      * ACC := (1/4)(I3 + I4) - (1/2)(I4)
           SACH :I4:,1       * I4 := (1/2)(I3 - I4)
*
*       calculate partial terms for R2, R4, I2, and I4
*
           LAC  :R1:,14      * ACC := (1/4)(R1)
           ADD  :R2:,14      * ACC := (1/4)(R1 + R2)
           SACH :R1:,1       * R1 := (1/2)(R1 + R2)
           SUB  :R2:,15      * ACC := (1/4)(R1 + R2) - (1/2)(R2)
           ADD  :I4:,15      * ACC := (1/4)[(R1 - R2) + (I3 - I4)]
```

FIGURE 12-36 Eight-point DIT FFT Code on TMS320C25. (*Continued*)

```
        SACH  :R2:              * R2 := (1/4)[(R1 - R2) + (T3 - I4)] = Re B2
        SUBH  :I4:              * ACC := (1/4)[(R1 - R2) - (I3 - I4)]
        DMOV  :R4:              * I4 := R4 = (1/2)(R3 - R4)
        SACH  :R4:              * R4 := (1/4)[(R1 - R2) - (I3 - I4)] = Re B4
        LAC   :I1:,14           * ACC := (1/4)(I1)
        ADD   :I2:,14           * ACC := (1/4)(I1 + I2)
        SACH  :I1:,1            * I1 := (1/2)(I1 + I2)
        SUB   :I2:,15           * ACC := (1/4)(I1 + I2) - (1/2)(I2)
        SUB   :I4:,15           * ACC := (1/4)[(I1 - I2) - (R3 - R4)]
        SACH  :I2:              * I2 := (1/4)[(I1 - I2) - (R3 - R4)] = Im B2
        ADDH  :I4:              * ACC := (1/4)[(I1 - I2) + (R3 - R4)]
        SACH  :I4:              * I4 := (1/4)[(I1 - I2) + (R3 - R4)] = Im B4
*
*       calculate partial terms for R1, R3, I1, and I3
*
        LAC   :R1:,15           * ACC := (1/4)(R1 + R2)
        ADD   :R3:,15           * ACC := (1/4)[(R1 + R2) + (R3 + R4)]
        SACH  :R1:              * R1 := (1/4)[(R1 + R2) + (R3 + R4)] = Re B1
        SUBH  :R3:              * ACC := (1/4)[(R1 + R2) - (R3 + R4)]
        SACH  :R3:              * R3 := (1/4)[(R1 + R2) - (R3 + R4)] = Re B3
        LAC   :I1:,15           * ACC := (1/4)(I1 + I2)
        ADD   :I3:,15           * ACC := (1/4)[(I1 + I2) + (I3 + I4)]
        SACH  :I1:              * I1 := (1/4)[(I1 + I2) + (I3 + I4)] = Im B1
        SUBH  :I3:              * ACC := (1/4)[(I1 + I2) - (I3 + I4)]
        SACH  :I3:              * I3 := (1/4)[(I1 + I2) - (I3 + I4)] = Im B3
        $END
*
* The following macro calculates a single butterfly with complex inputs P and Q
* under the condition of unity twiddle factor and k = 0
* PR is the real part of P ; PI is the imaginary part of P
* QR is the real part of Q ; QI is the imaginary part of Q
* The outputs are P + Q and P - Q.  A total of 10 instructions are used.
* Execution time is equal to 10 machine cycles
*
ZERO $MACRO            PR,PI,QR,QI
*
*       Calculate Real parts of (P + Q) and (P - Q)
*
        LAC   :PR:,15           * ACC := (1/2)(PR)
        ADD   :QR:,15           * ACC := (1/2)(PR + QR)
        SACH  :PR:              * PR := (1/2)(PR + QR) = Re P'
        SUBH  :QR:              * ACC := (1/2)(PR + QR) - (QR)
        SACH  :QR:              * QR := (1/2)(PR - QR) = Re Q'
*
*       Calculate Imaginary parts of (P + Q) and (P - Q)
*
        LAC   :PI:,15           * ACC := (1/2)(PI)
        ADD   :QI:,15           * ACC := (1/2)(PI + QI)
        SACH  :PI:              * PI := (1/2)(PI + QI) = Im P'
```

FIGURE 12-36 (*Continued*)

```
        SUBH  :QI:              * ACC := (1/2)(PI + QI) - (QI)
        SACH  :QI:              * QI := (1/2)(PI - QI) = Im Q'
        $END
*
* Butterfly computation for k = N/8 with complex inputs P and Q.
* PR is the real part of P ; PI is the imaginary part of P
* QR is the real part of Q ; QI is the imaginary part of Q
* This macro requires W to be the absolute value of cos (pi/4) and sin (pi/4).
* The outputs are P + Q*W and P - Q*W.  A total of 20 instructions are used.
* Execution time is equal to 20 machine cycles
*
PIBY4      $MACRO     PR,PI,QR,QI,W
*
        LT    :W:              * T-register := W = |cos (pi/4)| = |sin (pi/4)|
        LAC   :QI:,14          * ACC := (1/4)(QI)
        SUB   :QR:,14          * ACC := (1/4)(QI - QR)
        SACH  :QI:,1           * QI := (1/2)(QI - QR)
        ADD   :QR:,15          * ACC := (1/4)(QI + QR)
        SACH  :QR:,1           * QR := (1/2)(QI + QR)
        LAC   :PR:,14          * ACC := (1/4)(PR)
        MPY   :QR:             * P-register := (1/4)(QI + QR)*W
        APAC                   * ACC := (1/4)[PR + (QI + QR)*W]
        SACH  :PR:,1           * PR := (1/2)[PR + (QI + QR)*W] = Re P'
        SPAC                   * ACC := (1/4)(PR)
        SPAC                   * ACC := (1/4)[PR - (QI + QR)*W]
        SACH  :QR:,1           * QR := (1/2)[PR - (QI + QR)*W] = Re Q'
        LAC   :PI:,14          * ACC := (1/4)(PI)
        MPY   :QI:             * P-register := (1/4)(QI - QR)*W
        APAC                   * ACC := (1/4)[PI + (QI - QR)*W]
        SACH  :PI:,1           * PI := (1/2)[PI + (QI - QR)*W] = Im P'
        SPAC                   * ACC := (1/4)(PI)
        SPAC                   * ACC := (1/4)[PI - (QI - QR)*W]
        SACH  :QI:,1           * QI := (1/2)[PI - (QI - QR)*W] = Im Q'
        $END
*
* Butterfly computation for k = N/4 and twiddle factor of -j with complex
* inputs P and Q.
* PR is the real part of P ; PI is the imaginary part of P
* QR is the real part of Q ; QI is the imaginary part of Q
* The outputs are P - jQ and P + jQ
* This macro requires the input samples QR and QI to be in consecutive memory
* locations of ascending order.  A total of 11 instructions is used
* Execution time is equal to 11 machine cycles
*
PIBY2      $MACRO     PR,PI,QR,QI
*
*       Calculate Real parts of (P + jQ) and (P - jQ)
*
        LAC   :PI:,15          * ACC := (1/2)(PI)
```

FIGURE 12-36 Eight-point DIT FFT Code on TMS320C25. (*Continued*)

```
        SUB    :QR:,15           * ACC := (1/2)(PI - QR)
        SACH   :PI:              * PI := (1/2)(PI - QR) = Im P'
        ADDH   :QR:              * ACC := (1/2)(PI - QR) + (QR)
        SACH   :QR:              * QR := (1/2)(PI + QR) = Re Q'
*
*       Calculate Imaginary parts of (P + jQ) and (P - jQ)
*
        LAC    :PR:,15           * ACC := (1/2)(PR)
        ADD    :QI:,15           * ACC := (1/2)(PR + QI)
        SACH   :PR:              * PR := (1/2)(PR + QI) = Re P'
        SUBH   :QI:              * ACC := (1/2)(PR + QI) - (QI)
        DMOV   :QR:              * QR → QI
        SACH   :QR:              * QR := (1/2)(PR - QI) = Re Q'
        $END
*
* Butterfly computation for k = 3N/8 with complex inputs P and Q.
* PR is the real part of P ; PI is the imaginary part of P
* QR is the real part of Q ; QI is the imaginary part of Q
* This macro requires W to be the absolute value of cos (3*pi/4) and
* sin (3*pi/4).  The outputs are P + Q*W and P - Q*W
* A total of 20 instructions are used.
* Execution time is equal to 20 machine cycles
*
PI3BY4     $MACRO
*
        LT     :W:               * T-register := W = |cos(3*pi/4)| = |sin(3*pi/4)|
        LAC    :QI:,14           * ACC := (1/4)(QI)
        SUB    :QR:,14           * ACC := (1/4)(QI - QR)
        SACH   :QI:,1            * QI := (1/2)(QI - QR)
        ADD    :QR:,15           * ACC := (1/4)(QI + QR)
        SACH   :QR:,1            * QR := (1/2)(QI + QR)
        LAC    :PR:,14           * ACC := (1/4)(PR)
        MPY    :QI:              * P-register := (1/4)(QI - QR)*W
        APAC                     * ACC := (1/4)[PR + (QI - QR)*W]
        SACH   :PR:,1            * PR := (1/2)[PR + (QI - QR)*W] = Re P'
        SPAC                     * ACC := (1/4)(PR)
        SPAC                     * ACC := (1/4)[PR - (QI - QR)*W]
        MPY    :QR:              * P-register := (1/4)(QI + QR)*W
        SACH   :QR:,1            * QR := (1/2)[PR - (QI - QR)*W] = Re Q'
        LAC    :PI:,14           * ACC := (1/4)(PI)
        SPAC                     * ACC := (1/4)[PI - (QI + QR)*W]
        SACH   :PI:,1            * PI := (1/2)[PI - (QI + QR)*W] = Im P'
        APAC                     * ACC := (1/4)(PI)
        APAC                     * ACC := (1/4)[PI + (QI + QR)*W]
        SACH   :QI:,1            * QI := (1/2)[PI + (QI + QR)*W] = Im Q'
        $END
*
* FFT code with bit-reversed input samples
*
```

FIGURE 12-36 (*Continued*)

```
FFT8PT    BITREV    2,3,8,9
          BITREV    6,7,12,13
*
* First & second stages combined with divide-by-4 interstage scaling
*
      COMBO     X0R,X0I,X1R,X1I,X2R,X2I,X3R,X3I
      COMBO     X4R,X4I,X5R,X5I,X6R,X6I,X7R,X7I
*
* Third stage with divide-by-2 interstage scaling
*
      ZERO      X0R,X0I,X4R,X4I
      PIBY4     X1R,X1I,X5R,X5I,W
      PIBY2     X2R,X2I,X6R,X6I
      PI3BY4    X3R,X3I,X7R,X7I,W
```

FIGURE 12-36   (*Continued*)

```
/*     calling the _fft8 routine on DSP16A     */

main:     auc = 0x2
          pt = inputs
          r0 = Array

/*    First, the 16-element array, inputs (comprises real and imaginary
parts), is copied into the RAM array, Array.     */

          do 16      {
              y = *r0    x = *pt++
              *r0++ = x
          }

          call _fft8                    /*     call the subroutine */

pause:    nop
          goto pause

inputs:   int  0.707    0.0  1.99993    0.0
          int  0.707    0.0  -1.0       0.0
          int  -0.707   0.0  0.0        0.0
          int  -0.707   0.0  -1.0       0.0

/*    order of input data is: real1, imaginary1, real2, imaginary2, ...,
      real8, imaginary8     */
 .ram
Array:    16 * int
/*    Input array MUST start at location 0x0   */
 .endram
```

FIGURE 12-37   Eight-point DIT FFT code on DSP16A. (The code fragment is reproduced by permission of AT&T with additional annotations by the author.)

```
/*      source code from file "fft8.asm"

        Routine memory size       = 104 ROM locations
        Input/output data size    = 16 RAM locations

        Registers used:
            a0, a1, x, y, p, auc, r0, r1, r2, i, pt, pr, cache

        After completion of the routine, the output replaces the input data.
        The results are scaled down by a factor of 8 (due to 3 arithmetic
        left shifts)

        where W = pi / 4 * # of points, in this case # of points = 8

        The subroutine executes in 375 instruction cycles.    */

_fft8:      auc = 0x2
            i = -1
/*      First Level Butterfly Computation */
            r0 = 0x0
            r1 = 0x8
            r2 = 0x7
            pt = _fft8_w0                   /* point to twiddle factor table */
            a0 = *r0++                      /* set a0 to Re A = Re X[0] */
            a1 = *r2                        /* set a1 to Im B = Im X[N/2-1] */
            x = *pt++ y = *r1++
/* set x to cos W and y to Re B = Re X[N/2] */
            do 4 {
                p = x*y          x = *pt++i       y = *r1--

/*      compute p = Re B * cos W = Re X[k+N/2] * cos W,
        set x to - sin W, and y to Im B = Im X[k+N/2] */

                a1 = a0-p *r2++ = a1

/*      compute a1 = Re A - Re B * cos W = Re X[k] - Re X[k+N/2] * cos W
        and store in memory location pointed to by r2 */

                a0 = a0+p p = x*y

/*      compute a0 = Re A + Re B * cos W = Re X[k] + Re X[k+N/2] * cos W
        and p = -Im B * sin W = -Im X[k+N/2] * sin W */

                a0 = a0-p y = *r1++

/*      compute a0 = Re A + Re B * cos W + Im B * sin W
        which is the Re A' and set y to Re B = Re X[k+N/2] */

                a0 = a0>>1                  /* divide a0 by 2 for scaling */
                a1 = a1+p p = x*y   *r0m2:a0

/*      compute a1 = Re A - Re B * cos W - Im B * sin W
        which is the Re B', compute p = -Re B * sin W,
```

FIGURE 12-37 (*Continued*)

```
        and set temp = a0, a0 = *r0--, *r0++2 = temp by compound addressing
        this operation sets a0 to Im A and stores Re A' */

                x = *pt++ y = *r1++ /* set x to cos W and y to Im B */
                a1 = a1>>1              /* set a1 to a1/2 for scaling */
                a1 = a0-p *r2++ = a1

/*      compute a1 = Im A + Re B * sin W
        and store Re B' into memory location pointed to by r2 */

                a0 = a0+p p = x*y

/*      compute a0 = Im A - Re B * sin W
        and p = Im B * cos W */

                a0 = a0+p y = *r1++

/*      compute a0 = Im A - Re B * sin W + Im B * cos W
        which is the Im A', and set y to new Re B */

                a0 = a0>>1              /* set a0 to a0/2 for scaling */
                a1 = a1-p *r0m2:a0

/*      compute a1 = Im A + Re B * sin W - Im B * cos W
        which is the Im B', by compound addressing set a0 to Re A,
        and store Im A' into memory location pointed to by *r0++2 */

                a1 = a1>>1              /* set a1 to a1/2 for scaling */
            }
        *r2 = a1

/* store Im B' into memory location pointed to by r2 */

/*      Second Level Butterfly Computation      */

            r0 = 0x0                   /* top two butterfly computations */
            r1 = 0x4
            r2 = 0x3
            pt = _fft8_w0
/* point to twiddle factor table first entry set */
            a0 = *r0++                 /* set a0 to Re A = Re X[0] */
            a1 = *r2                   /* set a1 to Im B = Im X[N/2-1] */
            x = *pt++ y = *r1++
/* set x to cos W and y to Re B = Re X[N/2] */
            redo 2

/*      execute the current contents of cache (loaded with the previous "do 4 {"
instruction) within the cache 2 times */

            *r2 = a1

/* store Im B' into memory location pointed to by r2 */
```

FIGURE 12-37   Eight-point DIT FFT Code on DSP16A. (*Continued*)

```
        r0 = 0x8                        /* bottom two butterfly computations */
        r1 = 0xc
        r2 = 0xb
        pt = _fft8_w2
/* point to twiddle factor table second entry set */
        a0 = *r0++                      /* set a0 to Re A = Re X[0] */
        a1 = *r2                        /* set a1 to Im B = Im X[N/2-1] */
        x = *pt++ y = *r1++
/* set x to cos W and y to Re B = Re X[N/2] */
        redo 2
```

/*   execute the current contents of cache (loaded with the previous "do 4 {"
instruction) within the cache 2 times */

```
        *r2 = a1
```

/* store Im B' into memory location pointed to by r2 */

/*   divide all the input data by 2   */

```
        r0 = 0x0
        a0 = *r0++
        do 16 {
            a0 = a0>>1
            *r0m2:a0
```

/* store updated data values in a0 by compound addressing into memory
location pointed to by r0 */

```
        }
```

/*   Third Level Butterfly Computation */

```
        r0 = 0x0
        r1 = 0x2
        r2 = 0x2
        pt = _fft8_w0
/* point to twiddle factor table first entry set */
        a0 = *r0                        /* set a0 to Re A = Re X[k] */
        x = *pt++ y = *r1++
/* set x to cos W and y to Re B = Re X[k+1] */
        p = x*y            y = *r1-- x = *pt++i
```

/*   compute p = Re B * cos W = Re X[k+1] * cos W,
      set y to Im B = Im X[k+1], and set x to -sin W */

```
        j = 3
        do 4 {
            a1 = a0-p                   /* compute a1 = Re A - Re B * cos W */
            a0 = a0+p p = x*y
```

FIGURE 12-37 (*Continued*)

```
/*    compute a0 = Re A + Re B * cos W = Re X[k] + Re X[k+1] * cos W
      and p = -Im B * sin W = -Im X[k+1] * sin W */

            a0 = a0-p  y = *r1++

/*    compute a0 = Re A + Re B * cos W + Im B * sin W
      which is the Re A' and set y to Re B */

            a1 = a1+p  p = x*y          *r0++ = a0

/*    compute a1 = Re A - Re B * cos W - Im B * sin W
      which is the Re B', compute p = -Re B * sin W,
      and store Re A' from a0 into memory location pointed to by r0 */

            x = *pt++  y = *r1++j     /* set x to cos W and y to Im B */
            a0 = *r0                  /* set a0 to Im A = Im X[k] */
            a1 = a0-p  *r2++ = a1

/*    compute a1 = Im A + Re B *' sin W = Im X[k] + Re X[k+1] * sin W
      and store Re B' from a1 into memory location pointed to by r2 */

            a0 = a0+p  p = x*y        y = *r1         x = *pt++

/*    compute a0 = Im A - Re B * sin W,
      compute p = Im B * cos W, set y to Re B = Re X[k+2] and x to -sin W */

            a0 = a0+p  y = *r1++ x = *pt++

/*    compute a0 = Im A - Re B * sin W + Im B * cos W
      which is the Im A', set y to new Re B = Re X[k+3] and x to cos W */

            a1 = a1-p  p = x*y          *r0++j = a0

/*    compute a1 = Im A + Re B * sin W - Im B * cos W
      which is the Im B', compute p = Re B * cos W,
      and store Im A from a0 into memory location pointed to by r0 */

            y = *r1-- x = *pt++i

/*    set y to Im B = Im X[k+3] and x to - sin W */

            a0 = *r0                    /* set a0 to Re B = Re X[k+2] */
            *r2++j = a1

/*    store Im B' from a1 into memory location pointed to by r2 */
            }


/*    Bit Reversal */


            pt = _fft8_bit
            y = a0        x = *pt++
```

FIGURE 12-37 Eight-point DIT FFT Code on DSP16A. (*Continued*)

```
        do 2 {
              r0 = a0
              a1 = x
              r1 = a1
              y = *r0++          x = *pt++
              a0 = x
              a1 = *r0--
              *r1zp:y            x = *pt++
```

```
/*     set temp = y, y = *r1, *r1++ = temp i.e., interchange the contents of r1
       and y and set x to the proper bit address */
```

```
              *r1zp:a1
              *r0++ = y
              *r0 = a1
        }
```

```
        return
```

```
/*     twiddle factors table      */
```

```
/*_fft8_wN             cos W              -sin W        */
```

```
_fft8_w0: int  1.0,             0.0
_fft8_w2: int  0.0,             -1.0
_fft8_w1: int  0.707106781,     -0.707106781
_fft8_w3: int  -0.707106781,    -0.707106781
```

```
/*     bit reversal table  */
```

```
_fft8_bit:        int   0x2
                  int   0x8
                  int   0x6
                  int   0xc
```

FIGURE 12-37   (*Continued*)

The program in Figure 12-37 (contained in file "fft8.s") can be run by invoking the simulator **env16** as follows:

```
env16 -16 fft8.s
 . . .
```

[simulator messages appear here]

In the "env16" instruction, the option " − 16" signifies that the program will be run on WE DSP16. For WE DSP16A, we use the option " − 16a." Within the simulator, a breakpoint is set at the label "pause," and then the program is run:

```
sim: bpset pause
sim: run
breakpoint at 0x9 [pause]
```

The results are in a data array in RAM, which can be displayed as follows:

```
sim: ramArray:Array+15 .f
0.460205
-0.301758
0.252075
-0.153076
0.124939
0.033630
-0.075623
0.023682
-0.356812
0.051758
-0.325500
-0.273743
0.125061
0.216370
0.502197
0.402893
sim: quit
```

The flag ".f" in the simulator display command "ram" specifies to output the data as fixed point with "precision" bits (14 bits by default) to the right of the binary point.

12-7-3 FFT Implementation on DSP56000

We include in Figure 12-38 the code for a radix-2, in-place, decimation-in-time algorithm using triple nested DO loops and bit reverse addressing by reverse carry modifier. Although this code is easier to understand and very compact, it is not as fast as an optimized code that uses the symmetry of the FFT and the parallelism of the DSP.

Since the computation is in-place, the bit reversal process is needed. This is done by "Reverse Carry Modifier Addressing." Reverse carry is selected on 56001 by setting the corresponding modifier register Mn to zero. The address modification is performed in hardware by propagating the carry in the reverse direction, that is, from the MSB to the LSB. If the Rn + Nn addressing mode is used with this address modifier and Nn contains the value 2^{k-1}, this addressing modifier is equivalent to bit reversing the k LSBs of Rn, incrementing Rn by 1, and bit reversing the k LSBs of Rn again.

As an example, consider a 1024-point FFT with real data stored in the X memory and imaginary data stored in Y memory. Thus k = 10, modifier register Mn = 0, and offset register Nn = 512 for bit reversed addressing. The starting address at which FFT results can be held is 3072 (= 3 × 1024). Initially, the pointer register Rn contains 3072. Postincrementing by +Nn generates the address se-

```
; Radix-2, In-Place, Decimation-In-Time FFT Macro for DSP56000
;
fftr2a              macro               points,data,coef
;
;      points       number of points (2 ≤ points ≤ 32768, power of 2)
;      data         start of data buffer
;      coef         start of coefficient (sine-cosine) table
;
fftr2a              ident               1,1
;
;              real part of input/output data in X memory
;              imaginary part of input/output data in Y memory
;      Normally ordered input data
;      Bit reversed output data
;              cosine lookup table in X memory
;              sine lookup table in Y memory
;
; The macro call alters the following Data ALU Registers
;      x1      x0      y1      y0
;      a2      a1      a0      a
;      b2      b1      b0      b
;
; The macro call alters the following Address Registers
;      r0      n0      m0
;      r1      n1      m1
;              n2
;      r4      n4      m4
;      r5      n5      m5
;      r6      n6      m6
;
; The macro call alters the following Program Control Registers
;      pc      sr
;
; It uses 6 locations on System Stack
;
               move #points/2,n0    ; initialize butterflies per group
               move #1,n2           ; initialize groups per pass
               move #points/4,n6    ; initialize coefficient pointer offset
               move #-1,m0          ; initialize A and B data address modifiers
               move m0,m1           ; for linear addressing
               move m0,m4
               move m0,m5           ; initialize coefficient address modifier for
               move #0,m6           ; reverse carry addressing
;
; Perform all FFT passes with triple nested DO loop
;
               do   #@cvi(@log(points)/@log(2)+0.5),_end_pass
               move #data,r0        ; initialize A data input pointer
               move r0,r4           ; initialize A data output pointer
               lua  (r0)+n0,r1      ; initialize B data input pointer
                                    ; by loading updated address (r0)+n0
```

FIGURE 12-38 Radix-2 DIT FFT code on DSP56000. (The code fragment is reproduced by permission of Motorola, Inc., with additional annotations by the author.)

```
            lua   (r1)-,r5          ; initialize B data output pointer
                                    ; by loading updated address (r1)-
            move  #coef,r6          ; initialize coefficient table pointer
            move  n0,n1             ; initialize pointer offsets
            move  n0,n4
            move  n0,n5

            do    n2,_end_grp
            move  x:(r1),x1 y:(r6),y0
;
; load real part of B data input into "x1" and sine coefficient value into "y0"
;
            move  x:(r5),a   y:(r0),b
;
; load real part of B data output into "a" and imaginary part of A data input
; into "b"
;
            move  x:(r6)+n6,x0     ; load cosine coefficient value into "x0"
            do    n0,_end_bfy
;
;     Radix 2 DIT Butterfly Kernel to Compute
;
;     Re A' = Re A + Re B * cos W + Im B * sin W
;     Im A' = Im A + Im B * cos W - Re B * sin W
;     Re B' = Re A - Re B * cos W - Im B * sin W
;     Im B' = Im A - Im B * cos W + Re B * sin W
;
;     where W = pi / 4 * points
;
            mac   x1,y0,b                y:(r1)+,y1
;
; compute x1 * y0 + b (accumulate Re B * sin W + Im A), store result in "b",
; and load imaginary part of B data input into "y1"
;
            macr  -x0,y1,b  a,x:(r5)+ y:(r0),a
;
; compute -x0 * y1 + b and round (accumulate Im A + Re B * sin W - Im B * cos W),
; store result in "b", store real part of B data output from "a", postincrement
; B output pointer, and load imaginary part of A data into "a"
;
            subl  b,a       x:(r0),b  b,y:(r4)
;
; compute 2 * a - b (= Im A - Re B * sin W + Im B * cos W), store result in "a",
; load real part of A data input into "b", and store imaginary part of A data
; output from "b"
;
            mac   -x1,x0,b  x:(r0)+,a a,y:(r5)
;
; compute -x1 * x0 + b (accumulate Re A - Re B * cos W), store result in "b",
; load real part of A data input into "a", postincrement A input pointer, and
; store imaginary part of B data output from "a"
```

FIGURE 12-38 Radix-2 DIT FFT Code on DSP56000. (*Continued*)

```
;
             macr -y1,y0,b   x:(r1),x1
;
; compute -y1 * y0 + b and round (accumulate Re A - Re B * cos W - Im B * sin W),
; store result in "b", and load real part of B data input into "x1"
;
             subl b,a          b,x:(r4)+ y:(r0),b
;
; compute 2 * a - b (= Re A + Re B * cos W + Im B * sin W), store result in "a",
; store real part of A data output from "b", postincrement A output pointer, and
; load imaginary part of A data input into register "b"
;
_end_bfy
             move       a,x:(r5)+n5    y:(r1)+n1,y1
;
; store real part of B output data from "a", update B output pointer, load
; imaginary part of B data input into "y1", and update B input pointer
;
             move       x:(r0)+n0,x1   y:(r4)+n4,y1
;
; load real part of A data input into "x1", update A input pointer, load
; imaginary part of A data output into "y1", and update A output pointer
;
_end_grp
             move n0,b1             ; load butterflies per group into "b1"
             lsr  b          n2,a1
;
; divide butterflies per group by 2 by logical shift right on "b1" and load
; groups per pass into "a1"
;
             lsl  a          b1,n0
;
; multiply groups per pass by 2 by logical shift left on "a1" and move updated
; butterflies per pass into "n0"
;
             move a1,n2             ; move updated groups per pass into "n2"
_end_pass
             endm                  ; end of macro fftr2a
```

FIGURE 12-38 (*Continued*)

quence 0, 512, 256, 768, 128, 640, . . . , which corresponds to Rn contents to be 3072, 3584, 3328, ... This is exactly the correct sequence of addressing for in-place results.

12-7-4 FFT Implementation on WE DSP32C

The C program in Figure 12-39 that gets included in a file "fft.c" calls the **fft** function to perform a 16-point complex FFT [AT88] with real input data consisting of a 1-kHz tone (sinusoid) sampled at 8 kHz. The input data are stored in the **fftIO**. The output data replace the input data in the array.

```
#include <libap.h>

/*    FFT Routine for DSP32C

        fftIO is a pointer to array of floating point data.
        The data are in the order:
        real1, imaginary1, real2, imaginary2, ..., realN, imaginaryN
        The FFT is performed in-place, decimation-in-time.
        So this array will hold the output after completion   */

float fftIO[32] = {
      0.707,     0.0, 1.0,      0.0, 0.707,     0.0, 0.0, 0.0,
      -0.707,    0.0, -1.0,     0.0, -0.707,    0.0, 0.0, 0.0,
      0.707,     0.0, 1.0,      0.0, 0.707,     0.0, 0.0, 0.0,
      -0.707,    0.0, -1.0,     0.0, -0.707,    0.0, 0.0, 0.0,
};

main ()
{
      int i, N, M;

/*    Call fft routine for a 16-point complex FFT    */

      N = 16;    /*    Number of points (≤ 1024) of the complex FFT    */

      M = 4;     /*    log₂N, 1 ≤ M ≤ 10    */

      fft (N, M, fftIO);

/*    Print out the first 8 complex results in 4 lines    */

      for (i = 0; i < 16; i++) {
           if ((i + 1) % 4)
                 printf("%f, ", fftIO[i]);
           else
                 printf("%f NEWLINE", fftIO[i]);
      }
}
```

FIGURE 12-39 Sixteen-point complex FFT computation on DSP32C. (The code fragment is reproduced by permission of AT&T with additional annotations by the author.)

The expected results are

```
  0.000000   0.000000   0.000000 0.000000
  5.656426  -5.656428   0.000000 0.000000
  0.000000   0.000000   0.000000 0.000000
 -0.000427  -0.000427   0.000000 0.000000
```

The example code in Figure 12-39 can be compiled into the load module "fft" as follows:

```
d3cc fft.c -o fft - lap
```

It can be run either on the DSP32-DS or the software simulator as follows:

```
d3sim -c [-d1] fft
```

The "−d1" option is used to run the example program in real time with the DSP32-DS Development System. Otherwise, the program is run on the simulator.

A complete C language code implementing the Cooley–Tukey algorithm is included in Embree and Kimble [EM91].

12-7-5 FFT Implementation on DSP96002

In this section Figure 12-40 illustrates radix-2 DIT macro implemented on DSP96002.

```
;
;        Radix - 2 - Decimation - In - Time Fast Fourier Transform Macro on DSP96002
;
metr2a           macro              points,data,coef,coefsize
;
;    points      number of points (power of 2, 2 ≤ points ≤ 2,147,483,648)
;    data        start of data buffer
;    coef        start of 1/2 cycle sine/cosine coefficient table
;    coefsize    number of entries in sine/cosine table
;                = i * points / 2, i = 1, 2, ...
;                ≤ 2,147,483,648
;
metr2a           ident              1,4
;
;            real part of input/output data in X memory
;            imaginary part of input/output data in Y memory
;        Normally ordered complex input data
;        Bit reversed complex output data
;            cosine look - up (1/2 cycle) table in X memory
;            sine look - up (1/2 cycle) table in Y memory
;
; Affected Data ALU Registers
;    d0    d1    d2    d3    d4
;    d5    d6    d7    d8    d9
;
; Affected Address Registers
;    r0    n0    m0
;    r1    n1    m1
;    r2    n2    m2
;    r4    n4    m4
;    r5    n5    m5
;
             move #points,d1.l
             move #@cvi(@log(points)/@log(2)+0.5),n1
             move #data,r2
             move #coef,m2
             move #coefsize,d2.l
```

FIGURE 12-40 Radix-2 DIT FFT macro on DSP96002. (The code fragment is reproduced by permission of Motorola, Inc., with additional annotations by the author.)

```
;
            move    #0,m6                   ; set up for bit reverse addressing
            move    #-1,m0                  ; set up for linear addressing
            clr     d0          m0,m1       ; set d0 to 0 and set m1 to -1
            inc     d0          m0,m4       ; set d0 to 1 and m4 to -1
            lsr     d2          m0,m5       ; set d2 to coefsize/2 and m5 to -1
            move    d2.1,n6                 ; set n6 to coefsize/2
;
; 3 stage nested DO loop
;
            do      n1,_end_pass
            move    r2,r0                   ; set r0 to r2 and initially to data
            move    d0.1,n2
            lsr     d1          m2,r6       ; divide d1 by 2 and copy m2 to r6
            dec     d1          d1.1,n0     ; set d1 to d1 - 1 and n0 to d1
            move    d1.1,n1
; set n1 to d1 and initially to (points/2) - 1
            move    n0,n4
            move    n0,n5
            lea     (r0)+n0,r1
;
; load r1 with effective address pointed by r0 and n0 which initially is
; (data)+points/2
;
            lea     (r0)-,r4
;
; load r4 with effective address r0 which initially is (data) and post
; decrement r0
;
            lea     (r1)-,r5
;
; load r5 with effective address r1 which initially is (data)+points/2 and post
; decrement r1
;
            do      n2,_end_grp
            move                            x:(r6)+n6,d9.s y:,d8.s
;
; load cosine coefficient value in single precision into register d9 and
; corresponding single precision sine coefficient value into register d8
;
            move                            x:(r1)+,d6.s   y:,d7.s
;
; load single precision real part of B data into register d6 and
; corresponding single precision imaginary part of B data into register d7
;
        fmpy.s      d8,d7,d3                                y:(r5),d2.s
;
; multiply in floating point Im B * sin W,
; round to single precision and store result in register d3,
; load single precision imaginary part of A (output) into register d2
;
        fmpy.s      d9,d6,d0                                y:(r4),d5.s
;
; multiply in floating point Re B * cos W,
; round to single precision and store result in register d0,
; load single precision imaginary part of B (output) into register d5
```

FIGURE 12-40   Radix-2 DIT FFT Macro on DSP96002. (*Continued*)

```
;
        fmpy.s      d9,d7,d1                                    y:(r1),d7.s
;
; multiply in floating point Im B * cos W,
; round to single precision and store result in register d1,
; load single precision imaginary part of B into register d7
;
            do    n0,_end_bfy
        fmpy d8,d6,d2   fadd.s     d3,d0      x:(r0),d4.s    d2.s,y:(r5)+
;
; multiply in floating point Re B * sin W,
; round to single precision and store result in register d2,
; add in floating point Re B * cos W + Im B * sin W,
; round to single precision and store it in register d0,
; load single precision real part of A into register d4,
; store Im A (output) from register d2 into memory address pointed to by r5,
; and postincrement r5
;
        fmpy d8,d7,d3  faddsub.s d4,d0      x:(r1)+,d6.s   d5.s,y:(r4)+
;
; multiply in floating point Im B * sin W,
; round to single precision and store result in register d3,
; add in floating point Re A + (Re B * cos W + Im B * sin W),
; round to single precision and store Re A' in register d0,
; subtract in floating point Re A - (Re B * cos W + Im B * sin W),
; round to single precision and store Re B' in register d4,
; load single precision real part of B into register d6 and postincrement r1,
; store Im B (output) from register d5 into memory address pointed to by r4,
; and postincrement r4
;
        fmpy d9,d6,d0   fsub.s     d1,d2      d0.s,x:(r4)    y:(r0)+,d5.s
;
; multiply in floating point Re B * cos W,
; round to single precision and store result in register d0,
; subtract in floating point  Im B * cos W - Re B * sin W,
; round to single precision and store it in register d2,
; store Re A (output) from register d0 into memory address pointed to by r4,
; load single precision imaginary part of A into d5,
; and postincrement r0
;
        fmpy d9,d7,d1  faddsub.s d5,d2      d4.s,x:(r5)    y:(r1),d7.s
;
; multiply in floating point Im B * cos W,
; round to single precision and store result in register d1,
; add in floating point Im A + (Im B * cos W - Re B * sin W),
; round to single precision and store Im A' in register d2,
; subtract in floating point Im A - (Im B * cos W - Re B * sin W),
; round to single precision and store Im B' in register d5,
; store Re B (output) from register d4 into memory address pointed to by r5,
; and load single precision imaginary part of B into d7
;
_end_bfy
            move                                x:(r0)+n0,d7.s d2.s,y:(r5)+n5
;
; load single precision real part of A into register d7,
; update address of A data,
```

FIGURE 12-40 *(Continued)*

```
; store Im A (output) from register d2 into memory address pointed to by r5,
; and update A data output address
;
            move                        x:(r1)+n1,d7.s d5.s,y:(r4)+n4
;
; load single precision real part of B into register d7,
; update address of B data,
; store Im B (output) from register d5 into memory address pointed to by r4,
; and update B data output address
;
_end_grp
            move n2,d0.l
            lsl  d0          n0,d1.l
; multiply contents of n2 by 2 by logical left shift
_end_pass
        endm
```

FIGURE 12-40   Radix-2 DIT FFT Macro on DSP96002. (*Continued*)

12-7-6   FFT Implementation on TMS320C30 [TE91]

Details on decimation-in-time radix-2 FFT can be found in Burrus and Parks [BU85]. The symbols and notations used in the program example are also from the same reference. Figure 12-41 shows the code for a radix-2 algorithm that is relatively easy to understand. However, a radix-4 implementation can increase the speed of execution by reducing the number of arithmetic operations.

An example table with twiddle factors for a 64-point FFT is shown in Figure 12-42 to illustrate its format.

```
*
* Real, radix 2, in-place FFT computation on TMS320C30
*
* Details of the algorithm used in the program can be found in the paper by
* H. V. Sorensen et al., "Real-Valued Fast Fourier Transform Algorithms",
* Trans. on Acoustics, Speech, and Signal Processing, pp. 849-863, June
* 1987.
*
* Real data resides in internal memory.
* The bit reversal is done at the beginning of the program
* The outputs are in order
*
* The twiddle factors are supplied in a table included in a ".data" section in a
* separate file.
* The size of the FFT: N and log₂(N) are defined in a ".globl" directive
* and specified during linking.
* The twiddle factor table is of length:  N/4 + N/4 = N/2.
*
            .globl    FFT            ; entry point for program execution
            .globl    N              ; FFT size
            .globl    M              ; log₂(N)
```

FIGURE 12-41   Radix-2 FFT code on TMS320C30. (The code fragment is reproduced by permission of Texas Instruments, Inc., with additional annotations by the author.)

```
        .globl     SINE              ; address of sine table

        .bss       INP,1024          ; memory with input data

        .text

* Initialize

FFTSIZ    .word     N
LOGFFT    .word     M
SINTAB    .word     SINE
INPUT     .word     INP

FFT: LDP  FFTSIZ                     ; command to load data page printer

* Do the bit reversal at the beginning

        LDI   @FFTSIZ,RC             ; set RC to N
        SUBI  1,RC                   ; set RC = RC - 1
        LDI   @FFTSIZ,IR0
        LSH   -1,IR0                 ; set IR0 = N/2 by left shift
        LDI   @INPUT,AR0
        LDI   @INPUT,AR1

        RPTB  BITRV

* repeat the following block of instructions upto BITRV label N - 1 times
*
* Exchange locations only if AR0<AR1
*
        CMPI AR1,AR0                 ; integer comparison between AR0 and AR1
        BGE  CONT
* branch if AR1 is greater than or equal to AR0
        LDF  *AR0,R0
* interchange the floating-point contents of AR0 & AR1
||      LDF  *AR1,R1
        STF  R0,*AR1
||      STF  R1,*AR0
CONT NOP  *AR0++
BITRV      NOP  *AR1++(IR0)B
*
* Here the index register (IR0) is added to the auxiliary register (AR1).
* This addition is done with a reverse-carry propagation and is used to
* yield a bit-reversed (B) address.
*
*    Length 2 Butterflies (DO-60 loop)
*
        LDI  @INPUT,AR0              ; AR0 points to X[I]
        LDI  IR0,RC                  ; set repeat counter to N/2
        SUBI 1,RC                    ; decrement RC

        RPTB BLK1                    ; repeat until BLK1 (N/2) - 1 times
        ADDF3    *+AR0,*AR0++,R0
```

FIGURE 12-41   (*Continued*)

```
*                                   ; add in floating point R0 = X[I] + X[I+1]
        SUBF3     *AR0,*-AR0,R1
* subtract in floating point R1 = X[I] - X[I+1]
BLK1 STF  R0,*-AR0                  ; replace X[I] by X[I] + X[I+1] in memory
 ||  STF  R1,*AR0++                 ; replace X[I+1] by X[I] - X[I+1] in memory


* First pass of the DO-20 loop (for stage K = 2 in DO-10 loop)

        LDI  @INPUT,AR0             ; AR0 points to X[I]
        LDI  2,IR0                  ; set IR0 to 2 which is equal to N2
        LDI  @FFTSIZ,RC
        LSH  -2,RC
* repeat counter RC set to N/4 since N1 = 2*N2 = 4
        SUBI 1,RC                   ; RC should be set 1 less than desired
*                                   ; number of repetition

        RPTB BLK2                   ; repeat upto BLK2
        ADDF3  *+AR0(IR0),*AR0++(IR0),R0
* add in floating point R0 - X[I] + X[I+2]
        SUBF3  *AR0,*-AR0(IR0),R1
* subtract in floating point R1 = X[I] - X[I+2]
        NEGF *+AR0,R0               ; set R0 to -X[I+3]
 ||  STF  R0,*-AR0(IR0)             ; replace X[I] by X[I] + X[I+2]
BLK2 STF  R1,*AR0++(IR0)
*                                   ; replace X[I+2] by X[I] - X[I+2]
 ||  STF  R0,*+AR0                  ; replace X[I+3] by -X[I+3]


* Main loop (FFT stages DO-10 loop)

        LDI  @FFTSIZ,IR0
        LSH  -2,IR0
* IR0 holds the index for E = 6.283185307179586/N1
        LDI  3,R5                   ; R5 holds the current stage number K
        LDI  1,R4                   ; R4 = N4
        LDI  2,R3                   ; R3 = N2
LOOP LSH  -1,IR0                    ; IR0 = E/2
        LSH  1,R4                   ; R4 = 2 * N4 = N2
        LSH  1,R3                   ; R3 = 2 * N2 = N1


* Inner loop (DO-20 loop in the program)

        LDI  @INPUT,AR5            ; AR5 points to X[I]
INLOP   LDI  IR0,AR0              ; AR0 = IR0
        ADDI @SINTAB,AR0          ; AR0 points to SINE/COSINE table
        LDI  R4,IR1               ; IR1 = R4

        LDI  AR5,AR1              ; AR1 = AR5
        ADDI 1,AR1               ; AR1 points to X[I1] = X[I+J]
        LDI  AR1,AR3             ; AR3 = AR1
        ADDI R3,AR3             ; AR3 points to X[I3] = X[I+J+N2]
        LDI  AR3,AR2            ; AR2 = AR3
        SUBI 2,AR2             ; AR2 points to X[I2] = X[I-J+N2]
        ADDI3    R3,AR2,AR4    ; AR4 points to X[I4] = X[I-J+N1]

        LDF  *AR5++(IR1),R0
```

FIGURE 12-41 Radix-2 FFT code on TMS320C30. (*Continued*)

```
*                                       ; R0 = X[I]
      ADDF3        *+AR5(IR1),R0,R1
*                                       ; R1 = X[I] + X[I+N2]
      SUBF3        R0,*++AR5(IR1),R0
*                                       ; R0 = -X[I] + X[I+N2]
  ||  STF   R1,*-AR5(IR1)               ; replace X[I] by X[I] + X[I+N2]
      NEGF  R0                          ; R0 = X[I] - X[I+N2]
      NEGF  *++AR5(IR1),R1
*                                       ; R1 = -X[I+N4+N2]
  ||  STF   R0,*AR5                     ; X[I+N2] is replaced by X[I] - X[I+N2]
      STF   R1,*AR5                     ; replace X[I+N4+N2] by its negation


* Innermost loop (DO-30 loop)

      LDI   @FFTSIZ,IR1                 ; IR1 = N
      LSH   -2,IR1
* IR1 = separation between SINE/COSINE table = N/4
      LDI   R4,RC                       ; RC = R4
      SUBI  2,RC                        ; repeat N4 - 1 times

      RPTB  BLK3                        ; repeat upto BLK3
      MPYF3        *AR3,*+AR0(IR1),R0
*                                       ; R0 = X[I3] * cos A
      MPYF3        *AR4,*AR0,R1         ; R1 = X[I4] * sin A
      MPYF3        *AR4,*+AR0(IR1),R1
*                                       ; R1 = X[I4] * cos A
  ||  ADDF3        R0,R1,R2             ; R2 = X[I3] * cos A + X[I4] * sin A
      MPYF3        *AR3,*AR0++(IR0),R0
*                                       ; R0 = X[I3] * sin A
      SUBF3        R0,R1,R0             ; R0 = -X[I3] * sin A + X[I4] * cos A
      SUBF3        *AR2,R0,R1           ; R1 = -X[I2] + R0
      ADDF3        *AR2,R0,R1           ; R1 = X[I2] + R0
  ||  STF   R1,*AR3++                   ; replace X[I3] by -X[I2] + R0
      ADDF3        *AR1,R2,R1           ; R1 = X[I1] + R2
  ||  STF   R1,*AR4--                   ; replace X[I4] by X[I2] + R0
      SUBF3        R2,*AR1,R1           ; R1 = X[I1] - R2
  ||  STF   R1,*AR1++                   ; replace X[I1] by X[I1] + R2
BLK3  STF   R1,*AR2--                   ; replace X[I2] by X[I1] - R2

      SUBI  @INPUT,AR5
      ADDI  R3,AR5                      ; AR5 = I + N1
      CMPI  @FFTSIZ,AR5
      BLED  INLOP
* loop back (delayed) to the inner loop if AR5 less than or equal to N
      ADDI  @INPUT,AR5
      NOP
      NOP

      ADDI  1,R5
      CMPI  @LOGFFT,R5
      BLE   LOOP

END   BR    END                        ; branch to itself
      .end
```

FIGURE 12-41 (*Continued*)

```
*
* Table with Twiddle Factors for a 64-point FFT
*
* This file needs to be linked with the source code for a 64-point, radix-2 FFT
*
        .globl    SINE
        .globl    N
        .globl    M
N       .set 64
M       .set 6                      ; M = log₂(N)

        .data
SINE
        .float    0.000000
        .float    0.98017
        .
        .
        .
        .float    0.995185
COSINE
        .float    1.000000
        .float    0.995185
        .
        .
        .
        .float    0.995185
```

FIGURE 12-42   The 64-point FFT twiddle factors. (The figure is reproduced by permission of Texas Instruments, Inc., with additional annotations by the author.)

REFERENCES

[AG87] R. C. Agarwal and J. W. Cooley, Vectorized mixed radix discrete Fourier transform algorithms. *Proc. IEEE,* **75,** 1283–1292 (1987).

[AT88] AT&T Microelectronics, *WE DSP32 and DSP32C C Language Compiler Library Reference Manual,* AT&T Microelectronics, Allentown, PA, 1988.

[AT89a] AT&T Microelectronics, *WE DSP16 and DSP16A Digital Signal Processors Information Manual,* AT&T Microelectronics, Allentown, PA, 1989.

[AT89b] AT&T Microelectronics, *WE DSP16 and DSP16A Application Software Library Reference Manual,* AT&T Microelectronics, Allentown, PA, 1989.

[AT90] AT&T Microelectronics, *WE DSP32C Digital Signal Processors Information Manual,* AT&T Microelectronics, Allentown, PA, 1990.

[BA80] K. E. Batcher, Design of massively parallel processors, *IEEE Trans. Comput.* BC-29, 836–840 (1980).

[BU85] C. S. Burrus and T. W. Parks, *DFT/FFT and Convolution Algorithms,* Wiley, New York, 1985.

[EM91] P. M. Embree and B. Kimble, *C. Language Algorithms for Digital Signal Processing,* Prentice-Hall, Englewood Cliffs, NJ, 1991.

[HO81] R. W. Hockney and C. R. Jesshope, *Parallel Computers,* Adam Hilger Ltd., Bristol, UK, 1981.

[JA86] L. B. Jackson, *Digital Filters and Signal Processing*, Kluwer Academic Publishers, Boston, MA, 1986.

[KI83] H. Kikuchi, T. Inaba, Y. Kubono, H. Hambe, and T. Ikasawa, A 23K gate CMOS DSP with 100 ns multiplication, *IEEE Int. Solid-State Circuits Conf. Dig. Pap., 1983*, pp. 128–129 (1983).

[KO81] P. M. Kogge, *The Architecture of Pipelined Computers*, Hemisphere Publishing Co., Washington, DC, 1981.

[MO89] Motorola, Inc., *DSP96002 IEEE Floating-Point Dual-Port Processor User's Manual*, Motorola, Inc., Phoenix, AZ, 1989.

[MO90] Motorola, Inc., *DSP56000/DSP56001 Digital Signal Processor User's Manual*, Motorola, Inc., Phoenix, AZ, 1990.

[NI81] T. Nishitani, R. Maruta, Y. Kawakami, and H. Goto, A single-chip digital signal processor for telecommunication applications, *IEEE J. Solid-State Circuits* **SC-16,** 372–376 (1981).

[OP89] A. V. Oppenheim and R. W. Schafer, *Discrete-Time Signal Processing*, Prentice-Hall, Englewood Cliffs, NJ, 1989.

[SH88] D. Shear, EDN's DSP benchmarks. *EDN* **33,** 126 (1988).

[SI84] H. Simon, ed., *Supercomputer Vectorization and Optimization Guide*, Engineering Technology Applications Division, Boeing Computer Services, 1984.

[SU87a] W. Sung, Vector and multiprocessor implementation of digital filtering algorithms, Ph.D. dissertation, University of California, Santa Barbara, 1987.

[SU87b] W. Sung and S. K. Mitra, Implementation of digital filtering algorithms using pipelined vector processors, *Proc. IEEE* **75,** 1293–1303 (1987).

[SU90] W. Sung, S. K. Mitra, and B. Jeren, An efficient implementation of digital filtering algorithms using a multiprocessor system. *Proc. IEEE Int. Symp. Circuits Syst., New Orleans, Louisiana*, 1990, pp. 1426–1429 (1990).

[TE90] Texas Instruments, INc., *TMS320C2x User's Guide, Digital Signal Processing Products*, Texas Instruments Inc., Houston, TX, 1990.

[TE91] Texas Instruments, INc., *TMS320C3x User's Guide, Digital Signal Processing Products*, Texas Instruments Inc., Houston, TX, 1991.

13 Special Filter Designs

PHILLIP A. REGALIA

Department of Electronics and Communications
National Institute of Telecommunications
Evry, France

13-1 INTRODUCTION

In a variety of applications it is desired to process a signal using a filter that is not necessarily of the type discussed in Chapters 4 and 5, that is, of the linear passband/stopband type with piecewise constant magnitude responses. The natural questions are how to approach the design of such a filter, and whether the design procedures share some common features with conventional filter design procedures of the earlier chapters. The design of certain deterministic "special" filters, such as Hilbert transformers, integrators and differentiators, smoothing filters, nonintegral delay networks, and median filters, is the topic of this chapter. These various filters are briefly introduced in turn in the subsequent paragraphs.

The first of these is the design of Hilbert transformers. The classical approach to this problem is to approximate by some means a filter that passes all frequencies with a 90° phase shift, and at present numerous formulations exist for this approximation problem [AN87; CI70; GO70; JA75; ME84; PE88; RE87a; SC87]. The approach followed here deviates somewhat from the "conventional" approach and begins by presenting a simple derivation of the Hilbert transformer as it relates to signal processing. This derivation places in evidence a very strong connection between the Hilbert transformers and the classical "passband/stopband" filtering problems. This connection was pointed out for the FIR filters in Jackson [JA75] and was extended to the IIR filters independently in Ansari [AN87] and Regalia and Mitra [RE87a].

The advantages of following this approach are twofold. First, the conceptual connection promotes a clear understanding of how the Hilbert transformer design relates to conventional filter design, thus leading useful insight to the approximation problem. Such theoretical questions as realizability and optimality of design are then seen to be no different from those encountered in conventional filtering. Second, the design process is simplified immensely. For example, existing com-

Handbook for Digital Signal Processing, Edited by Sanjit K. Mitra and James F. Kaiser.
ISBN 0-471-61995-7 © 1993 John Wiley & Sons, Inc.

puter design routines [GR76; MC73] can be adapted almost trivially for the design of Hilbert transformers. By comparison, a separate reformulation of the approximation problem implies that a separate optimization routine must be developed for its solution. In some instances, this may require a significant modification of an existing computer design program and/or the addition of various subroutines. In practice, the modification of computer code is a time consuming and potentially error-prone procedure, as the modified program must be carefully debugged and reevaluated with respect to numerical accuracy if it is to be as reliable as its ''parent'' program.

We concentrate on the design of optimal (equiripple) Hilbert transformers; once the connection to conventional filter design is established, alternate approximation strategies are readily extended.

In Section 13-3 the design of digital differentiators and integrators is discussed. The idea here is to simulate an operation performed on an analog signal given only its sampled version. This problem can be understood as a special filter design problem in which the passband has a particular shape to be realized. Some connection to classical procedures for numerical integration is briefly discussed. One should keep in mind, however, that classical numerical integration procedures are derived by assuming a polynomial form for the function to be integrated. By comparison, in digital signal processing, one assumes that the sampled function is a bandlimited function. In this case, a frequency-domain approach provides superior performance to a ''polynomial'' approach. Once the frequency-domain formulation is developed, the design problem becomes fairly straightforward.

The design of smoothing filters is explored from both the time and frequency domains in Section 13-4. A time-domain approach is useful when waveform fidelity is of prime importance. We review an interesting approach to smoothing time sequences based on fitting successive polynomials to the data in a least-squares sense [SA64; ST72]. The resulting output sequence turns out to be a linear convolution of the input sequence. The observed properties of these smoothing filters suggest that frequency-domain formulations are also useful, particularly when spectral information is more important than waveform fidelity. The frequency domain formulation again suggests a simple connection to conventional linear-phase filter design.

The design of delay networks offering a signal delay not equal to an integral number of the sampling period is addressed in Section 13-5. An FIR design strategy proceeds quite systematically from the polyphase decomposition of an appropriately designed linear-phase lowpass filter. The technique leads to a delay network that implements a rational sample delay. An IIR design is accomplished by designing an allpass filter whose phase response has an arbitrary slope linear-phase characteristic. In this fashion, an arbitrary signal delay can be approximated over some well-defined frequency interval.

The chapter closes by exploring median filters in Section 13-6. The median operation yields a particularly effective smoothing filter for removing impulsive type noise from a signal while preserving edges, in contrast to linear filters, which can only suppress impulse noise by suppressing *all* high-frequency components. The median filter and its many generalizations have thus found uncommon success in signal restoration applications where waveform characteristics are of prime importance, such as image processing and biomedical signal processing.

The median operation though is nonlinear, which complicates conventional analysis techniques based on linear system theory. The notions of eigenfunctions and signal decompositions for linear systems are replaced by their median filter analogies of root signals and threshold decompositions. A root signal is invariant to median filtering and results if a median filter is applied repeatedly to a given signal. If the desired signal characteristics resemble those of a root signal, the median filter will succeed in its noise suppression task. Threshold decomposition lends itself to a superposition type of result for median filters and indeed suggests possible hardware implementations to replace the conventional sorting steps required in the median operation. Using these background concepts, the statistical properties of median filters are reviewed, with application to understanding the noise attenuation properties of median filters.

Median filters belong to a more general class of *stack order filters*, a natural extension following the threshold decomposition ideas. Stack order filters, in possible collaboration with linear filter substructures, permit greatly increased design flexibility. This allows the root signal(s) of the filter to be tailored to a specific application, resulting effectively in a waveform selective filter providing robust discrimination between a desired signal and corruptive noise. Numerous generalizations of stack order and hybrid linear/nonlinear filters have been proposed in the open literature; this chapter reviews some of the more general purpose modifications while briefly summarizing their attributes. The section concludes with implementation considerations, and application examples in image processing are included to demonstrate the performance of these filters.

13-2 DESIGN OF HILBERT TRANSFORMERS

The design of Hilbert transformers can be viewed as a particular case of half-band filter design. Although this is perhaps not the standard approach to this design problem, we feel it lends useful insight and simple design procedures, and so is pursued here.

13-2-1 Background

The spectrum of any real-valued signal exhibits Hermitian symmetry about the frequency $\Omega = 0$; that is, the negative frequency spectrum is simply the Hermitian mirror image of the positive frequency spectrum: $H(j\Omega) = H^*(-j\Omega)$. This suggests that there is some redundancy between the positive and negative portions of the frequency spectrum, and indeed the information of a real-valued time signal is completely contained within its positive frequency spectrum, save for a possible ambiguity at $\Omega = 0$ [OP75].

In certain applications it is desirable to discard the negative frequency spectrum of a real signal and process (or analyze) only the positive frequency spectrum. Examples include single sideband modulators in communications systems and frequency division multiplexers. In this section we consider the approximation theory and design of filters that remove the negative frequency components from a real-valued signal while leaving the positive frequency spectrum (ideally) unaltered. The derivation of this class of filters naturally leads us to the concept of a Hilbert transformer.

To begin, suppose we have a causal continuous-time function $u(t)$ and an associated Fourier transform $U(j\Omega)$, as exemplified in Figure 13-1. The reader will recall from Chapter 2 that, since the time signal $u(t)$ is causal, its transform $U(j\Omega)$ admits an analytic continuation into the complex s-plane. That is, we regard $U(j\Omega)$ as the values of an analytic function $U(s)$ evaluated along the imaginary axis $s = j\Omega$.

What happens if we impose "causality" in the frequency domain? Specifically, suppose we pass the signal $u(t)$ through a linear filter whose (ideal) frequency response is that of Figure 13-2, to obtain an output signal $y(t)$ whose spectrum vanishes for negative frequencies. By combining the duality property of the Fourier transform with the arguments from the preceding paragraph, we anticipate that "causality" of $Y(j\Omega)$ in the frequency domain should imply that $y(t)$ admit an analytic continuation throughout a complex time plane. Let us call this plane the τ-plane, with the understanding that the real axis of this plane is the familiar t-axis. We do not claim to offer a physical interpretation of the imaginary component of a complex time variable; rather, we regard the complex time plane as a conceptual artifice. Thus we can envisage an analytical function $y(\tau)$ in the complex time plane, which reduces to the complex-valued time signal $y(t)$ when evaluated along the real axis $\tau = t$. In this sense, a complex-valued time signal whose Fourier transform vanishes for negative frequencies is often termed an "analytic" signal.

Since $y(\tau)$ is an analytic function of τ, its real and imaginary parts along the real-time axis are related to each other via a pair of relations known as Hilbert transformations [HI59]. A signal processing derivation proceeds as follows. We begin by considering the real-valued continuous-time signal $u(t)$, which is passed through the linear filter $H(j\Omega)$ with the frequency response

$$H(j\Omega) = \begin{cases} 2, & \Omega > 0, \\ 0, & \Omega < 0. \end{cases} \tag{13.1}$$

This is simply a step function in the frequency domain; the factor 2 in the passband gain is for convenience. The impulse response corresponding to this filter can be

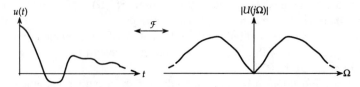

FIGURE 13-1   Real-valued signal $u(t)$ and its Fourier transform $U(j\Omega)$.

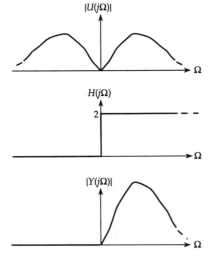

FIGURE 13-2 Removing negative frequency components from $U(j\Omega)$ to obtain $Y(j\Omega)$.

found with the aid of a Fourier transform pair table, leading to the result

$$h(t) = \delta(t) + \frac{j}{\pi t}. \tag{13.2}$$

The imaginary part of this impulse response is taken to be zero at $t = 0$. With $u(t)$ as the input to this filter, the output $y(t)$ is given as the convolution of the input $u(t)$ with the impulse response $h(t)$ as

$$y(t) = u(t) \circledast h(t) = u(t) + j[u(t) \circledast h_i(t)], \tag{13.3}$$

where $h_i(t)$ is the imaginary part of $h(t)$ in Eq. (13.2), as given by

$$h_i(t) = \frac{1}{\pi t}, \qquad t \neq 0. \tag{13.4}$$

Note that the real part of $y(t)$ is simply the original real signal $u(t)$, and that the imaginary part of $y(t)$ is obtained by passing $u(t)$ through a real linear filter whose impulse response is that of Eq. (13.4). Since the real and imaginary parts of $y(t)$ form a Hilbert transform pair, the linear filter with the impulse response of Eq. (13.4) is called the ideal Hilbert transformer. The frequency response of the ideal Hilbert transformer is found readily from Eq. (13.1) using the Fourier transform relation

$$\mathcal{F}\{h_i(t)\} = \frac{1}{2j}[H(j\Omega) - H^*(-j\Omega)]. \tag{13.5}$$

Using $H(\Omega)$ from Eq. (13.1), we obtain

$$\mathscr{F}\{h_i(t)\} = \begin{cases} -j, & \Omega > 0, \\ j, & \Omega < 0, \end{cases} \tag{13.6}$$

as the frequency response of the ideal Hilbert transformer. This frequency response is often interpreted as an allpass characteristic providing phase shift of $\pi/2$ radians at all frequencies.

It is seen from Eq. (13.4) that the impulse response of the ideal Hilbert transformer is two-sided and thus noncausal. The ideal Hilbert transformer, like the ideal lowpass filter, is to be regarded as a theoretical ideal, which can only be approximately realized in practice. Methods for obtaining such approximations are the subject of the remaining subsections.

The discrete-time Hilbert transformer can be derived in a manner analogous to that of its continuous-time counterpart. That is, we envisage a real-valued time sequence $u[n]$ with a convergent Fourier transform $U(e^{j\omega})$. We can imagine passing this signal through a complex linear filter $H(e^{j\omega})$ with the frequency response

$$H(e^{j\omega}) = \begin{cases} 2, & 0 < \omega < \pi, \\ 0, & -\pi < \omega < 0, \end{cases} \tag{13.7}$$

to obtain an output signal whose spectrum vanishes for negative frequencies. The inverse Fourier transform of Eq. (13.7) gives the impulse response

$$h[n] = \begin{cases} 1 & \text{for } n = 0, \\ \dfrac{j \cdot 2}{\pi n} & \text{for } n \text{ odd}, \\ 0 & \text{otherwise.} \end{cases} \tag{13.8}$$

The complex output sequence $y[n]$ of such a filter is

$$y[n] = u[n] \circledast h[n] = u[n] + j\{u[n] \circledast h_i[n]\}, \tag{13.9}$$

where $h_i[n]$ is the imaginary part of the impulse response of Eq. (13.8), as given by

$$h_i[n] = \begin{cases} \dfrac{2}{\pi n} & \text{for } n \text{ odd}, \\ 0 & \text{for } n \text{ even.} \end{cases} \tag{13.10}$$

Note from Eq. (13.9) that the real part of the complex sequence $y[n]$ is just the original sequence $u[n]$, and that the imaginary part of $y[n]$ is obtained by passing $u[n]$ through a linear filter with the impulse response of Eq. (13.10). In analogy with the continuous-time case, the filter of Eq. (13.10) is called the ideal discrete-time Hilbert transformer [CI70]. One should recognize, however, that because $y[n]$ is a discrete-time sequence, the notion of analyticity is not properly applicable, and hence it is somewhat misleading to imply that the real and imaginary parts of $y[n]$ should be related by a Hilbert transform operation. Nonetheless, $y[n]$ may be interpreted as a sampled version of an analytic continuous-time signal [GO70]. From this observation, a sequence whose Fourier transform vanishes for negative frequencies is often termed an ''analytic'' sequence.

The frequency response of the ideal discrete-time Hilbert transformer is found from the relation

$$\mathcal{F}\{h_i[n]\} = \frac{1}{2j}[H(e^{j\omega}) - H^*(e^{-j\omega})]. \tag{13.11}$$

The use of $H(e^{j\omega})$ from Eq. (13.7) yields

$$\mathcal{F}\{h_i[n]\} = \begin{cases} -j, & 0 < \omega < \pi, \\ j, & -\pi < \omega < 0. \end{cases} \tag{13.12}$$

As in the continuous-time case, the discrete-time Hilbert transformer can be regarded as an allpass filter providing a phase shift of $\pi/2$ radians at all frequencies.

13-2-2 Design of Hilbert Transformers

The derivation of the discrete-time Hilbert transformer suggests an attractive design strategy whose simplicity should not be overlooked. Specifically, suppose we shift the frequency response $H(e^{j\omega})$ of Eq. (13.7) by $\pi/2$ radians to obtain a new filter $G(e^{j\omega})$ whose frequency response is

$$G(e^{j\omega}) \triangleq H(e^{j[\omega + \pi/2]}) = \begin{cases} 2, & |\omega| < \pi/2, \\ 0, & \pi/2 \leq |\omega| < \pi. \end{cases} \tag{13.13}$$

Note that this is simply the ideal half-band lowpass filter, whose passband magnitude is scaled to the value 2. Thus if we have some satisfactory approximation to $G(e^{j\omega})$, we need only shift its frequency response by $\pi/2$ radians to obtain an equally good approximation to the frequency response of Eq. (13.7). For convenience, we call the filter $H(e^{j\omega})$ of Eq. (13.7) the ideal complex half-band filter.

Let $g[n]$ denote the inverse Fourier transform of $G(e^{j\omega})$. Upon shifting $G(e^{j\omega})$ by $\pi/2$ radians, we obtain an approximation to the ideal complex half-band filter. Recalling the modulation property of the Fourier transform, this operation is the time-domain equivalent of multiplying the sequence $g[n]$ by j^n to obtain the sequence $h[n]$:

$$h[n] = j^n g[n]. \tag{13.14}$$

Note that j^n produces the sequence $1, j, -1, -j, \ldots$, and thus the transformation of Eq. (13.14) is quite simple. In the following subsections, we explore the application of Eq. (13.14) to half-band lowpass filters of both the FIR and IIR types.

To summarize the discussion up to this point, a complex half-band filter can be realized using a Hilbert transform as the imaginary part of a complex filter. This is sketched in Figure 13-3(a), which shows the equivalent real filtering operations to obtain a complex-valued signal ($y(t)$ in the continuous-time case, or $y[n]$ in the discrete-time case) whose spectrum vanishes for negative frequencies. If the input signal u is real valued, the filter reduces to that of Figure 13-3(b).

FIR Linear-Phase Designs. We begin by reviewing some simple properties of FIR linear-phase half-band filters. A FIR half-band filter is often designed so that the passband attenuation is the complement of the stopband attenuation. Specifically, if $G(z)$ is a half-band lowpass filter of length $2N$, then $G(-z)$ becomes a half-band highpass filter of the same length, because the passband of $G(z)$ is the stopband of $G(-z)$, and vice versa. This complementary property between the passband and stopband of $G(z)$ may then be expressed mathematically as

$$G(z) - G(-z) = z^{-N}, \tag{13.15}$$

where N is the group delay of $G(e^{j\omega})$, which may be assumed to be an odd integer without loss of generality [VA87a]. Equation (13.15) then implies that $G(z)$ may be written as

$$G(z) = \frac{z^{-N} - E(z^2)}{2}. \tag{13.16}$$

The term $E(z^2)$ describes a linear-phase filter with a group delay of N samples. Since $E(z^2)$ can be written as a polynomial in z^{-2}, its inverse z-transform evidently yields a sequence whose odd-indexed samples are all zero. In the time domain,

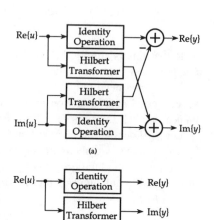

(a)

FIGURE 13-3 (a) Complex half-band filter realization using a Hilbert transformer. The scheme is applicable to both continuous-time and discrete-time signals. (b) Simplification of part (a) when the input signal is real valued.

(b)

Eq. (13.16) thus implies that every odd-indexed sample of $g[n]$ is zero save for the Nth, whose value is $\frac{1}{2}$. A quick design procedure for FIR linear-phase half-band filters is to note that $E(z)$ (which is the "decimated" version of $E(z^2)$) is a wideband filter whose single transmission zero lies at $z = -1$ [VA87a].[1]

By shifting the frequency response corresponding to $G(z)$ by $\pi/2$ radians, we can obtain a complex half-band filter according to

$$H(z) \triangleq j2G(-jz) = (-1)^{(N-1)/2}z^{-N} + jE(-z^2). \qquad (13.17)$$

Let us decompose $E(-z^2)$ as

$$E(-z^2) = z^{-N}E_0(-z^2), \qquad (13.18)$$

where $E_0(-z^2)$ is the zero-phase portion of $E(-z^2)$. If $(N + 1)/2$ is even, Eq. (13.17) may be written as

$$H(z) = z^{-N}[1 + jE_0(-z^2)]. \qquad (13.19)$$

(If $(N + 1)/2$ were odd, a factor of -1 would then appear before the z^{-N} term.) Since the stopband of this filter (ideally) occurs in the interval $-\pi < \omega < 0$, the output signal of this complex filter has nearly zero signal energy at negative frequencies. Thus we identify $E_0(-z^2)$ as a *zero-phase* approximation to the ideal Hilbert transformer, and $E(-z^2)$ becomes a *causal* approximation to the ideal Hilbert transformer. A realization corresponding to Eq. (13.18) is depicted in Figure 13-4, assuming the input sequence $\{u[n]\}$ to be real valued. These concepts are illustrated in greater detail in the following design example.

Design Example 13.1. We begin by designing a wideband FIR filter $E(z)$ to meet the following specifications:

$$|E(e^{j\omega})| \begin{cases} \approx 1, & 0 < \omega < \pi - \alpha, \\ = 0, & \omega = \pi. \end{cases} \qquad (13.20)$$

If the parameter α is small, the passband edge is fairly close to $\omega = \pi$. By choosing $E(z)$ as a symmetric filter of odd degree (i.e., a "Type II" impulse response in the terminology of Gold and Rader [GO69]), we obtain automatically a zero at $z = -1$ of odd multiplicity. We have used the program of McClellan et al. [MC73]

[1]See Section 4-10-2.

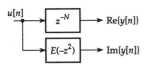

FIGURE 13-4 FIR realization of complex half-band filter.

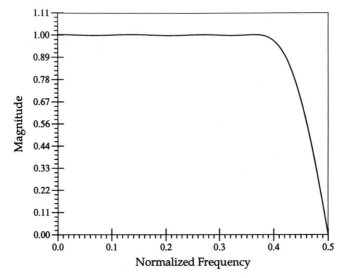

FIGURE 13-5 Frequency response of the wideband Type II FIR filter.

to design a filter $E(z)$ of degree 13 whose frequency response appears as Figure 13-5, satisfying $\alpha = 0.12\pi$. The impulse response is listed in Table 13-1. This sequence is then interpolated by a factor of 2 to obtain $E(z^2)$, to which we add the term z^{-13} to obtain a real half-band filter according to

$$2G(z) = z^{-13} + E(z^2). \tag{13.21}$$

A complex half-band filter is then obtained according to

$$H(z) = -2G(-jz) = z^{-13} - jE(-z^2), \tag{13.22}$$

TABLE 13-1 Impulse Response for Wideband Filter $E(z)$

	Coefficients of Polynomial $E(z)$	
n	$e[n]$	n
0	$-4.2106e^{-03}$	13
1	$1.13279e^{-02}$	12
2	$-2.53380e^{-02}$	11
3	$5.00879e^{-02}$	10
4	$-9.46106e^{-02}$	9
5	0.190787	8
6	-0.628044	7

TABLE 13-2   Impulse Response for Complex Half-Band Filter in Example 13.1

	Impulse Response $h[n]$	
n	$h_r[n]$	$jh_i[n]$
0	0.0	$-j4.21061e^{-03}$
2	0.0	$-j1.13279e^{-02}$
4	0.0	$-j2.53380e^{-02}$
6	0.0	$-j5.00879e^{-02}$
8	0.0	$-j9.46106e^{-02}$
10	0.0	$-j0.190787$
12	0.0	$-j0.628044$
13	1.0	$j0.0$
14	0.0	$j0.628044$
16	0.0	$j0.190787$
18	0.0	$j9.46106e^{-02}$
20	0.0	$j5.00879e^{-02}$
22	0.0	$j2.53380e^{-02}$
24	0.0	$j1.13279e^{-02}$
26	0.0	$j4.21061e^{-03}$

Note: Time indices that are not specified indicate a zero sample, for example $h[1] = h[3] = 0$, and so on.

whose impulse response is listed in Table 13-2. Note that the term $E(-z^2)$ has an antisymmetric impulse response. The frequency response corresponding to this term is shown in Figure 13-6. To summarize then, an FIR Hilbert transformer may be designed as follows:

1. Specifications: Approximate a Hilbert transformer over the frequency range $\omega_1 < |\omega| < \pi - \omega_1$. The transfer function approximates $-jz^{-N}$ for some integer N ($=$ the group delay of the filter). The maximum deviation from unity in the magnitude response we denote as δ.

2. Design a wideband FIR filter $E(z)$ of degree N, with N odd, whose impulse response is symmetric, using the program of McClellan et al. [MC73], for example. The passband of $E(z)$ should cover the range $0 \le \omega \le \pi - 2\omega_1$, and its magnitude response should satisfy

$$1 - \delta \le |E(e^{j\omega})| \le 1 + \delta, \quad 0 \le \omega \le \pi - 2\omega_1.$$

Since the filter is of odd degree and the impulse response is symmetric, we automatically obtain a zero at $\omega = \pi$.

3. The filter obtained as $E(-z^2)$ is the desired FIR Hilbert transformer with a group delay of N samples. The magnitude response of the resulting complex

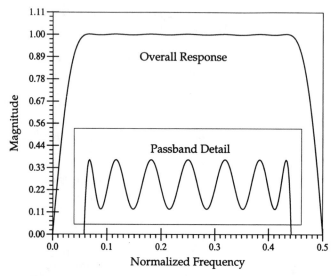

FIGURE 13-6 The designed wideband FIR Hilbert transformer.

half-band filter $z^{-N} + jE(-z^2)$ satisfies

$$|z^{-N} + jE(-z^2)|_{z=e^{j\omega}} \le \delta, \qquad -\pi + \omega_1 \le \omega \le -\omega_1,$$

so that the parameter δ functions as the stopband attenuation for negative frequencies.

The Hilbert transformer obtained in this fashion has an antisymmetric impulse response, which of course gives the desired factor j in the zero-phase response. The design procedure above in fact always yields a Type III design for the Hilbert transformer, so that the group delay is always an integral number of samples. One could conceivably develop design procedures for Type IV designs as well, since this case also represents an antisymmetric impulse response. Suppose we call this filter $F(z)$. One would simply force the magnitude response $|F(e^{j\omega})|$ of the design to approximate unity over a sufficiently large frequency range of interest, since the zero-phase response is automatically imaginary valued. Some caution is in order here though, because the group delay response of such a filter is no longer an integral number of samples. Rather, the group delay becomes $N + \frac{1}{2}$ samples, for some integer N. To obtain a complex half-band filter, one would then try to write

$$H(z) = z^{-[N+(1/2)]} + jF(z). \tag{13.23}$$

The immediate problem is realizing the fractional delay. One alternative is to replace the term $z^{-[N+(1/2)]}$ by a Type II linear-phase filter $D(z)$, say. (The Type II impulse response implies that the zero-phase response of $D(z)$ is real valued, and that its group delay has the desired "half-sample" property.) However, now the problem involves designing two filters instead of one. Moreover, even if $|D(e^{j\omega})|$ and $|F(e^{j\omega})|$ are both equiripple approximations to unity over some desired frequency range, this does not guarantee that the complex sum $D(z) + jF(z)$ has a frequency response that is equiripple. To obtain a proper equiripple response in this case, a filter design package must be properly modified to handle complex impulse response sequences [CR68]. In the interests of efficient designs, it is often preferable to use the complex half-band filter of Eq. (13.17), since one branch is simply a delay, while the other branch has every second impulse response coefficient as zero.

IIR Hilbert Transformer Design. Our approach here is to exploit the properties of a class of recursive half-band filters that may be written as the sum of two allpass functions. Included in this class are half-band Butterworth and elliptic filters, although nonclassical filter approximations are also quite useful, as shown below for the class of approximately linear-phase IIR filters. Our derivation will naturally lead to the concept of a $90°$ $(=\pi/2$ radians) phase splitter. Such a system is simply a pair of allpass functions whose phase responses differ by approximately $\pi/2$ radians over some well-defined frequency band. If we take the outputs of the two allpass filters as the real and imaginary parts of a complex signal, we find that the spectrum of such a signal nearly vanishes over much of the negative frequency interval. As such, the outputs of the two allpass filters are quite nearly a Hilbert transform pair.

To begin, we exploit the fact that a useful class of odd-ordered recursive half-band filters may be written as the sum of two allpass transfer functions [VA87b; WE79]:

$$G(z) = \tfrac{1}{2}[A_1(z^2) + z^{-1}A_2(z^2)]. \tag{13.24}$$

Here $G(z)$ is the assumed half-band filter approximation, while $A_1(z^2)$ and $A_2(z^2)$ are stable allpass transfer functions. For the sake of clarity, we point out that $A_1(z^2)$ is an even function of z, while $z^{-1}A_2(z^2)$ is an odd function of z. Thus in the time domain the even-indexed samples of the impulse response $g[n]$ come from the first allpass function, whereas the odd-indexed samples come from the second allpass function. This "even–odd" decomposition may be shown to be a consequence of the half-band property [VA87b; WE79].

It is now a simple matter to obtain a complex half-band filter according to

$$H(z) \triangleq 2G(-jz) = [A_1(-z^2) + jz^{-1}A_2(-z^2)]. \tag{13.25}$$

It is important to note that the terms $A_1(-z^2)$ and $A_2(-z^2)$ remain real allpass functions. Equation (13.25) thus expresses the complex half-band filter $H(z)$ as the complex sum of two real allpass functions, whose realization for a real-valued input sequence is depicted in Figure 13-7. We illustrate with a design example.

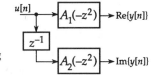

FIGURE 13-7 Complex half-band filter realization using allpass filters.

Design Example 13.2. Here we begin by designing a real half-band elliptic filter $G(z)$ whose frequency response satisfies

$$1 - \delta_1 \le |G(e^{j\omega})| \le 1 \quad \text{for } 0 \le \omega \le \omega_p, \tag{13.26}$$

$$|G(e^{j\omega})| \le \delta_2 \quad \text{for } \omega_s \le \omega \le \pi.$$

To obtain a half-band design, we choose $\omega_s + \omega_p = \pi$, and $(1 - \delta_1)^2 + \delta_2^2 = 1$. These two constraints ensure that the poles of the optimum elliptic design lie on the imaginary axis in the z-plane [VA87b]. Choosing, for example, $\omega_s = 0.55\pi$ and $\delta_2 = 0.01$ and using the design program of Gray and Markel [GR76], we obtain a seventh-order elliptic filter whose poles are

$$z = 0, \quad z = \pm j0.436688, \quad z = \pm j0.743707, \quad \text{and} \quad z = \pm j0.927758, \tag{13.27}$$

as sketched in Figure 13-8(a). The half-band filter $G(z)$ takes the form

$$G(z) = \tfrac{1}{2}[A_1(z^2) + z^{-1}A_2(z^2)], \tag{13.28a}$$

where

$$A_1(z^2) = \frac{z^{-2} + 0.190696}{1 + 0.190696z^{-2}} \cdot \frac{z^{-2} + 0.860735}{1 + 0.860735z^{-2}}, \tag{13.28b}$$

$$A_2(z^2) = \frac{z^{-2} + 0.553100}{1 + 0.553100z^{-2}}, \tag{13.28c}$$

which gives the frequency response of Figure 13-8(b). Note that the poles in Eq. (13.28) alternate between the two allpass functions $A_1(z^2)$ and $z^{-1}A_2(z^2)$, as depicted in Figure 13-8(a). This "alternating" property of pole assignment between the two allpass functions always holds for classical filter approximations [GA85]. The complex half-band filter $H(z)$ is obtained as

$$H(z) = 2G(-jz) = A_1(-z^2) + jz^{-1}A_2(-z^2), \tag{13.29a}$$

where now

$$A_1(-z^2) = \frac{z^{-2} - 0.190696}{1 - 0.190696z^{-2}} \cdot \frac{z^{-2} - 0.860735}{1 - 0.860735z^{-2}}, \tag{13.29b}$$

$$A_2(-z^2) = \frac{z^{-2} - 0.553100}{1 - 0.553100z^{-2}}. \tag{13.29c}$$

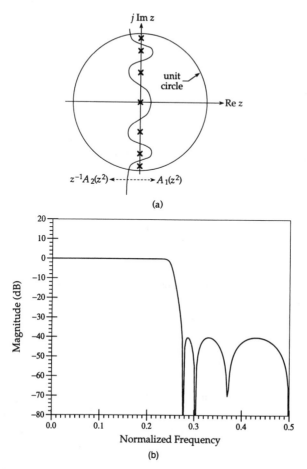

FIGURE 13-8 (a) Illustrating how the poles of the elliptic filter alternately distribute between the two allpass functions. (b) Frequency response of the seventh-order elliptic half-band filter.

The poles of these allpass filters now lie on the real axis in the z-plane. The frequency response corresponding to $H(z)$ appears in Figure 13-9. The phase difference between the two allpass functions $A_1(-z^2)$ and $z^{-1}A_2(-z^2)$ is shown in Figure 13-10. Note that the phase difference is automatically equiripple because we began with an elliptic filter prototype. One can show that the maximum deviation from phase quadrature, which we denote as $\Delta\phi_{\max}$, is related to the minimum stopband attenuation via

$$\text{stopband attenuation} = \sin(\Delta\phi_{\max}/2). \qquad (13.30)$$

For the example shown, the maximum deviation from phase quadrature is seen to be about 0.006π radians, corresponding to the negative frequencies being attenuated 40 dB relative to the passband amplitude.

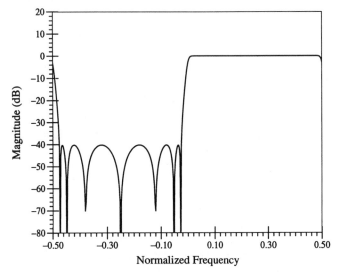

FIGURE 13-9   Magnitude response of the elliptic complex half-band filter (passband gain normalized to 0 dB).

The design procedure can be summarized as follows:

1. Specifications: Design two allpass functions whose phase responses are within $\Delta\phi_{max}$ of phase quadrature over the interval $\omega_1 \leq \omega \leq \pi - \omega_1$.
2. Design a half-band elliptic filter as per Eq. (13.26) of odd degree satisfying the following specifications (a design program can be found in Gray and

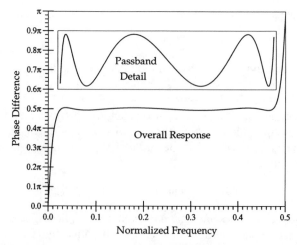

FIGURE 13-10   Phase difference between the allpass functions.

Markel [GR76]):

$$\text{stopband attenuation } \delta_2 = \sin (\Delta \phi_{max}/2),$$
$$\text{passband attenuation } \delta_1 = 1 - \sqrt{1 - \delta_2^2},$$
$$\text{passband edge frequency } \omega_p = \pi/2 - \omega_1,$$
$$\text{stopband edge frequency } \omega_s = \pi/2 + \omega_1.$$

As a check, the poles of the design should lie on the imaginary axis in the z-plane.

3. Rotate the poles 90° to lie on the real axis (i.e., use the transformation $z \rightarrow -jz$).

4. Alternately distribute the poles between the two allpass functions. (One allpass function will receive a pole at the origin, corresponding to a pure delay factor.)

The two allpass functions so formed will exhibit approximate phase quadrature over the specified frequency range.

Approximately Linear-Phase Designs. An interesting modification to the above design procedure results by starting with a class of half-band recursive filters that are known as "approximately linear-phase" filters [KI86; RE86]. When configured as half-band lowpass filters, these take the form

$$G(z) = \tfrac{1}{2}[z^{-N} + A(z^2)], \tag{13.31}$$

where the delay exponent N is assumed to be an odd integer. The allpass transfer function $A(z^2)$ is an even function of z, which is forced to be approximately in phase with the delay z^{-N} over the range $|\omega| < \pi/2$, and approximately out of phase with the delay over the range $\pi/2 < |\omega| < \pi$. The phasor sum in Eq. (13.31) thus exhibits approximately linear phase throughout the passband and stopband regions; hence the name. As above, a complex half-band filter may be obtained as

$$H(z) \triangleq 2jG(-jz) = [(-1)^{(N+1)/2}z^{-N} + jA(-z^2)]. \tag{13.32}$$

If $(N + 1)/2$ is an even integer, the output sequence of the allpass filter $A(-z^2)$ is the approximate Hilbert transformation of a delayed version of the input sequence. (If $(N + 1)/2$ is odd, the output from $A(-z^2)$ can be multiplied by -1 to obtain an approximate Hilbert transformation of a delayed version of the input.) As an example, we obtain a half-band filter designed from the program in Renfors and Saramäki [RE86] for the case $N = 17$. The pole locations of $A(z)$ (the "decimated" version of $A(z^2)$) are listed in Table 13-3 [RE87b]. Upon realizing $A(z)$ with a suitable structure, the term $A(z^2)$ is obtained by replacing each delay with a cascade of two delays. Upon adding the term $z^{-N} = z^{-17}$, the half-band

TABLE 13-3 Pole Locations for the Allpass Transfer Function $A(z)$ Used in the Approximately Linear-Phase Design

Pole Locations for $A(z)$
$z = -0.8699928078$
$0.491194141 \pm j0.183666529$
$0.252724179 \pm j0.463085544$
$-0.109950894 \pm j0.548611467$
$-0.447326028 \pm j0.356810323$

filter with the frequency response of Figure 13-11 results, which provides better than 49 dB of stopband attenuation. The phase difference between the terms z^{-17} and $A(-z^2)$ (the transformed version of $A(z^2)$) appears as Figure 13-12.

Even-Degree Case. The previous subsections have outlined the development of "phase splitters" derived from odd-ordered recursive half-band filters. We present here a similar development for even-ordered half-band filter prototypes. Our development relies on the fact that a useful class of even-ordered lowpass filters may be decomposed into the sum of two complex allpass functions according to [VA87b]

$$G(z) = \tfrac{1}{2}[A(z) + A^*(z^*)], \tag{13.33}$$

where $A(z)$ is now a complex allpass transfer function. The term $A^*(z^*)$ is obtained from $A(z)$ by replacing each constant coefficient with its complex conjugate. If

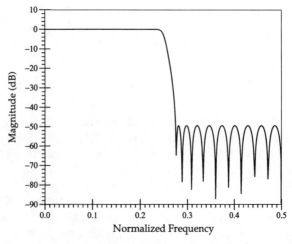

FIGURE 13-11 Frequency response of the approximately linear-phase half-band filter.

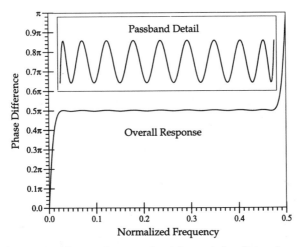

FIGURE 13-12   Phase difference between the delay and the allpass function for the approximately linear-phase Hilbert transform design.

$G(z)$ is a half-band filter, it can be shown [VA87b] that $A(z)$ then satisfies the relationship

$$A(-z) = \pm jA^*(z^*).$$ (13.34)

Specifically, if $G(z)$ is an even-ordered Butterworth or elliptic filter, one can show that the poles of $A(z)$ all lie on the imaginary axis, and that $A(z)$ takes the form

$$A(z) = e^{j\pi/4} \prod_{k=1}^{M} \frac{z^{-1} - j\alpha_k}{1 + j\alpha_k z^{-1}},$$ (13.35)

where $2M$ is the degree of $G(z)$, and α_k is real for all k. By applying the transformation $z \rightarrow -jz$, one obtains the complex half-band filter $H(z)$ according to

$$H(z) \triangleq 2G(-jz)$$

$$= \left[e^{j\pi/4} j^M \prod_{k=1}^{M} \frac{z^{-1} - \alpha_k}{1 - \alpha_k z^{-1}} + e^{-j\pi/4} j^M \prod_{k=1}^{M} \frac{z^{-1} + \alpha_k}{1 + \alpha_k z^{-1}} \right].$$ (13.36)

A factor $e^{j\pi/4} j^M$ is common to both transformed allpass functions and so may be omitted. The complex half-band filter may then be written as

$$G_1(z) = \frac{G(z)}{e^{j\pi/4} j^M} = B(z) + jB(-z)$$ (13.37a)

with

$$B(z) = \prod_{k=1}^{M} \frac{z^{-1} - \alpha_k}{1 - \alpha_k z^{-1}}. \tag{13.37b}$$

Note that $B(z)$ is simply the product of first-order allpass transfer functions. Again, we illustrate with a design example.

Design Example 13.3. The design procedure here is quite similar to that presented for the odd-ordered designs. Suppose we want two allpass transfer functions whose phase responses are within 0.0125π radians of phase quadrature over the range $0.05\pi \le \omega \le 0.95\pi$. Using Eq. (13.30), we find that the required negative frequency attenuation is 35 dB. Upon shifting the specifications by $\pi/2$ radians, we define a half-band lowpass filter according to the following specifications: minimum stopband attenuation = 35 dB; stopband edge frequency = 0.55π; maximum passband attenuation = 0.0316 dB; passband edge frequency = 0.45π. Using, for example, the design program of Gray and Markel [GR76], we obtain a sixth-order elliptic filter with the following pole locations:

$$z = \pm j0.261925, \quad z = \pm j0.674524, \quad \text{and} \quad z = \pm j0.912402. \tag{13.38}$$

Note that these poles lie on the imaginary axis as a consequence of the half-band property. Upon rotating these poles $90°$ to the real axis and alternately distributing them between two allpass functions, we obtain the complex half-band filter $G_1(z)$ according to

$$G_1(z) = B(z) + jB(-z), \tag{13.39a}$$

where

$$B(z) = \frac{z^{-1} + 0.261925}{1 + 0.261925z^{-1}} \cdot \frac{z^{-1} - 0.674524}{1 - 0.674524z^{-1}} \cdot \frac{z^{-1} + 0.912402}{1 + 0.912402z^{-1}}. \tag{13.39b}$$

The phase difference between $B(z)$ and $B(-z)$ is plotted in Figure 13-13.

Designs Nonsymmetric About $\omega = \pi/2$. Thus far the designs we have seen are obtained from real half-band filters using the transformation $z^{-1} \rightarrow jz^{-1}$. Since the frequency response of a real filter exhibits Hermitian symmetry about $\omega = 0$, the complex half-band filters designed above exhibit Hermitian symmetry about $\omega = \pi/2$. In some applications one may want a Hilbert transform system that is not necessarily centered about the frequency $\omega = \pi/2$; we treat that case here for the IIR designs.

Suppose we desire two allpass functions that exhibit approximate phase quadrature over the frequency range $\omega_1 \le \omega \le \omega_2$, where $\omega_1 + \omega_2 \ne \pi$. We can obtain this as a transformed version of a design that is symmetric about $\omega = \pi/2$. Spe-

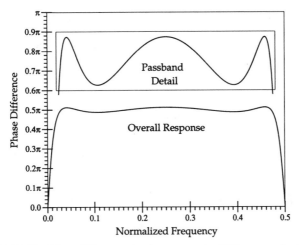

FIGURE 13-13   Phase difference for the even-order Hilbert transform system.

cifically, define two frequencies ω_3 and ω_4 by

$$\omega_3 = 2 \tan^{-1}\left[\left(\frac{\tan (\omega_1/2)}{\tan (\omega_2/2)}\right)^{1/2}\right],$$

$$\omega_4 = 2 \tan^{-1}\left[\left(\frac{\tan (\omega_2/2)}{\tan (\omega_1/2)}\right)^{1/2}\right]. \tag{13.40}$$

By construction we obtain $\omega_3 + \omega_4 = \pi$. Thus the desired specifications have been transformed to a frequency interval that is symmetric about $\pi/2$, and the two allpass functions can be designed using the methods described above. The resulting design can then be transformed back to meet the original specifications via

$$z^{-1} \rightarrow \gamma(z) = \frac{z^{-1} + \dfrac{\beta - 1}{\beta + 1}}{1 + z^{-1}\dfrac{\beta - 1}{\beta + 1}}, \tag{13.41a}$$

where

$$\beta = [\tan (\omega_1/2) \tan (\omega_2/2)]^{1/2}. \tag{13.41b}$$

We illustrate with an example.

Design Example 13.4.   Here we consider the design of two allpass functions that are within 0.0125π radians of phase quadrature over the frequency range $0.025\pi \le \omega \le 0.9\pi$. Setting $\omega_1 = 0.025\pi$ and $\omega_2 = 0.9\pi$, Eqs. (13.40) give $\omega_3 = 0.05\pi$

and $\omega_4 = 0.95\pi$, as the transformed frequency interval. This design problem was solved in Design Example 13.3, and hence we need only transform it back to the original specifications using Eq. (13.41). Specifically, we find from Eq. (13.41b) that

$$\beta = 0.498$$

and hence the transformation in Eq. (13.41a) looks like

$$z^{-1} \rightarrow \gamma(z) = \frac{z^{-1} - 0.334}{1 - 0.334z^{-1}}.$$

Thus the complex half-band filter becomes

$$H(z) = [B(z) + B(-z)]|_{z^{-1} \rightarrow \gamma(z)}, \tag{13.42}$$

where $B(z)$ was determined in Example 13.3. The pole locations obtained in Example 13.3 are transformed according to Eq. (13.42), with the results listed in Table 13-4. The phase difference between the transformed allpass functions obtained in Eq. (13.42) is plotted in Figure 13-14, and is seen to exhibit the desired phase quadrature over the interval $0.025\pi \leq \omega \leq 0.9\pi$.

Discussion. The complex half-band filters we have discussed above usually process real input sequences, in which case the filter becomes two branches, one of which produces the real portion of the output sequence, the other producing the imaginary portion. The design examples we have seen above indicate that FIR and IIR Hilbert transform systems appear to lend themselves to two classifications. Specifically, in the FIR designs we have seen the transfer functions of the two branches comprising the complex half-band filter exhibit perfect phase quadrature at all frequencies. The magnitude responses, however, are only approximately the same. By comparison, the transfer functions of the two branches of the IIR designs have magnitude responses that are identically the same, but the phase responses exhibit only approximate phase quadrature. Based on this observation, design books have traditionally classified Hilbert transform systems into two categories:

Type I. The two branches of the complex half-band filter exhibit perfect phase quadrature but only approximately the same magnitude.

TABLE 13-4   Pole Locations for the Transformed Design of Example 13.4

Poles for Transformed Design	
First Section	**Second Section**
−0.789848	−0.547986
−0.823089	0.439551
0.831925	−0.955286

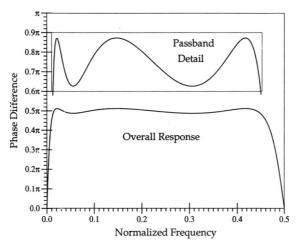

FIGURE 13-14 Phase difference for the transformed design.

Type II. The two branches of the complex half-band filter exhibit the exact same magnitude response (typically unity), but their phase responses are in only approximate phase quadrature.

In fact, the distinction above is artificial, as will now be shown. We use the fact that if $H(z)$ is any complex filter, its magnitude and group delay responses are totally unaffected if $H(z)$ is multiplied by $e^{j\theta}$, where θ is any real number. Thus let $H(z)$ denote some approximation to the ideal complex half-band filter, and decompose it as

$$H(z) = H_r(z) + jH_i(z), \tag{13.43}$$

where $H_r(z)$ and $H_i(z)$ are, respectively, the z-transforms of the real and imaginary components of the complex impulse response $h[n]$. Similarly, define $E(z) = e^{j\theta} H(z)$, decomposed as

$$E(z) = E_r(z) + jE_i(z), \tag{13.44}$$

again obtained as the z-transform of a "real–imaginary" decomposition in the time domain. If $\theta = \pi/4$, we find that

$$E_r(z) = \frac{1}{\sqrt{2}} [H_r(z) - H_i(z)],$$

$$E_i(z) = \frac{1}{\sqrt{2}} [H_r(z) + H_i(z)]. \tag{13.45}$$

From this it can be verified that if $H_r(e^{j\omega})$ and $H_i(e^{j\omega})$ have the same magnitude, then $E_r(e^{j\omega})$ and $E_i(e^{j\omega})$ exhibit perfect phase quadrature. Likewise, if $H_r(e^{j\omega})$

and $H_i(e^{j\omega})$ exhibit perfect phase quadrature, then $E_r(e^{j\omega})$ and $E_i(e^{j\omega})$ share the same magnitude response. Thus $H(z)$ represents a Type I Hilbert design if and only if $E(z)$ ($= e^{j\pi/4}H(z)$) represents a Type II design, and vice versa. Equation (13.45) is thus seen to give a simple recipe to convert a Type I design to a Type II design. One should also note that by choosing θ as something other than $\pi/4$, Hilbert transform systems can be generated that are neither Type I nor Type II designs.

The natural classifications of Hilbert transform systems, as with all digital filters, is thus into FIR and IIR designs. The FIR designs have the advantage of providing exactly linear phase, but at the expense of requiring a higher filter order compared to the IIR designs to achieve a given negative frequency attenuation level. A useful compromise between the FIR and IIR designs is the approximately linear-phase design, for which one branch of the complex filter is FIR and the other is IIR. This class of filters achieves very low phase distortion using a modest filter order.

Continuous-Time Hilbert Transformers. Continuous-time Hilbert transformers are typically realized using two continuous-time allpass functions $A_1(s)$ and $A_2(s)$ whose phase responses along the imaginary axis exhibit approximate phase quadrature over some specified frequency interval. Suppose we denote the endpoints of this interval by Ω_1 and Ω_2. One can define two transformed discrete-time frequencies ω_3 and ω_4 according to the frequency mapping

$$\omega_3 = 2 \tan^{-1} [\sqrt{\Omega_1/\Omega_2}],$$

$$\omega_4 = 2 \tan^{-1} [\sqrt{\Omega_2/\Omega_1}]. \tag{13.46}$$

Note that we obtain $\omega_3 + \omega_4 = \pi$. Thus we have transformed the design problem into an equivalent digital filter design problem of designing two allpass transfer functions to exhibit approximate phase quadrature over a frequency interval symmetric about $\omega = \pi/2$. One can then obtain two digital allpass transfer functions ($A_1(z^2)$ and $z^{-1}A_2(z^2)$ in the odd-degree case, or $B(z)$ and $B(-z)$ in the even-degree case), according to the design procedures outlined above. Let z_i, $i = 1, \ldots, N$, denote the set of poles of the two digital allpass functions so obtained, which we recall lie on the real axis in the z-plane, and distribute alternately between the two allpass transfer functions. Define a set of analog pole locations according to

$$s_i = \sqrt{\Omega_1 \Omega_2}\, \frac{z_i - 1}{z_i + 1}, \qquad i = 1, \ldots, N.$$

These poles will lie on the negative real axis in the s-plane, because the digital poles z_i obtained by the methods outlined above always lie on the real axis in the z-plane. These analog poles are then alternately distributed between two analog allpass transfer functions $A_1(s)$ and $A_2(s)$:

$$A_1(s) = \frac{s + s_1}{s - s_1} \cdot \frac{s + s_3}{s - s_3} \cdots, \qquad A_2(s) = \frac{s + s_2}{s - s_2} \cdot \frac{s + s_4}{s - s_4} \cdots,$$

where the poles are ordered such that $s_1 < s_2 < \cdots < s_n < 0$. The two analog allpass transfer functions so formed will have phase responses exhibiting the desired approximate phase quadrature over the specified frequency interval $[\Omega_1, \Omega_2]$.

13-3 DIFFERENTIATORS AND INTEGRATORS

Digital signal processing was originally approached as a method of simulating continuous-time systems using discrete-time computations. By now, of course, digital signals and systems have taken a significance in their own right. Nonetheless, some useful applications of analog simulation remain of interest, and in this section we treat some design aspects of two of these: digital differentiators and integrators.

13-3-1 Background

Suppose we have a continuous-time function $u(t)$ whose Fourier transform $U(j\Omega)$ is perfectly bandlimited; that is,

$$U(j\Omega) = 0, \qquad |\Omega| \geq \Omega_{\max}. \tag{13.47}$$

We know that the information of $u(t)$ is completely contained in its sampled version

$$\{u[n]\} \triangleq \{u(nT)\} \tag{13.48}$$

provided $T < 2\pi/\Omega_{\max}$. We are interested here in obtaining sampled versions of the following two functions:

$$y_1(t) = \int_{-\infty}^{t} u(v)\, dv \tag{13.49a}$$

and

$$y_2(t) = \frac{du(t)}{dt}. \tag{13.49b}$$

The frequency-domain versions of these two functions are (using the continuous-time Fourier transform)

$$y_1(t) \overset{\mathscr{F}}{\leftrightarrow} \frac{U(j\Omega)}{j\Omega} \tag{13.50a}$$

and

$$y_2(t) \overset{\mathcal{F}}{\leftrightarrow} j\Omega \, U(j\Omega).$$
(13.50b)

Note that because $U(j\Omega)$ is perfectly bandlimited according to Eq. (13.47), the transforms of $y_1(t)$ and $y_2(t)$ satisfy the same bandlimited condition. As such, both these time functions may also be sampled every T seconds with no loss of information. This suggests the possibility of directly processing the discrete sequence $\{u[n]\}$ to obtain sampled versions of the derivative and integral of the time signal $u(t)$. Specifically, if $u[n]$ has $U(e^{j\omega})$ as its discrete-time Fourier transform, we should obtain the correspondences

$$y_1[n] \overset{\mathcal{F}}{\leftrightarrow} \frac{U(e^{j\omega})}{j\omega}$$
(13.51a)

and

$$y_2[n] \overset{\mathcal{F}}{\leftrightarrow} j\omega U(e^{j\omega}),$$
(13.51b)

where $y_1[n]$ and $y_2[n]$ are the sampled versions of the two functions in Eq. (13.49), and $\leftrightarrow$ now denotes the discrete-time Fourier transform. For this reason, the filter with frequency response given by

$$H_1(e^{j\omega}) = \frac{1}{j\omega}$$
(13.52)

is called the *ideal digital integrator*, while the filter with the frequency response given by

$$H_2(e^{j\omega}) = j\omega$$
(13.53)

is called the *ideal digital differentiator*. The ideal differentiator is perhaps simpler to examine initially since there are no singularities (i.e., poles) in its frequency response.

13-3-2   Design of Differentiators

Let us recall the definition of the derivative operation:

$$\frac{du(t)}{dt} = \lim_{\Delta t \to 0} \frac{u(t) - u(t - \Delta t)}{\Delta t}.$$
(13.54)

If we ignore the limit operation and take $\Delta t = T$, we obtain a crude approximation of the differentiation operation as

$$y_2[n] = u[n] - u[n-1],$$
(13.55)

which corresponds to the transfer function

$$H_2(z) = \frac{Y_2(z)}{U(z)} = 1 - z^{-1}. \tag{13.56}$$

Along the unit circle, this transfer function becomes

$$H_2(e^{j\omega}) = 2je^{-j\omega/2} \sin(\omega/2). \tag{13.57}$$

The term $e^{-j\omega/2}$ indicates a half-sample delay through the system, whereas the factor j indicates a $\pi/2$ radian phase shift at all frequencies, which occurs because the impulse response corresponding to Eq. (13.56) is antisymmetric. If $\omega/2$ is small, one can use the approximation $\sin(\omega/2) \approx \omega/2$ to obtain

$$H_2(e^{j\omega}) \approx j\omega \cdot e^{-j\omega/2}, \qquad \omega/2 \text{ small}. \tag{13.58}$$

This crude filter approximates the behavior of the ideal differentiator for low-frequency signals, but at high frequencies the approximation becomes progressively worse, as illustrated by the plot of Figure 13-15. To obtain improved performance, we can cascade a correction filter $H_{cor}(z)$ whose frequency response is ideally

$$H_{cor}(e^{j\omega}) = e^{-jM\omega} \frac{\omega/2}{\sin(\omega/2)}, \tag{13.59}$$

where M is the delay in samples of the correction filter. In this fashion, the cascaded filter $H_2(z)H_{cor}(z)$ achieves the frequency response $j\omega$ times a linear-phase term. As a practical point, the impulse response of the product filter $H_2(z)H_{cor}(z)$

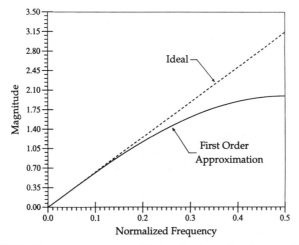

FIGURE 13-15 Frequency response of first-order approximation to a differentiator.

should be antisymmetric to obtain a factor j in the zero-phase response. Since $H_2(z)$ in Eq. (13.56) is already antisymmetric, it is clear that $H_{cor}(z)$ should have a symmetric impulse response.

To proceed, consider the following approximation problem. We want to minimize

$$\max_{\omega} |D(\omega) - H_{cor,0}(e^{j\omega})| \qquad (13.60)$$

over some suitable frequency interval, say $\omega \in [0, \pi]$ to obtain a wideband differentiator. Here $H_{cor,0}$ is the zero-phase response of the correction filter, and $D(\omega)$ is the desired zero-phase response of the correction filter, as given by

$$D(\omega) = \frac{\omega/2}{\sin(\omega/2)}. \qquad (13.61)$$

If we denote the cascade filter $H_c(z)$ as

$$H_c(z) = H_2(z)H_{cor}(z) = (1 - z^{-1})H_{cor}(z), \qquad (13.62)$$

then the minimization problem of Eq. (13.60) is equivalent to that of minimizing

$$\max_{\omega \in [0, \pi]} \left[\frac{2}{\sin(\omega/2)} |j\omega - H_{c,0}(e^{j\omega})| \right], \qquad (13.63)$$

where $H_{c,0}$ is the zero-phase response of H_c. The term $2/[\sin(\omega/2)]$ is seen to behave as a weighting function, which gives more weight to the error at low frequencies. This weighting of the error function is desirable since it ensures that the low-frequency portion of the filter is well behaved. Another commonly used weighting function for differentiator design is $1/\omega$ [MC73], again since it varies inversely proportional with frequency. In this case the design problem is that of minimizing

$$\max_{\omega \in [0, \pi]} \left[\frac{1}{\omega} |j\omega - H_{c,0}(e^{j\omega})| \right]. \qquad (13.64)$$

Note that a direct implementation of the minimization problem as posed in Eq. (13.63) or Eq. (13.64) is ill advised, because at low frequencies the weighting function blows up while the desired response goes to zero, somewhat akin to a "pole-zero" cancellation. This can cause potential numerical problems depending on the dynamic range and overflow behavior of the host computer. An alternate strategy, which avoids the overflow problem of the weighting function, is simply to design the correction filter $H_{cor}(z)$ according to Eq. (13.60), and then cascade this filter with the term $(1 - z^{-1})$. In this "transformed" approximation problem, the desired function $D(\omega)$ in Eq. (13.61) is bounded and well behaved at all frequencies, and the actual weighting function used is unity. As such, the approximation problem shows better numerical conditioning than either version Eq. (13.63) or Eq. (13.64).

Design Example 13.5. The design procedure consists essentially of designing a correction filter according to Eq. (13.60) and then cascading this with the term (1 − z^{-1}). To obtain a zero-phase response that is real valued, we must use either a Type I or a Type II filter (i.e., symmetric impulse response). A Type II filter always has a transmission zero at $\omega = \pi$, which is undesirable for wideband differentiator design. Thus we choose a Type I filter for $H_{cor}(z)$, that is, $H_{cor}(z)$ has even degree (and thus odd length). We have used the routine of McClellan et al. [MC73] to accomplish the minimization problem of Eq. (13.60). This routine nominally designs symmetric FIR filters, and the desired response and weighting functions can be supplied by the user. We have used a weighting function of unity, the desired response of Eq. (13.61), and have chosen the filter length as 65. The design so obtained is cascaded with the filter (1 − z^{-1}) to obtain an overall impulse response that is antisymmetric, as desired. The frequency of the overall filter is shown in Figure 13-16, which is very close to a straight line. The error function corresponding to this design is shown in Figure 13-17. Note that the envelope of the error function behaves as $\sin(\omega/2)$, since this is the effective weighting function for the overall design (cf. Eq. (13.63)). To summarize:

1. Design a Type I filter according to the minimization problem of Eq. (13.60). The program of McClellan et al. [MC73] can be used upon supplying a weighting function of unity and the desired response as

$$D(\omega) = \frac{\omega/2}{\sin(\omega/2)}.$$

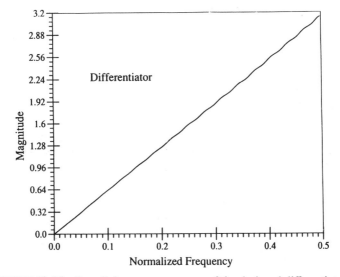

FIGURE 13-16 Overall frequency response of the designed differentiator.

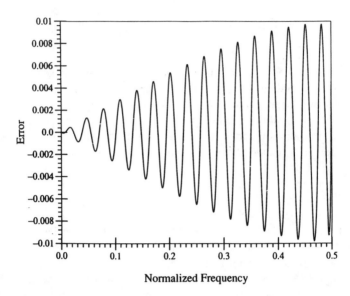

FIGURE 13-17   Error function corresponding to the design of Figure 13-16. Note that the envelope behaves as sin $(\omega/2)$, because this is the weighting function used.

2. Next, cascade this design with the transfer function $(1 - z^{-1})$ to obtain a wideband differentiator.

We should point out that many versions of the program of McClellan et al. [MC73] provide the option of directly designing a differentiator, using a weighting function of $1/\omega$. This corresponds to a direct optimization of the problem in Eq. (13.64), which, as pointed out above, is not as well conditioned numerically as the procedure in steps 1 and 2 above.

One other comment is in order. The differentiators designed according to steps 1 and 2 result effectively in a Type IV impulse response for the overall filter. As such the linear-phase portion of the filter exhibits the "half-sample" delay phenomenon. In some applications one may conceivably want sampled versions of a signal and its derivative corresponding to the *same* sampling instants. Clearly, this cannot be achieved if the differentiator has a half-sample delay in its linear-phase portion. The obvious fix is to modify step 1 in Design Example 13.5 by using a Type II filter for the correction filter. Upon cascading the term $(1 - z^{-1})$, the overall design is now a Type III filter, which has the desired integral sample delay. The filter obtained, however, necessarily has a transmission zero at $\omega = \pi$, and as such wideband performance is sacrificed. Accordingly, the frequency interval in the approximation problem of Eq. (13.60) should exclude $\omega = \pi$. Rather, one optimizes Eq. (13.60) over the frequency interval $\omega \in [0, \pi - \delta]$, where the selection of the parameter δ involves a trade-off between the bandwidth of the differentiator and the peak ripple in the error function.

13-3-3 Design of Integrators

One of the simplest approximations to an integral is the following formula:

$$y_1[n] = \sum_{m=-\infty}^{n} u[m], \tag{13.65}$$

which may be rewritten as the difference equation

$$y_1[n] = y_1[n-1] + u[n]. \tag{13.66}$$

Taking the z-transform of this equation leads to the transfer function

$$H(z) = \frac{Y_1(z)}{U(z)} = \frac{1}{1 - z^{-1}}, \qquad |z| > 1, \tag{13.67}$$

which is often called the *lossless digital integrator* [BR75]. Note that this filter contains a pole on the unit circle at $z = 1$. The filter is thus only marginally stable, and as such its frequency response blows up near $\omega = 0$. We can evaluate this function for some fixed radius r in the z-plane, with $r > 1$, and determine the limit as r approaches unity, leading to the result

$$\lim_{r \to 1} H(re^{j\omega}) = \frac{e^{j\omega/2}}{2j \sin(\omega/2)}. \tag{13.68}$$

Using the approximation $\sin(\omega/2) \approx \omega/2$ for small $\omega/2$ reveals that

$$H_1(e^{j\omega}) \approx \frac{e^{j\omega/2}}{j\omega}, \qquad \omega/2 \text{ small}. \tag{13.69}$$

Thus for low-frequency signals the filter behaves as an integrator with a linear-phase term corresponding to a one half-sample advance. At high frequencies though, the approximation is less accurate. Taking $H_1(e^{j\omega})$ as the frequency response of the ideal integrator from Eq. (13.52), we find that the ratio of ideal frequency response to actual frequency response is given by

$$\frac{H_1(e^{j\omega})}{H(e^{j\omega})} = \frac{1/j\omega}{e^{j\omega/2}/2j \sin(\omega/2)} = e^{-j\omega/2} \frac{\sin(\omega/2)}{\omega/2}. \tag{13.70}$$

Note that this is the frequency response of a zero-order hold (a common component in digital-to-analog converters). Thus if the signal $y_1[n]$ is to be converted to analog form, no additional processing is necessary; the additional frequency response correction is provided automatically by the zero-order hold portion in the D/A converter.

At that point it is interesting to consider some classical numerical integration schemes [BA77]. First of all, the simple scheme in Eq. (13.66) is recognized as the rectangular method of integration. Some other standard methods are mentioned here:

1. *The Trapezoidal Method.*

$$y_1[n] = y_1[n-1] + \tfrac{1}{2}(u[n] + u[n-1]),$$

which corresponds to the transfer function

$$H_1(z) = \frac{Y_1(z)}{U(z)} = \frac{1}{2} = \frac{1+z^{-1}}{1-z^{-1}}.$$

2. *Simpson's Method.* Here the integral is approximated by the $N+1$ term sum

$$y_1[n] = y_1[n-N] + \tfrac{1}{3}(u[n] + 4u[n-1] + 2u[n-2]$$

$$+ \cdots + 4u[n-N+1] + u[n-N]), \qquad N \text{ even.}$$

The inverse error function H/H_1 is plotted for the rectangular and trapezoidal methods in Figure 13-18. The plots show the reciprocal of the magnitude response necessary in each case for a postcorrection filter. Since the plot corresponding to the trapezoidal method has a zero at $\omega = \pi$, a postcorrection filter in this case would require a pole at $z = -1$.

One should keep in mind that these integration methods are designed from a polynomial perspective. For example, the trapezoidal rule is exact if the function

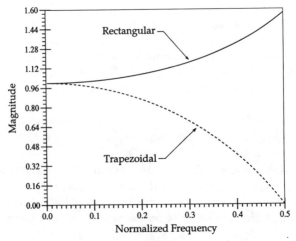

FIGURE 13-18   Ratio of actual to ideal response for the rectangular and trapezoidal integration methods.

to be integrated is a degree one polynomial. Likewise, Simpson's method is exact for a second degree polynomial, and schemes that are exact for progressively higher order polynomials can be developed at the expense of greater complexity [DH84]. In using a signal processing approach, however, we assume the signals to be integrated are *bandlimited*. In general, bandlimited time signals are not well approximated as polynomial functions of the time index, and this explains why the classical integration routines mentioned above do not appear particularly well suited to the present design problem.

Thus, let us again follow the "correction filter" approach. We assume a filter of the form

$$H_1(z) = \frac{N(z)}{1 - z^{-1}}, \tag{13.71}$$

whose zero-phase response $H_{1,0}(e^{j\omega})$ approximates the ideal integrator:

$$H_{1,0}(e^{j\omega}) \approx \frac{1}{j\omega}. \tag{13.72}$$

Note that the denominator $1 - z^{-1}$ gives the desired pole at $\omega = 0$ as well as the factor j in the zero-phase response. Hence we choose $N(z)$ to have a real-valued zero-phase response, which indicates that $N(z)$ should be a symmetric FIR filter. The desired zero-phase response for $N(z)$ is simply that of Eq. (13.70):

$$N_0(e^{j\omega}) = \frac{\sin(\omega/2)}{\omega/2}. \tag{13.73}$$

Again the program of McClellan et al. [MC73] can be used upon providing the right-hand side of Eq. (13.73) as the desired response.

Figure 13-19 shows the magnitude responses of two filters designed to approximate the right-hand side of Eq. (13.70). The curve labeled "15 point" corresponds to a length 15 symmetric FIR filter (Type I impulse response), while the curve labeled "16 point" corresponds to a Type II impulse response. The 15 point response allows a wideband approximation and provides a group delay corresponding to an integral number of samples. The denominator term $1 - z^{-1}$ contributes an additional half-sample advance to the group delay, so that the overall filter $N(z)/(1 - z^{-1})$ has the "half-sample" property in the group delay response. In some applications one may want sampled versions of the input signal and its integral at the same sampling instants. This feature is not provided by linear-phase filters with the half-sample property. The remedy in this case is to use a Type II impulse response for the filter $N(z)$. The half-sample factor in the group response of $N(z)$ cancels with the half-sample advance provided by the denominators $1 - z^{-1}$, resulting in an overall filter $N(z)/(1 - z^{-1})$ whose group delay response is an integral number of samples. The price paid for this change is that the Type II filter $N(z)$ necessarily has a transmission zero at $\omega = \pi$. For the filter design shown, the frequency response was optimized over the band $\omega \in [0, 0.8\pi]$, with

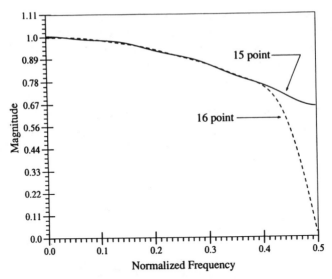

FIGURE 13-19   Frequency responses of two FIR correction filters used in the integrator design. The desired frequency response shape is $(\sin \omega/2)/(\omega/2)$.

the band from $(0.8\pi, \pi]$ left as a "don't care" band. To summarize briefly, if the half-sample property is tolerable in a given application, $N(z)$ should be chosen as a Type I filter to obtain wideband performance. Otherwise a Type II design must be used for $N(z)$, and the "integration bandwidth" must be sacrificed.

13-3-4 Section Summary

We have seen in this section that differentiators and integrators are linear-phase filters with a special frequency response shape. The special passband shape can be obtained using most standard design/optimization procedures; in this section we have concentrated on minimax designs. If a wideband approximation is described, the group delay response of the resulting filters exhibits the "half-sample" property. In certain applications, sampled versions of a function and its derivative or integral are required, corresponding to the same sampling instants. As pointed out, if the differentiator or integrator filter has a group delay response with the half-sample property, the output corresponds to the derivative or antiderivative of the underlying continuous time signal, but sampled at different time instants. When this phenomenon is intolerable, then a narrowband design must be used.

13-4 SMOOTHING FILTERS

Suppose we are given a sequence $\{u[n]\}$ that we would like to smooth in order to obtain information about the general trends of the signal without being distracted by its transient components. Applications for such processing include recovering long-term economic trends based on daily fluctuations, or seasonal temperature variations derived from day-to-day thermometer readings, and so on. We present in this section a method of least-squares smoothing described in Savitzky and Golay [SA64] and Steiner et al. [ST72].

The method begins by windowing a set of samples of the sequence $\{u[n]\}$. For convenience, we assume that the window contains an odd number of samples and is centered about the sample $u[0]$; that is, the window contains the samples $\{u[n]: -m \le n \le m\}$ for some integer m. (This can always be achieved by shifting the origin of the time axis if necessary.) To these points we attempt to fit a polynomial $x(t)$:

$$x(t) = a_0 + a_1 t + a_2 t^2 + \cdots + a_k t^k, \tag{13.74}$$

whose degree k is less than $2m + 1$. (In effect, the degree of the polynomial is less than the number of samples in the input window; the reason for this choice will become clear later.) A conceptual diagram is sketched in Figure 13-20. Our criterion for choosing the coefficients of the polynomial is to minimize the following error measure:

$$\epsilon = \sum_{n=-m}^{m} \{u[n] - x(n)\}^2, \tag{13.75}$$

where $x(n)$ denotes the polynomial $x(t)$ evaluated at $t = n$. Once this polynomial is obtained, we take the central value $x(0)$ of the polynomial as the output sample. The window of input samples is then adjusted one sample in the positive n direction and the process is repeated to obtain a new output sample, and so on. In this manner we obtain an output based on fitting "smooth" polynomials to the data, which in some sense should lead to an output that is a smoothed version of the input.

The mathematical construction proceeds as follows. We write the original samples as

$$u[n] = x(n) + e[n], \tag{13.76}$$

where $x(n)$ is the polynomial of Eq. (13.74) evaluated at $t = n$, and $e[n]$ is the resulting error in the polynomial approximation. For the sake of illustration, sup-

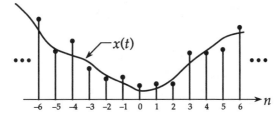

FIGURE 13-20   Fitting a polynomial $x(t)$ to a window of discrete-time samples.

pose we use a five-sample window of the sequence $\{u[n]\}$ centered about $n = 0$, and suppose furthermore that the degree of the polynomial $x(t)$ is three. We can write Eq. (13.76) for $n = -2, -1, \ldots, 2$ in matrix form as

$$
\begin{bmatrix}
u[2] \\
u[1] \\
u[0] \\
u[-1] \\
u[-2]
\end{bmatrix}
=
\begin{bmatrix}
1 & 2 & 2^2 & 2^3 \\
1 & 1 & 1^2 & 1^3 \\
1 & 0 & 0^2 & 0^3 \\
1 & -1 & (-1)^2 & (-1)^3 \\
1 & -2 & (-2)^2 & (-2)^3
\end{bmatrix}
\begin{bmatrix}
a_0 \\
a_1 \\
a_2 \\
a_3
\end{bmatrix}
+
\begin{bmatrix}
e[2] \\
e[1] \\
e[0] \\
e[-1] \\
e[-2]
\end{bmatrix}
$$

$$(13.77a)$$

or

$$\mathbf{u} = \mathbf{\Theta a} + \mathbf{e}. \qquad (13.77b)$$

Note that the free variables are the coefficients in the $\mathbf{a}$ vector. The optimum choice for these variables should minimize the norm of the error vector $\mathbf{e}$. The norm of this vector is of course the quantity ϵ in Eq. (13.75):

$$\epsilon = \sum_{n=-2}^{2} e^2[n] = \mathbf{e}'\mathbf{e}. \qquad (13.78)$$

This minimization is a standard least-squares problem, and the solution is found readily as

$$\mathbf{a} = (\mathbf{\Theta}'\mathbf{\Theta})^{-1}\mathbf{\Theta}'\mathbf{u}. \qquad (13.79)$$

It is seen that the coefficients in the vector $\mathbf{a}$ may be obtained as the appropriate linear combinations of the elements of the vector $\mathbf{u}$.

As stated above, our goal is to obtain the central value $x(0)$ of the fitted polynomial. We see from Eq. (13.74) that this is given by

$$x(0) = a_0, \qquad (13.80)$$

which is taken as the output sample at $n = 0$. At the next sample instant the only change in the formulation is that the vector $\mathbf{u}$ now consists of a shifted window of input samples; the term $(\mathbf{\Theta}'\mathbf{\Theta})^{-1}\mathbf{\Theta}'$ remains the same. Upon shifting the window of input samples, the new computed value of a_0 is taken as the output at time $n = 1$, and so on. It is thus seen that the output sequence defined by this method is essentially a linear convolution of the input samples, with the impulse response coefficients obtained from the first row of the matrix $(\mathbf{\Theta}'\mathbf{\Theta})^{-1}\mathbf{\Theta}'$.

Before we treat some practical aspects of this method, it is interesting to give some interpretation to the other elements of the **a** vector. Namely, we have

$$a_1 = \frac{dx(t)}{dt}\bigg|_{t=0} , \qquad 2a_2 = \frac{d^2x(t)}{dt^2}\bigg|_{t=0} , \quad \cdots . \qquad (13.81)$$

The remaining coefficients thus give information concerning the derivatives of the fitted curve at $t = 0$. The technique outlined above thus also yields the derivatives of the smoothed output function. For example, by using the computed value of a_1 as the output at each sample instant, we obtain the first derivative of the fitted curve as the output sequence.

Practical Details. We need to first establish under what conditions the matrix $(\Theta'\Theta)^{-1}$ in Eq. (13.79) exists. Fortunately, the answer becomes fairly simple in view of the special structure assumed by the matrix Θ. To this end, we take advantage of the following definition [GR83].

Definition. Let the $k \times n$ matrix **V** take the form

$$\mathbf{V} = \begin{bmatrix} 1 & x_1 & x_1^2 & \cdots & x_1^{k-1} \\ 1 & x_2 & x_2^2 & \cdots & x_2^{k-1} \\ \vdots & \vdots & \vdots & & \vdots \\ 1 & x_n & x_n^2 & \cdots & x_n^{k-1} \end{bmatrix} . \qquad (13.82)$$

Then **V** and **V'** are *Vandermonde* matrices.

The matrix Θ as defined implicitly in Eq. (13.77) is seen to be a Vandermonde matrix with the x_i's as successive integers. One can show [GR83] that the rank of a general $k \times n$ Vandermonde matrix **V** is given by

$$\text{rank } \mathbf{V} = \min(k, n, r), \qquad (13.83)$$

where r is the number of *distinct* values in the set $\{x_i\}_{i=1}^n$.

In Eq. (13.79) for example, the set $\{x_i\}$ consists of five distinct integers (namely $-2, -1, 0, 1,$ and $2,$ in the formula). Thus in Eq. (13.79) the matrix Θ has $r = 5$ and, of course, $k = 5$ and $n = 4$. Its rank is thus 4 ($=$ the number of columns). Since Θ has full column rank, the matrix $(\Theta'\Theta)^{-1}$ exists. In the general case Θ is, by construction, a Vandermonde matrix based on distinct integers. Thus, provided Θ has more rows than columns, it will have full column rank, and as such the solution to the **a** vector in Eq. (13.79) will exist and be unique.

Observe that the number of rows of Θ is precisely the number of samples contained in the input window, while the number of columns is one less than the degree of the polynomial $x(t)$. Let us investigate further this connection, starting first with the case in which the matrix Θ is square. In this case, since Θ has full rank, it can be directly inverted, so that one can obtain a vector **a** in Eq. (13.77) that results in the error vector **e** having zero norm. (Simply choose $\mathbf{a} = \boldsymbol{\Theta}^{-1}\mathbf{u}$.) A square $\boldsymbol{\Theta}$ matrix, of course, corresponds to the situation in which the degree of the polynomial $x(t)$ is one less than the number of samples in the input window. A well-known result of polynomial theory, summarized in the Lagrange interpolation formula [CH66], is that the coefficients of a polynomial of degree $n - 1$ can always be chosen so that the polynomial passes through n given distinct points. The central value of this polynomial is simply the center sample of the input window; the filtering function thus provided in this case is the identity operation.

If the degree of $x(t)$ is greater than or equal to the number of samples in the input window, then Eq. (13.77) presents too many unknowns, and an infinite number of solutions are possible, all of them again providing an identity filtering operation. We have no practical interest in this case. Instead, we shall concentrate on the case in which the degree of $x(t)$ is at least two less than the size of the input window. In this case the polynomial $x(t)$ in general cannot be chosen such that it passes through all the points in the input window; rather a unique least-squares fit exists, as we have previously justified.

We anticipate that the amount of smoothing is related to the discrepancy between the polynomial degree and the input window size. For example, a square $\boldsymbol{\Theta}$ matrix results in effectively no filtering of the input signal. From another perspective, we recall that the successive coefficients in the polynomial $x(t)$ are related to its derivatives (cf. Eq. (13.81)). By restricting the degree of $x(t)$, we restrict the number of nonzero derivatives of $x(t)$, which evidently exerts an influence on the smoothness of the output sequence.

Before presenting some computed examples, some words concerning the computation of $(\boldsymbol{\Theta}'\boldsymbol{\Theta})^{-1}\boldsymbol{\Theta}'$ are in order. First, it is generally *not* recommended to compute $(\boldsymbol{\Theta}'\boldsymbol{\Theta})^{-1}$ and then form the product $(\boldsymbol{\Theta}\boldsymbol{\Theta}')^{-1}\boldsymbol{\Theta}'$; one alternate approach is to solve the system of linear equations

$$\boldsymbol{\Theta}'\boldsymbol{\Theta}\mathbf{H} = \boldsymbol{\Theta}' \tag{13.84}$$

using standard linear algebra subroutines [DO79; ST73], and then identifying **H** as the desired result. Even this approach, however, requires the formation of the matrix $\boldsymbol{\Theta}'\boldsymbol{\Theta}$; this "squaring up" procedure needlessly degrades the numerical conditioning of the problem. A variety of routines are available for the reliable computation of least-squares problems [DO79], such as methods based on the Cholesky decomposition, the QR decomposition, and so on. For a given application, however, the result need be computed only once, since we are only interested in the elements from the appropriate row of $(\boldsymbol{\Theta}'\boldsymbol{\Theta})^{-1}\boldsymbol{\Theta}'$. Fortunately, the results for a wide number of choices of the parameters are tabulated in Savitzky and Golay [SA64] (with corrections subsequently published in Steiner et al. [ST72]).

In Figure 13-21 we have plotted the frequency responses corresponding to some different smoothing filters obtained from the method of this section. The degree of the interpolating polynomial is six in each case, and the window size varies from 13 to 17, as indicated in the figure. The corresponding impulse responses are sum-

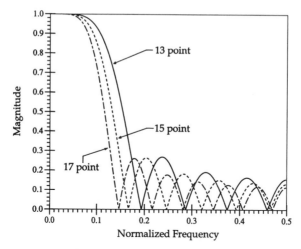

FIGURE 13-21 Frequency responses of the smoothing filter.

marized in Table 13-5, where it is seen that each filter is linear phase with a symmetric impulse response. Note that each curve has a lowpass type of characteristic, as expected from a smoothing filter.

Figure 13-22 shows the frequency responses of some derivative filters obtained from this formulation. Again the interpolation polynomial has degree six, and the window sizes match those used in Figure 13-21. The corresponding impulse re-

TABLE 13-5   Coefficients for the Smoothing Filters of Figure 13-21

	Coefficients of Smoothing Filters		
n	13 point	15 point	17 point
−8			0.046439628
−7		0.046439628	−0.046439628
−6	0.045248869	−0.061919505	−0.061919505
−5	−0.081447964	−0.063586568	−0.027863777
−4	−0.0555327	−0.003572279	0.032150512
−3	0.045248869	0.081296413	0.098833055
−2	0.1604278	0.162376323	0.157180281
−1	0.246812	0.219208037	0.196475351
0	0.2784862	0.239515902	0.210288164
1	0.246812	0.219208037	0.196475351
2	0.1604278	0.162376323	0.157180281
3	0.045248869	0.081296413	0.098833055
4	−0.0555327	−0.003572279	0.032150512
5	−0.081447964	−0.063586568	−0.027863777
6	0.045248869	−0.061919505	−0.061919505
7		0.046439628	−0.046439628
8			0.046439628

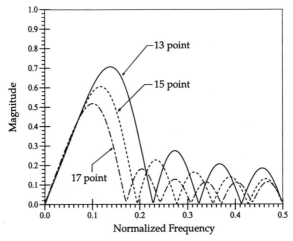

FIGURE 13-22   Frequency responses for the differentiation filters.

sponses appear in Table 13-6 and are antisymmetric. Each plot in Figure 13-22 is seen to behave as a differentiator over the passband of the corresponding plot in Figure 13-21. The attenuation at high frequencies is due to the smoothing operation, and hence these filters do not amplify high-frequency noise. This property can be desirable in applications where the high noise gain of a wideband differentiator cannot be tolerated.

TABLE 13-6   Coefficients for the Differentiating Filters of Figure 13-22

	Coefficients of Differentiating Filters		
n	13 point	15 point	17 point
−8			−0.028586171
−7		−0.030199071	0.048942208
−6	−0.0330977459	0.067404541	0.033101135
−5	0.0929529626	0.025890688	−0.0172084
−4	−0.0000411706	−0.051800429	−0.064114472
−3	−0.1149723814	−0.105739859	−0.087268794
−2	−0.1569321028	−0.111127649	−0.080342542
−1	−0.1076611658	−0.069510598	−0.047521235
0	−0.0	0.0	0.0
1	0.1076611658	0.069510598	0.047521235
2	0.1569321028	0.111127649	0.080342542
3	0.1149723814	0.105739859	0.087268794
4	0.0000411706	0.051800429	0.064114472
5	−0.0929529626	−0.025890688	0.0172084
6	0.0330977459	−0.067404541	−0.033101135
7		0.030199071	−0.048942208
8			0.028586171

The filters derived in this section are based on a time-domain fitting of the output to the input, and hence the waveform preservation properties are typically quite good [SA64]. The frequency responses are not generally optimal though, as indicated by the unequal ripples in the stopband responses of Figure 13-21 and 13-22. For designs requiring optimal frequency response characteristics, the design program of McClellan et al. [MC73] can be used. Figure 13-23 shows a comparison of two narrowband differentiators, both FIR filters of length 15. That labeled ''least squares'' is the corresponding response of Figure 13-21, whereas that labeled ''optimal'' has been designed using the program of McClellan et al. [MC73]. The desired function is ω over the passband, and 0 over the stopband. Likewise the weighting function is $1/\omega$ in the passband[2] and 5 in the stopband, and the bandedges between the passband and the stopband have been chosen approximately equal to those of the least-squares design. The main advantage of using the program of McClellan et al. [MC73] is that greater design flexibility is available for the frequency response shape.

In summary, the least-squares procedure developed here provides a somewhat heuristic method of smoothing and differentiating data that preserves waveform characteristics. The designs obtained by this method can be shown to always yield linear-phase FIR filters and can be considered an alternative to conventional optimal FIR filter designs. We should point out that a generalized procedure can be developed by using a weighted error measure in place of Eq. (13.75). That is, one

---

[2]To avoid the singularity in the weighting function $1/\omega$, the filter was actually designed using the ''compensating'' filter approach, as described in Section 13-3-2. The desired function and weighting function have been predistorted for the compensating filter to obtain the desired end results.

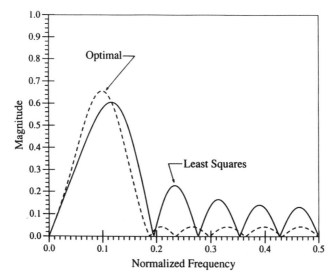

FIGURE 13-23 Comparison of optimal versus least-squares designs for a narrowband differentiator.

seeks to minimize the error measure

$$\epsilon = \sum_{n=-m}^{m} \beta_n \{u[n] - x(n)\}^2, \tag{13.85}$$

where $\{\beta_n\}$ represents a sequence of positive weights that control the relative importance of the errors at each sample in the approximation window of Figure 13-20. This remains a standard least-squares problem, and hence the solution follows with straightforward modification of that presented above. With various choices of weighting sequence $\{\beta_n\}$ the specific characteristics of the smoothing filters can be adjusted.

13-5 NONINTEGER DELAY FILTERS

A linear-phase shift or delay is easily achievable in digital systems by simply cascading an appropriate number of unit delays, provided the desired signal delay is an integral multiple of the sampling period. It has been noted [CR75] that in various applications, such as echo cancellation, phased array antenna systems, or pitch synchronous synthesis of speech, the desired signal delay may not equal an integral number of samples, in which case more elaborate processing schemes must be considered. Two such techniques are considered in this subsection, one based on FIR filtering, the other on IIR filtering.

The FIR technique exploits multirate processing to approximate a linear-phase characteristic providing a delay equal to a rational number of samples. Further simplifications of the method lead to a single rate system, in which the output sequence is obtained from a linear convolution of the input sequence with the appropriate decimated polyphase component of an Mth-band linear-phase lowpass filter. The problem then reduces to designing an appropriate linear-phase Mth-band lowpass filter.

The IIR technique exploits the fact that the phase response of an IIR filter is a continuous function of its pole locations. This suggests the idea of realizing an approximately linear-phase allpass filter providing an *arbitrary* signal delay over some usable signal bandwidth, which can be accomplished using available design programs [KI86; RE86].

13-5-1 FIR Noninteger Delay Filters

The method described in this section uses multirate techniques to delay a signal by a rational number of samples, as first proposed in Crochiere et al. [CR75]. Further simplification of the technique results in the output being a linear convolution of the input sequence with a linear-phase FIR filter whose response approximates an allpass transfer function with a nonintegral but rational delay.

The basic system is sketched in Figure 13-24. The first block increases the sampling rate by a factor of M, by introducing $M - 1$ zero-valued samples between

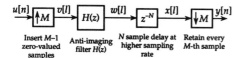

Insert M−1 zero-valued samples | Anti-imaging filter H(z) | N sample delay at higher sampling rate | Retain every M-th sample

FIGURE 13-24 Multirate system for approximating a rational delay line.

each input sample $u[n]$. The filter $H(z)$ is chosen as a linear-phase FIR lowpass filter, whose cutoff frequency (normalized to the higher sampling rate) is approximately π/M rad/s. The interpolated signal $w[l]$ is delayed by N samples with $N < M$, to form $x[l] = w[l - N]$, which is then decimated by a factor of M to form the output signal $y[n]$ whose sampling rate matches that of the input.

The operation can be understood in greater detail using standard multirate analysis techniques [CR83] (cf. Chapter 14). The signal $v[l]$ relates to the input signal $u[n]$ through

$$v[l] = \begin{cases} u[n] & \text{for } l = nM, \\ 0 & \text{otherwise,} \end{cases} \tag{13.86}$$

so that the spectrum $V(e^{j\omega})$ relates to that of $U(e^{j\omega})$ through

$$V(e^{j\omega}) = U(e^{jM\omega}). \tag{13.87}$$

In effect, the original spectrum $U(e^{j\omega})$ is compressed to the interval $\omega \in [0, \pi/M]$ and then replicated every π/M radians to obtain the spectrum $V(e^{j\omega})$. The portion of $V(e^{j\omega})$ lying in the range $\omega \in [0, \pi/M]$ contains all the information of the input signal $u[n]$; the replicas of the compressed input spectrum that lie in the range $\omega \in [\pi/M, \pi]$ are termed imaging components and constitute a form of redundancy in the spectrum $V(e^{j\omega})$. The anti-imaging filter $H(z)$ suppresses this spectral redundancy provided its frequency response satisfies

$$|H(e^{j\omega})| \approx \begin{cases} 1, & 0 \le \omega \le \pi/M, \\ 0, & \pi/M \le \omega \le \pi. \end{cases} \tag{13.88}$$

Typically $H(z)$ is chosen as a linear-phase FIR design to avoid any additional delay distortion. We shall let D denote the group delay response of $H(e^{j\omega})$. The output sequence $w[l]$ obtained from the filter then has a spectrum that satisfies

$$W(e^{j\omega}) \approx \begin{cases} e^{-jD\omega}V(e^{j\omega}) = e^{-jD\omega}U(e^{jM\omega}), & 0 \le \omega \le \pi/M, \\ 0, & \pi/M \le \omega \le \pi. \end{cases} \tag{13.89}$$

The sequence $x[l]$ is obtained by an N sample delay at the higher sampling frequency, where we assume $0 \le N \le M - 1$. The signal $y[n]$ is then obtained by retaining only every Mth sample of $x[l]$:

$$y[n] = x[nM]. \tag{13.90}$$

The spectrum $Y(e^{j\omega})$ is then given by

$$Y(e^{j\omega}) = \frac{1}{M} \sum_{k=0}^{M-1} X(e^{jk\omega/M}) \approx \frac{1}{M} \sum_{k=0}^{M-1} e^{-j(D+N)\omega/M} U(e^{jk\omega}). \quad (13.91)$$

To the extent that $H(z)$ gives a perfect lowpass response, there is negligible aliasing distortion introduced by the down-sampling operation, and the latter inequality in Eq. (13.91) becomes progressively more accurate. This results in

$$y[n] \approx u[n - (D + N)/M]. \quad (13.92)$$

Thus the technique makes it possible to approximate a rational sample delay, the delay in this case given by $(D + N)/M$ samples.

Since only every Mth sample is retained from the multirate system of Figure 13-24, the system as shown is not fully efficient, and indeed further simplifications are possible. To start, let us decompose $H(z)$ into its polyphase components:

$$H(z) = \sum_{k=0}^{M-1} z^{-k} H_k(z^M). \quad (13.93)$$

In effect, the term $H_0(z^M)$ receives the impulse response samples $h[0]$, $h[M]$, $h[2M]$, . . . , the term $z^{-1}H_1(z)$ receives the samples $h[1]$, $h[M + 1]$, . . . , and so on, up to $z^{-M+1}H_{M-1}(z^M)$, which receives the samples $h[M - 1]$, $h[2M - 1]$, ... The polyphase decomposition of $H(z)$ is indicated in Figure 13-25(a).

The next step is to pass the up-sampler through each polyphase component, while replacing z^M by z in each filter. This leads ultimately to the commutator model of Figure 13-25(b), in which the filters to the left of the commutator all operate at the (lower) sampling rate of the input sequence $u[n]$. The output sequence from the commutator remains at the higher sampling rate. Upon delaying this sequence by N samples and then discarding every Mth sample, one sees that only the output from the filter $H_{M-N}(z)$ ever passes through the decimator. [When $N = 0$, this corresponds to the output from the filter $H_0(z)$.] Hence the system reduces to that of Figure 13-25(c), involving a linear convolution between the input sequence and the appropriate decimated polyphase filter $H_{M-N}(z)$, operating at the output sequence sampling rate. Thus although multirate techniques were used to derive the system, the overall implementation is a single rate system. The derivation places in evidence the fact that the polyphase components of an appropriately designed lowpass filter are approximately allpass transfer functions with rational group delay responses.

To summarize the design procedure, we have:

1. Specifications: Design a rational delay network to provide a delay of $(D + N)/M$ samples, where $0 \le N < M$.

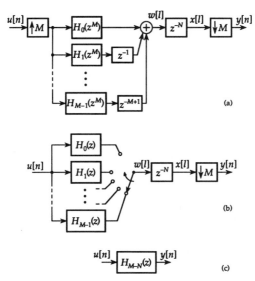

(a)

(b)

(c)

FIGURE 13-25 (a) Illustrating the polyphase decomposition of the lowpass filter $H(z)$. (b) The equivalent commutator model of Figure 13-25(a). (c) Simplified single rate implementation of Figure 13-25(b).

2. Design a linear-phase lowpass filter $H(z)$ of length $2D + 1$ (using, e.g., the program of McClellan et al. [MC73]), satisfying

$$1 - \delta_1 \le |H(e^{j\omega})| \le 1 + \delta_1, \qquad 0 \le \omega \le \pi/M - \omega_1,$$

$$|H(e^{j\omega})| \le \delta_2, \qquad\qquad \pi/M + \omega_1 \le \omega \le \pi,$$

where δ_1, δ_2, and ω_1 are small constants. The trade-off between the size of these constants depends on the choice of filter length $2D + 1$.

3. Decompose $H(z)$ into its polyphase components as per Eq. (13.93).

4. The decimated polyphase component $H_{M-N}(z)$ approximates an allpass transfer function over the frequency range $\omega \in [0, \pi - M\omega_1]$ with a group delay of $(D + N)/M$ samples. The deviation of the magnitude response from unity over this frequency range can be shown to be bounded by $(\delta_1 + \delta_2)$.

13-5-2 IIR Designs with Arbitrary Delay

The previous section described a FIR filtering technique to introduce a fractional delay N/M into the system, plus a delay offset D/M to account for the group delay response of the anti-imaging filter $H(z)$. By using instead an IIR allpass filter, irrational delays can be approximated, since the phase response of an allpass function is a continuous function of its parameters. Suppose we want to approximate a delay $z^{-\gamma}$, where γ is an arbitrary positive real number. Let M denote the smallest integer greater than γ, and begin with an allpass transfer function $A(z)$ of degree M. If we let $-\phi(\omega)$ denote its phase response, then

$$A(e^{j\omega}) = e^{-j\phi(\omega)}. \tag{13.94}$$

If $A(z)$ is stable, then the phase response $-\phi(\omega)$ is a monotonically decreasing function of ω that satisfies the boundary conditions

$$-\phi(0) = 0 \quad \text{and} \quad -\phi(\pi) = -M\pi. \tag{13.95}$$

Now define an approximate interval $\omega \in [0, \pi - \alpha]$, and consider the minimax approximation problem

$$\min \{ \max_{\omega \in [0, \pi - \alpha]} |\gamma\omega - \phi(\omega)| \}, \tag{13.96}$$

where the minimization is performed with respect to the coefficients of the allpass transfer function $A(z)$. The introduction of the parameter α is necessary since Eq. (13.95) shows that $\phi(\pi)$ is fixed, corresponding to an error of $M - \gamma$ radians in the phase approximation at $\omega = \pi$. By excluding the frequency $\omega = \pi$ from the frequency interval in Eq. (13.96), a more meaningful approximation problem results. Programs to design allpass transfer functions approximating a linear-phase response have been presented in Kim and Ansari [KI86] and Renfors and Saramäki [RE86] and can be adapted to the present problem by using $\gamma\omega$ as the desired phase response and $\omega \in [0, \pi - \alpha]$ as the approximation interval. The approximation problem so posed has M degrees of freedom, corresponding to the M poles of $A(z)$, and in general a smaller α will result in a larger deviation from the desired linear-phase response.

If, for a fixed α, the optimal phase response provided by an Mth-order allpass function is unsatisfactory, an improvement can by obtained by increasing the order of $A(z)$ to $M + N$, and replacing the approximation problem in Eq. (13.96) with

$$\min \{ \max_{\omega \in [0, \pi - \alpha]} |(\gamma + N)\omega - \phi(\omega)| \}. \tag{13.97}$$

The optimal allpass filter design then approximates a linear-phase system with a delay of $N + \gamma$ samples.

A design example was run using the program of Renfors and Saramäki [RE86], corresponding to $N + \gamma = 5 + \sqrt{2}$ samples as the desired signal delay, over a frequency range $\omega \in [0, 0.9\pi]$. The allpass transfer function order is chosen as seven (obtained by rounding $5 + \sqrt{2}$ to the next largest integer). The pole locations for the optimized design are listed in Table 13-7, corresponding to the phase response plotted in Figure 13-26. The phase response of the design so obtained is accurate to within 0.014 radians over the usable frequency range.

TABLE 13-7 Pole Locations for the Allpass Transfer Function $A(z)$ Used in the Irrational Delay Approximation Design

Pole Locations for $A(z)$
$z = -0.884555$
$0.4235216 \pm j0.2154549$
$0.0792957 \pm j0.4759793$
$-0.3514971 \pm j0.3771274$

13-6 MEDIAN FILTERS[3]

13-6-1 Definition and Basic Properties

Median filtering is a nonlinear filtering technique that is used for signal smoothing. Its performance is particularly good when the noise is of the impulsive type. The best known characteristic of a median filter is its ability to remove noise while retaining sudden changes in the signal. The median filter is robust and the magnitude of the error in a few of the input samples does not affect the error in the filter output.

The median filter is implemented by sliding a window of odd length over the input signal one sample at a time. At each window position the samples inside the filter window are sorted by magnitude and the mid-value (the median) is the filter

---

[3]This section was contributed by Petri Haavisto of the Department of Electrical Engineering, Tampere University of Technology, Tampere, Finland.

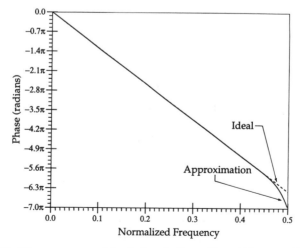

FIGURE 13-26 Phase response of the designed allpass filter $A(z)$ to provide an irrational delay.

output. We denote the filter length by N and since it is required to be odd it can be represented as $N = 2k + 1$. The output of the filter at any window position is then the $(k + 1)$th largest or smallest sample in the filter window. This filtering procedure is denoted here as

$$y[n] = \text{med}\ (x[n - k], \ldots, x[n], \ldots, x[n + k]), \qquad (13.98)$$

where $x[n]$ and $y[n]$ are the input and output sequences. The filter defined by Eq. (13.98) is often called the *running median*.

If the samples in the filter window are denoted as $x_1, x_2, \ldots, x_{2k+1}$, a notation that is commonly used for the sorted list of those same samples is $x_{(1)}, x_{(2)}, \ldots, x_{(2k+1)}$. Here, $x_{(1)}$ and $x_{(2k+1)}$ are the smallest and the greatest sample, respectively. By using this notation the median value is $x_{(k+1)}$. In median filtering, the definition of median for an even number of samples is of academic interest only. In the rare instances where it is needed, the median is usually defined as

$$\text{med}\ (x_1, \ldots, x_N) = \begin{cases} x_{(k+1)}, & N = 2k + 1, \\ \frac{1}{2}(x_{(k)} + x_{(k+1)}), & N = 2k. \end{cases} \qquad (13.99)$$

In most instances it is reasonable to assume that the signal is of finite length, say, L, consisting of samples from $x[0]$ to $x[L - 1]$. When the filter window is centered at the beginning or at the end of the input signal some values must be assigned to the empty window positions. Throughout this chapter the first and the last value carry-on appending strategy is used. This means that values from $x[-k]$ to $x[-1]$ are taken to be equal to $x[0]$, and the values from $x[L]$ to $x[L + k - 1]$ are equal to $x[L - 1]$. The filtering procedure is illustrated in Figure 13-27.

Important characteristic properties of median filters are shown by examples in

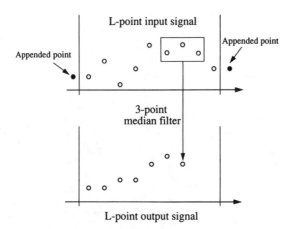

FIGURE 13-27 Example of median filtering with window width $N = 3$.

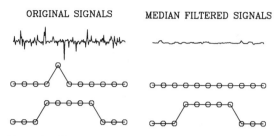

ORIGINAL SIGNALS   MEDIAN FILTERED SIGNALS

FIGURE 13-28 Examples illustrating characteristic properties of median filters. Median filter of length 7 was used.

Figure 13-28. The first example presents the median filter as a robust noise smoother. The level of smoothing achieved grows with the filter length. In the second example a single data point that differs from its neighboring values is completely removed. The median filter is able to remove aberrant data values when at most k such values are inside the filter window. The third example shows that a stepwise change in the signal level, an edge, is preserved without change in the filter output. The square pulse in question is an example of a nonconstant median filter invariant signal. If the length of the pulse is $k + 1$ or more, the median filter of length $2k + 1$ has no effect on it, and if it is less than $k + 1$, the filter removes it.

To analyze the median filters, new methods have been developed in recent years. Root signal analysis deals with signals that are invariant to median filtering and develops the concept of "passband" for median filters. The local monotone nature of these root signals was observed in Tyan [TY81] and Gallagher and Wise [GA81]. Statistical methods also give useful information on the median filter output [JU81].

It has been observed that threshold decomposition and the stacking property of the median filter can be used to decompose the input signal into several binary signals that are easier to analyze. The median filter can then be applied to the binary signals and the binary outputs can be added together to produce the final filter output [FI84].

The median filter itself is simple and allows no design options other than the window size, and in the multidimensional case, the window shape. During the time median filters have been studied numerous extensions and modifications have been proposed to the standard median filter. Only a few of these are described in this chapter.

13-6-1-1 Root Signals. Each median filter has a finite set of signals that pass the filter unaltered. For a median filter of length $N = 2k + 1$ this means that

$$x[n] = \text{med}\ (x[n - k], \dots, x[n], \dots, x[n + k]) \quad \text{for all } n. \quad (13.100)$$

If the above condition is satisfied, $x[n]$ is called a *root signal* of that particular median filter.

In Gallagher and Wise [GA81] some concepts that are useful in describing the root signals were defined:

A *constant neighborhood* is a region of at least $k + 1$ consecutive identically valued points.

An *edge* is a monotonically rising or falling set of points surrounded on both sides by constant neighborhoods.

An *impulse* is a set of at least one but less than $k + 1$ points whose values are different from the surrounding regions and whose surrounding regions are identically valued constant neighborhoods.

Note that the definition of these concepts is tied to the filter length. A constant neighborhood of a length N filter is also a constant neighborhood of all the filters whose length is less than N, but not generally of those whose length is more than N.

Using the concepts defined above the root signals of median filters can be completely characterized. *A signal is a root signal exactly when it consists of only constant neighborhoods and edges* [GA81; TY81]. The following relationship is immediately implied between the root signal sets of different median filters: if $x[n]$ is a root signal of the median filter of length N it is also a root signal of any median filter whose length is less than N.

These results have some interesting consequences. For a signal invariant to a median filter to contain both positive and negative slopes, these slopes must be separated by a constant neighborhood. Since the length of the constant neighborhood depends on the filter length, the longer the filter the farther apart the opposite slopes must be. However, the magnitude of the slope has no effect. This, in a way, says that the median filter preserves edges but filters out impulses and oscillations, and that longer filters attenuate impulses and oscillations more than shorter filters.

It is interesting to note that a root signal can always be generated by repeated median filtering of a finite length sequence. An upper bound, n_{max}, for the number of filterings needed can be proved to be

$$n_{max} = 3 \left\lceil \frac{(L - 2)}{2(k + 2)} \right\rceil . \tag{13.101}$$

Here, L is the length of the signal and $N = 2k + 1$ is the filter length. Usually, the required number of filterings is much lower than the upper bound given by Eq. (13.101). Each constant neighborhood effectively divides the input signal into shorter signals and the upper bound can then be applied to each of these separately.

Figure 13-29 gives an example where a binary input signal has been filtered to a root signal with three different median filters of lengths 3, 5, and 7, respectively. In this case the three-point filter produced the root signal in one filter pass, whereas the other two filters needed three passes.

Figure 13-30 shows an example where a root was searched with a median filter

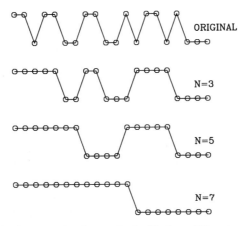

FIGURE 13-29 Root signals searched with three different median filters.

of length 15. Three filter passes were needed, whereas Eq. (13.101) would give 69 as the upper bound.

13-6-1-2 Median Filtering by Threshold Decomposition. The superposition property of linear filters cannot be applied to nonlinear filters. This is the main factor that makes the theoretical analysis of median filters somewhat difficult. However, another powerful technique, called *threshold decomposition*, allows us to divide the analysis problem into smaller parts. Essentially, using threshold decomposition, we can derive all properties of median filters by just studying their effect on binary signals.

Threshold decomposition of an M-valued signal $x[n]$, where the samples are integer valued, $0 \le x[n] < M$, means decomposing it into $M - 1$ binary signals, $x^1[n], x^2[n], \ldots, x^{M-1}[n]$, according to the following rule:

$$x^m[n] = \begin{cases} 1 & \text{if } x[n] \ge m, \\ 0 & \text{otherwise.} \end{cases} \tag{13.102}$$

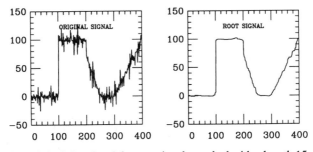

FIGURE 13-30 Original signal and the root signal searched with a length 15 median filter.

By replacing the sample values by their respective rank values this thresholding scheme can be applied also to noninteger signals or, in general, to all signals that are quantized to a finite number of arbitrary levels.

The original multivalued signal can be constructed from the binary signals by adding these together:

$$x[n] = \sum_{m=1}^{M-1} x^m[n]. \tag{13.103}$$

A very important property of median filters states that [FI84] applying a median filter to an M-valued signal is equivalent to decomposing the signal to $M - 1$ binary components using the algorithm in Eq. (13.102), filtering these components separately with the median filter, and then adding the binary signals together using the algorithm in Eq. (13.103).

An example of filtering by threshold decomposition is shown in Figure 13-31. The importance of the property arises from the fact that binary signals are much easier to analyze than multivalued signals. The median operation on binary samples reduces to a simple majority decision: if the number of zeros in the filter window is greater than the number of ones, then the median filter output at that filter position will also be zero, otherwise it will be one.

A binary median filter operating on a binary signal is essentially a Boolean function of the $N = 2k + 1$ variables inside the filter window. For example, the binary three-point median filter is defined by the Boolean function

$$y[n] = x[n-1] \cdot x[n] + x[n-1] \cdot x[n+1] + x[n] \cdot x[n+1], \tag{13.104}$$

where $\cdot$ and $+$ denote Boolean AND and OR operators. This binary function operates on each binary signal obtained by threshold decomposition, Eq. (13.102).

The approach of threshold decomposition and binary filtering has led to a new class of filters, called *stack filters* [WE86]. Stack filters are defined by the Boolean function used for filtering each threshold level of the multivalued signal. Only positive Boolean functions (PBF) that can be expressed as Boolean expressions

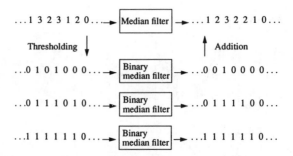

FIGURE 13-31 Median filtering of a four-valued sequence by threshold decomposition. $N = 3$.

containing no complements of input variables have the *stacking property* [WE86] and can be used to define stack filters. If Boolean function does not have the stacking property the addition process of Eq. (13.103) does not produce meaningful results.

The integer domain stack filter corresponding to a given PBF can be expressed by replacing Boolean AND and OR operations with MIN and MAX operations. For example, the three-point median filter can be expressed as

$$
y[n] = \text{med} \ (x[n-1], x[n], x[n+1])
$$
$$
= \max \ (\min \ (x[n-1], x[n]), \ \min \ (x[n-1], x[n+1]),
$$
$$
\min \ (x[n], x[n+1])). \tag{13.105}
$$

13-6-2 Statistical Properties

Median filters are used as robust noise smoothers. The filtering results are often evaluated in a qualitative way, which in many cases, like in image filtering, is the preferred method. The performance of the filter depends on how well it can suppress the undesired part of the signal and, equally importantly, how well the desired information is retained. There is no general numerical measure to combine these two properties and thus to evaluate the performance of the filter. With linear filters the two objectives can both be achieved if, and only if, the noise and the signal occupy different frequency bands.

The noise attenuation properties of a filter nevertheless are keys to the proper understanding of that filter. It is common, for example, for the desired signal to remain constant over a certain period and during this period the filter should be able to suppress the undesired variations in the signal. To get quantitative information on how much median filters reduce noise, statistical methods must be applied. Since the median operation is nonlinear, the effects of the filter on the noise and on the signal cannot be separated as with linear filters. The amount of smoothing also depends on the properties of the noise.

We restrict ourselves here to the case of white additive noise on a constant signal. This is usually meant when the noise attenuation of the median filter is discussed. In addition, we consider a case where the noise is embedded in a signal containing a step edge. This gives statistical background to the argument that median filters attenuate noise while retaining sharp changes in the signal.

13-6-2-1 Constant Signal and White Noise. White noise is here modeled by independent identically distributed random variables with mean μ and variance σ^2. The corresponding distribution and density functions are denoted as $F(x)$ and $f(x)$. The probability distribution $G(x)$ and the density $g(x)$ of the output of the median filter of length $N = 2k + 1$ are given by

$$
G(x) = \sum_{i=k+1}^{2k+1} \binom{2k+1}{i} F(x)^i (1 - F(x))^{2k+1-i} \tag{13.106}
$$

and

$$g(x) = \frac{(2k + 1)!}{k!k!} f(x)F(x)^k(1 - F(x))^k. \tag{13.107}$$

The above two equations provide the basis for a quantitative analysis of the noise attenuation of median filters. Usually, numerical methods must be used to apply these equations.

Under some very general assumptions the asymptotic distribution (for large N) of the median filter output, when the input is a white noise, is normal with mean μ_{med} and variance σ^2_{med},

$$\mu_{med} = x_{0.5},$$

$$\sigma^2_{med} = \frac{1}{4N[f(x_{0.5})]^2}, \tag{13.108}$$

where $x_{0.5}$ is defined by $F(x_{0.5}) = 0.5$. This result explains the robust noise smoothing properties of the median filter. Regardless of the input distribution, the median operation produces an unbiased estimate of the distribution median $x_{0.5}$. Moreover, the estimate is always consistent; that is, $\lim_{N \to \infty} \sigma^2_{med} = 0$. The output variance does not depend on the input variance directly, but on $f(x_{0.5})$ instead. With heavy-tailed, or impulsive, noise distributions the variance of the distribution grows with the amplitude of the impulses, whereas $f(x_{0.5})$ does not necessarily change. The sample mean, that is, the average value of $x_1, \ldots, x_N$, does not have this same property.

The asymptotic distribution of the sample mean is normal with mean μ_{ave} and variance σ^2_{ave}

$$\mu_{ave} = \mu,$$

$$\sigma^2_{ave} = \frac{\sigma^2}{N}, \tag{13.109}$$

where N is again the sample size.

Equations (13.108) and (13.109) provide a means to compare the asymptotic behavior of the sample mean and the sample median. Generally, the expected values $x_{0.5}$ and μ coincide and both operators give an unbiased estimate of the distribution mean. The ratio of their asymptotic variances is

$$\frac{\sigma^2_{ave}}{\sigma^2_{med}} = 4\sigma^2(f(x_{0.5}))^2. \tag{13.110}$$

Their respective performance as a noise smoother is therefore dependent on the input noise distribution and can easily be determined using Eq. (13.110).

TABLE 13-8 Asymptotic Output Variances of the Averaging and the Median Filters of Length N for Different Input Noise Distributions[a]

Noise Density	Average	Median
Uniform $$f(x) = \begin{cases} \dfrac{1}{2\sigma\sqrt{3}}, & -\sigma\sqrt{3} \le x \le \sigma\sqrt{3} \\ 0, & \text{otherwise} \end{cases}$$	$\dfrac{\sigma^2}{N}$	$\dfrac{3\sigma^2}{N+2}$
Gaussian $$f(x) = \frac{1}{\sigma\sqrt{2\pi}}\, e^{-x^2/2\sigma^2}$$	$\dfrac{\sigma^2}{N}$	$\dfrac{\pi\sigma^2}{2N}$
Laplacian $$f(x) = \frac{1}{\sigma\sqrt{2}}\, e^{-\sqrt{2}|x|/\sigma}$$	$\dfrac{\sigma^2}{N}$	$\dfrac{\sigma^2}{2N}$

[a]The noise is white with zero mean and σ^2 variance.

Table 13-8 lists asymptotic output variances for some common probability distributions. The good performance of the median filter with Laplacian distributed noise is due to an important optimality property of the median operation: *the sample median is the maximum likelihood estimate for the mean of the Laplacian distribution.* This property guarantees that the median filter is, at least asymptotically, optimal in the mean-square error sense for attenuating double exponentially distributed white noise. In the same way the sample mean is optimal for Gaussian distributed noise. This analogy results from alternate definitions for the average and the median of N values $x_1, \ldots, x_N$. Consider the expression

$$\Phi(\beta) = \sum_{i=1}^{N} |x_i - \beta|^{\gamma}. \tag{13.111}$$

The median of $x_1, \ldots, x_N$ can be defined as the value β minimizing $\Phi(\beta)$ in Eq. (13.111) when $\gamma = 1$. The definition always produces one of the samples x_i as a result. This can easily be seen from the example in Figure 13-32. The sample mean minimizes the same expression when $\gamma = 2$.

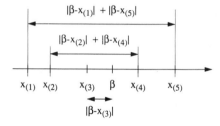

FIGURE 13-32 The sample median is the value β minimizing $\Phi(\beta)$ in Eq. (13.111) when $\gamma = 1$. In this example $N = 5$. The two sums $|\beta - x_{(1)}| + |\beta - x_{(5)}|$ and $|\beta - x_{(2)}| + |\beta - x_{(4)}|$ remain constant, and it is easy to see that the sum of distances, $\Phi(\beta)$, is minimized when $\beta = x_{(3)}$, that is, the median.

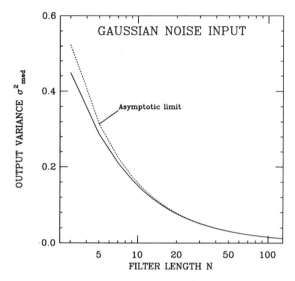

FIGURE 13-33   Output variance of the median filter when the input is Gaussian distributed white noise with unit variance. The figure also shows the asymptotic limit given by Eq. (13.108).

Figure 13-33 shows the output variance of the median filter when the input is a Gaussian distributed white noise. Asymptotically, the variance is 57% larger than that of the averaging filter of the same length. Note that with short windows the difference is smaller. For example, when $N = 3$ and $N = 5$ the respective percentages are 35 and 43.

The analysis for impulsive noise distributions depends on how the impulses are modeled. For example, are the impulses allowed to have different amplitude values? Here, we only consider a simple model where the signal value remains constant, say, 0, and where impulses of constant value e occur at a given probability p:

$$x[n] = \begin{cases} 0 & \text{with probability } 1 - p, \\ e & \text{with probability } p. \end{cases} \qquad (13.112)$$

Other models have been studied, for example, in Justusson [JU81]. In the model in Eq. (13.112) the output of the filter is either 0 or, if there are more than k impulses inside the filter window, the erroneous value e. Of interest is the probability of an error.

Table 13-9 lists error probabilities for different filter window sizes and impulse rates. With practical impulse rates the performance of the median filter is very good: single outliers are removed. Note, however, that if the desired signal is not constant, impulses cause some distortion in the filter output. This must be taken into account when reading the error rates in Table 13-9, especially with large values of N. In spite of that, impulse filtering is one of the areas where the advan-

TABLE 13-9 Error Probability in the Noise Model of Eq. (13.112)[a]

p	$N = 3$	$N = 5$	$N = 7$	$N = 9$	$N = 15$	$N = 25$
0.01	2.980×10^{-4}	9.851×10^{-6}	3.417×10^{-7}	1.219×10^{-8}	6.045×10^{-13}	4.650×10^{-20}
0.05	7.250×10^{-3}	1.158×10^{-3}	1.936×10^{-4}	3.322×10^{-5}	1.830×10^{-7}	3.591×10^{-11}
0.10	2.800×10^{-2}	8.560×10^{-3}	2.728×10^{-3}	8.909×10^{-4}	3.362×10^{-5}	1.621×10^{-7}
0.15	6.075×10^{-2}	2.661×10^{-2}	1.210×10^{-2}	5.629×10^{-3}	6.096×10^{-4}	1.689×10^{-5}
0.20	1.040×10^{-1}	5.792×10^{-2}	3.334×10^{-2}	1.958×10^{-2}	4.240×10^{-3}	3.690×10^{-4}
0.30	2.160×10^{-1}	1.631×10^{-1}	1.260×10^{-1}	9.881×10^{-2}	5.001×10^{-2}	1.747×10^{-2}
0.40	3.520×10^{-1}	3.174×10^{-1}	2.898×10^{-1}	2.666×10^{-1}	2.131×10^{-1}	1.538×10^{-1}
0.50	5.000×10^{-1}	5.000×10^{-1}	5.000×10^{-1}	5.000×10^{-1}	5.000×10^{-1}	5.000×10^{-1}

[a] N is the filter window size and p is the absolute probability of an impulse in the filter input.

tages of median filters are most striking. Image filtering examples of this appear later.

13-6-2-2   *Signal Edge and White Noise.* The median filters preserve ideal signal edges with no noise. When noise is added some amount of blurring occurs. To evaluate the behavior of the median filter consider a signal edge of height h represented by

$$x[n] = \begin{cases} \eta[n], & n < 0, \\ h + \eta[n], & n \geq 0, \end{cases} \tag{13.113}$$

where $\eta[n]$ is white noise. Obviously some bias error in the filter output will result from the signal edge. Because of symmetry, only the samples $y[n]$ where $n < 0$ need be considered. When an averaging filter of length $N = 2k + 1$ is used the bias error can be expressed as

$$E[y[n]] = \frac{k + 1 + n}{2k + 1} h, \qquad -k \leq n < 0, \tag{13.114}$$

and the corresponding mean-square error as

$$E[y[n]^2] = \left(\frac{k + 1 + n}{2k + 1}\right)^2 h^2 + \frac{\sigma^2}{N}, \qquad -k \leq n < 0. \tag{13.115}$$

For the median filter the corresponding result is not as readily available as for the averaging filter and the noise characteristics will affect the result. Here, only Gaussian distributed noise will be considered.

Figure 13-34 shows the expected edge responses of three different median and averaging filters when the edge is reasonably low, that is, only twice the noise deviation. The difference is rather small even though the averaging filter has a greater bias error. This edge height ($h = 2\sigma$) is considered a limit after which the median filter outperforms the average as a noise smoother [JU81]. When the edge height $h = 4.0$ (see Figure 13-35), the differences become clearer: the averaging filter turns the step into a ramp whereas the median filter preserves the step.

Note that both these edge heights are still quite low. In most practical applications the signal steps of interest could easily be higher than five times the noise deviation. As the edge height h grows, the error values of linear filters increase without limit. The error made by the median filter, however, has an asymptotic upper limit and increasing h has no further effect. Actually, the bias error has practically reached its maximum value at $h = 5.0$.

Finally, Figure 13-36 gives an example where the performance of the median filter on both constant areas and near an edge is shown. The original signal has additive noise components from several different probability distributions and the robustness of the median filter can be seen immediately.

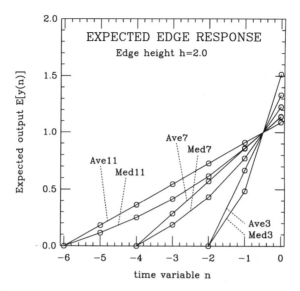

FIGURE 13-34 Edge responses of three median and averaging filters of different lengths (e.g., Med11 stands for a median filter of length 11). The edge height is 2.0 and the edge is contaminated by Gaussian distributed noise with variance 1.0. See Eq. (13.113).

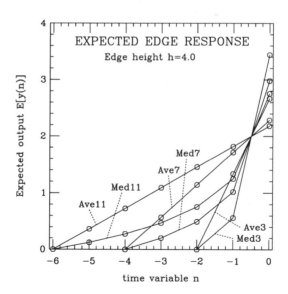

FIGURE 13-35 Edge responses of three median and averaging filters of different lengths. The edge height is 4.0 and the edge is contaminated by Gaussian distributed noise with variance 1.0. See Eq. (13.113).

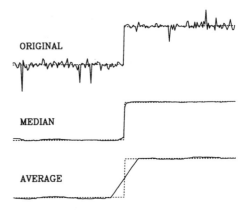

FIGURE 13-36 The median filter as a noise smoother. The original signal contains a step edge and is contaminated by white noise. The length of the signal is 200 samples. The dotted line shows the uncontaminated signal. The results when the signal is filtered with median and averaging filters of length $N = 25$ are shown.

13-6-3 Two-Dimensional Median Filtering

Extending the median filter to two dimensions involves using two-dimensional filter windows. Here, it is assumed that the filter window advances from right to left on every row, and from top to bottom for row advances. As in one-dimensional filtering the filter output is the median value inside a fixed size window. Examples of two-dimensional filter windows are shown in Figure 13-37. For example, the operation of the cross filter in a 3×3 window is described by the equation

$$y[m, n] = \text{med} \left(x[m, n-1], x[m, n], x[m, n+1], x[m-1, n], x[m+1, n]\right),$$

$$(13.116)$$

where in $x[m, n]$ the first index is the row and the second is the column. The appending strategy with two-dimensional signals is similar to that with one-di-

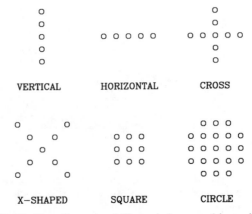

FIGURE 13-37 Two-dimensional filter windows used in median filtering.

mensional signals. The borders of the image can be filtered by duplicating the outmost values as many times as needed.

The deterministic properties of two-dimensional median filters are much more difficult to analyze than the corresponding properties of one-dimensional filters. The definition of monotonicity and concepts like the constant neighborhood are not clear for two-dimensional signals and the two-dimensional median filters can often have quite complicated root signals. Also, the variety of filter window shapes makes the formulation of general theorems difficult.

It is also known that the common two-dimensional median filters, that is, the cross, the X-shaped, and the square, do not necessarily convert the input to a root signal if repeatedly applied, but instead, there are signals that will oscillate. Because of these difficulties one common approach to study the effect of different filters on images is to consider their effect on some simple two-dimensional structures, like lines, line segments, or edges having different orientations. In most image processing tasks it is essential that certain image patterns not be disturbed by the filter.

Of the basic filter masks in Figure 13-37 the square mask is the least sensitive to image details. It filters out narrow lines and cuts corners of square-shaped objects. Of course, it also smooths the image more than the other filters. In image filtering the key factor is the trade-off between the amount of smoothing and the preservation of details.

It is easy to observe that the cross filter is able to preserve horizontal and vertical lines, whereas the X-shaped filter preserves only diagonal lines. For most applications the cross filter is preferred over the X-shaped filter since horizontal and vertical details are usually more important for a human observer than diagonal details.

The best method to evaluate the different filters is to apply them to real images. Figure 13-38 shows a rather critical image containing a lot of high-frequency details. The same image contaminated by impulsive noise is shown in Figure 13-39.

Figure 13-40 gives an example of why linear algorithms fail at trying to remove

FIGURE 13-38   Original image: 300 × 200 pixels, 8 bits per pixel.

FIGURE 13-39 Noisy image: 1% of the pixels have been randomly set black and 1% white.

the noise. The 3 × 3 moving average blurs the image badly and is still not able to remove the impulses. This results in a picture that appears "out-of-focus."

In Figure 13-41 the result of filtering with the 3 × 3 square median filter is shown. The impulses have been effectively eliminated but the contouring effects caused by streaking and the resulting loss of detail can clearly be seen if the image is compared with the original. The loss of narrow lines can also be seen especially in the background. The advantages over linear filters are obvious but the results can be further improved, for example, with the cross window.

13-6-4 Some Generalizations

Since the introduction of median filters, numerous generalizations and modifications to the standard median filter have been proposed. These modifications are often quite application specific and designed to improve some particular property

FIGURE 13-40 Noisy image filtered with the 3 × 3 moving average.

FIGURE 13-41 Noisy image filtered with the square 3 × 3 moving median.

of the median filter. For example, in an early application to speech processing [RA75] a cascade of a median filter and an averaging filter was proposed to reduce streaking caused by the median filter.

Many median-related filters are based on order statistics but it is difficult to say whether they are actually modifications of the median filter or if their development was only inspired by the success of the median filter. Filters that are often connected with the median filter include other rank order filters (e.g., min and max) [NO82], and the so-called α-trimmed mean filters [BE84], which provide a compromise between the moving average and the median. In Bovik et al. [BO83] the filter output was taken as a linear combination of different order statistics. Recently, an extension of the median operation to vector-valued signals has been proposed [AS90] with applications to processing of complex and color signals.

Three modifications that utilize the median operation are discussed here in greater depth: the recursive median filter, the weighted median filter, and the FIR–median hybrid filter. A good list of references for median-related filters can be found in Coyle et al. [CO89]. There are also good median filter reviews that can be consulted [AR86; JU81; PI90; TY81].

13-6-4-1 Recursive Median Filter. An early modification of the median filter uses the previous output samples also for determining the filter output. The output of a recursive median filter [NO82] of length $N = 2k + 1$ is defined by the equation:

$$y[n] = \text{med}\,(y[n - k], \ldots, y[n - 1], x[n], x[n + 1], \ldots, x[n + k]).$$

$$(13.117)$$

In recursive median filtering, the filtering operation is performed "in-place" so that the output of the filter replaces the old input value before the filter window is moved to the next position. The recursive median filter is more reluctant to change the output as a result when there is a change in the input signal than the nonrecursive filter. The output of the recursive filter is more correlated also than that of the nonrecursive filter. The practical use of recursive median filters is limited to rather short filters where this effect is not too dominating.

It is easy to observe that a signal is a root of a recursive median filter exactly when it is a root of the nonrecursive filter of the same length. An attractive property of the recursive median filter is that it produces the root signal just after one filtering operation. However, starting from an arbitrary signal, the recursive median filter does not generally produce the same root as repeated applications of the nonrecursive filter. This is illustrated in Figure 13-42.

13-6-4-2   Weighted Median Filters. In the standard median filter each sample inside the filter window has the same influence on the filter output. To achieve new filter properties one may want to give more weight to signal samples in specific filter window positions, say, to the center samples.

In the weighted median filter [JU81] each filter window position is assigned a weight w_i, $i = 1, \ldots, N$. We denote the sum of the weights as S_w:

$$S_w = \sum_{i=1}^{N} w_i. \qquad (13.118)$$

Like the median filter, the weighted median filter can be defined in two different ways that both produce the same result. The first definition can be used in the common case of positive integer weights: sort the samples inside the filter window and replicate each sample to the number of the corresponding weight w_i. Choose the median value from the new sequence of S_w points. Here, S_w should be odd.

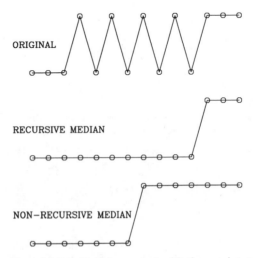

FIGURE 13-42 A binary signal filtered to a root with three-point recursive and nonrecursive median filters. The original signal oscillates between zero and one before settling to the higher value.

This filtering procedure can be represented by the following equation,

$$y[n] = \text{med}\,(w_1 \lozenge x[n - k], \ldots, w_{k+1} \lozenge x[n], \ldots, w_N \lozenge x[n + k]),$$

(13.119)

where the symbol $\lozenge$ is used to denote replication

$$n \lozenge x = \underbrace{x, \ldots, x.}_{n\,\text{times}}$$

(13.120)

The second definition of the weighted median operation also allows positive noninteger weights to be used: the weighted median of the sequence $x_1, \ldots, x_N$ is the value β minimizing the expression

$$\Phi(\beta) = \sum_{i=1}^{N} w_i |x_i - \beta|.$$

(13.121)

Here, β is guaranteed to be one of the samples x_i because $\Phi(\beta)$ is piecewise linear and convex if $w_i \geq 0$ for all i. Note that this definition reduces to the definition of the standard median if $w_i = 1$ for all i.

An interesting analogy exists between linear FIR filters and weighted median filters [YL91]. If in the expression

$$\Phi(\beta) = \sum_{i=1}^{N} w_i |x_i - \beta|^{\gamma}$$

(13.122)

the exponent is set to $\gamma = 2$, the minimizing value β can be expressed as

$$\beta = \frac{\sum\limits_{i=1}^{N} w_i x_i}{\sum\limits_{i=1}^{N} w_i},$$

(13.123)

which can be taken as a normalized FIR filter, that is, a weighted average. Likewise, if $\gamma = 1$, the value β minimizing Eq. (13.122) is

$$\beta = \text{med}\,(w_1 \lozenge x_1, \ldots, w_N \lozenge x_N),$$

(13.124)

that is, a weighted median.

All weighted median filters have the property that they are statistically unbiased in the sense of the distribution median. When the input to the filter is identically distributed white noise, the distribution median of the output is the same as that of the input.

13-6-4-3 FIR–Median Hybrid Filters. The principle of FIR–median hybrid (FMH) filters [HE87] is to divide the filter window into an odd number M of smaller subwindows and to use a linear-phase FIR filter in each of these subwindows. The filter output is taken as the median of the subfilter outputs. Let $H_1(z)$, . . . , $H_M(z)$ be linear-phase FIR filters. The output of the FMH filter is then

$$y[n] = \text{med} \ (y_1[n], y_2[n], \ . \ . \ . \ , y_M[n]), \tag{13.125}$$

where $y_i[n]$ is the output of the FIR filter $H_i(z)$.

The FMH filters exhibit many properties similar to those of the median filter. Due to the subfilters this filter type allows greater design flexibility than the median filter. Also, M, the number of subwindows, is quite small, typically, three or five, compared to the number of samples in these subwindows. As a result, the number of operations needed to find the median is significantly reduced. This advantage can be important in many signal processor environments where the FIR filters can be efficiently implemented and where the compare/swap operations needed to find the median are costly.

An important class of FMH filters of length $N = 2k + 1$ is considered here, and its properties are compared to those of the median filter of the same length. This filter class is defined by the following three substructures:

$$H_1(z) = \frac{1}{k} (z^k + z^{k-1} + \cdot \cdot \cdot + z),$$

$$H_2(z) = 1,$$

$$H_3(z) = \frac{1}{k} (z^{-1} + z^{-2} + \cdot \cdot \cdot + z^{-k}). \tag{13.126}$$

This filter will be called the basic FMH filter.

It is obvious that the basic FMH filter preserves signal edges as does the median filter. For a square pulse to pass through the basic FMH filter unaltered, its length must be at least $2k$. The critical length for the median filter is $k + 1$. The square pulse is an example of the differences between the FMH filter and the median filter roots: *a sufficient, but not necessary, condition for a finite length signal to be a root of the FMH filter of Eq. (13.126) is that it be a root of the median filter of length $4k - 1$.* Note, however, that these are not the only roots. The convergence to roots is also much slower than with median filters. In the theoretical case of infinite number of signal levels the convergence is actually only asymptotic.

In impulse filtering the performance of the FMH filter differs from that of median filtering. Whereas the median filter is able to filter out impulses of length k or less, the FMH filter can only remove single impulses. If several positive going impulses occur inside the filter window the output level will rise depending on the number of impulses. Figure 13-43 gives an example of another difference. An impulse near an edge causes edge jitter in the median filter output. There is no

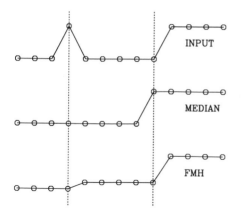

FIGURE 13-43 The effect of an impulse near an edge for median and FMH filtering. The median filter causes edge jitter whereas in the FMH filter output the signal level rises a little before the edge but the location of the edge does not change. In this example the filter length is 11.

edge jitter in the FMH filter output but the signal level before the edge changes slightly.

Figure 13-44 shows that the output distributions of the median filter and the FMH filter are practically identical when the input is Gaussian distributed white noise. Consequently, both filters have approximately the same noise attenuation for Gaussian distributed white noise. Near noisy signal edges the bias error of the basic FMH filter is significantly smaller than that of the median filter [HE87].

One problem of median filters, and also of the basic FMH filters, is that their noise smoothing ability is smaller on rising and falling parts of the signal than on constant parts. The substructures of FMH filters can be used to alleviate this prob-

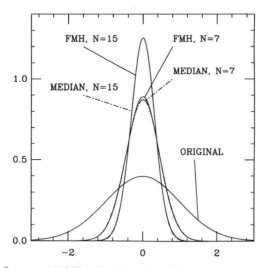

FIGURE 13-44 Output probability densities of median and FMH filters when the input is Gaussian distributed white noise with unit variance. The figure shows the original normal density and the densities of two median and FMH filters.

lem since first-order predictors can be added to the structure of Eq. (13.126). The substructures can also be made adaptive.

13-6-5   Implementation Considerations

A straightforward implementation of a median filter of length N requires sorting of the N elements inside the filter window for each input sample. Well-known sorting algorithms accomplish this task in $O(N \log N)$ time. Implementing long median filters using this approach can be quite time consuming.

In the one-dimensional running median filter only one new input sample enters the filter window at each time moment. This can easily be utilized by keeping the already sorted input samples in memory. As the new sample enters the window the oldest sample is deleted from the sorted list and the new one is added to the list.

A good way to implement this idea in software is to use a circular buffer of length N as the filter window. Each element in the buffer consists of the value of the input sample, and also of the current rank of that sample. When a new sample enters the window the ranks of all the other samples are modified according to the values of the new sample and the oldest sample now leaving the filter window. After the new ranks have been found the sample with the rank $(N + 1)/2$ is chosen to be the filter output.

The above algorithm is easy to code efficiently and operates in $O(N)$ time. The algorithm can be modified to implement other rank order filters and also weighted median filters. For median filters with very large N the use of the algorithm proposed in Astola and Campbell [AS89] can be considered. The samples are stored in a special data structure, called a double heap, which is modified for each new input sample. This algorithm enables median filters to operate with logarithmic time complexity, $O(\log N)$, but is typically somewhat difficult to implement.

In two-dimensional median filters there are more than one new sample entering and leaving the filter window at each time moment and the previous algorithms cannot be implemented directly. A simple and efficient way to implement two-dimensional median filters is to store a histogram of the samples inside the filter window in memory and modify it as new samples enter and old samples leave the window [HU79]. The filter output can be searched from the histogram quite fast when the search is initiated from the previous output. If the number of grey levels in the two-dimensional image is very large, or if the signal is real-valued, this histogram method is not practical.

The binary representation of data can often be utilized in implementing median filters. When the radix-2 representation is used the kth most significant bit of the median can be determined by inspecting the k most significant bits of the samples [AT80]. For example, if the majority of the most significant bits are equal to 1(0), then the most significant bit of the median is also equal to 1(0). With slight modifications the same reasoning can be continued for the other bits too.

The radix algorithm is suitable for hardware implementation and can be applied both to one- and two-dimensional signals. Many of the tasks can be done in parallel and the speed and hardware complexity of the algorithm do not depend on N, but on the wordlength of the data.

13-6-6 Applications

The use of the median operation in digital filtering was first suggested by Tukey [TU71; TU74] for smoothing statistical data. Later, Pratt used two-dimensional median filters for image smoothing [PR78]. As an edge preserving filtering technique, the median filter does not blur the image like linear filters. Images are very good examples of signals where the noise and the signal cannot be separated by conventional frequency selective methods since the high-frequency components form an important part of the information content of the image.

The impulse removing properties of the median filter have been exploited in speech processing [JA76]. Jayant studied its use in improving the quality of DPCM coded speech in the presence of transmission errors. He found that with small error rates the subjective results were better when the speech was not processed, but when the bit error rate grows the quality of speech could be improved by median filtering.

Another early application to speech processing was presented in Rabiner et al. [RA75]. According to these results simple linear smoothing routines generally fail to provide adequate smoothing for data that exhibit both local roughness and sharp discontinuities.

Many applications of median filters include removing of artifacts caused by imperfect data acquisition systems. For example, optical scanners often produce noise that is seen as horizontal stripes in the image. This kind of noise can be removed by vertical median filtering [WE74].

In digital subtraction angiography (DSA) the radiographic images suffer from noise. In these images spatial resolution and edge detail (outline of vessels) are critical for clinical usefulness. Linear filters introduce unwanted effects when applied to DSA images, but median filters have been used successfully [RI84].

Median filters are also used in many commercial tomographic scan systems. For example, in ultrasonic computed tomography Gaussian distributed noise is introduced in the analog signal processing chain while shotlike impulse noise is generated by mechanical stepping motors. Especially speed of sound imaging is susceptible to impulse noise which produces bright and dark streaks through the image. Median filters have been successfully used for reducing these artifacts [SC84].

Some other applications in biomedical signal processing are in the areas of EEG spike detection and in the analysis of blood pressure recording.

The use of median filters in commercial applications is growing rapidly. One good example is the future television where digital image processing is becoming commonplace also in commercial TV sets. It seems probable that median filters will be used in many advanced TV sets. Application areas in television include scan rate up-conversion, noise elimination, and coding. Median filters are suitable for TV applications because of their very good performance/cost ratio: median filters can be implemented at video rates and even the simplest filters have properties that are difficult to achieve with other methods.

13-7 CONCLUDING REMARKS

We have explored in this chapter some deterministic special filter design problems including such linear filters as Hilbert transformers, differentiation and integration filters, smoothing, filters, and delay networks, as well as nonlinear median-based filters. The design procedures for the linear filters become straightforward once the subtleties of the respective approximation problems are understood. The median filters were explored from the viewpoint of understanding their filtering influence on a signal.

The problem of designing Hilbert transformers was shown to be a particular case of half-band filter design. Thus methods of designing conventional real half-band filters can readily be tailored to the Hilbert transformer problem. In addition to the simplified design procedure, the resulting filters are often quite computationally efficient in view of the half-band characteristic. For example, the FIR Hilbert transformers designed in Example 13.1 have every other impulse response sample as zero. Similarly, the odd-ordered IIR designs consist of two allpass functions, each of which has every other impulse response sample as zero.

The integrators and differentiators were designed using a "correction" filter approach. In the differentiator case it was pointed out that this approach can provide better numerical conditioning of the design problem, compared to a direct formulation of the approximation problem. In the case of integrator design, it becomes necessary to factor a pole from the frequency response before the design procedure commences; in this case one is naturally faced with designing a correction filter.

A brief examination of some classical numerical integration schemes showed that, although these methods are quite good for low-frequency signals, the high-frequency performance deviates somewhat from the ideal. If the underlying analog signal to be integrated or differentiated is grossly oversampled, the resulting digital signal has most of its energy concentrated in the low-frequency portion of the spectrum, in which case the simple first-order approximations to differentiation and integration work quite well. This technique of oversampling corresponds to choosing Δt progressively smaller in Eq. (13.54); in the limit as $\Delta t \to 0$, the first-order filter of Eq. (13.55) becomes arbitrarily close to an ideal differentiator for the oversampled signal.

We also showed that wideband performance can be obtained only if the application can tolerate a half-sample term in the group delay of the integrator or differentiator. In practical applications (such as simulating digitally the solution to a differential equation), one may want sampled versions of the signal and its integral (or derivative) corresponding to the same sample instants. The filter designed to meet these applications has a Type III impulse response (i.e., antisymmetric with a group delay corresponding to an integral number of samples) and thus provides a transmission zero at $\omega = \pi$. In this case the designed filter cannot achieve full wideband performance. In practice, this restriction may not be severe if the underlying analog signal is oversampled.

A least-squares approach to smoothing was presented that provides reasonably good retention of the time-domain characteristics of the signal to be smoothed. The method leads to a linear convolution of the given sequence to be smoothed, and the frequency response corresponding to these filters not surprisingly takes a lowpass characteristic. The method also yields differentiation filters whose frequency response takes the form of a narrowband differentiator. From these observations it becomes clear that conventional linear-phase filter design can be used by supplying an appropriate frequency response to be approximated. An example was briefly illustrated for the case of an optimal design; the other approximation techniques presented in Chapter 4 can be applied as well.

The realization of delay networks providing a signal delay distinct from an integral multiple of the sampling period was addressed. If the desired signal delay is a rational number, a FIR design technique becomes straightforward, by exploiting the polyphase decomposition of an appropriately designed Mth-band linear-phase lowpass filter. For arbitrary signal delays, one can use an IIR allpass filter that provides approximately linear phase over a prescribed frequency range, and no magnitude distortion in the frequency response.

Median filtering is an effective technique for removing impulse-like noise while preserving edges and constant neighborhoods of a signal. Root signals are an important feature for one-dimensional median filters, providing a conceptual analogy with the passband of a linear filter. Two-dimensional median filters are more difficult to analyze as, for example, root signals may not always be applicable. Moreover, the statistical properties vary with the window shape.

Increased design flexibility is obtained with various generalizations of the median filter, including stack order filters and hybrid linear/median filters. The noise attenuation properties, as well as the root signals of the filter, can be tailored to specific applications. Some of the more widely applicable generalizations have been reviewed, with their salient features summarized. Median and related filters show great promise in image processing, advanced television systems, and biomedical signal processing because of their effective cost/performance ratio and their desirable signal restoration properties not afforded by linear filters.

ACKNOWLEDGMENTS

The author is indebted to Dr. Peter Steffen of the Universität Erlangen-Nürnberg, whose careful proofreading contributed to the accuracy of the material and to the consistency of the notation. In addition, Dr. Markku Renfors of the Department of Electrical Engineering, Tampere University of Technology, Tampere, Finland, is gratefully acknowledged for his assistance in obtaining the computer programmed designs used in the examples.

REFERENCES

[AN87] R. Ansari, IIR discrete-time Hilbert transformers. *IEEE Trans. Acoust., Speech, Signal Process.* **ASSP-35**, 1116–1119 (1987).

[AR86] G. R. Arce, N. C. Gallagher, Jr., and T. A. Nodes, Median filters: Theory for one- and two-dimensional filters. In *Advances in Computer Vision and Image Processing.* (T. S. Huang, ed.), 89–166, Vol. 2, JAI, Greenwich, CT; 1986.

[AS89] J. Astola and T. G. Campbell, On computation of the running median. *IEEE Trans. Acoust., Speech, Signal Process.* **ASSP-37**, 572–574 (1989).

[AS90] J. Astola, P. Haavisto, and Y. Neuvo, Vector median filters. *Proc. IEEE* **78**, 678–689 (1990).

[AT80] E. Ataman, V. K. Aatre, and K. M. Wong, A fast method for real-time median filtering. *IEEE Trans. Acoust., Speech, Signal Process.* **ASSP-28**, 415–421 (1980).

[BA77] J. Baranger, *Introduction à l'Analyse Numérique.* Hermann, Paris, 1977.

[BE84] J. B. Bednar and T. L. Watt, Alpha-trimmed means and their relationship to median filters. *IEEE Trans. Acoust., Speech, Signal Process.* **ASSP-32**, 145–153 (1984).

[BO83] A. C. Bovik, T. S. Huang, and D. C. Munson, Jr., A generalization of median filtering using linear combinations of order statistics. *IEEE Trans. Acoust., Speech, Signal Process.* **ASSP-31**, 1342–1350 (1983).

[BR75] L. T. Bruton, Low sensitivity digital ladder filters. *IEEE Trans. Circuits Syst.* **CAS-22**, 168–176 (1975).

[CH66] E. W. Cheney, *Introduction to Approximation Theory.* McGraw-Hill, New York, 1966.

[CI70] V. Cížek, Discrete Hilbert transform. *IEEE Trans. Audio Electroacoust.* **AU-18**, 340–343 (1970).

[CO89] E. J. Coyle, J.-H. Lin, and M. Gabbouj, Optimal stack filtering and the estimation and structural approaches to image processing. *IEEE Trans. Acoust., Speech, Signal Process.* **ASSP-37**, 2037–2066 (1989).

[CR68] T. H. Crystal and L. Ehrman, The design and applications of digital filters with complex coefficients. *IEEE Trans. Audio Electroacoust.* **AU-16**, 315–320 (1968).

[CR75] R. E. Crochiere, R. R. Rabiner, and R. R. Shively, A novel implementation of digital phase shifters. *Bell Syst. Tech. J.* **54**, 1497–1502 (1975).

[CR83] R. E. Crochiere and L. R. Rabiner, *Multirate Digital Signal Processing.* Prentice-Hall, Englewood Cliffs, NJ, 1983.

[DH84] G. Dhatt and G. Touzot, *Une Présentation de la Méthode des Eléments Finis.* Maloine, Paris, 1984.

[DO79] J. J. Dongarra et al., *LINPACK User's Guide.* SIAM, Philadelphia, PA, 1979.

[FI84] J. P. Fitch, E. J. Coyle, and N. C. Gallagher, Jr., Median filtering by threshold decomposition. *IEEE Trans. Acoust., Speech, Signal Process.* **ASSP-32**, 1183–1188 (1984).

[GA81] N. C. Gallagher, Jr. and G. L. Wise, A theoretical analysis of the properties of median filters. *IEEE Trans. Acoust., Speech, Signal Process.* **ASSP-29**, 1136–1141 (1981).

[GA85] L. Gazsi, Explicit formulas for lattice wave digital filters. *IEEE Trans. Circuits Syst.* **CAS-32**, 68–88 (1985).

[GO69] B. Gold and C. M. Rader, *Digital Processing of Signals*. Prentice-Hall, Englewood Cliffs, NJ, 1969.

[GO70] B. Gold, A. V. Oppenheim, and C. M. Rader, Theory and implementation of the discrete Hilbert transformer. *Proc. Symp. Computer Process. Commun.* **19**, (1970).

[GR76] A. H. Gray, Jr. and J. D. Markel, A computer program for designing digital elliptic filters. *IEEE Trans. Acoust., Speech, Signal Process.* **ASSP-24**, 529–534 (1976).

[GR83] F. A. Graybill, *Matrices with Application in Statistics*. Wadsworth, Belmont, CA, 1983.

[HE87] P. Heinonen and Y. Neuvo, FIR-median hybrid filters. *IEEE Trans. Acoust., Speech, Signal Process.* **ASSP-35**, 832–838 (1987).

[HI59] E. Hille, *Analytic Function Theory*. Ginn, Boston, MA, 1959.

[HU79] T. S. Huang, G. J. Yang, and G. Y. Tang, A fast two-dimensional median filtering algorithm. *IEEE Trans. Acoust., Speech, Signal Process.* **ASSP-27**, 13–18 (1979).

[JA75] L. B. Jackson, On the relationship between Hilbert transformers and certain low-pass filters. *IEEE Trans. Acoust., Speech, Signal Process.* **ASSP-23**, 381–383 (1975).

[JA76] N. S. Jayant, Average- and median-based smoothing techniques for improving digital speech quality in the presence of transmission errors. *IEEE Trans. Commun.* **COM-24**, 1043–1045 (1976).

[JU81] B. I. Justusson, Median filtering: Statistical properties. *Top. Appl. Phys.* **43**, 161–196 (1981).

[KI86] C. W. Kim and R. Ansari, Approximately linear phase IIR filters using all-pass sections. *Proc. IEEE Int. Symp. Circuits Syst., San Jose, California, 1986*, pp. 661–664 (1986).

[MC73] J. H. McClellan, T. W. Parks, and L. R. Rabiner, A computer program for designing optimum FIR linear phase digital filters. *IEEE Trans. Audio Electroacoust.* **AU-21**, 506–526 (1973).

[ME84] K. Meerkötter and M. Romeike, Wave digital Hilbert transformer. *Proc. IEEE Int. Symp. Circuits Syst., Montreal, Canada, 1984*, pp. 258–260 (1984).

[NO82] T. A. Nodes and N. C. Gallagher, Jr., Median filters: Some modifications and their properties. *IEEE Trans. Acoust., Speech, Signal Process.* **ASSP-30**, 739–746 (1982).

[OP75] A. V. Oppenheim and R. W. Schafer, *Digital Signal Processing*. Prentice-Hall, Englewood Cliffs, NJ, 1975.

[PE88] S.-C. Pei and J.-J. Shyu, Design of FIR Hilbert transformers and differentiators by eigenfilter. *IEEE Trans. Circuits Syst.*, **CAS-35**, 1457–1461 (1988).

[PI90] I. Pitas and A. N. Venetsanopoulos, *Nonlinear Digital Filters*. Kluwer, Boston, MA, Academic Publishers, 1990.

[PR78] W. K. Pratt, *Digital Image Processing*. Wiley, New York, 1978.

[RA75] L. R. Rabiner, M. R. Sambur, and C. E. Schmidt, Applications of a nonlinear smoothing algorithm to speech processing. *IEEE Trans. Acoust., Speech, Signal Process.* **ASSP-23**, 552–557 (1975).

[RE86] M. Renfors and T. Saramäki, A class of approximately linear phase digital filters composed of all-pass subfilters. *Proc. IEEE Int. Symp. Circuits Syst., San Jose, California, 1986*, pp. 678–681 (1986).

[RE87a] P. A. Regalia and S. K. Mitra, Quadrature mirror Hilbert transformers. In *Digital Signal Processing-87* (V. Cappellini and A. G. Constantinides, eds.). North-Holland Publ., Amsterdam, pp. 775–782, 1987.

[RE87b] M. Renfors, private communication (1987).

[RI84] E. R. Ritenour, T. R. Nelson, and U. Raff, Applications of the median filter to digital radiographic images. *Proc. Int. Conf. Acoust., Speech, Signal Process. San Diego, California, 1984*, pp. 23.1.1–23.1.4 (1984).

[SA64] A. Savitzky and M. J. E. Golay, Smoothing and differentiation of data by simplified least squares procedures. *Anal. Chem.* **36,** 1627–1639 (1964).

[SC84] R. M. Schmitt, C. R. Meyer, P. L. Carson, and B. I. Samuels, Median and spatial low-pass filtering in ultrasonic computed tomography. *Med. Phys.* **11,** 767–771 (1984).

[SC87] H. W. Schüssler and J. Weith, On the design of recursive Hilbert transformers. *Proc. IEEE Int. Conf. Acoust., Speech, Signal Process. Dallas, Texas, 1987,* 876–879 (1987).

[ST72] J. Steiner, Y. Termonia, and J. Deltour, Comments on "Smoothing and differentiation of data by simplified least squares procedures." *Anal. Chem.* **44,** 1906–1909 (1972).

[ST73] G. W. Stewart, *Introduction to Matrix Computations.* Academic Press, New York, 1973.

[TU71] J. W. Tukey, *Exploratory Data Analysis.* Addison-Wesley, Menlo Park, CA, 1971/ 1977.

[TU74] J. W. Tukey, Nonlinear (nonsuperposable) methods for smoothing data. *Cong. Rec. EASCON,* 673 (1974).

[TY81] S. G. Tyan, Median filtering: Deterministic properties. *Top. Appl. Phys.* **43,** 197–217 (1981).

[VA87a] P. P. Vaidyanathan and T. Q. Nguyen, A "trick" for the design of FIR halfband filters. *IEEE Trans. Circuits Syst.* **CAS-34,** 297–300 (1987).

[VA87b] P. P. Vaidyanathan, P. Regalia, and S. K. Mitra, Design of doubly complementary IIR digital filters using a single complex all-pass filter, with multirate applications. *IEEE Trans. Circuits Syst.* **CAS-34,** 378–389 (1987).

[WE74] G. W. Wecksung and K. Campbell, Digital image processing at EG&G. *Computer.* **7,** 63–71 (1974).

[WE79] W. Wegener, Wave digital directional filters with reduced number of multipliers and adders. *Arch. Elektr. Üebertragungstech.* **33,** 239–243 (1979).

[WE86] P. D. Wendt, E. J. Coyle, and N. C. Gallagher, Jr., Stack filters. *IEEE Trans. Acoust., Speech, Signal Process.* **ASSP-34,** 898–911 (1986).

[YL91] O. Yli-Harja, J. Astola, and Y. Neuvo, Analysis of the properties of median and weighted median filters using threshold logic and stack filter representation. *IEEE Trans. Signal Process* **SP-39,** 395–410, (1991).

14 Multirate Signal Processing

RASHID ANSARI

Bellcore
Morristown, New Jersey

BEDE LIU

Department of Electrical Engineering
Princeton University
Princeton, New Jersey

14-1 INTRODUCTION

In digital signal processing, there is no inherent reason to use only a single sampling frequency throughout a system. Indeed, multiple sampling rates may be unavoidable for some signal processing tasks or may be introduced intentionally to reduce the computational burden. The term multirate signal processing refers to the use of multiple sampling rates in a signal processing system. A large body of DSP methodologies has been accumulated since the early 1970s to deal with situations involving multiple sampling rates [CR83]. In order to see the benefits of multirate signal processing, a brief look is taken at some applications that are illustrative of the opportunities for exploiting multirate techniques.

A common task in digital signal processing is to pass a signal through a filter in which the widths of the passband and the transition band(s) are a small fraction of the sampling frequency. If a finite-duration impulse response (FIR) filter is used for this task, the order of the filter required to meet the specifications is usually very large, entailing a heavy computational burden. On the other hand, an infinite-duration impulse response (IIR) filter can be used to save computation, but the implementation of such a filter often suffers from rather severe quantization effects. Multirate techniques can significantly alleviate these problems, thus providing an alternative approach to narrowband filtering [MI78; RA75]. There are many other situations where multirate techniques furnish the only solution to the problem or help provide a more efficient solution. This chapter provides a review of the basic concepts and the theory of multirate signal processing, outlines the design techniques, and discusses several typical applications.

Handbook for Digital Signal Processing, Edited by Sanjit K. Mitra and James F. Kaiser.
ISBN 0-471-61995-7 © 1993 John Wiley & Sons, Inc.

An example of the application of multirate signal processing in telecommunications is the implementation of an interface between two multiplexing formats, time division multiplexing (TDM) and frequency division multiplexing (FDM). Due to the continued growth of digital transmission and digital switching, TDM is being increasingly used in telecommunication systems. However, FDM is still in use for long-haul transmission. Since both formats are used in existing systems, there is often a need for translation between the formats. The interface between the two formats is commonly known as a transmultiplexer. Consider a TDM format in which the signal stream consists of several interleaved signals each sampled at 8 kHz. An example of a TDM–FDM conversion is the combining of 12 TDM voice channel signals using single-sideband modulation to create an FDM signal composed of contiguous 4-kHz frequency bands and occupying the frequency band from 60 to 108 kHz. In an all-digital implementation of the format conversion, the large bandwidth FDM signal requires a much higher sampling rate than that of the individual channel signals. A conversion of one format to the other immediately opens up opportunities for exploiting multirate techniques [PU82; SC81].

A third application is the subband coding of speech [CR76b] and image [WO86] signals. Advantage can be taken of the nonuniform distribution of energy and of the different perceptual importance in the frequency domain to achieve data compression. The signal to be coded is applied to a bank of filters and the sampling rates of the filter outputs are reduced. These output signals are referred to as subband signals and are represented with a judicious and possibly adaptive allocation of an allowed budget of bits according to the perceptual significance of the subband signals. The objective is to be able to reconstruct the signal with a minimum of degradation at the prescribed bit rate. Two key operations in subband coding are the decomposition and synthesis of the signals. These operations are performed using suitably designed filter banks, which are discussed in Section 14-5.

Multirate signal processing can also be used in beamforming applications, where the outputs of an array of sensors are suitably combined to enhance the energy of a wavefront propagating in a particular direction relative to background noise and interference from other directions. Conventional beamforming is performed in the time domain with a delay-and-sum operation on the weighted outputs of the sensor array. Digital implementation of beamforming requires that the temporal sampling rate be at least equal to the Nyquist rate. The sampling period of the signal limits the possible set of "look" directions that can be achieved exactly. The beams for which exact look directions are possible are called synchronous beams. At an available sampling rate, the signal samples may not be dense enough to provide an adequate set of synchronous beam look directions. However, the sampled sensor outputs can be processed digitally to increase its sampling rate and provide a larger set of look directions. Efficient methods of achieving a denser grid of look directions using multirate techniques have been examined [PR79; SC84] and result in a significant reduction on the hardware.

Another application of multirate signal processing is in the area of spectrum estimation. The computational requirements for narrowband spectrum estimation based on discrete Fourier transform can be reduced significantly by using sampling rate changes [CO80; LI78]. The idea is to reduce the signal sampling rate in a way that preserves the signal information content in the frequency band in which the spectrum is being estimated while limiting the aliasing that is caused by nonideal filtering before the rate reduction. Multirate techniques can also be applied to produce a significant saving in computational requirements in autoregressive (AR) spectrum estimation [QU83] while maintaining high resolution. This procedure can be used to carry out a full-band AR spectrum estimation with a significant reduction in computations while producing a resolution comparable to the direct AR spectrum estimation procedure.

14-2 SAMPLING RATE CONVERSION

The choice of sampling rates in signal processing is an important determinant of system performance with regard to computation requirement. There may be a need for adjusting the sampling rate according to the frequency content of the relevant information in the signal at various stages in an algorithm. This adjustment may be from a low sampling rate to a high rate or vice versa. The process of increasing the sampling rate of a signal is commonly referred to as *interpolation* and that of decreasing the rate is referred to as *decimation*.

The problem of sampling rate conversion can be put in the proper perspective for digital signal processing by its interpretation in the frequency domain [SC73], which is discussed in this section. First the sampling theorem is reviewed.

14-2-1 Implications of the Sampling Theorem

Suppose a discrete-time signal $x[n]$ is obtained by uniformly sampling a continuous-time signal $x_c(t)$ as follows:

$$x[n] = x_c(nT), \tag{14.1}$$

where T is the sampling period. The discrete-time Fourier transform $X(e^{j\omega})$ of $x[n]$ and the continuous-time Fourier transform $X_c(j\Omega)$ of $x_c(t)$ are defined, respectively, as

$$X(e^{j\omega}) = \sum_{n=-\infty}^{\infty} x[n]e^{-j\omega n} \tag{14.2}$$

and

$$X_c(j\Omega) = \int_{-\infty}^{\infty} x_c(t)e^{-j\Omega t} \, dt. \tag{14.3}$$

It can be shown that $X(e^{j\omega})$ and $X_c(j\Omega)$ are related by[1]

$$X(e^{j\omega}) = \frac{1}{T} \sum_{k=-\infty}^{\infty} X_c\left(\frac{\omega - 2\pi k}{T}\right). \tag{14.4}$$

If the signal $x_c(t)$ is bandlimited so that

$$|X_c(j\Omega)| = 0, \qquad |\Omega| > \Omega_0 \tag{14.5}$$

with $\Omega_0 T \le \pi$, then the sampling theorem [OP89] states that $x_c(t)$ can be reconstructed from $x[n]$ using the well-known interpolation formula:

$$x_c(t) = \sum_{k=-\infty}^{\infty} x[k] \frac{\sin \dfrac{\pi}{T}(t - kT)}{\dfrac{\pi}{T}(t - kT)}. \tag{14.6}$$

Figures 14-1(a) and 14-1(b) show a plot of typical $X_c(j\Omega)$ (assumed real) and the corresponding plot of $X(e^{j\omega})$.

14-2-2 Interpolation by an Integer Factor

A bandlimited continuous-time signal $x_c(t)$ with its Fourier transform $X_c(j\Omega)$ equal to zero outside the interval $[-\Omega_0, \Omega_0]$ is sampled with a sampling period $T_0(T_0 \le$

[1]See Section 2-9 for the derivation.

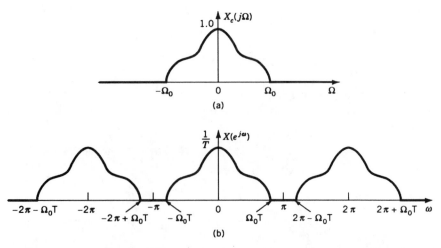

FIGURE 14-1 (a) Fourier transform of continuous-time signal $x_c(t)$. (b) Discrete-time Fourier transform of its sampled version $x[n]$.

π/Ω_0) to obtain a discrete-time signal $x[n]$. Suppose the same analog signal $x_c(t)$ is sampled at a different sampling period $T = T_0/L$, L a positive integer ≥ 2, to obtain another signal $y[n]$. An assumed waveform $x_c(t)$ is shown in Figure 14-2(a), and the two sampled signals $x[n]$ and $y[n]$ are shown in Figures 14-2(b) and 14-2(c). The problem discussed in this section is that of efficiently determining $y[n]$ from $x[n]$.

As a first step, consider the seemingly artificially constructed signal $x^u[n]$, obtained from $x[n]$ by inserting $L - 1$ zeros between existing neighboring samples, as illustrated in Figure 14-2(d) for $L = 3$. This process of inserting zeros is referred

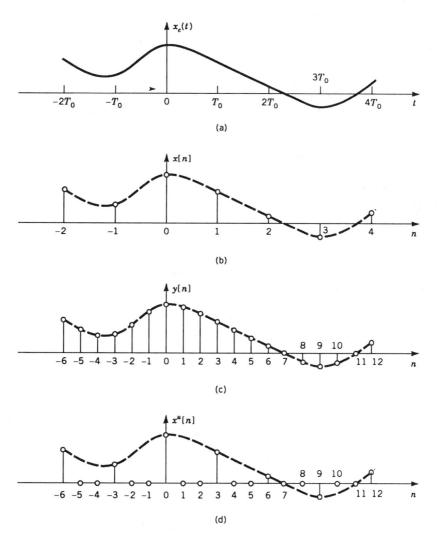

FIGURE 14-2 (a) Continuous-time signal $x_c(t)$. (b) Signal $x[n]$ sampled at low rate. (c) Signal $y[n]$ sampled at high rate. (d) Up-sampled version $x^u[n]$ ($L = 3$) of signal in (b).

to as *up-sampling* and $x^u[n]$ is called the up-sampled version of $x[n]$. The device used for inserting the zero-valued samples is called an *up-sampler*. In the frequency domain, $X(e^{j\omega})$ is given by the right side of Eq. (14.4) with T replaced by T_0. The Fourier transform $Y(e^{j\omega})$ is given by the same expression with T replaced by T_0/L. The two transforms $X^u(e^{j\omega})$ and $X(e^{j\omega})$ are related as follows:

$$X^u(e^{j\omega}) = \sum_{n=-\infty}^{\infty} x^u[n]e^{-j\omega n} = \sum_{m=-\infty}^{\infty} x[m]e^{-j\omega mL} = X(e^{j\omega L}). \quad (14.7)$$

The plot of $X_c(j\Omega)$ (assumed real) and the corresponding sketches of $X(e^{j\omega})$, $X^u(e^{j\omega})$, and $Y(e^{j\omega})$ are given in Figures 14-3(a)–(d) for $L = 3$. It is clear from Figure 14-3(c) and 14-3(d) that $y[n]$ can be obtained from $x^u[n]$ by (1) eliminating that spectral content of $X^u(e^{j\omega})$ in the frequency range $(-\pi, \pi)$ that is outside the interval $(-\pi/L, \pi/L)$, and (2) scaling the amplitude of the remaining spectrum by a factor L. In other words, the required operation is up-sampling followed by ideal lowpass filtering with a passband gain L as illustrated in Figure 14-3(e).

If the initial sampling rate $1/T_0 = \Omega_0/\pi\alpha$ is higher than the Nyquist rate, then as shown in Figure 14-3(f), $X(e^{j\omega})$ would be zero in the band $\alpha\pi \le |\omega| \le \pi$, $0 < \alpha < 1$. In such a case, $X^u(e^{j\omega})$ would be zero for $|\omega - (2k - 1)\pi/L| \le (1 - \alpha)\pi/L$ for integer values of k. The ideal filter response requirements can then be relaxed to allow for *don't care* bands. The required frequency response of the lowpass filter is $LH(e^{j\omega})$, where

$$H(e^{j\omega}) = \begin{cases} 1, & |\omega| \le \alpha\pi/L, \\ 0, & |\omega - 2k\pi/L| < \alpha\pi, \ k = 1, 2, \ldots, L - 1. \end{cases} \quad (14.8)$$

For this case, $X^u(e^{j\omega})$, $Y(e^{j\omega})$, and the required $H(e^{j\omega})$ are shown in Figures 14-3(g), 14-3(h), and 14-3(i). The shaded regions in Figure 14-3(i) represent the *don't care* bands.

The complete system for interpolation consisting of an up-sampler followed by a lowpass filter is shown in Figure 14-4. The lowpass filter suppresses images of baseband spectrum and is referred to as an anti-imaging filter.

14-2-3 Decimation by an Integer Factor

Given a signal $x[n]$ with sampling rate $1/T$, the problem considered in this section is that of constructing a signal $y[n]$ whose sampling rate, $1/T_0$, is lower than $1/T$ by an integer factor $D > 1$, that is, $T_0 = TD$. In order for $x[n]$ to be free of aliasing error, the analog signal from which $x[n]$ is obtained must be bandlimited to $-\pi/T < \Omega < \pi/T$. Similarly, in order for $y[n]$ to be free of aliasing error, it should contain only the information content in the analog signal within the band $-\pi/T_0 < \Omega < \pi/T_0$. Since π/T_0 is smaller than π/T, the energy in the spectrum of $X(e^{j\omega})$ beyond π/T_0 must be removed by filtering $x[n]$ before the sampling rate can be reduced.

Suppose $x[n]$ is first filtered. The output of the filter is denoted by $v[n]$ and its

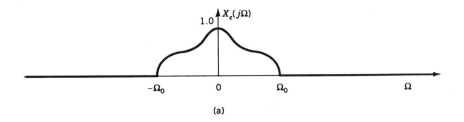

(a)

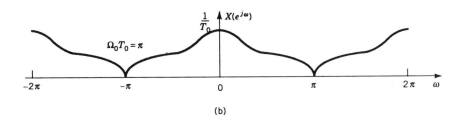

(b)

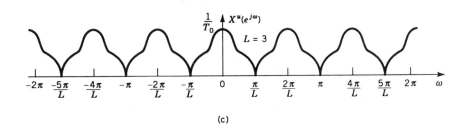

(c)

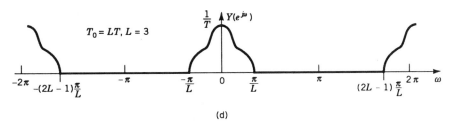

(d)

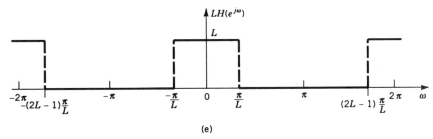

(e)

FIGURE 14-3 (a) Continuous-time Fourier transform $X_c(j\Omega)$. Discrete-time Fourier transforms with sampling rate $= \pi/\Omega_0$: (b) $X(e^{j\omega})$, (c) $X^u(e^{j\omega})$, and (d) $Y(e^{j\omega})$. (e) Ideal lowpass filter frequency response $LH(e^{j\omega})$. Discrete-time Fourier transforms with sampling rate $= \pi/\alpha\Omega_0$: (f) $X(e^{j\omega})$, (g) $X^u(e^{j\omega})$, and (h) $Y(e^{j\omega})$. (i) Ideal lowpass filter frequency response $LH(e^{j\omega})$ with don't care bands.

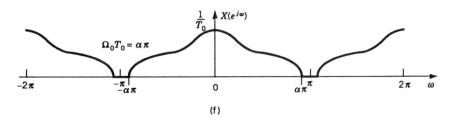

(f)

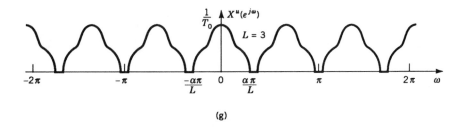

(g)

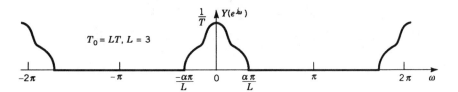

(h)

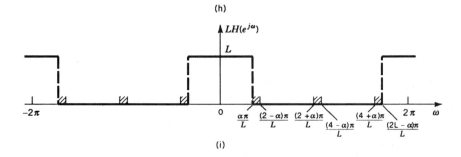

(i)

FIGURE 14-3   (*Continued*)

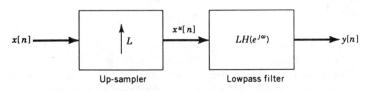

FIGURE 14-4   Block diagram of procedure for interpolation.

Fourier transform by $V(e^{j\omega})$. The signal $y[n]$ is obtained by simply removing $D - 1$ samples between neighboring samples of $v[n]$. The process is called *down-sampling* and the device employed to perform this operation is called a *down-sampler*. That is,

$$y[n] = v[Dn].\qquad(14.9)$$

It can be shown that $Y(e^{j\omega})$ is related to $V(e^{j\omega})$ via

$$Y(e^{j\omega}) = \frac{1}{D}\sum_{i=0}^{D-1} V(e^{j(\omega - 2\pi i)/D}).\qquad(14.10)$$

From Eq.(14.10), it is seen that $Y(e^{j\omega})$ is the sum of D frequency-scaled (stretched) shifted images of $V(e^{j\omega})$. Also, it is seen that any two adjacent *stretched* images of $V(e^{j\omega})$ that appear in $X(e^{j\omega})$ are separated by 2π. Any overlap of these terms results in aliasing. Thus in order to avoid aliasing, $V(e^{j\omega})$ should be zero for $\pi/D < |\omega| \le \pi$. If the Fourier transform $X(e^{j\omega})$ of the high rate signal $x[n]$ is nonzero almost everywhere in the interval $(-\pi, \pi)$ then $x[n]$ should first be processed by an ideal lowpass filter with frequency response $H(e^{j\omega})$ given by

$$H(e^{j\omega}) = \begin{cases} 1, & |\omega| < \pi/D, \\ 0, & \pi/D < |\omega| < \pi. \end{cases}\qquad(14.11)$$

The overall decimation system consisting of a lowpass filter referred to as an anti-aliasing filter, followed by a down-sampler is shown in Figure 14-5. The sketches of a typical $X(e^{j\omega})$ and the corresponding plots of $H(e^{j\omega})$, $V(e^{j\omega})$, and $Y(e^{j\omega})$ with $D = 2$ are given in Figure 14-6(a)–(d). Incidentally, if $x[n]$ is down-sampled without any prior filtering, then the Fourier transform $W(e^{j\omega})$ of the resulting signal would be that illustrated in Figure 14-6(e), which is clearly different from $Y(e^{j\omega})$ due to the effect of aliasing.

14-2-4 Computational Requirements in a Conventional Implementation

The computational requirements for sampling rate alteration depend on the type of filters used. For large conversion factors, the use of FIR filters turns out to require less computation than the use of IIR filters with *conventional* implementations of

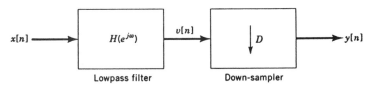

FIGURE 14-5 Block diagram of procedure for decimation.

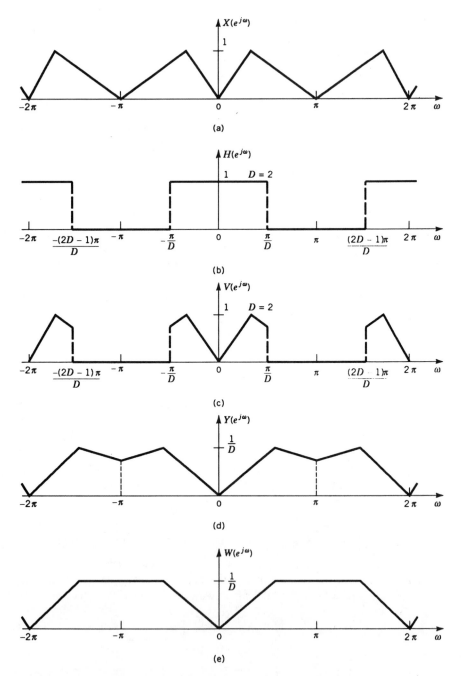

FIGURE 14-6   Fourier transforms of signals in the process of decimation. (a) $X(e^{j\omega})$, (b) $H(e^{j\omega})$, (c) $V(e^{j\omega})$, (d) $Y(e^{j\omega})$ (with filtering), and (e) $W(e^{j\omega})$ (without filtering).

the filters. The reasons are discussed in this section. However, a different approach, based on the use of a *polyphase* structure can make IIR filters computationally more efficient than FIR filters. This approach is discussed in Section 14-2-6.

For interpolation by a factor L, the system of Figure 14-4 can employ either a FIR or an IIR filter to approximate the ideal frequency response $H(e^{j\omega})$ given by Eq. (14.8). Consider first the use of a FIR filter of odd length N_F. The filter impulse response is assumed to have an even symmetry about $n = 0$ and is zero for $|n| \geq (N_F + 1)/2$. A typical impulse response $h[n]$ with $N_F = 17$ for $L = 3$ is shown in Figure 14-7,[2] where $N_F = 2KL - 1$, and filter gain is L.

The filter operates on the signal $x^u[n]$, which is obtained from $x[n]$ by upsampling. The output $y[n]$ can be expressed in terms of $x^u[n]$ as follows:

$$y[n] = \sum_{m = n-(N_F-1)/2}^{n+(N_F-1)/2} h[n - m]x^u[m]. \tag{14.12}$$

Since only one out of every L neighboring samples in $x^u[n]$ is possibly nonzero, roughly N_F/L multiplications are needed to compute each $y[n]$. This can be seen from the realization in Figure 14-8(a) where in practice only those multipliers are active for which the input samples are nonzero. The processing can be viewed as a time-varying filter in which only some of the filter coefficients are activated at one time [ME75]. For convenience a noncausal form of implementation is shown in Figure 14-8(a). This form does not take advantage of the symmetry of coefficients since inputs to multipliers with the same coefficients are not necessarily zero at the same time. However, the transpose of the structure of Figure 14-8(a), modified to share multipliers, as shown in Figure 14-8(b), can further reduce the complexity [NA78]. Note that in practice multipliers are active only when input samples are nonzero. Therefore, for a large value of L, the approximate average number

---

[2]In this figure, $h[0] = 1$ and $h[mL] = 0$, $m \neq 0$. These conditions are not necessary, but if they hold, then samples of the low rate input signal appear unaltered in the output signal.

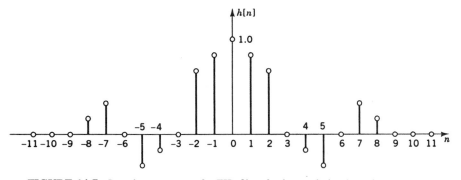

FIGURE 14-7 Impulse response of a FIR filter for interpolation by a factor $L = 3$.

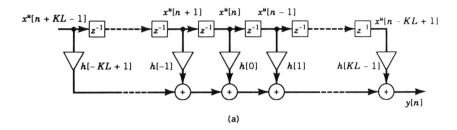

(a)

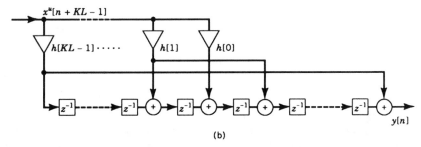

(b)

FIGURE 14-8 (a) A realization of interpolation filter. (b) Transpose of the structure in (a) for exploiting impulse response symmetry.

of *multipliers per output sample* for FIR filter, MPOS_{FIR}, is given by

$$\text{MPOS}_{\text{FIR}} \cong \frac{N_F}{2L}. \tag{14.13}$$

Consider now the use of a conventional IIR filter structure for performing the filtering. The filter transfer function is assumed to be of the form

$$H(z) = \frac{\displaystyle\sum_{k=0}^{N_I} q_k z^{-k}}{1 - \displaystyle\sum_{k=1}^{N_I} d_k z^{-k}}. \tag{14.14}$$

With $x^u[n]$ as the input, the output $y[n]$ in a direct form realization is computed by using

$$y[n] = \sum_{k=1}^{N_I} d_k y[n - k] + \sum_{k=0}^{N_I} q_k x^u[n - k]. \tag{14.15}$$

Even though the filter input $x^u[n]$ has $L - 1$ zeros among every L neighboring samples, the output $y[n]$ is not necessarily zero for any n. Therefore, from Eq. (14.15), the average number of *multiplies per output sample* for an IIR filter implementation, MPOS_{IIR}, is given by

$$\text{MPOS}_{\text{IIR}} \cong N_I + \frac{1 + N_I}{2L},\tag{14.16}$$

where it is assumed that $q_k = q_{N_I - k}$, $k = 0, 1, \ldots, N_I$. A comparison of Eq. (14.13) and (14.16) shows that if $2\ LN_I \geq N_F$, then symmetric FIR filters are preferable when *conventional* implementations are used. Symmetric FIR filters are also attractive for applications requiring a strictly linear-phase passband response.

Consider now the implementation of a *decimator* by a factor D using the scheme of Figure 14-5, where the lowpass filter is a symmetric FIR filter of length N_F. By exploiting coefficient symmetry, each output sample $v[n]$ in Figure 14-5 requires roughly $N_F/2$ multiplies on the average. However, since only one of every D samples of $v[n]$ is of interest, only those values of $v[n]$ corresponding to the final output $y[n]$ need to be computed. Therefore, the average number of *multiplies per output sample* for a FIR filter implementation is given by

$$\text{MPOS}_{\text{FIR}} = N_F/2.\tag{14.17}$$

If instead of FIR filters, an IIR filter of order N_I is used with transfer function $H(z)$ given by Eq. (14.14), then $y[n]$ can be obtained by first computing $v[n]$ from the input $x[n]$ by using

$$v[n] = \sum_{k=1}^{N_I} d_k v[n - k] + \sum_{k=0}^{N_I} q_k x[n - k].\tag{14.18}$$

In order to obtain the final (low rate) signal $y[n]$ given in terms of $v[n]$ by Eq. (14.9), $D - 1$ out of every D samples of $v[n]$ are discarded. However, all intermediate values of $v[n]$ must still be computed as they are needed in performing the recursion in Eq. (14.18). The average number of *multiplies per output sample* is therefore given by

$$\text{MPOS}_{\text{IIR}} \cong 1.5DN_I,\tag{14.19}$$

where it is assumed that $q_k = q_{N_I - k}$, $k = 0, 1, \ldots, N_I$. If $N_F \leq 3DN_I$, then FIR filters would be preferable in conventional implementations for performing decimation. This condition is true in many applications for large values of D. As a result, most of the initial work on interpolation and decimation was focused on FIR filters [CR72; SC73]. However, subsequent work shows that IIR filters with constrained structures can be computationally more efficient than FIR filters [AN83; MA79]. In order to arrive at these special structures, the problem of sampling rate alteration in the time domain is examined. This leads to a polyphase structure [BE76] that provides a unified framework for looking at IIR and FIR filters used in sampling rate conversion.

Consider an example of interpolation with $L = 3$ and a filter of length 17 ($=2KL - 1$, $K = 3$) with an impulse response as shown in Figure 14-7. The input and output signal waveforms of the corresponding interpolator are shown in Figure 14-9. Note that the input samples appear unaltered at the output. The rest of the required output samples, located between the retained input samples, are computed using $2K = 6$ adjacent input samples with a properly chosen set of coefficients of $h[n]$.

14-2-5 Sampling Rate Conversion by Rational Factors

The discussion so far has been restricted to the change of sampling rate by integer factors. It is straightforward to extend the results to the case of sampling rate alteration by a rational factor L/D, where L and D are integers. This can be achieved using the scheme of Figure 14-10(a), where *interpolation* by a factor L is performed first using a lowpass filter $LH^L(e^{j\omega})$, followed then with *decimation* by a factor D using a lowpass filter $H^D(e^{j\omega})$. The superscripts are used here to distinguish the filtering operations in interpolation and decimation. For the purpose of implementation, the lowpass filtering operations required in interpolation and decimation as shown in Figure 14-10(b) can be combined so that the overall filter

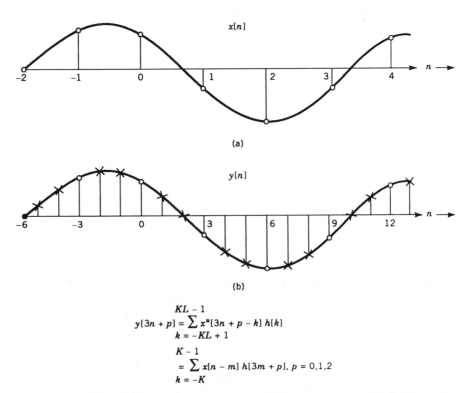

(a)

(b)

$$y[3n + p] = \sum_{k = -KL + 1}^{KL - 1} x^u[3n + p - k]\, h[k]$$

$$= \sum_{k = -K}^{K - 1} x[n - m]\, h[3m + p],\ p = 0,1,2$$

FIGURE 14-9 (a) Low-rate input signal. (b) Interpolated signal ($L = 3$).

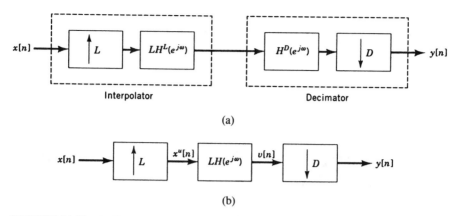

FIGURE 14-10 (a) Rate conversion by rational factor L/D using an interpolator followed by a decimator. (b) Rate conversion by combining the two filters in (a).

approximates the ideal frequency response:

$$
H(e^{j\omega}) = \begin{cases} 1, & |\omega| < \dfrac{\pi}{M}, \\[2ex] 0, & \dfrac{\pi}{M} < |\omega| \le \pi, \end{cases}
\tag{14.20}
$$

where $M = \max(L, D)$. The frequency-domain interpretation of the procedure is illustrated in Figure 14-11 for a change in sampling rate by a factor $\frac{2}{3}$. The Fourier transforms of the input signal $x[n]$, the up-sampled signal $x''[n]$, the filtered signal $v[n]$, and the output $y[n]$ are sketched in Figure 14-11(a), 14-11(b), 14-11(d), and 14-11(e), respectively. The filter response given by Eq. (14.20) with $D = 3$ and $L = 2$ is shown in Figure 14-11(c). If $D > L$, then signal information is lost in the band $(\pi L/D, \pi]$. It is to be noted that reversing the order of the operations of interpolation and decimation in Figure 14-10(a) would cause an unnecessary loss in the output of that part of the input signal information in the frequency range $(\pi/D, \pi L/M]$.

14-2-6 Polyphase Network Approach to Interpolation and Decimation

It has been shown in Section 14-2-3 and 14-2-4 that FIR filters offer a significant computational advantage for interpolation and decimation if the filters are implemented in the conventional manner. This section reexamines the problem of sampling rate conversion, but in the time domain. This leads to a general framework in which both the FIR and the IIR filters can be employed advantageously for multirate signal processing. The resulting digital filter structure is referred to as a polyphase network [BE76]. The polyphase network for the problem of interpolation by an integer factor L is first developed.

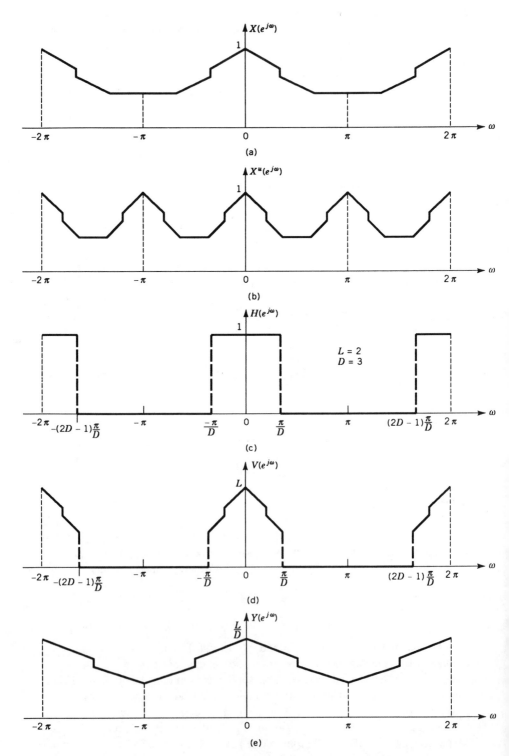

FIGURE 14-11 Fourier transforms of signals and frequency response of filter in rate conversion by a factor $\frac{2}{3}$: (a) $X(e^{j\omega})$ and (b) $X^u(e^{j\omega})$; (c) $H(e^{j\omega})$, (d) $V(e^{j\omega})$, and (e) $Y(e^{j\omega})$.

Suppose the signal $x_0[n]$ is obtained by sampling the continuous-time signal $x_c(t)$ with sampling period T_0:

$$x_0[n] = x_c(nT_0). \tag{14.21}$$

Assume that the Fourier transform $X_c(j\Omega)$ of $x_c(t)$ is zero outside the interval $[-\Omega_0, \Omega_0]$, where $\Omega_0 T_0 \leq \pi$. Let $y_c(t)$ be the signal obtained by shifting the signal $x_c(t)$ to the left by βT_0; that is,

$$y_c(t) = x_c(t + \beta T_0), \qquad 0 < \beta < 1. \tag{14.22}$$

It is desired to construct from $x_0[n]$ the signal $y[n]$, which is the sampled version of the signal $y_c(t)$:

$$y[n] = y_c(nT_0). \tag{14.23}$$

The signals $x_0[n]$ and $y[n]$ are shown in Figure 14-12(b) and 14-12(c) for an assumed waveform $x_c(t)$ shown in Figure 14-12(a). Note that

$$Y_c(j\Omega) = e^{j\Omega T_0 \beta} X_c(j\Omega). \tag{14.24}$$

From the sampling theorem

$$Y(e^{j\omega}) = \frac{1}{T} Y_c\left(\frac{j\omega}{T_0}\right), \qquad |\omega| < \pi. \tag{14.25}$$

So the Fourier transforms $X_0(e^{j\omega})$ and $Y(e^{j\omega})$ are related by

$$Y(e^{j\omega}) = e^{j\omega\beta} X_0(e^{j\omega}), \qquad |\omega| < \pi. \tag{14.26}$$

Therefore $y[n]$ can be obtained as shown in Figure 14-12(d) by applying $x_0[n]$ to a linear shift-invariant system with an ideal frequency response

$$H^d(e^{j\omega}) = e^{j\omega\beta}, \qquad |\omega| < \pi. \tag{14.27}$$

The ideal response is thus an allpass filter with a fractional group delay equal to $-\beta$.

Now visualize a factor-of-L interpolator as being composed of L such fractional delay filters whose outputs are interleaved to give the interpolated signal. This is shown in Figure 14-13 for $L = 3$. The filters ideally have a delay $-\beta_p$, where

$$\beta_p = p/L, \qquad p = 0, 1, \ldots, L - 1. \tag{14.28}$$

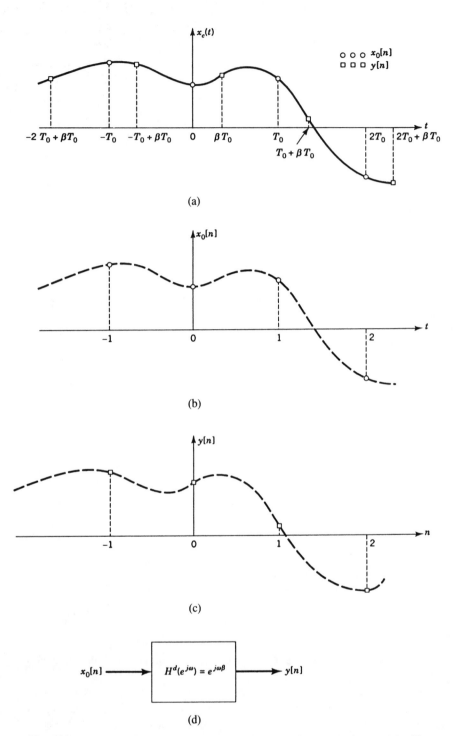

FIGURE 14-12 (a) Continuous-time signal $x_c(t)$. (b) Sampled waveform $x_0[n]$ with sampling period T_0. (c) Sampled waveform $y[n]$ offset by βT_0. (d) Ideal filter with fractional group delay equal to $-\beta$.

Denote by $H_p(z)$ the transfer function of a stable fractional delay filter whose frequency response $H_p(e^{j\omega})$ approximates the desired response $H_p^d(e^{j\omega})$ given by

$$H_p(e^{j\omega}) \cong H_p^d(e^{j\omega}) = e^{jp\omega/L}, \quad |\omega| < \pi. \tag{14.29}$$

Note that $H_0(z)$ just preserves the input samples, while, for $p = 1, 2, \ldots,$ $L - 1$, $H_p(z)$ produces the samples of $x_c(t)$ that are separated by pT_0/L units from the input samples. If $x_p[n]$ denotes the output of $H_p(z)$, then interleaving the signals $x_0[n], x_1[n], \ldots, x_{L-1}[n]$ yields the high-rate signal $x[n]$ as shown in Figure 14-13. The interleaving is performed by a counter-clockwise commutator seen on the right-hand side of Figure 14-13. This interleaving is represented very simply by:

1. up-sampling $x_p[n]$ to get $x_p^u[n]$, where

$$x_p^u[n] = \begin{cases} x_p(n/L) & \text{for } n = 0, \pm L, \pm 2L, \ldots, \\ 0 & \text{otherwise,} \end{cases} \tag{14.30}$$

2. shifting $x_p^u[n]$ to the right by p samples, and

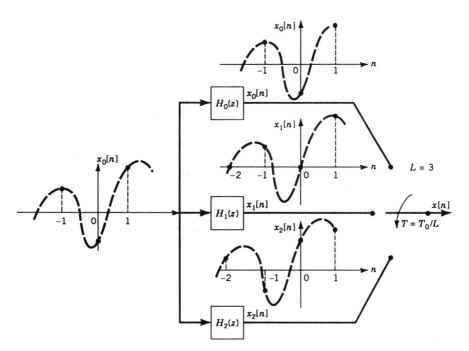

FIGURE 14-13 Input and output waveforms of filters in polyphase configuration used in interpolation by a factor $L = 3$.

3. adding the resulting L signals $x_p^u[n - p]$ to give the output signal $x[n]$:

$$x[n] = \sum_{p=0}^{L-1} x_p^u[n - p]. \qquad (14.31)$$

Also $x_p[n]$ is related to $x[n]$ as follows:

$$x_p[n] = x[nL + p]. \qquad (14.32)$$

The z-transform of $x_p^u[n]$ in Eq. (14.30) is given by

$$X_p^u(z) = X_p(z^L) = X_0(z^L)H_p(z^L) = X_0^u(z)H_p(z^L). \qquad (14.33)$$

Therefore $x_p^u[n]$ can be obtained by applying $x_0^u[n]$ to a filter with transfer function $H_p(z^L)$. Thus an alternate representation for the overall system for interpolation is shown in Figure 14-14(a) and 14-14(b). From Eq. (14.31) and (14.33) we see that

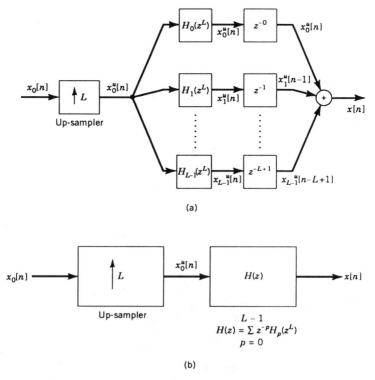

(a)

(b)

FIGURE 14-14   (a) Block diagram of polyphase network for interpolation. (b) Equivalent form as in Figure 14-4 (scaling factor included in transfer function).

$X(z)$ and $X_0^u(z)$ are related by

$$X(z) = X_0^u(z) \sum_{p=0}^{L-1} z^{-p} H_p(z^L). \tag{14.34}$$

The overall transfer function of the system shown in Figure 14-14(a) for computing $x[n]$ from $x_0^u[n]$ is given by

$$H(z) = \sum_{p=0}^{L-1} z^{-p} H_p(z^L). \tag{14.35}$$

The corresponding sample domain relationship is

$$h[n] = \sum_{p=0}^{L-1} h_p^u[n - p] \tag{14.36}$$

and

$$h_p[n] = h[nL + p]. \tag{14.37}$$

If each $H_p(z)$ has the ideal response given by $H_p^d(e^{j\omega})$ in Eq. (14.29), then the frequency response of $H(e^{j\omega})$ can be shown to be the same as that of the ideal lowpass filter given by Eq. (14.8) with a gain equal to L and $\alpha = 1$ [BE76]. Thus Figure 14-14(b) is equivalent to Figure 14-4, with the exception that the expression of $H(z)$ in Eq. (14.35) includes the scaling factor of L shown in Figure 14-4. The implementation of the lowpass filter of Figure 14-14(a) is called the *polyphase* configuration.

A different polyphase representation is obtained if the polyphase filters approximate an ideal frequency response $(H_p^d(e^{j\omega}))^*$, the conjugate of $H_p^d(e^{j\omega})$ in Eq. (14.29). Let $\overline{H}_p(z)$ be the transfer function of the filters that approximate a group delay response of $+\beta_p$ given by Eq. (14.28) and let $\overline{x}_p[n]$ be the output signal for an input $x_0[n]$. Note that $\overline{x}_0[n] = x_0[n]$. In this case the signals $\overline{x}_p[n]$ are interleaved to obtain the high rate signal $x[n]$ according to

$$\overline{x}_p[n] = x[nL - p]. \tag{14.38}$$

Also,

$$x[n] = \sum_{p=0}^{L-1} \overline{x}_p^u[n + p]. \tag{14.39}$$

Proceeding as before it follows that

$$\overline{H}(z) = \sum_{p=0}^{L-1} z^p \overline{H}_p(z^L). \tag{14.40}$$

If the filters $\overline{H}_p(z^L)$ are causal and the overall filter $H(z)$ is required to be causal, then this is ensured by adding a delay (z^{-L+1}) to the above filter to yield the following modified transfer function:

$$H(z) = \sum_{p=0}^{L-1} z^{-(L-1-p)}\overline{H}_p(z^L). \qquad (14.41)$$

The commutator sampling direction in this polyphase formulation is the reverse of the one described earlier. A bar has been used above to distinguish this representation. However, in the following the bar is dropped. It will be clear from the context which version is being used. In those cases where a distinction is needed a bar is inserted in the notation.

It should be noted that even though the overall filter order for $H(z)$ in Eq. (14.35) may be large, in practice one uses the implementation scheme of Figure 14-13, where in order to compute any one output sample, only one of the L filters $H_p(z)$ has to operate on the low-rate input signal. So the computations required per output sample are just the average of those needed for implementing the filters $H_p(z)$.

To perform decimation by a factor D, a lowpass decimation filter followed by a down-sampler is needed as shown in Figure 14-5. Now a filter with transfer function $H(z)$ in Eq. (14.35) with the structure shown in Figure 14-14(a), with L replaced by D, can be used to perform the lowpass filtering (aside from a scaling factor D). The input to the filter is available at a high rate $1/T$ whereas the output of the down-sampler is required at a low rate of $1/T_0$, $(T_0 = LT)$. Therefore, for each sample $x[n]$, only one out of every L filter output samples needs to be computed. This can be accomplished by using either a FIR or an IIR filter without computing intermediate samples. The reason for this is that each filter $H_p(z^L)$ in Figure 14-14(a) is a function of z^L, so that if any recursion is required, it can be performed using only those samples that are ultimately retained in the low-rate output signal. In practice therefore, the input signal is broken into L sequences of the form $\ldots , x[-L - p], x[-p], x[L - p], x[2L - p], \ldots$, which are applied to the filters $H_p(z)$ (not $H_p(z^L)$) operating at the low rate. This process is illustrated in Figure 14-15.

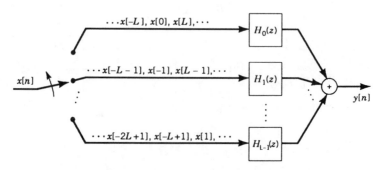

FIGURE 14-15 Processing with polyphase network for decimation.

If FIR filters are used in interpolation and decimation, the transfer function can also be expressed in the form given in Eq. (14.35). A symmetric impulse response and its polyphase decomposition for the case $L = 2$ is illustrated in Figure 14-16. The two polyphase components with impulse responses $h_0[n]$ and $h_1[n]$ are shown in Figure 14-16(b) and 14-16(c). It can be verified that

$$h[n] = h_0^u[n] + h_1^u[n-1], \tag{14.42}$$

where $h_p^u[n]$, $p = 0, 1$, are the up-sampled versions of $h_p[n]$, that is, with one zero inserted between successive samples. The transfer function is given by

$$\begin{aligned} H(z) &= H_0^u(z) + z^{-1}H_1^u(z) \\ &= H_0(z^2) + z^{-1}H_1(z^2), \end{aligned} \tag{14.43}$$

which is a special case of Eq. (14.35) with $L = 2$. The desired frequency response of $H_p(z)$, $p = 0, 1$, is given by Eq. (14.29), which is seen to be satisfied by $H_0(z)$. The frequency response of $H_1(z)$ is given by

$$H_1(e^{j\omega}) = e^{j\omega/2} \sum_{k=0}^{K-1} h_1[k] \cos{(2k + \tfrac{1}{2})\omega}, \qquad |\omega| < \pi, \tag{14.44}$$

The magnitude response is shown in Figure 14-16(d) for $K = 3$. By comparing Eq. (14.44) to Eq. (14.29), it is seen that the ideal passband phase response is achieved but the unit magnitude response is only approximated. The magnitude response of $H_1(e^{j\omega})$ in Figure 14-16(d) shows that the ideal response is well approximated for low frequencies only. One can require $H_0(z)$ and $H_1(z)$ to be allpass filters and approximate the ideal *phase* response for $p = 1$ in Eq. (14.29). An example of this is considered later.

The polyphase network proves to be useful for employing IIR filters in interpolation and decimation. In some applications, phase linearity of the response in Eq. (14.8) or Eq. (14.11) may not be important. In such a case, the transfer function $H_0(z)$ in Figure 14-13 is not constrained to be equal to unity. As a result, the input samples are not preserved in the output. Relaxing the constraint of $H_0(z) = 1$ when using IIR filters leads to a very significant reduction in computation. The polyphase transfer function in Eq. (14.35) can be alternately expressed as

$$H(z) = H_0(z^L)\left[1 + \sum_{p=1}^{L-1} z^{-p} \frac{H_p(z^L)}{H_0(z^L)}\right]. \tag{14.45}$$

Ideally, $H_0(z) = 1$, and the term in the brackets on the right-hand side has the lowpass response in Eq. (14.8) with a gain L and $\alpha = 1$. However, in practice the term in the brackets only approximates the lowpass response in Eq. (14.8). Also, the transfer function $H_0(z^L)$ is either allpass or close to allpass, and its (nonlinear) passband phase response is close to the overall passband phase response of the

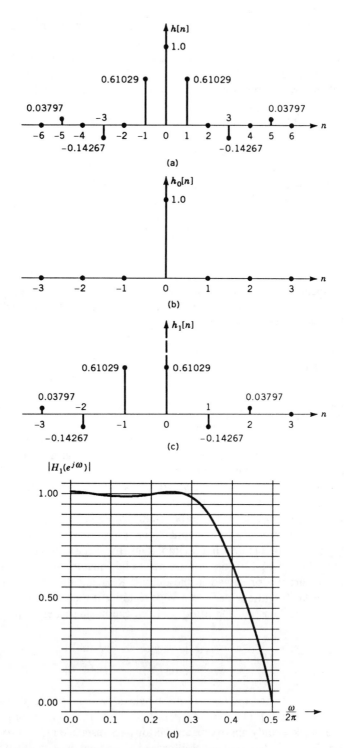

FIGURE 14-16 Polyphase decomposition of FIR filter impulse response: (a) $h[n]$, (b) $h_0[n]$, (c) $h_1[n]$, and (d) magnitude response of $H_1(e^{j\omega})$.

filter in Eq. (14.45). The frequency response of $H_p(z)/H_0(z)$ closely approximates the ideal response of the fractional delay filters $H_p^d(e^{j\omega})$ given by Eq. (14.29). As shown later in Section 14-3, the transfer function of an odd-order digital Butterworth interpolation filter can be expressed in the polyphase form of Eq. (14.35) for $L = 2$. For example, the transfer function of a fifth-order filter can be written as

$$H(z) = \frac{a_0 z^2 + 1}{z^2 + a_0} + z^{-1} \frac{a_1 z^2 + 1}{z^2 + a_1}, \qquad (14.46)$$

where

$$a_p = \tan^2 \frac{(p + 1)\pi}{10}, \qquad p = 0, 1. \qquad (14.47)$$

Equation (14.46) is precisely in the polyphase form of Eq. (14.35) with $L = 2$, where

$$H_p(z^2) = \frac{a_p + z^{-2}}{1 + a_p z^{-2}}, \qquad p = 0, 1. \qquad (14.48)$$

The magnitude response of $H(z)$ is 2.0 at $\omega = 0$ and monotonically decreases to 0.0 at $\omega = \pi$, with the 3-dB *cutoff* frequency at $\omega = \pi/2$. It approximates L times the ideal response in Eq. (14.8) with $L = 2$. The phase response $\phi(\omega)$ of $H_1(z)/H_0(z)$ approximates the ideal phase response $\omega/2$ for $|\omega| < \pi$ of $H_1^d(e^{j\omega})$ at low frequencies.

The filter with the transfer function given by Eq. (14.46) can be efficiently implemented in practice. Consider doubling the sampling rate of a signal $x_0[n]$. As shown in Figure 14-17(a), $x_0[n]$ is first up-sampled by inserting zeros to get the signal $x_0^u[n]$ at twice the rate. The signal $x_0^u[n]$ is then applied to a polyphase network for $L = 2$ with $H_p(z^2)$ given by Eq. (14.48). Consider the following method of computing $y_p^u[n]$, the output of the system with transfer function $H_p(z^2)$:

$$y_p^u[n] = a_p(x_0^u[n] - y_p^u[n - 2]) + x_0^u[n - 2], \qquad p = 0, 1. \quad (14.49)$$

Since the signal $x_0^u[n]$ is zero for n odd, $y_p^u[n]$ is zero for n odd. Therefore, in practice, only every other sample of $y_p[n]$ needs to be computed. The interpolated signal $x[n]$ is given by

$$x[n] = y_0^u[n] + y_1^u[n - 1]$$

$$= \begin{cases} y_0^u[n] & \text{for } n \text{ even,} \\ y_1^u[n - 1] & \text{for } n \text{ odd.} \end{cases} \qquad (14.50)$$

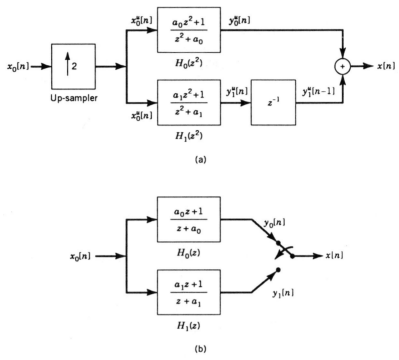

FIGURE 14-17 (a) Interpolation using IIR polyphase filter. (b) Efficient processing procedure used in practice.

Let $y_p[n]$, $p = 0, 1$, denote the down-sampled versions of $y_p^u[n]$:

$$y_p[n] = y_p^u[2n]. \tag{14.51}$$

Then, in practice, just $y_0[n]$ and $y_1[n]$ need to be computed directly from $x_0[n]$ at the low sampling rate as shown in Figure 14-17(b). The final output is obtained by interleaving the signals $y_0[n]$ and $y_1[n]$. One multiplication and two additions for computing each output sample in the direct form given by Eq. (14.49) are needed. Equation (14.49) describes just one possible realization, and other realizations may similarly be used.

The same transfer function can be used for an efficient implementation of decimation by a factor of $D = 2$, as shown in Figure 14-18(a). The signal $x[n]$ is first lowpass filtered and then the filtered signal $v[n]$ is down-sampled to produce the low-rate signal $y[n]$. The signal $v[n]$ is computed by adding $v_0'[n]$ and $v_1'[n - 1]$, where $v_p'[n]$ is the output of the system $H_p(z^2)$, $p = 0, 1$, with input $x[n]$. Since only every other sample of $v[n]$ is retained as $y[n]$, giving

$$\begin{aligned} y[n] &= v[2n] \\ &= v_0'[2n] + v_1'[2n - 1], \end{aligned} \tag{14.52}$$

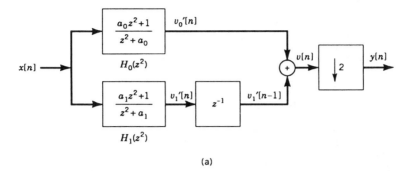

(a)

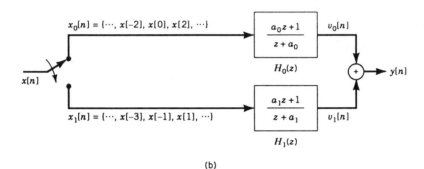

(b)

FIGURE 14-18 (a) Decimation using IIR polyphase filter. (b) Efficient processing procedure used in practice.

where

$$v_0'[2n] = a_0(x[2n] - v_0'[2n - 2]) + x[2n - 2] \qquad (14.53)$$

and

$$v_1'[2n - 1] = a_1(x[2n - 1] - v_1'[2n - 3]) + x[2n - 3]. \qquad (14.54)$$

It is clear from Eq. (14.52) through (14.54) that in order to compute $y[n]$, only $v_0'[n]$ for even values of n and $v_1'[n]$ for odd values of n are needed. Define, for $p = 0, 1,$

$$v_p[n] = v_p'[2n - p] \qquad (14.55)$$

and

$$x_p[n] = x[2n - p]. \qquad (14.56)$$

Then from Eq. (14.52) and (14.55),

$$y[n] = v_0[n] + v_1[n] \qquad (14.57)$$

and from Eq. (14.53) through (14.56),

$$v_p[n] = a_p(x_p[n] - v_p[n-1]) + x_p[n-1]. \qquad (14.58)$$

The actual processing is shown in Figure 14-18(b), where $x[n]$ is split into low-rate signals $x_0[n]$ and $x_1[n]$ according to Eq. (14.56). The low-rate outputs $v_0[n]$ and $v_1[n]$ are computed according to Eq. (14.58) and added to give $y[n]$ as in Eq. (14.57). Therefore, to compute each output sample, two multiplications and five additions are needed.

The procedures for $L = 2$ and $D = 2$ can be extended in a straightforward manner to arbitrary integer factors for interpolation and decimation.

14-2-7 Multistage Decimation and Interpolation

The basic procedure for interpolation by a factor L is shown in Figure 14-4. If L is factorable into a product of integers, then it is possible to increase the sampling rate in several stages. Similarly, if D is a product of integers, decimation by D can be accomplished in more than one stage. It turns out that such a multistage implementation can provide a very significant gain in computational efficiency [BE74; CR75; SH75].

First consider the multistage implementation of decimation by a ratio D with

$$D = \prod_{i=1}^{J} D_i, \qquad (14.59)$$

where D_i are integers.

Figure 14-19(a) shows the single-stage implementation of decimation using the scheme of Figure 14-5. In Figure 14-19(a), $H(z)$ is the transfer function of the lowpass filter and $h[n]$ is its impulse response. In Figure 14-19(b), an implementation is shown in J stages, where the decimation is carried out by a factor D_1 in the first stage, D_2 in the second stage, and so on. The lowpass filter transfer function for the ith stage, $i = 1, 2, \ldots, J$ is denoted by $H_i(z)$ and the corresponding impulse response by $h_i[n]$. It can be shown straightforwardly that Figure 14-19(b) is identical to Figure 14-19(c). The equivalent filter transfer function $H(z)$ can be expressed as

$$H(z) = H_1(z) H_2(z^{D_1}) \cdots H_J(z^{D_1 D_2 \cdots D_{J-1}}). \qquad (14.60)$$

The multistage implementation of interpolation by a factor L with

$$L = \prod_{i=1}^{J} L_i, \qquad (14.61)$$

where L_i are integers, can similarly be reduced to a single-stage form.

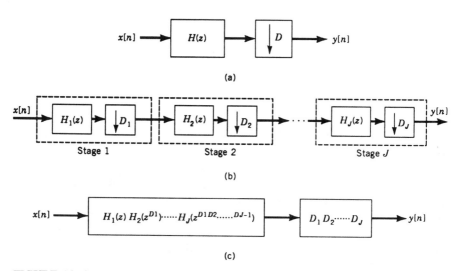

FIGURE 14-19 (a) Single-stage decimation by a factor $D = \Pi_{i=1}^{J} D_i$. (b) J-stage implementation of decimation. (c) Single-stage decimation equivalent to (b).

Figure 14-20(a) shows a single-stage implementation of interpolation consisting of an up-sampler followed by a lowpass filter with transfer function $H(z)$ where scaling by a factor L is assumed. Figure 14-20(b) shows a J-stage implementation in which interpolation is carried out by a factor L_J in the first stage, L_{J-1} in the second stage, and so on. The transfer function of the filter for the ith stage, $i = 1, 2, \ldots, J$, is denoted by $H_{J+1-i}(z)$. As in the case of decimation, it can be shown that the equivalent transfer function $H(z)$ is given by

$$H(z) = H_1(z)H_2(z^{L_1})H_3(z^{L_1 L_2}) \cdots H_J(z^{L_1 L_2 \cdots L_{J-1}}). \qquad (14.62)$$

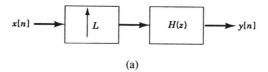

(a)

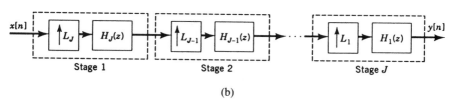

(b)

FIGURE 14-20 (a) Single-stage interpolation by a factor $L = \Pi_{i=1}^{J} L_i$. (b) J-stage implementation of interpolation.

The expression for $H(z)$ in Eq. (14.62) is similar to that given in Eq. (14.60). It should be noted that the subscripts of the filters in the J stages in Figures 14-19(b) and 14-20(b) appear in reverse order.

14-2-8 Bandpass Interpolation and Decimation

The subject of interpolating a signal $x[n]$ by a factor L to obtain a signal $y[n]$ has been discussed in Section 14-2-2. As shown in Figures 14-3(b) and 14-3(d), the information content of $X(e^{j\omega})$ in the interval $(-\pi, \pi)$ has been compressed to the band $(-\pi/L, \pi/L)$ to yield the output Fourier transform $Y(e^{j\omega})$. In the case of decimation, as illustrated in Figures 14-6(a) and 14-6(d), the input Fourier transform $X(e^{j\omega})$ in the interval $(-\pi/D, \pi/D)$ has been retained and stretched by frequency scaling to cover the period $(-\pi, \pi)$ of the output Fourier transform $Y(e^{j\omega})$. The operation yields the minimum sampling rate for the frequency components of interest in the band $(-\pi/D, \pi/D)$ of the input signal.

In both interpolation and decimation, the full frequency content of the low-rate signal (output signal in decimation and input signal in interpolation) corresponds to the *low*-frequency components in the corresponding high-rate signal. The operation can be generalized so that the frequency content of the low-rate signal corresponds to any arbitrary frequency band in the high-rate signal. The resulting operations are referred to as bandpass interpolation and decimation. A sketch of the Fourier transform $X(e^{j\omega})$ of a low-rate signal $x[n]$ and the Fourier transform $Y(e^{j\omega})$ of the corresponding bandpass interpolated signal $y[n]$ is shown in Figures 14-21(a) and 14-21(b), respectively, for $L = 3$. The lower edge of the designated band is ω_0, $0 \le \omega_0 \le (L - 1)\pi/L$, and the upper band edge is at $\omega_0 + \pi/L$. The

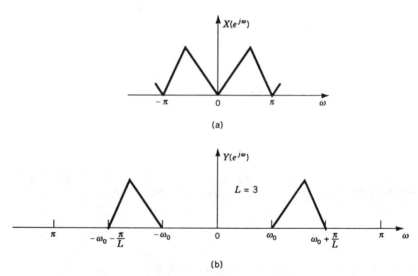

FIGURE 14-21   (a) Fourier transform $X(e^{j\omega})$ of low-rate signal $x[n]$. (b) Fourier transform $Y(e^{j\omega})$ of bandpass interpolated signal $y[n]$.

bandpass interpolation of a signal can be thought of as a process of inserting redundancy in sampling so that ideally all the input signal information appears in a designated frequency band of the interpolated signal. On the other hand, bandpass decimation consists of generating a signal at a minimum sampling rate in such a way that the input signal information in a designated band is preserved.

Performing bandpass interpolation by an integer factor L turns out to be particularly straightforward in the special case in which the designated frequency band containing (ideally) all the signal energy is of the form $(k\pi/L, (k + 1)\pi/L)$, for $k = 0, 1, \ldots, L - 1$. The bandpass interpolation procedure is shown in Figure 14-22. The procedure is a modification of the scheme of Figure 14-4 where the lowpass filter is replaced by a bandpass filter $LH_k(z)$ with an ideal frequency response:

$$H_k(e^{j\omega}) = \begin{cases} 1, & k\pi/L < |\omega| < (k + 1)\pi/L, \\ 0, & |\omega| < k\pi/L \text{ or } (k + 1)\pi/L < |\omega| < \pi. \end{cases}$$

$$(14.63)$$

The subscript k in H_k refers to the kth integer band. For odd values of k, the sign of every other input signal sample is reversed to ensure that the output signal contains information corresponding to the upper sideband. When $k = 0$ the procedure reduces to the familiar case of lowpass interpolation shown in Figure 14-4. Increase in sampling rate for the case in which the designated band is of the form $(k\pi/L, (k + 1)\pi/L)$ for $k = 0, 1, \ldots, L - 1$ is referred to as integer-band interpolation. The effect of the processing as interpreted in the frequency domain is illustrated in Figure 14-23 for an interpolation factor $L = 5$ and the designated band is specified by $k = 2$. The Fourier transforms $X(e^{j\omega})$, $X^u(e^{j\omega})$, and $Y(e^{j\omega})$, of the input signal $x[n]$, the up-sampled $x^u[n]$, and the bandpass filtered (interpolated) signal $y[n]$, respectively, are shown in Figures 14-23(a), 14-23(b), and 14-23(d). The frequency response $H_k(e^{j\omega})$ is sketched in Figure 14-23(c).

The counterpart of integer-band interpolation is referred to as integer-band decimation in which the sampling rate is reduced by an integer factor D. Here the sampling rate is reduced by preserving the information in the high-rate signal in a designated frequency band using a bandpass filter with passband $(k\pi/D, (k + 1)\pi/D)$ for $k = 0, 1, \ldots, D - 1$.

As has been seen in this section, bandpass interpolation and decimation can be performed in a straightforward manner if the frequency band of interest is of the

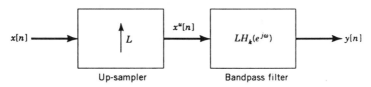

FIGURE 14-22 Block diagram of integer-band bandpass interpolation.

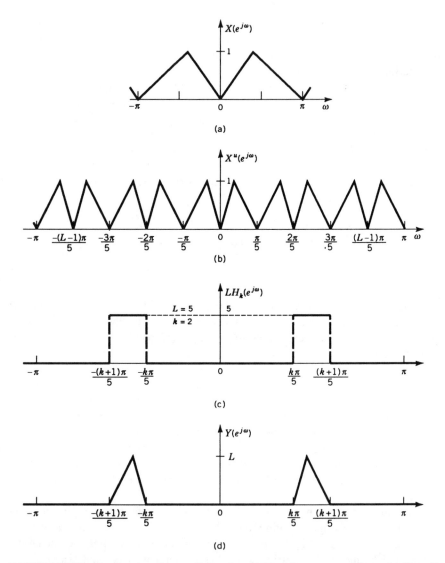

FIGURE 14-23 Fourier transforms of signals and filter frequency response in $(k = 2)$ integer-band interpolation for factor $L = 5$: (a) $X(e^{j\omega})$, (b) $X^u(e^{j\omega})$, (c) $H_k(e^{j\omega})$, and (d) $Y(e^{j\omega})$.

form $(k\pi/M, (k + 1)\pi/M)$, where M is a positive integer. Now examine the general case by considering bandpass interpolation by an integer factor L. Specifically, a procedure is developed in which the low-rate input signal $x[n]$ and the high-rate signal $y[n]$ are related in the frequency domain as shown by the sketches of the respective Fourier transforms $X(e^{j\omega})$ and $Y(e^{j\omega})$ in Figure 14-21. In order to obtain the signal $y[n]$, the Hilbert transform $\hat{x}[n]$ is first generated by applying the signal $x[n]$ to a system with frequency response $-j\,\text{Psgn}(e^{j\omega})$, where $\text{Psgn}(e^{j\omega})$

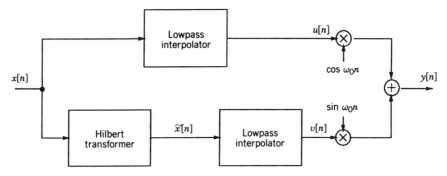

FIGURE 14-24   Bandpass interpolation using a Hilbert transformer.

is the periodic signum function given by

$$
\mathrm{Psgn}(e^{j\omega}) = \begin{cases} 1, & 0 < \omega < \pi, \\ 0, & \omega = 0, \pi, \\ -1, & -\pi < \omega < 0. \end{cases} \tag{14.64}
$$

The function is periodic in ω with period 2π. The signal $x[n]$ and $\hat{x}[n]$ are lowpass interpolated by a factor L to get the high-rate signals $u[n]$ and $v[n]$. It is straightforward to verify that

$$
y[n] = u[n] \cos(\omega_0 n) - v[n] \sin(\omega_0 n) \tag{14.65}
$$

is the required bandpass interpolated signal. The bandpass interpolation scheme is shown in Figure 14-24.

Decimation to preserve signal components in the band $(\omega_0, \omega_0 + \pi/D)$ can also be performed using the Hilbert transform operation together with bandpass filtering with an ideal passband of $(\omega_0, \omega_0 + \pi/D)$, followed by modulation and down-sampling. The actual processing is done at the low rate.

14-3   FILTER DESIGN FOR SAMPLING RATE ALTERATION

The preceding sections described the underlying principles of sampling rate conversion by integer and rational factors. The basic operations involved are up-sampling, lowpass filtering, and down-sampling. The filters are designed so that the aliasing and distortion in the frequency band of interest are within specified tolerances. A multirate system designer can exercise some control over the filtering operation and in some cases over the rate conversion factors. By tailoring the filter design and implementation techniques for multirate applications one can significantly reduce the computational complexity when compared with conventional techniques. This section discusses procedures that are especially suited for digital filter design to be used in sampling rate conversion. Though the procedures of

Chapters 4 and 5 can be used directly for this task, the design solution may not lead to a computationally efficient implementation, especially in the case of IIR filters. It turns out that imposition of certain constraints on the form of the transfer function of an IIR filter can provide significant advantages in implementation.

The design depends on several factors. Perhaps the most important one is the optimization criterion (frequency domain, time domain, or a combination; mean-square error, minimax error, etc.) The design can also depend on the nature of the signal, as would be the case if the signal spectrum is taken into account in the optimization. Another consideration is the possible inclusion of constraints on the filter transfer function. Finally, the solution depends on whether the rate conversion is carried out using a multistage implementation described in Section 14-2-7, in which case one needs to examine the different possibilities in implementing the multiple stages of rate conversion. This section summarizes several design procedures for FIR and IIR filters for a single-stage operation. Multistage implementation is discussed in Section 14-4.

14-3-1   Frequency-Domain Design Considerations

Some commonly given specifications in the frequency domain for interpolation and decimation are now briefly reviewed. The ideal frequency response $H(e^{j\omega})$ of the lowpass filter required in interpolation by a factor L is given by Eq. (14.8) and for decimation by a factor D is given by Eq. (14.11). Consider the specifications for the ideal filter $H(e^{j\omega})$ for sampling rate conversion (interpolation or decimation) by a factor M,

$$H(e^{j\omega}) = \begin{cases} 1, & |\omega| < \pi/M, \\ 0, & \pi/M \leq |\omega| \leq \pi, \end{cases} \tag{14.66}$$

where the scaling factor M needed for interpolation has been ignored. Since this ideal frequency response cannot be met exactly in practice, the usual allowances are made in the form of *don't care* bands and tolerances on passband and stopband ripple. Depending on the application, three types of specifications used in practice are sketched in Figure 14-25 for $M = 5$. The impulse response of the system is assumed to be real so that the allowed magnitude response is shown only for $[0, \pi]$. The first type of specification, shown in Figure 14-25(a), is used in interpolation when the Fourier transform of the signal to be interpolated is nonzero in the interval $[-\pi, \pi]$. This filter response ensures that the high-rate signal has all replicas of the baseband spectrum adequately suppressed in the frequency band $[\pi/M, \pi]$. In the case of decimation, this filter response ensures that the aliasing is negligible in the low-rate signal.

The response shown in Figure 14-25(b) is suitable for interpolation when the Fourier transform of the low-rate signal is zero or negligible in the range $[\alpha\pi, \pi]$ with $0 < \alpha < 1$. This case is discussed in Section 14-2-2 and illustrated in Figure 14-3(f)–(h). If this filter response is used in decimation it may cause significant aliasing to occur in the frequency band $[\alpha\pi, \pi]$ of the decimated signal. However,

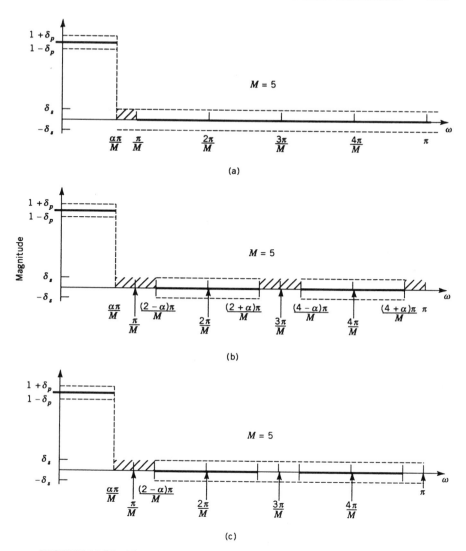

FIGURE 14-25 Three types of specifications for rate conversion by factor M.

this can be acceptable in some applications such as spectrum estimation in the band $[0, \alpha\pi]$. The specification of Figure 14-25(b) is often used in designing filters in a multistage implementation of decimation or interpolation. Finally, the response shown in Figure 14-25(c) is sometimes used in decimation when significant aliasing in the band $[\alpha\pi, \pi]$ is allowed but only from the interval $[\pi/M, (2 - \alpha)\pi/M]$ in the high-rate signal. The aliasing in the band $[\alpha\pi, \pi]$ in the case of the response shown in Figure 14-25(b) can be excessive when M is very large, posing problems in some applications even when some aliasing is accepted. In such a case the specifications of Figure 14-25(c), allowing aliasing only from the adjacent band, may be used.

The choice of the maximum allowable passband and stopband ripples in Figure 14-25 is dictated by the degree of distortion that is acceptable in a particular application. For example, in some communication applications of decimation one may have prescribed bounds on the permissible *cross-talk* caused by aliasing in the decimated signal. These bounds set the minimum attenuation requirement in the stopband. The next three sections discuss filter design procedures for FIR and IIR filters to meet given frequency-domain specifications for a single-stage implementation of decimation and interpolation filters.

14-3-2 Design of FIR Filters for Sampling Rate Conversion

Several FIR filter design techniques are briefly reviewed next.

14-3-2-1 Windowing Method. The ideal lowpass filter with the frequency response given by Eq. (14.66) has an impulse response

$$
h[n] = \begin{cases} \dfrac{1}{M} & n = 0, \\[2ex] \dfrac{\sin{(n\pi/M)}}{n\pi} & \text{for } n \neq 0. \end{cases} \tag{14.67}
$$

A simple approach to the design of a FIR filter of duration N is to window[3] the ideal impulse response in Eq. (14.67). This method is suitable for the specifications given in Figures 14-25(a) and 14-25(c). In this approach the samples of the ideal filter impulse response are multiplied by a suitable window function [OP89]. A variety of window functions that provide a reduced ripple at the cost of a wider transition band have been proposed. The windowing procedure is simple, but the resulting filter is not optimal according to the minimax criteria.

14-3-2-2 Chebyshev Approximation: Parks–McClellan Design Procedure. The Parks–McClellan technique[4] [MC73] is a numerical procedure that can be used to obtain linear-phase FIR filters for any of the three specifications in Figure 14-25. The frequency response of such a filter is optimum in the minimax or Chebyshev sense over the set of specified frequency bands shown in Figure 14-25. The filter can readily be designed using a widely available program, due to McClellan, Parks, and Rabiner, to be referred to as the MPR program. In this design program one specifies the following:

1. The number of frequency bands. Note that the number of bands is equal to 2, $\lfloor M/2 \rfloor + 1$, and 2 for the specifications in Figures 14-25(a), 14-25(b), and 14-25(c), respectively.[5]

---

[3]See Section 4-4 for a review of the window method of filter design and the commonly used window functions.
[4]See Section 4-8 for a review of the Parks–McClellan method of FIR filter design.
[5]$\lfloor x \rfloor$ denotes the integer part of x.

2. The edges of the specified frequency bands given in specification 1 as a fraction of the (high) sampling rate.

3. The desired magnitude in the care bands (1.0 in the passband and 0.0 in the stopbands).

4. The weight for frequency response deviation from the desired value for each band, for example, δ_p and δ_s in Figure 14-25

A good estimate of the filter length N for meeting a given set of specifications can be found in Chapter 4.

Note that the ideal impulse response given by Eq. (14.67) satisfies

$$h[n] = 0 \quad \text{for } n = \pm M, \pm 2M, \ldots . \tag{14.68}$$

If such a constraint is imposed on the FIR filter, then the resulting filters are referred to as the Mth band filter [MI82]. Under this constraint, the MPR program cannot readily be used. Also, the constraint implies certain restrictions on the frequency response, which makes the design suitable only for the specifications in Figure 14-25(b). A simple suboptimal approach to the Mth band filter design using the MPR program is described in Mintzer [MI82]. Another approach, to be reviewed later, is to decompose the design problem using the polyphase representation, and then use a modified MPR program to design the filters.

The MPR program cannot be used for designing FIR filters with constraints imposed on the transfer function and specialized techniques are needed. One such technique is applied in the design of linear-phase FIR filters with the transfer function constrained to be of the form $H(z)A(z^M)$ [SA84], where $A(z^M)$ is used to adjust the behavior in the passband while $H(z)$ acts to provide the stopband attenuation. The filters can be properly chosen using an iterative design [SA84].

14-3-2-3 Linear Programming Based METEOR Program. A numerical procedure based on linear programming that offers greater flexibility in design compared with the Parks–McClellan technique is described in Steiglitz and Parks [ST86]. A program based on the simplex method, called METEOR, is available to carry out the filter design. In addition to minimax approximation, this program allows the use of a wide variety of constraints such as the convexity of the response. For multirate applications, the focus is on linear-phase filters with even symmetric impulse response with odd duration of impulse response. The design procedure finds the filter of least order with a frequency response that remains within the prescribed limits of Figure 14-25 and maximizes the distance of this response from the prescribed constraints. In some cases one may desire that the response "hug" (have a zero distance from) the constraint. The design algorithm uses second derivatives in the set of constraints to allow convexity requirements to be incorporated in the design. The problem is formulated to choose the coefficients of the basis functions subject to inequality constraints.

14-3-2-4 Minimum Mean-Squared Error Design for Interpolation. The FIR filter design for interpolation by integer factors can be carried out by posing it as a problem of minimizing the mean-squared error between the output signal and the desired signal [OE75; OE79]. The problem can be solved either in the time domain or in the frequency domain, and the solution depends on the type of prior signal information that is assumed to be available. Consider here the problem of interpolation of a low-rate real-valued signal $x_0[n]$ by an integer factor M. The focus is on the case where $x[n]$ is a wide-sense stationary random process with autocorrelation R_x:

$$R_x[k] = E\{x[n]\, x[n - k]\}. \tag{14.69}$$

It is assumed that the following $2K$ observations of process x are available:

$$x_0[n] = x[nM] \quad \text{for } -K + 1 \leq n \leq K. \tag{14.70}$$

It is desired to obtain the linear minimum mean-squared error estimate $y[p]$ of $x[p], p = 1, 2, \ldots, K - 1$, where

$$y[p] = \sum_{n = -K + 1}^{K} g_p[n]\, x[nM]. \tag{14.71}$$

The object is to determine $g_p[n]$ such that the mean-squared error in the estimate is minimized. The application of the orthogonality principle (the error $e[p] = y[p] - x[p]$ is orthogonal to observations) results in

$$E\left\{ x[mM] \left(\sum_{n = -K + 1}^{K} g_p[n] x[nM] - x[p] \right) \right\} = 0, \quad -K + 1 \leq m \leq K. \tag{14.72}$$

Therefore

$$\sum_{n = -K + 1}^{K} g_p[n]\, R_x[(m - n)M] = R_x[mM - p]. \tag{14.73}$$

The processing with the weighting coefficients $\{g_p[n], p = 1, 2, \ldots, M - 1\}$ can be represented as filtering using polyphase networks. One can also use the minimum mean-squared error design for the case of deterministic signals [OE75; OE79]. The design can also be carried out using a Chebyshev criterion for the frequency-domain deviation of the polyphase filters [OE79]. It is observed that the overall frequency response is close to the value M at K distinct points. For the optimum choice of filters, it is also observed that the errors for different values of p are almost exactly proportional to each other over the frequency range $0 \leq \omega \leq \alpha\pi$. These observations are used in designing interpolation filters with a response that is close to equiripple and the design solution is suitable for the specifications in Figure 14-25(b). One can begin by designing the filter for $M = 2$ to get what is referred to as the FIR half-band solution. Methods for half-band filter design are described later. The zeros of this filter transfer function are determined and are then used to get the remaining polyphase components $h_p[n]$. An alternate procedure that avoids the root finding is described next.

14-3-2-5 Design of Fractional Delay Filters. Fractional delay filters are required in a polyphase network to get a lowpass response. These filters are also useful in other applications, as, for example, in implementing delays in beamforming for a given look direction. In this section an efficient design procedure is examined for approximating any fractional delay response $e^{-j\beta\omega}$, $0 \le \beta < 1$. This method produces results similar to Oetken [OE79] for the special case when $\beta = p/M$.

The procedure [PY87] is based on modifying the Parks–McClellan filter design program, which allows the design of close to a minimax solution of the fractional delay filters. Note that for $p = 0, 1, 2, \ldots, M - 1$, $H_p(e^{j\omega})$ approximates the ideal frequency response $H_p^d(e^{j\omega})$ given by

$$H_p^d(e^{j\omega}) = e^{-jp\omega/M}, \qquad |\omega| \le \pi. \tag{14.74}$$

The frequency response of the ideal filters $H_p^d(e^{j\omega})$ and $H_{M-p}^d(e^{j\omega})$ for $p = 1, 2, \ldots, M - 1$ are related by

$$H_p^d(e^{j\omega}) = e^{-j\omega} H_{M-p}^d(e^{-j\omega}). \tag{14.75}$$

Using the above relations, the expressions for the desired Fourier transforms of the sum $F_1^d(e^{j\omega})$ and the difference $F_2^d(e^{j\omega})$ of the filters $H_p^d(e^{j\omega})$ and $H_{M-p}^d(e^{j\omega})$ are

$$F_1^d(e^{j\omega}) = \tfrac{1}{2}(H_p^d(e^{j\omega}) + H_{M-p}^d(e^{j\omega}))$$

$$= e^{-j\omega/2} \cos\left(\left(\frac{1}{2} - \frac{p}{M}\right)\omega\right) \tag{14.76}$$

and

$$F_2^d(e^{j\omega}) = \tfrac{1}{2}(H_p^d(e^{j\omega}) - H_{M-p}^d(e^{j\omega}))$$

$$= je^{-j\omega/2} \sin\left(\left(\frac{1}{2} - \frac{p}{M}\right)\omega\right). \tag{14.77}$$

These two ideal functions $F_1^d(e^{j\omega})$ and $F_2^d(e^{j\omega})$ can be approximated in the minimax sense by using the MPR program, where they are labeled Case 2 and Case 4 [MC73]. Assume the filter duration to be even $(2K)$ and consider the approximation of $F_k^d(e^{j\omega})$ by $j^{k-1}e^{-j\omega/2}G_k(\omega)$, $k = 1, 2$, where

$$G_1(\omega) = \sum_{n=1}^{K} g_1[n] \cos\left(\omega(n - \tfrac{1}{2})\right)$$

$$= \cos(\omega/2) \sum_{n=0}^{K} g_1'[n] \cos(\omega n) \tag{14.78}$$

and

$$G_2(\omega) = \sum_{n=1}^{K} g_2[n] \sin \left(\omega \left(n - \tfrac{1}{2}\right)\right)$$

$$= \sin \left(\omega/2\right) \sum_{n=0}^{K} g_2'[n] \cos \left(\omega n\right) \tag{14.79}$$

where the relation between $g_k[n]$ and $g_k'[n]$ for $k = 1, 2$ is obtained by simple trigonometric manipulation. This approximation is performed over the interval $[0, \alpha \pi]$, $0 < \alpha < 1.0$. Once $g_k'[n]$ is obtained for approximating $F_1^d(e^{j\omega})$ and $F_2^d(e^{j\omega})$ in the Chebyshev sense by using the modified MPR program, the equiripple filter results can be combined in order to obtain an approximation to $H_p^d(e^{j\omega})$ and $H_{M-p}^d(e^{j\omega})$. The combination provides an approximation that is very close to equiripple. The response is nearly equiripple due to the fact that in practice for an even filter duration $(2K)$ the approximation error in $F_2^d(e^{j\omega})$ is almost two orders of magnitude less than that in $F_1^d(e^{j\omega})$. As a result, the magnitude of the complex deviation $(H_p(e^{j\omega}) - e^{-jp\omega/M})$ is dominated by the approximation error arising from $F_1^d(e^{j\omega})$, which is strictly equiripple. The absolute value of the complex deviation is shown in Figure 14-26 for $K = 5$, $\alpha = 0.8$ with $p = 1$ and $M = 3, 7, 10$. This behavior agrees with the result described in Oetken [OE79], which is achieved by a root finding procedure. The design in Pyfer and Ansari [PY87] is carried out by modifying the MPR program to define two related target functions in the program function EFF. This essentially requires two runs of the MPR program for each fractional delay filter design. When the different polyphase filters are combined, an approximation to the specification in Figure 14-25(b) is obtained.

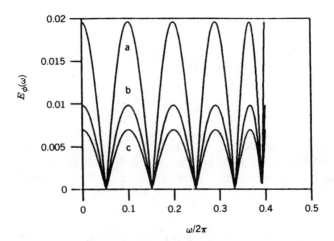

FIGURE 14-26 Approximation error in FIR fractional delay filter $H_1(e^{j\omega})$ for $K = 5$, $\alpha = 0.8$, and $M = 3, 7, 10$ labeled a, b, c, respectively. (Adapted from Pyfer and Ansari [PY87] with permission from IEEE.)

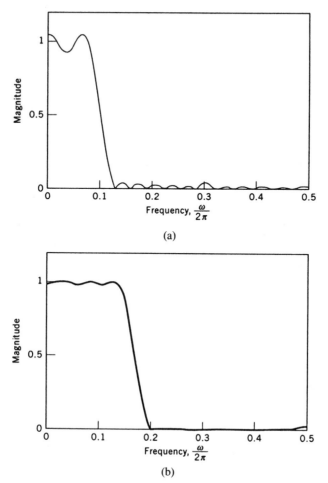

FIGURE 14-27   Magnitude response of Mth band filters for (a) $M = 5$, $K = 3$, and $\alpha = 0.8$, and (b) $M = 3$, $K = 6$, and $\alpha = 0.8$, designed with the modified MPR program.

The frequency response plotted in Figure 14-27(a) is obtained by using this procedure with $M = 5$, $K = 3$, and $\alpha = 0.8$.

The frequency response plotted in Figure 14-27(b) is obtained for $M = 3$, $K = 6$, and $\alpha = 0.8$. The order of the third-band filter is 35. For a desired passband magnitude of 1.0, the passband deviation is 0.0064 and the stopband deviation is 0.00514. Using the MPR program, a passband deviation equal to 0.00564 and a stopband deviation equal to 0.00564 are obtained. However, the filter in the latter case does not satisfy Eq. (14.68).

14-3-3   Design of IIR Filters for Sampling Rate Conversion

The design of IIR filters for meeting the lowpass specification of Figures 14-25(a) and 14-25(c) can be achieved by using well-known procedures, such as Butterworth and elliptic filter design. However, as mentioned in Section 14-2, these techniques do not provide the computational efficiency when compared with a specialized design, where the filter is assumed to be in a polyphase form with coefficients to be determined by suitably formulated design. The advantage of the polyphase structure representation accrues from the fact that the denominator polynomial of the polyphase path transfer functions contains only powers of z^M, where M is the sampling rate conversion factor. This allows the filtering in the polyphase path to be performed at the lower sampling rate. Martinez and Parks [MA79] have described a procedure in which the filter transfer is constrained to have a denominator polynomial containing only powers of z^M. An approach for sampling rate conversion in which the polyphase filters are constrained to be allpass is proposed in Ansari and Liu [AN81]. A transmultiplexer design using allpass filters is proposed in Taxen [TA81]. The choice of allpass filters is made in view of the allpass characteristic of the ideal polyphase filters. By constraining one of the polyphase paths to contain a pure delay, one can obtain an approximately linear-phase response [AN83].

It is observed that the choice of polyphase filters $H_p(z)$ with real poles leads to a low-order design solution. The filter transfer function is assumed to be of the form

$$H(z) = \sum_{p=0}^{M-1} z^{-p} H_p(z^M), \tag{14.80}$$

where

$$H_p(z^M) = z^{-k_p M} \prod_{k=1}^{J_p} \frac{\alpha_{pk} z^M + 1}{z^M + \alpha_{pk}}. \tag{14.81}$$

In Eq. (14.81), J_p and k_p are integers, and $\{\alpha_{pk}\}$ are constants to be determined. With the choice of allpass polyphase filters, one has certain restrictions on the control over the frequency response. For example, if $k_p = 0$ and $J_p = K$, both independent of p, then the magnitude response at the frequencies $\omega = m\pi/M$, $1 \le m \le M$, is given by [PU82, pp. 1569–1574]

$$|H(e^{jm\pi/M})| = \begin{cases} 0 & \text{for } m \text{ even,} \\ \operatorname{cosec} \dfrac{m\pi}{2M} & \text{for } m \text{ odd.} \end{cases} \tag{14.82}$$

Thus the filter is suitable for meeting the specifications of Figure 14-25(b). Several properties of and efficient design techniques for the filters given by Eqs. (14.80) and (14.81) are described in Drews and Gazsi [DR86] and Renfors and Saramäki [RE87a; RE87b]. For each ω, $0 \le \omega \le \pi$, define a sequence $\{F_\omega[k]\}_{k=0}^{M-1}$ as follows:

$$F_\omega[k] = H(e^{j(\omega + 2\pi k/M)})$$

$$= \sum_{p=0}^{M-1} e^{-j2\pi pk/M} H_p(e^{jM\omega}) e^{-jp\omega}, \qquad k = 0, 1, \ldots, M-1. \tag{14.83}$$

The last expression, which follows from Eq. (14.80), shows that $\{F_\omega[k]\}$ is the DFT of the sequence $\{f_\omega[p]\}$, where

$$f_\omega[p] = H_p(e^{jM\omega})e^{-jp\omega}, \quad p = 0, 1, \ldots, M - 1. \tag{14.84}$$

Therefore, by Parseval's relation,

$$\sum_{k=0}^{M-1} |F_\omega[k]|^2 = M \sum_{p=0}^{M-1} |f_\omega[p]|^2, \tag{14.85}$$

and since $H_p(z)$ is allpass the result is

$$\sum_{k=0}^{M-1} |H(e^{j(\omega + 2\pi k/M)})|^2 = M. \tag{14.86}$$

This implies that for each frequency ω_0 in the range for 0 to π/M, the sum of the squared magnitude response at frequencies in the range $[0, \pi)$ given by $\{2k\pi/M \pm \omega_0\}$ with k an integer, is equal to M. Due to the interrelated nature of the response at these frequencies, we allow symmetric *don't care* bands about the frequency $\omega_m = m\pi/M$ with m odd and $1 \le m \le M$. Also, the constraint Eq. (14.86) implies that if the squared magnitude response is approximately equal to M in a passband $[0, \alpha\pi/M]$, $0 < \alpha < 1$, then the magnitude response is automatically close to zero at frequencies that are in one of the stopband intervals $[(2k - \alpha)\pi/M, (2k + \alpha)\pi/M]$, $k = 1, \ldots, \lfloor M/2 \rfloor$. If the maximum stopband deviation is $\delta_s \ll 1$ then the passband deviation δ_p is guaranteed to be bounded by

$$\delta_p \le 1 - \sqrt{1 - (M - 1)\delta_s^2}. \tag{14.87}$$

The relation Eq. (14.86) can be used to obtain the largest number of attenuation zeros in the passband, where the magnitude is equal to M. It is found [RE87a] that with no consideration given to the phase response, the smallest stopband attenuation is obtained with $k_p = 0$ for all p and

$$J_p = \begin{cases} K + 1, & 0 \le p < P, \\ K, & P \le p \le M - 1. \end{cases} \tag{14.88}$$

where $0 \le P \le M - 1$. For approximately linear-phase design, $k_p = 0$ is chosen in all except the Pth path in which $k_P = K$ with $J_P = 0$. And

$$J_p = \begin{cases} K + 1, & 0 \le p < P, \\ K, & P \le p \le M - 1. \end{cases} \tag{14.89}$$

The final selection of the unknown α_{pk}'s is made using a Remez-type algorithm in which an initial guess of attenuation zeros is iteratively adjusted. As an example,

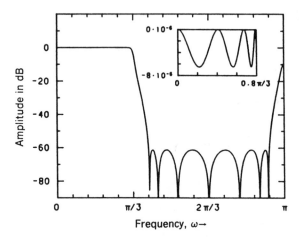

FIGURE 14-28   Magnitude response of a nonlinear-phase third-band IIR filter with $\omega_p = 0.8\pi/3$. (Adapted from Renfors and Saramäki [RE87a] with permission from IEEE.)

consider a rate conversion filter for $M = 3$ with $\alpha = 0.8$. The passband is $[0, 0.8\pi/3]$ and the stopband is $[1.2\pi/3, 2.8\pi/3]$. With no consideration given to the phase response, the magnitude response with $J_p = 2$ for $p = 0, 1, 2$ is shown in Figure 14-28. The minimum stopband attenuation is 61.24 dB and the maximum passband attenuation is 6.5×10^{-6} dB. The linear-phase design with $k_0 = J_1 = J_2 = 3$ and $k_1 = k_2 = J_0 = 0$ has its magnitude, phase and group delay response as shown in Figure 14-29. In this case the minimum stopband attenuation is 40.28 dB and the maximum passband attenuation is 7.1×10^{-4} dB. For the same stopband specifications, a FIR filter of length 31 is needed to get $\delta_p = \delta_s = 0.0106$. With a third-band filter of length 35 designed using the fractional delay design of Section 14-3-2, $\delta_p = 0.0064$ and $\delta_s = 0.00514$ is obtained.

One can also design hybrid IIR/FIR filters for sample rate conversion using the technique in Ramstad [RA82].

14-3-4   FIR and IIR Half-Band Filters

Half-band filters is a term generally used to describe special filters that are suitable for rate conversion by a factor of 2. The term *half-band frequency response* filters is used here specifically to refer to stable (possibly noncausal) filters with transfer functions of the form

$$H(z) = 1 + zT(z^2). \qquad (14.90)$$

If $H(z)$ is lowpass then $H(-z)$ is the transfer function of a highpass filter with a complementary frequency response since the sum of $H(z)$ and $H(-z)$ is 2. The expression for $H(z)$ in Eq. (14.90) is in the polyphase form for $M = 2$. The magnitude response $|H(e^{j\omega})|$ approximates an ideal lowpass filter response with a cutoff at $\omega = \pi/2$ and magnitude of 2. The assumed form of $H(z)$ implies that the

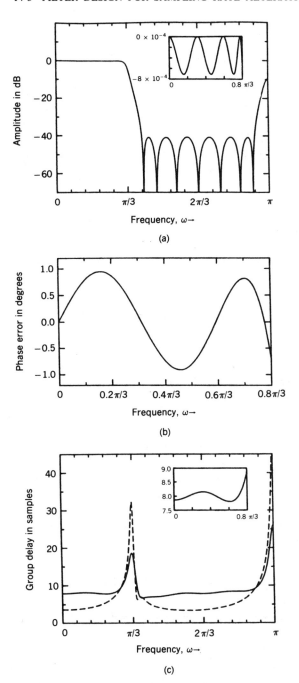

FIGURE 14-29 Approximately liner-phase third-band IIR filter with $\omega_p = 0.8\pi/3$. (a) Magnitude response. (b) Phase error. (c) Group delay response (broken line shows response of nonlinear-phase filter). (Adapted from Renfors and Saramäki [RE87a] with permission from IEEE.)

corresponding impulse response $h[n]$ satisfies the condition that

$$h[n] = 0, \qquad n = \pm 2, \pm 4, \ldots . \tag{14.91}$$

Originally, the term half-band filters was used for a FIR filter with transfer function given by Eq. (14.90) and with a symmetric impulse response [BE77]. The condition given by Eq. (14.91) can be extended to other rate conversion factors and the term Nth band filters is used in that context [MI82]. In this discussion the half-band filters may be FIR or IIR. The frequency response $H(z)$ in Eq. (14.90) satisfies

$$H(e^{j(\pi - \omega)}) - 1 = -\{H(e^{-j\omega}) - 1\}, \tag{14.92}$$

which can be expressed as

$$\text{Re } \{H(e^{j\omega}) - 1\} = -\text{ Re } \{H(e^{j(\pi - \omega)}) - 1\} \tag{14.93a}$$

and

$$\text{Im } \{H(e^{j\omega})\} = \text{Im } \{H(e^{j(\pi - \omega)})\}. \tag{14.93b}$$

From Eq. (14.92),

$$H(e^{j(\pi - \omega)}) = -\{H(e^{-j\omega}) - 2\}, \tag{14.94}$$

which implies that the complex deviations of $H(e^{j\omega})$ from the desired value of 2 and 0 in the passband and stopband are related, having the same absolute value. For symmetric FIR filters the frequency response is real so that the magnitude deviations in the passband $[0, \omega_p]$ and stopband $[\pi - \omega_p, \pi]$ are equal. However, for IIR filters, the magnitude deviations can be different. For a special useful class this relation between the magnitude deviation is discussed in Section 14-3-4-2.

14-3-4-1 Symmetric Half-Band FIR Filters.
In this case the filter transfer function is given by

$$H(z) = 1 + \sum_{n=0}^{K-1} h[2n + 1](z^{-2n-1} + z^{2n+1}). \tag{14.95}$$

The frequency response is purely real with the passband deviation δ_p in $[0, \omega_p]$ equal to the stopband deviation δ_s in $[\pi - \omega_p, \pi]$. The MPR program can be used to design equiripple filters directly by choosing an odd length filter and assigning equal weights to the passband deviation and stopband deviation. Also ω_s is set equal to $\pi - \omega_p$. The design can also be done by using the fractional delay design procedure described in Section 14-3-2. The MPR design program can also be used in this case to get the polyphase filter $h_1[n]$ given by

$$h_1[n] = t[n] = h[2n + 1], \tag{14.96}$$

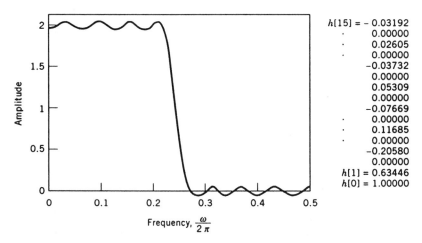

$$h[15] = -0.03192$$
$$\cdot \qquad 0.00000$$
$$\cdot \qquad 0.02605$$
$$\cdot \qquad 0.00000$$
$$\cdot \qquad -0.03732$$
$$0.00000$$
$$0.05309$$
$$0.00000$$
$$-0.07669$$
$$\cdot \qquad 0.00000$$
$$\cdot \qquad 0.11685$$
$$\cdot \qquad 0.00000$$
$$\cdot \qquad -0.20580$$
$$0.00000$$
$$h[1] = 0.63446$$
$$h[0] = 1.00000$$

FIGURE 14-30 Magnitude and impulse responses of a FIR half-band filter.

where $t[n]$ is the impulse response corresponding to $T(z)$. The filter with impulse response $h_1[n]$ is of even length $2K$ (as compared with $4K - 1$ in the previous case) and can be designed using the MPR program without any modification. A single *care* band $[0, 2\omega_p]$ is specified with a desired magnitude of 1.0 (and weight 1.0). The result of a design for $K = 8$ and $\omega_p = 0.45\pi$ is shown in Figure 14-30. The magnitude deviation is 0.0426 for a desired passband magnitude of 2.0 (i.e., $\delta_p = \delta_s = 2.13 \times 10^{-2}$).

Instead of an equiripple design, one can find filters with a maximally flat half-band response to $\omega = 0$ and π. A maximally flat half-band design results in a Lagrange interpolation filter [GU78] and the filter coefficients are given by

$$h_1[n] = \frac{(-1)^{n+K} \prod_{i=1}^{2K} (K + \tfrac{1}{2} - i)}{(K - n)!(K - 1 + n)!(\tfrac{1}{2} - n)}. \tag{14.97}$$

The coefficients $\tfrac{1}{2}h_1[n]$ are listed in Table 14-1 for $K = 1, 2, \ldots, 10$ and $n = 1, 2, \ldots, K$.

Note that many of the coefficients in Table 14-1 are integers that can be useful for simple implementation. A set of eight half-band filters with simple coefficients that are attractive for implementation [GO77] are listed in Table 14-2. Some of these filters are the same as Lagrange filters as seen from Table 14-2. Design rules for using these filters along with a comb filter in a multistage implementation of rate conversion are described in Goodman and Carey [GO77].

14-3-4-2 IIR Half-Band Filters. Here an analytic solution to the filter design is derived by assuming that $T(z^2)$ in Eq. (14.90) is allpass. The system may be non-causal and stable with a two-sided infinite-duration impulse response. This implies that some of the poles of $T(z^2)$ may be outside the unit circle. So $T(z^2)$ can be

TABLE 14-1 LaGrange Half-Band Filter Coefficients

K	$h_1[1]$	$h_1[2]$	$h_1[3]$	$h_1[4]$	$h_1[5]$	$h_1[6]$	$h_1[7]$	$h_1[8]$	$h_1[9]$	$h_1[10]$
1	$\dfrac{1}{4}$									
2	$\dfrac{9}{32}$	$\dfrac{-1}{32}$								
3	$\dfrac{150}{512}$	$\dfrac{-25}{512}$	$\dfrac{3}{512}$							
4	$\dfrac{1225}{4096}$	$\dfrac{-245}{4096}$	$\dfrac{49}{4096}$	$\dfrac{-5}{4096}$						
5	$\dfrac{19845}{2^{16}}$	$\dfrac{-2205}{2^{15}}$	$\dfrac{567}{2^{15}}$	$\dfrac{-405}{2^{17}}$	$\dfrac{35}{2^{17}}$					
6	$\dfrac{160083}{2^{19}}$	$\dfrac{-38115}{2^{19}}$	$\dfrac{22869}{2^{20}}$	$\dfrac{-5445}{2^{20}}$	$\dfrac{847}{2^{20}}$	$\dfrac{-63}{2^{20}}$				
7	$\dfrac{7(429)^2}{2^{22}}$	$\dfrac{-7(429)^2}{2^{24}}$	$\dfrac{21(143)^2}{2^{24}}$	$\dfrac{-3(143)^2}{2^{23}}$	$\dfrac{13013}{2^{23}}$	$\dfrac{-3549}{2^{24}}$	$\dfrac{231}{2^{24}}$			
8	$\dfrac{(6435)^2}{2^{27}}$	$\dfrac{-21(715)^2}{2^{27}}$	$\dfrac{21(429)^2}{2^{27}}$	$\dfrac{-33(195)^2}{2^{27}}$	$\dfrac{77(65)^2}{2^{27}}$	$\dfrac{-61425}{2^{27}}$	$\dfrac{7425}{2^{27}}$	$\dfrac{-429}{2^{27}}$		
9	$\dfrac{(36465)^2}{2^{32}}$	$\dfrac{-15(2431)^2}{2^{30}}$	$\dfrac{77(663)^2}{2^{30}}$	$\dfrac{-55(663)^2}{2^{31}}$	$\dfrac{1001(85)^2}{2^{31}}$	$\dfrac{-13(255)^2}{2^{30}}$	$\dfrac{55(51)^2}{2^{30}}$	$\dfrac{-123981}{2^{33}}$	$\dfrac{6435}{2^{33}}$	
10	$\dfrac{5(46189)^2}{2^{35}}$	$\dfrac{-165(4199)^2}{2^{35}}$	$\dfrac{33(4199)^2}{2^{34}}$	$\dfrac{-2145(323)^2}{2^{34}}$	$\dfrac{715(323)^2}{2^{34}}$	$\dfrac{-195(323)^2}{2^{34}}$	$\dfrac{165(323)^2}{2^{36}}$	$\dfrac{-7293(19)^2}{2^{36}}$	$\dfrac{715(19)^2}{2^{36}}$	$\dfrac{-12155}{2^{36}}$

TABLE 14-2 Filter Coefficients

Filter	Order	$h[0]$	$h[1]$	$h[3]$	$h[5]$	$h[7]$	$h[9]$
F1[a]	3	1	1				
F2	3	2	1				
F3	7	16	9	-1			
F4	7	32	19	-3			
F5	11	256	150	-25	3		
F6	11	346	208	-44	9		
F7	11	512	302	-53	7		
F8	15	802	490	-116	33	-6	
F9	19	8192	5042	-1277	429	-116	18

Adapted from Goodman and Carey with permission from IEEE.
[a]F1 in general is a comb filter of order N ($= 3$ above).

expressed as

$$T(z^2) = H_1(z^2)/H_0(z^2), \qquad (14.98)$$

where $H_0(z^2)$ and $H_1(z^2)$ are allpass with all poles inside the unit circle. A stable filter with the same magnitude response can then be obtained, but now with a right-sided response. The filter transfer function is given by

$$H'(z) = H_0(z^2) + zH_1(z^2). \qquad (14.99)$$

The filter with transfer function $H'(z)$ in Eq. (14.99), where $H_0(z^2)$ is allpass, is referred to as a half-band *magnitude* response filter to distinguish from the half-band *frequency* response filter in Eq. (14.90). Unlike $H(z)$ in Eq. (14.90) the sum of $H'(z)$ and $H'(-z)$ is not constant but has constant magnitude.

Consider $H(z)$ in Eq. (14.90) where $T(z^2)$ is allpass with poles on the imaginary axis:

$$T(z^2) = \prod_{i=1}^{L} \frac{a_i z^2 + 1}{z^2 + a_i}, \qquad (14.100)$$

where a_i are real and positive. The frequency response of $H(z)$ is given by

$$H(e^{j\omega}) = 1 + e^{j(\omega + \phi(2\omega))} \qquad (14.101)$$

and

$$H(e^{j(\pi - \omega)}) = 1 - e^{-j(\omega + \phi(2\omega))}, \qquad (14.102)$$

where $\phi(2\omega)$ is the phase response of $T(z^2)$. It is assumed that

$$\omega_p = \pi - \omega_s, \qquad 0 < \omega_p < \pi/2. \qquad (14.103)$$

It is easy to show that the passband magnitude deviation $2\delta'$ and the stopband magnitude deviation 2δ are related by

$$(1 - \delta')^2 + \delta^2 = 1. \tag{14.104}$$

For small values of δ,

$$\delta' \sim \tfrac{1}{2}\delta^2, \tag{14.105}$$

which makes the passband ripple much smaller than the stopband ripple.

Let δ_p and δ_s be given as the maximum deviations in the passband and stopband for specification assuming derived passband magnitude as 1.0. Define

$$\delta = \min \{\delta_s, (2\delta_p - \delta_p^2)^{1/2}\} \tag{14.106}$$

and use δ as the required stopband magnitude deviation.

For a maximally flat filter with a two-sided infinite-duration impulse response, the solution is a modified Butterworth filter [AN83]. The transfer function $T(z^2)$ in this case is given by Eq. (14.101), where the a_i are equal to

$$a_i = \cot^2\left(\frac{\pi i}{2L + 1}\right). \tag{14.107}$$

The Butterworth filter transfer with a right-sided response is given by Eq. (14.100), where

$$H_0(z) = \prod_{i=1}^{J} \frac{z^2\tilde{a}_i + 1}{z^2 + \tilde{a}_i} \tag{14.108}$$

with $J = \lfloor L/2 \rfloor$ and

$$\tilde{a}_i = \tan^2\left(\frac{\pi i}{2L + 1}\right). \tag{14.109}$$

And

$$H_1(z) = \prod_{i=J+1}^{L} \frac{z^2 a_i + 1}{z^2 + a_i}, \tag{14.110}$$

where a_i is given by Eq. (14.107).

Now consider the design of *equiripple* IIR half-band filter [AN81; AN85; GA85]. The design leads to a two-sided infinite-duration impulse response by using a special class of elliptic filters. Given δ_s, δ_p, ω_p, $\omega_s = \pi - \omega_p$, the procedure in Ansari [AN85] consists of the following steps:

1. Compute δ using Eq. (14.106).
2. Compute $k = \tan^2(\omega_p/2)$ and $k' = (1 - k^2)^{1/2}$.
3. Compute p and q from

$$p = \frac{1}{2}\frac{1 - \sqrt{k'}}{1 + \sqrt{k'}}, \tag{14.111}$$

$$q \cong p + 2p^5 + 15p^9 + 150p^{13}. \tag{14.112}$$

4. Estimate filter order by

$$2L + 1 = \left[\frac{2 \ln(4(1 - \delta^2)/\delta^2)}{-\ln q}\right]_O, \tag{14.113}$$

where $[\cdot]_O$ is the smallest odd integer equal to or exceeding the argument.
5. Compute the analog extremal frequencies Ω_i. Find

$$r_i = [(1 - k\Omega_i^2)(1 - \Omega_i^2/k)]^{1/2} \tag{14.114}$$

and determine

$$\cos\theta_i = \frac{(-1)^{i+1}r_i}{1 + \Omega_i^2}. \tag{14.115}$$

6. Compute $a_i = (1 - \cos\theta_i)/(1 + \cos\theta_i)$, $i = 1, 2, \ldots, L$.

Now substitute a_i in Eq. (14.101) to get the transfer function. The right-sided impulse response is obtained using the manipulations described before.

As an example, suppose $\delta_p = 0.0001$ and $\delta_s = 0.1$. The value of L for the design is 3 with $N = 7$. The values of a_i with $a_i < 1$ are 0.172636 and 0.843466. The value of $1/a_i$ with $a_i > 1$ is 0.519974. The actual deviations obtained are $\delta' = 0.245 \times 10^{-4}$. In this case the filter order is determined by the passband deviation requirement. The magnitude response is shown in Figure 14-31.

If a right-sided impulse response for the transfer function in Eq. (14.90) is required, that is, an approximately linear-phase causal equiripple filter is required, then an analytic solution is not known and methods described in Section 14-3-3 should be used.

14-4 MULTISTAGE IMPLEMENTATION OF RATE CONVERSION

Multistage rate conversion has been examined briefly in Section 14-2-7. As mentioned there, a multistage implementation has the potential for reducing the computational burden in sampling rate conversion when the rate conversion ratio is

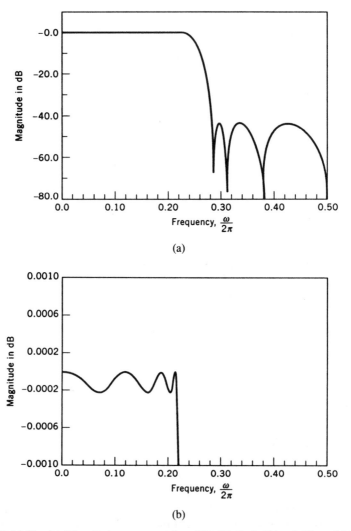

FIGURE 14-31 (a) Magnitude response of an IIR elliptic half-band filter. (b) Passband detail of the response in (a). (Adapted from Ansari [AN85] with permission from IEEE.)

factorable into smaller integers. In this section methods are described for designing filters for multistage implementation. It is assumed that $D = D_1 D_2 \ldots D_J$ is a product of J integers D_i, $i = 1, 2, \ldots, J$. Thus the rate conversion by a factor D is carried out in J stages with factors D_1 in the first stage, D_2 in the second stage, and so on. A J-stage decimator is illustrated in Figure 14-32(a), where the lowpass filter transfer function for the ith stage is denoted by $H_i(z)$, $i = 1, 2, \ldots, J$. The discussion here focuses on the design of decimators, although the technique is equally applicable to interpolation.

Consider the first two stages in Figure 14-32(a). Suppose the order of the op-

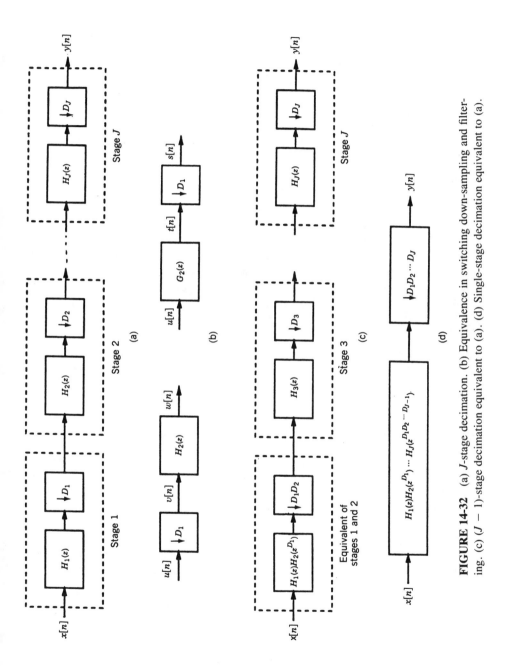

FIGURE 14-32 (a) J-stage decimation. (b) Equivalence in switching down-sampling and filtering. (c) $(J-1)$-stage decimation equivalent to (a). (d) Single-stage decimation equivalent to (a).

erations of down-sampling by D_1 and filtering by $H_2(z)$ is switched, as shown in Figure 14-32(b). It can be shown that the two systems are equivalent if

$$G_2(z) = H_2(z^{D_1}). \qquad (14.116)$$

Using this result, it is clear that the system of Figure 14-32(c) is equivalent to the system in Figure 14-32(a). Repeated application of this switching yields the equivalent single-stage system of Figure 14-32(d), where the single-stage filter has a transfer function $H(z)$ given by

$$H(z) = H_1(z)H_2(z^{D_1})H_3(z^{D_1D_2}) \cdots H_J(z^{D_1D_2 \cdots D_{J-1}}). \qquad (14.117)$$

Each filter $H_i(z)$, $i = 1, 2, \ldots, J$, is ideally a lowpass filter with a passband $(0, \pi/D_i)$ and stopband $(\pi/D_i, \pi)$. Note that $H_i(z^{D_1D_2 \cdots D_{i-1}})$, $i = 2, 3, \ldots, J$, have multiple passbands. However, with individual $H_i(z)$ properly designed, the overall frequency response of the right-hand side of Eq. (14.117) can be made to approximate a lowpass response with a passband $(0, \pi/D)$ and stopband $(\pi/D, \pi)$.

There may be several ways of expressing D as a product of J integers. In practice, however, each D_i is often limited to small integers such as 2, 3, or 4. Also, if $D = 2^J$, then half-band filters can be used in all except perhaps the last stage. It is often the case that using more than two stages does not yield a significant further reduction in complexity.

In Section 14-4-1, a simple method of choosing the filter for multistage implementation for a given factorization is reviewed. Then methods of jointly optimizing the different stages for FIR [CR75; CR76a; MI78; RA75; SA84] and IIR filters [CR83; RE87b] are presented.

14-4-1 Suboptimal Procedure for Multistage Rate Conversion

Recall the three commonly used specifications for the passbands and stopbands in sampling rate conversion by an integer factor M as illustrated in Figure 14-25 for $M = 5$, where the permissible passband and stopband deviations are denoted by δ_p and δ_s, respectively. In this section a procedure for J-stage decimation is described, and for convenience the specifications are shown again in Figure 14-33 for decimation by an integer factor D. These three types of specifications are referred to as Types A, B, and C. The Type A response has a single transition band $(\alpha\pi/D, \pi/D)$, $0 < \alpha < 1$. The Type B response contains multiple transition bands given by

$$((2k + \alpha)\pi/D, (2k + 2 - \alpha)\pi/D) \cap [0, \pi],$$

$$k = 0, 1, \ldots, \lfloor D/2 \rfloor - 1.$$

The Type C specification has a single transition band $(\alpha\pi/D, (2 - \alpha)\pi/D)$, which, unlike Type A, is centered at $\omega = \pi/D$. The reasons behind these choices of specifications are discussed in Section 14-3.

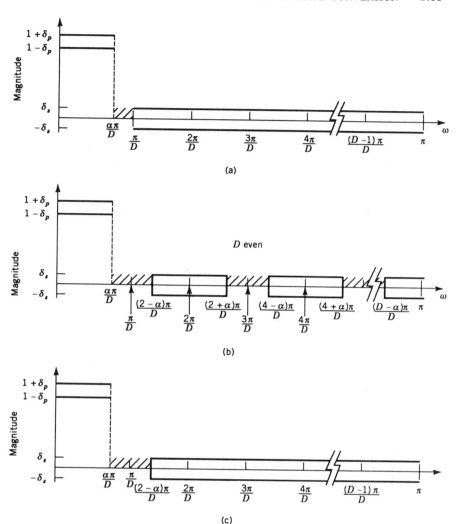

FIGURE 14-33 Three types of specification for decimation by a factor D, referred to as (a) Type A, (b) Type B, and (c) Type C.

In multistage systems, the maximum overall passband deviation is bounded by the sum of the deviations in each of the stages. The maximum overall stopband deviation is less than the largest stopband deviation among the multiple stages scaled by the product of the peak passband magnitudes of the different bands. Let the overall passband and stopband deviation limits be denoted by δ_p and δ_s, respectively. It is convenient to set all stages to have equal passband and stopband deviation limits. Let δ_{pJ} and δ_{sJ} be these limits. It is simple to see that the following choice satisfies the requirements:

$$\delta_{pJ} = (1 + \delta_p)^{1/J} - 1 \qquad (14.118)$$

and

$$\delta_{sJ} = \delta_s/(1 + \delta_p)^{(J-1)/J}. \qquad (14.119)$$

In practice, the number of stages J is small (<4) and $\delta_p \ll 1$, so that

$$\delta_{pJ} \cong \delta_p/J \quad \text{and} \quad \delta_{sJ} \cong \delta_s. \qquad (14.120)$$

If the magnitude response of the filters is constrained to be no greater than unity, then Eq. (14.120) defines the specifications for the individual stages. With these choices of deviation specifications, and for a given factorization of D, the design task is to choose the passbands and stopbands for each individual stage so that the transfer function Eq. (14.117) meets the overall specifications. Equation (14.117) can be expressed in the form

$$H(z) = \prod_{i=1}^{J} H_i(z^{\Delta_i}), \qquad (14.121)$$

where

$$\Delta_1 = 1 \quad \text{and} \quad \Delta_{i+1} = D_1 D_2 \cdots D_i, \quad i = 1, 2, \ldots, J. \qquad (14.122)$$

Note that $\Delta_{J+1} = D$ and that $\Delta_{i+1} = \Delta_i D_i$. The ideal stopband SB_i^d of the filter $H_i(z)$ is

$$\text{SB}_i^d = [\pi/D_i, \pi]. \qquad (14.123)$$

The ideal magnitude response of $H_i(z^{\Delta_i})$ has multiple stopband intervals due to frequency scaling by $1/\Delta_i$. The first stopband interval is $[\pi/D_i\Delta_i, \pi/\Delta_i] = [\pi/\Delta_{i+1}, \pi/\Delta_i]$. Since Δ_i increases with i, only the last stage provides attenuation in the band $[\pi/\Delta_{J+1}, \Delta_J]$ or $[\pi/D, D_J \pi/D]$. It turns out the last stage plays a key role in meeting the specifications. Consider the case of $D = 12$ with stopband intervals as shown in Figures 14-34(a)-(c). Suppose $J = 2$, $D_1 = 4$, and $D_2 = 3$. Focus now only on the choice of passband and stopband intervals. The frequency response of the last stage should be chosen as shown in Figures 14-34(d)-(f) corresponding to the three cases in Figures 14-34(a)-(c). Note that the specifications in Figures 14-34(d)-(f) are of Type A, B, and C, respectively, with the same α but with a conversion ratio of D_J. Note also that Figures 14-34(d)-(f) are obtained by stretching out the specifications of Figures 14-34(a)-(c) in the interval $[0, \pi D_J/D]$ of Figures 14-34(a)-(c) thereby widening the transition band.

For the stage preceding the last stage, it is observed that its ideal stopband $\text{SB}_{J-1}^d = [\pi/D_J, \pi]$ is scaled by $1/\Delta_{J-1}$ and maps to $[\pi/\Delta_J, \pi/\Delta_{J-1}] = [D_J\pi/D, D_{J-1}D_J\pi/D]$. None of the preceding $J - 2$ stages provides any stopband attenuation in the interval $[D_J\pi/D, D_{J-1}D_J\pi/D]$. Only the stopbands due to multiple images of the response of the last filter $H_J(z)$ and the stopband due to $H_{J-1}(z)$ appear in the interval $[D_J\pi/D, D_{J-1}D_J\pi/D]$. Now focus on the specifications in

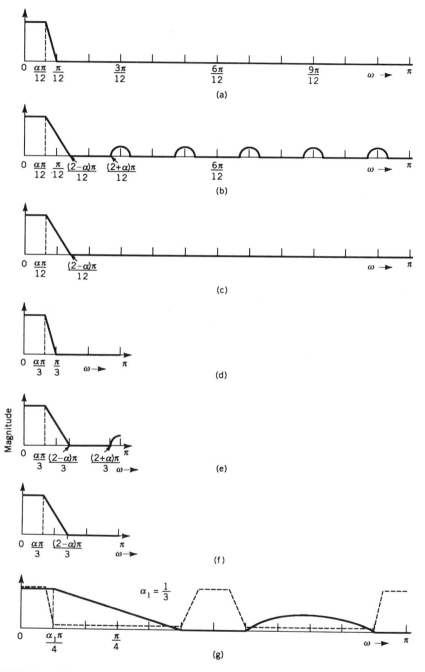

FIGURE 14-34 Frequency responses in two-stage decimation for $D = 12$, $D_1 = 4$, $D_2 = 3$. Overall responses: (a) Type A, (b) Type B, and (c) Type C. (d)–(f) Frequency response at last stage for Types A, B, and C, respectively. (g)–(i) Frequency response of first stage for Types A, B, and C, respectively.

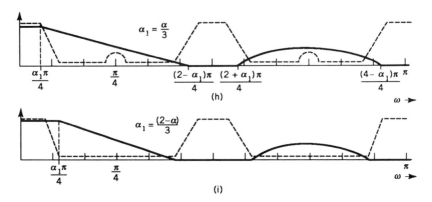

FIGURE 14-34 *(Continued)*

the interval $[D_J \pi/D, D_{J-1}D_J \pi/D]$ and examine the stopband requirements that are met due to the last stage filter $H_J(z^{\Delta_J})$ in this interval. For the purpose of illustration let $J = 2$, $D_1 = 4$, $D_2 = \Delta_2 = 3$. The attenuation provided by $H_J(z^{\Delta_J})$ using the specifications in Figures 14-34(d)–(f) is shown in Figures 14-34(g)–(i) with the dashed line. Examining the overall responses in Figures 14-34(a)–(c) it is clear that the specification for $H_{J-1}(z)$ should be chosen as shown by the solid lines in Figures 14-34(g)–(i). Note that in Figures 14-34(g) and 14-34(i) the specifications are of Type B with a conversion ratio D_1 and the passband interval specification can be chosen as $[0, \alpha'\pi/4]$ in all three cases where $\alpha' = \alpha/3$. However, the choice of the slightly wider passband in Figures 14-34(g) and 14-34(i) is used due to the possibility of using the D_{J-1}th band filters. Note that the specifications for $H_{J-1}(z)$ in all three cases are suitable for the use of D_{J-1}th band filters.

Once the filters for the last two stages are chosen, the procedure for each preceding stage can be repeated. The filters in the last two stages together meet the required specifications in $[0, D_{J-1}D_J \pi/D]$ as well as in other intervals due to periodicity. This means that one can now proceed with the equivalent single stage that combines the last two stages. This new last stage of a $(J-1)$ stage system performs decimation by a factor $D_{J-1}D_J$.

The requirements can be summarized as follows:

Last Stage (Jth Stage)
    Passband: $[0, \alpha\pi/D_J]$ in all three cases
    Stopband: Type A: $[\pi/D_J, \pi]$
                  Type B: $\bigcup_{k=1}^{\lfloor D_J/2 \rfloor} [(2k - \alpha)\pi/D_J, (2k + \alpha)\pi/D_J] \cap [0, \pi]$
                  Type C: $[(2 - \alpha)\pi/D_J, \pi]$
  Preceding Stages ($i = 1, 2, \ldots, J - 1$)

    The passbands can be defined as $[0, \alpha\pi\Delta_i/D]$ for all three types, with Δ_i as in Eq. (14.122). However, the following passband and stopband intervals are specified with the use of the D_ith band filter in mind. These intervals are specified in terms of a parameter α_i that is different in the three cases and is given by:

Type A: $\alpha_i = \Delta_{i+1}/D$
Type B: $\alpha_i = \alpha\Delta_{i+1}/D$
Type C: $\alpha_i = (2 - \alpha)\,\Delta_{i+1}/D$
Passband: $[0,\, \alpha_i\pi/D_i]$
Stopband: $\bigcup_{k=1}^{\lfloor D_i/2 \rfloor} [(2k - \alpha_i)\,\pi/D_i,\, (2k + \alpha_i)\pi/D_i] \cap [0, \pi]$

Consider now an example of decimation by a factor $D = 8$ with $J = 3$ and $D_1 = D_2 = D_3 = 2$. Let $\alpha = 0.8$ and $\delta_p = \delta_s = 0.1$. The frequency responses shown in Figure 14-35 are obtained using the METEOR program. The filter lengths are 5, 7, and 35 for the three stages giving a computation rate of $4 \times (5 + 1)/2 + 2 \times (7 + 1)/2 + (35 + 1)/2 = 38$ multiplies per output. In a single-stage implementation the specifications are met with a filter of length 109, requiring 55 multiplies per output. With larger conversion ratios and larger values of α the savings can be even higher.

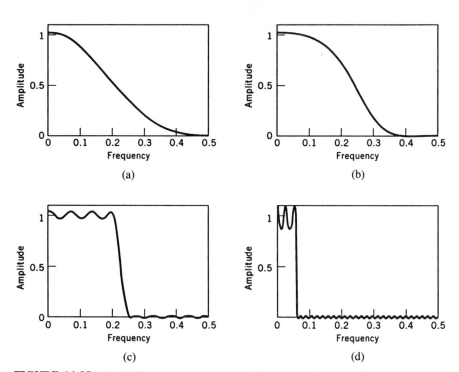

FIGURE 14-35 (a)–(c) Frequency responses of first, second, and third stages in three-stage decimation with $D = 8$, $D_i = 2$, $i = 1, 2, 3$. (d) Frequency response required to meet specifications in one-stage decimation.

14-4-2 Optimum Procedures for Multistage Rate Conversion

For a given number, J, of stages there may be several ways in which D can be partitioned into J factors and for each partition the actual order in which the decimation by these factors is carried out needs to be specified. The complexity of these implementations can be different and one needs to choose the scheme with the lowest complexity, as measured in terms of minimum computational requirements, minimum storage, and so on. In this section, some methods based on the complexity measure of the average multiplications required for computing each output sample are briefly described. It should be pointed out that the lowest multiplication rate may be accompanied by increased control complexity and these issues should be balanced in the selection of a system. Often at low sampling rates, such as for speech signals, these computational gains may not necessarily lead to reduced hardware if one uses the fast programmable digital signal processors currently available. However, at higher sampling rates the reduction in complexity can be significant.

A procedure for constructing an efficient system for rate conversion is to vary the number J of stages and in each case vary the choice of factors $D_1, D_2, \ldots ,$ D_J and search for the best solution [CR75; VI88]. A suitable objective function is set up for the optimization. Once the parameters $J, D_1, D_2, \ldots , D_J$ are chosen, the frequency response of the overall system can be further optimized. In this section some available procedures and the sources for carrying out multistage rate conversion using FIR and IIR filters are pointed out to the reader.

Multistage Rate Conversion Using FIR Filters. Consider the J-stage decimation using FIR filters. If the length of the ith filter is N_i, then the number of multiplications per output sample (MPOS$_{FIR}$) is given by

$$\text{MPOS}_{FIR} = \sum_{i=1}^{J} N_i D / \Delta_{i+1}, \tag{14.124}$$

where no symmetries in filter coefficients have been exploited. In order to optimize the overall system by choosing the best value for J and the best set of D_i's, a search procedure can be used by incorporating the filter length estimate in terms of filter specifications. Although in principle, stopband and passband deviations can also be optimized, this would add greatly to the complexity of the optimization. Experience has shown that in most cases the bulk of saving occurs in going from a one- to a two-stage implementation. A further optimization of the responses can be carried out by iteratively adjusting the response of individual stages [SA84].

Consider an example of decimation by a factor $L = 100$. A filter for the specifications of Type A with $\alpha = 0.95$ $\delta_p = 0.001$, and $\delta_s = 0.0001$ requires an estimated length of 15590 [CR75]. On the other hand, an implementation in three stages with $L_1 = 10$, $L_2 = 5$, and $L_3 = 2$ requires filters of lengths 32, 43, and 355, respectively. The average multiplies per output sample are 7.84 in a three-stage implementation as compared with 78 for a single stage.

Multistage Rate Conversion Using IIR Filters. A procedure similar to that described above can be employed for IIR filters using available estimates of filter orders [CR83]. Closed-form expressions are not available for filters designed using the polyphase expression of IIR filter transfer functions. However, for a small number of stages, one can carry out a joint optimization [RE87b] of the response for each choice of the set $\{D_i\}$ and pick the set with the lowest complexity. Note that allpass polyphase filters are used here. A correction filter can be used at the lowest rate to provide attenuation of the frequency components that alias into the transition band.

Consider an IIR filter design using allpass polyphase filters for sampling rate reduction by a factor $D = 20$. Type A specifications are to be met with $\alpha = 0.9$ and $\delta_p = 0.1$ and $\delta_s = 0.005$. For $J = 2$, the best design is obtained with $D_1 = 5$ and $D_2 = 4$ [RE87b]. Using this design, one obtains partial and overall responses as shown in Figure 14-36. The total number of filter coefficients for all the allpass sections in stage 1 is four and in stage 2 is nine. The correction filter is of order 3.

14-5 MULTIRATE FILTER BANKS

A common application of multirate techniques is encountered in the processing of signals using filter banks. In this section, the structures, design, and implementation of filter banks [BE76; CR83; MA87; RA84; VA87] are considered.

14-5-1 Issues in Filter Bank Design and Implementation

A filter bank refers to a processing unit comprised of a set of filters that usually have bandpass characteristics. The unit is typically employed either in an *analysis* mode with a single input and multiple outputs or in a *synthesis* mode with multiple inputs and a single output. An analysis filter bank is shown on the left side in Figure 14-37 and is used to decompose a given signal into N component signals, which are in most cases down-sampled before subsequent use. When the signal support straddles two integer bands, as discussed earlier, then the component signals are modulated before down-sampling. The down-sampling factors are M_0, M_1, ..., M_{N-1}. A synthesis filter bank, shown on the right side of Figure 14-37, enables the merging of several input signals into a higher rate signal. In principle, this operation consists of up-sampling the N channel signals to a common higher sampling rate and applying the up-sampled signals to suitable filters followed by an addition of the filtered signals.

The choice of filters in the analysis and synthesis filter banks depends on the application. There are several features of a filter bank that can vary with specific applications. These include the widths and spacings of the frequency bands, the extent of allowed overlap of the frequency bands, methods of implementation and the choice of modulation if used, the down-sampling and up-sampling factors, and the nature of reconstruction expected in a back-to-back connection of an analysis–

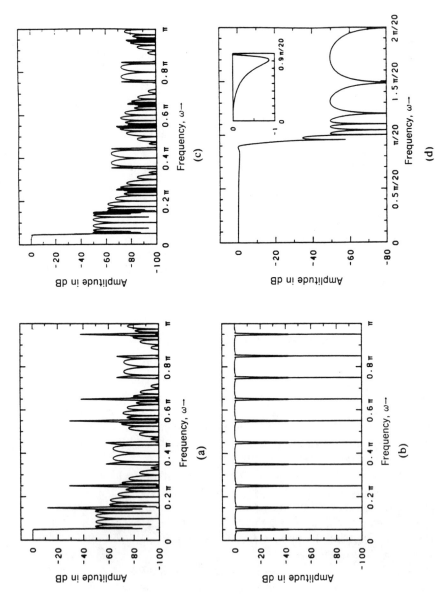

FIGURE 14-36 Magnitude responses of IIR filters used in two-stage decimation for $D = 20$. (a) Decimation stages without correction filter. (b) Correction filter. (c) Overall filter. (d) Passband and transition band detail of overall filter. (Adapted from Renfors and Saramäki [RE87b] with permission from IEEE.)

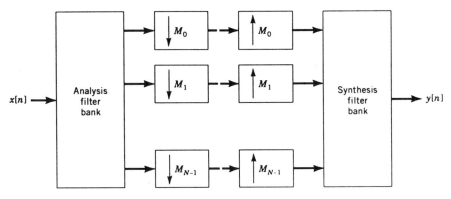

FIGURE 14-37 Analysis and synthesis filter banks.

synthesis filter bank pair. These issues are briefly explored here by focusing on two specific applications.

The first application is in the subband coding of speech and images, where the input signal is split into several bandpass signals whose energy is mainly concentrated in disjoint frequency bands. These bandpass signals are decimated to reduce redundant information. Each of the resulting subband signals is then compressed by exploiting the different subband signal characteristics. The subband signals, after data compression, can be recombined by a synthesis filter bank to yield a close approximation to the original signal. The exact original signal cannot be regenerated in practice due to the loss of information that is typically caused by data compression and arithmetic quantization. In the absence of these distortions, it is possible to carry out an exact reconstruction of the signal with a properly designed pair of analysis and synthesis filter banks.

A second application of filter banks is in a digital transmultiplexer, where an analysis filter bank is used to convert a frequency division multiplexed (FDM) signal into a time division multiplexed (TDM) signal. The reverse conversion is accomplished by a synthesis filter bank.

The design objectives for the construction of filter banks depend on the application requirements and the input and output signal characteristics. A key consideration is the choice of the passband widths of the filters, which is dictated by the application. Three types of frequency band partitions that are used in practice are shown in Figure 14-38. The first, shown in Figure 14-38(a), is a nonoverlapping partition of the frequency bands, where the different humps show the magnitude responses of the bandpass filters that constitute the filter bank. Such a frequency band partition is used in a transmultiplexer, where the bandpass response is essentially zero outside the designated bands so as to avoid aliasing when the filtered signal is down-sampled. This is done to limit the cross-talk caused by aliasing to levels below specified thresholds. However, there are many applications, such as subband coding and spectrum estimation, where it is permissible or even necessary to have an overlap of the frequency bands. An example of this partition is shown in Figure 14-38(b).

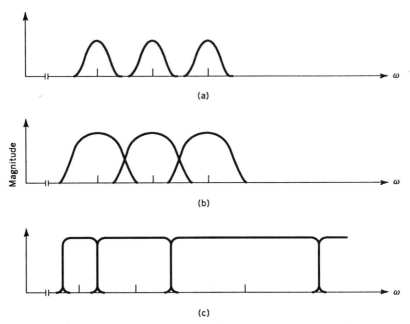

FIGURE 14-38   Three types of frequency band partitions in filter banks.

The frequency bands shown in Figures 14-38(a) and 14-38(b) have equal width and equal spacing and the filtered signals are all down-sampled by the same factor. This type of partition is referred to as a *uniform filter bank*. In some applications, the partition of the frequency band is nonuniform, leading to a *nonuniform filter bank*, as illustrated in Figure 14-38(c).

A uniform filter bank can be realized with a single suitably designed lowpass filter. The bandpass filtering is accomplished by either modulating the lowpass impulse response to obtain the bandpass filter response or modulating the signal to translate the information in the different bands down to the baseband for processing by the lowpass filter. These frequency translations are effected by multiplication using either complex exponential sequences or real sinusoids. The modulation of the different channels can be cast as a task of computing a discrete Fourier transform (DFT) and the actual implementation can be efficiently carried out using a fast algorithm for computing the DFT (see Section 8-2).

An arbitrary nonuniform filter bank does not, in general, lend itself to efficient implementations based on fast transform algorithms. However, some special cases do offer the possibility of efficient implementations. The frequency band partition shown in Figure 14-38(c) is characterized by the fact that the width of the highest frequency band is twice that of the next highest band. Thereafter, the lower frequency bands have a geometrically decreasing width. Such structures can be implemented using a tree-structured partition by a factor of 2 at each stage. The lower frequency band signal of each stage is further split into two frequency bands of equal width. These structures are discussed in detail later.

Another feature that leads to differences in filter banks is the choice of the sampling rate for the outputs of the different filters in the analysis bank. As shown in Figure 14-37, the outputs of a bank of N filters are down-sampled by factors $M_0, M_1, \ldots, M_{N-1}$. For uniform filter banks the down-sampling factors are equal ($M_k = M$, $0 \le k \le N - 1$). On the other hand, in the case of the filter bank in Figure 14-38(c), $M_{k+1} = 2M_k$ for $1 \le k \le N - 2$ and $M_1 = M_0$. Often the decimation factors are chosen such that the bandpass signals are sampled at rates that are together the lowest for recovering the original signal. For real input and output signals, if $\sum_{k=0}^{N-1} 1/M_k = 1$, then the filter bank is referred to as critically sampled or maximally decimated. Critical sampling or maximal decimation is often encountered in subband systems where an analysis–synthesis filter bank pair is designed to provide an output signal that very closely approximates or exactly reconstructs an arbitrary input signal. The topic of exact reconstruction filter banks is discussed later. If $\sum_{k=0}^{N-1} 1/M_k > 1$, then the filter bank is oversampled, since the total output sampling rate is greater than the input sampling rate. There are also situations of undersampling, such as a transmultiplexer operation where $\sum_{k=0}^{N-1} 1/M_k < 1$. In this case the input to the analysis filter bank is not arbitrary but is a FDM signal that is sampled at a slightly higher rate than the sum of the sampling rates of the multiplexed channels.

The filter banks that are used in a large number of applications are uniform; the next section is devoted to examining their structure and implementation. In Section 14-5-3, exact reconstruction filter banks are considered.

14-5-2   Uniform Filter Banks

Consider an analysis filter bank with N channels in which the frequency bands are of equal width and uniformly spaced. The centers of the bands are separated by $2\pi/K$, where K is referred to as the band separation factor. The input and output signals may be complex or real valued and the two cases are discussed separately.

14-5-2-1   Uniform Filter Banks for Complex Signals ($K = N$).   Consider the case where the number of bands N is equal to the band separation factor K, which defines the spacing between centers of adjacent bands to be equal to $2\pi/K$. For $K = 8$, three possible ideal frequency band partitions are shown in Figures 14-39(a)–(c). In each, the kth band interval is of width $2\pi/K$ and centered at the frequency

$$\omega_k = (k + k_0) \, 2\pi/K, \tag{14.125}$$

where $k = 0, 1, \ldots, K - 1$ and $-\frac{1}{2} < k < \frac{1}{2}$. When $k_0 = 0$ the bands are centered at frequencies that are even multiples of π/K, as shown in Figure 14-39(a). Such a filter bank is referred to as even type. With $k_0 = \frac{1}{2}$, the center frequencies of the bands are at odd multiples of π/K, as shown in Figure 14-39(b). This is the odd type. In both Figures 14-39(a) and 14-39(b), the width of the bands is equal to the spacing between bands. Finally, Figure 14-39(c) shows a filter bank with $K = 8$ and $k_0 = \frac{1}{4}$, and the bands are centered at ω_k given by Eq. (14.125).

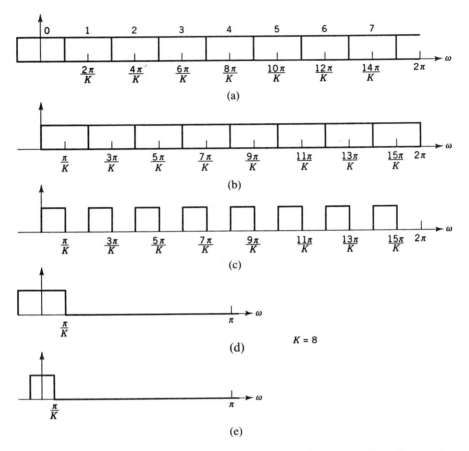

FIGURE 14-39   (a)–(c) Three typical frequency band partitions in uniform filter banks. (d) Prototype filter for filter banks in (a) and (b). (e) Prototype filter for filter bank in (c).

However, the width of the bands in this case is half that in the previous two cases considered. Now the bandpass filtering for creating the partitions in Figure 14-39(a) and 14-39(b) can be carried out using a lowpass filter with frequency response as shown in Figure 14-39(d), together with suitable modulation. For the band partition in Figure 14-39(c), one would use a lowpass filter with the frequency response shown in Figure 14-39(e).

Now consider methods for carrying out the bandpass filtering operation. One option is to use a lowpass filter of the type shown in Figure 14-39(d) and modulate the signal so that its content in the desired band is modulated down into the passband of the filter. Let $h[n]$ be the impulse response and $H(z)$ be the transfer function of a filter that approximates the required lowpass response. Then in order to perform the partition in Figure 14-39 the signal has to be modulated by

$$e^{-j\omega_k n} = W_K^{(k+k_0)n}, \tag{14.126}$$

where ω_k is given by Eq. (14.125) for $k = 0, 1, \ldots, K - 1$, and

$$W_K = e^{-j2\pi/K}. \tag{14.127}$$

Each of the K modulated signals $\tilde{x}_k[n]$ is applied to the filter with impulse response $h[n]$ as shown in Figure 14-40(a). The outputs are then down-sampled by a factor M to yield the output signals $S_k[m]$, which can be expressed as

$$S_k[m] = \sum_{n=-\infty}^{\infty} h[mM - n]x[n]W_K^{(k+k_0)n}, \qquad k = 0, 1, \ldots, K - 1. \tag{14.128}$$

Uppercase letter S is used to denote the output signal since it can be expressed as the generalized DFT of another finite-duration sequence.

The uniform synthesis filter bank has K input signals, each of which is first up-sampled by a factor M and filtered, as shown in Figure 14-40(b). The impulse response of the lowpass interpolation filter is $g[n]$ and may be different from $h[n]$. The interpolated signals are then translated in frequency and added to yield the output signal $y[n]$. The frequency shift is the reverse of that in the analysis filter bank and the modulating signal is

$$e^{j\omega_k n} = W_K^{-(k+k_0)n}. \tag{14.129}$$

If the input signals to the synthesis filter bank are the unmodified outputs $S_k[m]$ of the analysis filter bank, then

$$y[n] = \sum_{m=-\infty}^{\infty} g[n - mM] \sum_{k=0}^{K-1} S_k[m] W_K^{-(k+k_0)n}. \tag{14.130}$$

The filters $h[n]$ and $g[n]$ are chosen according to the requirements in the application. Equations (14.128) and (14.130) represent a procedure for carrying out the bandpass filtering based on the complex modulation of the signal $x[n]$. An alternate procedure consists of modulating the *impulse response* $h[n]$ of the lowpass filter to create K bandpass filters, which are used to process the input signal. The filtered outputs are modulated down to the baseband and decimated. The analysis procedure is based essentially on a different interpretation of Eq. (14.128), which can be written as

$$S_k[m] = W_K^{(k+k_0)mM} \sum_{n=-\infty}^{\infty} W_K^{-(mN-n)(k+k_0)} h[mM - n]x[n]$$

$$= W_K^{mM(k+k_0)} R_k[m], \tag{14.131}$$

where

$$R_k[m] = \tilde{h}_k[n] \circledast x[n]\big|_{n=mM}, \tag{14.132}$$

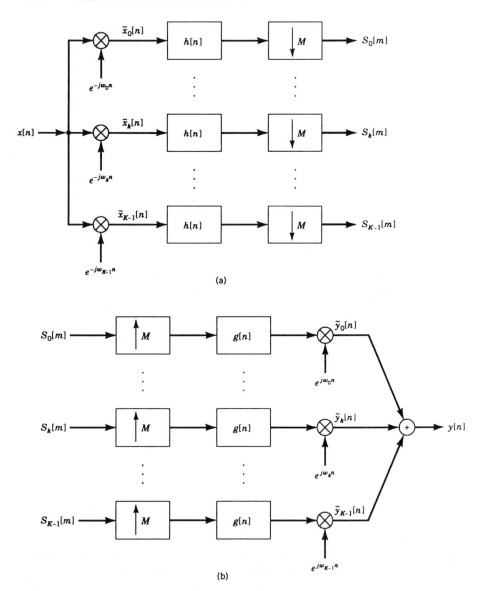

FIGURE 14-40 Processing in uniform filter banks with modulation of *signals*: (a) analysis bank and (b) synthesis bank.

and

$$\tilde{h}[n] = h[n] W_K^{-n(k + k_0)}. \tag{14.133}$$

The kth path in the analysis filter bank in Figure 14-40(a) is replaced by the processing shown in Figure 14-41(a). Now $\tilde{y}_k[n]$ in Figure 14-40(b) can be expressed

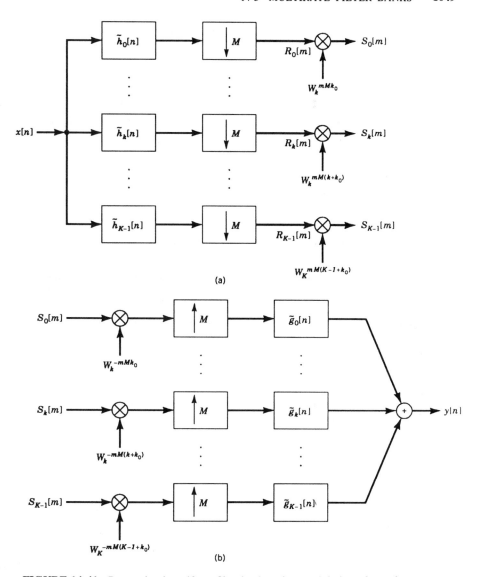

FIGURE 14-41 Processing in uniform filter banks using modulation of *impulse response*: (a) analysis bank and (b) synthesis bank.

as

$$\tilde{y}_k[n] = \sum_{m=-\infty}^{\infty} W_K^{-(k+k_0)mM} S_k[m] g[n - mM] W^{-(k+k_0)(n-mM)}. \quad (14.134)$$

Therefore the kth path in Figure 14-40(b) can be replaced by the processing shown in Figure 14-41(b), where

$$\tilde{g}_k[n] = g[n] W_K^{-(k+k_0)n}. \quad (14.135)$$

Figures 14-40 and 14-41 represent the basic operations used in the processing in filter banks. Efficient implementations based on DFT computations are now examined.

DFT Based Implementations. The modulating signals used in the filter banks bear a relationship that can be exploited to efficiently implement the filter banks. The expression in Eq. (14.132) for $R_k[m]$ will be examined (see Figure 14-41(a)), which is the kth intermediate output signal of the analysis filter bank. Note that $R_k[m]$ is given by

$$R_k[m] = \sum_{n=-\infty}^{\infty} x[mM - n]h[n]W_K^{-n(k+k_0)}.$$ (14.136)

For a fixed value of m, $\{R_k[m], k = 0, 1, \ldots, K - 1\}$ are samples of the z-transform of $x[mM - n]h[n]$ evaluated at K points on the unit circle, at $z = e^{j\omega_k}$, $k = 0, 1, \ldots, K - 1$, where ω_k is given by Eq. (14.125). If all nonzero samples of the product sequence $x[mM - n]h[n]$ are contained in the interval $0 \le n \le K - 1$, then $\{R_k[m], k = 0, 1, \ldots, K - 1\}$ is the DFT of $x[mM - n]h[n]W_K^{-nk_0}$. If the effective duration of $x[mM - n]h[n]$ is greater than K, then $R_k[m]$ is the DFT of the time-aliased version of the modulated product sequence. $R_k[m]$ can be expressed as

$$R_k[m] = \sum_{n=0}^{K-1} \sum_{i=-\infty}^{\infty} x[mM + iK + n]h[-iK - n]W_K^{-(k+k_0)(-iK-n)}$$

$$= \sum_{n=0}^{K-1} W_K^{(k+k_0)n} \sum_{i=-\infty}^{\infty} x[mM + iK + n]h[-iK - n]W_K^{ik_0K}.$$ (14.137)

Define $r_n[m]$ as the sequence

$$r_n[m] = \begin{cases} \sum_{i=-\infty}^{\infty} W_K^{ik_0K} x[mM + iK + n]h[-iK - n] & \text{for } 0 \le n \le K - 1, \\ 0 & \text{otherwise.} \end{cases}$$ (14.138)

Then for each fixed m, $R_k[m]$ is a special case of the K-point generalized DFT (GDFT) of $r_n[m]$ given by [CR83]

$$R_k[m] = \sum_{n=0}^{K-1} r_n[m]W_K^{(k+k_0)n}.$$ (14.139)

Note that in Eq. (14.138)

$$W_K^{ik_0K} = \begin{cases} 1, & k_0 = 0, \\ (j)^i, & k_0 = \frac{1}{4}, \\ (-1)^i, & k_0 = \frac{1}{2}. \end{cases}$$ (14.140)

Equations (14.138) and (14.139) suggest an implementation based on a weighted overlapping and summing of the input signal followed by a GDFT computation. For $k_0 = 0$, the regular DFT computation can be used. The synthesis filter bank can be obtained using the reverse operations. Among the operations required in the synthesis is the computation of the inverse GDFT of the sequence $R_k[m]$:

$$
\begin{aligned}
r_n[m] &= \frac{1}{K} \sum_{k=0}^{K-1} R_k[m] W_K^{-(k+k_0)n} \\
&= W_K^{-k_0 n} \frac{1}{K} \sum_{k=0}^{K-1} R_k[m] W_K^{-kn}.
\end{aligned}
\tag{14.141}
$$

From Eqs. (14.139) and (14.141) it is easy to verify that the GDFT relationships are valid. Note that $r_n[m] W_K^{k_0 n}$ and $R_k[m]$ form a DFT pair.

Polyphase Implementation for Critically Sampled Filter Banks. The overlap-add procedure described in the previous section is suited for a situation when $M \neq K$ and FIR lowpass filters are used so that the overlap-add is performed over a finite number of blocks. When $M = K$, a particularly efficient structure results when the polyphase representation for the lowpass filter is used. Consider the following polyphase representation of $h[n]$ as given by

$$
h[n] = \sum_{p=0}^{M-1} h_p^u[n - p]
\tag{14.142}
$$

and

$$
h_p[n] = h[nM + p].
\tag{14.143}
$$

Recall that h_p^u denotes the up-sampled version of h_p obtained with the insertion of $M - 1$ zeros between adjacent samples of h_p. The polyphase representation of $x[n]$ in terms of $\bar{x}_p[n]$ is defined as

$$
\bar{x}_p[n] = x[nM - p].
\tag{14.144}
$$

Now considering the filtering operation in the kth path of Figure 14-40(a), Eq. (14.128) can be rewritten as

$$
\begin{aligned}
S_k[m] &= h[n] \circledast \bar{x}_k[n]\big|_{n=mM} \\
&= \sum_{n=-\infty}^{\infty} x[mM - n] W_M^{(k+k_0)(mM-n)} h[n] \\
&= \sum_{p=0}^{M-1} \sum_{n=-\infty}^{\infty} x[mM - nM - p] W_M^{(k+k_0)(mM-nM-p)} h[nM + p] \\
&= W_M^{mMk_0} \sum_{p=0}^{M-1} W_M^{-p(k+k_0)} \sum_{n=-\infty}^{\infty} \bar{x}_p[m - n] h_p[n] W_M^{-nMk_0}.
\end{aligned}
\tag{14.145}
$$

Equation (14.145) suggests an implementation in which the polyphase components $\bar{x}_p$ of the input signal are applied to the filters with impulse response $h_p[n] W_M^{-nMk_0}$. The GDFT of the M output is computed for values of $k = 0, 1, \ldots, K - 1$ to obtain $S_k[m]$ at each fixed value of m. The outputs of the GDFT are modulated by $W_M^{mMk_0}$. The processing for the analysis filter bank is shown in Figure 14-42. The computational advantages in the implementation in Figure 14-42 over that in Figure 14-40 are clear from the fact that operations of the polyphase filters are shared by all channel filters, and the GDFT can be implemented by fast algorithms. The implementation of the synthesis filter bank can be carried out in a similar manner.

14-5-2-2 Uniform Filter Banks for Real Signals.
The techniques for processing complex signals described in Section 14-5-2 can be applied to real signals as well. However, if both the input and output signals are real, then the resulting symmetries in the frequency domain affect the choice and structure of filter banks. The required operations include bandpass interpolation or decimation, which can be interpreted in terms of single sideband modulation or demodulation.

Consider three specific cases of filter banks that can be used to process real signals. These are related to the three cases considered in Figures 14-39(a)–(c) for complex signals. Attention is focused first on the analysis bank for partitioning a real signal whose Fourier transform is sketched in Figure 14-43(a). Consider the ideal partition of the signal shown in Figures 14-43(b)–(d). Due to the restriction that the signals are real, some bands duplicate the information contained in other bands. So separate channel signals corresponding to the different bands in Figures 14-43(b)–(d) are not used. These signals can be paired when necessary. Let N

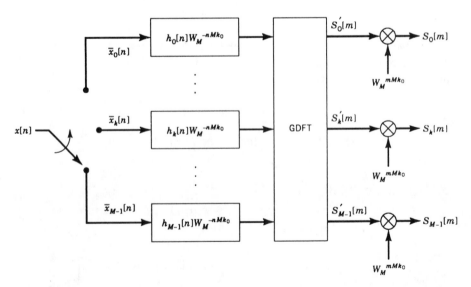

FIGURE 14-42 Efficient implementation of uniform analysis filter bank using GDFT and polyphase filters.

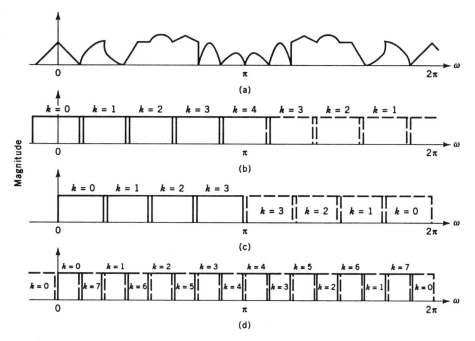

FIGURE 14-43 (a) Assumed Fourier transform of real-valued signal. (b)–(d) Three types of frequency band partitions in filter banks for real-valued signals.

denote the total number of bands. Assume that the centers of the bands are separated by $2\pi/K$, where as before K is referred to as the band separation factor. In dealing with complex signals, the number of bands and the band separation factors are considered equal. Here, in the case of real signals, a distinction between the two is made.

In Figure 14-43(b)–(d) the ideal bands are chosen symmetrically about $\omega = 0$ (or about $\omega = \pi$). The band center frequencies are $\pm \omega_k$, where

$$\omega_k = (k + k_0)2\pi/K \qquad (14.146)$$

and $k = 0, 1, \ldots, N - 1$. In this figure, $K = 8$. The choice of N is made in a manner that avoids duplication of information in the bands. In Figures 14-43(b)–(d) the values of k_0 are $0, \frac{1}{2}, \frac{1}{4}$, and the widths of the bands are $2\pi/K, 2\pi/K, \pi/K$, respectively. The filters for these bands can be obtained by a translation $\pm \omega_k$ of the responses shown in Figure 14-39(d) for the response in Figures 14-43(b) and 14-43(c) and of the response in Figure 14-39(e) for the partition in Figure 14-43(d).

In Figure 14-43(b) the number of bands N is $\lfloor (K + 2)/2 \rfloor$. When K is even the lowest and highest frequency channel signals have half the bandwidth of the remaining signals. This is because the shifts by $+\omega_k$ and $-\omega_k$ for $k = 0$ and $k = K/2$ give overlapping bands. The sampling rate for these two signals is lower than that of the remaining signals and is ideally $1/K$ times the input rate. The remaining signals need to be sampled at twice the rate of the lowest band signal. When K is odd, only the lowest frequency band is sampled at a lower rate compared with the remaining channels. The remaining $\lfloor K/2 \rfloor$ channel signals have equal bandwidth.

In Figure 14-43(c) the number of banks N is $\lfloor (K + 1)/2 \rfloor$. With K even, all channel signals have equal bandwidth and the channel down-sampling factors are equal. When K is odd, then the highest-frequency band has half the bandwidth of the remaining bands and the signal needs to be sampled at half the rate of the other channel signals. Finally, in Figure 14-43(d) the number of bands $N = K$. For both odd and even K, the bandwidths of the K signals are equal and the down-sampling factors are equal. Note that the gaps in the bands obtained by the shift ω_k, $k = 0$, $1, \ldots, K - 1$, are filled by the bands due to shifts by $-\omega_k$.

Once the choice of the frequency bands is made, the processing can be carried out as in the case of complex signals by sharing the filtering computations among the different channels and casting the modulation operations in terms of DFT computations, which can efficiently be performed using fast algorithms. An example of this processing is described in Section 14-6 for carrying out the transmultiplexer implementation.

14-5-3 Aliasing Cancellation and Exact Reconstruction Filter Banks

The discussion on the uniform filter banks considered in Section 14-5-2 focused on the efficient implementation of a set of filters using a single prototype lowpass filter. The underlying assumption is that the channel signals have negligible aliasing because the filter attenuation characteristics are chosen to ensure it. Consider now a situation where the channel signals may have aliasing distortion, but the analysis–synthesis filter bank pair is carefully designed to either mitigate or eliminate it. A two-channel filter bank is examined first and then the multichannel case is discussed.

14-5-3-1 Two-Channel Filter Bank Pair with Aliasing Cancellation. Consider the two-channel analysis–synthesis filter bank pair shown in Figure 14-44. The analysis filter bank splits the input signal into two channel signals by processing it with a lowpass filter $G_0(z)$ in one path and with a highpass filter $G_1(z)$ in the other. The filtered signals $u_0[n]$ and $u_1[n]$ are down-sampled by a factor of 2 to obtain the subband signals. These signals are then processed according to the application requirements. In subband coding the signals are subjected to data compression and encoded suitably for transmission or storage. For recovering the original signal, the channel signals are first decoded. The decoded signals suffer from possible distortions due to coding and quantization errors and channel impairments. In the discussion here, it is assumed that there is no coding or quantization degradation. These signals are up-sampled and processed by filters $F_0(z)$ and $F_1(z)$ as shown in Figure 14-44. The output signals $y_0[n]$ and $y_1[n]$ are then added to yield $y[n]$, which approximates the original input signal $x[n]$. Note that in a back-to-back connection of the analysis and synthesis banks, the signals $u_k[n]$

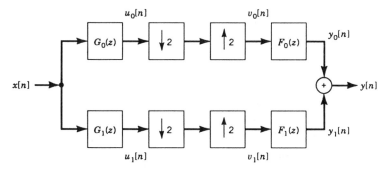

FIGURE 14-44 Two-channel analysis and synthesis filter banks.

and the up-sampled signals $v_k[n]$, $k = 0, 1$, are related by

$$v_k[n] = \begin{cases} u_k[n], & n \text{ even} \\ 0, & n \text{ odd} \end{cases}$$

$$= \tfrac{1}{2}(1 + (-1)^n)u_k[n]. \tag{14.147}$$

The corresponding z-transforms are related by

$$V_k(z) = \tfrac{1}{2}[U_k(z) + U_k(-z)]. \tag{14.148}$$

Looking at Figure 14-44 and using Eq. (14.148), the z-transform of the output $y[n]$ can be expressed as

$$Y(z) = \tfrac{1}{2}[F_0(z)G_0(z) + F_1(z)G_1(z)]X(z)$$

$$+ \tfrac{1}{2}[F_0(z)G_0(-z) + F_1(z)G_1(-z)]X(-z). \tag{14.149}$$

The first term on the right-hand side is proportional to $X(z)$ and the second term containing the aliasing component $X(-z)$ arises due to a frequency translation caused by down-sampling and up-sampling. The undesirable aliasing term vanishes if the filters are chosen to satisfy the condition

$$F_0(z)G_0(-z) + F_1(z)G_1(-z) = 0. \tag{14.150}$$

If in addition to aliasing cancellation, it is required that the highpass filter be obtained by a frequency translation of the lowpass filter to get a pair of mirror image responses, then

$$G_0(z) = G_1(-z) = H(z). \tag{14.151}$$

Under the restriction Eq. (14.151), Eq. (14.150) can be satisfied by choosing

$$F_0(z) = CH(z) \tag{14.152}$$

and

$$F_1(z) = -CH(-z). \tag{14.153}$$

The resulting filter bank is referred to as a *quadrature mirror filter* (QMF) bank [ES77; GA84]. The overall system with input $x[n]$ and output $y[n]$ now becomes a time-invariant system. If $C = 2$ in Eqs. (14.152) and (14.153), $Y(z)$ and $X(z)$ are related by

$$Y(z) = D(z)X(z), \tag{14.154}$$

where

$$D(z) = H^2(z) - H^2(-z). \tag{14.155}$$

If it is required that $Y(z) = z^{-n_0} X(z)$ for some integer n_0, then $D(z) = z^{-n_0}$. Suppose $H(z)$ is a linear-phase FIR filter. The filter is either of odd length $(2K + 1)$ whose (real) impulse response satisfies

$$h[n] = h[-n] \tag{14.156}$$

with $h[K] = h[-K] \neq 0$, or of even length $(2K)$ with

$$h[n] = h[1 - n] \tag{14.157}$$

and $h[K] = h[-K + 1] \neq 0$. If the filter length is odd, then $H^2(e^{j\omega})$ is non-negative with $H^2(e^{j\pi/2}) = H^2(-e^{j\pi/2})$. Since this forces $D(e^{j\pi/2}) = 0$, the choice of odd-length filters is ruled out. For even-length filters in Eq. (14.157) it is observed that $H^2(z)$ has linear phase with group delay of 1. The requirement is,

$$z^{n_0}[H^2(z) - H^2(-z)] = 1. \tag{14.158}$$

$H(z)$ can be expressed in its polyphase form with a scaling factor $\frac{1}{2}$ inserted for convenience; thus

$$H(z) = \tfrac{1}{2}[H_0(z^2) + z^{-1}H_1(z^2)] \tag{14.159}$$

so that the condition Eq. (14.158) can be rewritten as

$$z^{n_0-1}H_0(z^2)H_1(z^2) = 1. \tag{14.160}$$

The symmetry in Eq. (14.157) implies that the polyphase components are related by

$$H_1(z^2) = H_0(z^{-2}) \tag{14.161}$$

so that

$$z^{n_0 - 1} H_0(z^2) H_0(z^{-2}) = 1. \tag{14.162}$$

Since $H_0(z^2)$ is the transfer function of a FIR filter, Eq. (14.162) forces $H_0(z^2)$ to have only one nonzero sample, which is either $2h[K]$ or $2h[-K + 1]$ depending on whether K is even or odd, respectively. Also, $n_0 = 1$. Therefore the only linear-phase solution to satisfy Eq. (14.158) is [VA85]

$$h[n] = \begin{cases} \frac{1}{2}, & \text{for } n = K, -K + 1, \\ 0, & \text{otherwise.} \end{cases} \tag{14.163}$$

The choice of $K = 1$ in Eq. (14.163) yields a nominal lowpass response. In order to design QMF filters so that $H(e^{j\omega})$ provides a better approximation of an ideal lowpass response, one can relax the requirement for exactly meeting the condition Eq. (14.158) and use a computer-aided design procedure [GR88; JO80] by setting up a suitable error criterion. A lowpass response is desired for $H(e^{j\omega})$ together with a close approximation of the following condition due to Eq. (14.158),

$$|H(e^{j\omega})|^2 + |H(e^{j(\omega + \pi)})|^2 = 1. \tag{14.164}$$

The objective function for optimization can be chosen as a weighted error function E defined as [JO80]

$$E = E_r + \alpha E_s(\omega_s), \tag{14.165}$$

where

$$E_r = \int_0^\pi [|H(e^{j\omega})|^2 + |H(e^{j(\omega + \pi)})|^2 - 1]^2 \, d\omega \tag{14.166}$$

and

$$E_s(\omega_s) = \int_{\omega_s}^\pi |H(e^{j\omega})|^2 \, d\omega. \tag{14.167}$$

E_r represents the error in approximating the condition on the filter frequency response in Eq. (14.164) and E_s represents the stopband approximation error for a specified stopband edge frequency ω_s, $0.5\pi < \phi_s < \pi$. A search procedure is used to find the coefficient values for a given filter order $2K$. Tables 14-3 and 14-4 list results [JO80] for choices of $2K = 8, 12, 16, 24, 32, 48$, and 64 and for different

TABLE 14-3   One-Sided Coefficients from Center to End

8 TAP

0.48998080E-00
0.69428270E-01
−0.70651830E-01
0.93871500E-02

12 TAP(A)	12 TAP(B)
0.48438940E-00	0.48079620E-00
0.88469920E-01	0.98085220E-01
−0.84695940E-01	−0.91382500E-01
−0.27103260E-02	−0.75816400E-02
0.18856590E-01	0.27455390E-01
−0.38096990E-02	−0.64439770E-02

16 TAP(A)	16 TAP(B)	16 TAP(C)
0.48102840E-00	0.47734690E-00	0.47211220E-00
0.97798170E-01	0.10679870E-00	0.11786660E-00
−0.90392230E-01	−0.95302340E-01	−0.99295500E-01
−0.96663760E-02	−0.16118690E-01	−0.26275600E-01
0.27641400E-01	0.35968530E-01	0.46476840E-01
−0.25897560E-02	−0.19209360E-02	0.19911500E-02
−0.50545260E-02	−0.99722520E-02	−0.20487510E-01
0.10501670E-02	0.28981630E-02	0.65256660E-02

24 TAP(B)	24 TAP(C)	24 TAP(D)
0.47312890E-00	0.46864790E-00	4.6542880E-01
0.11603550E-00	0.12464520E-00	1.3011210E-01
−0.98297830E-01	−0.99878850E-01	−9.9844220E-02
−0.25615330E-01	−0.34641430E-01	−4.0892220E-02
0.44239760E-01	0.50881620E-01	5.4029850E-02
0.38915220E-02	0.10046210E-01	1.5473930E-02
−0.19019930E-01	−0.27551950E-01	−3.2958390E-02
0.14464610E-02	−0.65046690E-03	−4.0137810E-03
0.64858790E-02	0.13540120E-01	1.9763800E-02
−0.13738610E-02	−0.22731450E-02	−1.5714180E-03
−0.13929110E-02	−0.51829780E-02	−1.0614000E-02
0.38330960E-03	0.23292660E-02	4.6984260E-03

32 TAP(C)	32 TAP(D)	32 TAP(E)
0.46640530E-00	4.6367410E-01	0.45964550E-00
0.12855790E-00	1.3297250E-01	0.13876420E-00
−0.99802430E-01	−9.9338590E-02	−0.97683790E-01
−0.39348780E-01	−4.4524230E-02	−0.51382570E-01
0.52947450E-01	5.4812130E-02	0.55707210E-01
0.14568440E-01	1.9472180E-02	0.26624310E-01
−0.31238620E-01	−3.4964400E-02	−0.38306130E-01
−0.41874830E-02	−7.9617310E-03	−0.14569000E-01

TABLE 14-3 (*Continued*)

32 TAP(C)	32 TAP(D)	32 TAP(E)
0.17981450E-01	2.2704150E-02	0.28122590E-01
−0.13038590E-03	2.0694700E-03	0.73798860E-02
−0.94583180E-02	−1.4228990E-02	−0.21038230E-01
0.14142460E-02	8.4268330E-04	−0.26120410E-02
0.42341950E-02	8.1819410E-03	0.15680820E-01
−0.12683030E-02	−1.9696720E-03	−0.96245920E-03
−0.14037930E-02	−3.9711520E-03	−0.11275650E-01
0.69105790E-03	2.2451390E-03	0.51232280E-02

48 TAP(C)	48 TAP(D)	48 TAP(E)
0.46424160E-00	0.46139480E-00	0.45817950E-00
0.13207910E-00	0.13639810E-00	0.14082370E-00
−0.99384370E-01	−0.98437790E-01	−0.96727910E-01
−0.43596380E-01	−0.48731140E-01	−0.53990540E-01
0.54326010E-01	0.55379000E-01	0.55307280E-01
0.18809490E-01	0.24020070E-01	0.29675180E-01
−0.34090220E-01	−0.36906340E-01	−0.38442230E-01
−0.78016710E-02	−0.12422540E-01	−0.18039380E-01
0.21736090E-01	0.25813150E-01	0.28813790E-01
0.24626820E-02	0.60226430E-02	0.11160250E-01
−0.13441620E-01	−0.18121920E-01	−0.22285690E-01
−0.61169920E-04	−0.23574670E-02	−0.66961680E-02
0.78402940E-02	0.12465680E-01	0.17437190E-01
−0.75614990E-03	0.33292710E-03	0.36242470E-02
−0.42153860E-02	−0.82474350E-02	−0.13593290E-01
0.78333890E-03	0.63647700E-03	−0.15049380E-02
0.20340170E-02	0.51489700E-02	0.10453240E-01
−0.52055740E-03	−0.95592250E-03	0.25450090E-04
−0.85293900E-03	−0.29611340E-02	−0.78527550E-02
0.24225190E-03	0.89979030E-03	0.99124640E-03
0.30117270E-03	0.15016570E-02	0.56447870E-02
−0.56157570E-04	−0.66471290E-03	−0.16909870E-02
−0.92054790E-04	−0.61083240E-03	−0.37667050E-02
−0.14619070E-04	0.40829340E-03	0.25404290E-02

64 TAP(D)	64 TAP(E)
0.46009810E-00	0.45725790E-00
0.13823630E-00	0.14202200E-00
−0.97790960E-01	−0.96089540E-01
−0.50954870E-01	−0.55443150E-01
0.55432450E-01	0.54974440E-01
0.26447000E-01	0.31331460E-01
−0.37649730E-01	−0.38449870E-01
−0.14853970E-01	−0.19830160E-01
0.27160550E-01	0.29128720E-01
0.82875600E-02	0.13062040E-01

TABLE 14-3   One-Sided Coefficients from Center to End (*Continued*)

64 TAP(D)	64 TAP(E)
−0.19943650E-01	−0.22925640E-01
−0.43136740E-02	−0.86300040E-02
0.14593960E-01	0.18350770E-01
0.18947140E-02	0.55598240E-02
−0.10506890E-01	−0.14772120E-01
−0.48579350E-03	−0.33824410E-02
0.73671710E-02	0.11865500E-01
−0.25847670E-03	0.18165760E-02
−0.49891470E-02	−0.94456860E-02
0.57431590E-03	−0.70816120E-03
0.32358770E-02	0.74161550E-02
−0.62437240E-03	−0.57608340E-04
−0.19861770E-02	−0.57047680E-02
0.53085390E-03	0.55112530E-03
0.11382600E-02	0.42629060E-02
−0.38236310E-03	−0.84474820E-03
−0.59535630E-03	−0.30644720E-02
0.22984380E-03	0.98342750E-03
0.27902770E-03	0.20745920E-02
−0.11045870E-03	−0.10196710E-02
−0.11235150E-03	−0.12577780E-02
0.35961890E-04	0.10798060E-02

Adapted from Johnston [JO80] with permission from IEEE.

values of weighting parameter α and the transition width $(2\omega_s - \pi)/2\pi$. Figure 14-45 shows the magnitude and error plots for $2K = 48$, $\alpha = 2$ and $\omega_s = 0.586\pi$.

A choice of filters that allows cancellation of aliasing and eliminates magnitude distortion is a class of IIR filters $H(z)$ such that the relations Eqs. (14.151)–(14.153) hold with $C = 2$ [BA82; RE87c]. The IIR filters that are suitable for such filter banks have been considered before in Section 14-4, where $H(z)$ is given by Eq. (14.159) and $H_0(z)$ and $H_1(z)$ are allpass transfer functions with magnitude unity and with poles inside the unit circle. In this case

$$D(z) = z^{-1}H_0(z^2)H_1(z^2) \qquad (14.168)$$

and $|D(e^{j\omega})| = 1$. The filters given in Section 14-4 serve as solutions in this case. Issues in the selection of initial conditions in exact reconstruction IIR filter banks are discussed in Babic et al. [BA91] and Mitra et al. [MI92].

14-5-3-2   *Two-Channel Exact Reconstruction Filter Bank.*   Section 14-5-3 has examined two-channel filter banks that permit signal reconstruction without aliasing distortions in the absence of coding and quantization errors. It is also possible to exactly reconstruct the signal by satisfying the condition Eq. (14.150) together

TABLE 14-4 Specifications and Properties of QMF Coefficients in Table 14-3

Number of TAPS	Transition Band[a]	Stopband Weighting	Passband Ripple (dB)	Stopband Rejection (dB)	Ultimate Stopband Rejection (dB)
8			0.06	31	31
12	A	1	0.025	48	50
16			0.008	60	75
12			0.04	33	36
16	B	1	0.02	44	48
24			0.008	60	78
16		1	0.07	30	36
24		1	0.02	44	49
32	C	2	0.009	51	60
48		2	0.002	63	80
24		1	0.1	30	38
32		2	0.025	38	48
48	D	2	0.006	50	66
64		5	0.002	65	80
48	E	2	0.07	32	46
64		5	0.025	40	51

Adapted from Johnston [JO80] with permission from IEEE.
[a]The band code letters correspond to the transition bands as follows:

Transition Code Letter	Normalized Transition Band
A	0.14
B	0.10
C	0.0625
D	0.043
E	0.023

with

$$D(z) = \tfrac{1}{2}(F_0(z)G_0(z) + F_1(z)G_1(z)) = z^{-n_0}. \qquad (14.169)$$

For a given choice of stable filters with rational transfer functions $G_0(z)$ and $G_1(z)$, the requirement on $F_0(z)$ and $F_1(z)$ for exact reconstruction can be stated as

$$\begin{bmatrix} G_0(z) & G_1(z) \\ G_0(-z) & G_1(-z) \end{bmatrix} \begin{bmatrix} F_0(z) \\ F_1(z) \end{bmatrix} = \begin{bmatrix} 2 \\ 0 \end{bmatrix} z^{-n_0} \qquad (14.170)$$

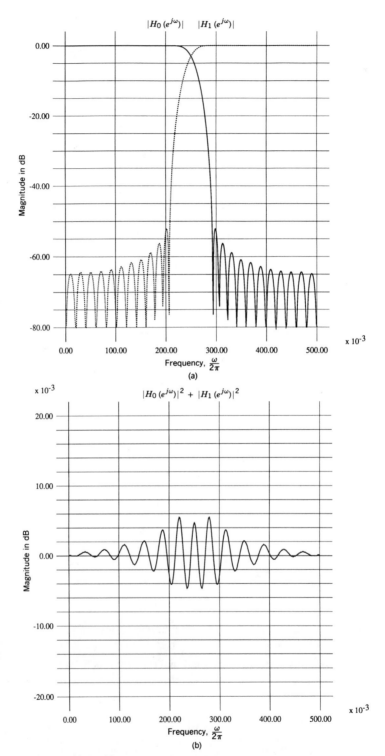

FIGURE 14-45 (a) Magnitude and (b) error plots for 48-tap (*D*) QMF filter.

or, using matrix and vector notation,

$$G(z)f(z) = 2z^{-n_0}e, \qquad (14.171)$$

where $G_{ij}(z) = G_{j-1}((-1)^{i-1}z)$, $i, j = 1, 2$, $f(z) = [F_0(z) \quad F_1(z)]^t$ and $e = [1 \quad 0]^t$, where superscript t indicates transpose. Assume that $\det G(z) = G_0(z)G_1(-z) - G_0(-z)G_1(z)$ is not zero for $|z| = 1$. Then BIBO stable, possibly noncausal, filters can be found that provide exact reconstruction. These filters are given by

$$f(z) = 2z^{-n_0}G^{-1}(z)e. \qquad (14.172)$$

First consider the use of IIR filters for constructing an exact reconstruction filter bank. In some applications such as image processing, it is permissible to use non-causal IIR filters. This allows an exact reconstruction filter bank pair to be constructed with mirror image magnitude response. Again, the analysis filters are expressed by Eq. (14.151) in terms of $H(z)$ given by Eq. (14.159), with $H_0(z^2) = 1$ and $H_1(z^2)$ allpass. The analysis filter transfer functions are given by

$$G_k(z) = \tfrac{1}{2}[1 + (-1)^k z^{-1}H_1(z^2)], \qquad k = 0, 1. \qquad (14.173)$$

It can be verified that the synthesis filter transfer functions for exact reconstruction are given by

$$F_k(z) = 1 + (-1)^k z H_1(z^{-2}), \qquad k = 0, 1. \qquad (14.174)$$

These filters can be used in tree-structured form [RE87c].

Now consider the use of FIR filters for analysis filters $G_k(z)$. If again $\det G(z) \neq 0$ for $|z| = 1$, then possibly noncausal IIR filters $F_k(z)$, $k = 0, 1$, can be found that are determined from Eq. (14.172). However, if both analysis and synthesis filters are constrained to be FIR, then there is a technique of deriving them from half-band filters [MI85; SM84]. Here constraints of Eqs. (14.151)–(14.153) that limit the possible solutions are removed.

Let $G_{HB}(z)$ be the transfer function of a (zero-phase) symmetric half-band FIR filter designed using the procedure described in Section 14-3-4, with $g_{HB}[n]$ denoting its inverse z-transform. Note that $G_{HB}(z) = G_{HB}(z^{-1})$ and $G_{HB}(z) + G_{HB}(-z) = 2g_{HB}(0)$. The filter length is $2K + 1$, with K odd, and an illustrative frequency response is sketched in Figure 14-46(a). The transfer function $G_{HB}(z)$ can be factorized as

$$G_{HB}(z) = \tfrac{1}{2}F_0(z)G_0(z), \qquad (14.175)$$

where the inverse z-transform of $G_0(z)$ is zero outside the range $0 \leq n \leq L$, and the inverse z-transform of $F_0(z)$ is equal to zero for $n \notin \{-K, -K + 1, \ldots,$

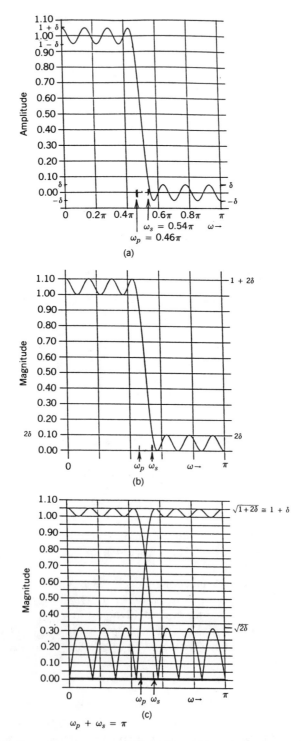

FIGURE 14-46 Frequency responses of (a) half-band filter, (b) half-band filter with non-negative frequency response, and (c) derived mirror image responses.

$K - L\}$. Let

$$G_1(z) = \tfrac{1}{2}z^{-K}F_0(-z) \qquad (14.176)$$

and

$$F_1(z) = 2z^K G_0(-z). \qquad (14.177)$$

Then

$$\tfrac{1}{2}(F_0(z)G_0(z) + F_1(z)G_1(z)) = G_{HB}(z) + G_{HB}(-z)$$

$$= 2\,g_{HB}(0) = 1. \qquad (14.178)$$

Also, the aliasing cancellation condition Eq. (14.150) is satisfied since K is odd.

Now consider the construction of a filter bank pair derived from a half-band filter such that the filters in each bank have mirror image magnitude response. The zeroth impulse response sample $g_{HB}(0)$ can be adjusted to make the frequency response nonnegative and this can further be scaled to get a modified frequency response $\tilde{G}(e^{j\omega})$ shown in Figure 14-46(b) such that $0 \le \tilde{G}_{HB}(e^{j\omega}) \le 1$. The modified transfer function can be factored as

$$\tilde{G}_{HB}(z) = G_0(z)G_0(z^{-1}), \qquad (14.179)$$

where $G_0(z) = \Sigma_{k=0}^{K} g_0[k]z^{-k}$. From Eqs. (14.170)–(14.172)

$$F_0(z) = 2G_0(1/z); \qquad G_1(z) = z^{-K}G_0(-1/z); \qquad F_1(z) = 2z^K G_0(-z).$$

$$(14.180)$$

The magnitude response of $G_0(e^{j\omega})$ is shown in Figure 14-46(c).

14-5-4 *M*-Channel Critically Sampled Filter Banks [VA87]

In studying a filter bank with M channels one can extend the ideas developed in Section 14-5-3. The aliasing cancellation procedure of a QMF bank can be extended to $M > 2$ channels. Under the assumption that aliasing into a channel occurs primarily from adjacent channels due to the fact that the filter stopband can be specified to cover the nonadjacent channels, one can focus on the cancellation of the aliasing due to adjacent bands. This problem is examined in Rothweiler [RO83], Nussbaumer [NU81] and Cox [CO86], and filter design procedures are suggested.

If perfect reconstruction is desired then one can proceed as in Section 14-5-3 to derive the expression for the output signal in terms of the input signal and its

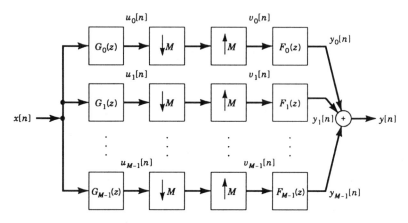

FIGURE 14-47 M-channel analysis and synthesis filter banks.

modulated versions [SM87; SW86; VE87]. Consider the directly connected analysis and synthesis filter banks in Figure 14-47. The output signal $u_k[n]$ from the analysis filter and the signal $v_k[n]$ obtained by down-sampling and up-sampling $u_k[n]$ by factors of M (equal to the number of channels) are related by

$$v_k[n] = \frac{1}{M} \left(\sum_{i=0}^{M-1} W_M^{-in} \right) u_k[n], \tag{14.181}$$

where $W_M = e^{-j2\pi/M}$ and $k = 0, 1, \ldots, M - 1$. The corresponding z-transforms are related by

$$V_k(z) = \frac{1}{M} \sum_{i=0}^{M-1} U_k(zW_M^i) \tag{14.182}$$

and $U_k(z) = G_k(z)X(z)$. The z-transform of the output y can be expressed as

$$Y(z) = \frac{1}{M} \sum_{k=0}^{M-1} F_k(z) \sum_{i=0}^{M-1} G_k(zW_M^i)X(zW_M^i). \tag{14.183}$$

This is compactly expressed as

$$Y(z) = \frac{1}{M} \mathbf{x}^t(z)\,\mathbf{G}(z)\,\mathbf{f}(z), \tag{14.183a}$$

where the element in the ith row and kth row of matrix $\mathbf{G}(z)$ is $G_k(zW_M^i)$, $k, i = 0, 1, \ldots, M - 1$, the kth element of M-vector $\mathbf{f}(z)$ is $F_k(z)$ and the ith element of M-vector $\mathbf{x}(z)$ is $X(zW_M^i)$. Note that the rows and column of matrices and elements of vectors are labeled beginning $0, 1, \ldots, M - 1$.

For perfect reconstruction, $Y(z) = z^{-n_0}X(z)$, which forces the following relationship on the filters:

$$\mathbf{G}(z)\mathbf{f}(z) = Mz^{-n_0}\mathbf{e}, \tag{14.184}$$

where $\mathbf{e}^t = [1 \ \ 0 \ \ 0 \ \ \cdots \ \ 0]^t$.

To get a better insight into the structure of the M-channel exact reconstruction filter bank [VA87], it is of interest to examine the polyphase representation of the filter $G_k(z)$ in the kth path of the analysis filter bank:

$$G_k(z) = \sum_{p=0}^{M-1} z^{-p} E_{kp}(z^M). \tag{14.185}$$

All $G_k(z)$ are assumed to be rational transfer functions and therefore so are $E_{kp}(z)$. The processing is shown in Figure 14-48(a), where $s_k[n]$ is the down-sampled signal in the kth channel and is referred to as the kth subband signal. The down-sampler can be moved forward in the system as shown in Figure 14-48(b) with $E_{kp}(z^M)$ replaced by $E_{kp}(z)$ and this does not affect the operations. Note that $x[n]$ can be written in polyphase form as

$$x[n] = \sum_{p=0}^{M-1} x_p^u[n+p]. \tag{14.186}$$

The input to the filter $E_{kp}(z)$ in Figure 14-48(b) is just

$$x_p[n] = x[nM - p]. \tag{14.187}$$

Note that to keep notation simple, the bar used earlier in the alternate polyphase representation of Eq. (14.38) is dropped. However, the proper commutator should be used. The processing in Figure 14-48(b) is redrawn in Figure 14-48(c), which represents the kth path of the overall analysis filter that is shown in Figure 14-48(d). $\mathbf{E}(z)$ is a matrix whose (k, p)th element is $E_{kp}(z)$. Let $S_k(z)$ denote the z-transform of $s_k[n]$. Denote by $\mathbf{x}$ and $\mathbf{s}$ the vectors

$$\mathbf{x}^t = [X_0(z) \ \ X_1(z) \ \ \cdots \ \ X_{M-1}(z)] \tag{14.188}$$

and

$$\mathbf{s}^t = [S_0(z) \ \ S_1(z) \ \ \cdots \ \ S_{M-1}(z)]. \tag{14.189}$$

Then

$$\mathbf{s} = \mathbf{E}(z)\mathbf{x}. \tag{14.190}$$

In the synthesis filter bank the signals $x_k[n]$, $k = 0, 1, \ldots, M - 1$, are to be recovered with a possible delay, given the signals $s_k[n]$, $k = 0, 1, \ldots, M - 1$.

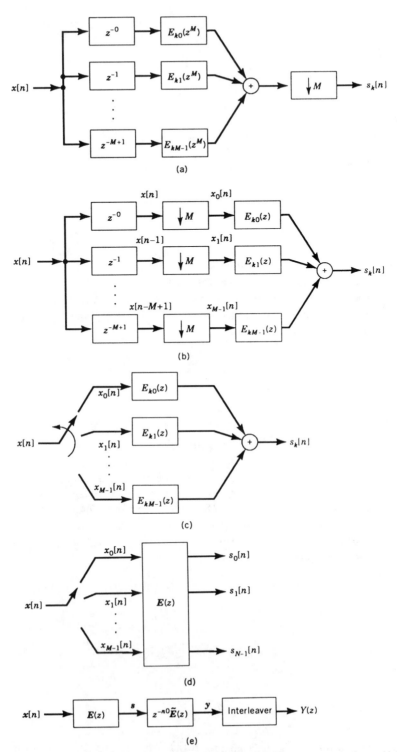

FIGURE 14-48 *M*-channel analysis filter bank and equivalent representations. (Adapted from Vaidyanathan [VA87] with permission from IEEE.)

Let $\tilde{\mathbf{E}}(z)$ be the transpose of the matrix $\mathbf{E}(z^{-1})$ with a conjugation of coefficients. If the system described by the matrix $z^{-n_0}\tilde{\mathbf{E}}(z)$ is used (see Figure 14-48(e)), then with the usual notation,

$$\mathbf{y} = z^{-n_0}\tilde{\mathbf{E}}(z)\,\mathbf{s} = z^{-n_0}\tilde{\mathbf{E}}(z)\,\mathbf{E}(z)\mathbf{x} = z^{-n_0}\mathbf{x}, \qquad (14.191)$$

where $\mathbf{y}^t = [Y_0(z)\quad Y_1(z)\quad Y_{M-1}(z)]$ with $Y_p(z)$ denoting the z-transform of the pth polyphase component of the signal $y[n]$. Therefore

$$\tilde{\mathbf{E}}(z)\mathbf{E}(z) = \mathbf{I} \qquad (14.192)$$

is required in a region $r_1 < |z| < r_2$, where $r_1 < 1 < r_2$ and $\mathbf{I}$ is an $M \times M$ identity matrix. Our requirement is met if $\mathbf{E}(z)$ is constrained to be unitary on the unit circle; that is,

$$[\mathbf{E}*(e^{j\omega})]^t\mathbf{E}(e^{j\omega}) = \mathbf{I}. \qquad (14.193)$$

By analytic continuation of Eq. (14.193), it is seen that Eq. (14.192) holds in a region $r_1 < |z| < r_2$ containing the unit circle, and $\mathbf{E}(z)$ is paraunitary.

Consider now possible choices for $\mathbf{E}(z)$ using allpass based analysis filters and FIR analysis and synthesis filters. If $G_k(z)$ are derived from a transfer function $H(z)$,

$$G_k(z) = H(W_M^k z) = \sum_{k=0}^{p} W_M^{-kp} z^{-p} H_p(z^M), \qquad (14.194)$$

then $\mathbf{E}(z)$ can be expressed as

$$\mathbf{E}(z) = \mathbf{\Omega}\mathbf{H}(z). \qquad (14.195)$$

where $\mathbf{\Omega}$ is the matrix whose (k, p)th element is $\Omega_{kp} = W_M^{-kp}$ and $\mathbf{H}$ is a diagonal matrix with (p, p)th diagonal element $H_p(z)$. If $H_p(z)$ is stable, causal allpass, then $H_p(z^{-1})$ would be stable but noncausal allpass. $\mathbf{E}(z)$ satisfies Eq. (14.193) and its implementation as well as that of $\tilde{\mathbf{E}}(z)$ would involve DFT computations

Suppose $\mathbf{E}(z)$ represents a FIR system. In order to design such a system with some control over the frequency responses of the channel filters, the filters must be suitably parametrized. A convenient method [VA89] is to express $\mathbf{E}(z)$ in a product form, where the factors are low-order FIR systems:

$$\mathbf{E}(z) = \mathbf{E}_K(z)\mathbf{E}_{K-1}(z)\cdots\mathbf{E}_1(z)\mathbf{E}_0, \qquad (14.196)$$

where $\mathbf{E}_0$ is an $M \times M$ unitary matrix and

$$\mathbf{E}_k(z) = \mathbf{I} - \mathbf{v}_k\mathbf{v}_k^{*t} + z^{-1}\mathbf{v}_k\mathbf{v}_k^{*t}, \qquad (14.197)$$

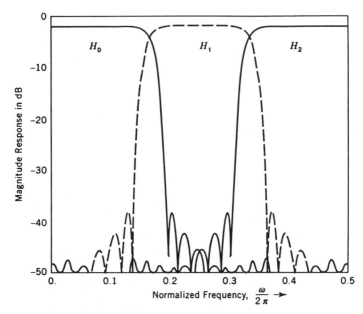

FIGURE 14-49   Magnitude response of filters in a three-channel exact reconstruction filter bank. (Adapted from Vaidyanathan [VA87] with permission from IEEE.)

where v_k is an M-vector with unit norm. With this parametrization the passband and stopband characteristics of the channel filters $G_k(z)$ can be optimized. For a paraunitary $E(z)$, the passband and stopband attenuations are related so that a criterion based only on stopband attenuation suffices. Figure 14-49 shows the analysis filter responses for $M = 3$ with filters of length 56.

14-6   APPLICATIONS

In this section three applications of multirate techniques are reviewed. These applications are FDM–TDM format conversion, efficient spectrum estimation, and subband image coding. There are several other applications such as subband speech coding [CR76b], conversion between delta modulation formats [GO71], and interpolation beamforming [PR79] that are not described here.

14-6-1   Transmultiplexer Design

The introduction of digital transmission systems and switches in predominantly analog telephone networks created a need for a translation between the digital format of time division multiplexing and the analog format of frequency division multiplexing. The interface between the two multiplexing formats is known as a transmultiplexer and its all-digital implementation using multirate techniques has received a great deal of attention. Early work on the application of multirate techniques in modulation and multiplexing is reported in Freeny et al. [FR71]. Several approaches to carrying out a transmultiplexer design using multirate techniques have since been proposed [AA78; PU82], and some of these techniques are summarized in Scheuermann and Gockler [SC81]. The basic principles of a 24-channel transmultiplexer used in the United States and Japan are briefly described here. The focus is on the TDM–FDM conversion of 12 digital signals out of the 24 channels into a sampled (digital) FDM signal. The overall system can be partitioned into the three subsystems as shown in Figure 14-50.

1. TDM Interface. Preliminary processing is carried out to extract the signal information and the pulse code modulated (PCM) input is converted from the 8-bit μ-law code to a linear code.

2. Digital Signal Processing (DSP) Unit. The front end of this part contains baseband filters to prevent disturbance from speech signals to signaling information. This is followed by the multiplexing of the proper sideband from the input signals to obtain the sampled version of the FDM signal.

3. FDM Interface. This unit filters out the FDM signal from the designated frequency band after the D/A conversion. Among the operations performed in the transmultiplexer, a key task is the bandpass filtering in the DSP unit in order to select a sideband of the input signal in the desired channel. This operation can be performed efficiently by using the polyphase network approach. A brief explanation of the procedure to accomplish this is first given. There are many variations to the procedure but the basic idea is captured here.

Consider now the placement of the sidebands of a set of 12 input signals sampled at 8 kHz, which together require a bandwidth of 48 kHz for single sideband frequency division multiplexing. Suppose the 12 sidebands are placed in a 60–108-kHz groupband. This can be accomplished by first placing the sidebands in the 4–52-kHz range of a digital signal sampled at 112 kHz. It is convenient to describe the process by adding to this set of 12 signals two more signals, identically zero, to be placed in the 0–4-kHz and 52–56-kHz bands. Thus consider the multiplexing of 14 signals, $x_m[n]$, $m = 0, 1, \ldots, 13$, where $x_0[n] = x_{13}[n] = 0$ for all n. The required 60–108-kHz groupband can be obtained by suitable bandpass filtering.

The extraction of the proper sideband in the mth channel is shown in Figure 14-51(a). In each path one needs bandpass interpolation, which is carried out as follows. Each of the signals $x_m[n]$ is first up-sampled by a factor of $N = 14$ by

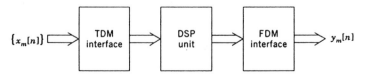

FIGURE 14-50 System for TDM–FDM conversion.

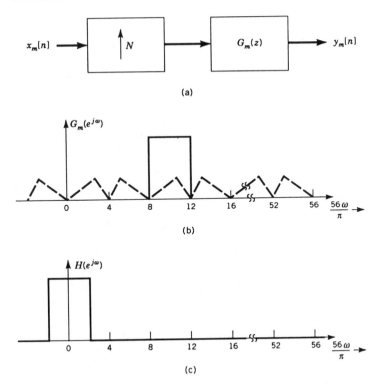

FIGURE 14-51 (a) Extraction of upper sideband in the mth channel. (b) Filter for extraction of signal in channel $m = 2$. (c) Ideal prototype filter response.

inserting zeros to get a sampling rate of 112 kHz. In the frequency domain this causes frequency-scaled images of the baseband spectrum to appear at even multiples of π/N in the discrete-time Fourier transform of the up-sampled signal $x_m^u[n]$. The upper (lower) sideband appears in the interval $I_m = (m\pi/N, (m + 1)\pi/N)$ for m even (odd). Therefore if the upper sideband for an interval I_m with m odd is to be retained, the frequencies in the baseband signal must be scrambled before the up-sampling operation. This is accomplished by reversing the sign of every other sample in the odd-numbered baseband signals. The up-sampled signal is then applied to a bandpass filter to extract the upper sideband signal from the appropriate interval. Let $G_m(z)$ be the transfer function of the filter that performs this operation. Its frequency response is shown in Figure 14-51(b), along with the Fourier transform of the up-sampled signal. The ideal filter $G_m(z)$ required for this purpose can be obtained from a prototype filter $H(z)$ whose response is shown for $m = 2$ in Figure 14-51(c), which is the ideal response of a lowpass interpolation filter for an interpolation factor of $M = 2N$. The transfer function $H(z)$ can be expressed in a polyphase network form as

$$H(z) = \sum_{k=0}^{2N-1} z^{-k} H_k(z^{2N}), \tag{14.198}$$

where N is the number of channels. $G_m(z)$ can be obtained by translating the prototype response by $\pm\omega_m$, where

$$\omega_m = \frac{(2m + 1)/\pi}{2N}.\qquad(14.199)$$

Note that $\pm\omega_m$ is the center of the frequency interval I_m. With this shift

$$G_m(z) = H(ze^{j\omega_m}) + H(ze^{-j\omega_m}).\qquad(14.200)$$

Using Eqs. (14.198)–(14.200) and noting that $(e^{j\omega_m})^{2N} = -1$, we find

$$G_m(z) = \sum_{k=0}^{2N-1} z^{-k}\, 2\cos(k\omega_m) H_k(-z^{2N}).\qquad(14.201)$$

Now the sampled version of the FDM signal $y[n]$ is obtained by summing the sideband signals extracted by $G_m(z)$. With the usual notation for the z-transform,

$$Y(z) = \sum_{m=0}^{N-1} X_m^u(z) G_m(z) = \sum_{m=0}^{N-1} X_m(z^N) G_m(z).\qquad(14.202)$$

Substituting Eq. (14.201) into Eq. (14.202) and reversing the order of summation yield

$$Y(z) = \sum_{k=0}^{2N-1} z^{-k} H_k(-z^{2N}) \left\{ \sum_{m=0}^{N-1} 2\cos(k\omega_m) X_m(z^N) \right\}.\qquad(14.203)$$

In this relation [MA81] the term in the braces is defined to be $V_k^u(z)$, which is the z-transform of the up-sampled version of the signal

$$v_k[n] = \sum_{m=0}^{N-1} 2\cos(k\omega_m) x_m[n], \qquad k = 0, 1, \ldots, 2N - 1.\qquad(14.204)$$

The processing thus reduces to the computation of the discrete cosine transform of the N input signal samples with time index m. The transform output is then processed by the bank of $2N$ filters $H_k(z)$ operating at the low rate of 8 kHz. The bulk of the computations is required in the filter bank, and the computation requirement is significantly reduced by the polyphase decomposition.

14-6-2 Spectrum Estimation

Estimation of the power spectrum of a discrete-time process from a finite record of observed samples finds applications in many areas. Early procedures are based on the use of the discrete Fourier transform computed with the fast Fourier transform algorithm. Recently, several model based procedures have become popular due to the improved resolution. Among these are the autoregressive (AR) spectrum estimation technique and the eigenanalysis based spectrum estimation techniques. In any spectrum estimation scheme, one is interested in the amount of computation required and the frequency resolution. Frequency resolution refers to the closest frequency separation of two sinusoids that can be distinguished. Two applications of decimation in these techniques are discussed: the reduction of computation in DFT and the improvement of frequency resolution in AR spectrum estimation.

Calculation of Narrowband Spectra Using FFT and Decimation. Computing the length N DFT of a real data sequence would produce a resolution of $1/N$. If one is interested only in a frequency band of (f_1, f_2) and the bandwidth $f_2 - f_1$ is only a small fraction of the sampling frequency, filtering and downsampling can reduce the computation burden while maintaining frequency resolution at $1/N$. The filter preserves the spectrum inside the band of interest and suppresses the aliasing caused by the subsequent down-sampling by D. The DFT of the decimated data sequence is then computed using the FFT algorithm. The multipliers $\{C_i\}$ are introduced to correct any distortion introduced by the filter in its passband. The simplest way to design the filter is to select a proper decimation factor and to choose the passband as (f_1, f_2) [LI78]. However, because of the postmultiplication by $\{C_i\}$, it is possible to use a much narrower passband, one that consists of a single point [CO80; QU83]. For very narrowband cases, the number of real multiplications per input data point is approximately 1.25 [QU83].

In the AR spectrum estimation approach [BU67; LA74; UL75; UL76], the data $\{y[n]\}$ are assumed to be a realization of an autoregressive random process of order M with coefficients γ_k excited by a white noise (innovations) $\{x[n]\}$.

$$y[n] = x[n] - \gamma_1 y[n-1] - \gamma_2 y[n-2] - \cdots - \gamma_M y[n-M].$$

$$(14.205)$$

The power density spectrum of $\{y[n]\}$ is

$$P_{AR}(\lambda) = \frac{T\sigma_x^2}{\left|1 + \sum_{k=1}^{M} \gamma_k e^{j2\pi k\lambda}\right|^2}, \qquad -0.5 \le \lambda < 0.5, \qquad (14.206)$$

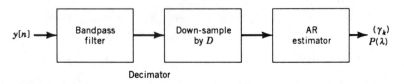

Decimator

FIGURE 14-52   Direct decimation system for AR spectrum analysis. (Adapted from Quirk and Liu [QU83] with permission from IEEE.)

where σ_x^2 is the variance of $x[n]$, $T = 1/f_s$ is the sampling interval, and $\lambda = f/f_s$ is the normalized digital frequency. The autoregressive coefficients $\{\gamma_k\}_1^M$ are related to the autocorrelation function ρ_i of the process $\{y[n]\}$ by the linear equations

$$\sum_{k=1}^{M} \gamma_k \rho_{i-k} = -\rho_i, \quad 1 \le i \le M.$$

There are many AR algorithms in literature that are used to obtain estimates of the autocorrelation or autocovariance function, which are then employed in solving for $\{\gamma_k\}_1^M$ [BU67; FO77; LA74; MA77; MO77: UL75; UL76]. Because of the form of Eqs. (14.205) and (14.206), the AR model is called an allpole model and this type of spectrum is also known as the linear prediction spectrum.

It has been shown that decimating the input data sequence can improve the frequency resolution [QU83]. Specifically, a model of order M/D applied to a signal decimated by D has the same resolution as an Mth order model applied to the undecimated signal. Moreover, decimation leads to more efficient computation.

The direct decimation technique is depicted in Figure 14-52. The bandpass filter preserves the spectrum inside the band of interest and suppresses the aliasing caused by the subsequent down-sampling by D. An AR spectrum is then computed from the decimated data sequence and the power spectrum so obtained has high resolution for the specified narrow frequency band. The high resolution is the consequence of two effects. The major effect is due to the down-sampling, which effectively expands the frequency scale by a factor of D, separating neighboring frequencies and thereby reducing the interference at one spectral peak caused by the proximity of the peak widths. Figure 14-53 shows three eight-order spectra for

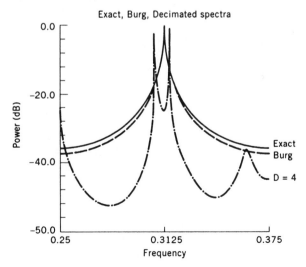

FIGURE 14-53 Eighth-order exact, Burg, and decimated ($D = 4$) spectra. (Adapted from Quirk and Liu [QU83] with permission from IEEE.)

two sine waves at frequencies separated by $\frac{1}{128}$, plus white Gaussian noise at 20 dB below the signal. Neither the exact AR nor the conventional Burg algorithm resolves the two sinusoids, but the spectrum of the decimated signal resolves the two components clearly and gives the correct frequencies. Figure 14-54 compares the conventional 32nd-order Burg spectrum to the 32nd-order decimated-by-four spectrum for a signal composed of five closely spaced sinusoids plus additive white Gaussian noise. Again, a dramatic increase in resolution results from this approach.

The benefit of using decimation in AR spectral estimation can be appreciated by looking at a typical plot of the resolution boundary shown in Figure 14-55. Resolution boundary is the minimum signal-to-noise ratio needed to resolve two equal-strength sinusoids at a given frequency separation for a given model order.

14-6-3   Subband Coding of Images

Recently, methods of subband coding of images have been proposed [GH86; WO86] and have spurred further research and application in still image and video coding [WO91]. The basic idea in subband coding is to take advantage of the nonuniform distribution of energy in the frequency domain in order to suitably allocate bits to the information content. In subband coding the image is decomposed into components that carry information from (ideally) nonoverlapping frequency bands in the original image. In view of the bandlimited nature of these signals they can be down-sampled to yield the *subband signals*. The lowest frequency subband signal is usually a lower resolution version of the original signal.

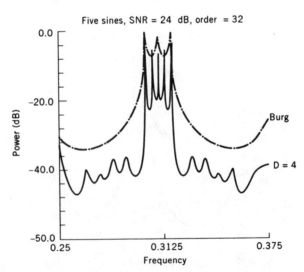

FIGURE 14-54   Decimated and undecimated spectra of five sine waves with solid line for the case of decimation by 4, and dashed line for undecimated Burg. (Adapted from Quirk and Liu [QU83] with permission from IEEE.)

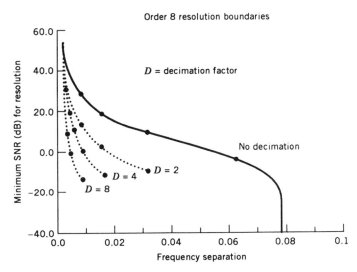

FIGURE 14-55 Resolution boundary curves for eighth-order AR spectra. (Adapted from Quirk and Liu [QU83] with permission from IEEE.)

It is sometimes convenient to use several stages of decomposition in a tree-structured manner leading to a hierarchy of resolutions. A related hierarchical coding procedure is the pyramidal coding procedure described in Burt and Adelson [BU83] in which at each level the signal is partitioned into a low- and a high-frequency component. However, while the low-frequency signal at each stage is down-sampled the high-frequency component is not. In the discussion here, all component signals in the subband decomposition are considered to be down-sampled.

An important task in subband coding is the construction of two sets of filters (analysis–synthesis filter banks), one of which decomposes the input signal into subband signals prior to coding and the other synthesizes the output signal from the coded subband signals. The use of quadrature mirror filter banks in two-dimensional subband decomposition is discussed in Vetterli [VE84]. In two-dimensional processing there is additional flexibility in choosing sampling grids. Filter banks using nonrectangular multirate techniques [DU85; ME83] have also been used in image subband coding [AN88]. Here, only separable methods of subband partition are considered.

A brief description of the subband coding procedure for images is given here. A four-band partition is shown in Figure 14-56(a). The input signal is applied to a bank of four filters whose outputs are down-sampled by a factor of 4 (by a factor of 2 in each of the vertical and horizontal directions). In practice, the filters are nonideal and would cause aliasing and in-band distortion. However, as discussed in Section 14-5, these filters can be designed to provide aliasing cancellation or even exact reconstruction. In practice, the decomposition can be carried out in two stages as shown in Figure 14-56(b). In the first stage the image, with $2M$ lines and $2N$ pixels/line, is applied to horizontal highpass and lowpass filters. The filter

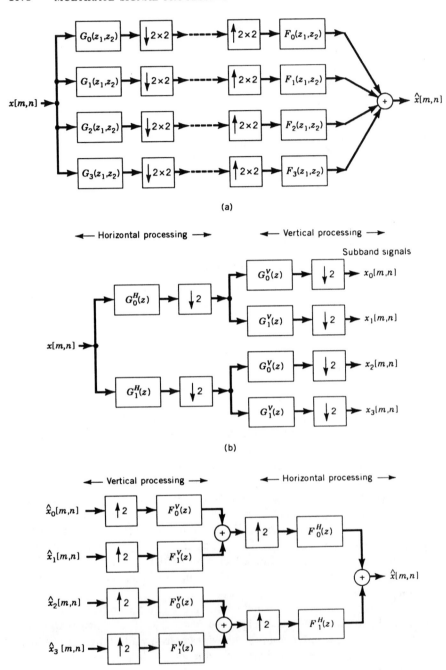

FIGURE 14-56 (a) Single-stage four-band subband analysis and synthesis filter banks. (b) Two-stage four-band subband analysis bank. (c) Two-stage four-band subband synthesis bank.

outputs are horizontally down-sampled by a factor 2 by dropping alternate columns of data to generate two $2M \times N$ arrays (L-band and H-band). Each of these signals is further split into two $M \times N$ subband signals by processing with vertical high-pass and lowpass filters. The synthesis is carried out as shown in Figure 14-56(c). In almost all commonly occurring natural images, most of the perceptually significant information resides in the lowest frequency (LOW–LOW or LL) band. The LL band signal is a lower resolution version of the original image, while the higher frequency band signals carry mostly information about the contours and edges and detail of texture. The four subband signals can be further split using the same decomposition procedure. In some cases only the lowest band is split [GH88] into four subbands giving a total of seven subband signals. The LL band can be coded either using differential pulse code modulation or discrete cosine transform coding. The high-frequency band signals can be coded efficiently with runlength and entropy coding [GH88]. After decoding, the subband signals are applied to a synthesis filter bank to recover the image.

14-7 SUMMARY

A review of the body of DSP methodologies dealing with situations involving multiple sampling rates is presented in this chapter. Multirate systems provide efficient implementation of signal processing tasks by the adjustment of the signal sampling rate according to the frequency content of the relevant information in the signal at various stages in an algorithm. Multirate processing is sometimes imposed by the requirements of a signal processing task or is intentionally introduced to reduce computational complexity. The tools required to exploit multirate processing are explained in this chapter.

The basic concepts and the theory of multirate signal processing are examined, relevant design techniques are outlined, and some applications are discussed. Early in the chapter techniques for interpolation and decimation by integer and rational factors, and rate conversion of bandpass signals are explained. Computational requirements in the implementation of FIR and IIR filters are investigated, and it is shown how polyphase network configurations allow efficient multirate processing using IIR filters. FIR and IIR filter designs for multirate processing are described with emphasis on special designs for filters in the polyphase configuration. Methods of multistage rate conversion and its advantages are discussed. The topic of filter banks is covered with focus on uniform filter banks and exact reconstruction filter banks. Two-dimensional multirate procedures are not discussed here, though an application is described. Finally, three applications of multirate processing are briefly explained.

REFERENCES

[AA78] M. R. Aaron and A. Lender, eds., Special issue on digital signal processing. *IEEE Trans. Commun.* **COM-26,** 697–741, (1978).

[AN81] R. Ansari and B. Liu, Interpolators and decimators as periodically time-varying filters. *Proc. IEEE Int. Symp. Circuits Syst., Chicago, 1981*, pp. 447–450. (1981).

[AN83] R. Ansari and B. Liu, Efficient sampling rate alteration using recursive (IIR) digital filters. *IEEE Trans. Acoust., Speech, Signal Process.* **ASSP-31**(6), 1366–1373, (1983).

[AN85] R. Ansari, Elliptic filter design for a class of generalized halfband filters. *IEEE Trans. Acoust., Speech, Signal Process.* **ASSP-33**(5), 1146–1150, (1985).

[AN88] R. Ansari, A. E. Cetin, and S. H. Lee, Subband coding of images using nonrectangular filter banks. *Proc. 32nd Annu. Int. Symp. Opt. Optoelectron. Appl. Sci. Eng., San Jose, California*, 315–323, (1988).

[BA82] T. P. Barnwell, III, Subband coder design incorporating recursive quadrature filters and ADPCM coders. *IEEE Trans. Acoust., Speech, Signal Process.* **ASSP-30**(5), 751–765, (1982).

[BA91] H. Babic, S. K. Mitra, C. Creusere, and A. Das, Perfect reconstruction recursive QMF banks for image subband coding. *Proc. 25th Ann. Asilomar Conf. Signals, Syst. Comput., Pacific Grove, California, 1991*, pp. 746–750, (1991).

[BE74] M. G. Bellanger, J. L. Daguet, and G. P. Lepagnol, Interpolation, extrapolation and reduction of computation speed in digital filters. *IEEE Trans. Acoust., Speech, Signal Process.* **ASSP-22**(4), 231–235, (1974).

[BE76] M. G. Bellanger, G. Bonnerot, and M. Coudreuse, Digital filtering by polyphase network: Application to sampling rate alteration and filter banks. *IEEE Trans. Acoust., Speech, Signal Process.* **ASSP-24**(2), 109–114, (1976).

[BE77] M. G. Bellanger, Computation rate and storage estimation in multirate digital filtering with half-band filters. *IEEE Trans. Acoust., Speech, Signal Process.* **ASSP-25**(4), 344–346, (1977).

[BU67] J. P. Burg, Maximum entropy spectral analysis. *Proc. Soc. Explor. Geophys. Meet., 37th, Oklahoma City, 1967*, reprinted in *Modern Spectral Analysis.* (D. G. Childers, ed.), IEEE Press, New York, pp. 34–41, (1978).

[BU83] P. J. Burt and E. H. Adelson, The Laplacian pyramid as a compact image code. *IEEE Trans. Commun.* **COM-31**(4), 532–540, (1983).

[CO80] J. W. Cooley and S. Winograd, A limited range discrete Fourier transform algorithm. *Proc. IEEE Conf. Acoust., Speech, Signal Process., 1980*, pp. 213–217, (1980).

[CO86] R. V. Cox, The design of uniformly and nonuniformly spaced pseudoquadrature mirror filters. *IEEE Trans. Acoust., Speech, Signal Process.* **ASSP-34**(5), 1090–1096, (1986).

[CR72] A. W. Crooke and J. W. Craig, Digital filters for sample-rate reduction. *IEEE Trans. Audio Electroacoust.* **AU-20**(4), 308–315, (1972).

[CR75] R. E. Crochiere and L. R. Rabiner, Optimum FIR digital filter implementations for decimation, interpolation and narrow-band filtering. *IEEE Trans. Acoust., Speech, Signal Process.* **ASSP-23**(5), 444–456, (1975).

[CR76a] R. E. Crochiere and L. R. Rabiner, Further considerations in the design of decimators and interpolators. *IEEE Trans. Acoust., Speech, Signal Process.* **ASSP-24**, 296–311, (1976).

[CR76b] R. E. Crochiere, S. A. Webber, and J. L. Flanagan, Digital coding of speech in subbands. *Bell Syst. Tech. J.* **55,** 1069–1085, (1976).

[CR83] R. E. Crochiere and L. R. Rabiner, *Multirate Digital Signal Processing*. Prentice-Hall, Englewood Cliffs, NJ, 1983.

[DR86] W. Drews and L. Gazsi, A new design method for polyphase filters using all-pass sections. *IEEE Trans. Circuits Syst.* **CAS-33**(3), 346–348, (1986).

[DU85] E. Dubois, The sampling and reconstruction of time-varying imagery with application in video systems. *Proc. IEEE.* **73**, 502–522, (1985).

[ES77] D. Esteban and C. R. Galand, Application of quadrature mirror filters to split-band voice coding schemes. *Proc. IEEE Conf. Acoust., Speech, Signal Process., Hartford, Connecticut, 1977*, pp. 191–195 (1977).

[FO77] P. F. Fougere, A solution to the problem of spontaneous line splitting in maximum entropy power spectrum analysis. *J. Geophys. Res.* **82**, 1051–1054, (1977).

[FR71] S. L. Freeny, R. B. Kieburtz, K. V. Mina, and S. K. Tewksbury, Design of digital filters for an all digital frequency division multiplex—time division multiplex translator. *IEEE Trans. Circuit Theory.* **CT-18**(6), 702–711, (1971).

[GA84] C. R. Galand and H. J. Nussbaumer, New quadrature mirror filter structures. *IEEE Trans. Acoust., Speech, Signal Process.* **ASSP-32**(3), 522–531, (1984).

[GA85] L. Gazsi, Explicit formulas for lattice wave digital filters. *IEEE Trans. Acoust., Speech, Signal Process.* **CAS-32**(1), 68–88, (1985).

[GH86] H. Gharavi and A. Tabatabai, Sub-band coding of digital images using two-dimensional quadrature mirror filtering. *Proc. SPIE Conf. Visual Commun. Image Process.* **707**, pp. 51–61, (1986).

[GH88] H. Gharavi and A. Tabatabai, Sub-band coding of monochrome and color images. *IEEE Trans. Circuits Syst.* **CAS-35**, 207–214, (1988).

[GO71] D. J. Goodman and J. L. Flanagan, Direct digital conversion between linear and adaptive delta modulation formats. *Proc. IEEE Int. Commun. Conf., Montreal, Canada, 1971.* (1971).

[GO77] D. J. Goodman and M. J. Carey, Nine digital filters for decimation and interpolation. *IEEE Trans. Acoust., Speech, Signal Process.* **ASSP-25**(2), 121–126, (1977).

[GR88] F. Grenez, Chebyshev design of filters for subband coders. *IEEE Trans. Acoust., Speech, Signal Process.* **ASSP-36**(2), 182–185, (1988).

[GU78] C. Gumacos, Weighting coefficients for certain maximally flat nonrecursive digital filters. *IEEE Trans. Circuits Syst.* **CAS-25**(4), 234–235, (1978).

[JO80] J. D. Johnston, A filter family designed for use in quadrature mirror filter banks. *Proc. IEEE Conf. Acoust., Speech, Signal Process., 1980*, pp. 291–294 (1980).

[LA74] R. T. Lacoss, Data adaptive spectral analysis methods. *Geophysics* **36**, 661–675, (1974).

[LI78] B. Liu and F. Mintzer, Calculation of narrow band spectra by direct decimation. *IEEE Trans. Acoust., Speech, Signal Process.* **ASSP-26**(6), 529–534, (1978).

[MA77] J. Makhoul, Stable and efficient lattice methods for linear prediction. *IEEE Trans. Acoust., Speech, Signal Process.* **ASSP-25**, 423–428, (1977).

[MA79] H. G. Martinez and T. W. Parks, A class of infinite-duration impulse response digital filters for sampling rate reduction. *IEEE Trans. Acoust., Speech, Signal Process.* **ASSP-27**(2), 154–162, (1979).

[MA81] T. G. Marshall, Jr., A multiple VLSI signal processor realization of a transmultiplexer. *Proc. Int. Conf. Commun., 1981*, pp. 7.7.1–7.7.5, (1981).

[MA87] T. G. Marshall, Jr., The polyphase transform and its application to block processing and filter bank structures. *Proc. IEEE Symp. Circuits Syst., Philadelphia, 1987*, pp. 1103–1109, (1987).

[MC73] J. H. McClellan and T. W. Parks, A unified approach to the design of optimum FIR linear-phase digital filters. *IEEE Trans. Circuit Theory.* **CT-20**(6), 697–701, (1973).

[ME75] R. A. Meyer and C. S. Burrus, A unified analysis of periodically time-varying digital filters. *IEEE Trans. Circuits Syst.* **CAS-22**(3), 162–168, (1975).

[ME83] R. M. Mersereau and T. C. Speake, The processing of periodically sampled multidimensional signals. *IEEE Trans. Acoust., Speech, Signal Process.* **ASSP-31**, 188–194, (1983).

[MI78] F. Mintzer and B. Liu, The design of optimal multirate bandpass and bandstop filters. *IEEE Trans. Acoust., Speech, Signal Process.* **ASSP-26**(6), 534–543, (1978).

[MI82] F. Mintzer, On half-band, third-band and nth-band FIR filters and their design. *IEEE Trans. Acoust., Speech, Signal Process.* **ASSP-30**(5), 734–738, (1982).

[MI85] F. Mintzer, Filters for distortion-free two-band multirate filter banks. *IEEE Trans. Acoust., Speech, Signal Process.* **ASSP-33**(3), 626–630, (1985).

[MI92] S. K. Mitra, C. D. Creseure, and H. Babic, A novel implementation of perfect reconstruction QMF banks using IIR filters for infinite length signals. *Proc. IEEE Int. Symp. Circuits & Systems, San Diego, California 1992*, pp. 2312–2315 (1992).

[MO77] M. Morf, B. Dickinson, T. Kailath, and A. Vieira, Efficient solution of covariance equations for linear prediction. *IEEE Trans. Acoust., Speech, Signal Process.* **ASSP-25,** 429–433, (1977).

[NA78] M. Narasimha and A. Peterson, On using the symmetry of FIR filters for digital interpolation. *IEEE Trans. Acoust., Speech, Signal Process.* **ASSP-26**(3), 267–268, (1978).

[NU81] H. J. Nussbaumer, Pseudo QMF filter bank. *IBM Technol. Disclosure Bull.* **24**(6), 3081–3087, (1981).

[OE75] G. Oetken, T. W. Parks, and H. W. Schuessler, New results in the design of digital interpolators. *IEEE Trans. Acoust., Speech, Signal Process.* **ASSP-23**(3), 301–309, (1975).

[OE79] G. Oetken, A new approach for the design of digital interpolating filters. *IEEE Trans. Acoust., Speech, Signal Process.* **ASSP-27**(6), 637–643, (1979).

[OP89] A. V. Oppenheim and R. W. Schafer, *Discrete-Time Signal Processing.* Prentice-Hall, Englewood Cliffs, NJ, 1989.

[PR79] R. G. Pridham and R. A. Mucci, Digital interpolation beam forming for lowpass and bandpass signals. *Proc. IEEE.* **67**(6), 904–919, (1979).

[PU82] C. M. Puckette and T. G. Marshall, eds., Special issue on transmultiplexers. *IEEE Trans. Commun.* **COM-30**(7), 1457–1655, (1982).

[PY87] M. F. Pyfer and R. Ansari, The design and application of optimal FIR fractional-slope phase filters. *Proc. IEEE Conf. Acoust., Speech, Signal Process., Dallas, Texas, 1987,* pp. 896–899, (1987).

[QU83] M. P. Quirk and B. Liu, Improving resolution for autoregressive spectral estimation by decimation. *IEEE Trans. Acoust., Speech, Signal Process.* **ASSP-31**(3), 630–637, (1983).

[RA75] L. R. Rabiner and R. E. Crochiere, A novel implementation for narrow-band FIR digital filters. *IEEE Trans. Acoust., Speech, Signal Process.* **ASSP-23**(5), 457–464, (1975).

[RA82] T. A. Ramstad, Digital two-rate IIR and hybrid IIR/FIR filters for sample rate conversion. *IEEE Trans. Commun.* **COM-30**(7), 1466–1476, (1982).

[RA84] T. A. Ramstad, Analysis/synthesis filter banks with critical sampling. *Proc. Int. Conf. Digital Signal Proc., Florence, Italy, 1984*, 130–134, (1984).

[RE87a] M. Renfors and T. Saramäki, Recursive N-th band digital filters. Part I. Design and properties. *IEEE Trans. Circuits Syst.* **CAS-34**(1), 24–39, (1987).

[RE87b] M. Renfors and T. Saramäki, Recursive N-th band digital filters. Part II. Design of multistage interpolators and decimators. *IEEE Trans. Circuits Syst.* **CAS-34**(1), 40–51, (1987).

[RE87c] P. A. Regalia, S. K. Mitra, P. P. Vaidyanathan, M. Renfors, and Y. Neuvo, Tree-structured complementary filter banks using all-pass sections. *IEEE Trans. Circuits Syst.* **CAS-34**, 1470–1484, (1987).

[RO83] J. H. Rothweiler, Polyphase quadrature filters—a new sub-band coding technique. *Proc. IEEE Conf. Acoust., Speech, Signal Process., 1983*, pp. 1280–1283, (1983).

[SA84] T. Saramäki, A class of linear-phase FIR filters for decimation, interpolation and narrow-band filtering. *IEEE Trans. Acoust., Speech, Signal Process.* **ASSP-32**(5), 1023–1036, (1984).

[SC73] R. W. Schafer and L. R. Rabiner, A digital signal processing approach to interpolation. *Proc. IEEE.* **61**(6), 692–702, (1973).

[SC81] H. Scheuermann and H. Gockler, A comprehensive survey of digital transmultiplexing methods. *Proc. IEEE.* **69**(11), 1419–1450, (1981).

[SC84] D. J. Scheibner and T. W. Parks, Slowness aliasing in the discrete radon transform: A multirate approach to beamforming. *IEEE Trans. Acoust., Speech, Signal Process.* **ASSP-32**(6), 1160–1165, (1984).

[SH75] R. R. Shively, On multistage FIR filters with decimation. *IEEE Trans. Acoust., Speech, Signal Process.* **ASSP-23**(4), 353–357, (1975).

[SM84] M. J. T. Smith and T. P. Barnwell, A procedure for designing exact reconstruction filter banks for tree structured subband coders. *Proc. IEEE Conf. Acoust., Speech, Signal Process., San Diego, California, 1984*, pp. 27.1.1–27.1.4 (1984).

[SM87] M. J. T. Smith and T. P. Barnwell, A new filter bank theory for time-frequency representation. *IEEE Trans. Acoust., Speech, Signal Process.* **ASSP-35**, 314–327, (1987).

[ST86] K. Steiglitz and T. W. Parks, What is the filter design problem. *Proc. Princeton Conf. Inf. Sci. Syst., Princeton, New Jersey, 1986*, pp. 604–609. (1986).

[SW86] K. Swaminathan and P. P. Vaidyanathan, Theory and design of uniform DFT, parallel quadrature mirror filter banks. *IEEE Trans. Circuits Syst.* **CAS-33**, 1170–1191, (1986).

[TA81] L. Taxen, Polyphase filter banks using wave digital filters. *IEEE Trans. Acoust., Speech, Signal Process.* **ASSP-29**(3), 423–428, (1981).

[UL75] T. J. Ulrych and T. N. Bishop, Maximum entropy spectral analysis and autoregressive decomposition. *Rev. Geophys. Space Phys.* **13**, 183–200, (1975).

[UL76] T. J. Ulrych and R. W. Clayton, Time series modeling and maximum entropy. *Phys. Earth Planet. Inter.* **12**, 188–200, (1976).

[VA85] P. P. Vaidyanathan, On power complementary FIR filters. *IEEE Trans. Circuits Syst.* **CAS-32**(12), 1308–1310, (1985).

[VA87] P. P. Vaidyanathan, Quadrature mirror filter banks, M-band extensions and perfect reconstruction techniques. *IEEE ASSP Mag.* **4**(3), 4–20, (1987).

[VA89] P. P. Vaidyanathan, T. Q. Nguyen, Z. Doganata, and T. Saramäki, Improved technique for design of perfect reconstruction FIR QMF filter banks with lossless polyphase matrices. *IEEE Trans. Acoust., Speech, Signal Process.* **ASSP-37**(7), 1042–1056, (1989).

[VE84] M. Vetterli, Multi-dimensional subband coding: Some theory and algorithms. *Signal Process.* **6,** 97–112, (1984).

[VE87] M. Vetterli, A theory of multirate filter banks. *IEEE Trans. Acoust., Speech, Signal Process.* **ASSP-35,** 356–372, (1987).

[VI88] E. Viscito and J. P. Allebach, On determining optimum multirate structures for narrow-band FIR filters. *IEEE Trans. Acoust., Speech, Signal Process.* **ASSP-36**(8), 1255–1271, (1988).

[WO86] J. W. Woods and S. D. O'Neil, Subband coding of images. *IEEE Trans. Acoust., Speech, Signal Process.* **ASSP-34**(5), 1278–1288, (1986).

[WO91] J. W. Woods, ed., *Subband Image Coding.* Kluwer Academic Publishers, Norwell, MA, 1991.

15 Adaptive Filtering

JOHN M. CIOFFI

Department of Electrical Engineering
Stanford University
Stanford, California

YOUN-SHIK BYUN

Department of Electronic Engineering
Incheon University, Korea

15-1 INTRODUCTION

Adaptive signal processing has found widespread practical applications in areas as diverse as voiceband data modems, antenna arrays, radar, sonar, digital satellite data transmission, mobile telephony, speech compression, voice echo cancellation, and spectral estimation to name just a few. The key reason for the widespread use of adaptive filters is their ability to optimize their own performance through recursive modification of internal parameters. Methods for performing the recursive modifications are called *adaptive algorithms*. This self-adjustment capability is especially advantageous when the application environment cannot be accurately described in advance to the system designer. A number of effective adaptive algorithms have been developed over the last three decades, and this chapter describes the most commonly used of these methods. These algorithms include "gradient" or LMS (Section 15-3), RLS (Section 15-4), and frequency-domain or block (Section 15-5) algorithms. In Section 15-2, we also outline the basic structural characteristics common to adaptive systems. Section 15-6 is an applications section, where we discuss data echo cancelers, equalizers for disk channels (a very recently developing application), and decision-feedback equalization of fading channels to illustrate a variety of the methods described.

Before beginning, we draw the reader's attention to a number of recent books (in alphabetical order) by Alexander [AL86], Bellanger [BE87], Cowan and Grant [CO85], Giordano and Hsu [GI85], Goodwin and Sin [GO84], Haykin [HA85; HA91], Honig and Messerschmitt [HO84], Monzingo and Miller [MO80], Mulgrew and Cowan [MU88], Orfanidis [OR88], Strobach [ST90], Treichler, John-

Handbook for Digital Signal Processing, Edited by Sanjit K. Mitra and James F. Kaiser.
ISBN 0-471-61995-7 © 1993 John Wiley & Sons, Inc.

son, and Larimore [TR87], and Widrow and Stearns [WI85], for more detailed developments.

15-2 ADAPTIVE FILTER BASICS

The *adaptive filter* differs from fixed filters in that the filter coefficients are varied with time as a function of the filter input(s). The basic concept is illustrated in Figure 15-1. There are two input time series: $x[k]$, the filter input, and $d[k]$, the desired response. The adaptive filter forms an N-term time-varying linear combination of $x[k], x[k - 1], \ldots, x[k - N + 1]$ to estimate $d[k]$. The coefficients in this combination are denoted $w_0[k], \ldots, w_{N-1}[k]$. An error sequence is formed by

$$\epsilon[k] = d[k] - \sum_{m=0}^{N-1} w_m[k]x[k - m] = d[k] - \mathbf{W}_k^t\mathbf{X}_k, \tag{15.1}$$

where $\mathbf{W}_k^t$ is the transpose of the column vector

$$\mathbf{W}_k = \begin{bmatrix} w_0[k] \\ w_1[k] \\ \vdots \\ w_{N-1}[k] \end{bmatrix} \tag{15.2}$$

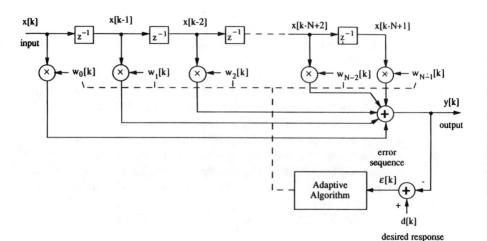

FIGURE 15-1 The adaptive filter.

and

$$
\mathbf{X}_k = \begin{bmatrix} x[k] \\ x[k-1] \\ \vdots \\ x[k-N+1] \end{bmatrix}
\tag{15.3}
$$

The error sequence and $\mathbf{X}_k$ are input to an adaptive algorithm that recursively determines $\mathbf{W}_{k+1}$.

More generally, $x[k]$ can be a $p \times 1$ vector and $d[k]$ a $q \times 1$ vector, yielding $w_m[k]$, a $q \times p$ matrix. This situation occurs frequently in practical applications, and we refer to it as a multichannel adaptive filter. A well-known special case of a multichannel adaptive filter is the *adaptive linear combiner* (ALC) that is shown in Figure 15-2(a) and corresponds to $N = 1$, $q = 1$, and $p > 1$. This configuration is common in adaptive antenna arrays [MO80], and we usually find $N \geq 1$ in these arrays. Another example is the fractionally spaced equalizer in Figure 15-2(b), where $q = 1$, $N > 1$, and $p = 2$ [QU85].[1] This fractionally spaced equalizer illustrates how the multichannel concept in adaptive filtering is used to accommodate applications when the two inputs, $\mathbf{x}[k]$ and $d[k]$ (in the fractionally spaced equalizer, $\mathbf{x}[k] = [x(kT) \quad x(kT - T/2)]^t$, while $d[k] = d(kT)$) are supplied at different rates. A third example is the "equation-error" adaptive infinite impulse response filter of Figure 15-2(c), where $N > 1$, $q = 1$, and $p = 2$; and the input to the adaptive filter is now

$$
\begin{bmatrix} x[k] \\ d[k-1] \end{bmatrix},
\tag{15.4}
$$

or, in other words, the desired response values are delayed and reused as inputs in a multichannel adaptive filter. In this last structure, one could write $w_m[k] = [\mathbf{a}_m[k] \; \vdots \; \mathbf{b}_m[k]]$, where $\mathbf{a}_m[k]$ and $\mathbf{b}_m[k]$ are the numerator (zero) and denominator (pole) coefficients, respectively, in a time-varying difference equation that describes a discrete infinite impulse response filter.

In practice, the filter output, $y[k]$, is often of most interest as opposed to $\mathbf{W}_k$, such as in adaptive equalization [QU85] or echo cancellation [SO80]. Thus there are many alternative structures to the transversal filter (or linear combiner) for the adaptive filter. The two most common alternatives are the lattice and the frequency-domain adaptive filters, the latter of which is considered in Section 15-5.

Although many performance criteria have been suggested for adaptive algorithms [GO84], in practice the minimum mean-square-error (MMSE) criterion is

---

[1]Actually, p can take on any rational value (>1) in the fractionally spaced equalizer, but 2 is most common.

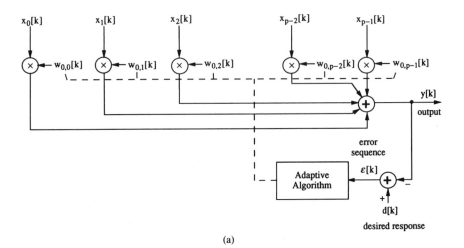

(a)

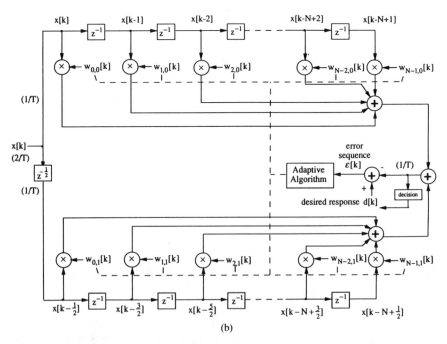

(b)

FIGURE 15-2 (a) The adaptive linear combiner. (b) The fractionally spaced equalizer. (c) The equation-error adaptive filter.

the most commonly used, and we therefore consider in this chapter only algorithms that try to approximate the MMSE performance.

The MMSE Criterion. The MMSE criterion determines the $\mathbf{W}_k$ that minimizes the mean-square error, $E(\epsilon^2[k])$, where E denotes statistical expectation; that is (let

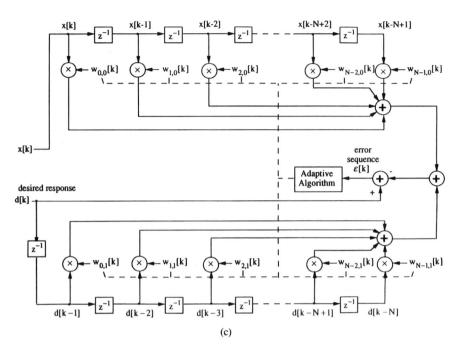

FIGURE 15-2 (*Continued*)

$q = 1$ and $p = 1$),

$$\text{MMSE} = \min_{\mathbf{W}_k} E(\epsilon^2[k]). \tag{15.5}$$

Equation (15.5) becomes, using Eq. (15.1),

$$\text{MMSE} = \min_{\mathbf{W}_k} E(d^2[k]) - 2\mathbf{P}_k^t\mathbf{W}_k + \mathbf{W}_k^t\mathbf{R}_k\mathbf{W}_k, \tag{15.6}$$

where

$$\mathbf{R}_k = E(\mathbf{X}_k\mathbf{X}_k^t)$$

$$= E \begin{bmatrix} x^2[k] & x[k]x[k-1] & \cdots & x[k]x[k-N+1] \\ x[k]x[k-1] & x^2[k-1] & \cdots & x[k-1]x[k-N] \\ \vdots & \vdots & \ddots & \vdots \\ x[k]x[k-N+1] & x[k-1]x[k-N] & \cdots & x^2[k-N+1] \end{bmatrix}$$

$$\tag{15.7}$$

and

$$\mathbf{P}_k = E(\mathbf{X}_k d[k]) = E \begin{bmatrix} d[k]x[k] \\ \vdots \\ d[k]x[k - N + 1] \end{bmatrix} \tag{15.8}$$

are the autocorrelation matrix and the cross-correlation vector, respectively. The solution is easily found as

$$\mathbf{W}_k^0 = \mathbf{R}_k^{-1} \mathbf{P}_k. \tag{15.9}$$

More generally, if $p > 1$ and $q > 1$, and the inputs are complex, one can show that Eq. (15.9) still holds if $\mathbf{P}_k = E[\mathbf{X}_k(d[k])^t]$ and a superscript of t denotes transpose in Eqs. (15.6), (15.7), and also in (15.1). If $x[k]$ and $d[k]$ are jointly stationary then $\mathbf{W}_k^0 = \mathbf{W}^0 = \mathbf{R}^{-1}\mathbf{P}$, since neither $\mathbf{R}_k$ nor $\mathbf{P}_k$ depend on k in this case. Often in practice $\mathbf{R}_k \neq \mathbf{R}$ (nonstationary inputs), but the product $\mathbf{R}_k^{-1}\mathbf{P}_k$ is constant, or nearly constant.

Statistical quantities such as $\mathbf{R}_k$ or $\mathbf{P}_k$ are usually not available, and the adaptive algorithm must estimate these quantities or their equivalents. There are three major categories of adaptive algorithms that try to approximate Eq. (15.9) by making use of only the observed time series $x[k]$ and $d[k]$. These categories are the stochastic-gradient (LMS, least-mean-squares) algorithms, the recursive-least-squares (RLS) algorithms, and the block adaptive algorithms. Each of these methods are discussed in the sections to follow.

Although there are many applications of adaptive filters, it is often convenient to consider these applications as belonging to one of two classes. The first class is illustrated in Figure 15-3 and is loosely termed as *system identification*. The distinguishing feature in the system-identification application is the absence of "noise" in the adaptive-filter input. This means $d[k]$ is a signal such that

$$d[k] = u[k] + (\mathbf{W}^0)^t \mathbf{X}_k, \tag{15.10}$$

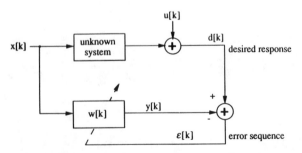

FIGURE 15-3 System identification configuration.

where $E(u[k]\mathbf{X}_k) = 0$; that is, $d[k]$ is a linear combination of current and past adaptive-filter inputs $x[k]$ and an additive uncorrelated (with $x[k]$) noise component $u[k]$. The MMSE adaptive algorithms converge to the setting $\mathbf{W}^0$ in this class of applications; that is, the unknown system is identified ($\mathbf{W}^0$ is found, either implicitly or explicitly), even though the actual quantity of interest in a particular application may be the adaptive-filter output, $y[k]$, or the error signal, $\epsilon[k]$. Applications in this category include echo cancellation [SO80], noise canceling [WI75b], plant or channel modeling [CI86; LJ83], and linear prediction of (autoregressively modeled) speech [MA76]. Of course, $\mathbf{W}_k^0$ can vary with time, and the MMSE methods will track this variation.

The second category is loosely termed *system inversion* and is illustrated in Figure 15-4. Here, the roles of $x[k]$ and $d[k]$ have essentially been reversed, and we write

$$x[k] = \mathbf{H}^t \begin{bmatrix} d[k + \overline{M}] \\ \vdots \\ d[k - M] \end{bmatrix} + \nu[k], \qquad (15.11)$$

where $\mathbf{H}$ is a vector corresponding to some system's response characteristic and $\nu[k]$ is again an additive uncorrelated (with $d[k]$) noise. Note $x[k]$ is presumed "noncausal" as a function of $\overline{M}$ future, 1 current, and M past samples of $d[k]$. In this type of application, $\mathbf{W}_k$ becomes, under a MMSE criterion, a biased estimate of the *inverse* of the *system* $\mathbf{H}$. The magnitude of the bias (deviation from the inverse of $\mathbf{H}$) increases with the variance of $\nu[k]$. Again, another quantity such as the adaptive-filter output or $\nu[k]$ may be of interest in a specific system-inversion application, and the approximate system inversion is implicit in the adaptive filters. Specific system-inversion applications include adaptive equalization [QU85], adaptive arrays [MO80], deconvolution, inverse control [GO77], and adaptive line enhancement [TR77].

In the remainder of this chapter, we discuss in more detail the various least-squares adaptive algorithms, their performance, and their implementation costs and constraints. Section 15-3 is concerned with the stochastic-gradient methods.

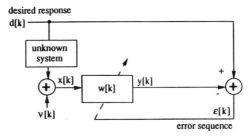

FIGURE 15-4 System-inversion configuration.

15-3 STOCHASTIC-GRADIENT (LMS) ADAPTIVE ALGORITHMS

The MMSE criterion in Eq. (15.6) is quadratic in the adaptive-filter coefficients $\mathbf{W}_k$, which ensures a unique MMSE optimum if $\mathbf{R}_k$ is nonsingular. Stochastic-gradient algorithms have their origin in the gradient-search procedures [LU73; RO51] that abound in nonlinear programming. The essence of such gradient-search methods is to update the quantity $\mathbf{W}_k$ along the direction of the negative gradient of $E(\epsilon^2[k])$ that corresponds to the direction of steepest descent of the mean-square-error surface in Eq. (15.6):

$$\mathbf{W}_{k+1} = \mathbf{W}_k - \frac{\mu}{2} \nabla E[\epsilon^2[k]], \tag{15.12}$$

where $\mu/2$ is a "step-size" constant that determines the magnitude of the update, and ∇ denotes gradient. For the adaptive filter,

$$\nabla E(\epsilon^2[k]) = 2E(\epsilon[k](-\mathbf{X}_k^t)) = -2(\mathbf{P}_k^t - \mathbf{R}_k\mathbf{W}_k^t), \tag{15.13}$$

which depends on the unknown quantities $\mathbf{P}_k$ and $\mathbf{R}_k$. Stochastic-gradient methods [RO51] approximate the quantity $E(\epsilon[k]\mathbf{X}_k^t)$ by its sample average value

$$E(\epsilon[k]\mathbf{X}_k^t) \cong \frac{1}{l} \sum_{m=0}^{l-1} \epsilon[k-m]\mathbf{X}_{k-m}^t. \tag{15.14}$$

The most important adaptive-filtering algorithm is the LMS (least-mean square) method that was introduced to the engineering community in a landmark paper by Widrow and Hoff [WI60]. It sets $l = 1$ in Eq. (15.14), reducing Eq. (15.12) to

$$\mathbf{W}_{k+1} = \mathbf{W}_k + \mu\epsilon[k]\mathbf{X}_k. \tag{15.15}$$

A commonly used extension, when $\mathbf{X}_k$, $d[k]$, and $\mathbf{W}_k$ are allowed to be complex, is [WI75a]

$$\mathbf{W}_{k+1} = \mathbf{W}_k + \mu\epsilon[k]\mathbf{X}_k^*, \tag{15.16}$$

where the superscript * denotes complex conjugate, and the mean-square magnitude of the complex error is to be minimized.

The computational requirements of the LMS algorithm are relatively low, requiring $2N + 1$ multiply-accumulate operations per input data pair $(x[k], d[k])$, or per iteration. The algorithm is most often implemented digitally, at signal bandwidths of tens to hundreds of kilohertz, with a single serial digital signal processor, such as the AT&T DSP32, Motorola 56000, or Texas Instruments TMS 320 families of DSP chips (see Chapters 11 and 12). As such, the implementation is essentially a computer program; no implementation diagram is necessary. At higher signal bandwidths, the update in Eq. (15.15) or Eq. (15.16) need not be executed at each time instant k and can be executed at arbitrary or periodic intervals (leading to an inevitable reduction in the tracking or learning rate of the adaptive filter).

15-3-1 Steady-State Performance of the Stochastic-Gradient LMS Algorithm

As the adaptive filter can only process a finite amount of data, it estimates the solution in Eq. (15.9), $\mathbf{R}_k^{-1}\mathbf{P}_k$. In the LMS algorithm, the parameter μ, often called the "stepsize," determines the proximity of $\mathbf{W}_k$ to $\mathbf{R}_k^{-1}\mathbf{P}_k$. If $\mathbf{R}_k = \mathbf{R}$ and $\mathbf{P}_k = \mathbf{P}$ (constant with time),[2] then smaller μ leads to a solution closer[3] to $\mathbf{W}^0 = \mathbf{R}^{-1}\mathbf{P}$. However, in a time-varying application or when the adaptive algorithm is initially converging μ cannot be too small as the adaptive filter must "adapt."

There are many analyses of the LMS algorithm's performance under a variety of conditions. In this handbook, we list only results that are useful to the system designer as rules of thumb for expected performance, and we assume $q = p = 1$. When Eq. (15.9) is substituted in Eq. (15.6), one obtains the MMSE (minimum mean-square error)

$$\sigma_{\text{MMSE}}^2 = E(d^2[k]) - \mathbf{P}_k^t \mathbf{R}_k^{-1} \mathbf{P}_k \tag{15.17}$$

(usually a roughly constant quantity), an idealized performance level to be approached by the adaptive filter. In the LMS algorithm, one can show (under the "independence assumptions" and the assumptions that $\mathbf{R}_k^{-1}\mathbf{P}_k$ varies very slowly with time k) that the following three relations hold after a time period sufficiently long for the adaptive filter to have converged:

$$E(\mathbf{W}_k) = \mathbf{W}^0, \tag{15.18}$$

$$E(\epsilon[k]) = 0, \tag{15.19}$$

and

$$\sigma_{\epsilon}^2 \triangleq E(\epsilon^2[k]) = \sigma_{\text{MMSE}}^2 + \frac{\mu N \sigma_x^2}{2 - \mu N \sigma_x^2} \sigma_{\text{MMSE}}^2, \tag{15.20}$$

where $\sigma_x^2 = E(x^2[k])$ and $d[k]$ and $x[k]$ are implicitly assumed to be of zero mean. Equations (15.18) and (15.19) indicate the desirable property that, on the average, the LMS adaptive filter converges to the true least-squares solution. Equation (15.20) is a measure of the fluctuation about this ideal average. Often cited quantities related to Eq. (15.20) are the *excess mean-square error* (EMSE),

$$\sigma_{\text{EMSE}}^2 \triangleq \sigma_{\epsilon}^2 - \sigma_{\text{MMSE}}^2 = \left(\frac{\mu N \sigma_x^2}{2 - \mu N \sigma_x^2} \right) \sigma_{\text{MMSE}}^2, \tag{15.21}$$

---

[2]Actually, only the matrix–vector product $\mathbf{R}_k^{-1}\mathbf{P}_k = \mathbf{W}^0$ needs to be constant.
[3]This presumes infinite precision; numerical effects are considered in Section 15-3-5.

and the *misadjustment factor*,

$$m \triangleq \frac{\sigma_{\text{EMSE}}^2}{\sigma_{\text{MMSE}}^2} = \frac{\mu N \sigma_x^2}{2 - \mu N \sigma_x^2}. \tag{15.22}$$

The reader is cautioned that the latter quantity is usually only of interest in those applications where $\sigma_{\text{MMSE}}^2 \geq \sigma_x^2$ (often not true) or, equivalently, the error is the desired output of the adaptive filter, in which case m is the inverse of the quantity often called "SNR" or signal-to-noise ratio in communication applications. In some applications, such as adaptive equalization, σ_{MMSE}^2 and σ_{EMSE}^2 can be very small, and as a result m effectively becomes indeterminate. Here σ_{EMSE}^2, or possibly $\sigma_{\text{EMSE}}^2/\sigma_x^2$, is a much more meaningful steady-state performance measure.

The adaptive filter designer needs to interpret the formula given by Eq. (15.21). First, note that

$$0 < \mu < \frac{2}{N \sigma_x^2} \tag{15.23}$$

in order for σ_{EMSE}^2 to be positive. If μ falls outside this range, the adaptive filter can be shown to diverge [HA91]. Second, as μ decreases, the performance improves for fixed N, σ_x^2, and σ_{MMSE}^2. If N, σ_x^2, or σ_{MMSE}^2 increases at a fixed μ, performance degrades, or μ must be chosen smaller (slower convergence and tracking) to maintain acceptable performance. If μ is chosen below $1/10N\sigma_x^2$, then the approximation

$$\sigma_{\text{EMSE}}^2 \cong \left(\frac{\mu}{2} N \sigma_x^2 \right) \sigma_{\text{MMSE}}^2 \tag{15.24}$$

is often used, or when $p > 1$ (but $q = 1$)

$$\sigma_{\text{EMSE}}^2 \cong \frac{\mu}{2} \text{ trace } (\mathbf{R}) \, \sigma_{\text{MMSE}}^2. \tag{15.25}$$

The performance relations in Eqs. (15.20)–(15.24) are only rough approximations because of a wide variety of approximate assumptions made in their derivations. Nevertheless, the dependencies discussed in the previous paragraph on μ, N, σ_x^2, and σ_{MMSE}^2 usually do hold until the onset of limited-precision effects, which are discussed in Section 15-3-5. Thus in real application, the system designer is strongly encouraged to experiment with the parameter μ according to these rules to obtain the best performance.

15-3-2   Initial Convergence of the LMS Adaptive Filter

Equations (15.18)–(15.25) represent the steady-state performance of the LMS algorithm after all learning transients have abated. Another measure of adaptive-filter performance is the duration of the initial learning period or equivalently the rate of convergence toward the steady-state values in Eqs. (15.18)–(15.25). This decay can be shown to be exponential for the LMS algorithms [HA91]. The quantities in Eqs. (15.18) and (15.19) converge as a sum of N geometrically decaying components with geometric decay ratios

$$(1 - \mu q_i), \quad i = 1, \dots, N, \tag{15.26}$$

where $\{q_i\}$ are the eigenvalues of $\mathbf{R}$.

This convergence of $E(\mathbf{W}_k)$ to $\mathbf{W}^0$ and $E(\epsilon[k])$ to 0 is called "convergence of the mean." Note that when Eq. (15.23) is satisfied, the geometric ratios in Eq. (15.26) are all of magnitude less than one, ensuring convergence in the mean. The quantity of σ_ϵ^2 also converges exponentially to its steady-state value in Eq. (15.20), but more slowly (the geometric ratios are larger). This second decay usually dominates the LMS convergence time. These slower geometric ratios are not readily expressed in simple form but can be found to be the eigenvalues of the following matrix [HA91; UN72]:

$$\begin{bmatrix} (1 - \mu q_1)^2 & \mu^2 q_1 q_2 & \cdots & \mu^2 q_1 q_N \\ \mu^2 q_1 q_2 & (1 - \mu q_2)^2 & \cdots & \mu^2 q_2 q_N \\ \vdots & \vdots & \ddots & \vdots \\ \mu^2 q_1 q_N & \mu^2 q_2 q_N & \cdots & (1 - \mu q_N)^2 \end{bmatrix}. \tag{15.27}$$

which also have magnitude less than unity when Eq. (15.23) is satisfied.

A lower bound (fastest rate) for the geometric convergence of the LMS algorithm is [CI85c]

$$\sigma_{\text{EMSE}}^2[k] \geq \sigma_{\text{EMSE}}^2[0] \left(1 - \frac{1}{2N}\right)^k. \tag{15.28}$$

The equality in Eq. (15.28) is achieved if the input $x[k]$ is unit power "white" or uncorrelated ($E(x[k]x[l]) = 0$ for $k \neq l$), and $\mu = 1/N$. $\sigma_{\text{EMSE}}^2[k]$ is the excess mean-square error at time k in Eq. (15.28).

15-3-3 Tracking Lag Error

In the above convergence expressions (Eqs. (15.26)–(15.28)), the quantity $\mathbf{W}_k^0 = \mathbf{R}_k^{-1}\mathbf{P}_k$ is assumed to vary so slowly that it is essentially constant with respect to the convergence rate of the adaptive filter. This assumption is usually a good one. Nevertheless, there are applications, such as the equalization of multipath fading channels, where this assumption is not satisfied. In such applications, a first-order Markov process (vector-random walk) is often used to model the time variation of

$\mathbf{W}_k^0$, that is,

$$\mathbf{W}_{k+1}^0 = \mathbf{W}_k^0 + \nu_k, \tag{15.29}$$

where ν_k is an uncorrelated "noise" vector such that $E(\nu[k]\nu[l]) = 0$ for $k \neq l$. In this case the excess mean-square error is given by

$$\sigma_{\text{EMSE}}^2 = \left(\frac{\mu N \sigma_x^2}{2 - \mu N \sigma_x^2} \right) \sigma_{\text{MMSE}}^2 + \left(\frac{N}{2\mu} \right) \sigma_\nu^2, \tag{15.30}$$

where the second term is called the "lag" excess mean-square error. Note that decrease of μ improves (decreases) the first term but degrades (increases) the lag term. This trade-off is classic and characteristic of all adaptive algorithms; that is, if too much data are averaged (small μ) then variations in $\mathbf{W}_k^0$ cannot be adequately tracked, while averaging too little data (larger μ) may cause a large fluctuation about the desired optimal solution.

15-3-4 Normalized LMS Algorithm

As the geometric ratios in Eq. (15.26) and implicit in Eq. (15.27) generally get "smaller" with larger μ, improved versions of LMS introduce some time variation in the step-size μ. Initially a good choice is on the order of $\mu = 1/N\sigma_x^2$, but the corresponding MMSE, according to Eq. (15.20), is

$$\sigma_\epsilon^2 = \sigma_{\text{MMSE}}^2 + \sigma_{\text{MMSE}}^2 = 2\sigma_{\text{MMSE}}^2, \tag{15.31}$$

which is unacceptably high in some applications. Then, after relatively swift convergence to the level in Eq. (15.31), μ can be decreased to further reduce σ_{EMSE}^2. This technique is often called *gear-shifting*.

A continuous variation of μ is often better in improving the excess error versus convergence-and-tracking trade-off. The most common time variation of μ is called the *normalized LMS algorithm*; see Bershad [BE86], for instance, where μ is replaced by

$$\mu[k] = \frac{\mu[0]N}{\mathbf{X}_k^t \mathbf{X}_k}, \tag{15.32}$$

effectively providing an instantaneous automatic gain control on the input. If $\mathbf{X}_k^t \mathbf{X}_k \approx 0$, then no update should be performed. It is often not implemented directly in real-time because of the division. Rather, μ is changed according to some look-up table with an input that specifies the range of $\mathbf{X}_k^t \mathbf{X}_k$. Expressions for σ_{EMSE}^2 are complicated [BE86] and are subject to many assumptions, making an application-independent expression intractable for the purposes of this handbook. An example of the improvement in convergence for a stationary system identification example is illustrated in Figure 15-5 for $N = 10$, $\sigma_x^2 = 1$, and $\sigma_{\text{MMSE}}^2 = 0.001$. The plot

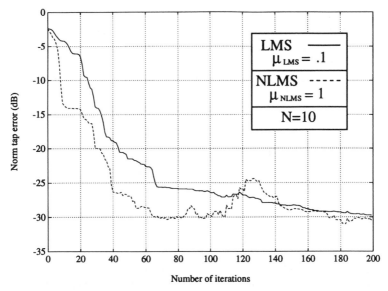

FIGURE 15-5 Normalized LMS convergence improvement.

of $\|\mathbf{W}_k - \mathbf{W}^0\|^2$ versus time is a standard form of comparison for the convergence of adaptive algorithms.[4]

15-3-5 Numerical Stability and Leakage

Up to this point, the discussion and performance measures have been for an infinite-precision implementation. In practice, all algorithmic and filter variables are quantized to a fixed limited precision. This quantization has a critical effect on the performance of the LMS algorithm, and we complete our discussions of the LMS algorithm by focusing on numerical effects.

The LMS algorithm is succinctly restated with the following two recursive equations, which are executed at each time instant k,

$$\epsilon[k] = d[k] - \mathbf{W}_k^t \mathbf{X}_k, \tag{15.33}$$

$$\mathbf{W}_{k+1} = \mathbf{W}_k + \mu \epsilon[k] \mathbf{X}_k. \tag{15.34}$$

Numerically, it is the second recursion Eq. (15.34) where difficulty arises in the LMS algorithm, as the scaled gradient estimate has been computed with quantization errors that can propagate and grow to an appreciable magnitude with time. To illustrate this growth, consider the gradient estimate (second term on the right in Eq. (15.34)):

$$\mathbf{G}_k = \mu \epsilon[k] \mathbf{X}_k, \tag{15.35}$$

---

[4] $\|x\|$ denotes the Euclidean norm or length of x.

which in the cost-conscious implementation should have the product of the two scalars, μ and $\epsilon[k]$, computed first. Assuming 2's-complement arithmetic, this product will be in double precision and must be rounded or truncated to the original precision, thus introducing a quantization error. The second scalar times vector product will also have quantization error. The limited-precision gradient estimate then becomes

$$\hat{\mathbf{G}}_k = (\mu(\epsilon[k] + \Delta\epsilon[k]) + \Delta_{1,k})(\mathbf{X}_k + \Delta\mathbf{X}_k) + \Delta_{2,k}, \qquad (15.36)$$

where $\Delta_{1,k}$ and $\Delta_{2,k}$ are the quantization errors for the first and second products, respectively. The individual nature of these quantization errors is application dependent. Nevertheless, in order to make the analysis tractable and still meaningful, assume that the quantity

$$\mathbf{b}_k = \hat{\mathbf{G}}_k - \mathbf{G}_k, \qquad (15.37)$$

is statistically characterized by N independent and identically distributed random quantization errors, with some nonzero mean in a cost-conscious implementation. One can then use this quantity to extract a considerable amount of information about the limited-precision performance of the LMS algorithm.

We begin by specifying the deviation from the infinite-precision performance in terms of the mean and mean-square values for the components of $b[k]$. Equation (15.34) can then be rewritten with the quantization vector included as

$$\mathbf{W}_{k+1} = \mathbf{W}_k + \mathbf{G}_k + \mathbf{b}_k. \qquad (15.38)$$

In other words, $\mathbf{b}_k$ is some quantization error vector introduced at time k. The mean value of Eq. (15.38) is (using Eq. (15.34))

$$E(\mathbf{W}_{k+1}) = [\mathbf{I} - \mu\mathbf{R}_k]E(\mathbf{W}_k) + \mu\mathbf{P}_k + E(\mathbf{b}_k), \qquad (15.39)$$

where $\mathbf{P}_k = E(\mathbf{X}_k d[k])$ and $\mathbf{I}$ is an identity matrix. Assuming all statistical quantities to be stationary (or independent of k), one determines in steady state, by setting $E(\mathbf{W}_{k+1}) = E(\mathbf{W}_k)$, that

$$E(\mathbf{W}_{\text{inf}}) = \mathbf{R}^{-1}\mathbf{P} + \frac{1}{\mu}\mathbf{R}^{-1}\mathbf{b}, \qquad (15.40)$$

where $\mathbf{b}$ is the mean of $\mathbf{b}_k$; that is, $\mathbf{b} = E(\mathbf{b}_k)$. The first term in Eq. (15.40) represents the infinite-precision solution, and the second term represents a mean deviation from the ideal. There is much that can be inferred from the second term.

Note that the deviation in Eq. (15.40) is inversely proportional to the step size μ. Decreasing the step size in the infinite-precision LMS algorithm (with very slow or no change in the relationship between $x[k]$ and $d[k]$) should improve its performance according to Eq. (15.21). Nevertheless, Eq. (15.40) shows that a decrease of the step size increases the deviation from the infinite-precision performance, an unexpected result to the designer not familiar with the effects of limited precision in the LMS algorithm. Thus the step size μ can only be decreased in practice to a level at which the effects of the second term become significant, which will of course be a function of the wordlength of the processor used to implement the algorithm.

Also note that the second term can be decomposed in an eigenvalue expansion as

$$\frac{1}{\mu} \mathbf{R}^{-1} \mathbf{b} = \frac{1}{\mu} \sum_{i=1}^{N} q_i^{-1} v_i(v_i^t \mathbf{b}), \qquad (15.41)$$

where $\{q_i\}$ and $\{v_i\}$ are the eigenvalues and eigenvectors of $\mathbf{R}$, respectively. The terms in Eq. (15.41) with the smallest eigenvalues will dominate, especially in the likely event that the vector $\mathbf{b}$ has a nonzero component along the eigenvectors with small eigenvalues. In a time-varying environment, the above analysis approximately applies. The magnitude of the second term in Eq. (15.40) can be enormous, even when large precision is used and $\mathbf{b}$ has relatively small component values. Thus finite-precision requirements increase when the eigenvalue spread is large.

In most theoretical and simulated studies of the LMS algorithm, the offset in Eq. (15.40) is not immediately evident. Remembering that the quantity in Eq. (15.40) grows over time from the accumulation of very small quantization errors, it can take on the order of tens of millions of iterations before the effects become noticeable, much longer than is often used in computer simulation studies. Nevertheless, even at modest voiceband sampling rates, like 8 or 9.6 kHz, the effect in Eq. (15.40) can grow to overflow proportions in real-time implementations within a period of minutes, and much faster in higher bandwidth (larger number of updates per second) applications.

A more heuristic, but basically equivalent, explanation for the "catastrophic overflow" in the unabetted LMS algorithm is obtained in the frequency domain. An input with little or no energy content over a certain frequency band is an example of a signal with a small minimum eigenvalue in its autocorrelation matrix (roughly speaking). An example of an application in which this minimum eigenvalue is very small is the so-called fractionally spaced equalizer [UN76].

In Figure 15-6, we plot the energy spectrum of such an input time series $x[k]$. If the rate of change of $\mathbf{W}_k$ is sufficiently low with respect to the signal bandwidths, one can think of the adaptive filter as having a frequency response $W(e^{j\omega})$, given by the Fourier transform of the filter coefficients. Let the input time series $x[k]$ have a frequency response

$$X(e^{j\omega}) = \begin{cases} \text{nonzero}, & |\omega - \omega_0| < \omega_z, \\ 0, & |\omega - \omega_0| \geq \omega_z. \end{cases} \qquad (15.42)$$

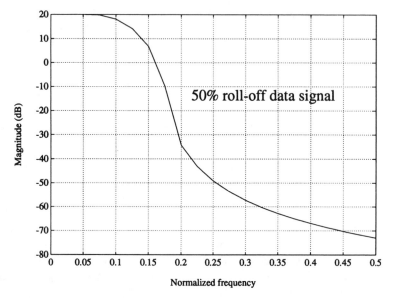

FIGURE 15-6 Input with high eigenvalue spread.

Then it is also true that

$$W(e^{j\omega})X(e^{j\omega}) = \begin{cases} \text{nonzero}, & |\omega - \omega_0| < \omega_z, \\ 0, & |\omega - \omega_0| \geq \omega_z \end{cases} \tag{15.43}$$

Over the frequency range $|\omega - \omega_0| \geq \omega_z$, the adaptive-filter frequency response $W(e^{j\omega})$ can take very large values with little or no effect on the filter output $y[k]$ or on the error $\epsilon[k]$. What happens in practice is that the accumulation of quantization errors toward the amount in Eq. (15.40) leads to a growth in $W(e^{j\omega})$ over this frequency range where it does not affect the filter error or the mean-square error, and thus the adaptive algorithm does not compensate this growth, as it is "hidden" from the LMS updating mechanism. If the implementation had an infinite dynamic range, this growth would not be a problem. In practice, especially in fixed-point implementations, the growth leads to an overflow in the registers storing $\mathbf{W}_k$, resulting in an adaptive-filter outage, as the overflow results in a sudden large increase in the mean-square error and the adaptive filter must reconverge. This effect is common in the aforementioned fractionally spaced equalizer [GI82; UN76], in adaptive arrays [JA86], and in adaptive line enhancers [TR77]. In Figure 15-7, we have plotted the $W(e^{j\omega})$ for a real-time simulation of a voiceband echo canceller after about 15 min of real-time adaptation using the LMS algorithm at a sampling rate of 9600 samples per second. Also plotted is the same characteristic just after convergence before quantization errors can accumulate appreciably.

In Figure 15-7, one notes that $W(e^{j\omega})$ is slowly rising with respect to its earlier

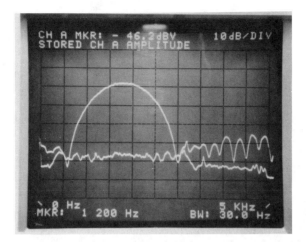

FIGURE 15-7 Photo of adaptive-filter input spectrum and the corresponding tap-error vector spectrum. Note the unconstrained growth of the tap-error spectrum where the input has little or no energy.

zero value in the low- and high-frequency ranges where the input has no energy. After waiting approximately another 15 min, the adaptive-filter setting had overflown the register size (12 bits, in this simulation), and the adaptive-filter had not reconverged. Sometimes it has been noted that after such a catastrophic overflow, the adaptive filter will not reconverge unless $\mathbf{W}_k$ is also reset to zero.

It is also possible to determine the component of the accumulation quantization error that is evident in the output mean-square error of the adaptive filter (at any point of steady-state operation prior to the catastrophic overflow). Caraiscos and Liu have determined a general formula for the extra quantization error power in the total output mean-square error as [CA84]

$$(\sigma^2_{\text{MMSE}} + \sigma^2_q) + \frac{\mu N \sigma^2_x}{2 - \mu N \sigma^2_x}(\sigma^2_u + \sigma^2_q)$$

$$+ \left[\frac{\|W^0\|^2}{2\mu} + \frac{\sigma^2_{\text{MMSE}} N}{4} \right] \sigma^2_{q'} + \left[2 \sum_{k=1}^{N} r[k](w^0[k])^2 + C \right] \sigma^2_q, \quad (15.44)$$

where σ^2_q and $\sigma^2_{q'}$ are the quantization noises associated with quantization of the ADCs and of the internal quantities, respectively. Note that it has been assumed that $\mathbf{b} = \mathbf{0}$ to focus on effects separate from the steady accumulation of errors toward catastrophic overflow or, alternatively, if all average biases have been carefully removed. C is a constant that Caraiscos and Liu have determined as

$$C = \sum_{j=1}^{N-1} E \left(\sum_{i=0}^{j} x[k - i] W^0[i] \right)^2. \quad (15.45)$$

Also in Eq. (15.44), $\mathbf{W}^0 = \mathbf{R}^{-1}\mathbf{P}$ and $W^0[i]$ is its ith component. $(\sigma_u^2 + \sigma_q^2)$ is the MMSE and contains an extra σ_q^2 term corresponding to possible quantization noise in the desired response, $d[k]$. Note that decreasing the step size or increasing N increases the third term in the excess error, while increasing μ increases the second more traditional *excess error* term, indicative of a classic trade-off in the value of μ. Quite often, this trade-off is far more important to adaptive-filter performance than is the other classic trade-off between averaging and tracking discussed in Section 15-3-3.

Of course, there are many assumptions made in the derivation of such a result to make the analysis tractable, and thus Eq. (15.44) only roughly characterizes the effects of the various parameters, such as the step size and eigenvalue disparity on the performance. One may observe deviations from this formula in a particular application. Cioffi and Werner [CI85a] characterize the distortion for the particular application of echo cancellation, which is thus more accurate in that application, but does not apply in general. Alexander [AL87] characterizes the deviations during the difficult convergence period as well. Nevertheless, the important results for the designer to keep in mind when making specifications for a particular application is that the step size cannot be chosen too small, even in a stationary application, despite infinite-precision theory specifying that being the best setting when little tracking is necessary. Second, the more ill-conditioned the input time series $x[k]$, the more pronounced will be the numerical problems, as was evident in Eq. (15.41).

Leaky LMS Solution. In the case of the LMS algorithm, there is a technique known generally as *leakage* that can be used to prevent the catastrophic overflow, at the expense of an increase in the cost of implementation and at the expense of a small degradation in performance away from the infinite-precision performance. That technique is summarized here and was introduced by Zahm in the early 1970s [ZA73], although it has been observed independently by others [GI82; UN76].

The leakage methods all essentially augment the performance criterion by a term that bounds or contains the energy in the adaptive-filter response. The resulting adaptive algorithms then try to find a compromise setting between minimizing the true mean-square error and containing the energy of the filter settings in a limited-precision implementation. Our own favorite view is that the performance criterion becomes

$$(\epsilon[k])^2 + \alpha \|\mathbf{W}_k\|^2. \tag{15.46}$$

The stochastic gradient then is

$$-2(\epsilon[k])^2\mathbf{X}_k + 2\alpha\mathbf{W}_k, \tag{15.47}$$

that is, the original gradient offset by the gradient of the energy-control term. The equivalent of Eq. (15.34) is then

$$\mathbf{W}_{k+1} = \mathbf{W}_k(1 - \mu\alpha) + \mu \cdot \epsilon[k] \cdot \mathbf{X}_k. \tag{15.48}$$

The resulting algorithm is essentially the LMS algorithm, except for the leakage factor $(1 - \mu\alpha)$ in the first term, which can be precomputed. This factor is the origin of the names *leaky LMS algorithm* [TR77] and *tap leakage algorithm* [GI82]. Further improvements can be made by allowing α to become a diagonal matrix, with the entries along the diagonal chosen to weight more heavily against unconstrained growth in regions where the input $x[k]$ is energy deficient.

It is very important to note that the presence of the leakage factor increases the cost of implementation by at most an additional N multiplies per iteration.[5] Other less costly increases have been investigated by Ungerboeck [UN76] with the fractionally spaced equalizer as the specific application. Ad hoc leakage methods may only leak one tap coefficient on each iteration, but cycle through all N tap coefficients on N successive iterations.

Leakage also costs in terms of performance deviations from the true infinite-precision performance (though, with leakage, these deviations remain bounded and catastrophic overflow is eliminated). One infers from Eq. (15.46) that the autocorrelation matrix of the input $x[k]$, as far as the adaptive algorithm is concerned, now becomes

$$\mathbf{R} + \alpha \mathbf{I}. \tag{15.49}$$

The smallest eigenvalue of this augmented autocorrelation matrix now must be greater than α, thereby favorably conditioning the input. Using standard matrix-inversion techniques, the offset in the converged filter setting $\mathbf{W}_k$ from its known statistically (stationary) value of $\mathbf{R}^{-1}\mathbf{P}$ is given by

$$\Delta\mathbf{W} = -\alpha\mathbf{R}^{-1}(\mathbf{I} + \alpha\mathbf{R}^{-1})^{-1}\mathbf{R}^{-1}\mathbf{P}. \tag{15.50}$$

The simple analysis of the standard excess mean-square error formula(s) that appear in Widrow and Stearns [WI85] and Haykin [HA91] for LMS can be approximately applied to leaky LMS, if $\mathbf{R}$ is replaced by $\mathbf{R} + \alpha\mathbf{I}$ in those formulas, leaving an excess error around the biased solution in Eq. (15.50) as (single-channel case)

$$\sigma^2_{\text{EMSE, bias}} = \mu N \frac{(\sigma_x^2 + \alpha)}{2 - \mu N(\sigma_x^2 + \alpha)} \sigma^2_{\text{MMSE, bias}}, \tag{15.51}$$

where σ_x^2 is again the power in the input $x[k]$, and $\sigma^2_{\text{MMSE, bias}}$ represents the minimum mean-square error corresponding to the solution $\mathbf{W} = (\mathbf{R} + \alpha\mathbf{I})^{-1}\mathbf{P}$, given by

$$\sigma^2_{\text{MMSE, bias}} = E(d^2) - \mathbf{P}^t(\mathbf{R} + \alpha\mathbf{I})^{-1}\mathbf{P}. \tag{15.52}$$

---

[5]Leakage need not be implemented on every update, as long as all taps see some leakage at regular intervals.

The total excess mean-square error, however, must also reflect the offset from the desired behavior that is given in Eq. (15.50). Summing the two effects, one obtains

$$\sigma^2_{\text{EMSE, total}} = \sigma^2_{\text{EMSE, bias}} + \Delta\mathbf{W}\mathbf{R}\,\Delta\mathbf{W}, \qquad (15.53)$$

where Eqs. (15.51) and (15.50) can be used to express the $\sigma^2_{\text{EMSE, bias}}$ in terms of the known quantities σ_x^2, $\mathbf{R}$, $\mathbf{P}$, μ, N, and α. One can then note from Eqs. (15.50)–(15.53) and (15.21) that, if α is chosen small, little performance degradation will be evident. However, α cannot be chosen too small or the above catastrophic overflow can occur. The proper trade-off between performance and numerical stability in the use of the leaky LMS can then be evaluated from the preceding analysis, given a particular application and implementation precision.

A nearly equivalent technique that can be used to stabilize LMS (without explicitly using leaky LMS or incurring the algorithmic cost increase) is to add a small (of power $= \alpha$) "white" or uncorrelated noise component to the input $x[k]$. This method has essentially the same performance as the leaky LMS algorithm. However, the generation of this extra component may often be more costly in practice than the extra cost of using the leaky LMS and therefore is not usually as attractive an option as it might initially seem.

The important numerical facts about the LMS algorithm are that it is sensitive to the accumulation of quantization errors over long periods of time, and the sensitivity is magnified by small step sizes or spectrally incomplete input time series $x[k]$. Nevertheless, at the expense of a decrease in performance (with respect to infinite precision) as given above and a maximum of 50% implementation-cost increase, the LMS algorithm can circumvent the sensitivity using the leakage techniques.

15-4 RECURSIVE LEAST-SQUARES ADAPTIVE ALGORITHMS

Recursive least-squares (RLS) algorithms are high-performance adaptive-filtering methods that often converge and track significantly faster than stochastic-gradient methods. The trade-off for the improved performance is usually an increase in the cost of implementation, although RLS methods sometimes are less expensive, depending on the details of the implementation. RLS adaptive filters approximate the MMSE criterion of Eq. (15.5) by the sum (sample average) of squared errors:

$$\min_{\mathbf{W}_k} E(|\epsilon(k)|^2) \cong \min_{\mathbf{W}_k} \frac{1}{k+1} \sum_{m=0}^{k} \|d[m] - \mathbf{W}_k^t \mathbf{X}_m\|^2. \qquad (15.54)$$

The factor $1/(k+1)$ does not affect the minimization and is often omitted. Also, as k increases, it is desirable to reduce the relative influence of terms in the remote past ($m \ll k$). There are two ways of *windowing* the sum of squared errors that are commonly used. The first method, called the *exponential windowing* approach, modifies Eq. (15.54) to

$$\min_{\mathbf{W}_k} \sum_{m=0}^{k} \|d[m] - \mathbf{W}_k^t \mathbf{X}_m\|^2 \lambda^{k-m}, \tag{15.55}$$

where λ is the important convergence factor,

$$0 < \lambda \le 1, \tag{15.56}$$

and determines the trade-off between the tracking and the excess error. The second method, called the *sliding window* approach, minimizes

$$\sum_{m=k-l+1}^{k} \|d[m] - \mathbf{W}_k^t \mathbf{X}_m\|^2, \tag{15.57}$$

where l is a constant that determines the window length, and thus the trade-off between the tracking and the excess errors.

The solution to Eq. (15.55) is easily determined as

$$\mathbf{W}_k = \left(\sum_{m=0}^{k} \mathbf{X}_m \mathbf{X}_m^t \lambda^{k-m} \right)^{-1} \left(\sum_{m=0}^{k} \mathbf{X}_m d[m] \lambda^{k-m} \right), \tag{15.58}$$

and for Eq. (15.57) as

$$\mathbf{W}_k = \left(\sum_{m=k-l+1}^{k} \mathbf{X}_m \mathbf{X}_m^t \right)^{-1} \left(\sum_{m=k-l+1}^{k} \mathbf{X}_m d[m] \right). \tag{15.59}$$

There are essentially three basic methods for recursively performing the matrix inversion in the above equations, all of which require $O(N^2)$ computations[6] per sample period to compute $\mathbf{W}_{k+1}$ (or its equivalent) from $\mathbf{W}_k$. Because of the special *shift-invariance* structure of the adaptive-filter input $x[k]$, these computational requirements can be reduced to $O(N)$ in many situations (see Section 15-4-6). The three methods are based on (1) direct propagation of the inverted matrix, (2) propagation of the $\mathbf{UDU}^t$ (upper-diagonal-lower) factors of the matrix to be inverted, and (3) propagation of the QR (orthogonal-triangular) factors of the matrix.

15-4-1 Performance of RLS Adaptive Algorithms

It is important to realize that as long as the RLS problem is exactly solved, its performance is independent of the matrix inversion method used (in infinite precision). We further note that there are really two errors that can be computed in

---

[6]$O(x)$ is a function increasing no faster than linearly with x.

an adaptive filter:

$$\epsilon[k] = d[k] - \mathbf{W}_k^t \mathbf{X}_k \qquad (15.60)$$

or

$$\epsilon^P[k] = d[k] - \mathbf{W}_{k-1}^t \mathbf{X}_k, \qquad (15.61)$$

differing only in the use of the old $\mathbf{W}_{k-1}$ (not updated) or new $\mathbf{W}_k$ (updated) filter settings. One should note that the LMS algorithm uses the old setting. Usually, the two errors are close in the steady state, and the excess mean-square error is of the same magnitude.[7] We will use the *old* error throughout the analysis, as it is most often used in practice.

For exponential RLS, Eqs. (15.18) and (15.19) also hold. The equivalent of Eq. (15.20) is

$$\sigma_\epsilon^2 = \sigma_{\text{MMSE}}^2 + \left(N \frac{1 - \lambda}{1 + \lambda} \right) \sigma_{\text{MMSE}}^2, \qquad (15.62)$$

and the second term on the right is the excess mean-square error, which is independent of σ_x^2. Thus choosing $\lambda \to 1$ yields the best performance if $\mathbf{W}_k^0 \equiv \mathbf{W}^0$. Note the factor $(1 - \lambda)$ has a role equivalent to μ in the LMS algorithm. For the sliding-window RLS algorithm,

$$\sigma_\epsilon^2 = \sigma_{\text{MMSE}}^2 + \frac{N}{l} \sigma_{\text{MMSE}}^2, \qquad (15.63)$$

where $1/l$, $(1 - \lambda)/(1 + \lambda)$, and μ are analogous quantities in determining the trade-off between convergence and noise averaging.

15-4-2 Convergence of RLS Versus LMS

The relations in Eqs. (15.62) and (15.63) are not complete without specifying the associated convergence times. We can then specify when a significant performance advantage exists with respect to the LMS algorithm. There has been considerable controversy concerning this subject and it still remains a sensitive and unsettled subject for some; see, for instance, Widrow and Walach [WI84] and Friedlander [FR82] for a sample of the differing viewpoints. There is little doubt that RLS performs at least as well as LMS, but whether RLS exhibits significant advantage or not remains the disputed issue. Actually, the answer depends on the application. We separate the range of applications into two types, for purposes of this com-

---

[7]An often unnoticed important fact is that the σ_{EMSE}^2 for the RLS method is always negative when the new taps are used, and positive (but statistically about the same magnitude) when the old taps are used.

parison.[8] These types are low noise and high noise. For an application to be low noise,

$$\sigma^2_{MMSE} < 0.01\sigma^2_d, \tag{15.64}$$

and for high noise

$$\sigma^2_{MMSE} > 0.01\sigma^2_d, \tag{15.65}$$

where σ^2_d is the desired-response power.

15-4-2-1 Low-Noise Case. For the LMS algorithm from Eq. (15.28), $\sigma^2_{EMSE}[k]$ decays toward its steady-state value no faster than exponentially as

$$\sigma^2_{EMSE}[k] \geq \left(1 - \frac{1}{2N}\right)^k (\sigma^2_d - \sigma^2_u) + \sigma^2_u \tag{15.66}$$

(with equality if $x[k]$ is white), where k is a time index on σ^2_{EMSE}. Equation (15.66) results from the *optimal* step-size choice of $\mu = 1/2N$ (for unit-power white $x[k]$). The corresponding steady-state excess error variance for $\mu = 1/2N$ is $\sigma^2_{EMSE} = \sigma^2_u$, which corresponds to $\sigma^2_{MMSE} = 2\sigma^2_u$.

For any unit-power input in the RLS algorithm, it can be shown [CI84] that for $k > N - 1$, or for N data points processed, that

$$0 \leq \sigma^2_{EMSE} \leq \sigma^2_u, \quad k \geq N. \tag{15.67}$$

Since σ^2_u is small, Eq. (15.67) states that in low-noise applications, the RLS adaptive algorithm converges in N steps. This is heuristically interpreted by noting that when $\sigma^2_u \cong 0$, the desired response is an N-parameter deterministic function of the input time series $x[k]$. One requires N independent equations in the N parameters to solve for these parameters, and the RLS provides and solves those N independent equations as soon as N independent data points have been acquired by the adaptive filter.

Although the LMS adaptive filter according to Eq. (15.66) theoretically takes infinitely long to reach $\sigma^2_{EMSE} = \sigma^2_u$ ($\sigma^2_{MMSE} = 2\sigma^2_u$), let us assume that $\sigma^2_{EMSE} = 2\sigma^2_u$ ($\sigma^2_{MMSE} = 3\sigma^2_u$) is *close enough* since σ^2_u is small (actually this is a best-case LMS performance assumption with respect to RLS for the sake of comparison). Also, let the constant g be defined by

$$\sigma^2_d = g\sigma^2_u, \quad g \geq 100. \tag{15.68}$$

---

[8]These types are for performance analysis only and differ from the two categories in Section 15-2, system identification and system inversion.

Then the LMS will have converged when $\sigma_{EMSE}^2 = 2\sigma_u^2$ in Eq. (15.66), such that

$$\left(1 - \frac{1}{2N}\right)^k g\sigma_u^2 = \sigma_u^2 \qquad (15.69)$$

or

$$k = \frac{-\log(g)}{\log\left(1 - \frac{1}{2N}\right)}. \qquad (15.70)$$

The advantage (ratio of convergence times) of RLS in this *low-noise case* is

$$\text{RLS } \textit{low-noise advantage} \geq -\frac{\log(g)}{N \log\left(1 - \frac{1}{2N}\right)}, \qquad g \geq 100. \quad (15.71)$$

For $N = 20$ and $g = 1000$ (numbers typical of adaptive equalizers and echo cancelers in data transmission, as well as other applications), this advantage is about 13 (or roughly an order of magnitude). This case is illustrated in Figures 15-8(a) and 15-8(b), where NTE (*norm tap error*) is plotted versus discrete time (number of iterations). The approximate factor of 10 convergence-speed increase is evident in Figure 15-8(a). One can easily show (using $\ln(1 - \delta) \cong \delta$, for small δ) that the RLS advantage approaches (for $N \geq 10$)

$$\text{RLS } \textit{low-noise advantage} \geq 4.6 \log(g), \qquad g \geq 100, \quad N \geq 10, \quad (15.72)$$

or $2 \ln(g)$, where "ln" is the natural log. A minor qualification on Eq. (15.72) is that $u[k]$ is not the quantity being estimated (most such noise-canceling applications would intentionally silence $u[k]$ during training if it were present, in any case), as one may then desire $\sigma_{EMSE}^2[k] < \sigma_u^2/100$, which really makes the application fall into the high-noise category.

15-4-2-2 High-Noise Case. In the high-noise case, the factor g in Eq. (15.68) is typically less than 100, and good performance is achieved when $|\sigma_{EMSE}^2| \leq \sigma_u^2/10$. The adaptive filter must average the noise ($u[k]$) in this case and N-iteration convergence is not possible with either the RLS or the LMS algorithm. In a stationary application, that is, $\mathbf{W}_k^0 \equiv \mathbf{W}^0$, either the LMS or the RLS method can achieve the desired σ_{EMSE}^2 level (say, $\sigma_u^2/10$), but there can be differences in the convergence time required to attain the prescribed σ_{EMSE}^2 level. This convergence time is measured by an exponential time constant, τ, where σ_{EMSE}^2 decays exponentially as

$$\sigma_{EMSE}^2[k] \geq e^{-k/\tau}\sigma_{EMSE}^2[0]. \qquad (15.73)$$

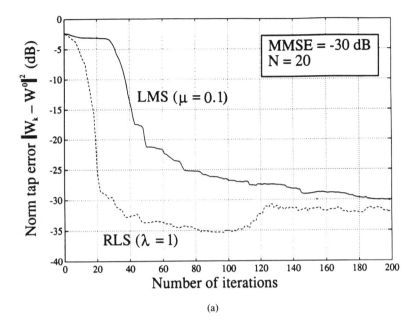

(a)

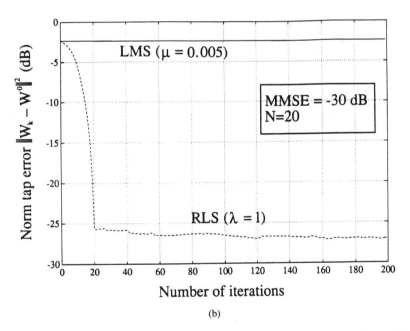

(b)

FIGURE 15-8 (a) Low-noise, low-eigenvalue spread, convergence comparison of LMS and RLS. (b) Low-noise, high-eigenvalue spread, convergence comparison of LMS and RLS.

To compare the RLS and the LMS methods, we make extensive use of some earlier results of [EL86]. For the LMS algorithm,

$$\tau_{LMS} \cong \frac{1}{\mu q_{min}}, \tag{15.74}$$

where q_{min} is the minimum eigenvalue[9] of $\mathbf{R}_k$ when $\sigma_x^2 = 1$. For the RLS algorithm [EL86],

$$\tau_{RLS} \cong \frac{1}{1 - \lambda}. \tag{15.75}$$

One notes that as μ or $(1 - \lambda)$ decreases in the corresponding formulas, σ_{EMSE}^2 improves at the expense of a slower (larger) time constant. This trade-off between σ_{EMSE}^2 and convergence time is essential in a comparison of the RLS and LMS methods to determine which algorithm provides a better trade-off. The answer depends on q_{min}. If we set the RLS excess error in Eq. (15.62) equal to the LMS excess error in Eq. (15.21), we obtain

$$\mu \cong (1 - \lambda) \tag{15.76}$$

for the equivalent steady-state σ_{EMSE}^2. However, use of Eq. (15.76) in Eq. (15.74) yields

$$\tau_{LMS} \cong \frac{1}{(1 - \lambda)q_{min}} \cong \frac{1}{q_{min}} \tau_{RLS}. \tag{15.77}$$

For the assumed unit-power input in the above analysis, $q_{min}^{-1} \geq 1$. Thus $\tau_{LMS} > \tau_{RLS}$ such that LMS cannot do better than RLS. If $q_{min} = 1$, the input sequence $x[k]$ is broadband or white, and in this high-noise case, there is no improvement of RLS over LMS. If $q_{min} < 1$, there is an approximate RLS convergence-time improvement of q_{min}^{-1}.

In Figures 15-9(a) and 15-9(b) we illustrate both situations ($q_{min} \cong 1$ and $q_{min} < 1$) for $\sigma_u^2 = 0.1\sigma_d^2$ or $g = 10$. In Figures 15-9(a) and 15-9(b) one should first note the great difference in the horizontal scale with respect to Figure 15-8, indicative of the dominance of the noise averaging of $u[k]$. Note in Figure 15-9(a) the RLS improvement when $q_{min} < 1$ (actually $q_{min} = 0.01$), but in Figure 15-9(b) there is no improvement since $q_{min} = 1$. Finally, we should note that all the analyses in this subsection (high-noise case) apply to the previous subsection (low-noise case), but in the low-noise case, it is acceptable to have $\sigma_{EMSE}^2 \approx \sigma_u^2$ (since σ_u^2 is very small with respect to the other signals of interest), which occurs for $k = N$ in RLS, but much later in LMS. It is for this practical concern that the two cases have been distinguished here.

---

[9]Strictly, ϵ_{min} is the smallest eigenvalue for which $\nu_{min}^t \mathbf{P}_k \neq 0$, and ν_{min} is the corresponding eigenvector.

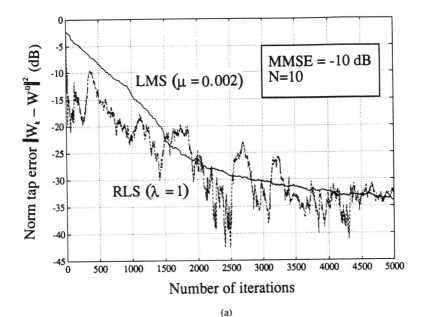

(a)

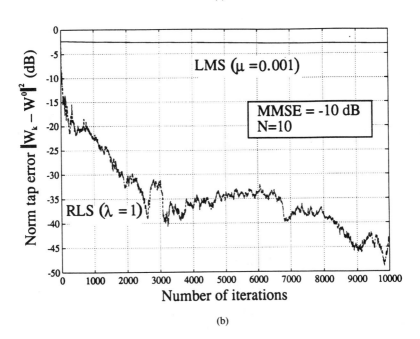

(b)

FIGURE 15-9 (a) High-noise, low-eigenvalue spread, convergence comparison of LMS and RLS. (b) High-noise, high-eigenvalue spread convergence comparison between LMS and RLS.

15-4-2-3 Tracking Comparison. As discussed in Section 15-3-3, the optimal setting $\mathbf{W}_k^0$ varies with time. The adaptive filter will attempt to follow or "track" this time variation of $\mathbf{W}_k^0$ with $\mathbf{W}_k$. However, the adaptive algorithm's memory of previous data, which is used to average the noise $u[k]$, hinders this tracking ability, and $\mathbf{W}_k$ will *lag* $\mathbf{W}_k^0$ somewhat. This classic trade-off between noise immunity and tracking is also characteristic of the RLS algorithm. Greater noise immunity implies lesser tracking speed (longer *lag*) and vice versa. The value of the parameter μ or λ controls the balance between noise immunity and tracking for LMS or RLS, respectively. For any adaptive algorithm, σ_{EMSE}^2 can be decomposed into two components,

$$\sigma_{EMSE}^2 = \sigma_{noise}^2 + \sigma_{lag}^2. \tag{15.78}$$

We now investigate the balance between these two terms for the LMS and RLS methods, relying heavily on the work of Elefthériou and Falconer [EL86].

The specific model for the time variation of $\mathbf{W}_k^0$ that this comparison uses is that each component of $\mathbf{W}_k$ varies according to a random dither of power σ_ν^2 at each point in time; that is,

$$\mathbf{W}_k^0 = \mathbf{W}_{k-1}^0 + \begin{bmatrix} \nu_1[k] \\ \nu_2[k] \\ \vdots \\ \nu_N[k] \end{bmatrix}, \tag{15.79}$$

where $\nu_i[k]$ are i.i.d. white-noise random processes with mean zero and variance σ_ν^2. For the LMS algorithm, σ_{EMSE}^2 is given by Eq. (15.30) as (note $\sigma_x^2 = 1$ here)

$$\sigma_{EMSE,LMS}^2 = \frac{\mu}{2 - \mu N} N\sigma_{MMSE}^2 + \frac{N}{2\mu} \sigma_\nu^2, \tag{15.80}$$

while for the RLS algorithm [EL86]

$$\sigma_{EMSE,RLS}^2 \cong N\left(\frac{1-\lambda}{1+\lambda}\right)\sigma_{MMSE}^2 + \frac{N}{2(1-\lambda)} \sigma_\nu^2. \tag{15.81}$$

Note in Eq. (15.80) or (15.81) that as μ or $(1 - \lambda)$ decreases, the noise component decreases while the lag increases. An increase in μ or $(1 - \lambda)$ improves the lag component but makes the noise component worse. By optimizing (taking the derivative with respect to μ or λ and setting the result to zero in) Eq. (15.80) or (15.81), one determines

$$\sigma_{EMSE,LMS}^2 = N\sigma_u\sigma_\nu + \frac{N^2}{2} \sigma_\nu^2. \tag{15.82}$$

If we let

$$h \triangleq \frac{\sigma_v}{\sigma_u}, \qquad (15.83)$$

Equation (15.82) becomes

$$\sigma^2_{\text{EMSE, LMS}} = Nh\sigma^2_u \left(1 + \frac{Nh}{2} \right). \qquad (15.84)$$

For RLS, one similarly finds

$$\sigma^2_{\text{EMSE, RLS}} \cong \begin{cases} Nh\sigma^2_{\text{MMSE}} \left(1 + \dfrac{h}{4} \right), & h \leq 2.5, \\ N\sigma^2_{\text{MMSE}}(0.56 + 0.56h^2), & h \geq 2.5. \end{cases} \qquad (15.85)$$

One notes that if h is small, implying that the time variation of $\mathbf{W}^0_k$ is insignificant with respect to noise power, then the LMS and RLS algorithms are equivalent and have

$$\sigma^2_{\text{EMSE}} = \sigma^2_{\text{EMSE, LMS}} = \sigma^2_{\text{EMSE, RLS}} = Nh\sigma^2_{\text{MMSE}}. \qquad (15.86)$$

Thus there is no advantage for the RLS method, as has also been determined previously in Widrow and Walach [WI84]. However, if h is relatively large, the advantage can be significant. For example, if $h = 1$ and $N = 10$, the advantage is about a factor of 5 (because of a reduction in the best attainable σ^2_{EMSE}). The advantage is plotted in Figure 15-10 for $h = 0.001, 0.01, 0.1,$ and 1 versus N. There one sees perhaps an unexpected result that for h small (less than 0.001), as is typical of most applications, the RLS algorithm has no advantage. If there is significant time variation and the noise is not also large, then the RLS method can have a significant advantage. For instance, there are applications such as the multipath-fade tracking in adaptive equalization [LI85] where conditions have been such that h is large, and significant performance improvement over LMS has been observed (see also Section 15-6). Analyses that show the RLS approach to be equivalent to the LMS approach in tracking performance are for small h, even if not explicitly stated. Also, the first-order walk model of time variation may not be appropriate, but most analyses find the RLS method tracking better as time variation becomes the dominant effect, no matter the model.

15-4-2-4 Performance Summary. We conclude our comparison of the RLS and LMS algorithms by restating the results of our analysis. A convergence-time improvement of RLS with respect to LMS occurs if:

1. The noise power σ^2_{MMSE} is small in relation to the other signal powers of interest in the application, in which case the advantage (in reduced conver-

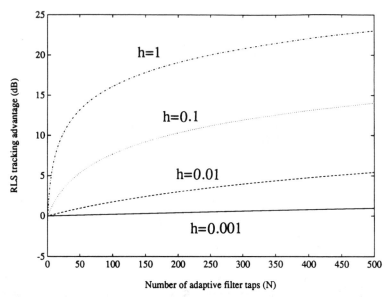

FIGURE 15-10   Tracking comparison of LMS and RLS for various ratios of time-variation to noise variance (h).

gence time) is at least $2 \ln(g)$, where g is essentially the signal-to-noise ratio for the desired response; or otherwise if,

2. The smallest eigenvalue of **R** is less than the input power (nonwhite input $x[k]$), in which case the convergence-time improvement is inversely proportional to this smallest (relevant) eigenvalue.

A convergence-time improvement will not occur if the noise power is comparable to the other relevant signal powers and the input spectrum is flat or nearly flat, in which case both algorithms perform the same.

For tracking *slow* statistical variations (as is the usual case), there is *no tracking advantage* of the RLS method over the LMS method. However, as the time variations increase in magnitude with respect to the noise power, the RLS algorithm does show an advantage, as is indicated in Figure 15-10.

We do not analyze the sliding window here as results are not available. Nevertheless, Cioffi and Kailath [CI85b] list empirical results that verify the natural conclusion that sliding windows perform better when $\mathbf{W}_k^0$ exhibits infrequent sudden changes in its coefficient values.

15-4-3   Direct RLS Algorithm

The exponential RLS solution in Eq. (15.58) can be written in terms of the sample autocorrelation matrix

$$\mathbf{R}_k = \sum_{m=0}^{k} \mathbf{X}_k \mathbf{X}_k^{t} \lambda^{k-m} \qquad (15.87)$$

and the sample cross-correlation vector

$$\mathbf{P}_k = \sum_{m=p}^{k} \mathbf{X}_k d[k] \lambda^{k-m} \tag{15.88}$$

and $\mathbf{W}_k = \mathbf{R}_k^{-1} \mathbf{P}_k$. Using the recursive relations

$$\mathbf{R}_k = \lambda \mathbf{R}_{k-1} + \mathbf{X}_k \mathbf{X}_k^t, \qquad \mathbf{P}_k = \lambda \mathbf{P}_{k-1} + \mathbf{X}_k d[k], \tag{15.89}$$

and a well-known matrix-inversion lemma, we obtain the exponentially windowed RLS algorithm.

Exponentially Windowed Direct RLS Algorithm

$$\epsilon[k] = d[k] - \mathbf{W}_{k-1}^t \mathbf{X}_k, \tag{15.90}$$

$$\mathbf{C}_k = \frac{-\mathbf{R}_{k-1}^{-1} \mathbf{X}_k}{\lambda + \mathbf{X}_k^t \mathbf{R}_{k-1}^{-1} \mathbf{X}_k}, \tag{15.91}$$

$$\mathbf{W}_k = \mathbf{W}_{k-1} - \epsilon[k] \mathbf{C}_k, \tag{15.92}$$

$$\mathbf{R}_k^{-1} = \frac{1}{\lambda} [\mathbf{R}_{k-1}^{-1} + \mathbf{C}_k \mathbf{X}_k^t \mathbf{R}_{k-1}^{-1}]. \tag{15.93}$$

The direct RLS requires about $2N^2 + 5N$ operations per recursive iteration (exploiting symmetry of $\mathbf{R}_k$ in Eq. (15.93)). The matrix $\mathbf{R}_0$ should be initially set to some small diagonal matrix, say, $0.001 \sigma_x^2 \mathbf{I}$, to avoid division by zero in the first iteration of this algorithm.

For the sliding window RLS algorithm, we define

$$\mathbf{R}_{l,k} = \sum_{m=k-l+1}^{k} \mathbf{X}_m \mathbf{X}_m^t \tag{15.94}$$

and

$$\mathbf{P}_{l,k} = \sum_{m=k-l+1}^{k} \mathbf{X}_m d[m], \tag{15.95}$$

which can be written in recursive form as

$$\mathbf{R}_{l,k} = \mathbf{R}_{l,k-1} + \mathbf{X}_k \mathbf{X}_k^t - \mathbf{X}_{k-l} \mathbf{X}_{k-l}^t \tag{15.96}$$

and

$$\mathbf{P}_{l,k} = \mathbf{P}_{l,k-1} + \mathbf{X}_k d[k] - \mathbf{X}_{k-l} d[k-l]. \tag{15.97}$$

This leads to the following algorithm.

Sliding Window Direct RLS Algorithm

$$\epsilon[k] = d[k] - \mathbf{W}_{l-1,k-1}^t \mathbf{X}_k, \tag{15.98}$$

$$\mathbf{C}_{l,k} = \frac{-\mathbf{R}_{l-1,k-1}^{-1} \mathbf{X}_k}{1 + \mathbf{X}_k^t \mathbf{R}_{l-1,k-1}^{-1} \mathbf{X}_k}, \tag{15.99}$$

$$\mathbf{W}_k = \mathbf{W}_{l-1,k-1} - \epsilon[k] \mathbf{C}_{l,k}, \tag{15.100}$$

$$\mathbf{R}_{l,k}^{-1} = \mathbf{R}_{l-1,k-1}^{-1} + \mathbf{C}_{l,k} \mathbf{X}_k^t \mathbf{R}_{l-1,k-1}^{-1}, \tag{15.101}$$

$$\nu_l[k] = d[k - l + 1] - \mathbf{W}_k^t \mathbf{X}_{k-l+1}, \tag{15.102}$$

$$\mathbf{D}_{l,k} = -\mathbf{R}_{l,k}^{-1} \mathbf{X}_{k-l+1}, \tag{15.103}$$

$$\delta_l[k] = 1 + \mathbf{D}_{l,k} \mathbf{X}_{k-l+1}, \tag{15.104a}$$

$$\mathbf{A}_{l,k} = \frac{\mathbf{D}_{l,k}}{\delta_l[k]}, \tag{15.104b}$$

$$\mathbf{W}_k = \mathbf{W}_{l-1,k} - \left(\frac{\nu_l[k]}{\delta_l[k]}\right) \mathbf{D}_{l,k} = \mathbf{W}_{l-1,k} - \nu_l[k] \mathbf{A}_{l,k}, \tag{15.105}$$

$$\mathbf{R}_{l,k}^{-1} = \mathbf{R}_{l-1,k}^{-1} - \frac{\mathbf{D}_{l,k} \mathbf{D}_{l,k}^t}{\delta_l[k]} = \mathbf{R}_{l-1,k}^{-1} - \mathbf{A}_{l,k} \mathbf{D}_{l,k}^t, \tag{15.106}$$

which requires approximately $3N^2 + 10N$ operations per iteration. Again, $\mathbf{R}_{l,0} = 0.001\sigma_x^2 \mathbf{I}$ (or some small value) to avoid division-by-zero problems. The variable $\mathbf{D}_{l,k}$ is sometimes termed the "Kalman gain vector." Equations (15.90)–(15.93) and (15.98)–(15.106) are sometimes called "Kalman" algorithms in the literature.

15-4-4 Cholesky RLS Algorithm

The additions in Eqs. (15.93) and (15.101) can lead to an erroneous matrix $\mathbf{R}_k$ that is no longer positive-definite in limited-precision applications [CI87; HS82]. This undesirable event leads to unacceptable numerical accuracy problems, and the direct RLS algorithms are restricted to applications where the number of updates is fixed and relatively small (100,000 iterations or less). This accuracy problem is circumvented by the $\mathbf{UDU}^t$ forms of the RLS algorithm, which do not propagate $\mathbf{R}_k$ directly but instead propagate its $\mathbf{UDU}^t$ factorization [BI77]

$$\mathbf{R}_k = \mathbf{U}_k \mathbf{D}_k \mathbf{U}_k^t, \tag{15.107}$$

where $\mathbf{U}_k$ is an upper triangular and $\mathbf{D}_k$ is a diagonal matrix, respectively. We note that

$$\mathbf{R}_k \mathbf{W}_k = \mathbf{P}_k = \mathbf{U}_k \mathbf{D}_k \mathbf{U}_k^t \mathbf{W}_k, \tag{15.108}$$

and

$$\mathbf{V}_k = \mathbf{U}_k^{-1}\mathbf{P}_k \qquad (15.109)$$

is a vector that can be found by back substitution since $\mathbf{U}_k$ is triangular. Then

$$\overline{\mathbf{V}}_k = \mathbf{D}_k^{-1}\mathbf{V}_k \qquad (15.110)$$

is trivially found, and finally

$$\mathbf{W}_k = (\mathbf{U}_k^t)^{-1}\overline{\mathbf{V}}_k, \qquad (15.111)$$

obtained by back substitution.

One can also directly propagate $\mathbf{U}_k^{-1}$ and $\mathbf{V}_k$ as is done in the recursive modified Gram–Schmidt methods of Ling and Proakis [LI86] or in Strobach's "Porla" methods [ST90]. These methods can be very robust in limited-precision implementations and also very efficient, despite a computation count (per iteration) that is proportional to the square of the number of parameters (N^2). Strobach's recent excellent text [ST90] discusses many of the trade-offs and algorithms in "Cholesky RLS" adaptive filters. In particular, the troublesome subtraction (addition) in Eqs. (15.93) and (15.101) can be avoided in the update of the corresponding $\mathbf{D}_k$ entries using what are often called "Schur ratios." For more details, see Strobach [ST90].

15-4-5 QR-RLS Algorithm

The third method for implementation of the RLS adaptive filter is based on QR (orthogonal-triangular) factorization of the underlying data matrix in the RLS adaptive filter. This is a special form of the Cholesky RLS algorithm that has many desirable features. One notes that the RLS problem in Eq. (15.55) can be interpreted as determining the $\mathbf{W}_k$ that minimizes the norm of the vector

$$\varepsilon_k = \begin{bmatrix} d[k] \\ d[k-1] \\ \vdots \\ d[0] \end{bmatrix}$$

$$- \begin{bmatrix} x[k] & x[k-1] & \cdots & x[k-N+1] \\ x[k-1] & x[k-2] & \cdots & x[k-N] \\ \vdots & \vdots & \ddots & \vdots \\ x[0] & x[-1] & \cdots & x[-N+1] \end{bmatrix} \mathbf{W}_k. \qquad (15.112)$$

The QR method notes that premultiplying by an orthogonal matrix $\mathbf{Q}$ ($\mathbf{QQ^t} = \mathbf{Q^tQ} = \mathbf{I}$) does not change the (minimized or otherwise) norm of ε_k. Assume that an orthogonal $(k+1) \times (k+1)$ matrix $\mathbf{Q}_k$ can be chosen to *triangularize* the matrix in Eq. (15.112), that is,

$$\mathbf{Q}_k \epsilon_k = \mathbf{Q}_k \mathbf{d}_k - \begin{bmatrix} 0 \\ \bar{\mathbf{u}}_{N,k} \end{bmatrix} \mathbf{W}_k, \tag{15.113}$$

where $\bar{\mathbf{u}}_{N,k}$ is an $N \times N$ triangular matrix that can be expressed as

$$\bar{\mathbf{u}}_{N,k} = \begin{bmatrix} \bar{u}_{11}[k] & \bar{u}_{12}[k] & \cdots & \bar{u}_{1N}[k] \\ \bar{u}_{21}[k] & \bar{u}_{22}[k] & \cdots & 0 \\ \vdots & \vdots & \ddots & \vdots \\ \bar{u}_{N1}[k] & 0 & \cdots & 0 \end{bmatrix}.$$

Therefore

$$\mathbf{Q}_k \epsilon_k = \mathbf{Q}_k \mathbf{d}_k - \begin{bmatrix} 0 & 0 & \cdots & 0 \\ \vdots & \vdots & \ddots & \vdots \\ 0 & 0 & \cdots & 0 \\ \bar{u}_{11}[k] & \bar{u}_{12}[k] & \cdots & \bar{u}_{1N}[k] \\ \bar{u}_{21}[k] & \bar{u}_{22}[k] & \cdots & 0 \\ \vdots & \vdots & \ddots & \vdots \\ \bar{u}_{N1}[k] & 0 & \cdots & 0 \end{bmatrix} \mathbf{W}_k. \tag{15.114}$$

The minimizing $\mathbf{W}_k$ can then be determined easily by back-substitution. This procedure is very stable numerically as long as N is not so large so as to cause significant error propagation in the back-substitution process. The least-squares problem can be solved easily in order N^2 computations recursively by noting that

$$\begin{bmatrix} 1 & 0 \\ 0 & \mathbf{Q}_{k-1} \end{bmatrix} \varepsilon_k = \begin{bmatrix} d[k] \\ \mathbf{Q}_{k-1}\mathbf{d}_{k-1} \end{bmatrix}$$

$$- \begin{bmatrix} x[k] & x[k-1] & \cdots & x[k-N+1] \\ 0 & 0 & \cdots & 0 \\ \vdots & \vdots & \ddots & \vdots \\ 0 & 0 & \cdots & 0 \\ \bar{u}_{11}[k-1] & \bar{u}_{12}[k-1] & \cdots & \bar{u}_{1N}[k-1] \\ \bar{u}_{21}[k-1] & \bar{u}_{22}[k-1] & \cdots & 0 \\ \vdots & \vdots & \ddots & \vdots \\ \bar{u}_{N1}[k-1] & 0 & \cdots & 0 \end{bmatrix} \mathbf{W}_k.$$

$$\tag{15.115}$$

The new data matrix on the right side of Eq. (15.115) can easily be triangularized by N successive Givens rotations [MC83], requiring on the order of N^2 computations. The triangularizing matrix, call it $\mathbf{Q}_k$, is given by

$$\mathbf{Q}_k = \hat{\mathbf{Q}}_k \mathbf{Q}_{k-1}. \tag{15.116}$$

We can write $\hat{\mathbf{Q}}_k$ as

$$\hat{\mathbf{Q}}_k = \begin{bmatrix} \cos \theta_{N,k} & \cdots & 0 & \cdots & -\sin \theta_{N,k} \\ \mathbf{0} & & \mathbf{I} & & \mathbf{0} \\ \sin \theta_{N,k} & \cdots & 0 & \cdots & \cos \theta_{N,k} \end{bmatrix} \cdots$$

$$\begin{bmatrix} \cos \theta_{1,k} & \cdots & -\sin \theta_{1,k} & \mathbf{0} \\ \vdots & \mathbf{I} & \vdots & \vdots \\ \sin \theta_{1,k} & \cdots & \cos \theta_{1,k} & \vdots \\ \mathbf{0} & \cdots & \cdots & \mathbf{I} \end{bmatrix}, \tag{15.117}$$

which reduces to

$$\hat{\mathbf{Q}}_k = \begin{bmatrix} \sqrt{\gamma[k]} & \mathbf{0} & \mathbf{b}_k^t \\ \mathbf{0} & \mathbf{I} & \mathbf{0} \\ \bar{\sigma}_{N,k} & \mathbf{0} & \mathbf{M}_k \end{bmatrix}, \tag{15.118}$$

where $\sqrt{\gamma[k]}$ is a real-valued scaling factor, $\mathbf{M}_k$ is an $N \times N$ matrix, $\mathbf{b}_k$ and $\bar{\sigma}_{N,k}$ are $N \times 1$ vectors, and $\mathbf{I}$ denotes the $(k - N - 1) \times (k - N - 1)$ identity matrix [HA91; MC83]. If the adaptive-filter designer is only interested in $\epsilon[k]$, it can be computed by knowing only the top row of $\hat{\mathbf{Q}}_k$ or

$$\hat{\mathbf{Q}}_k \begin{bmatrix} 1 \\ 0 \\ \vdots \\ 0 \end{bmatrix} = \begin{bmatrix} \sqrt{\gamma[k]} \\ 0 \\ \vdots \\ 0 \\ \bar{\sigma}_{N,k} \end{bmatrix}, \tag{15.119}$$

where

$$\sqrt{\gamma[k]} = \prod_{n=1}^{N} \cos \theta_{n,k}, \tag{15.120}$$

and

$$\bar{\sigma}_{N,k} = \left[-\sin \theta_N, \ \sin \theta_{N-1} \cos \theta_N, \ -\sin \theta_{N-2} \cos \theta_N \cos \theta_{N-1}, \ \cdots, \right.$$
$$\left. -\sin \theta_1 \left(\prod_{n=2}^{N} \cos \theta_n \right) \right] \qquad (15.121)$$

(where the k subscript has been dropped for notational convenience). For a more detailed derivation, see Bellanger [BE89]. Then, $\epsilon[k]$ is computed as

$$\epsilon[k] = \sqrt{\gamma[k]} (\sqrt{\gamma[k]} d[k] + \bar{\mathbf{d}}_{N,k}^t \bar{\sigma}_{N,k}), \qquad (15.122)$$

and $\bar{d}_{N,k}$ is recursively computed as

$$\mathbf{Q}_k \mathbf{d}_k = \begin{bmatrix} \tilde{\mathbf{d}}_{N,k} \\ \bar{\mathbf{d}}_{N,k} \end{bmatrix} = \hat{\mathbf{Q}}_{k+1} \begin{bmatrix} d[k] \\ \tilde{\mathbf{d}}_{N,k-1} \\ \bar{\mathbf{d}}_{N,k-1} \end{bmatrix}. \qquad (15.123)$$

The quantity in parentheses in Eq. (15.122) is actually computed as the top entry in

$$\hat{\mathbf{Q}}_k \begin{bmatrix} d[k] \\ \tilde{\mathbf{d}}_{N,k-1} \\ \bar{\mathbf{d}}_{N,k-1} \end{bmatrix}. \qquad (15.124)$$

The computational requirements (for exponential windowing) are $2.5N^2 + 6N$. A systolic implementation is straightforward [GE81; MC83] and is shown in Figure 15-11, where the individual computation units are detailed in Figures 15-12(a) and 15-12(b).

15-4-6   Fast RLS Algorithms

Fast versions, requiring only $O(N)$ computations/update of all three types of RLS algorithms (standard, $\mathbf{UDU}^t$, QR) have been derived under the names of "fast transversal filters" [CI84; CI85b] and in references therein, "fast lattice" [FR82] and the references therein, and "fast QR-RLS" [CI90a], respectively. The books by Haykin [HA91] and Alexander [AL86] have good developments of these methods.

It should be noted that the "fast transversal" methods, as noted in Cioffi and Kailath [CI84], had severe numerical problems that magnified the above-mentioned difficulties of the standard RLS algorithm. Slock's recent work [SL91] presents a practical and efficient method for eliminating the numerical problems.

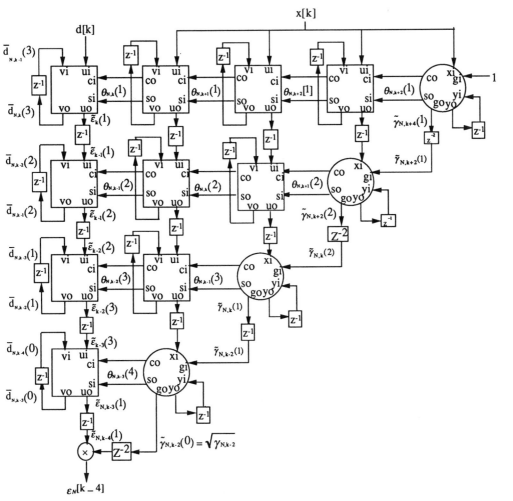

FIGURE 15-11 Processor array for $N = 4$ QR-RLS adaptive filter.

15-5 FREQUENCY-DOMAIN AND BLOCK ADAPTIVE FILTERS

There has been considerable interest in block adaptive filters, as they are easily pipelined and have excellent numerical accuracy. The block adaptive filter is illustrated in Figure 15-13. The input $x[k]$ is converted from a serial input sequence into successive blocks of l inputs by the serial-to-parallel converter (s/p). Each block is processed as a single vector input sample and a block error signal is generated for use in updating the adaptive-filter setting for the next block of inputs. The updating and implementation of the adaptive filter can be in the time domain or the frequency domain. These updates can be gradient-like (block LMS) or least-squares (recursive block least squares). We begin with the gradient algorithms and then proceed to least-squares methods.

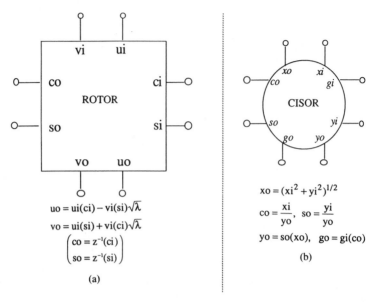

FIGURE 15-12   Processors for QR adaptive filters: (a) rotation processor and (b) cosine/sine processor.

15-5-1   Block LMS Algorithm

In terms of our previous notation, we can write the filter-error blocks as follows (returning to $\mathbf{W}_{N,l,k}$, a column vector), where the first subscript denotes the number of taps, the second denotes the block length, and the third indicates the time index:

$$\boldsymbol{\varepsilon}_{N,l,k} = \mathbf{d}_{l,k} - \mathbf{X}_{N,l,k}\mathbf{W}_{N,l,k}, \tag{15.125}$$

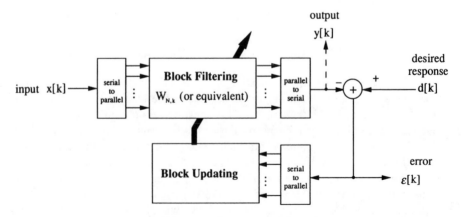

FIGURE 15-13   The block adaptive filter.

remembering now that l is independent of k and that we are interested in successive nonoverlapping blocks, $\varepsilon_{N,l,k}$, $\varepsilon_{N,l,k-l}$, $\varepsilon_{N,l,k-2l}$, ... For any block, the block-average mean-square error is

$$\sigma_{\text{BMSE}}^2 = \frac{1}{l} E(\varepsilon_{N,l,k}^t \varepsilon_{N,l,k}), \tag{15.126}$$

or

$$\sigma_{\text{BMSE}}^2 = \tfrac{1}{2}\{E(\mathbf{d}_{l,k}^t \mathbf{d}_{l,k}) - 2E(\mathbf{d}_{l,k}^t \mathbf{X}_{N,l,k})\mathbf{W}_{N,l,k}$$
$$+ \mathbf{W}_{N,l,k}^t E(\mathbf{X}_{N,l,k}^t \mathbf{X}_{N,l,k})\mathbf{W}_{N,l,k}\}, \tag{15.127}$$

with minimizing solution

$$\mathbf{W}_{N,l,k} = \mathbf{R}^{-1}\mathbf{P}, \tag{15.128}$$

where

$$\mathbf{R} = E(\mathbf{X}_{N,l,k}^t \mathbf{X}_{N,l,k}) \tag{15.129}$$

and

$$\mathbf{P} = E(\mathbf{X}_{N,l,k}^t \mathbf{d}_{l,k}). \tag{15.130}$$

The corresponding MBMSE (minimum block mean-square error) is

$$\sigma_{\text{MBMSE}}^2 = \sigma_{\text{MMSE}}^2 = E(d^2[k]) - \mathbf{P}^t\mathbf{R}^{-1}\mathbf{P}. \tag{15.131}$$

The block LMS algorithm [CL81] moves a small amount in the direction of the negative gradient of $(1/l)[\varepsilon_{N,l,k}^t \varepsilon_{N,l,k}]$:

$$\mathbf{W}_{N,l,k+l} = \mathbf{W}_{N,l,k} - \frac{\mu_B}{2} l \nabla \left(\frac{1}{l} \varepsilon_{N,l,k}^t \varepsilon_{N,l,k} \right), \tag{15.132}$$

or

$$\mathbf{W}_{N,l,k+l} = \mathbf{W}_{N,l,k} + \mu_B \mathbf{X}_{N,l,k}^t \varepsilon_{N,l,k}. \tag{15.133}$$

Note the new time subscript is $k + l$, not $k + 1$, to emphasize the block updating. Actually, the blocks can be either nonsuccessive or overlapping if so desired. The error vector $\varepsilon_{N,l,k}$ is computed using $\mathbf{W}_{N,l,k}$ for all its entries. The reader will note that the block LMS would be just l LMS updates in succession, were it not for the same $\mathbf{W}_{N,l,k}$ used for all errors in a block. Clark, Mitra, and Parker [CL81] have shown that for $\mathbf{W}_k^0 = \mathbf{W}^0$, the excess error and time constants for BLMS and LMS are the same when $\mu_B = \mu$. This result can be improved if μ_B is not constant over a block and the resultant filtering and updating are performed in the frequency domain. Note, at this point, the computational complexities of LMS and BLMS are essentially the same (although BLMS requires much more storage).

15-5-2 Frequency-Domain Adaptive Filter

There are essentially two types of frequency-domain adaptive filters. One that follows naturally from BLMS and uses successive data blocks, and another that uses blocks that overlap in $l - 1$ of l data points.

The first type was introduced simultaneously by Ferrara [FE80] and by Clark, Mitra, and Parker [CL81]. This method makes use of the "overlap-save" DFT-convolution methods. For this, let $L = l + N - 1$ and cyclically extend $\mathbf{X}_{N,l,k}$ to get the following $L \times L$ cyclic matrix

$$\mathbf{X}_{N,L,k}^c = \begin{bmatrix} x^t[k] & x^t[k-1] & \cdots & x^t[k-L+1] \\ x^t[k-1] & x^t[k-2] & \cdots & x^t[k] \\ \vdots & \vdots & \cdots & \vdots \\ x^t[k-L+2] & x^t[k-L+1] & \cdots & x^t[k-L+3] \\ x^t[k-L+1] & x^t[k] & \cdots & x^t[k-L+2] \end{bmatrix}. \tag{15.134}$$

Then note that the first l entries in

$$\epsilon_{N,l,k} = \begin{bmatrix} d[k] \\ 0 \\ \vdots \\ 0 \end{bmatrix} - \mathbf{X}_{N,L,k}^c \begin{bmatrix} \mathbf{W}_{N,l,k} \\ 0 \\ \vdots \\ 0 \end{bmatrix} \tag{15.135}$$

are the desired error signals and that the cyclic convolution on the right-hand side of Eq. (15.135) can be computed in the frequency domain by multiplying the L-point discrete Fourier transforms of the sequences $x[k-L+1], \ldots, x[k]$, and $W_0, \ldots, W_{N-1}, 0, \ldots, 0$; that is,

$$\mathcal{Y}_n = \mathcal{X}_n \cdot \mathcal{W}_n, \tag{15.136}$$

where $\mathcal{Y}_n$, $\mathcal{X}_n$, and $\mathcal{W}_n$ are the discrete Fourier transforms of the block filter outputs, $\mathbf{X}_{L,k}$, and $\mathbf{W}_{N,l,k}$, respectively. That is, for example,

$$\mathcal{X}_n = \sum_{m=k}^{k-L+1} x_m e^{-j(2\pi/L)mn}. \tag{15.137}$$

Let $\mathbf{P}_{l/L}$ be the projection matrix (identity in upper left l entries, zero elsewhere) corresponding to retaining the upper l error components. Then

$$\varepsilon_{N,l,k} = \mathbf{P}_{l/L} \left(\begin{bmatrix} d_l[k] \\ 0 \\ \vdots \\ 0 \end{bmatrix} - \mathbf{X}^{c}_{N,L,k} \begin{bmatrix} \mathbf{W}_{N,l,k} \\ 0 \\ \vdots \\ 0 \end{bmatrix} \right). \qquad (15.138)$$

We also note that the update $\mathbf{X}^{t}_{N,l,k}\varepsilon_{N,l,k}$ can be written as the first N terms of

$$(\mathbf{X}^{c}_{N,L,k})^{t} \begin{bmatrix} \varepsilon_{N,l,k} \\ 0 \\ \vdots \\ 0 \end{bmatrix}, \qquad (15.139)$$

which is actually a cyclic convolution of the time reverse of the series $x[k - L + 1], \cdots, x[k]$, and the sequence $0, \cdots, 0, \epsilon_{N,l}[k - L + 1], \cdots, \epsilon_{N,l}[k]$. Again it can be computed in the frequency domain as

$$\mathcal{G}_n \triangleq \mathcal{X}^{*}_{-n}\mathcal{E}_n, \qquad (15.140)$$

where $\mathcal{E}_n$ is the DFT of the zero-extended $\varepsilon_{N,l,k}$. An individual one-tap LMS algorithm is thus applied to each of the frequency bins

$$\mathbf{W}_{N,l,k+1} = \mathbf{W}_{N,l,k} + \mu_B \mathbf{P}_{N/L} \cdot \mathbf{g}_{N,L,k}, \qquad (15.141)$$

where $\mathbf{g}_{N,L,k}$ is a vector that contains the inverse DFT entries for $\mathcal{G}_n$ in Eq. (15.140).

Equations (15.134)–(15.141) are best summarized in Figure 15-14. It is very important to note at this point that, to the extent that any finite-length transform (such as the DFT here) can decouple a signal into independent or orthogonal components (frequency bins), the gradient in Eq. (15.140) corresponds to the updates of L independent LMS filters. As such, μ_B can be incorporated into the frequency update and can be frequency dependent: $\mu_B \to \mu_{B,n}$. Such schemes can have fast convergence and tracking properties that rival RLS methods. In fact, it is not difficult to show that if $x[k]$ and $d[k]$ are periodic with the same period $L(= l)$, the RLS and block frequency-domain methods are identical (if $\mu_{B,n}$ is appropriately chosen and varies from block to block); see Cioffi [CI86]. Of course, this type of periodicity is too restrictive to impose in practice, so RLS usually outperforms frequency-domain methods, but the advantage can be insignificant in certain applications, such as fast-starting adaptive equalizers with periodic training sequences. One also notes that for the length l data block, if $\log_2 L \ll N$, there is

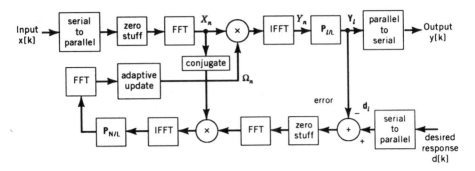

FIGURE 15-14   Overlap-save frequency-domain adaptive filter (see Clark et al. [CL81]).

a computational savings, as the configuration in Figure 15-14 requires only $O(L \log_2 L)$ computations for a block (if FFTs are used), which is less than $O(lN)$ in the previous LMS and fast RLS methods. Nevertheless, because of the five FFTs required, we must have $N > 50$ for a realization of this savings [CL81]. A hybrid time/frequency approach, such as that introduced by Cioffi and Bingham [CI93], can have significant advantages in data transmission applications.

15-5-3   Frequency-Domain Approximations

There are a number of approximations to the frequency-domain adaptive filter of Figure 15-14, [DE78; NA83]. These methods make a variety of approximations that essentially amount to ignoring some or all circular convolution effects when implementing in the time domain. The number of FFTs can correspondingly be reduced. Such methods work well when $l > 10N$ in practice.

An alternative structure appears in Figure 15-15, which, for a shift-invariant filter, passes the input through a "bank" of bandpass filters that "decouple" the input. This is equivalent to overlapping blocks in $l - 1$ points. A separate adaptive filter (a single weight) is used to process each bandpass filter output. These outputs are summed and compared to the desired response. For reasons of document length, we will not pursue this type of adaptive filter further, although it may be of equal importance to some of the previously described algorithms.

15-5-4   Limited-Precision Errors in Block LMS

Basically, the sum of gradient estimates in the block LMS algorithm will have l times greater variance than that of the LMS algorithm. Of course, since block LMS updates at only $(1/l)$th the rate, the infinite-precision performance is the same in real time. Recall, however, from Section 15-3-4 that μ in the LMS algorithm could not be made too small without incurring a significant degradation in adaptive-filter performance caused by the accumulation of roundoff errors. In the block LMS, we

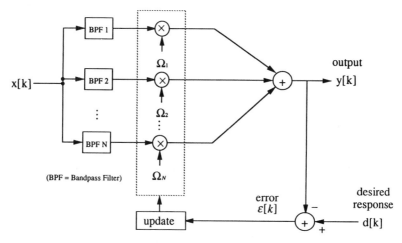

FIGURE 15-15 Bank-of-filters frequency-domain adaptive filter.

choose

$$\mu_B = \frac{\mu}{l},$$
(15.142)

which will slow the tracking rate of the block LMS algorithm by a factor of l with respect to the LMS algorithm. However, the choice in Eq. (15.35) is equivalent to dividing the sum of gradient estimates in Eq. (15.34) by l and then using μ as the step size, say,

$$\overline{\mathbf{G}}_{k+L-1} = \frac{\mu \mathbf{X}_{N,l,k}\varepsilon_{N,l,k}}{l}.$$
(15.143)

If the finite-precision errors, β_{k+i}, $i = 0, \ldots, l-1$, which correspond to the quantization errors in computing Eq. (15.143), are assumed to be independent and identically distributed, then the mean bias will be the same as the LMS algorithm. Thus we need to ensure that multiplier rounding forces this mean bias to zero. Nevertheless, the variance of the gradient-roundoff error in block LMS will be l times smaller than LMS, over l iterations,

$$\sigma^2_{q,\text{BLMS}} = \frac{\sigma^2_{q,\text{LMS}}}{l}.$$
(15.144)

The result was first observed by Panda et al. [PA87].

To interpret the result in Eq. (15.144), note that a reduction in the coefficient wordlength by 1 bit increases σ^2_q by a factor of 4. To keep accumulation of finite-precision errors from enlarging with respect to the original wordlength, we increase l by a factor of 4. Actually, this means $\mu_B = \mu/4$, thereby also reducing the tracking speed by a factor of 4. To reduce precision by 4 bits, we need

$$l = 4^4 = 256. \tag{15.145}$$

15-5-5 Block Least-Squares Method

Block least-squares methods, as they are presented here, are strictly time domain, although there are frequency-domain methods; see Hertz et al. [HE86] and Lee and Mitra [LE87]. The number of computations per block-update is proportional to lN. As with time-domain block LMS, their primary advantage with respect to FFT adaptive filters is the variability of block length and their applicability to nonsuccessive blocks as described in Cioffi [CI86] (no overlap-save necessary).

The general situation is illustrated in Figure 15-16. The two inputs are acquired in blocks by the DMA (dynamic memory access), and a microprocessor processes the data, typically at rates significantly below the incoming data rate. An update is formed and programmed into the transversal filter. The BFTF algorithm is used to solve the "soft-constrained" least-squares problem

$$\mathbf{W}_{N,l,k}^{\min} \mu_k \|\mathbf{W}_{N,l,k} - \mathbf{W}_0\|^2$$

$$+ \sum_{m=k-l+1}^{k} (d[m] + \mathbf{W}_{N,l,k}\mathbf{X}_{N,m})^t (d[m] + \mathbf{W}_{N,l,k}\mathbf{X}_{N,m}), \tag{15.146}$$

where $\mathbf{W}_0$ is some initial setting for $\mathbf{W}_{N,l,k}$ prior to processing the data from $k - l + 1$ to k. The factor μ_k permits control of the influence of the soft constraint with respect to the squared-error sum; reasonable choices for μ_k are discussed in Cioffi

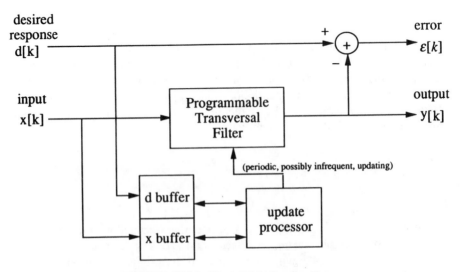

FIGURE 15-16 Block RLS adaptive filter.

[CI86]. $\mathbf{W}_0$ could be the solution from a previous use of the BFTF algorithm, thus providing a means by which to transfer the useful information from previous data blocks to the current data block. For more information, see Cioffi [CI86], [CI88] and Yu and He [YU88].

15-6 APPLICATIONS

In this section, we consider three different applications of adaptive filters as case studies. The first application will be that of an adaptive data echo canceler found in CCITT high-speed duplex voiceband data modems. The numbers that we use are typical of what are known as V.32 or V.32bis modems. Only the basic features are shown, but the multirate structure that can be viewed as several adaptive filters in parallel is characteristic of the use of adaptive filters in quadrature modulated data transmission. The algorithm chosen here will be the block LMS of Section 15-5-1 for reasons of numerical accuracy. We then look at the use of a transversal equalizer in the application [CI90c] of increasing density in disk storage devices. Here signed-LMS is used for reasons of implementation simplicity. Finally, we investigate the use of an adaptive decision-feedback equalizer contemplated for use in a packet digital mobile context, where both a frequency-domain and a block algorithm are suggested.

It is worthy of note that while the LMS or RLS methods are the basis for adaptive filters, it is usually some modified version of these that is used in practice.

15-6-1 The Data Echo Canceler

The data echo canceler is illustrated in Figure 15-17. The hybrid, H, couples the four-wire modem system to the associated two-wire subscriber loop. The hybrid is often imperfect, as the exact channel termination cannot be implemented, and a near-end echo passes (undesirably) from the transmitter to the receiver. Another far-end echo can be generated by a second hybrid located at the far-end of the transmission path. Both echoes are undesirable and must be canceled by the echo canceler. While it would seem that the adaptive filter directly applies with x_k as the transmitter output and d_k as the near-end hybrid output, further improvement and computational reduction are possible as we now show.

T/3 Canceler Architecture. The hybrid output signal in Figure 15-17 can be expressed [WE85] as

$$d(t) = \mathrm{Re}\left[\sum_m A_m h_{\mathrm{bb}}(t - mT)e^{j\omega_c t}\right] + n(t), \qquad (15.147)$$

where A_m is the complex data signal

$$A_m = a_m + jb_m \qquad (15.148)$$

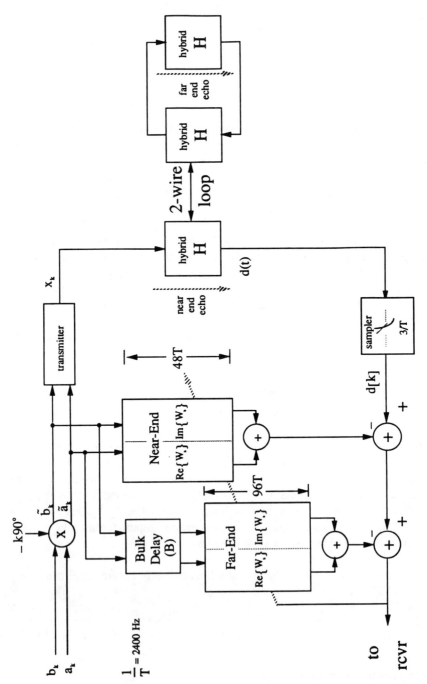

FIGURE 15-17 Voiceband data-driven echo cancellation.

formed by taking a_m, the real part, as the horizontal component of a quadrature modulated data signal, and b_m, the imaginary part, as the vertical component. $h_{bb}(t)$ is the baseband-equivalent pulse response (at $1/T = 2400$) [PR83] of the combined echo-pulse shaping path, or, equivalently, the pulse response of the combined pulse-shaping-echo path is

$$h(t) = \text{Re } [h_{bb}(t)e^{j\omega_c t}]. \tag{15.149}$$

Also in Eq. (15.147), $n(t)$ is a noise signal that may consist of both channel noise, a far-end double talking data signal, and other nonideal channel impairments; $\omega_c = 2\pi(1800)$ is the carrier frequency in radians, and t denotes the continuous-time variable.

To see the equivalence of Eq. (15.147) and Figure 15-17 [CI90b; WE84], we write

$$d(t) = \text{Re } \left\{ \sum_m (A_m e^{j\omega_c mT})h_{bb}(t - mT)e^{j\omega_c(t - mT)} \right\} + n(t), \tag{15.150}$$

which can be reduced to

$$d(t) = \text{Re } \left\{ \sum_k \tilde{A}_m H(t - mT) \right\} + n(t), \tag{15.151}$$

where

$$\tilde{A}_m = A_m e^{j\omega_c mT} \tag{15.152}$$

and

$$H(t) = h_{bb}(t)e^{j\omega_c t}. \tag{15.153}$$

Note that $H(t)$ can also be written in terms of $h(t)$ in Eq. (15.149) and $\hat{h}(t)$ (the Hilbert transform of $h(t)$) as

$$H(t) = h(t) + j\hat{h}(t). \tag{15.154}$$

Thus $d(t)$ becomes

$$d(t) = \sum_m [\tilde{a}_m h(t - mT) - \tilde{b}_m \hat{h}(t - mT)] + n(t), \tag{15.155}$$

which can be estimated using only two adaptive filters at the rate $1/T_1 = 3/T = 7200$ Hz (a rate higher than twice the highest passband frequency) and in parallel with the rotated in-phase and quadrature inputs, $\tilde{a}_m$ and $\tilde{b}_m$, respectively. The two outputs are summed to form an estimate of the echo, which is then subtracted from the received signal. There is no need for a Hilbert transform or phase splitter in the $T/3$ canceler, nor is there need for a modulator or demodulator, other than the trivial rotations in Eq. (15.152), since $\omega_c T = 3\pi/2$ (or $-90°$), that is, a simple signed interchange of the in-phase and quadrature data.

Further reductions in computations are possible using a performance and mathematically equivalent subcanceler configuration [WE85] (actually a special case of the polyphase systems in Crochiere and Rabiner [CR83]), as illustrated in Figure 15-18. If $d(t)$ is sampled at rate $1/T_1 = 3/T$, we can write

$$d(kT_1) = \sum_m [\tilde{a}_m h(kT_1 - mT) - \tilde{b}_m \hat{h}(kT_1 - mT)] + n(kT_1), \quad (15.156)$$

or by letting $kT_1 = lT + pT_1$, where $p = 0, 1,$ or 2,

$$d(lT + pT_1) = \sum_m [\hat{a}_m h(lT - mT + pT_1)$$
$$- \tilde{b}_m \hat{h}(lT - mT + pT_1)] + n(lT + pT_1). \quad (15.157)$$

In Eq. (15.157) there are three phases (one for each value of p) of $d(kT_1)$ for each symbol interval T, each of which can be construed as a convolution of the same input, $\tilde{a}_m$, with a "subecho" channel indexed by the corresponding value of p. The echo estimate can similarly be configured, leaving only one in-phase and one quadrature subcanceler to be implemented and updated at each sampling instant $(T/3)$, resulting in a factor of 3 reduction in the computation required to implement the echo canceler.

Also, as there is often a long delay in the far-end echo, its contribution to $h(t)$ is nonoverlapping with that of the near-end echo, so the bulk delay shift register queue is inserted between the near-end and far-end cancelers. Note there are actually six smaller (real or in-phase) echo cancelers in Figure 15-18 (and 12 overall, counting the imaginary and quadrature components). The computational requirements are significantly reduced by a factor of 33% with respect to a system employing a single adaptive filter at the same rate. This system has two other important advantages with respect to a single adaptive filter: first, the input to the three subcancelers (near- or far-end, in-phase or quadrature components) in each group is the same, reducing the input memory by 3 in length, and second, because the inputs are 1–2-bit quantities, they are stored exactly with far less precision than quantized transmitter outputs. There is therefore a large memory reduction for the adaptive filter input. Finally, as the far-end data signal is often 30–40 dB below the near-end echo, and a prereceiver signal-to-(echo-plus-noise)-ratio of 25 dB is desirable, 60 dB of echo rejection is necessary. This extremely high level of cancellation would be difficult to attain if quantized transmitter outputs, rather than exactly stored data signals, were input to the cancelers. Also, the transmit filter input is white in practice, leaving a much lower eigenvalue spread in this data-driven echo canceler. This further improves performance as described in Section 15-3-5. The echo canceler coefficients, call them $\mathbf{W}_{p,k}$ for the pth subcanceler,

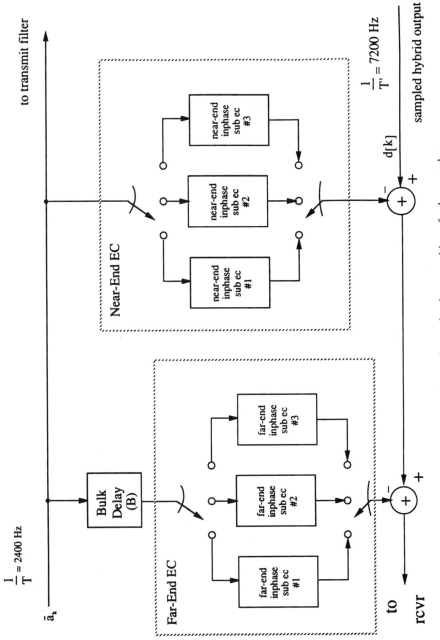

FIGURE 15-18 In-phase subcanceler decomposition of echo canceler.

1133

could be updated (on phase p of T) according to the stochastic-gradient or LMS algorithm:

$$\epsilon_p = d_p[k] - \text{Re } [\mathbf{W}_{p,k+1}]\tilde{a}[k] - \text{Im } [\mathbf{W}_{p,k}]\tilde{b}[k], \quad (15.158)$$

$$\text{Re } [\mathbf{W}_{p,k+1}] = \text{Re } [\mathbf{W}_{p,k}] + \mu\epsilon_p[k]\tilde{a}[k], \quad (15.159)$$

$$\text{Im } [\mathbf{W}_{p,k+1}] = \text{Im } [\mathbf{W}_{p,k}] + \mu\epsilon_p[k]\tilde{b}[k], \quad (15.160)$$

where $\tilde{a}[k]$ and $\tilde{b}[k]$ are the vectors of inputs corresponding to the echo canceler coefficients. Thus the echo canceler continues to adapt through the entire call, tracking slow variation of the echo response over the duration of the call. However, the LMS method in Eqs. (15.158) and (15.159) is notorious for its large coefficient-memory requirement, requiring at least 20 bits-per-coefficient for 60 dB of near-end cancellation (and 16 bits-per-coefficient in the far-end cancelers). As in Section 15-5-4, this near-end 20-bit requirement can be reduced to 16 bits by using a block LMS algorithm with block size 256 (and thereby slowing the tracking by 256) for implementation on the common 16-bit digital signal processing chips that are widely available commercially. Initial convergence of such a slow block algorithm can be improved by using periodic training sequences, as discussed in Cioffi [CI90b].

15-6-2   Equalization of a Magnetic Disk Channel

The adaptive equalizer for a magnetic disk channel is illustrated in Figure 15-19(a). This system, unlike the previous data modem, is inherently baseband and two-level at the read-channel input (which corresponds to the head and ensuing circuitry in a disk system) because of the inherent two-state (two-directional flux) nature (hysteresis) of the previously recorded and magnetically stored data. The readback channel undesirably attenuates low and high frequencies, and the equalizer is a filter that tries to reverse this attenuation to some degree. The readback channel varies with the movement of the head, and also dramatically with the radius of the track on which the data are stored. Thus an adaptive equalizer is desirable in high-density recording to correspond to the variation with track. The head signal can be quantized by a 6–8-bit analog-to-digital converter (ADC), and this is the input, $x[k]$, for the adaptive filter. A typical ADC may run at 54 MHz, which is $9/8$ or 11% higher than the 48-MHz (6 Mbytes/s) data rate output of the head. A special rate $8/9$ coder forces transitions between flux changes to occur at least every 4 clock periods to ensure robust operation of the adaptive equalizer and associated phase-locking circuitry. To generate $d[k]$, the desired response, one would need the stored data (which, of course, is what we are trying to estimate). In this case, the desired response is estimated using decision-directed training, also illustrated in Figure 15-19(a). As the channel passes no dc, a desirable shaping is

$$d[k] = \hat{a}[k - \Delta] - \hat{a}[k - \Delta - 2], \quad (15.161)$$

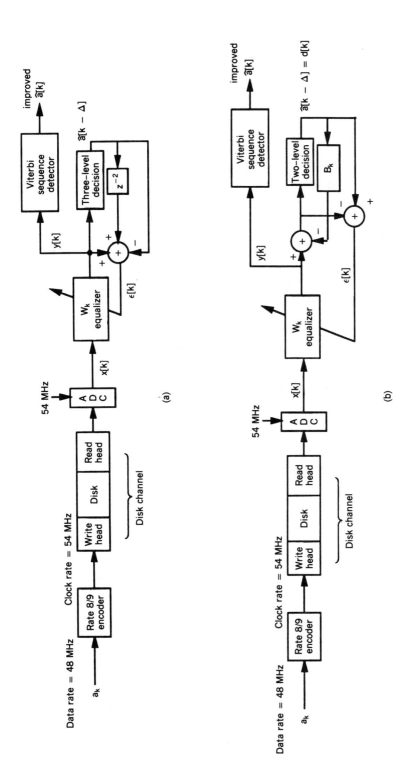

FIGURE 15-19 (a) Partial-response maximum likelihood (PRML) with adaptive equalizer. (b) Adaptive decision-feedback equalizer for disk recording.

1135

sometimes referred to as "class IV partial-response" signaling. (Δ is a delay that is specific to the disk and is roughly equal to the sum of the channel group delay and the length of W_k, the adaptive filter weights.) The input is differentially encoded; $\hat{a}[k - \Delta] = 1$ if the current and last bit differ and $\hat{a}[k - \Delta] = -1$ if they are the same. Then a noise-free output decision on the three possible outputs corresponding to Eq. (15.161) is ± 2 decode to -1 and 0 decodes to $+1$. The error signal between the actual equalizer output and the constructed desired response is used to update the equalizer using the LMS algorithm, typically. The remaining intersymbol interference in Eq. (15.161) can be eliminated with a sequence (Viterbi) detector as described in Cioffi et al. [CI90c]. This method is popularly called the *partial response maximum likelihood* (PRML) approach in the literature.

An alternative, effective, and increasingly popular structure for disk-channel equalization is shown in Figure 15-19(b). In this *decision-feedback equalizer* (DFE), a second adaptive filter B_k is used to reconstruct interference from previously estimated data. It has N_B taps. This DFE is a form of a multichannel ($p = 2$) adaptive filter. The adaptive-filter inputs are $x[k]$ and $\hat{a}[k - \Delta - 1]$, and the filter outputs are summed to estimate a single desired response $d[k] = \hat{a}[k - \Delta]$. As multiplication is expensive at the high sampling rates, a "signed LMS" algorithm is often used (decisions are assumed correct here):

$$\epsilon[k] = d[k] - \mathbf{W}_k^t \mathbf{X}_k - \mathbf{B}_k^t \mathbf{A}_{k-\Delta-1}, \tag{15.162}$$

$$\mathbf{W}_{k+1} = \mathbf{W}_k + \mu \operatorname{sgn}(\epsilon[k]) \mathbf{X}_k, \tag{15.163}$$

where

$$\mathbf{B}_{k+1} = \mathbf{B}_k + \mu \operatorname{sgn}(\epsilon[k]) \mathbf{A}_{k-\Delta-1},$$

$$\operatorname{sgn}(x) = \begin{cases} +1 & \text{for } x > 0, \\ 0 & \text{for } x = 0, \\ -1 & \text{for } x < 0. \end{cases} \tag{15.164}$$

This eliminates multiplications in the updating, which need not be performed at high speed and may be easier at lower speeds. Of course, some initial training is usually necessary with a known training sequence before decision-direction performs well. This is accomplished by storing a short training sequence in the header information on each track, and the equalizer is "retrained" every time the head moves to a different track. In practice, the feedback section can be replaced by a table look-up because there are only 2^{N_B} possible inputs. A full description of this popular approach for the disk application appears in Fisher et al. [FI91].

15-6-3 Decision-Feedback Equalization for Digital Wideband Packet Radio Networks

An application where convergence and tracking speed are of importance to the adaptive filter is digital wideband mobile radio. In such an application, at least one of the two ends of the channel is physically moving and the channel characteristic (often dominated by spectral nulls induced by multipath fading and gain variations caused by shadowing) varies rapidly with time. A typical situation, such as approximately occurs in the new IS-54 North American standard for digital mobile cellular telephony, might be a symbol rate of 24 kHz using two quadrature-modulated bit streams for an overall bit rate of 48 kbps. Here the decision-feedback equalizer shown in Figure 15-20 (again, a special case of the equation-error adaptive filter in Figure 15-2(c), as discussed for disk storage in Section 15-6-2) is used to mitigate the effect of any moving spectral null. Experimental evidence has suggested that the channel can only be assumed constant for short periods of time, roughly corresponding to the packet size in the network, as long as the multipath is not too severe. Use of the LMS algorithm is an order of magnitude too slow in converging; in fact, an entire packet is often not sufficient just to converge the equalizer with LMS, as generically illustrated in Figure 15-8(a) for this case of a low-noise application. Furthermore, the usual fast-start methods, [CH87; QU85] are not applicable as the requirement of a periodic (at least *two* periods) training sequence unnecessarily lengthens the initialization period. For the packet digital radio application, various special forms of the RLS algorithm that compute a single equalizer for the entire packet can be used. Usually this equalizer is computed by identifying the channel and then computing the equalizer for a known embedded training sequence. This is a form of the block adaptive filter that was discussed in Section 15-5. Then LMS can be used to update for the rest of the packet (again using decision-aided training).

During training the adaptive-filter input is the known training sequence (the entire packet is presumed to have been received and stored prior to adaptive filtering). The desired response during training is the sampled (and demodulated) antenna output. The adaptive filter computes as estimate of the channel. Since the

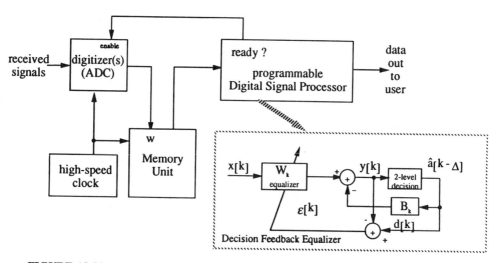

FIGURE 15-20 Packet-based equalization for digital cellular telephony as a block adaptive filter.

adaptive-filter input corresponds to white input data, no eigenvalue problems exist, although the application is still low-noise so that RLS is very much preferred. The channel estimate can be converted into a decision-feedback equalizer in a variety of ways [BE79].

As additional data are estimated by the DFE, these data can be used in a LMS adaptive filter to continue updating the channel estimate, and the DFE coefficients can periodically be recomputed.

15-7   CONCLUSION

In this chapter, we have focused on describing basic principles. Many of the most widely used adaptive algorithms have been described, listed, and analyzed. In most applications, the choice of the adaptive-filter structure and algorithm will depend on two, often-opposing, factors: performance and cost of a numerically stable implementation. We have assumed all signal and filter coefficients to be real valued in this chapter. In the general case of complex signals and filters, the superscript t should be replaced with the superscript †.

An important area of adaptive filtering not discussed here is that of adaptive IIR (infinite impulse response) filtering, although equation-error techniques were discussed earlier in the first section. Equation-error methods are included implicitly in the discussion here using the multichannel adaptive filters introduced in the first section. However, there is another class of adaptive IIR filters that attempt to minimize the true output MSE, rather than the equation error. The reader is referred to Chapter 4 in Cowan and Grant [CO85]. Also, the increasingly important applications without a true desired response are not covered here, but the "constant modulus" adaptive algorithms are often used in these applications. The reader is deferred to Chapter 6 in Treichler et al. [TR87]. Commercial markets for products using the adaptive filters described in general in this chapter currently generate worldwide revenues of tens of billions of dollars with an order of magnitude increase on the horizon. Defense applications also generate a similarly large revenue base of adaptive-filter applications. With such enormous impact, there has been an explosion of interest in this field, and the area will continue to be dynamic in many aspects for the foreseeable future.

ACKNOWLEDGMENTS

The authors thank Professor John J. Shynk of the University of California at Santa Barbara for his critical review of the chapter. Thanks are also due to Ms. Renee Leach who typed the final version of this chapter.

REFERENCES

[AL86] S. T. Alexander, *Adaptive Signal Processing: Theory and Applications*. Springer-Verlag, New York, 1986.

[AL87] S. T. Alexander, Transient weight misadjustment properties for the finite-precision LMS algorithm. *IEEE Trans. Acoust., Speech, Signal Process.* **ASSP-35**(9), 1250–1258 (1987).

[BE79] C. A. Belfiore and J. H. Park, Jr., Decision feedback equalization. *Proc. IEEE* **67**, 1143–1156 (1979).

[BE86] N. J. Bershad, Analysis on the normalized LMS algorithm with Gaussian inputs. *IEEE Trans. Acoust., Speech, Signal Process.* **ASSP-34**(4), 793–806 (1986).

[BE87] M. G. Bellanger, *Adaptive Digital Filters and Signal Analysis.* Dekker, New York, 1987.

[BE89] M. G. Bellanger, The FLS-QR algorithm for adaptive filtering. *Signal Process.* **19**(4), 291–304 (1989).

[BI77] G. J. Bierman, *Factorization Methods for Discrete Sequential Estimation.* Academic Press, New York, 1977.

[CA84] C. Caraiscos and B. Liu, A round-off analysis of the LMS adaptive algorithm. *IEEE Trans. Acoust., Speech, Signal Process.* **ASSP-32**(1), 34–41 (1984).

[CH87] P. R. Chevillat, D. Maiwald, and G. Ungerboeck, Rapid training of a voiceband data-modem employing an equalizer with fractional-T spaced coefficients. *IEEE Trans. Commun.* **35**(9), 869–876 (1987).

[CI84] J. M. Cioffi and T. Kailath, Fast, RLS, transversal filters for adaptive filtering. *IEEE Trans. Acoust., Speech, Signal Process.* **ASSP-32**(2), 304–337 (1984).

[CI85a] J. M. Cioffi and J. J. Werner, The tap-drifting problem in digitally implemented data-driven echo cancelers. *AT&T Bell Lab. Tech. J.* **64**(1), 115–138 (1985).

[CI85b] J. M. Cioffi and T. Kailath, Windowed FTF adaptive algorithms with normalization. *IEEE Trans. Acoust., Speech, Signal Process.* **ASSP-33**(3), 607–625 (1985).

[CI85c] J. M. Cioffi, When do I use an RLS adaptive filter? *Proc. Asilomar Conf. Signals, Syst., Comput., Pacific Grove, California, 1985*, pp. 636–639 (1985).

[CI86] J. M. Cioffi, Least-squares storage channel identification. *IBM J. Res. Dev.* **30**(3), 310–320 (1986).

[CI87] J. M. Cioffi, Limited precision effects in adaptive filtering. *IEEE Trans. Circuits Syst.* **CAS-34**(7), 821–833 (1987).

[CI88] J. M. Cioffi, Precision-efficient use of block adaptive algorithms in data-driven echo cancelers. *IEEE Int. Conf. Commun. Rec. Philadelphia, 1988*, Pap. No. 12.1, pp. 355–359, 1988.

[CI90a] J. M. Cioffi, The fast adaptive rotors RLS algorithm. *IEEE Trans. Acoust., Speech, Signal Process.* **ASSP-38**(4), 631–653 (1990).

[CI90b] J. M. Cioffi, A fast echo canceler initialization method for the CCITT V.32 modem. *IEEE Trans. Commun.* **38**(5), 629–638 (1990).

[CI90c] J. M. Cioffi, W. L. Abbott, H. K. Thapar, C. M. Melas, and F. O. Fisher, Adaptive equalization in magnetic-disk storage channels. *IEEE Commun. Mag.*, **28**(2) February, pp. 14–29 (1990).

[CI93] J. M. Cioffi and J. Bingham, A data-driven multitone echo canceler. *IEEE Trans. Commun.*, (1993), to appear.

[CL81] G. A. Clark, S. K. Mitra, and S. R. Parker, Block implementation of adaptive digital filters. *IEEE Trans. Acoust., Speech, Signal Process.* **ASSP-29**(3), 744–752 (1981).

[CO85] C. F. N. Cowan and P. M. Grant, *Adaptive Filters.* Prentice-Hall, Englewood Cliffs, NJ, 1985.

[CR83] R. E. Crochiere and L. R. Rabiner, *Multirate Digital Signal Processing*. Prentice-Hall, Englewood Cliffs, NJ, 1983.

[DE78] M. Dentino, J. McCool, and B. Widrow, Adaptive filtering in the frequency domain. *Proc. IEEE* **66**(2), 1658–1659 (1978).

[EL86] E. Eleisthériou and D. Falconer, Tracking properties and steady-sate performance of RLS adaptive filter algorithms. *IEEE Trans. Acoust., Speech, Signal Process.* **ASSP-34**(5), 1097–1110 (1986).

[FE80] E. R. Ferrara, Fast implementation of LMS adaptive filters. *IEEE Trans. Acoust., Speech, Signal Process.* **ASSP-28**(4), 474–475 (1980).

[FI91] K. D. Fisher, J. M. Cioffi, W. A. Abbot, P. Bednarz, and C. M. Melas, The RAM-DFE for disk storage channels. *IEEE Trans. Commun.* **39**(11), 1559–1568 (1991).

[FR82] B. Friedlander, Lattice methods for spectral estimation. *Proc. IEEE* **70**(9), 990–1017 (1982).

[GE81] W. M. Gentleman and H. T. Kung, Matrix triangularization by systolic arrays. *Proc. SPIE—Int. Soc. Opt. Eng.* **298,** 19–26 (1981).

[GI82] R. D. Gitlin, H. C. Meadows, Jr., and S. B. Weinstein, The tap-leakage algorithm: An algorithm for the stable operation of a digitally implemented, fractionally adaptive spaced equalizer. *Bell Syst. Tech. J.* **61**(8), 1817–1839 (1982).

[GI85] A. A. Giordano and F. M. Hsu, *Least Squares Estimation with Applications to Digital Signal Processing*. Wiley, New York, 1985.

[GO77] G. C. Goodman and R. L. Payne, *Dynamic System Identification*. Academic Press, New York, 1977.

[GO84] G. C. Goodwin and K. S. Sin, *Adaptive Filtering, Prediction, and Control*. Prentice-Hall, Englewood Cliffs, NJ, 1984.

[HA85] S. Haykin, ed., *Array Signal Processing*. Prentice-Hall, Englewood Cliffs, NJ, 1985.

[HA91] S. Haykin, *Adaptive Filter Theory*, 2nd ed. Prentice Hall, Englewood Cliffs, NJ, 1991.

[HE86] D. Hertz, D. Mansour, and I. Engel, On least square frequency domain adaptive filters. *IEEE Trans. Circuits Syst.* **CAS-33,** 335–337 (1986).

[HO84] M. L. Honig and D. G. Messerschmitt, *Adaptive Filters*. Kluwer Academic Publishers, Boston, MA, 1984.

[HS82] F. Hsu, Square-root Kalman filtering for high-speed data received over fading dispersive HF channels. *IEEE Trans. Inf. Theory.* **IT-28**(5), 753–763 (1982).

[JA86] N. Jablon, Adaptive beamforming with the generalized sidelobe canceler in the presence of array imperfections. *IEEE Trans. Antennas Propag.* **34**(8), 996–1012 (1986).

[LE87] J. C. Lee and S. K. Mitra, On frequency domain least square adaptive algorithm. *Proc. IEEE Int. Conf. Acoust., Speech, Signal Process., Dallas, Texas, 1987*, pp. 411–414 (1987).

[LI85] F. Ling and J. G. Proakis, Lattice decision-feedback equalizers and their performance and application to time-variant multipath channels. *IEEE Trans. Commun.* **33**(4), 348–356 (1985).

[LI86] F. Ling and J. G. Proakis, A recursive modified Gramm-Schmidt algorithm with applications to least square estimation and adaptive filtering. *IEEE Trans. Acoust., Speech, Signal Process.* **ASSP-34**(4), 829–836 (1986).

[LJ83] L. Ljung and G. T. Söderström, *Theory and Practice of Recursive Identification.* MIT Press, Cambridge, MA, 1983.

[LU73] D. G. Luenberger, *Introduction to Linear and Nonlinear Programming.* Addison-Wesley, Reading, MA, 1973.

[MA76] J. D. Markel and A. H. Gray, Jr., *Linear Prediction of Speech.* Springer-Verlag, New York, 1976.

[MC83] J. G. McWhirter, RLS minimization using a systolic array. *Proc. SPIE—Int. Soc. Opt. Eng.* **431**, 105–112 (1983).

[MO80] R. A. Monzingo and T. W. Miller, *Introduction to Adaptive Arrays.* Wiley, New York, 1980.

[MU88] B. Mulgrew and C. F. N. Cowan, *Adaptive Filters and Equalizers.* Kluwer Academic Publishers, Boston, MA, 1988.

[NA83] S. S. Narayan, A. M. Peterson, and M. J. Narasimha, Transform domain LMS algorithm. *IEEE Trans. Acoust., Speech, Signal Process.* **ASSP-31**(3), 609–615 (1983).

[OR88] S. J. Orfanidis, *Optimum Signal Processing: An Introduction*, 2nd ed. McGraw-Hill, New York, 1988.

[PA87] G. Panda, C. F. N. Cowan, and P. M. Grant, Assessment of finite-precision limitations in LMS and BLMS adaptive algorithms. *Proc. Int. Conf. Acoust., Speech, Signal Process., Dallas, Texas, 1987*, pp. 137–140 (1987).

[PR83] J. G. Proakis, *Digital Communications.* McGraw-Hill, New York, 1983.

[QU85] S. Qureshi, Adaptive equalization. *Proc. IEEE.* **73**(9), 1349–1387 (1985).

[RO51] H. Robbins and S. Monro, A stochastic approximation method. *Ann. Math. Stat.* **22**(3), 400–407 (1951).

[SL91] D. T. M. Slock and T. Kailath, Numerically stable fast transversal filters for recursive least square adaptive filtering. *IEEE Trans. Signal Process.* **SP-30**(1), 92–114 (1991).

[SO80] M. M. Sondhi and D. A. Berkley, Silencing echoes on the telephone network. *Proc. IEEE* **68**(8), 948–963 (1980).

[ST90] P. Strobach, *Linear Prediction Theory.* Springer-Verlag, New York, 1990.

[TR77] J. R. Treichler, The spectral line enhancer. Ph.D. dissertation, Stanford University, Stanford, CA, 1977.

[TR87] J. R. Treichler, C. R. Johnson, Jr., and M. G. Larimore, *Theory and Design of Adaptive Filters.* Wiley (Interscience), New York, 1987.

[UN72] G. Ungerboeck, Theory on the speed of convergence in adaptive equalizer for digital communication. *IBM J. Res. Dev.* **16**(5), 546–555 (1972).

[UN76] G. Ungerboeck, Fractional tap-spacing equalizer and consequences for clock recovery in data modems. *IEEE Trans. Commun.* **24**(8), 856–864 (1976).

[WE84] J. J. Werner, An echo-cancellation-based 4800 bit/s full-duplex DDD modem. *IEEE J. Select. Areas Commun.* **2**(5), 722–730 (1984).

[WE85] J. J. Werner, Effects of channel impairments on the performance of an In-band data-driven echo canceler. *AT&T Bell Lab. Tech. J.* **64**(1), 91–113 (1985).

[WI60] B. Widrow and M. E. Hoff, Jr., Adaptive switching circuits. *IRE WESCON Conv. Rec., 1960*, pp. 96–104 (1960).

[WI75a] B. Widrow, J. McCool, and M. Ball, The complex LMS algorithm. *Proc. IEEE.* **63**(4), 719–720 (1975).

[WI75b] B. Widrow et al., Adaptive noise cancelling: Principles and applications. *Proc. IEEE* **63**(12), 1692–1716 (1975).

[WI84] B. Widrow and E. Walach, On the statistical efficiency of the LMS algorithm with nonstationary inputs. *IEEE Trans. Inf. Theory* **IT-30**(2), 211–221 (1984).

[WI85] B. Widrow and S. D. Stearns, *Adaptive Signal Processing*. Prentice-Hall, Englewood Cliffs, NJ, 1985.

[YU88] X-H. Yu and Z-Y. He, Efficient implementation of exact sequential least-squares problems. *IEEE Trans. Acoust., Speech, Signal Process.* **ASSP-36**(3), 392–399 (1988).

[ZA73] C. L. Zahm, Applications of adaptive arrays to suppress strong jammers in the presence of weak signals. *IEEE Trans. Aerosp. Electron. Syst.* **AES-9**(2), 260–271 (1973).

16 Spectral Analysis

RAMDAS KUMARESAN

Department of Electrical Engineering
The University of Rhode Island, Kingston

16-1 INTRODUCTION

Spectral analysis is concerned with characterizing the frequency content of a signal. It has applications in a wide variety of fields including radar signal processing, underwater acoustics, seismology, and spectrometry. A large number of spectral analysis techniques are available in the literature. These can be broadly classified into nonparametric or Fourier analysis based methods and parametric model based methods. In this chapter, we provide an overview of the various spectral analysis methods currently available. A large number of textbooks and research monographs covering the topics in spectral analysis already exist. Some of the books are written from the point of view of statistical time-series analysis [BL58; BR80a; JE68; KO74], whereas others are oriented toward an engineering audience [CH78; HA82, HA83; KA87; KE86; MA87].

In 1807, Fourier proposed that any finite duration signal, even a signal with discontinuities, can be expressed as an infinite summation of harmonically related sinusoidal components; that is,

$$x(t) = \sum_{k=1}^{\infty} (A_k \cos k\Omega_0 t + B_k \sin k\Omega_0 t),$$

where A_k and B_k are the Fourier coefficients and Ω_0 is the fundamental angular frequency. Fourier analysis has since become an indispensable tool in a large number of disciplines. An important step in the application of Fourier analysis to spectral analysis was made by Schuster [SC1898]. Schuster called his numerical spectrum estimation method the *periodogram*. Schuster computed the periodogram $P(e^{j\omega})$ using discrete signal samples $x[0], x[1], x[2], \ldots, x[N-1]$ as follows:

$$P(e^{j\omega}) = \frac{1}{N} \left| \sum_{n=0}^{N-1} x[n] e^{-j\omega n} \right|^2.$$

Handbook for Digital Signal Processing, Edited by Sanjit K. Mitra and James F. Kaiser.
ISBN 0-471-61995-7 © 1993 John Wiley & Sons, Inc.

If the signal samples, $x[n]$, consisted of a sinusoid of frequency ω_1 then the periodogram showed a peak at $\omega = \omega_1$. Thus the peaks in the periodogram appeared to show the location of the frequencies of the underlying sinusoidal signals. Schuster successfully applied the periodogram to processing sunspot numbers. However, other empirical time-series data observed in nature yielded periodograms with erratic behavior not exhibiting any dominant peaks. Because of this fact many researchers at that time were disenchanted with the periodogram. However, the periodogram remained the only numerical tool for spectral analysis of time-series data, until the late 1920s.

In 1930, in a classic paper, Wiener established the framework for spectral analysis of random processes using the continuous Fourier transform [WI30]. Specifically, he defined the autocorrelation function of a random signal and showed its relationship to the power spectral density via the Fourier transform. This result, that the autocorrelation function and the power spectral density are Fourier transform pairs, is often known as the *Wiener–Khintchine theorem*. It paved the way in 1958 for the Blackman–Tukey [BL58] implementation of a power spectrum estimation procedure by first computing the estimate of the autocorrelation sequence from the given data sequence and then Fourier transforming the windowed autocorrelation sequence.

The next major advance in spectral analysis appeared in the form of an efficient computational algorithm suitable for calculation of the discrete Fourier transform using a digital computer. This algorithm, known as the fast Fourier transform, was proposed by Cooley and Tukey [CO65] in 1965. Although there have been many contributions to the development of fast Fourier transform algorithms (see Chapter 8), the Cooley–Tukey work was instrumental in popularizing digital spectral analysis. This algorithm also rekindled interest in the periodogram approach to spectral analysis. A modified version of Schuster's periodogram is currently the most popular approach to spectral analysis.

The erratic behavior of the Schuster periodogram when used with random data motivated Yule to search for alternatives. In 1927, he proposed a finite parametric model for a random process [YU27]. Yule observed that the movement of a simple harmonic pendulum with damping, discretized in time, can be represented by the homogeneous difference equation

$$s[n] + a_1 s[n - 1] + a_2 s[n - 2] = 0,$$

where $s[n]$ is the amplitude of the pendulum swing at the integer time instant n. The solution of this difference equation gives the impulse response of the pendulum, which is a damped sinusoid. However, the measurements of the pendulum swing have errors. Instead of modeling the deviations as superposed errors on the errorless measurements of the pendulum swing, Yule modeled the measurements using a driving function. That is, the pendulum motion is now described by the following nonhomogeneous difference equation

$$s[n] + a_1 s[n - 1] + a_2 s[n - 2] = e[n],$$

where $e[n]$ is the white noise input driving the pendulum. The pendulum is no longer supposed to move in a damped oscillatory motion but its amplitude and phase vary continuously depending on the driving sequence $e[n]$. Yule thus proposed a finite-order all-pole model characterized by only two parameters and excited by a white noise sequence. Then, given a sequence $s[n]$, he went to find the model parameters by using regression analysis. Since $s[n]$ regresses on its own past, his model is called an *autoregressive* model. The power spectrum is then determined from the model parameters and the variance of the input noise sequence. Yule's work initiated the study of model based spectral analysis but was not popular till the late 1960s. The above homogeneous equations used by Yule were also known to Prony [PR1795] as early as 1795. Prony was not interested in spectrum analysis but in curve fitting a sum of damped exponentials to some observed data. Also, he did not use the least-squares technique to obtain a fit to the data.

Wold [WO38] proposed a fundamental theorem showing that any time series may be decomposed into two parts: one a deterministic or predictable part and the other a random process part that may be modeled as an all-zero filter excited by a white noise sequence. The Kolmogorov [KO41], Wiener [WI49], and Levinson [LE47] theories of linear prediction are also closely connected to many of the model based spectral analysis methods. A current resurgence of widespread interest in spectral analysis may partly be attributed to the work of J. P. Burg. Burg [BU67], in his pioneering work, introduced a new philosophy in spectral analysis based on general variational principles and in particular based on the maximum entropy formalism. The maximum entropy spectrum he proposed turned out to be closely related to the autoregressive spectrum analysis methods proposed earlier by Parzen [PA64] and Bartlett [BA48], who extended the original work of Yule. The elegance of Burg's work caused more attention to be focused on model based methods. More detailed histories of spectrum analysis may be found in the works of Robinson [RO82] and Marple [MA87].

As pointed out by Robinson [RO82], the signals observed during a natural phenomenon may be broadly classified into noise-like processes and signal-plus-noise processes. A noise-like process possesses an underlying statistical structure but is different in detail in each of its realizations. For example, an unvoiced speech signal observed when a letter such as "f" or "s" is uttered may be regarded as a noise-like process. On the other hand, signal-plus-noise processes may appear essentially the same from record to record; examples include seismic records [RO67; SC84] and nuclear magnetic resonance signals [ST88]. For noise-like processes one may be interested in estimating the power of the process as a function of frequency, that is, its power spectral density. The nonparametric or Fourier analysis based methods have been most successful in such cases. On the other hand, for signal-plus-noise processes one may be interested in answering more specific questions, such as "Does this region in the frequency domain consist of one or two or more closely spaced tones?" Such questions are more readily answered by parametric model based methods. This is particularly true if the data record is short. Therefore basic issues in applying spectral analysis involve the proper choice of the spectral estimation method based on the understanding of the underlying physical phenomena, the choice of the particular nonparametric or parametric modeling algorithm, and the questions one wants answered.

In Sections 16-2 and 16-3 we summarize the classical methods of spectral analysis based on the Fourier transform. In Section 16-4 we describe some of the more successful parametric model based methods for analyzing segments of stationary random processes. In Section 16-5 we focus on specific techniques that are suitable for processing sinusoidal signals in noise. In Section 16-6 we briefly describe methods for processing spatial time-series data.

16-2 FOURIER TRANSFORM OF FINITE-TIME SIGNALS

Consider a sampled signal $x[n]$, $-\infty \leq n \leq \infty$. Assume that the signal has finite energy; that is, $\sum_{n=-\infty}^{\infty} |x[n]|^2 < \infty$. The discrete-time Fourier transform (DTFT) of $x[n]$ is

$$X(e^{j\omega}) = \sum_{n=-\infty}^{\infty} x[n]e^{-j\omega n}. \tag{16.1}$$

By Parseval's relation, the energy E is

$$E = \sum_{n=-\infty}^{\infty} |x[n]|^2 = \frac{1}{2\pi} \int_{-\pi}^{\pi} |X(e^{j\omega})|^2 \, d\omega. \tag{16.2}$$

The energy spectral density (ESD) of a finite-energy signal is defined as

$$S_{xx}(e^{j\omega}) = |X(e^{j\omega})|^2. \tag{16.3}$$

The ESD is the squared magnitude of the Fourier transform of the sequence $x[n]$. It gives the concentration of the energy of a signal as a function of frequency ω. An equivalent expression for the ESD is obtained by defining it as the Fourier transform of the autocorrelation sequence $r_{xx}[k]$:

$$S_{xx}(e^{j\omega}) = \sum_{k=-\infty}^{\infty} r_{xx}[k]e^{-j\omega k}, \tag{16.4}$$

where the autocorrelation sequence is given by

$$r_{xx}[k] = \sum_{n=-\infty}^{\infty} x^*[n]x[k+n]. \tag{16.5}$$

The equivalence of Eqs. (16.3) and (16.4) can readily be seen by substituting the expression for $r_{xx}[k]$ into Eq. (16.4). If the sequence $x[n]$ is of finite duration, being nonzero only for $n = 0, 1, \ldots, N - 1$, then the energy spectral density is given by

$$S_{xx}(e^{j\omega}) = \left| \sum_{n=0}^{N-1} x[n]e^{-j\omega n} \right|^2. \tag{16.6}$$

Since the ESD is just the square of the magnitude of the Fourier transform of $x[n]$, computing the ESD involves essentially computing the DTFT, $X(e^{j\omega})$, of the finite-time sequence $x[n]$. Using a digital computer, however, $X(e^{j\omega})$ can be evaluated at only a finite number of points. Since $X(e^{j\omega})$ is periodic with period 2π, it can be conveniently evaluated at N uniformly spaced points in this interval. The values at the sample points on the frequency axis are given by the discrete Fourier transform (DFT):

$$X[k] = \sum_{n=0}^{N-1} x[n]e^{-j2\pi kn/N}, \qquad k = 0, 1, \ldots, N - 1. \tag{16.7}$$

If N is a power of 2 or is a product of relatively prime integers, then the DFT can be efficiently computed using a fast Fourier transform (FFT) algorithm (see Chapter 8).

In practice, evaluation of the ESD at these N points may not be sufficient. One may need to know the energy spectrum at L points, $L > N$; that is, one may need a finer frequency spacing. This is accomplished by zero-padding the data sequence with $L - N$ zeros, and computing an L-point DFT of the zero-padded sequence. That is, the sample points are now computed as

$$X[k] = \sum_{n=0}^{L-1} x[n]e^{-j2\pi kn/L}, \qquad k = 0, 1, \ldots, L - 1, \tag{16.8}$$

where the samples $x[N]$ to $x[L - 1]$ are set equal to zero. Note that the effective frequency spacing is now $2\pi/L$ ($<2\pi/N$). It should be emphasized that zero-padding does not increase the resolution of the energy spectrum but merely interpolates the values of the energy spectrum at more values.

Example 16.1. Consider the ESD of a signal composed of two sinusoidal signals with frequencies 0.135 and $0.135 + \Delta f$. Let $x[n]$ be given by

$$x[n] = \cos[2\pi(0.135)n - \pi/2] + \cos[2\pi(0.135 + \Delta f)n],$$

$$n = 0, 1, \ldots, 15. \tag{16.9}$$

Let us consider the cases for which $\Delta f = 0.06$ and $\Delta f = 0.01$. Figure 16-1(a) illustrates the ESD for the first case, with $L = N = 16$. Since the signal is real-valued, only normalized frequencies up to $f = 0.5$ are plotted. From this figure it is not possible to conclude whether the two large-amplitude sample points belong

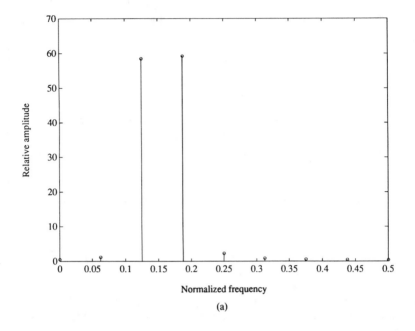

(a)

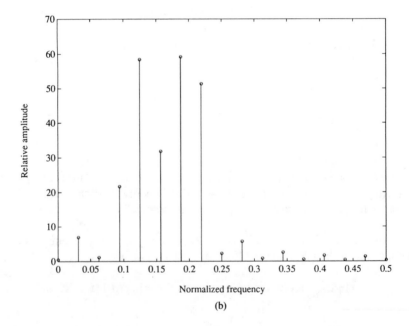

(b)

FIGURE 16-1   Spectrum/energy spectral density of a signal composed of two sinusoidal signals with frequencies 0.135 and 0.195 Hz. N is the number of signal samples. L is the number of samples after appending zeros. (a) $N = L = 16$; (b) $N = 16$, $L = 32$; (c) $N = 16$, $L = 64$, and (d) $N = 16$, $L = 128$.

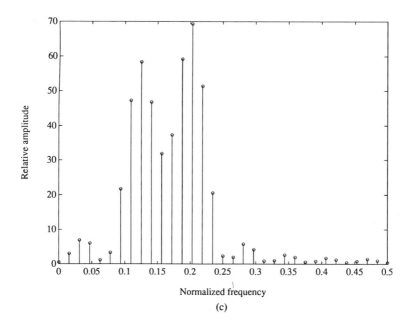

(c)

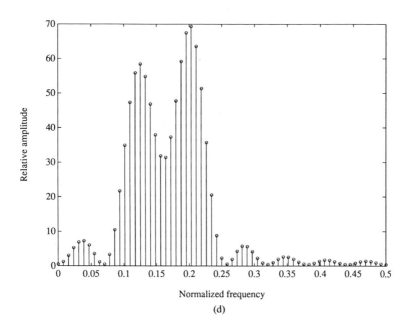

Normalized frequency

(d)

FIGURE 16-1 (*Continued*)

to one broad peak or are points on two distinct peaks. To resolve this ambiguity, we need to evaluate the samples at finer intervals. Figures 16-1(b)–(d) show the ESD for $L = 32$, 64, and 128. For these L values, 16, 48, and 112 zeros were appended to the original 16-point sequence $x[n]$ before the DFT was computed. Clearly, with increasing L values we get a better picture of the computed ESD. In Figure 16-2 we have plotted the ESD for the second case in which the frequency spacing was closer but for identical values of L and N. Observe that zero-padding has not increased the resolution, which is our ability to distinguish between two line components closely spaced in frequency. Resolution depends on the number of original data points, N, available for processing. More about this is discussed in Section 16-2-1.

If a signal $x[n]$ exists over the entire interval $(-\infty, \infty)$ and its energy is infinite (e.g., sinusoidal or periodic signals), it is convenient to define the power P of such a signal as follows:

$$P = \lim_{N \to \infty} \frac{1}{2N + 1} \sum_{n=-N}^{N} |x[n]|^2.$$

If P is finite, then the power spectral density, the distribution of power of the signal as a function of frequency, is defined as

$$\Phi_{xx}(e^{j\omega}) = \lim_{N \to \infty} \frac{1}{2N + 1} \left| \sum_{n=-N}^{N} x[n]e^{-j\omega n} \right|^2 \tag{16.10}$$

because $P = 1/2\pi \int_{-\pi}^{\pi} \Phi_{xx}(e^{j\omega})d\omega$.

16-2-1 Effects of Data Windowing: Resolution Loss and Spectral Leakage

A central problem we consider in this chapter is that of estimating the energy or power spectral density of a signal from its finite-time observation. The observation time is often limited because the signal remains stationary only over a short interval. The finite observation time, unfortunately, limits the quality of the energy or power spectral density estimate that one can obtain. The effects of the finite observation time are easily understood by studying them using sinusoidal signals. Such sinusoidal signals would correspond to impulses in the frequency domain if the signal were known for all time. But in practice, since the signals are known only over a finite interval, it is of interest to know what effect this has on the spectra of such signals.

It is convenient to view a finite-duration signal as a product of a window function and the underlying infinite-duration signal. For example, the observed signal $s[n]$ can be expressed as a product of a rectangular window

$$w_R[n] = \begin{cases} 1 & \text{for } 0 \leq n \leq N - 1, \\ 0 & \text{otherwise,} \end{cases} \tag{16.11}$$

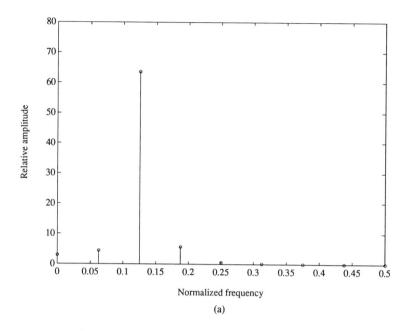

(a)

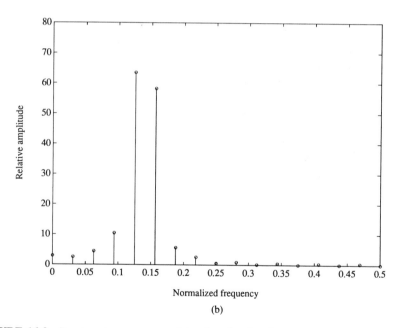

(b)

FIGURE 16-2 Spectrum/energy spectral density of a signal composed of two sinusoidal signals with frequencies 0.135 and 0.145 Hz. Note that the frequency spacing $\Delta f = 0.01$ is much smaller than in Figure 16-1. N is the number of signal samples. L is the number of samples after appending zeros. (a) $N = L = 16$; (b) $N = 16$, $L = 32$; (c) $N = 16$, $L = 64$; and (d) $N = 16$, $L = 128$.

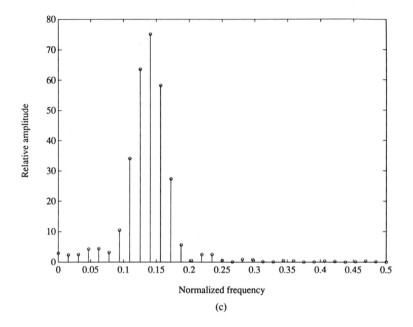

(c)

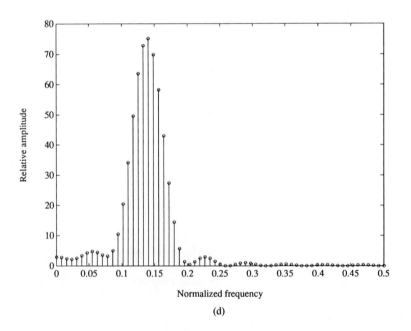

(d)

FIGURE 16-2   (*Continued*)

and the infinite-duration signal $x[n]$, as

$$s[n] = x[n] \cdot w_R[n]. \tag{16.12}$$

Of course, we have made the tacit assumption that the signal $s[n]$ is zero outside the observation window. This may be an unrealistic one. To obtain the signal spectrum $S(e^{j\omega})$, we use the property of the Fourier transform that multiplication in the time domain corresponds to convolution in the frequency domain. Hence

$$S(e^{j\omega}) = \frac{1}{2\pi} X(e^{j\omega}) \circledast W_R(e^{j\omega}), \tag{16.13}$$

where $W_R(e^{j\omega})$ is the discrete Fourier transform of the rectangular window $w_R[n]$ given by

$$W_R(e^{j\omega}) = \frac{\sin(N\omega/2)}{\sin(\omega/2)} e^{-j\omega(N-1)/2}. \tag{16.14}$$

Thus the spectrum of the observed signal $s[n]$ is a distorted version of that of the infinite-duration signal $x[n]$ because of the convolution with the discrete sinc function $W_R(e^{j\omega})$.

Example 16.2. To see this distortion more clearly, consider the spectrum of a truncated complex sine wave of frequency f_0, $x[n] = e^{j2\pi f_0 n}$, $n = 0, 1, 2, \ldots, 9$. The spectrum of the infinite-duration signal would show an impulse at f_0. However, truncation in the time domain or, equivalently, convolution with the window transform in the frequency domain causes this spectrum to be smeared. The computed spectrum now shows a sinc function centered at f_0. This is illustrated in Figure 16-3(a) for a complex sine wave of normalized frequency 0.3. The frequency axis in Figure 16-3 has been normalized with respect to the sampling frequency f_s. Note that the narrowest frequency component that can now be observed is limited by the main lobe width of the window transform, which in this case is given by Eq. (16.14). The only way to reduce the mainlobe width is to make the window length large. This in effect means that we have to observe the signal for a longer duration. In practice, it may not be possible to make measurements for arbitrarily long time intervals. Hence one cannot avoid this inherent limitation when using conventional Fourier techniques for spectral analysis.

The smearing of spectra discussed above can become quite serious when we attempt to compute the spectrum of a signal with many sinusoidal components, especially if some of them are closely spaced in frequency. Again, consider the case of a signal with two sinusoidal components. For the moment, we assume that their strengths are equal. Figure 16-3(b) shows an example in which the truncated signal has two complex sine waves of normalized frequencies 0.3 and 0.7, re-

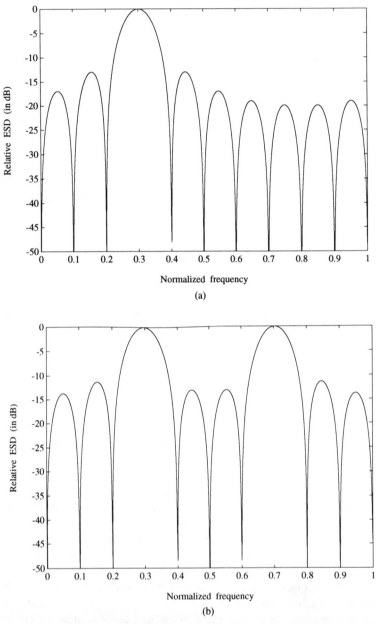

FIGURE 16-3 (a) Energy spectral density of a single complex sinusoid located at frequency $f_0 = 0.3$ Hz. (b) Energy spectral density of two complex sinusoids located at frequencies $f_0 = 0.3$ Hz and $f_1 = 0.7$ Hz. They have equal amplitudes. (c) Energy spectral density of two closely spaced complex sinusoids located at frequencies $f_0 = 0.3$ Hz and $f_1 = 0.32$ Hz with equal amplitudes. (d) Energy spectral density of two complex sinusoids located at frequencies $f_0 = 0.3$ Hz and $f_1 = 0.7$ Hz but with unequal amplitudes. The peak amplitude of the weaker sinusoid at f_1 is 20 dB below that of the stronger one at f_0.

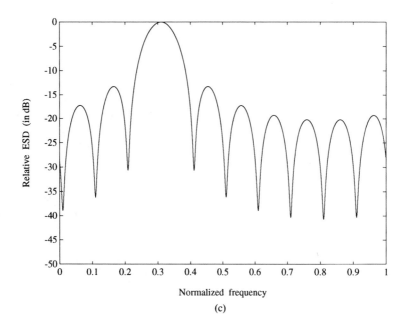

(c)

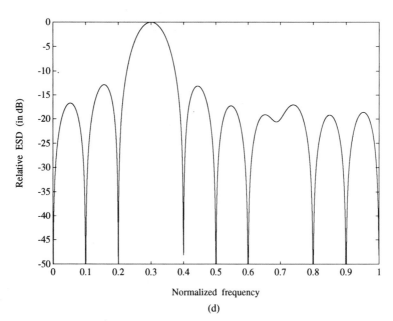

(d)

FIGURE 16-3 (*Continued*)

spectively. Since these two frequencies are spaced far apart, one can still distinguish the two signals despite the broadening of their respective impulsive components. But as the frequency spacing decreases, the two peaks of the respective sinc functions come closer, and beyond a certain point it becomes impossible to distinguish the two peaks. Therefore, in such cases, just by observing the spectrum we cannot conclude whether the data consist of one component or two closely spaced components. This situation is shown in Figure 16-3(c) for the case of a signal having two sinusoidal components of normalized frequencies 0.3 and 0.32, respectively. We thus encounter a serious loss in resolution as a result of dealing with a finite segment of data.

The resolution problem is considerably worsened when the strengths of the two components are quite different. Consider now the case when the amplitude of the second frequency component is significantly less than the first one. In such cases, even if the frequencies are quite widely separated, it is now possible to observe clearly the weaker signal. Figure 16-3(d) shows the spectrum of two sinusoids of normalized frequencies 0.3 and 0.7 respectively, but not the peak amplitude of the second sinusoid being 20 dB less than the first one. We are barely able to see the peak corresponding to the weaker signal. The reason for this is that the sinc ripples of the stronger signal mask or submerge the response of the lower amplitude signal. This energy spreading in the spectral domain is called *leakage*. Even in the cases in which we are able to clearly discern two closely spaced peaks, the sidelobe leakage may cause the peaks to be displaced from their true frequency locations. Also, since the discrete-time Fourier transform is periodic, sidelobe aliasing from adjacent spectral periods may contribute additional bias, worsening the problem.

The effects of sidelobe leakage can be mitigated if the energy in the sidelobes can be made small. Simple truncation or, equivalently, rectangular windowing causes maximum sidelobe interference because of the window's abrupt transitions. The problem of minimizing energy leakage into the sidelobe of a window function has received considerable attention in the literature [HA78]. The solution is to allow for a more gradual transition in the window function, which results in lesser sidelobe level. Some of the commonly used window functions are as follows[1]:

Triangular (Bartlett) Window:

$$
w[n] = \begin{cases} \dfrac{2n}{N-1}, & 0 \le n \le \dfrac{N-1}{2}, \\[2ex] 2 - \dfrac{2n}{N-1}, & \dfrac{N-1}{2} < n \le N-1. \end{cases} \tag{16.15}
$$

[1]See Section 4-4 for a review of window functions.

Hann Window:

$$w[n] = \frac{1}{2}\left[1 - \cos\left(\frac{2\pi n}{N-1}\right)\right], \quad 0 \leq n \leq N-1. \quad (16.16)$$

Hamming Window:

$$w[n] = 0.54 - 0.46\cos\left(\frac{2\pi n}{N-1}\right), \quad 0 \leq n \leq N-1. \quad (16.17)$$

Blackman Window:

$$w[n] = 0.42 - 0.5\cos\left(\frac{2\pi n}{N-1}\right) + 0.08\cos\left(\frac{4\pi n}{N-1}\right),$$

$$0 \leq n \leq N-1. \quad (16.18)$$

Kaiser Window:

$$w[n] = \frac{I_0\left[\omega_a\sqrt{\left(\frac{N-1}{2}\right)^2 - \left[n - \left(\frac{N-1}{2}\right)\right]^2}\right]}{I_0\left[\omega_a\left(\frac{N-1}{2}\right)\right]}, \quad (16.19)$$

where $I_0(\cdot)$ is the modified zeroth-order Bessel function of the first kind. Although these windows have lesser sidelobe levels than a simple rectangular window, they have increased mainlobe width and hence poorer resolution. Among these windows, all except the Kaiser window have fixed mainlobe width and peak sidelobe level. The Kaiser window offers flexible design, in that, one can trade off mainlobe width for peak sidelobe amplitude using the parameter ω_a. The strategy for window selection is dictated by the specific application on hand. For filter design applications, peak sidelobe level will be of primary concern, while in spectral analysis applications the asymptotic roll-off will be the deciding factor. For various trade-offs with window functions and their usage, we refer to Harris [HA78] (see also Section 4-4).

16-3 FOURIER ANALYSIS OF RANDOM SIGNALS

In the previous section we considered signals with finite energy and used the energy spectral density to characterize such signals in the frequency domain. Another class of signals that are observed during many physical phenomena are random signals or processes. These signals have an underlying statistical structure. An important subclass among these are the so-called wide-sense stationary random processes.[2] Such processes have average values that are constant and whose autocorrelation functions depend only on time differences. For simplicity, let us assume that the observed sampled random process $x[n]$ has zero mean. Its autocorrelation sequence is defined by

$$r_{xx}[m] = \mathcal{E}\{x^*[n]x[m + n]\}, \tag{16.20}$$

where $\mathcal{E}\{\cdot\}$ denotes the ensemble average. The power spectral density (PSD) $\Phi_{xx}(e^{j\omega})$ of the random process is given by the discrete-time Fourier transform of $r_{xx}[m]$:

$$\Phi_{xx}(e^{j\omega}) = \sum_{m=-\infty}^{\infty} r_{xx}[m]e^{-j\omega m}. \tag{16.21}$$

This relationship between the autocorrelation sequence and the PSD is known as the *Wiener–Khintchine theorem*. An alternate definition of the PSD is given by the following:

$$\Phi_{xx}(e^{j\omega}) = \lim_{N\to\infty} \mathcal{E}\left\{\frac{1}{2N + 1}\left|\sum_{n=-N}^{N} x[n]e^{-j\omega n}\right|^2\right\}. \tag{16.22}$$

That is, the PSD $\Phi_{xx}(e^{j\omega})$ is obtained by first computing the DTFT of $x[n]$, letting the number of points grow to infinity and scaling the magnitude squared of the DTFT by the number of data samples. Since for a random signal the quantity within the brackets in Eq. (16.22) is a random variable, its ensemble average has to be obtained to compute the PSD. It is instructive to show that the definitions of the PSD in Eqs. (16.21) and (16.22) are equivalent. Observe that the expression in Eq. (16.22) may be rewritten as follows:

$$\Phi_{xx}(e^{j\omega}) = \lim_{N\to\infty} \mathcal{E}\left\{\frac{1}{2N + 1}\sum_{m=-N}^{N}\sum_{n=-N}^{N} x[m]x^*[n]e^{-j\omega(m - n)}\right\}. \tag{16.23}$$

Interchanging the order of expectation operation and summation we have

$$\Phi_{xx}(e^{j\omega}) = \lim_{N\to\infty} \frac{1}{2N + 1}\sum_{m=-N}^{N}\sum_{n=-N}^{N} r_{xx}[m - n]e^{-j\omega(m - n)}. \tag{16.24}$$

Furthermore, the double summation in the above equation may be converted into a single summation using the following identity:

$$\sum_{n=-N}^{N}\sum_{m=-N}^{N} r_{xx}[m - n]e^{-j\omega(m-n)} = \sum_{n=-2N}^{2N} (2N + 1 - |n|)\, r_{xx}[n]e^{-j\omega n}. \tag{16.25}$$

[2]See Section 2-14 for a review of random processes.

Therefore

$$\Phi_{xx}(e^{j\omega}) = \lim_{N \to \infty} \sum_{n=-2N}^{2N} \left[1 - \frac{|n|}{2N+1} \right] r_{xx}[n] e^{-j\omega n}. \tag{16.26}$$

However, if the autocorrelation sequence $r_{xx}[n]$ decays rapidly enough, then the second term in the summation tends to zero. This establishes the equivalence of the two definitions of the PSD in the formulas of Eqs. (16.21) and (16.22). In fact, the two definitions of the PSD point us to two different methods of estimating the power spectral density when only a finite segment of the random process $x[n]$ is available for processing. Equation (16.21) shows that the autocorrelation values may first be computed from the data and then Fourier transformed to obtain the power spectral density. On the other hand, Eq. (16.22) shows that we may directly compute the DFT of the given data, the magnitude squared of which gives an estimate of the PSD (i.e., Schuster's periodogram). We discuss these two basic approaches in the sequel. Detailed discussions including the properties of various estimators of the power spectral density are to be found in Blackman and Tukey [BL58], Jenkins and Watts [JE68], Koopmans [KO74], Kay [KA87], and Marple [MA87].

16-3-1   The Periodogram

Let us consider the problem of estimating the power spectral density $\Phi_{xx}(e^{j\omega})$ of a random signal sequence $x[n]$. Although, in theory, the sequence $x[n]$ exists for all time, often only a finite segment of the sequence, say, $x[n]$, $n = 0, 1, \ldots, L - 1$, is available for processing. As before, we can view the truncated data as the product with a window function $w[n]$. Let $S(e^{j\omega})$ be the DTFT of $x[n] \cdot w[n]$; that is,

$$S(e^{j\omega}) = \sum_{n=0}^{L-1} x[n] \cdot w[n] e^{j\omega n}, \tag{16.27}$$

where $w[n]$ is nonzero for $n = 0, 1, \ldots, L - 1$. We now consider an estimate of the power spectral density given by

$$\check{I}_{xx}(e^{j\omega}) = \frac{1}{LU} |S(e^{j\omega})|^2. \tag{16.28}$$

The constant U is a normalizing factor to remove any bias in the spectral estimate that might be caused by the window $w[n]$. It is defined as

$$U = \frac{1}{L} \sum_{n=0}^{L-1} |w[n]|^2. \tag{16.29}$$

$\check{I}_{xx}(e^{j\omega})$ is called the *periodogram* if $w[n]$ is the simple rectangular window function. Otherwise, it is called the *modified periodogram*. As noted in the previous section, the window length determines the resolving capabilities of the periodogram. Using a digital computer, the estimated periodogram can be computed only at discrete frequencies, $\omega_k = 2\pi k/N$, $k = 0, 1, \ldots, N - 1$. If we denote the frequency ω_k by the index k, we have

$$\check{I}_{xx}[k] = \frac{1}{LU} |S[k]|^2. \tag{16.30}$$

In general, N can be chosen greater than L, in which case the sequence $x[n]\,w[n]$ is zero-padded with $N - L$ zero sample values, before computing the DFT.

Since $x[n]$ is a random signal sequence the estimate of the PSD, $\check{I}_{xx}(e^{j\omega})$, at each value of ω is a random variable. Its mean value and variance determine the quality of the PSD estimate at the frequency ω. It is easily shown that the mean value of the PSD estimate $\check{I}_{xx}(\omega)$ is [JE68]

$$\mathcal{E}\{\check{I}_{xx}(e^{j\omega})\} = \frac{1}{2\pi LU} \int_{-\pi}^{\pi} \Phi_{xx}(e^{ju})|W(e^{j(\omega - u)})|^2 \, du, \tag{16.31}$$

where $\Phi_{xx}(e^{ju})$ is the true PSD of the sequence $x[n]$ and $W(e^{j\omega})$ is the Fourier transform of the window $w[n]$. Equation (16.31) shows that the PSD estimate $\check{I}_{xx}(e^{j\omega})$ is biased. The bias is caused by the convolution of the true power spectrum with the magnitude squared of the Fourier transform of the data window. If the window length, L, is increased, then $|W(e^{j\omega})|^2$ asymptotically tends to a periodic impulse train. Therefore it should be clear from the integral in Eq. (16.31) that the bias approaches zero asymptotically as the window length increases. Computation of the variance of the periodogram is difficult. However, Jenkins and Watts [JE68] show that in general as the window length increases the variance of the periodogram approximately equals $\Phi_{xx}^2(e^{j\omega})$. That is, the variance is of the same order as the power spectrum we are estimating. This point is illustrated in the following example.

Example 16.3. Figures 16-4(a)–(c) show the periodogram estimate of a computer generated pseudorandom white noise sequence $x[n]$. The noise samples $x[n]$ were drawn from a zero mean and unit-variance Gaussian distribution. Since the noise sequence has negligible sample-to-sample correlation, its power spectral density is expected to be flat over the Nyquist interval. The power spectral density estimates were obtained using Eq. (16.30) for $L = 64$, 256, and 1024. $w[n]$ was chosen to be a rectangular window. In Figure 16-4(a) we have plotted the estimated spectrum for $L = 64$. Figures 16-4(b) and 16-4(c) show the estimates for $L = 256$ and 1024, respectively. It is seen that with increasing window length, instead of converging to a flat power spectral density, the spectrum shows more rapid fluctuations. Some appreciation of this fact may be gained be rewriting the periodo-

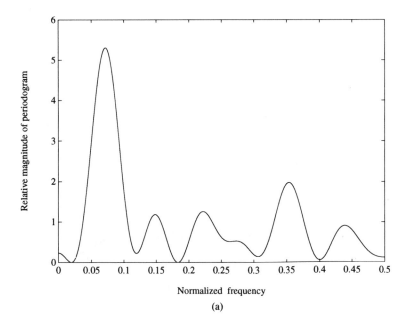

(a)

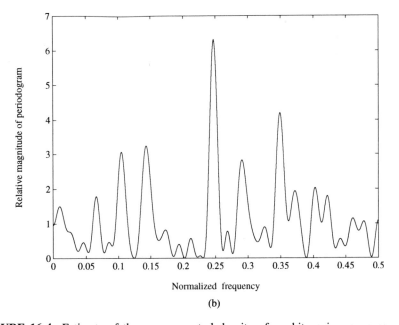

(b)

FIGURE 16-4 Estimate of the power spectral density of a white noise sequence with different data lengths. (a) Periodogram with $L = 64$, (b) periodogram with $L = 256$, and (c) periodogram with $L = 1024$. (d) Welch's averaged periodogram estimate with $Q = 1024$, $P = 7$, $S = 128$, and $L = 256$.

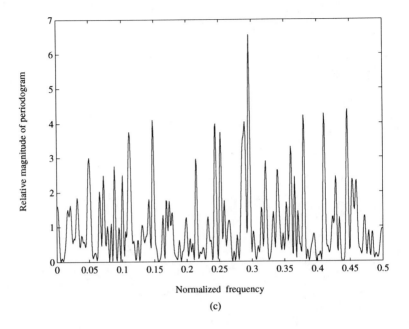

Normalized frequency

(c)

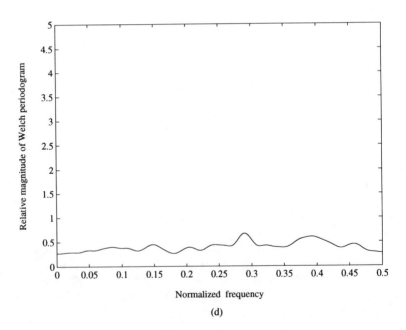

Normalized frequency

(d)

FIGURE 16-4   (*Continued*)

gram as follows (ignoring the window $w[n]$):

$$
\begin{aligned}
\check{I}_{xx}(e^{j\omega}) &= \frac{1}{L} \sum_{m=0}^{L-1} \sum_{l=0}^{L-1} x[l]x^*[m]e^{-j\omega(l-m)} \\
&= \sum_{k=-(L-1)}^{L-1} \left(\frac{1}{L} \sum_{m=0}^{L-1-|k|} x[m+k]x^*[m] \right) e^{-j\omega k} \\
&= \sum_{k=-(L-1)}^{L-1} \hat{r}_{xx}[k]e^{-j\omega k}.
\end{aligned}
\tag{16.32}
$$

Therefore the periodogram is actually the Fourier transform of the periodic correlation $\hat{r}_{xx}[k]$ of the finite-length sequence $x[n]$. The $\hat{r}_{xx}[k]$ are the estimates of the true autocorrelation values $r_{xx}[k]$. However, note that when k is close to L, only a few samples of $x[n]$ are involved in the computation of $\hat{r}_{xx}[k]$, resulting in poor estimates of $r_{xx}[k]$. These cause the wild fluctuations in the spectral estimates $\check{I}_{xx}(e^{j\omega})$, which were observed by Schuster in 1898. To get smoother PSD estimates, many independent periodogram estimates have to be averaged. The averaging of periodograms was first studied by Bartlett [BA48]. Later Welch [WE67] developed methods for averaging modified periodograms. These averaging methods are discussed next.

16-3-2 Periodogram Averaging: Welch and Bartlett Methods

Welch's method computes the periodograms of overlapping segments of data and averages them. Let us assume that Q contiguous samples of $x[n]$ are available for processing. We divide this data record into P segments of L samples each. Let S be the shift between successive segments. These parameters are related by the formula $Q = (P-1)S + L$. The weighted pth segment is

$$
x^{(p)}[n] = w[n] \cdot x[n+pS]
\tag{16.33}
$$

for $0 \leq n \leq L-1$ and $0 \leq p \leq P-1$. The periodogram of the pth segment is

$$
\check{I}_{xx}^{(p)}(e^{j\omega}) = \frac{1}{LU} |S^{(p)}(e^{j\omega})|^2,
\tag{16.34}
$$

where $S^{(p)}(e^{j\omega})$ is the discrete-time Fourier transform of the pth windowed data. The average of these individual segments yields the Welch estimate

$$
\hat{I}_{xx}^{W}(e^{j\omega}) = \frac{1}{P} \sum_{p=0}^{P-1} \check{I}_{xx}^{(p)}(e^{j\omega}).
\tag{16.35}
$$

The expected value of $\hat{I}_{xx}^W(e^{j\omega})$ can be shown to be [JE68]

$$\mathcal{E}\{\hat{I}_{xx}^W(e^{j\omega})\} = \frac{1}{2\pi}\,\Phi_{xx}\,\circledast\,|W(e^{j\omega})|^2,\qquad(16.36)$$

where

$$|W(e^{j\omega})|^2 = \frac{1}{LU}\left|\sum_{n=0}^{L-1}w[n]e^{-j\omega n}\right|^2.\qquad(16.37)$$

For the rectangular window, the expected value of $\hat{I}_{xx}^W(e^{j\omega})$ is the convolution of the true power spectrum $\Phi_{xx}(e^{j\omega})$ with a sinc-squared function. Thus the Welch estimate of the PSD is biased. Assuming that the P periodogram estimates are essentially independent of each other, the variance of $\hat{I}_{xx}^W(e^{j\omega})$ is approximately reduced by a factor of P due to averaging [JE68]:

$$\mathrm{Var}\,\{\hat{I}_{xx}^W(e^{j\omega})\} \approx \frac{1}{P}\,\Phi_{xx}^2(e^{j\omega}).\qquad(16.38)$$

In fact, for the Welch estimate of the PSD the bias and the variance asymptotically approach zero as Q, the number of samples, increases. However, for a given value of Q, the decrease in variance is achieved at the cost of poorer resolution. To obtain maximum resolution we should choose L as large as possible; that is, $L = Q - 1$. But this results in the standard periodogram, which has a large variance. For variance reduction, we should choose L small, so that the number of segments averaged, P, will be large. Since these are conflicting requirements, we have to exchange resolution or bias for variance by choosing a suitable value of P. Figure 16-4(d) shows the Welch estimate for the power spectrum using the white noise sequence used earlier. Clearly, the estimate is much smoother than the corresponding periodogram in Figure 16-4(c).

The Bartlett periodogram is the forerunner to the Welch method and can be considered a special case of it. In this method, the data record is divided into nonoverlapping segments and each segment's periodogram estimate is obtained. Only a rectangular data window $w[n]$ is used for each segment. The final PSD estimate is obtained by averaging the periodograms of all the segments. As before, there is a decrease in the variance but the spectrum is biased. The usual trade-off between bias and variance for a fixed Q applies to this estimator as well. However, the Welch estimate, which allows overlapping, tends to have less variance.

Example 16.4. To highlight the effects of the trade-off between variance and resolution we computed the Welch PSD estimate of some 2048 complex-valued samples recorded from a Bruker nuclear magnetic resonance (NMR) spectrometer using a chemical (1-octanol) sample. In NMR spectroscopy, the location of the spectral lines and their intensities are used to identify the structure of the chemical under study. Figures 16-5(a) and 16-5(b) show the real and imaginary parts of the NMR data record. Typically, NMR data records tend to decay with time but it may not

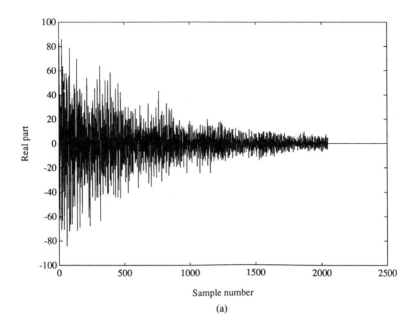

(a)

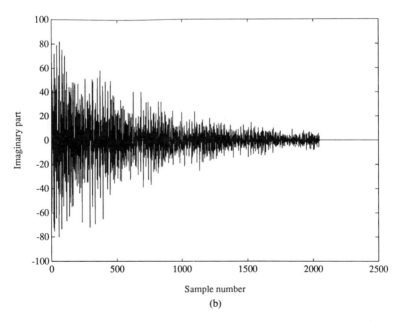

(b)

FIGURE 16-5 (a) Real part of the NMR data samples. (b) Imaginary part of the NMR data samples. (c) Periodogram of NMR data. $N = 2048$ complex data samples. No averaging was used. Note the presence of two very closely spaced lines around $0.85f_s$. (d) Welch's averaged periodogram estimate. $L = 256$ and $S = 128$. These parameters correspond to averaging over 15 overlapping segments. Note that the two distinct peaks at $f \approx 0.85f_s$ have merged together, whereas the spectrum shows much less variability elsewhere.

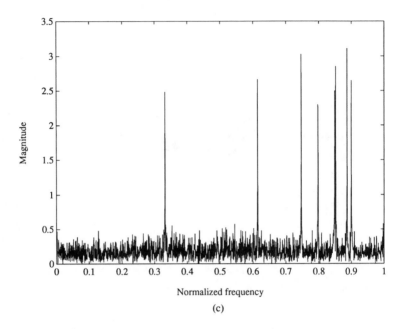

Normalized frequency

(c)

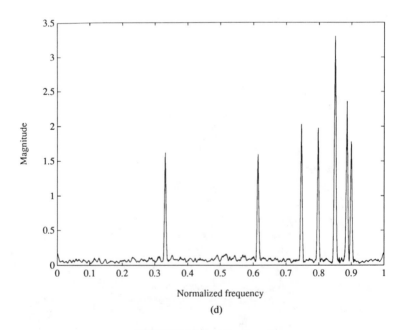

Normalized frequency

(d)

FIGURE 16-5 (*Continued*)

be possible to collect all the data samples until the signals decay to zero. Figure 16-5(c) shows the raw periodogram estimate without any averaging. Note that the spectral lines are sharp but there is significant variability due to the measurement noise in regions where the spectral lines are absent. To reduce this variability we segmented the data into segments of $L = 256$ samples each. The shift between segments, S, was set equal to 128. This corresponds to averaging over 15 segments. Then the Welch PSD estimate was computed as in Eq. (16.35) and the result is shown in Figure 16-5(d). Clearly, the variability is reduced but the lines show a distinct broadening. In particular, the two closely spaced frequencies around $f \approx 0.85 f_s$ are no longer distinguishable due to the decreased resolution.

16-3-3 PSD Estimation Using Correlation Sequence

In the previous subsection we computed the periodogram estimates directly by Fourier transforming the segments of the data sequence $x[n]$ and averaging them to obtain the PSD estimate. Another approach to spectral estimation, as suggested by Eq. (16.21), is to first estimate the autocorrelation sequence $r_{xx}[m]$ of $x[n]$ and then Fourier transform it to obtain the PSD estimate [BL58].

Consider the following autocorrelation estimate at lag m, obtained from the given finite data record $x[n]$, $n = 0, 1, \ldots, Q - 1$:

$$\hat{r}_{xx}[m] = \frac{1}{Q} \sum_{n=0}^{Q-|m|-1} x^*[n]x[n+m], \qquad |m| < Q - 1. \tag{16.39}$$

The expected value of $\hat{r}_{xx}[m]$ is

$$\mathcal{E}\{\hat{r}_{xx}[m]\} = \left(\frac{Q - |m|}{Q}\right) r_{xx}[m], \qquad |m| < Q - 1. \tag{16.40}$$

That is, $\hat{r}_{xx}[m]$ is a biased estimate of $r_{xx}[m]$ for finite Q. We can also define an unbiased estimate of the autocorrelation sequence by changing the scale factor in Eq. (16.39) from $1/Q$ to $1/(Q - |m|)$. However, this estimator may not always yield valid autocorrelation sequences. More importantly, for a finite value of Q, as m approaches Q the variance of the autocorrelation estimate $\hat{r}_{xx}[m]$ increases significantly. This is because for large lag indices, only a few samples of $x[n]$ enter into the computation of $\hat{r}_{xx}[k]$, resulting in very unreliable estimates. Since the periodogram $\bar{I}_{xx}(e^{j\omega})$ is the DTFT of the sequence $\hat{r}_{xx}[m]$ (see Eq. (16.32)), the increased variance in the estimates of the autocorrelation at large lags results in large variability in the periodogram. The effect of the large variance at large lags can be mitigated by weighting the correlation estimates by a window function, which deemphasizes the autocorrelation values at large lags. The Blackman–Tukey PSD estimate does exactly this. It is defined as

$$\check{\Phi}_{xx}(e^{j\omega}) = \sum_{k=-(R-1)}^{R-1} \hat{r}_{xx}[k]w_c[k]e^{-j\omega k}, \tag{16.41}$$

where only the lag values up to R (typically much less than Q) are used. A maximum value of $R = Q/5$ is often recommended. Note that the autocorrelation window $w_c[k]$ must have an even symmetry to produce a PSD estimate that is a real and even function of ω. Since the Blackman–Tukey estimate is the DTFT of the product of the autocorrelation sequence estimate $\hat{r}_{xx}[m]$ and the autocorrelation window $w_c[m]$, we may write it in the frequency domain as follows:

$$\check{\Phi}_{xx}(e^{j\omega}) = \frac{1}{2\pi} \int_{-\pi}^{\pi} \check{I}_{xx}(e^{ju}) W_c(e^{j(w-u)}) \, du. \tag{16.42}$$

That is, the Blackman–Tukey estimate of the PSD is the convolution of the periodogram with the Fourier transform of the window function, that is, $W_c(e^{j\omega})$. This has the effect of smoothing the fluctuations in the periodogram. Note that to ensure that $\check{\Phi}_{xx}(e^{j\omega})$ is nonnegative, $W_c(e^{j\omega})$ has to be nonnegative. The Bartlett window in Eq. (16.15) automatically satisfies this property. However, some other windows such as Hamming and Hann do not.

The expected value of the Blackman–Tukey estimate is

$$\mathcal{E}\{\check{\Phi}_{xx}(e^{j\omega})\} = \frac{1}{2\pi} \int_{-\pi}^{\pi} \mathcal{E}\{\check{I}_{xx}(e^{ju})\} W_c(e^{j(\omega-u)}) \, du. \tag{16.43}$$

The expression for $\mathcal{E}\{\check{I}_{xx}(e^{j\omega})\}$ is given in Eq. (16.31). As Q increases $\mathcal{E}\{\check{I}_{xx}(e^{j\omega})\}$ tends to the true PSD. Hence, even for large data records, the mean value of the Blackman–Tukey estimate is biased due to autocorrelation windowing. Its variance can be shown to be [JE68]

$$\text{Var}\{\check{\Phi}_{xx}(e^{j\omega})\} \approx \frac{\Phi_{xx}^2(e^{j\omega})}{Q} \int_{-\pi}^{\pi} W_c^2(e^{j\omega}) \, d\omega. \tag{16.44}$$

The factor $(1/Q)\int_{-\pi}^{\pi} W_c^2(e^{j\omega})$ is of course less than one and hence the variance of the Blackman–Tukey estimate is reduced by this factor in comparison with the periodogram. For rectangular and Bartlett windows ($w_c[n]$), this factor is $2R/Q$ and $2R/3Q$, respectively. As in the Welch method, the resolution/variance trade-off is evident in Eqs. (16.43) and (16.44). For higher resolution, we need R to be large so that the window $W_c(e^{j\omega})$ is closer to an impulse function. But for small variance Eq. (16.44) demands that R be a small fraction of Q.

We now briefly summarize the computational load involved in the Blackman–Tukey, Bartlett, and Welch methods. All three methods utilize the DFT in their computation, which in practice is done via the FFT. This results in these methods being more computationally efficient than any of the parametric methods described in subsequent sections. Among these three methods, there are slight computational differences. The computation of the autocorrelation estimates required for the Blackman–Tukey method can be efficiently done using the FFT if the number of lag values needed is large. But if only a few lag estimates are required, then direct computation will be faster. The only difference between Bartlett and Welch methods is the increased number of segments in the latter due to overlap between successive segments. Hence the Welch estimator requires a greater number of FFT computations. For the NMR data in Example 16.4 considered earlier, the Welch method required fifteen 256-point FFT computations for 50% overlap. On the other hand, the Bartlett method would require eight such FFT computations since there are no overlaps. However, it should be noted that on the whole, there is little difference in the computational requirements among these three methods.

16-4 PARAMETRIC SPECTRUM ANALSYIS OF RANDOM SIGNALS

The Fourier analysis based, nonparametric methods discussed in the previous two sections are attractive because they are easy to compute, robust, and their statistical properties have been well studied. However, when the data record is short, their spectral resolution is severely limited by spectral leakage from one region in the frequency domain to its neighborhood. Also, for a data record length N, the nonparametric methods implicitly assume that the autocorrelation values $r_{xx}[m]$ for lag values $m > N$ are zero. This is an unrealistic assumption. The model based methods that we discuss in this section extrapolate the autocorrelation sequence to lag values beyond N. This is made possible by assuming a model for signal generation. The signal generation model may be a function of a number of parameters. These parameters in turn are estimated using the observed signal sequence $x[n]$, $n = 0$, $1, \ldots, N - 1$. Once the model parameters are estimated then the power spectral density implied by the model is computed.

16-4-1 Models for Random Processes

The parametric spectrum analysis methods described in this section are based on modeling the given data sequence $x[n]$ as the output of a linear, time-invariant, causal filter $H(z)$ whose input is a white noise sequence $e[n]$, with mean value zero and variance σ_e^2. The most general form for the transfer function, $H(z)$, is

$$
H(z) = \frac{A(z)}{B(z)}
$$

$$
= \frac{\sum\limits_{k=0}^{q} a_k z^{-k}}{1 + \sum\limits_{k=1}^{p} b_k z^{-k}}
$$

$$
= \sum\limits_{n=0}^{\infty} h[n] z^{-n}. \tag{16.45}
$$

The denominator polynomial $B(z)$ is assumed to have all its roots inside the unit circle in the z plane. $h[n]$ is the impulse response of the filter. The difference equation relating the input and output of the filter is

$$x[n] = -\sum_{k=1}^{p} b_k x[n-k] + \sum_{k=0}^{q} a_k e[n-k]. \tag{16.46}$$

The output sequence $x[n]$ is a wide-sense stationary process with power spectral density $\Phi_{xx}(e^{j\omega})$:

$$\begin{aligned} \Phi_{xx}(e^{j\omega}) &= \Phi_e(e^{j\omega})|H(e^{j\omega})|^2 \\ &= \sigma_e^2 \left|\frac{A(e^{j\omega})}{B(e^{j\omega})}\right|^2 \end{aligned} \tag{16.47}$$

where

$$A(e^{j\omega}) = \sum_{k=0}^{q} a_k e^{-j\omega k}, \qquad B(e^{j\omega}) = 1 + \sum_{k=1}^{q} b_k e^{-j\omega k}.$$

The random process $x[n]$ generated by the pole-zero filter is called an autoregressive moving average (ARMA) process of order (p, q). If $A(z) = 1$ $(q = 0)$, then $x[n]$ is an allpole or autoregressive (AR) process of order p. If $B(z) = 1$ $(p = 0)$, then $x[n]$ is generated by a filter

$$H(z) = \sum_{k=0}^{q} a_k z^{-k}$$

and is called a moving average (MA) process of order q. For the AR and MA processes the respective power spectral densities are as follows.

$$\text{MA:} \quad \Phi_x(e^{j\omega}) = \sigma_e^2 |A(e^{j\omega})|^2; \qquad \text{AR:} \quad \Phi_x(e^{j\omega}) = \frac{\sigma_e^2}{|B(e^{j\omega})|^2}. \tag{16.48}$$

The parametric methods of spectral analysis involve the following steps: selecting an appropriate model (AR, MA, or ARMA), selecting the orders of the filter $H(z)$ (i.e., p and/or q), and estimating the coefficients of $H(z)$ from the available data samples $x[n]$, $n = 0, 1, \ldots, N-1$, and evaluating the appropriate expression for $\Phi_{xx}(e^{j\omega})$ (Eq. (16.47) or (16.48)).

Selecting the model for a given application is difficult if nothing is known about the signal. If some knowledge about the way the signal is generated is available, then an informed choice can be made. For example, in speech signal analysis several resonant peaks are to be expected in the signal spectrum. In this case an allpole (AR) or pole-zero (ARMA) model is a good choice. On the other hand, if the true spectrum is likely to have "narrow valleys," then a MA or an ARMA model is more appropriate. Even when a wrong model choice is made, for example, an AR model is chosen instead of the correct ARMA model, a sufficiently high-order AR model can approximate an ARMA model. However, to estimate the large-order AR model coefficients we need a larger number of data samples.

The AR model is the most often invoked model. There are at least three reasons for this popularity. First, many methods for estimating the AR parameters involve solving only linear simultaneous equations. For this purpose, computationally efficient algorithms are available. Also, these algorithms compute the parameters of all AR models starting from order 1, 2, . . . , to p in a recursive manner. This helps us in choosing the model order that best explains the data. Second, many methods for estimating the AR parameters often result in a denominator polynomial $B(z)$, which has roots inside the unit circle. Although this property for $B(z)$ is not essential in spectral analysis, it is useful in other applications. Lastly, the AR spectral analysis methods are related to maximum entropy (ME) spectral analysis. The principle behind ME spectral analysis states that any method that attempts to estimate the power spectral density from a finite data record should do so such that the PSD estimate is consistent with the measured data but noncommital about the unobserved part of the data. This is in contrast with the Fourier methods discussed in the previous section, which assume that the data outside the observed interval are either periodic or zero. Therefore, it is intuitively satisfying that AR spectral analysis has at least a theoretical link to the maximum entropy principle of spectral analysis. Among the three models, the ARMA model is the most parsimonious. But unfortunately, the simultaneous estimation of the numerator and the denominator coefficients of an ARMA model requires the solution of nonlinear equations.

16-4-2 Relationships Between the Autocorrelation Sequence and the Model Filter Coefficients

The parametric methods described in this section rely on the relationships between the autocorrelation sequence and the coefficients of the model filter $H(z)$. The following paragraphs enlist these relationships. Let us first consider the general ARMA model. Multiplying Eq. (16.46) by $x^*[n - m]$ and taking expected values, we have

$$r_{xx}[m] = -\sum_{k=1}^{p} b_k r_{xx}[m - k] + \sum_{k=0}^{q} a_k r_{ex}[m - k], \qquad -\infty < m < +\infty,$$

(16.49)

where $r_{xx}[m] = \mathcal{E}(x^*[n]x[n + m])$ and $r_{ex}[m] = \mathcal{E}(x^*[n]e[n + m])$ is the cross-correlation between $x[n]$ and $e[n]$. Since $x[n] = e[n] \circledast h[n]$, $r_{ex}[m]$ may be expressed as

$$r_{ex}[m] = \sum_{k=0}^{\infty} h^*[k]\mathcal{E}(e^*[n - k]e[n + m])$$

$$= \sigma_e^2 h^*[-m],$$

(16.50)

where in the second step we have used the fact that $e[n]$ is a white noise sequence with variance σ_e^2.

Since $h[m]$ is a causal sequence, $r_{ex}[m]$ is zero for $m > 0$. Using this result in Eq. (16.50), we have the following relations for an ARMA model for various values of the index m:

$$
r_{xx}[m] = \begin{cases}
-\sum_{k=1}^{p} b_k r_{xx}[m-k] & \text{for } m > q, \\[2mm]
-\sum_{k=1}^{p} b_k r_{xx}[m-k] + \sigma_e^2 \sum_{k=0}^{q-m} h^*[k] a_{k+m} & \text{for } 0 \le m \le q, \\[2mm]
r_{xx}^*[-m] & \text{for } m < 0.
\end{cases}
$$

(16.51)

In words, the autocorrelation values $r_{xx}[m]$ for $|m| \ge q$ are completely determined by the coefficients of the denominator polynomial $B(z)$ and the autocorrelation values $r_{xx}[0]$, $r_{xx}[1]$, $\ldots$, $r_{xx}[q]$. Thus, when we invoke a model, the autocorrelation sequence $r_{xx}[m]$ is automatically extrapolated for all lag values.

If we restrict our model to an AR model then q is zero. Then we have

$$
r_{xx}[m] = \begin{cases}
-\sum_{k=1}^{p} b_k r_{xx}[m-k] & \text{for } m > 0, \\[2mm]
-\sum_{k=1}^{p} b_k r_{xx}[m-k] + \sigma_e^2 & \text{for } m = 0, \\[2mm]
r_{xx}^*[-m] & \text{for } m < 0.
\end{cases}
$$

(16.52)

In this case, if the autocorrelation values $r_{xx}[0]$, $r_{xx}[1]$, $\ldots$, $r_{xx}[p]$ are known, then the AR model filter coefficients can be obtained by solving the p linear equations corresponding to the indices $m = 1, 2, \ldots, p$ in Eq. (16.52).

$$
\begin{bmatrix}
r_{xx}[0] & r_{xx}[-1] & \cdots & r_{xx}[-p+1] \\
r_{xx}[1] & r_{xx}[0] & \cdots & r_{xx}[-p+2] \\
\vdots & \vdots & \ddots & \vdots \\
r_{xx}[p-1] & r_{xx}[p-2] & \cdots & r_{xx}[0]
\end{bmatrix}
\begin{bmatrix}
b_1 \\ b_2 \\ \vdots \\ b_p
\end{bmatrix}
= -
\begin{bmatrix}
r_{xx}[1] \\ r_{xx}[2] \\ \vdots \\ r_{xx}[p]
\end{bmatrix}.
$$

(16.53)

Note that all the filter coefficients are determined by knowing only $p + 1$ correlation values. Also, once the coefficients are found, the autocorrelation values can be extended to any index m. The variance σ_e^2 can also be obtained from Eq. (16.52) corresponding to $m = 0$:

$$
\sigma_e^2 = r_{xx}[0] + \sum_{k=1}^{p} b_k r_{xx}[-k].
$$

(16.54)

Combining Eqs. (16.53) and (16.54) into a single matrix equation, we have

$$
\begin{bmatrix}
r_{xx}[0] & r_{xx}[-1] & \cdots & r_{xx}[-p] \\
r_{xx}[1] & r_{xx}[0] & \cdots & r_{xx}[-p+1] \\
\vdots & \vdots & \ddots & \vdots \\
r_{xx}[p] & r_{xx}[p-1] & \cdots & r_{xx}[0]
\end{bmatrix}
\begin{bmatrix}
1 \\
b_1 \\
b_2 \\
\vdots \\
b_p
\end{bmatrix}
=
\begin{bmatrix}
\sigma_e^2 \\
0 \\
0 \\
\vdots \\
0
\end{bmatrix}. \tag{16.55}
$$

These equations are known as the *Yule–Walker* equations. The matrix on the left is a Toeplitz matrix, that is, a matrix that has the same entry along each diagonal.

Finally, if $H(z) = \sum_{k=0}^{q} a_k z^{-k}$ (i.e., $p = 0$), then we have the MA model. In this case,

$$
r_{xx}[m] = \begin{cases}
0 & \text{for } |m| > q, \\[2mm]
\sigma_e^2 \displaystyle\sum_{k=0}^{q-m} a_k^* a_{k+m} & \text{for } 0 \le m \le q, \\[2mm]
r_{xx}^*[-m] & \text{for } -q \le m < 0.
\end{cases} \tag{16.56}
$$

This follows from Eq. (16.51) by letting $b_k = 0$ for $k = 1, 2, \ldots, p$ and noting that $h[k] = a_k$ for $k = 0, 1, \ldots, q$. The above relationships between the auto-correlation values and the model filter coefficients are exploited by the parametric model based methods of spectrum analysis, which are discussed in Section 16-4-4. But first we derive an important recursive relationship that exists between the AR model coefficients and the autocorrelation values.

16-4-3 Levinson–Durbin Recursions

In the case of an AR model, if the true autocorrelation values $r_{xx}[0]$, $r_{xx}[1]$, $\ldots$, $r_{xx}[p]$ are known, the model coefficients $1, b_1, b_2, \ldots, b_p$ can be calculated from the Yule–Walker equations in Eq. (16.55). The Yule–Walker equations can be solved by using the standard techniques, such as Gaussian elimination, which require a number of operations proportional to p^3. An operation is defined as a complex multiplication plus a complex addition. However, a fast algorithm that makes use of the regular structure in Eq. (16.55) to compute the solutions using a number of operations proportional to only p^2 is available [DU73; LE47]. We briefly describe this algorithm in this section. The key feature of this algorithm is the recursive manner in which the coefficients of an AR model filter are calculated from those of the lower-order model.

Let us start with model order $p = 1$. In this case we have to determine only one coefficient $b_1^{(1)}$. The subscript corresponds to the coefficient number, while the superscript corresponds to the current model order. For $p = 1$ (see Eq. (16.53)),

$$
r_{xx}[0] b_1^{(1)} = -r_{xx}[1]. \tag{16.57}
$$

We will also denote the last coefficient of the jth order filter by K_j, that is, $K_j = b_j^{(j)}$. K_j is also called the jth reflection coefficient. For $p = 1$, b_1^1 is the last and only coefficient. Therefore

$$K_1 = b_1^{(1)} = \frac{r_{xx}[1]}{r_{xx}[0]}. \tag{16.58}$$

Note that $|K_1| \le 1$. We may also define a convenient set of intermediate quantities called prediction errors, $P_1, P_2, \ldots$. P_1 is defined as

$$P_1 = r_{xx}[0] + b_1^{(1)} r_{xx}^*[1]$$

$$= r_{xx}[0](1 - |K_1|^2) \tag{16.59}$$

since $r_{xx}[-m] = r_{xx}^*[m]$. Now consider $p = 2$. For this case the equations in Eq. (16.53) are as follows:

$$\begin{bmatrix} r_{xx}[0] & r_{xx}^*[1] \\ r_{xx}[1] & r_{xx}[0] \end{bmatrix} \begin{bmatrix} b_1^{(2)} \\ b_2^{(2)} \end{bmatrix} = - \begin{bmatrix} r_{xx}[1] \\ r_{xx}[2] \end{bmatrix} \tag{16.60}$$

$$\mathbf{R}_2 \mathbf{b}_2 = -\mathbf{r}_2. \tag{16.61}$$

Since $r_{xx}[1] = k_1 r_{xx}[0]$, $\mathbf{R}_2^{-1}$ is

$$\mathbf{R}_2^{-1} = \frac{1}{r_{xx}[0](1 - |K_1|^2)} \begin{bmatrix} 1 & b_1^{(1)} \\ (b_1^{(1)})^* & 1 \end{bmatrix}. \tag{16.62}$$

Also,

$$\mathbf{R}_2^{-1} = \frac{1}{P_1} \begin{bmatrix} 1 & b_1^{(1)} \\ (b_1^{(1)})^* & 1 \end{bmatrix}. \tag{16.63}$$

Using Eq. (16.63) in Eq. (16.60) we can compute K_2:

$$K_2 = b_2^{(2)} = \frac{-(r_{xx}[1] + b_1^{(1)})}{P_1}. \tag{16.64}$$

Now we can express the coefficients of the second-order model in terms of the first-order model and the second reflection coefficient, K_2, as follows:

$$\mathbf{b}_2 = \begin{bmatrix} b_1^{(2)} \\ b_2^{(2)} \end{bmatrix} = \begin{bmatrix} b_1^{(1)} \\ 0 \end{bmatrix} + \begin{bmatrix} K_2 (b_1^{(1)})^* \\ K_2 \end{bmatrix}. \tag{16.65}$$

In general, the coefficients of the pth order filter may be expressed in terms of the $(p - 1)$th order filter and the reflection coefficient K_p as

$$\mathbf{b}_p = \begin{bmatrix} b_1^{(p)} \\ b_2^{(p)} \\ \vdots \\ b_p^{(p)} \end{bmatrix} = \begin{bmatrix} \mathbf{b}_{p-1} \\ \hline 0 \end{bmatrix} + \begin{bmatrix} K_p \mathbf{c}_{p-1} \\ \hline K_p \end{bmatrix}. \tag{16.66}$$

The elements of $\mathbf{b}_{p-1}$ are the coefficients of the $(p-1)$th order filter. Assume that the filter vector $\mathbf{b}_{p-1}$ and the minimum prediction error P_{p-1} have already been computed. To compute the pth order filter we need K_p and the $(p - 1)$-element vector $\mathbf{c}_{p-1}$. Let the $p \times p$ matrix $\mathbf{R}_p$ in Eq. (16.55) be partitioned as follows:

$$\mathbf{R}_p = \begin{bmatrix} \mathbf{R}_{p-1} & \mathbf{J}_{p-1}\mathbf{r}_{p-1}^* \\ \mathbf{r}_{p-1}^t\mathbf{J}_{p-1} & r_{xx}[0] \end{bmatrix}, \tag{16.67}$$

where $\mathbf{J}_{p-1}$ is a $(p - 1) \times (p - 1)$ exchange matrix. An exchange matrix has ones along the cross-diagonal and zeros elsewhere:

$$\mathbf{J}_{p-1} = \begin{bmatrix} 0 & 0 & \cdots & 1 \\ 0 & 0 & \cdots & 0 \\ \vdots & \vdots & & \vdots \\ 0 & 1 & \cdots & \vdots \\ 1 & 0 & \cdots & 0 \end{bmatrix}. \tag{16.68}$$

Note that premultiplying a vector by $\mathbf{J}_{p-1}$ reorders its elements in the reverse order; that is, $\mathbf{J}_{p-1}(x_1, x_2, \ldots, x_{p-1})^t = (x_{p-1}, x_{p-2}, \ldots, x_2, x_1)^t$. Plugging Eqs. (16.67) and (16.66) into Eq. (16.53), we have

$$\begin{bmatrix} \mathbf{R}_{p-1} & \mathbf{J}_{p-1}\mathbf{r}_{p-1}^* \\ \mathbf{r}_{p-1}^t\mathbf{J}_{p-1} & r_{xx}[0] \end{bmatrix} \left\{ \begin{bmatrix} \mathbf{b}_{p-1} \\ 0 \end{bmatrix} + \begin{bmatrix} K_p\mathbf{c}_{p-1} \\ K_p \end{bmatrix} \right\} = -\begin{bmatrix} \mathbf{r}_{p-1} \\ r_{xx}[p] \end{bmatrix} = -\mathbf{r}_p. \tag{16.69}$$

Performing the matrix-vector multiplication on the left side of Eq. (16.69) and comparing the result with the right side, we have the following two equations:

$$\mathbf{R}_{p-1}\mathbf{b}_{p-1} + K_p\mathbf{R}_{p-1}\mathbf{c}_{p-1} + K_p\mathbf{J}_{p-1}\mathbf{r}_{p-1}^* = -\mathbf{r}_{p-1}, \tag{16.70}$$

$$\mathbf{r}_{p-1}^t\mathbf{J}_{p-1}\mathbf{b}_{p-1} + K_p r_{xx}[0] + K_p\mathbf{r}_{p-1}^t\mathbf{J}_{p-1}\mathbf{c}_{p-1} = -r_{xx}[p].$$

Since $\mathbf{R}_{p-1}\mathbf{b}_{p-1} = -\mathbf{r}_{p-1}$ from the first equation above, we have

$$\mathbf{R}_{p-1}\mathbf{c}_{p-1} = -\mathbf{J}_{p-1}\mathbf{r}_{p-1}^*. \tag{16.71}$$

Let $\mathbf{I}_{p-1}$ denote the $(p-1) \times (p-1)$ identity matrix. Note that $\mathbf{J}_{p-1}\mathbf{J}_{p-1} = \mathbf{I}_{p-1}$. Also $\mathbf{J}_{p-1}\mathbf{R}_{p-1}\mathbf{J}_{p-1} = \mathbf{R}_{p-1}^*$. Using these two relationships in Eq. (16.71), we observe that

$$\mathbf{c}_{p-1} = \mathbf{J}_{p-1}\mathbf{b}_{p-1}^*. \tag{16.72}$$

That is, the vector $\mathbf{c}_{p-1}$ is the $\mathbf{b}_{p-1}$ vector with its elements reversed in order and complex conjugated. Substituting Eq. (16.72) in the second equation in Eq. (16.70) and noting that the prediction error for model order $p-1 = r_{xx}[0] + \mathbf{r}_{p-1}^t\mathbf{b}_{p-1}^*$, we have

$$K_p = b_p^p = -\frac{r_{xx}[p] + \mathbf{r}_{p-1}^t\mathbf{J}_{p-1}\mathbf{b}_{p-1}}{P_{p-1}}. \tag{16.73}$$

Note that again K_p depends only on the known quantities, namely, P_{p-1} and $\mathbf{b}_{p-1}$. Substituting for $\mathbf{c}_{p-1}$ in Eq. (16.66) we can write the pth order model coefficient vector $\mathbf{b}_p$ in terms of the known quantities as follows:

$$[\mathbf{b}_p] = \left[\begin{array}{c} \mathbf{b}_{p-1} \\ \text{-----} \\ 0 \end{array}\right] + \left[\begin{array}{c} K_p\mathbf{J}_{p-1}\mathbf{b}_{p-1} \\ \text{---------------} \\ K_p \end{array}\right]. \tag{16.74}$$

Then the only quantity left is P_p. We can also express P_p in terms of P_{p-1}. Since $P_p = r_{xx}[0] + \mathbf{r}_p^t\mathbf{b}_p^*$, using Eqs. (16.73) and (16.74) we can express P_p as follows:

$$P_p = P_{p-1}(1 - |K_p|^2). \tag{16.75}$$

Summarizing, the steps involved in Levinson–Durbin recursions are:

1. Initialize the recursions with $b_1^1 = K_1 = r_{xx}[1]/r_{xx}[0]$ and $P_1 = r_{xx}[0](1 - |K_1|^2)$.
2. For $i > 0$, compute the $(i + 1)$th reflection coefficient using Eq. (16.73) (with p replaced by $i + 1$), which in expanded form is

$$K_{i+1} = b_{i+1}^{i+1} = \frac{r_{xx}[i+1] + \sum_{j=1}^{i} b_j^i r_{xx}[i+1-j]}{P_i}. \tag{16.76}$$

3. Compute the jth coefficient ($j \leq i$) of the $(i + 1)$th order model using Eq. (16.74) (with p replaced by $i + 1$):

$$b_j^{i+1} = b_j^i + K_{i+1}(b_{i+1-j}^i)^*.$$ (16.77)

4. Compute the prediction error P_{i+1} using the formula

$$P_{i+1} = P_i(1 - |K_i|^2).$$ (16.78)

5. Repeat steps 2 to 4 until the coefficient vector of the desired model order is computed.

The Levinson recursions lead to a number of important properties of the resulting filter $B(z)$. First, if the autocorrelation values $r_{xx}[0]$, $r_{xx}[1]$, ... correspond to a pth order AR process, then the recursions terminate with an order p. That is, if we seek a filter of order higher than p, the higher order reflection coefficients $K_{p+1}, K_{p+2}, \ldots$ are equal to zero and hence $P_{p+1} = P_{p+2} = \cdots = P_p$. However, if the autocorrelation values do not correspond to a strictly autoregressive process of order p but to a MA or an ARMA process, then the Levinson recursions continue indefinitely. The prediction error power tends to zero as $p \rightarrow \infty$. Second, the structure of the recursive equations in the Levinson algorithm leads to a special form of digital filter implementation called the "lattice structure" [FR82]. Finally, it can be shown that the magnitude of the reflection coefficients K_i is always less than unity. This causes the roots of the polynomial $B(z)$ to always lie inside the unit circle [PA83].

16-4-4 Estimation of Model Parameters and Power Spectral Densities

Once a model is invoked the estimation of the power spectral density involves estimating the parameters of the model filter coefficients $H(z)$ from the given data samples, $x[n]$, $n = 0, 1, \ldots, N - 1$. This problem can be approached from the point of view of standard statistical procedures such as maximum-likelihood estimation by assuming a probability density function for the observations. However, the resulting estimation techniques generally require solution of nonlinear equations. In this section, we describe simple and effective, although suboptimal, power spectrum estimation methods using the models described above, which have proved useful.

16-4-4-1 AR Modeling: Yule–Walker Method and Linear Prediction. A popular method for estimating the parameters of an AR model is the Yule–Walker method. It makes use of the estimates of the autocorrelation sequence computed from the data samples. The autocorrelation value at the kth lag is estimated from the given data samples $x[0], x[1], \ldots, x[N - 1]$ as follows:

$$\hat{r}_{xx}(k) = \frac{1}{N} \sum_{n=0}^{N-1-|k|} x^*[n]x[n + k], \qquad k = 0, 1, \ldots, N - 1. \quad (16.79)$$

These autocorrelation estimates are then used in Eq. (16.53) in place of the true autocorrelation values, which are often unavailable, resulting in the following linear equations:

$$
\begin{bmatrix}
\hat{r}_{xx}[0] & \hat{r}_{xx}[-1] & \cdots & \hat{r}_{xx}[-p+1] \\
\hat{r}_{xx}[1] & \hat{r}_{xx}[0] & \cdots & \hat{r}_{xx}[-p+2] \\
\vdots & \vdots & \ddots & \vdots \\
\hat{r}_{xx}[p-1] & \hat{r}_{xx}[p-2] & \cdots & \hat{r}_{xx}[0]
\end{bmatrix}
\begin{bmatrix}
\hat{b}_1 \\
\hat{b}_2 \\
\vdots \\
\hat{b}_p
\end{bmatrix}
= -
\begin{bmatrix}
\hat{r}_{xx}[1] \\
\hat{r}_{xx}[2] \\
\vdots \\
\hat{r}_{xx}[p]
\end{bmatrix}.
$$

(16.80)

The $\hat{b}_j$ terms represent the estimated AR filter coefficients. Since the structure of the equations in Eq. (16.80) is the same as in Eq. (16.53), the Levinson recursions described in Section 16-4-3 can be used to solve these equations. If we denote the corresponding prediction error for a pth order filter as $\hat{P}_p$, then the corresponding power spectrum estimate is

$$
\hat{\Phi}_{xx}[e^{j\omega}] = \frac{\hat{P}_p}{\left| 1 + \displaystyle\sum_{k=1}^{p} \hat{b}_k e^{-j\omega k} \right|^2},
$$

(16.81)

where

$$
\hat{P}_p = \hat{r}_{xx}[0] \prod_{i=1}^{p} (1 - |\hat{K}_i|^2)
$$

(see Eq. (78)).

We now show that the same linear equations (as in Eq. (16.80)) result, if we solve a related one-step linear prediction problem. Consider the given sequence of samples $x[0], x[1], \ldots, x[N-1]$ and assume that zeros are appended before the first sample $x[0]$ and after the last sample $x[N-1]$. We shall attempt to predict the nth sample as a linear combination of the previous p consecutive samples. Denote the predicted value of $x[n]$ by $\hat{x}[n]$ and let $e_f[n]$ be the error in the predicted value of the nth sample. Then

$$
\hat{x}[n] = -\sum_{s=1}^{p} \hat{b}_s x[n-s]
$$

$$
= x[n] - e_f[n].
$$

(16.82)

The following equation shows all the nontrivial prediction equations for the given data sequence:

$$
x[n] + \sum_{s=1}^{p} x[n-s]\hat{b}_s = e_f[n], \qquad n = 0, 1, \ldots, N+p-1. \quad (16.83)
$$

We can minimize the sum of squares of the prediction errors $(1/N)\sum_{n=0}^{N+p-1} |e_f[n]|^2$ by finding the best linear prediction coefficients, $\hat{b}_s$. Standard least-squares minimization leads to the following set of linear simultaneous (normal) equations:

$$\frac{1}{N} \sum_{s=1}^{p} \left\{ \sum_{n=0}^{N-1-|l-s|} x^*[n]x[n+l-s] \right\} \hat{b}_s$$

$$= -\frac{1}{N} \sum_{l=0}^{N-1-l} x[n]x^*[n+l], \qquad l = 1, 2, \ldots, p. \qquad (16.84)$$

One may easily verify that these equations are the same as those in Eq. (16.53). In other words, the linear prediction filter that minimizes the prediction error is also the AR model filter obtained by solving the Yule–Walker equations using the autocorrelation estimates in Eq. (16.53). The minimum value of the linear prediction error can be expressed as

$$\left\{ \frac{1}{N} \sum_{n=0}^{N+p-1} |e_f[n]|^2 \right\}_{\min} = \hat{r}_{xx}[0] + \sum_{s=1}^{p} \hat{b}_s \hat{r}_{xx}^*[s]. \qquad (16.85)$$

This quantity is equal to $\hat{P}_p$ and is an estimate of σ_e^2. In speech signal processing literature this method is called the "autocorrelation" method of linear prediction [MA75]. A significant advantage of this method in speech applications is the fact that the filter $\hat{B}(z) = 1 + \hat{b}_1 z^{-1} + \hat{b}_2 z^{-2} + \cdots + \hat{b}_p z^{-p}$ has all its roots inside the unit circle in the z plane; that is, the AR model implied by $1/\hat{B}(z)$ is always stable. Another advantage is the computational savings resulting from the use of Levinson's recursions. However, if N is much larger than p, then the load involved in computing the autocorrelation estimates in Eq. (16.79) will dominate.

A number of other related linear prediction methods are also available for AR modeling. In the "covariance" method of linear prediction [MA75] the prediction error is minimized over the indices $n = p$ to $N - 1$; that is, $\sum_{n=p}^{N-1} |e_f[n]|^2$ is minimized, where $e_f[n]$ is defined in Eq. (16.83). In the "modified covariance" method [NU76; UL76] the sum of squared forward and backward prediction errors are minimized. That is, we minimize

$$\sum_{n=p}^{N-1} (|e_f[n]|^2 + |e_b[n]|^2), \qquad (16.86)$$

where the first summation consists of the forward prediction error samples as before and the second term consists of the backward prediction error samples defined as follows:

$$e_b[n] = x^*[n-p] + \sum_{s=1}^{p} x^*[n+s-p]\hat{b}_s, \qquad n = p, p+1, \ldots, N-1. \qquad (16.87)$$

The backward prediction error samples are obtained by performing linear prediction on time-reversed, complex-conjugated data samples. These linear prediction methods are closely related to Prony's method and are discussed in the next section. Fast Levinson-like algorithms for solving the normal equations that result when using the covariance or modified covariance method are also available in the literature [MA80; MC88; MO77].

16-4-4-2 Maximum Entropy Spectral Analysis and Burg's Method.
Maximum entropy spectral analysis is based on extrapolating a known segment of autocorrelation values $r_{xx}[0]$, $r_{xx}[1]$, ..., $r_{xx}[p]$ of a sequence $x[n]$ to indices for which the autocorrelation values are not known, that is, $r_{xx}[p + 1]$, $r_{xx}[p + 2]$, ... Of course, the Fourier transform of the extrapolated autocorrelation sequence should be nonnegative so that it can be a legitimate power spectral density. Given the first $p + 1$ autocorrelation sequence values, there are an infinite number of possible extrapolations that will yield a valid power spectral density function. In maximum entropy spectral analysis it is argued that the extrapolation of the autocorrelation sequence is such that it does not add any new information arbitrarily to the underlying sequence. Information is measured in terms of Shannon's entropy. In particular, for a Gaussian random process $x[n]$, the entropy rate or entropy per sample is proportional to

$$\int_{-\pi}^{\pi} \ln \left(\Phi_{xx}(e^{j\omega}) \right) d\omega. \tag{16.88}$$

The maximum entropy spectrum maximizes the entropy rate subject to the constraint that

$$\frac{1}{2\pi} \int_{-\pi}^{\pi} \Phi_{xx}(e^{j\omega}) e^{j\omega n} d\omega = r_{xx}[n], \qquad |n| \le p. \tag{16.89}$$

These constraints ensure that the first $p + 1$ autocorrelation values corresponding to the maximum entropy spectrum coincide with the given autocorrelation values $r_{xx}[0]$, $r_{xx}[1]$, ..., $r_{xx}[p]$. This constrained maximization problem can be solved using Lagrange multipliers resulting in the following solution [JA82; RO82]:

$$\Phi_{xx}(e^{j\omega}) = \frac{P_p}{\left| 1 + \sum_{k=1}^{p} b_k e^{-j\omega k} \right|^2}, \tag{16.90}$$

where the coefficients b_k, $k = 1, 2, \ldots, p$, and P_p are found by solving the Yule–Walker equations in Eq. (16.52) using the known autocorrelation values $r_{xx}[0]$, $r_{xx}[1]$, ..., $r_{xx}[p]$. Therefore it turns out that the maximum entropy spectrum is indeed identical to the AR spectrum.

The promise of maximum entropy spectral analysis is invariably never delivered in practice because the true autocorrelation values may not be known, even if we assume that the data are stationary, which may not be realistic. If we substitute the estimated autocorrelation values, we implicitly assume that the data are zero outside the given interval. In an attempt to circumvent this dilemma, Burg devised a clever method [BU67] that does not require the explicit computation of the autocorrelation estimates of the data. In Burg's method the sum of the squared forward and backward prediction errors (with an ith order linear prediction filter),

$$P_i = \sum_{n=i}^{N-1} (|e_f^i[n]|^2 + |e_b^i[n]|^2), \tag{16.91}$$

is minimized with respect to the prediction coefficients. The following are the expressions for the forward and backward prediction errors for an ith order filter:

$$e_f^i[n] = x[n] + \sum_{s=1}^{i} \hat{b}_s^i x[n-s], \tag{16.92}$$

$$e_b^i[n] = x^*[n-i] + \sum_{s=1}^{i} \hat{b}_s^i x^*[n+s-i]. \tag{16.93}$$

Note that these errors are the same as the ones encountered in the "modified covariance" method in Eq. (16.86). However, unlike the "modified covariance" method, Burg's method requires that the prediction coefficients satisfy the Levinson recursions (see Eq. (16.77)) given by

$$\hat{b}_j^i = \hat{b}_j^{i-1} + \hat{K}_i (\hat{b}_{i-j}^{i-1})^*, \quad \text{for } 1 \le j \le i-1 \text{ and } 1 \le i \le p. \tag{16.94}$$

The hat ($\hat{}$) indicates estimated prediction and reflection coefficients. At each value of i ranging from 0 to p the error P_i is minimized with respect to the complex-valued reflection coefficients $\hat{K}_i$. This results in the following expression for $\hat{K}_i$:

$$\hat{K}_i = \frac{-2 \sum\limits_{n=i}^{N-1} e_f^i[n](e_b^i[n-1])^*}{\sum\limits_{n=i}^{N-1} (|e_f^i[n]|^2 + |e_b^i[n-1]|^2)}. \tag{16.95}$$

Note that the numerator corresponds to the cross-correlation between the forward and the backward errors and the denominator is the sum of the error energies. It is clear that $|\hat{K}_i| < 1$ and hence the AR model implied by Burg's method is always stable. Burg's algorithm is initialized with $e_f^0[n] = x[n]$ and $e_b^0[n] = x[n]$. Also, if we denote the minimum error at each i as $\hat{P}_i$, then $\hat{P}_i$ is initialized with

$$\hat{P}_0 = \frac{1}{N} \sum_{n=0}^{N-1} |x[n]|^2.$$

Then for all i between 1 and p the $\hat{K}_i$ are computed using Eq. (16.95). Note that by definition $\hat{K}_i = \hat{b}_i^i$. All other coefficients of the ith order filter are computed using the Levinson recursions in Eq. (16.94). The $\hat{P}_i$ are updated using $\hat{P}_i = (1 - |\hat{K}_i|^2)\hat{P}_{i-1}$. Furthermore, the forward and the backward errors can be written in order recursively as follows:

$$e_f^i[n] = e_f^{i-1}[n] + \hat{K}_i e_b^{i-1}[n-1],$$
$$e_b^i[n] = e_b^{i-1}[n-1] + \hat{K}_i^* e_f^{i-1}[n].$$

(16.96)

Once the coefficients of the pth order filter are computed, Burg's power spectrum estimate is given by

$$\hat{\Phi}_{xx}(e^{j\omega}) = \frac{\hat{P}_p}{\left|1 + \sum\limits_{k=1}^{p} \hat{b}_k^p e^{-j\omega k}\right|^2}.$$

(16.97)

16-4-4-3 AR Modeling: Model Order Selection. The best choice of the AR filter order is usually not known ahead of time. Whereas a low model order produces a smooth spectral estimate, too high a model order tends to introduce spurious detail in the spectral estimate. Increasing the AR model order causes the prediction error power to decrease. Unfortunately, the prediction error power decreases monotonically with respect to the model order and hence it does not clearly indicate when to stop the search procedure.

Several criteria are available for AR model selection. Akaike [AK69; AK74] has provided two different criteria. The first criteria is called the *final prediction error* (FPE) criterion. The FPE for an AR process is defined as

$$\text{FPE}_{\hat{p}} = \hat{P}_{\hat{p}}\left(\frac{N + \hat{p} - 1}{N - \hat{p} - 1}\right),$$

(16.98)

where N is the number of data samples, $\hat{p}$ is the AR filter order, and $\hat{P}_{\hat{p}}$ is the corresponding prediction error power. Since $\hat{P}_{\hat{p}}$ decreases with increasing p and the other term increases with increasing $\hat{p}$, the $\text{FPE}_{\hat{p}}$ will reach a minimum at some $\hat{p}$. This order is chosen as the best estimate of the true AR filter order p, using the FPE criterion.

The second criterion due to Akaike is based on the minimization of the log-likelihood of the prediction error power as a function of the filter order $\hat{p}$. This criterion, called the *Akaike information theoretic criterion* (AIC) is defined by

$$\text{AIC}_{\hat{p}} = \ln(\hat{P}_{\hat{p}}) + \frac{2\hat{p}}{N}.$$

(16.99)

The second term in the $\text{AIC}_{\hat{p}}$ may be thought of as the penalty for increasing the order of the AR filter, whereas the first term again decreases monotonically with increasing $\hat{p}$. As N tends to infinity the $\text{AIC}_{\hat{p}}$ and $\text{FPE}_{\hat{p}}$ are equivalent.

A third criterion is due to Rissanen [RI83]. It is based on selecting the order that minimizes the description length (MDL), where MDL is defined as

$$\text{MDL}_{\hat{p}} = N \ln \hat{P}_{\hat{p}} + \hat{p} \ln N. \tag{16.100}$$

Yet another criterion is called CAT or *criterion autoregressive transfer* function [PA74] and is defined by

$$\text{CAT}_{\hat{p}} = \left(\frac{1}{N} \sum_{j=1}^{\hat{p}} \frac{1}{\hat{P}_j} - \frac{1}{\hat{P}_p} \right), \tag{16.101}$$

where

$$\hat{P}_j = \frac{N}{N-j} P_j.$$

Again $\hat{p}$ is chosen to minimize $\text{CAT}_{\hat{p}}$.

None of the order selection criteria appear to be well suited for determining the model order in all cases. For short data segments, filter orders between $N/3$ and $N/2$ appear to produce satisfactory results. Unfortunately, subjective judgment is required in the order selection process especially in the absence of additional information about the physics of signal generation.

16-4-4-4 MA Modeling: Estimating the MA Filter Parameters. The power spectral density corresponding to a MA (q) process may be written as

$$\Phi_{xx}(e^{j\omega}) = \sigma_e^2 \left| \sum_{k=0}^{q} a_k e^{-j\omega k} \right|^2. \tag{16.102}$$

This is also equal to

$$\Phi_{xx}(e^{j\omega}) = \left| \sum_{k=-q}^{q} r_{xx}[k] e^{-j\omega k} \right|^2. \tag{16.103}$$

An obvious estimate of the MA power spectral density is

$$\hat{\Phi}_{xx}(e^{j\omega}) = \left| \sum_{k=-q}^{q} \hat{r}_{xx}[k] e^{-j\omega k} \right|^2. \tag{16.104}$$

where $\hat{r}_{xx}[k]$ is the usual autocorrelation estimate (biased or unbiased) computed from the data samples. Note that the above estimate of the power spectral density does not require an estimate of the MA filter coefficients. Also note that the MA power spectral density estimate above closely resembles the Blackman–Tukey power spectrum estimate given in Eq. (16.41).

Durbin [DU59] has developed an approximate maximum-likelihood method for estimating the MA filter coefficients from the given data samples. See also Kay [KA88]. Durbin's method relies on the fact that the MA filter $A(z) = a_0 + a_1 z^{-1} + a_2 z^{-2} + \cdots + a_q z^{-q}$ may be approximated by an AR filter of large order L. That is,

$$A(z) \approx \frac{1}{B(z)}.$$

Durbin's method consists of two steps. In the first step, Yule–Walker equations (see Eq. (16.80)) are used to estimate the coefficients of the long AR filter $B(z)$ from the data samples. Of course this assumes that the data length N is much larger than L. Let us call the estimates of the coefficients $1, \hat{b}_1, \hat{b}_2, \ldots, \hat{b}_L$. In the second step, using this sequence of coefficients as data samples, the Yule–Walker equations are again solved to obtain an AR(q) model. q is usually far less than L. The coefficients of this AR(q) model are the estimates of the desired MA parameters. These parameters may be used in Eq. (16.102) along with the prediction error estimate obtained in the first step to estimate the MA power spectrum.

16-4-4-5 ARMA Modeling: Estimating the ARMA Filter Parameters.

In the case of an ARMA(p, q) model the power spectral density estimate is defined as follows:

$$\hat{\Phi}_{xx}(e^{j\omega}) = \hat{\sigma}_e^2 \frac{\left| 1 + \sum_{k=1}^{q} \hat{a}_k e^{-j\omega k} \right|^2}{\left| 1 + \sum_{k=1}^{p} \hat{b}_k e^{-j\omega k} \right|^2}, \tag{16.105}$$

where the parameters $\hat{a}_k$, $\hat{b}_k$, and $\hat{\sigma}_e^2$ have to be estimated from the given data samples. Approximate maximum-likelihood methods for estimating these parameters have been described by Kay [KA88]. But these methods involve solution of nonlinear equations using iterative methods.

A simple but suboptimal approach is to make use of the relationships between the autocorrelation values and the assumed model parameters described in Eq. (16.51). Note that from Eq. (16.51)

$$-\sum_{k=1}^{p} b_k r_{xx}[m - k] = r_{xx}[m], \qquad m > q, \tag{16.106}$$

or, in matrix notation,

$$
\begin{bmatrix}
r_{xx}[q] & r_{xx}[q-1] & \cdots & r_{xx}[q-p+1] \\
r_{xx}[q+1] & r_{xx}[q] & \cdots & r_{xx}[q-p+2] \\
\vdots & \vdots & \ddots & \vdots \\
r_{xx}[q+p-1] & r_{xx}[q+p-2] & \cdots & r_{xx}[q]
\end{bmatrix}
\begin{bmatrix}
b_1 \\
b_2 \\
\vdots \\
b_p
\end{bmatrix}
$$

$$
= - \begin{bmatrix}
r_{xx}[q+1] \\
r_{xx}[q+2] \\
\vdots \\
r_{xx}[q+p]
\end{bmatrix}. \tag{16.107}
$$

These p equations have been selected from Eq. (16.51) in such a way that they do not involve the numerator coefficients of the ARMA model. These are sometimes known as the *modified Yule–Walker equations*. The estimates of the denominator coefficients of the ARMA model ($\hat{b}_k$'s) may be obtained by solving the above equations after using the estimated autocorrelation values in place of the true auto-correlation values.

Often, better estimates of the $\hat{b}_k$ values may be obtained by solving an over-determined set of linear equations in the least-square sense as follows. Let $\hat{r}_{xx}[m]$, $m = q + 1, q + 2, \ldots, M$, be the autocorrelation estimates obtained from the data samples. M is assumed less than N, the number of data samples, but greater than $q + p$. Then, analogous to Eq. (16.107), we have the following set of equations:

$$
\begin{bmatrix}
\hat{r}_{xx}[q] & \hat{r}_{xx}[q-1] & \cdots & \hat{r}_{xx}[q-p+1] \\
\hat{r}_{xx}[q+1] & \hat{r}_{xx}[q] & \cdots & \hat{r}_{xx}[q-p+2] \\
\vdots & \vdots & \ddots & \vdots \\
\hat{r}_{xx}[M-1] & \hat{r}_{xx}[M-2] & \cdots & \hat{r}_{xx}[M-p-1]
\end{bmatrix}
\begin{bmatrix}
\hat{b}_1 \\
\hat{b}_2 \\
\vdots \\
\hat{b}_p
\end{bmatrix}
$$

$$
\approx - \begin{bmatrix}
\hat{r}_{xx}[q+1] \\
\hat{r}_{xx}[q+2] \\
\vdots \\
\hat{r}_{xx}[M]
\end{bmatrix}. \tag{16.108}
$$

If we call the $(M - q) \times (p)$ matrix on the left $\mathbf{R}'$, and the vector on the right $\mathbf{r}'$, then the least-squares solution for the coefficient vector $\hat{\mathbf{b}}$ is given by $\hat{\mathbf{b}} = (\mathbf{R}'^\dagger \mathbf{R}')^{-1} \mathbf{R}'^\dagger \mathbf{r}'$. This method is sometimes called the least-squares modified Yule–

Walker method. M is chosen to be greater than $q + p$ but significantly less than N to eliminate unreliable autocorrelation lag estimates. Also, Eq. (16.108) may be premultiplied by a diagonal matrix to deemphasize the estimates for large lag indices. Once the denominator coefficients are estimated, the sequence $x[n]$ is filtered by the FIR filter $\hat{B}(z) = 1 + \hat{b}_1 z^{-1} + \hat{b}_2 z^{-2} + \cdots + \hat{b}_p z^{-p}$. If we call the resulting sequence $x'[n]$, then $x'[n]$ is essentially a MA process of order q, assuming that the orders p and q have been correctly chosen. Then the MA spectrum is obtained by computing the autocorrelation estimates $\hat{r}_{x'x'}[m]$, $m = 0, 1, \ldots, q$, of the sequence $x'[n]$. An autocorrelation window (such as Bartlett's window) may be used to deemphasize the correlation estimates at large lags. Finally, the ARMA power spectrum estimate is

$$\hat{\Phi}_{xx}(e^{j\omega}) = \frac{\displaystyle\sum_{m=-q}^{q} \hat{r}_{x'x'}[m] e^{-j\omega m}}{\left| 1 + \displaystyle\sum_{k=1}^{p} \hat{b}_k e^{-j\omega k} \right|^2}. \tag{16.109}$$

In the above method one need not estimate the numerator parameters and σ_e^2 explicitly. Alternatively, Durbin's method [DU59; KA88] described briefly in the previous subsection may be used to model the sequence $x'[n]$ by a qth order MA model. The order selection for the ARMA(p, q) model has been addressed by Bruzzone and Kaveh [BR80b] and Chow [CH72]. For an ARMA process the Akaike information criterion (AIC) given in Eq. (16.99) may be modified to include the numerator order as follows:

$$\text{AIC}_{(\hat{p},\hat{q})} = \ln(\hat{P}_{(\hat{p},\hat{q})}) + 2 \frac{(\hat{p} + \hat{q})}{N}. \tag{16.110}$$

The minimum value of $\text{AIC}_{(\hat{p},\hat{q})}$ gives the estimates of the numerator and denominator orders.

Example 16.5. In this example we apply various parametric spectral analysis methods described in this section to some computer generated data. The data samples $x[n]$ were generated by exciting a pole-zero filter $H(z)$ by a real-valued, white Gaussian noise sequence, $e[n]$, with variance σ_e^2. The filter system function $H(z)$ is

$$H(z) = \frac{a_0 + a_1 z^{-1} + a_2 z^{-2}}{1 + b_1 z^{-1} + b_2 z^{-2} + b_3 z^{-3} + b_4 z^{-4}}.$$

The numerator coefficients are $a_0 = 1$, $a_1 = 0.618034$, and $a_2 = 1$ and the denominator coefficients b_1, b_2, b_3, and b_4 are -1.991603, 2.760220, -1.912735, and 0.922368, respectively. The filter $H(z)$ has two zeros on the unit circle at frequencies 0.3 and -0.3 and poles at frequencies 0.2, -0.2, 0.125, and -0.125, all having a radius of 0.98. For each method, in each trial, 300 data samples of $x[n]$ are made available for processing. The number of trials is 10. For each trial the input sequence $e[n]$ is different. The true power spectral density $S_{xx}(\omega)$ is

$$S_{xx}(e^{j\omega}) = \sigma_e^2 \left| \frac{a_0 + a_1 e^{-j\omega} + a_2 e^{-j2\omega}}{1 + b_1 e^{-j\omega} + b_2 e^{-j2\omega} + b_3 e^{-j3\omega} + b_4 e^{-j4\omega}} \right|^2,$$

which is shown in Figure 16-6(a). For Burg, autocorrelation/Yule–Walker, and modified covariance methods, model orders of 4 and 16 were used. The estimated PSDs are shown in Figures 16-6(b)–(g). Clearly, the spectral peaks are well estimated but the spectral nulls are not. As the model order is increased from 4 to 16, the spectral nulls tend to appear in the estimated spectrum. However, there is increased variability in the spectrum when the model order is high because the

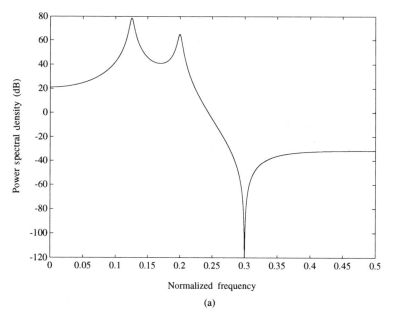

(a)

FIGURE 16-6 (a) The true power spectral density (PSD) $S_{xx}(\omega)$ of the sequence $x[n]$ output by the filter $H(z)$ when excited by a white noise sequence $e[n]$. The estimated PSDs obtained by using various methods are given in (b)–(h). In all cases 300 samples of $x[n]$ were used to estimate the PSD. (b) Burg's method (see Eq. (16.97)), model order is 4. (c) Burg's method (see Eq. (16.97)), model order is 16. (d) Autocorrelation/Yule–Walker method (see Eq. (16.81)), model order is 4. (e) Autocorrelation/Yule–Walker method (see Eq. (16.81)), model order is 16. (f) Modified covariance method. In this method both forward and backward linear prediction errors are minimized as in Eq. (16.86). The model order is 4. (g) Modified covariance method. The model order is 16. (h) Least-squares modified Yule–Walker method (see Eq. (16.109)). The denominator model order is 4 and the numerator order is 2. Unlike in Eq. (16.109), Durbin's method was used to determine the numerator.

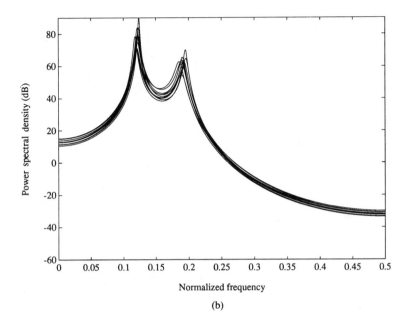

(b)

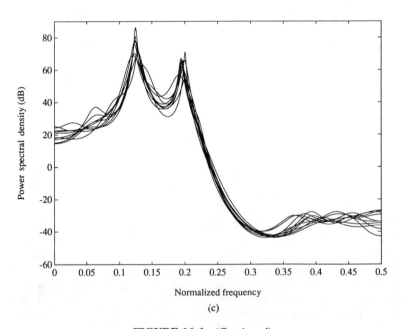

(c)

FIGURE 16-6   (*Continued*)

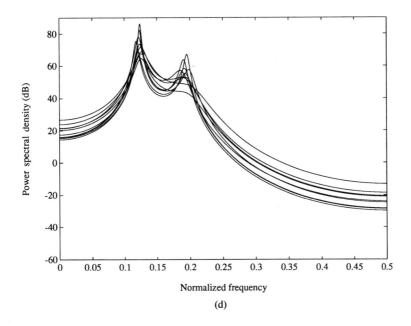

(d)

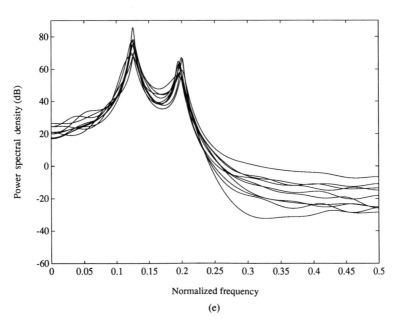

(e)

FIGURE 16-6 (*Continued*)

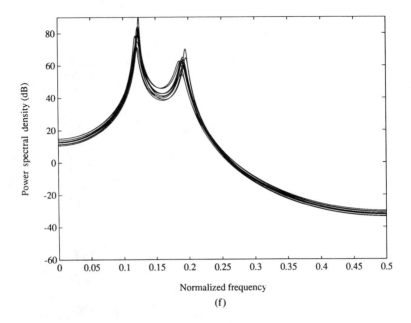

(f)

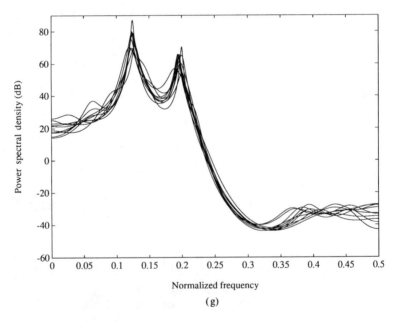

(g)

FIGURE 16-6   (*Continued*)

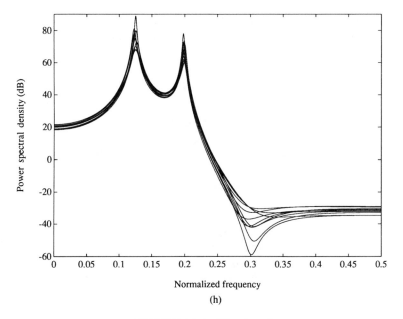

FIGURE 16-6 (*Continued*)

higher lag autocorrelation estimates are estimated with poor accuracy. The PSD estimate using the least-squares modified Yule–Walker method in Figure 16-6(h) clearly shows the spectral null. In general, if the model type, AR or ARMA or MA, coincides with the way the signal has been generated, the PSD estimate is quite satisfactory.

16-5 PARAMETRIC SPECTRUM ANALYSIS OF SINUSOIDAL SIGNALS

In many practical applications of spectrum analysis one is required to analyze data consisting of deterministic signals observed in the presence of random noise perturbations. These include problems such as determination of direction of arrival of plane waves at an equispaced line array of sensors encountered in radar and communication applications, high-resolution estimation of Doppler frequencies, identifying a linear time-invariant system from its impulse response measurements, and processing nuclear magnetic resonance signals in order to identify the molecular structure of a chemical sample. In many such applications it is essential that one is able to resolve signals that are closely spaced in frequency. But, often the number of data samples that can be observed is rather limited. When the data record is short, as discussed in Section 16-2, simple Fourier transform processing suffers from spectral leakage and loss of resolution and hence may not be suitable. In such cases, considering the way the signal is generated, it is appropriate to fit a parametric signal model to the observed data samples $x[n]$, $n = 1, 1, \ldots, N - 1$. A useful signal model, $h[n]$, is a sum of exponential or sinusoidal signals. That is,

$$h[n] = \sum_{k=1}^{p} A_k e^{s_k n}, \qquad n = 0, 1, \ldots, N - 1, \qquad (16.111)$$

where A_k are the complex amplitudes and e^{s_k} are the complex frequencies (or poles) of the underlying signal (or system), $s_k = \alpha_k + j2\pi f_k$, while α_k and f_k are the damping factors and frequencies, respectively. In some problems such as Doppler frequency estimation, since the signal is expected to be a sum of undamped sinusoids the damping factors may be assumed to be zero. Note that the frequencies f_k lie in the normalized range of 0 to 1 and are not necessarily multiples of a common fundamental frequency. If the data are real valued, then A_k and s_k may be assumed to occur in complex conjugate pairs. If such a model is successfully fit to the data, the signal model can be used to extrapolate the data to any desired length, thereby overcoming the resolution limit dictated by traditional Fourier transform processing. However, the limiting factor in this case would be the inaccuracies in the model fit caused by the noise in the data samples and deviations from the sinusoidal model assumption.

Basically, there are two major approaches to this modeling problem. The first is the direct approach in which one attempts to minimize the model fitting error E,

$$E = \sum_{n=0}^{N-1} |x[n] - h[n]|^2, \qquad (16.112)$$

by choosing the model parameters A_k and e^{s_k}. Also, one may have to choose p, the number of signal components, if it is not known. If the noise samples corrupting the underlying signal are independent and Gaussian, then this least-squares criterion is also optimum in the maximum-likelihood sense [LA73]. But this is a nonlinear minimization problem involving a multidimensional search or an iterative algorithm. Some methods that attempt to minimize E are discussed in Section 16-5-2.

In the second major approach, called the indirect approach, one attempts to extract the signal parameters of interest, the A_k's and e^{s_k}'s, as accurately as possible without explicitly attempting to minimize the error criterion in Eq. (16.112). Generally, these methods are computationally less expensive than the direct approaches. We discuss the indirect approaches first.

Strictly speaking, the methods we discuss in this section are not spectral estimation methods in the sense that the outcome of the method is not a power spectral density estimate. But if the assumed signal model truly reflects the underlying physics of signal generation, then these methods can be used to accurately model the data with a parsimonious set of signal parameters. Using the estimated signal and noise parameters, if desired, one may reconstruct a power spectral density estimate. These methods are especially effective when the data record is short but the SNR is sufficiently high.

16-5-1 Indirect Approaches

As mentioned, the indirect methods do not explicitly minimize the error E in Eq. (16.112) but attempt to estimate the underlying signal parameters computationally efficiently but at the cost of suboptimal performance. Prony's method is the earliest of the indirect methods.

16-5-1-1 Prony's Method. Prony's method [PR1795] is a technique for determining the parameters of exponential/sinusoidal signals, which are supposedly contained in a sequence of noiseless signal samples. Presently, let us assume that the data samples $x[n]$ are noiseless and consist of p exponential signals. That is,

$$x[n] = \sum_{k=1}^{p} A_k e^{s_k n}, \qquad n = 0, 1, \ldots, N - 1. \qquad (16.113)$$

The problem addressed by Prony [PR1795] was to determine the unknown A_k's and s_k's from the N given samples. Since there are $2p$ complex-valued unknowns in Eq. (16.113), at least $2p$ samples of $x[n]$ are needed to determine them. An apparent difficulty in determining the A_k's and e^{s_k}'s is that they enter bilinearly in Eq. (16.113). An important observation, due to Prony, was that the determination of the A_k's and e^{s_k}'s can be decoupled and that they can be determined by solving two sets of linear simultaneous equations and a polynomial root finding.

A standard derivation of Prony's method can be found, for example, in Hildebrand [HI56]. However, we shall derive Prony's difference equations by a slightly different route in this section. This derivation leads to the prefiltering methods discussed in the subsequent subsections. Consider the polynomial $X(z)$, the z-transform of the given finite-duration sequence, $x[n]$:

$$X(z) = \sum_{n=0}^{N-1} \sum_{k=1}^{p} A_k e^{s_k n} z^{-n}. \qquad (16.114)$$

Interchanging the summations and summing the geometric series, we have

$$X(z) = \sum_{k=1}^{p} A_k \frac{1 - e^{s_k N} z^{-N}}{1 - e^{s_k} z^{-1}}$$

$$= \frac{\sum_{k=1}^{p} \left[A_k (1 - e^{s_k N} z^{-N}) \left(\prod_{(j=1, j \neq k)}^{p} (1 - e^{s_j} z^{-1}) \right) \right]}{\prod_{k=1}^{p} (1 - e^{s_k} z^{-1})}. \qquad (16.115)$$

$X(z)$ can be written as a ratio of two polynomials as follows:

$$X(z) = \frac{C(z)}{B(z)}. \qquad (16.116)$$

Comparing Eqs. (16.115) and (16.116), we observe that $B(z)$ is a pth degree polynomial with roots at $e^{s_1}, e^{s_2}, \ldots, e^{s_p}$.

$$B(z) = 1 + \sum_{k=1}^{p} b_k z^{-k} = \prod_{k=1}^{p} (1 - e^{s_k} z^{-1}).$$

After some simplification of the numerator in Eq. (16.115), we note that $C(z)$ is a polynomial of degree $(N + p - 1)$, defined as

$$C(z) = \sum_{k=0}^{p-1} c_k z^{-k} + \sum_{k=0}^{p-1} c_{N+k} z^{-(N+k)}.$$

Note that although some coefficients of $C(z)$, specifically c_0 to c_{p-1} and c_N to c_{N+p-1}, depend on the unknown amplitudes and poles (or frequencies), the other coefficients $c_p, c_{p+1}, \ldots, c_{N-1}$ are identically zero. From Eq. (16.116), we have

$$X(z)B(z) = C(z). \tag{16.117}$$

Therefore inverse z-transforming both sides of Eq. (16.117), we get

$$x[n] \circledast b_n = c_n. \tag{16.118}$$

Writing Eq. (16.118) explicitly we have

$$x[n] + \sum_{k=1}^{p} x[n-k] b_k = c_n, \qquad n = 0, 1, \ldots, N + p - 1. \tag{16.119}$$

Since $c_p, c_{p+1}, \ldots, c_{N-1}$ are zero, the middle $N - p$ equations corresponding to the zero entries on the right side can be extracted and written as follows:

$$\begin{bmatrix} x[p] & x[p-1] & \cdots & x[0] \\ x[p+1] & x[p] & \cdots & x[1] \\ \vdots & \vdots & \ddots & \vdots \\ x[N-1] & x[N-2] & \cdots & x[N-p-1] \end{bmatrix} \begin{bmatrix} 1 \\ b_1 \\ b_2 \\ \vdots \\ b_p \end{bmatrix} = \begin{bmatrix} 0 \\ 0 \\ 0 \\ \vdots \\ 0 \end{bmatrix}. \tag{16.120}$$

Or, in matrix notation,

$$X_f b = 0,$$

where the (i, j)th element of X_f is $X_f(i, j) = x[p + i - j]$ for $i = 1, 2, \ldots, N - p$ and $j = 1, 2, \ldots, p + 1$ and $b = [1, b_1, b_2, \ldots, b_p]^T$. If $N = 2p$, then we have just enough equations to solve for the unknowns, $b_1, b_2, \ldots, b_p$. Since the first entry in **b** is unity, the first column of $\mathbf{X}_f$ is moved to the right side before solving the equations for the unknowns $b_1, b_2, \ldots, b_p$. These difference equations were first used by Prony. Once the b_k's are determined, then the polynomial $B(z)$ $= 1 + b_1 z^{-1} + b_2 z^{-2} + \cdots + b_p z^{-p}$ is formed and its roots $e^{s_1}, e^{s_2}, \ldots,$ e^{s_p} are computed. The A_k's are then determined by solving the following p linear equations:

$$\sum_{k=1}^{p} A_k e^{s_k n} = x[n], \qquad n = 0, 1, \ldots, p - 1. \tag{16.121}$$

Therefore the key idea in Prony's method is the way the determination of the A_k's and e^{s_k}'s is decoupled. Summarizing, the steps involved in Prony's method are:

1. Solve for the coefficients of $B(z)$ using Eq. (16.120).
2. Root the polynomial $B(z)$ to find $e^{s_1}, e^{s_1}, \ldots, e^{s_p}$.
3. Determine the amplitudes $A_1, A_2, \ldots, A_p$ using Eq. (16.121).

16-5-1-2 Least-Squares Prony's Methods. The original Prony's method assumed that the signal samples are noise free. Unfortunately, often the observed signals are corrupted by measurement errors. However, the number of observed data samples, N, is likely to be greater than the minimum required, which is $2p$. In this case, we can formulate $N - p$ difference equations, instead of the minimum p in Eq. (16.120). Taking the first column of $\mathbf{X}_f$ to the right side, we have

$$\begin{bmatrix} x[p-1] & x[p-2] & \cdots & x[0] \\ x[p] & x[p-1] & \cdots & x[1] \\ \vdots & \vdots & & \vdots \\ x[N-1] & x[N-2] & \cdots & x[N-p-1] \end{bmatrix} \begin{bmatrix} \hat{b}_1 \\ \hat{b}_2 \\ \vdots \\ \hat{b}_p \end{bmatrix}$$

$$= -\begin{bmatrix} x[p] \\ x[p+1] \\ \vdots \\ x[N-1] \end{bmatrix} + \begin{bmatrix} e_f[1] \\ e_f[2] \\ \vdots \\ e_f[N-p] \end{bmatrix} \tag{16.122}$$

or in matrix notation,

$$\mathbf{X}_f' \hat{\mathbf{b}}' = -\mathbf{x}_f' + \mathbf{e}_f,$$

where $e_f(i)$ is the error in the ith equation caused by the noise in the data. $\mathbf{e}_f$ is the vector of these equation errors and $\hat{\mathbf{b}} = [\hat{b}_1, \hat{b}_2, \ldots, \hat{b}_p]'$. The standard least-squares solution of $\hat{\mathbf{b}}'$, which minimizes the euclidean length of $\mathbf{e}_f$, is given by

$$\hat{\mathbf{b}}' = -(\mathbf{X}_f'^\dagger \mathbf{X}_f')^{-1} \mathbf{X}_f'^\dagger \mathbf{x}_f, \qquad (16.123)$$

where $\dagger$ denotes complex conjugate transpose. This is known as the least-squares Prony's method [HI56]. Note that the equation errors in Eq.(16.122) are identical to the equation errors corresponding to the covariance method of linear prediction (see Eq. (16.83)). Although computation of $\hat{\mathbf{b}}'$ via Eq. (16.123) can be achieved using standard least-squares methods such as Cholesky decomposition, a fast Levinson-like algorithm that requires less computation is available [MO77]. McClellan [MC88] gives a readable account of this algorithm. Once $\hat{\mathbf{b}}'$ is computed, the polynomial $\hat{B}(z)$ may be formed and its roots determined.

If the data are known to be composed of undamped sine waves, then time reversing and complex conjugating the data samples does not alter the characteristics of the signal. Only the phases of the sine waves are affected. In this case we can use the forward and the backward difference equations together, resulting in twice as many equations as in Eq. (16.122), to determine $\hat{\mathbf{b}}'$. That is, $\hat{\mathbf{b}}'$ is computed as follows:

$$\hat{\mathbf{b}}' = -(\mathbf{X}_{fb}'^\dagger \mathbf{X}_{fb}')^{-1} \mathbf{X}_{fb}'^\dagger \mathbf{x}_{fb}. \qquad (16.124)$$

where $\mathbf{X}_{fb}$ is the $(2(N - p) \times (p + 1))$ forward–backward data matrix formed by concatenating $\mathbf{X}_f$ and $\mathbf{X}_b$ as

$$[\mathbf{X}_{fb}] = \begin{bmatrix} \mathbf{X}_f \\ \mathbf{X}_b \end{bmatrix}. \qquad (16.125)$$

Here $\mathbf{x}_{fb}$ is the first column of $\mathbf{X}_{fb}$ and $\mathbf{X}_{fb}'$ contains the remaining columns of $\mathbf{X}_{fb}$. The elements of $\mathbf{X}_f$ are defined in Eq. (16.120). The (i, j)th element of the $(N - p) \times (p + 1)$ backward data matrix $\mathbf{X}_b$ is $x^*[N - 1 - p + i - j]$. A row of the $\mathbf{X}_b$ matrix, when reversed and complex conjugated, is identical to a row of $\mathbf{X}_f$. This method is identical to the modified covariance method mentioned earlier. A fast Levinson-like algorithm is available for this method as well [MA80].

Once the roots of $\hat{B}(z)$ ($e^{\hat{s}_k}$, $k = 1, 2, \ldots, p$), which are the estimates of the true parameters, are computed, the estimates of the amplitudes $\hat{A}_k$, $k = 1, 2, \ldots, p$, can be obtained by minimizing the following error criterion with respect to the $\hat{A}_k$'s:

$$\sum_{n=0}^{N-1} \left| x[n] - \sum_{k=1}^{p} \hat{A}_k e^{\hat{s}_k n} \right|^2. \qquad (16.126)$$

This is a linear least-squares problem. The amplitudes estimate vector $\hat{\mathbf{a}} = [\hat{A}_1, \hat{A}_2, \ldots, \hat{A}_p]'$, which minimizes the above error, is

$$\hat{\mathbf{a}} = [\hat{\mathbf{T}}^\dagger \hat{\mathbf{T}}]^{-1} \hat{\mathbf{T}}^\dagger \mathbf{x}, \qquad (16.127)$$

where $\mathbf{x}$ is the data vector $[x[0], x[1], \ldots, x[N-1]]'$ and $\hat{\mathbf{T}}$ is a $N \times p$ matrix with (i, j)th element defined as follows:

$$\hat{T}(i, j) = e^{\hat{s}_j(i-1)}, \quad i = 1, 2, \ldots, N \text{ and } j = 1, 2, \ldots, p. \quad (16.128)$$

We shall use a computer simulation to evaluate the performance of the least-squares Prony's method.

Example 16.6. Figure 16-7(a) shows the zeros of the polynomial $\hat{B}(z)$ estimated from computer generated data for 100 different trials. The data set consisted of 32 complex-valued samples for each trial. It was generated by using the following formula:

$$x[n] = \sum_{k=1}^{5} A_k e^{j2\pi f_k n} + u[n], \quad n = 0, 1, \ldots, 31, \quad (16.129)$$

where $u[n]$ is a complex white Gaussian noise sequence with real and imaginary parts having equal variance, equal to σ^2. The amplitudes of the five undamped complex sine waves were chosen as follows: $A_1 = 1$, $A_2 = 1$, $A_3 = 0.9$, $A_4 = 1$, $A_5 = 1$. The normalized frequencies (sampling frequency $= 1$) were $f_1 = 0.2, f_2 = 0.25, f_3 = 0.36, f_4 = 0.38,$ and $f_5 = 0.4$. The signal-to-noise power ratio (SNR) of the kth sine wave is defined as $10 \log_{10} |A_k|^2/2\sigma^2$. In this example, σ^2 was chosen as 5×10^{-3}. Therefore the SNRs of the complex sine waves, in order, are 20, 20, 19.08, 20, and 20 dB. For each of the 100 trials the signal part of the data remained the same, whereas the noise sequence $u[n]$ was changed. Figure 16-7(b)

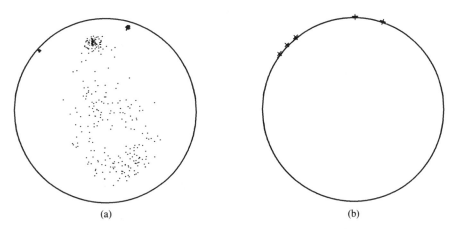

(a)   (b)

FIGURE 16-7 (a) The plot of zeros of the polynomial $B(z)$ obtained in 100 trials using the least-squares Prony's method. The zeros are shown relative to the unit circle $|z| = 1$. The angular location of the zeros is used to estimate the frequencies of the sine waves. The signal-to-noise ratio for each sine wave is about 20 dB. (b) The location of the zeros of $B(z)$ when the data are noiseless.

shows the true locations of frequencies (i.e., at angles of $2\pi f_k$, $k = 1, 2, \ldots, 5$ in radians relative to the real axis) by crosses with reference to the unit circle $|z| = 1$. Since the sine waves were all undamped, the data were used in both the forward and backward directions simultaneously (i.e., Eq. (16.124) was used) to estimate the coefficients of the polynomial $\hat{B}(z)$. We assumed that the value of p is known to be equal to 5. The polynomial roots for each trial were determined and denoted by dots in Figure 16-7(a). Clearly, the least-squares Prony's method has difficulty is estimating the frequencies of the complex sinusoidal signals. The zeros of $\hat{B}(z)$ tend to cluster around regions where the signal energy is significant. However, none of the frequencies can be determined accurately from the root locations. If we use only the forward-difference equations (as in Eq. (16.123)) the results are worse. This computer simulation indicates that in spite of using more than the minimum number of samples required, the performance of the least-squares Prony's method is not satisfactory, even at relatively high signal-to-noise ratio values. In the following subsections methods that improve the performance of Prony's method are presented.

16-5-1-3 Least-Squares Prony's Method Using a Prefilter. In many instances, some prior knowledge regarding the signal such as the approximate regions in the frequency domain where the signals are concentrated is available. A prime example is the case of nuclear magnetic resonance (NMR) signals. In NMR signal processing [ST88], in many situations, the general region in the frequency domain where line components are expected to lie are known. Only the relative locations of the components known as *chemical shifts* are unknown. Such prior information could also be obtained by a preliminary Fourier analysis of the data. For example, an examination of the magnitude of the DFT of the data used in the Example 16.6, shown in Figure 16-8(a), indicates the regions of energy concentration. The simple method described in this subsection can be useful if such prior information is available.

Prony's method extracts the signal parameters accurately only if the data consist of noiseless exponential signals. Therefore, intuitively, one might expect that the performance of Prony's method, when the data are noisy, can be improved if we prefilter the data to emphasize the regions in the frequency domain where the signal components are known to be dominant. However, we must ensure that this prefiltering does not interfere with the basic idea of Prony's method. Specifically, since Prony's method depends on the annihilation property of the polynomial $B(z)$ (as in Eq. (16.118)), any prefiltering operation should ensure that this basic property is not violated.

Consider a finite impulse response (FIR) prefilter, $W(z)$, defined as follows:

$$W(z) = 1 + w[1]z^{-1} + w[2]z^{-2} + \cdots + w[q]z^{-q}. \qquad (16.130)$$

We wish to use this filter to prefilter the data. Let us convolve $w(n)$, the impulse response of $W(z)$, with both sides of Eq. (16.118):

$$w[n] \circledast x[n] \circledast b_n = w[n] \circledast c_n. \qquad (16.131)$$

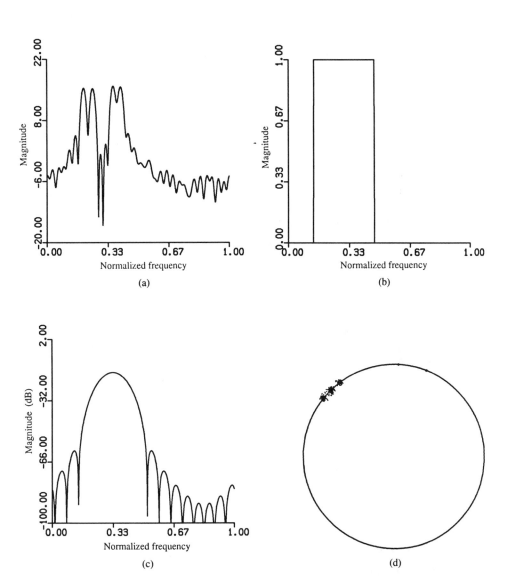

FIGURE 16-8   (a) Magnitude of the DFT of a typical realization of $x[n]$ consisting of five sinusoidal signals and noise given in example 16.6. (b) The desired magnitude response of the prefilter. (c) The prefilter magnitude response. $q = 15$. (d) The roots of $B(z)$ after using the prefilter. Compare with Figure 16-7(a).

Let $x'[n] = w[n] \circledast x[n]$ and $d_n = w[n] \circledast c_n$. Then we have the following linear equations in the noiseless case:

$$
\begin{bmatrix}
x'[0] & 0 & \cdots & 0 \\
x'[1] & x'[0] & \cdots & \cdots \\
x'[2] & x'[1] & \cdots & 0 \\
\vdots & x'[2] & \cdots & x'[0] \\
& & & \\
\vdots & \vdots & \vdots & \vdots \\
x'[p+q-1] & x'[p+q-2] & \cdots & x'[q-1] \\
\hdashline
x'[p+q] & x'[p+q-1] & \cdots & x'[q] \\
\vdots & \vdots & \vdots & \vdots \\
x'[N-1] & x'[N-2] & \cdots & x'[N-p-1] \\
\hdashline
x'[N] & x'[N-1] & \cdots & x'[N-p] \\
\vdots & \vdots & \vdots & \vdots \\
x'[N+q-1] & \cdots & \cdots & \cdots \\
0 & x'[N+q-1] & \cdots & \cdots \\
\vdots & \vdots & \vdots & \vdots \\
0 & \cdots & 0 & x'[N+q-1]
\end{bmatrix}
\begin{bmatrix}
1 \\
b_1 \\
b_2 \\
\vdots \\
b_p
\end{bmatrix}
=
\begin{bmatrix}
d_0 \\
d_1 \\
d_2 \\
\vdots \\
\vdots \\
d_{p+q-1} \\
\hdashline
0 \\
\vdots \\
0 \\
\hdashline
d_N \\
d_{N+1} \\
d_{N+2} \\
\vdots \\
\vdots \\
d_{N+p+q-1}
\end{bmatrix}
$$

$$(16.132)$$

The convolution of $w[n]$ and c_n has resulted in a sequence d_n that has *q fewer zeros in the middle*, compared to c_n, due to the smearing caused by the convolution operation. We have assumed that q is small enough that this convolution has not resulted in the elimination of all the zeros in the middle of the right side vector. Extracting those $N = (p + q)$ equations corresponding to the zero entries on the right side, we have

$$
\begin{bmatrix}
x'[p+q] & x'[p+q-1] & \cdots & x'[q] \\
x'[p+q+1] & x'[p+q] & \cdots & x'[q+1] \\
\vdots & \vdots & \ddots & \vdots \\
x'[N-1] & x'[N-2] & \cdots & x'[N-p-1]
\end{bmatrix}
\begin{bmatrix}
1 \\
b_1 \\
b_2 \\
\vdots \\
b_p
\end{bmatrix}
=
\begin{bmatrix}
0 \\
0 \\
\vdots \\
0
\end{bmatrix}.
$$

$$(16.133)$$

Since we have to determine the p unknown coefficients of $B(z)$ from Eq. (16.133), we should have at least p equations. Thus $N - (p + q)$ should be greater than or equal to p. Therefore q, the order of the prefilter, has to obey the following inequality:

$$0 \leq q \leq N - 2p. \tag{16.134}$$

Simply, Eq. (16.133) shows that the annihilation property (see Eq. (16.120)) on which Prony's method depends is still valid with a prefilter as long as the transients due to prefiltering on the right side of Eq. (16.132) die out fast enough. If an IIR filter is chosen as the prefilter, then, since the transient effects persist for a long time, this property is violated. This leads to a deterministic bias in the frequency estimates [KA84]. When the data are noisy, as before, Eq. (16.133) is solved in the least-squares sense to estimate the coefficients of $B(z)$.

Figure 16-8(a) shows the DFT magnitude of the 32 samples of $x[n]$ in Eq. (16.129) for a particular noise realization. In Figure 16-8(b) we have chosen a desired prefilter magnitude response after examining Figure 16-8(a). A prefilter with this response will have an infinitely long impulse response and hence is unsuitable since it violates Eq. (16.134). A 15th order FIR prefilter (i.e., q is 15), which approximates this desired response, was designed with a Hamming window (bandwidth $\Delta f = 0.15, f_c = 0.33$) (see Chapter 5 [OP75]). Figure 16-8(c) shows the magnitude response of the FIR prefilter. Once $w[n]$ is computed, then the equations in Eq. (16.133) (and also the corresponding backward-difference equations) were formulated and solved simultaneously in the least-squares sense to estimate the coefficients $b_1, b_2, \ldots, b_p$. p was assumed 5. Then the fifth degree polynomial $\hat{B}(z)$ was formed and factored to find the roots. The roots give the estimates of the frequencies of the signals. Figure 16-8(d) shows the roots of the estimated polynomial for 100 trials with respect to the unit circle $|z| = 1$. Clearly, there is significant improvement compared to the results of the least-squares Prony's method shown in Figure 16-7(a), because the frequency estimates are clearly much more accurate in Figure 16-8(d). Although in the above example we had used only one prefilter, it is possible to use several such filters for different regions in the spectrum and combine the results. Summarizing, the steps involved in the least-squares Prony's method with prefiltering are:

1. Identify from the DFT magnitude or prior information a desired prefilter response.
2. Determine a FIR filter of order $q(\leq N - 2p)$ such that its magnitude response is close to the desired response. Typically, q is between $N/3$ and $3N/4$.
3. Formulate the equations in Eq. (16.133) (and also the corresponding backward-difference equations if the signals are known to be undamped sinusoidal signals).
4. Solve for the estimates of $b_1, b_2, \ldots, b_p$ in the least-squares sense.
5. Root the polynomial to estimate the pole locations/frequencies and determine the associated amplitudes.

16-5-1-4 Least-Squares Prony's Method with a Data-Adaptive Prefilter. A disadvantage of the previous method is that the prefilter $W(z)$ has to be determined off-line before proceeding to estimate the pole locations or frequencies. In this method the prefilter $W(z)$ is computed iteratively from the data. The rationale for this method is as follows. Consider Eq. (16.131). To start with, let us assume that $W(z) = 1$. Then $B(z)$ is determined by solving Eq. (16.133) in the least-squares sense. Since $W(z) = 1$, this is simply the least-squares Prony's method. Since the coefficient vector of $B(z)$ attempts to annihilate the signal, it tends to have nulls or zeros close to regions of signal energy concentration although the frequency/pole estimates themselves are inaccurate. This can be observed in Figure 16-6(a). Therefore $1/B(z)$, when evaluated around the unit circle, tends to have peaks of those locations. Therefore we may determine the prefilter, $W(z)$, for the next iteration such that it approximates $1/B(z)$. In other words, $W(z)$ may be chosen as the inverse filter of $B(z)$ computed in the previous iteration. That is, choose $W(z)$ such that

$$1/B(z) \approx W(z). \tag{16.135}$$

Equivalently, we may write

$$B(z)W(z) = 1 + E(z), \tag{16.136}$$

where $E(z)$ is the transform of an "error" sequence $e[n]$. In the time domain

$$b_n \circledast w[n] = \delta[n] + e[n], \tag{16.137}$$

where $\delta[n] = 1$ for $n = 0$ and $\delta[n] = 0$ for $n \neq 0$. Since b_n and $w[n]$ are $p + 1$ and $q + 1$ long sequences, respectively, $e[n]$ is a $q + p + 1$ long sequence. Therefore to determine the filter $W(z)$ we now minimize

$$\sum_{n=0}^{p+q} |e[n]|^2 = \sum_{n=0}^{p+q} |\delta[n] - b_n \circledast w[n]|^2 \tag{16.138}$$

by choosing $w[1], w[2], \ldots, w[q]$. Note that $w[0] = 1$. Using standard least-squares minimization, we have the following set of linear equations whose solution gives the $w[n]$'s:

$$\sum_{j=1}^{q} r_{bb}[i - j]w[j] = -r_{bb}[i], \qquad i = 1, 2, \ldots, p, \tag{16.139}$$

where

$$r_{bb}[m] = \sum_{k=0}^{p-|m|} b_n^* b_{n+|m|}, \qquad b_0 = 1. \tag{16.140}$$

$r_{bb}[m]$ are the autocorrelation values of the filter coefficients 1, b_1, b_2, ... , b_p when it is regarded as a sequence of $p + 1$ samples. Note that the equations in (16.139) are indeed the Yule–Walker equations encountered in Section 16-4-3. Therefore they can be solved efficiently by means of the Levinson–Durbin recursive algorithm described in Section 16-4-3. This also causes the filter $W(z)$ to be minimum phase. Once the prefilter $W(z)$ is computed, we go back and filter the original data sequence (resulting in $x'[n]$) and form the equations in Eq. (16.131) and calculate a new set of coefficients 1, b_1, b_2, ... , b_p. This is followed by another estimate of the prefilter $W(z)$. We repeat the procedure until the coefficients 1, b_1, b_2, ... , b_p do not change appreciably from iteration to iteration. Figure 16-9(a) shows the zero locations for the polynomial $B(z)$ after six iterations in all 100 trials. Clearly, the estimates of the frequencies are far superior, when compared to the least-squares Prony's method (in Figure 16-7(a)). Figure 16-9(b) shows the zeros of the prefilter $W(z)$ for the 100 trials. Figure 16-8(c) shows $|W(e^{j\omega})|$ for a typical trial. Note that this data-adaptive prefilter essentially has the same passband as the off-line prefilter shown in Figure 16-8(c). Summarizing, the steps involved in the least-squares Prony's method with a data-adaptive prefilter are:

1. Set $w[0] = 1$ and $w[n] = 0$ for $n = 1, 2, ... , q$, for some $q(\leq N - 2p)$. Typically, q is chosen between $N/3$ and $3N/4$.
2. Formulate Eq. (16.133) (and the corresponding backward-difference equations if the signals are known to be undamped sinusoids) after filtering $x[n]$ by $W(z)$.
3. Solve for the estimates of b_1, b_2, ... , b_p in the least-squares sense.
4. Solve for new $w[n]$, $n = 1, 2, ... , q$ ($w[0] = 1$), using Eq. (16.139).
5. Go to step 2 and repeat until the b_k's do not change appreciably (i.e., convergence is reached).
6. Root the polynomial and estimate the frequencies/pole locations and associated amplitudes.

16-5-1-5 Least-Squares Prony's Method Using Singular Value Decomposition (SVD).

Prefiltering of the data to improve the signal-to-noise ratio can also be achieved noniteratively, by using singular value decomposition (SVD) of the data matrix, thereby improving the performance of Prony's method. This is accomplished by exploiting a certain rank property of the data matrix.

Consider Eq. (16.118) with noiseless data. Since the convolution operation is commutative, we can rewrite Eq. (16.118) as follows:

$$x[n] \circledast w[n] \circledast b_n = w[n] \circledast c_n. \qquad (16.141)$$

Let $v[n] = w[n] \circledast b_n$. Then $V(z) = W(z)B(z)$. $v(n)$ is a $p + q + 1$ long sequence with $v[0] = 1$. Since the right side of the above equation is unaltered, we can once again extract the homogeneous part (i.e., $N - p - q$) of the above equations. Let

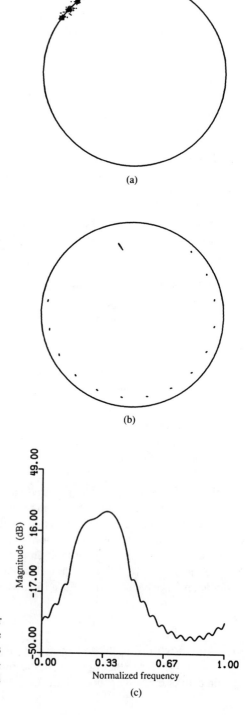

(a)

(b)

(c)

FIGURE 16-9 (a) The roots of $B(z)$ after using the data-adaptive prefilter. (b) The roots of the prefilter $W(z)$ for the 100 trials after convergence. (c) The magnitude response of the prefilter $W(z)$ for a typical trial. Compare with Figure 16-8(c).

$L = p + q$. Then

$$
\begin{bmatrix}
x[L] & x[L-1] & \cdots & x[1] & x[0] \\
x[L+1] & x[L] & \cdots & x[2] & x[1] \\
\vdots & \vdots & \ddots & \vdots & \vdots \\
x[N] & x[N-1] & \cdots & x[N-L-2] & x[N-L-1]
\end{bmatrix}
\begin{bmatrix}
1 \\
v[1] \\
v[2] \\
\vdots \\
v[L]
\end{bmatrix}
$$

$$
=
\begin{bmatrix}
0 \\
0 \\
\vdots \\
\vdots \\
0
\end{bmatrix}
$$

or, in matrix notation,

$$\mathbf{Y}_f \boldsymbol{v} = \mathbf{0}. \tag{16.142}$$

Since $L = p + q$ and q has to satisfy Eq. (16.134), L has to obey the following inequality:

$$p \le L \le N - p. \tag{16.143}$$

L has to satisfy Eq. (16.142) if $V(z)$ (i.e., its factor $B(z)$) has to have roots at the locations e^{s_k}, $k = 1, 2, \ldots, M$. Equations (16.142) and (16.143) imply that for all values of L in the above range, the rank of $\mathbf{Y}_f$ is p. This rank property is peculiar to Toeplitz (and Hankel) matrices $\mathbf{Y}_f$ (and $\mathbf{Y}_b$, see Eq. (16.125)) embedded with data samples composed of exponential signals. This property is exploited by the SVD based methods [HE81; KU81, KU82; TU82b].

As before, we can solve Eq. (16.142) for the $v[k]$'s in the least-squares sense if the data are noisy. The resulting method is, of course, the covariance/least-squares Prony's method, except that the number of signals is overestimated to be L instead of p. Therefore, in effect, when the number of signals is overestimated by q, the covariance/least-squares Prony's method is equivalent to using a prefilter $W(z)$ of order q. However, unlike the methods in the previous two subsections, the prefilter is adaptively determined from the data, along with $B(z)$ in one step, in the form of $V(z)$. Since with noisy data the prefilter is different for each realization of the data, the q zeros of $V(z)$ corresponding to the polynomial factor $W(z)$ tend to fall all over the plane about the unit circle $|z| = 1$, as Figure 16-10(a) shows. However, note that the zeros corresponding to the factor $B(z)$, that is, signal zeros, tend to fall close to the true signal zero locations. But the presence of the spurious or extraneous zeros makes it difficult to distinguish the true signal-related zeros from the extraneous ones. The SVD based prefiltering [KU81, KU82, KU83c, TU82b], can be used to overcome this problem.

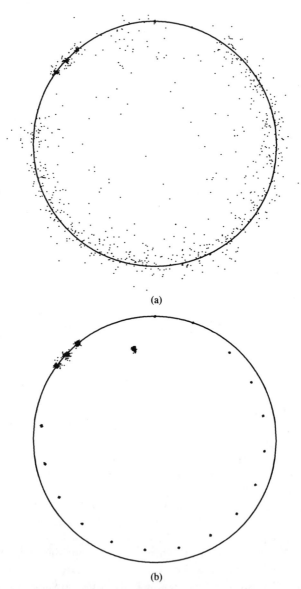

(a)

(b)

FIGURE 16-10   (a) The roots of the polynomial $V(z)$ for the 100 trials. The coefficients of the polynomial $v(1)$ to $v(20)$ were obtained by solving Eq. (16.142) using least-squares. This amounts to the least-squares Prony's method, where the order is overestimated to be 20 instead of 5. Note that the signal zeros are close to the true locations although the extraneous ones are all over the plane. (b) The roots of the polynomial $V(z)$ for 100 trials using SVD. The matrix $\mathbf{Y}_{fb}$ is replaced by its rank 5 approximant before the coefficients of the polynomial $v(1)$ to $v(20)$ are computed. (c) The singular values of the matrix $\mathbf{Y}_{fb}$ with noiseless data. Note the sudden drop in the singular value after 5. This information can be used to determine the number of signal components in the data. However, if the data consist of a large number of signal components of varying strength and are noisy, then singular values may not exhibit such an abrupt change.

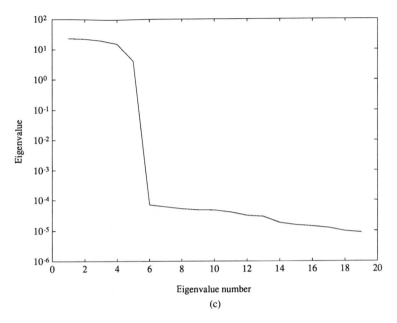

Eigenvalue number

(c)

FIGURE 16-10 (*Continued*)

The SVD of $\mathbf{Y}_f$ (or for that matter any matrix) is a product of three matrices [LA74]:

$$\mathbf{Y}_f = \mathbf{U}_f \mathbf{\Sigma}_f \mathbf{V}_f^\dagger. \qquad (16.144)$$

The matrices $\mathbf{U}_f$ and $\mathbf{V}_f$ are unitary matrices of dimension $(N - L) \times (N - L)$ and $L \times L$, respectively. $\mathbf{\Sigma}_f$ is a $(N - L) \times (L)$ diagonal matrix with real nonnegative diagonal entries. The diagonal entries $\sigma_1, \sigma_2, \ldots$, called the singular values of $\mathbf{Y}_f$, are conventionally ordered with the largest value in the upper left-hand corner. The singular values of $\mathbf{Y}_f$ are the nonnegative square roots of the eigenvalues of $\mathbf{Y}_f^\dagger \mathbf{Y}_f$ or $\mathbf{Y}_f \mathbf{Y}_f^\dagger$.

If the data are noiseless, then as shown in Eq. (16.142) the rank of $\mathbf{Y}_f$ is p and only the singular values σ_1 to σ_p are nonzero. If the data are noisy, then all singular values may be nonzero. Motivated by this fact, when the data samples are noise corrupted, before solving for v, $\mathbf{Y}_f$ is replaced by its pth rank approximant, $\hat{\mathbf{Y}}_f$. That is, $\hat{\mathbf{Y}}_f$ is the rank p matrix such that

$$\sum_{i=1}^{N-L} \sum_{j=1}^{L} |\mathbf{Y}_f(i, j) - \hat{\mathbf{Y}}_f(i, j)|^2$$

is minimum. The pth rank approximant $\hat{\mathbf{Y}}_f$ of $\mathbf{Y}_f$ is obtained from the SVD in Eq. (16.144) by replacing $\mathbf{\Sigma}_f$ by $\hat{\mathbf{\Sigma}}_f$, where $\hat{\mathbf{\Sigma}}_f$ is the same matrix as $\mathbf{\Sigma}_f$ except that the singular values $\sigma_{p+1}, \sigma_{p+2}, \ldots$ are set to zero. This approximation has the effect of increasing the signal-to-noise ratio in the data. That is, if the SNR in the data is sufficiently high, the approximant $\hat{\mathbf{Y}}_f$ is much closer to the signal-only data matrix [TU82a]. Once this is done, v is computed by solving

$$\hat{\mathbf{Y}}_f v = \mathbf{0}. \tag{16.145}$$

Since the rank of $\hat{\mathbf{Y}}_f$ is less than full, the above equation has many solutions. But the vector v that has the minimum euclidean length given by

$$\sum_{n=0}^{L} |v[n]|^2, \quad \text{where } v[0] = 1, \tag{16.146}$$

is picked as the unique solution vector. The resulting Lth $(L = p + q)$ degree polynomial is $V(z) = 1 + \sum_{k=1}^{L} v(k)z^{-k} = B(z)W(z)$. Figure 16-10(b) shows the zeros of $V(z)$ for the 100 trials. When the data are noiseless the effect of the constraint in Eq. (16.146) is to cause the factor $W(z)$ to be minimum phase [KU83a]. This turns out to be the case in Figure 16-10(b) although the data are noisy. Figure 16-10(c) shows the 12 singular values of the matrix $\mathbf{Y}_{fb}$ when the data are noiseless. Note the abrupt drop in the size of the sixth and higher singular values. Often this information can be used to guess the number of signal components in the data.

If the data consist of damped complex sinusoids, we may use the backward data matrix $\mathbf{Y}_b$ [KU82]. In this case the signal zeros corresponding to the damped sinusoids will fall outside the unit circle, whereas the extraneous roots corresponding to $W(z)$ will remain inside the unit circle [KU83a]. This fact can be used to separate the signal zeros from the rest. If the data consist of undamped signals, then $\mathbf{Y}_{fb} = [\mathbf{Y}_f, \mathbf{Y}_b]'$ may be used, as was the case in Figure 16-9(b).

Summarizing, the steps involved in the least-squares Prony's method using SVD are:

1. Choose the value of $L(= p + q)$. Often, L is chosen between $N/3$ and $3N/4$. But L has to obey the inequality in Eq. (16.143).
2. Form the data matrix $\mathbf{Y}_{fb}$ (if the data are known to consist of undamped sine waves) or $\mathbf{Y}_b$ (if it consists of damped sine waves).
3. Compute the SVD of $\mathbf{Y}_{fb}$ (or $\mathbf{Y}_b$) and compute $\hat{\mathbf{Y}}_{fb}$ (or $\hat{\mathbf{Y}}_b$).
4. Root the polynomial $V(z)$. If undamped signals are expected, the frequency estimates may be obtained from the p roots closest to the unit circle. If the data consisted of damped signals, the p roots outside the unit circle (reflected inside) give the desired estimates [KU82].

The accuracy of the parameter estimates obtained by using SVD based methods have been the topic of many recent reports [HU88; PO87; RA88; VA87]. Generally, the parameter estimation accuracy achieved by the SVD based methods are comparable to the maximum-likelihood/direct methods discussed below at high

SNR. The advantage of the SVD based methods is that they are noniterative. But they are computationally expensive.

16-5-2 Direct Approaches

In the direct approach one attempts to minimize the model fitting error E,

$$E = \sum_{n=0}^{N-1} |y[n] - h[n]|^2 = \sum_{n=0}^{N-1} \left| y[n] - \sum_{k=1}^{p} A_k e^{s_k} \right|^2, \qquad (16.147)$$

by choosing the model parameters e^{s_k} and A_k. The relevant references where similar direct approaches are used include McDonnough and Huggins [MC68], van den Boss [VA80], Tuttle [TU58], Rife and Boorstyn [RI76], and Aigrain and Williams [AI49] . The above criterion is optimum in the maximum-likelihood sense if the noise samples corrupting the underlying signal are independent and Gaussian. Setting the derivatives of E to zero with respect to the real and imaginary parts of the s_k's and A_k's leads to a complicated set of nonlinear simultaneous equations. General iterative methods such as the Newton–Raphson method may be used to solve these equations. But these methods are computationally expensive and require a good initial guess for the unknown parameters. We now discuss some simpler methods that attempt to minimize E. These methods take advantage of the special structure of the signal model $h[n]$.

16-5-2-1 Multidimensional Search for Minimizing **E.** We can rewrite the error E as follows in matrix-vector notation:

$$\mathbf{E} = \mathbf{e}^\dagger \mathbf{e}, \qquad (16.148)$$

where

$$\mathbf{e} = \mathbf{x} - \mathbf{Ta} \qquad (16.149)$$

and $\mathbf{x} = (x[0], x[1], \ldots, x[N-1])^t$, $\mathbf{a} = (A_1, A_2, \ldots, A_p)^t$, and $\mathbf{T}$ is a $N \times p$ matrix defined as follows:

$$\mathbf{T} = \begin{bmatrix} 1 & 1 & \cdots & 1 \\ e^{s_1} & e^{s_2} & \cdots & e^{s_p} \\ e^{2s_1} & e^{2s_2} & \cdots & e^{2s_p} \\ \vdots & \vdots & \ddots & \vdots \\ e^{(N-1)s_1} & e^{(N-1)s_2} & \cdots & e^{(N-1)s_p} \end{bmatrix}. \qquad (16.150)$$

Theoretically at least, it is possible to search the entire parameter space spanned by the A_k's and the e^{s_k}'s to find the minimum of the error E, although this would be extremely time consuming. However, because of the bilinear nature of the signal model it is possible to eliminate [GO73; TU80] the explicit appearance of the amplitude parameters in the error expression in Eq. (16.148) and hence reduce the parameter space. To show this, let us assume for a moment that the e^{s_k}'s are known. Then we can find the best A_k's by minimizing E using the standard linear least-squares method. For the assumed e^{s_k}'s, the best A_k's are given by

$$\mathbf{a} = (\mathbf{T}^\dagger \mathbf{T})^{-1} \mathbf{T}^\dagger \mathbf{x}. \tag{16.151}$$

Substituting this $\mathbf{a}$ in the above error expression, we note that the error E can be rewritten entirely in terms of the $\mathbf{T}$ matrix as follows:

$$E = \mathbf{x}^\dagger (\mathbf{I} - \mathbf{T}(\mathbf{T}^\dagger \mathbf{T})^{-1} \mathbf{T}^\dagger) \mathbf{x} = \mathbf{x}^\dagger (\mathbf{I} - \mathbf{P_T}) \mathbf{x}, \tag{16.152}$$

where $\mathbf{P_T}$ is called the projection matrix of $\mathbf{T}$ [RA71]. Note that E is also minimized by maximizing $\mathbf{x}^\dagger \mathbf{P_T} \mathbf{x}$. This approach to minimizing E is not unreasonable in the special case when the data are known to consist of undamped sine waves and noise. Because the s_k's are purely imaginary numbers, one needs to search for the maximum of $\mathbf{x}^\dagger \mathbf{P_T} \mathbf{x}$ only in the interval of 0–2π in p dimensions. We may also point out that when the data consist of one complex undamped sine wave then $\mathbf{x}^\dagger \mathbf{P_T} \mathbf{x}$ is simply

$$\frac{1}{N} \left| \sum_{n=0}^{N-1} x[n] e^{-j\omega_1 n} \right|^2 ;$$

that is, the best estimate of the frequency parameter ω_1 is the location of the global peak of the DTFT magnitude of the data. However, if the data are composed of two or more closely spaced sine waves, as we noted before, the DTFT magnitude of the data cannot be used to determine the frequencies of the sinusoidal signals even in the noiseless case because of spectral leakage. Figure 16-11 shows the three-dimensional plot of the expression $\mathbf{x}^\dagger \mathbf{P_T} \mathbf{x}$ with respect to the independent variables ω_1 and ω_2. In this example [TU82b], the data consist of two closely spaced undamped sinusoidal signals in additive noise. Note that the plot is symmetrical about the diagonal line because the variable names ω_1 and ω_2 are not unique and can be interchanged. The global peak location in the three-dimensional plot gives the estimates of the two frequencies of the sine waves, one along each axis. A coarse search followed by a hill-climbing algorithm may be used to locate the global peak. A preliminary Fourier analysis can be used to locate the regions where the search is to be concentrated. The amplitudes may subsequently be determined by a linear least-squares fit to the data.

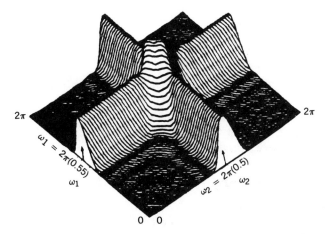

FIGURE 16-11 Plot of the error surface as a function of the candidate frequency estimates $\hat{\omega}_1$ and $\hat{\omega}_2$. In this example the frequencies of the two sinusoids were $\omega_1 = 2\pi \times 0.55$ and $\omega_2 = 2\pi \times 0.5$. They had equal amplitudes. The number of data samples used is 12. SNR = 20 dB. Note that the error surface is multimodal. However, if the number of signal components is small, a search procedure to find the global peak can be devised.

16-5-2-2 Iterative Approach to Minimizing **E.** The above method required a multidimensional search to determine the signal parameters. This approach may be computationally expensive if the data consisted of several damped sinusoidal signals. Alternatively, the error E can be reformulated in terms of a different set of parameters. This leads to a somewhat simpler iterative algorithm. This also shows the connection between the direct approaches and the indirect approaches discussed in the previous section.

Consider a polynomial

$$B(z) = 1 + \sum_{k=1}^{p} b_k z^{-k}.$$

Let its roots be e^{s_k}, $k = 1, 2, \ldots, p$. By definition, $B(e^{s_k}) = 0$, $k = 1, 2, \ldots, p$. Expressing this fact in a matrix-vector form, we have

$$\begin{bmatrix} b_p & b_{p-1} & \cdots & \cdots & \cdots & \cdots & b_1 & 1 & 0 & \cdots & 0 \\ 0 & b_p & b_{p-1} & \cdots & \cdots & \cdots & \cdots & b_1 & 1 & \cdots & 0 \\ \vdots & \vdots & \vdots & \vdots & \vdots & \vdots & \vdots & \vdots & \vdots & \vdots & \vdots \\ 0 & \cdots & \cdots & b_p & b_{p-1} & \cdots & \cdots & \cdots & \cdots & b_1 & 1 \end{bmatrix} [T] = [0]$$

$$(16.153)$$

or, equivalently,

$$\mathbf{BT} = \mathbf{0}, \tag{16.154}$$

where $\mathbf{B}$ is the $(N - p) \times N$ matrix on the left and $\mathbf{0}$ is a $(N - p) \times p$ matrix with zero entries. That is, $\mathbf{B}^\dagger$, the matrix of coefficients of $B^*(z)$, is the orthogonal complement of the $\mathbf{T}$ matrix in Eq. (16.150). Then, using a result from linear algebra [RA71] that says that the projection matrices of $\mathbf{T}$ and $\mathbf{B}^\dagger$ add up to an identity matrix, we have

$$\mathbf{I} - \mathbf{P_T} = \mathbf{P_{B^\dagger}}, \quad \text{where} \quad \mathbf{P_{B^\dagger}} = \mathbf{B}^\dagger (\mathbf{BB}^\dagger)^{-1} \mathbf{B}. \tag{16.155}$$

Using Eq. (16.155) in Eq. (16.152), we can rewrite the error in terms of the coefficients of $B(z)$ as follows:

$$E = \mathbf{x}^\dagger \mathbf{B}^\dagger (\mathbf{BB}^\dagger)^{-1} \mathbf{Bx}. \tag{16.156}$$

Furthermore, due to the special structure of the $\mathbf{B}$ matrix, $\mathbf{Bx}$ can also be written as $\mathbf{X}_f \mathbf{b}$, where $\mathbf{b} = (1, b_1, b_2, \ldots, b_p)'$ and $\mathbf{X}_f$ is the $(N - p) \times (p + 1)$ forward data matrix encountered in Eq. (16.120). Its (i, j)th element $\mathbf{X}_f(i, j) = x[p + i - j]$. Therefore the error E can be further written as

$$E = \mathbf{b}^\dagger \mathbf{X}_f^\dagger (\mathbf{BB}^\dagger)^{-1} \mathbf{X}_f \mathbf{b}. \tag{16.157}$$

Because of the matrix $(\mathbf{BB}^\dagger)^{-1}$ in the middle of the error expression, minimization of E with respect to the coefficients of $B(z)$ is still a nonlinear problem. However, an iterative algorithm that linearizes the minimization at each iteration is available [BR86; EV73; KU86]. In this algorithm the quantity

$$E^{(i)} = \mathbf{b}^{\dagger(i)} \mathbf{X}_f^\dagger (\mathbf{B}^{(i-1)} \mathbf{B}^{\dagger(i-1)})^{-1} \mathbf{X}_f \mathbf{b}^{(i)} \tag{16.158}$$

is minimized at the ith iteration, as indicated by the superscripts (i). During the first iteration ($i = 1$), $(\mathbf{B}^{(0)} \mathbf{B}^{(0)\dagger})^{-1}$ is chosen as the identity matrix. Then the coefficient vector $\mathbf{b}^{(1)} = (1, b_1^{(1)}, b_2^{(1)}, \ldots, b_p^{(1)})'$ is determined by minimizing $E^{(1)} = \mathbf{b}^{\dagger(1)} \mathbf{X}_f^\dagger \mathbf{X}_f \mathbf{b}^{(1)}$. A little thought will convince the reader that the result is the same as in the least-squares Prony's method (see Eq. (16.123)). For $i = 2$, $\mathbf{B}^{(1)}$ is formed (in the same format as $\mathbf{B}$ in Eq. (16.153)) using the coefficients $1, b_1^{(1)}, b_2^{(1)}, \ldots, b_p^{(1)}$ obtained from the first iteration. Then $E^{(2)} = \mathbf{b}^{\dagger(2)} \mathbf{X}_f^\dagger (\mathbf{B}^{(1)} \mathbf{B}^{(1)\dagger})^{-1} \mathbf{X}_f \mathbf{b}^{(2)}$ is minimized with respect to the coefficients $b_1^{(2)}, b_2^{(2)}, \ldots, b_p^{(2)}$. This also leads to linear simultaneous equations as in the least-squares Prony's method except that these equations are now weighted by the square root of the matrix $(\mathbf{B}^{(1)} \mathbf{B}^{(1)\dagger})^{-1}$.

This process is repeated until convergence is reached. Once the coefficients of $B(z)$ are estimated, the roots of $B(z)$ are computed to obtain the estimates of e^{s_k}'s. The amplitude estimates are subsequently estimated by solving a linear least-squares problem. Some simulation results using this method are available in Evans and Fischl [EV73], Kumaresan et al. [KU86], and Bressler and Macovski [BR86]. If the signals contained in the data are known to be undamped sinusoids, it may be advantageous to constrain the coefficients of $B(z)$ to be symmetric, in order to attempt to force its roots to lie on the unit circle [BR86; KU86].

The error E can be rewritten in yet another equivalent form. Let us define $H(z)$ such that $H(z) = 1/B(z)$. In the time domain $h[n] \circledast b_n = \delta[n]$. That is, in matrix-vector form we have

$$
\begin{bmatrix}
b_p & b_{p-1} & \cdots & & \cdots & & \cdots & b_1 & 1 & 0 & \cdots & 0 \\
0 & b_p & b_{p-1} & \cdots & & \cdots & & \cdots & b_1 & 1 & \cdots & 0 \\
\vdots & \vdots & \vdots & \vdots & \vdots & \vdots & \vdots & \vdots & \vdots & \vdots & & \vdots \\
0 & \cdots & & \cdots & b_p & b_{p-1} & \cdots & & \cdots & & b_1 & 1
\end{bmatrix}
$$

$$
\cdot
\begin{bmatrix}
h[0] & 0 & \cdots & & 0 \\
h[1] & h[0] & \cdots & & \cdots \\
h[2] & h[1] & \cdots & & \cdots \\
\cdots & \cdots & \cdots & & 0 \\
\cdots & \cdots & \cdots & & h[0] \\
\cdots & \cdots & \cdots & & h[1] \\
\vdots & \vdots & \vdots & & \vdots \\
h[N-1] & \cdots & & \cdots & h[N-p]
\end{bmatrix}
= [0]
\qquad (16.159)
$$

or, equivalently,

$$\mathbf{BH} = \mathbf{0}. \qquad (16.160)$$

This shows that the matrix $\mathbf{B}^\dagger$ is also the orthogonal complement of $\mathbf{H}$; that is, $\mathbf{T}$ and $\mathbf{H}$ span the same space. Again using the properties of the projection matrices $\mathbf{I} - \mathbf{P_H} = \mathbf{P_{B^\dagger}}$. Therefore the error E may also be written as

$$E = \mathbf{x}^\dagger (\mathbf{I} - \mathbf{P_H})\mathbf{x}. \qquad (16.161)$$

In the above form the error E is not amenable to an iterative minimization algorithm. We may rewrite it further in a slightly different form using the following observation. Let us filter the data $x[n]$ by the filter $1/B(z)$. Let the filtered sequence be called $x'[n]$. That is,

$$X'(z) = \frac{1}{B(z)} X(z)$$

or

$$X(z) = B(z)X'(z). \tag{16.162}$$

$$
\begin{bmatrix}
x[0] \\
x[1] \\
x[2] \\
\vdots \\
\vdots \\
\vdots \\
\vdots \\
x[N-1]
\end{bmatrix}
=
\begin{bmatrix}
x'[0] & 0 & \cdots & 0 \\
x'[1] & x'[0] & \cdots & \cdots \\
x'[2] & x'[1] & \cdots & \cdots \\
\cdots & \cdots & \cdots & 0 \\
\cdots & \cdots & \cdots & x'[0] \\
\cdots & \cdots & \cdots & x'[1] \\
\vdots & \vdots & \vdots & \vdots \\
x'[N-1] & \cdots & \cdots & x'[N-p]
\end{bmatrix}
\begin{bmatrix}
1 \\
b_1 \\
b_2 \\
\vdots \\
b_p
\end{bmatrix}
\tag{16.163}
$$

$$\mathbf{x} = \mathbf{X'b}. \tag{16.164}$$

Substituting Eq. (16.164) in Eq. (16.161), we have

$$E = \mathbf{b}^\dagger \mathbf{X'}^\dagger (\mathbf{I} - \mathbf{P_H}) \mathbf{X'b}. \tag{16.165}$$

The error E in the above form may be minimized iteratively. At the ith iteration the following error is minimized:

$$E^{(i)} = \mathbf{b}^{(i)\dagger} \mathbf{X'}^{(i-1)\dagger} (\mathbf{I} - \mathbf{P}_{\mathbf{H}^{(i-1)}}) \mathbf{X'}^{(i-1)} \mathbf{b}^{(i)}. \tag{16.166}$$

Each step in the iteration will involve only the solution of linear equations. The error in Eq. (16.166) is indeed identical to the one in Eq. (16.158) [MC90]. This simply follows from the fact that $\mathbf{I} - \mathbf{P}_{\mathbf{H}^{(i-1)}} = \mathbf{P}_{\mathbf{B}^{(i-1)\dagger}}$ and that $\mathbf{B}^{(i-1)}\mathbf{X}'^{(i-1)} = \mathbf{X}_f$. However, the error in Eq. (16.166) may be computationally simpler to minimize since it needs inversion of only a $p \times p$ matrix $(\mathbf{H}^{(i-1)\dagger}\mathbf{H}^{(i-1)})$ [PA87], whereas the one in Eq. (16.158) involves inversion of a possibly large $(N - p) \times (N - p)$ matrix, although the algorithm given in Kumaresan et al. [KU86] can be used to reduce the computational load. An improved version of this algorithm is given by Clark and Scharf [CL92].

Another iterative method that attempts to minimize E, which predates the above algorithms, is due to Steiglitz and McBride [ST65, ST77]. In fact, McClellan and Lee [MC90] show that the Steiglitz–McBride algorithm is theoretically identical to the above algorithm. In this method, making use of the fact that our signal model, a linear combination of exponentials, may be expressed as the impulse response of a pole/zero filter $C(z)/B(z)$, the error E is rewritten as follows:

$$E = \sum_{n=0}^{N-1} \left| x[n] - \frac{C(z)}{B(z)} \{\delta[n]\} \right|^2, \qquad (16.167)$$

where $\delta[n]$ is the impulse sequence and $\{\cdot\}$ denotes an "operator." That is, for example, $(1 + 2z^{-1})\{x[n]\}$ reads "$(1 + 2z^{-1})$ operating on $x[n]$" and the result is $x[n] + 2x[n - 1]$.

$$C(z) = \sum_{k=0}^{p-1} c_k z^{-k}$$

and

$$B(z) = 1 + \sum_{k=1}^{p} b_k z^{-k},$$

so that $[C(z)/B(z)]\{\delta[n]\}$ represents the impulse response of the model filter $C(z)/B(z)$. Now one wishes to minimize E with respect to the coefficients of $C(z)$ and $B(z)$. This problem is still nonlinear. Instead, Kalman [KA58] suggested that the following error E' be minimized:

$$E' = \sum_{n=0}^{N-1} |B(z)\{x[n]\} - C(z)\{\delta[n]\}|^2. \qquad (16.168)$$

The advantage of this modified error criterion is that it can be minimized with respect to the coefficients of $C(z)$ and $B(z)$ by solving the following linear equations in the least-squares sense:

$$
\begin{bmatrix}
x[0] & 0 & \cdots & \cdots & 0 & -1 & 0 & \cdots & \cdots & 0 \\
x[1] & x[0] & 0 & \cdots & \cdots & 0 & -1 & 0 & \cdots & \cdots \\
x[2] & x[1] & x[0] & \cdots & \cdots & \cdots & 0 & -1 & \cdots & \cdots \\
\cdots & \cdots & \cdots & \cdots & \cdots & \cdots & \cdots & \cdots & \cdots & 0 \\
\cdots & \cdots & \cdots & \cdots & 0 & \cdots & \cdots & \cdots & 0 & -1 \\
\cdots & \cdots & \cdots & \cdots & x[0] & 0 & \cdots & \cdots & \cdots & 0 \\
\cdots & \cdots & \cdots & \cdots & x[1] & \cdots & \cdots & \cdots & \cdots & 0 \\
\vdots & \vdots & \vdots & \vdots & \vdots & \vdots & \vdots & \vdots & \vdots & \vdots \\
x[N-1] & \cdots & \cdots & \cdots & x[N-p-1] & 0 & \cdots & \cdots & \cdots & 0
\end{bmatrix}
$$

$$
\cdot
\begin{bmatrix}
1 \\ b_1 \\ b_2 \\ \vdots \\ b_p \\ c_0 \\ c_1 \\ \vdots \\ c_{p-1}
\end{bmatrix}
=
\begin{bmatrix}
e[0] \\ e[1] \\ \vdots \\ \vdots \\ \vdots \\ \vdots \\ \vdots \\ \vdots \\ \vdots \\ e[N-1]
\end{bmatrix}
. \qquad (16.169)
$$

That is, after moving the first column to the right side (since the first element of the coefficient vector is unity), the rest of the coefficients may be determined by minimizing $\Sigma_{k=0}^{N-1} |e[k]|^2$. But in Eq. (16.169) the bottom $N - p$ equations do not depend on the coefficients c_0 to c_{p-1}. Therefore one has to minimize $\Sigma_{k=p}^{N-1} |e[k]|^2$ with respect to the b_k's only and the errors $e[0]$ to $e[p-1]$ can subsequently be made identically zero by choosing the c_k's. Therefore the above step is really the least-squares Prony's method discussed earlier, in disguise, in which the numerator coefficients are also computed.

Unfortunately, the error E' is not the same error as E. Steiglitz and McBride improved on this idea by minimizing the following error iteratively. In the ith iteration, the minimized

$$E^{(i)} = \sum_{n=0}^{(N-1)} \left| \frac{B^{(i)}(z)\{e[n]\} - C^{(i)}(z)\{\delta[n]\}}{B^{(i-1)}(z)} \right|^2 \tag{16.170}$$

with respect to $C^{(i)}(z)$ and $B^{(i)}(z)$ with $B^{(i-1)}(z)$ known from the previous iteration. The rationale behind this method is that if the iterations converge, that is $B^{(i-1)} = B^{(i)}$, then the error $E^{(i)}$ coincides with the true error E. The factor $1/B^{(i-1)}(z)$ may be regarded as a prefilter through which the data samples are filtered before a new $B^{(i)}(z)$ is determined. The resulting linear equations are as follows:

$$
\begin{bmatrix}
x'[0] & 0 & \cdots & & 0 & -h^{(i-1)}[0] & 0 & \cdots & 0 \\
x'[1] & x'[0] & \cdots & & \cdots & -h^{(i-1)}[1] & -h^{(i-1)}[0] & \cdots & \cdots \\
x'(2) & x'[1] & \cdots & & \cdots & -h^{(i-1)}[2] & -h^{(i-1)}[1] & \cdots & \cdots \\
\cdots & \cdots & & \cdots & & \cdots & \cdots & \cdots & 0 \\
\cdots & \cdots\cdots & \cdots & 0 & & \cdots & & \cdots & -h^{(i-1)}[0] \\
\cdots & \cdots\cdots & \cdots & x'[0] & & \cdots & & \cdots & -h^{(i-1)}[1] \\
\cdots & \cdots\cdots & \cdots & x'[1] & & \cdots & & \cdots & -h^{(i-1)}[2] \\
\vdots & \vdots & \vdots & \vdots & & \vdots & \vdots & \vdots & \vdots \\
x'[N-1] & \cdots & \cdots & x'[N-p-1] & -h^{(i-1)}[N-1] & & \cdots & \cdots & -h^{(i-1)}[N-p]
\end{bmatrix}
$$

$$
\cdot
\begin{bmatrix}
1 \\
b_1^{(i)} \\
b_2^{(i)} \\
\vdots \\
b_p^{(i)} \\
c_0^{(i)} \\
c_1^{(i)} \\
\vdots \\
c_{p-1}^{(i)}
\end{bmatrix}
=
\begin{bmatrix}
e^{(i)}[0] \\
e^{(i)}[1] \\
\vdots \\
\vdots \\
\vdots \\
\vdots \\
\vdots \\
\vdots \\
e^{(i)}[N-1]
\end{bmatrix}
\cdot
\tag{16.171}
$$

In the above, $h^{(i-1)}[n]$ is the impulse response of the prefilter $1/B^{(i-1)}(z)$. Note that we need only the first N samples of this impulse response. $x'[n] = x[n]$ ⊛

$h^{(i-1)}[n]$. In more compact notation we can write the above equation as follows:

$$\mathbf{X}'^{(i-1)}\mathbf{b}^{(i)} - \mathbf{H}^{(i-1)}\mathbf{c}^{(i)} = \mathbf{e}^{(i)},$$

where $\mathbf{b}^{(i)} = [1, b_1^{(i)}, \dots, b_p^{(i)}]'$ and $\mathbf{c}^{(i)} = [c_0^{(i)}, c_1^{(i)}, \cdots, c_{p-1}^{(i)}]'$. The error $E^{(i)} = \|\mathbf{e}^{(i)}\|^2$. After some manipulations resulting in the elimination of the numerator coefficients, it can be shown [MC90] that the error $E^{(i)}$ coincides with the expression in Eq. (16.166); that is, all the above iterative algorithms minimize the same error with respect to the denominator coefficients $\mathbf{b}^{(i)} = [1, b_1^{(i)}, \dots, b_p^{(i)}]'$. However, they differ computationally. Furthermore, if $1/B^{(i-1)}(z)$ is approximated by FIR filter $W^{(i-1)}(z)$ of order q before the minimization of $E^{(i)}$ in Eq. (16.170), then this method coincides with the data-adaptive prefiltering method described in the previous section. Summarizing, the steps involved in the iterative prefiltering algorithm are:

1. For $i = 1$, solve for $\mathbf{b}^{(1)}$ using the least-squares Prony's method (see Eqs. (16.122) and (16.123)). This is identical to minimizing $E^{(1)}$ in Eq. (16.166) with respect to $\mathbf{b}^{(1)}$.
2. For $i > 1$, filter the data $x[n]$ by the filter $1/B^{(i-1)}(z)$ to obtain the first N samples of the sequence $x'^{(i-1)}[n]$.
3. Form the matrices $\mathbf{X}'^{(i-1)}$ and $\mathbf{H}^{(i-1)}$ as in Eq. (16.171).
4. Compute the matrix $\mathbf{I} - \mathbf{P}_{\mathbf{H}^{(i-1)}}$ after inverting the $p \times p$ matrix $(\mathbf{H}^{(i-1)\dagger}\mathbf{H}^{(i-1)})$.
5. Minimize the euclidean length of the vector $(\mathbf{I} - \mathbf{P}_{\mathbf{H}^{(i-1)}})\mathbf{X}'^{(i-1)}\mathbf{b}^{(i)}$ by finding the coefficients $\mathbf{b}^{(i)} = [1, b_1^{(i)}, \dots, b_p^{(i)}]'$. Since $\mathbf{I} - \mathbf{P}_{\mathbf{H}^{(i-1)}}$ is an idempotent matrix, this amounts to minimizing the error in Eq. (16.166).
6. Go back to step 2 and repeat the algorithm until convergence is reached.

Unfortunately, the above algorithm is not guaranteed to converge. Even if it converges it may not reach the minimum value of E in the parameter space. However, if it converges, the corresponding parameters may be used as the initial estimates in a Newton–Raphson-like procedure to attain at least a local minimum of E.

Sometimes it is advantageous to achieve the prefiltering in the frequency domain. For example, if the noise perturbations in the data at certain frequencies are known to be relatively stronger than at other frequencies, one may wish to weight the error terms at different frequencies appropriately. Let us transform the data and the signal model samples $x[n]$ and $h[n]$ in Eq. (16.147) into the frequency domain using the discrete Fourier transform (DFT). Let W_N denote the complex Nth roots of unity,

$$W_N = e^{j(2\pi/N)}. \tag{16.172}$$

Let us denote the discrete Fourier transform samples of the sequences $x[n]$ and $h[n]$ by $X[m]$ and $H[m]$, respectively:

$$X[m] = \sum_{n=0}^{N-1} x[n] W_N^{-nm}, \quad m = 0, 1, 2, \ldots, N-1; \quad (16.173)$$

$$H[m] = \sum_{n=0}^{N-1} h[n] W_N^{-nm}, \quad m = 0, 1, 2, \ldots, N-1. \quad (16.174)$$

Our signal model being a sum of exponentials, $h[n] = \sum_{k=1}^{p} A_k e^{s_k n}$, substituting in Eq. (16.174) and summing the geometric series we have

$$H[m] = \sum_{k=1}^{p} A_k \frac{1 - e^{s_k N}}{1 - e^{s_k} W_N^m}. \quad (16.175)$$

After some simplifications we can write $H[m]$ as follows:

$$H[m] = \frac{C(W_N^m)}{B(W_N^m)}, \quad (16.176)$$

where

$$C(W_N^m) = c_0 + c_1 W_N^{-m} + c_2 W_N^{-2m} + \cdots + c_{p-1} W_N^{-(p-1)m}, \quad (16.177)$$

$$B(W_N^m) = 1 + b_1 W_N^{-m} + b_2 W_N^{-2m} + \cdots + b_p W_N^{-pm}. \quad (16.178)$$

Using Parseval's relation the error E in Eq. (16.147), neglecting a scale factor, can be written in terms of the frequency-domain samples as follows:

$$E = \sum_{m=0}^{N-1} |X[m] - H[m]|^2 \quad (16.179)$$

or

$$E = \sum_{m=0}^{N-1} \left| X[m] - \frac{C(W_N^m)}{B(W_N^m)} \right|^2. \quad (16.180)$$

Again, to linearize the minimization of E, we can use an iterative procedure similar to the one discussed above. In fact, such a frequency-domain iterative procedure already exists [SA63]. That is, in the ith iteration we minimize the error $E^{(i)}$,

$$E^{(i)} = \sum_{m=0}^{N-1} \left| \frac{B^{(i)}(W_N^m) X[m] - C^{(i)}(W_N^m)}{B^{(i-1)}(W_N^m)} \right|^2, \quad (16.181)$$

with respect to the coefficients $c_0^{(i)}$, $c_1^{(i)}$, ..., $c_{p-1}^{(i)}$ and $b_1^{(i)}$, $b_2^{(i)}$, ..., $b_p^{(i)}$. The corresponding linear equations are

$$[B]^{-1} \begin{bmatrix} X[0] & X[0] & X[0] & 1 & 1 & 1 \\ X[1] & X[1]W_N^{-1} & X[1]W_N^{-p} & 1 & W_N^{-1} & W_N^{-(p-1)} \\ X[2] & X[2]W_N^{-2} & X[2]W_N^{-2p} & 1 & W_N^{-2} & W_N^{-2(p-1)} \\ \vdots & \vdots & \vdots & \vdots & \vdots & \vdots \\ X[N-1] & X[N-1]W_N^{-(N-1)} & X[N-1]W_N^{-(N-1)p} & 1 & W_N^{-(N-1)} & W_N^{-(N-1)(p-1)} \end{bmatrix}$$

$$\cdot \begin{bmatrix} 1 \\ b_1^{(i)} \\ b_2^{(i)} \\ \vdots \\ b_p^{(i)} \\ c_0^{(i)} \\ c_1^{(i)} \\ \vdots \\ c_{p-1}^{(i)} \end{bmatrix} = \begin{bmatrix} e^{(i)}[0] \\ e^{(i)}[1] \\ \vdots \\ e^{(i)}[N-1] \end{bmatrix}, \tag{16.182}$$

where **B** is the $N \times N$ diagonal matrix with diagonal entries $B_{i-1}(W_N^m)$, $m = 0$, $1, \ldots, N - 1$. Any desired frequency-dependent weighting of the error may be applied to the left side of the above equation. Again, the coefficients $c_0^{(i)}$, $c_1^{(i)}$, ..., $c_{p-1}^{(i)}$ and $b_1^{(i)}$, $b_2^{(i)}$, ..., $b_p^{(i)}$ are determined by minimizing $\sum_{m=0}^{N-1}|e^{(i)}(m)|^2$. The iterations are repeated until convergence is reached. Again, note that if convergence is reached then $E^{(i)} = E$.

Example 16.7. We now present an example using NMR data consisting of a relatively large number of signal components. The data consisted of 710 samples of complex data collected in an experimental study of plasma blood samples using ^{1}H-NMR spectroscopy [DE88]. Roughly, a NMR experiment amounts to monitoring the impulse response of nuclei in a chemical sample. A typical impulse response may comprise up to about 100 damped complex sinusoids having different frequencies, damping factors, amplitudes, and phases. It is often necessary to quantify the parameters of the sinusoids. The signal parameters can be useful in the analysis of the chemical sample. However, since only a finite number of samples can be collected, often the signal is truncated before it decays to zero. Figure 16-12(a) shows a portion of the real part of the DFT spectrum, after discarding the first four samples, which were known to be corrupted during the measurement process. In NMR practice, often the real part of the DFT of the data is plotted because it has many advantages. Since the signal consists of damped sinusoids, the real part of the DFT produces narrower peaks than the magnitude of the DFT, thereby offering more "resolution." Also, it can be shown that for a single damped complex sinusoid the area under its peak gives the amplitude of the sinusoid and the width at half height is related to the damping factor. However, if there are many signal components it is difficult to estimate the amplitude and the damping factor this way. Figure 16-12(a) shows several peaks; the sinc wiggles caused by the rectangular window are also obvious.

First, we applied the SVD based Prony method on these data. Since the data are known to comprise damped exponentials, we used a backward data matrix Y_b (see Eq. (16.125)) of size 456×251 using the 706 data samples. This corresponds to an $L = p + q$ value of 250. Figure 16-12(b) shows the singular values of the matrix Y_b. We truncated the singular values at 35 and calculated the minimum norm solution to the equation $\hat{Y}_b v = 0$. The roots of the 250th degree polynomial $V(z)$ were computed. There were 26 roots outside the unit circle, which when reflected inside the circle gave the estimates of the poles. The amplitudes/residues were then obtained by a least-squares fit to the data. Using the residues and the poles (after accounting for the first four samples that were discarded) a pole-zero model was constructed and evaluated around the unit circle, the real part of which is shown in Figure 16-12(c). The sinc wiggles due to data truncation seen in Figure 16-12(a) are of course absent in Figure 16-12(c) because the assumed model has essentially extrapolated the data to infinity. The fitting error defined as

$$\frac{\sum\limits_{4}^{710} |x[n] - h[n]|^2}{\sum\limits_{4}^{710} |x[n]|^2}$$

was 4.5×10^{-4}. The model samples $h[n]$ agreed with the data samples to the second or third decimal place. Note that in Figure 16-12(c) some features in the 200-Hz region are not well resolved. In Le Beer et al. [DE88], using the same data set and the SVD based Prony method the authors show that they can reproduce all the features in the DFT spectrum. However, they used a larger order polynomial $V(z)$.

Unfortunately, the iterative approaches described in this section did not converge to a solution for these data for a model order $p = 35$. It is believed that this is due to the poor initial estimates provided by the first iteration (least-squares Prony's method) and the need to prefilter the data by a filter $1/B^{(i-1)}$, which may be unstable. However, an ad hoc procedure described below, which updates one pole at a time but all amplitudes together in one iteration, gave reasonable results.

Recall the frequency-domain model fitting method described earlier. In that method we fitted a signal model in the frequency domain, $H[m]$, to the DFT of the data, $X[m]$, by minimizing $E = \sum_{m=0}^{N-1} |X[m] - H[m]|^2$. A disadvantage of this method, apart from the fact that it did not converge, is that to obtain the estimates of the poles one must root a polynomial $B(z)$. Subsequently, another set of linear equations must be solved to estimate the amplitudes and phases. Alter-

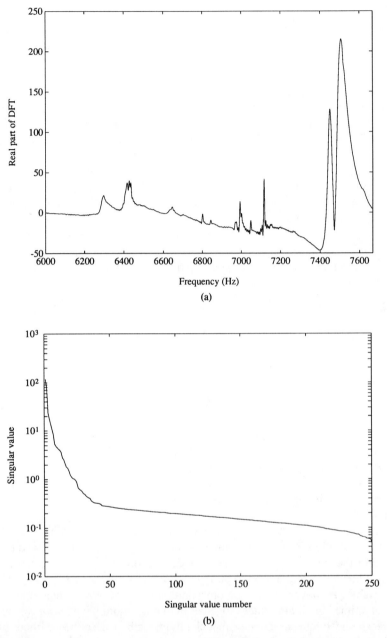

FIGURE 16-12 (a) Real part of the DFT of the NMR data of a blood sample: 706 samples of data were used. Note the sinc wiggles caused by the rectangular window. (b) Singular values of the 456×251 backward data matrix $\mathbf{X}_b$. The singular values were truncated at 35. (c) Real part of the DFT of the model fit obtained by using SVD based Prony's method. (d) Real part of the DFT of the model fit obtained by using a pole-by-pole iterative method.

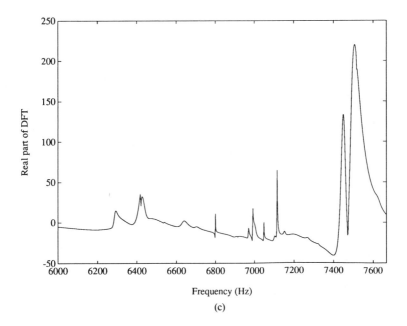

(c)

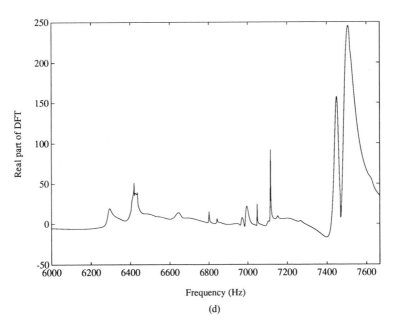

(d)

FIGURE 16-12 (*Continued*)

natively, we may fit the same model but by using a pole-by-pole iterative algorithm. By "pole-by-pole" we mean that in the iterative algorithm in each iteration the decay factor and frequency of only one signal component are updated. But all amplitudes and phases are simultaneously updated.

The error function we minimize in the ith iteration (we have renamed e^{s_k} as p_k, denoting the kth pole, for simplicity):

$$E_s^{(i)} = \sum_{m=0}^{N-1} \left| X[m] \frac{1 - p_s^{(i)} e^{-j2\pi m/N}}{1 - p_s^{(i-1)} e^{-j2\pi m/N}} - \sum_{k=1}^{p} A_k^{(i)} \frac{1 - (p_k^{(i-1)})^N}{1 - p_k^{(i-1)} e^{-j2\pi m/N}} \right|^2$$

(16.183)

for $= 1, 2, \ldots, M$. Rewriting the above in matrix notation, we have

$$\begin{bmatrix} \dfrac{X[0]e^{-j2\pi 0/N}}{1 - p_s^{(i-1)} e^{-j2\pi 0/N}} & \dfrac{1 - (p_1^{(i-1)})^N}{1 - p_1^{(i-1)} e^{-j2\pi 0/N}} & \cdots & \dfrac{1 - (p_M^{(i-1)})^N}{1 - p_M^{(i-1)} e^{-j2\pi 0/N}} \\[4mm] \dfrac{X[1]e^{-j2\pi 1/N}}{1 - p_s^{(i-1)} e^{-j2\pi 1/N}} & \dfrac{1 - (p_1^{(i-1)})^N}{1 - p_1^{(i-1)} e^{-j2\pi 1/N}} & \cdots & \dfrac{1 - (p_p^{(i-1)})^N}{1 - p_p^{(i-1)} e^{-j2\pi 1/N}} \\[4mm] \vdots & \vdots & \ddots & \vdots \\[4mm] \dfrac{X[N-1]e^{-j2\pi(N-1)/N}}{1 - p_s^{(i-1)} e^{-j2\pi(N-1)/N}} & \dfrac{1 - (p_1^{(i-1)})^N}{1 - p_1^{(i-1)} e^{-j2\pi(N-1)/N}} & \cdots & \dfrac{1 - (p_p^{(i-1)})^N}{1 - p_p^{(i-1)} e^{-j2\pi(N-1)/N}} \end{bmatrix}$$

$$\cdot \begin{bmatrix} p_s^i \\ A_1^i \\ A_2^i \\ \vdots \\ A_p^i \end{bmatrix} \approx \begin{bmatrix} \dfrac{X[0]}{1 - p_s^{(i-1)} e^{j2\pi 0/N}} \\[4mm] \dfrac{X[1]}{1 - p_s^{(i-1)} e^{-j2\pi 1/N}} \\[4mm] \vdots \\[4mm] \dfrac{X[N-1]}{1 - p_s^{(i-1)} e^{-j2\pi(N-1)/N}} \end{bmatrix}$$

(16.184)

These equations are solved using the least-squares method. That is, at each iteration we update *one pole* (i.e., the sth pole is chosen randomly among the p in each iteration) and *all amplitudes*. As before, if there is convergence, $p_s^{(i)} \approx p_s^{(i-1)}$ and hence $E_s^{(i)} \approx E$, for $s = 1, 2, \ldots, M$, resulting in the true error we wished to minimize. The advantage of the above formulation is that we can pick the initial pole locations in a convenient manner. For example, the DFT of the given data will give us a clue to where the actual poles are located and one can initialize the poles based on these rough initial estimates (rather than choosing them arbitrarily), thereby minimizing convergence problems. In this example we chose 35 poles uniformly distributed around the unit circle at radius 0.9. Figure 16-12(d) shows the corresponding model fit for $p = 35$. The corresponding fitting error was 1.3 $\times 10^{-4}$, slightly smaller than for the SVD based method.

We have some comments on the sinusoidal model fitting methods discussed in this section. First, the iterative methods are not guaranteed to converge and they quite often fail to converge to reasonable estimates especially when the number of signal components are unknown and large. Generally, SVD based methods are more dependable at high signal-to-noise ratios, but when the number of signal components is large their computational needs are significant. For example, Gesmar and Led [GE88] have applied the SVD based method described above to a NMR data set, consisting of about 8000 complex-valued data samples. They have decomposed a 5760×2400 data matrix $(\mathbf{Y}_b)$, retained 237 significant singular values, and obtained estimates of the corresponding signal parameters. They show that the signal estimate reconstructed from these parameters explains the observed data quite well. They report that the above calculations consumed 90 hours of CPU time on a minicomputer capable of 0.03 Mflops. In NMR applications, one encounters even more complex signal parameter estimation problems. In these cases it may be preferable to first decompose the whole Nyquist range into manageable regions by using filters on the raw data and then model each region separately. It is also possible to use the results of the above methods to initialize nonlinear optimization methods, such as Newton's method, near the globally optimum combination of parameters [PA85].

16-6 SPECTRUM ANALYSIS FOR SENSOR ARRAY PROCESSING

16-6-1 Introduction

In the previous sections we considered processing a segment of a time series to characterize its frequency content. Another important application of spectral analysis is in processing spatial data gathered by an array of sensors. An important problem in sensor array signal processing is to determine the angular location of certain sources that are radiating or reflecting energy. This is often called the bearing or direction of arrival (DOA) estimation problem [HA85; OW85]. Contemporary developments in spectral analysis have led to the development of many array processing methods that are promising in resolving signal sources that are closely spaced in their angular location. Such methods are of interest in applications including radar, sonar, and communications.

Consider an equally spaced linear array of N sensors shown in Figure 16-13. Suppose that plane waves propagating from M point sources from distinct direc-

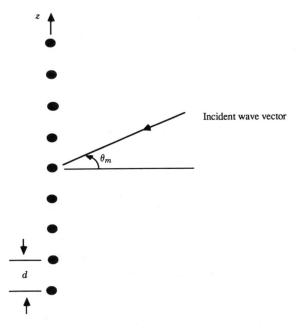

FIGURE 16-13   Geometry of linear array showing a ray of incident plane waves.

tions impinge on the array. The received signal is often correlated in both time and space, that is, from sensor to sensor. Usually this signal is processed in the spatial and temporal dimensions separately. The received signal is first narrowband filtered and sampled simultaneously at the N sensors. At the kth instant in time, N such samples form a vector $\mathbf{x_k}$, called the "snapshot" vector, which is defined as follows:

$$\mathbf{x_k} = [x_k[0] \quad x_k[1] \quad \dots \quad x_k[N-1]]'. \qquad (16.185)$$

K such snapshot vectors are observed at equally spaced time intervals. The observed signals are composed of signals from the M point sources, namely, $\mathbf{s_k} = [s_k[0], s_k[1], \dots, s_k[N-1]]'$ and noise $\mathbf{w_k} = [w_k[0], w_k[1], \dots, w_k[N-1]]'$. That is,

$$\mathbf{x_k} = \mathbf{s_k} + \mathbf{w_k}, \qquad k = 1, 2, \dots, K. \qquad (16.186)$$

The source signals, depending on the angles of arrivals, arrive at each sensor with different time delays. See Figure 16-13. Therefore the signal amplitude samples (after translation to baseband) at the nth sensor element for the kth snapshot may be modeled as follows:

$$s_k[n] = \sum_{m=1}^{M} A_{mk} e^{j((2\pi d/\lambda)\sin\theta_m)n + j\alpha_{mk}}, \qquad n = 0, 1, \dots, N-1. \qquad (16.187)$$

Note that the arrival time delays from sensor to sensor now appear as phase progression in the amplitude of the signals at each sensor. M is the number of point sources, d is the spacing between the sensor elements, λ is the radiation wavelength, N is the number of elements in the array $(N > 2M)$, μ_m is the bearing angle in relation to the array normal, while A_{mk} and α_{mk} are the amplitude and phase of the signal induced by the mth source at the kth snapshot. In somewhat simplified notation the signal samples are rewritten as follows:

$$s_k[n] = \sum_{m=1}^{M} a_{mk} e^{j\omega_m n}, \qquad (16.188)$$

where ω_m and a_{mk} are defined as

$$\omega_m = \frac{2\pi d}{\lambda} \sin \theta_m$$

and

$$a_{mk} = A_{mk} e^{j\alpha_{mk}}.$$

$w_k[n]$, the noise sample at the nth sensor element at the kth time instant, represents the observation noise and noise introduced by the medium. The noise is assumed uncorrelated with the signal. Furthermore, it is assumed that the number of snapshots used for processing K, is small enough that the bearings do not change during data acquisition. Clearly, the spatial frequencies of the sinusoids $e^{j\omega_m}$ yield the direction of arrival of the plane waves assuming that the wavelength λ is known. Therefore it should be clear as to why spectral analysis plays a central role in sensor array signal processing. The problem considered here is limited to a linear equispaced array and the signals are assumed to be narrowband signals. In principle, all the methods discussed in the previous section can be adapted to processing data gathered by such a sensor array. However, some of the methods discussed in this section may also be adapted for processing data gathered by an array having arbitrary shape.

Using Eq. (16.187) in Eq. (16.186), x_k can be written as

$$x_k = Ta_k + w_k, \qquad k = 1, 2, \ldots, K, \qquad (16.189)$$

where T is a $N \times M$ Vandermonde matrix,

$$T = \begin{bmatrix} 1 & 1 & \cdots & 1 \\ e^{j\omega_1} & e^{j\omega_2} & \cdots & e^{j\omega_M} \\ \vdots & \vdots & \ddots & \vdots \\ e^{j\omega_1(N-1)} & e^{j\omega_2(N-1)} & \cdots & e^{j\omega_M(N-1)} \end{bmatrix},$$

and

$$\mathbf{a_k} = [a_{1k} \quad a_{2k} \quad \ldots \quad a_{Mk}]^t.$$

The elements of the vector $\mathbf{a_k}$, which are complex amplitudes, may be modeled as samples of a stationary random process because of the unpredictability of the behavior of the sources. The spatial correlation matrix of the observed random vector $\mathbf{x}$ is

$$\mathbf{R} = \mathcal{E}[\mathbf{x_k x_k^\dagger}], \tag{16.190}$$

where $\mathcal{E}$ is the expectation operation. Since the signal and noise are assumed uncorrelated,

$$\begin{aligned} \mathbf{R} &= \mathbf{R}_s + \mathbf{R}_w \\ &= \mathbf{TR_a T}^\dagger + \mathbf{R}_w, \end{aligned} \tag{16.191}$$

where $\mathbf{R}_s$ is the signal correlation matrix, $\mathbf{R}_a$ is the $M \times M$ correlation matrix of the vector of complex amplitudes $\mathbf{a_k}$, and $\mathbf{R}_w$ is the noise correlation matrix. If the noise is uncorrelated from sensor to sensor and the noise samples, $w_k[n]$, have equal variance σ_n^2, then $\mathbf{R}_w = \sigma_n^2 \mathbf{I}$. For simplicity, if we assume that the source signals are incoherent or uncorrelated, then the matrix $\mathbf{R}_a$ is a diagonal matrix with diagonal elements σ_i^2, $i = 1, 2, \ldots, M$, which denote the signal powers corresponding to the M source signals.

16-6-2 Beamforming

Consider a snapshot vector $\mathbf{x_k} = [x_k[0], x_k[1], \ldots, x_k[N-1]]^t$ observed at the output of the array in Figure 16-14. Analogous to the discrete-time Fourier transform for finite-time signals defined in Section 16-2, we may define the discrete spatial Fourier transform of the spatial series $x_k[0], x_k[1], \ldots, x_k[N-1]$ as

$$X_k(\omega) = \sum_{n=0}^{N-1} x_k[n] e^{-j\omega n}, \tag{16.192}$$

where ω denotes the spatial frequency variable. $|X_k(\omega)|^2$ is the periodogram discussed in Section 16-3-1.

First, let us consider the special case of one source located in the direction θ_m. Assume for simplicity that the signal at the sensors is observed without noise. Then $x_k[n] = a_{mk} e^{j\omega_m n}$, $n = 0, 1, \ldots, N-1$. Substituting $x_k[n]$ in Eq. (16.192) and summing the geometric series, we have

$$\begin{aligned} X_k(\omega) &= a_{mk} \frac{\sin N(\omega - \omega_m)/2}{\sin (\omega - \omega_m)/2} \\ &= a_{mk} W(\omega - \omega_m), \end{aligned} \tag{16.193}$$

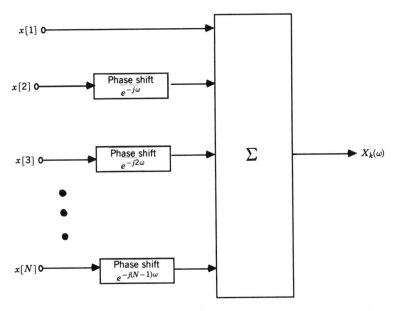

FIGURE 16-14 Beamformer network for computing the angle spectrum $X_k(\omega)$.

where

$$W(\omega) = \frac{\sin N\omega/2}{\sin \omega/2}.$$

The angular spectrum $X_k(\omega)$ may be implemented using the beamformer structure shown in Figure 16-14, which consists of $N - 1$ phase shifters and a summer. If the phase shifters are adjusted such that they correspond to phase shifts 1, $e^{-j\omega_m}$, $e^{-j2\omega_m}$, ..., $e^{-j(N-1)\omega_m}$, then the angular spectrum $X_k(\omega)$ has its largest peak located at ω_m. For this reason the vector of phase shifts in Figure 16-14, namely,

$$\mathbf{e}(\omega) = [1, e^{j\omega}, e^{j2\omega}, \ldots, e^{j(N-1)\omega}]^t, \qquad (16.194)$$

is referred to as the beamsteering vector. Therefore we may determine the direction of the single signal source by locating the dominant spectral peak in the magnitude of the angular spectrum $X_k(\omega)$. The function $W(\omega)$, shown in Figure 16-15 is the beam pattern of a linear array, which is steered to an angle of zero degrees. The pattern exhibits the familiar mainlobe and several sidelobes. The beamwidth, which is the separation between the peak of the magnitude and the nearest null, is $2\pi/N$ radians. Unfortunately, the sidelobes introduce uncertainty by indicating false arrival directions. The effects of the sidelobe may be reduced by applying one of the many weighting/tapering functions (discussed in Section 16-2-1) to the array aperture (i.e., modify the vector of phase shifts in Eq. (16.194) with weights) but only at the cost of increasing the width of the mainlobe.

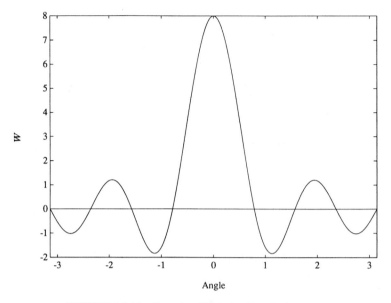

FIGURE 16-15   Function $W(\omega)$ for $N = 8$ elements.

When signals from M sources are incident on the array, then using superposition we have

$$X_k(\omega) = \sum_{m=1}^{M} a_{mk} W(\omega - \omega_m).$$

When the sources are separated in angle by more than a beamwidth, the angular spectrum $|X_k(\omega)|$ shows M distinct peaks located at or close to their angular locations determined by their directions. However, if two source directions are separated by less than one beamwidth, the angular spectrum may or may not exhibit distinct peaks depending on the complex amplitudes associated with the two source signals. This effect is the spatial analog of spectral leakage and resultant loss of spectral resolution observed in processing time-series data in Section 16-2. In this case the beamforming method fails to resolve the sources. Resolution of closely spaced sources can be achieved by increasing the number of elements in the array. But this is of limited value because of the costs involved in increasing the size of an array. For these reasons high-resolution spectrum analysis methods which attempt to overcome the resolution limit of the beamformer are currently being studied. Some of them are discussed below.

The ensemble average of the angular spectrum, called $P_{BF}(\omega)$, is

$$P_{BF}(\omega) = E\{|X_k(\omega)|^2\}$$

$$= \mathbf{e}^\dagger(\omega)\,\mathbf{Re}(\omega), \tag{16.195}$$

where **R** is the true spatial correlation matrix. In practice the true correlation matrix is not known. Then **R** is replaced by its estimate by averaging the K output products formed with the snapshot vectors $\mathbf{x}_k$. That is, in place of **R**

$$\hat{\mathbf{R}} = \frac{1}{K} \sum_{k=1}^{K} \mathbf{x}_k \mathbf{x}_k^\dagger$$

is used. This method is often called the Bartlett estimate of the angular spectrum because it is similar to the power spectrum estimation procedure described in Section 16-4.

16-6-3 Capon's Maximum-Likelihood Method

One of the more common high-resolution array processing methods is due to Capon [CA69; CA70] and is called the maximum-likelihood method (MLM). In spite of its name this method does not correspond to the standard maximum-likelihood approach used in statistics. In this method the angular power spectrum estimation is posed as a linearly constrained quadratic minimization problem.

Let **a** denote a complex-valued steering vector used to process the observed data:

$$\mathbf{a} = [a[0], a[1], \ldots, a[N-1]]^t.$$

The snapshot vector $\mathbf{x_k}$ when processed by the steering vector **a** produces the output

$$\mathbf{a}^\dagger \mathbf{x}_k = \sum_{n=0}^{N-1} a^*[n] x_k[n]. \tag{16.196}$$

We desire that the response of the beam when steered by the vector **a** to the plane wave arriving from a direction corresponding to the frequency variable ω by unity, that is,

$$\mathbf{a}^\dagger \mathbf{e}(\omega) = \sum_{n=0}^{N-1} a^*[n] e^{j\omega n} = 1,$$

and also the power due to plane waves and noise arriving from all other directions, that is,

$$E[|\mathbf{a}^\dagger \mathbf{x}_k|^2] = \mathbf{a}^\dagger \mathbf{R} \mathbf{a}, \tag{16.197}$$

be minimum. This constrained minimization problem may be stated as follows.

Minimize with respect to **a**, the power $\mathbf{a}^\dagger \mathbf{R} \mathbf{a}$ subject to the constraint $\mathbf{a}^\dagger \mathbf{e}(\omega) = 1$. This optimization problem may be solved by the technique of Lagrange multipliers [JO82]. We minimize the quantity Q:

$$Q = \mathbf{a}^\dagger \mathbf{R} \mathbf{a} + \lambda [\mathbf{a}^\dagger \mathbf{e}(\omega) - 1], \tag{16.198}$$

where λ is the Lagrange multiplier. Since the vector $\mathbf{a}$ is complex valued, we may set the derivatives of Q with respect to the real and imaginary parts of $\mathbf{a}$ to be zero. This results in the solution $\mathbf{a} = -\frac{1}{2}\lambda \mathbf{R}^{-1} \mathbf{e}(\omega)$. Using this value of $\mathbf{a}$ in the constraint equation $\mathbf{a}^\dagger \mathbf{e}(\omega) = 1$, we get

$$\mathbf{a} = \frac{\mathbf{R}^{-1} \mathbf{e}(\omega)}{\mathbf{e}^\dagger(\omega) \mathbf{R}^{-1} \mathbf{e}(\omega)}. \tag{16.199}$$

Inserting this value of $\mathbf{a}$ in Eq. (16.197) we have the angular power spectrum given by MLM:

$$P_{\mathrm{MLM}}(\omega) = \frac{1}{\mathbf{e}^\dagger(\omega) \mathbf{R}^{-1} \mathbf{e}(\omega)}. \tag{16.200}$$

Since the true special correlation matrix $\mathbf{R}$ is unavailable, one may use the estimate $\hat{\mathbf{R}}$ defined in the previous section.

16-6-4   High-Resolution Methods Based on Eigendecomposition of the Correlation Matrix

Recently, a class of angular power spectrum estimation methods based on the eigenvalue–eigenvector decomposition of the spatial correlation matrix $\mathbf{R}$ have been developed [BI79; KU83b; PI73; RE79; SC79]. These methods rely on certain algebraic properties of the eigenvectors of the spatial correlation matrix $\mathbf{R}$. The eigenvectors $\boldsymbol{v}_i$ of the correlation matrix $\mathbf{R}$ are characterized by the following property:

$$\mathbf{R} \boldsymbol{v}_i = \lambda_i \boldsymbol{v}_i, \quad i = 1, 2, \ldots, N, \tag{16.201}$$

where λ_i is the eigenvalue corresponding to the eigenvector $\boldsymbol{v}_i$. The matrix $\mathbf{R}$ is Hermitian symmetric and therefore has a complete set of orthonormal eigenvectors. If we assume that the signals are received from only one source, the correlation matrix $\mathbf{R}$ may be written as

$$\mathbf{R} = \mathbf{R}_s + \sigma_n^2 \mathbf{I}$$
$$= \sigma_1^2 \mathbf{t}_1 \mathbf{t}_1^\dagger + \sigma_n^2 \mathbf{I}. \tag{16.202}$$

$\mathbf{t}_1$ is the direction vector of a plane wave source (the first column of the matrix $\mathbf{T}$ in Eq. (16.189)). It is easily verified that the correlation matrix $\mathbf{R}$ has one unique eigenvalue $\lambda_1 = N\sigma_1^2 + \sigma_n^2$ and that the other eigenvalues $\lambda_2, \lambda_3, \ldots, \lambda_N$ are all equal to σ_n^2. The eigenvector $\boldsymbol{v}_1$ corresponding to λ_1 is

$$\boldsymbol{v}_1 = \frac{1}{\sqrt{N}} [1, e^{j\omega_1}, e^{j2\omega_1}, \ldots, e^{j(N-1)\omega_1}]^t.$$

That is, the eigenvector corresponding to the largest eigenvalue λ_1 spans the space of the direction vector $\mathbf{T}_1$ or it is said to span the "signal subspace." The other eigenvectors $\boldsymbol{v}_2, \boldsymbol{v}_3, \ldots, \boldsymbol{v}_N$ are orthogonal to $\boldsymbol{v}_1$ and they span the "noise-subspace." Since $\boldsymbol{v}_1^\dagger \boldsymbol{v}_i = 0$ for $i = 2, 3, \ldots, N$, it follows that the inner product between the noise subspace eigenvectors and the steering vector $\mathbf{e}(\omega)$ (defined in Eq. (16.194)), $\mathbf{e}^\dagger(\omega)\boldsymbol{v}_i$, is zero at the angular frequency $\omega = \omega_1$. This important property is exploited by the eigendecomposition based methods.

When more than one source is present, the correlation matrix $\mathbf{R}$ is given by

$$\mathbf{R} = \mathbf{R}_s = \sigma_n^2 \mathbf{I}$$

$$= \sum_{i=1}^{M} \sigma_i^2 \mathbf{t}_i \mathbf{t}_i^\dagger = \sigma_n^2 \mathbf{I}. \tag{16.203}$$

In this case the signal subspace is spanned by the eigenvectors corresponding to the M largest eigenvalues of $\mathbf{R}$. Although the individual eigenvectors $\boldsymbol{v}_1, \boldsymbol{v}_2, \ldots, \boldsymbol{v}_M$ may not lie alongside the direction vectors (the columns of $\mathbf{T}$ in Eq. (16.189)) the space spanned by them coincides with the column space of the matrix $\mathbf{T}$. Therefore, as before, the inner product of the noise-subspace eigenvectors $\boldsymbol{v}_{M+1}, \boldsymbol{v}_{M+2}, \ldots, \boldsymbol{v}_N$ with $\mathbf{e}(\omega)$ is zero at $\omega = \omega_1, \omega_2, \ldots, \omega_M$.

Schmidt [SC79] proposed the so-called MUSIC (multiple signal classification) algorithm, where he used the above properties of the eigendecomposition to calculate an angular spectrum as follows:

$$P_{\text{MUSIC}}(\omega) = \frac{1}{\displaystyle\sum_{i=M+1}^{N} |\mathbf{e}^\dagger(\omega)\boldsymbol{v}_i|^2}. \tag{16.204}$$

Since $\mathbf{e}^\dagger(\omega)\boldsymbol{v}_i$ for $i = M + 1, M + 2, \ldots, N$ is ideally equal to zero, $P_{\text{MUSIC}}(\omega)$ tends to have large peaks at the angular frequencies that correspond to the plane wave arrival directions.

Another popular method that uses the eigenvalue decomposition of $\mathbf{R}$ is called the minimum-norm method [KU83b; RE79]. In this method, a single vector $\mathbf{d} = [d[0], d[1], \ldots, d[N-1]]^t$, which spans the noise subspace (i.e., a linear combination of the noise-subspace eigenvectors $\boldsymbol{v}_{M+1}, \boldsymbol{v}_{M+2}, \ldots, \boldsymbol{v}_N$) is computed with the constraint that $d[0] = 1$ and $\sum_{k=1}^{N-1} |d[k]|^2$ is minimum—hence the name.

Simple calculations show that the vector **d** is given by

$$\mathbf{d} = \begin{bmatrix} 1 \\ v_N' \mathbf{C}^\dagger \\ \mathbf{cc}^\dagger \end{bmatrix} \qquad (16.205)$$

where $\mathbf{V}_N$ is the matrix formed with the noise-subspace eigenvectors,

$$\mathbf{V}_N = [\mathbf{v}_{M+1} \quad \mathbf{v}_{M+2} \quad \cdots \quad \mathbf{v}_N] = \begin{bmatrix} \mathbf{c} \\ \mathbf{V}_N' \end{bmatrix},$$

and $\mathbf{V}_N'$ consists of the last $N - 1$ rows of $\mathbf{V}_N$ and $\mathbf{c}$ is the first row of $\mathbf{V}_N$. Once the vector **d** is computed, the corresponding angular spectrum is $P_{\text{min-norm}}(\omega)$:

$$P_{\text{min-norm}}(\omega) = \frac{1}{|\mathbf{e}^\dagger(\omega)\mathbf{d}|^2}. \qquad (16.206)$$

As before, since the true spatial correlation matrix **R** is unavailable, one may use an estimate of **R** and use its eigendecomposition in Eqs. (16.204) and (16.206). Again, since all the noise-subspace eigenvectors tend to have nulls at the angular frequencies corresponding to the plane waves, $1/|\mathbf{e}^\dagger(\omega)\mathbf{d}|^2$ tends to have large peaks at those frequencies. Generally, the effect of the minimum-norm constraint is that it tends to cause fewer false peaks in the angular spectrum. A number of comments are in order regarding the methods discussed in this section, but first a word of caution regarding the angular spectra in Eqs. (16.204) and (16.206). They do not correspond to the traditional power spectral density in that the areas under the peaks do not correspond to the power of the sources. Their utility is largely in locating the angular locations of the sources. The power in each source has to be determined by other methods.

We now compare the performance of these different methods by applying them to the output samples of a set of sensors excited by two sources. The simulation model has two plane waves from two sources incident on a line array consisting of $N = 8$ equispaced sensors. The number of snapshots, K, equals 10. The two incoherent signals and the additive noise can be modeled as

$$x_k[n] = a_1 e^{j(n\omega_1 + \phi_{1k})} + a_2 e^{j(n\omega_2 + \phi_{2k})} + w_k[n], \qquad n = 1, 2, \ldots, 8,$$

$$(16.207)$$

where $w_k[n]$ is a complex Gaussian white noise sequence having unit variance. a_1 and a_2 are the amplitudes of the voltages induced and ϕ_{1k} and ϕ_{2k} are independent, uniformly distributed phase angles in the interval $(-\pi, \pi)$, $\omega_i = 2\pi(d \sin \theta_i)/\lambda$, $d = \lambda/2$, $\theta_1 = 338°$, and $\theta_2 = 342°$. The parameters d and θ correspond to that given in Figure 16-14 and are related to the sensor spacing, frequency of the incident wave, and its angle of arrival. Since the noise has unit variance, the SNR was varied by changing the amplitudes a_1 and a_2. $a_1 = a_2 = 31.62$ for 30-dB SNR and 3.162 for 10-dB SNR. Figure 16-16 shows the spectrum obtained by the two eigendecomposition methods described above and the minimum-variance estimator of Capon, for a SNR of 30 dB. The MUSIC spectrum was computed using the expression given in Eq. (16.204), while the minimum-norm spectrum was calculated using Eq. (16.206). Capon's estimator used Eq. (16.200). At this high SNR, these methods pick the two closely spaced peaks without any difficulty. However, for the lower SNR case, that is, for SNR = 10 dB, the minimum-variance estimator is unable to distinguish between the two closely spaced signals, while the performance is only slightly better for the MUSIC estimator. In contrast, the minimum-norm method gives the best resolution among these three methods.

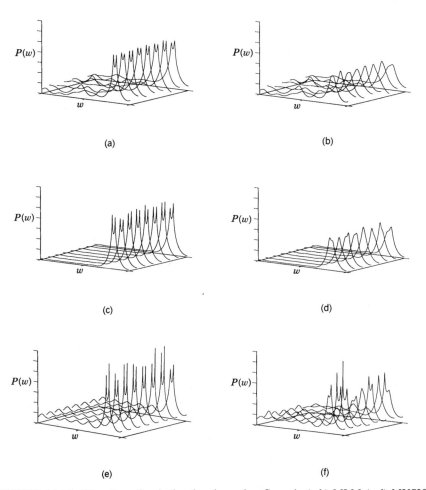

(a)   (b)

(c)   (d)

(e)   (f)

FIGURE 16-16   Direction of arrival estimation using Capon's (a,b) MLM (c,d) MUSIC algorithm, and (e,f) the MIN-NORM method. Parts (a), (c), and (e) correspond to 30-dB SNR and parts (b), (d), and (f) correspond to 10-dB SNR.

16-6-5 Two-Dimensional Spectral Analysis

In many applications including image analysis, sonar, and synthetic aperture radar data processing, one has to process a two-dimensional array of data samples $x[m, n]$, $m = 0, 1, \ldots, M - 1$ and $n = 0, 1, \ldots, N - 1$. The reader may consult the text by Dudgeon and Mersereau [DU84] and the review paper by McClellan [MC82] for details on multidimensional digital signal processing and spectral analysis. Kay [KA87] and Marple [MA87] give an overview of parametric spectral analysis of two-dimensional signals as well.

The Fourier methods discussed in earlier sections, that is, Fourier transforming the windowed autocorrelation function and averaging the windowed periodogram, are directly extendable to processing two-dimensional data. The unbiased two-dimensional autocorrelation estimate at lag (k, l) is

$$
r_{xx}(k, l) = \begin{cases}
\dfrac{1}{(M - k)(N - 1)} \displaystyle\sum_{m=0}^{M-1-k} \sum_{n=0}^{N-1-l} x[m + k, n + l]x^*[m, n], \\
\qquad\qquad\qquad\qquad\qquad\qquad k \geq 0, l \geq 0, \\[2mm]
\dfrac{1}{(M - k)(N - 1)} \displaystyle\sum_{m=0}^{M-1-k} \sum_{n=-l}^{N-1} x[m + k, n + l]x^*[m, n], \\
\qquad\qquad\qquad\qquad\qquad\qquad k \geq 0, l \leq 0, \\[2mm]
r_{xx}^*[-k, -l], \\
\qquad\qquad\qquad\qquad\qquad\qquad k \leq 0, \text{ any } l.
\end{cases}
\qquad (16.208)
$$

$r_{xx}[k, l]$ is defined over a range of $|k| < p$ and $|l| < q$ where p and q are limited to at most $M - 1$ and $N - 1$, respectively. The power spectral estimate is then given by

$$
\hat{P}(\omega_1, \omega_2) = T_1 T_2 \sum_{k=-p}^{p} \sum_{l=-q}^{q} w[k, l]\hat{r}_{xx}[k, l]e^{-j\omega_1 kT_1 - j\omega_2 lT_2}, \qquad (16.209)
$$

where $w[k, l]$ is the window used to trade off resolution and bias. T_1 and T_2 are the sampling intervals in the two domains. The two-dimensional windows are often obtained from the one-dimensional window by forming an outer product, that is, $w[k, l] = w[k]w[l]$ for $|k| \leq p$ and $|l| \leq q$. Windows that possess other than rectangular symmetry are also available [HU72].

The two-dimensional periodogram is defined analogous to the one-dimensional periodogram as follows:

$$
\hat{P}(\omega_1, \omega_2) = \frac{T_1 T_2}{MN} \left| \sum_{m=0}^{M-1} \sum_{n=0}^{N-1} w[m, n]x[m, n]e^{-j\omega_1 mT_1 - j\omega_2 nT_2} \right|^2, \qquad (16.210)
$$

where $w[m, n]$ is a suitable data window. As in Section 16-4 the variance of the periodogram may be reduced by partitioning the data into smaller possibly overlapping two-dimensional arrays and averaging their periodograms. However, in many applications the data are limited in extent and hence reduction in variance without significant loss of resolution is not possible.

16-7 SUMMARY

In this chapter we have reviewed some of the common spectral analysis methods. In Sections 16-2 and 16-3 we presented the traditional Fourier analysis based methods. Section 16-4 summarizes model based methods for estimating the power spectral density of a random process given a finite number of observations. In Section 16-5 we have outlined several methods for determining the parameters of sinusoidal signals, which are embedded in noise. In Section 16-6 we pointed out the application of spectral analysis methods to spatial data processing.

ACKNOWLEDGMENTS

This work was supported in part by the United States Office of Naval Research, Air Force Office of Scientific Research and the Alexander von Humboldt Foundation of the Federal Republic of Germany. I thank C. S. Ramalingam, Y. Feng and M. Muzzio for help with computer simulations, Dr. D. van Ormondt of Delft University and Dr. D. Traficante of the Chemistry Department at the University of Rhode Island for providing me with NMR data.

REFERENCES

[AI49] P. R. Aigrain and E. M. Williams, Synthesis of n-reactance networks for desired transient response. *J. Appl. Phys.* **20,** 597–600 (1949).

[AK69] H. Akaike, Power spectrum estimation through autoregressive model fitting. *Ann. Inst. Stat. Math.* **21,** 407–419 (1969).

[AK74] H. Akaike, A new look at the statistical model identification. *IEEE Trans. Autom. Control* **AC-19,** 716–723 (1984).

[BA48] M. S. Bartlett, Smoothing periodograms from time series with continuous spectra. *Nature (London)* **161,** 686–687 (1948).

[BI79] G. Bienvenu, Influence of the spatial coherence of the background noise on high resolution passive methods. *Proc. IEEE Int. Conf. Acoust., Speech, Signal Process., Washington, D.C., 1979,* pp. 306–309 (1979).

[BL58] R. B. Blackman and J. W. Tukey, *The Measurement of Power Spectra.* Dover, New York, 1958.

[BR80a] D. R. Brillinger, *Time Series: Data Analysis and Theory,* rev. ed. Holden-Day, Oakland, CA, 1980.

[BR80b] S. P. Bruzzone and M. Kaveh, On some sub-optimal ARMA spectral estimators. *IEEE Trans. Acoust., Speech Signal Process.* **ASSP-28,** 735–755 (1980).

[BR86] Y. Bressler and A. Macovski, Exact maximum likelihood estimation of super-imposed exponential signals in noise. *IEEE Trans. Acoust., Speech, Signal Process.* **ASSP-34,** 1081–1089 (1986).

[BU67] J. P. Burg, Maximum entropy spectral analysis. *Proc. 37th Meet. Soc. Explor. Geophys., Oklahoma City, 1967,* reprinted in Ref. [CH78].

[CA69] J. Capon, High resolution frequency wavenumber spectrum analysis. *Proc. IEEE* **57,** 1408–1418 (1969).

[CA70] J. Capon and N. R. Goodman, Probability distribution for estimation of the frequency wavenumber spectrum. *Proc. IEEE* **58,** 1785–1786 (1970).

[CH72] J. C. Chow, On estimating the orders of an autoregressive-moving average process with uncertain observations. *IEEE Trans. Autom. Control* **AC-17,** 707–709 (1972).

[CH78] D. G. Childers, ed., *Modern Spectrum Analysis.* IEEE Press, New York, 1978.

[CL92] M. P. Clark and L. L. Scharf, Reducing the complexity of parametric estimators for deterministic modal analysis. *IEEE Trans. Acoust., Speech, Signal Process., 1992* (submitted for publication).

[CO65] J. W. Cooley and J. W. Tukey, An algorithm for machine calculation of complex Fourier series. *Math. Comput.* **19,** 297–301 (1965).

[DE88] R. de Beer, W. M. M. J. Bovée *et al.,* Retrieval of signal parameters from magnetic resonance signals via SVD. In *SVD and Signal Processing Algorithms* (E. F. Deprettre, ed.). Elsevier North-Holland Publ., Amsterdam, 1988.

[DU59] J. Durbin, Efficient estimation of parameters in moving average models. *Biometrika.* **46,** 306–316 (1959).

[DU73] J. Durbin, Efficient estimation of parameters in moving average models. *Biometrika.* **60,** 255–265 (1973).

[DU84] D. E. Dudgeon and R. M. Mersereau, *Multidimensional Digital Signal Processing,* Prentice-Hall, Englewood Cliffs, NJ, 1984.

[EV73] A. G. Evans and R. Fischl, Optimal least squares time domain synthesis of recursive digital filters. *IEEE Trans. Audio Electroacoust.* **AU-21**(1), 61–65 (1973).

[FR82] B. Friedlander, Recursive lattice forms for spectral estimation. *IEEE Trans. Acoust., Speech, Signal Process.* **ASSP-30,** 920–930 (1982).

[GE88] H. Gesmar and J. J. Led, Spectral estimation of complex time-domain NMR signals by linear prediction. *J. Magn. Reson.* **76,** 183–192 (1988).

[GO73] G. H. Golub and V. Pereyra, The difference of pseudo-inverses and non-linear least squares problems whose variables separate. *SIAM. Numer. Anal.* **10,** 413–432 (1973).

[HA78] F. J. Harris, On the use of windows for harmonic analysis with the discrete Fourier transform. *Proc. IEEE* **66,** 51–83 (1978).

[HA82] S. Haykin and J. Cadzow, ed., *Proc. IEEE, Spec. Issue Spectral Estim.* **70** (1982).

[HA83] S. Haykin, ed., *Non-Linear Methods of Spectrum Analysis,* 2nd ed. Springer-Verlag, New York, 1983.

[HA85] S. Haykin, Radar array processing for angle of arrival estimation. In *Array Signal Processing* (S. Haykin *et al.,* eds.) Prentice-Hall, Englewood Cliffs, NJ, 1985.

[HE81] T. L. Henderson, Geometric methods for determining system poles from transient response. *IEEE Trans. Acoust., Speech, Signal Process.* **29,** 982–988 (1981).

[HI56] F. B. Hildebrand, *Introduction to Numerical Analysis*, Chapter 9. McGraw-Hill, New York, 1956.

[HU72] T. S. Huang, Two dimensional windows. *IEEE Trans. Audio Electroacoust.* **AU-20,** 88–90 (1972).

[HU88] Y. B. Hua and T. K. Sarkar, Perturbation analysis of TK method for harmonics retrieval. *IEEE Trans. Acoust., Speech, Signal Process.* **ASSP-36,** 228–240 (1988).

[JA82] E. T. Jaynes, On the rationale of maximum entropy methods. *Proc. IEEE* **70,** 939–952 (1982).

[JE68] G. M. Jenkins and D. G. Watts, *Spectral Analysis and Its Applications.* Holden-Day, Oakland, CA, 1968.

[JO82] D. H. Johnson, The application of spectral estimation methods to bearing estimation problems. *Proc. IEEE* **70,** 1018–1028 (1982).

[KA58] R. E. Kalman, Design of a self-optimizing control system. *Trans. ASME* **80,** 468–478 (1958).

[KA84] S. Kay, Accurate frequency estimation at low signal-to-noise ratio. *IEEE Trans. Acoust., Speech, Signal Process.* **32,** 540–547 (1984).

[KA87] S. M. Kay, *Modern Spectrum Analysis.* Prentice-Hall, New York, 1987.

[KA88] S. M. Kay, Spectral estimation. In *Advanced Topics in Signal Processing* (J. S. Lim and A. V. Oppenheim, eds.). Prentice-Hall, New York, 1988.

[KE86] S. Kesler, ed., *Modern Spectrum Analysis II.* IEEE Press, New York, 1986.

[KO41] A. N. Kolmogorov, Interpolation und extrapolation von stationaren zufalligen folgen. *Bull. Acad. Sci. USSR., Math. Ser.* **5,** 3–14 (1941).

[KO74] L. H. Koopmans, *The Spectral Analysis of Time Series.* Academic Press, New York, 1974.

[KU81] R. Kumaresan and D. W. Tufts, Singular value decomposition and spectral analysis. *Proc. IEEE Sponsored 1st Workshop Spectral Estim., Hamilton, Ontario, 1981*, pp. 6.4.1–6.4.12, (1981).

[KU82] R. Kumaresan and D. W. Tufts, Estimating the parameters of exponentially damped sinusoids and pole-zero modeling in noise. *IEEE Trans. Acoust., Speech, Signal Process.* **ASSP-30,** 833–840 (1982).

[KU83a] R. Kumaresan, On the zeros of linear prediction-error filter for deterministic signals. *IEEE Trans. Acoust., Speech, Signal Process.* **ASSP-31,** 217–220 (1983).

[KU83b] R. Kumaresan and D. W. Tufts, Estimating the angles of arrival of multiple plane waves. *IEEE Trans. Aerosp. Electron. Syst.* **AES-19,** 134–138 (1983).

[KU83c] S. Y. Kung *et al.*, State-space and singular value decomposition based approximation method for the harmonic retrieval problem. *J. Opt. Soc. Am.* **73,** 1799–1811 (1983).

[KU86] R. Kumaresan, L. L. Scharf, and A. K. Shaw, An algorithm for pole-zero modeling and spectral analysis. *IEEE Trans. Acoust., Speech, Signal Process.* **ASSP-34,** 637–640 (1986).

[LA73] H. J. Larson, *Introduction to the Theory of Statistics.* Wiley, New York, 1973.

[LA74] C. L. Lawson, and R. J. Hanson, *Solving Least Squares Problems.* Prentice-Hall, Englewood Cliffs, NJ, 1974.

[LE47] N. Levinson, The Weiner RMS error criterion in filter design and prediction. *J. Math. Phys.* **25,** 261–278 (1947).

[MA75] J. Makhoul, Linear prediction: A tutorial review. *Proc. IEEE* **63,** 561–580 (1975).

[MA80] S. L. Marple, A new autoregressive spectrum analysis algorithm. *IEEE Trans. Acoust., Speech, Signal Process.* **28**, 441–454 (1980).

[MA87] L. Marple, *Digital Spectrum Analysis with Applications*. Prentice-Hall, New York, 1987.

[MC68] R. N. McDonnough and W. H. Huggins, Best least squares representation of signals by exponentials. *IEEE Trans. Autom. Control* **20**, 408 (1968).

[MC82] J. H. McClellan, Multidimensional spectral estimation. *Proc. IEEE* **70**, 1029–1039 (1982).

[MC88] J. H. McClellan, Parametric signal modeling. In *Advanced Topics in Signal Process* (J. S. Lim and A. V. Oppenheim, eds.). Prentice-Hall, New York, pp. 1–57, 1988.

[MC90] J. H. McClellan and D. Lee, Exact equivalence of the Steiglitz-McBride iteration and IQML. *IEEE Trans. Acoust., Speech, Signal Process.* **39**, 509–512 (1991).

[MO77] M. Morf, B. Dickinson, T. Kailath, and A. Vieira, Efficient solution of covariance equations for linear prediction. *IEEE Trans. Acoust., Speech, Signal Process.* **25**, 429–433 (1977).

[NU76] A. H. Nuttall, Spectral analysis of a univariate process with bad data points via maximum entropy and linear predictive techniques. *Nav. Underwater Syst. Cent. (NUSC) Sci. Eng. Stud., Spectral Estim., NUSC,* New London, Connecticut, 1976.

[OP75] A. V. Oppenheim and R. W. Schafer, *Digital Signal Processing*. Prentice-Hall, Englewood Cliffs, NJ, 1975.

[OW85] N. L. Owsley, Sonar array processing. In *Array Signal Processing* (S. Haykin *et al.*, eds.), Chapter 3. Prentice-Hall, Englewood Cliffs, NJ, 1985.

[PA64] E. Parzen, An approach to empirical time series analysis. *J. Nat. Res. Bur. Stand., Sect. B* **68B**, 937–951 (1964).

[PA74] E. Parzen, On consistent estimates of the spectrum of a stationary time series. *Ann. Math. Stat.* **28**, 329–348 (1974).

[PA83] L. Pakula and S. Kay, New and simple proofs for the minimum phase property. *IEEE Trans. Acoust., Speech, Signal Process.* **31**, 501–502 (1983).

[PA85] S. Parthasarathy and D. W. Tufts, Maximum-likelihood estimation of parameters of exponentially damped sinusoids. *Proc. IEEE* **73**, 1528–1530 (1985).

[PA87] S. Park and J. T. Cordaro, Maximum likelihood estimation of poles from impulse response data in noise. *Proc. Int. Conf. Acoust., Speech, Signal Process., Dallas, Texas, 1987,* pp. 1501–1504 (1987).

[PI73] V. F. Pisarenko, The retrieval of harmonics from a covariance function. *Geophys. J. R. Astron. Soc.* **33**, 347–366 (1973).

[PO87] B. Porat and B. Friedlander, On the accuracy of the Kumaresan-Tufts method for estimating complex damped exponentials. *IEEE Trans. Acoust., Speech, Signal Process.* **ASSP-35**, 231–234 (1987).

[PR1795] R. Prony, *Essai expérimental et analytique*, Vol. 1, No. 2, pp. 24–76. L'Ecole Polytechnique, Paris, 1795.

[RA71] C. R. Rao and S. K. Mitra, *Generalized Inverses of Matrices and Their Applications*. Wiley, New York, 1971.

[RA88] B. D. Rao, Perturbation analysis of SVD-based linear prediction method for estimating the frequencies of multiple sinusoids. *IEEE Trans. Acoust., Speech, Signal Process.* **ASSP-36**, 1026–1035 (1988).

[RE79] S. S. Reddi, Multiple source location: A digital approach. *IEEE Trans. Aerosp. Electron. Syst.* **AES-15**, 95–105 (1979).

[RI76] D. C. Rife and R. R. Boorstyn, Multiple tone parameter estimation from discrete time observations. *Bell Syst. Tech. J.* **55**, 1389–1410 (1976).

[RI83] J. Rissanen, A Universal prior for the integers and estimation by minimum description length. *Ann. Stat.* **11**, 417–431 (1983).

[RO67] E. A. Robinson, Predictive decomposition of time-series with applications to seismic exploration. *Geophysics* **32**, 418–484 (1967).

[RO82] E. A. Robinson, A historical perspective of spectrum estimation. *Proc. IEEE* **70**, 885–907 (1982).

[SA63] C. K. Sanathan and J. Koerner, Transfer function synthesis as a ratio of two complex polynomials. *IEEE Trans. Autom. Control* **AC-8**, 56 (1963).

[SC79] R. Schmidt, Multiple emitter location and signal parameter estimation. *Proc. Rome Air Dev. Cent. Spectral Estim. Workshop*, Rome, New York, pp. 243–258 (1979).

[SC84] M. Schoenberger, ed., *Proc. IEEE Spec. Issue Seismic Signal Process.* **72**, 1233–1424, (1984).

[SC1898] A. Schuster, On the investigation of hidden periodicities with application to a supposed 26 days period of meteorological phenomenon. *Terr. Magn.* **3**, 13–41 (1898).

[ST65] K. Steiglitz and L. E. McBride, A technique for identification of linear systems. *IEEE Trans. Autom. Control* **AC-10**, 461–464 (1965).

[ST77] K. Steiglitz, On the simultaneous estimation of poles and zeros in speech analysis. *IEEE Trans. Acoust., Speech, Signal Process.* **25**, 229–234 (1977).

[ST88] D. S. Stephenson, Linear prediction and maximum entropy methods in NMR spectroscopy. *Prog. Nucl. Magn. Reson.* **20**, 515–526 (1988).

[TU58] D. F. Tuttle, *Network Synthesis*, Vol. 1. Wiley, New York, 1958.

[TU80] D. W. Tufts and R. Kumaresan, Improved spectral resolution II. *Proc. IEEE Conf. Acoust., Speech, Signal Process., Denver, CO, 1980*, pp. 392–397 (1980).

[TU82a] D. W. Tufts, R. Kumaresan, and I. Kirsteins, Data adaptive signal estimation by singular value decomposition of a data matrix. *Proc. IEEE* **70**, 684–685 (1982).

[TU82b] D. W. Tufts and R. Kumaresan, Frequency estimation of multiple sinusoids: Making linear prediction perform like maximum likelihood. *Proc. IEEE, Spec. Issue Spectral Estim., 1982*, Vol. 70, pp. 975–989 (1982).

[UL76] T. J. Ulrych and R. W. Clayton, Time series modeling and maximum entropy. *Phys. Earth Planet. Inter.* **12**, 188–200 (1976).

[VA80] A. van den Boss, A class of small sample nonlinear least squares problems. *Automatica* **16**, 487–490 (1980).

[VA87] R. J. Vaccaro and A. Kot, A perturbation theory for the analysis of SVD based algorithms. *Proc. Int. Conf. Acoust., Speech, Signal Process., Dallas, Texas, 1987*, pp. 1613–1616 (1987).

[WE67] P. D. Welch, The use of fast Fourier transform for the estimation of power spectra: A method based on time averaging over short modified periodograms. *IEEE Trans. Audio Electroacoust.* **AU-15**, 70–73 (1967).

[WI30] N. Wiener, Generalized harmonic analysis. *Acta Math.* **55**, 117–258 (1930).

[WI49] N. Wiener, *Extrapolation, Interpolation and Smoothing of Stationary Time Series with Engineering Applications*. Wiley, New York, 1949.

[WO38] H. Wold, *A Study in the Analysis of Stationary Time Series*. Almqvist & Wiksell, Uppsala, Sweden, 1938.

[YU27] G. U. Yule, On a method of investigating periodicities in disturbed series with special reference to Wolfer's sun-spot numbers. *Philos. Trans. R. Soc. London, Ser. A* **226,** 276–298 (1927).

INDEX